SECOND EDITION

ENVIRONMENTAL CONTAMINANTS IN BIOTA

Interpreting Tissue Concentrations

SECOND EDITION

ENVIRONMENTAL CONTAMINANTS IN BIOTA

Interpreting Tissue Concentrations

Edited by

W. Nelson Beyer
James P. Meador

CRC Press
Taylor & Francis Group
Boca Raton   London   New York

CRC Press is an imprint of the
Taylor & Francis Group, an **informa** business

CRC Press
Taylor & Francis Group
6000 Broken Sound Parkway NW, Suite 300
Boca Raton, FL 33487-2742

© 2011 by Taylor and Francis Group, LLC
CRC Press is an imprint of Taylor & Francis Group, an Informa business

No claim to original U.S. Government works

Printed in the United States of America on acid-free paper
10 9 8 7 6 5 4 3 2 1

International Standard Book Number: 978-1-4200-8405-4 (Hardback)

Visit the Taylor & Francis Web site at
http://www.taylorandfrancis.com

and the CRC Press Web site at
http://www.crcpress.com

Contents

Editors......vii

Authors......ix

Reviewers...... xiii

Introduction......1

W. Nelson Beyer and James P. Meador

Chapter 1   History of Wildlife Toxicology and the Interpretation of Contaminant Concentrations in Tissues......9

Barnett A. Rattner, Anton M. Scheuhammer, and John E. Elliott

Chapter 2   DDT and Other Organohalogen Pesticides in Aquatic Organisms......47

Nancy Beckvar and Guilherme R. Lotufo

Chapter 3   Dioxins, PCBs, and PBDEs in Aquatic Organisms......103

Richard J. Wenning, Linda Martello, and Anne Prusak-Daniel

Chapter 4   Methylmercury in Freshwater Fish: Recent Advances in Assessing Toxicity of Environmentally Relevant Exposures......169

Mark B. Sandheinrich and James G. Wiener

Chapter 5   Selenium Accumulation and Toxicity in Freshwater Fishes......193

David K. DeForest and William J. Adams

Chapter 6   Trace Metals in Aquatic Invertebrates......231

Philip S. Rainbow and Samuel N. Luoma

Chapter 7   Organotins in Aquatic Biota: Occurrence in Tissue and Toxicological Significance......255

James P. Meador

Chapter 8   Active Pharmaceutical Ingredients and Aquatic Organisms......287

Christian G. Daughton and Bryan W. Brooks

Chapter 9   Organic Contaminants in Marine Mammals: Concepts in Exposure, Toxicity, and Management......349

Lisa L. Loseto and Peter S. Ross

Chapter 10 Select Elements and Potential Adverse Effects in Cetaceans and Pinnipeds 377

              Todd O'Hara, Takashi Kunito, Victoria Woshner, and Shinsuke Tanabe

Chapter 11 Toxicological Significance of Pesticide Residues in Aquatic Animals 409

              Michael J. Lydy, Jason B. Belden, Jing You, and Amanda D. Harwood

Chapter 12 DDT, DDD, and DDE in Birds 425

              Lawrence J. Blus

Chapter 13 Cyclodiene and Other Organochlorine Pesticides in Birds 447

              John E. Elliott and Christine A. Bishop

Chapter 14 Effects of Polychlorinated Biphenyls, Dibenzo-*p*-Dioxins and
              Dibenzofurans, and Polybrominated Diphenyl Ethers in Wild Birds 477

              Megan L. Harris and John E. Elliott

Chapter 15 Toxicological Implications of PCBs, PCDDs, and PCDFs in Mammals 531

              Matthew Zwiernik, Frouke Vermeulen, and Steven Bursian

Chapter 16 Lead in Birds 563

              J. Christian Franson and Deborah J. Pain

Chapter 17 Lead in Mammals 595

              Wei-Chun Ma

Chapter 18 Mercury in Nonmarine Birds and Mammals 609

              Richard F. Shore, M. Glória Pereira, Lee A. Walker, and David R. Thompson

Chapter 19 Cadmium in Small Mammals 627

              John A. Cooke

Chapter 20 Cadmium in Birds 645

              Mark Wayland and Anton M. Scheuhammer

Chapter 21 Selenium in Birds 669

              Harry M. Ohlendorf and Gary H. Heinz

Chapter 22 Radionuclides in Biota 703

              Bradley E. Sample and Cameron Irvine

Index 733

Editors

Dr. W. Nelson Beyer is an ecotoxicologist at the Patuxent Wildlife Research Center, in the U.S. Geological Survey. He earned a doctorate in terrestrial ecology from Cornell University in 1976, studying the natural history of slugs, and remains curious about all manner of living things. With help from colleagues in the U.S. Fish and Wildlife Service, he has examined contaminated sites, studying phytotoxic injury to the chestnut oak forest adjacent to the Palmerton zinc smelters in Pennsylvania, lead poisoning of swans and other waterfowl in the Coeur d'Alene River Basin in Idaho, and zinc poisoning of wild birds at the Tri-State Mining District in Oklahoma, Kansas, and Missouri. Many of his studies have documented the movement of environmental contaminants from soil and sediments into wildlife and through food chains, often showing how ingestion of soil or sediment is the most important route of exposure of lead and related metals to wildlife. Recently he has turned his eye toward sediment ingestion by bottom-feeding fish. He believes in using tissue residues as a means to emphasize injury to animals, basic to wildlife toxicology. His work is usually conducted to support Natural Resource Damage Assessments. He balances his down-to-earth approach to wildlife toxicology, however, with an inordinate fondness for clouds. His daily cycling inspires him, and his wife, Mary, keeps his laboratory running.

Dr. James P. Meador is an environmental toxicologist with the National Marine Fisheries Service in Seattle, WA, U.S., which is part of the National Oceanic and Atmospheric Administration (NOAA). Jim has a Ph.D. in aquatic toxicology from the University of Washington and has more than 30 years experience in the field. He has been with NOAA for 20 years and has previously worked at the Scripps Institution of Oceanography and the Naval Oceans Systems Center in San Diego. For many years he has studied the environmental factors that control the bioavailability of contaminants and their role in bioaccumulation with the goal of developing relationships for predicting tissue concentrations, especially for aquatic species in the field. This work was supplemented with toxicokinetic studies that are useful for predicting bioaccumulation and highlighting species-specific differences in toxic responses. Gradually, these studies evolved to include analysis of toxic effects as a function of bioaccumulated tissue concentrations. In 2007, Jim organized and chaired a Pellston workshop, sponsored by the Society of Environmental Toxicology and Chemistry, to review the tissue residue approach for toxicity assessment. His current research is focused on alterations to growth and energetics in fish resulting from low-dose exposure to a variety of environmental contaminants. Jim has been married for 30 years to Susan, a recently retired NOAA statistician, who keeps him busy in the garden and serves as his buddy on their tropical-therapy SCUBA diving trips. He is also an avid bicycle rider/commuter and logs far more miles on his bike than on his car.

Authors

William J. Adams
Rio Tinto
Lake Point, UT

Nancy Beckvar
National Oceanic and Atmospheric
    Administration
Office of Response and Restoration
Assessment and Restoration Division
Seattle, WA

Jason B. Belden
Department of Zoology
Oklahoma State University
Stillwater, OK

W. Nelson Beyer
U.S. Geological Survey
Patuxent Wildlife Research Center
Laurel, MD

Christine A. Bishop
Environment Canada
Science and Technology Branch
Pacific Wildlife Research Centre
Delta, Canada

Lawrence J. Blus (retired)
U.S. Geological Survey
Forest and Rangeland Ecosystem
    Science Center
Corvallis, OR

Bryan W. Brooks
Department of Environmental Science
Center for Reservoir and Aquatic
    Systems Research
Baylor University
Waco, TX

Steven Bursian
Department of Animal Science
Michigan State University
East Lansing, MI

John A. Cooke
School of Biological and Conservation Sciences
University of KwaZulu-Natal
Durban, South Africa

Christian G. Daughton
U.S. Environmental Protection Agency
Office of Research and Development
Environmental Sciences Division
National Exposure Research Laboratory
Las Vegas, NV

David K. DeForest
Windward Environmental
Seattle, WA

John E. Elliott
Environment Canada
Science and Technology Branch
Pacific Wildlife Research Centre
Delta, Canada

J. Christian Franson
U.S. Geological Survey
National Wildlife Health Center
Madison, WI

Amanda D. Harwood
Fisheries and Illinois Aquaculture Center
Department of Zoology
Southern Illinois University
Carbondale, IL

Megan L. Harris
Lorax Environmental
Bonshaw, Canada

Gary H. Heinz
U.S. Geological Survey
Patuxent Wildlife Research Center
Laurel, MD

Cameron Irvine
CH2M HILL
Sacramento, CA

Takashi Kunito
Department of Environmental Sciences
Shinshu University
Matsumoto, Nagano, Japan

Lisa L. Loseto
Fisheries and Oceans Canada
Freshwater Institute
Winnipeg, Canada

Guilherme R. Lotufo
U.S. Army Corps of Engineers
Engineer Research and Development Center
Vicksburg, MS

Samuel N. Luoma
John Muir Institute of the Environment
University of California at Davis
Davis, CA

Michael J. Lydy
Fisheries and Illinois Aquaculture Center
Department of Zoology
Southern Illinois University
Carbondale, IL

Wei-Chun Ma
Environmental Sciences Group
Wageningen University and Research Centre
Wageningen, The Netherlands

Linda Martello
ENVIRON International Corporation
San Francisco, CA

James P. Meador
National Oceanic and Atmospheric
  Administration
Northwest Fisheries Science Center
Ecotoxicology and Environmental Fish
  Health Program
Seattle, WA

Todd O'Hara
Institute of Arctic Biology
University of Alaska Fairbanks
Fairbanks, AK

Harry M. Ohlendorf
CH2M HILL
Sacramento, CA

Deborah J. Pain
Wildfowl & Wetlands Trust
Slimbridge
Gloucestershire, UK

M. Glória Pereira
Centre for Ecology & Hydrology
Lancaster Environment Centre
Lancaster, UK

Anne Prusak-Daniel
ENVIRON International Corporation
Portland, ME

Philip S. Rainbow
Department of Zoology
The Natural History Museum
London, UK

Barnett A. Rattner
U.S. Geological Survey
Patuxent Wildlife Research Center
Laurel, MD

Peter S. Ross
Fisheries and Oceans Canada
Institute of Ocean Sciences
Sidney, Canada

Bradley E. Sample
Principal Scientist
Ecological Risk, Inc.
Rancho Murieta, CA

Mark B. Sandheinrich
University of Wisconsin–La Crosse
River Studies Center
La Crosse, WI

Anton M. Scheuhammer
Environment Canada
National Wildlife Research Centre
Ottawa, Canada

Richard F. Shore
Centre for Ecology & Hydrology
Lancaster Environment Centre
Lancaster, UK

Shinsuke Tanabe
Center for Marine Environmental Studies
Ehime University
Bunkyo-cho, Matsuyama, Japan

David R. Thompson
National Institute of Water and Atmospheric
  Research Ltd.
Wellington, New Zealand

Frouke Vermeulen
Department of Animal Science
Michigan State University
East Lansing, MI

Lee A. Walker
Centre for Ecology & Hydrology
Lancaster Environment Centre
Lancaster, UK

Mark Wayland
Canadian Wildlife Service
Environment Canada
Saskatoon, Canada

Richard J. Wenning
ENVIRON International Corporation
San Francisco, CA

James G. Wiener
University of Wisconsin–La Crosse
River Studies Center
La Crosse, WI

Victoria Woshner
Independent Consultant
Pittsburg, PA

Jing You
State Key Laboratory of Organic Geochemistry
Guangzhou Institute of Geochemistry
Chinese Academy of Sciences
Guangzhou, China

Matthew Zwiernik
Department of Animal Science
Michigan State University
East Lansing, MI

Reviewers

The authors and editors thank our distinguished reviewers for their assistance.

Mehran Alaee
Aquatic Ecosystem Protection
    Research Division
Water Science and Technology Directorate
Science and Technology Branch
Environment Canada
Burlington, Canada

Gerald Ankley
U.S. Environmental Protection Agency
Mid-continent Ecology Division
Duluth, MN

Jill Awkerman
Biological Effects & Population Responses
U.S. EPA—Gulf Ecology Division
Gulf Breeze, FL

Mace G. Barron
Biological Effects & Population Responses
U.S. EPA—Gulf Ecology Division
Gulf Breeze, FL

Nancy Beckvar
National Oceanic and Atmospheric
    Administration
Office of Response and Restoration
Assessment and Restoration Division
Seattle, WA

Ronny Blust
Department of Biology
University of Antwerp
Antwerp, Belgium

Birgit Braune
Environment Canada
National Wildlife Research Centre
Ottawa, Canada

Jeremy Buck
U.S. Fish and Wildlife Service
Oregon State Fish and Wildlife Office
Portland, OR

Christine M. Custer
U.S. Geological Survey
LaCrosse, WI

Thomas W. Custer
U.S. Geological Survey
LaCrosse, WI

Damon Delistraty
Washington State Dept. of Ecology
Spokane, WA

Paul Drevnick
INRS-ETE, Université du Québec
Québec, Canada

David Evers
BioDiversity Research Institute
Gorham, ME

Patricia Fair
National Oceanic and Atmospheric
    Administration
Center for Coastal Environmental Health and
    Biomolecular Research at Charleston
Charleston, SC

Antonio Juan Garcia Fernandez
Department of Sociosanitary Sciences
University of Murcia
Murcia, Spain

Michael Fry
Conservation Advocacy American Bird
    Conservancy
Washington, DC

Robert A. Grove
U.S. Geological Survey
Forest and Rangeland Ecosystem
  Science Center
Corvallis, OR

Ailsa J. Hall
Scottish Oceans Institute
University of St Andrews
St. Andrews, Scotland

Chad R. Hammerschmidt
Department of Earth & Environmental
  Sciences
Wright State University
Dayton, OH

Dr. Thomas Heberer
Lower Saxony Federal State Office of
  Consumer Protection and Food Safety
  (LAVES)
Oldenburg, Germany

Gary H. Heinz
U.S. Geological Survey
Patuxent Wildlife Research Center
Laurel, MD

David Hoffman
U.S. Geological Survey
Patuxent Wildlife Research Center
Laurel, MD

Michael J. Hooper
U.S. Geological Survey
Columbia Environmental Research Center
Columbia, MO

Lisa Loseto
Fisheries and Oceans Canada
Freshwater Institute
Winnipeg, Canada

Rafael Mateo
Instituto de Investigación en Recursos
  Cinegéticos (IREC)
Ciudad Real, Spain

Chris Mebane
U.S. Geological Survey
Boise, ID

Joseph S. Meyer
ARCADIS
Lakewood, CO

Steve Mihok
Environmental Risk Assessment Division
Canadian Nuclear Safety Commission
Ottawa, Canada

John Newsted
ENTRIX
Okemos, MI

Lisa Nowell
Chemist, Pesticides National Synthesis Project
National Water Quality Assessment Program
  (NAWQA)
U.S. Geological Survey
Sacramento, CA

Oliver H. Pattee
U.S. Geological Survey
Patuxent Wildlife Research Center
Laurel, MD

Kevin D. Reynolds
U.S. Fish and Wildlife Service
Arizona Ecological Services Field Office
Phoenix, AZ

Peter S. Ross
Institute of Ocean Sciences (Fisheries and
  Oceans Canada)
Sidney, Canada

Michael Salazar
Applied Biomonitoring
Kirkland, WA

Lance J. Schuler
School of Biological Sciences
Louisiana Tech University
Ruston, LA

Jeffery Steevens
U.S. Army Corps of Engineers
Waterways Experimental Station
Vicksburg, MS

White-Tailed Sea Eagles

By G. Mutzel, from *The Royal Natural History,* edited by Richard Lydekker, Frederick
Warne & Co., London, 1893–94.

Introduction

W. Nelson Beyer and James P. Meador

Ecotoxicology is the study of the movement of environmental contaminants through ecosystems and their effects on plants and animals. Examining tissue residues of these contaminants in biota is basic to ecotoxicology, both for understanding the movement of contaminants within organisms and through food chains, and for understanding and quantifying injuries to organisms and their communities. This book provides guidance on interpreting tissue concentrations of environmental contaminants.

Tissue concentrations have long been used both to identify the cause of toxicity in animals and as a measure of the severity of toxicity. More recently, they have been incorporated into environmental models, tying together exposure, kinetics, and toxic effects. Measuring tissue concentrations is basic to studies on the kinetics of contaminants, which entails characterizing the rates of uptake and elimination in organisms, as well as redistribution (organs, lipid, and plasma) within them. Tissue concentrations are also used in ecological studies examining the movement of contaminants between organisms and within biological communities.

In monitoring programs, tissue concentrations tell us about the geographical distribution of contaminants and how they change through time. Measuring contaminants in tissue can also be important for defining the background, or the uncontaminated condition, as well as identification of hot spots and gradients from point sources. Although analyses of soils and sediments also provide information on the distribution of contaminants, analyses of tissues provide information that is more meaningful to ecotoxicologists. In some instances, chemical analyses of tissues gave the first hint of the global dispersion of chemicals. The environmental importance of polychlorinated biphenyls, tributyltin, and perfluorooctanesulfonic acid was not recognized until these compounds were found in tissues of widely distributed animals. Sometimes knowing simply that a contaminant is present in an organism is useful. For example, if an avian die-off has occurred and brain tissue shows greatly reduced activity of cholinesterase, then documenting the presence of an organophosphate or carbamate pesticide in the carcasses may be all that is required to find the cause of that die-off (Mineau and Tucker 2002). When pathologists examine toxicological cases, tissue analyses are usually essential to making a diagnosis. For the most part, however, this book provides guidance on relating tissue concentrations quantitatively to injury, which lies at the core of ecotoxicology. Thousands of research papers reporting tissue concentrations are published each year, and their value depends on ecotoxicologists being able to interpret the toxicological consequences of those concentrations.

The logic for relying on tissue residues in wildlife toxicology was put forth by Bill and Lucille Stickel (1973), who explained how tissue concentrations may best be used in diagnosing poisoning of birds by organochlorine pesticides. Biologists had suspected that birds were being poisoned by applications of pesticides, but differences among species, the physiological condition of the birds, and extraneous factors made it difficult to establish the cause of death. Analyzing the contents of the digestive tract for the presence of a pesticide, the usual means of diagnosing poisoning in humans, failed because all of the birds in a sprayed area had some exposure to the pesticide. Live birds collected at the site often had whole-body concentrations of pesticides that exceeded those of birds found dead. In a series of controlled studies on birds dosed with various organochlorine pesticides, the Stickels and colleagues demonstrated that because the lipids that store the pesticides are metabolized when a bird stops feeding (due to sickness caused by exposure to these pesticides), those

1

pesticides in the body may be mobilized and then rapidly become lethal. Although various organs could be used to indicate sublethal exposure to pesticides, analyzing the brain was key to identifying those birds that had lethal residues. Unlike concentrations in other organs, lethal concentrations in the brain were remarkably consistent, even in different species and in birds with different exposures. The brain is the logical organ to analyze because the organochlorine pesticides were neurotoxic, but the decision to rely on brain residues was a practical one. The reader is referred to Keith (1996) and to the following chapter in this book for a history of the use of tissue residues in evaluating hazards of contaminants to wildlife.

Aquatic toxicologists also rely on tissue concentrations when interpreting hazards, although much of this research has been relatively recent. The underpinnings of tissue residue toxicity were considered in the early 1900s by Meyer and Overton, who addressed the narcotic effect of organic compounds (Lipnick 1995) and by researchers who measured pesticides and metals in fish (Ferguson 1939, McCarty et al. in press). As these sporadic papers touted the virtues of using tissue residues to assess toxic responses, they were largely ignored by aquatic toxicologists, who emphasized exposure to contaminants in water and sediment. It was not until the early 1990s that a more in-depth analysis of tissue residue toxicity for a variety of chemicals and modes of action was considered (McCarty 1991, McCarty and Mackay 1993). After that, a flurry of research papers explored this topic in greater detail. These include published works on PAHs and other compounds at narcotic concentrations (Di Toro et al. 2000), chlorophenols (Kukkonen 2002), PCBs in salmonids (Meador et al. 2002a), tributyltin (Meador 2000, Meador et al. 2002b), mercury and DDT in fish (Beckvar et al. 2005), dioxins in fish (Steevens et al. 2005), and general reviews from Barron et al. (2002), Meador (2006), and Meador et al. (2008). At a Pellston workshop in 2007, 40 of the world's leading experts conducted a critical review of the tissue residue approach for toxicity assessment (see Integrated Environmental Assessment and Monitoring Jan. 2011).

The wide assortment of terms used in the field illustrates how researchers have evolved different ways of thinking about tissue concentrations. We begin this discussion with terms based on a mechanistic approach, originally defined by a work group on metals (Norberg 1976), although applicable to other contaminants as well. Several definitions are relevant here. The work group defines the "critical concentration" for a cell as the concentration at which undesirable functional changes, reversible or irreversible, occur in the cell. The "critical organ concentration" is defined as the mean concentration in the organ at the time any of its cells reaches critical concentration. The "critical organ" is that organ that first attains the critical concentration of a metal under specified circumstances of exposure and for a given population (Nordberg 1976). This approach is precise, assuming cause and effect. Cadmium's well-known effect on renal function seems to fit well into this framework. In practice, however, this approach does not work well for many environmental contaminants. A toxicant, such as lead, may affect many organs and systems simultaneously, and the signs and lesions observed among lead-poisoned individuals may vary substantially. Because organochlorine compounds are stored in lipids, throughout the body, they are not associated with a single organ. Nor is identifying "that organ that first attains the critical concentration" as simple as it sounds. A histopathologist using electron microscopy may detect lesions not visible using light microscopy. Drawing on more sensitive measures, such as those used in genomics, a toxicologist may detect alterations at lower tissue concentrations and exacerbate the difficulty in differentiating a harmless response from an adverse response. Risk assessors try to select endpoints that they consider meaningful to an assessment, which is not always the same as selecting the most sensitive endpoint.

The expression "critical concentration" is often useful when generalizing about tissue concentrations applicable to a taxon, as long as the effect and the circumstances are made clear. For example, based on studies conducted on several species ingesting lead shot, we might identify a critical concentration in livers of waterfowl expected to be associated with death. The term "threshold" means the concentration at which an effect is first observable.

The terminology of tissue concentrations used commonly by aquatic toxicologists is based on the traditional toxicological expressions of exposure—LCp or LDp (lethal) and ECp or EDp (sublethal) values, where C is the external concentration, D is the administered dose, and p is the percentage responding. In many cases aquatic toxicologists use LRp or ERp, where "R" denotes tissue residue (Meador 1997). There is a distinction between toxicity metrics that are expressed in terms of the amount of a toxicant that is delivered or administered to the organism and the actual tissue concentrations associated with the response. The dose is generally expressed as µg or µmol toxicant/gram body weight/day or as single-dose µg/g or µmol/g and is usually administered by feeding, injection, gavage, or bolus to determine the LD50 or other measures of toxicity. The acquired dose (tissue residue) is used to characterize adverse effects as a function of the measured or predicted tissue concentration, such as an LR50. The administered dose, as it is metabolized and excreted, may be very different from the tissue concentration associated with toxicity (Meador 2006). For aquatic toxicologists, "critical body residue" (CBR) is a general term often implying a whole-body concentration that is related to an adverse effect. A CBR can be characterized by any one of a number of toxicity metrics (e.g., LR50, ER10, or LOER) depending on the application. These values are best expressed as a molar concentration, especially when comparing among toxicants.

The terms "diagnosis" and "diagnostic" have well-established uses in veterinary science, and these terms can be applied in some instances to aquatic and wildlife toxicological studies. A diagnosis is a determination of the cause of an illness from its signs and lesions, through an examination by a trained diagnostician or pathologist. Making a diagnosis implies not only identifying a cause but also ruling out other potential causes of the observed signs and lesions. Consequently, ecotoxicologists may determine that the probable cause of death is a contaminant, but they are not making a "diagnosis" unless other causes are ruled out. A diagnostic residue is a concentration in tissue that supports a diagnosis of poisoning if the signs and lesions observed in the animal are consistent with the poison in question. A diagnostician starts with observed effects and reasons back to a cause, establishing the diagnosis, whereas an ecotoxicologist usually starts with an exposure or tissue concentration and tries to deduce possible toxic effects.

The need for screening values in ecological risk assessment has led to the use of "hazardous concentrations." For example, Aldenberg and Slob (1993) described a statistical method to calculate the lower confidence limit based on a percentile of a distribution of no-effect or lowest effect levels measured in different species within a taxonomic group. When calculated at the fifth percentile, the value is meant to be protective for 95% of the species or focal group. This threshold, or protection value, is lower than those derived from central tendency values (e.g., mean or median) that will protect far fewer organisms. The calculation of the HC_5 usually requires a large database from comparable studies.

For a critical concentration to be credible, it must be based on substantial evidence. Well-designed, controlled toxicological studies establish a cause-and-effect relation between the administration of a poison and an effect. Some controlled studies also establish a cause-and-effect relation between whole-body or specific tissue concentrations and an effect on an organism or that specific tissue. More often, however, the relation between a tissue concentration and an effect is a correlation. If an observed relationship between tissue concentration and injury holds true in other experiments and is consistent with observations in the field, then the correlation becomes credible and useful. In some instances the relation cannot possibly be based on cause and effect. For example, in the classic toxicological example in which researchers related DDE residues in raptor eggs to eggshell thinning, the DDE that caused the eggshell thinning was in the female that laid the egg. The DDE in the egg could not have caused the thinning. The important point is that the relation was found to be consistently reliable and was based on well-designed studies conducted under both controlled and field conditions. Because the DDE in the egg was correlated at some level to DDE in the adult, the concentration in egg became a useful surrogate. In some cases such as these where the mechanism is known, ancillary correlations may be useful as surrogate measures for the actual biologically effective dose at the receptor.

The more evidence collected under variant conditions, the more credible the argument. Critical concentrations are least reliable when based on few data, when they are applied to species that are not closely related, and when the timing or route of exposure is different from the conditions in the study used as a reference. Whenever animals are dosed under experimental conditions, concentrations in many organs will increase and be correlated with each other and with effects, but most of those correlations will fail to be robust. Critical concentrations may be derived from field studies, but will be in error if an observed effect is incorrectly attributed to the contaminant or if the animals were subjected to additional stressors, lowering their sensitivity. Extrapolating from tissue concentrations to effects on populations or ecosystems is especially tenuous. To establish a credible relation between a contaminant and a population requires extensive work on several populations, as described by Ohlendorf and Heinz in Chapter 21 on selenium in this book.

Tissue concentrations of some contaminants are especially challenging to interpret. Concentrations of polycyclic aromatic hydrocarbons (PAHs), for example, are difficult to interpret in higher animals because they tend to be rapidly metabolized and excreted (Eisler 2000). However, a recent study correlated the administered dietary dose of PAHs with biliary metabolites in fish (Meador et al. 2008). Even though the biliary metabolites are not tissue concentrations, these values do represent an internal dose that can be correlated to toxic effects and measured in field collected animals in a similar fashion to assess harm. Further, elements that are homeostatically regulated in an organism pose another difficulty. Sometimes a target organ, however, may be identified that does show a sharp increase in tissue concentrations as toxicity is approached, even though concentrations are still regulated in most tissues. For some other elements, such as mercury, the total concentration of the element may be misleading, because the element's toxicity is so dependent on its chemical form.

The large number of poorly studied manufactured and natural chemicals is daunting. These industrial compounds, elements, pharmaceuticals, personal care products, pesticides, and others, are often best considered as chemical classes because of their overwhelming numbers. As shown for many toxicants, grouping chemicals by class and mode of action often results in similar toxicity metrics among several species and higher taxa, which is immensely helpful in our quest to characterize toxicity and quantify the concentrations likely to result in adverse responses. The authors of the book chapters adeptly address the challenges. With patience, the relations between tissue concentrations and toxicity are becoming better understood and their use in ecotoxicology gradually refined.

The study of tissue concentrations rests entirely on the validity of the chemical analyses supporting them. In general, the ability of today's analytical chemists to provide reliable analyses of most important environmental contaminants surpasses the ability of ecotoxicologists to interpret those concentrations. There is a perception that some poisons leave no traces, especially among mystery readers. Consider, for instance: *"I am assured that there are many poisons known only to a few chemists in the world, a single grain of which is sufficient to destroy the strongest man and leave not the slightest trace behind. If the poisoner be sufficiently accomplished he can pursue his calling without the faintest risk of detection." Mr. Sabin sipped his wine thoughtfully* (from E. Phillips Oppenheim, 1903, *The Yellow Crayon*). Now, however, concentrations of almost all important contaminants or their metabolites may be detected in wine and in tissues, and, most importantly, they may be interpreted. Although uncovering the relation between concentration and effect requires considerable research and careful interpretation, the results are worthy, as the chapters of this book prove.

We are excited to present this second edition. Many of the chapters in this book address chemical classes that were explored in the first edition, which the authors have painstakingly updated with current data and, in some cases, with new ways of analyzing those data. We are also fortunate to have chapters that address tissue concentrations of some toxicants that have not been considered previously. Lastly, our second edition is illustrated with eighteenth-century engravings of fish, wildlife, and invertebrates, to remind us of what ecotoxicology is about. They are taken from *The Royal Natural History*, edited by British naturalist Richard Lydekker, and published in six volumes by Frederick Warne, 1893–1894.

ACKNOWLEDGMENTS

We thank Randy Brehm, Jennifer Smith, David Fausel, and Soniya Ashok of Taylor & Francis for their expert help in publishing this book. Mary Beyer of the Patuxent Wildlife Research Center provided editorial assistance and Jaime Rodríguez Estival of the Instituto de Investigación en Recursos Cinegéticos, Ciudad Real, Spain, prepared the figures of the engravings.

LITERATURE CITED

Aldenberg, T., and W. Slob. 1993. Confidence limits for hazardous concentrations based on logistically distributed NOEC toxicity data. *Ecotoxicol. Environ. Saf.* 25:48–63.

Barron, M. G., J. A. Hansen, and J. Lipton. 2002. Association between contaminant tissue residues and effects in aquatic organisms. *Rev. Environ. Contam. Toxicol.* 173:1–37.

Beckvar, N., T. Dillon, and L. Reed. 2005. Approaches for linking whole-body fish tissue residues of mercury or DDT to biological effects thresholds. *Environ. Toxicol. Chem.* 8:2094–2105.

Buekers, J., E. Steen Redeker, and E. Smolders. 2009. Lead toxicity to wildlife: derivation of a critical blood concentration for wildlife monitoring based on literature data. *Sci. Total Environ.* 407:3431–3438.

Di Toro, D. M., J. A. McGrath, and D. J. Hansen. 2000. Technical basis for narcotic chemicals and polycyclic aromatic hydrocarbon criteria I. water and tissue. *Environ. Toxicol. Chem.* 19:1951–1970.

Eisler, R. 2000. Polycyclic aromatic hydrocarbons. In *Handbook of chemical risk assessment, Vol. 2, organics,* pp. 1343–1411, chapter 25. Boca Raton, FL: Lewis Publishers.

Ferguson, J. 1939. The use of chemical potentials as indices of toxicity. *Proc. Roy. Soc. London* 127B:387–404.

Keith, J. O. 1996. Residue analyses: how they were used to assess the hazards of contaminants to wildlife. In *Environmental contaminants in wildlife: Interpreting tissue concentrations,* eds. W. N. Beyer, G. H. Heinz, and A. W. Redmon-Norwood, pp. 1–47. Boca Raton, FL: CRC Press.

Kukkonen, J. V. K. 2002. Lethal body residue of chlorophenols and mixtures of chlorophenols in benthic organisms. *Arch. Environ. Contam. Toxicol.* 43:214–220.

Lipnick, R. L. 1995. Structure-activity relationships. In *Fundamentals of aquatic toxicology II: Effects, environmental fate, and risk assessment,* ed. G. M. Rand, pp. 609–665, chapter 20. Bristol, PA: Taylor and Francis.

McCarty, L. S. 1991. Toxicant body residues: implications for aquatic bioassays with some organic chemicals. In *Aquatic toxicology and risk assessment: Fourteenth volume STP,* eds. M. A. Mayes and M. G. Barron, pp. 183–192. Philadelphia, PA: American Society for Testing and Materials.

McCarty, L. S., and D. Mackay. 1993. Enhancing ecotoxicological modeling and assessment. *Environ. Sci. Technol.* 27:1719–1728.

McCarty, L. S., et al. A review of tissue residue dosimetry in environmental toxicology. *Integr. Environ. Assess. Manag.* In press.

Meador, J. P. 2006. Rationale and procedures for using the tissue-residue approach for toxicity assessment and determination of tissue, water, and sediment quality guidelines for aquatic organisms. *Hum. Ecol. Risk Asses.* 12:1018–1073.

Meador, J. P., J. Buzitis, and C. Bravo. 2008. Using fluorescent aromatic compounds (FACs) in bile from juvenile salmonids to determine exposure to polycyclic aromatic hydrocarbons. *Environ. Toxicol. Chem.* 27:845–853.

Meador, J. P. 2000. Predicting the fate and effects of tributyltin in marine systems. *Rev. Environ. Contam. Toxicol.* 166:1–48.

Meador, J. P., T. K. Collier, and J. E. Stein. 2002a. Use of tissue and sediment based threshold concentrations of polychlorinated biphenyls (PCBs) to protect juvenile salmonids listed under the U.S. Endangered Species Act. *Aquat. Conserv.: Mar. Freshwat. Ecosyst.* 12:493–516.

Meador, J. P., T. K. Collier, and J. E. Stein. 2002b. Determination of a tissue and sediment threshold for tributyltin to protect prey species for juvenile salmonids listed by the U.S. Endangered Species Act. *Aquat. Conserv.: Mar. Freshwat. Ecosyst.* 12:539–551.

Meador, J. P., L. S. McCarty, B. I. Escher, and W. J. Adams. 2008. The tissue-residue approach for toxicity assessment: concepts, issues, application, and recommendations. *J. Environ. Mon.* 10:1486–1498.

Mineau, P., and K. R. Tucker. 2002. Improving detection of pesticide poisoning in birds. *J. Wildlife Rehab.* 25(2):4–13.

Nordberg, G. F. 1976. Effects and dose-response relationships of toxic metals: a report from an international meeting. *Scand. J. Work Environ. Health* 2:37–43.

Oppenheim, E. P. 1903. *The yellow crayon.* New York: Mckinlay, Stone and Mackenzie.

Reynolds, K., M. S. Schwarz, C. A. McFarland, T. McBride, and B. Adair. 2006. Northern pocket gophers (*Thomomys talpoides*) as bioindicators of environmental metal contamination. *Environ. Toxicol. Chem.* 25:458–469.

Steevens, J. A., M. R. Reiss, and A. V. Pawlisz. 2005. A methodology for deriving tissue residue benchmarks for aquatic biota: a case study for fish exposed to 2,3,7,8-tetrachlorodibenzo-*p*-dioxin and equivalents. *Integr. Environ. Assess. Manage.* 1:142–51.

Stickel, L. F. 1973. Pesticide residues in birds and mammals. In *Environmental pollution by pesticides*, ed. C. A. Edwards, pp. 254–312. London: Plenum Press.

Peregrine Falcon

By G. Mutzel, from *The Royal Natural History*, edited by Richard Lydekker, Frederick Warne & Co., London, 1893–94.

1 History of Wildlife Toxicology and the Interpretation of Contaminant Concentrations in Tissues

Barnett A. Rattner
Anton M. Scheuhammer
John E. Elliott

CONTENTS

1.1 Introduction 9
1.2 The Beginnings of Wildlife Toxicology 10
1.3 Synthetic Pesticides and Poisoning of Wildlife... 11
1.4 "Silent Spring" and Population Level Effects 12
1.5 Advances in Measurement Endpoints of Contaminant Exposure 14
1.6 Interpreting Exposure Using New Molecular and Modeling Techniques 16
1.7 Case History: Mercury in Wild Birds... 17
1.8 Case History: Dioxins and PCBs in Wild Birds... 21
Conclusion 27
References..28

1.1 INTRODUCTION

The detection and interpretation of contaminants in tissues of wildlife belongs to the field of toxicology, a scientific discipline with a long, intriguing, and illustrious history (reviewed by Hayes 1991, Gallo 2001, Gilbert and Hayes 2006, Wax 2006). We review its history briefly, to provide a context for understanding the use of tissue residues in toxicology, and to explain how their use has developed over time. Because so much work has been conducted on mercury, and dioxins and polychlorinated biphenyls (PCBs), separate case histories are included that describe the evolution of the use of tissue concentrations to assess exposure and effects of these two groups of contaminants in wildlife.

The roots of toxicology date back to early man, who used plant and animal extracts as poisons for hunting and warfare. The Ebers papyrus (Egypt ~1550 BC) contains formulations for hemlock, aconite (arrow poison), opium, and various metals used as poisons. Hippocrates (~400 BC) is sometimes credited with proposing the treatment of poisoning by decreasing absorption and using antidotes (Lane and Borzelleca 2007). Chanakya (350–283 BC), Indian advisor of the Maurya Emperor Chandragupta (340–293 BC), urged the use of food tasters as a precaution against poisoning, and the Roman emperor Claudius may have even been poisoned by his taster Halotus in 54 AD. Moses ben Maimon (1135–1204), author of a treatise on poisoning, noted that dairy products could delay absorption of some poisons. Paracelsus (1493–1541) shaped the field of toxicology with his corollaries that experimentation is essential to examining the response, that therapeutic properties should be distinguished from toxic properties, that chemicals have specific modes of action, and that the dose makes the poison. The art of concocting and using poisons reached its "zenith" during the Italian Renaissance, eventually culminating in its commercialization by Catherine Deshayes (a.k.a., La Voisine, 1640–1680) in France.

One of the first to suggest a chemical method for the detection of a poison in modern times was Herman Boerhaave (1668–1738), a physician and botanist, who, according to Jurgen Thorwald (*The Century of the Detective*), placed the suspected poison on red-hot coals, and tested for odors. The Spanish physician Orfila (1787–1853) served in the French court, and was the first toxicologist to systematically use autopsy and chemical analysis to prove poisoning. He has been credited with developing and refining techniques to detect arsenic poisoning. Other historic accounts include extraction of alkaloids from postmortem specimens (Jean Servais Stas ~1851) as evidence in a nicotine poisoning case (Levine 2003). The chemical analysis of organs and tissues became the basis for establishing poisoning. Much of the early history of toxicology addressed whether someone had been poisoned and how to treat poisoning.

1.2   THE BEGINNINGS OF WILDLIFE TOXICOLOGY

Wildlife toxicology has generally dealt with environmental contamination and the unintentional poisoning of amphibians, reptiles, birds, and mammals (Rattner 2009). Concern over poisoning of wildlife began in the late nineteenth century, and initially focused more on identifying environmental problems than determining contaminant concentrations in tissues. Reports of pheasant (*Phasianus colchicus*) and waterfowl mortality related to ingestion of spent lead shot appeared in the popular literature (Calvert 1876, Grinell 1894). Once recognized, it was considered a common occurrence in waterfowl (Phillips and Lincoln 1930). Wetmore (1919) described postmortem signs of intoxication in waterfowl that contained shot in the gizzard and other portions of the digestive tract. Poisoning of waterfowl from lead mining wastes dumped into the Spring River in Kansas was described in 1923 (Phillips and Lincoln 1930). A report of arsenic-related mortality of fallow deer (*Dama dama*) near factories processing metal ores in Freiberg, Germany, made its way into the popular press in 1887 (Newman 1979). Controlled exposure studies with mercury, strychnine, and arsenic were conducted in domestic and wild fowl (Gallagher 1918, Whitehead 1934), including measurement of arsenic in tissues of dosed chickens (*Gallus gallus*) (Whitehead 1934). Alkali poisoning of thousands of eared grebes (*Podiceps nigricollis*) and shovelers (*Anas clypeata*) was documented in California in 1891 (Fisher 1893). Similar cases were subsequently described in many locations in the western United States, and alkali poisoning was even experimentally duplicated by Wetmore using captive birds, which were administered chlorides of calcium and magnesium (Phillips and Lincoln 1930). The hazard of ingested phosphorus from military munitions by waterfowl and swans was first recognized in 1923, and emerged as a problem on several occasions decades later (Phillips and Lincoln 1930). With the expansion of oil production and its use for marine propulsion after World War I, oiling of waterbirds and numerous die-offs occurred along the coast of the United States (over 35 incidents documented in Phillips and Lincoln 1930). In the aforementioned mortality incidents (i.e., lead shot, arsenic, alkali, phosphorus, and oil), the source and the presence of the toxicant were usually readily apparent (e.g., recovery of ingested shot, alkali salts or oil in or on birds).

Qualitative and quantitative determination of presence of lead in stomach and caeca of waterfowl was described as early as 1919 (Wetmore 1919, Magath 1931). Traces of arsenic were reported in the liver of dead deer following application of calcium arsenate for forest insect control in 1926 (Danckwortt and Pfau cited by Keith 1996). In the detailed description of poisoning of nontarget wildlife by thallium baits used to control ground squirrels, Linsdale (1931) mentioned the use of qualitative spectroscopic methods to detect thallium in tissues of dead geese, and quantitative methods to determine concentrations in edible tissues. Before 1940, the presence or actual concentrations principally served as evidence of exposure. Quantitative methods were used in analyzing lead-poisoned geese (*Branta canadensis*) in the early 1940s, and "some correlation" between the number of shot found in the gizzard with lead content of the liver and kidney (but not leg bone) was suggested (Adler 1944).

Environmental contaminant studies with captive waterfowl began at the Patuxent Research Refuge in the late 1940s. Through controlled exposure studies with captive birds, evidence of adverse effects (histopathological lesions, impaired reproduction, and lethality) on individuals began to be generated. In toxicity studies of white phosphorus used in military munitions, Coburn and coworkers (1950) found statistically significant changes in concentrations of elemental phosphorus in heart, liver, and kidney of mallards (*Anas platyrhynchos*) and black ducks (*Anas rubripes*) that were dosed with various quantities of white phosphorus. These data were then used to interpret phosphorus concentration in tissues of redhead ducks (*Aythya americana*) collected from northern Chesapeake Bay, which led to the conclusion that "it appears probable" that the birds had been killed from ingestion of elemental phosphorus (Coburn et al. 1950). In order to interpret quantities of lead that produce toxic signs in waterfowl, mallards were dosed with lead nitrate (soluble and readily absorbable form of lead) for several weeks (Coburn et al. 1951). Anemia, emaciation, and a number of pathological lesions were consistently noted, and lead concentration in bone and liver was 7 and 40 times greater than that found in control birds. The critical lead intake level was suggested to be between 6 and 8 mg/kg body weight/day, and it was stated that bone, liver, or soft tissues could be used to chemically verify lead poisoning in field samples.

1.3   SYNTHETIC PESTICIDES AND POISONING OF WILDLIFE

By the 1930s, a total of about 30 pesticides were in use in the United Kingdom, United States and elsewhere, including plant derivatives (e.g., pyrethrum and nicotine), inorganic compounds (e.g., calcium arsenate and lead), mercurial fungicides, and the synthetic weed killer dinitro-ortho cresol (Sheail 1985). Aerial application of pesticides became a common practice in the 1930s, and potential adverse effects of pesticides to wildlife were acknowledged at the Third North American Wildlife Conference (Strong 1938).

The discovery of the insecticidal properties of dichlorodiphenyltrichloroethane (DDT) in 1939, the development, production, and use of other organochlorine (e.g., hexachlorocyclohexane), organophosphorus pesticides (e.g., schradan), and rodenticides (e.g., Compound 1080), increased dramatically during and after World War II (Hayes 1991). Concerns about potential damaging effects of DDT on wildlife appeared in The Atlantic Monthly (Wigglesworth 1945), which prompted experimental studies. Field studies of DDT effects on wildlife were undertaken in Maryland, Pennsylvania, and Texas, and reduced numbers of some avian species and dead birds were noted at application concentrations of 4.4 and 5 pounds per acre (Hotchkiss and Pough 1946, George and Stickel 1949, Robbins and Stewart 1949). Parathion poisoning of geese attributed to spray drift was also reported at this time (Livingston 1952). In the United Kingdom, large numbers of wildlife poisonings (e.g., passerines, game birds, mammals) occurred in the early and mid-1950s related to the use of aldrin and dieldrin as seed dressings, and application of schradan for control of aphids (Sheail 1985). Many other organochlorine insecticides (e.g., chlordane, heptachlor, and toxaphene) came into use in agricultural and forest settings in the 1950s, and wildlife mortality was noted (Peterle 1991).

Reports of wildlife mortality from pesticide use were controversial, pitting scientists associated with agriculture and chemical companies against environmental scientists. Biologists relied on tissue analyses not just to understand environmental hazards associated with pesticides, but also to provide more definitive evidence of exposure and even adverse effect. This controversy also served as an impetus to conduct controlled studies, which often relied on tissue concentrations as a measurement endpoint.

Data on concentrations of organic pesticides in tissues of wildlife began to appear in the early 1950s. Analysis of liver tissue from dead and intoxicated pheasants collected at pesticide-treated orchards revealed elevated concentrations of DDT (up to 326 µg/g), while parathion was detected in only a few birds (up to 5 µg/g) (Barnett 1950). Chronic dietary DDT and parathion feeding trials in pheasants demonstrated that the kinetics of the two pesticides were quite different (Barnett 1950). Substantial quantities of DDT were detected in fat (up to 8104 µg/g) and liver (up to 94 µg/g), but usually only trace amounts of parathion were found in liver. In a songbird study evaluating DDT (applied at 3 pounds per acre), Mitchell et al. (1953) found that whole body DDT concentrations in dead nestling songbirds were variable (up to 77 µg/g) and tended to be greater in dead nestlings in the sprayed area compared to the reference site. However, the overall songbird population was not affected (Mitchell et al. 1953). During this period, aldrin, dieldrin, and heptachlor were detected in tissues of dead birds and mammals following their field application (Post 1952, Clawson and Baker 1959, Scott et al. 1959, Rosene 1965). As aptly pointed out by Keith (1996), during this era investigators documented pesticide exposure in tissues of dead birds, but were often hesitant to conclude that the cause of death was pesticide-related.

Acute and chronic exposure studies were conducted using captive game birds that described signs of intoxication, lethality, and accumulation of residues of organochlorine pesticides in tissues (Dahlen and Haugen 1954, DeWitt 1955, 1956, DeWitt et al. 1955). The hazard of toxic chemicals to wildlife was frequently investigated using a combined laboratory-field approach (viz., determining the tissue concentrations of the compound and/or metabolites present in intoxicated or dead wild animals, and then comparing those values to concentrations in experimentally dosed animals exhibiting toxicological signs or effects) (Peakall 1992, Keith 1996). This approach worked well for organochlorine contaminants that readily bioaccumulated in tissues and exerted their lethal effects through neurotoxic mechanisms. For example, dietary feeding studies with captive quail and pheasants demonstrated that the concentrations of DDT in breast muscle were related to the severity of intoxication, with 34 µg/g in adult bobwhite (*Colinus virginianus*) and 22 µg/g in adult pheasants being associated with death (DeWitt et al. 1955). Barker (1958) reported brain concentrations of DDT and DDE (dichlorodiphenyldichloroethane) in robins (*Turdus migratorius*) and other passerines that succumbed following DDT application for Dutch elm disease. Based on this field study it was concluded that "the brain, being a suspected site of action, was considered to be best as an indicator of toxicity," with greater than 60 µg/g indicative of death in robins (Barker 1958). Other investigators made similar conclusions on the toxic concentration of DDT in brain tissue (>50 µg/g) in several species of birds, and extended findings by considering the sum of metabolites (Bernard 1963, Wurster et al. 1965, Stickel et al. 1966). This approach was used for many organochlorine compounds, including chlordane, heptachlor, dieldrin, and Aroclor 1254 (DeWitt et al. 1960, Stickel et al. 1969, 1984, and reviewed by Hoffman et al. 1996, Peakall 1996, Wiemeyer 1996). To improve diagnostic capabilities for free-ranging wildlife, the effects of body condition, lipid reserves, cessation of feeding, cold, and other stressors on tissue distribution and mobilization of organochlorines were examined in both controlled exposure and field studies (Harvey 1967, Stickel et al. 1970, Van Velzen et al. 1972, Heinz and Johnson 1981).

1.4 "SILENT SPRING" AND POPULATION LEVEL EFFECTS

With the publication of Rachel Carson's *Silent Spring* (1962), issues such as adverse effects of pesticides on nontarget organisms, ecological imbalances, chemical persistence, pesticide resistance, and human safety were publicized and debated not only among scientists, but also in all sectors of society. Eventually, some pesticides and environmental issues were addressed through testimony before government entities and courtroom litigation. This environmental movement sparked new legislation (e.g., in the United States, Resource Conservation and Recovery Act in 1965, National Environmental

Policy Act in 1970, and Toxic Substances Control Act in 1976; reviewed by Fairbrother 2009) and the establishment of distinct governmental agencies to deal with environmental pollution (e.g., U.S. Environmental Protection Agency in 1970). Research programs related to pesticides and industrial chemicals expanded in North America, Europe, and elsewhere. In the United Kingdom, Monks Wood Experimental Station was established to investigate effects of chemicals on animals and their supporting habitat. Chemical screening programs were initiated to examine toxicity, repellency and potential hazard of chemicals to birds and mammals (Heath et al. 1972, Schafer et al. 1983), and long-term environmental contaminant monitoring programs were established.

Population declines observed in many species of fish-eating and raptorial birds were of great concern to biologists. Following the discovery of an increased frequency of broken peregrine falcon (*Falco peregrinus*) eggs in England, Moore and Ratcliffe (1962) and many other investigators (reviewed in Sheail 1985 and Keith 1996) detected organochlorine pesticide residues (e.g., DDE, dieldrin, and lindane) in eggs. In a classic paper, Ratcliffe (1967) reported that weights of raptor eggshells fell markedly and rapidly after DDT use was instituted, and Hickey and Anderson (1968) used correlation analysis to demonstrate that shell thickness was inversely related to the concentration of DDE in eggs. Controlled exposure studies followed that proved DDE caused eggshell thinning and impaired reproduction (Heath et al. 1969). Similar relationships have been demonstrated in a number of predatory avian species (Hickey and Anderson 1968, Blus et al. 1972, Blus 1996), although some species are considerably more sensitive (e.g., brown pelican, *Pelecanus occidentalis*) than others. Concerns over the effects of moisture loss related to incubation stage, particularly in addled eggs, resulted in the development of concentration correction factors (Stickel et al. 1973). Adverse effects of organochlorine pesticides were also described in wild mammals, most notably bats (reviewed in Clark and Shore 2001), and the first reports of organochlorine contaminants and mercury appeared in marine mammals in the 1960s (reviewed in O'Shea and Tanabe 2003). The use of tissue residues has evolved from merely explaining the cause of local wildlife die-offs to its use in the investigation of the status of wildlife populations, and in some cases the possible fate of species.

Advances in chemical analysis and instrumentation enhanced detection capabilities and revealed some unsuspected problems. In 1966, Swedish scientist Soren Jensen reported several unknown peaks in a gas chromatogram that interfered with the quantification of DDT in environmental samples (Jensen 1966). These unknown peaks were subsequently identified as PCBs, which raised the possibility that previously reported DDT and metabolite values may have been falsely elevated by these interfering peaks. It was quickly recognized that PCBs were present in biota on a global scale, with perhaps the highest concentrations in fish-eating birds (up to 14,000 µg/g) (Risebrough et al. 1968, Wasserman et al. 1979). Quantification of these complex mixtures was based in part on matching chlorinated biphenyl patterns to the commercial Aroclor formulations, classified on the percentage of chlorination of the biphenyl. In the environment, these complex mixtures changed substantially due to natural weathering and biological processes, and Aroclor pattern recognition techniques were used to quantify total PCB concentrations in free-ranging wildlife. Toxicity studies of PCBs were undertaken in birds and mammals. It was realized that poor reproduction in ranch mink (*Mustela vison*) was due to the presence of PCBs in their food source, Great Lakes coho salmon (*Oncorhynchus kistutch*) (Aulerich and Ringer 1977). In the 1970s, studies focused on commercial mixtures (e.g., Aroclor 1254, Clophen A60), and concentration thresholds associated with lethality and embryotoxicity (e.g., <10 µg/g in eggs; reviewed by Hoffman et al. 1996). Concentrations of PCBs in liver and whole bodies were found to be indicative of exposure, but of limited diagnostic value in explaining mortality events (i.e., extremely high concentrations of PCBs are necessary to evoke mortality in adult birds).

Studies in the 1960s indicating a possible link between contaminant exposure (mainly organochlorines) and reproductive dysfunction in Great Lakes colonial nesting waterbirds (Hickey et al. 1966, Keith 1966, Gilbertson 1974, 1975, Gilman et al. 1977) led to a long-term research and monitoring program (Peakall and Fox 1987) using herring gulls (*Larus argentatis*) as bioindicators.

This herring gull monitoring program still continues today as an integral part of a multifacetted bi-national program to evaluate Great Lakes ecosystem health (Shear et al. 2003).

The toxicity of metals during this era focused primarily on mercury and lead. By the late 1950s, waterfowl poisoning by ingestion of spent lead shot, and effects on populations were further characterized (Bellrose 1959). Studies with captive Canada geese (Cook and Trainer 1966) and mallards (Locke et al. 1966, Barrett and Karstad 1971) dosed with lead shot reported concentrations of lead in liver and blood, associated pathological lesions, signs of intoxication, and death. However, in a lead study by Longcore and coworkers (1974) ranges indicative of exposure were proposed (>3 µg/g wet weight in brain, 6–20 µg/g in liver or kidney, and 10 µg/g in blood), that when combined with necropsy findings (presence of lead fragments in digestive tract), case history and histopathological lesions, could be used to make a definitive diagnosis of lead poisoning. In time and with the acquisition of additional data, these ranges became a more formalized criteria, including categories of lead exposure (e.g., background), and levels of injury or effect (e.g., subclinical, clinical, and severe poisoning) (Friend 1985, Franson 1996, Pain 1996).

1.5 ADVANCES IN MEASUREMENT ENDPOINTS OF CONTAMINANT EXPOSURE

By the 1970s, restrictions were placed on the use of some organochlorine compounds, including DDT and PCBs, although to some scientists the decision on DDT was misguised (Roberts et al. 2010). Controlled exposure studies in wild birds and mammals began to focus on sublethal biochemical, physiological, and behavioral effects of organic compounds and metals.

Use of organophosphorus and carbamate pesticides for farm crops, mosquito abatement, and for control of forest insect pests (e.g., spruce budworm in Canada described as "The Thirty Years' War," Burnett 1999) increased dramatically. Although these anticholinesterase pesticides had short environmental half-lives, they were not without adverse effects to birds and other nontarget organisms (Mineau 1991, Kendall and Lacher 1994, Grue et al. 1997). Because these compounds are rapidly metabolized, laboratory studies focused on enzymatic indicators (cholinesterase and other esterases) in blood and tissues of exposed birds (Bunyan et al. 1968a, 1968b, Ludke et al. 1975). Detection of organophosphate poisoning in wildlife quickly evolved to include the combination of inhibition of cholinesterase activity in brain tissue (~50%) along with the presence of organophosphorus or carbamate parent compounds or metabolites in tissues or ingesta (Hill and Fleming 1982). Many direct poisoning cases, and intriguing incidents involving secondary poisoning, are described in the peer-reviewed literature (Henny et al. 1985, Mineau et al. 1999). Refinements over time included the development of extensive libraries of reference values for unexposed animals and cholinesterase reactivation assays (Fairbrother et al. 1991). The combination of reduced cholinesterase activity and detection of residues or metabolites in tissues for diagnosis of poisoning has remained steadfast (Hill 2003).

Although a longstanding problem, major petroleum spills resulting in large bird kills (e.g., *Torrey Canyon* in 1967, Union Oil drilling platform in 1969, *Arrow* tanker in 1970) heightened public awareness and concern. From both an historical (Phillips and Lincoln 1930) and modern day perspective, evidence of exposure of wildlife following major oil spills is usually apparent by visual inspection and petroleum odor of the integument (feathers or fur) of suspect animals. In the 1960s and 1970s, numerous controlled exposure studies were undertaken that focused on characterizing the effects of crude petroleum oil and refined petroleum products on wildlife (Holmes 1984, Jessup and Leighton 1996). Despite the development of analytical methods (e.g., Gay et al. 1980), tissue concentrations and related measures (e.g., total resolved hydrocarbons; presence of aromatic, high molecular weight hydrocarbons and odd-numbered hydrocarbons; and ratios of pristine to n-C17 and phytane to n-C18; Hall and Coon 1988) are only occasionally measured in wildlife following oil spill events. In time it became recognized that (1) the composition of crude and refined petroleum varies considerably, (2) the chemical and physical properties of petroleum change through weathering and volatilization following a spill, (3) ingested petroleum compounds are often rapidly metabolized, and (4) there are substantial differences in toxicity following external exposure and ingestion of various crude or refined petroleum products (Jessup and Leighton 1996, Albers 2003). Accordingly, tissue concentrations of aliphatic and aromatic components of petroleum oil that are associated with adverse effects have not been developed. Instead, measurements are used to document exposure, most commonly for purposes of natural resource damage assessments following a spill event. However, confirmation of the presence of petroleum oil on the integument does not necessarily indicate that oil was the cause of death (Jessup and Leighton 1996). In the 1980s, enzyme-linked immunosorbent assays for detection of oil, and detailed fingerprinting for matching oil on exposed animals with its source were developed, and are now commonly used (Peters et al. 2005).

By the end of the 1980s, the use of lead shot for hunting waterfowl and coots was banned in the United States, and restrictions were placed on the use of lead fishing tackle in the United Kingdom due to the unintentional poisoning of mute swans (*Cygnus olor*) (Pattee and Pain 2003). Investigations on effects of heavy metals (e.g., lead, cadmium, and zinc) at industrial, mining, and hazardous waste sites examined exposure and responses at the individual and population levels of biological organization. Selenium became a significant environmental issue in the early 1980s when dramatic effects, including death and embryonic deformity of birds, were observed at the Kesterson National Wildlife Refuge in California (Ohlendorf and Hothem 1995, Ohlendorf 2003). In response to findings of selenosis and waterbird death at the Kesterson Reservoir, numerous field and feeding studies of birds were undertaken to establish toxicity thresholds. Using various statistical models (logit, probit, Weibull functions), much emphasis was placed on determining the toxicity of selenium in bird eggs. The probability of teratogenesis in black-necked stilts (*Himantopus mexicanus*) increased when selenium concentrations exceeded 37 µg/g dry weight (i.e., EC10, estimate of concentration affecting 10% of the population), while the EC10 for teratogenesis in mallards and in American avocets (*Recurvirostra americana*) was estimated to be 23 and 74 µg/g, respectively (Skorupa 1998a, 1998b). The threshold for reduced egg hatchability, a more sensitive measure of selenosis, was estimated to be 6–7 µg/g in stilt eggs, but avocets were found to be considerably more tolerant with hatchability effects at 60 µg/g. There has been considerable debate on the selenium threshold concentration for impaired hatchability in waterfowl, with an EC10 ranging from 12.5 to 16 µg/g.

The development and use of biomarkers of contaminant exposure and adverse effect expanded dramatically in the 1990s (McCarthy and Shugart 1990, Huggett et al. 1992, Peakall and Shugart 1993). The impetus was multifold. Organic contaminant and elemental analysis of tissues was, and continues to be, highly quantitative but costly and time consuming. Some biochemical measurements were amenable to rapid screening of samples, and a few were rather specific for certain contaminants and linked to the mechanism of toxicity (e.g., δ-aminolevunic acid dehydratase inhibition and protoporphyrin accumulation in blood of lead-poisoned birds and mammals). In some instances, rapid metabolism does not permit detection of parent compounds or metabolites in tissues, and thus enzymatic and other biochemical assays are utilized in place of tradition analytical methods (e.g., neurotoxic esterase activity for organophosphorus-induced delayed neuropathy caused by tri-*o*-tolyl phosphate and leptophos; Ecobichon 1996). It was quickly recognized that other biochemical measurements (e.g., changes in plasma transaminase and lactate dehydrogenase activities) were sensitive generalized responses that were precursors or indicators of cellular damage, although such measurements lack toxicant specificity. In time, other biomarkers (cytochrome P450, metallothionein, heat stress proteins, DNA damage, and measures of oxidative stress) were utilized as indicators of exposure and/or adverse effects in wildlife, and several exhibited dose-response relationships. Endpoint measurements were characterized for sensitivity, specificity, variability, clarity of interpretation, validity, and applicability to field sampling (McCarthy and Shugart 1990, Huggett et al. 1992, Peakall and Shugart 1993). Although biochemical markers are of tremendous value in ecotoxicology, only a few have gained widespread acceptance for risk assessments and natural resource damage assessments. Often these endpoints are most valuable as ancillary measures used in combination with contaminant concentration and other endpoints.

1.6 INTERPRETING EXPOSURE USING NEW MOLECULAR AND MODELING TECHNIQUES

On a grand scale, high prevalence of embryonic deformity was observed in some populations of colonial nesting fish-eating birds in the Great Lakes (Great Lakes Embryo Mortality, Edema, and Deformities Syndrome; Gilbertson et al. 1991). Such epidemic-like events and catastrophes including the Chernobyl nuclear reactor meltdown, and the Exxon Valdez and Gulf War oil spills, greatly expanded ecotoxicological research worldwide. Exposure studies with captive wild birds and mammals, in parallel with modeling efforts, were used to estimate no adverse effect level (NOAEL) and lowest observable adverse effect level (LOAEL) for PCBs in diets and drinking water (e.g., Heaton et al. 1995, U.S. EPA 1995, Forsyth 2001). Perhaps more germane to this text, the tissue concentrations of PCBs that correspond to the dietary NOAEL and LOAEL were also estimated (Heaton et al. 1995, Forsyth 2001). The realization of extreme toxicity of dioxin and dioxin-like coplanar PCB congeners in laboratory rodents (Poland and Knutson 1982), chicken eggs, and cell culture systems (Safe 1984, 1990) led to measurement (Kubiak et al. 1989) and toxicity testing (reviewed by Hoffman et al. 1996) of individual congeners in wildlife. The use of mammalian toxic equivalency factors to estimate dioxin equivalents of coplanar PCB congeners was applied to wild bird eggs. Potency estimates for dioxin-like PCB congeners (toxic equivalents, TEQs) were subsequently compiled (reviewed in Hoffman et al. 1996), and along with dioxins and dibenzofurans were eventually formalized at a World Health Organization workshop in 1997 (Van den Berg et al. 1998).

The use of nondestructive and minimally or noninvasive sampling techniques became more common in the 1990s. The rationale arose from the desire to use samples that did not entail the sacrifice of animals for ethical or scientific reasons (species status as threatened or endangered) and the sampling of animals repeatedly at a site where only a few individuals were found (Fossi and Leonzio 1994). Much of the analysis of such samples has focused on biochemical endpoints. Concentrations of organochlorine pesticides and metabolites, PCBs, and metals (lead, mercury, cadmium, and vanadium) in blood, milk, feathers, hair, and excreta are often correlated with levels found in historically used tissues (e.g., liver and kidney), and thus critical concentration values associated with harm were developed for some contaminants in these matrices (Fossi and Leonzio 1994).

Pesticide hazards to migratory species were highlighted by the death of some 20,000 Swainson's hawks (*Buteo swainsoni*) from monocrotophos poisoning during their winter migration to Argentina (Hooper et al. 2003). Monitoring and forensic studies documented anticoagulant rodenticide exposure and secondary poisoning in raptors (e.g., Stone et al. 1999, 2003), and restrictions were placed on the use of some of these compounds (US EPA 2008). Investigation of wildlife die-offs at industrial and mining sites continued (e.g., Hill and Henry 1996, Henny 2003), and in some instances metal concentration thresholds in tissues associated with toxicity were established (e.g., vanadium, Rattner et al. 2006). Studies of forest birds exposed to the organic-arsenical pesticide MSMA (monosodium methanearsonate) used for suppression of the mountain pine beetle in British Columbia revealed a significant hazard to woodpeckers (*Picoides* spp.), and findings led to the removal of MSMA from the marketplace (Morrissey et al. 2007, Albert et al. 2008). With reports of feminization of alligators (*Alligator mississippiensis*) in Lake Apopka, Florida (Guillette et al. 1994) and the publication of *Our Stolen Future* (Colborn et al. 1996), laboratory and field investigations were launched that focused on endocrine-disruptive effects of pollutants on wildlife. Despite extensive research, widespread effects of pollutants on endocrine function of free-ranging wildlife have been difficult to demonstrate; however, effects on the gonadal subsystem of wild fish seem to be pronounced (Jobling et al. 1998). Ecotoxicological research and monitoring of amphibians greatly expanded in response to worldwide declines of their populations, and the realization that some pesticides might be responsible for limb and other structural deformities (Sparling et al. 2000).

Emerging contaminant issues in the twenty-first century have included the global detection of perfluoroalkyl surfactants in wildlife (Giesy and Kannon 2001), and the dramatic increase in concentrations of polybrominated diphenyl ether (PBDE) flame retardants in eggs and tissues of birds (Norstrom et al. 2002). The population crash of *Gyps* vultures in the Indian subcontinent resulted in a remarkable forensic investigation documenting secondary poisoning (renal failure) of vultures that fed on carcasses of cattle that had been treated with the nonsteroidal anti-inflammatory drug diclofenac (Oaks et al. 2004). Old World vultures were found to be quite sensitive to diclofenac (LD50 of 0.1 to 0.2 mg/kg body weight), with concentrations in kidney and liver of affected birds being <1 µg/g wet weight (Oaks et al. 2004, Shultz et al. 2004, Swan et al. 2006). This catastrophic event was the first time that a veterinary drug resulted in species endangerment. Notably, New World vultures do not seem to be sensitive to diclofenac (Rattner et al. 2008).

The use of stable isotopes to identify the environmental source of a metal is a recent development in the field of wildlife toxicology. In a study of California condors (*Gymnogyps californianus*), lead concentrations and stable isotope ratios demonstrated that the source of exposure was a combination of background environmental lead, and ingested spent lead ammunition that has a distinctly lower ^{207}Pb to ^{206}Pb ratio (Church et al. 2006).

New models have been developed to estimate tissue concentration and distribution of legacy and emerging contaminants in wildlife. For example, a toxicokinetic model in the developing herring gull embryo predicts lipid mass balance and distribution of PCBs between the embryo, and yolk and albumen compartment (Drouillard et al. 2003). The model predicts that greatest PCB concentrations in the embryo occur during pipping, or shortly thereafter, when yolk lipids have been completely absorbed into the embryo, which is consistent with empirical data. Retention and elimination half-lives have also been modeled for numerous PCB and PBDE congeners in juvenile and adult American kestrels (*Falco sparverius*) (Drouillard et al. 2001, 2007). A bioenergetic-based model for tree swallow (*Tachycineta bicolor*) nestlings has been used to quantitatively examine factors (weight-normalized food consumption) and processes (growth dilution) that influence PCB bioaccumulation (Nichols et al. 1995, 2004). Several kinetic models have been developed for mercury in birds, and one such model with a bioenergetics-based component has been used to predict blood mercury concentration as a function of food intake, food mercury content, body mass, and mercury absorption and elimination in common loon (*Gavia immer*) chicks (Karasov et al. 2007). Work has been initiated on physiological-based pharmacokinetic models for some chlorinated hydrocarbons, methylmercury, and anticoagulant rodenticides in wild birds. Such models permit calculation of tissue concentrations (internal dose) of contaminants for a variety of administered doses, and support interspecific extrapolations for risk assessments. Application of uncertainty factors in estimating toxicity reference values have become well-accepted, and are now used to estimate adverse effect concentrations for toxicant intake (e.g., milligrams per kilogram body weight per day), concentrations in media, and tissue-based toxicant concentrations (micrograms per day) (USACHPPM 2000). Using this approach, predicted no effect concentrations for perfluorooctane sulfonate (PFOS) in the diet (i.e., 0.013 mg PFOS/kg body weight/day) and in the liver, serum, and egg yolk (0.08 µg PFOS/g wet weight, 0.15 µg PFOS/mL, and 1 µg PFOS/mL, respectively) of a generic female top-level avian predatory species have been generated (Newsted et al. 2005). Statistical techniques are now being used to derive tissue concentrations associated with toxicological benchmarks. Buekers et al. (2009) have recently calculated the fifth percentile hazard concentration (HC5) of blood lead levels associated with a no observed effect concentration (NOEC) in bird and mammals. Theoretically, at blood lead concentrations below the HC5, 95% of all higher vertebrates will be protected. As these examples illustrate, tissue concentrations are being used to answer increasingly more complex questions. Although tissue concentrations are still used to examine the fate of a particular organism, they are also used to elucidate contaminant hazards to populations and to ecosystems.

1.7 CASE HISTORY: MERCURY IN WILD BIRDS

Mercury (Hg) exposure has long been considered a potentially serious threat to the health of both humans and wildlife. The ecotoxicological literature on Hg is substantial, having evolved over many years, and serves to illustrate some of the problems faced by ecotoxicologists when trying to evaluate injury based on tissue contaminant concentrations. The fate of Hg in the environment is complex. For example, Hg originates from both natural sources and industrial processes; it may be released into the environment in a number of different chemical forms; it may be chemically interconverted within the abiotic environment; and it may be metabolized by microorganisms to form either methyl Hg or inorganic Hg, which differ considerably in dietary absorption, tissue distribution, and toxicity in exposed wildlife. Furthermore, the chemical forms of Hg may be changed within some organs (e.g., demethylation by liver), further complicating interpretation of wildlife tissue concentrations. The principles learned from the literature on Hg are applicable to other less well-studied environmental contaminants.

Mercury first received attention as a toxicological issue for wildlife in the 1950s and 1960s when elevated Hg concentrations and poisonings were reported in a wide variety of seed-eating birds and small mammals, and their predators. In these cases, the ultimate source of Hg exposure in small granivores was the consumption of agricultural seeds (mainly wheat, barley, and oats) coated with alkyl Hg fungicide compounds (commonly methylmercury dicyandiamide). Predators, including raptorial birds and various carnivorous mammals, were in turn poisoned after feeding on Hg-poisoned prey. Although poisoning of wildlife from this Hg source occurred in a number of different countries, the most comprehensive report of the phenomenon is probably that of Borg et al. (1969) who described the Swedish experience and concluded that the extent of Hg poisoning was great enough to have caused population reductions of some affected species. Although Borg et al. (1969) did not indicate specific threshold tissue concentrations for assessing Hg poisoning in wildlife, they were among the first to suggest that tissue-Hg (or alkyl Hg) concentrations, together with supporting evidence such as behavioral signs and/or characteristic histopathology, were the primary criteria for diagnosing Hg poisoning in wildlife.

In the 1960s, Hg poisoning of scavenging and fish-eating birds in Japan was related to the industrial release of methyl Hg, the most notable effects occurring in Minamata Bay (Doi et al. 1984). Other major point-sources of Hg to the environment during the 1960s and 1970s were effluents from pulp mills and chloralkali plants. Aquatic wildlife, especially fish-eating species, sampled from environments affected by these industrial emissions, commonly demonstrated elevated tissue-Hg concentrations (e.g., Fimreite 1974). Occasionally, overt intoxication and mortality of fish-eating wildlife (e.g., wild mink, Wobeser and Swift 1976; and wild otter, *Lutra canadensis,* Wren 1985) were attributed to Hg exposure from such sources. In addition, reproductive impairment in wild fish-eating birds was linked to elevated Hg exposure in such environments (Fimreite 1974, Barr 1986). These early studies examined mainly gross toxicological endpoints such as overt neurotoxicity, reproductive failure, and outright mortality, but sometimes also included histopathological examination for lesions at the cellular level (Tejning 1967, Borg et al. 1969, 1970, Fimreite 1971, Fimreite and Karstad 1971, Aulerich et al. 1974, Heinz 1974, Pass et al. 1975, Heinz and Locke 1976, Wobeser et al. 1976, Finley and Stendell 1978, Finley et al. 1979, Heinz 1979).

A combination of field studies of methyl Hg-exposed animals, and controlled dosing studies using captive animals, helped elucidate tissue and dietary levels of methyl Hg that were associated with overt toxicity or reproductive impairment (reviewed by Wren et al. 1986, Eisler 1987, Scheuhammer 1987, and more recent reviews by Heinz 1996, Thompson 1996, Burger and Gochfeld 1997, Wolfe et al. 1998). In these studies, it was common to measure and report only total Hg concentrations in tissues, with the implicit assumption that because exposure was known to be primarily to methyl Hg, tissue levels of total Hg and methyl Hg would be essentially identical. This assumption was probably valid for most field and lab studies conducted in the 1960s and 1970s because dietary methyl Hg exposures in these studies tended to be high, and the duration of exposure was generally fairly brief, certainly not more than a few months. However, the assumption is not valid for scenarios involving chronic, lower-level dietary methyl Hg exposure. A review of Hg concentrations in liver and kidney tissue of wildlife that died from Hg poisoning during the methyl

Hg-treated grains era, indicated that total Hg concentrations >20 µg/g wet weight represent potentially lethal exposures. But in other unrelated studies, much higher concentrations of total Hg (in some cases, several hundred µg/g) in liver of apparently healthy wild animals were reported (e.g., ringed seal, *Pusa hispida* and bearded seal, *Erignathus barbatus,* Smith and Armstrong 1975; striped dolphins, *Stenella coeruleoalba,* Itano et al. 1984; polar bears, *Ursus maritimus,* Norstrom et al. 1986). An apparently healthy wandering albatross (*Diomedea exulans*) had liver concentrations >1000 µg Hg/g dry weight (Thompson and Furness 1989). How could such dramatically elevated Hg concentrations fail to be accompanied by signs of severe toxicity? When studied further, Hg in livers of these and other species was shown to contain variable proportions of organic (methyl) and inorganic forms, with a generally decreasing organic fraction as total Hg concentrations increased. After absorption from the diet, some methyl Hg is apparently demethylated in certain tissues in response to increasing methyl Hg accumulation. Thompson and Furness (1989) demonstrated this phenomenon in a number of seabird species, suggesting that long-lived species with relatively slow molt cycles might be slow to eliminate methyl Hg through new feather growth and that, therefore, demethylation of methyl Hg might be an important additional mechanism to reduce the body burden of toxic methyl Hg. More recent studies have addressed apparent species differences in demethylation efficiency among different wild avian species (e.g., Scheuhammer et al. 2008, Eagles-Smith et al. 2009). Taken together, these studies demonstrated that Hg in liver cannot be assumed to be present primarily as methyl Hg, even though wildlife are exposed primarily to dietary methyl Hg in fish and other prey. Inorganic Hg resulting from demethylation in liver is often found in close association with selenium (Se), especially at higher Hg concentrations (e.g., Koeman et al. 1975, Thompson and Furness 1989, Dietz et al. 1990, Scheuhammer et al. 1998a). Further discussion of the biological Hg–Se relationship may be found in accompanying chapters by O'Hara et al. (Chapter 10 of this volume) and Shore et al. (Chapter 18 of this volume).

A major lesson for wildlife toxicologists from the published literature on apparent demethylation of methyl Hg and accumulation of relatively nontoxic Hg–Se complexes is that toxicological assessments should not rely solely on total Hg concentration measurements in typically analyzed tissues such as liver. This is especially true for long-lived piscivores and other aquatic predators for which years of chronic low-level dietary methyl Hg exposure may be occurring. In such cases, a high proportion of liver Hg may be present as inorganic Hg bound with Se. Scheuhammer et al. (1998a) suggested that total Hg, organic (methyl) Hg, and Se should be analyzed rather than total Hg alone, when using liver, kidney, and/or brain tissue for toxicological assessments. Wiener et al. (2003) recommended that, when only total Hg measurements were available, Hg in skeletal muscle should be analyzed in addition to liver, as almost all of the Hg in muscle remains methylated. Total Hg concentrations in liver are not by themselves sufficiently informative to make confident toxicological judgments.

By the 1980s, the use of mercurial seed dressings had been abandoned, and releases of Hg from the chloralkali and pulp industries had been eliminated or drastically curtailed. At least in North America, some other sources of previously significant environmental Hg releases (e.g., its use in gold mining) had already been phased out by the early 1900s (Eisler 2000). However, some of these older sources of environmental Hg contamination can still cause substantial exposure in wildlife today. For example, waterbirds nesting in the Carson River basin, contaminated with Hg from gold refining operations during the late 1800s, continue to be exposed to substantially elevated dietary methyl Hg concentrations that are of toxicological concern, especially with respect to egg hatchability and health of young (Henny et al. 2002, Hill et al. 2008). Similarly, predatory fish and fish-eating wildlife such as bald eagles (*Haliaeetus leucocephalus*) continue to experience elevated Hg exposure near a former Hg mine in central British Columbia, Canada (Weech et al. 2004, 2006). However, in addition to locations experiencing continued Hg contamination from past point-source emissions, a growing recognition evolved during the 1980s that environments remote from such releases could also contain fish (and consequently fish-eating wildlife) with elevated Hg concentrations. Predominant among such remote "Hg-sensitive" environments were acid-impacted lakes

(Björklund et al. 1984, Scheuhammer 1991, Spry and Wiener 1991, Scheuhammer and Blancher 1994), and reservoirs and lakes created by flooding of vegetated land where environmental Hg methylation and food chain transfer of Hg are enhanced (Bodaly et al. 1984, Johnson et al. 1991, Hall et al. 2005). Furthermore, in some very remote environments, temporal investigations indicated that Hg levels in wildlife were increasing near the end of the twentieth century (Monteiro and Furness 1997, Braune et al. 2005), whereas levels were declining in some more industrialized areas (Koster et al. 1996).

In response to the recognition of elevated Hg concentrations in food chains of certain remote ecosystems, studies of the effects of environmental Hg exposure in wildlife continued through the 1990s and beyond. A number of field studies sought to characterize exposure and effects in fish-eating wildlife in Hg-sensitive habitats, and to better understand the relation between elevated Hg in wildlife and their prey (Wren et al. 1987a, 1987b, Meyer et al. 1995, DesGranges et al. 1998, Evers et al. 1998, 2003, Scheuhammer et al. 1998b, 2001, Burgess et al. 2005, Champoux et al. 2006). In more recent toxicological studies, emphasis has been placed on clarifying the effects of environmentally realistic dietary methyl Hg exposures at the molecular (Spalding et al. 2000, Heath and Frederick 2005, Basu et al. 2005, 2006, 2007, 2009, Kenow et al. 2008, Scheuhammer et al. 2008), organismal (reproductive endpoints) (Wren et al. 1987a, 1987b, Heinz and Hoffman 1998, 2003, Dansereau et al. 1999, Albers et al. 2007), and population (Meyer et al. 1998, Moore et al. 1999, Sample and Suter 1999, Evers et al. 2005, 2008, Burgess and Meyer 2008) levels of organization in wild birds and mammals. In addition, species differences in methyl Hg toxicity *in ovo* have begun to be explicitly addressed through avian egg injection studies (Heinz et al. 2006, 2009). Contemporary reviews of the ecological impacts and toxicology of methyl Hg in wildlife have explicitly recognized that current levels of Hg exposure for some wildlife species in some environments are sufficiently high to be of toxicological concern (Wiener et al. 2003, Scheuhammer et al. 2007, Wolfe et al. 2007). In addition, tissue-Hg concentrations recognized to be harmful have gradually decreased as increasingly sensitive cellular and biochemical effects have been identified. For example, significant correlations between brain Hg concentrations and the density of some neurotransmitter receptors in mink have been observed well below the previously estimated lowest observable effect concentration (LOEC) for Hg in mink or otter brain (Scheuhammer et al. 2007). Shore et al. (2010), using species sensitivity distributions, have established egg-Hg concentrations that are protective of 95% (HC5) of avian species.

Although fish-eating wildlife generally exhibit substantially higher exposure to dietary methyl Hg than terrestrial animals, recent research has identified certain terrestrial food chains in forest habitats that appear to concentrate methyl Hg. Some forest songbird species feeding in such food chains can experience dietary methyl Hg exposure at least as high as fish-eating birds. For example, blood Hg concentrations exceeding 4 μg/mL were reported in red-eyed vireos (*Vireo olivaceus*) and Carolina wrens (*Thryothorus ludovicianus*) (Cristol et al. 2008); these levels exceed the estimated threshold for reproductive impairment in common loons (2.87 μg/mL in breeding females; Scheuhammer et al. 2007). Spiders (order Araneae), which had methyl Hg concentrations similar to fish preyed upon by belted kingfishers (*Megaceryle alcyon*), were found to be a major dietary source of methyl Hg for these terrestrial birds (Cristol et al. 2008). Additional studies are required to better understand the environmental conditions that lead to elevated methyl Hg concentrations in these terrestrial food webs, and to determine if reproductive or other impairments accompany elevated Hg exposure in the most at-risk wildlife species.

In summary, Hg in wildlife has been studied for more than 50 years, and much has been learned regarding its food chain transfer, accumulation, and toxic effects. Recent studies have begun to document subtle, yet important effects of Hg on behavior, neurochemistry, and endocrine function in wildlife at currently relevant levels of environmental exposure. Insofar as substantial global anthropogenic Hg emissions will likely continue into the foreseeable future, there will undoubtedly be a need for continued research on ecotoxicology of Hg, and a revisiting of tissue concentration effect thresholds, well into the twenty-first century.

1.8 CASE HISTORY: DIOXINS AND PCBs IN WILD BIRDS

PCBs, polychlorinated dibenzo-*p*-dioxins (PCDDs), and polychlorinated dibenzofurans (PCDFs) are structurally similar, persistent, and lipophilic chemicals, which have widely contaminated environmental media, where they have the potential to cause toxicological effects in wild birds. PCBs are anthropogenic in origin and were manufactured and widely used until the latter decades of the twentieth century. PCDDs and PCDFs were produced as by-products of industrial processes and combustion, especially of plastic wastes. The chemistry, environmental fate, and toxicology of these chemicals are complex, and hence controversial.

The word "dioxin" became known to the scientific community, and eventually part of the public lexicon, as a result of the death of millions of broiler chickens during the 1950s in parts of the eastern and southwestern United States. The condition was labeled "chick edema disease" as it was characterized by excessive fluid in the pericardial sac and abdominal cavity, subcutaneous edema, liver necrosis, and death beginning at about 3 weeks of age (Friedman et al. 1959). Investigators quickly traced the source to toxic factors present in fatty acid feed supplements obtained from "fleshing greases" produced as a by-product of the hide tanning industry (Wootton and Alexander 1959). Several years of toxicological and chemical research eventually implicated the use of chlorophenolic biocides as hide preservatives, and the identification of PCDDs, particularly 1,2,3,7,8,9-hexachlorodibenzo-*p*-dioxin as the main chick edema factor (Higginbotham et al. 1968, Firestone 1973). Verrett, Flick and coworkers dosed both chicks and chick embryos with individual PCDDs and PCDFs providing the first data of potential value for interpreting tissue concentrations (Verrett 1970, Flick et al. 1973).

Concerned over PCDD contaminants in chlorophenolic pesticides, some researchers began to investigate food chain contamination in areas of intensive use. During the 1960s, an estimated 400 kg of 2,3,7,8-tetrachlorodibenzo-*p*-dioxin (TCDD) was sprayed by the United States military onto the forests of Indochina as a contaminant in the 20 million kg of the herbicide 2,4,5-T, a component of Agent Orange, used as a chemical warfare agent (Huff and Wassom 1974). A 1970 survey of Vietnamese rivers found that whole body samples of catfish (Siluridae), for example, from the Dong Nai River had mean 2,3,7,8-TCDD concentrations of 810 pg/g wet weight (Baughman and Meselson 1973). There appears to have been no published attempt to extrapolate that finding to wildlife; however, assuming that the reported concentrations in fish were accurate within an order of magnitude, and using the biomagnification factor from fish to fish-eating birds for 2,3,7,8-TCDD of 32 (Braune and Norstrom 1989), aquatic birds feeding in that system would have accumulated sufficient TCDD alone to cause overt toxicity, even in less sensitive species.

While a number of laboratories were investigating environmental contamination by dioxins, in 1966 during gas chromatographic analysis for DDT, Jensen identified a series of PCB compounds (Jensen 1966, Jensen et al. 1969). Riseborough and coworkers (1968) soon reported that birds from the remotest areas of the globe were contaminated by PCBs. Studies of PCB toxicity to birds, particularly chickens, soon followed (Chapter 14 of this volume, reviewed by Eisler 1986, Bosveld and Van den Berg 1994, Barron et al. 1995). Compared to many of the organochlorine insecticides in wide use at that time, acute toxicity of PCBs was low and also varied according to the degree of chlorination of the Aroclor mixture (Hill et al. 1975). In cases of experimentally caused mortality, the brain was the most reliable diagnostic tissue for determining lethal concentrations of PCBs (e.g., brain: 300–400 µg/g wet weight in pheasants, Dahlgren et al. 1972; 76–445 µg/g in fish-eating birds, Koeman et al. 1973; and 310 µg/g in passerines, Stickel et al. 1984; liver: 70–697 µg/g in Bengalese finch, *Lonchura striata*, Prestt et al. 1970). A study of lethal effects on the great cormorant (*Phalacrocorax carbo*) reported a lower brain threshold, and attempted to determine if that species was more sensitive or whether results were confounded by furan contamination from the Clophen A60 dosing mixture (Koeman et al. 1973). Mortality and residue analyses of ring-billed gulls (*Larus dalawarensis*) in the early 1970s on the Great Lakes also implicated PCBs as a possible causative factor (Sileo et al. 1977).

During the late 1960s, scientists began to assess the effects of PCB mixtures on avian reproduction, principally focusing on chickens, but also including other galliform and nongalliform species. It soon became evident that, as with dioxins, chickens were more sensitive than other tested species. Some researchers employed the egg injection technique, and findings varied according to factors such as the Aroclor mixture, injection site, and dosing vehicle. For example, 5 μg/g of Aroclor 1242 injected into chicken eggs on day zero of incubation caused a variety of malformations in embryos, and other effects in hatched chicks (Carlson and Duby 1973). Peakall and Peakall (1973) conducted a feeding study with Aroclor 1254 in the ring dove (*Streptopelia risoria*), and showed that embryonic mortality increased with egg concentrations. In subsequent experiments with artificially incubated eggs, they found that the embryonic mortality was caused by altered parental behavior, specifically reduced nest attentiveness.

By the late 1970s, surveys of PCB contamination revealed, not surprisingly, that wildlife in heavily industrialized ecosystems, such as the North American Great Lakes and the Baltic region, were particularly contaminated, and thus those regions became foci for investigating the effects of environmental contaminants on birds (Gilman et al. 1977, Falandysz 1980) (Table 1.1). As early as 1970, colonies of gulls (*Larus* spp.) and terns (*Sterna* spp.) nesting on Lakes Ontario and Michigan, were exhibiting high rates of nest failure associated with embryotoxicity and various deformities among hatched birds (Gilbertson 1974, 1975). Mean PCB concentrations (as Aroclor 1254:1260) in herring gull eggs were 142 μg/g wet weight at a colony in Lake Ontario and 92 μg/g at a Lake Michigan colony (Gilman et al. 1977). There were extensive field and laboratory investigations of the Great Lakes avifauna; however, establishing cause–effect linkages, and thus critical concentrations of specific compounds proved problematic. During the period when signs of toxicity were overt, fish-eating bird eggs contained elevated concentrations of a complex mixture of halogenated aromatic contaminants in addition to PCBs, including DDTs, mirex, hexachlorobenzene, and TCDD (Peakall and Fox 1987).

The early research and monitoring of wildlife contamination by PCBs and dioxins was hindered by limitations in analytical chemistry. PCB quantification was based on one or two peaks resolved by packed column gas chromatography (GC), ineffective at separating most individual congeners. With the introduction and widespread use of fused-silica capillary GC columns, greater resolution of compounds was possible, but identification of many peaks remained problematic. In the early 1980s, Mullin et al. (1984) reported the synthesis and relative retention times of all 209 PCB congeners, which allowed researchers to comprehensively assess the patterns of PCB congeners present in various environmental media. By comparing patterns of congener peaks among sediment, forage fish, and birds, Norstrom (1988) showed which congeners were more resistant to metabolic degradation and therefore, tended to bioaccumulate, and he suggested some general structural properties governing bioaccumulation in birds.

Given the findings of widespread embryotoxicity, including deformities, during the late 1960s and early 1970s, the presence of 2,3,7,8-TCDD in the Great Lakes food chain had been hypothesized, but could not be established with analytical methods employed at that time (Bowes et al. 1973). By the early 1980s, the availability of high resolution mass spectrometry (MS) combined with GC/MS enabled the quantification of PCDDs and PCDFs in tissue samples at <10 pg/g. A new GC/MS analytical method was developed and applied to a spatial survey of the Great Lakes, and to a temporal survey made possible by retrospective analysis of herring gull egg samples archived in the Canadian Wildlife Service National Specimen Bank (Elliott et al. 1988). The results showed that eggs from a colony in Lake Ontario contained mean concentrations of 2, 3, 7, 8-TCDD that were greater than 1000 pg/g in 1971, and which had decreased to about 100 pg/g by 1980 (Stalling et al. 1985). The GC/MS method enabled examination of PCDD and PCDF patterns in environmental media and biota, demonstrating that chlorine substitution at the 2, 3, 7, and 8 carbon positions conferred resistance to metabolic breakdown, indicating therefore that those compounds tended to bioaccumulate (Stalling et al. 1985).

TABLE 1.1
Some Representative Studies of PCBs and Dioxins in Wild Birds

Species	Location	Contaminants	Study Type	Reference
Western grebe, *Aechmophorus occidentalis*	British Columbia	PCBs, PCDDs, PCDFs, OCs	Monitoring	Elliott and Martin 1998
Black-footed albatross, *Phoebastria nigripes*	North Pacific	PCBs, OCs	Monitoring	Auman et al. 1996
Northern gannet, *Morus bassanus*	Eastern Canada	PCBs, DDE, OCs	Poor reproductive success	Elliott et al. 1988
Great cormorant, *Phalacrocorax carbo*	Netherlands, Japan	PCBs	Reports of mortality, monitoring	Koeman et al. 1973, Guruge and Tanabe 1994
Double-crested cormorant, *Phalacrocorax auritus*	Great Lakes	PCBs, OCs	Deformities, variable reproductive success	Larson et al. 1996, Powell et al. 1998, Custer et al. 1999
Black-crowned night heron, *Nycticorax nycticorax*	Eastern U.S.	PCBs, OCs, PCDDs, PCDFs	Monitoring study	Rattner et al. 1997, 2000, 2001
Great blue heron, *Ardea herodias*	British Columbia	PCDDs, PCDFs	Reproductive failure	Elliott et al. 1989, 2001a, Bellward et al. 1990
Wood duck, *Aix sponsa*	Arkansas	TCDD, TCDF	Poor reproductive success, egg injection	White and Seginak 1994, Augspurger et al. 2008
Herring gull, *Larus argentatus*	Great Lakes	PCBs, OCs	Poor reproductive success	Gilman et al. 1977
Glaucous gull, *Larus hyperboreus*	Norway	PCBs, OCs	Monitoring study	Bustnes et al. 2001
Common tern, *Sterna hirundo*	Great Lakes, Netherlands	PCBs, PCDDs, PCDFs	Deformities, poor reproductive success	Gilbertson et al. 1976, Bosveld et al. 1995
Forster's tern, *Sterna forsteri*	Great Lakes	PCBs, TCDD	Reproductive problems	Kubiak et al. 1989, Harris et al. 1993
Caspian tern, *Hydroprogne caspia*	Great Lakes	PCBs, OCs	Monitoring	Struger and Weseloh 1985
Atlantic puffin, *Fratercula arctica*	Great Britain	PCBs	Toxicological field experiment	Harris and Osborne 1981
White-tailed sea-eagle, *Haliaeetus albicilla*	Sweden	PCBs, PCDDs, PCDFs, OCs	Poor reproductive success	Helander et al. 2002
Bald eagle, *Haliaeetus leucocephalus*	United States, British Columbia	PCBs, OCs, PCDDs, PCDFs	Poor reproductive success	Wiemeyer et al. 1993, Elliott and Harris 2001
Osprey, *Pandion haliaetus*	Pacific Northwest, Wisconsin, Ontario	PCBs, PCDDs, PCDFs	Monitoring, industrial site assessment	Elliott et al. 2001a, Henny et al. 2009, Woodford et al. 1998, DeSolla and Martin 2009
American kestrel, *Falco sparverius*	Lab study	CB-126, CB-77, Aroclor mixture	Egg injection study, feeding study	Hoffman et al. 1998, Fernie et al. 2001, 2003

continued

TABLE 1.1 (continued)
Some Representative Studies of PCBs and Dioxins in Wild Birds

Species	Location	Contaminants	Study Type	Reference
Northern bobwhite, *Colinus virginianus*	Lab study	CB-126, CB-77	Egg injection study	Hoffman et al. 1998
Ring-necked pheasant, *Phasianus colchicus*	Lab study	2,3,7,8-TCDD	Feeding study, egg injection	Nosek et al. 1992, 1993
Great horned owl, *Bubo virginianus*	Kalamazoo River, Michigan	PCBs	Contaminated site assessment	Strause et al. 2007a, 2007b
Tree swallow, *Tachycineta bicolor*	New York, Massachusetts, Rhode Island	PCBs, TCDD	Contaminated site assessment	McCarty and Secord 1999a, 1999b, Custer et al. 2003, 2005
Eastern bluebird, *Sialia sialis*	Wisconsin, Michigan	TCDD	Lab study, contaminated site assessment	Thiel et al. 1988, Neigh et al. 2007
American robin, *Turdus migratorius*	Massachusetts	PCBs	Contaminated site assessment	Henning et al. 2003
Eurasian dipper, *Cinclus cinclus*	Wales	PCBs	Point source assessment	Ormerod et al. 2000
American dipper, *Cinclus mexicanus*	British Columbia coastal watershed	PCBs, OCs	Source determination	Morrissey et al. 2005
Starling, *Sturnus vulgaris*	Illinois	PCBs	Contaminated site assessment	Arenal et al. 2004

Availability of a full range of compounds for toxicity testing advanced understanding of the structure–activity relationships of the 17 various 2,3,7,8-substituted PCDDs and PCDFs and the structurally similar non-*ortho* and mono-*ortho* PCB congeners. That similarity in structure and effects furthered the theory that there was a common mechanism of action that hinged on the binding to the cytosolic aryl hydrocarbon (Ah) receptor protein, translocation into the nucleus and induction of gene transcription and corresponding proteins (Poland and Knutson 1982). Ranking of potencies for individual congeners relative to 2,3,7,8-TCCD resulted in development of the toxic equivalence factor (TEF) scheme, whereby the toxicity of complex mixtures could be estimated by multiplying each congener concentration in a given sample by its TEF and summing the results of the multiple congeners to obtain the TCDD TEQ concentration of the sample (Safe 1984, 1990). An expert panel recommended avian-specific TEFs, now in wide usage (Van den Berg et al. 1998); however, recent experiments have reported that 2,3,7,8-TCDF may be more toxic than 2,3,7,8-TCDD in some bird species, while 2,3,4,7,8-pentaCDF may be tenfold or more toxic to Japanese quail (*Coturnix japonica*), than TCDD, requiring a reassessment of avian TEFs (Cohen-Barnhouse et al. 2008).

Brunström and coworkers (Brunström 1988, 1990, Brunström and Andersson 1988, Brunström and Lund 1988) conducted a series of egg injection experiments using chickens and other avian species. Those and other studies provided avian-specific data on the relative potencies of various PCB congeners, and further demonstrated that the chicken was in a class of its own in relative sensitivity to dioxin-like compounds, while the pheasant and the turkey (*Melleagris gallopavo*) were intermediate in sensitivity, with other species such as ducks and gulls being much less sensitive. Meanwhile, Nosek et al. (1992, 1993) studied the toxicology of TCDD in more depth using the pheasant as a model species.

Congener-specific analytical techniques were employed in field and laboratory investigations of ongoing health problems including poor reproductive success of bird populations in the Great Lakes. Given the similarity between the apparent syndrome in that region and chick edema disease,

Gilbertson and coworkers (1983, 1991) developed the concept of the Great Lakes Embryo Mortality, Edema and Deformities Syndrome (GLEMEDS). By the late 1980s, the non-*ortho* PCBs were suggested as the likely causative factor, because of their wide distribution, greater environmental concentrations, and dioxin-like toxicity. Kubiak et al. (1989) investigated a Forster's tern (*Sterna forsteri*) colony in the Green Bay region of Lake Michigan that was exhibiting what was later considered GLEMEDS-type signs of toxicity. Their egg-swap studies in particular pointed toward a parental behavioral mechanism to explain lower productivity, rather than embryotoxicity. A subsequent study of those birds suggested NOAELs for PCBs and for hatching success (Harris et al. 1993). Studies of common terns (*Sterna hirundo*) in North America (e.g., Hoffman et al. 1993), and in Europe (Becker et al. 1993, Bosveld et al. 2000), reported some sublethal effects on chick growth and development, but no clear evidence of PCB effects on hatching success. Reproduction and contaminants, particularly elevated PCB concentrations were investigated by Struger and Weseloh (1985) in Great Lakes Caspian terns (*Sterna caspia*), and despite relatively high concentrations of PCBs, there were no apparent effects on productivity.

Of particular concern to many researchers were the continued reports during the early 1990s of deformed nestlings in the Great Lakes, considered by some to be a clear diagnostic of poor fitness in wildlife. Clustered incidences of bill deformities among double-crested cormorants (*Phalacrocorax auritus*) nesting at Lake Michigan colonies were regularly reported. Ludwig et al. (1996) and Giesy et al. (1994) considered that the correlative evidence was sufficient to implicate PCBs as a chemical driver of deformities in Great Lakes cormorants. It has, however, proven difficult to conclusively establish cause and effect between the observed deformities and PCB concentrations in the field, given inconsistency in laboratory results, and potential confounding factors such as disease and genetics, which are discussed more thoroughly in Chapter 14.

In the mid-1980s, eggs of aquatic birds from the Pacific coast of Canada were found with high concentrations and an unusual pattern of PCDDs and PCDFs (Elliott et al. 1989). Work with great blue herons (*Ardea herodias*) explored possible links between colony failures and increasing PCDD and PCDF contamination from forest industry sources. Throughout the 1990s, field work was conducted on a variety of potentially vulnerable species, including herons, cormorants, bald eagles, osprey (*Pandion haliaetus*), tree swallows, and various waterfowl species, which described spatial and temporal patterns in contamination, and successfully established linkages with specific forest industry sources (Elliott and Martin 1994, Elliott et al. 1996a, 2001a, Harris and Elliott 2000, Harris et al. 2003). A complementary series of laboratory studies employing artificial incubation and egg injection explored the toxicological aspects in more depth, and generated data useful for recommending criteria for interpreting tissue concentrations of PCDDs in a number of avian species (Bellward et al. 1990, Sanderson et al. 1994a, 1994b, Sanderson and Bellward 1995, Elliott et al. 1996b, 2001b, Janz and Bellward 1996). The contamination and potential effects of PCDDs and PCDFs from the pulp and paper industry were also studied in fish-eating and insectivorous birds from other locations in North America (Champoux 1996, Wayland et al. 1998, Woodford et al. 1998, Custer et al. 2002).

Given its conservation status until the mid-1990s as federally endangered in the United States and in some Canadian provinces, and its position as a top predator, the role of contaminants in the decline of the bald eagle was widely investigated. Nests in many regions of North America were visited regularly to document reproduction and to salvage unhatched eggs for contaminant analysis. Statistically significant negative associations were found between productivity and various contaminants, including PCBs, while DDE effects on shell quality was identified as the main determinant (Wiemeyer et al. 1993). In the Great Lakes region, Best et al. (2010) reported that associations continued between PCBs and productivity into at least the late 1990s. To improve the quantity of data obtained, the salvaged egg metric was enhanced by measuring contaminant burdens in blood samples from nestling bald eagles (Bowerman et al. 1995, 1998), an approach also applied in Sweden to the white-tailed sea-eagle (*Haliaeetus albicilla*). Meanwhile, improved analytical techniques also made it possible to measure PCDDs, PCDFs, and non-*ortho* PCBs in eagle nestlings (Elliott and Norstrom 1998). Tentative critical concentrations for PCBs and DDE in nestling eagle blood samples were derived (Elliott and Norstrom 1998) based on regressions between nestling blood and egg samples, which were later modified by Elliott and Harris (2001) and Strause et al. (2007a). As exposure to legacy contaminants such as PCBs was declining in many jurisdictions during the 1990s, some researchers began to directly investigate ecological factors, particularly the relative role of food supply and weather that may interact with contaminant exposure and effects (Elliott et al. 1998, 2005, Dykstra et al. 1998, Gill and Elliott 2003, Hoff et al. 2004, Elliott et al. 2005). Similarly, Helander et al. (2002) studied the ecotoxicology of the congeneric white-tailed sea-eagle for many decades in Sweden, in the process developing critical tissue values for PCBs in eggs.

Field research on the effects of PCBs and other contaminants on productivity and other parameters of birds nesting in the Great Lakes continued through the 1990s and into the present century (e.g., Tillitt et al. 1992, Giesy et al. 1995, Fox et al. 1998, Ryckman et al. 1998, Custer et al. 1999). As with bald eagles, the role of ecological variables was increasingly factored into understanding sources and dynamics of contaminants (Hebert and Weseloh 2006), eventually incorporating tools such as stable isotopes and fatty acid profiles (Hebert et al. 2008). Drouillard and Norstrom (2000) made valuable progress in understanding the pharmokinetics of PCBs in birds, and applied those collective advances to develop a bioenergetics model for contaminant dynamics in wild birds (Norstrom et al. 2007). The ongoing concerns about contamination of birds in the Great Lakes and elsewhere, led to more comparative egg injection studies with TCDD and non-*ortho* PCBs. These studies furthered the understanding of species variation in sensitivity and provided critical egg concentrations for the double-crested cormorant (Powell et al. 1998), and common tern and American kestrel (Hoffman et al. 1998). Several investigators (Fernie et al. 2001, 2003, Fisher et al. 2001, Smits et al. 2002) carried out a feeding study of an Aroclor mixture with the American kestrel as a laboratory model of predatory and fish-eating birds, which has yielded valuable data on a wide range of reproductive and physiological endpoints.

Currently, widespread restrictions on use of PCBs and the need to regulate dioxin releases have been in place for at least 30 years. There remain, however, numerous point sources of those compounds, associated primarily with waste dumps, and soil and sediment contamination at former manufacturing and storage sites. Birds have been used to determine the exposure and evaluate impacts to wildlife in Canada (Bishop et al. 1999, Harris and Elliott 2000, Ormerod et al. 2000, Kocan et al. 2001, Kuzyk et al. 2003, Jaspers et al. 2006), and particularly in the United States, where investigations of contamination of wild birds have been conducted as part of Natural Resource Damage Assessments. Researchers and risk assessors have looked principally at fish-eating birds and raptors (Williams et al. 1995, Hart et al. 2003, Strause et al. 2007a, 2007b), and increasingly at cavity-nesting passerines (Custer et al. 1998, McCarty and Secord 1999a, 1999b, Arenal et al. 2004). Custer and colleagues (1998, 1999, 2002, 2003, 2005) in particular have made effective use of the tree swallow and provided data on various endpoints useful for determining critical tissue concentrations.

Outside of specific hotspot areas, long-term monitoring of PCBs and other contaminants in avian indicator species has continued in some regions, such as the Great Lakes (Norstrom and Hebert 2006), and other North American aquatic environments (Rattner et al. 2004, Toschik et al. 2005, Henny et al. 2009), various marine systems including the Arctic (Barrett et al. 1996, Braune and Simon 2003), the Pacific, and Atlantic coasts of Canada (Elliott et al. 1992, 2001a, Harris et al. 2003), and the Baltic (Bignert et al. 1995). The Arctic has been a focus of ongoing study as biologists from Scandinavia, Canada, and Alaska have investigated the exposure and potential effects of PCBs and other persistent organic pollutants in high trophic-level marine birds, particularly the glaucous gull (*Larus hyperboreus*), a species, which often preys on other marine birds (Henriksen et al. 1998, Sagerup et al. 2000, 2002, Bustnes et al. 2001, Verreault et al. 2006a, 2007).

Monitoring of PCBs and dioxins in wildlife has been complemented by the use of biomarkers, often measured nondestructively in blood or by bioassay methods, to assess relationships between exposure and various endpoints such as hepatic cytochrome P450-associated monooxygenase activities (Fossi et al. 1986, Rattner et al. 1993, 1994, 1997, 2000, Sanderson et al. 1994a, 1994b, Davis et al. 1997, Custer et al. 1998, Feyk et al. 2000, Kennedy et al. 2003, Fox et al. 2007a), gene mutations (Stapleton et al. 2001), porphyrin metabolism (Fox et al. 1988, Kennedy et al. 1998), immune system responses (Grasman and Fox 2001, Grasman et al. 1996, Bustnes et al. 2004, Fox et al. 2007b), thyroid hormone levels (Smits et al. 2002, McNabb and Fox 2003, Saita et al. 2004), retinoids (Spear et al. 1990, Elliott et al. 1996b, 2001b, Kuzyk et al. 2003, Murvoll et al. 2006), sex steroids (Verreault et al. 2006b), stress hormones (Martinovic et al. 2003), disease (Hario et al. 2000), and behavior (McCarty and Secord 1999b). Further developments of analytical methodology have led to surveys of PCB and other organochlorine (OC) metabolites in wild birds and some examination of relations with biomarkers (Fangstrom et al. 2005, McKinney et al. 2006).

Most recently, advances in molecular biology have furthered understanding of the mechanisms of dioxin-like toxicity to birds and of the basis for variation in species sensitivity. The cytochrome P450 response of birds exposed to Ah receptor ligands has been shown to be unique, with birds having two distinct CYP1A isoforms (Gilday et al. 1996, Mahajan and Rifkind 1999). Kennedy et al. (1996) developed avian *in vitro* assays and showed that the magnitude of *in vitro* response to CYP1A induction may be predictive of species differences in embryotoxicity *in ovo*. Application of molecular techniques examined the interspecific variation of response to TCDD-like exposure, and showed that sensitivity is closely associated with differences in the molecular structure of the Ah receptor and to differences in preferential induction of CYP1A isoforms (Head 2006, Karchner et al. 2006, Head and Kennedy 2007, Yasui et al. 2007). Sensitivity to dioxin-like compounds among avian species varies according to amino acid differences in the Ah receptor ligand binding domain (Head et al. 2008). Consistent with previous toxicological data, chickens exhibit high sensitivity, while of particular interest some upland game birds, passerines, and an albatross exhibit moderate sensitivity. All other species tested to date, including raptors, waterbirds, and waterfowl, appear relatively insensitive to dioxin-like toxicity (Head et al. 2008).

In summary, beginning with the identification of the chick edema factor, the collaboration among biologist, chemists, and toxicologists over the past 50 years has successfully investigated many aspects of the exposure and toxicology of PCBs, PCDDs, and PCDFs in birds. Field studies have made correlative links between dioxin-like chemicals and alterations in the metabolic, endocrine, and immune functions of populations of avian top predators and aquatic insectivores. In some instances, reproductive success has been significantly affected, although it has often proved difficult to separate causal factors, including other contaminants and cumulative anthropogenic and natural stressors. The particular sensitivity of the chicken to these chemicals has now been linked to the structure of its Ah receptor. One or two changes in the amino acids of the receptor's binding domain causes greatly reduced sensitivity to dioxins. That likely explains in large part the findings that, despite widespread exposure to PCBs, dioxins, and furans at concentrations that would severely compromise reproduction of chickens, there is limited evidence of a significant impact on populations of wild birds. That contrasts to the population declines associated with DDT and dieldrin, and more recently the veterinarian pharmaceutical, diclofenac. Nevertheless, given the ongoing problems posed by numerous contaminated sites, and the global nature particularly of PCB contamination, we can expect these chemicals to remain an issue for sometime into the current century.

CONCLUSION

As pointed out in a recent review, the field of wildlife toxicology has been shaped by chemical use and misuse, ecological mishaps, and research in the allied field of human toxicology (Rattner 2009). The development and use of new chemicals, and unexpected and unpredicted contamination problems continue to drive this discipline. In some instances, environmental release of toxicants could have resulted in species extinction (e.g., bald eagle, sparrowhawk, *Accipiter nisus*, and California condor) had not regulatory and remedial actions been undertaken. Dramatic advances in analytical technology over the past 50 years now permit routine detection and measurement of minute quantities of chemicals in a myriad of matrices. However, our greatest challenge remains the extrapolation of exposure data from laboratory and field studies to effects in diverse species and free-ranging populations, which are often subject to multiple environmental and toxicological stressors.

REFERENCES

Adler, F. W. E. 1944. Chemical analysis of organs from lead-poisoned Canada geese. *J. Wildl. Manage.* 8:83–85.

Albers, P. H. 2003. Petroleum and individual polycyclic aromatic hydrocarbons. In *Handbook of ecotoxicology*, 2nd edition, eds. D. J. Hoffman, B. A. Rattner, G. A. Burton Jr., and J. Cairns Jr., pp. 341–371. Boca Raton, FL: Lewis Publishing Inc.

Albers, P. H., M. T. Koterba, R. Rossman, W. A. Link, J. B. French, R. S. Bennett, and W. C. Bauer. 2007. Effects of methylmercury on reproduction in American kestrels. *Environ. Toxicol. Chem.* 26:1856–1866.

Albert, C. A., T. D. Williams, W. R. Cullen, V. Lai, C. A. Morrissey, and J. E. Elliott. 2008. Tissue uptake, mortality and sub-lethal effects of monomethylarsonic acid (MMA (V)) in nestling Zebra Finches (*Taeniopygia guttata*). *J. Toxicol. Environ. Health A* 71:353–360.

Arenal, C. A., R. S. Halbrook, and M. J. Woodruff. 2004. European starling (*Sturnus vulgaris*): avian model and monitor of polychlorinated biphenyl contamination at a superfund site in southern Illinois, USA. *Environ. Toxicol. Chem.* 23:93–104.

Augspurger, T. P., D. E. Tillitt, S. J. Bursian, S. D. Fitzgerald, D. E. Hinton, and R. T. Di Giulio. 2008. Embryo toxicity of 2,3,7,8-tetrachlorodibenzo-*p*-dioxin to the wood duck (*Aix sponsa*). *Arch. Environ. Contam. Toxicol.* 55:659–669.

Aulerich, R. J., and R. K. Ringer. 1977. Current status of PCB toxicity to mink and effect on their reproduction. *Arch. Environ. Contam. Toxicol.* 6:279–292.

Aulerich, R. J., R. K. Ringer, and J. Iwamoto. 1974. Effects of dietary mercury on mink. *Arch. Environ. Contam. Toxicol.* 2:43–51.

Auman, H. J., J. P. Ludwig, C. L. Summer, D. A. Verbrugge, K. L. Froese, T. Colborn, and J. P. Giesy. 1996. PCBS, DDE, DDT, and TCDD-EQ in two species of albatross on Sand Island, Midway Atoll, North Pacific Ocean. *Environ. Toxicol. Chem.* 16:498–504.

Barker, R. J. 1958. Notes on some ecological effects of DDT sprayed on elms. *J. Wildl. Manage.* 22:269–274.

Barnett, D. C. 1950. The effect of some insecticide sprays on wildlife. Proceedings of the Thirtieth Annual Conference of Western Association of State Game and Fish Commissioners, pp. 125–134.

Barr, J. F. 1986. Population dynamics of the common loon (*Gavia immer*) associated with mercury-contaminated waters in north-western Ontario, Can. Wildl. Ser. Occasional Paper 56, Ottawa, Canada, 25 pp.

Barrett, M. W., and L. H. Karstad. 1971. A fluorescent erythrocyte test for lead poisoning in waterfowl. *J. Wildl. Manage.* 35:109–119.

Barrett, R. T., J. U. Skaare, and G. A. Gabrielsen. 1996. Recent changes in levels of persistent organochlorines and mercury in eggs of seabirds from the Barents Sea. *Environ. Pollut.* 92:13–18.

Barron, M. G., H. Galbraith, and D. Beltman. 1995. Comparative reproductive and developmental toxicology of PCBs in birds. *Comp. Biochem. Physiol. Part C Comp. Biochem. Toxicol.* 112:1–14.

Basu, N., K. Klenavic, M. Gamberg, M. O'Brien, R. D. Evans, A. M. Scheuhammer, and H. M. Chan. 2005. Effects of mercury on neurochemical receptor binding characteristics in wild mink. *Environ. Toxicol. Chem.* 24:1444–1450.

Basu, N., A. M. Scheuhammer, K. Rouvinen-Watt, N. Grochowina, R. D. Evans, M. O'Brien, and H. M. Chan. 2007. Decreased N-methyl-D-aspartic acid (NMDA) receptor levels are associated with mercury exposure in wild and captive mink. *Neurotoxicology* 28:587–593.

Basu, N., A. M. Scheuhammer, K. Rouvinen-Watt, N. Grochowina, K. Klenavic, R. D. Evans, and H. M. Chan. 2006. Methylmercury impairs components of the cholinergic system in captive mink (*Mustela vison*). *Toxicol. Sci.* 91:202–209.

Basu, N., A. M. Scheuhammer, C. Sonne, R. J. Letcher, E. W. Born, and R. Dietz. 2009. Is dietary mercury of neurotoxicological concern to wild polar bears (*Ursus maritimus*)? *Environ. Toxicol. Chem.* 28:133–140.

Baughman, R. W., and M. S. Meselson. 1973. An analytical method for detecting TCDD. *Environ. Health Perspect.* 5:27–35.

Becker, P. H., S. Schuhmanns, and C. Koepff. 1993. Hatching failure in common terns (*Sterna hirundo*) in relation to environmental chemicals. *Environ. Pollut.* 79:207–213.

Bellrose, F. C. 1959. Lead poisoning as a mortality factor in waterfowl populations. *Ill. Nat. History Survey Bull.* 27:235–288.

Bellward, G. D., et al. 1990. Comparison of dibenzodioxin (PCDD) and dibenzofuran (PCDF) levels with mixed-function oxidase induction in Great Blue Herons. *J. Toxicol. Environ. Health* 30:33–52.

Bernard, R. F. 1963. Studies of the effects of DDT on birds. *Michigan State University Biological Series* 2:155–192.

Best, D. A., et al. 2010. Productivity, embryo and eggshell characteristics and contaminants in bald eagles from the Great Lakes, USA, 1986–2000. *Environ. Toxicol. Chem.* 29:1581–1592.

Bignert, A., K. Litzen, T. Odsjo, M. Olsson, W. Persson, and L. Reutergårdh. 1995. Time-related factors influence the concentrations of sDDT, PCBs and shell parameters in eggs of Baltic guillemot (*Uria aalge*), 1861–1989. *Environ. Pollut.* 89:27–36.

Bishop, C. A., N. A. Mahony, S. Trudeau, and K. E. Pettit. 1999. Reproductive success and biochemical effects in tree swallows (*Tachycineta bicolor*) exposed to chlorinated hydrocarbon contaminants in wetlands of the Great Lakes and St. Lawrence River basin, USA and Canada. *Environ. Toxicol. Chem.* 18:263–271.

Björklund, I., H. Borg, and K. Johansson. 1984. Mercury in Swedish lakes—its regional distribution and causes. *Ambio* 13:118–121.

Blus, L. J. 1996. DDT, DDD, and DDE in birds. In *Environmental contaminants in wildlife: interpreting tissue concentrations,* eds. W. N. Beyer, G. H. Heinz, and A. W. Redmon-Norwood, pp. 49–71. Boca Raton, FL: Lewis Publishing Inc.

Blus, L. J., C. D. Gish, A. A. Belisle, and R. M. Prouty. 1972. Logarithmic relationship of DDE residues to eggshell thinning. *Nature* 235:376–377.

Bodaly, R. A., R. E. Hecky, and R. J. P. Fudge. 1984. Increases in fish mercury levels in lakes flooded by the Churchill River Diversion, northern Manitoba. *Can. J. Fish. Aquat. Sci.* 41:682–691.

Borg, K., K. Erne, E. Hanko, and H. Wanntorp. 1970. Experimental secondary methylmercury poisoning in the goshawk (*Accipiter g. gentiles L.*). *Environ. Pollut.* 1:91–104.

Borg, K., H. Wanntorp, K. Erne, and E. Hanko. 1969. Alkyl mercury poisoning in terrestrial Swedish wildlife. *Viltrevy* 6:301–379.

Bosveld, A. T. C., et al. 1995. Effects of PCBs, PCDDs and PCDFs in common tern (*Sterna hirundo*) breeding in estuarine and coastal colonies in the Netherlands and Belgium. *Environ. Toxicol. Chem.* 14:99–115.

Bosveld, A. T. C., et al. 2000. Biochemical and developmental effects of dietary exposure to polychlorinated biphenyls 126 and 153 in common tern chicks (*Sterna hirundo*). *Environ. Toxicol. Chem.* 19:719–730.

Bosveld, A. T. C., and M. Van den Berg. 1994. Effects of polychlorinated biphenyls, dibenzo-*p*-dioxins, and dibenzofurans on fish-eating birds. *Environ. Rev.* 2:147–166.

Bowerman, W. W., J. P. Giesy, D. A. Best, and K. J. Kramer. 1995. A review of factors affecting productivity of bald eagles in the Great Lakes region: implications for recovery. *Environ. Health Perspec.* 103(Suppl 4):51–59.

Bowerman, W. W., D. A. Best, T. G. Grubb, G. M. Zimmerman, and J. P. Giesy. 1998. Trends of contaminants and effects in bald eagles of the Great Lakes basin. *Environ. Monit. Assess.* 53:197–212.

Bowes, G. W., B. R. Simoneit, A. L. Burlingame, B. W. de Lappe, and R. W. Risebrough. 1973. The search for chlorinated dibenzofurans and chlorinated dioxins in wildlife populations showing elevated levels of embryonic death. *Environ. Health Perspect.* 5:191–198.

Braune, B. M., and R. J. Norstrom. 1989. Dynamics of organochlorine compounds in herring gulls: III. Tissue distribution and bioaccumulation in Lake Ontario gulls. *Environ. Toxicol. Chem.* 8:957–968.

Braune, B. M., and M. Simon. 2003. Dioxins, furans, and non-ortho PCBs in Canadian Arctic seabirds. *Environ. Sci. Technol.* 37:3071–3077.

Braune, B. M., et al. 2005. Persistent organic pollutants and mercury in marine biota of the Canadian Arctic: an overview of spatial and temporal trends. *Sci. Total Environ.* 351–352:4–56.

Brunström, B. 1988. Sensitivity of embryos from duck, goose, herring gull, and various chicken breeds to 3,3′,4,4′-tetrachlorobiphenyl. *Poult. Sci.* 67:52–57.

Brunström, B. 1990. Mono-ortho-chlorinated chlorobiphenyls: toxicity and induction of 7-ethoxyresorufin *O*-deethylase (EROD) activity in chick embryos. *Arch. Toxicol.* 64:1188–1192.

Brunström, B., and L. Andersson. 1988. Toxicity and 7-ethoxyresorufin *O*-deethylase induction potency of coplanar polychlorinated biphenyls (PCBs) in chick embryos. *Arch. Toxicol.* 62:263–266.

Brunström, B., and J. Lund. 1988. Differences between chick and turkey embryos in sensitivity to 3,3′4,4′-tetrachlorobiphenyl and in concentration/affinity of the hepatic receptor for 2,3,7,8-tetra-chlorodibenzo-*p*-dioxin. *Comp. Biochem. Physiol. C. Comp. Pharmacol. Toxicol.* 91:507–512.

Buekers, J., E. S., Redeker, and E. Smolders. 2009. Lead toxicity to wildlife: Derivation of a critical blood concentration for wildlife monitoring based on literature data. *Sci. Total Environ.* 407:3431–3438.

Bunyan, P. J., D. M. Jennings, and A. Taylor. 1968a. Organophosphorus poisoning, some properties of avian esterases. *J. Agric. Food Chem.* 16:326–331.

Bunyan, P. J., D. M. Jennings, and A. Taylor. 1968b. Organophosphorus poisoning, diagnosis of poisoning in pheasants owing to a number of common pesticides. *J. Agric. Food Chem.* 16:332–339.

Burger, J., and M. Gochfeld. 1997. Risk, mercury levels, and birds: relating adverse laboratory effects to field biomonitoring. *Environ. Res.* 75:160–172.

Burgess, N. M., and M. W. Meyer. 2008. Methylmercury exposure associated with reduced productivity in common loons. *Ecotoxicology* 17:83–91.

Burgess, N. M., D. C. Evers, and J.D. Kaplan. 2005. Mercury and other contaminants in Common Loons breeding in Atlantic Canada. *Ecotoxicology* 14:241–252.

Burnett, J. A. 1999. A passion for wildlife: a history of the Canadian Wildlife Service, 1947–1997. Chapter 8. Wildlife toxicology. *Can. Field Nat.* 113:121–136.

Bustnes, J. O., V. Bakken, K. E. Erikstad, F. Mehlum, and J. U. Skaare. 2001. Patterns of incubation and nest-site attentiveness in relation to organochlorine (PCB) contamination in glaucous gulls. *J. Appl. Ecol.* 38:791–801.

Bustnes, J. O., S. A. Hanssen, I. Folstad, K. E. Erikstad, D. Hasselquist, and J. U. Skaare. 2004. Immune function and organochlorine pollutants in arctic breeding glaucous gulls. *Arch. Environ. Contam. Toxicol.* 47:530–541.

Calvert, J. H. 1876. Pheasant poisoning by swallowing shot. *The Field* 47 No. 1208, Feb 19, p. 189.

Carlson, R. W., and R. T. Duby. 1973. Embryonic effects of three PCBs in the chicken. *Bull. Environ. Contam. Toxicol.* 9:261–266.

Carson, R. L. 1962. *Silent spring.* Boston, MA: Houghton Mifflin. 368 pp.

Champoux, L. 1996. PCBs, dioxins and furans in Hooded Merganser (*Lophodytes cucullatus*), Common Merganser (*Mergus merganser*) and mink (*Mustela vison*) collected along the St.Maurice River near La Tuque, Quebec. *Environ. Pollut.* 92:147–153.

Champoux, L., D. Masse, D. Evers, O. Lane, M. Plante, and S. T. A. Timmerman. 2006. Assessment of mercury exposure and potential effects on Common Loons (*Gavia immer*) in Quebec. *Hydrobiologia* 567:263–274.

Church, M. E., R. Gwiazda, R. W. Risebrough, K. Sorenson, C. P. Chamberlain, S. Farry, W. Heinrich, B. A. Rideout, and D. R. Smith. 2006. Ammunition is the principal source of lead accumulated by California condors re-introduced to the wild. *Environ. Sci. Technol.* 40:6143–6150.

Clark, D. R. Jr., and R. F. Shore. 2001. Chiroptera. In *Ecotoxicology of wild mammals*, eds. R. F. Shore and B. A. Rattner, pp. 159–214. Chichester, UK: John Wiley and Sons, Ltd.

Clawson, S. G., and M. F. Baker. 1959. Immediate effects of dieldrin and heptachlor on bobwhites. *J. Wildl. Manage.* 23:215–219.

Coburn, D. R., J. B. DeWitt, J. V. Derby Jr., and E. Ediger. 1950. Phosphorus poisoning in waterfowl. *J. Amer. Pharm. Assoc.* 39:151–158.

Coburn, D. R., D. W. Metzler, and R. Treichler. 1951. A study of absorption and retention of lead in waterfowl in relation to clinical evidence of lead poisoning. *J. Wildl. Manage.* 15:186–192.

Cohen-Barnhouse, A., et al. 2008. Effects of TCDD, TCDF and PeCDF injected into the air cell of Japanese quail (*Coturnix japonica*) prior to incubation. Abstracts of the 29th Society of Environmental Toxicology and Chemistry North America Annual Meeting. Abstract WP210.

Colborn, T., D. Dumandski, and J. P. Myers. 1996. *Our stolen future: are we threatening our fertility, intelligence and survival? A scientific detective story.* New York: Dutton Publishing, pp. 306.

Cook, R. S., and D. O. Trainer. 1966. Experimental lead poisoning of Canada geese. *J. Wildl. Manage.* 30:1–8.

Cristol, D. A., et al. 2008. The movement of aquatic mercury through terrestrial food webs. *Science.* 320:335.

Custer, C. M., T. W. Custer, P. D. Allen, K. L. Stromborg, and M. J. Melancon. 1998. Reproduction and environmental contamination in tree swallows nesting in the Fox River drainage and Green Bay, Wisconsin, USA. *Environ. Toxicol. Chem.* 17:1786–1798.

Custer, C. M., T. W. Custer, P. M. Dummer, and K. L. Munney. 2003. Exposure and effects of chemical contaminants on tree swallows nesting along the Housatonic River, Berkshire County, Massachusetts, USA, 1998–2000. *Environ. Toxicol. Chem.* 22:1605–1621.

Custer, T. W., C. M. Custer, and R. K. Hines. 2002. Dioxins and congener-specific polychlorinated biphenyls in three avian species from the Wisconsin River, Wisconsin. *Environ. Pollut.* 119:323–332.

Custer, T. W., et al. 1999. Organochlorine contaminants and reproductive success of double-crested cormorants from Green Bay, Wisconsin, USA. *Environ. Toxicol. Chem.* 18:1209–1217.

Custer, C. M., T. W. Custer, C. J. Rosiu, and M. J. Melancon. 2005. Exposure and effects of 2,3,7,8-tetrachlo-rodibenzo-*p*-dioxin in tree swallows (*Tachycineta bicolor*) nesting along the Woonasquatucket River, Rhode Island, USA. *Environ. Toxicol. Chem.* 24:93–109.

Dahlen, J. H., and A. O. Haugen. 1954. Acute toxicity of certain insecticides to the bobwhite quail and mourning dove. *J. Wildl. Manage.* 18:477–481.

Dahlgren, R. B., R. J. Bury, R. L. Linder, and R. F. Reidinger Jr. 1972. Residue levels and histopathology in pheasants given polychlorinated biphenyls. *J. Wildl. Manage.* 36:524–533.

Dansereau, M., N. Lariviere, D. D. Tremblay, and D. Belanger. 1999. Reproductive performance of two generations of female semidomesticated mink fed diets containing organic mercury contaminated freshwater fish. *Arch. Environ. Contam. Toxicol.* 36:221–226.

Davis, J. A., D. M. Fry, and B. W. Wilson. 1997. Hepatic ethoxyresorufin-*O*-deethylase activity and inducibility in wild populations of double-crested cormorants (*Phalacrocorax auritus*). *Environ. Toxicol. Chem.* 16:1441–1449.

DesGranges, J.-L., J. Rodrigue, and M. Laperle. 1998. Mercury accumulation and biomagnification in ospreys (*Pandion haliaetus*) in the James Bay and Hudson Bay regions of Quebec. *Arch. Environ. Contam. Toxicol.* 35:330–341.

DeSolla, S. R., and P. A. Martin. 2009. PCB accumulation in osprey exposed to local sources in lake sediment. *Environ. Pollut.* 157:347–355.

DeWitt, J. B. 1955. Effects of chlorinated hydrocarbon insecticides upon quail and pheasants. *J. Agric. Food. Chem.* 3:672–676.

DeWitt, J. B. 1956. Chronic toxicity to quail and pheasants of some chlorinated insecticides. *J. Agric. Food. Chem.* 4:863–866.

DeWitt, J. B., J. V. Derby Jr., and G. F. Mangan Jr. 1955. DDT vs. wildlife. Relationships between quantities ingested, toxic effects and tissue storage. *J. Amer. Pharm. Assoc.* 44:22–24.

DeWitt, J. B., C. M. Menzie, V. A. Adomaitis, and W. L. Reichel. 1960. Pesticidal residues in animal tissues. *Trans. N. Amer. Wildl. Conf.* 25:277–285.

Dietz, R., C. O. Nielsen, M. M. Hansen, and C. T. Hansen. 1990. Organic mercury in Greenland birds and mammals. *Sci. Total Environ.* 95:41–51.

Doi, R., H. Ohno, and M. Harada. 1984 Mercury in feathers of wild birds from the mercury-polluted area along the shore of the Shiranui Sea. *Sci. Total Environ.* 40:155–167.

Drouillard, K. G., and R. J. Norstrom. 2000. Dietary absorption efficiencies and toxicokinetics of polychlorinated biphenyls in ring doves following exposure to Aroclor mixtures. *Environ. Toxicol. Chem.* 19:2707–2714.

Drouillard, K. G., et al. 2007. Bioaccumulation and biotransformation of 61 polychlorinated biphenyl and four polybrominated diphenyl ether congeners in juvenile American kestrels (*Falco sparverius*). *Environ. Toxicol. Chem.* 26:313–324.

Drouillard, K. G., K. J. Fernie, J. E. Smits, G. R. Bortolotti, D. M. Bird, and R. J. Norstrom. 2001. Bioaccumulation and biotransformation of 42 polychlorinated biphenyl in American kestrels (*Falco sparverius*). *Environ. Toxicol. Chem.* 20:2514–2522.

Drouillard, K. G., R. J. Norstrom, G. A. Fox, A. Gilman, and D. B. Peakall. 2003. Development and validation of a herring gull embryo toxicokinetic model for PCBs. *Ecotoxicology* 12:55–68.

Dykstra, C. R., M. W. Meyer, D. K. Warnke, W. H. Karasov, D. E. Andersen, W. W. Bowerman IV, et al. 1998. Low reproductive rates of Lake Superior bald eagles: low food delivery rates or environmental contaminants? *J. Great Lakes Res.* 24:32–44.

Eagles-Smith, C. A., J. T. Ackerman, J. Yee, and T. L. Adelsbach. 2009. Mercury demethylation in waterbird livers: dose-response thresholds and differences among species. *Environ. Toxicol. Chem.* 28:568–577.

Ecobichon, D. J. 1996. Toxic effects of pesticides. In *Casarett and Doull's Toxicology The Basic Science of Poisons*, 5th edition, eds. C. D. Klaassen, M. O. Amdur, and J. Doull, pp. 643–689. New York, NY: McGraw-Hill.

Eisler, R. 1986. Polychlorinated biphenyl hazards to fish, wildlife, and invertebrates: a synoptic review. *U.S. Fish Wildl. Ser. Biol. Rep.* 85(1.7):72.

Eisler, R. 1987. Mercury hazards to fish, wildlife, and invertebrates: a synoptic review. *U.S. Fish Wildl. Ser. Biol. Rep.* 85(1.10):90.

Eisler, R. 2000. *Biogeochemical, health, and ecotoxicological perspectives on gold and gold mining.* Boca Raton, FL: CRC Press, 355 pp.

Elliott, J. E., and M. L. Harris. 2001. An ecotoxicological assessment of chlorinated hydrocarbon effects on bald eagle populations. *Rev. Toxicol.* 4:1–60.

Elliott, J. E., and P. A. Martin. 1994. Chlorinated hydrocarbons and shell thinning in eggs of *Accipiter* hawks in Ontario, 1986–1989. *Environ. Pollut.* 86:189–200.

Elliott, J. E., and P. A. Martin. 1998. Chlorinated hydrocarbon contaminants in grebes and seaducks wintering on the coast of British Columbia, Canada: 1988–1993. *Environ. Mon. Assess.* 53:337–362.

Elliott, J. E., and R. J. Norstrom. 1998. Chlorinated hydrocarbon contaminants and productivity of bald eagle populations on the Pacific coast of Canada. *Environ. Toxicol. Chem.* 17:1142–1153.

Elliott, J. E., R. W. Butler, R. J. Norstrom, and P. E. Whitehead. 1989. Environmental contaminants and reproductive success of great blue herons *Ardea herodias* in British Columbia, 1986–87. *Environ. Pollut.* 59:91–114.

Elliott, J. E., M. L. Harris, L. K. Wilson, P. E. Whitehead, and R. J. Norstrom. 2001a. Monitoring temporal and spatial trends in polychlorinated dibenzo-*p*-dioxins (PCDDs) and dibenzofurans (PCDFs) in eggs of great blue heron (*Ardea herodias*) on the coast of British Columbia, Canada, 1983–1998. *Ambio* 30:416–428.

Elliott, J. E., I. E. Moul, and K. M. Cheng. 1998. Variable reproductive success of bald eagles on the British Columbia coast. *J. Wildl. Manage.* 62:518–529.

Elliott, J. E., R. J. Norstrom, S. K. Kennedy, and G. A. Fox. 1988. Trends and effects of environmental contaminants determined from analysis of archived wildlife samples. In *Progress in environmental specimen banking*, eds. S. A. Wise, and G. M. Zeisler. U.S. National Bureau of Standards Special Publication No. 740:131–142.

Elliott, J. E., et al. 1992. Patterns and trends of organic contaminants in Canadian seabirds, 1968–1990. In *Persistent Pollutants in the Marine Environment*, eds. C. H. Walker, and D. R. Livingston, pp. 181–194, Oxford: Pergamon Press.

Elliott, J. E., R. J. Norstrom, and G. E. J. Smith. 1996a. Patterns, trends and toxicological significance of chlorinated hydrocarbons and mercury in bald eagle eggs. *Archiv. Environ. Contam. Toxicol.* 31:354–367.

Elliott, J. E., et al. 1996b. Biological effects of polychlorinated dibenzo-*p*-dioxins, dibenzofurans, and biphenyls in bald eagle (*Haliaeetus leucocephalus*) chicks. *Environ. Toxicol. Chem.* 15:782–793.

Elliott, J. E., et al. 2001b. Assessment of biological effects of chlorinated hydrocarbons in osprey chicks. *Environ. Toxicol. Chem.* 20:866–879.

Elliott, K. H., C. E. Gill, and J. E. Elliott. 2005. Influence of tides and weather on bald eagle provisioning rates. *J. Raptor Res.* 39:99–108.

Evers, D. C., et al. 2005. Patterns and interpretation of mercury exposure in freshwater avian communities in northeastern North America. *Ecotoxicology* 14:193–221.

Evers, D. C., et al. 1998. Geographic trend in mercury measured in Common Loon feathers and blood. *Environ. Toxicol. Chem.* 17:173–183.

Evers, D. C., et al. 2008. Adverse effects from environmental mercury loads on breeding common loons. *Ecotoxicology* 17:69–81.

Evers, D. C., K. M. Taylor, A. Major, R. J. Taylor, R. H. Poppenga, and A. M. Scheuhammer. 2003. Common loon eggs as indicators of methylmercury availability in North America. *Ecotoxicology* 12:69–81.

Fairbrother, A. 2009. Federal environmental legislation in the U.S. for protection of wildlife and regulation of environmental chemicals. *Ecotoxicology* 18:784–790.

Fairbrother, A., B. T. Marden, J. K. Bennett, and M. J. Hooper. 1991. Methods used in determination of cholinesterase activity. In *Cholinesterase-inhibiting insecticides: their impact on wildlife and the environment. Vol. 2 Chemicals in agriculture*, ed. P. Mineau, pp. 35–71. New York: Elsevier.

Falandysz, J. 1980. Chlorinated hydrocarbons in gulls from the Baltic south coast. *Mar. Pollut. Bull.* 11:75–80.

Fangstrom, B., M. Athanasiadou, I. Athanassiadis, P. Weihe, and A. Bergman. 2005. Hydroxylated PCB metabolites in nonhatched Fulmar eggs from the Faroe Islands. *Ambio* 2005. 34:184–187.

Fernie, K., J. Smits, and G. Bortolotti. 2003. Developmental toxicity of *in ovo* exposure to polychlorinated biphenyls: I. Immediate and subsequent effects on first-generation nestling American kestrels (*Falco sparverius*). *Environ. Toxicol. Chem.* 22:554–560.

Fernie, K. J., J. E. Smits, G. R. Bortolotti, and D. M. Bird. 2001. Reproduction success of American kestrels exposed to dietary polychlorinated biphenyls. *Environ. Toxicol. Chem.* 20:776–781.

Feyk, L. A., J. P. Giesy, A. T. C. Bosveld, and M. van den Berg. 2000. Changes in cytochrome P4501A activity during development in common tern chicks fed polychlorinated biphenyls, as measured by the caffeine breath test. *Environ. Toxicol. Chem.* 19:712–718.

Fimreite, N. 1971. Effects of dietary methylmercury on ring-necked pheasants. *Can. Wildl. Ser. Occasional Paper 9*, Ottawa, Canada, 39 pp.

Fimreite, N. 1974. Mercury contamination of aquatic birds in northwestern Ontario. *J. Wildl. Manage.* 38:120–131.

Fimreite, N., and L. Karstad. 1971. Effects of dietary methyl mercury on red-tailed hawks. *J. Wildl. Manage.* 35:293–300.

Finley, M. T., and R. C. Stendell. 1978. Survival and reproductive success of black ducks fed methylmercury. *Environ. Pollut.* 16:51–64.

Finley, M. T., W. H. Stickel, and R. E. Christensen. 1979. Mercury residues in tissues of dead and surviving birds fed methylmercury. *Bull. Environ. Contam. Toxicol.* 21:105–110.

Firestone, D. 1973. Etiology of chick-edema disease. *Environ. Health Perspect.* 5:59–66.

Fisher, A. K. 1893. Report on the ornithology of the Death Valley expedition of 1891, comprising notes on the birds observed in southern California, southern Nevada, and parts of Arizona and Utah. In *Part II. The Death Valley Expedition: A biological survey of parts of California, Nevada, Arizona, and Utah*, A. K. Fisher, et al., 7-158 North American Fauna No. 7. U.S. Department of Agriculture, Division of Ornithology and Mammalogy, Washington, DC.

Fisher, S. A., et al. 2001. Courtship behavior of captive American kestrels (*Falco sparverius*) exposed to polychlorinated biphenyls. *Arch. Environ. Contam. Toxicol.* 41:215–220.

Flick, D. F., D. Firestone, J. Ress, and J. R. Allen. 1973. Studies of the chick edema disease. *Poulty Sci.* 52:1637–1641.

Forsyth, D. J. 2001. Extrapolation of laboratory tests to field populations. In *Ecotoxicology of wild mammals. ecological and environmental toxicology series,* eds. R. F. Shore and B. A. Rattner, pp. 577–634. New York: John Wiley and Sons, Ltd.

Fossi, M. C., and C. Leonzio. 1994. *Nondestructive biomarkers in vertebrates.* Boca Raton, FL: Lewis Publishing, Inc. pp. 345.

Fossi, C., C. Leonzio, and S. Focardi. 1986. Mixed function oxidase activity and cytochrome P-450 forms in black-headed gulls feeding in different areas. *Mar. Pollut. Bull.* 17:546–548.

Fox, G. A., S. W. Kennedy, R. J. Norstrom, and D. C. Wigfield. 1988. Porphyria in herring gulls: a biochemical response to chemical contamination in Great Lakes food chains. *Environ. Toxicol. Chem.* 7:831–839.

Fox, G. A., S. Trudeau, H. Won, and K. A. Grasman. 1998. Monitoring the elimination of persistent toxic substances from the Great Lakes: chemical and physiological evidence from adult herring gulls. *Environ. Monit. Assess.* 53:147–168.

Fox, G. A., D. A. Jeffrey, K. S. Williams, S. W. Kennedy, and K. A. Grasman. 2007a. Health of herring gulls (*Larus argentatus*) in relation to breeding location in the early 1990s. I Biochemical measures. *J. Toxicol. Environ. Health A* 70:1443–1470.

Fox, G. A., K. A. Grasman, and G. D. Campbell. 2007b. Health of herring gulls (*Larus argentatus*) in relation to breeding location in the early 1990s. II Cellular and histopathological measures. *J. Toxicol. Environ. Health* A 70:1471–1491.

Franson, J. C. 1996. Interpretation of tissue lead residues in birds other than waterfowl. In *Environmental contaminants in wildlife: interpreting tissue concentrations,* eds. W. N. Beyer, G. H. Heinz, and A. W. Redmon-Norwood, pp. 265–279. Boca Raton, FL: Lewis Publishing Inc.

Friedman, L., D. Firestone, W. Horwitz, D. Banes, M. Anstead, and G. Shue. 1959. Studies of the chick edema disease factor. *J. Ass. Offic. Agr. Chem* 42:129–140.

Friend, M. 1985. Interpretation of criteria commonly used to determine lead poisoning problem areas. *U.S. Fish Wildl. Ser.*, Fish and Wildlife Leaflet 2, Washington, DC.

Gallagher, B. A. 1918. Experiments in avian toxicology. *J. Amer. Vet. Med. Assoc.* 54:337–356.

Gallo, M. A. 2001. History and scope of toxicology. In *Casarett and Doull's toxicology: The basic science of poisons*, 6th edition, ed. C. D. Klaassen, pp. 3–10. New York: McGraw-Hill.

Gay, M. L., A. A. Belisle, and J. F. Patton. 1980. Quantification of petroleum-type hydrocarbons in avian tissue. *J. Chromat.* 187:153–160.

George, J. L., and W. H. Stickel. 1949. Wildlife effects of DDT dust used for tick control on a Texas prairie. *Am. Midland Nat.* 42:228–237.

Giesy, J. P., and K. Kannan. 2001. Global distribution of perfluorooctane sulfonate in wildlife. *Environ. Sci. Technol.* 35:1339–1342.

Giesy, J. P., et al. 1995. Contaminants in fishes from the Great Lakes-influenced sections and above dams of three Michigan Rivers: III. Implications for health of bald eagles. *Arch. Environ. Contam. Toxicol.* 29:309–321.

Giesy J. P., J. P. Ludwig, and D. E. Tillitt. 1994. Dioxins, dibenzofurans, PCBs and colonial, fish-eating water birds. In *Dioxins and health*, ed. A. Schecter, pp. 249–307. New York: Plenum Press.

Gilbert, S. G., and A. Hayes. 2006. Milestones in toxicology. http://www.asmalldoseof.org/historyoftox/Milestones.poster.02.03.06.pdf (accessed September 2, 2008).

Gilbertson, M. 1974. Pollutants in breeding herring gulls in the lower Great Lakes. *Can. Field. Nat.* 88:273–280.

Gilbertson, M. 1975. A Great Lakes tragedy. *Nature Canada* 4:22–25.

Gilbertson, M, 1983. Etiology of chick edema disease in herring gulls in the lower Great Lakes. *Chemosphere* 12:357–370.

Gilbertson, M., T. Kubiak, J. Ludwig, and G. Fox. 1991. Great Lakes embryo mortality edema, and deformities syndrome (GLEMEDS) in colonial fish-eating birds: similarity to chick edema disease. *J. Toxicol. Environ. Health* 33:455–520.

Gilbertson, M., R. D. Morris, and R. A. Hunter. 1976. Abnormal chicks and PCB residue levels in eggs of colonial birds in the lower Great Lakes (1971–1973). *Auk* 93:434–434.

Gilday, D., M. Gannon, K. Yutzey, D. Bader, and A. B. Rifkind. 1996. Molecular cloning and expression of two novel avian cytochrome P450 1A enzymes induced by 2,3,7,8-tetrachlorodibenzo-*p*-dioxin. *J. Biol. Chem.* 271:33054–33059.

Gill, C. E., and J. E. Elliott. 2003. Influence of food supply and chlorinated hydrocarbon contaminants on breeding success of bald eagles. *Ecotoxicology* 12:95–111.

Gilman, A. P., G. A. Fox, D. B. Peakall, S. M. Teeple, T. R. Carroll, and G. T. Haymes. 1977. Reproductive parameters and egg contaminant levels of Great Lakes herring gulls. *J. Wildl. Manage.* 41:458–468.

Grasman, K. A., and G. A. Fox. 2001. Associations between altered immune function and organochlorine contamination in young Caspian terns (*Sterna caspia*) from Lake Huron, 1997–1999. *Ecotoxicology* 10:101–114.

Grasman, K. A., G. A. Fox, P. F. Scanlon, and J. P. Ludwig. 1996. Organochlorine-associated immunosuppression in prefledgling Caspian terns and herring gulls from the Great Lakes: an ecoepidemiological study. *Environ. Health. Perspect.* 104(Suppl4):829–842.

Grinell, G. B. 1894. Lead poisoning. *Forest and Stream* 42(6):117–118.

Grue, C. E., P. L. Gibert PL, and M. E. Seeley. 1997. Neurophysiological and behavior changes on non-target wildlife exposed to organophosphorus and carbamate pesticides: thermoregulation, food consumption, and reproduction. *Amer. Zool.* 37:369–388.

Guillette, L. J. Jr., T. S. Gross, G. R. Masson, J. M. Matter, H. F. Percival, and A. R. Woodward. 1994. Developmental abnormalities of the gonad and abnormal sex hormone concentrations in juvenile alligators from contaminated and control lakes in Florida. *Environ. Health Perspect.* 102:680–688.

Guruge, K. S., and S. Tanabe. 1997. Congener specific accumulation and toxic assessment of polychlorinated biphenyls in common cormorants, *Phalacrocorax carbo*, from Lake Biwa, Japan. *Environ. Pollut.* 96:425–433.

Hall, B. D., et al. 2005. Impacts of reservoir creation on the biogeochemical cycling of methyl mercury and total mercury in boreal upland forests. *Ecosystems*, 8:248–266.

Hall, R. J., and N. C. Coon. 1988. Interpreting residues of petroleum hydrocarbons in wildlife tissues. Biological Report 88(15). *Fish Wildl. Ser.*, U.S. Department of the Interior, Washington, DC. 7 pp.

Hario, M., K. Himberg, T. Hollmén, and E. Rudbäck. 2000. Polychlorinated biphenyls in diseased lesser black-backed gull (*Larus fuscus fuscus*) chicks from the Gulf of Finland. *Environ. Pollut.* 107:53–60.

Harris, H. J., T. C. Erdman, G. T. Ankley, and K. B. Lodge. 1993. Measures of reproductive success and PCB residues in eggs and chicks of Forster's tern on Green Bay, Lake Michigan—1988. *Arch. Environ. Contam. Toxicol.* 25:304–314.

Harris, M. L., and J. E. Elliott. 2000. Reproductive success and chlorinated hydrocarbon contamination in tree swallows (*Tachycineta bicolor*) nesting along rivers receiving pulp and paper mill effluent discharges. *Environ. Pollut.* 110:307–320.

Harris, M. P., and D. Osborn. 1981. Effect of a polychlorinated biphenyl on the survival and breeding of puffins. *J. Appl. Ecol.* 18:471–479.

Harris, M. L., L. K. Wilson, R. J. Norstrom, and J. E. Elliott. 2003. Egg concentrations of polychlorinated dibenzo-*p*-dioxins and dibenzofurans in double-crested (*Phalacrocorax auritus*) and pelagic (*P. pelagicus*) cormorants from the Strait of Georgia, Canada, 1973–1998. *Environ. Sci. Technol.* 37:822–831.

Hart, C. A., I. C. T. Nisbet, S. W. Kennedy, and M. E. Hahn. 2003. Gonadal feminization and halogenated environmental contaminants in common terns (*Sterna hirundo*): evidence that ovotestes in male embryos do not persist to the prefledgling stage. *Ecotoxicology* 12:125–140.

Harvey, J. M. 1967. Excretion of DDT by migratory birds. *Can. J. Zool.* 45:629–633.

Hayes, W. J. Jr. 1991. Introduction. In *Handbook of pesticide toxicology, Volume 1 General Principles,* eds. W. J. Hayes Jr., and E. R. Laws, Jr., pp. 1–37. New York: Academic Press, Inc.

Head, J. A. 2006. Variation in the cytochrome P4501A response to dioxin-like compounds in avian species. PhD Thesis, University of Ottawa, Ottawa, ON.

Head, J. A., and S. W. Kennedy. 2007. Differential expression, induction, and stability of CYP1A4 and CYP1A5 mRNA in chicken and herring gull embryo hepatocytes. *Comp Biochem Physiol Part C Comp. Biochem. Toxicol.* 145:617–624.

Head, J. A., M. E. Hahn, and S. W. Kennedy. 2008. Key amino acids in the aryl hydrocarbon receptor predict dioxin sensitivity in avian species. *Environ. Sci. Technol.* 42:7535–7541.

Heath, J. A., and P. C. Frederick. 2005. Relationships among mercury concentrations, hormones, and nesting effort of white ibises (*Eudocimus albus*) in the Florida Everglades. *Auk* 122:255–267.

Heath, R. G., J. W. Spann, and J. F. Kreitzer.1969. Marked DDE impairment of mallard reproduction in controlled studies. *Nature* 224:47–48.

Heath, R. G., J. W. Spann, E. F. Hill, and J. F. Kreitzer. 1972. Comparative dietary toxicities of pesticides to birds. *U.S. Fish Wildl. Ser.*, Special Scientific Report-Wildlife 152:57.

Heaton, S. N., et al. 1995. Dietary exposure of mink to carp from Saginaw Bay, Michigan. 1. Effects on reproduction and survival, and the potential risks to wild mink populations. *Arch. Environ. Contam. Toxicol.* 28:334–343.

Hebert, C. E., and D. V. C. Weseloh. 2006. Adjusting for temporal change in trophic position results in reduced rates of contaminant decline. *Environ. Sci. Technol.* 40:5624–5628.

Hebert, C. E., et al. 2008. Restoring picivorous fish populations in the Laurentian Great Lakes causes seabird dietary change. *Ecology* 89:891–897.

Heinz, G. 1974. Effects of low dietary levels of methyl mercury on mallard reproduction. *Bull. Environ. Contam. Toxicol.* 11:386–392.

Heinz, G. H. 1979. Methylmercury: reproductive and behavioral effects on three generations of mallard ducks. *J. Wildl. Manage.* 43:94–401.

Heinz, G. H. 1996. Mercury poisoning in wildlife. In *Noninfectious diseases of wildlife*, 2nd edition, eds. A. Fairbrother, L. N. Locke, and G. L. Hoff, pp. 118–127. Ames, IA: Iowa State University Press.

Heinz, G. H., and D. J. Hoffman. 1998. Methylmercury chloride and selenomethionine interactions on health and reproduction in mallards. *Environ. Toxicol. Chem.* 17:139–145.

Heinz, G. H., and D. J. Hoffman. 2003. Embryotoxic thresholds of mercury: estimates from individual mallard eggs. *Arch. Environ. Contam. Toxicol.* 44:257–264.

Heinz, G. H., and R. W. Johnson. 1981. Diagnostic brain residues of dieldrin: some new insights. In *Avian and mammalian wildlife toxicology: second conference*, eds. D. W. Lamb and E. E. Kenega, pp. 72–92. Philadelphia: ASTM STP 757.

Heinz, G. H., and L. N. Locke. 1976. Brain lesions in mallard ducklings from parents fed methylmercury. *Avian Dis.* 20:9–17.

Heinz, G. H., D. J. Hoffman, J. D. Klimstra, K. R. Stebbins, S. L. Kondrad, and C. A. Erwin. 2009. Species differences in the sensitivity of avian embryos to methylmercury. *Arch. Environ. Contam. Toxicol.* 56:129–138.

Heinz, G. H., D. J. Hoffman, S. L. Kondrad, and C. A. Erwin. 2006. Factors affecting the toxicity of methylmercury injected into eggs. *Arch. Environ. Contam. Toxicol.* 50:264–279.

Helander, B., A. Olsson, A. Bignert, L. Asplund, and K. Litzén. 2002. The role of DDE, PCB, coplanar PCB and eggshell parameters for reproduction in the white-tailed sea eagle (*Haliaeetus albicilla*) in Sweden. *Ambio* 31:386–403.

Henning, M. H., S. K. Robinson, K. J. McKay, J. P. Sullivan, and H. Brucker. 2003. Productivity of American robins exposed to polychlorinated biphenyls, Housatonic River, Massachusetts, USA. *Environ. Toxicol. Chem.* 22:2783–2788.

Henny, C. J. 2003. Effects of mining lead on birds: a case history at Coeur d'Alene Basin, Idaho. In *Handbook of Ecotoxicology*, 2nd edition, eds. D. J. Hoffman, B. A. Rattner, G. A. Burton, Jr., and J. Cairns, Jr., pp. 755–766. Boca Raton, FL: Lewis Publishing Inc.

Henny, C. J., L. J. Blus, E. J. Kolbe, and R. E. Fitzner. 1985. Organophosphate insecticide (famphur) topically applied to cattle kills magpies and hawks. *J. Wildl. Manage.* 49:648–658.

Henny, C. J., E. F. Hill, D. J. Hoffman, M. G. Spalding, and R. A. Grove. 2002. Nineteenth century mercury: Hazard to wading birds and cormorants of the Carson River, Nevada. *Ecotoxicology* 11:213–231.

Henny, C. J., J. L. Kaiser, and R. A. Grove. 2009. PCDDs, PCDFs, PCBs, OC pesticides and mercury in fish and osprey eggs from Willamette River, Oregon (1993, 2001 and 2006) with calculated biomagnification factors. *Ecotoxicology* 18:151–173.

Henriksen, E. O., G. W. Gabrielsen, J. U. Skaare, N. Skjegstad, and B. M. Jenssen. 1998. Relationships between PCB levels, hepatic EROD activity and plasma retinol in glaucous gulls. *Mar. Environ. Res.* 46:45–49.

Hickey, J. J., and D. W. Anderson. 1968. Chlorinated hydrocarbons and eggshell changes in raptorial and fish-eating birds. *Science* 162:271–273.

Hickey, J. J., J. A. Keith, and F. B. Coon. 1966. An exploration of pesticides in a Lake Michigan ecosystem. *J. Appl. Ecol.* 3(Suppl.):141–153.

Higginbotham, G. R., A. Huang, D. Firestone, J. Verrett, J. Ress, and A. D. Campbell. 1968. Chemical and toxicological evaluations of isolated and synthetic chloro derivatives of dibenzo-*p*-dioxin. *Nature* 220:702–703.

Hill, E. F. 2003. Wildlife toxicology of organophosphorus and carbamate pesticides. In *Handbook of ecotoxicology*, 2nd edition, eds. D. J. Hoffman, B. A. Rattner, G. A. Burton Jr., and J. Cairns Jr., 281–312. Boca Raton, FL: Lewis Publishing Inc.

Hill, E. F., and W. J. Fleming. 1982. Anticholinesterase poisoning of birds: field monitoring and diagnosis of acute poisoning. *Environ. Toxicol. Chem.* 1:27–38.

Hill, E. F., R. C. Heath, J. W. Spann, and J. D. Williams. 1975. Lethal dietary toxicities of environmental pollutants to birds. *U.S. Fish Wildl. Ser.*, Special Scientific Report-Wildlife No. 191, Washington, DC., 61 pp.

Hill, E. F., and P. F. P. Henry. 1996. Cyanide. In *Noninfectious diseases of wildlife*, 2nd edition, eds. A. Fairbrother, L. N. Locke, and G. L. Hoff, pp. 99–107. Ames, IA: Iowa State University Press.

Hill, E. F., C. J. Henny, and R. A. Grove. 2008. Mercury and drought along the lower Carson River, Nevada: II. Snowy egret and black-crowned night-heron reproduction on Lahontan Reservoir, 1997–2006. *Ecotoxicology* 17:117–131.

Hoff, M. H., M. W. Meyer, J. Van Stappen, and T. W. Fratt. 2004. Relationships between bald eagle productivity and dynamics of fish populations and fisheries in the Wisconsin waters of Lake Superior, 1983–1999. *J. Great Lakes Res.* 30:434–442.

Hoffman, D. J., C. L. Rice, and T. J. Kubiak. 1996. PCBs and dioxins in birds. In *Environmental contaminants in wildlife: interpreting tissue concentrations*, eds. W. N. Beyer, G. H. Heinz, and A. W. Redmon-Norwood, pp. 165–207. SETAC Special Publication Series. Boca Raton, FL: Lewis Publishing Inc.

Hoffman, D. J., M. J. Melancon, P. N. Klein, J. D. Eisemann, and J. W. Spann. 1998. Comparative developmental toxicity of planar polychlorinated biphenyl congeners in chickens, American kestrels, and common terns. *Environ. Toxicol. Chem.* 17:747–757.

Hoffman, D. J., G. L. Smith, and B. A. Rattner. 1993. Biomarkers of contaminant exposure in common terns and black-crowned night herons in the Great Lakes. *Environ. Toxicol. Chem.* 12:1095–1103.

Holmes, W. N. 1984. Petroleum pollutants in the marine environment and their possible effects on seabirds. In *Reviews in Environmental Toxicology Vol. 1.*, ed. E. Hodgson, pp. 252–317. Amsterdam: Elsevier Science Publishers.

Hooper, M. J., P. Mineau, M. E. Zaccagnini, and B. Woodbridge. 2003. Pesticides and international migratory bird conservation. In *Handbook of ecotoxicology*, 2nd edition, eds. D. J. Hoffman, B. A. Rattner, G. A. Burton Jr., and J. Cairns Jr., pp. 737–754. Boca Raton, FL: Lewis Publishing Inc.

Hotchkiss, N., and R. H. Pough. 1946. Effect on forest birds of DDT used for gypsy moth control in Pennsylvania. *J. Wildl. Manage.* 10:202–207.

Huff, J. E., and J. S. Wassom. 1974. Health hazards from chemical impurities: chlorinated dibenzodioxins and chlorinated dibenzofurans. *Inter. J. Environ. Studies* 6:1–17.

Huggett, R. J., R. A. Kimerle, P. M. Mehrle Jr., and H. L. Bergman. 1992 *Biomarkers. Biochemical, physiological, and histological markers of anthropogenic stress*. SETAC Special Publication Series. Boca Raton, FL: Lewis Publishing, Inc., pp. 347.

Itano, K., S. Kawai, N. Miyazaki, R. Tatsukawa, and T. Fujiyama. 1984. Mercury and selenium levels in striped dolphins caught off the Pacific coast of Japan. *Agric. Biol. Chem.* 48:1109–1116.

Janz, D. M., and G. D. Bellward. 1996. *In ovo* 2,3,7,8-tetrachlorodibenzo-*p*-dioxin exposure in three avian species. I. Effects on thyroid hormones and growth during the perinatal period. *Toxicol. Appl. Pharmacol.* 139:281–291.

Jaspers, V. L. B., A. Covaci, S. Voorspoels, T. Dauwe, M. Eens, and P. Schepens. 2006. Brominated flame retardants and organochlorine pollutants in aquatic and terrestrial predatory birds of Belgium: levels, patterns, tissue distribution and condition factors. *Environ. Pollut.* 139:340–352.

Jensen, S. 1966. Report of a new chemical hazard. *New Scientist* 32:612.

Jensen, S., A. G. Johnels, M. Olsson, and G. Otterlind. 1969. DDT and PCB in marine animals from Swedish waters. *Nature* 224:247–250.

Jessup, D. A., and F. A. Leighton. 1996. Oil pollution and petroleum toxicity to wildlife. 1996. In *Noninfectious Diseases of Wildlife*, eds. A. Fairbrother, L. N. Locke, and G. L. Hoff, pp. 141–156. Ames, IA: Iowa State University Press.

Jobling, S., M. Nolan, C. R. Tyler, G. Brighty, and J. P. Sumpter. 1998. Widespread sexual disruption in wild fish. *Environ. Sci. Technol.* 32:2498–2506.

Johnson, T. A., R. A. Bodaly, and J. A. Mathias. 1991. Predicting fish mercury levels from physical characteristics of boreal reservoirs. *Can. J. Fish. Aquat. Sci.* 48:1468–1475.

Karasov, W. H., K. P. Kenow, M. W. Meyer, and F. Fournier. 2007. Bioenergetic and pharmacokinetic model for exposure of common loon (*Gavia immer*) chicks to methylymercury. *Environ. Toxicol. Chem.* 26:677–685.

Karchner, S. I., D. G. Franks, S. W. Kennedy, and M. E. Hahn. 2006. The molecular basis for differential dioxin sensitivity in birds: role of the aryl hydrocarbon receptor. *Proc. Nat. Acad. Sci. USA* 103:6252–6257.

Keith, J. A. 1966. Reproduction in a population of herring gulls (*Larus argentatus*) contaminated by DDT. *J. Appl. Ecol.* 3(Suppl.):57–70.

Keith, J. O. 1996. Residue analyses: how they were used to assess hazards of contaminants to wildlife. In *Environmental contaminants in wildlife: interpreting tissue concentrations*, eds. W. N. Beyer, G. H. Heinz, and A. W. Redmon-Norwood, pp. 1–46. SETAC Special Publication Series. Boca Raton, FL: Lewis Publishing Inc.

Kendall, R. J., and T. E. Lacher Jr. 1994. *Wildlife toxicology and population modeling: integrated studies of agroecosystems.* Boca Raton, FL: Lewis Publishers Inc., 576 pp.

Kennedy, S. W., G. A. Fox, S. P. Jones, and S. F. Trudeau. 2003. Hepatic EROD activity is not a useful biomarker of polychlorinated biphenyl exposure in the adult herring gull (*Larus argentatus*). *Ecotoxicology* 12:153–161.

Kennedy, S. W., G. A. Fox, S. Trudeau, L. J. Bastien, and S. P. Jones. 1998. Highly carboxylated porphyrin concentration: a biochemical marker of PCB exposure in herring gulls. *Mar. Environ. Res.* 46:65–69.

Kennedy, S. W., A. Lorenzen, S. P. Jones, M. E. Hahn, and J. J. Stegeman. 1996. Cytochrome P4501A induction in avian hepatocyte cultures: a promising approach for predicting the sensitivity of avian species to toxic effects of halogenated aromatic hydrocarbons. *Toxicol. Appl. Pharmacol.* 141:214–230.

Kenow, K. P., et al. 2008. Effects of methylmercury exposure on glutathione metabolism, oxidative stress, and chromosomal damage in captive-reared common loon (*Gavia immer*) chicks. *Environ. Pollut.* 156:732–738.

Kocan, A., J. Petrik, S. Jursa, J. Chovancova, and B. Drobna. 2001. Environmental contamination with polychlorinated biphenyls in the area of their former manufacture in Slovakia. *Chemosphere* 43:595–600.

Koeman, J. H., H. C. van Velzen-Blad, R. de Vries, and J. G. Vos. 1973. Effects of PCBs and DDE in cormorants and evaluation of PCB residues from an experimental study. *J. Reprod. Fertil. Suppl.* 19:353–364.

Koeman, J. H., W. S. M. van de Ven, J. J. M. de Goeij, P. S. Tjioe, and J. L. van Haaften. 1975. Mercury and selenium in marine mammals and birds. *Sci. Tot. Environ.* 3:279–287.

Koster, M. D., D. P. Ryckman, D. V. C. Weseloh, and J. Struger. 1996. Mercury levels in Great Lakes herring gull (*Larus argentatus*) eggs, 1972–1992. *Environ. Pollut.* 93:261–270.

Kubiak, T. J., H. J. Harris, L. M. Smith, T. R. Schwartz, J. A. Stalling, L. Sileo, D. E. Docherty, and T. C. Erdman. 1989. Microcontaminants and reproductive impairment of the Forster's tern on Green Bay, Lake Michigan—1983. *Arch. Environ. Contam. Toxicol.* 18:706–727.

Kuzyk, Z. Z. A., N. M. Burgess, J. P. Stow, and G. A. Fox. 2003. Biological effects of marine PCB contamination on black guillemot nestlings at Saglek, Labrador: liver biomarkers. *Ecotoxicology* 12:183–197.

Lane, R. W., and J. F. Borzelleca. 2007. Harming and helping through time: the history of toxicology. In *Principles and Methods of Toxicology*, ed. A. W. Hayes, pp. 3–43. New York: Informa Healthcare USA, Inc.

Larson, J. M., et al. 1996. Reproductive success, developmental anomalies, and environmental contaminants in double-crested cormorants (*Phalacrocorax auritus*). *Environ. Toxicol. Chem.* 15:553–559.

Levine, B. 2003. Postmortem forensic toxicology. In *Principles of forensic toxicology*, ed. B. Levine, pp. 3–13. Washington, DC. American Association for Clinical Chemistry.

Linsdale, J. M. 1931. Facts concerning the use of thallium in California to poison rodents—its destructiveness to game birds, song birds and other valuable wild life. *Condor* 33:92–106.

Livingston, M. L. 1952. Parathion poisoning in geese. *J. Amer. Vet. Med. Assoc.* 120:27.

Locke, L. N., G. E. Bagley, and H. D. Irby 1966. Acid-fast intranuclear inclusion bodies in the kidneys of mallards fed lead shot. *Bull. Wildl. Dis. Assoc.* 2:127–131.

Longcore, J. R., L. N. Locke, G. E. Bagley, and R. Andrews. 1974. Significance of lead residues in mallard tissues. *U.S. Fish Wildl. Ser.*, Special Scientific Report—Wildlife No. 182, 24 pp.

Ludke, J. L., E. F. Hill, and M. P. Dieter. 1975. Cholinesterase (ChE) response and related mortality among birds fed ChE inhibitors *Arch. Environ. Contam. Toxicol.* 3:1–21.

Ludwig, J. P., et al. 1996. Deformities, PCBs and TCDD-equivalents in double-crested cormorants (*Phalacrocorax auritus*) and Caspian terns (*Hydroprogne caspia*) of the upper Great Lakes 1986–1991: testing a cause-effect hypothesis. *J. Great Lakes Res.* 22:172–197.

Magath, T. B. 1931. Lead poisoning in wild ducks. *Proc. Staff Mtg. Mayo Clinic* 6:749–752.

Mahajan, S. S., and A. B. Rifkind. 1999. Transcriptional activation of avian CYP1A4 and CYP1A5 by 2,3,7,8-tetrachlorodibenzo-*p*-dioxin: differences in gene expression and regulation compared to mammalian CYP1A1 and CYP1A2. *Toxicol. Appl. Pharmacol.* 155:96–106.

Martinovic, B., D. R. S. Lean, C. A. Bishop, E. Birmingham, A. Secord, and K. Jock. 2003. Health of tree swallow (*Tachycineta bicolor*) nestlings exposed to chlorinated hydrocarbons in the St. Lawrence River Basin. Part II. Basal and stress plasma corticosterone concentrations. *J. Toxicol. Environ. Health A* 66:2015–2029.

McCarthy, J. F., and L. R. Shugart. 1990. *Biomarkers of environmental contamination.* Boca Raton, FL: Lewis Publishers, Inc.

McCarty, J. P., and A. L. Secord. 1999a. Reproductive ecology of tree swallows (*Tachycineta bicolor*) with high levels of polychlorinated biphenyl contamination. *Environ. Toxicol. Chem.* 18:1433–1439.

McCarty, J. P., and A. L. Secord. 1999b. Nest-building behavior in PCB-contaminated tree swallows. *Auk* 116:55–63.

McKinney, M. A., L. S. Cesh, J. E. Elliott, T. D. Williams, D. K. Garcelon, and R. J. Letcher. 2006. Novel brominated and chlorinated contaminants and hydroxylated analogues among North American west coast populations of bald eagles (*Haliaeetus leucocephalus*). *Environ. Sci. Techol.* 40:6275–6281.

McNabb, F. M. A., and G. A. Fox. 2003. Avian thyroid development in chemically contaminated environments: is there evidence of alterations in thyroid function and development? *Evolut. Devel.* 5:76–82.

Meyer, M. W., D. C. Evers, T. Saulton, and W. E. Braselton. 1995. Common loons (*Gavia immer*) nesting on low pH lakes in northern Wisconsin have elevated blood mercury content. *Water Air Soil Pollut.* 80:871–880.

Meyer, M. W., D. C. Evers, J. J. Hartigan, and P. S. Rasmussen. 1998. Patterns of common loon (*Gavia immer*) mercury exposure, reproduction, and survival in Wisconsin, USA. *Environ. Toxicol. Chem.* 17:184–190.

Mineau, P. 1991. *Cholinesterase-inhibiting insecticides: their impact on wildlife and the environment. Vol 2 Chemicals in agriculture.* New York: Elsevier, 348 pp.

Mineau, P., et al. 1999. Poisoning of raptors with organophosphorus and carbamate pesticides with emphasis on Canada, U.S. and U.K. *J. Raptor Res.* 33:1–35.

Mitchell, R. T., H. P. Blagbrough, and R. C. VanEtten. 1953. The effects of DDT upon the survival and growth of nestling songbirds. *J. Wildl. Manage.* 17:45–54.

Monteiro, L. R., and R. W. Furness. 1997. Accelerated increase in mercury contamination in North Atlantic mesopelagic food chains as indicated by time series of seabird feathers. *Environ. Toxicol. Chem.* 16:2489–2493.

Moore, N. W., and D. A. Ratcliffe. 1962. Chlorinated hydrocarbon residues in the egg of a peregrine falcon (*Falco peregrinus*) from Perthshire. *Bird Study* 9:242–244.

Moore, D. R. J., B. E. Sample, G. W. Suter, B. R. Parkhurst, and R. S. Teed. 1999. A probabilistic risk assessment of the effects of methylmercury and PCBs on mink and kingfishers along East Fork Poplar Creek, Oak Ridge, Tennessee, USA. *Environ. Toxicol. Chem.* 18:2941–2953.

Morrissey, C. A., L. I. Bendell-Young, and J. E. Elliott. 2005. Identifying sources and biomagnification of persistent organic contaminants in biota from mountain streams of southwestern British Columbia, Canada. *Environ. Sci. Technol.* 39:8090–8098.

Morrissey, C. A., C. A. Albert, P. L. Dods, W. R. Cullen, V. Lai, and J. E. Elliott. 2007. Arsenic accumulation in bark beetles and forest birds occupying mountain pine beetle infested stands treated with monosodium methanearsonate (MSMA). *Environ. Sci. Technol.* 41:1494–1500.

Mullin, M. D., C. M. Pochini, S. McCrindle, M. Romkes, S. H. Safe, and L. M. Safe. 1984. High resolution PCB analysis: synthesis and chromatographic properties of all 209 PCB congeners. *Environ. Sci. Technol.* 18:468–476.

Murvoll, K. M., J. U. Skaare, E. Anderssen, and B. M. Jenssen. 2006. Exposure and effects of persistent organic pollutants in European shag (*Phalacrocorax aristotelis*) hatchlings from the coast of Norway. *Environ. Toxicol. Chem.* 25:190–198.

Neigh, A. M., et al. 2007. Reproductive success of passerines exposed to polychlorinated biphenyls through the terrestrial food web of the Kalamazoo River. *Ecotoxicol. Environ. Safety* 66:107–118.

Newman, J. R. 1979. Effects of industrial air pollution on wildlife. *Biol. Conserv.* 15:181–190.

Newsted, J. L., P. D. Jones, K. Coady, and J. P. Geisy. 2005. Avian toxicity reference values for perfluorooctane sulfonate. *Environ. Sci. Toxicol.* 39:9357–9362.

Nichols, J. W., C. P. Larsen, M. E. McDonald, G. J. Niemi, and G. T. Ankley. 1995. Bioenergetics-based model for accumulation of polychlorinated biphenyls by nestling tree swallows. *Tachycineta bicolor. Environ. Sci. Technol.* 29:604–612.

Nichols, J. W., K. R. Echols, D. E. Tillitt, A. L. Secord, and J. P. McCarty. 2004. Bioenergetics-based modeling of individual PCB congeners in nestling tree swallows from two contaminated sites on the upper Hudson River, New York. *Environ. Sci. Technol.* 38:6234–6239.

Norstrom, R. J. 1988. Bioaccumulation of polychlorinated biphenyls in Canadian wildlife. In *Hazards, decontamination, and replacement of PCB, A comprehensive guide*, ed. J.-P. Crine, pp. 85–100. New York: Plenum Publishers.

Norstrom, R. J., and C. E. Hebert. 2006. Comprehensive re-analysis of archived herring gull eggs reconstructs historical temporal trends in chlorinated hydrocarbon contamination in Lake Ontario and Green Bay, Lake Michigan, 1971–1982. *J. Environ. Monit.* 8:835–847.

Norstrom, R. J., T. P. Clark, M. Enright, B. Leung, K. G. Drouillard, and C. R. Macdonald. 2007. ABAM, a model for bioaccumulation of POPs in birds: validation for adult herring gulls and their eggs in Lake Ontario. *Environ. Sci. Technol.* 41:4339–4347.

Norstrom, R. J., R. E. Schweinsberg, and B. T. Collins. 1986. Heavy metals and essential elements in livers of the polar bear (*Ursus maritimus*). *Sci. Tot. Environ.* 48:195–212.

Norstrom, R. J., M. Simon, J. Moisey, B. Wakeford, and D. V. Weseloh. 2002. Geographical distribution (2000) and temporal trends (1981–2000) of brominated diphenyl ethers in Great Lakes Herring gull eggs. *Environ. Sci. Technol.* 36:4783–4789.

Nosek, J. A., S. R. Craven, J. R. Sullivan, S. S. Hurley, and R. E. Peterson. 1992. Toxicity and reproductive effects of 2,3,7,8-tetrachlorodibenzo-*p*-dioxin in ring-necked pheasant hens. *J. Toxicol. Environ. Health* 35:187–198.

Nosek, J. A., J. R. Sullivan, S. R. Craven, A. Gendron-Fitzpatrick, and R. E. Peterson. 1993. Embryotoxicity of 2,3,7,8-tetrachlorodibenzo-*p*-dioxin in the ring-necked pheasant. *Environ. Toxicol. Chem.* 12:1215–1222.

Oaks, J. L., et al. 2004. Diclofenac residues as the cause of vulture population decline in Pakistan. *Nature* 427:630–633.

Ohlendorf, H. M. 2002. The birds of Kesterson Reservoir: a historical perspective. *Aquatic Toxicol.* 57:1–10.

Ohlendorf, H. M. 2003. Ecotoxicology of selenium. In *Handbook of ecotoxicology*, 2nd edition, eds. D. J. Hoffman, B. A. Rattner, G. A. Burton Jr., and J. Cairns Jr., pp. 465–500. Boca Raton, FL: Lewis Publishing Inc.

Ohlendorf, H. M., and R. L. Hothem. 1995. Agricultural drainwater effects on wildlife in central California. In *Handbook of ecotoxicology*, eds. D. J. Hoffman, B. A. Rattner, G. A. Burton Jr., and J. Cairns Jr., pp. 577–595. Boca Raton, FL: Lewis Publishing Inc.

Ormerod, S. J., S. J. Tyler, and I. Juttner. 2000. Effects of point-source PCB contamination on breeding performance and post-fledging survival in the dipper *Cinclus cinclus*. *Environ. Pollut.* 110:505–513.

O'Shea, T. J., and S. Tanabe. 2003. Persistent ocean contaminants and marine mammals: a retrospective overview. In *Toxicology of marine mammals*, eds. J. G. Vos, G. D. Bossart, M. Fournier, and T. J. O'Shea, pp. 99–134. New York, NY: Taylor and Francis.

Pain, D. J. 1996. Lead in waterfowl. In *Environmental contaminants in wildlife: interpreting tissue concentrations*, eds. W. N. Beyer, G. H. Heinz, and A. W. Redmon-Norwood, pp. 251–264. SETAC Special Publication Series. Boca Raton, FL: Lewis Publishing Inc.

Pass, D. A., P. B. Little, and L. H. Karstad. 1975. The pathology of subacute and chronic methylmercury poisoning of mallard ducks (*Anas platyrhynchos*). *J. Comp. Pathol.* 85:7–21.

Pattee, O. H., and D. J. Pain. 2003. Lead in the environment. In *Handbook of ecotoxicology*, 2nd edition, eds. D. J. Hoffman, B. A. Rattner, G. A. Burton Jr., and J. Cairns Jr., pp. 373–408. Boca Raton, FL: Lewis Publishing Inc.

Peakall, D. B. 1992. *Animal biomarkers as population indicators*. New York: Chapman and Hall, 291 pp.

Peakall, D. B. 1996. Dieldrin and other cyclodiene pesticides in wildlife. In *Environmental contaminants in wildlife: interpreting tissue concentrations*, eds. W. N. Beyer, G. H. Heinz, and A. W. Redmon-Norwood, pp. 73–97. Boca Raton, FL: Lewis Publishing Inc.

Peakall, D. B., and G. A. Fox. 1987. Toxicological investigations of pollutant-related effects in Great Lakes gulls. *Environ. Health Perspec.* 71:187–193.

Peakall, D. B., and M. L. Peakall. 1973. Effects of a polychlorinated biphenyl on the reproduction of artificially and naturally incubated dove eggs. *J. Appl. Ecol.* 10:863–868.

Peakall, D. B., and L. R. Shugart. 1993. *Biomarker-research and applications in the assessment of environmental health. NATO ASI Series, Ser H, Cell Biology, Vol. 68*. Berlin: Springer-Verlag, 119 pp.

Peterle, T. J. 1991. *Wildlife toxicology*. New York: Van Nostrand Reinhold.

Peters, K. E., C. C. Walters, and J. M. Moldwan. 2005. *The Biomarker Guide. I. Biomarkers and isotopes in the environment and human history*, 2nd edition. Cambridge: Cambridge University Press, 473 pp.

Phillips, J. C., and F. C. Lincoln. 1930. *American waterfowl, their present situation and the outlook for their future*. New York: Houghton Mifflin Co., 312 pp.

Poland, A., and J. C. Knutson. 1982. 2,3,7,8-Tetrachlorodibenzo-*p*-dioxin and related halogenated aromatic hydrocarbons: examination of the mechanism of action. *Ann. Rev. Pharmacol. Toxicol.* 22:517–554.

Post, G. 1952. The effects of aldrin on birds. *J. Wildl. Manage.* 16:492–497.

Powell, D. C., et al. 1998. Effects of 3,3′,4,4′,5-pentachlorobiphenyl and 2,3,7,8-tetrachlorodibenzo-*p*-dioxin injected into the yolks of double-crested cormorant (*Phalacrocorax auritus*) eggs prior to incubation. *Environ. Toxicol. Chem.* 17:2035–2040.

Prestt, I., D. J. Jefferies, and M. W. Moore. 1970. Polychlorinated biphenyls in wild birds in Britain and their avian toxicity. *Environ. Pollut.* 1:3–26.

Ratcliffe, D. A. 1967. Decrease in eggshell weight in certain birds of prey. *Nature* 215:208–210.

Rattner, B. A. 2009. History of wildlife toxicology. *Ecotoxicology* 18:773–783.

Rattner, B. A., J. S. Hatfield, M. J. Melancon, T. W. Custer, and D. E. Tillitt. 1994. Relation among cytochrome P450, A*h*-active PCB congeners and dioxin equivalents in pipping black-crowned night-heron embryos. *Environ. Toxicol. Chem.* 13:1805–1812.

Rattner, B. A., D. J. Hoffman, M. J. Melancon, G. H. Olsen, S. R. Schmidt, and K. C. Parsons. 2000. Organochlorine and metal contaminant exposure and effects in hatching black-crowned night herons (*Nycticorax nycticorax*) in Delaware Bay. *Arch. Environ. Contam. Toxicol.* 39:38–45.

Rattner, B. A., et al. 2004. Contaminant exposure and reproductive success of ospreys (*Pandion haliaetus*) nesting in Chesapeake Bay regions of concern. *Arch. Environ. Contam. Toxicol.* 47:126–140.

Rattner, B. A., P. C. McGowan, J. S. Hatfield, C.-S. Hong, and S. G. Chu. 2001. Organochlorine contaminant exposure and reproductive success of black-crowned night herons (*Nycticorax nycticorax*) nesting in Baltimore Harbor, Maryland. *Arch. Environ. Contam. Toxicol.* 41:73–82.

Rattner, B. A., et al. 2006. Toxicity and hazard of vanadium to mallard ducks (*Anas platyrhynchos)* and Canada geese (*Branta canadensis*). *J. Toxicol. Environ. Health A* 69:331–351.

Rattner, B. A., et al. 1993. Biomonitoring environmental contamination in pipping black-crowned night heron embryos: induction of cytochrome P450. *Eviron. Toxicol. Chem.* 12:1719–1732.

Rattner, B. A., M. J. Melancon, C. P. Rice, W. Riley Jr., J. Eisemann, and R. K. Hines. 1997. Cytochrome P450 and organochlorine contaminants in black-crowned night-herons from the Chesapeake Bay Region, USA. *Eviron. Toxicol. Chem.* 16:2315–2322.

Rattner, B. A., et al. 2008. Apparent tolerance of turkey vultures (*Cathartes aura*) to the non-steroidal anti-inflammatory drug diclofenac. *Environ. Toxicol. Chem.* 27:2341–2345.

Risebrough, R. W., P. Rieche, D. W. Peakall, S. G. Herman, and M. N. Kirven. 1968. Polychlorinated biphenyls in the global ecosystem. *Nature* 220:1098–1102.

Robbins, C. S., and R. E. Stewart. 1949. Effects of DDT on bird population of scrub forest. *J. Wildl. Manage.* 13:11–16.

Roberts, D., R. Tren, R. Bate, and J. Zambone. 2010. *The excellent powder DDT's political and scientific history.* Indianapolis, IN. Dog Ear Publishing, 432 pp.

Rosene, W. Jr. 1965. Effects of field application of heptachlor on bobwhite quail and other wild animals. *J. Wildl. Manage.* 29:554–580.

Ryckman, D. P., et al. 1998. Spatial and temporal trends in organochlorine contamination and bill deformities in double-crested cormorants (*Phalacrocorax auritus*) from the Canadian Great Lakes. *Environ. Monit. Assess.* 53:169–195.

Safe, S. 1984. Polychlorinated biphenyls (PCBs) and polybrominated biphenyls (PBBs): biochemistry, toxicology, and mechanism of action. *Crit. Rev. Toxicol.* 13:319–393.

Safe, S. 1990. Polychlorinated biphenyls (PCBs), dibenzo-*p*-dioxins (PCDDs), dibenzofurans (PCDFs), and related compounds: environmental and mechanistic considerations which support the development of toxic equivalency factors (TEFs). *Crit. Rev. Toxicol.* 21:51–58.

Sagerup, K., E. O. Henriksen, J. U. Skaare, and G. W. Gabrielsen. 2002. Intraspecific variation in trophic feeding levels and organochlorine concentrations in glaucous gulls (*Larus hyperboreus*) from Bjørnøya, the Barents Sea. *Ecotoxicology* 11:119–125.

Sagerup, K., E. O. Henriksen, A. Skorping, J. U. Skaare, and G. W. Gabrielsen. 2000. Intensity of parasitic nematodes increases with organochlorine levels in the glaucous gull. *J. Appl. Ecol.* 37:532–539.

Saita, E., S. Hayama, H. Kajigaya, K. Yoneda, G. Watanabe, and K. Taya. 2004. Histologic changes in thyroid glands from great cormorant (*Phalacrocorax carbo*) in Tokyo Bay, Japan: possible association with environmental contaminants. *J. Wildl. Dis.* 40:763–768.

Sample, B. E., and G. W. Suter. 1999. Ecological risk assessment in a large river-reservoir: 4. Piscivorous wildlife. *Environ. Toxicol. Chem.* 18:610–620.

Sanderson, J. T., and G. D. Bellward. 1995. Hepatic microsomal ethoxyresorufin *O*-deethylase-inducing potency *in ovo* and cytosolic Ah receptor binding affinity of 2,3,7,8-tetrachlorodibenzo-*p*-dioxin: comparison of four avian species. *Toxicol. Appl. Pharmacol.* 132:131–145.

Sanderson, J. T., J. E. Elliott, R. J. Norstrom, P. E. Whitehead, L. E. Hart, K. M. Cheng, and G. D. Bellward. 1994a. Monitoring biological effects of polychlorinated dibenzo-p-dioxins, dibenzofurans and biphenyls in great blue heron chicks. *J. Toxicol. Environ. Health* 41:435–450.

Sanderson, J. T., R. J. Norstrom, J. E. Elliott, L. E. Hart, K. M. Cheng, and G. D. Bellward. 1994b. Biological effects of polychlorinated dibenzo-*p*-dioxins, dibenzofurans, and biphenyls in double-crested cormorant chicks (*Phalacrocorax auritus*). *J. Toxicol. Environ. Health* 41:247–265.

Schafer, E. W. Jr., W. A. Bowles Jr., and J. Hurlbut. 1983. The acute oral toxicity, repellency, and hazard potential of 998 chemicals to one or more species of wild and domestic birds. *Arch. Environ. Contam. Toxicol.* 12:355–382.

Scheuhammer, A. M. 1987. The chronic toxicity of aluminum, cadmium, mercury, and lead in birds: a review. *Environ. Pollut.* 46:263–295.

Scheuhammer, A. M. 1991. Effects of acidification on the availability of toxic metals and calcium to wild birds and mammals. *Environ. Pollut.* 71:329–375.

Scheuhammer, A. M., and P. J. Blancher. 1994. Potential risk to common loons (*Gavia immer*) from methylmercury exposure in acidified lakes. *Hydrobiologia* 278/280:445–455.

Scheuhammer, A. M., C. M. Atchison, A. H. K. Wong, and D. C. Evers. 1998b. Mercury exposure in breeding common loons (*Gavia immer*) in central Ontario, Canada. *Environ. Toxicol. Chem.* 17:191–196.

Scheuhammer, A. M., et al. 2008. Relationships among mercury, selenium, and neurochemical parameters in common loons (*Gavia immer*) and bald eagles (*Haliaeetus leucocephalus*). *Ecotoxicology* 17:93–101.

Scheuhammer, A. M., M. W Meyer, M. B. Sandheinrich, and M. W. Murray. 2007. Effects of environmental methylmercury on the health of wild birds, mammals, and fish. *Ambio* 36:12–19.

Scheuhammer, A. M., J. A. Perrault, and D. E. Bond. 2001. Mercury, methylmercury, and selenium concentrations in eggs of common loons (*Gavia immer*) from Canada. *Environ. Monit. Assess.* 72:79–94.

Scheuhammer, A. M., A. H. K. Wong, and D. E. Bond. 1998a. Mercury and selenium accumulation in common loons (*Gavia immer*) and common mergansers (*Mergus merganser*) from eastern Canada. *Environ. Toxicol. Chem.* 17:197–201.

Scott, T. G., Y. L. Willis, and J. A. Ellis 1959. Some effects of a field application of dieldrin on wildlife. *J. Wildl. Manage.* 23:409–427.

Sheail, J. 1985. *Pesticides and nation conservation: The British experience, 1950–1975.* Oxford: Clarendon Press, 276 pp.

Shear, H., N. Stadler-Salt, P. Bertram, and P. Horvatin. 2003. The development and implementation of indicators of ecosystem health in the Great Lakes basin. *Environ. Monit. Assess.* 88:119–152.

Shultz, S., et al. 2004. Diclofenac poisoning is widespread in declining vulture populations across the Indian subcontinent. *Proc. Royal Soc. Lond. B* 271(Suppl.):S458–S460.

Sileo, L., L. Karstad, R. Frank, M. V. H. Holdrinet, E. Addison, and H. E. Braun. 1977. Organochlorine poisoning of ring-billed gulls in southern Ontario. *J. Wildl. Dis.* 3:313–322.

Skorupa, J. P. 1998a. Selenium poisoning of fish and wildlife in nature: lessons from twelve real world experiences. In *Environmental chemistry of selenium,* eds. W. T. Frankenberger, Jr. and R. A. Engberg, pp. 315–354. New York, NY: Marcel Dekker.

Skorupa, J. P. 1998b. Risk assessment for the biota database of the National Irrigation Water Quality Program. Prepared for the National Irrigation Water Quality Program, U.S. Department of the Interior, Washington, DC.

Smith, T. G., and F. A. J. Armstrong. 1975. Mercury in seals, terrestrial carnivores, and principal food items of the Inuit from Holman, N. W. T. *J. Fish. Res. Board Can.* 32:795–801.

Smits, J. E., K. J. Fernie, G. R. Bortolotti, and T. A. Marchant. 2002. Thyroid hormone suppression and cell-mediated immunomodulation in American kestrels (*Falco sparverius*) exposed to PCBs. *Arch. Environ. Contam. Toxicol.* 43:338–344.

Spalding, M. G., et al. 2000. Histologic, neurologic, and immunologic effects of methylmercury in captive great egrets. *J. Wildl. Dis.* 36:423–435.

Sparling, D. W., G. Linder, and C. A. Bishop. 2000. *Ecotoxicology of amphibians and reptiles.* Pensacola, FL: SETAC Press, 877 pp.

Spear, P. A., D. H. Bourbonnais, R. J. Norstrom, and T. W. Moon. 1990. Yolk retinoids (vitamin A) in eggs of the herring gull and correlations with polychlorinated dibenzo-*p*-dioxins and dibenzofurans. *Environ. Toxicol. Chem.* 9:1053–1061.

Spry, D. J., and J. G. Wiener. 1991. Metal bioavailability and toxicity to fish in low-alkalinity lakes: a critical review. *Environ. Pollut.* 71:243–304.

Stalling, D. L., R. J. Norstrom, L. M. Smith, and M. Simon. 1985. Patterns of PCDD, PCDF and PCB contamination in Great Lakes fish and birds and their characterization by principal components analysis. *Chemosphere* 14:627–643.

Stapleton, M., P. O. Dunn, J. McCarty, A. Secord, and L. A. Whittingham. 2001. Polychlorinated biphenyl contamination and minisatellite DNA mutation rates of tree swallows. *Environ. Toxicol. Chem.* 20:2263–2267.

Stickel, L. F., W. H. Stickel, and R. Christensen. 1966. Residues of DDT in brains and bodies of birds that died on dosage and in survivors. *Science* 151:1549–1551.

Stickel, L. F., W. H. Stickel, and J. W. Spann. 1969. Tissue residues of dieldrin in relation to mortality in birds and mammals. In *Chemical fallout: Current research on persistent pesticides*, eds. M. W. Miller and G. G. Berg, pp. 174–204, Springfield, IL: Charles C. Thomas.

Stickel, L. F., S. N. Wiemeyer, and L. J. Blus. 1973. Pesticide residues in eggs of wild birds: adjustment for loss of moisture and lipid. *Bull. Environ. Contam. Toxicol.* 9:193–196.

Stickel, W. H., L. F. Stickel, and F. B. Coon. 1970. DDE and DDD residues correlated with mortality of experimental birds. In *Pesticide symposia*, ed. W. P. Deichmann, pp. 287–294. Miami, FL: Helios and Associates.

Stickel, W. H., L. F. Stickel, R. A. Dyrland, and D. L. Hughes. 1984. Aroclor 1254® residues in birds: lethal levels and loss rates. *Arch. Environ. Contam. Toxicol.* 13:7–13.

Stone, W. B., J. C. Okoniewski, and J. R. Stedelin. 1999. Poisoning of wildlife with anticoagulant rodenticides in New York. *J. Wildl. Dis.* 35:187–193.

Stone, W. B., J. C. Okoniewski, and J. R. Stedelin. 2003. Anticoagulant rodenticides and raptors: recent findings from New York, 1998–2001. *Bull. Environ. Cont. Toxicol.* 70:34–40.

Strause, K. D., et al. 2007a. Plasma to egg conversion factor for evaluating polychlorinated biphenyl and DDT exposures in great horned owls and bald eagles. *Environ. Toxicol. Chem.* 26:1399–1409.

Strause, K. D., et al. 2007b. Risk assessment of great horned owls (*Bubo virginianus*) exposed to polychlorinated biphenyls and DDT along the Kalamazoo River, Michigan, USA. *Environ. Toxicol. Chem.* 26:1386–1398.

Strong, L. 1938. Insect and pest control in relation to wildlife. *Trans 3rd N. Am. Wildl. Conf.*, American Wildlife Institute, Washington, DC, USA, pp. 543–547.

Struger, J., and D. V. Weseloh. 1985. Great Lakes caspian terns: egg contaminants and biological implications. *Col. Waterbirds* 8:142–149.

Swan, G. E., et al. 2006. Toxicity of diclofenac to *Gyps* vultures. *Biol. Letters* 2:279–282.

Tejning, S. 1967. Biological effects of methyl mercury dicyandiamide-treated grain in the domestic fowl *Gallus gallus* L. *Oikos* (Suppl 8):116.

Thiel, D. A., S. G. Martin, J. W. Duncan, M. J. Lemke, W. R. Lance, and R. E. Peterson. 1988. Evaluation of the effects of dioxin-contaminated sludges on wild birds. In: Proceedings of the 1988 Technical Association of Pulp and Paper Environmental Conference, Charleston, SC, pp. 145–148.

Thompson, D. R. 1996. Mercury in birds and terrestrial mammals. In *Environmental contaminants in wildlife: Interpreting tissue concentration*, eds. W. N. Beyer, G. H. Heinz, and A. W. Redmond-Norwood, pp. 341–356. Boca Raton, FL: Lewis Publishers Inc.

Thompson, D. R., and R. W. Furness. 1989. The chemical form of mercury stored in South Atlantic seabirds. *Environ. Pollut.* 60:305–317.

Tillitt, D. E., et al. 1992. Polychlorinated biphenyl residues and egg mortality in double-crested cormorants from the Great Lakes. *Environ. Toxicol. Chem.* 11:1281–1288.

Toschik, P. C., et al. 2005. Effects of contaminant exposure on reproductive success of ospreys (*Pandion haliaetus*) nesting in Delaware River and Bay, USA. *Environ. Toxicol. Chem.* 24:617–628.

[USACHPPM] US Army Center for Health Promotion and Preventive Medicine. 2000. Standard practice for wildlife toxicity reference values. Technical Guide No. 254. http://usaphcapps.amedd.army.mil/erawg/tox/tg254(Oct00final).pdf

[U.S. EPA] U.S. Environmental Protection Agency. 1995. Final water quality guidance for the Great Lakes system; final rule. *Federal Register* 60:15366–15425.

[U.S. EPA] U.S. Environmental Protection Agency. 2008. Final risk mitigation decision for ten rodenticides. http://www.epa.gov/pesticides/reregistration/rodenticides/finalriskdecision.htm (accessed June 20, 2008).

Van den Berg, M., et al. 1998. Toxic equivalency factors (TEFs) for PCBs, PCDDs, PCDFs for humans and wildlife. *Environ. Health Perspect.* 106:775–792.

Van Velzen, A. C., W. B. Stiles, and L. F. Stickel. 1972. Lethal mobilization of DDT by cowbirds. *J. Wildl. Manage.* 36:733–739.

Verreault, J., R. J. Letcher, E. Ropstad, E. Dahl, and G. W. Gabrielsen. 2006b. Organohalogen contaminants and reproductive hormones in incubating glaucous gulls (*Larus hyperboreus*) from the Norwegian Arctic. *Environ. Toxicol. Chem.* 25:2990–2996.

Verreault, J., S. Shahmiri, G. W. Gabrielsen, and R. J. Letcher. 2007. Organohalogen and metabolically-derived contaminants and associations with whole body constituents in Norwegian Arctic glaucous gulls. *Environ. Internat.* 33:823–830.

Verreault, J., R. A. Villa, G. W. Gabrielsen, J. U. Skaare, and R. J. Letcher. 2006a. Maternal transfer of organo-halogen contaminants and metabolites to eggs of Arctic-breeding glaucous gulls. *Environ. Pollut.* 144:1053–1060.

Verrett, M. J. 1970. Effects of 2,4,-T on man and the environment. In: Hearings before the Subcommittee on Energy, Natural Resources and the Environment of the Committee on Commerce, US Senate, Serial 91–60. US Government Printing Office, Washington DC.

Wasserman, M., D. Wasserman, S. Cucos, and H. J. Miller. 1979. World PCBs map: storage and effects in man and his biologic environment in the 1970's. *Ann. N. Y. Acad. Sci.* 320:69–124.

Wax, P. M. 2006. Historical principles and perspectives. In *Goldfrank's toxicological emergencies*, 8th edition, eds. L.R. Goldfrank, N. Flomenbaum, R.S. Hoffman, M. A., N. A. Lewin, and L. S. Nelson, pp. 1–17. New York, NY: McGraw-Hill.

Wayland, M., S. Trudeau, T. Marchant, D. Parker, and K. A. Hobson. 1998. The effect of pulp and paper mill effluent on an insectivorous bird, the tree swallow. *Ecotoxicology* 7:237–251.

Weech, S. A., A. M. Scheuhammer, J. E. Elliott, and K. M. Cheng. 2004. Mercury in fish from the Pinchi Lake region, British Columbia, Canada. *Environ. Pollut.* 131:275–286.

Weech, S. A., A. M. Scheuhammer, and J. E. Elliott. 2006. Mercury exposure and reproduction in fish-eating birds breeding in the Pinchi Lake region, British Columbia, Canada. *Environ. Toxicol. Chem.* 25:1433–1440.

Wetmore, A. 1919. Lead poisoning in waterfowl. U.S. Department of Agriculture, Bulletin 793, 12 pp.

White, D. H., and J. T. Seginak. 1994. Dioxins and furans linked to reproductive impairment in wood ducks. *J. Wildl. Manage.* 58:100–106.

Whitehead, F. E. 1934. The effect of arsenic, as used in poisoning grasshoppers, upon birds. Oklahoma Agricultural and Mechanical College Agriculture Experiment Station. Experiment Station Bulletin Number 218, 54 pp.

Wiemeyer, S. N. 1996. Other organochlorine pesticides in birds. In: *Environmental contaminants in wildlife: interpreting tissue concentrations,* eds. W. N. Beyer, G. H. Heinz, and A. W. Redmon-Norwood, pp. 99–115. SETAC Special Publication Series. Boca Raton, FL: Lewis Publishing Inc.

Wiemeyer, S. N., C. M. Bunck, and C. J. Stafford. 1993. Environmental contaminants in bald eagle eggs—1980–1984—and further interpretations of relationships to productivity and shell thickness. *Arch. Environ. Contam. Toxicol.* 24:213–227.

Wiener, J. G., D. P. Krabbenhoft, G. H. Heinz, and A. M. Scheuhammer. 2003. Ecotoxicology of mercury. In *Handbook of Ecotoxicology*, 2nd edition, eds. Hoffman, D. J., B. A. Rattner, G. A. Burton, and J. Cairns, pp. 409–463. Boca Raton, FL, CRC Press.

Wigglesworth, V. B. 1945. DDT and the balance of nature. *The Atlantic Monthly* 176:107–113.

Williams, L. L., J. P. Giesy, D. A. Verbrugge, S. Jurzysta, and K. Stromborg. 1995. Polychlorinated biphenyls and 2,3,7,8-tetrachlorodibenzo-*p*-dioxin equivalents in eggs of double-crested cormorants from a colony near Green Bay, Wisconsin, USA *Arch. Environ. Contam. Toxicol.* 29:327–333.

Wobeser, G., and M. Swift. 1976. Mercury poisoning in a wild mink. *J. Wildl. Dis.* 12:335–340.

Wobeser, G., N. O. Nielsen, and B. Schiefer. 1976. Mercury and mink. II. Experimental methyl mercury intox-ication. *Can. J. Comp. Med.* 40:34–45.

Wolfe, M. F., S. Schwarzbach, and R. A. Sulaiman. 1998. The effects of mercury on wildlife: A comprehensive review. *Environ. Toxicol. Chem.* 17:146–160.

Wolfe, M. F., et al. 2007. Wildlife Indicators. In *Ecosystem response to mercury contamination: indica-tors of change*, eds. R. Harris, D. P. Krabbenhoft, R. Mason, M. W. Murray, R. Reash and T. Saltman, pp. 123–189. Webster, New York, NY: CRC Press.

Woodford, J. E., W. H. Karasov, M. W. Meyer, and L. Chambers. 1998. Impact of 2,3,7,8-TCDD exposure on survival, growth, and behaviour of ospreys breeding in Wisconsin, USA. *Environ. Toxicol. Chem.* 17:1323–1331.

Wootton, J. C., and J. C. Alexander. 1959. Some chemical characteristics of the chicken edema disease factor. *J Assoc Offic Agr Chem* 42:141–148.

Wren, C. D. 1985. Probable case of mercury poisoning in a wild otter, *Lutra canadensis*, in northwestern Ontario. *Can. Field Nat.* 99:112–114.

Wren, C. D. 1986. A review of metal accumulation and toxicity in wild mammals. I. Mercury. *Environ. Res.* 40:210–244.

Wren, C. D., D. B. Hunter, J. F. Leatherland, and P. M. Stokes. 1987a. The effects of polychlorinated biphenyls and methylmercury, singly and in combination, on mink. I: Uptake and toxic responses. *Arch. Environ. Contam. Toxicol* 16:441–447.

Wren, C. D., D. B. Hunter, J. F. Leatherland, and P. M. Stokes. 1987b. The effects of polychlorinated biphenyls and methylmercury, singly and in combination on mink. II. Reproduction and kit development. *Arch. Environ. Contam. Toxicol.* 16:449–454.

Wren, C. D., P. M. Stokes, and K. L. Fischer. 1986. Mercury levels in Ontario mink and otter relative to food levels and environmental acidification. *Can. J. Zool.* 64:2854–2859.

Wurster, D. H., C. J. Wurster Jr., and W. N. Strickland. 1965. Bird mortality following DDT spray for Dutch elm disease. *Ecology* 46:488–499.

Yasui, T., et al. 2007. Functional characterization and evolutionary history of two aryl hydrocarbon receptor isoforms (AhR1 and AhR2) from avian species. *Toxicol. Sci.* 99:101–117.

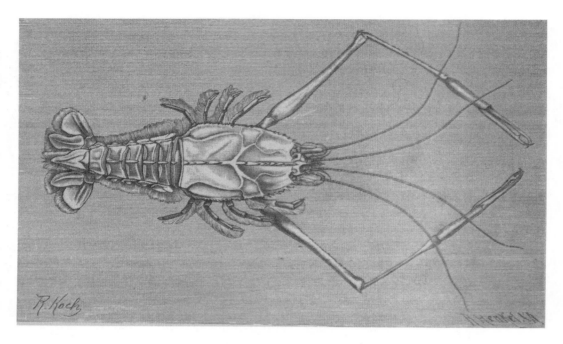

Slender-clawed crayfish

By R. Koch, from *The Royal Natural History,* edited by Richard Lydekker, Frederick
Warne & Co., London, 1893–94.

2 DDT and Other Organohalogen Pesticides in Aquatic Organisms

Nancy Beckvar
Guilherme R. Lotufo

CONTENTS

2.1 Introduction..48
2.2 Analytical Method Considerations..50
2.3 Environmental Occurrence 52
    2.3.1 Water, Sediment, and Air 52
    2.3.2 Tissues...53
2.4 Bioaccumulation 56
    2.4.1 Biomagnification..60
    2.4.2 Lipids 60
    2.4.3 Maternal Transfer 61
2.5 Toxicity 61
    2.5.1 Modes of Action 61
2.6 Tissue-Residue Effects—Fish and Invertebrates.................................63
    2.6.1 DDT 63
        2.6.1.1 p,p'-DDT 63
        2.6.1.2 o,p'-DDT 67
        2.6.1.3 o,p'-DDE, p,p'-DDE..68
        2.6.1.4 o,p'-DDD, p,p'-DDD...69
    2.6.2 DDT Discussion..70
        2.6.2.1 Fish 70
        2.6.2.2 Invertebrates...71
    2.6.3 Cyclodienes..72
        2.6.3.1 Aldrin and Dieldrin 72
        2.6.3.2 Endrin 74
        2.6.3.3 Endosulfan 75
        2.6.3.4 Chlordane...76
        2.6.3.5 Heptachlor and Heptachlor Epoxide..............................77
    2.6.4 Cyclodiene Discussion...78
        2.6.4.1 Fish 78
        2.6.4.2 Invertebrates...79
    2.6.5 Mirex, Lindane, and Toxaphene 80
        2.6.5.1 Mirex...80

        2.6.5.2   Lindane 81

        2.6.5.3   Toxaphene 83

    2.6.6   Mirex, Lindane, and Toxaphene Discussion 84

        2.6.6.1   Fish 84

        2.6.6.2   Invertebrates 84

2.7   Considerations 85

    2.7.1   Mixtures 85

    2.7.2   Resistance/Pre-exposure 86

    2.7.3   Life Stage/Tissue Type 87

Summary 88

References 91

2.1   INTRODUCTION

Organohalogen (OH) compounds are persistent hydrocarbon compounds containing a halogen group, often chlorine or bromine, that substitutes for hydrogen atoms in different positions in the hydrocarbon. They may occur naturally, but this chapter's focus is on synthetically produced compounds, mainly organochlorines, that were produced for use as pesticides. Nine OH compounds (aldrin, chlordane, dichlorodiphenyltrichloroethane [DDT], dieldrin, endrin, heptachlor, hexachlorobenzene, mirex, and toxaphene) are in the top 12 list of particularly toxic and persistent organic pollutants (POPs) identified by the Stockholm Convention treaty implemented in 2004 under the United Nations Environment Program (UNEP). More than 90 countries have signed on to this treaty as Parties. These chemicals became classified as POPs because they may remain in the environment for decades following their use, they accumulate in fatty tissues of exposed organisms, they have a variety of toxic endpoints, and they travel long distances from source areas through atmospheric or aqueous transport.

Synthetic broad-spectrum OH pesticides such as DDT (1,1,1-trichloro-2,2-bis(p-chlorophenyl) ethane) became widely used in agriculture beginning in the 1940s. The term total DDT or ΣDDT refers to the sum of DDT and metabolites: 1,1'-(2,2-dichlor-ethenylidene)-bis[4-chlorobenzene] (DDE); and 1,1-dichloro-2,2-bis(4-chlorophenyl)ethane (DDD, also referred to as TDE); and their ortho para (o,p' or 2, 4) and para para (p,p', or 4, 4) isomers. Use of these compounds generally consisted of wide-scale spraying and initially was supported by their effectiveness in controlling pest vectors and their apparent low acute toxicity to humans and other mammals. Estimated application amount in 1959 exceeds 450,000 metric tons applied to 5% of the land area in the United States, with cotton growing areas of the southeast United States having high DDT, toxaphene, and lindane applications (Johnson 1968). Rothane, a metabolite of DDT known as DDD (p,p' dichlorodiphenyldichloroethane), or TDE (1,1-dichloro-2,2-bis(p-chlorophenyl)ethane) also was applied as a pesticide in the United States (Schmitt et al. 1990, ATSDR 2002).

DDT's acute toxicity to nontarget aquatic organisms was noted early (Ellis et al. 1944). Effects to fish and invertebrates from aerial spraying were observed in the 1950s with dramatic die-offs from pesticide application (Cope 1961). Shortly after DDT spraying, major fish and benthic invertebrate kills were noted in various locations; such as in the Yellowstone River, Wyoming; in New Brunswick hatchery fish after spruce bud worm spraying; and around Lake George, New York after spraying to control gypsy moths. The persistence in tissue, much higher concentrations in tissue compared to water, development of resistance, and an ability to travel long distances in aquatic systems also were noted early in their use. Public outcry and environmental investigations ensued after the publication of Rachael Carson's book *Silent Spring* (Carson 1962), eventually leading to either a complete ban, or restricted use, in a number of countries. DDT use was banned in Canada and Sweden in 1970, in the United States in 1972 (37 FR 13369, July 7, 1972) and in western European countries in the early 1970s. DDT production continued in some countries even after use was banned. Use of DDT for agriculture was banned in India in 1989, but DDT continues to be manufactured and have restricted use for public health. Aldrin and dieldrin were banned in the United States in 1974 (39 FR 37246, October 18, 1974), toxaphene in 1983, and chlordane in 1988. Current use of DDT, supported by the World Health Organization since 2006, includes indoor residual spraying of DDT in dwellings as part of a comprehensive program to control malaria epidemics and transmission in African countries such as Zambia, Uganda, Zimbabwe, and Kenya, and some Asian countries. India has used indoor spraying to reduce disease transmission. Recent global DDT use was estimated at 4000–5000 metric tons per year and production occurred in India, China, and the Democratic People's Republic of Korea (UNEP 2008).

After DDT, aldrin, and dieldrin were the next most heavily used pesticides in the United States during the 1960s–1970s. Use was concentrated on corn crops in mid-western states. Dieldrin was used heavily in the northeast United States on a variety of crops, but was also used for pest control such as termites in the southern states. Aldrin also was used on large scale sugar cane crops in Brazil and Australia. Aldrin use continues in some countries; however, once released to the environment, aldrin readily degrades to dieldrin, which is very persistent. After DDT, dieldrin, and aldrin use were restricted, compounds such as endosulfan and toxaphene were used as replacement pesticides.

The National Pesticide Monitoring Program (NPMP) was established in the United States in 1964 to assess regional and national contaminant trends for these widely used compounds. The National Contaminant Biomonitoring Program (NCBP) originated as part of NPMP and was operated by the U.S. Fish and Wildlife Service (USFWS). This program was eventually expanded into the Biomonitoring of Environmental Status and Trends (BEST) program and transferred to the U.S. Geological Survey (USGS) in 1996. Trends in contamination generally have been downward in the United States but continued presence in tissue highlights the extreme persistence of some of these compounds which remain a threat in some areas. Southern California, home to one of the largest DDT manufacturers in the world, still had an estimated 156 metric tons of DDT in shelf sediments. Sampling in 1994 measured DDT in all Pacific and longfin sanddab and Dover sole collected in the area (Schiff and Allen 2000). The Huntsville Spring Branch in Huntsville, Alabama, received over 400 metric tons of DDT from manufacturing at the Redstone Arsenal from 1947 to 1971. More than 35 years after its ban in the United States, the number of DDT fish advisories continues to increase (U.S. EPA 2006). p,p'-DDT is ranked 12th on the U.S. Environmental Protection Agency's (EPA) 2007 list of priority hazardous substances, dieldrin is 17th and chlordane is 20th (ATSDR 2007).

Many OH compounds are semivolatile and spread locally and globally via the atmosphere. The wide geographic reach of OH pesticide contamination is highlighted by their elevated concentrations in deep sea fish (Looser et al. 2000) and in higher trophic Arctic organisms (AMAP Assessment 2002). Transport to remote Arctic areas may occur through what is known as "grasshopping," chemicals volatilize and condense, resulting in fractionation, as they are transported atmospherically in step-wise fashion to remote areas. Ocean circulation also transports OH compounds to the Arctic, but more slowly than atmospheric transport for most OH compounds (Lohman et al. 2007).

The high molecular weight chlorinated insecticides (e.g., DDT and dieldrin) are particularly concerning due to their ability to bioaccumulate and persist in tissue. Early studies to understand the significance of tissue residues began with these lipophilic and persistent OH pesticides. Reproductive failure noted in New York lake trout populations as a result of DDT spraying was among the first studies to relate the tissue concentration in egg with mortality (Burdick et al. 1964). The challenge of determining sublethal no-effect concentrations in aquatic organisms was noted early. Researchers at the U.S. EPA Gulf Breeze laboratory (currently known as the Gulf Ecology Division Laboratory) were among the first to conduct laboratory studies measuring tissue-residue concentrations associated with effects, or critical body residues (CBR).

Organochlorine pesticides are grouped into categories, the diphenyl aliphatics, the cyclodienes, hexachlorocyclohexane (HCH), and polychloroterpenes. Properties and uses are summarized briefly in Table 2.1. DDT belongs to the diphenyl aliphatic group. The cyclodienes include aldrin and dieldrin, endrin, chlordane, heptachlor, heptachlor epoxide, and endosulfan. Lindane is the gamma

TABLE 2.1
Organohalogen Compound Properties, Products and Uses

Chemical	Formula	Mol. Wt. g/mole	Metabolites	Isomers	Log K_{ow}[a]	Example Uses
DDT[b]	$C_{14}H_9C_{l5}$	354.49	DDE, DDD	o,p', p,p'	6.79*	Agriculture, disease vectors
DDD[b]	$C_{14}H_{10}C_{l4}$	320.05		o,p', p,p'	5.87*	Agriculture
DDE[b]	$C_{14}H_8CL_4$	318.03		o,p', p,p'	6.0*	
Aldrin	$C_{12}H_8CL_6$	364.92	Dieldrin		6.75	Agriculture (corn), soil insects (termites)
Chlordane (pure)	$C_{10}H_6CL_8$	409.78	Oxychlordane, heptachlor, chlordene epoxide	Cis, trans	6.22	Corn, citrus, household pests (termites)
Dieldrin	$C_{12}H_8CL_6O$	380.91			5.45	Agriculture, disease vectors
Endrin	$C_{12}H_8CL_6O$	380.91		Dieldrin	5.45	Agriculture, cotton
Endosulfan	$C_9H_6CL_6O_3S$	406.92	Endosulfan sulfate	Alpha, beta	3.5	Agriculture
Heptachlor	$C_{10}H_5CL_7$	373.32	Heptachlor epoxide		5.86	Agriculture, termites only since 1983 in US
Hexachloro-cycohexane (HCH)	$C_6H_6CL_6$	290.83		Lindane (γ-HCH), 7 others	4.26	Ornamentals, soil pests, head lice
Mirex	$C_{10}CL_{12}$	545.55			7.01	Fire ants, flame retardant
Toxaphene (camphechlor)	$C_{10}H_{10}CL_8$ (average of chlorines)	413.82 (component average)	Mixture of polychlorinated terpenes	Numerous	6.79	Agriculture (cotton), fish eradication

[a] Log K_{ow} estimated in EPA EPI Suite KOWWIN v1.67.
[b] Kow for p,p' isomers.

isomer of HCH. Toxaphene, a mixture of more than 670 chemicals, is a polychloroterpene and mirex has a caged structure. Several OH pesticides were produced as technical mixtures in addition to pure compounds.

This chapter reviews the tissue residue-effect data available for OH compounds measured frequently in tissues of fish and invertebrates. First we review some issues pertinent to interpreting tissue-residue data such as analytical methods, which have changed over time and are continuing to evolve. We also briefly review factors important to the accumulation and elimination of OH compounds (toxicokinetics), modes of action relevant to the tissue-residue literature, and the importance of lipids and maternal transfer in understanding their toxicodynamics. Most sections of this chapter are divided into separate subsections for fish and invertebrates.

2.2 ANALYTICAL METHOD CONSIDERATIONS

To accurately separate, identify, and quantify the variety of chlorinated hydrocarbon concentrations in tissue, several analytical challenges have to be overcome. These are reviewed here to help understand considerations needed to evaluate both older and recent data.

Due to their persistence and widespread occurrence, OH compounds generally occur as mixtures in the tissues of fish and invertebrates. In addition, environmental degradation and/or transformation within the organism results in various metabolites and isomers of the original forms within the tissue. Measured concentrations often are presented as sums of parent and metabolite compounds, and may represent a complex assortment of related compounds. Chlordane and toxaphene contain many components making quantification very difficult.

OH pesticides have characteristics such as high hydrophobicity and consequently tend to bind to, and concentrate in fatty tissues. The efficient and complete extraction of fat-soluble substances from tissue is therefore an important analytical step. Interference with other compounds and accurate quantification methods can present substantial difficulties potentially causing inaccurate measurements. Compound separation achieved by paper chromatography, an early analytical method, was not always sufficient to estimate individual compound concentrations (Burdick et al. 1964), so measured concentrations may have been biased high. The development of gas–liquid chromatography in the early 1950s was one of the most significant advances to improve the quantification of OH pesticides. When coupled with electron capture detection, developed late in the 1950s, researchers became able to more reliably measure low concentrations of individual OH compounds.

Current analytical methods include lipid extraction with solvents, followed by cleanup techniques to remove coextracted compounds. Lipids typically are removed with liquid-liquid partitioning and Florisil adsorption chromatography. Additional adsorption chromatography columns are used for further cleanup, and supplemental cleanup steps may be included. Analytical standards such as standard reference material should be used to estimate measurement errors.

Past use of packed columns could have resulted in coelution of some additional components and less sensitivity than measurements from capillary columns. For example, differentiating between DDD and *o,p'*-DDT was not always possible with packed columns (Buhler et al. 1969, Jarvinen et al. 1976). Sample cleanup with acid and alkaline treatment overcame some packed column problems (Brevik et al. 1996). Comparison of the two column methods showed similar results even without extra cleanup, although the packed column results were higher for the DDT metabolite. Although the results from packed columns produce different concentrations than capillary columns, the differences may fall within experimental uncertainty (Brevik et al. 1996).

Toxaphene quantification can still be problematic due to the large number of components (Muir et al. 2006). Schmitt (2002a) considers toxaphene residue measurement as approximations, even with the analysis techniques used in this decade.

To circumvent restrictions posed by the typical high costs of analytical chemistry and the detection limits that demand exposure to high concentration in the exposure media, investigations of invertebrate and some fish bioaccumulation and CBRs have resorted to the use of radioactive tracer techniques. The exposure compound radioactivity is measured using liquid scintillation counting (LSC) and its specific activity (Pawlisz and Peters 1993). To identify the compounds providing the source of radioactivity being measured, the thin layer chromatographic (TLC) method typically is used (e.g., Guarino et al. 1974, Lotufo et al. 2000a). All studies reporting DDT invertebrate CBRs derived from radioactivity data (Johnson et al. 1971, Lotufo et al. 2000a, 2000b, 2001a, 2001b), except for Mulsow and Landrum (1995), employed TLC as a compound identity confirmatory tool. The radiotracer technique was also employed for deriving CBRs for DDE (Fisher et al. 1999, Hwang et al. 2004) and endrin (Keilty et al. 1988a, 1988b). Because those compounds are not expected to biotransform in invertebrate tissues, all radioactivity was assumed to be associated with the parent compound.

The cost associated with OH measurement is a major obstacle for understanding distribution and risks in countries that continue to use, or have recently banned these products. Muir and Sverko (2006) review best practice approaches and new advances as they relate to needs in countries with emerging economies.

2.3 ENVIRONMENTAL OCCURRENCE

2.3.1 Water, Sediment, and Air

In the environment, OH pesticides have low aqueous solubility; tend to bind to particulates and organic matter; and bind strongly to sediment where they have long residence times. Consequently, many monitoring programs measure concentrations in sediment and biota for assessing OH contaminant trends. Although dissolved concentrations typically are very low, aqueous measurements do provide OH compound transport and flux information. Several recent studies report aqueous concentrations in areas where OH pesticide use is on-going or recent. For example, in South Africa where DDT is used indoors for malarial control, aqueous ΣDDT concentrations in areas downstream of spraying ranged from below detection (<0.05 µg/L) to 1.1–7.0 µg/L (Barnhoorn et al. 2009). Aqueous ΣDDT ranged from 0.02 to 5.2 ng/L in the Daliao River estuary in China (Tan et al. 2009). Aqueous concentrations for other OH pesticides in this region are also provided in this study.

Surface water sampled for 19 organochlorine pesticides in the Bering and Chukchi Seas measured HCH at the highest concentration (Strachan et al. 2001). ΣDDT in this 1993 study ranged from 0.17 to 0.26 ng/L.

Water column DDT concentrations measured in the United States include locations near former discharges of DDTs such as the Palos Verde shelf in Southern California. Aqueous phase DDT concentrations were measured using solid-phase microextraction (SPME) in 2003–2004. Water column concentrations of DDT ranged from <0.073 to 2.58 ng/L for p,p'-DDE and from <0.043 to 0.264 ng/L for o,p'-DDE (Zeng et al. 2005). Water column ΣDDT concentrations measured in 1997 from this area ranged from 2.3 to 14.5 ng/L (Zeng and Venkatesan 1999). In San Francisco Bay, aqueous DDT concentrations ranged from 0.16 to 0.657 ng/L; chlordanes from 0.062 to 0.136 ng/L; and dieldrin from 0.028 to 0.067 ng/L (Connor et al. 2007). Water column concentrations of p,p'-DDE and o,p'-DDE in other parts of the world are summarized in Zeng et al. (2005) and OH compounds in Fowler (1990) for older data.

In their review of toxaphene concentrations in the Great Lakes, Muir et al. (2006) report surface water concentrations are highest in Lake Superior (0.91–1.12 ng/L) and lowest in Lake Ontario and Lake Erie (0.081–0.23 ng/L). In addition to water, toxaphene and HCH are important OH compounds in Great Lakes air and precipitation, especially in late summer and early fall (Muir et al. 2006).

Aqueous concentrations of two current use pesticides (endosulfan and γ-HCH) were studied in Arctic seawater to understand transport patterns and processes (Weber et al. 2006). Air monitoring also is a useful tool for understanding long-range transport and spatial OH compound variation in the Arctic (e.g., Su et al. 2008).

The sea surface microlayer, the thin surface film enriched with organic matter, is another area of study because it may play an important role in OH transport and transfer (García-Flor et al. 2005). Concentrations of organochlorine compounds were enriched in the surface microlayer, sometimes by up to 7 times, compared to concentrations in underlying water.

The extreme persistence of many OH pesticide compounds in freshwater sediments is demonstrated by their continued detection in most U.S. streams measured as part of the USGS NAWQA program (Gilliom et al. 2006). DDE is the metabolite detected most frequently in those areas.

In the marine environment, sediment DDT concentrations on the Palos Verdes shelf from 0.014 to 12.5 mg/kg dw were composed of 92–95% DDE (Zeng and Venkatesan 1999). DDT metabolites may form from aerobic degradation, photochemical decomposition, or abiotic hydrolytic dehydrochlorination. DDE is degraded by reductive dechlorination to DDMU [1,1-dichloro-2,2-bis(4-chlorophenyl)ethylene] under both sulfidogenic and methanogenic conditions. In a study by Eganhouse et al. (2000), DDE concentrations generally comprised 60–70% of the ΣDDT concentrations on the Palos Verde shelf. DDE is very resistant to degradation and only 9–23% of the DDE inventory on the Palos Verdes shelf has been converted to DDMU since releases began in the 1930s (Eganhouse et al. 2000). Therefore, decreases in DDE concentrations are due mainly to remobilization, not degradation.

Sarkar et al. (2008) report on sediment DDT and HCH concentrations in estuaries and bays of some Asian countries where OH pesticide use has been more recent. DDT was prevalent in sediments along the coast of India and the ratio of p,p'-DDT to ΣDDT ranged from 0.36 to 0.75 (Sarkar et al. 2008), consistent with recent use. In China, DDT remains prevalent in coastal sediments, in part due to its use in antifouling paints (Lin et al. 2009). ΣDDT sediment concentrations ranged from 0.009 to 7.35 mg/kg dw in coastal fishing harbors of China with p,p'-DDD comprising 64% of the ΣDDT (Lin et al. 2009). Comparison to DDT sediment concentrations in other harbors of the world are provided in Lin et al. (2009).

2.3.2 Tissues

OH compounds in fish tissue have similar patterns of detection compared to sediment, but typically are detected more frequently because they accumulate to higher concentrations (Gilliom et al. 2006). An early study on the prevalence and distribution of DDT in fish tissue from a variety of widely distributed Wisconsin fish measured DDT (0.021–16.2 mg/kg ww or 0.22–534.6 mg/kg lipid) in every fish sampled in that state (Kleinert et al. 1968). Results from this study were consistent with similar programs in other U.S. states for frequency of detection and range of concentrations. More recently, DDE and DDT were detected in fish from 90% and 30%, respectively, of agricultural streams sampled in the United States (Gilliom et al. 2006). Many local, regional, and national programs monitor OH pesticide concentrations in fish and invertebrate tissue. A number of different State and National monitoring programs in the United States generally have measured decreasing concentrations of DDTs and other OH pesticides in fish and invertebrates from many areas during the past several decades since their ban (e.g., Nowell et al. 1999, Gilliom et al. 2006, O'Connor and Lauenstein 2006). In contrast, monitoring programs in countries where DDT has been used more recently have continued to measure stable or increasing concentrations (Wiktelius and Edwards 1997, Tanabe et al. 2000; Ramu et al. 2007). DDT in coastal mussels of Asia were highest in Hong Kong (maximum 0.58 mg/kg), followed by Vietnam (maximum 0.4 mg/kg) and China (maximum 0.18 mg/kg) in the study by Ramu et al. (2007). This study also reports concentrations for HCH and chlordane in green mussels from these areas.

Table 2.2 provides a small subset of DDT residues in marine/estuarine and freshwater organisms from various localities around the world to provide some examples of concentrations measured in a variety of aquatic organisms. Table 2.2 also presents historical and recent DDT concentrations in fish and invertebrates from locations near DDT manufacturers. DDT biota concentrations measured in the past near DDT manufacturing sometimes reached >100 mg/kg. Locations near former DDT manufacturers continue to have elevated DDT concentrations in organisms (Bettinetti et al. 2006, NOAA and EPA 2007, Hinck et al. 2009).

In the marine environment, fish surveys indicated that benthic feeders have higher concentrations of DDT than pelagic feeders (ATSDR 2002, NOAA and EPA 2007). The reverse was observed in a freshwater lake (Kidd et al. 2001).

The USGS National Water-Quality Assessment program measured the following ratios of DDT metabolites and isomers in whole-body field-collected freshwater fish in the United States (1992–1995): 84.5% p,p'-DDE, 9.6% p,p'-DDD, 4.6% p,p'-DDT, 0.6% o,p'-DDD, 0.4% o,p'-DDE, and 0.3% o,p'-DDT (Wong et al. 2000). After uptake, p,p'-DDT is metabolized to DDD and to DDE, and therefore their ratios change over time.

For compounds resistant to degradation, environmental half-lives, or the time for the average tissue concentration to decrease by 50%, ranged from 9 to 17.7 years for DDT, 7.7–9.4 years for dieldrin and 1–1.9 years for toxaphene in Great Lakes biota (Hickey et al. 2006). DDT in fish from a Norwegian Lake took 5–7 years for 50% reduction (Brevik et al. 1996).

TABLE 2.2
Environmental DDT Concentrations in Aquatic Organisms from Various Locations Globally and in Locations Near Where DDT Was Manufactured

Species	Tissue	Form	Conc. (Mean/ Median or Range) mg/kg ww	Location	Year(s)	Reference
			Estuarine/Marine			
Fish						
Anchovy	Fillet	DDE	0.009	Adriatic Sea	1997	Bayarri et al. 2001
Mackerel	Fillet	DDE	0.0254	Adriatic Sea	1997	Bayarri et al. 2001
Red mullet	Fillet	DDE	0.0088	Adriatic Sea	1997	Bayarri et al. 2001
Herring	WB	ΣDDT	0.01–0.04	N. Baltic Sea	1991	Strandberg et al. 1998
Herring	WB	ΣDDT	0.116	Gulf of Gdansk	1992	Strandberg et al. 1998
Mullet	Muscle	ΣDDT	0.013[a]	Portugal	2001	Ferreira et al. 2004
Mullet	WB	ΣDDT	0.064	China	2001	Nakata et al. 2005
Goatfish	WB	ΣDDT	0.276	Midway Atoll		Hope and Scatolini 2005
Flounder	Muscle	ΣDDT	0.003[a]	Portugal	2001	Ferreira et al. 2004
Perch	WB	ΣDDT	0.0023–0.0064	N. Baltic Sea		Strandberg et al. 1998
Snapper	WB	ΣDDT	0.37–3.1	Suez Canal	2003	Said and Hamed 2005
Rabbitfish	WB	ΣDDT	0.16–0.56	Suez Canal	2003	Said and Hamed 2005
Fish (19 species)	WB	ΣDDT	0.0003–0.04	Indonesia	2003	Sudaryanto et al. 2007
African lungfish	Muscle	ΣDDT	0.7	Uganda		Ssebugere et al. 2009
Fish (17 species)	WB	ΣDDT	0.0018–0.287	China	2004	Qiu et al. 2009
Dog fish	Muscle	ΣDDT	0.033	India		Pandit et al. 2006
Sturgeon	Muscle	ΣDDT	0.012–0.44	Caspian Sea	2001–02	Kajiwara et al. 2003
Catfish & bass	Fillet	ΣDDT	0.00003–0.002	Gulf of Mexico	1996	Lewis et al. 2002
Chinook salmon	WB	ΣDDT	0.0005–0.041	Pacific NW	1996–01	Johnson et al. 2007a
Fish (7 species)		ΣDDT	0.005–0.061	San Francisco Bay	1997	Greenfield et al. 2005
Invertebrates						
Mussels & Oyster		ΣDDT	0.0026[a]	United States	2003	O'Connor and Lauenstein 2006
Oysters			0.0012[a]	Mexico	2000	Carvalho et al. 2009
Clam		DDE	0.0008	Adriatic Sea	1997	Bayarri et al. 2001
Sea urchin		ΣDDT	0.002	Midway Atoll		Hope and Scatolini 2005
Green mussel		ΣDDT	0.24	China	2001	Monirith et al. 2003
Green mussel		ΣDDT	0.12	Hong Kong	1998	Monirith et al. 2003
Green mussel		ΣDDT	0.0042	India	1998	Monirith et al. 2003
Green mussel		ΣDDT	0.004–0.507	China	2005	Guo et al. 2007
Blue mussel		ΣDDT	0.0035	Japan	1994	Monirith et al. 2003
Green mussel		ΣDDT	0.04	Vietnam	1997	Monirith et al. 2003
Squid		DDE	0.0038	Adriatic Sea	1997	Bayarri et al. 2001
Norway lobster		DDE	0.0012	Adriatic Sea	1997	Bayarri et al. 2001
Prawn		ΣDDT	0.002–0.014	China	2005	Guo et al. 2007
Shrimp	WB	ΣDDT	0.014	China	2001	Nakata et al. 2005
Crab	WB	ΣDDT	0.055	China	2001	Nakata et al. 2005

TABLE 2.2 (continued)
Environmental DDT Concentrations in Aquatic Organisms from Various Locations Globally and in Locations Near Where DDT Was Manufactured

Species	Tissue	Form	Conc. (Mean/ Median or Range) mg/kg ww	Location	Year(s)	Reference
			Freshwater			
Fish						
Predator fish	Fillet	ΣDDT	0.008–1.48	500 U.S. Lakes	2000–03	Stahl et al. 2009
Bottom fish	WB	ΣDDT	0.008–1.76	500 U.S. Lakes	2000–03	Stahl et al. 2009
Game and bottom fish	WB + fillets	*p,p'*-DDE	0.056	U.S. Background	1987	U.S. EPA 1992
Largemouth bass	WB	*p,p'*-DDE	2.7	Gila River, AZ	2003	Hinck et al. 2006a
Largemouth bass	WB	ΣDDT	0.059–0.072	Alabama	2004	Hinck et al. 2009
Walleye	WB	ΣDDT	0.076–0.15	Lake Erie	1991–96	Hickey et al. 2006
Carp	WB	*p,p'*-DDE	0.16–0.58	Idaho	1998	Hinck et al. 2006b
Carp	WB	ΣDDT	0.055	China	2000	Nakata et al. 2005
Lake trout	WB	ΣDDT	0.76–1.6	Lake Michigan	1991–98	Hickey et al. 2006
Fish (4 species)	Muscle	ΣDDT	0.00017–0.0026	Thailand	1997	Kumblad et al. 2001
Trahira	Muscle	ΣDDT	0.027–0.074[a]	Brazil	2005	Miranda et al. 2008
Nile Tilapia	Muscle	ΣDDT	0.051	Uganda		Ssebugere et al. 2009
Tilapia	WB	ΣDDT	0.0781	Indonesia	1998, 2003	Sudaryanto et al. 2007
Catfish	Muscle	ΣDDT	0.0009	Uganda		Ssebugere et al. 2009
Lake trout	Muscle	ΣDDT	0.0615	Arctic	2003	Ryan et al. 2005
Invertebrates						
Zooplankton		ΣDDT	0.00013	NW Territories	1995	Kidd et al. 1998
Chironominae		ΣDDT	0.0015	NW Territories	1995	Kidd et al. 1998
Prawn		ΣDDT	0.0001–0.052	China	2005	Guo et al. 2007
			Near Former DDT Manufacturing			
Fish						
Dover sole	Flesh	ΣDDT	39.7	Palos Verdes, CA	1977	Mearns et al. 1991
White croaker	Flesh	ΣDDT	39.17	Palos Verdes, CA	1975	Mearns et al. 1991
Spiny dogfish	Flesh	ΣDDT	81.2	Palos Verdes, CA	1981	Matta et al. 1986
White croaker	Fillet	ΣDDT	3.18	Palos Verdes, CA	2004	NOAA and EPA 2007
Northern anchovy	WB	ΣDDT	0.061	Southern CA Bight	2004	Jarvis et al. 2007
Largemouth bass	WB		0.24–225	Tennessee River	1978	U.S. EPA 2004
Channel catfish	WB		Up to 411.6	Tennessee River	1978	U.S. EPA 2004
Largemouth bass	WB	ΣDDT	0.42–49.8	Tombigbee River, AL	2004	Hinck et al. 2009

continued

TABLE 2.2 (continued)
Environmental DDT Concentrations in Aquatic Organisms from Various Locations Globally and in Locations Near Where DDT Was Manufactured

Species	Tissue	Form	Conc. (Mean/ Median or Range) mg/kg ww	Location	Year(s)	Reference
			Near Former DDT Manufacturing			
Landlocked shad		p,p'-DDT +DDE	0.458	Lake Maggiori, Italy	1998	Bettinetti et al. 2006
Barbel	Muscle	ΣDDT	0.997	Cinca River, Spain	2002	de la Cal et al. 2008
Bleak	WB	ΣDDT	0.840	Cinca River, Spain	2002	de la Cal et al. 2008
Invertebrates						
Coastal mussels		ΣDDT	3.24	Palos Verdes, CA	1969	Matta et al. 1986
Mussels		ΣDDT	0.092[a]	Palos Verdes, CA	2005	Kimbrough et al. 2008
Mussels		ΣDDT	0.013–0.035	Lake Maggiore, Italy	2003	Binelli and Provini 2003
Zooplankton		ΣDDT	0.387	Palos Verdes, CA	1969	Matta et al. 1986
Penaid shrimp		ΣDDT	4.49	Palos Verdes, CA	1969	Matta et al. 1986
Lobster		ΣDDT	0.562	Palos Verdes, CA	1976	Mearns et al. 1991

[a] Converted from dry-weight using 80% moisture.

2.4 BIOACCUMULATION

OH compounds readily bioaccumulate in aquatic organisms, especially in lipid-rich tissues. OH uptake occurs through passive diffusion across gills, dermal surfaces, and digestive organs. Uptake and elimination kinetics (toxicokinetics) and related topics are briefly reviewed in this section to help interpret tissue-residue toxicity.

The tendency of OH compounds to partition into lipid is an important factor influencing toxicokinetics and the degree of hydrophobicity is expressed as the n-octanol (octanol) water partition coefficient (K_{ow}). K_{ow} is the ratio of the solute concentration in octanol to water, where octanol acts as a surrogate for lipid, and the K_{ow} gives an indication about the ability or tendency of the compound to transfer between water-lipid phases. K_{ow} is generally expressed as a logarithm (log K_{ow}), and different techniques exist for measuring log K_{ow} so a range of values exist for any one compound (Pontolillo and Eganhouse 2001, Shen and Wania 2005). The OH compounds covered in this chapter and their log K_{ow} predicted using the U.S. EPA EPIweb software are listed in Table 2.1. Arnot and Gobas (2006) review both data availability and considerations for bioconcentration factors (BCFs) and bioaccumulation factors (BAFs) of organic chemicals in aquatic organisms.

Quantitative structure–activity relationship (QSAR) models relate BCFs and log K_{ow} using linear regression (see discussion in Arnot and Gobas 2006). OH compounds generally follow QSAR model predictions for bioaccumulation. However, BCFs predicted from $\log K_{ow}$ may be over-estimated when the biotransformation rates are high (Oliver and Niimi 1985).

Bioaccumulation is the net result of competing process of chemical uptake into the organism from food and water, and elimination from the organism into the environment. The uptake and elimination of OH compounds is influenced by numerous factors, including exposure temperature and organism size (e.g., Landrum et al. 1992, Lotufo et al. 2000a), therefore influencing net bioaccumulation in an organism. The time to steady-state bioaccumulation is directly proportional to the rate of elimination of a chemical from the organism, the slower the rate the longer the time to steady state.

Elimination of OH compounds occurs through passive diffusion through dermal surfaces, excretion, egestion of feces, and biotransformation of the parent compound. The elimination rate (k_e) and time to approximate steady-state body residues ($3/k_e$) of DDT and its major transformation products varied substantially across compounds and invertebrate species (Table 2.3). Comparison of the elimination of DDT, DDD, and DDE was reported for four invertebrate species. The slowest

TABLE 2.3
Elimination Rate Coefficient (k_e) and Corresponding Time to Achieve 95% of Steady-State Body Residue (95% SS) in Invertebrates Exposed to OH Compounds in Water

Species	k_e (d^{-1})	Time to 95% SS (d)	Reference
Nereis virens			
DDT	0.024	125	Haya and Burridge 1988
DDT	0.029	102	Kennedy et al. 2010
DDD	0.033	90	Kennedy et al. 2010
DDE	0.073	41	Kennedy et al. 2010
Dieldrin	0.024	125	Haya and Burridge 1988
Endosulfan	0.432	7	Haya and Burridge 1988
Macoma nasuta			
DDT	0.041	73	Boese et al. 1997
DDD	0.078	38	Boese et al. 1997
DDE	0.030	100	Boese et al. 1997
Dieldrin	0.054	56	Boese et al. 1997
Hyalella azteca			
DDT	0.113	27	Lotufo et al. 2000a
DDD	0.180	17	Lotufo et al. 2000a
DDE	0.386	8	Lotufo et al. 2000a
Leptocheirus plumulosus			
DDT	0.48	6.25	Lotufo et al. unpublished
Diporeia spp.			
DDT	0.017	179	Lotufo et al. 2000a
DDD	0.012	250	Lotufo et al. 2000a
DDE	0.014	208	Lotufo et al. 2000a
Heteromastus filiformis			
DDT	0.103	29	Mulsow and Landrum 1995
Acartia erythraea			
DDT	0.01 – 0.05	300–500	Wang and Wang 2005

elimination rates for DDT, DDD, and DDE, and therefore longest time for steady-state, were reported for the Great Lakes amphipods of the genus *Diporeia*, while the fastest were reported for the more widely distributed amphipod *Hyalella azteca* (Lotufo et al. 2000a). The slowest elimination rate for DDT was reported for the copepod *Acartia erythraea*. Few reports on toxicokinetics of other OH pesticides were found for invertebrates. The elimination of endosulfan was much faster than DDT or dieldrin in the polychaete *Nereis virens* (Haya and Burridge 1988), and the elimination rate of dieldrin in the clam *Macoma nasuta* was in the same range as those reported for DDT, DDD, and DDE in the same study (Boese et al. 1997). For fish, the DDT elimination rate constant for the mangrove snapper (*Lutjanus argentimaculatus*) was 0.028 d^{-1} and 0.002 d^{-1} for aqueous and dietary exposure, respectively (Wang and Wang 2005). Elimination of DDT from fish tissue varied by tissue type and was lowest in viscera compared to gills and carcass after aqueous or dietary exposure (Kwong et al. 2008).

Exposure route (aqueous or dietary) may influence tissue distribution and transformation processes (Kwong et al. 2008) but did not influence DDT absorption efficiency (Wang and Wang 2005). The OH compound concentration in the diet may influence assimilation efficiency. DDT contaminant uptake efficiency by fish reported from studies using part per million (mg/kg) range concentrations in diet were 20% in rainbow trout (*Oncorhynchus mykiss*) (Macek et al. 1970), 12–25% in chinook salmon (*O. tshawytscha*), 38–68% in coho salmon (*O. kisutch*) (Buhler et al. 1969) and 17–27% in Atlantic menhaden (*Brevoortia tyrannus*) (Warlen et al. 1977). DDT uptake efficiency by Asian seabass (*Lates calcarifer*) from food containing DDT in the part per billion (μg/kg) range was 98% (Bayen et al. 2005). Assimilation efficiency also may be influenced by prey type, as DDT assimilation in mangrove snappers (*Lutjanus argentimaculatus*) was 72% when clams were the DDT food source and 99% when copepods were the DDT source (Wang and Wang 2005).

Biotransformation is typically an important route of parent compound elimination especially if it results in products that are more water soluble. Metabolic products of endosulfan include endosulfan sulfate, which is not excreted, and alcohol, lactone, and ether, which are excreted. Fifty percent of accumulated dietary endosulfan was estimated to be eliminated by transformation (Berntssen et al. 2008). Endosulfan taken up by striped mullet (*Mugil cephalus*) was metabolized to endosulfan sulfate after 28-day (d) aqueous exposure (Schimmel et al. 1977a). Chlordane metabolizes to oxychlordane and also can be dehydrochlorinated to heptachlor. For DDT, metabolism (chemical reduction) results in reduction of the parent DDT compound in fish and an increase in the more resistant DDE metabolite over the more readily excreted DDD metabolite (Kwong et al. 2008) (Table 2.4). The result is an enrichment of the DDE metabolite in fish tissue after longer exposures. Biotransformation rate data in fish are lacking for most compounds. A QSAR was developed to predict screening level whole-body biotransformation of organic compounds in fish (Arnot et al. 2009).

In invertebrates, the degree and product of the transformation of *p,p'*-DDT varies among species (Table 2.4). Following a 3-d exposure of seven species to DDT in water, the parent compound was present at higher concentration than the sum of the transformation products in five species, and DDE was present at higher concentrations than DDD, which was below detection limit in the tissues of five species (Johnson et al. 1971). The amphipods *Diporeia* spp., exposed for 28 d, transformed only a very small fraction of the DDT entering the tissues to DDD, while most DDT was transformed to DDE in *H. azteca* (Lotufo et al. 2000a). Biotransformation considerations will improve kinetics modeling for compounds subject to transformation, and is especially relevant when parent compound and different metabolites vary in toxic potency.

Variable bioaccumulation of DDT in different tissues and organs is a function of differences in lipid content and biotransformation activity. In catfish (*Heteropneustes fossilis*), concentrations were highest in the liver followed by ovary and brain (Singh and Singh 2007). Holden (1966) noted that DDT distribution in salmon differed under different exposure durations. After acute exposure salmon had higher concentrations in the liver, spleen, and fat, while chronic exposures resulted in liver concentrations similar or lower than muscle concentrations. Organ to whole-body

TABLE 2.4
DDT and Transformation Products Expressed as Percentage of Total Body Residue in Fish and Invertebrates Exposed to Either Technical- or p,p'-DDT

Species	DDT	DDE	DDD	Other	Exposure (Days)	Reference
Fish						
Fathead minnow	7.8	80.3	13.4	0	266	Jarvinen et al. 1976
Brook Trout	45.7	14.8	39.5	0	156	Macek 1968
Goldfish	60.2	34.9	4.8	0	4	Davy et al. 1972
Invertebrates						
Daphnia magna	73.4	19.7	6.6	0	3	Johnson et al. 1971
Gamarus fasciatus	79.1	20.9	0	0	3	Johnson et al. 1971
Palaemonetes kadiakensis	50.9	13.2	7.2	28.7 (DTMC[a], DBP[b])	3	Johnson et al. 1971
Hexagenia bilineata	14.9	85	0	0	3	Johnson et al. 1971
Ishnura vericalis	39.2	60.2	0	0	3	Johnson et al. 1971
Libellula sp.	56.3	28.4	0	15 (DTMC)	3	Johnson et al. 1971
Chironomus sp.	80.8	19.1	0	0	3	Johnson et al. 1971
Neanthes arenaceodentata	63.5	6.3	4.5	25.7 (polar)	28	Lotufo et al. 2000b
Hyalella azteca	34.4	64.4	0	1.2 (polar)	10	Lotufo et al. 2000a
Diporeia spp.	95.7		4	0.3 (polar)	28	Lotufo et al. 2000a
Leptocheirus plumulosus	83.6	9.7	0	6.7 (polar)	2	Lotufo et al. unpublished

Note: All organisms were exposed in water to p,p'-DDT, except brook trout which were exposed to technical DDT and *Neanthes arenaceodentata*, which were exposed to p,p'-DDT-spiked sediment.

[a] 1,1,1-trichoro-2,2,-bis(p-chlorophenyl)ethanol.

[b] 4,4'-dichlorobenzophenone.

relationships can be difficult to predict and may vary among species, between genders, and by age. DDT concentrations in a variety of whole fish were 4–12 times higher and averaged 10-times higher than the skin-off fillet concentrations (NOAA and EPA 2007). White croaker DDT whole-body to fillet ratio was 7–8 in this study.

The fate of DDT and its transformation products in different organs and tissues was examined in crustaceans. In aqueous exposure to DDT, Crosby and Tucker (1971) reported that a significant proportion of the total body residue in daphnids was adsorbed externally by the exoskeleton. A long-term exposure (56 d) of the pink shrimp, *Farfantepenaeus duorarum* (formerly *Penaeus duorarum*), to spiked water resulted in residues of DDT in the hepatopancreas exceeding those in other tissues by wide factors (e.g., 30, 275, and 550 when compared to residues in the gills, digestive tract, and tail muscle, respectively) (Nimmo et al. 1970). For all organs investigated, DDT transformation products were only approximately 10% of the total residue. Similar findings were reported for the lobster *Homarus americanus* (Guarino et al. 1974). Seven days following a 7-d water exposure, approximately 91% of the administered radioactivity was found in the hepatopancreas, 1% in the gill, 4% in the intestine, 3.2% in the egg masses, and 1% in the tail muscle. Similar distribution among organs and tissues was obtained 7 d following a single dose injection of DDT. The hepatic organ contained large amounts of lipids, 50–60% of the wet weight, explaining in part the high concentration relative to other organs. The identity of the radioactivity in the hepatopancreas, determined 48 h after DDT injection, was 91% parent compound, 6.8% DDD, and 2.6% DDE.

2.4.1 Biomagnification

Biomagnification occurs when the thermodynamic activity of the chemical in an organism exceeds that of its diet (Arnot and Gobas 2006). It is expressed as a factor (biomagnification factor or BMF), which for hydrophobic organics is the ratio of the lipid-normalized residue in the predator divided by the lipid-normalized prey residue. Use of stable isotopes to establish trophic relationships has enabled improved quantification and understanding of factors affecting biomagnification. In addition to lipid content, organism size, dietary sources, and intrinsic toxicokinetic parameters have a role in biomagnification (Banas et al. 2009). Biomagnification was reported for DDT (e.g., Fisk et al. 1998, Kidd et al. 2001, Hu et al. 2010), mirex (Fisk et al. 1998), p,p' DDE and chlordane (Strandberg et al. 1998, Ruus et al. 1999), and endosulfan (Berntssen et al. 2008).

Concentrations of some OH compounds biomagnify for specific metabolites. For example, DDT concentrations, in the form of p,p' DDE, and chlordane in the form of oxychlordane and trans-nonachlor increase in higher trophic predator species compared to concentrations in prey (Strandberg et al. 1998, Ruus et al. 1999). β-Endosulfan had higher biomagnification than α-endosulfan (Berntssen et al. 2008).

BMFs were determined for mirex and 3 toxaphene congeners (Fisk et al. 1998). BMFs in rainbow trout were 1.8 and 2.9 for mirex at two concentrations in the diet. For the higher chlorinated toxaphene congeners, BMFs ranged from 2.1 to 4.9, depending on concentration in the diet. In this study of laboratory-derived parameters relating to bioaccumulation, Fisk et al. (1998) observed a curvilinear relationship between log K_{ow} and BMFs. BMFs were highest for compounds with high log K_{ow} and low transformation. The BMF for p,p'-DDT was 1.7 in a freshwater food web in China (Hu et al. 2010). Trophic position was a key determinant, along with lipid, of DDT body residue in a freshwater food web (Kidd et al. 2001). The relationship between lipid and DDT bioaccumulation is explored in more detail in the next section.

2.4.2 Lipids

OH compounds that have low solubility in water (hydrophobic) will partition into lipids within an organism. Lipid content and lipid dynamics are important considerations for understanding OH tissue-residue effect data. Accumulation, transfer, and toxicity of OH compounds can be affected by lipid quantity and type. Contaminants tend to partition into lipids with similar polarity. Polar lipids are associated with cell membranes, which are the site of baseline toxicity. Nonpolar lipids such as triglycerides, cholesterol, and wax esters serve as energy storage. Triglycerides are an important storage lipid in fish. Contaminants can become associated with these storage lipids and are thereby removed from sites of toxic action. Periods of stress, starvation, migration, and/or spawning can draw on and deplete those lipid reserves. As lipid stores are depleted, organochlorine compounds can move to sites of toxic action within the organism where they become available to exert toxic effects (Jørgensen et al. 2006).

The OH–lipid relationship is complicated by a number of factors, but for many compounds and species, lipid normalization can reduce variability in measured tissue residues. For example, lipid normalization reduced the variability of aqueous lindane toxicity across multiple fish species and toxicity to lindane decreased with increasing lipid content (Geyer et al. 1994). Lipid normalization also reduced DDT and DDD lethal CBR variability between two freshwater amphipods (Lotufo et al. 2000a). However, difference in lipid content did not account for major differences in DDT uptake and elimination kinetics between males and females of the polychaete *Neanthes arenaceodenta*, as steady-state lipid-normalized body residues were three times higher in females than in males (Lotufo et al. 2000b).

Understanding how best to interpret the residue-effect literature, and whether to express residue-effects as wet-weight or lipid-normalized (ww or dw residue divided by % tissue lipid) is still being evaluated and the best approach may differ for different compounds, species, and tissue types. For hydrophobic compounds, normalization to a specific lipid pool such as storage lipid, may improve the toxicity relationship (Delbeke et al. 1995). Residue effect studies often do not report lipid concentrations, and lipid concentrations vary seasonally, with life stage and gender. Different measurement techniques and solvents used for lipid analysis can also yield different results on the same samples (Randall et al. 1991, 1998). Understanding lipid dynamics is a key issue for OH residue transfer and toxicity, especially as it relates to maternal transfer and dose to developing sensitive life stages. Elskus et al. (2005) review many of the important factors affecting persistent contaminant toxicity in fish in the context of lipids.

2.4.3 MATERNAL TRANSFER

The maternal transfer of accumulated OH contaminants to developing offspring is a significant and critical pathway for early life-stage exposure and potential for toxicity to this sensitive life stage. Highly lipophilic compounds may be mobilized from parental stores and transferred into the developing eggs as lipids move from adult storage tissues during oogenesis. Lipophilic OH pesticides tend to concentrate in lipid-rich gonads where they are available to exert toxic effects during early development. During oogenesis, estrogen stimulates the formation of the egg yolk precursor protein known as vitellogenin that serves as nutrient reserves to the developing oocyte in the fish ovary. OH compounds can bind to vitellogenin and other lipoproteins and be transported into the developing oocytes (Ungerer and Thomas 1996). Fish early developmental stages are a period of enhanced sensitivity during the formation of the body's major systems. Maternally transferred OH pesticides may have a stronger influence on toxic effects to fish larvae during the first few weeks of feeding than dietary sources (Westin et al. 1985). Bioaccumulation in fish at early life stages is higher than in juvenile or adult fish because of lower metabolic capacity. In addition, enzyme systems may not be sufficiently developed to reduce the toxicity of accumulated compounds.

2.5 TOXICITY

2.5.1 MODES OF ACTION

Early studies on the effects of OH compounds on aquatic organisms focused primarily on mortality endpoints during short-term exposures. Many researchers noted the rapid death of sensitive individuals at relatively low tissue concentrations and the ability of remaining "resistant" organisms to accumulate high residues. Death from central nervous system (CNS) disruption was considered the primary mode of action for these compounds. The focus of recent literature for DDT and some other OHs has shifted toward understanding sublethal effects, in particular endocrine system disruption in fish. The potency of OH pesticides to act as endocrine disruptors in fish and invertebrates is generally low compared to natural hormones, however environmental OH concentrations may be high enough alone, or in combination with other endocrine disrupting substances, to cause effects. The cumulative effect of endocrine disruptors in the environment is an emerging area of research. The next paragraphs provide a brief overview for several modes or mechanisms of action to help interpret residue-effect endpoints frequently encountered in the fish and invertebrate OH tissue-residue literature.

DDT has been described as a neurotoxin (Bloomquist 1996). Accumulation of OH pesticides within the nerve tissue is the site of toxic action for CNS disruption. DDT and other organochlorines act on the CNS via nerve cell membrane sodium channels. Molecules of OH pesticides bind to the lipid-rich membrane-sheaths of nerve axons where they may interfere with sodium and potassium ion permeability. The passage of "action potential" is disrupted and results in uncontrolled spontaneous discharges along the nerve rather than normal responses to stimulation. Endpoints resulting from this mode of action include seizures and behavioral abnormalities that can result in mortality. Cyclodiene pesticides may act on the nervous system through a different mechanism.

For example, lindane, dieldrin, and endosulfan alter chloride ion flux via the gamma-aminobutyric acid (GABA) receptor leading to excitation and hyperactivity, and afterward, suppression (Narahashi 2000). Reduced activity, also known as hypoactivity or lethargy, may result from this mechanism of action (Ballesteros et al. 2009).

For the neurotoxic mode of action, DDT is expected to promote mortality when whole-body residues reach levels well below those observed for organic compounds that act by general narcosis, or baseline toxicity (McCarty and Mackay 1993). However, given the wide range of lethal CBRs in fish and invertebrates, DDT may have a strong neurotoxic action in some organisms, but only a weak action in others; it may produce mortality mostly by a baseline toxicity mechanism. Neurotoxic effects of DDT were noted as spasms, lack of coordination or swimming ability or immobilization in daphnids (*Daphnia magna*) (Crosby and Tucker 1971), crabs (*Cancer irroratus*) (Neufeld and Pritchard 1979), and amphipods (Lotufo et al. 2000a), and hyperactivity in bluegill (*Lepomis macrochirus*) (Ellgaard et al. 1977). It was noted that the amphipods *Diporeia* spp. exposed to DDT and DDD became sluggish and increasingly immobilized long before death occurred (Lotufo et al. 2000a).

Although DDT and DDD are expected to promote neurotoxicity to varying degrees in invertebrates, that mode of action may be weak or not-exist for the transformation product DDE, which produced mortality at levels considered typical for compounds with a baseline toxicity mode of action (Lotufo et al. 2000a, Hwang et al. 2004).

Some OH pesticides are also endocrine disruptors. The endocrine system controls the activity of hormones in the body and substances that interfere with the modulation of these hormones are known as endocrine disruptors. The endocrine (estrogenic) effects of DDT exposure were identified early when cockerels were noted to have decreased testes size and reduced reproduction (Burlington and Lindeman 1950). The similarity in the molecular structure of DDT to synthetic estrogens was noted at that time, but the field of endocrinology was not well developed. Since about the mid-1990s, endocrine disruption from organochlorine exposure has been a major research area for fish and other vertebrates.

Hormone activity can be disturbed by different mechanisms such as OH interaction with hormone receptors, or alteration of steroid synthesis and metabolism (Garcia-Reyero et al. 2006). Endocrine disruptors can interfere with hormone synthesis in the thyroid gland via a number of mechanisms that are still not well-understood. The thyroid controls a number of physiological functions in fish such as reproductive status and embryogenesis. Sex-steroid hormones control sex-specific gonad differentiation activities such as gonadal recrudescence. Substances that mimic sex hormones are known as estrogens or xenoestrogens if they can bind to estrogen receptors and regulate the activity of estrogen responsive genes. Examples of xenoestrogenic substances are: *o,p'* DDD; *p,p'* DDT and *o,p'* DDT; and endosulfan, as described later. Compounds that interfere with male sex hormones are androgenic. Examples include *p,p'* DDE and *p,p'* DDT. Since the endocrine system controls many reproductive functions, disruption can impact the number and health of offspring produced.

During fish sexual maturation, estrogen stimulates the formation of the egg yolk precursor protein, known as vitellogenin, in the liver of female fish. Female fish livers contain high concentrations of estrogen receptors and can synthesize large quantities of vitellogenin, especially during their reproductive season. Vitellogenin serves as nutrient reserves to the developing oocyte in the fish ovary and concentrations increase dramatically during the reproductive phase to enable female fish to form the many, often thousands, of eggs. In contrast, male fish produce no or extremely small quantities of vitellogenin (Sumpter and Jobling 1995). The concentration of plasma vitellogenin indicates estrogen stimulation (Donohoe and Curtis 1996) and is one way to measure exposure to endocrine disrupting compounds in fish. A number of studies in this chapter report vitellogenin and steroid hormone measurements as an endpoint. Induction of vitellogenin was initially considered a biomarker for exposure to endocrine disrupting compounds; however, its connection to reproductive impairment, especially in female fish is becoming supported by more data (Cheek et al. 2001, Thorpe et al. 2007). Therefore, it will be the focus for endocrine disruptive effects reported here.

Studies investigating altered gene expression connected with hormone synthesis from OH exposure is an active area of research but residues are typically not reported (e.g., Zhang and Hu 2008).

A number of studies have examined endocrine responses from exposure to DDT isomers to fish; however, few have measured tissue residues associated with these effects. The naturally occurring estrogen, 17β-estradiol (E_2) is often used as a positive control to compare the responsiveness of potential estrogenic substances. Studies reporting the dose administered to fish egg/embryo via nanoinjection are included in the DDT residue section below.

Endocrine disruption in invertebrates is an area of increasing research but relationship with residues has not been reported to date. Depledge and Billinghurst (1999) suggested that DDT, endrin, toxaphene, and endosulfan may cause endocrine disruption in invertebrates. Exposure of midges to DDE promoted decreased fecundity in association with body residues 2 orders of magnitude lower than those found to promote mortality (Hwang et al. 2004). The DDE effect on insect fecundity may be attributable to endocrine disruption, although the onset of endocrine disruption of DDE may be mechanistically different in insects and fish (Hwang et al. 2004).

2.6 TISSUE-RESIDUE EFFECTS—FISH AND INVERTEBRATES

Studies reporting organismal-level effects associated with a whole-body residue, including appropriate control treatments, were identified with the aid of the Army Corps of Engineer Environmental Residue Effects Database (ERED) as well as library and internet search engines. For some compounds with little or no whole-body effect data, studies that reported fish organ concentrations were also selected. No-effect and low-effect whole-body residue concentrations and the associated effect were identified from each study. CBR data are reported individually in tables or in text descriptions. CBR data are combined and summarized when data are sufficient. The following sections, organized by OH compound, present CBR data for fish and invertebrates. Unless specified, all residue concentrations are wet-weight.

2.6.1 DDT

The DDT pesticide was applied as a technical mixture. Technical DDT is composed of a mixture of DDT forms, approximately 77% p,p'-DDT, 15% o,p'-DDT, 4% p,p'-DDE, and 0.4% DDD (U.S. EPA 1980). The less persistent DDT analogue methoxychlor was once widely used as a pesticide but is not covered in this review.

The ratios of o,p'-isomers to total DDT in the environment can provide information about the DDT source since the form applied as an insecticide was technical DDT enriched in the p,p' DDT form. When the o,p' DDT to DDT ratio is greater than 20%, the source likely is not insecticidal and could be from either industrial DDT production or storage (Nowell et al. 1999). Early studies with DDT primarily dosed fish with the technical mixture, with a high percentage of p,p'-DDT.

The following sections review the residue effect literature for DDT in fish and invertebrates. Sections are grouped by DDT isomer and include literature that has focused on endocrine disruption in fish. Invertebrate CBRs for p,p'-DDT, -DDD, and -DDE from specific studies are presented in Table 2.5. ΣDDT fish CBR ranges and medians are summarized in Table 2.6, and invertebrate CBR ranges and medians for DDT isomers are summarized in Table 2.7.

2.6.1.1 p,p'-DDT

2.6.1.1.1 Fish

In a previous analysis, Beckvar et al. (2005) examined different approaches for developing protective whole-body DDT residues for fish. Using scientific literature primarily consisting of mortality studies and technical or p,p'-DDT exposure, four approaches were explored for analyzing no- and low-observed effect residues from different fish species. Results from the different approaches

TABLE 2.5
Toxicity Associated with DDT, DDD, and DDE in Tissues of Aquatic Invertebrates (mg/kg ww)

Species	No Effect mg/kg	Low Effect mg/kg	Exposure	Effect	Reference
			p,p′-DDT		
Callinectes sapidus	—	0.95	14 d in diet	Mortality	Butler 1969
Callinectes sapidus	0.13	1.0	21 d in diet	Mortality	Leffler 1975
Callinectes sapidus	—	0.30	Single oral dose	Behavior	Neufeld and Pritchard 1979
Penaeus sp.	—	0.26	15 d in diet	Mortality	Butler 1969
Farfantepenaeus duorarum	0.1	0.15	22 d in water	Mortality	Nimmo et al.1970
Hyalella azteca	1.0	2.1	10 d in water	Mortality	Lotufo et al. 2000a
Hyalella azteca	0.5	2.8	10 d in sediment	Mortality	Lotufo et al. 2001b
Diporeia spp.	9	15.6	28 d in water	Mortality	Lotufo et al. 2000a
Diporeia spp.	—	5.9	28 d in sediment	Mortality	Lotufo et al. 2001b
Daphnia magna	128	1150	26 h in water	Mortality	Crosby and Tucker 1971
Leptocheirus plumulosus	1.2	2.7	28 d in sediment	Mortality	Lotufo et al. 2001a
Heteromastus filiformis	3.4	5.9	28 d in sediment	Feeding rate	Mulsow and Landrum 1995
Neanthes arenaceodentata	28	35	28 d in diet	Growth	Lotufo et al. 2000b
			p,p′-DDD		
Hyalella azteca	6.5	15.0	10 d in water	Mortality	Lotufo et al. 2000a
Diporeia spp.	—	93.1	28 d in water	Mortality	Lotufo et al. 2000a
			p,p′-DDE		
Hyalella azteca	30	126.6	10 d in water	Mortality	Lotufo et al. 2000a
Hyalella azteca	—	116.8	10 d in water	Mortality	Landrum et al. 2005
Chironomus riparius	—	60	From 2nd instar to adults in diet	Mortality	Hwang 2000
Chironomus riparius	—	0.8	From 2nd instar to adults in diet	Development delayed, fecundity	Hwang et al. 2004
Lumbriculus variegatus	79.2	178.4	35 d in diet	Mortality	Fisher et al. 1999
Daphnia pulex	—	14	5 d in water	Feeding	Bengtsson et al. 2004

— Not available.

were compared to the mean of the control concentrations from the included studies, and ambient DDT concentrations to determine the reasonableness of the calculated no-effect threshold residues. A concentration of 0.6 mg/kg DDT was calculated as the concentration below which effects (primarily mortality) to juvenile and adult fish were expected to be unlikely. For early life stages a threshold residue of 0.7 mg/kg DDT was calculated from the available literature. Data used in this analysis were mostly older studies reporting lethal effects so the calculated thresholds were considered provisional. Sublethal residue-effect data and data correlating isomer-specific residues with an effect

TABLE 2.6
Summary of CBR Ranges and Medians for OH Compounds in Whole-Body Fish (mg/kg ww)

Chemical	Response	Life Stage	Range	Median[a]	n
ΣDDT[b]	Lethal	Juv/adult	0.29–113	2.38	6
ΣDDT[b]	Lethal	ELS[c]	0.89–24	1.27	9
ΣDDT	Sublethal	Egg	0.005–91	0.045	4
p,p'-DDE	↓ Sex steroids	Adult	0.38–0.59		2
o,p'-DDT	Sublethal	Egg	0.02–91	0.07	3
Dieldrin	Lethal	Juv	0.2–5.9		2
Endrin	Lethal	Juv/adult	0.012–1.7	0.88	9
Endosulfan	Lethal	Juv/adult	0.03–0.36	0.27	3
Chlordane	Lethal	Adult	0.7–16.6	11.4	3
Heptachlor	Lethal	Adult	0.33–34	4.1	6
Heptachlor epoxide	Lethal	Adult	0.23–11	1.0	6
Lindane	Lethal	Adult	5.2–79	10.8	4
Toxaphene	Lethal	Adult	1.9–6.1	3.25	4
Toxaphene	Lethal	Fry/juveniles	24.7–46.6	34	4
Toxaphene	Sublethal	Adult	0.4–5.9	1.95	4
Toxaphene	Sublethal	ELS[c]	0.4–10	0.95	6

[a] Up to 100% mortality included for lethal effects.
[b] From studies reporting lethality in Beckvar et al. (2005); primarily technical or p,p'-DDT.
[c] ELS, early life stages; egg, embryo, and fry.

TABLE 2.7
Summary of CBR Ranges and Medians for OH Compounds in Invertebrates (mg/kg ww)

Chemical	Response	CBR			n
		Min	Max	Median	
p,p'-DDT	Lethal	0.15	1150	2.4	10
p,p'-DDD	Lethal	15	93	—	2
p,p'-DDE	Lethal	60	178	122	4
Dieldrin	Lethal	0.08	2.1	1.05	4
Endrin	Lethal	0.03	358	0.61	7
Endosulfan	Lethal	<0.01	0.21	—	2
Chlordane	Lethal	1.7	9.1	—	2
Heptachlor	Lethal	0.02	3.5	—	2
Mirex	Lethal	0.02	10.4	0.27	11
Lindane	Lethal	0.03	5.2	—	2
Toxaphene	Lethal	0.54	2.7	—	2

were insufficient for analysis. No recent studies using technical or p,p'-DDT were found to add to this previously compiled dataset. Using the studies identified in Beckvar et al. (2005), the median effect concentrations for different life stages were calculated and are reported in Table 2.6 to compare with median CBRs from other OH pesticides.

2.6.1.1.2 Invertebrates

The lowest reported lethal CBR (0.15 mg/kg) for p,p'-DDT was for the pink shrimp, *F. duorarum*, exposed for 22 d to 0.14 µg/L in water in a bioaccumulation experiment (Nimmo et al. 1970) (Table 2.5). The authors stated that this concentration is lethal to those shrimp after 22 d, but did not report percent mortality. Close to 100% mortality likely occurred by day 28, as bioaccumulation data for that sampling period were not provided. A lethal CBR (0.26 mg/kg) was reported for shrimp (*Penaeus* sp.) fed oysters exposed to p,p'-DDT (Butler 1969). After 15 d, mortality was 57% (no statistical comparison provided) and whole-body residue in dead shrimp was 0.26 mg/kg. Mortality in both control shrimp (13%) and DDT treatment groups was partially attributed to cannibalism. The same study reported a lethal CBR of 0.95 mg/kg for crabs (*Callinectes sapidus*) fed DDT-exposed oysters (Butler 1969). After 14 d, mortality (54%) was lower than in control crabs (no statistical comparison provided), where high mortality (38%) was also explained by cannibalism. A later study reported three weekly feedings of 3.2 µg DDT resulting in crab mortality at a body residue of 1 mg/kg (Leffler 1975). The percent mortality associated with this lethal residue was not reported. The study also reported that slightly lower concentration (0.82 mg/kg) caused a significant decrease in ability to regenerate limb and doubled metabolic rate. No effects were associated with a body residue of 0.13 mg/kg. Sublethal effects for adult *C. sapidus* injected with DDT (isomer not specified) targeting 0.3 mg/kg body weight were convulsions and lack of coordination, which disappeared within 24 h (Neufeld and Pritchard 1979).

The lethal CBRs of p,p'-DDT for two amphipods commonly used in toxicity testing, the estuarine *Leptocheirus plumulosus* (Lotufo et al. 2001a) and the freshwater *H. azteca* (Lotufo et al. 2000a), were similar and higher than those reported for decapod crustaceans (Table 2.5). The median lethal CBR for *L. plumulosus* (2.7 mg/kg) and *H. azteca* (2.8 mg/kg), measured after a 28-d exposure to spiked sediment, were almost identical. A slightly lower median lethal CBR (2.1 mg/kg) was observed after a 10-d exposure to water only (Lotufo et al. 2000a). In a 10-d sediment exposure, where no supplemental food was provided, exposure to sublethal DDT concentrations significantly enhanced growth in *L. plumulosus* (Lotufo et al. 2001a). No sublethal effects were reported for *L. plumulosus* (reproduction or decreased growth) and *H. azteca* (decreased growth) in the latter studies.

Freshwater amphipods, *Diporeia* spp., which are ecologically relevant benthic organisms in the Great Lakes, were more tolerant to p,p'-DDT exposure than *H. azteca* and *L. plumulosus*. In 28-d aqueous exposures, the median lethal CBR was 15.6 mg/kg (0.682 µmol/g lipid) (Lotufo et al. 2000a). Using inability to actively swim on contact stimulus as the effect endpoint for lethargy, a substantially lower median CBR of 4.5 mg/kg was reported for the same exposure. In a 28-d sediment exposure, a lower median lethal CBR was reported (5.9 mg/kg) (Lotufo et al. 2001b), but mortality was reported as unexpectedly high and attributed to low lipid content and higher sensitivity of the batch of field organisms used in the experiment.

The CBRs reported for the water flea, *D. magna* (Crosby and Tucker 1971), were orders of magnitude higher than those for other crustaceans. When exposed to p,p'-DDT in water for 26 h, the median lethal CBR was 1170 mg/kg. A substantially lower CBR of 128 mg/kg caused only 4% mortality but promoted immobilization in 64% of the exposed cladocerans. Full recovery of those lethargic animals occurred upon transfer to clean water.

Polychaete worms (*Heteromastus filiformis*) were exposed to increasing spiked sediment concentrations of radiolabeled p,p'-DDT (Mulsow and Landrum 1995). Although no significant mortality was reported, the feeding rate was estimated at various time periods as the quantity of fecal pellets produced per individual per unit time. Significantly decreased feeding was associated with mean CBRs of 5.9 mg/kg obtained in the highest sediment treatment group. The growth of another species of polychaete, *Neanthes arenoceodentata*, exposed to p,p'-DDT-spiked food (fish flakes), sediment or both was significantly decreased at CBRs of 35 mg/kg (Lotufo et al. 2000b). Body residues as high as approximately 140 mg/kg did not elicit lethal effects in those invertebrates, a no-lethal-effect body residue similar to that reported for *D. magna* (Crosby and Tucker 1971).

The range and median CBRs for p,p'-DDT in invertebrate species are summarized in Table 2.7. Several studies reporting body residues of invertebrates exposed to p,p'-DDT in bioaccumulation exposures did not report any associated mortality or sublethal effects. For bivalves, body residue as high as 0.88 mg/kg in *Mya arenaria* and 0.13 mg/kg in *Mercenaria mercenaria* did not impact their survival or feeding behavior (Butler 1971). When freshwater invertebrates were exposed to DDT in water for 3 d for the determination of BCFs, highest reported body residues were 0.014 mg/kg for the dragonfly (*Libellula* sp.) 0.047 mg/kg for the crayfish (*Orconectes nais*) 0.075 mg/kg for the damselfly (*Ischnura verticalis*) 0.1 mg/kg for the grass shrimp, *Palaemonetes kadiakensis*, 0.336 mg/kg for the mayfly, *Hexagenia bilineata*, and 0.440 mg/kg for the midge, *Chironomus* sp., (Johnson et al. 1971). No effects were reported at a body residue of 6 mg/kg in the mayfly (*Ephemera danica*) exposed in the lab (Sodergren and Svensson 1973). Longer exposure to DDT would likely have resulted in higher body residues in the above species and until further investigation is conducted, their maximum tolerance to DDT bioaccumulation is unknown. Ingersoll et al. (2003) reported a p,p'-DDT body residue of 1.9 mg/kg for the oligochaete, *Lumbriculus variegatus*, exposed in the lab. They did not observe mortality, but did observe less activity and smaller size in DDT-exposed oligochaetes compared to control.

2.6.1.2 *o,p'*-DDT

2.6.1.2.1 Fish

Several recent studies examine the effect of maternally transferred DDT isomers using egg-nanoinjection techniques. For fish, DDT concentrations in the eggs were not measured during most of these experiments. Instead, egg residues were estimated based on the injected dose and the approximate weight of the egg/embryo. In the Edmunds et al. (2000) study described below, DDT embryo concentrations measured by gas chromatography were within 20% of the estimated embryo dose. Therefore, doses reported in nanoinjection studies likely estimate actual residues in fish embryos.

Edmunds et al. (2000) injected five doses of *o,p'*-DDT into d-rR strain medaka (*Oryzias latipes*) embryos 6–8 h after fertilization. Triolein injected and uninjected embryos served as controls and estradiol-injected eggs served as positive estrogenic controls. DDT doses were injected into the yolk of the embryos, not the oil globule, unlike studies with DDE described later. Percent hatch, survival to 14 d posthatch and sex-reversal were measured and followed by breeding trials. Survival decreased with increasing DDT dose with a least squares calculated LD_{50} of 511 ng/egg (or 511 mg/kg based on average egg mass of 1 mg). Male to female sex-reversal occurred in 6 of 7 genetic males injected with 227 ng/egg (or 227 mg/kg) DDT. Breeding success of these sex-reversed fish was about 50% and similar to controls. The authors observed that the DDT oil droplet remained intact throughout hatching, so the amount of DDT exposure is uncertain and may not be related to the injected dose.

Adult female medaka were exposed to aqueous concentrations of *o,p'*-DDT to study maternal transfer and effects to offspring (Metcalfe et al. 2000). Adult females were exposed to a single sublethal concentration of 2.5 µg/L *o,p'*-DDT for 2 weeks. During exposure, medaka maternal tissue residues varied from a maximum of 109.6 mg/kg during exposure and decreased to 0.28 mg/kg after 23 weeks. Lipid-weight (lw) DDT concentrations were similar between eggs and maternal tissues. Median time to hatch was significantly longer in treated groups compared to controls, and ovarian development was increased in offspring of DDT-treated fish. The mean DDT concentration in the treated 3- and 6-week-old eggs was 91.2 mg/kg. In the same study, posthatch medaka treated with aqueous exposures of *o,p'*-DDT experienced altered sex ratios and males with testes-ova (intersex). Tissue residues were not measured but testes-ova induction occurred at exposure to about 2 µg/L mean aqueous concentration, and statistically significant difference in sex ratio occurred at about 5 µg/L mean concentration.

Lowered gonadosomatic index (GSI, gonad weight/body weight × 100) was measured in male-summer flounder (*Paralichthys dentatus*) injected subcutaneously with *o,p'*-DDT using a slow-release solvent (Mills et al. 2001). Liver concentrations were highly variable within a treatment group but increased in a dose–responsive manner with decreasing GSI after 8 weeks. Gonad alterations were confirmed by histological analysis. Plasma testosterone concentration decreased in a dose–dependent manner. Liver concentrations in o,p'-DDT treated fish ranged from 188.0 to 521.3 mg/kg compared to 0.3 mg/kg in control flounder after injection with 30, 60, and 120 mg/kg body weight. No effects on these endpoints were observed when fish were injected with p,p'-DDE. Study duration was important as a subset of fish that were followed for 15 weeks showed effects (elevated plasma estradiol) not apparent when the study was concluded at 8 weeks.

Faulk et al. (1999) exposed adult Atlantic croaker (*Micropogonias undulatus*) to o,p'-DDT in the diet and tested offspring to behavioral assays representing behaviors needed for survival such as feeding and predator avoidance. Growth differences were not observed between control and treated fish but behavioral alterations such as burst speed were reduced in larvae of treated fish. DDT egg residues were 0.07 mg/kg in the low treatment group and 0.2 mg/kg in the high treatment group and effects were dose-responsive.

2.6.1.2.2 Invertebrates

Studies reporting CBRs for o,p'-DDT for invertebrates were not found in the available literature. Insufficient data for o,p'-DDT preclude toxicity comparison with p,p'-DDT based on exposure water concentration.

2.6.1.3 o,p'-DDE, p,p'-DDE

2.6.1.3.1 Fish

Villalobos et al. (2003) injected o,p'-DDE into fertilized d-rR strain medaka embryos using a triolein carrier solvent with 4 doses separated by tenfold dilution (0.0005–0.5 ng/egg or mg/kg based on 1 mg medaka egg weight). o,p'-DDE was injected directly into the oil globule of early gastrula embryos. The oil globule is used by the developing fish before hatching. Uninjected eggs and carrier-only injected eggs served as controls. Survival was followed for 2 months. Mortality was statistically different in the highest treatment group with 56% mortality. Mortality occurred most frequently after early development near the time of hatching. The authors note that the morphological defects noted at death were consistent with effects observed from organochlorine exposure in field populations. The survival-based no-observed adverse effect level (NOAEL) and low-observed adverse effect level (LOAEL) egg residues from these dosing intervals were 0.05 and 0.5 mg/kg, respectively.

Papoulias et al. (2003) injected fertilized embryos of d-rR strain medaka using a triolein carrier solvent with three doses of o,p'-DDE (0.005, 0.05, and 0.5 ng/egg, or mg/kg). After injection and hatching, fish were reared until sexual maturity. Reproduction was assessed by quantifying GSI in male and female fish. GSI was significantly reduced in all DDE treatments compared to triolein-injected controls. Histopathic analyses revealed developmental alterations in oocyte formation in the ovaries for all treated females, and reduction of testes size in males at the highest treatment (0.5 ng/egg or mg/kg). Fish in the highest treatment group also weighed significantly more and had the highest mortality (56%). The lowest CBRs for different endpoints from this study were 0.005 mg/kg for reduction in male and female GSI and 0.5 mg/kg for growth (increased) and mortality.

Carlson et al. (2000) injected 21-d postfertilization embryos of rainbow trout and coho salmon using menhaden oil as a carrier. Prior to sexual differentiation and hatching of the fish, but after organ formation, yolks were injected with o,p'-DDE and p,p'-DDE individually or as a mixture. Both oil-injected and noninjected rainbow trout embryos served as control treatments. The ratio of males to females was statistically different in fish injected with 80 or 160 mg/kg o,p'-DDE in one of the experiments, but not in the others. Mortality in oil-injected control rainbow trout was high in two of the experiments and was statistically higher than noninjected control fry in doses greater than 40 mg/kg. Fish that were grown out for several years after injection had normal gonads and spawned successfully. Residues measured in fat at the end of the study were higher in fish injected with the p,p' isomer by a factor of 10 compared to the o,p' isomer.

Milston et al. (2003) report effects from o,p'-DDE exposure on immune function in fall chinook salmon. Fertilized eggs were exposed to aqueous nominal concentrations of 10 and 100 mg/L o,p'-DDE for 1 h after fertilization and an additional 2 h after hatch. After treatment, fish were reared in the hatchery for 1 year when the effect on gonad histology and humoral immunity was assessed. Fry with 0.02 mg/kg o,p'-DDE whole-body residue had significantly reduced ability to respond to a bacterial antigen. These fry also weighed significantly less than controls at the first sampling event. The authors believed that alteration of gonadal steroids during early development likely caused the immune suppression observed 1 year after exposure, although they could not definitely rule out direct toxicity to o,p'-DDE.

Muller et al. (2004) exposed 2-year-old largemouth bass to p,p'-DDE spiked food for 30 and 50 d and noted statistically reduced sex-steroid concentrations, 17β-estradiol and 11-ketotestosterone (E_2 and 11-KT), in both female and male fish at whole-body concentrations of about 0.375 mg/kg DDE. No statistically significant difference in GSI of treated fish compared to control fish was noted in these sexually immature fish.

Johnson et al. (2007b) measured sex-steroid concentrations in 2-year-old largemouth bass exposed to p,p'-DDE in food for 120 d. Concentrations of E_2 in female bass were significantly reduced compared to controls at 0.589 mg/kg p,p'-DDE in female carcass (ovary not included, Johnson 2005). The authors were not able to fit a Hill slope dose–response relationship using the dose in the food (not the tissue residues), and did not report whether the data could be fit to other dose–response models.

Other studies have measured endocrine activity from exposure to p,p'-DDE; but did not measure associated residues. For example, mature male guppies exposed to p,p'-DDE in food for 30 d demonstrated impacts at cellular through organism level with reduced sperm cell counts, loss in sexual coloration and changes in courtship behavior, and reduction in clutch size (Baatrup and Junge 2001). The sperm count of males fed p,p'-DDE at the lowest dose had a hormetic effect, that is, was higher than control fish. Juvenile male guppies exposed to the same concentration in food experienced delayed development time, a skewed sex ratio, reduced sperm cell counts, reduced growth, and altered sexual display coloration (Bayley et al. 2002). Male Japanese medaka exposed to p,p'-DDE had reduced GSI and the highest exposure group had intersex (Zhang and Hu 2008).

2.6.1.3.2 Invertebrates

Lethal CBRs for p,p'-DDE were similar for a variety of freshwater invertebrates (Table 2.5). Median lethal CBRs were 126.6 mg/kg for *H. azteca* exposed in water for 10 d (Lotufo et al. 2000a), 116.8 mg/kg at day 10 for *H. azteca* exposed in water for 28 d (Landrum et al. 2005), and 178.4 mg/kg for *L. variegatus* exposed via feeding on algae grown in spiked medium (Fisher et al. 1999). For the midge *Chironomus riparius* fed algae exposed to p,p'-DDE, a lethal body residue of 60 mg/kg was reported (Hwang 2000). Sublethal effects associated with DDE, were significantly delayed development time in female midges and decreased fecundity occurring at approximately 0.8 mg/kg (Hwang et al. 2004). For midge *Chironomus dilutus* (formerly *C. tentans*), body residues ranging from 10 to 60 mg/kg were reported as causing developmental delay (Derr and Zabik 1972). A body residue of 14 mg/kg for sublethal effect was reported for the cladoceran *D. magna* associated with decreased feeding rate (Bengtsson et al. 2004). No significant mortality occurred in *Diporeia* spp. exposed for 28 d to a water concentration approaching the DDE solubility limit and producing body residues as high as 429 mg/kg (Lotufo et al. 2000a).

Studies reporting CBRs for o,p'-DDE for invertebrates were not found in the available literature.

2.6.1.4 o,p'-DDD, p,p'-DDD

2.6.1.4.1 Fish

o,p'-DDD is often present in tissue either as a metabolite of DDT or from the technical mixture. A high DDD residue may indicate past use of the insecticide rothane. DDD acts on adrenal steroidogenesis in fish affecting cortisol production but effects have not been associated with residues. No dosing studies could be found that exposed fish to the DDD metabolite and measured residue-effects.

2.6.1.4.2 Invertebrates

The lethal CBR of p,p'-DDD was reported for freshwater amphipods exposed to spiked water (Lotufo et al. 2000a) (Table 2.5). Median lethal CBRs were 15 mg/kg for *H. azteca* exposed for 10 d and 93.1 mg/kg for *Diporeia* spp. exposed for 28 d. In that study, a mean body residue of 46.7 mg/kg was associated with paralysis characterized by inability to actively swim on contact stimulus. All amphipods in the lowest exposure treatment were affected. No adverse effects were reported in association with a measured body residue of 60 mg/kg in *L. variegatus* (Ingersoll et al. 2003).

Studies reporting CBRs for o,p'-DDD for invertebrates were not found in the available literature.

2.6.2 DDT DISCUSSION

2.6.2.1 Fish

Most of the older literature reporting DDT residues associated with effects to fish focused on lethal body residues from exposure to p,p'-DDT alone, or as the predominant form in the technical formulation (Beckvar et al. 2005). Lethal body residues spanned several orders of magnitude for a variety of fish species with CNS the mode of action studied (Table 2.6). Many authors noted behavioral disorders preceded death, early life stages were more sensitive than adults, and smaller, leaner, or more sensitive fish died more quickly. Few studies reported residues associated with sublethal effects during these early investigations.

Recent investigations on effects to fish have focused on effects from specific DDT isomers, mostly related to endocrine disruption. These studies observed effects at concentrations similar to and much lower than those reported in the older literature (0.005–91 mg/kg). The variety of effect endpoints, dosing techniques (nanoinjection), and life stages preclude detailed data analyses using these recent investigations. However some general insights about sublethal effect endpoints and effect ranges for the different isomers can be summarized from these more recent studies.

The studies dosing with o,p'-DDT measured endocrine-related effects and mortality at very high egg residues. Doses of o,p'-DDT injected into medaka embryos resulted in sex-reversal in males at 227 mg/kg, and survival was reduced by 50% at 511 mg/kg (Edmunds et al. 2000). This study differed from the DDE nanoinjection studies because o,p'-DDT was injected into the yolk, not the oil globule of the embryo. o,p'-DDT fed to Atlantic croaker was found primarily in the triglyceride-rich oil globule in oocytes (Ungerer and Thomas 1996). The way lipids are used from different parts of the fish egg during early embryonic development may impact the ability of DDT to reach sites of toxic action. DDT and lipid movement within developing embryos were correlated in brook trout (Atchison 1976). Comparison of Edmunds et al. (2000) with the studies that injected DDT into other parts of the embryo may not be appropriate.

Two studies using maternal transfer and o,p'-DDT measured effects for different endpoints at very different egg concentrations. Eggs from aqueous-exposed adult female medaka contained a mean concentration of 91.2 mg/kg o,p'-DDT, and median time to hatch was significantly longer for these offsprings (Metcalfe et al. 2000). Eggs from adult female Atlantic croaker exposed to dietary o,p'-DDT had residues of 0.07 mg/kg. At this much lower concentration, larvae demonstrated behavioral alterations. The studies used different species, exposure routes, and effect endpoints, and the relative contribution of these variables on the three order of magnitude difference in egg-effect residues is unknown.

Alterations of sex-steroid hormones were observed in adult fish at residue concentrations from 0.3 to 0.6 mg/kg p,p'-DDE (Muller et al. 2004, Johnson 2005). Fish were exposed as adults in these studies, not during sensitive early life stages, so the effect of a life-time exposure is unknown.

Effects associated with *o,p'*-DDE exposure were reported as doses injected into fish embryos. A reasonable assumption is that the injected embryo dose is a good estimate of the embryo residue concentration (Edmunds et al. 2000). Dose–response relationships also support this assumption. Medaka embryo CBRs associated with increased mortality were 0.5 mg/kg (Papoulias et al. 2003, Villalobos et al. 2003), and an embryo dose/residue of 0.005 mg/kg was associated with reduced gonad weights (GSI) in males and females (Papoulias et al. 2003). Immune response in Chinook salmon fry 1 year after aqueous egg exposure to *o,p'*-DDE was reduced at a measured whole-body CBR of 0.02 mg/kg, with growth significantly reduced at the first sampling event (Milston et al. 2003). CBRs for the DDE isomers in fish were less variable than for the DDT isomer, and the lowest effect concentrations for fish embryos in Atlantic croaker for DDT (Faulk et al. 1999) were similar to that observed for DDE in Chinook salmon (Milston et al. 2003).

The results from different isomer experiments report a variable ability for DDT isomers to affect endocrine activity. This variability may be a function of life-stage differences, and the timing and method of dosing in relation to development and sexual maturation of the organism. Species, gender, and other variables among the reviewed studies make direct comparisons about the strength of different DDT isomers as endocrine disruptors uncertain. For the studies reviewed here, the *o,p'*-DDT isomer appeared to require higher concentrations compared to the DDE isomers for endocrine disruption. This result contrasts with effects in rats where the *o,p'*-DDT isomer was observed to have higher potency for endocrine disruption (ATSDR 2002).

The lowest CBR for fish egg/embryo from the isomer-specific DDT studies reviewed were associated with reproduction and behavior endpoints (0.005 mg/kg *o,p'*-DDE and 0.07 mg/kg *o,p'*-DDT, respectively). Residues associated with mortality in fish eggs were 0.5 mg/kg *o,p'*-DDE. Intersex or sex-reversal endpoints were associated with much higher egg DDT CBRs (80–90 mg/kg *o,p'*-DDT).

Reduction in immune function response in 1-year-old fish exposed as eggs (Milston et al. 2003) was the endpoint associated with the lowest whole-body fry residue (0.02 mg/kg *o,p'*-DDE). Residues associated with reductions in sex-steroid concentrations in adult fish (0.3–0.6 mg/kg) were measured at concentrations comparable to residues associated with mortality reported by Beckvar et al. (2005). However, for the sex-steroid endpoints (Muller et al. 2004, Johnson et al. 2007b), DDE exposure was limited to only the adult life stage. Fish exposed throughout their life-time may experience effects at lower residue concentrations.

Few studies are available to compare summary CBR data for DDT isomers in fish (Table 2.6). Generally, median CBRs for fish egg life stage and sublethal effects are one to two orders of magnitude lower than CBRs for fish exposed during older life stages or lethal effects.

Several studies report reproductive effects from DDT exposure associated with concentrations in fish organs. Hose et al. (1989) observed complete absence of spawning in white croaker (*Genyonemus lineatus*) at concentrations greater than 3.8 mg/kg DDT in ovary, and reduction in spawning at ovary concentrations of 2.2 mg/kg. Ovary concentration >0.2 mg/kg and dieldrin >0.1 mg/kg impaired reproductive success in North Sea whiting (*Merlangius merlangus*) (von Westernhagen et al. 1989). These concentrations are within the range of CBRs observed in the DDT isomer effect studies.

2.6.2.2 Invertebrates

Lethal CBRs reported for invertebrates span a wide range, from 0.15 mg/kg for pink shrimp, to 1150 mg/kg for daphnia with a median 2.4 mg/kg (Table 2.7). While most invertebrates investigated appear to be sensitive to the neurotoxic effects of DDT; some such as daphnia and polychaete worms, are apparently mostly tolerant to DDT specific mode of action. Lethal CBRs for the *p,p'* isomers of DDT, DDD, and DDE were compared using two species of freshwater amphipods. Striking differences in the lethal CBRs for those compounds suggest that they may cause mortality via different modes of action. While DDT, and to a lesser degree DDD, likely act via neurotoxicity impairing the normal functioning of voltage-sensitive sodium channels, DDE likely caused amphipod mortality by a combination of baseline toxicity and a weak specific, yet unknown, mode of action

(Lotufo et al. 2000a). Because the lethal CBR for DDE is much higher (lower toxicity) than for DDT in amphipods, biotransformation acts as a protective mechanism against mortality. The DDE lethal body residues in midges and oligochaetes were similar to those for amphipods, suggesting similar mode of action across taxa. Unfortunately the lethal CBRs of DDT and DDD are unknown for those species, and overall broad comparisons on modes of action for parent and transformation products cannot be established. Based on toxicity comparison using water concentrations, DDD and DDE are more toxic than the parent compound to freshwater planarians (Bonner and Wells 1987), suggesting DDT metabolites may also act by specific modes of action associated with low body residues in some invertebrates.

Lethal body residues for amphipods are among the lowest CBRs reported for invertebrates; growth and reproductive effects were not significant for amphipods (Lotufo et al. 2000a, 2001a). Delayed development in female chironomids and reduced fecundity occurred at a relatively low CBR (0.8 mg/kg, Hwang et al. 2004). Sublethal effects were manifested in polychaete worms with relatively high tolerance to the lethal effect of DDT, as both growth (Lotufo et al. 2000b) and feeding rate (Mulsow et al. 2002). These data suggest that for species tolerant to the neurotoxic effects of DDT, other specific modes of action may cause sublethal effects.

2.6.3 CYCLODIENES

2.6.3.1 Aldrin and Dieldrin

Aldrin and dieldrin exist as pure compounds or technical mixtures. The technical mixture of aldrin contains not less than 85.5% pure aldrin. Dieldrin technical mixture contains not less than 80.75% pure dieldrin. The toxicity of the pure compound is greater than the formulations in both cases. Aldrin and dieldrin were used extensively in the United States until the 1970s and continue to be used elsewhere to control insects such as termites and tsetse fly (Table 2.1). Aldrin rapidly metabolizes to dieldrin in the environment, so even though toxic, tissue residues are generally measured as dieldrin. Review of the residue literature for aldrin therefore primarily deals with dieldrin in tissue.

2.6.3.1.1 Fish

Even though dieldrin is highly accumulated in fish tissue and widespread, studies for both lethal and sublethal residue-effects are limited (Table 2.6). A dieldrin concentration of 1.21 mg/kg in winter flounder (*Pseudopleuronectes americanus*) eggs was associated with 100% mortality (Smith and Cole 1973). These eggs also contained 0.76 mg/kg DDT. Juvenile rainbow trout exposed to dieldrin in water for 16 weeks had whole-body residues of 0.2 mg/kg and increased mortality compared to control fish, but no statistics were provided (Shubat and Curtis 1986). Juvenile rainbow trout had estimated whole-body dieldrin residues of 5.9 mg/kg in the treatment group with 100% mortality after dietary exposure (Shubat and Curtis 1986). Behavioral effects in goldfish and bluegill (*Lepomis macrochirus*) were noted at whole-body concentrations of approximately 3.7 mg/kg (Gakstatter and Weiss 1967). One study observed 65% mortality at 62.4 mg/kg whole-body dieldrin in sheepshead minnow (*Cyprinodon variegatus*) (Parrish et al. 1974), but residues in the control minnows were 1.1 mg/kg. Therefore, CBRs associated with mortality in sheepshead minnow from pre-exposed fish in this study may be biased high.

Studies reporting sublethal effects observed histological changes and endocrine effects. A whole-body dieldrin residue in spot (*Leiostomus xanthurus*) of 2.9 mg/kg was associated with histological aberrations of lamellae in gill and mucosal epithelium in small intestine after aqueous exposure for 4 d. Circulating sex-steroid levels of E_2 and 11-KT were reduced in largemouth bass (*Micropterus salmoides floridanus*) fed dieldrin in diet compared to control bass (Muller et al. 2004). GSI was not significantly different between treated and control bass; however, according to the authors, bass were sexually immature, possibly due to handling stress. Residues in whole-body bass were approximately 0.1 mg/kg dieldrin in the treated bass, with fish retaining 30–35% of the 50-d total administered dose. In a similar study, Johnson et al. (2007b, Johnson 2005) exposed 2-year-old male and female largemouth bass to dietary dieldrin for 30 and 120 d. The longer exposure was needed to measure reductions in sex-steroid concentrations in all the treated bass. Whole-body residues at 120 d were similar in control and the lowest treatment group of fish, but sex steroids were significantly reduced in all the treatment groups compared to control fish. Bass had average whole-body dieldrin residues of 0.2 mg/kg in the lowest treatment and control groups. No difference in GSI was noted after 120 d exposure.

2.6.3.1.2 Invertebrates

Invertebrate CBRs for aldrin and dieldrin are presented in Table 2.8 and summarized in Table 2.7. For dieldrin, invertebrate lethal body residue was lowest, 0.08 mg/kg, for the pink shrimp, *F. duorarum* (Parrish et al. 1974) exposed in water for 96 h (44% mortality, full survival in the control). Lethal body residues for other invertebrates were higher and within the narrow range of 1–2.1 mg/kg. For the midge, *C. riparius*, mortality (percent not reported) after 1 d exposure was

TABLE 2.8
Toxicity Associated with Cyclodienes in Tissues of Aquatic Invertebrates (mg/kg ww)

Species	No Effect mg/kg	Low Effect mg/kg	Exposure	Effect	Reference
			Aldrin		
Chlamydotheca arcuata	—	1	96 h in water	Mortality	Kawatski and Schmulbach 1971
			Dieldrin		
Chironomus riparius	—	1.1	24 h in water	Mortality	Estenik and Collins 1979
Chlamydotheca arcuata	—	1	96 h in water	Mortality	Kawatski and Schmulbach 1971
Palaemonetes pugio	0.09	2.1	96 h in water	Mortality	Parrish et al. 1974
Farfantepenaeus duorarum	0.016	0.08	96 h in water	Mortality	Parrish et al. 1974
Crassostrea virginica	13.9	20	96 h in water	Growth	Parrish et al. 1974
			Endrin		
Palaemonetes pugio	0.18	0.61	96 h in water	Mortality	Tyler-Schroeder 1979
Palaemonetes pugio	0.02	0.19	96 h in water	Mortality	Schimmel et al. 1975
Farfantepenaeus duorarum	0.01	0.025	96 h in water	Mortality	Schimmel et al. 1975
Pteronarcys dorsata	0.03	0.07	96 h in water	Mortality	Anderson and DeFoe 1980
Crassostrea virginica	0.26	16.4	68 h in water	Mortality	Mason and Rowe 1976
Crassostrea virginica	—	5.8	96 h in water	Growth	Schimmel et al. 1975
Limnodrilus hoffmeisteri	58	358	43 d in sediment	Mortality	Keilty et al. 1988b
Limnodrilus hoffmeisteri	58	148	43 d in sediment	Growth	Keilty et al. 1988b
Limnodrilus hoffmeisteri	58	0.003	43 d in sediment	Sediment reworking	Keilty et al. 1988b
Stylodrilus heringianus	62	118	43 d in sediment	Mortality	Keilty et al. 1988a
Stylodrilus heringianus	—	1.5	43 d in sediment	Sediment reworking	Keilty et al. 1988a

continued

TABLE 2.8 (continued)
Toxicity Associated with Cyclodienes in Tissues of Aquatic Invertebrates (mg/kg ww)

Species	No Effect mg/kg	Low Effect mg/kg	Exposure	Effect	Reference
			Endosulfan		
Palaemonetes pugio	0.065	0.21	96 h in water	Mortality	Schimmel et al. 1977a
Farfantepenaeus duorarum	<0.01	<0.01	96 h in water	Mortality	Schimmel et al. 1977a
Mytilus edulis	1.9	8.1	112 d in water	Spawning	Roberts 1972
			Chlordane		
Palaemonetes pugio	4.8	9.1	96 h in water	Mortality	Parrish et al. 1976
Farfantepenaeus duorarum	0.71	1.7	96 h in water	Mortality	Parrish et al. 1976
Crassostrea virginica	<0.01	27	96 h in water	Growth	Parrish et al. 1976
			Heptachlor (heptachlor and heptachlor epoxide)		
Farfantepenaeus duorarum	—	0.068[a]	96 h in water	Mortality	Schimmel et al. 1976
Farfantepenaeus duorarum	—	0.016[b]	96 h in water	Mortality	Schimmel et al. 1976
Farfantepenaeus duorarum	—	0.21[c]	96 h in water	Mortality	Schimmel et al. 1976
Palaemonetes vulgaris	0.32	3.5[c]	96 h in water	Mortality	Schimmel et al. 1976
Crassostrea virginica	—	8.4[c]	96 h in water	Growth	Schimmel et al. 1976

[a] Sum heptachlor and heptachlor epoxide (exposure to analytical grade heptachlor).
[b] Heptachlor epoxide (exposure to heptachlor epoxide).
[c] Sum heptachlor and heptachlor epoxide (exposure to technical grade heptachlor).
— Not available.

associated with a body residue of 1.1 mg/kg (Estenik and Collins 1979). Mortality (% not reported) in the ostracod, *Chlamydotheca arcuata*, exposed for 48–96 h was associated with body residues less than 1.0 mg/kg (Kawatski and Schmulbach 1971). Parrish et al. (1974) reported a lethal CBR of 2.1 mg/kg in the grass shrimp, the *Palaemonetes pugio*, displaying 20% mortality (full control survival) in a 96-h exposure to dieldrin in water. Dieldrin body residues ranging from 14 to 107 mg/kg were reported as nonlethal to the Eastern oyster, *Crassostrea virginica* (Parrish et al. 1974, Mason and Rowe 1976, Emanuelsen 1978), while this bivalve displayed decreased growth (24%, no decrease reported for control oysters) at 13.9 mg/kg (Parrish et al. 1974). Statistical comparisons for mortality data were not reported in the above studies.

The only lethal CBR for aldrin was reported for *C. arcuata* (Kawatski and Schmulbach 1971) as less than 1 mg/kg. Highest no-effect body residues in invertebrates exposed to aldrin in water ranged from 0.13 to 2.3 mg/kg (Butler 1971, Johnson et al. 1971). The later studies did not indicate whether aldrin was transformed to dieldrin in those invertebrates. The reported body residues are assumed to correspond to the aldrin-only body residue.

2.6.3.2 Endrin

Endrin is the stereoisomer of dieldrin. The technical formulation contains not less than 92% pure endrin (WHO 1992). Endrin was introduced in 1951 and was used to control cotton crop pests in the United States (Table 2.1). Endrin use is banned or highly restricted in many countries. Although similar in structure to dieldrin, endrin is more rapidly metabolized and accumulates to a lesser degree in lipids.

2.6.3.2.1 Fish

Endrin has high acute aqueous toxicity to fish with a relatively narrow LC50 range. Lethal CBRs for seven fish species ranged from 0.012 mg/kg in largemouth bass fingerlings (Fabacher and Chambers 1976) to 1.7 mg/kg in sailfin mollies (*Poecilia latipinna*) (Schimmel et al. 1975), and 0.9 mg/kg median lethal CBR (Table 2.6). Lethal CBR for fingerling channel catfish (100% mortality) was 1.0 mg/kg (Argyle et al. 1973) and bluegill sunfish LR50 was 0.3 mg/kg (Bennett and Day 1970). Fathead minnow fry CBR (89% mortality) was 0.24 mg/kg (Jarvinen and Tyo 1978). Adult fathead minnow lethal CBR was 1.1 mg/kg (60% mortality), similar to adult sheepshead minnow lethal CBRs (0.88–1.5 mg/kg; Schimmel et al. 1975, Hansen et al. 1977). One study reported hyperactivity in golden shiner (*Notemigonus crysoleucas*) at 0.21 mg/kg sublethal CBR (Ludke et al. 1968). Exposures were acute and chronic duration, and all but one were aqueous exposures. Immune suppression has been noted from endrin exposure, but CBRs were not reported.

2.6.3.2.2 Invertebrates

Invertebrate CBRs for endrin are presented in Table 2.8. Endrin lethal CBRs varied much more widely than those for dieldrin, ranging from 0.025 to 358 mg/kg for eight species. As for dieldrin, the lowest invertebrate lethal CBR, 0.025 mg/kg, was reported for the pink shrimp, *F. duorarum* (Schimmel et al. 1975) exposed to technical grade endrin in water for 96 h (25% mortality, full survival in the control). In that same study, the grass shrimp, *P. pugio* was more tolerant, accumulating 0.19 mg/kg when mortality was 20% (full control survival). The 28-d median lethal CBR for giant black stonefly *Pteronarcys dorsata* (Anderson and DeFoe 1980) was 0.07 mg/kg. Other invertebrates investigated were substantially more tolerant to endrin. High mortality (90%) of the eastern oyster *C. virginica* occurred at a body residue of 16.4 mg/kg after a 7-d exposure (Mason and Rowe 1976). For the latter species, a lower reported residue of 5.8 mg/kg resulted in 40% decrease in growth of juvenile oysters relative to control organisms after 96 h (Schimmel et al. 1975). Statistical comparisons of mortality data were not reported in the above studies.

Keilty et al. (1988a) exposed the tubificid oligochaete, *Limnodrilus hoffmeisteri*, to sediment spiked with endrin to investigate several toxicity endpoints as well as bioaccumulation. In contrast to CBR of 358 mg/kg for significantly decreased survival (44%; 12% in the control) and 148 mg/kg for decreased growth, feeding activity, measured as sediment reworking rate, was significantly decreased at 0.003 mg/kg. Similarly, Keilty et al. (1988b) reported a significant decrease in feeding activity in the oligochaete, *Stylodrilus heringianus*, associated with a CBR (1.5 mg/kg), much lower than the reported lethal body residue (118 mg/kg) following a 43-d exposure to spiked sediment. Endrin body residues as high as 0.62 mg/kg in the bivalves *Mya arenaria* and 0.24 mg/kg in *Mercenaria mercenaria* did not impact their survival or feeding behavior (Butler 1971).

2.6.3.3 Endosulfan

Endosulfan consists of two stereoisomers, alpha and beta, also known as endosulfan I and endosulfan II. The technical grade mixture has alpha and beta isomers in an approximately 70:30 ratio (WHO 1984a), and the isomers have different toxicity. Aquatic toxicity to endosulfan I is generally greater than toxicity from the other isomers, but the sulfate metabolite can have similar or greater toxicity in some species (Knauf and Schulze 1973). Endosulfan is highly toxic to fish and invertebrates and has resulted in kills following aerial spraying (Naqvi and Vaishnavi 1993). Fairly persistent in water, soil, and sediment, endosulfan fish tissue residues declined rapidly after exposure ceased (Schimmel et al. 1977a). Endosulfan use in the United States was reregistered by the U.S. EPA in 2001 (U.S. EPA 2002) but concentrations in fish were not analyzed by the USGS under its National

Water-Quality Assessment (NAWQA) program. Endosulfan use in other countries such as China and India also continues (PAN North American 2009).

2.6.3.3.1 Fish

Few studies have measured whole-body residues and toxicity. After uptake, endosulfan is oxidized to the metabolite endosulfan sulfate. One study reported lethal CBRs after a 96-h exposure to endosulfan. Spot, pinfish (*Lagodon rhomboids*), and striped mullet experienced 35–40% mortality at whole-body total endosulfan concentrations of 0.031, 0.272, and 0.36 mg/kg, respectively (Schimmel et al. 1977a). Most of the endosulfan was endosulfan sulfate. Adult zebra fish exposed to endosulfan accumulated different concentrations of the individual isomers with the alpha isomer measured at the highest whole-body concentration (Toledo and Jonsson 1992). Maximum total endosulfan residue after the 27-d exposure was 0.9 mg/kg and was associated with alterations in gill lamella and zonal necrosis, and lipid accumulation in liver. In Atlantic salmon, β-endosulfan was accumulated more readily and depurated more slowly than α-endosulfan (Berntssen et al. 2008).

Matthiessen et al. (1982) measured whole-body residues in fish that died within 3 d after aerial spraying with 35% endosulfan emulsifiable concentrate for tsetse fly control in Botswana. Concentrations in juvenile fish ranged from 0.07 mg/kg whole body in African catfish (*Clarias* sp.) to 1.08 mg/kg in African cichlid (*Haplochromis* sp.). Adult southern mouthbrooder (*Pseudocrenilabrus* sp.) that died had 1.46 mg/kg whole body. After 12 months, field fish still had detectable residues in their tissue. In lab studies, Matthiessen et al. (1982) measured concentrations in organs associated with mortality and noted that lean fish died more rapidly.

Endosulfan has been reported to have estrogenic properties, although study results are sometimes conflicting (e.g., Smeets et al. 1999, Hemmer et al. 2001, Balasubramani and Pandian 2008). Reduction in cortisol levels in fish has been reported from *in-vitro* studies (Leblond et al. 2001). CBRs associated with endocrine disruptive effects have not been reported.

2.6.3.3.2 Invertebrates

In the only study reporting endosulfan CBRs for invertebrates (Table 2.8), Schimmel et al. (1977a) observed lethal effects (90% mortality, 100% survival in the control) in pink shrimp, *F. duorarum* at water concentrations of 0.076 μg/L resulting in nondetectable residues of endosulfan I and II and endosulfan sulfate (0.01 mg/kg detection limit) after 96 h. The same study reported a lethal residue of 0.21 mg/kg for the grass shrimp, *P. pugio* (35% mortality, 100% control survival), that was 63% endosulfan sulfate, 28% endosulfan I, and 9% endosulfan II (Schimmel et al. 1977a). The same study reported decreased growth (78%), higher than in the control (10%) for *C. virginica*, and no lethal effects were associated with a body residue of 19.9 mg/kg. Statistical comparisons for mortality data were not reported in that study. Endosulfan body residues as high as 8.1 mg/kg in the mussel *Mytilus edulis* did not impact their survival although timing of spawning was affected (Roberts 1972).

2.6.3.4 Chlordane

Chlordane has been used on agricultural crops and for termite control. Pure chlordane consists of cis- and trans-isomers. In its technical form it is a complex mixture of more than 140 related compounds, including isomers, other chlorinated hydrocarbons, and byproducts. The approximate composition of the technical grade chlordane manufactured by Velsicol is 24% trans-chlordane, 19% cis-chlordane, 10% heptachlor, 20.5% chlordenes, 5% trans-nonachlor, and 2.8% cis-nonachlor (Cardwell et al. 1977). The most persistent forms often measured in tissue are trans-nonachlor and cis-chlordane (Schmitt 2002b). Chlordane can be metabolized in the organism to form a number of different metabolites (Table 2.1). The metabolic products such as heptachlor and its metabolite heptachlor epoxide can be more toxic than the parent compound. Chlordane may be reported as the sum of chlordane-related compounds (cis- and trans-chlordane, nonachlors, oxychlordane, and heptachlor epoxide) (Schmitt 2002b). Chlordane use is either banned or severely restricted in many countries. Use in the United States has been prohibited since 1983.

2.6.3.4.1 Fish

Tissue-residue concentrations of chlordane associated with effects to fish come from one study that used technical chlordane and two studies that used technical heptachlor for dosing. Pinfish lethal CBR was 16.6 mg/kg and sheepshead minnow fry CBR for behavior was 87 mg/kg (Parrish et al. 1976). When exposed to technical grade heptachlor in chronic exposures, the sum of cis- and trans-chlordane was measured in whole body, and heptachlor and heptachlor epoxide were also present in tissues. Chlordane CBRs were 0.71 mg/kg for mortality in whole-body spot, and 3.85 mg/kg for decreased swimming, and 11.1 mg/kg for mortality in whole-body sheepshead minnow (Schimmel et al. 1976, Goodman et al. 1977a).

2.6.3.4.2 Invertebrates

Invertebrate CBRs for chlordane are summarized in Table 2.8. Parrish et al. (1976) reported CBRs of chlordane associated with toxic effects in invertebrates. Lethal CBRs were lower for pink shrimp, *F. duorarum* (1.7 mg/kg associated with 55% mortality), than for grass shrimp, *P. pugio* (9.1 mg/kg associated with 45% mortality), exposed in water to technical grade chlordane; full control survival was observed for both species. For *C. virginica*, decreased growth (41%), higher than in the control (10%), and no lethal effects, were associated with a CBR of 27 mg/kg. Statistical comparisons for mortality data were not reported in that study.

2.6.3.5 Heptachlor and Heptachlor Epoxide

Pure heptachlor was originally isolated from technical chlordane. The technical formulation for heptachlor contains about 65% heptachlor, 22% trans-chlordane, 2% cis-chlordane, and 2% non-achlor (Schimmel et al. 1976). Tissue residues in organisms exposed to technical heptachlor contain these other compounds. Heptachlor epoxide is a metabolic product of analytical heptachlor and is typically measured in tissue residues after exposure.

2.6.3.5.1 Fish

Residue-effect data for four fish species primarily were mortality endpoints from exposure to technical heptachlor (Table 2.9). The lowest CBR was 0.23 mg/kg whole body for heptachlor epoxide in bluegill resulting in death of almost all the fish in the pond after a 16-h exposure (Andrews et al. 1966). Heptachlor was also present at 2.8 mg/kg for a total heptachlor residue of 3.03 mg/kg. Spot were sensitive to heptachlor with a lethal CBR of 2.08 mg/kg for total heptachlor (Schimmel et al. 1976). Low concentrations of trans- and cis-chlordane also occurred in these fish (0.55 and 0.16 mg/kg, respectively). Sheepshead minnow sublethal CBR for embryos and fry exposed to technical heptachlor occurred at a residue three times lower than the residue associated with adult mortality after acute exposure, and two times lower than the residue associated with mortality after chronic exposure of embryo/fry (Table 2.9). Sheepshead minnow exposed to heptachlor as embryos and juveniles had twice the mortality as minnows exposed during only the juvenile stage (Goodman et al. 1977b).

The ratio of heptachlor epoxide: heptachlor in the tissues is higher when the exposure duration is longer (Table 2.9). After acute exposure, heptachlor residues in sheepshead minnow were about 4 times higher than heptachlor epoxide residues, probably reflecting less time for metabolites to form (Schimmel et al. 1976). Heptachlor and heptachlor epoxide CBRs in sheepshead minnow whole bodies were similar to each other after chronic exposure to the same technical heptachlor aqueous concentration (Goodman et al. 1977b).

2.6.3.5.2 Invertebrates

Only one study (Schimmel et al. 1976) reported CBRs of heptachlor for invertebrates (Table 2.8). For the pink shrimp, *F. duorarum*, exposure to analytical grade compound resulted in 40% mortality

TABLE 2.9
Toxicity Associated with Heptachlor and Heptachlor Epoxide in Whole-Body Fish (mg/kg ww) after Exposure to Technical Heptachlor

Fish	Heptachlor		Heptachlor Epoxide		Effect, Exposure	Reference
	No Effect	Low Effect	No Effect	Low Effect		
Sheepshead minnow[a]	4.8	10.4	4.2	8.0	Mortality, 25 d	Goodman et al. 1977b
Spot	0.01	1.5	0.016	0.58	Mortality, 96 h	Schimmel et al. 1976
Spot	1.7	5.3	0.64	1.4	Mortality,[b] 96 h	Schimmel et al. 1976
Sheepshead minnow	0.022	20	0.02	6.7	Mortality, 96 h	Schimmel et al. 1976
Pinfish	5.7	34	3.2	11	Mortality, 96 h	Schimmel et al. 1976
Bluegill		2.8		0.23	Mortality, 16 h	Andrews et al. 1966
Bluegill		0.33–6.7		0.23–5.0	Mortality, 84 d	Andrews et al. 1966
Sheepshead minnow[a]	0.038	4.5	0.056	3.6	Behavior, 25 d	Goodman et al. 1977b

[a] Exposed as embryo/fry.
[b] Exposed to analytical grade heptachlor.
  No statistically significant differences reported in Schimmel et al. 1976.

and a total heptachlor CBR of 0.068 mg/kg (0.058 mg/kg heptachlor epoxide and 0.010 mg/kg heptachlor), and exposure to heptachlor epoxide resulted in 65% mortality and 0.016 mg/kg CBR. Exposure to technical grade heptachlor resulted in 82% mortality associated with a whole-body CBR of 0.21 mg/kg (0.18 mg/kg heptachlor epoxide and 0.03 mg/kg heptachlor). The same study reported 3.5 mg/kg of total heptachlor (2.5 mg/kg heptachlor epoxide and 1 mg/kg heptachlor) associated with 70% mortality in the grass shrimp, *P. vulgaris*, exposed to technical grade heptachlor. The above exposures were 96 h and did not cause mortality in the control. Sublethal effects in the Eastern oyster (*C. virginica*) exposed to technical grade heptachlor was 33% decreased growth associated with a tissue concentration of 8.4 mg/kg total heptachlor (7.7 mg/kg heptachlor and 0.8 mg/kg heptachlor epoxide, in addition to 6.5 mg/kg trans-chlordane and 0.8 mg/kg cis-chlordane), indicating the higher tolerance of this species. Statistical comparisons for mortality data were not reported.

2.6.4   Cyclodiene Discussion

2.6.4.1   Fish

Dieldrin is frequently measured in tissues of field-collected fish tissue, yet few studies report residues associated with effects investigated in laboratory studies. The lowest whole-body CBR was 0.1 mg/kg for reduced sex steroids in adult largemouth bass (Muller et al. 2004), and 0.2 mg/kg reported for mortality in juvenile rainbow trout (Shubat and Curtis 1986). Reduced production of steroid hormones can lead to reduced reproductive output, but the reviewed dieldrin studies did not provide that link. African catfish exposed to aqueous concentrations greater than 1.5 µg/L dieldrin did not produce eggs, but residues were not measured after the 2-month exposure (Lamai et al. 1999). The dieldrin-exposed fish had significantly less adipose fat reserves than nonexposed fish, which the authors speculate may be related to failure to produce eggs.

Almost all endrin residue-effect studies report fish mortality from aqueous exposures, likely due to its high acute aqueous toxicity. Largemouth bass was the most sensitive species tested with 40% mortality at 0.012 mg/kg CBR. Figure 2.1 compares the trends among fish species for cyclodiene and other OH compounds. Fathead and sheepshead minnow had similar endrin sensitivity.

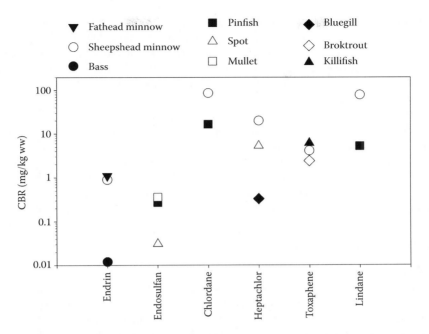

FIGURE 2.1 Comparison of fish critical body residues of select organochlorine pesticides for different species (data reported in text). Mortality endpoints (25–85% mortality) for all species except for sheepshead minnow exposed to chlordane (behavior).

Sublethal endrin CBRs, which would theoretically occur at lower residues, are lacking, possibly as a result of the high acute toxicity of the compound. Only one study reported a sublethal effect for behavior in golden shiner at 0.2 mg/kg (Ludke et al. 1968).

Most of the CBRs for endosulfan are from a single study where residue-effect concentrations ranged over an order of magnitude, and spot were the most sensitive species (Schimmel et al. 1977a) (Figure 2.1). A larger range (0.07–1.5 mg/kg) in species sensitivity was observed during a field application in Africa with African catfish the most sensitive species for mortality (Matthiessen et al. 1982). Species-sensitivity variability in this study was reduced to less than an order of magnitude by lipid normalization (7.4–47.1 mg/kg lipid). Species-sensitivity variability in aquatic toxicity tests for endosulfan ranged up to 4 orders of magnitude (18,000) in saltwater species (Chapman 1983), so the variability in toxicity was reduced when based on residue concentrations.

Fish CBRs from chlordane are too few to make conclusions about species sensitivity or effect ranges, but OH CBRs are highest for this compound (Figure 2.1). Only one study reports residue-effects associated with technical chlordane exposure (Parrish et al. 1976). When present along with heptachlor and heptachlor epoxide, spot lethal chlordane CBRs were close to an order of magnitude lower than sheepshead minnow sublethal and lethal CBRs (Schimmel et al. 1976).

Fish exposed to technical heptachlor had heptachlor, heptachlor epoxide, and often chlordane compounds in their tissue. Most of the studies for heptachlor report mortality and the median lethal CBR for heptachlor was 4.1 mg/kg compared to 1.0 mg/kg median lethal CBR for heptachlor epoxide (Table 2.6).

2.6.4.2 Invertebrates

Similar 96-h water exposures were conducted to assess the effects of dieldrin, endrin, chlordane, and heptachlor on survival of pink shrimp and grass shrimp and (for all but endosulfan) on growth of the Eastern oyster (Parrish et al. 1974, Schimmel et al. 1975, Parrish et al. 1976, Schimmel et al. 1977a).

Table 2.7 reports lowest, highest, and median lethal CBR, and Figure 2.2 compares the trends among invertebrate species for cyclodiene and other OH compounds. The pink shrimp was consistently the species with the lowest lethal CBR, followed by grass shrimp, while much higher residues failed to promote mortality in the Eastern oyster. Endrin and heptachlor were overall the most toxic and chlordane the least toxic cyclodiene for those three marine species.

Endrin lethal CBRs are available for seven species. Freshwater oligochaetes were substantially more tolerant than crustaceans and the sole insect investigated. Therefore, similar to DDT, endrin may be a potent toxicant causing mortality via specific modes of action in some invertebrates, such as pink shrimp and stonefly nymph, while lethal at residues approaching the range typical for nonspecific, baseline-toxicity-type toxicants for other invertebrates, such as oligochaetes. In contrast, endrin sublethal CBRs associated with decreased sediment reworking rates were orders of magnitude lower than lethal body residues for oligochaetes (Keilty et al. 1988a, 1988b), suggesting that physiological impairment by that compound is caused by toxic action other than baseline toxicity.

2.6.5 MIREX, LINDANE, AND TOXAPHENE

2.6.5.1 Mirex

Mirex was used mainly for fire ant control in the southeastern U.S., but also as a fire retardant (dechlorane). Application was concentrated in the southern U.S., but manufacturing of mirex resulted in contamination in the Niagara and Oswego Rivers as well as Lake Ontario and the St. Lawrence River. Use of mirex-containing products was cancelled in 1977–1978 in the United States. The technical grade mixture contains about 95% mirex and less than 3% chlordecone (kepone). Due to lower acute toxicity, mirex replaced heptachlor for some uses.

2.6.5.1.1 Fish

In short-term aqueous toxicity tests, mirex did not cause fish mortality even at high concentrations (Tagatz et al. 1975, 1976, Skea et al. 1981). Mirex was not acutely toxic to fathead minnows (*Pimephales promelas*), and percent hatchability and growth of 30-d fry was higher in mirex-exposed

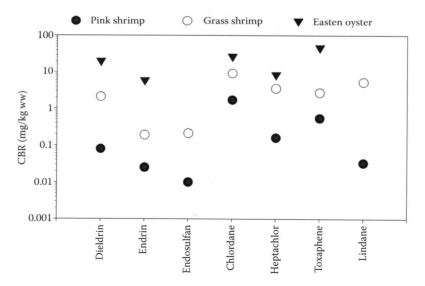

FIGURE 2.2 Comparison of invertebrate critical body residues of select organochlorine pesticides for different species (data reported in Tables 2.7, 2.8, and 2.10). Mortality endpoints for both shrimp species (20–90% mortality), and growth endpoints for Eastern oyster (24–78% reduction in growth).

fish compared to controls (Buckler et al. 1981). The number of spawns by fathead minnow was reduced at a whole-body CBR of 63 mg/kg (Buckler et al. 1981). Histological changes in gill lamellae were observed at whole-body CBR of 0.35–0.94 mg/kg in sheepshead minnow exposed to mirex leached from fire ant bait (Tagatz et al. 1975). Catfish (*Ictalurus punctatus*) survival was reduced in outdoor ponds at 0.015 mg/kg in muscle (Hyde et al. 1974).

Contamination of control fish was a problem in a study of bluegill and goldfish (van Valin et al. 1968). Bluegill fed mirex had reduced growth at whole-body residues that ranged from approximately 15–70 mg/kg from 30 to 160 d, but mirex in control fish ranged from 0.7 to 1.0 mg/kg.

Young juvenile striped mullet exposed to aqueous concentrations of mirex for 96 h experienced 6.4%, 27%, and 32% mortality at whole-body concentrations of 0.18, 0.82, and 3.5 mg/kg, respectively (Lee et al. 1975). The highest-dosed fish, with residues of 22.5 mg/kg experienced no mortality, which the authors could not explain. Older juvenile and adult mullet did not experience mortality after exposure to aqueous concentrations of mirex.

Leatherland and Sonstegard (1979) exposed coho salmon to 50 mg/kg mirex in diet. A mirex concentration in carcass of 9.6 mg/kg was associated with a significant reduction in the lipid content of the carcass after 3 months exposure. The treatment group exposed to 5 mg/kg mirex in diet had 1.6 mg/kg mirex in carcass and had carcass lipid content similar to the control salmon. When PCBs and mirex were combined in the diet, mirex appeared to inhibit the uptake of PCBs by coho salmon.

2.6.5.1.2   Invertebrates

Invertebrate CBRs for mirex are summarized in Table 2.10. Mirex lethal CBRs were reported for a variety of estuarine invertebrates in multispecies exposures (Tagatz et al. 1975, 1976). After a 28-d exposure to mirex released to water from ant bait, body residues associated with significant mortality were lowest for the blue crab, *C. sapidus* (0.02 mg/kg), followed by the pink shrimp, *F. duorarum* (0.12 mg/kg), and the grass shrimp, *P. pugio* (0.27 mg/kg) (Tagatz et al. 1975). In a similar experiment conducted for 70 d, lethal body residues were similar for the mud crab, *Panapeus herbstii* (0.22 mg/kg), and the hermit crab, *Clibanarius vittatus* (0.23 mg/kg) (Tagatz et al. 1976). Lethal CBRs were reported for crab by Bookhout and Costlow (1975) in exposure to water throughout its development (0.07 mg/kg) and by Leffler (1975) for a 28-d dietary exposure (0.42 mg/kg). The lowest CBR reported for crayfish juveniles, *Procambarus blandingii*, was 1.45 mg/kg following a 54-h exposure to water contaminated via mirex released from ant bait (Ludke et al. 1971). Various freshwater invertebrates tested in bioaccumulation experiments experienced 100% mortality after 7-d exposures to technical mirex (de la Cruz and Naqvi 1973). Body residues measured after 2 d were 1.06 mg/kg in the crayfish *Orconectes mississippiensis*, 2.56 mg/kg in the amphipod, *H. azteca*, and 10.37 mg/kg in the dragonfly naiad, *Macromia* sp. The highest mirex nonlethal body residues were 2 mg/kg for the ribbed mussel, *Modiolus demissus* and 2.8 mg/kg for the Eastern oyster (Tagatz et al. 1976), as well as 1.92, 5.07, and 4.91 mg/kg for the leeches, *Erpobdella puctata*, *Placobdella rugosa*, and *Glossiphonia* sp., respectively (de la Cruz and Naqvi 1973).

2.6.5.2   Lindane

The compound 1,2,3,4,5,6-HCH has eight isomers and the gamma (γ) isomer is lindane (Table 2.1). Another name used for lindane is benzene hexachloride (BHC). Lindane is not less than 99% pure γ-isomer of HCH (BHC), and is the active insecticide. Product composition varies from different manufacturers. Technical HCH is 65–70% α-HCH, 7–10% β-HCH, 14–15% γ-HCH, and smaller amounts of some additional compounds (WHO 1991).

2.6.5.2.1   Fish

The lindane residue-effect literature reports over an order of magnitude difference in species sensitivity from lethal residue data for fish tested under the same conditions. Lethal whole-body CBRs

TABLE 2.10

Toxicity Associated with Mirex, Lindane, and Toxaphene in Tissues of Aquatic Invertebrates (mg/kg ww)

Species	No Effect mg/kg	Low Effect mg/kg	Exposure	Effect	Reference
			Mirex		
Procambarus blandingii	—	1.45	54 h in water	Mortality	Ludke et al. 1971
Callinectes sapidus	—	0.42	28 d in diet	Mortality	Leffer 1975
Callinectes sapidus	0.03	0.07	Throughout development in water	Mortality	Bookhout and Costlow 1975
Callinectes sapidus	—	0.02	28 d in water	Mortality	Tagatz et al. 1975
Palaemonetes pugio	—	0.27	28 d in water	Mortality	Tagatz et al. 1975
Farfantepenaeus duorarum	—	0.12	28 d in water	Mortality	Tagatz et al. 1975
Clibanarius vittatus	—	0.23	70 d in water	Mortality	Tagatz et al. 1976
Panopeus herbstii	—	0.22	70 d in water	Mortality	Tagatz et al. 1976
Macromia sp.	—	10.37	7 d in water	Mortality	de la Cruz and Naqvi 1973
Orconectes mississippiensis	—	1.06	7 d in water	Mortality	de la Cruz and Naqvi 1973
Hyalella azteca	—	2.56	7 d in water	Mortality	de la Cruz and Naqvi 1973
			Lindane		
Farfantepenaeus duorarum	—	0.033	96 h in water	Mortality	Schimmel et al. 1977b
Palaemonetes pugio	—	5.2	96 h in water	Mortality	Schimmel et al. 1977b
Mytilus edulis	—	0.014	7 d in suspended sediment	Feeding	Hermsen et al. 1994
			1,2,3,4,5,6-Hexachlorocyclohexane		
Farfantepenaeus duorarum	—	0.028	96 h in water	Mortality	Schimmel et al. 1977b
			Toxaphene		
Farfantepenaeus duorarum	—	0.54	96 h in water	Mortality	Schimmel et al. 1977c
Palaemonetes pugio	—	2.7	96 h in water	Mortality	Schimmel et al. 1977c
Crassostrea virginica	—	47	96 h in water	Growth	Schimmel et al. 1977c

— Not available.

associated with LC50 data (LR50) after a 96-h exposure were 79 mg/kg for sheepshead minnow and 5.2 mg/kg for pinfish (Schimmel et al. 1977b). Fathead minnow exposed to lindane for 18 months had about 30% reduced survival at 9.5 mg/kg in carcass (Macek et al. 1976). Threespine stickleback mortality was associated with 12 mg/kg ww estimated from whole-body dry-weight (using water content provided in the chapter) (Hansen 1980). In this chapter Hansen (1980) reported that stickleback assimilated about 50% of the lindane in food.

A lindane concentration of 1.07 mg/kg in gudgeon muscle tissue (*Gobio gobio*) was associated with approximately 50% mortality after a 96-h exposure (Marcelle and Thome 1983).

Eyed eggs of rainbow trout were exposed to lindane in sediment for 27 d. Lindane residues in eggs and fry of rainbow trout did not change appreciably during the 27 d continuous exposure.

Sublethal effect residues for fry associated with altered behavior (lethargy) were 0.055–0.1 mg/kg (Ramamoorthy 1985). Rainbow trout egg residues averaged 0.048–0.08 mg/kg in 7-d-old embryos for those fry.

Reduced growth and reproduction in brook trout after exposure to aqueous lindane for 18 months were associated with residues in muscle of 1.2 mg/kg (Macek et al. 1976). Recent sublethal effect studies have measured residues in fat (Pesce et al. 2008) and organs (González de Canales et al. 2009) but not in whole bodies.

Immunosuppression by lindane has also been observed, but residues were not reported. The β-HCH and γ-HCH isomers have been reported to have estrogenic properties (Wester and Canton 1986, Singh and Singh 2007).

2.6.5.2.2 Invertebrates

Lethal effects of lindane (Table 2.10) were reported at body residue of 0.033 mg/kg for the pink shrimp, *F. duorarum*, and a much higher lethal body residue, 5.2 mg/kg, for the grass shrimp, *P. pugio*, following 96 h (Schimmel et al. 1977b). The percent mortality associated with the latter residues was higher than 50%. Lethality to pink shrimp yielded a body residue (0.028 mg/kg) for a mixture of HCH isomers that was similar to the lethal body residue reported for lindane alone in the same study (Schimmel et al. 1977b). Significant sublethal decrease (20%) of the feeding activity in the mussel, *M. edulis*, was associated with a lindane body residue of 0.014 mg/kg (Hermsen et al.1994) (Table 2.10).

2.6.5.3 Toxaphene

Toxaphene (also known as camphechlor) is a complex mixture of compounds and is the technical grade of chlorinated camphene. It contains about 67–69% chlorine (WHO 1984b). Toxaphene was used extensively on cotton crops in the southeastern United States and replaced DDT after DDT use was banned. Toxaphene also has been found in areas where it was never used such as in the Great Lakes (Schmitt et al. 1990) and in the Arctic. Toxaphene use was banned for most purposes in the United States and Canada in 1982. Toxaphene can be difficult to analyze because of coeluting PCBs and other compounds, but use of capillary columns provides good estimates of concentrations (Schmitt 2002a).

2.6.5.3.1 Fish

Toxaphene CBR data exists for six fish species from laboratory studies. Residues in fish contain the more chlorinated toxaphene congeners. Fry and adults had similar accumulation factors.

Toxaphene in adult fish and fry was associated with reduced growth, mortality, and reduced egg viability. The highest CBRs were associated with mortality in fry and juveniles. Lethal body residues ranged from 24.7 to 46.4 mg/kg for longnose killifish (*Fundulus similis*) and sheepshead minnow and were associated with 35–90% mortality. Brook trout and pinfish were more sensitive than both fathead minnow and sheepshead minnow to toxaphene residues but were not represented in the fry and juvenile dataset. Lethal body residues associated in adult fish ranged from 1.9 mg/kg from acute exposure for pinfish to 6.1 mg/kg from chronic exposure for killifish (Schimmel et al. 1977c). The authors note that adult killifish were more sensitive than juveniles. They speculate that the increased sensitivity of adult fish may have been due to their reproductive status, because exposure occurred during their spawning period. Mayer et al. (1975) also noted that adult mortality in brook trout occurred just prior to spawning.

Sublethal effects in adult fish ranged from 0.4 mg/kg for reduced reproduction in brook trout (Mayer et al. 1975) to 5.9 mg/kg for reduced growth in fathead minnow (Mehrle and Mayer 1975a). Sublethal effects in early life-stage fish ranged from 0.4 mg/kg for reduced growth in brook trout fry to 10 mg/kg for abnormal behavior in sheepshead minnow fry (Goodman et al. 1977b).

The lowest CBRs were for sublethal effects to brook trout. Brook trout whole-body residues of 0.4 mg/kg were associated with reduced growth in fry, and in adult tissue, reduced viability of spawned eggs (Mayer et al. 1975, Mehrle and Mayer 1975b). Fathead minnow fry had reduced growth at 1.0 mg/kg or 17 mg/kg lw, and adult growth was reduced at 3.3 mg/kg or 46.5 mg/kg lw (Mayer et al. 1977). In addition, atrophy and degeneration of cells and tissues in liver, pancreas, and kidney were seen during histological analysis.

Sublethal effects in early life stages were associated with lower whole-body median residues compared to similar effects observed in adult fish (Table 2.6). The opposite was observed for adult residues; median CBRs associated with mortality were lower (3.25 mg/kg) in adult fish compared to median lethal CBRs for juveniles and fry (35.0 mg/kg) (Table 2.6).

Residue effects of toxaphene were examined in a field-based study where wild lake trout and white sucker were injected with a single dose of toxaphene, and along with control fish, monitored for 5 years to assess effects on survival, growth and reproduction (Delorme et al. 1999). White sucker were more sensitive than lake trout to effects of toxaphene when spawning success, using the field-deployed fish, was tested in the laboratory. Adult carcass concentrations of 0.859 mg/kg were associated with reduced fertilization, viability, and survival of embryo and sac fry 3 years after white sucker were injected with toxaphene. White sucker egg concentrations of 0.029 mg/kg experienced an 80% decrease in fertilization success.

2.6.5.3.2   Invertebrates

In the only study reporting toxaphene CBRs for invertebrates (Table 2.10), Schimmel et al. (1977c) reported a 20% decrease in survival associated with 0.54 mg/kg in the pink shrimp, *F. duorarum*, and a 25% decrease in survival at 2.7 mg/kg in the grass shrimp, *P. pugio* (full survival in the control for both species). Growth decreased 34% in the Eastern oyster, *C. virginica*, but no significant effect on survival was observed at a much higher whole-body CBR, 47 mg/kg.

2.6.6   Mirex, Lindane, and Toxaphene Discussion

2.6.6.1   Fish

Mirex is readily accumulated by fish but residue-effects data are insufficient to compare species sensitivity or lethal and sublethal effect concentrations. Mirex is not included in Table 2.6 because endpoints or tissues measured were too dissimilar to combine. The lowest whole-body CBR was 0.82 mg/kg for young juvenile mullet associated with increased mortality after acute exposure (Lee et al. 1975). Statistically significant reduction in catfish survival was associated with 0.015 mg/kg muscle concentration (Hyde et al. 1974) and sheepshead minnow had gill alterations at whole-body residues from 0.35 to 0.94 mg/kg.

The median lethal lindane CBR for four adult fish species was 10.8 mg/kg (Table 2.6). Pinfish were the most sensitive species (Figure 2.1). Altered behavior was also noted in adult fish with high residues. Early life stage sublethal effects were the most sensitive endpoints and were four orders of magnitude lower (0.055–0.1 mg/kg) than the highest LR_{50} adult residues.

Toxaphene has sufficient data to summarize lethal and sublethal CBRs by life stage (Table 2.6). Data within each summarized group is limited to at most three species and therefore likely does not capture the full range of species sensitivities. The early life stages were twice as sensitive to the sublethal effects of toxaphene residues as adult fish. The reverse was true for mortality, as adult fish were more sensitive than juvenile fish. This observation may reflect the species used in the experiments: adult lethal studies included sensitive species such as pinfish and brook trout, early life stage studies included killifish and sheepshead minnow, species typically more tolerant to OH compounds (Figure 2.1).

2.6.6.2   Invertebrates

Five marine and estuarine crustacean species had similar sensitivity to the lethal effects of mirex, with lethal CBRs ranging from 0.02 to 0.27 mg/kg (Tagatz et al. 1975, 1976), while four species of freshwater invertebrates were more tolerant, with lethal CBRs ranging from 1.06 to 10.4 mg/kg (Ludke et al. 1971, de la Cruz and Naqvi 1973). The relatively narrow range of lethal CBRs for mirex, lindane, and toxaphene established with the few existing studies (Table 2.10) suggest less across-species variability for these compounds compared to that for other OH pesticides. Pink shrimp, the most sensitive species for most OH pesticides reviewed in this chapter, was similarly sensitive to the three compounds, while grass shrimp was more sensitive to mirex. Similar to CBRS reported for other OH compounds, the Eastern oyster appears to be more tolerant to other OH pesticides as well (Figure 2.2). The range of invertebrate lethal CBRs for mirex, lindane, and toxaphene is similar to the range associated with potent neurotoxic effects speculated for DDT.

2.7 CONSIDERATIONS

Interpretation of the OH tissue-residue literature has a number of considerations and complexities. Most of the studies were not designed to assess effects from residue concentrations. The differences with species and life stages tested, dosing techniques, dosing concentrations and intervals, and the variety of endpoints increases the difficulty in comparing and contrasting studies. Application of this data to field assessments also has a number of considerations. Single isomer or single chemical exposures rarely occur in field organisms, fish will contain variable amounts of the different OH compounds, their metabolites and isomers, along with other persistent compounds. Aquatic organisms living constantly in contaminated environments typically have life-time exposures. Fish used in laboratory experiments typically have short life spans, and other life-history characteristics such as short time to reproduction, increased frequency but decreased duration of reproduction, and other strategies that may contrast sharply with wild fish living in the field that live much longer, mature more slowly, and possibly stay reproductive for years. Laboratory studies may include effects to life-stages that do not generally have tissue-residue measurement in the field such as fish eggs. Exposure through maternal transfer is an important pathway in the environment, but infrequently considered in laboratory studies. Therefore, applying data from laboratory exposures to concentrations measured in the field has many challenges. These topics are explored further to help with interpretation of the residue data.

2.7.1 MIXTURES

One of the greatest challenges to the application of CBRs derived for DDT and other individual OH pesticides is how to interpret residues in field-collected organisms where their concentrations typically co-occur as mixtures. In addition, OH pesticides in the environment typically co-occur in tissues with other hydrophobic compounds such as PCBs and PBDEs. Fifty percent of the streams receiving agricultural or urban runoff had five or more pesticide compounds in fish tissue measured by the USGS NAWQA program (USGS 2008).

Contaminants in a mixture can interact during uptake and excretion, during distribution and metabolism and during action at the receptor site. Interpreting the toxicological consequences of a mixture, its magnitude and nature (noninteraction and synergistic or antagonistic interaction) is an evolving science. The two most common models for noninteraction are dose addition and response addition (Borgert 2007). Response addition assumes that the compounds involved have independent mechanisms of action, in contrast to dose addition, which assumes that the compounds have different potencies with similar mechanism of action. Response addition is typically assumed for risk assessments (Borgert 2007).

The toxic unit (TU) approach calculates the mixture toxicity according to the degree of toxicity for each individual compound. Typically, aqueous mixture TUs are calculated as each compound's aqueous concentration divided by its LC_{50} and summed. When the sum is >1 the TU mixture would predict mortality, or the effect endpoint under consideration. This method assumes response addition, or that the joint toxicity is represented by addition of the individual toxicities.

The use of the TU approach to predict the toxicity of mixtures of DDT, DDD, and DDE in the tissues was validated using freshwater amphipods (Lotufo et al. 2001b). Because substantial differences in lethal body residues occur among those compounds, predicting mortality for a mixture using the toxicity data for a single compound (e.g., DDT) could under- or over-predict mortality. The use of the TU approach assumes that CBRs are independent of the exposure route and that the effects of DDT, DDD, and DDE on survival are additive. Fifty percent mortality was associated with sum TUs of 1.1 for *H. azteca* and 0.53 for *Diporeia* spp., close to the expected value of 1, demonstrating that the assumption of additive effects holds true.

von Westernhagen et al. (1989) used a contamination factor (CF) to scale the relative toxicities of different organochlorine compounds measured in field-collected fish in the mixture, similar to a TU approach. The CFs were developed based on aqueous 96 h LC50 toxicity data and their ratio to PCBs. They scaled each contaminant's LC50 to PCBs because all fish contained PCBs. Residue concentrations for each contaminant were multiplied by its CF and then all the concentrations were summed for each sample. Their CF for endrin was highest and next was α-endosulfan and endosulfan sulfate. Lindane toxicity was lowest of the chlorinated hydrocarbons and closest to PCBs.

Koenig (1977) measured the combined effects of mirex and DDT to the estuarine cyprinodont, *Adinia xenica*. Adult female fish were exposed to either *p,p'*-DDT or mirex separately, or combined at different ratios in diet for 9 d. Mortality of embryo and larval offspring was followed. Embryo concentrations of mirex up to 27.9 mg/kg were not associated with increased mortality. An average *p,p'*-DDT embryo concentration of 8.6 mg/kg had 22% mortality. Embryos exposed to both compounds had 47% mortality with average DDT residue of 8.4 and 2.82 mg/kg mirex. Koenig (1977) estimated that the toxicity of DDT and mirex was more than additive. Another study using mixtures observed that effects in red-eared slider turtle were more than additive when low doses of trans-nonachlor, chlordane and *p,p'*-DDE were combined (Willingham 2004). A mixture of estrogenic compounds (*o,p'* DDT, nonylphenol, octylphenol, arochlor 1221, and bisphenol A) resulted in a higher response (increased vitellogenin synthesis) in male rainbow trout cultured hepatocytes than any of the compounds alone (Sumpter and Jobling 1995).

Monod (1985) measured DDT and PCB concentrations in eggs in Lake Geneva char. PCB and DDT concentrations in eggs were significantly correlated with each other and lipid-normalized concentrations of both were correlated with egg mortality in laboratory rearings. Most DDT measured was DDE, and median concentrations of 20 mg/kg lw DDT and 44.5 mg/kg lw PBCs in eggs correlated with greater than 50% mortality.

Models used to deal with aqueous mixture toxicity may also provide useful concepts for other potential approaches to address residue mixture toxicity (Belden et al. 2007). A more complete discussion about mixture considerations for the tissue-residue approach is available in the discussion in Meador (2006).

2.7.2 RESISTANCE/PRE-EXPOSURE

Resistance, that is, the ability to survive or be less affected by exposure to concentrations that were toxic to naive or earlier generations, has been observed for many invertebrate and fish species exposed to OH pesticides. The development of resistance may be due either to an organism's ability to physiologically acclimate or to genetically adapt. In either case, organisms that develop resistance through short- or long-term processes may respond differently to chemical exposure than naive organisms. For neuroactive OH compounds that target sodium and chloride channels, slight substitutions at the target site can confer resistance to other compounds in the same chemical class, resulting in what is known as cross-resistance (Casida and Quistad 1998).

Resistant fish tolerated aqueous concentrations of toxaphene, aldrin, dieldrin, and endrin that were 36–70 times higher than fish that had not been pre-exposed (Ferguson et al. 1964). Mosquito fish became resistant to insecticides used and the 300-fold resistance to aqueous concentrations persisted for several generations of fish maintained in pesticide-free water (Boyd and Ferguson 1964).

Resistance in mosquito fish pre-exposed to endrin appeared to be related to slower uptake in brain tissues where endrin residues were about half the brain endrin concentration in mosquito fish not resistant to endrin toxicity (Fabacher and Chambers 1976).

Development of resistance has a number of implications. Laboratory studies often employ species with short life spans that have a higher likelihood to develop resistance. OH compounds are present in almost all environments and habitats, whether lab-reared, or field-collected, and some of the studies reviewed for this chapter report elevated OH concentrations in control organisms, indicating pre-exposure. The influence on residue-effect studies is unknown. The potential effect of cross-resistance to other compounds is also unknown. Results from OH studies using field-collected or laboratory-reared organisms may be biased by this pre-exposure. Especially for the older literature reviewed in this chapter, results potentially biased by use of resistant organisms cannot be identified or evaluated. Most researchers have reported an increased ability to withstand exposure which would indicate that tissue residues that used resistant organisms could be biased high. Although recent aqueous toxicity studies continue to report effects resulting from pre-exposure (Brausch and Smith 2009), environmental OH concentrations in the United States have been declining during the past decades so the bias resulting from resistance may be more pronounced in older studies.

2.7.3 Life Stage/Tissue Type

Most of the OH pesticides reviewed here had the lowest effect residues in the egg/embryo and fry stages of fish. These life stages are infrequently measured as part of assessment or monitoring programs. The reproductive cycle in fish may exert important control over organochlorine concentrations in these early and sensitive life stages, and these life stages may provide the best tool for evaluating tissue-residue effects to fish. Several researchers have reported on the transfer of hydrophobic contaminants and noted a relationship between concentration in fish eggs and maternal tissues. A variety of fish species have been shown to have a correlation between organochlorine concentrations in eggs and muscle (Miller 1993, Miller and Amrhein 1995, Fisk and Johnston 1998) and also between liver and oocytes (Serrano et al. 2008). Macek (1968) noted that DDT concentrations in fry were related to the maternal dose. When egg residues are not available for comparison to laboratory studies the following information may provide some guidance on how to assess effects to early life stage when adult tissue has been measured.

Niimi (1983) noted that the percentage of maternal organic contaminants transferred to eggs ranged from about 5% to 30% and that the percentage of lipid in the fish and the percent of that total lipid in the egg significantly influenced contaminant transfer in the five species examined. Russell et al. (1999) reported that the ratio of lipid-normalized organochlorines egg concentrations to lipid-normalized muscle concentrations was approximately one in a variety of fish, with 95% of all the measured ratios within a factor of 2 of the mean ratio (0.56–2.51). Metcalfe et al. (2000) found that the lipid-normalized o,p'-DDT concentration in the eggs and the lipid-normalized adult concentration were nearly the same. Serano et al. (2008) found that less than half the contaminant concentration in the liver of sea bream was measured in oocytes.

Residues measured in fish whole body or muscle may be converted to an egg concentration if enough data are collected to make a conversion. Lipid concentration in adult and early life stage tissue is needed to use the relationship explored by Russell et al. (1999).

Burdick et al. (1964) concluded that there was a relationship between maternal and egg DDT concentrations, however, variability in this relationship could be explained by a maternal diet component. Female fish could continue to take up DDT after egg deposition and thereby have a different relationship to the egg concentration than when the oils in the eggs were deposited.

In addition to diet, gender, age, and breeding strategy, influence the uptake, disposition, and elimination of organochlorine compounds. Male pike had significantly higher organochlorine concentrations than the females (Larrson et al. 1993). While oogenesis results in contaminant movement into the ovaries, spawning can result in contaminant loss. Miller (1994) estimated that spawning eliminated about 28–39% of the *p,p'*-DDE concentration in Lake Michigan Chinook salmon. A negative correlation between age and organochlorine concentration in female pike was explained by the yearly loss of contaminant via deposition of roe (Larsson et al. 1993). For invertebrates, female midges lost 11.6–30.9% of their DDE burden during egg deposition (Derr and Zabik 1972). Therefore, the seasonal reproductive cycle should be a factor considered when sampling tissues for OH compounds.

Summary

OH pesticides remain important global contaminants due to extreme persistence in environmental media, continued use in some countries, and ability to accumulate in tissues of biota. Their widespread occurrence in biota, especially for the hydrophic OH compounds, and their ability to biomagnify, highlights the need for understanding the significance of residue concentrations.

Key processes affecting residue concentrations are elimination rates and biotransformation of parent compounds. DDT and transformation products have slow elimination rates and long time-frames for achieving steady-state concentrations. For some compounds, metabolites may have different toxicities than parent compounds. For instance, the toxicity of DDE to amphipods was much lower than the parent DDT compound, and biotransformation protected against mortality.

For fish, CBR data for OH compounds are supported by a number of studies for DDT, endrin, heptachlor, heptachlor epoxide, and toxaphene (Table 2.6). However, dieldrin and chlordane, which are very persistent in fish tissue, have few studies for discerning residue-toxicity relationships. The same is true for mirex and lindane. More studies are needed to understand the significance of these compounds in fish tissue.

For invertebrates, tissue residue-effect data for OH compounds are supported by a number of studies for *p,p'*-DDT, endrin, and mirex (Table 2.7). Data for chlordane, lindane, toxaphene, heptachlor, and endosulfan are too few to determine residue-toxicity relationships in invertebrates.

Even for OH compounds with multiple studies, data for sublethal effects, long-term exposures, or effects to early life stages are often limited for both fish and invertebrates. Although residue-toxicity relationships have been reported for sublethal effects for some OH compounds, data are generally too few to allow comparison of the same sublethal endpoint for a single compound across different species.

For similar endpoints, species-sensitivity differences sometimes accounted for a one to two order of magnitude difference in CBRs for both fish and invertebrate species (Figures 2.1 and 2.2). Species tended to have similar sensitivity across the different groups of pesticides, but substantial differences in sensitivity among species were observed for most OH pesticides. For fish, sheepshead minnow, fathead minnow, and killifish were less sensitive than bass, spot, pinfish, and bluegill. For invertebrates, pink and grass shrimp were the most sensitive species.

Life stage exposed was an important factor influencing CBRs. The lowest CBRs were reported for egg and embryo life stages in fish, and early development in invertebrates. Sublethal endpoints for early life stages were typically associated with the lowest CBRs.

Comparison between fish and invertebrate CBRs for DDT and endrin are possible with the available datasets (Figures 2.3 and 2.4). DDT lethal residues in fish are within the range of lethal residues for invertebrates with only a slightly narrower range compared to invertebrates (Figure 2.3). In addition, the median lethal *p,p'*-DDT concentrations are about the same for fish and invertebrates (Tables 2.6 and 2.7). The median lethal residues for fish and invertebrates can be compared for the *p,p'*-DDT form since the ΣDDT fish data came primarily from exposure to the *p,p'*-DDT form (Beckvar et al. 2005).

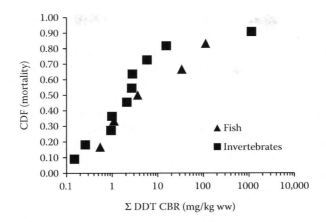

FIGURE 2.3 DDT cumulative distribution function (CDF) for fish and invertebrate lethal CBRs. Solid triangles represent fish species, solid squares invertebrate species. Data for fish from mortality endpoints (35–85% mortality) from adult fish reported in Beckvar et al. (2005). Invertebrate data are mostly LR50 endpoints. Data primarily for *p,p'*-DDT.

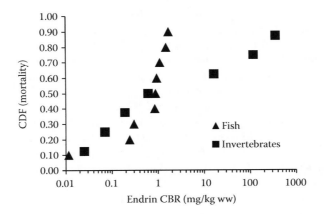

FIGURE 2.4 Endrin cumulative distribution function (CDF) for fish and invertebrate lethal CBRs. Solid triangles represent fish species (data reported in text; 20–100% mortality), solid squares invertebrate species (Table 2.8; 20–90% mortality).

DDT metabolites differed in toxicity for invertebrates. *p,p'*-DDT was clearly more toxic than p,p-DDE, and the two metabolites should be considered separately when metabolite data are available. For fish, lethal residue-effect studies are very limited for the DDE form, so comparing toxicity between metabolites is hampered. CBR for mortality from *o,p'*-DDE tissue residues in medaka eggs (0.5 mg/kg) from one study was comparable to the value derived for fish ELS mortality (0.7 mg/kg) from studies primarily based on *p,p'*-DDT and mortality endpoints (Beckvar et al. 2005). In adult fish, *p,p*-DDE residues (0.4–0.6) were associated with reduced sex-steroid production at concentrations close to the tissue-residue concentration of 0.6 mg/kg derived from mortality studies, primarily the *p,p'*-DDT form, in Beckvar et al. (2005). Sublethal effects from DDE residues in fish may be within the effects range for DDT, but more data are needed to adequately assess that comparison.

For DDT sublethal effects in fish, the median sublethal residue-effect for *o,p'*-DDT in fish eggs was 0.07 mg/kg which is an order of magnitude lower than the threshold no-effect residue concentration of 0.7 mg/kg derived for fish ELS mortality in Beckvar et al. (2005). More studies reporting sublethal effects for this life stage are needed to develop a protective and robust CBR.

DDT sublethal effects reported for invertebrates were limited. For sensitive invertebrate species, strong neurotoxicity appeared to be the primary cause of mortality. Loss of swimming ability or immobilization occurred for both tolerant and sensitive species. Decreased feeding was observed in polychaete species tolerant to the lethal effects of DDT. DDE caused reproductive impacts in midges (0.8 mg/kg), suggesting a potential effect from endocrine disruption. This contrasts with data for other invertebrates which reported no reproductive effects from exposure to the DDT metabolite.

For endrin, variability in fish lethal CBRs is much lower than for invertebrates (Figure 2.4). Mortality of fish at relatively low endrin residues and low variability (range of 0.2–1.7 mg/kg for all but one species) is consistent with toxicity acting via a specific mode of action. Neurotoxic effects in fish may occur through inhibition of the neurotransmitter gamma-aminobutyric acid (GABA) similar to the mechanism in higher vertebrates. In contrast, invertebrate CBRs for endrin were variable, indicating they may experience toxicity through multiple modes of action. Crustaceans, more sensitive (0.025–0.61 mg/kg CBR range) than either bivalves or oligochaetes, had CBRs similar to fish, suggesting that toxicity in crustaceans may also act via a specific mechanism. Feeding reduction in oligochaete occurred at residues similar to lethal endrin CBRs in sensitive invertebrate species. This behavioral alteration may have been caused by a single specific mechanism, such as inhibition of GABA receptor function. Mortality at elevated CBRs in oligochaetes is consistent with baseline toxicity. The combination of rapid metabolism of endrin by fish and sensitivity to neurotoxicity may explain reduced CBR variability in fish compared to invertebrate species.

Limited and inconsistent data preclude robust CBR comparisons between fish and invertebrates for other OH pesticides. However some general observations can be made. Invertebrate lethal CBRs for dieldrin span 2 orders of magnitude (0.08–2.1 mg/kg) and were lower overall than for fish (0.2–5.9 mg/kg). Endosulfan residue effect data were low for both fish (0.03–0.36 mg/kg) and invertebrates (<0.01–0.21 mg/kg) in the results reported from a single study (Tables 2.6 and 2.7).

Chlordane CBRs also span a similar range in both fish (0.7–16.6 mg/kg) and invertebrates (0.7–4.8 mg/kg) for the few species investigated. Heptachlor and the metabolite heptachlor epoxide CBRs span a similar range in fish (0.3–34 mg/kg and 0.2–11 mg/kg, respectively) but the metabolite heptachlor epoxide is more potent than the parent compound. Invertebrate CBRs for heptachlor (0.07–3.5 g/kg) were an order of magnitude lower than CBRs for fish, so invertebrates appear more sensitive.

Fish appear to be less sensitive to the lethal effects of mirex than invertebrates, but fish data are limited and variable. Several studies report no fish mortality from exposure to mirex (van Valin et al. 1968, Tagatz et al. 1975, 1976, Buckler et al. 1981), but increased mortality was observed in catfish at 0.015 mg/kg muscle (Hyde et al. 1974) and in young juvenile striped bass at 0.82 mg/kg (Lee et al. 1975). Invertebrate lethal CBRs for mirex span 2 orders of magnitude (0.03–10.0 mg/kg) with a median lethal residue <0.5 mg/kg. Various crab and shrimp species were very sensitive to the lethal effects of mirex with lethal CBRs for blue crab as low as 0.02 mg/kg.

The lowest invertebrate CBR for lindane (0.033 mg/kg), derived for a sensitive species (pink shrimp) and mortality, is comparable to the lowest fish CBR (0.055 mg/kg) for a sensitive fish species (rainbow trout), life stage (fry), and sublethal endpoint (behavior). The highest fish lindane CBR, an LR50 (79 mg/kg) for insensitive sheepshead minnow was much higher than the highest invertebrate CBR for grass shrimp (5.2 mg/kg), typically a sensitive species.

Toxaphene studies for invertebrates were very limited compared to data available for fish. The range of toxaphene lethal CBRs were narrower in invertebrates (0.54–2.7 mg/kg) compared to adult fish (1.9–6.1 mg/kg) and compared to the range for sublethal effects in adult fish (0.4–5.9 mg/kg). The reason for the apparent reduced toxaphene sensitivity of juvenile compared to adult fish is not known.

While some species are sensitive to the specific mode of action of OH pesticides (e.g., interference with sodium and potassium ion permeability in axons, endocrine disruption), others (some invertebrates) are affected mostly by the baseline toxicity effect of those compounds. DDT's potency as an endocrine disruptor for invoking complete sex-reversal is weak compared to natural estrogens and would likely require higher residues than typically measured in field-collected fish. DDTs potency for endocrine disruption for other effect endpoints such as reduced reproduction and immune system and behavior alterations is within the range of environmental tissue concentrations. If combined with other endocrine-active compounds, DDT may act additively or synergistically to cause effects. Investigations on mode of action of these compounds using biochemical tools such as gene expression quantification (toxicogenomics) are warranted.

Other important considerations for OH pesticide residues include laboratory to field extrapolation and methods for assessing mixture effects. The use of a TU approach is recommended for DDT compounds in invertebrates. Sublethal toxicity to early life stages through maternal transfer, especially for fish species, should also be considered.

The OH pesticide residue data overall capture a range of species sensitivities and lethal and sublethal effects. Some species were consistently more or less sensitive to OH compounds. For example, within the invertebrate species represented, growth in oysters was typically the least sensitive endpoint and species. Pink and grass shrimp tended to be the most sensitive species for the invertebrates. For fish, sheepshead minnow tended to be one of the least sensitive species. Data for specific compounds often do not have enough species and endpoints represented to fully capture the range of species-sensitivity differences and endpoint differences. If sensitive species or effect endpoints are not represented in a specific OH data set, CBRs developed from that dataset may not be protective of aquatic receptors.

REFERENCES

Anderson, R. L. and D. L. DeFoe. 1980. Toxicity and bioaccumulation of endrin and methoxychlor in aquatic invertebrates and fish. *Environ. Pollut. (Series A)* 22:111–121.

Andrews, A. K., C. C. van Valin, and B. E. Stebbings. 1966. Some effects of heptachlor on bluegills. *Trans. Am. Fish Soc.* 65:297–309.

Arctic Monitoring and Assessment Programme (AMAP). 2002. *Arctic pollution 2002: Persistent organic pollutants, heavy metals, radioactivity, human health, changing pathways.* Arctic Monitoring and Assessment Programme. Oslo, Norway: xii+112.

Argyle, R. L., G. C. Williams, and H. K. Dupree. 1973. Endrin uptake and release by fingerling channel catfish (*Ictalurus punctatus*). *J. Fish. Res. Board. Can.* 30:1743–1744.

Arnot, J. A., et al. 2009. A quantitative structure-activity relationship for predicting metabolic biotransformation rates for organic chemicals in fish. *Environ. Toxicol. Chem.* 28:1168–1177.

Arnot, J. A., and F. A. Gobas. 2006. A review of bioconcentration factor (BCF) and bioaccumulation factor (BAF) assessments for organic chemicals in aquatic organisms. *Environ. Rev.* 14:257–297.

Atchison, G. J. 1976. The dynamics of lipids and DDT in developing brook trout eggs and fry. *J. Great Lakes Res.* 2:13–19.

ATSDR 2007. Agency for Toxic Substances and Disease Registry. 2007 CERCLA Priority List of Hazardous Substances. http://www.atsdr.cdc.gov/cercla/07list.html (accessed September 30, 2008).

ATSDR 2002. Agency for Toxic Substances and Disease Registry. *Toxicological profile for DDT, DDD, DDE update.* Atlanta, GA: U.S. Department of Health and Human Services, ATSDR.

Baatrup, E., and M. Junge. 2001. Antiandrogenic pesticides disrupt sexual characteristics in the adult male guppy (*Poecilia reticulata*). *Environ. Health. Perspect.* 109:1063–1070.

Balasubramani, A., and T. J. Pandian. 2008. Endosulfan suppresses growth and reproduction in zebrafish. *Curr. Sci.* 94:883–890.

Ballesteros, M. L., G. E. Bianchi, M. Carranza, and M. A. Bistoni. 2007. Endosulfan acute toxicity and histomorphological alterations in *Jenynsia multidentata* (Anablepidae, Cyprinodontiformes). *J. Environ. Sci. Health B* 42:351–357.

Banas, D., et al. 2009. Can we use stable isotopes for ecotoxicological studies? Effect of DDT on isotopic fractionation in *Perca fluviatilis*. *Chemosphere* 76:734–739.

Barnhoorn, I. E. J., M. S. Bornman, C. Jansen van Rensburg, and H. Bouwman. 2009. DDT residues in water, sediment, domestic and indigenous biota from a currently DDT-sprayed area. *Chemosphere* 77:1236–1241.

Bayarri, S., L. T. Baldassarri, N. Lacovella, F. Ferrara, and A. di Domenico. 2001. PCDDS, PCDFs, PCBs and DDE in edible marine species from the Adriatic Sea. *Chemosphere* 43:601–610.

Bayen, S., P. Giusti, H. K. Lee, P. J. Barlow, and J. P. Obard. 2005. Bioaccumulation of DDT pesticide in cultured Asian seabass following dietary exposure. *J. Toxicol. Environ. Health, Part A—Curr. Iss.* 68:51–65.

Bayley, M., M. Junge, and E. Baatrup. 2002. Exposure of juvenile guppies to three antiandrogens causes demasculinization and a reduced sperm count in adult males. *Aquat. Toxicol.* 56:227–239.

Beckvar, N., T. M. Dillon, and L. B. Read. 2005. Approaches for linking whole-body fish tissue residues of mercury or DDT to biological effects thresholds. *Environ. Toxicol. Chem.* 24:2094–2105.

Belden, J. B., R. J. Gilliom, and M. J. Lydy. 2007. How well can we predict the toxicity of pesticide mixtures to aquatic life? *Integr. Environ. Assess. Manag.* 3:364–372.

Bengtsson, G., L. A. Hansson, and K. Montenegro. 2004. Reduced grazing rates in *Daphnia pulex* caused by contaminants: Implications for trophic cascades. *Environ. Toxicol. Chem.* 23:2641–2648.

Bennett, H. J., and J. W. Day. 1970. Absorption of endrin by bluegill sunfish, *Lepomis macrochirus*. *Pestic. Monit. J.* 3:201–203.

Berntssen, M. H. G., C. N. Glover, D. H. F. Robb, J. V. Jakobsen, and D. Petri. 2008. Accumulation and elimination kinetics of dietary endosulfan in Atlantic salmon (*Salmo salar*). *Aquat. Toxicol.* 86:104–111.

Bettinetti, R., V. Croce, S. Galassi, and P. Volta. 2006. *p,p*'-DDT and *pp*'DDE accumulation in a food chain of Lake Maggiore (northern Italy): Testing steady-state condition. *Environ. Sci. Poll. Res.* 13:59–66.

Binelli, A., and A. Provini. 2003. DDT is still a problem in developed countries: The heavy pollution of Lake Maggiore. *Chemosphere* 52:717–723.

Bloomquist, J. R. 1996. Ion channels as targets for insecticides. *Annu. Rev. Entomol.* 41:163–190.

Boese, B. L., H. Lee, and S. Echols. 1997. Evaluation of a first-order model for the prediction of the bio-accumulation of PCBs and DDT from sediment into the marine deposit-feeding clam *Macoma nasuta*. *Environ. Toxicol. Chem.* 16:1545–1553.

Bonner, J. C., and M. R. Wells. 1987. Comparative acute toxicity of DDT metabolites among American and European species of planarians. *Comp. Biochem. Physiol. Pharmacol. Toxicol. Endocrinol.* 87:437–438.

Bookhout, C. G., and J. D. Costlow Jr. 1975. Effects of mirex on the larval development of blue crab. *Water Air Soil Poll.* 4:113–126.

Borgert, C. J. 2007. Predicting interactions from mechanistic information: Can omic data validate theories? *Toxicol. Appl. Pharmacol.* 223:114–120.

Boyd, C. E., and D. E. Ferguson. 1964. Susceptibility and resistance of mosquito fish to several insecticides. *J. Econ. Entomol.* 57:430–431.

Brausch, J. M., and P. N. Smith. 2009. Pesticide resistance from historical agricultural chemical exposure in *Thamnocephalus platyurus* (Crustacea: Anostraca). *Environ. Pollut.* 157:481–487.

Brevik, E. M., M. Grande, J. Knutzen, A. Polder, and J. U. Skaare. 1996. DDT contamination of fish and sediments from Lake Orsjoen, Southern Norway—Comparison of data from 1975 and 1994. *Chemosphere* 33:2189–2200.

Buckler, D. R., A. Witt, F. L. Mayer, and J. N. Huckins. 1981. Acute and chronic effects of Kepone and mirex on the fathead minnow. *Trans. Am. Fish Soc.* 110:270–280.

Buhler, D. R., M. E. Rasmusson, and W. E. Shanks. 1969. Chronic oral DDT toxicity in juvenile coho and chinook salmon. *Toxicol. Appl. Pharmacol.* 14:535–555.

Burdick, G. E., E. J. Harris, H. J. Dean, T. M. Walker, J. Skea, and D. Colby. 1964. The accumulation of DDT in lake trout and the effect on reproduction. *Trans. Am. Fish Soc.* 93:127–136.

Burlington, H., and V. F. Lindeman. 1950. Effect of DDT on testes and secondary sex characters of white leghorn cockerels. *Proc. Soc. Exp. Biol. Med.* 74:48–51.

Butler, P. A. 1969. The significance of DDT residues in estuarine fauna. In: *Chemical fallout: Current research on persistent pesticides,* ed. M. W. Miller and G. G. Berg, pp. 205–220. Springfield, IL: C.C. Thomas.

Butler, P. A. 1971. Influence of pesticides on marine ecosystems. *Proc. R Soc. Lond. B Biol. Sci.* 177:321–329.

Cardwell, R. D., D. G. Foreman, T. R. Payne, and D. J. Wilbur. 1977. Acute and chronic toxicity of chlordane to fish and invertebrates. Duluth, MN: U.S. Environmental Protection Agency.

Carlson, D. B., L. R. Curtis, and D. E. Williams. 2000. Salmonid sexual development is not consistently altered by embryonic exposure to endocrine-active chemicals. *Environ. Health Perspect.* 108:249–255.

Carson, R. 1962. *Silent Spring.* Boston, MA: Houghton Mifflin.

Carvalho, F. P., J. P. Villeneuve, C. Cattini, J. Rendón, and J. Mota de Oliveira. 2009. Pesticide and PCB residues in the aquatic ecosystems of Laguna de Terminos, a protected area of the coast of Campeche, Mexico. *Chemosphere* 74:988–995.

Casida, J. E., and G. B. Quistad. 1998. Golden age of insecticide research: Past, present, or future? *Annu. Rev. Entomol.* 43:1–16.

Chapman, G. A. 1983. Do organisms in laboratory toxicity tests respond like organisms in nature? In: *Aquatic Toxicology and Hazard Assessment: Sixth Symposium,* eds. W. E. Bishop, R. D. Cardwell, and B. B. Heidolph, pp. 315–327. Philadelphia, PA: American Society of Testing Materials STP 802.

Cheek, A. O., T. H. Brouwer, S. Carroll, S. Manning, J. A. McLachlan, and M. Brouwer. 2001. Experimental evaluation of vitellogenin as a predictive biomarker for reproductive disruption. *Environ. Health Perspect.* 109:681–690.

Connor, M. S., et al. 2007. The slow recovery of San Francisco Bay from the legacy of organochlorine pesticides. *Environ. Res.* 105:87–100.

Cope, O. B. 1961. Effects of DDT spraying for spruce budworm on fish in the Yellowstone River system. *Trans. Am. Fish Soc.* 90:251.

Crosby, D. G., and R. K. Tucker. 1971. Accumulation of DDT by *Daphnia magna. Environ. Sci. Technol.* 5:714–716.

Davy, F. B., H. Kleerekoper, and P. Gensler. 1972. Effects of exposure to sublethal DDT on the locomotor behavior of the goldfish (*Carassius auratus*). *J. Fish Res. Board. Can.* 29:1333–1336.

de la Cal, A., E. Eljarrat, D. Raldúa, C. Durán, and D. Barceló. 2008. Spatial variation of DDT and its metabolites in fish and sediment from Cinca River, a tributary of Ebro River (Spain). *Chemosphere* 70:1182–1189.

de la Cruz, A. A., and S. M. Naqvi. 1973. Mirex incorporation in the environment: Uptake in aquatic organisms and effects on the rates of photosynthesis and respiration. *Arch. Environ. Contam. Toxicol.* 1:255–264.

Delbeke, K., T. Teklemariam, E. de la Cruz, and P. Sorgeloos. 1995. Reducing variability in pollution data— the use of lipid classes for normalization of pollution data in marine biota. *Int. J. Environ. Anal. Chem.* 58:147–162.

Delorme, P. D., W. L. Lockhart, K. H. Mills, and D. C. G. Muir. 1999. Long-term effects of toxaphene and depuration in lake trout and white sucker in a natural ecosystem. *Environ. Toxicol. Chem.* 18:1992–2000.

Depledge, M. H., and Z. Billinghurst. 1999. Ecological significance of endocrine disruption in marine invertebrates. *Mar. Poll. Bull.* 39:32–38.

Derr, S. K., and M. J. Zabik. 1972. Biologically active compounds in the aquatic environment: The uptake and distribution of [1,1-dichloro-2,2-bis(*p*-chlorophenyl)ethylene], DDE by *Chironomus tentans* Fabricius (Diptera: *Chironomidae*). *Trans. Am. Fish Soc.* 101:323–329.

Donohoe, R. M., and L. R. Curtis. 1996. Estrogenic activity of chlordecone, *o,p'*-DDT and *o,p'*-DDE in juvenile rainbow trout: Induction of vitellogenesis and interaction with hepatic estrogen binding sites. *Aquat. Toxicol.* 36:31–52.

Edmunds, J. S. G., R. A. McCarthy, and J. S. Ramsdell. 2000. Permanent and functional male-to-female sex reversal in d-rR strain medaka (*Oryzias latipes*) following egg microinjection of *o,p'*-DDT. *Environ. Health Perspect.* 108:219–224.

Eganhouse, R. P., J. Pontolillo, and T. J. Leiker. 2000. Diagenetic fate of organic contaminants on the Palos Verdes Shelf, California. *Mar. Chem.* 70:289–315.

Ellgaard, E. G., J. C. Ochsner, and J. K. Cox. 1977. Locomotor hyperactivity induced in the bluegill sunfish, *Lepomis macrochirus,* by sublethal concentrations of DDT. *Can. J. Zool.* 55: 1077–1081.

Ellis, M. M., B. A. Westfall, and M. D. Ellis. 1944. Toxicity of dichloro-diphenyl-trichlorethane (DDT) to goldfish and frogs. *Science* 100:477.

Elskus, A. A., T. K. Collier, and E. Monosson. 2005. Interactions between lipids and persistent organic pollutants in fish. *Environ. Toxicol.* 6:119–152.

Emanuelsen, M., J. L. Lincer, and E. Rifkin. 1978. The residue uptake and histology of American oysters (*Crassostrea virginica* Gmelin) exposed to dieldrin. *Bull. Environ. Contam. Toxicol.* 19:121–129.

Estenik, J. F., and W. J. Collins. 1979. *In Vivo* and *in vitro* studies of mixed-function oxidase in an aquatic insect, Chironomus riparius. In: *Pesticide and Xenobiotic Metabolism in Aquatic Organisms*, eds. M. A. Q. Khan, J. J. Lech, and J. J. Menn, pp. 349–370. Washington, DC: American Chemical Society.

Fabacher, D. L. 1976. Toxicity of endrin and an endrin-methyl parathion formulation to largemouth bass fingerlings. *Bull. Environ. Contam. Toxicol.* 16:376–378.

Fabacher, D. L., and H. Chambers. 1976. Uptake and storage of ^{14}C-labeled endrin by livers and brains of pesticide-susceptible and resistant mosquitofish. *Bull. Environ. Contam. Toxicol.* 16:203–207.

Faulk, C. K., L. A. Fuiman, and P. Thomas. 1999. Parental exposure to ortho,para-dichlorodiphenyltrichloroethane Impairs survival skills of Atlantic croaker (*Micropogonias undulatus*) larvae. *Environ. Toxicol. Chem.* 18:254–262.

Ferguson, D. E., D. D. Culley, W. D. Cotton, and R. P. Dodds. 1964. Resistance to chlorinated hydrocarbon insecticides in three species of freshwater fish. *Bioscience* 14:43–44.

Ferreira, M., P. Antunes, O. Gil, C. Vale, and M. A. Reis-Henriques. 2004. Organochlorine contaminants in flounder (*Platichthys flesus*) and mullet (*Mugil cephalus*) from Douro Estuary, and their use as sentinel species for environmental monitoring. *Aquat. Toxicol.* 69:347–357.

Fisher, S. W., S. W. Chordas, and P. F. Landrum. 1999. Lethal and sublethal body residues for PCB intoxication in the oligochaete, *Lumbriculus variegatus. Aquat. Toxicol.* 45:115–126.

Fisk, A. T., and T. A. Johnston. 1998. Maternal transfer of organochlorines to eggs of walleye (*Stizostedion vitreum*) in Lake Manitoba and Western Lake Superior. *J. Gt Lakes Res.* 24:917–928.

Fisk, A. T., R. J. Norstrom, C. D. Cymbalisty, and D. C. G. Muir. 1998. Dietary accumulation and depuration of hydrophobic organochlorines: Bioaccumulation parameters and their relationship with the octanol/water partition coefficient. *Environ. Toxicol. Chem.* 17:951–961.

Fowler, S. W. 1990. Critical review of selected heavy-metal and chlorinated-hydrocarbon concentrations in the marine environment. *Mar. Environ. Res.* 29:1–64.

Gakstatter, J. H., and C. M. Weiss. 1967. The elimination of DDT-C^{14}, dieldrin-C^{14}, and lindane-C^{14} from fish following a single sublethal exposure in aquaria. *Trans. Am. Fish Soc.* 96:301–307.

García-Flor, N., et al. 2005. Enrichment of organochlorine contaminants in the sea surface microlayer: An organic carbon-driven process. *Mar. Chem.* 96:331–345.

Garcia-Reyero, N., et al. 2006. Dietary exposure of largemouth bass to OCPs changes expression of genes important for reproduction. *Aquat. Toxicol.* 78:358–369.

Geyer, H. J., et al. 1994. The relevance of aquatic organisms' lipid content to the toxicity of lipophilic chemicals: toxicity of lindane to different fish species. *Ecotoxicol. Environ. Saf.* 28:53–70.

Gilliom, R. J., et al. 2006. The quality of our Nation's waters: Pesticides in the Nation's streams and ground water, 1992–2001. Reston, VA: National Water-Quality Assessment Program, U.S. Geological Survey Circular 1291.

González de Canales, M. L., M. Oliva, and C. Garrido. 2009. Toxicity of lindane (γ-hexachlorocyclohexane) in *Sparus aurata*, *Crassostrea angulata* and *Scrobicularia plana*. *J. Environ. Sci. Health Part B Pestic. Food Contam. Agric. Wastes* 44:95–105.

Goodman, L. R., D. J. Hansen, J. A. Couch, and J. Forester. 1977a. Organochlorine insecticide residues in salmonid fish. *J. Appl. Ecol.* 3:42–53.

Goodman, L. R., D. J. Hansen, J. A. Couch and J. Forester. 1977b. Effects of heptachlor and toxaphene on laboratory-reared embryos and fry of the sheepshead minnow. Proceedings 30th Annual Conference of the SE Fish and Wildlife Agencies, Gulf Breeze Environmental Research Laboratory.

Greenfield, B. K., et al. 2005. Seasonal, interannual, and long-term variation in sport fish contamination, San Francisco Bay. *Sci. Total Environ.* 336:25–43.

Guarino, A. M., J. B. Pritchard, J. B. Anderson, and D. P. Rall. 1974. Tissue distribution of ^{14}C-DDT in the lobster after administration via intravascular or oral routes or after exposure from ambient sea water. *Toxicol. Appl. Pharmacol.* 29:277–288.

Guo, J.-Y., et al. 2007. Organochlorine pesticides in seafood products from southern China and health risk assessment. *Environ. Toxicol. Chem.* 26:1109–1115.

Hansen, P. D. 1980. Uptake and transfer of the chlorinated hydrocarbon lindane (γ-BHC) in a laboratory freshwater food chain. *Environ. Poll. (Series A)* 21:97–108.

Hansen, D. J., S. C. Schimmel, and J. Forester. 1977. Endrin: effects on the entire life cycle of a saltwater fish, *Cyprinodon variegatus. J. Toxicol. Environ. Health* 3:721–733.

Haya, K., and L. E. Burridge. 1988. Uptake and excretion of organochlorine pesticides by *Nereis virens* under normoxic and hypoxic conditions. *Bull. Environ. Contam. Toxicol.* 40:170–177.

Hemmer, M. J., et al. 2001. Effects of p-nonylphenol, methoxychlor, and endosulfan on vitellogenin induction and expression in sheepshead minnow (*Cyprinodon variegatus*). *Environ. Toxicol. Chem.* 20:336–343.

Hermsen, W., I. Sims, and M. Crane. 1994. The bioavailability and toxicity to *Mytilus edulis* L. of two organochlorine pesticides adsorbed to suspended solids. *Mar. Environ. Res.* 38:61–69.

Hickey, J. P., S. A. Batterman, and S. M. Chernyak. 2006. Trends of chlorinated organic contaminants in Great Lakes trout and walleye from 1970 to 1998. *Arch. Environ. Contam. Toxicol.* 50:97–110.

Hinck, J. E., R. J. Norstrom, C. E. Orazio, C. J. Schmitt, and D. E. Tillitt. 2009. Persistence of organochlorine chemical residues in fish from the Tombigbee River (Alabama, USA): continuing risk to wildlife from a former DDT manufacturing facility. *Environ. Pollut.* 157:582–591.

Hinck, J. E., et al. 2006a. *Biomonitoring of Environmental Status And Trends (BEST) Program: Environmental Contaminants, Health Indicators, and Reproductive Biomarkers in Fish from the Colorado River Basin.* 119. Reston, VA: U.S. Geological Survey. Scientific Investigation Report 2006–5163.

Hinck, J. E., et al. 2006b. Environmental contaminants and biomarker responses in fish from the Columbia River and its tributaries: spatial and temporal trends. *Sci. Total Environ.* 366:549–578.

Holden, A. V. 1966. Organochlorine insecticide residues in salmonid fish. *J. Appl. Ecol.* 3:45–53.

Hope, B. K., and S. Scatolini. 2005. DDT, DDD, and DDE in abiotic media and near-shore marine biota from Sand Island, Midway Atoll, North Pacific ocean. *Bull. Environ. Contam. Toxicol.* 75:554–560.

Hose, J. E., J. N. Cross, S. G. Smith, and D. Diehl. 1989. Reproductive impairment in a fish inhabiting a contaminated coastal environment off Southern California. *Environ. Pollut.* 57:139–148.

Hu, G., et al. 2010. Concentrations and accumulation features of organochlorine pesticides in the Baiyangdian Lake freshwater food web of North China. *Arch. Environ. Contam. Toxicol.* 58:700–710.

Hwang, H. 2000. Contaminant body residues in *Chironomus riparius* to assess acute and chronic toxic effects of environmental contaminants. PhD dissertation, Ohio State University.

Hwang, H., S. W. Fisher, K. Kim, and P. F. Landrum. 2004. Comparison of the toxicity using body residues of DDE and select PCB congeners to the midge, *Chironomus riparius*, in partial-life cycle tests. *Arch. Environ. Contam. Toxicol.* 46:32–42.

Hyde, K. M., S. Stokes, J. F. Fowler, J. B. Graves, and F. L. Bonner. 1974. The effect of mirex on channel catfish production. *Trans. Am. Fish Soc.* 103:366–369.

Ingersoll, C. G., et al. 2003. Uptake and depuration of nonionic organic contaminants from sediment by the oligochaete, *Lumbriculus variegatus*. *Environ. Toxicol. Chem.* 22:872–885.

Jarvinen, A. W., M. J. Hoffman, and T. W. Thorslund. 1976. Toxicity of DDT food and water exposure to fathead minnows. Duluth, MN: U.S. Environmental Protection Agency 600/3–76–114.

Jarvinen, A., and R. Tyo. 1978. Toxicity to fathead minnows of endrin in food and water. *Arch. Environ. Contam. Toxicol.* 7:409–421.

Jarvis, E., K. Schiff, L. Sabin, and M. J. Allen. 2007. Chlorinated hydrocarbons in pelagic forage fishes and squid of the Southern California Bight. *Environ. Toxicol. Chem.* 26:2290–2298.

Johnson, B. T., C. R. Saunders, H. O. Sanders, and R. S. Campbell. 1971. Biological magnification and degradation of DDT and aldrin by freshwater invertebrates. *Fish Res. Board Can.* 28:705–709.

Johnson, D. W. 1968. Pesticides and fishes—a review of selected literature. *Trans. Am. Fish Soc.* 97:398–424.

Johnson, K. G. 2005. Dietary exposure to organochlorine pesticides *p,p′*-DDE and dieldrin and their effects on steroidogenesis and reproductive success in Florida largemouth bass (*Micropoterus salmoides* floridanus). M.S. Thesis, Gainsville, University of Florida.

Johnson, L. L., et al. 2007a. Contaminant exposure in outmigrant juvenile salmon from Pacific Northwest estuaries of the United States. *Environ. Monit. Assess.* 124:167–194.

Johnson, K. G., et al. 2007b. Influence of seasonality and exposure on the accumulation and reproductive effects of *p,p′*-dichlorodiphenyldichloroethane and dieldrin in largemouth bass. *Environ. Toxicol. Chem.* 26:927–934.

Jørgensen, E. H., M. M. Vijayan, J. E. A. Killie, N. Aluru, O. Aas-Hansen, and A. Maule. 2006. Toxicokinetics and effects of PCBs in Arctic fish: A review of studies on Arctic charr. *J. Toxicol. Environ. Health A* 69:37–52.

Kajiwara, N., et al. 2003. Contamination by organochlorine compounds in sturgeons from Caspian Sea during 2001 and 2002. *Mar. Pollut. Bull.* 46:741–747.

Kawatski, J. A., and J. C. Schmulbach. 1971. Accumulation of insecticide in freshwater ostracods exposed continuously to sublethal concentrations of aldrin or dieldrin. *Trans. Am. Fish Soc.* 100:565–567.

Keilty, T. J., D. S. White, and P. F. Landrum. 1988a. Sublethal responses to endrin in sediment by *Limnodrilus hoffmeisteri (Tubificidae)*, and in mixed-culture with *Stylodrilus heringianus (Lumbriculidae)*. *Aquat. Toxicol.* 13:227–249.

Keilty, T. J., D. S. White, and P. F. Landrum. 1988b. Sublethal responses to endrin in sediment by *Stylodrilus heringianus* (Lumbriculidae) as measured by a 137 cesium marker layer technique. *Aquat. Toxicol.* 13:251–270.

Kennedy, A. J., G. R. Lotufo, J. A. Steevens, and T. S. Bridges. 2010. Determination of steady state tissue residues for invertebrates in contaminated sediment. US Army Engineer Research and Development Center. Vicksburg, MS. ERDC/EL TR-03-2.

Kidd, K. A., R. H. Hesslein, B. J. Ross, K. Koczanski, G. R. Stephens, and D. C. G. Muir. 1998. Bioaccumulation of organochlorines through a remote freshwater food web in the Canadian Arctic. *Environ. Pollut.* 102:91–103.

Kidd, K. A., H. A. Bootsma, R. H. Hesslein, D. C. G. Muir, and R. E. Hecky. 2001. Biomagnification of DDT through the benthic and pelagic food webs of Lake Malawi, East Africa: Importance of trophic level and carbon source. *Environ. Sci. Technol.* 35:14–20.

Kimbrough, K. L., W. E. Johnson, G. G. Lauenstein, J. D. Christensen, and D. A. Apeti. 2008. An assessment of two decades of contaminant monitoring in the Nation's coastal zone. 105 pp. Silver Spring, MD: NOAA Technical Memorandum NOS NCCOS 74.

Kleinert, S. J, P. E. Degurse, and T. L. Wirth. 1968. Occurrence and significance of DDT and dieldrin residues in Wisconsin fish. 43 pp. Madison, WI: Technical Bulletin Number 41, Department of Natural Resources.

Knauf, W., and E. F. Schulze. 1973. New findings on the toxicity of endosulfan and its metabolites to aquatic organisms. Submitted in fulfillment of Pesticide Registration Notice 38: 717–732.

Koenig, C. C. 1977. The effects of DDT and mirex alone and in combination on the reproduction of a salt marsh cyprinodont fish, *Adinia xenica*. In *Physiological Responses of Marine Biota to Pollutants*, eds. F. J. Vernberg, A. Calabrese, F. P. Thurberg, and W. B. Vernberg, pp. 357–376. New York, NY: Academic Press.

Kumblad, L., A. Olsson, V. Koutny, and H. Berg. 2001. Distribution of DDT residues in fish from the Songkhla Lake, Thailand. *Environ. Pollut.* 112:193–200.

Kwong, R. W. M., P. K. N. Yu, P. K. S. Lam, and W.-X. Wang. 2008. Uptake, elimination, and biotransformation of aqueous and dietary DDT in marine fish. *Environ. Toxicol. Chem.* 27:2053–2063.

Lamai, S. L., G. F. Warner, and C. H. Walker. 1999. Effects of dieldrin on life stages of the African catfish, *Clarias gariepinus* (Burchell). *Ecotoxicol. Environ. Saf.* 42:22–29.

Landrum, P. F., H. Lee, and M. J. Lydy. 1992. Toxicokinetics in aquatic systems: Model comparisons and use in hazard assessment. *Environ. Toxicol. Chem.* 11:1709–1725.

Landrum, P. F., J. A. Steevens, M. McElroy, D. C. Gossiaux, J. S. Lewis, and S. D. Robinson. 2005. Time-dependent toxicity of dichlorodiphenyldichloroethylene to *Hyalella azteca*. *Environ. Toxicol. Chem.* 24:211–218.

Larsson, P., L. Okla, and L. Collvin. 1993. Reproductive status and lipid content as factors in PCB, DDT and HCH contamination of a population of pike (*Esox lucius* L). *Environ. Toxicol. Chem.* 12:855–861.

Leatherland, J. F., and R. A. Sonstegard. 1979. Effect of dietary mirex and PCB (Aroclor 1254) on thyroid activity and lipid reserves in rainbow trout *Salmo gairdneri* Richardson. *J. Fish Dis.* 2:43–48.

Leblond, V. S., M. Bisson, and A. Hontela. 2001. Inhibition of cortisol secretion in dispersed head kidney cells of rainbow trout (*Oncorhynchus mykiss*) by endosulfan, an organochlorine pesticide. *Gen. Comp. Endocrinol.* 121:48–56.

Lee, J. H., J. R. Sylverster, and C. E. Nash. 1975. Effects of mirex and methoxychlor on juvenile and adult striped mullet, *Mugil cephalus* (L.). *Bull. Environ. Contam. Toxicol.* 14:180–186.

Leffler, C. W. 1975. Effects of ingested mirex and DDT on juvenile *Callinectes sapidus* Rathbun. *Environ. Pollut.* 8:283–300.

Lewis, M. A., et al. 2002. Fish tissue quality in near-coastal areas of the Gulf of Mexico receiving point source discharges. *Sci. Total Environ.* 284:249–261.

Lin, T., et al. 2009. Levels and mass burden of DDTs in sediments from fishing harbors: The importance of DDT-containing antifouling paint to the coastal environment of China. *Environ. Sci. Technol.* 43:8033–8038.

Lohmann, R., K. Breivik, J. Dachs, and D. Muir. 2007. Global fate of POPs: Current and future research directions. *Environ. Pollut.* 150:150–165.

Looser, R., O. Froescheis, G. M. Cailliet, W. M. Jarman, and K. Ballschmiter. 2000. The deep-sea as a final global sink of semivolatile persistent organic pollutants? Part II: Organochlorine pesticides in surface and deep-sea dwelling fish of the North and South Atlantic and the Monterey Bay Canyon (California). *Chemosphere* 40:661–670.

Lotufo, G.R., M.D. Duke, and T.S. Bridges. Unpublished. Bioaccumulation of sediment-associated DDT in the amphipod *Leptocheirus plumulosus*. Manuscript in preparation.

Lotufo, G. R., P. F. Landrum, M. L. Gedeon, E. A. Tigue, and L. R. Herche. 2000a. Comparative toxicity and toxicokinetics of DDT and its major metabolites in freshwater amphipods. *Environ. Toxicol. Chem.* 19:368–379.

Lotufo, G. R., J. D. Farrar, and T. S. Bridges. 2000b. Effects of exposure source, worm density, and sex on DDT bioaccumulation and toxicity in the marine polychaete *Neanthes arenaceodentata*. *Environ. Toxicol. Chem.* 19:472–484.

Lotufo, G. R., J. D. Farrar, B. M. Duke, and T. S. Bridges. 2001a. DDT toxicity and critical body residue in the amphipod *Leptocheirus plumulosus* in exposures to spiked sediment. *Arch. Environ. Contam. Toxicol.* 41:142–150.

Lotufo, G. R., Landrum, P. F., and Gedeon, M. L. 2001b. Toxicity and bioaccumulation of DDT in freshwater amphipods in exposures to spiked sediments. *Environ. Toxicol. Chem.* 20:810–825.

Ludke, J. L., D. E. Ferguson, and W. D. Burke. 1968. Some endrin relationships in resistant and susceptible populations of golden shiners, *Notemigonus crysoleucas*. *Trans. Am. Fish Soc.* 97:260–263.

Ludke, J. L., M. T. Finley, and C. Lusk. 1971. Toxicity of mirex to crayfish, *Procambarus blandingi*. *Bull. Environ. Contam. Toxicol.* 6:89–96.

Macek, K. G. 1968. Reproduction in the brook trout (*Salvelinus fontinalis*) fed sublethal concentration of DDT. *Fish Res. Board Can.* 25:1787–1796.

Macek, K. J., C. R. Rodgers, D. L. Stalling, and S. Korn. 1970. The uptake, distribution and elimination of dietary ^{14}C-DDT and ^{14}C-dieldrin in rainbow trout. *Trans. Am. Fish Soc.* 99:689–695.

Macek, K. J., K. S. Buxton, S. K. Derr, J. W. Dean, and S. Sauter. 1976. Chronic toxicity of lindane to selected aquatic invertebrates and fishes. U.S. Environmental Protection Agency, EPA—600/3-76-046.

Marcelle, C., and J. P. Thome.1983. Acute toxicity and bioaccumulation of lindane in gudgeon, *Gobio gobio* (L.). *Bull. Environ. Contam. Toxicol.* 31:453–458.

Mason, J. W., and D. R. Rowe. 1976. The accumulation and loss of dieldrin and endrin in the Eastern oyster. *Arch. Environ. Contam. Toxicol.* 4:349–360.

Matta, M. B., A. J. Mearns, and M. F. Buchman. 1986. *Trends in DDT and PCBs in U.S. West Coast fish and invertebrates*. Seattle, WA: Ocean Assessments Division, National Ocean Service, NOAA.

Matthiessen, P., P. J. Fox, R. J. Douthwaite, and A. B. Wood. 1982. Accumulation of endosulfan residues in fish and their predators after aerial spraying for the control of testse fly in Botswana. *Pestic. Sci.* 13:39–48.

Mayer, F. L. J., P. M. J. Mehrle, and W. P. Dwyer. 1977. Toxaphene: Chronic Toxicity to Fathead Minnows and Channel Catfish. *Ecological Research Series*, 39. Duluth, MN: U.S. Environmental Protection Agency 600/3-77-069.

Mayer, F. L., Jr., P. M. Mehrle, Jr., and W. P. Dwyer. 1975. Toxaphene effects on reproduction, growth, and mortality of brook trout. Duluth, MN: U.S. Environmental Protection Agency 600/3-75/013.

McCarty, L. S., and D. Mackay. 1993. Enhancing ecotoxicological modeling and assessment. *Environ. Sci. Technol.* 27:1718–1728.

Meador, J. 2006. Rationale and procedures for using the tissue-residue approach for toxicity assessment and determination of tissue, water, and sediment quality guidelines for aquatic organisms. *Hum. Ecol. Risk Assess.* 12:1018–1073.

Mearns, A. J., et al. 1991. Contaminant trends in the Southern California bight: Inventory and assessment. In: *NOAA Technical Memorandum NOS ORCA 62*. Seattle, WA: Office of Ocean Resources Conservation and Assessment, National Ocean Service.

Mehrle, P. M., and F. L. Mayer. 1975a. Toxaphene effects of growth and bone composition of fathead minnows, *Pimephales promelas*. *Fish Res. Board Can.* 32:593–598.

Mehrle, P. M., and F. L. Mayer. 1975b. Toxaphene effects on growth and development of brook trout (*Salvelinus fontinalis*). *Fish Res. Board Can.* 32:609–613.

Metcalfe, T. L., C. D. Metcalfe, Y. Kiparissis, A. J. Niimi, C. M. Foran, and W. H. Benson. 2000. Development and endocrine responses in Japanese medaka (*Oryzias latipes*) exposed to *o,p′*-DDT in water or through maternal transfer. *Environ. Toxicol. Chem.* 19:1893–1900.

Miller, M. A. 1993. Maternal transfer of organochlorine compounds in salmonines to their eggs. *Can. J. Fish Aquat. Sci.* 50:1405–1413.

Miller, M. A. 1994. Organochlorine concentration dynamics in Lake Michigan chinook salmon (*Oncorhynchus tshawytscha*). *Arch. Environ. Contam. Toxicol.* 27:367–374.

Miller, M. A., and J. F. Amrhein. 1995. Maternal transfer of organochlorine compounds in Lake-Superior Siscowet (*Salvelinus namaycush siscowet*) to their eggs. *Bull. Environ. Contam. Toxicol.* 55:96–103.

Mills, L. J., et al. 2001. Effects of estrogenic (*o,p′*-DDT; octylphenol) and anti-androgenic (*p,p′*-DDE) chemicals on indicators of endocrine status in juvenile male summer flounder (*Paralichthys dentatus*). *Aquat. Toxicol.* 52:157–176.

Milston, R. H., et al. 2003. Short-term exposure of chinook salmon (*Oncoryhnchus tshawytscha*) to *o,p'*-DDE or DMSO during early life-history stages causes long-term humoral immunosuppression. *Environ. Health Perspect.* 111:1601–1607.

Miranda, A. L., H. Roche, M. A. F. Randi, M. L. Menezes, and C. A. O. Ribeiro. 2008. Bioaccumulation of chlorinated pesticides and PCBs in the tropical freshwater fish *Hoplias malabaricus*: Histopathological, physiological, and immunological findings. *Environ. Int.* 34:939–949.

Monirith, I., et al. 2003. Asia-Pacific Mussel Watch: Monitoring contamination of persistent organochlorine compounds in coastal waters of Asian Countries. *Mar. Pollut. Bull.* 46:281–300.

Monod, G. 1985. Egg mortality of Lake Geneva charr (*Salvelinus alpinus* l.) contaminated by PCB and DDT derivatives. *Bull. Environ. Contam. Toxicol.* 35:531–536.

Muir, D., and E. Sverko. 2006. Analytical methods for PCBs and organochlorine pesticides in environmental monitoring and surveillance: A critical appraisal. *Anal. Bioanal. Chem.* 386:769–789.

Muir, D., D. Swackhamer, T. Bidleman, and L. Jantunen. 2006. Toxaphene in the Great Lakes. In *Persistent Organic Pollutants in the Great Lakes*, ed. R. A. Hites, pp. 201–265. Berlin Heidelberg: Springer-Verlag.

Muller, J. K., K. G. Johnson, M. S. Sepulveda, C. J. Borgert, and T. S. Gross. 2004. Accumulation of dietary DDE and dieldrin by largemouth bass, *Micropterus salmoides floridanus*. *Bull. Environ. Contam. Toxicol.* 73:1078–1085.

Mulsow, S., P. F. Landrum, and J. A. Robbins. 2002. Biological mixing responses to sublethal concentrations of DDT in sediments by *Heteromastus filiformis* using a 137-Cs marker layer technique. *Mar. Ecol. Prog. Ser.* 239:181–191.

Mulsow, S. G., and P. F. Landrum. 1995. Bioaccumulation of DDT in a marine polychaete, the conveyor-belt deposit feeder *Heteromastus filiformis* (Claparede). *Chemosphere* 31:3141–3152.

Nakata, H., et al. 2005. Concentrations and compositions of organochlorine contaminants in sediments, soils, crustaceans, fishes and birds collected from Lake Tai, Hangzhou bay and Shanghai city region, China. *Environ. Pollut.* 133:415–429.

Naqvi, S. M., and C. Vaishnavi. 1993. Bioaccumulative potential and toxicity of endosulfan insecticide to non-target animals. *Comp. Biochem. Physiol. C: Pharmacol. Toxicol.* 105:347–361.

Narahashi, T. 2000. Neuroreceptors and ion channels as the basis for drug action: Past, present, and future. *J. Pharmacol. Exp. Ther.* 294:1–26.

National Oceanic and Atmospheric Administration (NOAA) and U.S. Environmental Protection Agency (EPA). 2007. 2002–2004 Southern California coastal marine fish contaminants survey. 91 pp. Long Beach, California: U.S. Department of Commerce, National Oceanic and Atmospheric Administration, on behalf of the Natural Resource Trustees, and U.S. Environmental Protection Agency–Region IX.

Neufeld, G. J., and J. B. Pritchard. 1979. An assessment of DDT toxicity on osmoregulation and gill Na, K-ATPase activity in the blue crab. In *Aquatic Toxicology*, eds. L. L. Marking and R. A. Kimerle, pp. 23–34. Philadelphia, PA: American Society for Testing and Materials STP 667.

Niimi, A. J. 1983. Biological and toxicological effects of environmental contaminants in fish and their eggs. *Can. J. Fish Aquat. Sci.* 40:306–312.

Nimmo, D. R., A. J. Wilson, and R. R. Blackman. 1970. Localization of DDT in the body organs of pink and white shrimp. *Bull. Environ. Contam. Toxicol.* 5:333–341.

Nowell, L. H., P. D. Capel, and P. D. Dileanis. 1999. Pesticides in stream sediment and aquatic biota: Distribution, trends, and governing factors. In: *Pesticides in the Hydrologic System Vol. 4*. Boca Raton, FL: Lewis Publishers, CRC Press.

O'Connor, T. P., and G. G. Lauenstein. 2006. Trends in chemical concentrations in mussels and oysters collected along the US coast: Update to 2003. *Mar. Environ. Res.* 62:261–285.

Oliver, B. G., and A. J. Niimi. 1985. Bioconcentration factors of some halogenated organics for rainbow trout: limitations in their use for prediction of environmental residues. *Environ. Sci. Technol.* 19:842–849.

Pandit, G. G., S. K. Sahu, S. Sharma, and V. D. Puranik. 2006. Distribution and fate of persistent organochlorine pesticides in coastal marine environment of Mumbai. *Environ. Int.* 32:240–243.

PAN North American. 2009. http://www.panna.org/campaigns/endosulfan (accessed October 13, 2009).

Papoulias, D. M., S. A. Villalobos, J. Meadows, D. B. Noltie, J. P. Giesy, and D. E. Tillitt. 2003. *In ovo* exposure to *o,p'*-DDE affects sexual development but not sexual differentiation in Japanese medaka (*Oryzias latipes*). *Environ. Health Perspect.* 111:29–32.

Parrish, P. R., J. A. Couch, J. Forester, J. M. Patrick, and G. H. Cook. 1974. Dieldrin: effects on several estuarine organisms. *Proc. 28th Ann. Conf. Southeast Assoc. Game Fish Comm.* 427–434.

Parrish, P. R., S. C. Schimmel, D. J. Hansen, J. M. Patrick, and J. Forester. 1976. Chlordane: Effects on several estuarine organisms. *J. Toxicol. Environ. Health Part A* 1:485–494.

Pawlisz, A. V., and R. H. Peters. 1993. A radioactive tracer technique for the study of lethal body burdens of narcotic organic chemicals in *Daphnia magna*. *Environ. Sci. Technol.* 27:2795–2800.

Pesce, S. F., J. Cazenave, M. V. Monferrán, S. Frede, and D. A. Wunderlin. 2008. Integrated survey on toxic effects of lindane on neotropical fish: *Corydoras paleatus* and *Jenynsia multidentata*. *Environ. Pollut.* 156:775–783.

Pontolillo, J., and R. P. Eganhouse. 2001. The search for reliable aqueous solubility (S_w) and octanol-water partition coefficient (k_{ow}) data for hydrophobic organic compounds: DDT and DDE as case study. 61 pp. Reston, VA: U.S. Geological Survey Water-Resources Investigations Report 01–4201.

Qiu, Y. W., et al. 2009. Current status and historical trends of organochlorine pesticides in the ecosystem of Deep Bay, South China. *Estuar. Coast Shelf Sci.* 85:265–272.

Ramamoorthy, S. 1985. Competition of fate processes in the bioconcentration of lindane. *Bull. Environ. Contam. Toxicol.* 34:349–358.

Ramu, K., et al. 2007. Asian mussel watch program: Contamination status of polybrominated diphenyl ethers and organochlorines in coastal waters of Asian countries. *Environ. Sci. Technol.* 41:4580–4586.

Randall, R. C., H. Lee II, R. J. Ozretich, J. L. Lake, and R. J. Pruell. 1991. Evaluation of selected lipid methods for normalizing pollutant bioaccumulation. *Environ. Toxicol. Chem.* 10:1431–1436.

Randall, R. C., D. R. Young, H. Lee II, and S. F. Echols. 1998. Lipid methodology and pollutant normalization relationships for neutral nonpolar organic pollutants. *Environ. Toxicol. Chem.* 17:788–791.

Roberts, D. 1972. The assimilation and chronic effects of sub-lethal concentrations of endosulfan on condition and spawning in the common mussel. *Mar. Biol.* 16:119–125.

Russell, R. W., F. A. P. C. Gobas, and G. D. Haffner. 1999. Maternal transfer and *in ovo* exposure of organochlorines in oviparous organisms: A model and field verification. *Environ. Sci. Technol.* 33:416–420.

Ruus, A., K. I. Ugland, O. Espeland, and J. U. Skaare. 1999. Organochlorine contaminants in a local marine food chain from Jarfjord, Northern Norway. *Mar. Environ. Res.* 48:131–146.

Ryan, M. J., et al. 2005. Temporal trends of organochlorine contaminants in burbot and lake trout from three selected Yukon lakes. *Sci. Total Environ.* 351–352:501–522.

Said, T. O., and M. A. F. Hamed. 2005. Distribution of chlorinated pesticides in surface water and fish of El Temsah and Bitter Lakes, Suez Canal. *Egyptian J. Aquat. Res.* 31:200–213

Sarkar, S. K., et al. 2008. Occurrence, distribution and possible sources of organochlorine pesticide residues in tropical coastal environment of India: An overview. *Environ. Int.* 34:1062–1071.

Schiff, K., and M. J. Allen. 2000. Chlorinated hydrocarbons in flatfishes from the Southern California, USA, Bight. *Environ. Toxicol. Chem.* 19:1559–1565.

Schimmel, S. C., P. R. Parrish, D. J. Hansen, J. M. Jr. Patrick, and J. Forester. 1975. Endrin: effects on several estuarine organisms. *Proc. 28th Ann. Conf. Southeast Assoc. Game Fish Comm.* 187–194.

Schimmel, S. C., J. M. Jr. Patrick, and J. Forester. 1976. Heptachlor: Toxicity to and uptake by several estuarine organisms. *J. Toxicol. Environ. Health* 1:955–965.

Schimmel, S. C., J. M. Jr. Patrick, and A. J. Wilson Jr. 1977a. Acute toxicity to and bioconcentration of endosulfan by estuarine animals. In: *Aquatic Toxicology and Hazard Evaluation*, eds. F. L. Mayer and J. L. Hamelink, pp. 241–252. Philadelphia, PA: American Society for Testing and Materials STP 634.

Schimmel, S. C., J. M. Jr. Patrick, and J. Forester. 1977b. Toxicity and bioaccumulation of BHC and lindane in selected estuarine animals. *Arch. Environ. Contam. Toxicol.* 6:355–363.

Schimmel, S. C., J. M. Patrick, and J. Forester. 1977c. Uptake toxicity of toxaphene in several estuarine organisms. *Arch. Environ. Contam. Toxicol.* 5:353–367.

Schmitt, C. J. 2002a. Organochlorine chemical residues in fish from the Mississippi River basin, 1995. *Arch. Environ. Contam. Toxicol.* 43:81–97.

Schmitt, C. J. 2002b. Biomonitoring of environmental status and trends (BEST) program: environmental contaminants and their effects on fish in the Mississippi River Basin. USGS/BRD/BSR-2002–0004.

Schmitt, C. J., J. L. Zajicek, and P. H. Peterman. 1990. National contaminant biomonitoring program: residues of organochlorine chemicals in U.S. freshwater fish, 1976–1984. *Arch. Environ. Contam. Toxicol.* 19:748–781.

Serrano, R., M. A. Blanes, and F. J. López. 2008. Maternal transfer of organochlorine compounds to oocytes in wild and farmed gilthead sea bream (*Sparus aurata*). *Chemosphere* 70:561–566.

Shen, L., and F. Wania. 2005. Compilation, evaluation, and selection of physical-chemical property data for organochlorine pesticides. *J. Chem. Eng. Data* 50:742–768.

Shubat, P. J., and L. R. Curtis.1986. Ration and toxicant preexposure influence dieldrin accumulation by rainbow trout (*Salmo gairdneri*). *Environ. Toxicol. Chem.* 5:69–77.

Singh, P. B., and V. Singh. 2007. Exposure recovery response of isomers of HCH, metabolites of DDT and estradiol-17 beta in the female catfish, *Heteropneustes fossilis*. *Environ. Toxicol. Pharmacol.* 24:245–251.

Skea, J. C., H. J. Simonin, S. Jackling, and J. Symula. 1981. Accumulation and retention of mirex by brook trout fed a contaminated diet. *Bull. Environ. Contam. Toxicol.* 27:79–83.

Smeets, J. M. W., I. van Holsteijn, J. P. Giesy, W. Seinen, and M. van den Berg. 1999. Estrogenic potencies of several environmental pollutants, as determined by vitellogenin induction in a carp hepatocyte assay. *Toxicol. Sci.* 50:206–213.

Smith, R. M., and C. F. Cole. 1973. Effects of egg concentration of DDT and dieldrin on development in winter flounder (*Pseudopleuronects americanus*). *Fish Res. Board Can.* 30:1894–1898.

Sodergren, A., and B. J. Svensson. 1973. Uptake and accumulation of DDT and PCB by *Ephermera danica* (Ephemeroptera) in continuous-flow systems. *Bull. Environ. Contam. Toxicol.* 9:345–350.

Ssebugere, P., B. T. Kiremire, M. Kishimba, S. O. Wandiga, S. A. Nyanzi, and J. Wasswa. 2009. DDT and metabolites in fish from Lake Edward, Uganda. *Chemosphere* 76:212–215.

Stahl, L., B. Snyder, A. Olsen, and J. Pitt. 2009. Contaminants in fish tissue from US lakes and reservoirs: A National probabilistic study. *Environ. Monit. Assess.* 150:3–19.

Strachan, W. M. J., D. A. Burniston, M. Williamson, and H. Bohdanowicz. 2001. Spatial differences in persistent organochlorine pollutant concentrations between the Bering and Chukchi Seas (1993). *Mar. Pollut. Bull.* 43:132–142.

Strandberg, B., et al. 1998. Concentrations and spatial variations of cyclodienes and other organochlorines in herring and perch from the Baltic Sea. *Sci. Total Environ.* 215:69–83.

Su, Y., et al. 2008. A circumpolar perspective of atmospheric organochlorine pesticides (OCPs): Results from six Arctic monitoring stations in 2000–2003. *Atmos. Environ.* 42:4682–4698.

Sudaryanto, A., et al. 2007. Levels and distribution of organochlorines in fish from Indonesia. *Environ. Int.* 33:750–758.

Sumpter, J. P., and S. Jobling. 1995. Vitellogenesis as a biomarker for estrogenic contamination of the aquatic environment. *Environ. Health Perspect.* 103:173–178.

Tagatz, M. E., P. W. Borhtwick, and J. Forester. 1975. Seasonal effects of leached mirex of selected estuarine animals. *Arch. Environ. Contam. Toxicol.* 3:371–383.

Tagatz, M. E., P. W. Borthwick, J. M. Ivey, and J. Knight. 1976. Effects of leached mirex on experimental communities of estuarine animals. *Arch. Environ. Contam. Toxicol.* 4:435–442.

Tan, L., M. He, B. Men, and C. Lin. 2009. Distribution and sources of organochlorine pesticides in water and sediments from Daliao River estuary of Liaodong Bay, Bohai Sea (China). *Estuar. Coast Shelf Sci.* 84:119–127.

Tanabe, S., M. S. Prudente, S. Kan-Atireklap, and A. Subramanian. 2000. Mussel watch: Marine pollution monitoring of butyltins and organochlorines in coastal waters of Thailand, Philippines and India. *Ocean Coast Manag.* 43:819–839.

Thorpe, K. L., R. Benstead, T. H. Hutchinson, and C. R. Tyler. 2007. Associations between altered vitellogenin concentrations and adverse health effects in fathead minnow (*Pimephales promelas*). *Aquat. Toxicol.* 85:176–183.

Toledo, M. C. F., and C. M. Jonsson. 1992. Bioaccumulation and elimination of endosulfan in zebra fish (*Brachydanio rerio*). *Pest. Sci.* 36:207–211.

Tyler-Schroeder, D. B. 1979. Use of the grass shrimp (*Palaemonetes pugio*) in a life cycle toxicity test. In: *Aquatic Toxicology*, eds. L. L. Marking and R. A. Kimerle, pp. 159–170. Philadelphia, PA: American Society for Testing and Materials STP 667.

Ungerer, J. R., and P. Thomas. 1996. Role of very low density lipoproteins in the accumulation of *o,p'*-DDT in fish ovaries during gonadal recrudescence. *Aquat. Toxicol.* 35:183–195.

United Nations Environment Programme (UNEP). 2008. Global status of DDT and its alternatives for use in vector control to prevent disease. Stockholm convention of persistent organic pollutants. Geneva, Henk van den Berg, Laboratory of Entomology, Wageningen University and Research Centre, Netherlands: 31.

U.S. Geological Survey (USGS). 2008. Pesticide National Synthesis Project, National Water-Quality Assessment Program, Frequency of Mixtures. http://water.usgs.gov/nawqa/pnsp/pubs/circ1291/show_description. php?chapter=5&figure=6/ (accessed October 26, 2009).

U.S. Environmental Protection Agency (U.S. EPA). 1980. Ambient water quality criteria for DDT. Springfield, VA. U.S. Environmental Protection Agency 440/5-80-038.

U.S. EPA. 1992. National study of chemical residues in fish. Washington, DC: Office of Science and Technology EPA-823-R-92–008.

U.S. EPA. 2002. Reregistration eligibility decision for endosulfan, U. S. Enviromental Protection Agency 738-F-02–012.

U.S. EPA. 2004. Five-year review report. Third five-year review report for Triana/Tennessee River site, Triana, Madison Country, Alabama. 14 pp + 17 attachments. Atlanta, GA: U.S. Environmental Protection Agency.

U.S. EPA. 2006. National Listing of Fish Advisories. General Fact Sheet: 2005/06 National Listing. http://www.epa.gov/waterscience/fish/advisories/2006/index.html (accessed September 30, 2008).

Ungerer, J. R., and P. Thomas. 1996. Role of very low density lipoproteins in the accumulation of *o,p'*-DDT in fish ovaries during gonadal recrudescence. *Aquat. Toxicol.* 35:183–195.

van Valin, C. C., A. K. Andrews, and L. L. Eller. 1968. Some effects of mirex on two warm-water fishes. *Trans. Am. Fish Soc.* 97:185–196.

Villalobos, S. A., et al. 2003. Toxicity of *o,p'*-DDE to medaka d-rR strain after a one-time embryonic exposure by *in ovo* nanoinjection: An early through juvenile life cycle assessment. *Chemosphere* 53:819–826.

von Westernhagen, H., P. Cameron, V. Dethlefsen, and D. Janseen. 1989. Chlorinated hydrocarbons in North Sea whiting (*Merlangius merlangus* L.), and effects on reproduction. 1. Tissue burden and hatching success. *Helgol. Meeresunters* 43:45–60.

Wang, X. H., and W. X. Wang. 2005. Uptake, absorption efficiency and elimination of DDT in marine phytoplankton, copepods and fish. *Environ. Pollut.* 136:453–464.

Warlen, S. M., Wolfe, D. A., Lewis, C. W., and Colby, D. R. 1977. Accumulation and retention of dietary ^{14}C-DDT by Atlantic menhaden. *Trans. Am. Fish Soc.* 106:95–104.

Weber, J., et al. 2006. Endosulfan and γ-HCH in the Arctic: An assessment of surface seawater concentrations and air-sea exchange. *Environ. Sci. Technol.* 40:7570–7576.

Wester, P. W., and J. H. Canton. 1986. Histopathological study of *Oryzias latipes* (medaka) after long-term beta-hexachlorocyclohexane exposure. *Aquat. Toxicol.* 9:21–45.

Westin, D. T., C. E. Olney, and B. A. Rogers. 1985. Effects of parental and dietary organochlorines on survival and body burdens of striped bass larvae. *Trans. Am. Fish Soc.* 114:125–136.

Wiktelius, S., and C. A. Edwards. 1997. Organochlorine insecticide residues in African fauna: 1971–1995. *Rev. Environ. Contam. Toxicol.* 151:1–37.

Willingham, E. 2004. Endocrine-disrupting compounds and mixtures: unexpected dose-response. *Arch. Environ. Contam. Toxicol.* 46:265–269.

Wong, C. S., P. D. Capel, and L. H. Nowell. 2000. Organochlorine pesticides and PCBs in stream sediment and aquatic biota—initial results from the National Water Quality Assessment Program, 1992–1995. Sacramento, CA: Water-Resources Investigations Report USGS 00–4053.

World Health Organization (WHO). 1984a. Environmental health criteria 40—Endosulfan. Published under the joint sponsorship of the United Nations Environment Programme, the International Labour Organisation, and the World Health Organization. Geneva: World Health Organization, International programme on Chemcal Safety.

World Health Organization (WHO). 1984b. Environmental health criteria 45—Camphechlor. Published under the joint sponsorship of the United Nations Environment Programme, the International Labour Organisation, and the World Health Organization. Geneva: World Health Organization, International programme on Chemical Safety.

World Health Organization (WHO). 1991. Environmental health criteria 124—Lindane. Published under the joint sponsorship of the United Nations Environment Programme, the International Labour Organisation, and the World Health Organization. Geneva: World Health Organization, International Programme on Chemical Safety.

World Health Organization (WHO). 1992. Environmental health criteria 130—Endrin. Published under the joint sponsorship of the United Nations Environment Programme, the International Labour Organisation, and the World Health Organization. Geneva: World Health Organization, International Programme on Chemical Safety.

Zeng, E. Y., and M. I. Venkatesan. 1999. Dispersion of sediment DDTs in the coastal ocean off southern California. *Sci. Total Environ.* 229:195–208.

Zeng, E. Y., et al. 2005. Distribution and mass inventory of total dichlorodiphenyldichlorothylene in the water column of the southern California Bight. *Environ. Sci. Technol.* 39:8170–8176.

Zhang, Z. B., and J. Y. Hu. 2008. Effects of *p,p'*-DDE exposure on gonadal development and gene expression in Japanese medaka (*Oryzias latipes*). *J. Environ. Sci. (China)* 20:347–352.

Flying Fish

From *The Royal Natural History,* edited by Richard Lydekker, Frederick Warne & Co.,
London, 1893–94.

3 Dioxins, PCBs, and PBDEs in Aquatic Organisms

Richard J. Wenning
Linda Martello
Anne Prusak-Daniel

CONTENTS

3.1   Introduction 103
3.2   Characteristics of Dioxins, PCBs, and PBDEs 104
    3.2.1   Dioxins 104
    3.2.2   PCBs 105
    3.2.3   PBDEs 107
3.3   Environmental Fate and Exposure 107
    3.3.1   Atmospheric Fate Considerations 107
    3.3.2   Aquatic Fate Considerations 108
    3.3.3   Bioaccumulation and Food Chain Transfers 109
3.4   Toxic Equivalency Factors 111
3.5   Tissue Residues in Aquatic Organisms 113
    3.5.1   Dioxin Tissue Residues 113
    3.5.2   PCB Tissue Residues 120
    3.5.3   PBDE Tissue Residues 123
3.6   Ecotoxicological Effects 127
    3.6.1   Dioxin Toxicity 148
    3.6.2   PCB Toxicity 152
    3.6.3   PBDE Toxicity 155
Summary 156
References 157

3.1   INTRODUCTION

Polychlorinated dibenzo-*p*-dioxins (PCDDs) and polychlorinated dibenzofurans (PCDFs; collectively referred to as dioxins) and polychlorinated biphenyls (PCBs) are members of a superfamily of compounds classified as polyhalogenated persistent organic pollutants (POPs). Polybrominated diphenyl ethers (PBDEs) are included along with dioxins and PCBs in a closely related group of contaminants identified in the Stockholm Convention as persistent, bioaccumulative, and toxic (PBT) substances (Davies 1999). Although not identified formally as a POP, the behavior of PBDEs in the environment fulfills the screening criteria in Annex D of the Stockholm Convention. POPs and PBT contaminants are generally acknowledged as long-lasting substances that can accumulate in fish, reaching levels that have the potential to affect the health of people and wildlife.

Dioxins, PCBs, and PBDEs share some—but not all—environmental fate and ecotoxicological properties. Their half-lives are on the order of 1–20 years or more, depending on the degree of chlorination or bromination (IUPAC 1989, Ritter et al. 1995). Their shared chemical characteristics predispose these substances to long environmental persistence and facilitate long-range transport to and accumulation in the polar region, often far removed from any source of use. Dioxins, PCBs, and PBDEs resist degradation and sorb tightly to organic matter in the aquatic environment. They are lipophilic, bioaccumulative, and have been shown to transfer from lower to higher trophic levels in aquatic food chains, thereby aiding their distribution throughout the environment. In general, dioxin levels occur at approximately 1–2 orders of magnitude lower in the environment than PCBs and PBDEs. Typically, environmental levels of PBDE are lower than that of PCBs. However, there are occasions where environmental samples contain PBDEs at levels 1–2 orders of magnitude higher than PCBs.

This chapter summarizes current knowledge on tissue residues and ecotoxicological effects typically observed in aquatic organisms exposed to dioxins, PCBs, and PBDEs. The chapter is organized into seven sections. The second section summarizes the physical and chemical properties of these compounds that are important to understanding environmental fate, bioaccumulation, distribution in the environment, and the significance of tissue residues in aquatic organisms. The third section summarizes what is currently understood about the environmental fate of these compounds, with an emphasis on factors that influence exposure and toxicity. The fourth section includes an overview of toxic equivalency factors (TEFs) assigned to different dioxins and certain dioxin-like PCBs used to evaluate the significance of environmental exposures, and a discussion of recent TEF schemes proposed to aid PBDE risk assessment. The fifth section summarizes available monitoring data describing tissue levels in various aquatic species, and comments on emerging trends in fish body burdens reported in different parts of the world. The sixth section reviews what is suspected or known about predominant acute and chronic ecotoxicological effects in different aquatic taxonomic groups. The final section is a summary of current gaps in knowledge about environmental fate, levels in aquatic organisms, and ecotoxicological effects.

3.2 CHARACTERISTICS OF DIOXINS, PCBs, AND PBDEs

3.2.1 DIOXINS

Dioxins are known to occur naturally from the incomplete combustion of organic material such as during forest fires or volcanic activity and are also produced by human activities. Dioxins are unintentional by-products of industrial, municipal, and domestic incineration, uncontrolled burning and certain industrial processes often used in incinerators, metal smelters, cement kilns, the manufacture of chlorinated organics, and coal-burning power plants (U.S. EPA 2005). Dioxins occur most notably as a contaminant in the manufacturing process of certain chlorinated organic chemicals such as some commercial herbicides and chlorinated phenols. The most widely studied sources of dioxins are the manufacture of 2,4-dichlorophenol (2,4-D) and 2,4,5-trichlorophenol (2,4,5-TCP), the manufacturing of the microbicide hexachlorphene, and the chlorine bleaching of wood pulp. 2,4,5-TCP was used to produce hexachlorophene and the herbicide, 2,4,5-trichlorophenoxyacetic acid (2,4,5-T). The manufacturing process for 2,4,5-T and related chlorinated herbicides typically results in contamination with trace amounts of 2,3,7,8-tetrachlorodibenzo-p-dioxin (2,3,7,8-TCDD). Several other manufacturing processes such as those involved in the manufacture of pentachlorophenol, chlorobenzenes, chlorobiphenyls, and polyvinyl chloride also are known to contain trace levels of dioxins and furans. Owing to improvements in production chemistry, manufacturing processes and chemical products contaminated with dioxins have been greatly reduced over the past two decades.

Dioxin is also produced by nonindustrial sources such as residential wood burning, backyard burning of household trash, oil heating, and emissions from automobiles using leaded or unleaded gasoline, and diesel fuel (U.S. EPA 2005). Burning of many materials that may contain chlorine, such as plastics, wood treated with PCP, pesticide-treated wastes, other polychlorinated chemicals, and even bleached paper are known or suspected to generate dioxins. Although emissions from these sources can vary greatly depending on practices and the technology, all of these sources are now considered by the U.S. Environmental Protection Agency (U.S. EPA) to represent the largest combined source of dioxins in the U.S. environment (U.S. EPA 2006a).

Strictly defined, dioxins are a class of halogenated aryl hydrocarbon compounds consisting of eight homologue groups encompassing 75 PCDD (dioxin) and 135 PCDF (furan) chemicals, each typically referred to as a dioxin or furan congener. The PCDD molecule consists of two phenyl rings joined by two oxygen bridges. The PCDF molecule comprises two phenyl rings joined by one oxygen bridge and one single bond (Figure 3.1). The degree of chlorination and the pattern of chlorine substitution on the two phenyl rings affect the stereochemistry of the congener, and are responsible for intercongener differences in environmental behavior and toxicity. The 17 dioxin and furan compounds substituted only at the 2,3,7, or 8-positions are widely regarded as the most toxic to humans and biota.

Increasingly, the term "dioxin-like" is used to describe compounds that share structural and biochemical similarities to PCDDs and PCDFs and share a common mode of action whereby they exert similar affects in humans and biota. At present, the term is most often used to describe 4 nonortho and 8 mono-ortho substituted PCBs (Figure 3.1). Including the coplanar PCBs, there are 29 dioxin-like compounds (17 dioxin and furan congeners plus 12 PCB congeners) that are often referred to simply as "dioxins." Table 3.1 provides information on the physical and chemical properties of dioxins and dioxin-like compounds.

3.2.2 PCBs

PCBs were introduced in the United States in 1929 and banned by the U.S. EPA nearly 50 years later due to concerns about toxicity to wildlife and human carcinogenicity. Unlike the dioxins, PCBs

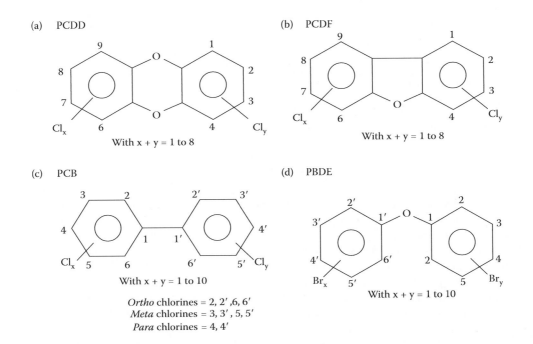

FIGURE 3.1 Polychlorinated dibenzo-*p*-dioxins (dioxins), polychlorinated dibenzofurans (furans), polychlorinated biphenyls (PCBs) and Polybrominated Diphenyl Ethers (PBDEs) with "dioxin-like" toxicity.

TABLE 3.1

Important Physical and Chemical Properties of Polychlorinated Dibenzo-*p*-Dioxins (PCDDs), Dibenzofurans (PCDFs), Biphenyls (PCBs), and Polybrominated Diphenyl Ethers (PBDEs)

Property	PCDD[a]	PCDF[a]	PCB[b]	PBDE
Octanol/water partition Coefficient (log K_{ow})	Between 6 and 12 for the 75 congeners	Between 6 and 10 for the 135 congeners	Between 4.5 and 8.5 for the 209 congeners	6.57[d] (pentaBDE) 8.35–8.90[e] (octaBDE) 9.97[e] (decaBDE)
Vapor pressure (mmHg)	Between 1.5×10^{-9} and 3.4×10^{-5} for the 75 congeners	9.21×10^{-7}	Between 7.7×10^{-5} and 4.1×10^{-3} for the 209 congeners	4.7×10^{-5f} (pentaBDE) Between 1.6×10^{-6} and 4.7×10^{-7} for the hexa- and hepta-BDEs 2.95×10^{-9h} (decaBDE)
Water solubility (µg/L)	0.019	0.692	Between 2.7 and 590 for the 209 congeners	13.3[i] (pentaBDE) 0.5[j] (octaBDE) <0.1[j] (decaBDE)
Henry's law constant (atm m³/mol)	Between 1.6×10^{-5} and 1.0×10^{-4} for the 75 congeners	1.48×10^{-5}	Between 5.2×10^{-4} and 2.0×10^{-3} for the 209 congeners	11[k] (pentaBDE) 10.6[l] (octaBDE) >44[m] (decaBDE)
BAF or BCF	130,000	61,000	0.2–101,000[c]	220,000–1,400,000[n]

Sources:

[a] U.S. EPA (2002).

[b] WHO/IPCS (1993) for selected Aroclors; water solubility, vapor pressure, and Henry's law measured at 25°C.

[c] PCB BCF range includes values for fish, crustaceans and mollusks. Values obtained from ECOTOX (U.S. EPA 2009a).

[d] MacGregor and Nixon (1997).

[e] Watanabe and Tatsukawa (1990).

[f] Stenzel and Nixon (1997).

[g] Tittlemier et al. (2002).

[h] Estimated value by Wania and Dugani (2003).

[i] Stenzel and Markley (1997).

[j] CMABFRIP (1997).

[k] European Communities (2001).

[l] European Communities (2003).

[m] European Communities (2002).

[n] Gustafsson et al. (1999).

were manufactured purposely because of their resistance to high temperatures, and widely used in fire prevention, insulation materials, and the manufacture of transformers, capacitors, electromagnets, circuit breakers, voltage regulators, and switches. Fluids containing PCBs were used in hydraulic systems, and as plasticizers and additives in lubricating and cutting oils. As the number of chlorines in a PCB mixture increases, the flash point rises and the mixture becomes less combustible (more stable) and, therefore, more highly favored in a wide range of heat-resistant and lubricating uses (Erickson 1997).

PCBs are a class of compounds consisting of 10 homologue groups encompassing 209 individual chlorinated congeners, including the 12 dioxin-like congeners (Erickson 1997, Van den Berg et al. 2006). The PCB molecule consists of two phenyl rings joined by a single bond (Figure 3.1). The individual PCB congeners differ in the number and position of the chlorine atoms on each ring. The toxicity of PCBs is typically determined by the mixture of homologue groups and, more specifically, the chlorination pattern of the biphenyl structure. Congeners with chlorines in both para positions (4 and 4′) and at least 2 chlorines at the meta positions (3, 5, 3′, 5′) on the biphenyl structure are considered "dioxin like" (Van den Berg et al. 2006). When there is just 1 or no substitution in the ortho position, the atoms of the congener are able to line up in a single plane (sometimes referred to as coplanar). The planar or "flat" configuration is regarded as the most potent configuration. Table 3.1 provides information on the physical and chemical properties of PCBs.

3.2.3 PBDEs

PBDEs are included in the class of manufactured chemicals referred to as brominated flame retardants (BFRs). On the basis of their use in the chemical industry, BFRs are classified as either reactive or additive. Reactive BFRs such as the tetrabromobisphenol A (TBBPA) are covalently bound into a polymer matrix. Additive BFRs are dissolved in the matrix and weakly associated with the polymer. The additive BFRs such as the PBDEs, polybrominated biphenyls (PBBs), and hexabromocyclododecane (HBCD) are more likely to be released into the environment more readily than reactive BFRs (Alaee and Wenning 2002, de Wit 2002).

Until regulatory-mandated and voluntary phase-out by industry began in about 2004, three technical formulations containing different mixtures of PBDE congeners were used by the industry: pentabromodiphenyl ether (Penta-PBDE), octabromodiphenyl ether (Octa-PBDE), and decabromodiphenyl ether (Deca-PBDE). Penta-PBDE was mostly used in polyurethane foams and Octa-PBDE was mainly used in rigid plastics such as ABS and high-impact polystyrene. The Deca-PBDE formulation was used in a wide range of polymers including textiles, resins, and rigid plastics (Alaee and Wenning 2002, de Wit 2002).

The PBDE molecule is structurally similar to PCBs. In addition to the bromine substitution, the major difference between PBDEs and PCBs is the presence of an ether group linking the two phenyl rings (Figure 3.1). Table 3.1 provides information on the physical and chemical properties of PBDEs. Although there are 209 possible congeners, analyses of congeners present in commercial PBDEs, organisms, including humans, and environmental matrices reveal a much smaller number of congeners tend to be found in the environment (U.S. EPA 2006b). PBDEs are numbered according to the same IUPAC system used for numbering PCBs.

3.3 ENVIRONMENTAL FATE AND EXPOSURE

A brief overview of the important chemical properties influencing the environmental fate of these compounds and the levels found in the aquatic environment is provided here. In general, dioxin, PCB, and PBDE congeners vary widely in their physical and chemical properties because of their differing degrees and patterns of chlorination or bromination. Although there is general agreement on the overall trends and relative magnitude of different properties, it is not uncommon to find disagreement among published literature values. Detailed reviews of the environmental fate of dioxins, PCBs, and PBDEs are available elsewhere (Rahman et al. 2001, de Wit 2002, Voorspoels et al. 2003, Wang et al. 2007, Beyer and Biziuk 2009).

3.3.1 ATMOSPHERIC FATE CONSIDERATIONS

Volatility (as measured by vapor pressure) affects environmental fate in two ways: by controlling the rate of partitioning between the vapor and the particle phases, and by controlling, together with water solubility (expressed as Henry's law constant), the rate of partitioning between the vapor phase in the atmosphere and the dissolved phase in water. Dioxins and PCBs have a wide range of volatilities according to the degree of chlorination. In general, the higher chlorinated congeners are less volatile than the lower chlorinated congeners (Davies 1999). PBDEs have low vapor pressures, generally lower than dioxins and PCBs. Experimentally determined vapor pressures for several

BDE congeners were found to be lower than for comparably chlorinated PCBs, and decreased with increasing number of bromines (Tittelmier and Tomy 2000). Halogen substitution pattern influences vapor pressure such that congeners with bromine substitution in the ortho positions to the ether bond tend to have higher vapor pressures (Wong et al. 2001).

Several fate and transport studies have demonstrated that the most important pathway for removal of these compounds from the atmosphere, and the main route through which these compounds enter the aquatic environment, is by gravitational settling and washout in rain (Raff and Hites 2007). Dioxins and PCBs attached to particulate matter will tend to settle out under gravity, with larger, coarser particles deposited more rapidly and closer to emission sources than smaller particles (Raff and Hites 2007). Bound to fine particulate and in gaseous form, these compounds are more prone to long-range atmospheric transport. These particles will tend to be deposited by rain and snow, although they may have traveled far from the emissions source before they are eventually removed. For PBDEs, however, experiments by Hale et al. (2002) suggest that fragments from the disintegration of polyurethane foam may be a mechanism by which PBDEs diffuse into the atmosphere.

3.3.2 Aquatic Fate Considerations

Dioxins, PCBs, and PBDEs are generally ubiquitous in the aquatic environment, particularly in sediments, and are transported to and recycled within aquatic systems. These compounds enter aquatic environments from wet and dry deposition, river inflows, groundwater flow, and direct and indirect discharges from industrial facilities. Dry and wet deposition may be the most important sources to water bodies such as lakes and seas with large surface areas. Long-term or temporary sequestration in aquatic systems can occur when bound to particles that settle as sediment, or volatilized across the air–water interface, or by chemical and biological transformations. The latter two processes are possible, but generally considered less significant than sedimentation; studies on transformation and decomposition of dioxins, PCBs, and PBDEs have not resolved the significance of these processes in the environment (Rahman et al. 2001, Hardy 2002).

Dioxins, PCBs, and PBDEs are highly lipophilic and hydrophobic compounds; these compounds are nearly insoluble in water and generally have low Henry's Law constant (H) values. The hydrophobicity of these compounds can be ascertained from their octanol-water partition coefficients (K_{ow}), which provides insight on the relationship between bioconcentration/bioaccumulation and toxicity. Dioxins are super-hydrophobic because experimentally determined log K_{ow}s are typically greater than 6 and as high as 12 (Gobas et al. 1989, Table 3.1). The log K_{ow}s of PCBs also are typically greater than 5, ranging from approximately 4.5 to 8.5 (Hawker and Connell 1988). The log K_{ow}s of PBDEs are in the range of 5.9–6.2 for tetraBDEs, 6.5–7.0 for pentaBDEs, 8.4–8.9 for octaBDEs, and 10 for decaBDE (Watanabe and Tatsukawa 1990, MacGregor and Nixon 1997). As a consequence of low vapor pressure, low water solubility, and high log K_{ow} values, dioxins, PCBs, and PBDEs entering the aquatic environment have a high affinity for the organic fraction of suspended particulate matter and sediments.

In general, dioxins and PBDEs partition weakly, more so than PCBs, between particulate and dissolved phases but generally remain bound to particulates when deposited or resuspended in the water column (Cetin and Odabasi 2009, Kitamura et al. 2009). The range of water-particle distribution coefficients (Kd), which represents the ratio of the concentration of a chemical in the particulate phase to its concentration in the dissolved phase is relatively low for dioxins in comparison to PCBs and PBDEs. The Kd value for the representative dioxin congener—2,3,7,8-TCDD—is 2.5E + 05. For PCBs and PBDEs, Kd values generally range by as much as six orders of magnitude (Hornbuckle et al. 2006); although for PBDEs, much of what is known is inferred from structure–activity relationships and comparisons to PCBs (Hardy 2002, Zeng et al. 2005). Regardless, the dioxins, PCBs, and PBDEs associate strongly with organic matter; hence, the degradation and mobility of organic carbon in sediment and, possibly, the influence of combustion-derived black carbon are significant factors contributing to their mobility in the aquatic environment (Lohmann et al. 2005). Suspended sediment in surface waters may be incorporated into the sediment bed and recycled at or near the sediment-water interface. Unless conditions occur where resuspension and bottom currents are not strong enough to cause further transport, sediments act as an important sink for these chemicals.

3.3.3 BIOACCUMULATION AND FOOD CHAIN TRANSFERS

It is widely accepted that the bioavailability of contaminants in sediments is governed by three factors: the composition and characteristics of the natural particles, the physical and chemical properties of the contaminants, and the behavior and physiological characteristics of the aquatic organisms (Guerrero 2007). Owing to their hydrophobic nature, the majority of dioxins, PCBs, and PBDEs released into aquatic systems ultimately become associated with the organic fraction of suspended and/or bed sediments and lipid-rich tissues of aquatic organisms. Lipid normalization may provide insight into species differences in contaminant concentrations. However, data is available which suggests that the relationship between lipophilic contaminants and lipid content of fish is not clear cut or well understood. In fact, there is just as much variation in lipid weight values as wet weight values.

It is important to distinguish between two mechanisms of uptake by aquatic organisms—bioaccumulation and bioconcentration—and understand the uncertainties associated with the consideration of both pathways in risk assessment (Arnot and Gobas 2006). Bioconcentration involves direct uptake from water (e.g., across the gill membrane in fish) and is distinct from bioaccumulation, which also includes dietary uptake and transfer from sediments. Bioconcentration factors (BCFs) are determined from the ratio of chemical concentration in the organism, assuming steady state metabolism and excretion, to that in the surrounding water column (U.S. EPA 1992). Bioaccumulation factors (BAFs) are based on lipid-normalized concentration of the chemical in the organism with respect to the concentration of the bioavailable fraction of chemical in water (Burkhard et al. 2003). Biota-sediment accumulation factors (BSAFs) are based on the lipid-normalized concentration in the organism with respect to organic carbon-normalized concentration in the sediments (Burkhard et al. 2003). Specific definitions for each of these terms are provided in Table 3.2.

Dioxins, PCBs, and PBDEs are known to bioaccumulate in fish and a wide variety of aquatic organisms. Field studies and models demonstrate that fish body burdens are largely due to dietary uptake rather than membrane transport across the gill surface (Saloranta et al. 2006a, Micheletti et al. 2008). Environmental monitoring data generally indicate higher body burdens in benthic organisms and bottom-dwelling fish than in pelagic fish residing in the surface water. The highest levels tend to occur in top predator species in food chains where successive stages of bioaccumulation at each trophic level results in significant biomagnification (a cumulative increase in the concentration of a persistent substance in successively higher trophic levels of the food chain). The differences are generally attributed to the close association of sediment-dwelling organisms with sediments, and the generally low opportunity for exposure to these substances in the water column. Differences in BCFs and BAFs between species are often explained by different feeding strategies and, perhaps, by different rates of biotransformation or excretion. For example, metabolic transformations of certain dioxin congeners have been suggested as an important factor in explaining low BCFs and BAFs in some fish species (Wan et al. 2005).

Results from dioxin measurements in fish and fish-eating birds from the North American Great Lakes indicate that 2,3,7,8-substituted congeners preferentially accumulate to a higher degree than non-2,3,7,8-substituted congeners in the food chain (Davies 1999, Kannan et al. 2001, Gauthier et al. 2008). Further, there is some evidence in fish suggesting that 2,3,7,8-substituted congeners have longer half-lives (on the order 50–100 days) than other dioxins (on the order of several weeks or less) (Davies 1999, Geyer et al. 2002).

TABLE 3.2

Bioaccumulation, Bioconcentration and Biota-Sediment Accumulation Factor Definition of Terms

BCF	Bioconcentration factor	The ratio (in L/kg) of a substance's concentration in tissue of an aquatic organism to its concentration in the ambient water, in situations where the organism is exposed through the water only and the ratio does not change substantially over time.
BAF	Bioaccumulation factor	The ratio (in L/kg) of a substance's concentration in tissue of an aquatic organism to its concentration in the ambient water, in situations where both the organism and its food are exposed and the ratio does not change substantially over time.
BSAF	Biota-sediment accumulation factor	The ratio (in kg of organic carbon/kg of lipid) of a substance's lipid-normalized concentration in tissue of an aquatic organism to its organic carbon-normalized concentration in surface sediment, in situations where the ratio does not change substantially over time, both the organism and its food are exposed, and the surface sediment is representative of average surface sediment in the vicinity of the organism.

Source: U.S. EPA (2009c).

In the absence of sufficient field data for all but a few PBDE congeners, the similarities in structure and physicochemical properties between PBDE and PCB congeners suggest that environmental processes may affect their fate comparably (Burreau et al. 2006). However, bioaccumulation behavior may differ based on the nature of the chemical structures (an ether bridge for PBDEs; a phenyl bridge for dioxins and PCBs) and properties of bromine compared to chlorine (deBruyn et al. 2009). Several studies suggest PBDEs are less biomagnified than PCBs (Kelly et al. 2008, Mizukawa et al. 2009, Wu et al. 2009). PBDEs have been found to exhibit a wider range of molecular size and hydrophobicity than PCBs, with measured and calculated K_{ow} values exceeding 10^{10} or greater (Palm et al. 2002, Braekevelt et al. 2003).

Furthermore, PBDEs are reported to undergo metabolic debromination more readily than either dioxin or PCB dechlorination, suggesting that bioaccumulation of higher brominated congeners might be less than that of PCBs (Stapleton et al. 2004a, 2004b, 2006). In PBDE metabolism studies investigated by feeding trout DeBDE spiked food, Kierkegaard et al. (1999) observed short-term increases in levels of hexa, hepta, octa, and nona-BDE congeners; after depuration, DeBDE rapidly decreased and the concentrations of lower BDE congeners were unaffected, concluding that DeBDE was metabolized by the trout and metabolic processes did not produce tetraBDE and pentaBDE congeners.

Overall, the higher chlorinated dioxins and PCBs tend to have lower BAFs and BCFs than less chlorinated compounds (Table 3.3). Similarly, the higher brominated PBDEs tend to have lower BCFs and, possibly, BAFs than less brominated compounds. This has been attributed to factors such as differences in membrane transport, larger molecular sizes, lower solubilities, and the possibility for preferential metabolism of certain congeners (Opperheizun 1986). Published BCFs for dioxins have been shown to vary over 4 orders of magnitude; for example, studies in the U.S. EPA (2009a) ECOTOX database report BCFs ranging from 0.08 to 86,000 in fish. The recommended default fish BCF for dioxins used by the California Office of Environmental Health Hazard Assessment (OEHHA 2000) is 19,000 based on mean BCFs reported in six studies cited by Hsieh et al. (1994) ranging from 2.7 to 64,000.

BCFs for PCBs reported in ECOTOX database also varied over 6 orders of magnitude in fish from 0.2 to 355,000 (U.S. EPA 2009a). At present, few studies report BCFs for PBDEs. Gustafsson

TABLE 3.3
Bioconcentration Factors (BCFs) Reported for the 2,3,7,8-TCDD and Aroclors

Species Group	BCF	
	Range	Mean
2,3,7,8-TCDD		
Crustaceans	49–7125	2883
Fish	0.08–86,000	11,225
Insects/Spiders	2846–9222	5689
Mollusks	720–3775	1839
Amphibians	0.12–3.78	1.04
PCBs (Aroclors)		
Crustaceans	160–108,000	11,374
Fish	0.2–355,210	51,117
Invertebrates	14.8–17,000	6962
Mollusks	2–101,000	4424
Polychaetes	—	8.6
Insects/Spiders	19.5–24,800	5444

Notes: Experimental concentrations used to derive BCFs varied widely by study.
2,3,7,8-TCDD: Amphibian data reported in dry weight and converted to wet weight: dry-weight concentration = 5 * wet-weight concentration.
PCBs: Includes Aroclors 1016, 1242, 1254, and 1260.
Source: U.S. EPA 2009a—ECOTOX Database (accessed May and December 2009).

et al. (1999) report BCFs in blue mussels (*Mytilus edulis*) of 1.4×10^6 and 1.3×10^6 for penta- and tetra-BDE congeners, respectively, and 2.2×10^5 for hexa-BDE, almost one order of magnitude lower, after 44-d exposure experiments. The high log K_{ow} values for PBDEs suggest high rates of bioaccumulation and biomagnification in aquatic biota, particularly for tetra- and penta-brominated PBDEs (Alaee and Wenning 2002, de Wit 2002; Law et al. 2006).

Similar wide range was found for dioxin BSAF values. Using lake trout and surficial (0–2 cm) sediment samples from southern Lake Michigan, Burkhard et al. (2004) calculated BSAFs ranging from <0.1 to 18 for PCBs and from <0.001 to 0.32 for dioxins. PCBs with zero or one chlorine in an ortho position had lower BSAFs than other PCBs. Dioxins with chlorines at the 2,3,7,8-positions had higher BSAFs than other congeners.

3.4 TOXIC EQUIVALENCY FACTORS

It is generally understood that different congeners of dioxin, PCB, and PBDE are not equally toxic to biota, as defined by their ability to elicit specific adverse effects in animals. Therefore, to assess the likely toxicological effect of a particular mixture of congeners, the World Health Organization (WHO) and U.S. EPA developed TEFs as early as the mid-1980s as a means to assign different degrees of toxicity based on chemical structure and relative potency of the different congeners. Since that time, different TEF schemes have been proposed in the United States and other countries for dioxins and certain PCBs exhibiting dioxin-like behavior relative to the potency of 2,3,7,8-TCDD, which is widely recognized as the most toxic to humans and wildlife among the dioxins. After 1998, the WHO and U.S. EPA reached scientific consensus on the use of separate sets of TEFs applicable to risk assessments for mammals, birds, and fish in response to the observed differences in sensitivities and toxicity of certain congeners among different taxa (van den Berg et al. 1998, 2006, U.S. EPA 2008). The TEF scheme, endorsed by WHO in 2005 for humans and mammals and in 1998 for birds and fish, is presented in Table 3.4.

TEF schemes have been proposed for mixtures of PCBs, though regulatory agencies have not formally adopted any scheme (Hornung et al. 1996, Simon et al. 2007, Bhavsar et al. 2008a, Burkhard and Lukasewycz 2008). While binding of PBDEs to the Ah receptor and estrogen receptors has been reported, the current knowledge of the toxicological behavior of PBDEs is rather limited and,

TABLE 3.4
World Health Organization (WHO) Toxic Equivalent Factors (TEFs) for Mammals, Birds, and Fish

Congeners	Mammals[a]	Birds[b]	Fish[b]
Polychlorinated dibenzo-p-dioxins			
2,3,7,8-tetraCDD	1	1	1
1,2,3,7,8-pentaCDD	1	1	1
1,2,3,4,7,8-hexaCDD	0.1	0.05	0.5
1,2,3,6,7,8-hexaCDD	0.1	0.01	0.01
1,2,3,7,8,9-hexaCDD	0.1	0.1	0.01
1,2,3,4,6,7,8-heptaCDD	0.01	<0.001	0.001
OctaCDD	0.0003	0.0001	<0.0001
Polychlorinated dibenzofurans			
2,3,7,8-tetraCDF	0.1	1	0.05
1,2,3,7,8-pentaCDF	0.03	0.1	0.05
2,3,4,7,8-pentaCDF	0.3	1	0.5
1,2,3,4,7,8-hexaCDF	0.1	0.1	0.1
1,2,3,6,7,8-hexaCDF	0.1	0.1	0.1
1,2,3,7,8,9-hexaCDF	0.1	0.1	0.1
2,3,4,6,7,8-hexaCDF	0.1	0.1	0.1
1,2,3,4,6,7,8-heptaCDF	0.01	0.01	0.01
1,2,3,4,7,8,9-heptaCDF	0.01	0.01	0.01
OctaCDF	0.0003	0.0001	<0.0001
Nonortho polychlorinated biphenyl congeners IUPAC#			
CB#77	0.0001	0.05	0.0001
PCB#81	0.0003	0.1	0.0005
PCB#126	0.1	0.1	0.005
PCB#169	0.03	0.001	0.00005
Mono-ortho polychlorinated biphenyl congeners IUPAC#			
PCB#105	0.00003	0.0001	<0.000005
PCB#114	0.00003	0.0001	<0.000005
PCB#118	0.00003	0.00001	<0.000005
PCB#123	0.00003	0.00001	<0.000005
PCB#156	0.00003	0.0001	<0.000005
PCB#157	0.00003	0.0001	<0.000005
PCB#167	0.00003	0.00001	<0.000005
PCB#189	0.00003	0.00001	<0.000005

Sources:

[a] Van den Berg, M. et al., *Toxicol. Sci.*, 93, 223–241, 2006.

[b] Van den Berg, M. et al., *Environ. Health Perspect.*, 106, 775–792, 1998. With permission.

therefore, TEFs have not been proposed by WHO or environmental regulatory agencies at this time.

Throughout this chapter, dioxins and furans are presented as 2,3,7,8-TCDD toxic equivalents (TEQs) using congener specific mammalian TEFs, following the approach described by the WHO International Program on Chemical Safety as part of the 2005 reevaluation of PCDD/F toxicity (Van den Berg et al. 2006). The use of mammalian TEFs is consistent with approaches used by the Ontario Ministry of the Environment (MOE) and U.S. EPA to report fish tissue residue data and to facilitate comparisons to comparable fish tissue data reported elsewhere (e.g., Saloranta et al. 2006a, 2006b, Bhavsar et al. 2007, 2008b, 2008c, Wintermyer and Cooper 2003, FDA 2009, Stahl et al. 2009, U.S. EPA 2009b). However, it should be noted that the evaluation of PCDD/F residues in aquatic organisms should include the use of fish and/or bird TEFs, depending on the use of the data for risk assessment purposes (e.g., Cook et al. 2003, Steevens et al. 2005).

The TEF scheme has significant implications for regulatory programs tasked with monitoring and setting exposure limits, despite the assumptions and uncertainties inherent in this approach (Dyke and Stratford 2002, Finley et al. 2003, Bhavsar et al. 2008b). The underlying premise for using a TEF scheme is twofold: first, the mode of action of all 2,3,7,8-substituted dioxins and dioxin-like congeners is the same (i.e., AhR mediated); and, second, the combined effects of the individual dioxin congeners are dose-additive. Additivity is an important prerequisite of the TEF concept. There is evidence supporting and contradicting both of these assumptions for dioxins and dioxin-like compounds. There is considerable evidence that the relative toxicity of different congeners varies significantly among different taxonomic groups (e.g., birds, fish, and mammals); differences may be even broader as more information becomes known regarding ecotoxicity to invertebrates, reptiles, and amphibians.

Further, the assumption that the combined effects of the congeners are additive may not be true in all cases. Some studies provide evidence that nondioxin-like AhR agonists and antagonists are able to increase or decrease the toxicity of 2,3,7,8-substituted compounds. In addition, there are natural nonchlorinated AhR agonists in the diets of many animals, and some studies have suggested that the potential effects of these may be significant.

The application of TEFs to obtain TEQs represent a fairly accurate estimate of the toxic potential of a complex mixture of persistent Ah-receptor agonists, such as PCDDs, PCDFs, and coplanar PCBs, but completely ignores any other components in a mixture that may elicit a similar toxicity via a non-Ah receptor pathway. This is important to keep in mind when attempts are made to associate toxic effects observed in the environment with exposures to dioxin-like compounds or TEQs.

3.5 TISSUE RESIDUES IN AQUATIC ORGANISMS

Numerous global, regional, and national surveys of dioxins, PCBs, PBDEs, and other POPs have been conducted during the past three decades and are available in the literature (e.g., de Boer and Denneman 1998, Alaee et al. 1999, Ikonomou et al. 2000, 2002, Luross et al. 2002, Hale et al. 2003, Holden et al. 2003, Kiviranta et al. 2003, AMAP 2004, Chiuchiolo et al. 2004, Hites et al. 2004, Rayne et al. 2004, Evans et al. 2005, Gómara et al. 2005, Johnson and Olson 2001, Brown et al. 2006, Isosaari et al. 2006, Fair et al. 2007, Mathews et al. 2008, Pandelova et al. 2008, Gewurtz et al. 2009, Szlinder-Richert et al. 2009, Yogui and Sericano 2009).

3.5.1 DIOXIN TISSUE RESIDUES

The U.S. EPA's National Study of Chemical Residues in Fish, previously referred to as the National Bioaccumulation Study, was released in 1992 and, at the time, was among the first comprehensive screening investigations of the prevalence of selected bioaccumulative substances in fish anywhere in the world (U.S. EPA 1992). One or more of the seventeen 2,3,7,8-substituted PCDDs and

PCDFs were detected at nearly all of the 388 locations surveyed throughout U.S.* In general, the frequency of detection was highest in the east and northeast and lowest in the west and southwest regions of the United States. Levels were higher in bottom-feeding fish than in pelagic fish. The congeners most frequently detected in fish included four PCDDs compounds (2,3,7,8-TCDD; 1,2,3,7,8-PeCDD; 1,2,3,6,7,8-HxCDD; and 1,2,3,4,6,7,8-HpCDD) and 3 PCDFs (2,3,7,8-TCDF; 2,3,4,7,8-PeCDF; and 1,2,3,4,6,7,8-HpCDF). The most frequently detected dioxins (1,2,3,4,6,7,8-HpCDD and 2,3,7,8-TCDF) were also detected at the highest concentrations in whole fish tissues: 249 ng/kg and 404 ng/kg, respectively. 2,3,7,8-TCDD was detected at 70% of the surveyed locations; the maximum concentration was 204 ng/kg and arithmetic mean concentration was 7 ng/kg.

Since 1992, nearly 1000 international, national, and regional surveys of dioxins in freshwater and marine fish have been conducted and published worldwide. In North America, the most comprehensive surveys conducted to date are those conducted by the U.S. EPA in U.S. lakes and reservoirs, and by U.S. and Canada scientists in the Great Lakes. Much of what is known about dioxin levels in fish in North America is reported in two studies: U.S. EPA's National Study of Chemical Residues in Lake Fish Tissue in 500 lakes randomly selected from 147,000 target water bodies in the lower 48 states (Olsen et al. 2009, Stahl et al. 2009, U.S. EPA 2009b) and the Ontario MOE on-going Great Lakes sport fish monitoring program (MOE 2009).

The Ontario MOE Sport Fish Contaminant Monitoring Program, which started in 1976, is the largest program of its kind in North America. Fish are tested annually from up to 1860 locations in Ontario's inland lakes and rivers and the Canadian waters of the Great Lakes for a variety of substances, including mercury, PCBs, mirex, DDT, and dioxins. The 2009–2010 Ontario sport fish consumption guide incorporates approximately 120,000 additional test results performed on 12,000 samples and 100 more locations than the previous 2007–2008 edition (MOE 2009).

The U.S. EPA National Study of Chemical Residues in Lake Fish Tissue is a 4-year national screening-level study of PBT substances in freshwater fish; the sampling work was conducted between 2000 and 2003. It is the U.S. EPA's first national fish tissue survey based on a probabilistic (random) sampling design, which is intended to support national estimates of the mean concentrations of 268 PBT substances in fish tissue from lakes and reservoirs in the lower 48 states (U.S. EPA 2009b).

Figures 3.2 through 3.5 summarize the arithmetic mean dioxin TEQ in fish calculated by combining the data available in both the U.S. EPA 500 Lakes Study (Figures 3.2 through 3.4) and the Ontario MOE Sport Fish Monitoring Program (Figure 3.5) based on geography, species, and the type of fish tissue sample collected and analyzed. The 982 fish tissue samples in the U.S. EPA data set included 442 whole-body samples and 540 fillet with skin samples. Whole-body fish samples were grouped by family: *Catostomidae* (18 species of suckers and buffalo fish; $n = 154$), *Cyprinidae* (3 species of carp and goldfish; $n = 119$), and *Ictaluridae* (9 species of catfish and bullhead; $n = 169$). Fillet with skin fish samples also were grouped by family: *Centrarchidae* (10 species of bass, crappies, and sunfish; $n = 352$), *Esocidae* (2 species of pike and pickerel; $n = 40$), *Percidae* (5 species of walleye, perch, sauger, saugeye, and pumpkinseed; $n = 84$), and *Salmonidae* (10 species of salmon, trout, and mountain and lake whitefish; $n = 64$). The 443 fish tissue samples from the Great Lakes included in the Ontario MOE data set represented fillet without skin samples. These data were similarly grouped by family: *Catostomidae* (1 species of white sucker; $n = 6$), *Cyprinidae* (1 species of carp; $n = 57$), *Esocidae* (1 species of pike; $n = 7$), *Ictaluridae* (2 species of catfish and bullhead; $n = 69$), *Percidae* (2 species of walleye and perch; $n = 35$), and *Salmonidae* (6 species of chinook and coho salmon, and brown, rainbow and lake trout, and Siscowet; $n = 269$).

Fish tissue data were analyzed using one-way analysis of variance (ANOVA), if the data satisfied the requirements of normality with or without log-transformation (Levene 1960, Shapiro and Wilk 1965). Data sets that did not meet the requirements of the ANOVA, were evaluated using a Kruskal–Wallis test (Kruskal 1952, Kruskal and Wallis 1952). If the ANOVA or Kruskal–Wallis

---

* As mentioned above, dioxins are presented as 2,3,7,8-TCDD TEQs, following the approach described by Van den Berg et al. 2006.

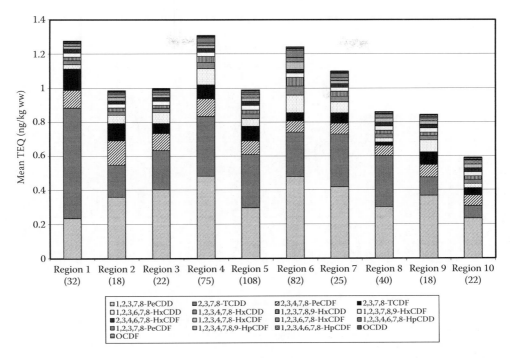

FIGURE 3.2   Mean dioxin total TEQ in fish tissue (whole body) from U.S. EPA regions 1–10 (2000–2003).

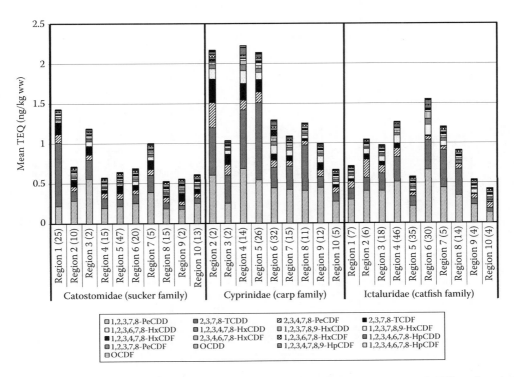

FIGURE 3.3   Mean dioxin total TEQ in fish tissue (whole body) by fish family from U.S. EPA regions 1–10 (2000–2003).

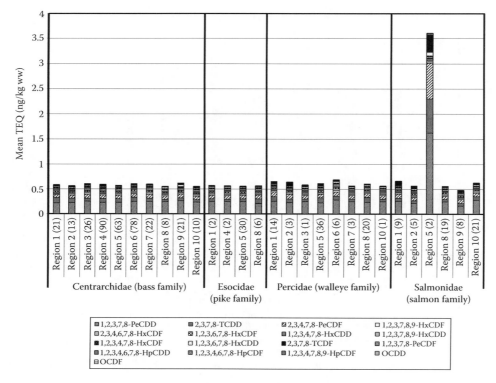

FIGURE 3.4 Mean dioxin total TEQ in fish tissue (fillet with skin) from U.S. EPA regions 1–10 (2000–2003).

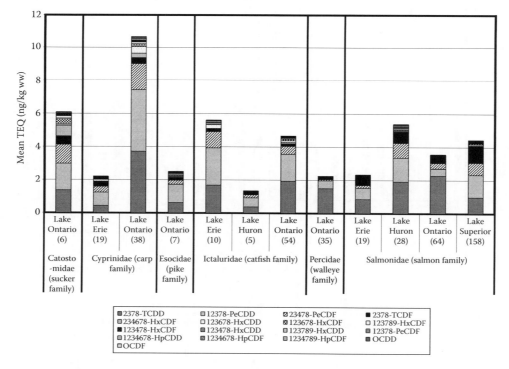

FIGURE 3.5 Mean dioxin total TEQ in fish tissue (fillet without skin) from the Great Lakes (2000–2008).

test was significant, pair-wise comparisons of the exposed groups to control were made least squares means and the Tukey–Kramer (Kramer 1956) comparison test or a Mann–Whitney U test (Mann and Whitney 1947), respectively.

The mean and range of total dioxin TEQ in fish from different lakes and U.S. EPA regions, regardless of species or type of sample (filet or whole body) were calculated. Total dioxin TEQ among all fish groups ranged from 0.4 to 2.2 ng/kg for whole-body samples and 0.48–3.6 ng/kg for fillet with skin samples. Figure 3.2 presents combined whole-body data for three fish families (*Catostomidae, Cyprinidae,* and *Ictaluridae*) to show regional differences in fish TEQ across the United States. When the data from the three fish families are combined in each region, the highest mean dioxin TEQ (1.3 ng/kg) occurs in fish in U.S. EPA region 4; the lowest mean dioxin TEQ (0.59 ng/kg) occurs in U.S. EPA region 10. Mean whole-body fish dioxin TEQ in U.S. EPA region 4 is significantly higher ($p < .05$) than in U.S. EPA regions 1, 5, 6, 8, and 10. In general, the whole-body fish tissue TEQ in the western U.S. (regions 9–10) is lower (0.72 ng/kg) than the mean total dioxin TEQ (1.11 ng/kg) in the eastern U.S. (regions 1–5).

Figure 3.3 summarizes mean total dioxin TEQ in the United States for different taxonomic fish families. The highest mean dioxin TEQs are found in *Cyprinidae* as compared to *Ictaluridae* and *Catostomidae*. Mean total dioxin TEQs in *Cyprinidae* are higher in U.S. EPA region 4 (2.22 ng/kg), region 2 (2.16 ng/kg), and region 5 (2.13 ng/kg) than in other regions. TEQ levels in all other fish families were lower in all ten regions (ranging between 0.43 and 1.55 ng/kg). There were no significant differences in mean TEQ values across the United States within the *Catostomidae* family. However, significant differences ($p < .5$) exist across the United States within the *Cyprinidae* and *Ictaluridae* families. Within the *Cyprinidae* family of fishes, U.S. EPA regions 4 and 5 have the highest mean dioxin TEQ, significantly higher than regions 6, 8, and 10; region 10 has the lowest mean dioxin TEQ (0.67 ng/kg). Within the *Ictaluridae* family of fishes, U.S. EPA region 6 has the highest mean dioxin TEQ (1.55 ng/kg), significantly greater than regions 2 and 4, and region 10 has the lowest mean dioxin TEQ (0.43 ng/kg).

Figures 3.4 and 3.5 summarize mean total dioxin TEQ according to the type of fish tissue sample—either fillet with or fillet without skin, respectively. In general, mean total dioxin TEQs in fish fillet with skin samples collected from U.S. lakes (ranging between 0.48 and 3.61 ng/kg; Figure 3.4) were lower than fish fillet without skin samples collected from the Great Lakes (ranging between 1.15 and 9.36 ng/kg; Figure 3.5). Among the six families of fish collected from the Great Lakes, only two, *Cyprinidae* and *Salmonidae,* showed significant ($p < .05$) differences in mean total dioxin TEQ. In *Cyprinidae,* the mean TEQ is significantly ($p = .01$) greater in Lake Ontario (10.6 ng/kg) than in Lake Erie (2.20 ng/kg). In *Salmonidae,* the mean total dioxin TEQ is significantly greater in Lake Huron (5.38 ng/kg) than in Lake Erie (2.34 ng/kg) and Lake Superior (4.44 ng/kg). In addition, mean dioxin TEQ in Lake Superior is significantly greater and in Lake Ontario (3.58 ng/kg).

On the basis of the combined fish tissue residue data from the U.S. EPA 500 Lakes Study and the Ontario MOE Sport Fish Monitoring Program, 4 of the 2,3,7,8-substituted dioxin congeners are most prevalent in fish and contribute most to total TEQ: 1,2,3,7,8-PentaCDD; 2,3,4,7,8-PentaDCF; 2,3,7,8-TCDD; and 2,3,4,7,8-PeCDF.

Figure 3.6 summarizes mean total dioxin TEQs reported in fish in other countries. Interpretation of the available data is challenging, at best, because of large differences in the analytical methodologies used to measure dioxins in fish and often failure of any explicit indication of the type of tissues analyzed (i.e., whole body, fillet, or fillet with and without skin). Moreover, typically only a mean concentration or TEQ value is reported in the literature, which limits meaningful statistical comparisons. Mean total dioxin TEQ in fish from all countries ranged from below detection to 155 ng/kg (Figure 3.6). No clear trends are discernable because TEQs in fish representing different feeding guilds or habitats vary widely by country. Dioxin TEQs in fish from the United States generally fall within the mid-range of dioxin TEQs reported in fish from other countries. Overall, mean total dioxin TEQs in fish from Asia (with the exception of Korea), Europe, and the United States are similar.

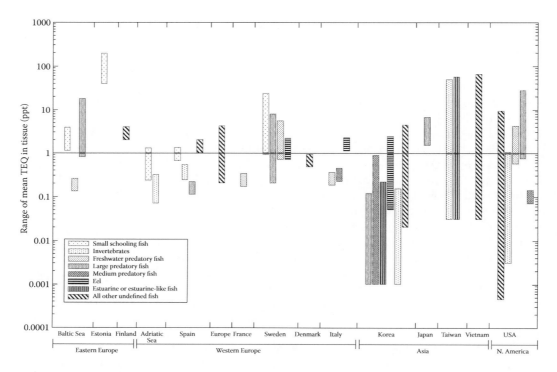

FIGURE 3.6 Ranges of mean dioxin TEQs reported in the edible portion of fish and aquatic invertebrates worldwide.

The significance of dioxin tissue residues in aquatic organisms is evident using the Environmental Residue-Effects Database (ERED) compiled jointly by the U.S. Army Corps of Engineers and U.S. EPA (U.S. EPA/ERED 2009). The ERED database is a compilation of data from the literature, where biological effects (e.g., reduced survival and growth) and tissue contaminant concentrations are simultaneously measured in the same organism. The database is limited to those instances where biological effects observed in an organism can be linked to a specific contaminant.

Table 3.5 summarizes no observable effect tissue residues (NOERs) associated with dioxins in tissues and lethal tissue residue to 50% of test organisms (LR50) reported in fish and other aquatic organisms reported in ERED. Lake trout are reported to have the lowest NOER indicating that this species may be particularly sensitive. Mosquito fish, invertebrates including crayfish and aquatic insects, and aquatic plants appear to be the least sensitive of the species included in the database as tissue residue concentrations between 0.4 and 1 mg/kg wet weight (ww) result in no observable adverse effect. Of the congeners evaluated, 1,2,3,4,7,8-HCDD appears to be the most toxic congener (to rainbow trout); however, the concentrations affecting 50% (LC50) of test organisms fall within the lower range of toxic tissue residues for 2,3,7,8-TCDD. Figure 3.7 illustrates the variation in sensitivities of a variety of aquatic organisms to TCDD body burdens using LR50s using a species sensitivity distribution (SSD). Note that early life stages of fish are more sensitive than that of juvenile and adult life stages or invertebrates.

Steevens et al. (2005) proposed a methodology for searching, reviewing, and analyzing linked, tissue residue effect data to derive benchmark distributions. The approach is demonstrated for contaminants having a dioxin-like mechanism of toxic action and is based on residue-effects data for 2,3,7,8-TCDD and equivalents in early life stage fish. The tissue-residue benchmark for TCDD TEQ for fish (based on the geometric means of NOERs and LOER values used in their study) could range from 0.057 to 0.699 ng TCDD TEQ/g lipid, depending on the level of protection needed (99% and 90% of fish species protected, respectively).

TABLE 3.5
**No Observable Effect Tissue Residue (NOER) Associated with Dioxins in Tissues and
Lethal Tissue Residue to 50% of Test Organisms (LR50) Reported in Fish and Other
Aquatic Organisms**

		Tissue Concentration	
Chemical	Species	Range	Mean
No Observable Effects Tissue Residue (NOERs)			
TCDD (2,3,7,8-TCDD)	Mosquito fish	440–900	670
CAS# 1746-01-6			
	Trout—Rainbow	0.1–5	1.6
	Trout—Lake	0.023–0.044	0.034
	Trout—Brook	—	1.2
	Yellow perch	0.143–1	0.5715
	Catfish—Channel	140–190	165
	Salmon—coho	—	0.125
	Black Bullhead	—	1
	Bluegill	—	1
	Largemouth Bass	—	1
	Crayfish	—	0.3
	Snail	9.7–1020	420
	Water flea	8.6–2080	580
	Midge	—	470
	Algae—Green	—	980
	Least Duckweed	—	12
Lethal Tissue Residue to 50% of the Test Group (LR50s)			
TCDD (2,3,7,8-TCDD)	Trout—Lake	0.034–65	3.10
CAS# 1746-01-6			
	Trout—Rainbow	0.17–10	1.42
	Common carp	2–3	2.73
	Yellow perch	—	3
	Black Bullhead	—	5
	Largemouth Bass	—	11
	Bluegill	—	16
1,2,3,4,7,8-HCDD	Trout—Rainbow	—	0.14
CAS # 39227-28-6			
1,2,3,4,7,8-HxCDF		—	0.99
Cas # 70648-26-9			
1,2,3,7,8-PcCDD		—	0.566
CAS # 40321-76-4			
1,2,3,7,8-PeCDF		—	7.34
CAS # 57117-41-6			
2,3,7,8-TCDF		—	8.08
CAS # 51207-31-9			
2,3,4,7,8-PCDF		—	0.7
57117-31-4			

Notes: All data in ng/g wet weight. Exposure routes include absorption, injection and/or ingestion or in combination.
Source: U.S. EPA/U.S. ACE ERED Database (accessed April 2009; http://el.erdc.usace.army.mil/ered/).

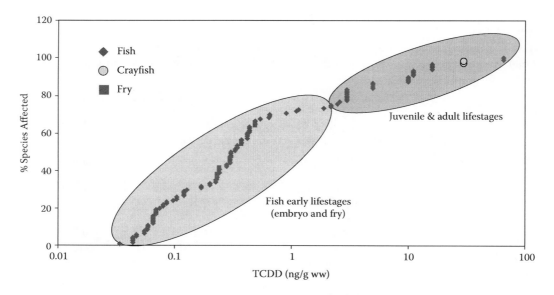

FIGURE 3.7   TCDD tissue residue LR50 species sensitivity distribution for a variety of aquatic organisms. (From U.S. EPA/U.S. ACE ERED Database, http://el.erdc.usace.army.mil/ered/)

3.5.2   PCB Tissue Residues

Data from the U.S. EPA's National Study of Chemical Residues in Lake Fish Tissue (U.S. EPA 2009b), the Ontario MOE Sport Fish Monitoring Program (MOE 2009), the Great Lakes National Program Office (GLNPO), and the Great Lakes Fish Monitoring Program (GLFMP) were compiled and analyzed to understand PCB residue in fish tissue from U.S. lakes and waterways and in the U.S.–Canadian Great Lakes. The GLFMP consists of two separate programs, the Open Lakes Trend Monitoring Program, which examines the health of fish and fish-consuming wildlife through trend analysis, and the Game Fish Fillet Monitoring Program, designed to monitor potential human exposure to contaminants through consumption of popular sport fish species (GLNPO 2009).

Figures 3.8–3.10 summarize the available data based on geography, species, and the type of fish-tissue sample collected and analyzed. All data are reported on a wet weight basis. The data were treated in the same manner as described for the dioxins. Figure 3.8 summarizes mean total PCB in whole-body fish from each of the ten U.S. EPA regions. In general, the highest mean total PCB concentrations were found in U.S. EPA region 4 (0.14 mg/kg) and the lowest in U.S. EPA region 8 (0.012 mg/kg). Consistent with the trend observed for dioxins, mean total PCB in fish from the western U.S. (0.03 mg/kg; regions 6–10) were substantially lower than in fish from the eastern U.S. (0.09 mg/kg). Similar to the dioxins, the higher mean total PCBs are found in *Cyprinidae* than in either *Ictaluridae* or *Catostomidae* (Figure 3.9a). Mean total PCBs in *Cyprinidae* are highest in U.S. EPA regions 2, 3, 4 (ranging from 0.26 to 0.32 mg/kg) and lowest in U.S. EPA regions 8 and 10 (0.018 and 0.017 mg/kg, respectively). Mean total PCBs in *Cyprinidae* in U.S. EPA region 8 are statistically lower ($p < .05$) than other regions, with the exception of U.S. EPA region 10. For *Ictaluridae*, the highest mean total PCBs are found in U.S. EPA regions 3 and 4 (0.09 and 0.12 mg/kg, respectively) and the lowest are found in U.S. EPA region 10 (0.001 mg/kg). Mean total PCBs reported for *Catostomidae* are highest in U.S. EPA region 2 (0.05 mg/kg) and lowest in U.S. EPA region 8 (0.006 mg/kg).

For fish collected in the Great Lakes between 2000 and 2005 (Figure 3.9b), *Salmonidae* from Lake Erie and Lake Superior had the lowest mean total PCB concentrations among the five Great Lakes (0.44 mg/kg in both lakes). Mean total PCB in fish was statistically higher ($p < .003$) in Lake Ontario (0.97 mg/kg) than in the other Great Lakes.

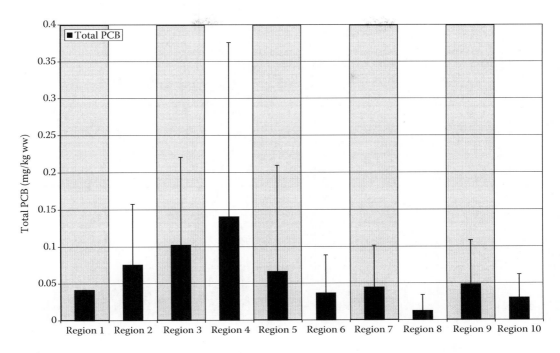

FIGURE 3.8   Total PCB in fish tissue (whole body) from U.S. EPA regions 1–10 (2000–2003).

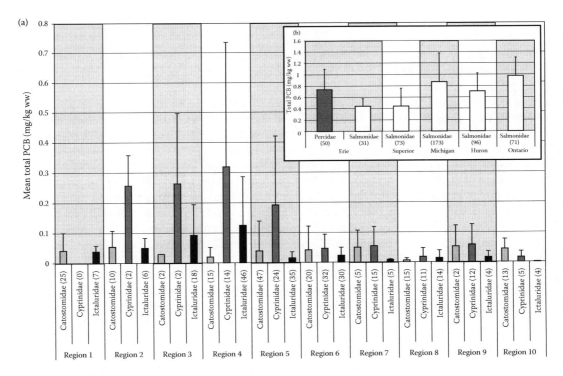

FIGURE 3.9   (a) Total PCB concentration in fish tissue (whole body) from U.S. EPA regions 1–10 (2000–2003), (b) comparison of total PCB concentration in fish tissue (whole body) from the Great Lakes (2000–2005).

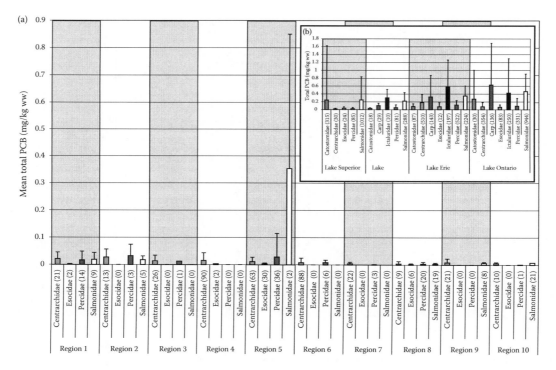

FIGURE 3.10 (a) Total PCB in fish tissue (fillet with skin) from U.S. EPA Regions 1–10 (2000–2003), (b) comparison of total PCB (fillet without skin) from the Great Lakes (2000–2008).

In general, mean total PCB levels in fish fillet with skin samples collected from U.S. lakes (ranging between 0.001 and 0.35 mg/kg) (Figure 3.10a) were lower than fish fillet without skin samples collected from the Great Lakes (ranging between 0.02 and 0.64 mg/kg; Figure 3.10b). Among the fish families collected from the U.S. lakes, mean total PCBs in *Centrarchidae* (fillet with skin) collected from region 8 (0.004 mg/kg) were statistically lower ($p < .05$) than all other regions; the highest mean total PCB concentrations were measured in U.S. EPA regions 1 and 2 (0.021 and 0.027 mg/kg, respectively). When the data from all fish families are combined, mean total PCB levels in fish (fillet with skin) in U.S. EPA region 8 are statistically lower ($p < .05$) than other regions with the exception of U.S. EPA region 7. In the Great Lakes, mean total PCBs in all fish (fillet without skin) are highest in Lake Ontario (0.33 mg/kg, $p < .001$) and lowest in Lake Huron (0.18 mg/kg, $p < .01$). This also is evident in the *Salmonidae* family (Figure 3.10b); mean total PCBs are highest in Lake Ontario (0.47 mg/kg, $p < .001$) and lowest in Lake Huron (0.22 mg/kg, $p < .001$). For the *Percidae* family, the highest mean total PCBs are found in Lake Erie (0.12 mg/kg, $p < .001$) and lowest in Lake Superior (0.03 mg/kg, $p < .001$) (Figure 3.10b).

Figure 3.11 summarizes mean total PCBs and dioxin-like PCB TEQs reported in fish from other countries. Similar to the dioxins, the interpretation of data reported in studies conducted outside North America is challenging, at best, because of large differences in the analytical methodologies used to measure PCBs and the lack of information on the type of tissues analyzed. Mean total PCB TEQ in fish from all countries ranged from 0.008 to 55 ng/kg. Like dioxins, trends are difficult to discern as TEQs in fish representing different feeding guilds or habitats vary widely by country.

Table 3.6 summarizes NOERs associated with PCBs in tissues and LR50s reported in fish and other aquatic organisms reported in ERED. NOER concentrations vary widely across the spectrum of species and tested congeners. Interestingly, the lowest and highest NOERs are observed for the same two PCB congeners. Among common carp, NOERs for PCB 153 and 180 were 0.00004 and

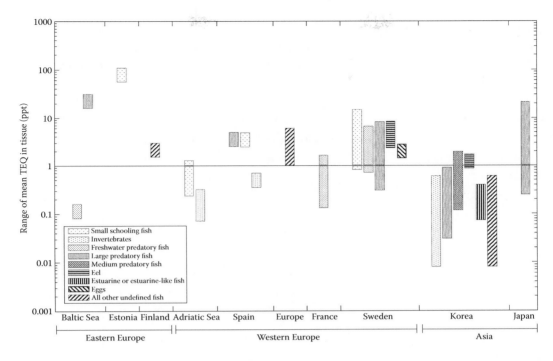

FIGURE 3.11   Ranges of mean dioxin-like PCB TEQs reported in the edible portion of fish and aquatic invertebrates worldwide.

0.00006 mg/kg (wet weight), respectively, while the same two congeners have the highest NOERs for fathead minnows, with concentrations averaging 6030 and 4565 mg/kg, respectively. Common carp were also particularly sensitive to body burdens of PCB 52, with a reported NOER of 0.00003 mg/kg. Figure 3.12 illustrates the variation in sensitivities of a variety of aquatic organisms to PCB body burdens using LR50s. Note that, like dioxins, early life stages of fish are more sensitive than that of juvenile and adult lifestages or invertebrates.

3.5.3   PBDE TISSUE RESIDUES

Various synonyms and abbreviations used to describe individual BDE congeners and homologues groups, as well as the technical formulations have lead to considerable unintentional misrepresentation of environmental levels in the scientific literature (Wenning 2002). Environmental monitoring data reported in the scientific literature typically report the concentrations of total PBDEs based on some, but not all, of the individual congeners, or the concentrations only of certain BDE congeners (Alaee and Wenning 2002, Wenning 2002). Interpretation of the available environmental data is further complicated by the fact that much more extensive analysis has been performed for only certain BDE congeners. Data reported in this manner hinder attempts to develop reliable estimates of environmental levels and trends or plausible human exposure. To date, there have been few, if any, studies strictly correlating the occurrence of specific BDE congeners with the use of technical formulations.

PBDEs were first detected in 1979 in sludge and soil samples collected in the vicinity of textile manufacturing plants, and have since been detected in abiotic and biotic samples throughout the world. Nearly every environmental monitoring program conducted during the past decade has shown increasing levels of some BDE congeners in wildlife, particularly in Nordic countries where this trend contrasts with a general decline in the occurrence of dioxins, PCBs, and chlorinated pesticides in marine mammals and aquatic wildlife (Hooper and MacDonald 2000).

TABLE 3.6
No Observable Effects Tissue Residue (NOERs) Associated with PCBs in Tissues and Lethal Tissue Residue to 50% of Test Organisms (LR50) Reported in Fish and Other Aquatic Organisms

Congener and CAS #	Species Common Name	Tissue Concentration	
		Range	Mean
	No Observable Effect Tissue Residues (NOERs)		
PCB 101 CAS #37680-73-2	Fathead minnow	1115–4690	2902
PCBs CAS #1336-36-3	Clam—Bent nose	—	1.7
	Crab—Blue	—	23
	Eastern oyster	33–425	229
	Mayfly	—	1.5
	Shrimp—Pink	—	1.3
	Starfish	—	19.3
	Catfish—Channel	1.2–14.6[a]	9
	Pinfish	17–170	66
	Salmon—Atlantic	—	44
	Sheepshead minnow	0.56–230[a]	66
	Spot	—	27
	Trout—Rainbow	—	0.09
	Trout—Lake	1.8–2.4	2.1
	Zebra Danio	—	0.14
PCB 138 CAS #35065-28-2	Fathead minnow	946–3980	2463
PCB 153 CAS #35065-27-1	Fathead minnow	1120–10,940	6030
	Common carp	—	4.00E-05
PCB 52 CAS # 35693-99-3	Amphipod—Freshwater	—	54
	Common carp	—	3.00E-05
	Fathead minnow	—	2179
	Trout—Rainbow	1–20[a]	10
PCB 118 CAS #31508-00-6	Starfish	—	3.3
PCB 180 CAS#35065-29-3	Fathead minnow	1210–7920	4565
	Common carp	—	6.00E-05
PCB 126 CAS # 57465-28-8	Trout—Rainbow	—	2
	Starfish	—	3.26
	Killifishes	—	1
PCB 77 CAS # 32598-13-3	Trout—Rainbow	—	2
Aroclor 1242 CAS # 53469-21-9	Amphipod—Freshwater	—	30
	Catfish—Channel	0.23–2.3[a]	1.7
Aroclor 1254 CAS # 11097-69-1	African Clawed Frog	—	114.09
	Amphipod Gammarus sp.	—	7.8
	Crayfish	—	0.04
	Daphnia magna	—	10.4
	Shrimp—Grass	0.4–17	8.7
	Fathead minnow	—	105
	Pinfish	—	17
	Salmon—Chinook	1–60	30
	Salmon—Coho	—	54
	Sheepshead minnow	—	5.4
	Trout—Brook	0.5[b]–71[c]	—
	Trout—Rainbow	2.3–103	22.4
	Trout—Lake	156–206	181

TABLE 3.6 (continued)

No Observable Effects Tissue Residue (NOERs) Associated with PCBs in Tissues and Lethal Tissue Residue to 50% of Test Organisms (LR50) Reported in Fish and Other Aquatic Organisms

Chemical	Species Common Name	Tissue Concentration Range	Mean
	Lethal Tissue Residues to 50% of the Test Group (LR50s)		
PCBs (PCB Mixture) CAS # 1336-36-3	Goldfish	250–324	287
	Pinfish	30–205	83
	Spot croaker	46	—
PCB 153 CAS # 35065-27-1	Midge	206	—
PCB 47 CAS # 2437-79-8	Amphipod—Marine	161–394	270
PCB 52 CAS # 35693-99-3	Amphipod—Freshwater	29	—
2,4,6,2'-tetrachlorobiphenyl CAS # 62796-65-0	Trout—Lake	9.2	—
Aroclor 1254 CAS # 11097-69-1	Trout—Brook	284	—
PCB 126 CAS # 57465-28-8	Trout—Rainbow	0.074	
	Trout—Lake	0.000054–0.029	0.015
Aroclor or PCB 1242 CAS # 53469-21-9	Amphipod—Freshwater	157	—
PCB 77 CAS # 32598-13-3	Trout—Rainbow	1.35	—

Notes: All data in µg/g wet weight. Exposure routes include absorption, injection, and/or ingestion, or in combination.

[a] Concentrations for various tissues and life stages.

[b] Egg.

[c] Embryo.

Source: U.S. EPA/U.S. ACE ERED Database (accessed April 2009) (http://el.erdc.usace.army.mil/ered/).

Few data are available on contamination pathways in aquaculture systems such as that for farmed fish. In the United States, for example, recent studies have shown the edible portions of farm-raised fish containing higher levels of PCDD/Fs, PCBs, and PBDEs than in wild fish (Brown et al. 2006, Hites et al. 2004). For PBDEs, a recent study by Pena-Abaurrea et al. (2009) of farmed and wild bluefin tuna (*Thunnus thynnus*) caught in the Mediterranean Sea indicated that total PBDE concentrations were similar in farmed and wild tuna (17–149 ng/g lipid weight, lw and 25–219 ng/g w, respectively). However, higher concentrations of naturally produced organobromines, such as methoxylated PBDEs (MeO-PBDEs) and polybrominated hexahydroxanthene derivatives (PBHDs), were observed only in wild tuna, suggesting that natural sources of brominated compounds can be significant.

A summary of total PBDE tissue levels reported in studies of fish and other aquatic organisms conducted for the most part after 1995 is presented in Tables 3.7–3.10. In general, 10–12 individual BDE congeners are commonly reported (although not consistently) by different researchers involved in environmental studies. The predominant PBDEs in ambient air, soil, sediment, and biota are BDE-28 (a triBDE), BDE-47 (a tetraBDE), BDE-66 (a tetraBDE), BDE-85 (a pentaBDE), BDE-99 (a pentaBDE), BDE-100 (a pentaBDE), BDE-153 (a hexaBDE), BDE-154 (a hexaBDE), BDE-183

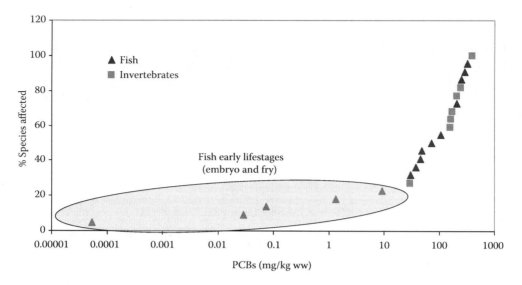

FIGURE 3.12 PCB and PCB mixture tissue residue LR50 species sensitivity distribution for a variety of aquatic organisms. (From U.S. EPA/U.S. ACE ERED Database, http://el.erdc.usace.army.mil/ered/)

(a hepta BDE), and BDE-209 (decaBDE). The predominant BDEs reported in fish—BDE-47, BDE-99, BDE-100, BDE-153, and BDE-154.

According to Yogui and Sericano (2009), PBDEs are ubiquitous in surface water, sediment and biota in U.S. marine and estuarine environments. PBDEs occur at higher levels in urbanized regions and, in some cases, at concentrations among the highest found in the world (Holden et al. 2003, She et al. 2004, Stapleton et al. 2006, Fair et al. 2007, Kannan et al. 2007). Seasonal variations in concentrations have been observed (Sellstrom et al. 1993), for example, lipid-normalized concentrations in herring were higher in the spring than fall probably due to a lower fat content in spring.

In the United States marine environment, BDEs 28, 47, 49, 66, 99, 100, 153, 154, and 155 were detected in more than 75% of the studies in which these congeners were analyzed. BDEs 47, 99, and 100 typically dominate the composition of PBDEs in most samples and exhibit high concentrations in several matrices. Higher congener levels in the U.S. marine environment as compared to elsewhere may be due, in part, to the predominant use of the Penta-BDE technical formulation in the United States. For example, BDE47, BDE99, and BDE100 typically dominate the composition of PBDEs in most samples. BDE17, BDE28, BDE33, BDE49, BDE153, BDE154, and BDE155 are also of concern since they are known to be present in a minor proportion in the Penta-BDE products (La Guardia et al. 2006). Mono through tri-brominated congeners such as BDE2, BDE8, BDE15, and BDE30 have been detected in U.S. marine organisms (Quakenbush 2007) even though these congeners do not occur at detectable levels in the technical formulations.

It has been speculated that the presence of lower brominated BDEs reflects debromination of the higher brominated BDEs to less brominated congeners is occurring (Hua et al. 2003, Eriksson et al. 2004, Soderstrom et al. 2004). However, BDE206, BDE207, BDE208, and BDE209, which occur in the Deca-BDE technical formulation, are not as prevalent in marine organisms in U.S. waters. Sharks are an exception as they have been found to contain high levels of BDE-209 (Johnson-Restrepo et al. 2005).

From the available U.S. studies, total PBDEs in either whole or fillet fish tissues range from nondetect to approximately 1300 ng/g ww in U.S. waterways; the highest levels tend to occur in the mid-Atlanta and southeastern regions and the Great Lakes. Levels in fish in the United States are comparable to those reported in Europe, with the exception of the Nordic countries, and higher than levels reported in Canada. By comparison, total PBDEs in either whole or fillet fish

TABLE 3.7

No Observable Effect Tissue Residues (NOERs) Associated with PBDEs in Tissues Reported in Fish and Other Aquatic Organisms

Chemical and CAS #	Species Common Name	Range	Mean	Effect
PBDE-47	Fathead minnow	16–51	42.4	Physiological
40088-47-9		51–61	47.4	Reproductive
PBDEs	Salmon—Atlantic	180–565	373.3	Biochemical
32534-81-9				
PBDEs	Oligochaete	22–24.3	23.4	Behavior
32534-81-9		22–24.3	23.4	Growth
		24.3	—	Mortality
PBDEs	African Clawed Frog	99.3	—	Development
32534-81-9		1030	—	Physiological
PBDE	Trout—Lake	0.06–0.2	0.13	Physiological
101-84-8				
PBDE	Common carp (Juvenile)	1.00E-07–6.00E-04	6.89E-05	Physiological
101-84-8				

Notes: All data in µg/g wet weight. Exposure routes include absorption, injection, and/or ingestion, or in combination.
Source: U.S. EPA/U.S. ACE ERED Database (accessed April 2009) (http://el.erdc.usace.army.mil/ered/).

tissues ranged as high as approximately 10000 ng/g ww in Sweden, including fish from the Baltic Sea. However, data are generally limited, and the congeners, measurement units, and nomenclature used to report results vary widely, making direct comparisons difficult.

With the exception of differences attributed to reporting "total PBDEs" based on some, but not all, congeners in whole body or certain tissues, the wide range is likely attributed to differences in feeding habits and trophic level. Among salmon, two studies found differences between Chinook and the other four salmon species that are likely due to differences in feeding habits (Hites et al. 2004). Coho, chum, pink, and sockeye salmon diets are based on zooplankton and invertebrates, while Chinook salmon tends to feed on higher trophic level organisms. In addition, Chinook salmon caught in coastal Oregon waters had higher PBDE levels than the same species caught in coastal Alaskan waters.

In bivalves, not all congeners are readily bioavailable since only BDEs 47, 99, and 100 have been found in bivalve tissues in both east and west U.S. coasts at levels as much as three orders of magnitude higher than detected in mussels from Europe, Asia, and Greenland (Yogui and Sericano 2009). According to Sericano et al. (2003) mean concentrations were highest in the Atlantic coast (39 ng/g dry wt; range: nd–145 ng/g) and Pacific coast (22 ng/g dry wt; range: nd–53 ng/g) than in the Gulf of Mexico (5.3 ng/g dry wt; range: 0.6–366 ng/g).

3.6 ECOTOXICOLOGICAL EFFECTS

Detailed reviews of the ecotoxicological effects of dioxins, PCBs, and PBDEs in aquatic organisms are available elsewhere (Niimi 1996, de Boer and Denneman 1998, U.S. EPA 2001, 2002, Giesy et al. 2003). Briefly, ecotoxicology studies conducted over the past two decades suggest that the mode of action among dioxins and dioxin-like PCBs is broadly the same, at least among vertebrate animals. The mode of action for PBDEs is the subject of considerable debate, at present. Most studies have been conducted on laboratory rodents and primates, which have traditionally been used as models to extrapolate the results to studies of potential human health effects. Laboratory

TABLE 3.8
PBDE Levels Reported in Studies of Fish and Other Aquatic Biota Published after 2000

Species	Survey Years	Location	Tissue Type	N	Mean Concentration Σ PBDE (ng/g)	Reference
Bivalve Invertebrates (wet wt.)						
Bivalves (species not specified)	1999	Gulf of Mexico	Soft tissue	38	5.26	Sericano et al. (2003)
Bivalves (species not specified) Blue mussel (*Mytilus edulis*)	1999	Atlantic coast and Massachusetts Bay	Soft tissue	58	38–212	Sericano et al. (2003); Yogui and Sericano (2009)
Bivalves (species not specified) California mussel (*Mytilus californianus*) Pacific oyster (*Crassostrea gigas*) Asian clam (*Corbicula fluminea*)	1999, 2002	Pacific coast and San Francisco Bay	Soft tissue	44	22–95	Oros et al. (2005); Sericano et al. (2003)
Crab Invertebrates (wet wt.)						
Crab	1992–2000	British Columbia, Canada coast	Whole body	5	4–480	Ikonomou et al. (2002)
Teleost Fishes (wet wt.)						
Chum salmon (*Oncorhynchus keta*) Chinook salmon (*O. tshawytscha*)	2000	Alaskan coast	Whole body	2	0.04, 0.5	Easton et al. (2002)
Pink salmon (*Oncorhynchus gorbuscha*) Chum salmon (*O. keta*) Sockeye salmon (*O. nerka*) Coho salmon (*O. kisutch*) Chinook salmon (*O. tshawytscha*)	2001–2002	Alaskan coast	Muscle & skin	27	0.05–0.5	Hites et al. (2004)
Sole	1992–2000	British Columbia, Canada coast	Whole body	10	12–340	Ikonomou et al. (2002)
Chinook salmon (*Oncorhynchus tshawytscha*)	2001–2002	West coast	Muscle & skin	3	2.1	Hites et al. (2004)
Rainbow trout Mountain whitefish Channel catfish Carp Largescale sucker	1994–1999	WA state rivers	Whole body	10	1.5–297	Johnson and Olson (2001)
Jacksmelt (*Atherinopsis californiensis*) Halibut	2002	West coast San Francisco Bay	Edible parts	21	6–44	Holden et al. (2003)

Species	Year	Location	Tissue			Reference
Striped bass (*morone saxatilis*)						
Perch						
King fish						
Speckled sanddab (*Citharichthys stigmaeus*)	2000–2001	California coast	Muscle & skin	23	0.04–7	Brown et al. (2006)
Kelp rockfish (*Sebastes atrovirens*)						
Rainbow surf perch						
Black rockfish (*Sebastes melanops*)						
White surf perch						
Canary rockfish (*Sebastes pinniger*)						
Kelp bass (*Paralabrax clathratus*)						
Striped bass (*Morone saxatilis*)						
Pacific mackerel (*Scomber japonicus*)						
Spotted sand bass (*Paralabrax maculatofasciatus*)						
White croaker (*Genyonemus lineatus*)						
Shiner surf perch (*Cymatogaster aggregata*)	2000–2001	California coast	Edible parts	18	6	Brown et al. (2006)
Jacksmelt (*Atherinopsis californiensis*)						
Salmon	2000	Lake Michigan	Whole body	21	47–148	Manchester-Neesvig et al. (2001)
Large mouth bass	2001	Detroit River, MI	Whole body	12	5.3–5.4	Rice et al. (2002)
Carp						
Carp	2001	Des Plaines River, IA	Whole body	14	7–14	Rice et al. (2002)
Silver perch (*Bairdiella chrysoura*)	2004	Eastern coast of Florida	Muscle	38	0.1–0.8	Johnson–Restrepo et al. (2005)
Striped mullet (*Mugil cephalus*)						
Spotted seatrout (*Cynoscion nebulosus*)						
Red drum (*Sciaenops ocellatus*)						
Hardhead catfish (*Arius felis*)						
Yellowtail tuna	2000	Japan coast	Edible parts	18	0.2–1.7	Ohta et al. (2002)
Mackerel						
Salmon						
Cartilaginous Fishes (wet wt.)						
Leopard shark (*Triakis semifasciata*)	2002	California coast	Edible parts	1	1.5	Holden et al. (2003)
Atlantic stingray (*Dasyatis sabina*)	2004	Eastern coast of Florida	Muscle	30	0.2–4	Johnson–Restrepo et al. (2005)
Spiny dogfish (*Squalus acanthias*)						
Atlantic sharpnose shark (*Rhizoprionodon terraenovae*)						
Bull shark (*Carcharhinus leucas*)						

TABLE 3.9
PBDE Levels in Fish from the United States and Abroad

Country	Compound	Concentration	Percent Lipid (Where Available)	Reference	Note
Laurentian Great Lakes, USA		**(ng/g wet)**	**(%)**		
Lake Ontario lake trout	BDE-47	58 ± 15		Luross et al. (2002)	Values are mean concentrations in freshwater fish and all samples were whole fish and ground
	BDE-66	1.3 ± 0.38			
	BDE-99	14 ± 3.5			
	BDE-100	5.7 ± 1.1			
	BDE-153	4.9 ± 1.6			
Lake Erie lake trout	BDE-47	16 ± 4.2		Luross et al. (2002)	Values are mean concentrations in freshwater fish and all samples were whole fish and ground
	BDE-66	0.18 ± 0.12			
	BDE-99	2.0 ± 0.48			
	BDE-100	2.5 ± 0.89			
	BDE-153	0.89 ± 0.14	—		
Lake Huron lake trout	BDE-47	27 ± 8.6		Luross et al. (2002)	Values are mean concentrations in freshwater fish and all samples were whole fish and ground
	BDE-66	0.82 ± 0.38			
	BDE-99	7.7 ± 3.8			
	BDE-100	3.8 ± 1.8			
	BDE-153	2.3 ± 0.98			
Lake Superior lake trout	BDE-47	29 ± 9.8		Luross et al. (2002)	Values are mean concentrations in freshwater fish and all samples were whole fish and ground
	BDE-66	0.89 ± 0.31			
	BDE-99	12 ± 5.9			
	BDE-100	4.1 ± 1.3			
	BDE-153	1.5 ± 0.54			
Lake Michigan, USA		**Range (ng/g wet)**			
Salmon	BDE-47	26.0–95.1		Manchester-Neesvig et al. (2001)	A sample number of 21 coho and chinook salmon were used. The average % lipid was 3.89 with a range between 1.83 and 7.19. A 100 g "steak" was tested which included skin, muscle, bone, and organ tissues and then blended—the steak was extracted from in front of the dorsal fin
	BDE-66	1.2–2.5			
	BDE-99	5.9–18.9			
	BDE-100	5.2–18.8	—		
	BDE-153	1.8–4.8			

		Mean (ng/g wet)	Reference	Notes
	BDE-154	2.8–8.5		
	Total PBDE	44.6–148.0		A sample number of 21 coho and chinook salmon were used. The average % lipid was 3.89 with a range between 1.83 and 7.19. A 100 g "steak" was tested which included skin, muscle, bone, and organ tissues and then blended—the steak was extracted from in front of the dorsal fin
	Total PBDE	773.0–8120.0 ng/g lipid	Manchester-Neesvig et al. (2001)	
	BDE-47	52.1		
	BDE-66	1.7		
	BDE-99	9.3		
	BDE-100	9.7		
	BDE-153	2.7		
	BDE-154	4.5		
	Total PBDE	80.1		
	Total PBDE	2440 ng/g lipid		

		Range (% of total PBDEs)	Reference	Notes
Virginia, USA				
	BDE-47	45–74		Values were extracted and approximated from a graph. Five species of fish were analyzed (channel catfish, flathead catfish, carp, striped bass, and white bass. Values were ranged across species—there had similar distribution within each species despite large variations in total PBDEs in fish between sampling sites. Number of samples: Catfish $n = 15$; flathead $n = 20$; carp $n = 5$; stiped $n = 24$; white $n = 6$
	BDE-49	2–12.0		
	BDE-99	0–27		
	BDE-100	16–24		
	BDE-153	0–7	Hale et al. (2001)	
	BDE-154	3.0–5.0		

		(μg/kg wet)	% lipid	Reference	Notes
Washington State, USA					
RI—Rainbow trout (whole)	Total PBDE	Nondetectable	5.2		Quantitation limit was approximately 5 μg/kg unless indicated values were on a lipid basis Samples were collected at different times between 9/94 and 7/99, RI = Rock Island Creek; DC = Douglas Creek; SR = Spokane River; SoR = Soleduck River; SnR = Snake River; YR = Yakima River; CR = Columbia River; Total PBDEs can be broken down into TBDEs, PeBDEs, and HxBDEs
RI—Rainbow trout (whole)	Total PBDE	Nondetectable	4.9		
SR—Rainbow trout (whole)	Total PBDE	297.0	2.1		
DC—Rainbow trout (whole)	Total PBDE	1.4	4.8		
DC—Rainbow trout (split)	Total PBDE	1.5	3.8	Johnson and Olson (2001)	
DC—Rainbow trout (split)	Total PBDE	1.4	NA		
SR—Rainbow trout (fillet)	Total PBDE	20.0	1.3		
SR—Rainbow trout (fillet)	Total PBDE	119.0	3.7		
SR—Rainbow trout (fillet)	Total PBDE	166.0	4		

continued

TABLE 3.9 (continued)
PBDE Levels in Fish from the United States and Abroad

Country	Compound	Concentration	Percent Lipid (Where Available)	Reference	Note
SR—Rainbow trout (fillet)	Total PBDE	174.0	0.9		
SoR—Mountain whitefish (fillet)	Total PBDE	Nondetectable	5.7		
SR—Mountain whitefish (whole)	Total PBDE	1250.0	14.9		
SR—Large scale sucker (fillet)	Total PBDE	120.0	2.6		
SR—Large scale sucker (whole)	Total PBDE	105.0	4.4		
YR—Large scale sucker (whole)	Total PBDE	64.0	4.2		
YR—Carp (fillet)	Total PBDE	22.0	2.3		
SnR—Channel catfish (fillet)	Total PBDE	8.0	4.3		
CR—Starry flounder (whole)	Total PBDE	30.0	4.3		
Hadley Lake—white crappie and bluegill	BDE-47	13 ± 2	3.2 ± 0.4		
	BDE-99	16 ± 2			
	BDE-100	7.4 ± 1			
	BDE-153	15 ± 2			
	BDE 154	13 ± 1			
	BDE -190	<0.007			
	BDE-209	<1.4			
	Total PBDE	65 ± 8			
	Total PBDE	2400 ± 600 ng/g lipid		Dodder et al. (2002)	This lake is in close proximity (1.3 km) to a suspected PBDE manufacturing facility
Carp	BDE-47	3.20	0.8		
	BDE-99	0.07			
	BDE-100	0.89			
	BDE-153	0.10			
	BDE-154	2.10			
	BDE-190	<0.007			
	BDE-209	<1.4			
	Total PBDE	6.2			
	Total PBDE	760 ng/g lipid			

Sample	Congener	Value		Reference	Note
Carp	BDE-47	9.80			
	BDE-99	0.12			
	BDE-100	3.60			
	BDE-153	0.04			
	BDE-154	6.90			
	BDE-190	<0.006			
	BDE-209	<1.3			
	Total PBDE	20.0	0.8	Dodder et al. (2002)	This lake is in close proximity (1.3 km) to a suspected PBDE manufacturing facility
	Total PBDE	2500 ng/g lipid			
Lake of the Ozarks—white crappie and bluegill	BDE-47	3.5 ± 0.7			
	BDE-99	1.9 ± 0.4			
	BDE-100	1.1 ± 0.2			
	BDE-153	0.27 ± 0.11			
	BDE-154	0.2 ± 0.05			
	BDE-190	<0.007			
	BDE-209	<1.4			
	Total PBDE	6.9 ± 1.4	3 ± 0.9	Dodder et al. (2002)	This lake is in close proximity (1.3 km) to a suspected PBDE manufacturing facility
	Total PBDE	300 ± 80 ng/g lipid			
Lake Superior—Smelt	BDE-47	5.7 ± 0.3			
	BDE-99	1.8 ± 0.2			
	BDE-100	0.98 ± 0.09			
	BDE-153	0.2 ± 0.02			
	BDE 154	0.45 ± 0.03			
	BDE -190	<0.007			
	BDE-209	<1.5			
	Total PBDE	9.1 ± 0.6	6.2 ± 0.5	Dodder et al. (2002)	This lake is in close proximity (1.3 km) to a suspected PBDE manufacturing facility
	Total PBDE	150 ± 9 ng/g lipid			
Lake Ontario—Smelt	BDE-47	10 ± 1			
	BDE-99	5.3 ± 0.7			
	BDE-100	1.6 ± 0.1			
	BDE-153	0.49 ± 0.02			
	BDE-154	0.9 ± 0.05			
	BDE-190	<0.007			
	BDE-209	<1.6			
	Total PBDE	18 ± 1	7.8 ± 1.0	Dodder et al. (2002)	This lake is in close proximity (1.3 km) to a suspected PBDE manufacturing facility
	Total PBDE	240 ± 30 ng/g lipid			

continued

TABLE 3.9 (continued)
PBDE Levels in Fish from the United States and Abroad

Country	Compound	Concentration Mean (ng/g wet)	Percent Lipid (Where Available)	Reference	Note
Michigan, USA—Detroit River			—		
Large mouth bass	BDE-47	2.80		Rice et al. (2002)	$n = 12$; composite subsamples of whole fish
	BDE-99	0.48			
	BDE-100	0.45			
	BDE-153	0.44			
	BDE 154	0.43			
	BDE-181	0.26			
	BDE-183	0.26			
	Total PBDE	5.25			
	Total PBDE	163 ng/g lipid			
Carp	BDE-47	3.00		Rice et al. (2002)	$n = 10$; composite subsamples of whole fish
	BDE-99	0.50			
	BDE-100	0.48			
	BDE-153	0.47			
	BDE-154	0.45			
	BDE-181	0.24			
	BDE-183	0.25			
	BDE-190	n/a			
	Total PBDE	5.39			
	Total PBDE	40.70			
Illinois, USA—Des Plaines River (lower)		**Mean (ng/g wet)**			
Carp	BDE-47	2.54	—	Rice et al. (2002)	It is speculated that these numbers are due to the manufacture. Discharges that use products containing these substances are from waste facilities along the river; $n = 10$; composite subsamples of whole fish
	BDE-99	0.50			
	BDE-100	0.44			
	BDE-153	1.01			
	BDE 154	1.89			
	BDE-181	3.28			
	BDE-183	2.99			
	BDE-190	1.75			
	Total PBDE	14.40			
	Total PBDE	281.00			

continued

		Mean (ng/g wet)	
Illinois, USA—Des Plaines River (upper)			
Carp	BDE-47	1.34	It is speculated that these numbers are due to the manufacture. Discharges that use products containing these substances are from waste facilities along the river; $n = 4$; composite subsamples of whole fish
	BDE-99	0.50	
	BDE-100	0.49	
	BDE-153	0.66	
	BDE 154	0.98	
	BDE-181	1.44	
	BDE-183	1.31	
	BDE-190	0.97	
	Total PBDE	7.68	
	Total PBDE	78.3	Rice et al. (2002)
Buffalo, New York, USA			
Carp Muscle	Total PBDE	18.7	Loganathan et al. (1995) as cited in Manchester-Neesvig et al. (2001)
Klosterfjorden, Sweden			
Sea Trout	Total PBDE	15.0	Andersson and Blomkvist (1981) as cited in Manchester-Neesvig et al. (2001)
Viskan River, Sweden			
Pike muscle	Total PBDE	124.0	Andersson et al. (1981) as cited in Manchester-Neesvig et al. (2001)
Pike liver	Total PBDE	9680.0	
Baltic Sea			
Salmon muscle	Total PBDE	14.0	Asplund et al. (1999) as cited in Manchester-Neesvig et al. (2001)
Salmon egg	Total PBDE	9.0	
Salmon blood	Total PBDE	6.0	
Wakayama, Japan			
Sardine	Total PBDE	0.8	Watanabe and Tatsukawa (1987) as cited in Manchester-Neesvig et al. (2001)
North Sea			
North—cod liver	Total PBDE	26.0	de Boer (1989) as cited in Manchester-Neesvig et al. (2001)
Central—cod liver	Total PBDE	54.0	
South—cod liver	Total PBDE	170.0	

TABLE 3.9 (continued)
PBDE Levels in Fish from the United States and Abroad

Country	Compound	Concentration (pg/g wet)	Percent Lipid (Where Available)	Reference	Note
Canada					
Salmon feed	BDE-27/33	61.0; 67.0			Two samples were taken at different locations. Both are listed; For individual concentrations of congeners (41 cong. were detected), NDR = Peak detected, but did not meet quantification criteria; ND = Not detected
	BDE-47	840; 1100			
	BDE-49	190; 170			
	BDE-99	130; 180			
	BDE-100	170; 230		Easton et al. (2002)	
	BDE-153	65; 38			
	BDE-154	170; 48			
	BDE-155	140; 19			
	Total PBDE	1875.2; 1902.4			
Farmed Salmon	BDE-27/33	36; 110			Two samples were taken at different locations. Both are listed; For individual concentrations of congeners (41 cong. were detected), NDR = Peak detected, but did not meet quantification criteria; ND = Not detected
	BDE-47	690; 2600			
	BDE-49	110; 210			
	BDE-99	140; 390	—		
	BDE-100	130; 470		Easton et al. (2002)	
	BDE-153	NDR; 80			
	BDE-154	41; 130			
	BDE-155	16; 67			
	Total PBDE	1187.9; 4147.4			
Wild Salmon	BDE-27/33	NDR19.0			Four samples were taken at different locations. Ranges are listed; For individual concentrations of congeners (41 cong. were detected), see Table 6 in article; NDR = Peak detected, but did not meet quantification criteria; ND = Not detected
	BDE-47	29–280			
	BDE-49	NDR-29.0			
	BDE-99	NDR-97.0			
	BDE-100	4.2–43		Easton et al. (2002)	
	BDE-153	ND-3.0			
	BDE-154	ND-5.2			
	BDE-155	NDR-3.2			
	Total PBDE	38.7–485.2			

				Reference	Comment
Baltic Sea		**(ng/g lipid)**			
Salmon muscle	BDE-47	200.0		Bergman et al., Cambridge Isotope Labs (no date)	Concentrations were found in Salmon muscle
	BDE-99	54.0			
	BDE-100	47.0			
Sweden—Baltic Sea		**Mean (pmol/g lipid)**			
Salmon	BDE-47	410	—	Bergman et al. (1999)	
	BDE-99	96			
		Range (pmol/g lipid)			
	BDE-47	210–840	—		
	BDE-99	46–130			
		Homogenate (pmol/g lipid)			
Herring	BDE-47	170	—		
	BDE-99	48			
Great Britain					
Tees Bay—Plaice, flounder and dab	BDE-47	520–9500		Allchin et al. (1999) as cited in de Wit (2002)	
	BDE-99	83–370			
	DE-71	920–1200			
	DE-79	500–1200			
Lune/Wyre (off River Calder)—Flounder	BDE-47	400		Allchin et al. (1999) as cited in de Wit (2002)	
	BDE-99	54			
	DE-71	100			
	DE-79	120			
Nith estuary—Flounder	BDE-47	73–120		Allchin et al. (1999) as cited in de Wit (2002)	
	BDE-99	nd-19			
	DE-71	47–120			
	DE-79	nd-83			

continued

TABLE 3.9 (continued)
PBDE Levels in Fish from the United States and Abroad

Country	Compound	Concentration	Percent Lipid (Where Available)	Reference	Note
Bideford Bay (off Avonmouth)— Paice, flounder, and dab	BDE-47	ND-370		Allchin et al. (1999) as cited in de Wit (2002)	
	BDE-99	ND-100			
	DE-71	94–120			
	DE-79	ND-970			
The Wash (off Great Ouse)—Dab	BDE-47	380		Allchin et al. (1999) as cited in de Wit (2002)	
	BDE-99	74			
	DE-71	110			
	DE-79	58			
Off River Humber	BDE-47	1600		Allchin et al. (1999) as cited in de Wit (2002)	
	BDE-99	160			
	DE-71	110			
	DE-79	900			
Sweden					
Herring muscle	BDE-47	82			Approximations based on graphed data Bothnian Sea, November 1986
	BDE-47	450	—		Baltic Proper, June 1987
	BDE-47	59			Skagerrak, April 1987
	BDE-47	12			Fladen, November 1987
	BDE-47	38			Utlangan, September 1987
	BDE-47	35			Landsort, October 1987
	BDE-47	27			Angskarsklubb, September–October 1987
	BDE-47	19		Sellstrom et al. (1993)	Harufjarden, October 1987
Bream muscle	BDE-47	250			River Viskan, Spring 1987
Bream muscle	BDE-47	750			River Viskan, Spring 1987
Pike muscle	BDE-47	6500			River Haggen, Spring 1987
Pike muscle	BDE-47	2000			River Viskan, Spring 1987
Perch muscle	BDE-47	24000			River Viskan, Spring 1987
Perch muscle	BDE-47	2200			River Viskan, Spring 1987
Trout muscle	BDE-47	460			Kesnacksalven, Fall 1988
Trout muscle	BDE-47	120			Bengtsbroholjen, Fall 1988

continued

Tissue	Compound	Value	Location, Date	Reference
Trout muscle	BDE-47	140	Kesnacksalven, Fall 1988	
Trout muscle	BDE-47	250	Bengtsbroholjen, Fall 1988	
Trout muscle	BDE-47	190	Skifors, Fall 1988	
Pike muscle	BDE-47	98	Kesnacksalven, Fall 1988	
Pike muscle	BDE-47	94	Bengtsfors, Fall 1988	
Arctic char muscle	BDE-47	400	Lake Vattern, May 1987	
Whitefish muscle	BDE-47	15	Lake Storvindeln, November 1986	
			Approximations based on graphed data	
Herring muscle	BDE-99	27	Bothnian Sea, November 1986	
	BDE-99	46	Baltic Proper, June 1987	
	BDE-99	9.8	Skagerrak, April 1987	
	BDE-99	3.4	Fladen, November 1987	
	BDE-99	17	Utlangan, September 1987	
	BDE-99	9.2	Landsort, October 1987	
	BDE-99	17	Angskarsklubb, September–October 1987	
	BDE-99	7.8	Harufjarden, October 1987	
Bream muscle	BDE-99	2.3	River Viskan, Spring 1987	
Bream muscle	BDE-99	2.4	River Viskan, Spring 1987	Sellstrom et al. (1993)
Pike muscle	BDE-99	1100	River Haggen, Spring 1987	
Pike muscle	BDE-99	78	River Viskan, Spring 1987	
Perch muscle	BDE-99	9400	River Viskan, Spring 1987	
Perch muscle	BDE-99	380	River Viskan, Spring 1987	
Trout muscle	BDE-99	590	Kesnacksalven, Fall 1988	
Trout muscle	BDE-99	130	Bengtsbroholjen, Fall 1988	
Trout muscle	BDE-99	130	Kesnacksalven, Fall 1988	
Trout muscle	BDE-99	220	Bengtsbroholjen, Fall 1988	
Trout muscle	BDE-99	64	Skifors, Fall 1988	
Pike muscle	BDE-99	79	Kesnacksalven, Fall 1988	
Pike muscle	BDE-99	60	Bengtsfors, Fall 1988	
Arctic char muscle	BDE-99	64	Lake Vattern, May 1987	
Whitefish muscle	BDE-99	7.2	Lake Storvindeln, November 1986	

TABLE 3.9 (continued)
PBDE Levels in Fish from the United States and Abroad

Country	Compound	Concentration (ng/g lipid)	Percent Lipid (Where Available)	Reference	Note % lipid
Swedish Area					
Whitefish	BDE-47	15			0.66
	BDE-99	7.2			
Arctic char	BDE-47	400		Jansson et al. (1993)	5.3
	BDE-99	64			
Herring	BDE-47	82	—		5.4
	BDE-99	27			
Herring	BDE-47	450		Jansson et al. (1993)	4.4
	BDE-99	46			
Herring	BDE-47	59			3.2
	BDE-99	9.8			
North Sea		ng/g lipid weight			
Herring fillet—Mean	BDE-28	1.9			
	BDE-47	37			
	BDE-99	12			
	BDE-100	9.2			
	BDE-153	0.9			
	BDE-154	1.5	—	Boon et al. (2002)	
Range	BDE-28	1.2–2.4			
	BDE-47	23–47			
	BDE-99	9.9–17			
	BDE-100	6.3–12			
	BDE-153	0.6–1.3			
	BDE-154	1.3–1.9			
Cod fillet—Mean	BDE-28	2.7			
	BDE-47	43	—	Boon et al. (2002)	
	BDE-99	6.3			
	BDE-100	13			

continued

			Reference
Range	BDE-153	<LOD	
	BDE-154	3.9	
	BDE-28	1.5–4.5	
	BDE-47	26–74	
	BDE-99	3.1–16	
	BDE-100	5.9–21	
	BDE-153	<LOD	
	BDE-154	3.9–3.9	
Whiting fillet—Mean	BDE-28	1.8	
	BDE-47	26	
	BDE-99	9	
	BDE-100	8.6	
	BDE-153	<LOD	
	BDE-154	3.3	Boon et al. (2002)
Range	BDE-28	1.3–2.4	
	BDE-47	7.1–40	
	BDE-99	5.3–14	
	BDE-100	4.2–12	
	BDE-153	<LOD	
	BDE-154	2.2–4.4	
Herring liver—Mean	BDE-28	2.1	
	BDE-47	30	
	BDE-99	13	
	BDE-100	9.1	
	BDE-153	2.1	
	BDE-154	2.6	Boon et al. (2002)
Range	BDE-28	1.6–2.5	
	BDE-47	19–52	
	BDE-99	8.0–21	
	BDE-100	5.6–17	
	BDE-153	1.1–3.9	
	BDE-154	1.5–4.4	

TABLE 3.9 (continued)
PBDE Levels in Fish from the United States and Abroad

Country	Compound	Concentration	Percent Lipid (Where Available)	Reference	Note
Cod liver—Mean	BDE-28	6.7		Boon et al. (2002)	
	BDE-47	133			
	BDE-99	15			
	BDE-100	40			
	BDE-153	0.7			
	BDE-154	6.4			
Range	BDE-28	2.0–12			
	BDE-47	63–307			
	BDE-99	1.4–53			
	BDE-100	18–93			
	BDE-153	0.5–1.3			
	BDE-154	4.3–12			
Whiting liver—Mean	BDE-28	3.6	—	Boon et al. (2002)	
	BDE-47	70			
	BDE-99	15			
	BDE-100	16			
	BDE-153	1.4			
	BDE-154	4.5			
Range	BDE-28	0.7–6.3			
	BDE-47	7.6–132			
	BDE-99	1.9–34			
	BDE-100	1.7–31			
	BDE-153	0.3–3.1			
	BDE-154	0.6–11			
British Columbia, Canada		(ng/g lipid)	% lipid		
Sole (Bamfield)	Total PBDE	22	10.1	Ikonomou et al. (2002)	Samples were collected between 1992 and 2000; Total PBDE consists of congen. BDE-15,17,28/33,47,49,66,75,99,100,119,15 3,154,155, and other (BDE-25,30,32,77,140); Crab ranged from 4.2 to 480 ng/g lipid; Sole ranged from 12 to 340 ng/g lipid; Porpoise ranged from 350 to 2300 ng/g lipid
Sole (Crofton)	Total PBDE	82	6.6		
Sole (Crofton)	Total PBDE	67	7.5		
Sole (Crofton)	Total PBDE	60	5.5		
Sole (Crofton)	Total PBDE	12	14.5		
Sole (Howe Sound)	Total PBDE	140	11.8		

			(pg/g fresh wt.)			Reference	Notes
Sole (Kitimat)	Total PBDE	66		8.8			Samples were collected between 1992 and 2000; Total PBDE consists of congen. BDE-15,17,28/33,47,49,66,75,99,100,119,153,154,155, and other (BDE-25,30,32,77,140); Crab ranged from 4.2 to 480 ng/g lipid; Sole ranged from 12 to 340 ng/g lipid; Porpoise ranged from 350 to 2300 ng/g lipid
Sole (Kitimat)	Total PBDE	70		6.9			
Sole (Kitimat)	Total PBDE	64		7.3			
Sole (Mill Bay)	Total PBDE	37		14.1			
Sole (Trincomali)	Total PBDE	110		6			
Sole (Trincomali)	Total PBDE	110		6.7			
Sole (Vancouver)	Total PBDE	340		6.5			
Sole (Vancouver)	Total PBDE	280		5.4		Ikonomou et al. (2002)	
Japan							
Wakayama prefecture (Young Yellowtail)	BDE-153		30				These are approximations based on graphed data
	BDE-154		190				
	BDE-99		125				
	BDE-100		240				
	BDE-47		1000				
	244'/2'34				—		
	Total		1650			Ohta et al. (2002)	
Ehime prefecture	BDE-153		30				These are approximations based on graphed data
	BDE-154		170				
	BDE-99		110				
	BDE-100		240				
	BDE-47		925				
	244'/2'34				—		
	Total		1580			Ohta et al. (2002)	
Iwate prefecture	BDE-153		70				These are approximations based on graphed data
	BDE-154		170				
	BDE-99		190				
	BDE-100		260				
	BDE-47		1000				
	244'/2'34						
	Total		1720			Ohta et al. (2002)	

continued

TABLE 3.9 (continued)
PBDE Levels in Fish from the United States and Abroad

Country	Compound	Concentration	Percent Lipid (Where Available)	Reference	Note
China Sea	BDE-153	50			
	BDE-154	140			
	BDE-99	190			
	BDE-100	250		Ohta et al. (2002)	These are approximations based on graphed data
	BDE-47	925			
	244′/2′34				
	Total	1620			
Iwate prefecture (Mackerel)	BDE-153	60			
	BDE-154	220			
	BDE-99	140			
	BDE-100	230		Ohta et al. (2002)	These are approximations based on graphed data
	BDE-47	750			
	244′/2′34				
	Total	1550			
Chiba prefecture	BDE-153	80			
	BDE-154	60			
	BDE-99	140			
	BDE-100	220		Ohta et al. (2002)	These are approximations based on graphed data
	BDE-47	700			
	244′/2′34				
	Total	1400			
Shizuoka prefecture	BDE-153	30			
	BDE-154	60			
	BDE-99	350			
	BDE-100	230		Ohta et al. (2002)	These are approximations based on graphed data
	BDE-47	825			
	244′/2′34				
	Total	1540			

continued

Barents Sea (Northern Europe)	BDE-153	30		
	BDE-154	125		
	BDE-99	310		Ohta et al. (2002)
	BDE-100	190		These are approximations based on graphed data
	BDE-47	700		
	244'/2'34		—	
	Total	1280		
Iwate prefecture (Yellowtail)	BDE-153	60		
	BDE-154	40		
	BDE-99	190		Ohta et al. (2002)
	BDE-100	190		These are approximations based on graphed data
	BDE-47	740		
	244'/2'34			
	Total	1320		
Shizuoka prefecture	BDE-153	40		
	BDE-154	125		
	BDE-99	125		Ohta et al. (2002)
	BDE-100	190		These are approximations based on graphed data
	BDE-47	600		
	244'/2'34			
	Total	985		
Miyagi prefecture (Salmon)	BDE-153	40		
	BDE-154	125		
	BDE-99	70		Ohta et al. (2002)
	BDE-100	125		These are approximations based on graphed data
	BDE-47	600		
	244'/2'34		—	
	Total	1040		
Hokkaido prefecture	BDE-153	50		
	BDE-154	125		
	BDE-99	70		Ohta et al. (2002)
	BDE-100	125		These are approximations based on graphed data
	BDE-47	500		
	244'/2'34			
	Total	897		

Environmental Contaminants in Biota

TABLE 3.9 (continued)
PBDE Levels in Fish from the United States and Abroad

Country	Compound	Concentration	Percent Lipid (Where Available)	Reference	Note
Iwate prefecture	BDE-153	50			
	BDE-154	90			
	BDE-99	125			These are approximations based on graphed data
	BDE-100	110		Ohta et al. (2002)	
	BDE-47	375			
	244′/2′34				
	Total	813			
Hokkaido prefecture	BDE-153	30			
	BDE-154	90			
	BDE-99	70			These are approximations based on graphed data
	BDE-100	60		Ohta et al. (2002)	
	BDE-47	250			
	244′/2′34				
	Total	593			
Indian Ocean (Yellow Tuna)	BDE-153	Unreadable			
	BDE-154	Unreadable			
	BDE-99	Unreadable			These are approximations based on graphed data
	BDE-100	Unreadable		Ohta et al. (2002)	
	BDE-47	Unreadable			
	244′/2′34	Unreadable			
	Total	18.5			
Western Pacific Ocean	BDE-153	Unreadable			
	BDE-154	Unreadable			
	BDE-99	Unreadable			These are approximations based on graphed data
	BDE-100	Unreadable		Ohta et al. (2002)	
	BDE-47	Unreadable			
	244′/2′34	Unreadable			
	Total	17.7			

TABLE 3.10
PBDE Levels in Aquatic Invertebrates

Country	Compound	Concentration (ng/g lipid)	Percent Lipid (Where Available) (%) Levels in Biota	Reference	Note
British Columbia, Canada					
Crab (Vancouver)	Total PBDE	350	8.1	Ikonomou et al. (2002)	Samples were collected between 1992 and 2000; Total PBDE consists of congen. BDE-15,17,28/33,47,49,66,75,99,100,119,15 3,154,155, and other (BDE-25,30,32,77,140); Crab ranged from 4.2 to 480 ng/g lipid
Crab (Cowichan)	Total PBDE	320	7.6		
Crab (Gardener)	Total PBDE	4.2	14.7		
Crab (Sechelt)	Total PBDE	320	16.3		
Crab (Prince Rupert)	Total PBDE	480	5.9		
Crab (Victoria)	Total PBDE	200	15.4		
Sweden					
Viskan River eel muscle	Total PBDE	215	—	Sellstrom et al. (1993) Andersson et al. (1981) as cited in Manchester-Neesvig et al. (2001)	% lipid not reported
Osaka, Japan					
mussel	Total PBDE	17.4	—	Wantanabe et al. (1987) as cited in Manchester-Neesvig et al. (2001)	
Shizuoka prefecture (Short-necked clam)	BDE-153	unreadable		Ohta et al. (2002)	These are approximations based on graphed data
	BDE-154	unreadable			
	BDE-99	unreadable			
	BDE-100	unreadable	—		
	BDE-47	unreadable			
	244′/2′34	unreadable			
	Total	61.3			
Shimane prefecture (Short-necked clam)	BDE-153	unreadable		Ohta et al. (2002)	These are approximations based on graphed data
	BDE-154	unreadable			
	BDE-99	unreadable			
	BDE-100	unreadable	—		
	BDE-47	unreadable			
	244′/2′34	unreadable			
	Total	43.5			

studies involving fish indicate similar modes of action, though the number of studies is significantly fewer; there is little laboratory ecotoxicological information for invertebrates, amphibians, and other aquatic species. It has not been firmly established that the mode of action is the same among all vertebrate species or in invertebrates.

3.6.1 DIOXIN TOXICITY

The mechanism of action for dioxin (and dioxin-like compounds) is generally accepted to function by binding to a vertebrate cytosolic protein known as the aryl hydrocarbon receptor (AhR) and inducing transcriptional upregulation of the AhR gene battery (Nebert et al. 2000, Mandal 2005). Although subject to some debate, invertebrates do not possess this receptor and are generally considered less sensitive to dioxins (Butler et al. 2001). The genes known to be regulated through the AhR include drug-metabolizing enzymes, of which cytochrome P450 or CYP1A are best known. This signal transduction pathway from AhR binding to gene expression in fish and the role of these genes in the toxicity induced by dioxins and dioxin-like PCBs has been extensively studied in the past decade. Fish, like mammals possess monooxygenase enzymes (such as CYP1A) designed to detoxify chemicals; however, the phase I metabolites of some contaminants may be more toxic than the parent compound (Guengerich and Liebler 1985). Induction of CYP1A is mediated through the binding of chemicals to AhR. AhR ligands generally have isoteric configurations and are similar in structure to 2,3,7,8-TCDD, a model CYP1A inducer. Receptor binding is followed by a series of molecular events leading to the expression of several genes (including CYP1A) known as the "Ah-gene battery" (Nebert et al. 1993). The toxic effects of dioxin and structurally similar compounds are thought to be mediated through the AhR, with induced proteins causing alterations in cellular homeostasis (DeVito et al. 1994). In mammals, these effects include wasting syndrome, tumor promotion, and thymic atrophy (Poland and Knutson 1982). In fish, early life stages appear to be particularly sensitive to AhR ligands (Mehrle et al. 1988, Walker and Peterson 1991), and recent evidence indicates the involvement of CYP1A enzymes specifically in this toxic response (Cantrell et al. 1996).

A wide range of effects have been observed in fish, aquatic invertebrates, benthic organisms exposed to dioxins in laboratory and field studies (Grimwood and Dobbs 1995, Boening 1998). The early developmental stages of most vertebrates are susceptible to the effects of dioxins, and the developing embryos of oviparous vertebrates are particularly prone to these adverse effects (Zabel et al. 1995, Zabel and Petersen 1996). Symptoms of dioxin exposure in fish embryos and fry include edema of the yolk sac and pericardium, hemorrhaging in the head and tail regions, craniofacial deformities, and wasting syndrome (Giesy et al. 2002, Heiden et al. 2005, Carney et al. 2006). The cardiovascular system, and in particular the vascular endothelium of the developing embryo, has been identified as uniquely sensitive to cytochrome P4501A enzyme induction in early life stages of freshwater fish such as lake trout, and marine fish such as seabreem (Guiney et al. 1997, Yamauchi et al. 2006).

Adverse impacts to fish tend to be greater when exposure involves tetra-, penta-, and hexachlorinated congeners. Lower chlorinated congeners tend to be more rapidly metabolized and eliminated, while higher chlorinated compounds have limited membrane permeability and bioavailability. It is theorized that the elevated exposure risk that fish experience through water, sediment, and dietary sources of dioxin has led to reduced and, in some cases, failed recruitment of young fish into breeding populations. For example, populations of lake trout in the Great Lakes were thought to be limited by dioxin-like contaminants in the 1960s–1970s through recruitment failure (Cook et al. 2003).

Aquatic toxicity data for organisms exposed in water to dioxins and furans are summarized in Tables 3.11 and 3.12, respectively. Figure 3.13 illustrates variations in the sensitivities of aquatic organisms exposed in water to 2,3,7,8-TCDD. As was shown in the tissue residue SSDs, early life stages of aquatic organisms are more sensitive to dioxins than adult life stages; however, sensitivities vary over approximately three orders of magnitude. Clear differences in the sensitivities of

TABLE 3.11
Summary of Available Ecotoxicity Data Describing Effects Concentrations Based on Water Exposure for Polychlorinated Dibenzo-p-dioxins in Fish and Other Aquatic Organisms

	Congener	Endpoint[a]	Species	Exposure Duration	Concentration (µg/L)	Effects
Reproduction	2,3,7,8-TCDD	NOEL	Brook Trout	105	0.1622–2.3	General reproductive effects
		NR[b]	Aquatic insects	35	3–30	
			Molluscs	36–55	0.2	
Physiological	2,3,7,8-TCDD	LOEC	Rainbow trout	50–200	0.09	Gap charge balance, osmolality in female trout
		EC50	Medaka	3–6	0.012–0.0056	Swim bladder inflation
		NOEL	Mummichog	27	0.000012–0.0001	General physiology
		EC50	Mummichog	27	2.03	General physiology and assimilation efficiency
		EC4.5–37	Trout/Minnow	30	105–110	Assimilation efficiency
Histology	2,3,7,8-TCDD	EC50	Medaka	Until hatch	0.006–19	Lesions
	1,2,3,4,7,8-HexaCDD		Medaka	Until hatch	0.2–1.1	
	1,2,3,7,8-PentaCDD		Medaka	Until hatch	0.01–0.03	
	2,3,7,8-TCDD	NOEL	crayfish	40–60	30	General histological changes
Growth	1,2,3,4,6,7,8-HeptaCDD	NOEC	Rainbow trout	30	109	Relative growth rate
	1,2,3,4,7,8-HexaCDD		Rainbow trout	30	109.3	
	1,2,3,4,7-PentaCDD		Rainbow trout	30	104.7	
	2,3,7,8-TCDD		Rainbow trout	30	109.8	
			Rainbow trout	2	3.7	Growth weight
	1,2,3,4,7,8-HexaCDD		Rainbow trout	2	7.9–10	
Genetic	2,3,7,8-TCDD	EC50	Medaka	3	3500–14,000	Abnormal growth
	2,3,7,8-TCDD	LOEC	Medaka	1–13	0.1–10	Gene expression
		NOEC	Medaka	1–13	0.1–10	
Biochemical	2,3,7,8-TCDD	LOEC	Rainbow trout	50–200	0.09	General biochemical effect
		EC50	Rainbow trout	7	0.17–0.79	
			trout	3–21	0.007–1.05	
			carp	7–14	0.048–0.17	
			Medaka	3	0.0012	
			flounder	10	1.62	

continued

TABLE 3.11 (continued)
Summary of Available Ecotoxicity Data Describing Effects Concentrations Based on Water Exposure for Polychlorinated Dibenzo-*p*-dioxins in Fish and Other Aquatic Organisms

	Congener	Endpoint[a]	Species	Exposure Duration	Concentration (µg/L)	Effects
Cyp450 Enzyme Activity	2,3,7,8-TCDD	LOEC	Carp	7–14 d	0.03–0.17	General enzyme activity
			Trout	7–14 d	0.1–0.5	
	1,2,3,7,8-PentaCDD		Carp	14-d	0.002–1.53	
	1,2,3,6,7,8-HexaCDD		Carp	14-d	0.31–1.53	
Mortality	2,3,7,8-TCDD	LC50	Medaka	3	0.0081–0.01	Early life-stage mortality
			Rainbow Trout	7	0.23–0.5	
			Rainbow Trout	30	0.24–0.41	
			Black bullhead	80	5	Mortality
			common carp	80	3	
			Bluegill	80	16	
			Largemouth bass	80	11	
			Rainbow trout	80	10	
			Yellow perch	80	3	
			Signal crayfish	45	30	
		LC76–80	Midge	35	30–3000	Mortality–Hatch
		LOEC	Rainbow trout	342	0.0018	Mortality in female Rainbow trout
			Lake trout	NR[b]	0.04	Early life-stage mortality
		NOEC	Rainbow trout	56–342	0.000038–0.0018	Mortality
			Lake trout	NR	0.034	Early life-stage mortality
			Water flea	14	1000	Mortality
			Bullfrog	35	500	Mortality
Histology	2,3,7,8-Tetrachlorodibenzofuran	EC50	Medaka, high-eyes	3–72	0.005–0.028	Lesions

Notes:

[a] Concentration at which effect is seen in 50% of the study population (EC50); concentration that is lethal to 50% of the study population (LC50); No observable effects concentration (NOEC); lowest observable effects concentration (LOEC).

[b] Not reported.

Source: U.S. EPA ECOTOX Database (accessed April 2009; http://cfpub.epa.gov/ecotox).

TABLE 3.12
Summary of Available Ecotoxicity Data Describing Effects Concentrations Based on Water Exposure for Polychlorinated Dibenzofurans in Fish and Other Aquatic Organisms

	Congener	Endpoint[a]	Species	Exposure Duration (days)	Concentration (µg/L)	Effects
Growth and Development	1,2,3,4,7,8-Hexachlorodibenzofuran	NOEC	Rainbow trout	2-16	806	Weight and general growth
	1,2,3,7,8-Pentachlorodibenzofuran			2-16	25	
	2,3,4,7,8-Pentachlorodibenzofuran	NOEC		2-56	12	
	2,3,7,8-Tetrachlorodibenzofuran				0.00041	
Enzyme	1,2,3,4,5,6,7,8-Octachlorodibenzofuran	NOEC	common carp	14	44.07	General changes in enzyme activity or function primarily to cytochrome P450 enzymes
	1,2,3,4,7,8-Hexachlorodibenzofuran	LOEC	Rainbow trout	2-16	0.61	
	1,2,3,7,8-Pentachlorodibenzofuran	LOEC	Rainbow trout	2-16	0.82	
	2,3,4,7,8-Pentachlorodibenzofuran	LOEC	common carp	14	0.07-0.36	
			Rainbow trout	2-16	0.8	
		NOEC	common carp	14	0.01-0.07	
	2,3,7,8-Tetrachlorodibenzofuran	EC50	Medaka, high-eyes	3	0.0026	
		LOEC	common carp	14	10.51	
			Rainbow trout	2-16	0.12	
		NOEC	common carp	14	2.1	
Mortality	1,2,3,4,5,6,7,8-Octachlorodibenzofuran	LC50	Zebra danio	3.5	0.008-0.149	Hatching
	1,2,3,4,7,8-Hexabromodibenzofuran		Rainbow trout	50-60	10.3-330,000	
	1,2,3,4,7,8-Hexachlorodibenzofuran	LC50		7	0.632-1.237	Mortality
	1,2,3,7,8-Pentabromodibenzofuran		Rainbow trout	50-60	5.12-15.7	
	2,3,4,7,8-Pentabromodibenzofuran		Rainbow trout	50-60	1.29-6.75	
			Medaka, high-eyes	3	0.0039	Hatching, mortality
	2,3,7,8-Tetrachlorodibenzofuran	EC50		3	0.012-0.022	Mortality
			Rainbow trout	50-60	0.871-1.94	
		NOEC	Rainbow trout	56	0.00179	

Notes:

[a] Concentration at which effect is seen in 50% of the study population (EC50); concentration that is lethal to 50% of the study population (LC50); No observable effects concentration (NOEC); lowest observable effects concentration (LOEC).

Source: U.S. EPA ECOTOX Database (accessed April 2009; http://cfpub.epa.gov/ecotox).

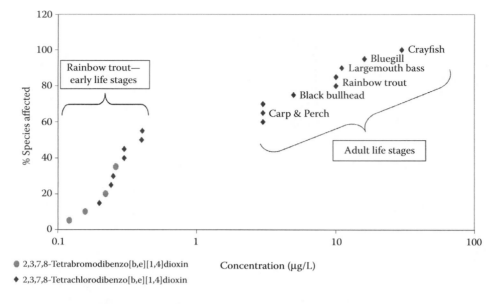

FIGURE 3.13 2,3,7,8-TCCD and TBDD chronic water exposure LC50 species sensitivity distribution (exposure duration: 30–80 days). (From U.S. EPA ECOTOX Database, accessed April 2009; http://cfpub.epa. gov/ecotox)

various fish species are evident between medaka and trout. Amphibians and crustaceans appear to be less sensitive to dioxins than fish; although the available data are too sparse to generate meaningful interspecies comparisons (Loonen et al. 1996). Studies have shown that many invertebrates do not possess a functional AhR, which may account for the apparent lack of susceptibility to dioxins reported in most invertebrate studies, though effects have been observed in a few species (Ashley et al. 1996, West et al. 1997, Wu et al. 2001). No significant mortality or morphological abnormalities have been observed in eggs, tadpoles, and frogs, which tend to eliminate dioxin relatively quickly, and appear to be about 100- to 1000-fold less sensitive to the deleterious effects of dioxin than early life stages of fish (Davies 1999, Karchner et al. 2000). Shorter birth cycles (3–6 days in most amphibians compared to 60 days or longer in fish) and reliance by fish on a yolk sac containing maternal lipids for 120 days after hatching are plausible explanations (Iguchi et al. 2001).

3.6.2 PCB Toxicity

Less so than in the 1980s and 1990s, ecotoxicological work in aquatic organisms continues to focus on the toxicity of commercial PCB formulations such as Aroclor, Clophen, Kanechlor, Chlorofen, Sovol, Delor, Phenoclor, and Chinese PCB mixtures (Lerner et al. 2007, Burkhard and Lukasewycz 2008). Toxicity from exposure to mixtures and individual PCB congeners has been demonstrated in several species of fish, aquatic invertebrates, benthic organisms, and avian and mammalian species; however, there is far less information on the chronic effects of PCBs on the full life cycles of aquatic animals. Considerable work has been done to understand cause–effect relationships between PCB levels and congener mixtures in sediment and sediment toxicity to benthic organisms (Fuchsman et al. 2006, Beckert and Ginn 2008).

Adverse effects to aquatic organisms include mortality, impaired growth and reproduction, disruption of the endocrine and immune systems, biochemical changes, behavioral alteration, and mutagenicity (Safe 1994, Fisk et al. 2005, Jorgensen et al. 2006). It is generally believed that the non- and mono-ortho substituted or planar congeners are more toxic than nonplanar PCBs due to their ability to interact with the Ah receptor (Giesy and Kannan 1998, Simon et al. 2007).

TABLE 3.13

Summary of Available Ecotoxicity Data Describing Effects Concentrations Based on Water Exposure for Aroclor 1254 in Fish and Other Aquatic Organisms

Effect	Endpoint	Species	Exposure Duration	Concentration (μg/L)	Measurements
Reproduction	EC50	Water flea	14	1.1–25	General Reproductive effects to adults
	EC50	Water flea	21	1.3–28	General Reproductive effects to adults
Behavioral	EC50	Common (Blue) mussel	1–2	2200–3000	Ability to detach from substrate
Physiological	LOEC	Mediterranean mussel	4	20	Tissue permeability
Genetic	LOEC	Oriental weatherfish	7	500	Micronuclei
Mortality					
				Amphibians	
	LC50	American toad	4	2.02	
	LC50	Fowler's toad	4	3.74	Mortality
	LC50	Leopard frog	4	1.03	
				Invertebrates	
	LC50	Daggerblade grass shrimp	4	6.1–7.8	Early life stages mortality
	NOEC	Hydra	1	10,000	
	LC50	Crayfish	1–7	80–550	
	LC50	Bay shrimp, Sand shrimp	1.25–4	12–21	
	LC50	Amphipod	2–4	40–98	
	LC50	Cockle	2	10,000	
	LC50	White shrimp	2	1640[a]	
	LC50	Amphipod	2–4	40–98	
	LC50	Opossum shrimp	2–4	57–109	
	LC50	Amphipod	4	9–40	Mortality
	LC50	Fiddler crab	4	10	
	LC50	Polychaete	4	1000	
	LC50	Daggerblade grass shrimp	4	86	
	LC50	Common shrimp	4	3000–10,000[b]	
	LC50	Amphipod	4	2400[c]	
	LC50	Crayfish	4	100[d]	
	LC50	Copepod	4	2400	
	LC50	Grass shrimp, freshwater prawn	7	3	

continued

TABLE 3.13 (continued)

Summary of Available Ecotoxicity Data Describing Effects Concentrations Based on Water Exposure for Aroclor 1254 in Fish and Other Aquatic Organisms

Endpoint	Species	Exposure Duration	Concentration (μg/L)	Measurements
			Fish	
LC50	Goldfish	4	1.18	
LC50	Redear sunfish	4	0.53	
LC50	Rainbow trout	4	0.32	
LC50	Rainbow trout	12–16	20.4–39.4	Early life stages mortality
LC50	Silver salmon	14	32.2	
LC50	Channel catfish	4	1.76	
LC50	Zebra danio	2.6–13.3	160–8000	Hatchability and survivorship
LC50	Yellow perch	1–4	150	
LC50	Cutthroat trout	1–4	42,500–44,900	
LC50	Harlequinfish, red rasbora	1–4	1100–6200	
LC50	Bluegill	1–15	200–16,000	
LC50	Channel catfish	1–15	200–12,000	
LC50	Hooknose	2	10,000	
LC50	Fathead minnow	4	7.7–33	
LC50	Bloater	4–5	3200–10,000	
LC50	Rainbow trout	4–16	8–1,500,000	
LC50	Western mosquitofish	4–6.5	305–460	Mortality
NOEC	Rainbow trout	5	0.081	
LOEC	Rainbow trout	7	500	
NOEC	Rainbow trout	7	100	
LC50	Striped bass	14	18–24	
LC50	Bluegill	20–30	54–260	
LC50	Channel catfish	20–30	113–300	
NOEC	Sheepshead minnow	28	5.6	
LC50	Sheepshead minnow	28	6100	

Note: All data refer to Aroclor 1254 obtained from U.S. EPA 2009a-ECOTOX database (accessed May 2009; http://cfpub.epa.gov/ecotox).

Sources:

a   Reyes et al. (2003).
b   Portmann et al. (1971).
c   Mayer et al. (1977).
d   Johnson and Finley (1980).

PCB toxicity in fish has been shown during embryo development and early larval stages, with exposure occurring through maternal transfer of PCBs to eggs during oogenesis (Table 3.13; Dethlefsen et al. 1996, Niimi 1996). PCBs have been associated with reduced egg hatchability and fry mortality at water concentrations orders of magnitude less than concentrations causing adult mortality (Eisler and Belisle 1986). Reproductive anomalies include inhibition of spermatogenesis and various testicular abnormalities as well as disruption of reproductive endocrine function (Sangalang et al. 1981, Vethaak et al. 2002, Khan and Thomas 2006). Interspecies differences in sensitivity and potency are evident in fish (Rankouhi et al. 2004). Similar to the dioxins, invertebrates are comparatively insensitive to PCBs, although some effects are evident for some congeners at high concentrations (Leney et al. 2006, Jofre and Karasov 2008). Similar to fish, maternal transfer of PCB to eggs has been shown in snapping turtles (*Chelydra serpentina*) (Eisenreich et al. 2009). Although maternal exposure of PCBs in *C. serpentina* has been shown not to affect embryonic development or hatching success, high rates of mortality are evident in individuals exposed maternally to PCBs beginning approximately 8 months after hatching (Eisenreich et al. 2009).

Figure 3.14 illustrates variations in the sensitivities of aquatic organisms exposed in water to PCBs. Similar to the dioxins, the early life stages of aquatic organisms are more sensitive to PCBs than adult life stages, and sensitivities vary among aquatic organisms over several orders of magnitude. While invertebrates are shown to be less sensitive to dioxin exposure, the sensitivities of various aquatic organisms to PCB exposure is less clear, particularly during the earlier stages of development. Unlike the dioxins, invertebrates show greater sensitivity to chronic PCB exposure than fish; however, toxicity testing methods may account for some of the differences.

3.6.3 PBDE Toxicity

A comprehensive overview on the toxicological effects of PBDEs is available elsewhere (e.g., McDonald 2002, Darnerud 2003, 2008, Legler and Brouwer 2003, Hamers et al. 2006). Despite the presence of PBDEs in numerous aquatic habitats and their detection in aquatic organisms collected in aquatic environments, little information is available, at present, on the toxicity of these compounds

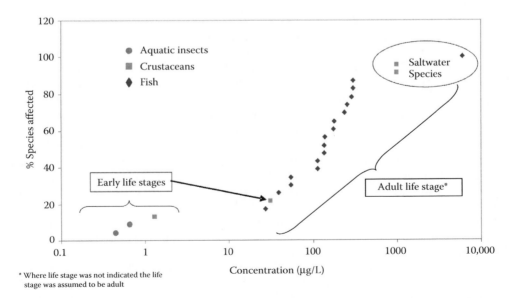

* Where life stage was not indicated the life
 stage was assumed to be adult

FIGURE 3.14 Aroclor 1254 chronic water exposure LC50 species sensitivity distribution (exposure duration: 20–35 days). (From U.S. EPA ECOTOX Database, accessed May 2009; http://cfpub.epa.gov/ecotox)

to aquatic organisms. PBDEs are suspected to be toxic to aquatic organisms, but dose–response relationships have yet to be developed (Ross et al. 2009).

Results from the few available laboratory studies indicate that PBDEs disrupt thyroid hormones in rodents and may have endocrine-disrupting effects on aquatic organisms (de Wit et al. 2006, Darnerud 2008, Talsness 2008). Studies on effects of PBDEs in fish species have revealed changes in hematocrit and blood glucose as well as reduction in spawning success (de Wit 2002). Plasma thyroxine levels decreased in juvenile rainbow trout (*Salvelinus namaycush*) fed a mixture of 13 PBDE congeners after a 56-day exposure period (Tomy et al. 2004). Male fathead minnows (*Pimephales promelas*) show decreased sperm production after oral intake of the tetra-brominated congener, BDE-47 (Muirhead et al. 2006). Behavioral effects such as altered larval swimming activity, predation rates and learning skills were observed in killifish (*Fundulus heteroclitus*) larvae exposed during embryonic development to as little as 0.001 μg/L commercial PBDE mixture (Timme-Laragay et al. 2006). Due to their structural similarity with PCBs and their affinity for the same cellular receptor (Luthe et al. 2008), it is possible that PBDE ecotoxicity will be found comparable to PCBs.

Summary

Although scientific interest regarding the ecological effects of dioxins, PCBs, and PBDEs stems from our knowledge of human health, much work remains regarding assessing risks to aquatic organisms. In fact, current information is insufficient to provide a thorough understanding of the potential for exposure and associated risks. Consequently, the few available environmental quality guidelines for these compounds are based on limited ecotoxicity data. While the most sensitive and ecologically important endpoints for mammals and birds are associated with reproduction, as shown in the ecotoxicity tables provided, there is a lack of reproduction bioassays and toxicokinetic information, for most aquatic species, which is needed to establish well-defined dose–response relationships. Outside the laboratory, it has not often been possible to demonstrate a clear cause/effect relationship between biological response and exposure to dioxins, PCBs, or PBDEs.

Furthermore, it is generally acknowledged that assessing the probability of an individual organism experiencing harm (e.g., reproductive impairment or mortality) is not useful for ecological assessment; risk is more appropriately assessed at the population, rather than the individual, level (Stark et al. 2004, Dearfield et al. 2005). Population endpoints, however, tend to be difficult to assess, requiring the use of dynamic population models covering effects on survival, breeding success, and immigration. In general, well validated population models do not yet exist for the majority of species, and it is difficult to estimate the extent of mortality or reproductive failure that could be incurred. Since population models are rarely available, it is more common for ecological risk assessments to define the NOEL for endpoints such as mortality or reproductive effects in individuals, and to assume that these may be used to set levels that will protect the whole population. Still, few NOEL values for dioxins, PCBs, and PBDEs are available, and those that are available are associated with high degrees of uncertainty (Bhavsar et al. 2007, 2008b).

Reflecting on the current state of knowledge regarding dioxins, PCBs, and PBDEs in aquatic organism, the available data in the published literature and results of the two largest studies in North America suggest the following trends:

- The frequency of detection of dioxins in the Continental United States is highest in the east and northeast and lowest in the west and southwest regions of the United States.
- Consistent with the trend observed for dioxins, mean total PCBs in fish from the western U.S. are lower than in fish from the eastern U.S.; in the Great Lakes, mean total PCBs in fish are highest in Lake Ontario and lowest in Lake Huron.

- Dioxin, PCB, and PBDE levels are generally higher in bottom-feeding fish than in pelagic fish.
- The most frequently detected dioxin congeners in fish include four PCDD compounds (2,3,7,8-TCDD; 1,2,3,7,8-PeCDD; 1,2,3,6,7,8-HxCDD; and 1,2,3,4,6,7,8-HpCDD) and 3 PCDF compounds (2,3,7,8-TCDF; 2,3,4,7,8-PeCDF; and 1,2,3,4,6,7,8-HpCDF).
- The results from the U.S. EPA and Ontario MOE studies indicate that four of the 2,3,7,8-substituted dioxin congeners are most prevalent in fish and contribute most to total TEQ (1,2,3,7,8-PeCDD; 2,3,4,7,8-PeCDF; 2,3,7,8-TCDD; and 2,3,4,7,8-PeCDF).
- Generally, early life stages of aquatic organisms are more sensitive to dioxins, PCBs, and PBDEs than adult life stages; however, the available ecotoxicology data indicates that sensitivities vary over approximately three orders of magnitude.
- Amphibians and crustaceans appear to be less sensitive to dioxins than fish; although the available data on invertebrates are too sparse to generate meaningful interspecies comparisons.
- Evaluating and predicting dioxin, PCB, and PBDE toxicity to aquatic organisms remains difficult; effect levels correlating to tissue residues vary over several orders of magnitude depending on species and life stage. Site-specific studies remain, at present, the best means for determining injuries to aquatic life and impairment of ecosystems.

REFERENCES

Alaee, M., and R. J. Wenning. 2002. The significance of brominated flame retardants in the environment: current understanding, issues and challenges. *Chemosphere* 46:579–582.

Alaee, M., J. Luross, D. B. Sergeant, D. C. G. Muir, D. M. Whittle, and K. Solomon. 1999. Distribution of polybrominated diphenyl ethers in the Canadian environment. *Organohalogen Compounds.* 40:347–350.

Allchin, C. R., R. J. Law, and S. Morris. 1999. Polybrominated diphenylethers in sediments and biota downstream of potential sources in the UK. *Environ. Pollut.* 105:197–207.

Andersson, O., and G. Blomkvist. 1981. Polybrominated aromatic pollutants found in fish in Sweden. *Chemosphere* 10:1051–1060.

Arctic Monitoring and Assessment Programme. (AMAP). 2004. AMAP Assessment 2002: Persistent Organic Pollutants in the Arctic. Arctic Monitoring and Assessment Programme (AMAP), Oslo, Norway. xvi +310 pp.

Arnot, J. A., and F. A. P. C. Gobas. 2006. A review of bioconcentration factor (BCF) and bioaccumulation factor (BAF) assessments for organic chemicals in aquatic organisms. *Environ. Rev.* 14:257–297.

Ashley, C. M., M. G. Simpson, D. M. Holdich, and D. R Bell. 1996. 2,3,7,8-Tetrachloro-Dibenzo-*p*-Dioxin is a Potent Toxin and Induces Cytochrome P450 in the Crayfish, *Pacifastacus leniusculus. Aquat. Toxicol.* 35:157–169.

Asplund, L., M. Athanasiadou, A. Sjodin, A. Bergman, and H. Bdrjeson. 1999. Organohalogen substances in muscle, egg and blood from M74 syndrome affected and non-affected Baltic salmon (Salmo salar). *Ambio* 28:67–76.

Bayarri, S., L. T. Baldassarri, N. Iacovella, F. Ferrara, and A. di Domenico. 2001. PCDDs, PCDFs, PCBs and DDE in edible marine species from the Adriatic Sea. *Chemosphere* 43:601–610.

Beckert, D. S., and T. C. Ginn. 2008. Critical evaluation of the sediment effect concentrations for polychlorinated biphenyls. *Integr. Environ. Assess. Manag.* 4:156–170.

Bergman, A., L. Asplund, M. Athanasiadou, E. Jakobsson, and E. K. Wehler. 1999. A Strategy for Chemical Analysis of Halogenated Environmental Pollutants. *Organohalogen Compounds.* 40:1–4.

Bergman, A., M. Athanasiadou, G. Marsh, A. Sjödin, and W. M. Grim III. No date Polybrominated Diphenyl Ethers—Environmental Contaminants of Concern. Cambridge Isotope Laboratories, Inc., Andover, MA. Application Note ENV 10.

Beyer, A., and M. Biziuk. 2009. Environmental fate and global distribution of polychlorinated biphenyls. In: *Reviews of Environmental Contamination and Toxicology, Vol. 201,* ed. D. M. Whitacre, pp. 137–158. New York: Springer.

Bhavsar, S. P., A. Hayton, E. J. Reiner, and D. A. Jackson. 2007. Estimating dioxin-like polychlorinated biphenyl toxic equivalents from total polychlorinated biphenyl measurements in fish. *Environ. Toxicol. Chem.* 26:1622–1628.

Bhavsar, S. P., E. J. Reiner, A. Hayton, R. Fletcher, and K. MacPherson. 2008a. Converting toxic equivalents (TEQ) of dioxins and dioxin-like compounds in fish from one toxic equivalency factor (TEF) scheme to another. *Environ. Int.* 34:915–921.

Bhavsar, S. P., A. Hayton, and D. A. Jackson. 2008b. Uncertainty analysis of dioxin-like polychlorinated biphenyls-related toxic equivalents in fish. *Environ. Toxicol. Chem.* 27:997–1005.

Bhavsar, S. P., E. Awad, R. Fletcher, A. Hayton, K. M. Somers, T. Kloic, K. MacPherson, and E. J. Reiner. 2008c. Temporal trends and spatial distribution of dioxins and furans in lake trout or lake whitefish from the Canadian Great Lakes. *Chemosphere* 73(1):158–165.

Boening, D. W. 1998. Toxicity of 2,3,7,8-tetrachlorodibenzo-*p*-dioxin to several ecological receptor groups: a short review. *Ecotoxicol. Environ. Saf.* 39:155–163.

Boon, J. P., W. E. Lewis, M. R. Tjoen-A-Choy, C. R. Allchin, R. J. Law, J. De Boer, C. C. Ten Hallers-Tjabbes, and B. N. Zegers. 2002. Levels of polybrominated diphenyl ether (PBDE) flame retardants in animals representing different trophic levels of the North Sea food Web. *Environ. Sci. Technol.* 36:4025–4032.

Braekevelt, E., S. A. Tittlemiert, and G. T. Tomy. 2003. Direct measurement of octanol-water partition coefficients of some environmentally relevant brominated diphenyl ether congeners. *Chemosphere* 51:563–567.

Brown, F. R., J. Winkler, P. Visita, J. Dhaliwal, and M. Petreas. 2006. Levels of PBDEs, PCDDs, PCDFs, and coplanar PCBs in edible fish from California coastal waters. *Chemosphere.* 64:276–286.

Burkhard, L. P., and M. T. Lukasewycz. 2008. Toxicity equivalency values for polychlorinated biphenyl mixtures. *Environ. Toxicol. Sci.* 27:529–534.

Burkhard, L. P., D. D. Endicott, P. M. Cook, K. G. Sappington, and E. L. Winchester. 2003. Evaluation of two methods for prediction of bioaccumulation factors. *Environ. Sci. Technol.* 37:4626–4634.

Burkhard, L. P., P. M. Cook, and M. T. Lukasewycz. 2004. Biota-sediment accumulation factors for polychlorinated biphenyls, dibenzo-p-dioxins, and dibenzofurans in southern Lake Michigan lake trout (Salvelinus namaycush). *Environ. Sci. Technol.* 38:5297–5305.

Burreau, S., Y. Zebühr, D. Broman, and R. Ishaq. 2006. Biomagnification of PBDEs and PCBs in food webs from the Baltic Sea and northern Atlantic Ocean. *Sci. Total Environ.* 366:659–672.

Butler, R. A., M. L. Kelley, W. H. Powell, M. E. Hahn, and R. J. Van Beneden. 2001. An aryl hydrocarbon receptor (AHR) homologue from the soft-shell clam, *Mya arenaria*: evidence that invertebrate AHR homologues lack 2,3,7,8-tetrachlorodibenzo-p-dioxin and beta-naphthoflavone binding. *Gene.* 278:223–234.

Cantrell, S. M., L. H. Lutz, D. E. Tillitt, and M. Hannink. 1996. Embryotoxicity of 2,3,7,8-tetrachlorodibenzo-p-dioxin (TCDD): the embryonic vasculature is a physiological target for TCDD-induced DNA damage and apoptotic cell death in Medaka (*Orizias latipes*). *Toxicol. Appl. Pharmacol.* 141(1):23–34.

Carney, S. A., A. L. Prasch, W. Heideman, and R. E. Peterson. 2006. Understanding dioxin developmental toxicity using the zebrafish model. *Birth Defects Res. A Clin. Mol. Teratol.* 76:7–18.

Cetin, B., and M. Odabasi. 2009. Air–water exchange and dry deposition of polybrominated diphenyl ethers at a coastal site in Izmir Bay, Turkey. *Environ. Sci. Technol.* 41:785–791.

Chiuchiolo, A. L., R. M. Dickhut, M. A. Cochran, and H. W. Ducklow. 2004. Persistent organic pollutants at the base of the Antarctic marine food web. *Environ. Sci. Technol.* 38:3551–3557.

CMABFRIP (Chemical Manufacturers Association Brominated Flame Retardant Industry Panel). 1997. Octabromodiphenyl Oxide (OBDPO): Determination of the water solubility. Wildlife International, Ltd. Project No. 439C-110.

Cook, P. M., et al. 2003. Effects of aryl hydrocarbon receptor-mediated early life stage toxicity on lake trout populations in Lake Ontario during the 20th century. *Environ. Sci. Technol.* 37(17):3864–3877.

Darnerud, P. O. 2003. Toxic effects of brominated flame retardants in man and in wildlife. *Environ. Int.* 29:841–853.

Darnerud, P. O. 2008. Brominated flame retardants as possible endocrine disruptors. *Int. J. Androl.* 31:152–160.

Davies, M. 1999. Compilation of EU Dioxin Exposure and Health Data Task 7—Ecotoxicology. European Commission DG Environment; UK Department of the Environment, Transport and the Regions. (DETR). 97/322/3040/DEB/E1.

de Boer, J., and M. Denneman. 1998. Polychlorinated diphenyl ethers: origin, analysis, distribution, and toxicity in the marine environment. *Rev. Environ. Contam. Toxicol.* 157:131–144.

DeVito, M. J., X. Ma, J. G. Babish, M. Menache, and L. S. Birnbaum. 1994. Dose-response relationships in mice following subchronic exposure to 2,3,7,8-tetrachlorodibenzo-p-dioxin: CYP1A1, CYP1A2, estrogen receptor, and protein tyrosine phosphorylation. *Toxicol. Appl. Pharmacol.* 124(1):82–90.

de Wit, C. A. 2002. An overview of brominated flame retardants in the environment. *Chemosphere* 46:583–624.

de Wit, C. A., M. Alaee, and D. C. G. Muir. 2006. Levels and trends of brominated flame retardants in the Arctic. *Chemosphere* 64:209–233.

Dearfield, K. L., et al. 2005. Ecological risk assessment issues identified during the U.S. Environmental Protection Agency's examination of risk assessment practices. *Integr. Environ. Assess. Manag.* 1:73–76.

deBruyn, A. M. H., L. M. Meloche, and C. J. Lowe. 2009. Patterns of bioaccumulation of polybrominated diphenyl ethers and polychlorinated biphenyl congeners in marine mussels. *Environ. Sci. Technol.* 43:3700–3704.

Dethlefsen, V., H. Von Westernhagen, and P. Cameron. 1996. Malformations in North sea pelagic fish embryos during the period 1984–1995. *ICES J. Mar. Sci.* 53:1024–1035.

Dodder, N. G., B. Strandberg, and R. A. Hites. 2002. Concentrations and spatial variations of polybrominated diphenyl ethers and several organochlorine compounds in fishes from the northeastern United States. *Environ. Sci. Technol.* 36:146–151.

Dyke, P. H., and J. Stratford. 2002. Changes to the TEF schemes can have significant impacts on regulation and management of PCDD/F and PCB. *Chemosphere* 47:103–116.

Easton, M. D. L., D. Luszniak, and E. Von der Geest. 2002. Preliminary examination of contaminant loadings in farmed salmon, wild salmon and commercial salmon feed. *Chemosphere* 46:1053–1074.

Eisenreich, K., S. M. Kelly, and C. L. Rowe. 2009. Latent mortality of juvenile snapping turtles from the Upper Hudson River, New York, exposed maternally and via the diet to polychlorinated biphenyls (PCBs). *Environ. Sci. Technol.* 43:6052–6057.

Eisler, R., and A. A. Belisle. 1986. *Planar PCB Hazards to Fish, Wildlife, and Invertebrates: A Synoptic Review.* Biological Report No. 31. Patuxent Wildlife Research Center, Laurel, MD.

Erickson, M. D. 1997. *Analytical Chemistry of PCBs*, 2nd edition. Boca Raton, FL: CRC Press.

Eriksson, J., N. Green, G. Marsh, and A. Bergman. 2004. Photochemical decomposition of 15 polybrominated diphenyl ether congeners in methanol/water. *Environ. Sci. Technol.* 38:3119–3125.

European Communities. 2001. European Union Risk Assessment Report. Diphenyl ether, Pentabromo Derivative (Pentabromodiphenyl Ether). CAS No.: 32534-81-9. EINECS No.: 251-084-2. Risk assessment. Final report, August 2000. United Kingdom on behalf of the European Union.

European Communities. 2002. European Union Risk Assessment Report. Bis(Pentabromophenyl) Ether. CAS No.: 1163-19-5. EINECS No.: 214-604-9. Risk assessment. Final report, 2002. France and United Kingdom on behalf of the European Union.

European Communities. 2003. *European Union Risk Assessment Report. Diphenyl Ether, Octabromo Derivative. CAS No.: 32536-52-0.* EINECS No.: 251-087-9. Risk assessment. Final Report. France and United Kingdom on behalf of the European Union.

Evans, M. S., D. Muir, W. L. Lockhart, G. Stern, M. Ryan, and P. Roach. 2005. Persistent organic pollutants and metals in the freshwater biota of the Canadian Subarctic and Arctic: an overview. *Sci. Total Environ.* 351–352: 94–147.

Fair, P. A., et al. 2007. Polybrominated diphenyl ethers (PBDEs) in blubber of free-ranging bottlenose dolphins (*Tursiops truncatus*) from two southeast Atlantic estuarine areas. *Arch. Environ. Contam. Toxicol.* 53:483–494.

Finley, B. L., K. T. Connor, and P. K. Scott. 2003. The use of toxic equivalency factor distributions in probabilistic risk assessments for dioxins, furans, and PCBs. *J. Toxicol. Environ. Health A.* 66:533–550.

Fisk, A. T., et al. 2005. An assessment of the toxicological significance of anthropogenic contaminants in Canadian arctic wildlife. *Sci. Total Environ.* 351–352:57–93.

Food and Drug Administration (FDA). 2009. Dioxin Analysis Results/Exposure Estimate. http://www.fda.gov/Food/FoodSafety/FoodContaminantsAdulteration/ChemicalContaminants/DioxinsPCBs/ucm077444.htm

Fuchsman, P. C., T. R. Barber, J. C. Lawton, and K. B. Leigh. 2006. An evaluation of cause-effect relationships between polychlorinated biphenyl concentrations and sediment toxicity to benthic invertebrates. *Environ. Toxicol. Chem.* 25:2601–2612.

Gauthier, L. T., C. E. Hebert, D. V. C. Weseloh, and R. J. Letcher. 2008. Dramatic changes in the temporal trends of polybrominated diphenyl ethers (PBDEs) in herring gull eggs from the Laurentian Great Lakes: 1982–2006. *Environ. Sci. Technol.* 42:1524–1530.

Gewurtz, S. B., et al. 2009. Factors influencing trends of polychlorinated naphthalenes and other dioxin-like compounds in lake trout (Salvelinus namaycush) from Lake Ontario, North America (1979–2004). *Environ. Toxicol. Chem.* 28:921–930.

Geyer, H. J., et al. 2002. Half-lives of tetra-, penta-, hexa-, hepta-, and octachlorodibenzo-*p*-dioxin in rats, monkeys, and humans: A critical review. *Chemosphere* 48(6):631–644.

Giesy, J. P., and K. Kannan. 1998. Dioxin like and non-dioxin-like toxic effects of polychlorinated biphenyls (PCBs): implications for risk assessment. *Crit. Rev. Toxicol.* 28:511–569.

Giesy, J. P., L. A. Feyk, P. D. Jones, K. Kannan, and T. Sanderson. 2003. Topic 4.8. Review of endocrine-disrupting chemicals in birds. *Pure Appl. Chem.* 75:2287–2303.

Giesy, J. P., P. D. Jones, K. Kannan, J. L. Newsted, D. E. Tillitt, and L. L. Williams. 2002. Effects of chronic dietary exposure to environmentally relevant concentrations to 2,3,7,8-tetrachlorodibenzo-*p*-dioxin on survival, growth, reproduction and biochemical responses of female rainbow trout (*Oncorhynchus mykiss*). *Aquat. Toxicol.* 59:35–53.

Gobas, F. A. P. C., K. E. Clark, W. Y. Shiu, and D. Mackay. 1989. Bioconcentration of polybrominated benzenes and biphenyls and related superhydrophobic chemicals in fish: Role of bioavailability and elimination in the feces. *Environ. Toxicol. Chem.* 8:231–245.

Gómara, B., et al. 2005. Levels and trends of polychlorinated dibenzo-*p*-dioxins/furans (PCDD/Fs) and dioxin-like polychlorinated biphenyls (PCBs) in Spanish commercial fish and shellfish products, 1995–2003. *J. Agric. Food Chem.* 53:8406–8413.

Great Lakes National Program Office (GLNPO). 2009. Great Lakes Fish Monitoring Program. http://www.epa.gov/glnpo/monitoring/fish/index.html (accessed August 11, 2009).

Grimwood, M. J., and T. J. Dobbs. 1995. A review of the aquatic ecotoxicology of polychlorinated dibenzo-*p*-dioxins and dibenzofurans. *Environ. Toxicol. Water Qual.* 10:57–75.

Guengerich, F. P., and D. C. Liebler. 1985. Enzymatic activation of chemicals to toxic metabolites. *Crit. Rev. Toxicol.* 14(3):259–307.

Guerrero, N. R. V. 2007. Predicting the uptake and bioaccumulation of organic pollutants from natural sediments. In: *Environmental Pollution: New Research,* ed. R. H. Plattenberg, pp. 141–185. New York, NY: Nova Science Pub.

Guiney, P. D., R. M. Smolowitz, R. E. Peterson, and J. J. Stegeman. 1997. Correlation of 2,3,7,8-tetrachlorodibenzo-*p*-dioxin induction of cytochrome P4501A in vascular endothelium with toxicity in early life stages of lake trout. *Toxicol. Appl. Pharmacol.* 143:256–273.

Gustafsson, K., M. Björk, S. Burreau, and M. Gilek. 1999. Bioaccumulation kinetics of brominated flame retardants (polybrominated diphenyl ethers) in blue mussel (Mytilus edulis). *Environ. Toxicol. Chem.* 18:1218–1224.

Hale, R. C., M. Alaee, J. B. Manchester-Neesvig, H. M. Stapleton, and M. G. Ikonomou. 2003. Polybrominated diphenyl ether flame retardants in the North American environment. *Environ. Int.* 29:771–779.

Hale, R. C., M. J. La Guardia, E. Harvey, and T. M. Mainor. 2002. Potential role of fire retardant-treated polyurethane foam as a source of brominated diphenyl ethers to the US environment. *Chemosphere* 46:729–735.

Hale, R. C., M. J. La Guardia, E. P. Harvey, T. M. Mainor, W. H. Duff, and M. O. Gaylor. 2001. Polybrominated diphenyl ether flame retardants in Virginia freshwater fishes (USA). *Environ. Sci. Technol.* 35:4585–4591.

Hamers, T., et al. 2006. *In vitro* profiling of the endocrine-disrupting potency of brominated flame retardants. *Toxicol. Sci.* 92:157–173.

Hardy, M. L. 2002. A comparison of the properties of the major commercial PBDPO/PBDE product to those of major PBB and PCB products. *Chemosphere* 46:717–729.

Hawker, D. W., and D. W. Connell. 1988. Octanol-water partition coefficients of polychlorinated biphenyl congeners. *Environ. Sci. Technol.* 22:382–387.

Heiden, T. K., R. J. Hutz, and M. J. Carvan. 2005. Accumulation, tissue distribution, and maternal transfer of dietary 2,3,7,8-tetrachlorodibenzo-*p*-dioxin: impacts on reproductive success of zebrafish. *Toxicol. Sci.* 87:497–507.

Hites, R. A., J. A. Foran, S. J. Schwager, B. A. Knuth, M. C. Hamilton, and D. O. Carpenter. 2004. Global assessment of polybrominated diphenyl ethers in farmed and wild salmon. *Environ. Sci. Technol.* 38:4945–4949.

Holden, A., J. She, M. Tanner, S. Lunder, R. Sharp, and K. Hooper. 2003. PBDEs in the San Francisco Bay Area: Measurements in fish. *Organohalogen Compounds.* 61:255–258.

Hooper, K., and T. A. McDonald. 2000. The BDEs: An emerging environmental challenge and another reason for breast-milk monitoring programs. *Environ. Health Perspect.* 108:387–392.

Hornbuckle, K. C., D. L. Carlson, D. L. Swackhammer, J. E. Baker, and S. J. Eisenreich. 2006. Polychlorinated biphenyls in the Great Lakes. In: *Environmental Pollution: New Research,* ed. R. A. Hites, pp. 13–71. Germany: Springer.

Hornung, M. W., E. W. Zabel, and R. E. Peterson. 1996. Toxic equivalency factors of polybrominated dibenzo-*p*-dioxin, dibenzofuran, biphenyl, and polyhalogenated diphenyl ether congeners based on rainbow trout early life stage mortality. *Toxicol. Appl. Pharmacol.* 140:227–234.

Hsieh, D. P. H., T. E. McKone, F. Chiao, R. C. Currie, and L. Kleinschmidt. 1994. Final Draft Report: Intermedia Transfer Factors for Contaminants Found at Hazardous Waste Sites. November. Prepared for the Office of Scientific Affairs, Department of Toxic Substances Control, California Environmental Protection Agency, Sacramento.

Hua, I., N. Kang, C. T. Jafvert, and J. R. Fábrega-Duque. 2003. Heterogeneous photochemical reactions of decabromodiphenyl ether. *Environ. Toxicol. Chem.* 22:798–804.

Iguchi, T., H. Watanabe, and Y. Katsu. 2001. Developmental effects of estrogenic agents on mice, fish, and frogs: A mini review. *Horm Behav.* 40(2):248–251.

Ikonomou, M. G., S. Rayne, and R. F. Addison. 2000. Exponential increases of the brominated flame retardants, polybrominated diphenyl ethers, in the Canadian Arctic from 1981 to 2000. *Environ. Sci. Technol.* 36:1886–1892.

Ikonomou, M. G., S. Rayne, and R. F. Addison. 2002. Exponential increases of the brominated flame retardants, polybrominated diphenyl ethers, in the Canadian Arctic from 1981 to 2000. *Environ. Sci. Technol.* 36:1886–1892.

International Union of Pure and Applied Chemistry (IUPAC). 1989. No. 209.

Isosaari, P., A. Hallikainen, H. Kiviranta, P. J. Vuorinen, R. Parmanne, J. Koistinen, and T. Vartiainen. 2006. Polychlorinated dibenzo-*p*-dioxins, dibenzofurans, biphenyls, naphthalenes and polybrominated diphenyl ethers in the edible fish caught from the Baltic Sea and lakes in Finland. *Environ. Pollut.* 141:213–225.

Jansson, B., R. Andersson, A. Lillemor, et al. 1993. Chlorinated and brominated persistent organic compounds in biological samples from the environment. *Environ. Toxicol. Chem.* 12(7):1163–1174.

Jofre, M. B., and W. H. Karasov. 2008. Effect of mono-ortho and di-ortho substituted polychlorinated biphenyl (PCB) congeners on leopard frog survival and sexual development. *Chemosphere* 70:1609–1619.

Johnson, A., and N. Olson. 2001. Analysis and occurrence of polybrominated diphenyl ethers in Washington State freshwater fish. *Arch. Environ. Contam. Toxicol.* 41:339–344.

Johnson, W. W., and M. T. Finley. 1980. *Handbook of Acute Toxicity of Chemicals to Fish and Aquatic Invertebrates. Summaries of toxicity Tests Conducted at Columbia National Fisheries Research Laboratory. 1965-78.* U.S. Fish and Wildlife Service, Washington, DC. Resource Publication 137.

Johnson-Restrepo, B., K. Kannan, R. Addink, and D. H. Adams. 2005. Polybrominated diphenyl ethers and polychlorinated biphenyls in a marine food web of coastal Florida. *Environ. Sci. Technol.* 39:8243–8250.

Jorgensen, E. H., M. M. Vijayan, J. E-A. Killie, N. Aluru, O. Aas-Hansen, and A. Maule. 2006. Toxicokinetics and effects of PCBs in Arctic fish: A review of studies on Arctic charr. *J. Toxicol. Environ. Health.* 69:37–52.

Kannan, K., E. Perrotta, N. J. Thomas, and K. M. Aldous. 2007. A comparative analysis of polybrominated diphenyl ethers and polychlorinated biphenyls in southern sea otters that died of infectious diseases and noninfectious causes. *Arch. Environ. Contam. Toxicol.* 53:293–302.

Kannan, K., K. Hilscherova, T. Imagawa, N. Yamashita, L. L. Williams, and J. P. Giesy. 2001. Polychlorinated naphthalenes, -biphenyls, -dibenzo-*p*-dioxins, and -dibenzofurans in double-crested cormorants and herring gulls from Michigan waters of the Great Lakes. *Environ. Sci. Technol.* 35:441–447.

Karchner, S. I., S. W. Kennedy, S. Trudeau, and M. E. Hahn. 2000. Towards molecular understanding of species differences in dioxin sensitivity: Initial characterization of Ah receptor cDNAs in birds and an amphibian. *Mar. Environ. Res.* 50:51–56.

Kelly, B. C., M. G. Ikonomou, J. D. Blair, and F. A. Gobas. 2008. Bioaccumulation behaviour of polybrominated diphenyl ethers (PBDEs) in a Canadian Arctic marine food web. *Sci. Total Environ.* 401:60–72.

Khan, I. A. and P. Thomas. 2006. PCB congener-specific disruption of reproductive neuroendocrine function in Atlantic croaker. *Mar. Environ. Res.* 62:S25–S28.

Kierkegaard, A., L. Balk, U. Tjarnlund, C. de Wit, and B. Jansson. 1999. Dietary uptake and biological effects of decabromo biphenyl ether in rainbow trout (*Oncorhynchus myskiss*). *Environ. Sci. Technol.* 33:1612–1617.

Kitamura, K., T. Sakurai, J-W. Choi, J. Kobayashi, Y. Imaizumi, N. Suzuki, and M. Morita. 2009. Particle-size-fractioned transfer of dioxins from sediments to water columns by resuspension process. *Environ. Pollut.* 157:2159–2165.

Kiviranta, H., T. Vartiainen, R. Parmanne, A. Hallikainen, and J. Koistinen. 2003. PCDD/Fs and PCBs in Baltic herring during the 1990s. *Chemosphere* 50:1201–1216.

Kramer, C. Y. 1956. Extension of multiple range tests to group means with unequal numbers of replications. *Biometrics* 12:307–310.

Kruskal, W. H. 1952. A nonparametric test for the several sample problem. *Ann. Math. Stat.* 23:525–540.

Kruskal, W. H., and W. A. Wallis. 1952. Use of ranks in one-criterion variance analysis. *J. Am. Stat. Assoc.* 47:583–621.

La Guardia, M. J., R. C. Hale, and E. Harvey. 2006. Detailed polybrominated diphenyl ether (PBDE) congener composition of the widely used penta-, octa-, and deca-PBDE technical flame-retardant mixtures. *Environ. Sci. Technol.* 40:6247–6254.

Law, K., et al. 2006. Bioaccumulation and trophic transfer of some brominated flame retardants in a Lake Winnipeg (Canada) food web. *Environ. Toxicol. Chem.* 25:2177–2186.

Legler, J., and A. Brouwer. 2003. Are brominated flame retardants endocrine disruptors? *Environ. Int.* 29:879–885.

Leney, J. L., K. G. Drouillard, and G. D. Haffner. 2006. Metamorphosis increases biotransformation of poly-chlorinated biphenyls: a comparative study of polychlorinated biphenyl metabolism in green frogs (*Rana clamitans*) and leopard frogs (*Rana pipiens*) at various life stages. *Environ. Toxicol. Chem.* 25:2971–2980.

Lerner, D. T., B. T. Bjornsson, and S. D. McCormick. 2007. Effects of aqueous exposure to polychlorinated biphenyls (Aroclor 1254) on physiology and behavior of smolt development of Atlantic salmon. *Aquat. Toxicol.* 81:329–336.

Levene, H. 1960. Robust tests for the equality of variance. In *Contributions to Probability and Statistics,* ed. I. Olkin, pp. 278–292. Palo Alto, CA: Stanford University Press.

Loganathan, B. G., et al. 1995. *Environ. Sci. Technol.* 29:1832–1838.

Lohmann, R., J. K. MacFarlane, and P. M. Gshwend. 2005. Importance of black carbon to sorption of native PAHs, PCBs, and PCDDs in Boston and New York harbor sediments. *Environ. Sci. Technol.* 39: 141–148.

Loonen, H., C. van de Guchte, J. R. Parsons, P. De Voogt, and H. A. Govers. 1996. Ecological hazard assess-ment of dioxins: hazards to organisms at different levels of aquatic food webs (fish-eating birds and mammals, fish and invertebrates). *Sci. Total Environ.* 182:93–103.

Luross, J. M., et al. 2002. Spatial and temporal distribution of polybrominated diphenyl ethers and polychlori-nated biphenyls in lake trout from the Laurentian Great Lakes. *Chemosphere.* 46:665–673.

Luthe, G., J. A. Jacobus, and L. W. Robertson. 2008. Receptor interactions by polybrominated diphenyl ethers versus polychlorinated biphenyls: A theoretical structure-activity assessment. *Environ. Toxicol. Pharmacol.* 25:202–210.

MacGregor, J. A., and W. B. Nixon. 1997. Pentabromodiphenyl oxide (PeBDPO): Determination of n-octanol/water partition coefficient. Wildlife International, Ltd. Project No. 439C-108.

Manchester-Neesvig, J. B., K. Valters, and W. C. Sonzogni. 2001. Comparison of polybrominated diphe-nyl ethers (PBDEs) and polychlorinated biphenyls (PCBs) in Lake Michigan salmonids. *Environ. Sci. Technol.* 35:1072–1077.

Mandal, P. K. 2005. Dioxin: a review of its environmental effects and its aryl hydrocarbon receptor biology. *J. Comp. Physiol.* 175:221–230.

Mann, H. B., and D. R. Whitney. 1947. On a test of whether one of two random variables is stochastically larger than the other. *Ann. Math. Stat.* 18:50–60.

Matthews, V., O. Päpke, and C. Gaus. 2008. PCDD/Fs and PCBs in seafood species from Moreton Bay, Queensland, Australia. *Mar. Pollut. Bull.* 57:392–402.

Mayer, F. L., P. M. Mehrle, and H. O. Sanders. 1977. Residue dynamics and biological effects of polychlori-nated biphenyls in aquatic organisms. *Arch. Environ. Contam. Toxicol.* 5:501–511.

McDonald, T. A. 2002. A perspective on the potential health risks of PBDEs. *Chemosphere* 46:745–755.

Mehrle, W., Hampp, R., Zimmermann, U., and Schwan, H. P. 1988. Mapping of the field distribution around dielectrophoretically aligned cells by means of small particles as field probes. *Biochimica et Biophysica Acta (BBA)-Biomembranes* 939(3):561–568.

Micheletti, C., T. Lovato, A. Critto, R. Pastres, and A. Marcomini. 2008. Spatially distributed ecological risk for fish of a coastal food web exposed to dioxins. *Environ. Toxicol. Chem.* 27:1217–1225.

Mizukawa, K., H. Takada, I. Takeuchi, T. Ikemoto, K. Omori, and K. Tsuchiya. 2009. Bioconcentration and biomagnification of polybrominated diphenyl ethers (PBDEs) through lower-trophic-level coastal marine food web. *Mar. Pollut. Bull.* 58:1217–1224.

Muirhead, E. K., A. D. Skillman, S. E. Hook, and I. R. Schultz. 2006. Oral exposure of PBDE-47 in fish: toxicokinetics and reproductive effects in Japanese Medaka (*Oryzias latipes*) and fathead minnows (*Pimephales promelas*). *Environ. Sci. Toxicol.* 40:523–528.

Nebert, D. W., Puga, A., and Vasiliou, V. 1993. Role of the Ah receptor and the dioxin-inducible [Ah] gene bat-tery in toxicity, cancer, and signal transduction. *Ann. N Y Acad. Sci.* June 23;685:624–640.

Nebert, D. W., A. L. Roe, M. Z. Dieter, W. A. Solis, Y. Yang, and T. P. Dalton. 2000. Role of the aromatic hydro-carbon receptor and [Ah] gene battery in the oxidative stress response, cell cycle control, and apoptosis. *Biochem. Pharmacol.* 59:65–85.

Niimi, A. J. 1996. PCBs in aquatic organisms. In: *Environmental Contaminants in Wildlife. Interpreting Tissue Concentrations*, eds. W. N. Beyer, G. H. Heinz, and A. W. Redmon-Norwood, pp. 117–152. Boca Raton, FL: Lewis Publishers.

Office of Environmental Health Hazard Assessment (OEHHA). 2000. Technical Support Document for Exposure Assessment and Stochastic Analysis. September. Appendix H. Fish Bioconcentration Factors. Department of Toxic Substances Control, California Environmental Protection Agency, Sacramento.

Ohta, S., et al. 2002. Comparison of polybrominated diphenyl ethers in fish, vegetables, and meats and levels in human milk of nursing women in Japan. *Chemosphere* 46:689–696.

Olsen, A. R., B. D. Snyder, L. L Stahl, and J. L. Pitt. 2009. Study design for lakes and reservoirs in the United States to assess contaminants in fish tissue. *Environ. Monit. Assess.* 150:91–100.

Ontario Ministry of the Environment (MOE). 2009. Sport Fish Contaminant Monitoring Program, Environmental Monitoring and Reporting Branch, Toronto, Ontario, Canada. http://www.ene.gov.on.ca/en/water/fish-guide/index.php (accessed August 11, 2009).

Opperheizun, A. 1986. Bioconcentration of hydrophobic chemicals in fish. In *Aquatic Toxicology and Environmental Fate. Ninth Volume, ASTM STP 921,* eds. T. M. Poston and R. Purdy, pp. 304–315. Philadelphia, PA: American Society for Testing and Materials.

Oros, D. R., D. Hoover, F. Rodigari, D. Crane, and J. Sericano. 2005. Levels and distribution of polybrominated diphenyl ethers in water, surface sediments, and bivalves from the San Francisco Estuary. *Environ. Sci. Technol.* 39:33–41.

Palm, A., I. T. Cousins, D. Mackay, M. Tysklind, C. Metcalfe, and M. Alaee. 2002. Assessing the environmental fate of chemicals of emerging concern: a case study of the polybrominated diphenyl ethers. *Environ. Pollut.* 117:195–213.

Pandelova, M., B. Henkelmann, O. Roots, M. Simm, L. Järv, E. Benfenati, and K. W. Schramm. 2008. Levels of PCDD/F and dioxin-like PCB in Baltic fish of different age and gender. *Chemosphere* 71:369–378.

Pena-Abaurrea, M., et al. 2009. Anthropogenic and naturally-produced organobrominated compounds in bluefin tuna from the Mediterranean Sea. *Chemosphere* 76:1477–1482.

Poland, A., and J. C. Knutson. 1982. 2,3,7,8-tetrachlorodibenzo-*p*-dioxin and related halogenated aromatic hydrocarbons: examination of the mechanism of toxicity. *Annu. Rev. Pharmacol. Toxicol.* 22:517–554.

Portmann, J. E., and K. W. Wilson. 1971. *The Toxicity of 140 Substances to the Brown Shrimp and Other Marine Animals.* Ministry Agriculture Fisheries & Food Laboratories, England. Shellfish Information Leaflet No 22.

Quakenbush, L. T. 2007. Polybrominated diphenyl ether compounds in ringed, bearded, spotted, and ribbon seals from the Alaskan Bering Sea. *Mar. Pollut. Bull.* 54:232–236.

Raff, J. D., and R. A. Hites. 2007. Deposition versus photochemical removal of PBDEs from Lake Superior Air. *Environ. Sci. Technol.* 41:6725–6731.

Rahman, F., K. H. Langford, M. D. Srimshaw, and J. N. Lester. 2001. Polybrominated diphenyl ether (PBDE) flame retardants. *Sci. Total Environ.* 275:1–17.

Rankouhi, T. R., J. T. Sanderson, I. van Holsteijn, C. Van Leeuwen, A. D. Vethaak, and M. Van den Berg. 2004. Effects of natural and synthetic estrogens and various environmental contaminants on vitellogenesis in fish primary hepatocytes: comparison of bream (*Abramis brama*) and carp (*Cyprinus carpio*). *Toxicol. Sci.* 81:90–102.

Rayne, S., M. G. Ikonomou, P. S. Ross, G. M. Ellis, and L. G. Barrett-Lennard. 2004. PBDEs, PBBs, and PCNs in three communities of free-ranging killer whales (*Orcinus orca*) from the northeastern Pacific Ocean. *Environ. Sci. Technol.* 38:4293–4299.

Reyes, G. G., J. M. Verdugo, D. Cassin, and R. Carvajal. 2003. Pollution by polychlorinated biphenyls in an estuary of the Gulf of California. Their toxicity and bioaccumulation in shrimp Litopenaeus vannamei. *Mar. Pollut. Bull.* 46:959–963.

Rice, C. P., S. M. Chernyak, L. Begnoche, R. Quintal, and J. Hickey. 2002. Comparison of PBDE composition and concentration in fish collected from the Detroit River, MI and Des Plaines River, IL. *Chemosphere* 49:731–737.

Ritter, L., K. R. Solomon, and J. Forget. 1995. An Assessment Report on DDT, Aldrin, Dieldrin, Endrin, Chlordane, Heptachlor, Hexachlorobenzene, Mirex, Toxaphene, Polychlorinated Biphenyls, Dioxins, and Furans for the International Program on Chemical Safety, Stockholm Convention on Persistent Organic Pollutants. Submitted to the Second Meeting of the ISG, Intergovernmental Forum on Chemical Safety, ISG/96.5B.

Ross, P. S., C. M. Couillard, M. G. Ikonomou, S. C. Johannessen, M. Lebeuf, R. W. Macdonald, and G. T. Tomy. 2009. Large and growing environmental reservoirs of Deca-BDE present an emerging health risk for fish and marine mammals. *Mar. Pollut. Bull.* 58:7–10.

Safe, S. H. 1994. Polychlorinated biphenyls (PCBs): Environmental impact, biochemical and toxic responses, and implications for risk assessment. *Crit. Rev. Toxicol.* 24:87–149.

Saloranta, T. M., T. Andersen, and K. Naes. 2006a. Flows of dioxins and furans in coastal food webs: inverse modeling, sensitivity analysis, and applications of linear system theory. *Environ. Toxicol. Chem.* 25:253–264.

Saloranta, T., J. Armitage, K. Næs, I. Cousins, and D. Barton. 2006b. SEDFLEX-Tool Multimedia Model Package: Model Code Description and Application Examples from the Grenland Fjords. NIVA-report 5216; Norwegian Institute for Water Research: Oslo, Norway, 2006.

Sangalang, G. B., H. C. Freeman, and R. Crowell. 1981. Testicular abnormalities in cod (*Gadus morhua*) fed Aroclor 1254. *Arch. Environ. Contam. Toxicol.* 10:617–626.

Sellstrom, U., B. Jansson, A. Kierkegaard, C. de Wit, T. Odsjö, and M. Olsson. 1993. Polybrominated diphenyl ethers (PBDE) in biological samples form the Swedish environment. *Chemosphere* 26:1703–1718.

Sericano, J. L., T. Wade, G. Denoux, Y. Qian, S. Sweet, and G. Wolff. 2003. NOAA's status and trends "MusselWatch" Program: PBDEs in bivalves from U.S. coastal areas. Abstract from the SETAC 24th Annual Meeting in North America. Pensacola: Society of Environmental Toxicology and Chemistry.

Shapiro, S. S., and M. B. Wilk. 1965. EDF statistics for goodness of fit and some comparisons. *J. Am. Stat. Assoc.* 69:730–737.

She, J., A. Holden, M. Tanner, M. Sharp, T. Adelsbach, and K. Hooper. 2004. Highest PBDE levels (max 63 ppm) yet found in biota measured in seabird eggs from San Francisco Bay. *Organohalogen Compounds.* 66:3939–3944.

Simon, T., J. K. Britt, and R. C. James. 2007. Development of a neurotoxic equivalence scheme of relative potency for assessing the risk of PCB mixtures. *Regul. Toxicol. Pharmacol.* 48:148–170.

Soderstrom, G., U. Sellstrom, C. A. de Wit, and M. Tysklind. 2004. Photolytic debromination of decabromodiphenyl ether (BDE 209). *Environ. Sci. Technol.* 38:127–132.

Stahl, L. L., B. D. Snyder, A. R. Olsen, and J. L. Pitt. 2009. Contaminants in fish tissue from US lakes and reservoirs: a national probabilistic study. *Environ. Monit. Assess.* 150:3–19.

Stapleton, H.M., R.J. Letcher, and J.E. Baker. 2004a. Debromination of polybrominated diphenyl ether congeners BDE 99 and BDE 183 in the intestinal tract of the common carp (*Cyprinus carpio*). *Environ. Sci. Technol.* 38:1054–1061.

Stapleton, H.M., R.J. Letcher, and J.E. Baker. 2004b. Debromination of the flame retardant decabromodiphenyl ether by juvenile carp (*Cyprinus carpio*). *Environ. Sci. Technol.* 38:112–119.

Stapleton, H.M., et al. 2006. *In vivo* and *in vitro* debromination of decabromodiphenyl ether (BDE 209) by juvenile rainbow trout and common carp. *Environ. Sci. Technol.* 40:4653–4658.

Stark, J. D., J. E. Banks, and R. Vargas. 2004. How risky is risk assessment: The role that life history strategies play in susceptibility of species to stress. *Proc. Nat. Acad. Sci.* 101:732–736.

Stenzel, J. I., and B. J. Markley. 1997. Pentabromodiphenyl Oxide: Determination of the Water Solubility. Wildlife International, Ltd. Project No. 439C-109.

Stenzel, J. I., and W. B Nixon. 1997. Octabromobiphenyl Oxide (OBDO): Determination of the Vapour Pressure using a Spinning Rotor gauge. Wildlife International Ltd. Project No. 439C-114.

Steevens, J. A., Reiss, M. R., and Pawlisz, A. V. 2005. A methodology for Deriving Tissue Residue Benchmarks for Aquatic Biota: A case study for fish exposed to 2,3,7,8-tetrachlorodibenzo-*p*-dioxin and equivalents. *Integr. Environ. Assess. Manag.* 1(2):142–151.

Szlinder-Richert, J., I. Barska, Z. Usydus, W. Ruczyńska, and R. Grabic. 2009. Investigation of PCDD/Fs and dl-PCBs in fish from the southern Baltic Sea during the 2002–2006 period. *Chemosphere* 74:1509–1515.

Talsness, C. E. 2008. Overview of toxicological aspects of polybrominated diphenyl ethers: A flame-retardant additive in several consumer products. *Environ. Res.* 108:158–167.

Timme-Laragy, A. R., E. D. Levin, and R. T. Di Giulio. 2006. Developmental and behavioral effects of embryonic exposure to the polybrominated diphenyl ether mixture DE-71 in the killifish (*Fundulus heteroclitus*). *Chemosphere* 62:1097–1104.

Tittelmier, S. A., and G. T. Tomy. 2000. Vapor pressures of six brominated diphenyl ether congeners. *Organohalogen Compounds.* 47:206–209.

Tittlemier, S. A., T. Halldorson, G. A. Stern, and G. T. Tomy. 2002. Vapor pressures, aqueous solubilities and Henry's law constants of some brominated flame retardants. *Environ. Toxicol. Chem.* 21:1804–1810.

Tomy, G. T., et al. 2004. Bioaccumulation, biotransformation, and biochemical effects of brominated diphenyl ethers in juvenile lake trout (*Salvenlinus namaycush*). *Environ. Sci. Technol.* 38:1496–1504.

U.S. EPA (United States Environmental Protection Agency). 1992. National Study of Chemical Residues in Fish. Volume 1. Office of Science and Technology, Washington, DC. EPA-823-R-92-008a. http://yosemite.epa.gov/water/owrcCatalog.nsf/9da204a4b4406ef885256ae0007a79c7/86f0f7d93d78a80185256d83004fd834

U.S. EPA (United States Environmental Protection Agency). 2001. Critical Review and Assessment of Published Research on Dioxins and Related Compounds in Avian Wildlife—Field Studies. External Review Draft. National Center for Environmental Assessment, Office of Research and Development, Cincinnati, OH.

U.S. EPA (United States Environmental Protection Agency). 2002. Dose-Response Assessment from Published Research of the Toxicity of 2,3,7,8-Tetrachlorodibenzo-*p*-dioxin and Related Compounds to Aquatic Wildlife—Laboratory Studies. National Center for Environmental Assessment, Office of Research and Development, Cincinnati, OH. EPA/600/R-02/095.

U.S. EPA (United States Environmental Protection Agency). 2005. The Inventory of Sources and Environmental Releases of Dioxin-Like Compounds in the United States: The Year 2000 Update. External Review Draft. National Center for Environmental Assessment, Washington, DC. EPA/600/p-03/002A.

U.S. EPA (United States Environmental Protection Agency). 2006a. An inventory of sources and environmental releases of dioxin-like compounds in the United States for the years 1987, 1995, and 2000. National Center for Environmental Assessment, Washington, DC. EPA/600/P-03/002F.

U.S. EPA (United States Environmental Protection Agency). 2006b. Polybrominated Diphenyl Ethers (PBDEs) Project Plan. http://www.epa.gov/ong/kg/pbde (accessed August 11, 2009).

U.S. EPA (United States Environmental Protection Agency). 2008. Framework for Application of the Toxicity Equivalence Methodology for Polychlorinated Dioxins, Furans, and Biphenyls in Ecological Risk Assessment. Office of the Science Advisor Risk Assessment Forum. EPA/100/R-08/004.

U.S. EPA (United States Environmental Protection Agency). 2009a. The ECOTOX (ECOTOXicology) database, Release 4.0. http://cfpub.epa.gov/ecotox/ (accessed August 11, 2009).

U.S. EPA (United States Environmental Protection Agency). 2009b. National Lake Fish Tissue Study. http://www.epa.gov/waterscience/fish/study/index.htm (accessed August 11, 2009).

U.S. EPA. 2009c. 40 CFR: Protection of the Environment. http://www.epa.gov/lawsregs/search/40cfr.html (accessed December 12, 2009).

U.S. EPA/U.S. ACE (United States Environmental Protection Agency/United States Army Corps of Engineers). Environmental Residue-Effects Database (ERED). 2009. http://el.erdc.usace.army.mil/ered/ (accessed August 2009).

Van den Berg, M., et al. 1998. Toxic Equivalency Factors (TEFs) for PCBs, PCDDs, PCDFs for Humans and Wildlife. *Environ. Health Perspect.* 106:775–792.

Van den Berg, M., et al. 2006. The 2005 World Health Organization re-evaluation of human and mammalian toxic equivalency factors for dioxins and dioxin-like compounds. *Toxicol. Sci.* 93:223–241.

Vethaak, A. D., J. Lahr, R. V. Kuiper, G. C. M. Grinwis, T. R. Rankouhi, J. P. Giesy, and A. Gerritsen. 2002. Estrogenic effects in fish in The Netherlands: Some preliminary results. *Toxicol.* 181–182:147–150.

Voorspoels, S., A. Covaci, and P. Schepens. 2003. Polybrominated diphenyl ethers in marine species from the Belgian North Sea and the Western Scheldt Estuary: Levels, profiles, and distribution. *Environ. Sci. Technol.* 37:4348–4357.

Walker, M. K., and Peterson, R. E. 1991. Potencies of polychlorinated dibenzo-*p*-dioxin, dibenzofuran, and biphenyl congeners, relative to 2,3,7,8-tetrachlorodibenzo-*p*-dioxin, for producing early life stage mortality in rainbow trout (*Oncorhynchus mykiss*). *Aqua Toxicol* 21(3–4):219–237.

Wan, Y., et al. 2005. Characterization of trophic transfer for polychlorinated dibenzo-*p*-dioxins, dibenzofurans, non- and mono-ortho polychlorinated biphenyls in the marine food web of Bohai Bay, North China. *Environ. Sci. Technol.* 39:2417–2425.

Wang, Y., G. Jiang, P. K. S. Lam, and A. Li. 2007. Polybrominated diphenyl ether in the East Asian environment: A critical review. *Environ. Int.* 33:963–973.

Wania, F., and C. B. Dugani. 2003. Assessing the long-range transport potential of polybrominated diphenyl ethers: a comparison of four multimedia models. *Environ. Toxicol. Chem.* 22:1252–1261.

Watanabe, I., and R. Tatsukawa. 1987. Formation of brominated dibenzofurans from the photolysis of flame retardant decabromobiphenyl ether in hexane solution by UV and sunlight. *Bull. Environ. Contam. Toxicol.* 39:953–959.

Watanabe, I., and R. Tatsukawa. 1990. Anthropogenic brominated aromatics in the Japanese environment. In: *Proceedings of Workshop on Brominated Aromatic Flame Retardants,* Skokloster, Sweden, October 24–26, 1989. Swedish National Chemicals Inspectorate, KEMI, Solna, Sweden, pp. 63–71.

Wenning, R. J. 2002. Uncertainties and data needs in risk assessment of the three commercial polybrominated diphenyl ethers: Probabilistic exposure analysis and comparison with European Commission results. *Chemosphere* 46:779–796.

West, C. W., G. T. Ankley, J. W. Nichols, G. E. Elonen, and D. E. Nessa. 1997. Toxicity and bioaccumulation of 2,3,7,8-tetrachlorodibenzo-*p*-dioxin in long-term tests with the freshwater benthic invertebrates *Chironomus tentans* and *Lumbriculus variegates*. *Environ. Toxicol. Chem.* 16:1287–1294.

WHO/IPCS (World Health Organization/ International Programme on Chemical Safety). 1993. Polychlorinated biphenyls and terphenyls, second edition (environmental health criteria 140). World Health Organization, Geneva.

Wintermyer, M. L., and Cooper, K. R. 2003. Dioxin/Furan and polychlorinated biphenyl concentrations in oyster (Crassostera virginica, gmelin) tissues and the effects on fertilization and development. *J. Shellfish Res*. 22(3)737–746.

Wong, A., Y. D. Lei, M. Alaee, and F. Wania. 2001. Vapor pressures of the polybrominated diphenyl ethers. *J. Chem. Eng. Data* 46:239–242.

Wu, J. P., et al. 2009. Biomagnification of polybrominated diphenyl ethers (PBDEs) and polychlorinated biphenyls in a highly contaminated freshwater food web from South China. *Environ. Pollut*. 157:904–909.

Wu, W. Z., W. Li, Y. Xu, and J. W. Wang. 2001. Long-term toxic impact of 2,3,7,8-tetrachlorodibenzo-*p*-dioxin on the reproduction, sexual differentiation, and development of different life stages of *Gobiocypris rarus* and *Daphnia magna*. *Ecotoxicol. Environ. Saf*. 48:293–300.

Yamauchi, M., E. Y. Kim, H. Iwata, Y. Shima, and S. Tanabe. 2006. Toxic effects of 2,3,7,8-tetrachlorodibenzo-*p*-dioxin (TCDD) in developing red seabream (*Pagrus major*) embryo: an association of morphological deformities with AHR1, AHR2 and CYP1A expressions. *Aquat. Toxicol*. 80:166–179.

Yogui, G. T., and J. L. Sericano. 2009. Polybrominated diphenyl ether flame retardants in the U.S. marine environment: A review. *Environ. Internat*. 35:655–666.

Zabel, E. W., and R. E. Peterson. 1996. TCDD-like activity of 2,3,6,7-tetrachloroxanthene in rainbow trout early life stages and in a rainbow trout gonadal cell line (RTG2). *Environ. Toxicol. Chem*. 15:2305–2309.

Zabel, E. W., M. K. Walker, M. W. Hornung, M. K. Clayton, and R. E. Peterson. 1995. Interactions of polychlorinated dibenzo-*p*-dioxin, dibenzofuran, and biphenyl congeners for producing rainbow trout early life stage mortality. *Toxicol. Appl. Pharmacol*. 134:204–213.

Zeng, X., P. K. Freeman, Y. V. Vasil'ev, V. G. Voinov, S. L. Simonich, and D. F. Barofsky. 2005. Theoretical calculation of thermodynamic properties of polybrominated diphenyl ethers. *J. Chem. Eng. Data*. 50:1548–1556.

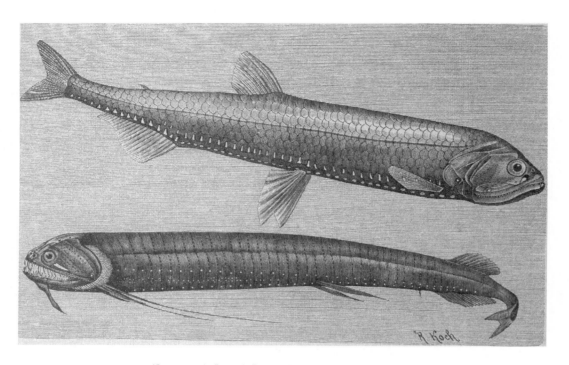

Silvery Light-Fish and Barbed Hedgehog

By R. Koch, from *The Royal Natural History,* edited by Richard Lydekker, Frederick
Warne & Co., London, 1893–94.

4 Methylmercury in Freshwater Fish

Recent Advances in Assessing Toxicity of Environmentally Relevant Exposures

Mark B. Sandheinrich
James G. Wiener

CONTENTS

4.1 Introduction 170
4.2 Methylmercury Bioaccumulation in Freshwater Fish 171
4.3 Toxicological Effects of Methylmercury Exposure on Freshwater Fish 172
    4.3.1 Effects on Survival and Growth 172
    4.3.2 Effects on Behavior 178
    4.3.3 Effects on Reproduction 178
    4.3.4 Effects on Tissue Histology and Biochemistry 180
4.4 Concentrations in Tissues Associated with Toxic Effects 182
4.5 Concentrations of Mercury in Freshwater Fish 184
Summary 185
Acknowledgments 186
References 186

Fish became contaminated with methylmercury to levels dangerous to consumers. There is little evidence to show what levels of mercury contamination in the fish may be lethal to the creatures themselves.

F. A. J. Armstrong (1979)

...scant progress has been made in defining fish-tissue residues of mercury associated with toxic effects and in assessing the toxicological significance of methylmercury at environmentally relevant exposure levels. Armstrong's statement...remains applicable 15 years later.

J. G. Wiener and D. J. Spry (1996)

4.1 INTRODUCTION

Scientific understanding of the adverse effects of methylmercury exposure on freshwater fish has advanced substantially since the first-edition chapter on this topic was prepared by Wiener and Spry (1996). Sublethal and reproductive effects of methylmercury in freshwater fish have been observed at environmentally relevant exposure levels in a number of recent field and laboratory studies. Moreover, the fish-tissue concentrations associated with adverse effects in many of these studies are much lower than those summarized by Wiener and Spry (1996).

Methylmercury contamination of fish remains a geographically widespread problem that has diminished the recreational, economic, and nutritional benefits derived from fishery resources in many fresh waters. In 2008, more than 68,000 km^2 of lake area and 2,019,000 km of rivers were under fish-consumption advisory for mercury in the United States, with mercury accounting for 80% of the advisories in the nation (U.S. EPA 2009). In Canada, 97% of fish-consumption advisories are attributed to mercury (U.S. EPA 2001). Atmospheric transport and deposition of mercury from anthropogenic sources has contaminated landscapes, surface waters, and aquatic food webs supporting freshwater fish production on a global scale. Analyses of sediment cores from remote lakes in both hemispheres, for example, show that net mercury deposition has increased about three-fold since preindustrial times because of anthropogenic emissions (Bindler et al. 2001, Lamborg et al. 2002, Lindberg et al. 2007).

Many fresh waters contain fish with high mercury concentrations. Early work on mercury-contaminated fish focused on waters polluted by industrial and mining sources. At such sites, mercury concentrations in fish have declined but commonly remain elevated for decades after peak pollution and cessation of operations (Turner and Southworth 1999, Wiener et al. 2003, Kinghorn et al. 2007). Mercury levels in fish also are elevated in aquatic systems that have enhanced entry of methylmercury into the base of the food web, as a result of significant influxes of methylmercury from external sources, high *in situ* rates of methylmercury production in bed sediments or anoxic hypolimnia, or a combination of these and other biogeochemical, biological, trophic, and human factors (Wiener et al. 2003, Munthe et al. 2007, Chasar et al. 2009). In newly flooded reservoirs, for example, concentrations of methylmercury in water, zooplankton, and fish increase rapidly after flooding of vegetated wetland or upland terrestrial habitats, in response to greatly increased rates of microbial methylation of inorganic Hg(II) present in the inundated landscapes (Kelly et al. 1997, Paterson et al. 1998, Bodaly and Fudge 1999, Bodaly et al. 2004, Hall and St. Louis 2004). Concentrations of methylmercury in piscivorous fish can increase as much as 10-fold in new reservoirs relative to preflood or reference values, and remain elevated for 20–30 years after flooding (Bodaly et al. 2007). Mercury concentrations in fish also increase in response to fluctuating water levels (Sorensen et al. 2005). Wetlands are sites of active mercury methylation, and water and biota in wetland-influenced lakes and streams can contain elevated concentration of methylmercury (Hurley et al. 1995, Branfireun et al. 2005, Wiener et al. 2006, Brigham et al. 2009, Chasar et al. 2009).

We summarize recent findings on the ecotoxicology of methylmercury in freshwater fish, with emphasis on assessing the toxicological significance of mercury concentrations in fish tissues. Our primary objectives are (1) to briefly summarize the state of scientific knowledge regarding bioaccumulation of mercury in fish; (2) to review recent advances in the state of scientific knowledge regarding toxicological effects of methylmercury in freshwater fish; and (3) to identify mercury concentrations in tissue associated with toxicological effects on fish. We do not revisit earlier studies of mercury-contaminated fish at grossly polluted sites, such as Minamata Bay (Japan) and Clay Lake (Ontario), which were reviewed in the first-edition chapter by Wiener and Spry (1996).

4.2 METHYLMERCURY BIOACCUMULATION IN FRESHWATER FISH

We begin with a brief review of mercury bioaccumulation in fish, as a prelude to the following sections on toxic effects. This summary is based on a number of reviews (Wiener and Spry 1996, Wiener et al. 2003, Crump and Trudeau 2009), as well as the original reports cited.

Nearly all of the mercury in fish is methylmercury (Grieb et al. 1990, Bloom 1992, Hammerschmidt et al. 1999), a highly toxic compound that readily crosses biological membranes, accumulates in exposed organisms, and can biomagnify to high concentrations in fish and wildlife atop aquatic food webs (Wiener et al. 2003, Scheuhammer et al. 2007). Methylmercury accumulates in fish to concentrations that exceed those in surface waters as much as 10^6 to 10^7 fold (Wiener et al. 2003). Most of the mercury in air, atmospheric deposition, watersheds, and surface waters exists as inorganic forms. However, inorganic mercury—in marked contrast to methylmercury—is not readily transferred through successive trophic levels and does not biomagnify (Watras et al. 1998, Pickhardt et al. 2002). Fish accumulate very little inorganic mercury (Oliveira Ribeiro et al. 1999), even in surface waters and ecosystems containing unusually high concentrations of inorganic mercury (Southworth et al. 1995, Kuwabara et al. 2007, Suchanek et al. 2008). Mercury in fish muscle is typically quantified by determination of total mercury, which requires less effort and cost than determination of methylmercury. Unless specifically indicated, the fish-mercury concentrations reported herein were from determination of total mercury.

Processes that affect the mass of methylmercury in aquatic ecosystems or its concentration at the base of the aquatic food web strongly affect its concentration in all trophic levels, including fish (Paterson et al. 1998, Wiener et al. 2003, Munthe et al. 2007, Chasar et al. 2009). Such processes include the production of methylmercury via the microbial methylation of inorganic Hg(II) and the destruction of methylmercury by microbial demethylation and photodemethylation. Concentrations of methylmercury in fish also can be influenced by biodilution of methylmercury at the base of the food web by algal blooms or high algal biomass (Pickhardt et al. 2002, 2005, Chen and Folt 2005).

Wild fish obtain methylmercury mostly from food (≥90% of total uptake) and—to a much lesser extent—from water passed over the gills (Rodgers 1994, Wiener and Spry 1996, Hall et al. 1997, Harris and Bodaly 1998). Most of the methylmercury in ingested prey is assimilated across the gut, with assimilation efficiencies of 65–80% estimated by bioenergetics-based modeling (Rodgers 1994) and efficiencies of about 90% estimated in recent laboratory experiments (Pickhardt et al. 2006). Uptake of methylmercury from water passing over the gills is less efficient, with assimilation of about 10% estimated by Phillips and Buhler (1978) and Rodgers and Beamish (1981). In the laboratory, fish can accumulate high concentrations of methylmercury via direct uptake from water if exposed to waterborne concentrations that greatly exceed those in toxic surface waters, which typically contain less than 1.0 ng Hg/L as methylmercury (Wiener and Spry 1996).

Methylmercury in digested food is rapidly assimilated into the intestine and is transferred to the blood, where most of it enters red blood cells and circulates throughout the body (Oliveira Ribeiro et al. 1999). The passage of methylmercury from the red blood cells to the tissues and organs involves the intermediate transfer from the red blood cells to the plasma, which is probably a rate-limiting step in the systemic transfer of methylmercury because of the low concentration of small mobile sulfhydryl ligands in the plasma (Oliveira Ribeiro et al. 1999).

Concentrations of methylmercury in fish are typically greatest in the blood and the highly perfused spleen, kidney, and liver. Assimilated methylmercury in experimentally exposed fish is redistributed among the tissues and organs within a few weeks; the concentrations and masses in the blood and visceral organs (spleen, kidney, and liver) decrease after experimental exposure to waterborne or dietary methylmercury ceases, and much of the redistributed methylmercury accumulates in skeletal muscle (Oliveira Ribeiro et al. 1999)—bound to cysteine in protein (Harris et al. 2003, Kuwabara et al. 2007). In experimentally exposed fish, the route of uptake (i.e., from food or water) has little effect on the internal distribution of methylmercury among the organs and tissues (Wiener and Spry 1996), except that concentrations in the gills are much greater after waterborne (than dietary) exposure, and concentrations in the intestines are greater after dietary exposure (McKim et al. 1976, Huckabee et al. 1979, Boudou and Ribeyre 1983, Harrison et al. 1990). Fish eliminate methylmercury very slowly (Trudel and Rasmussen 1997, Van Walleghem et al. 2007).

Concentrations of methylmercury in the skeletal muscle of individual fish in a given water body typically increase with increasing size or age (Phillips and Buhler 1978, Huckabee et al. 1979, Lange et al. 1993), partly because of the extremely slow rate of elimination of methylmercury relative to its rapid and efficient dietary uptake. In addition, the methylmercury content of the diet of some fishes, particularly those that are piscivorous as adults, increases as the fish grow larger (MacCrimmon et al. 1983, Mathers and Johansen 1985). The rate of methylmercury accumulation in lake trout (*Salvelinus namaycush*), for example, increases greatly when the fish become large enough to switch from a diet of invertebrates to prey fish (MacCrimmon et al. 1983).

Fish seem to have few defenses against methylmercury, which readily crosses the gills, intestines, and internal cellular membranes and is neither effectively excreted nor bound to metallothioneins (Wiener and Spry 1996). In mammals, demethylation in the liver is a key step in the elimination of mercury accumulated as methylmercury. Wiener and Spry (1996, p. 312) concluded that there was no compelling evidence of *in vivo* demethylation of methylmercury by freshwater fish. However, Drevnick et al. (2008), who analyzed northern pike (*Esox lucius*) from lakes in Isle Royale National Park (a remote archipelago in Lake Superior), found that methylmercury constituted most of the mercury in livers with total mercury less than 0.5 µg/g wet weight, but only 28–51% of the total mercury in livers with total concentrations exceeding 0.5 µg/g. Drevnick et al. (2008) hypothesized that the low percent methylmercury observed in livers of northern pike from Isle Royale was due to the accumulation of inorganic mercury, rather than to hepatic demethylation of methylmercury. Yet Evans et al. (2000), who reported similar patterns in the methylmercury fraction in livers and kidneys of wild river otter (*Lutra canadensis*) from south-central Ontario, interpreted variations in percent methylmercury among individual otters as evidence of demethylation. We believe that the question of demethylation of methylmercury in fish remains unresolved and merits focused research.

4.3 TOXICOLOGICAL EFFECTS OF METHYLMERCURY EXPOSURE ON FRESHWATER FISH

4.3.1 Effects on Survival and Growth

Survival, growth, and reproduction are endpoints commonly measured in toxicity tests with fish (Organization for Economic Cooperation and Development 1992a, 1992b, U.S. EPA 2002a, 2002b) and frequently used to estimate threshold concentrations of contaminants for regulatory purposes. Direct mortality due to methylmercury is observed only at unusually high tissue concentrations (6–20 µg/g wet weight in muscle) in cases involving extreme, gross pollution (Wiener and Spry 1996). The effects of methylmercury on growth of fish vary between laboratory experiments in which fish were fed contaminated diets (Table 4.1) and field studies that assessed the relation between fish condition and mercury concentrations in the tissues (Table 4.2). Hammerschmidt et al. (2002), for example, reported that fathead minnows (*Pimephales promelas*) fed a methylmercury-contaminated diet were larger than control fish. Wet weights of male and female minnows were weakly, positively correlated with mercury burdens in the fish, and females with 5.6 µg/g wet weight in the carcass were about 30% larger than control fish containing 0.12 µg Hg/g (wet concentrations estimated from dry-weight values in Table 2 of Hammerschmidt et al. 2002). In contrast, juvenile male walleye (*Sander vitreus*) with a mean concentration of 2.4 µg Hg/g wet weight in the carcass (minus viscera), after 6 months on a methylmercury-contaminated diet containing 1.0 µg Hg/g wet weight, had slower growth than male controls with 0.25 µg Hg/g (Friedmann et al. 1996). Houck and Cech (2004), however, found no alteration in bioenergetics or growth of juvenile Sacramento

TABLE 4.1
Laboratory Studies Reporting Total Mercury Concentrations in Tissues of Fish Exhibiting Symptoms of Methylmercury Toxicity

Species	Exposure Concentration and Duration (Days)		Life Stage, Age or Sex of Fish	Total Hg Concentration (µg/g ww)					Toxic Effect(s)[a]	Reference
	Water (µg/L)	Diet (µg/g Dry Weight)		Whole Body	Muscle	Brain	Liver	Other		
Medaka *Oryzias latipes*	2.5 (8)		10-month old		0.03	0.18	3.63	0.51 (gonad)	↓ Sum of cholinesterase activity in liver, brain, gill, and muscle	Liao et al. (2006)
Walleye		0.1[c] (180)	Age 1 ♂	0.25 ± 0.02 (minus viscera)					↓ Plasma cortisol (indicative of impaired immune function) ↑Gonadal atrophy	Friedmann et al. (1996)
Grayling	0.8 (10)		Embryo	0.27					↓ Feeding efficiency	Fjeld et al. (1998)
Golden shiner		0.96 (90)	Adult	0.54 ± 0.15		1.12 ± 0.20			↓ Shoaling behavior	Webber and Haines (2003)
Fathead minnow		0.88 (195)	Adult	♂ 0.71 (0.53–0.97)[d] ♀ 0.85 (0.56–1.14)[d] (minus gonads)					↓ Number of fish spawning ↑ Time required to first spawn ↓ Egg production ↑ Instantaneous rate of reproduction ↓ Female gonadosomatic index	Hammerschmidt et al. (2002)
Fathead minnow		0.87 (>250)	Adult ♂	0.71 ± 0.044					↓ Reproductive behavior	Sandheinrich and Miller (2006)

continued

TABLE 4.1 (continued)
Laboratory Studies Reporting Total Mercury Concentrations in Tissues of Fish Exhibiting Symptoms of Methylmercury Toxicity

Species	Exposure Concentration and Duration (Days)		Life Stage, Age or Sex of Fish	Total Hg Concentration (µg/g ww)					Toxic Effect(s)[a]	Reference
	Water (µg/L)	Diet (µg/g Dry Weight)		Whole Body	Muscle	Brain	Liver	Other		
Fathead minnow		0.87 (250)	Adult	♂ 0.86 ± 0.04 ♀ 0.92 ± 0.03 (minus gonads)					↓ Number of spawning fish ↓ Sex hormones (T, E2) ↑ Time to first spawn ↓ Female gonadal somatic index ↑ Ovarian follicular apoptosis	Drevnick and Sandheinrich (2003); Drevnick et al. (2006)
Wolffish		0.075 µg/g body wet weight (14 doses over 70 d)	Adult		1.45		1.069		↑ Melano-macrophage centers and other histological changes in liver and kidney	Mela et al. (2007)
Atlantic salmon		5 (160)	Parr			1.16 ± 0.29	4.06 ± 1.11	2.00 ± 0.21 (kidney)	↑ Antioxidant enzyme activity	Berntssen et al. (2003)
Zebra fish		5 (21)	Adult ♂		2[bd]	3.75[bd]	3.0[bd]		↓↑ Gene transcription	Gonzalez et al. (2005)
Atlantic salmon		10 (120)	Parr			0.68 ± 0.62	6.69 ± 1.02	3.05 ± 0.68 (kidney)	↓ Antioxidant enzyme activity ↑ Histological change in brain ↑ Lipid peroxidative products ↓ Neural enzyme activity ↓ Feeding behavior	Berntssen et al. (2003)

Species				Effect	Reference
Walleye	1[c] (180)	Age 1 ♂	2.37 ± 0.09 (minus viscera)	↓ Gonadal somatic index ↑ Gonadal atrophy ↓ Growth	Friedmann et al. (1996)
Zebra fish	13 (25)	Adult ♂	5.08 ± 1.0[d]	↓ Mitochondrial energy metabolism	Cambier et al. (2009)
Zebra fish	13.5 (63)	Adult ♂	8.55 ± 0.36[d]	↑ Histological changes in skeletal muscle	Oliveira Ribeiro et al. (2008)
Sacramento blackfish	22.2 (70)	Juvenile	15[b]	↓ Growth	Houck and Cech (2004)
Medaka	40 (16)	10-month old	24.9 28.7 54.5 38.7 (gonad)	↑ Histological changes in liver, gill, and testes	Liao et al. (2006)

Note: Total mercury concentrations in fish are the mean ± SE or SD (range). Fish were either fed naturally contaminated diets or diets spiked with methylmercury chloride or received aqueous exposure to methylmercury chloride.

[a] An increase is indicated by ↑; a decrease is indicated by ↓.

[b] Concentration estimated from figure in original reference.

[c] Wet weight.

[d] Wet-weight concentration estimated from reported dry-weight concentration and assuming tissue with 75% water content.

TABLE 4.2
Field Studies Reporting Total Mercury Concentrations in Tissues of Fish Exhibiting Symptoms of Methylmercury Toxicity

Species	Location	Life Stage, Age or Sex of Fish	Total Hg Concentration (μg/g ww)			Toxic Effect(s)[a]	Reference
			Whole Body	Muscle	Liver		
Cutthroat trout	Wilcox Lake, Washington, USA	Juvenile	0.055[b]			↓↑ Gene transcription	Moran et al. (2007)
Brook trout	Western U.S. National Parks	1–13 years	(0.05–0.29)[b]			↑ Histological change in spleen and kidney	Schwindt et al. (2008)
White sturgeon	Columbia River, Oregon, USA	Sexually immature ♂		(0.040–0.520)[b]	(0.013–0.780)[b]	↓ Plasma testosterone ↓ Gonadal somatic index	Webb et al. (2006)
White sturgeon	Columbia River, Oregon, USA	Sexually immature ♂ & ♀			(0.020–0.780)[b]	↓ Plasma estradiol ↓ Condition factor and relative weight	Webb et al. (2006)
White sturgeon	Columbia River, Oregon, USA	Sexually immature ♂ & ♀		(0.040–0.520)[b]		↓ Plasma testosterone ↓ 11-Ketotestosterone	Webb et al. (2006)
Yellow perch	Lake Desjardins-East, Quebec, Canada	Not reported			(0.035–0.220)[b]	↓ Glutathione peroxidase selenium-dependent activity and glutathione S-transferase activity	Larose et al. (2008)

Species	Location	Age			Effect	Reference
Northern pike	Isle Royale, Michigan, USA	Adult	(0.069–0.622)	(0.048–3.074)	↑Amount of lipofuscin in liver ↓Condition factor	Drevnick et al. (2008)
Walleye	Upper Columbia River, Washington, USA	Age 2	(0.20–0.375)[b]		↓Condition factor	Munn and Short (1997)
Northern pike	Oder River, Germany	Adult	(0.22–0.54)[b]	(0.05–0.45)[b]	↑Histological change in liver, spleen, and kidney	Meinelt et al. (1997)
Walleye	Lake Malartic, Quebec Canada	1–8 years		(0.100–3.40)[b]	↓Hepatosomatic index	Larose et al. (2008)
Striped bass	Lake Mead, Arizona and Nevada, USA	Various	0.309 (0.063–1.058)	0.531 (0.056–4.450)	↓Condition factor	Cizdziel et al. (2003)
Bleak	Cinca River, Spain	Not reported	(0.321–2.362)		↑Histological change in liver	Raldúa et al. (2007)
Barbel	Cinca River, Spain	<1–8 years	1.48 ± 0.17 (0.57–2.433)	1.78 ± 0.24 (0.747–3.641)	↑Histological change in liver	Raldúa et al. (2007)
Largemouth bass	Atlantic City Reservoir, New Jersey, USA	Adult	5.42 ± 0.56		↑11-Ketotestosterone	Friedmann et al. (2002)

Note: Total mercury concentrations are the mean (range).

[a] An increase is indicated by ↑; a decrease is indicated by ↓.

[b] Concentration estimated from figure in original reference.

blackfish (*Orthodon microlepidotus*) that had accumulated about 0.75 µg/g wet weight in axial muscle from a methylmercury-contaminated diet containing 0.52 µg Hg/g dry weight and administered for 70 days. Growth rates were reduced in Sacramento blackfish fed diets with 22.1 and 55.5 µg Hg/g dry weight and accumulated mercury concentrations in axial muscle of about 15 µg/g wet weight or greater (Houck and Cech 2004); however, the methylmercury levels in these two treatments vastly exceed those that would be typically encountered by wild freshwater fish.

In field studies, an inverse relation between mercury concentration in tissue and condition factor (an index of fish weight relative to length; Anderson and Neumann 1996) has been reported for several fishes, including northern pike from Isle Royale National Park, Michigan (axial muscle, 0.07–0.62 µg Hg/g wet weight; Drevnick et al. 2008), walleye from the upper Columbia River (axial muscle, 0.21–0.39 µg Hg/g wet weight; Munn and Short 1997), white sturgeon (*Acipenser transmontanus*) from the lower Columbia River (liver, 0.02–0.78 µg Hg/g wet weight; Webb et al. 2006), and striped bass (*Morone saxatilis*) from Lake Mead (axial muscle, 0.06–1.06 µg Hg/g wet weight; Cizdziel et al. 2003). In some field studies, the relation between methylmercury exposure and fish condition may be confounded and complicated by other, co-occurring contaminants or by changes in condition factor with fish age.

4.3.2 Effects on Behavior

Fish behavior is a sensitive and ecologically important indicator of contaminant stress. Critical reviews indicate that swimming (Little and Finger 1990) and feeding behaviors (Sandheinrich and Atchison 1990) of many fishes can be disrupted by chemicals at concentrations that subsequently reduced growth. Similarly, dietary methylmercury altered the predator-evasion behavior of adult golden shiners (*Notemigonus crysoleucas*) that were fed diets containing 0.01–0.96 µg Hg/g dry weight for 90 days and subsequently tested with a model avian predator (Webber and Haines 2003). Brain acetylcholinesterase did not differ among the treatment groups, yet golden shiners with whole-body concentrations of 0.52 µg Hg/g wet weight after being fed the high-methylmercury diet (0.96 µg Hg/g) were hyperactive and had altered shoaling behavior relative to fish fed control or low-methylmercury diets. Mercury concentrations in fish with altered behavior in the laboratory were within the range of concentrations measured in wild golden shiners inhabiting lakes in the northern United States, leading Webber and Haines (2003) to conclude that methylmercury exposure would also alter predator-avoidance behavior of wild golden shiners and possibly increase their vulnerability to predation. Fjeld et al. (1998) increased the concentrations of methylmercury in grayling (*Thymallus thymallus*) by exposing eggs to four treatments of aqueous methylmercury ranging from 0.16 to 20 µg Hg/L for 10 days. Concentrations of mercury ranged from 0.09 to 3.8 µg/g wet weight in newly hatched fry of grayling, which were raised for 3 years without supplemental methylmercury in either water or diet and then tested in foraging experiments with *Daphia magna* as prey. In tests with 3-year-old grayling (length 13.8 ± 0.8 cm), fish from treatment groups with mercury concentrations of 0.27 µg/g wet weight or greater as newly hatched fry had impaired feeding efficiencies and reduced competitive abilities relative to controls (0.01 µg Hg/g wet weight as fry) and would be expected to grow at a slower rate in the wild.

The consequences of methylmercury exposure may be more severe for wild fish than for fish in laboratory tests in which fish are provided ample food (Wiener and Spry 1996). The neurotoxic effects of methylmercury may severely impede the ability of wild fish to locate, capture, handle, and ingest prey and to avoid predation.

4.3.3 Effects on Reproduction

There is compelling and consistent evidence from laboratory and field studies indicating that methylmercury impairs reproduction of fish at environmentally relevant concentrations (Scheuhammer et al. 2007, Crump and Trudeau 2009, Tan et al. 2009). Moreover, it impairs reproduction, in part, by disruption of the hypothalamic-pituitary-gonadal axis and is an endocrine disrupter (Colborn et al. 1993, Tan et al. 2009). In separate experiments, Hammerschmidt et al. (2002) and Drevnick and Sandheinrich (2003) fed fathead minnows from the juvenile stage through sexual maturity with diets contaminated with methylmercury chloride. At sexual maturity, pairs of male and female fathead minnows were provided with spawning substrates and allowed to reproduce. Dietary methylmercury markedly altered several reproductive endpoints, and results were consistent between studies. For example, the spawning success of pairs of fish with mean carcass (i.e., whole body minus gonads) concentrations of 0.71–0.92 μg Hg/g wet weight was about 60% or less than that of fish fed control diets. Spawning of methylmercury-contaminated fish was delayed by an average of 5 days in fish that spawned (Drevnick and Sandheinrich 2003). Daily reproductive effort (eggs laid per gram of female carcass per day) and instantaneous rate of reproduction were suppressed; the reproductive potential of female controls was calculated to be 1.4 times that of females with 0.85 μg Hg/g wet weight (Hammerschmidt et al. 2002). Male reproductive behavior was also affected. Control males spent about 5% of their time spawning and were inactive only 8% of the time, whereas males with mean carcass concentrations of 0.71 μg Hg/g wet weight spent 0.5% of their time spawning and were inactive 19% of the time (Sandheinrich and Miller 2006). Altered reproduction was associated with suppression of plasma estradiol and testosterone. Plasma testosterone concentrations in male fathead minnows fed control diets were 20% and 106% greater, respectively, than those in fish fed methylmercury-contaminated diets and with 0.86 and 3.6 μg Hg/g wet weight in the carcass. Female controls (0.08 μg Hg/g wet weight in carcass) had estradiol concentrations 149% and 402% greater than those in fish with 0.92 and 3.8 μg Hg/g wet weight in the carcass (Drevnick and Sandheinrich 2003). Expression of genes related to endocrine function also was altered in these fish (Klaper et al. 2006). Tan et al. (2009) have critically reviewed literature on the effects of mercury on the endocrine systems of humans and wildlife, including fish.

Field studies also have documented altered sex hormones in fish relative to mercury burden. Webb et al. (2006) found a significant negative correlation between concentrations of plasma testosterone, 11-ketotestosterone, and muscle mercury of male white sturgeon in the lower Columbia River. Mean mercury concentrations in muscle, liver, and gonads were 0.17, 0.14, and 0.027 μg Hg/g wet weight, respectively. There also was a negative correlation between mercury in the liver (range 0.02–0.78 μg Hg/g wet weight) and estradiol in the plasma of female fish. For male fish, Webb et al. (2006) suggested that total mercury concentrations of 0.19 μg/g wet weight in muscle, 0.09 μg/g in liver, and 0.07 μg/g in gonads represented threshold concentrations affecting steroidogenesis; no male sturgeon with these or greater levels of mercury had plasma testosterone exceeding 4 ng/mL. A positive correlation between mercury in the axial fillet and 11-ketotestosterone in serum was reported for largemouth bass (*Micropterus salmoides*) from three reservoirs in New Jersey (Friedmann et al. 2002); however, mercury in muscle of the fish (range 0.30–5.42 μg Hg/g wet weight) was unrelated to serum testosterone (Friedmann et al. 2002).

Methylmercury appears to affect multiple sites of the reproductive axis in fish (hypothalamus, pituitary, gonads; Crump and Trudeau 2009), and altered concentrations of sex hormones may be caused by the cellular effects of methylmercury on the gonads. Drevnick et al. (2006) proposed that methylmercury induces apoptosis in steroidogenic cells in the gonads of fish. They observed an increase in the number of apoptotic follicular cells in the gonads of female fathead minnows that had impaired reproduction due to chronic methylmercury exposure (as previously reported by Drevnick and Sandheinrich 2003). There was a significant inverse relation between the number of apoptotic cells in the ovarian follicle, the relative size of the gonads (gonadal somatic index), and the concentration of estradiol in the serum of female fish. Female fish with whole-body mercury of about 0.90 μg Hg/g wet weight had significantly more apoptotic follicular cells than controls with 0.08 μg Hg/g weight. Similarly, Webb et al. (2006) found an inverse relation between gonadal somatic index, testosterone, and mercury in immature male sturgeon with 0.04–0.52 μg Hg/g wet weight in cheek muscle.

Dietary methylmercury also suppressed gonadal development in walleye. Friedmann et al. (1996) observed gonadal atrophy in male walleye with whole-body (minus viscera) concentrations of 0.25 µg Hg/g wet weight. Histological examination of the testes indicated that methylmercury caused multifocal cell atrophy and hypertrophy of cells adjacent to atrophied cells.

4.3.4 Effects on Tissue Histology and Biochemistry

Disruption of cell structure and changes in biochemistry associated with oxidative stress have been reported at tissue concentrations of mercury less than those affecting reproduction. In a coordinated set of experiments, adult *Hoplias malabaricus* (a neotropical fish) were fed prey fish (*Astyanax* sp.) every 5 days for 70 days. Prey fish were injected with either distilled water (controls) or with methylmercuric chloride equivalent to a dose of 0.075 µg methylmercury/g wet weight for the predator *H. malabaricus*. Methylmercury exposure increased the number of red blood cells, leukocytes, neutrophils, and monocytes in *H. malabaricus*. In addition, hemoglobin concentration, hematocrit, and mean corpuscular volume were greater in fish receiving the methylmercury-contaminated diets than in controls (Oliveira Ribeiro et al. 2006).

Dietary methylmercury also damaged liver cells and suppressed plasma δ-aminolevulinic acid dehydratase and cholinergic activity in muscle of *H. malabaricus* (Alves Costa et al. 2007). The livers of the fish fed methylmercury-treated prey had prenecrotic lesions (leukocyte infiltration) and necrotic areas, as well as greater numbers of melano-macrophage centers, abnormal cells, phagocytic areas and intercellular spaces than the controls (Mela et al. 2007). The head kidneys of treated fish had increased numbers of leukocytes, dead and atypical cells, and necrotic regions as well as increased number of melano-macrophage centers—indicating increased phagocytic activity. Mela et al. (2007) suggested that methylmercury caused oxidative stress, which contributed to the development of necrotic tissues. The concentrations of mercury in the livers of *H. malabaricus* fed the control and treated diet were 0.60 and 1.07 µg/g wet weight, respectively, and corresponding concentrations in muscle were 0.67 and 1.45 µg/g wet weight. Pre-existing concentrations of mercury in the prey fish and predator fish used in these studies were not reported.

Laboratory and field studies have shown that methylmercury exposure causes oxidative stress in fish tissues through the formation of radical oxygen species and lipid peroxidation. Berntssen et al. (2003), who fed juvenile Atlantic salmon (*Salmo salar*) diets containing 0.03 (control), 4.35 (medium), or 8.48 (high) µg Hg/g dry weight for 4 months, observed no effect of dietary methylmercury on growth or condition of test fish. However, the medium diet of methylmercury induced a defense response in the brain, indicated by a significant increase of the activity of super oxide dismutase. Brains of fish fed the high-methylmercury diet had suppressed super oxide dismutase and glutathione peroxidase activity, but a marked increase in thiobarbituric acid-reactive substances, products of lipid peroxidation. Brain monoamine oxidase activity also was inhibited by the high-methylmercury diet, suggesting disruption of the monaminergic system; the activity of these fish also was suppressed after feeding, but the behavior of fish fed the medium diet of methylmercury was not assessed. Brains of fish fed the medium and high diets of methylmercury had mean tissue concentrations of 1.16 and 0.68 µg/g wet weight, respectively, and exhibited severe vacuolization and cell necrosis. Livers of fish fed the high-methylmercury diet also had increased activity of super oxide dismutase and glutathione peroxidase. Berntssen et al. (2003) concluded that the medium diet of methylmercury induced redox defenses, but that such defenses were overcome—causing injury—in fish receiving the high-methylmercury diet.

Larose et al. (2008) examined the relation between methylmercury and the glutathione system in walleye and yellow perch (*Perca flavescens*) from four boreal lakes in eastern Canada. The size-standardized concentration of total mercury in axial muscle of 35-cm walleyes ranged from 0.3 to 0.79 µg/g wet weight in the four lakes. Mean concentrations of methylmercury in livers of walleye were greater in Lake Malartic (≈0.38 µg/g wet weight) than in the other three lakes (≈0.18–0.20 µg/g), which did not differ (mean values estimated from Fig. 1 in Larose et al. 2008). In Lake Malartic, the hepatosomatic index (size of the liver relative to total body mass) in walleye was inversely related to hepatic methylmercury concentration. The activities of glutathione reductase and glutathione S-transferase were positively correlated with liver size (and by inference, negatively correlated with methylmercury concentration). In Lake Desjardins-East, which had the greatest mean concentration of hepatic methylmercury in yellow perch, the activities of glutathione S-transferase and selenium-dependent glutathione peroxidase were negatively correlated with methylmercury concentration in the liver. Larose et al. (2008) concluded that environmentally relevant concentrations of methylmercury altered cell metabolism and physiology in these fishes.

Oxidative stress associated with methylmercury exposure also was documented in a study of four salmonid species (lake trout, brook trout [*Salvelinus fontinalis*], cutthroat trout [*Oncorhynchus clarkii*], rainbow trout [*O. mykiss*]) from 14 lakes in eight national parks or preserves in the western United States and Alaska (Schwindt et al. 2008). Concentrations of total mercury in whole fish and macrophage aggregates in kidney and spleen were quantified. Macrophage aggregates are groupings of macrophages within tissues that collect components of damaged cells, including cells damaged by oxidation and lipid peroxidation. Increases in the number of macrophage aggregates in the kidney and spleen were associated with concentrations of mercury in whole fish. Although the number of macrophage aggregates can increase with age, Schwindt et al. (2008) demonstrated that mercury affected macrophage aggregates independent of fish age. Other contaminants (including polychlorinated biphenyls and polybrominated diphenyl ethers) also were present in these fish; however, mercury alone explained 36% of the variation in macrophage aggregates in the spleens of brook trout—the species with the largest sample size and most geographically extensive data set in the study. Whole-body concentrations of total mercury in individual brook trout ranged from about 0.03 to 0.29 µg/g wet weight (values estimated from Fig. 3 in Schwindt et al. 2008).

Drevnick et al. (2008) examined mercury and livers in northern pike from eight inland lakes in Isle Royale National Park. Quantitative analysis of pigment in livers showed that color (absorbance at 400 nm) was positively related to total mercury in the organ. Concentrations of total mercury in liver (range 0.048–3.074 µg/g wet weight) and skin-on fillets (range 0.069–0.62 µg/g wet weight) were positively correlated. Lipofuscin, the pigment identified as responsible for altered liver color, is formed as a result of lipid peroxidation of membranous organelles and also is found frequently in macrophage aggregates. Condition factor of northern pike was negatively correlated with total mercury concentration in the liver, indicating that methylmercury exposure adversely affected fish health. Raldúa et al. (2007) also observed greater concentrations of mercury and a greater incidence of liver pathologies—including macrophage aggregates and lipofuscin—in fish downstream from a chlor-alkali plant (an industrial source of mercury) than in fish collected upstream from the plant. Total mercury in the livers of the fish downstream from the chlor-alkali plant ranged from 0.32 to 1.96 µg/g wet weight (Raldúa et al. 2007).

Induction of oxidation defenses by methylmercury exposure also can be detected by alterations in gene transcription. Gonzalez et al. (2005) fed diets containing 0.08 (control), 5, and 13.5 µg methylmercury/g dry weight to adult male zebrafish (*Danio rerio*) for 63 days. Dietary concentrations of 5 µg Hg/g dry weight approximate those in the diets of some piscivorous fish inhabiting natural lakes and reservoirs in North and South America (Hammerschmidt et al. 2002, Gonzalez et al. 2005). The expression of genes associated with mitochondrial metabolism, apoptosis, and oxidative stress (including super oxide dismutase) was upregulated in the skeletal muscle and liver after 21–63 days of dietary exposure to the 5-µg/g dry weight diet, whereas genes associated with DNA repair were downregulated in the skeletal muscle. After 63 days, methylmercury in the skeletal muscle of these fish was 15 µg/g dry weight (≈3 µg/g wet weight, assuming a water content of 80%). Expression of the 13 genes evaluated in the study did not change in the brains of the exposed fish, although methylmercury in the brain exceeded 20 µg/g dry weight. Experimental results highlighted by Gonzalez et al. (2005) include (1) lack of effect on gene expression in the brain, particularly those associated with antioxidant defense, and (2) effects of dietary methylmercury on gene expression in skeletal muscle. The lack of effect on gene expression in the brain may partly explain the neurotoxicity of methylmercury.

A subsequent study by Oliveira Ribeiro et al. (2008) demonstrated that dietary methylmercury disrupted muscle fibers and mitochondria of skeletal muscle in zebrafish that were fed a diet containing 13.5 µg methylmercury/g dry weight for 63 days. The mitochondria and spaces between muscle fibers of skeletal muscle from fish fed methylmercury were smaller than those fed a control diet. On the basis of the prior work of Gonzalez et al. (2005) and their own results, Oliveira Ribeiro et al. (2008) concluded that methylmercury damaged red muscle tissue and that modification of the mitochondria probably decreased adenosine triphosphate production by the muscle cells. Wiener and Spry (1996, p. 313) hypothesized that storage of methylmercury in the muscle may function as a detoxification mechanism for methylmercury in fish, given that the binding of assimilated methylmercury to proteins in the skeletal muscle reduces the exposure of the brain to methylmercury. Their hypothesis is not supported by the subsequent work of Gonzalez et al. (2005) and Oliveira Ribeiro et al. (2008), who reported adverse effects of methylmercury on muscle cells and tissue in zebrafish exposed experimentally to methylmercury.

Moran et al. (2007) compared gene expression in livers of cutthroat trout that differed in mercury concentrations in two high-altitude lakes in the northwestern United States. Mean concentrations in cutthroat trout were about 0.016 µg/g wet weight in Skymo Lake and 0.054 µg/g in Wilcox Lake (values estimated from Fig. 3 in Moran et al. 2007). Expression of 45 of the 147 genes evaluated differed between the two lakes. Genes upregulated in the more contaminated fish (Wilcox Lake) included those associated with stress response (including glutathione peroxidase), intermediary metabolism, and endocrine response. Differences in transcriptional responses of fish were attributed to differences in mercury contamination between the two lakes. Organic contaminants (DDE, total PCB) in these fish were determined, but their concentrations were low (median sum of all organochlorines detected in tissues was 12.6 ng/g wet weight) and differed little between lakes.

4.4 CONCENTRATIONS IN TISSUES ASSOCIATED WITH TOXIC EFFECTS

Recent laboratory experiments and field studies demonstrate that freshwater fish are adversely affected at tissue concentrations of methylmercury well below 1.0 µg Hg/g wet weight (Tables 4.1 and 4.2). We conclude that changes in biochemical processes, damage to cells and tissues, and reduced reproduction in fish occur at methylmercury concentrations of about 0.3–0.7 µg Hg/g wet weight in the whole body and about 0.5–1.2 µg Hg/g wet weight in axial muscle (Figure 4.1). These concentration ranges are based on the studies summarized in the preceding text and in Tables 4.1 and 4.2, and were derived in part by interconversion of concentrations between whole fish and axial muscle tissue with the formula provided in Figure 2 of Peterson et al. (2007).

Threshold tissue concentrations of methylmercury associated with toxic effects in fish are not readily defined in field studies. However, the statistical correlations between mercury concentrations in fish tissues and biomarkers of effects in field studies (Table 4.2) agree well with the results of toxicological experiments with fish (Table 4.1). Moreover, correlations indicative of adverse effects in wild fishes have been reported for multiple recent field studies in which maximal tissue concentrations were less than 1.0 µg Hg/g wet weight (Webb et al. 2006, Moran et al. 2007, Drevnick et al. 2008, Larose et al. 2008, Schwindt et al. 2008).

The skeletal muscle contains most or much of the methylmercury accumulated in fish, present in the form of methylmercury cysteine (Harris et al. 2003, Kuwabara et al. 2007). Harris et al. (2003) have suggested that the toxicity of methylmercury is overestimated in laboratory studies that use methylmercury chloride (CH_3HgCl) as a model compound for methylmercury in fish, partly because methylmercury chloride readily crosses biological membranes and is thus expected to be more toxic than methylmercury cysteine. Our synthesis, however, shows that the results of recent laboratory studies—most of which used methylmercury chloride—are generally consistent with

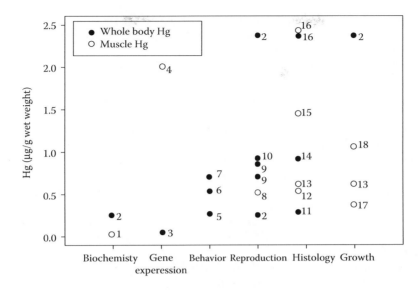

FIGURE 4.1 Tissue concentrations of mercury affecting fish biochemistry, gene transcription, behavior, reproduction, histology, and growth. Values are either mean effects concentrations from laboratory studies or maximum effects concentrations from correlative field studies (see Tables 4.1 and 4.2). Effects concentrations exceeding 2.5 µg/g wet weight were not included on the graph. Numbers on the graph indicate the source of the data, as follows: (1) Liao et al. 2006; (2) Friedmann et al. 1996; (3) Moran et al. 2007; (4) Gonzalez et al. 2005; (5) Fjeld et al. 1998; (6) Webber and Haines 2003; (7) Sandheinrich and Miller 2006; (8) Webb et al. 2006; (9) Hammerschmidt et al. 2002; (10) Drevnick and Sandheinrich 2003; (11) Schwindt et al. 2008; (12) Meinelt et al. 1997; (13) Drevnick et al. 2008; (14) Drevnick et al. 2006; (15) Mela et al. 2007; (16) Raldúa et al. 2007; (17) Munn and Short 1997; and (18) Cizdziel et al. 2003.

the results of mensurative field studies of freshwater fish. Many of the cited field studies (Table 4.2) analyzed piscivorous fish (e.g., largemouth bass, northern pike, striped bass, and walleye) that were presumably exposed to methylmercury in the form of methylmercury cysteine via the ingestion of prey fish. Moreover, adverse effects in some field studies were observed at tissue concentrations lower than those observed in laboratory studies.

Experimental studies providing food contaminated with methylmercury bioaccumulated in aquatic food webs have also produced results similar to experimental studies exposing fish to methylmercury chloride in food or water. For example, walleyes fed skeletal muscle of contaminated northern pike (mercury presumably present as methylmercury cysteine) from an industrially contaminated lake became emaciated; exhibited greatly diminished locomotor activity, coordination, and escape behavior; and had high mortality (Scherer et al. 1975). These same biotic effects have been reported at similar tissue concentrations in laboratory studies with fish exposed to methylmercury chloride (reviewed by Wiener and Spry 1996).

Species differences in sensitivity to methylmercury exposure in fishes have not been assessed, but may be considerable given the variation in avian sensitivity reported by Heinz et al. (2009). Consequently, methylmercury-sensitive fishes may be adversely affected at concentrations significantly lower than those summarized here.

The tissue concentrations associated with adverse effects in recent studies are substantially lower than those estimated by Wiener and Spry (1996), whose evaluation was based on published information available in the mid-1990s. Wiener and Spry (1996) reported that concentrations of total mercury (present as methylmercury) associated with *sublethal and lethal* effects in freshwater fish range from 5 to 20 µg/g wet weight in axial muscle and from 5 to 10 µg/g wet weight in whole fish. On the basis of these values, they inferred that tissue concentrations of mercury associated with overt toxic effects in freshwater fish would, therefore, not be expected except in surface waters grossly contaminated by point-source industrial pollution. Such an inference is no longer tenable, given that the tissue concentrations associated with adverse sublethal and reproductive effects in recent studies occur in many freshwater fishes and geographic regions of North America and elsewhere.

4.5   CONCENTRATIONS OF MERCURY IN FRESHWATER FISH

Concentrations of mercury in recreational fish have been widely surveyed and monitored in the United States and Canada, to provide information for issuance of fish-consumption advice to persons who consume wild-caught fish. Most monitoring and surveillance programs quantify concentrations of total mercury in axial muscle or skin-on fillets of fish, providing reliable estimates of methylmercury in edible fish flesh (Wiener et al. 2007). Concentrations of total mercury in axial muscle of freshwater fish are strongly correlated with those in blood (Schmitt and Brumbaugh 2007). Thus, concentrations of total mercury in the axial muscle or fillets of freshwater fish, which have been quantified in many fresh waters, provide a toxicologically relevant indicator of methylmercury exposure in adult fish.

Concentrations of methylmercury in freshwater fish have been characterized for a number of geographic regions. Kamman et al. (2005) analyzed a large dataset of mercury concentrations in fish from northeastern North America (Figure 4.2). Mean concentrations of mercury in the axial fillets from eight piscivorous species—including smallmouth bass (*Micropterus dolomieu*), largemouth bass, chain pickerel (*Esox niger*), northern pike, walleye, lake trout, white perch (*Morone americana*), and muskellunge (*Esox masquinongy*)—ranged from 0.54 to 0.98 µg/g wet weight (Kamman et al. 2005). A survey of fish from 775 lakes and rivers in the province of Quebec, Canada, reported that mean concentrations in the muscle of standard-length walleye (40–50 cm), northern pike (55–70 cm), and lake trout (55–70 cm) ranged from 0.64 to 0.75 µg/g wet weight; mean concentrations in fish longer than 50 cm (walleye) or 70 cm (northern pike and lake trout) exceeded 1.0 µg Hg/g (Laliberté 2004).

Concentrations of methylmercury in fish from many streams and rivers of the western United States are generally less than those from the northeastern region of North America. Peterson et al. (2007), who analyzed more than 2,700 large fish from 626 stream and riverine sites in 12 states in the western United States, reported mean whole-body concentrations of 0.26 µg/g wet weight in piscivorous fish—equivalent to 0.43 µg/g in the fillets. Moreover, concentrations of mercury in piscivorous fish exceeded the fillet equivalent of 0.30 µg/g in 57% of the assessed stream reaches. In California, concentrations of mercury have recently been quantified in sport fishes from the state's

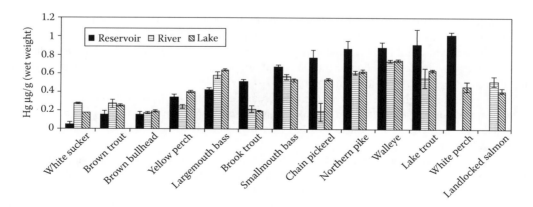

FIGURE 4.2   Mean Hg concentrations in fillets of standard-length freshwater fish of various species from across northeastern North America. (From Kamman et al., *Ecotoxicology*, 14, 163–180. With permission.)

lakes and reservoirs (Davis et al. 2009) and from major rivers of the Sacramento-San Joaquin Delta basin (Melwani et al. 2009). Mercury concentrations in axial muscle of fish exceeded 0.44 µg/g wet weight in 23% of 50 randomly selected lakes and reservoirs and in 26% of 102 targeted lakes and reservoirs in the state (Davis et al. 2009). Concentrations of mercury in axial muscle of largemouth bass exceeded 0.40 µg/g wet weight in many of the riverine sites sampled in the Sacramento-San Joaquin basin (Melwani et al. 2009).

In considering potential toxic effects associated with mercury concentrations in fish, it is evident that concentrations exceeding 0.3 Hg µg/g wet weight in whole fish and 0.5 Hg µg/g wet weight in axial muscle are common in piscivorous fish throughout North America. Moreover, many piscivorous fish inhabiting fresh waters in the midwestern and eastern United States and the eastern half of Canada contain concentrations exceeding 1.0 µg Hg/g wet weight in axial muscle. Concentrations also exceed 1.0 µg Hg/g in piscivorous fish inhabiting some waters in the western United States that have been contaminated with mercury from mining activities (Brumbaugh et al. 2001, Kuwabara et al. 2007, Davis et al. 2008).

Reduced methylmercury exposures of fish-eating humans and wildlife are generally considered to be principal benefits of regulatory actions to reduce anthropogenic emissions of mercury to the environment, given that consumption of fish is the dominant pathway for exposure of humans and many wildlife species to methylmercury (National Research Council 2000, Mergler et al. 2007, Scheuhammer et al. 2007, Swain et al. 2007). It is reasonable to infer—based on the synthesis presented in this chapter—that a reduction of methylmercury concentrations in fish would also be beneficial to the health of fishes inhabiting many fresh waters.

Summary

We summarize recent advances in the state of scientific knowledge regarding the toxicity of methylmercury in freshwater fish, with emphasis on assessing the toxicological significance of methylmercury concentrations in fish tissues. Sublethal and reproductive effects have been documented at environmentally relevant exposure levels in both field and laboratory studies, and the fish-tissue concentrations associated with adverse effects in recent studies are much lower than those presented in the first-edition chapter by Wiener and Spry (1996). There is compelling evidence that methylmercury exposure of fish causes oxidative stress through the formation of radical oxygen species and lipid peroxidation that cause biochemical and structural changes in tissues. Alterations in gene transcription and glutathione metabolism as well as increases in macrophage aggregates have been documented. Thus, effects of methylmercury on fish are not only limited to neurotoxicity, but also include histological changes in the spleen, kidney, liver, and gonads. These effects have been observed in multiple species of freshwater fish at tissue concentrations of methylmercury well below 1.0 µg Hg/g wet weight. Effects on biochemical processes, damage to cells and tissues, and reduced reproduction in fish have been documented at methylmercury concentrations of about 0.3–0.7 µg Hg/g wet weight in the whole body and about 0.5–1.2 µg Hg/g wet weight in axial muscle.

Concentrations of total mercury in axial muscle or fillets have been quantified for freshwater fishes from many surface waters and geographic regions of North America and elsewhere, providing both reliable estimates of methylmercury concentrations in edible fish flesh and a toxicologically relevant indicator of methylmercury exposure in adult fish. Piscivorous fish in many freshwaters and geographic regions bioaccumulate methylmercury to concentrations exceeding those associated with sublethal and reproductive effects in toxicological studies. Aquatic systems characterized by high-methylmercury levels in piscivorous fish include low-pH and highly organic waters, water bodies with fluctuating water levels, newly flooded impoundments, and waters contaminated by mercury from industrial and mining sources. We conclude that the principal effects of methylmercury on fish populations at existing exposure levels in

North American freshwaters would be sublethal damage to tissues and depressed reproduction. Furthermore, we infer that a reduction of methylmercury concentrations in fish would be beneficial to the health of fish inhabiting many freshwaters.

ACKNOWLEDGMENTS

During preparation of this manuscript, the lead author (MBS) was supported by the University of Wisconsin-La Crosse (UW-L), and the second author (JGW) was supported by the University of Wisconsin System Distinguished Professors Program and the UW-L Foundation. We thank James P. Meador and two anonymous reviewers for constructive comments on an earlier draft.

REFERENCES

Alves Costa, J. R. M., et al. 2007. Enzymatic inhibition and morphological changes in *Hoplias malabaricus* from dietary exposure to lead(II) or methylmercury. *Ecotoxicol. Environ. Safety* 67:82–88.

Anderson, R. O., and R. M. Neumann. 1996. Length, weight, and associated structural indices. In *Fisheries techniques*, 2nd edition, ed. B. R. Murphy and D. W. Willis, pp. 447–482. Bethesda, MD: American Fisheries Society.

Armstrong, F. A. J. 1979. Effects of mercury compounds on fish. In *Biogeochemistry of mercury in the environment*, ed. J. O. Nriagu, pp. 657–670. New York, NY: Elsevier/North-Holland Biomedical Press.

Berntssen, M. H. G., A. Aatland, and R. D. Handy. 2003. Chronic dietary mercury exposure causes oxidative stress, brain lesions, and altered behavior in Atlantic salmon (*Salmo salar*) parr. *Aquat. Toxicol.* 65:55–72.

Bindler, R., I. Renberg, P. G. Appleby, N. J. Anderson, and N. L. Rose. 2001. Mercury accumulation rates and spatial patterns in lake sediments from West Greenland: a coast to ice margin transect. *Environ. Sci. Technol.* 35:1736–1741.

Bloom, N. S. 1992. On the chemical form of mercury in edible fish and marine invertebrate tissue. *Can. J. Fish. Aquat. Sci.* 49:1010–1017.

Bodaly, R. A., and R. J. P. Fudge. 1999. Uptake of mercury by fish in an experimental boreal reservoir. *Arch. Environ. Contam. Toxicol.* 37:103–109.

Bodaly, R. A., et al. 2004. Experimenting with hydroelectric reservoirs. *Environ. Sci. Technol.* 38:346A–352A.

Bodaly, R. A., et al. 2007. Postimpoundment time course of increased mercury concentrations in fish in hydroelectric reservoirs of northern Manitoba, Canada. *Arch. Environ. Contam. Toxicol.* 53:379–389.

Boudou, A., and F. Ribeyre. 1983. Contamination of aquatic biocenoses by mercury compounds: an experimental ecotoxicological approach. In *Aquatic Toxicology*, ed. J. O. Nriagu, pp. 73–116. New York, NY: John Wiley and Sons.

Branfireun, B. A., D. P. Krabbenhoft, H. Hintelmann, R. J. Hunt, J. P. Hurley, and J. W. M. Rudd. 2005. Speciation and transport of newly deposited mercury in a boreal forest wetland: a stable mercury isotope approach. *Water Resour. Res.* 41:W06016, doi:10.1029/2004WR003219.

Brigham, M. E., D. A. Wentz, G. R. Aiken, and D. P. Krabbenhoft. 2009. Mercury cycling in stream ecosystems. 1. Water column chemistry and transport. *Environ. Sci. Technol.* 43:2720–2725.

Brumbaugh, W. G., D. P. Krabbenhoft, D. R. Helsel, J. G. Wiener, and K. R. Echols. 2001. A national pilot study of mercury contamination of aquatic ecosystems along multiple gradients: bioaccumulation in fish. U.S. Geological Survey, Biological Science Report USGS/BRD/BSR-2001–0009. http://www.cerc.usgs.gov/pubs/center/pdfdocs/BSR2001–0009.pdf (accessed May 20, 2009).

Cambier, S., et al. 2009. At environmental doses, dietary methylmercury inhibits mitochondrial energy metabolism in skeletal muscles of the zebra fish (*Danio rerio*). *Internat. J. Biochem. Cell Biol.* 41:791–799.

Chasar, L. C., B. C. Scudder, A. R. Stewart, A. H. Bell, and G. R. Aiken. 2009. Mercury cycling in stream ecosystems. 3. Trophic dynamics and methylmercury bioaccumulation. *Environ. Sci. Technol.* 43:2733–2739.

Chen, C. Y., and C. L. Folt. 2005. High plankton densities reduce mercury biomagnification. *Environ. Sci. Technol.* 39:115–121.

Cizdziel, J., T. Hinners, C. Cross, and J. Pollard. 2003. Distribution of mercury in the tissues of five species of freshwater fish from Lake Mead, USA. *J. Environ. Monitoring* 5:802–807.

Colborn, T., F. S. vom Saal, and A. M. Soto. 1993. Developmental effects of endocrine disrupting chemicals in wildlife and humans. *Environ. Health Perspect.* 101:378–384.

Crump, K. L., and V. L. Trudeau. 2009. Mercury-induced reproductive impairment in fish. *Environ. Toxicol. Chem.* 28:895–907.

Davis, J. A., B. K. Greenfield, G. Ichikawa, and M. Stephenson. 2008. Mercury in sport fish from the Sacramento-San Joaquin Delta region, California, USA. *Sci. Total Environ.* 391:66–75.

Davis, J. A., et al. 2009. Contaminants in fish from California lakes and reservoirs: Technical report on year one of a two-year screening survey. Report of the Surface Water Ambient Monitoring Program (SWAMP). Sacramento, CA: California State Water Resources Control Board.

Drevnick, P. E., and M. B. Sandheinrich. 2003. Effects of dietary methylmercury on reproductive endocrinology of fathead minnows. *Environ. Sci. Technol.* 37:4390–4396.

Drevnick, P. E., A. P. Roberts, R. R. Otter, C. R. Hammerschmidt, R. Klaper, and J. T. Oris. 2008. Mercury toxicity in livers of northern pike (*Esox lucius*) from Isle Royale, USA. *Comp. Biochem. Physiol. Part C* 147:331–338.

Drevnick, P. E., M. B. Sandheinrich, and J. T. Oris. 2006. Increased ovarian follicular apoptosis in fathead minnows (*Pimephales promelas*) exposed to dietary methylmercury. *Aquat. Toxicol.* 79:49–54.

Evans, R. D., E. M. Addison, J. Y. Villeneuve, K. S. MacDonald, and D. G. Joachim. 2000. Distribution of inorganic and methylmercury among tissues in mink (*Mustela vison*) and otter (*Lutra canadensis*). *Environ. Res.* (Section A) 84:133–139.

Fjeld, E., T. O. Haugen, and L. A. Vøllestad. 1998. Permanent impairment in the feeding behavior of grayling (*Thymallus thymallus*) exposed to methylmercury during embryogenesis. *Sci. Total Environ.* 213:247–254.

Friedmann, A. S., E. K. Costain, D. L. MacLatchy, W. Stansley, and E. J. Washuta. 2002. Effect of mercury on general and reproductive health of largemouth bass (*Micropterus salmoides*) from three lakes in New Jersey. *Ecotoxicol. Environ. Safety* 52:117–122.

Friedmann A. S., M. C. Watzin, T. Brinck-Johnsen, and L. C. Leiter. 1996. Low levels of methylmercury inhibit growth and gonadal development in juvenile walleye (*Stizostedion vitreum*). *Aquat. Toxicol.* 35:265–278.

Gonazalez, P., Y. Dominique, J. C. Massabuau, A. Boudou, and J. P. Bourdineaud. 2005. Comparative effects of dietary methylmercury on gene expression in liver, skeletal muscle, and brain of the zebrafish (*Danio rerio*). *Environ. Sci. Technol.* 39:3972–3980.

Grieb, T. M., C. T. Driscoll, S. P. Gloss, C. L. Schofield, G. L. Bowie, and D. B. Porcella. 1990. Factors affecting mercury accumulation in fish in the upper Michigan peninsula. *Environ. Toxicol. Chem.* 9:919–930.

Hall, B. D., and V. L. St. Louis. 2004. Methylmercury and total mercury in plant litter decomposing in upland forests and flooded landscapes. *Environ. Sci. Technol.* 38:5010–5021.

Hall, B. D., R. A. Bodaly, R. J. P. Fudge, J. W. M. Rudd, and D. M. Rosenberg. 1997. Food as the dominant pathway of methylmercury uptake by fish. *Water Air Soil Pollut.* 100:13–24.

Hammerschmidt, C. R., M. B. Sandheinrich, J. G. Wiener, and R. G. Rada. 2002. Effects of dietary methylmercury on reproduction of fathead minnows. *Environ. Sci. Technol.* 36:877–883.

Hammerschmidt, C. R., J. G. Wiener, B. E. Frazier, and R. G. Rada. 1999. Methylmercury content of eggs in yellow perch related to maternal exposure in four Wisconsin lakes. *Environ. Sci. Technol.* 33:999–1003.

Harris, R. C., and R. A. Bodaly. 1998. Temperature, growth and dietary effects on fish mercury dynamics in two Ontario lakes. *Biogeochemistry* 40:175–187.

Harris, H. H., I. J. Pickering, and G. N. George. 2003. The chemical form of mercury in fish. *Science* 301:1203.

Harrison, S. E., J. F. Klaverkamp, and R. H. Hesslein. 1990. Fates of metal radiotracers added to a whole lake: accumulation in fathead minnow (*Pimephales promelas*) and lake trout (*Salvelinus namaycush*). *Water Air Soil Pollut.* 52:277–293.

Heinz, G. H., D. J. Hoffman, J. D. Klimstra, K. R. Stebbins, S. L. Kondrad, and C. A. Erwin. 2009. Species differences in the sensitivity of avian embryos to methylmercury. *Arch. Environ. Contam. Toxicol.* 56:129–138.

Houck, A., and J. J. Cech, Jr. 2004. Effects of dietary methylmercury on juvenile Sacramento blackfish bioenergetics. *Aquat. Toxicol.* 69:107–123.

Huckabee, J. W., J. W. Elwood, and S. G. Hildebrand. 1979. Accumulation of mercury in freshwater biota. In *Biogeochemistry of mercury in the environment*, ed. J. O. Nriagu, pp. 277–302. New York, NY: Elsevier/North-Holland Biomedical Press.

Hurley, J. P., et al. 1995. Influences of watershed characteristics on mercury levels in Wisconsin rivers. *Environ. Sci. Technol.* 29:1867–1875.

Kamman, N. C., et al. 2005. Mercury in freshwater fish of northeast North America—a geographic perspective based on fish tissue monitoring databases. *Ecotoxicology* 14:163–180.

Kelly, C. A., et al. 1997. Increases in fluxes of greenhouse gases and methyl mercury following flooding of an experimental reservoir. *Environ. Sci. Technol.* 31:1334–1344.

Kinghorn, A., P. Solomon, and H. M. Chan. 2007. Temporal and spatial trends of mercury in fish collected in the English-Wabigoon river system in Ontario, Canada. *Sci. Total Environ.* 372:615–623.

Klaper, R., C. B. Rees, P. Drevnick, D. Weber, M. Sandheinrich, and M. J. Caravan. 2006. Gene expression changes related to endocrine function and decline in reproduction in fathead minnow (*Pimephales promelas*) after dietary methylmercury exposure. *Environ. Health Perspect.* 11:1337–1343.

Kuwabara, J. S., Y. Arai, B. R. Topping, I. J. Pickering, and G. N. George. 2007. Mercury speciation in piscivorous fish from mining-impacted reservoirs. *Environ. Sci. Technol.* 41:2745–2749.

Laliberté, D. 2004. Répertoire des données sur les teneurs en mercure dans la chair des poissons du Québec pour la période de 1976 à 1999 inclusivement. Envirodoq n° ENV/2004/0375, collection n° QE/153. Direction du suivi de l'état de l'environnement, ministère de l'Environnement, QC, Canada.

Lamborg, C. H., W. F. Fitzgerald, A. W. H. Damman, J. M. Benoit, P. H. Balcom, and D. R. Engstrom. 2002. Modern and historic atmospheric mercury fluxes in both hemispheres: global and regional mercury cycling implications. *Global Biogeochem. Cycles* 16(0): doi:10.1029/2001GB001847.

Lange, T. R., H. E. Royals, and L. L. Connor. 1993. Influence of water chemistry on mercury concentration in largemouth bass from Florida lakes. *Trans. Am. Fish. Soc.* 122:74–84.

Larose, C., R. Canuel, M. Lucotte, and R. T. Di Giulio. 2008. Toxicological effects of methylmercury on walleye (*Sander vitreus*) and perch (*Perca flavescens*) from lakes of the boreal forest. *Comp. Biochem. Physiol. Part C* 147:139–149.

Liao, C. Y., et al. 2006. Methylmercury accumulation, histopathology effects, and cholinesterase activity alterations in medaka (*Oryzias latipes*) following sublethal exposure to methylmercury chloride. *Environ. Toxicol. Pharm.* 22:225–233.

Lindberg, S., et al. 2007. A synthesis of progress and uncertainties in attributing the sources of mercury in deposition. *Ambio* 36:19–32.

Little, E. E., and S. E. Finger. 1990. Swimming behavior as an indicator of sublethal toxicity in fish. *Environ. Toxicol. Chem.* 9:13–20.

MacCrimmon, H. R., C. D. Wren, and B. L. Gots. 1983. Mercury uptake by lake trout, *Salvelinus namaycush*, relative to age, growth, and diet in Tadenac Lake with comparative data from other Precambrian Shield lakes. *Can. J. Fish. Aquat. Sci.* 40:114–120.

Mathers, R. A. and P. H. Johansen. 1985. The effects of feeding ecology on mercury accumulation in walleye (*Stizostedion vitreum*) and pike (*Esox lucius*) in Lake Simcoe. *Can. J. Zool.* 62:2006–2012.

McKim, J. M., G. F. Olson, G. W. Holcombe, and E. P. Hunt. 1976. Long-term effects of methylmercuric chloride on three generations of brook trout (*Salvelinus fontinalis*): toxicity, accumulation, distribution, and elimination. *J. Fish. Res. Board Can.* 33:2726–2739.

Meinelt, T., R. Krüger, M. Pietrock, R. Osten, and C. Steinberg. 1997. Mercury pollution and macrophage centres in pike (*Esox lucius*) tissues. *Environ. Sci. Pollut. Res.* 4:32–36.

Mela, M., M. A. F. Rand, D. F. Ventura, C. E. V. Carvalho, E. Pelletier, and C. A. Oiveira Ribeiro. 2007. Effects of dietary methylmercury on liver and kidney histology in the neotropical fish *Hoplia malabaricus*. *Ecotoxicol. Environ. Safety* 6:426–435.

Melwani, A. R., et al. 2009. Spatial trends and impairment assessment of mercury in sport fish in the Sacramento-San Joaquin Delta watershed. *Environ. Pollut.* 157:3137–3149.

Mergler, D., et al. 2007. Methylmercury exposure and health effects in humans: a worldwide concern. *Ambio* 36:3–11.

Moran, P., N. Alur, R. W. Black, and M. M. Vijayan. 2007. Tissue contaminants and associated transcriptional response in trout liver from high elevation lakes of Washington. *Environ. Sci. Technol.* 41:6591–6597.

Munn, M. D., and T. M. Short. 1997. Spatial heterogeneity of mercury bioaccumulation by walleye in Franklin D. Roosevelt Lake and the upper Columbia River, Washington. *Trans. Am. Fish. Soc.* 126:477–487.

Munthe, J., et al. 2007. Recovery of mercury-contaminated fisheries. *Ambio* 36:33–44.

National Research Council. 2000. *Toxicological effects of methylmercury*. Committee on the Toxicological Effects of Methylmercury. Washington, DC: National Academy Press.

Oliveira Ribeiro, C. A., C. Rouleau, E. Pelletier, C. Audet, and H. Tjälve. 1999. Distribution kinetics of dietary methylmercury in the artic charr (*Salvelinus alpinus*). *Environ. Sci. Technol.* 33:902–907.

Oliveira Ribeiro, C. A., et al. 2006. Hematological findings in neotropical fish *Hoplias malabaricus* exposed to subchronic and dietary doses of methylmercury, inorganic lead, and tributyl tin chloride. *Environ. Res.* 101:74–80.

Oliveira Ribeiro, C. A., et al. 2008. Effects of dietary methylmercury on zebrafish skeletal muscle fibers. *Environ. Toxicol. Pharmacol.* 25:304–309.

Organization for Economic Cooperation and Development. 1992a. Fish, acute toxicity test. OECD Guideline for Testing Chemicals, Guideline no. 203, OECD, Paris. http://puck.sourceoecd.org/vl=1195175/cl=14/nw=1/rpsv/ij/oecdjournals/1607310x/v1n2/s4/p1 (accessed May 9, 2009).

Organization for Economic Cooperation and Development. 1992b. Fish, early life-stage toxicity test. OECD Guideline for Testing Chemicals, Guideline no. 210, OECD, Paris. http://puck.sourceoecd.org/vl=1195175/cl=14/nw=1/rpsv/ij/oecdjournals/1607310x/v1n2/s11/p1 (accessed May 9, 2009).

Paterson, M. J., J. W. M. Rudd, and V. St. Louis. 1998. Increases in total and methylmercury in zooplankton following flooding of a peatland reservoir. *Environ. Sci. Technol.* 32:3868–3874.

Peterson, S. A., J. Van Sickle, A. T. Herlihy, and R. M. Hughes. 2007. Mercury concentrations in fish from streams and rivers throughout the western United States. *Environ. Sci. Technol.* 41:58–65.

Phillips, G. R., and D. R. Buhler. 1978. The relative contributions of methylmercury from food or water to rainbow trout (*Salmo gairdneri*) in a controlled laboratory environment. *Trans. Am. Fish. Soc.* 107:853–861.

Pickhardt, P. C., C. L. Folt, C. Y. Chen, B. Klaue, and J. D. Blum. 2002. Algal blooms reduce the uptake of toxic methylmercury in freshwater food webs. *Proc. Natl. Acad. Sci.* 99:4419–4423.

Pickhardt, P. C., C. L. Folt, C. Y. Chen, B. Klaue, and J. D. Blum. 2005. Impacts of zooplankton composition and algal enrichment on the accumulation of mercury in an experimental freshwater food web. *Sci. Total Environ.* 339:89–101.

Pickhardt, P. C., M. Stepanova, and N. S. Fisher. 2006. Contrasting uptake routes and tissue distributions of inorganic and methylmercury in mosquitofish (*Gambusia affinis*) and redear sunfish (*Lepomis microlophus*). *Environ. Toxicol. Chem.* 25:2132–2142.

Raldúa, D., S. Diez, J. M. Bayona, and D. Barceló. 2007. Mercury levels and liver pathology in feral fish living in the vicinity of a mercury cell chlor-alkali factory. *Chemosphere* 66:1217–1225.

Rodgers, D. W. 1994. You are what you eat and a little bit more: bioenergetics-based models of methylmercury accumulation in fish revisited. In *Mercury pollution: Integration and synthesis*, ed. C. J. Watras and J. W. Huckabee, 427–439. Boca Raton, FL: Lewis Publishers.

Rodgers, D. W., and F. W. H. Beamish. 1981. Uptake of waterborne methylmercury by rainbow trout (*Salmo gairdneri*) in relation to oxygen consumption and methylmercury concentration. *Can. J. Fish. Aquat. Sci.* 38:1309–1315.

Sandheinrich, M. B., and G. J. Atchison. 1990. Sublethal toxicant effects on fish foraging behavior: empirical vs. mechanistic approaches. *Environ. Toxicol. Chem.* 9:107–119.

Sandheinrich, M. B., and K. M. Miller. 2006. Effects of dietary methylmercury on reproductive behavior of fathead minnows (*Pimephales promelas*). *Environ. Toxicol. Chem.* 25:3053–3057.

Scherer, E., F. A. J. Armstrong, and S. H. Nowak. 1975. Effects of mercury-contaminated diet upon walleyes, *Stizostedion vitreum vitreum* (Mitchill). Can. Fish. Mar. Serv. Res. Dev. Branch, Winnipeg, Tech. Rep. No. 597. 21 pp.

Scheuhammer, A. M., M. W. Meyer, M. B. Sandheinrich, and M. W. Murray. 2007. Effects of environmental methylmercury on the health of wild birds, mammals, and fish. *Ambio* 36:12–18.

Schmitt, C. J., and W. G. Brumbaugh. 2007. Evaluation of potentially nonlethal sampling methods for monitoring mercury concentrations in smallmouth bass (*Micropterus dolomieu*). *Arch. Environ. Contam. Toxicol.* 53:84–95.

Schwindt, A. R., J. W. Fournie, D. H. Landers, C. B. Schreck, and M. L. Kent. 2008. Mercury concentrations in salmonids from western U.S. national parks and relationships with age and macrophage aggregates. *Environ. Sci. Technol.* 42:1365–1370.

Sorensen, J. A., L. W. Kallemeyn, and M. Sydor. 2005. Relationship between mercury accumulation in young of the year yellow perch and water level fluctuations. *Environ. Sci. Technol.* 39:9237–9243.

Southworth, G. R., R. R. Turner, M. J. Peterson, and M. A. Bogle. 1995. Form of mercury in stream fish exposed to high concentrations of dissolved inorganic mercury. *Chemosphere* 30:779–787.

Suchanek, T. H., et al. 2008. Spatiotemporal trends in fish mercury from a mine-dominated ecosystem: Clear Lake, California. *Ecol. Appl.* 18(Suppl. 8):A177–A195.

Swain, E., P. Jakus, G. Rice, F. Lupi, P. Maxson, J. Pacyna, A. Penn, S. Spiegel, and M. Veiga. 2007. Socioeconomic consequences of mercury use and pollution. *Ambio* 36:45–61.

Tan, S. W., J. C. Meiller, and K. R. Mahaffey. 2009. The endocrine effects of mercury in humans and wildlife. *Crit. Rev. Toxicol.* 39:228–269.

Trudel, M., and J. B. Rasmussen. 1997. Modeling the elimination of mercury by fish. *Environ. Sci. Technol.* 31:1716–1722.

Turner, R. R., and G. W. Southworth. 1999. Mercury-contaminated industrial and mining sites in North America: an overview with selected case studies. In *Mercury contaminated sites*, eds. R. Ebinghaus, R. R. Turner, L. D. Lacerda, O. Vasiliev, and W. Salomons, pp. 89–112. Berlin, MA: Springer.

U.S. Environmental Protection Agency (U.S. EPA). 2001. Update: national listing of fish and wildlife advisories. Fact Sheet EPA-823-F-01–010. Office of Water, Washington, DC.

U.S. EPA. 2002a. Methods for estimating the acute toxicity of effluents and receiving waters to freshwater and marine organisms, 5th edition. Report No. EPA-821-R-02–012, http://www.epa.gov/waterscience/methods/wet/disk2/ (accessed May 9, 2009).

U.S. EPA. 2002b. Short-term methods for estimating the chronic toxicity of effluents and receiving waters to freshwater organisms, 4th edition. Report No. EPA 821/R-02–013, http://www.epa.gov/waterscience/methods/wet/disk3/ (accessed May 9, 2009).

U.S. EPA. 2009. 2008 Biennial national listing of fish advisories. Fact Sheet EPA-823-F-09-007, Office of Water, Washington, DC. http://water.epa.gov/scitech/swguidance/fishshellfish/fishadvisories/upload/2009_09_16_fish_advisories_tech2008.pdf (accessed September 14, 2010).

Van Walleghem, J. L. A., P. L. Blanchfield, and H. Hintelmann. 2007. Elimination of mercury by yellow perch in the wild. *Environ. Sci. Technol.* 41:5895–5901.

Watras, C. J., R. C. Back, S. Halvorsen, R. J. M. Hudson, K. A. Morrison, and S. P. Wente. 1998. Bioaccumulation of mercury in pelagic freshwater food webs. *Sci. Total Environ.* 219:183–208.

Webb, M. A. H., et al. 2006. Mercury concentrations in gonad, liver, and muscle of white sturgeon *Acipenser transmontanus* in the lower Columbia River. *Arch. Environ. Contam. Toxicol.* 50:443–451.

Webber, H. M., and T. A. Haines. 2003. Mercury effects on predator avoidance behavior of a forage fish, golden shiner (*Notemigonus crysoleucas*). *Environ. Toxicol. Chem.* 22:1556–1561.

Wiener, J. G., and D. J. Spry. 1996. Toxicological significance of mercury in freshwater fish. In *Environmental contaminants in wildlife: Interpreting tissue concentrations*, eds. W. N. Beyer, G. H. Heinz, and A. W. Redmon-Norwood, pp. 297–339. Boca Raton, FL: CRC Press.

Wiener, J. G., D. P. Krabbenhoft, G. H. Heinz, and A. M. Scheuhammer. 2003. Ecotoxicology of mercury. In *Handbook of ecotoxicology*, 2nd edition, eds. D. J. Hoffman, B. A. Rattner, G. A. Burton, Jr., and J. Cairns, Jr., pp. 409–463. Boca Raton, FL: CRC Press.

Wiener, J. G., et al. 2006. Mercury in soils, lakes, and fish in Voyageurs National Park (Minnesota): importance of atmospheric deposition and ecosystem factors. *Environ. Sci. Technol.* 40:6261–6268.

Wiener, J. G., et al. 2007. Monitoring and evaluating trends in methylmercury accumulation in aquatic biota. In *Ecosystem responses to mercury contamination: Indicators of change*, eds. R. C. Harris, D. P. Krabbenhoft, R. P. Mason, M. W. Murray, R. J. Reash, and T. Saltman, pp. 87–122. Boca Raton, FL: Taylor and Francis.

Common Bass, Sea-Perch, and Stone-Bass

From *The Royal Natural History,* edited by Richard Lydekker, Frederick Warne & Co.,
London, 1893–94.

5 Selenium Accumulation and Toxicity in Freshwater Fishes

David K. DeForest
William J. Adams

CONTENTS

5.1 Introduction 194
5.2 Field Examples of Se Poisoning in Fish 195
5.3 Se Bioaccumulation in Fish 195
    5.3.1 Overview of Se in the Aquatic Food Chain... 195
    5.3.2 Effects of Age, Size, and Trophic Position on Se Bioaccumulation in Fish............. 196
    5.3.3 Influence of Exposure Route and Form on Se Bioaccumulation............................ 196
    5.3.4 Se Bioaccumulation in Different Tissues 197
5.4 Se Toxicity in Fish 198
    5.4.1 Maternal Transfer Se Exposures: Effects on Embryo/Larval Development and Survival... 200
        5.4.1.1 Bluegill 201
        5.4.1.2 Brook Trout.. 204
        5.4.1.3 Brown Trout... 205
        5.4.1.4 Cutthroat Trout... 205
        5.4.1.5 Dolly Varden 206
        5.4.1.6 Fathead Minnow 206
        5.4.1.7 Largemouth Bass 207
        5.4.1.8 Mosquitofish.. 207
        5.4.1.9 Northern Pike... 208
        5.4.1.10 Rainbow Trout 208
        5.4.1.11 Razorback Sucker 208
        5.4.1.12 White Sucker... 209
    5.4.2 Direct Dietary Se Exposures: Effects on Larval/Juvenile Growth and Survival.....209
        5.4.2.1 Bluegill 209
        5.4.2.2 Chinook Salmon 211
        5.4.2.3 Fathead Minnow 212
        5.4.2.4 Rainbow Trout 212
        5.4.2.5 Razorback Sucker 213
        5.4.2.6 White Sturgeon 215
5.5 Recommendation for a Fish Tissue Se Threshold 215
    5.5.1 Life Stage Considerations for a Se Fish Tissue Threshold 216
        5.5.1.1 Relative Sensitivity of Maternal Transfer and Juvenile Toxicity Studies.... 216

           5.5.1.2   Consistency of Concentration-Response Relationships 218
           5.5.1.3   Considerations for Developing and Applying Tissue-Based Toxicity
                     Thresholds 219
      5.5.2   Recommendation of a Fish Tissue Se Threshold 219
           5.5.2.1   Egg Se Threshold 220
           5.5.2.2   Ovary Se Threshold 220
           5.5.2.3   Whole-Body Se Threshold 221
           5.5.2.4   Summary of Fish Tissue Se Thresholds 222
      5.5.3   Comparisons of Se Fish Tissue Thresholds 223
Summary 225
Acknowledgment 225
References 225

5.1   INTRODUCTION

Selenium (Se) is a naturally occurring element that is essential to fish, but at sufficiently high concentrations can cause large-scale reproductive failure in fish populations (Lemly and Smith 1987). Many of the early studies that evaluated the chronic toxicity of Se to fish focused on aqueous exposures to inorganic selenite (e.g., Goettl and Davies 1977; Hodson et al. 1980). However, these studies were not able to duplicate the Se toxicity observed in the field (e.g., Belews Lake, NC, U.S.) at similar aqueous Se concentrations. Subsequent studies demonstrated that the critical exposure pathway for fish is dietary exposure to organic Se (Bryson et al. 1984, 1985a, 1985b, Finley 1985, Woock et al. 1987). The biogeochemistry of Se is rather complex. Human-caused releases of Se to the aquatic environment are typically in the form of inorganic selenite (e.g., coal power plant discharges) or selenate (e.g., mobilization of Se from marine shales via groundwater leaching) (Lemly and Smith 1987). Once in the aquatic environment, entry of Se into the food web depends on the Se form released (selenite is more bioaccumulative than selenate [Besser et al. 1993]) and site-specific properties that influence the reduction of Se to organic forms (such as redox conditions and biological productivity [Lemly and Smith 1987]). For example, at a given Se concentration in surface water, Se concentrations in food webs tend to be much higher in standing water (e.g., a pond or wetland) than in flowing waters (e.g., streams) (Lillebo et al. 1988, Orr et al. 2006). Accordingly, aqueous Se concentrations resulting in toxicity to fish can be highly variable between sites and, therefore, tissue Se concentrations are a better indicator of Se toxicity to fish.

The U.S. EPA has developed a draft whole-body fish tissue-based ambient water quality criterion for Se of 7.91 µg/g dry wt., which includes a summer/fall monitoring trigger of 5.85 µg/g dry wt. (U.S. EPA 2004). This draft criterion is based on a toxicity study in which juvenile bluegill sunfish (*Lepomis macrochirus*) were exposed to aqueous and dietary Se under simulated winter conditions (i.e., reduced temperature and photoperiod) (Lemly 1993a). This study, which will be summarized below in more detail, was expanded upon by a separate laboratory in 2007 with somewhat different results. At the time this book went to press, the final draft Se criterion for fish tissue was still pending. There has been some disagreement in the literature on appropriate Se tissue thresholds for fish (Lemly 1993b, 1996, DeForest et al. 1999, Hamilton 2002), with differences in interpretation largely due to different opinions on the environmental relevance of different Se toxicity study types. Lemly (1993b, 1996) and Hamilton (2002), for example, recommended a whole-body Se threshold of 4 µg/g dry wt., while DeForest et al. (1999) recommended whole-body Se thresholds of 9 µg/g dry wt. for warm water fish and 6 µg/g dry wt. for coldwater anadromous fish. In addition, Lemly (1993b, 1996) recommended an ovary/egg Se threshold of 10 µg/g dry wt., while DeForest et al. (1999) recommended an ovary Se threshold of 17 µg/g dry wt. Although the differences in these values may seem small in the field of aquatic toxicology, the differences can be important due to the often narrow range between background levels of Se and toxicity.

This chapter first provides background information on Se toxicity by summarizing field examples of Se impacts on fish populations and discussing Se bioaccumulation in fish. Mechanisms of Se toxicity in fish are then summarized, followed by a detailed review of Se toxicity studies in which Se concentrations were measured in fish tissue. Finally, fish tissue-based Se thresholds are recommended.

5.2 FIELD EXAMPLES OF Se POISONING IN FISH

The most clearly documented case of Se poisoning in fish occurred in Belews Lake (NC, U.S.), a cooling plant reservoir for a coal power plant. Operation of the power plant began in 1974 and by 1976 no centrarchids in the 0–64 mm size range were found and only 8 black and flat bullheads were found in the projected young-of-the-year size range (Cumbie and Van Horn 1978). Similarly, Hyco Reservoir (NC, U.S.) served as a cooling water impoundment for a large coal-fired electric power plant (Skorupa 1998a). In four separate coves of the reservoir, 38–75% reductions in densities of adult fish were observed between 1979 and 1980, and greater than 95% reductions in larval fish densities were observed along three transects. These episodes at Belews Lake and Hyco Reservoir provided the impetus for several of the Se toxicity studies summarized later in this chapter.

Skorupa (1998a) identified 12 real-world examples of Se poisoning, five of which were associated with Se poisoning of fish populations (the other seven examples were for Se poisoning of aquatic bird populations). The five examples with fish included Belews Lake and Hyco Reservoir, and also Martin Lake, TX (U.S.), Sweitzer Lake, CO (U.S.), and the Swedish Lakes Project (Sweden). Martin Lake received unauthorized Se discharges of two fly ash settling ponds containing 2200–2700 μg Se/L. After a die-off of fish in the reservoir was detected in the late 1970s, measured Se concentrations in fish muscle ranged from 2.0 to 9.1 μg/g wet wt. (or 8.0–37.6 μg/g dry wt. assuming 75% moisture) (Garrett and Inman 1984). Sweitzer Lake was created near Delta, Colorado in 1954 for recreational purposes. As cited in Skorupa (1998a), Barnhart (1957) reported that mortality of seven stocked game fishes was attributed to excessive dietary intake of Se. Butler et al. (1991) state that the presence of large green sunfish and carp populations in the lake suggest they may have adapted to Se in the lake, but populations of species once present, such as suckers, are minimal. In addition, channel catfish have been stocked in the lake, but there is no evidence of catfish reproduction in the lake (although it is unclear if suitable spawning habitat was available for catfish, which is not unusual in man-made reservoirs [Skorupa 1998a]). The minimum, median, and maximum whole-body Se concentrations in fish collected in 1988–1989 were approximately 15, 30, and 50 μg/g dry wt., respectively; levels sufficiently high to suggest Se would be toxic. In the Swedish Lakes Project, a leachable rubber matrix containing sodium selenite was suspended in a sack in 11 lakes as a means to mitigate high mercury (Hg) levels. The lakes initially had a Se concentration of about 0.1 μg/L and the target Se concentration was 3–5 μg/L (although up to 25–35 μg Se/L was measured within 100 m of the leach sacks) (Skorupa 1998a). Prior to Se treatment, muscle Se concentrations in yellow perch averaged 0.8–2.8 μg/g dry wt. After the first year of treatment the muscle Se concentrations averaged 6–36 μg/g dry wt. and by the second year perch populations collapsed in several lakes. Ultimately, the researchers could not clearly establish the cause of the perch population collapse, but the possibility of Se poisoning was not ruled out (Skorupa 1998a).

5.3 Se BIOACCUMULATION IN FISH

5.3.1 Overview of Se in the Aquatic Food Chain

In nature, juvenile and adult fish may be simultaneously exposed to inorganic and organic Se forms in both the water and the diet; however, in general, it is exposure to organic Se forms in the diet that is most important at environmentally relevant exposure levels (Bertram and Brooks 1986, Woock et al. 1987, Besser et al. 1993, Coyle et al. 1993). Field-based Se bioaccumulation factors (BAFs) for fish, defined as the ratio of the whole-body Se concentration in fish to the Se concentration in colocated surface water, are generally greater than 1000 (Brix et al. 2005), inversely related to exposure concentration, and can be highly variable between sites and depending on the site conditions. In terms of the BAF, most of the Se bioaccumulation in the aquatic food chain occurs at the level of primary producers and microorganisms. A number of studies have identified primary producers and microorganisms as being the critical link where Se enters the aquatic food chain (e.g., Riedel et al. 1991, Maier and Knight 1994, Sanders and Gilmour 1994, Williams et al. 1994, Malchow et al. 1995, Riedel and Sanders 1996). Although it is mostly organic Se forms that are transferred in the aquatic food chain, Se typically enters aquatic systems, either naturally or anthropogenically, in an inorganic form (e.g., selenate, selenite). Primary producers can rapidly bioconcentrate and incorporate inorganic Se into selenoamino acids, such as selenocysteine or selenomethionine (Ogle et al. 1988). Once in the aquatic food chain, Se trophic transfer factors (TTFs), defined as the Se concentration in an organism's tissue to its concentration in the organism's food item, typically range from <1 to 2 for fish in laboratory studies (DeForest et al. 2007, Luoma and Presser 2009).

5.3.2 Effects of Age, Size, and Trophic Position on Se Bioaccumulation in Fish

Unlike chemicals that truly biomagnify up successive levels of the food chain and with concentrations increasing in older organisms, such as that observed for methyl mercury, Se concentrations in fish do not appear to increase with trophic level or organism age. May et al. (2001, 2008) reported whole-body Se concentrations in various fish species of different sizes collected from the Republican River Basin (Colorado, Nebraska, Kansas—U.S.) and the Solomon River Basin (Kansas—U.S.) and did not observe a consistent pattern of whole-body Se concentrations as a function of fish size. In both basins, whole-body Se concentration in channel catfish tended to be highest in the smallest fish collected, while for other species (common carp, green sunfish, bluegill sunfish, and largemouth bass) there was no apparent trend in Se concentration as a function of size. Muscatello et al. (2008) measured Se concentrations in aquatic food chains from two Se exposure sites downstream from uranium mining and milling operations in northern Saskatchewan, Canada. They observed that Se concentrations increased by approximately 1.5 to 6-fold between plankton, invertebrates, and forage fish; however, no biomagnification was observed between forage fish and predatory fish. Overall, the data from laboratory and field studies suggest that Se concentrations in fish are not size-, age-, or trophic level-dependent. However, work by Stewart et al. (2004) has shown using stable isotope techniques that Se concentrations can increase across trophic levels in specific food webs.

5.3.3 Influence of Exposure Route and Form on Se Bioaccumulation

As summarized in Luoma and Presser (2009), the uptake of dissolved Se by aquatic animals is slow, regardless of form, and therefore makes little or no direct contribution to bioaccumulation and toxicity in animals. This observation is supported by a variety of aquatic Se studies. Lemly (1985), for example, stated that the results of waterborne Se toxicity studies do not give a true indication of the relative sensitivity of fish to Se on a long-term basis in the environment and that it is the dietary component that is responsible for the largest portion of Se present in aquatic animals in natural waters. As concluded in a recent expert Se workshop on Se in aquatic environments, the dietary route of exposure generally dominates Se bioaccumulation processes and traditional approaches for predicting bioaccumulation based on dissolved concentrations does not work for Se (Chapman et al. 2009). Further, organic Se is the critical Se form in aquatic food webs. As summarized in Palace et al. (2004), oviparous vertebrates are the most sensitive organisms to Se, as these organisms efficiently transfer selenomethionine to their eggs, and that selenomethionine is the dominant Se form in fish and bird eggs. These observations are further supported by the studies summarized below.

The Se bioaccumulation and toxicity data for bluegill reported in Cleveland et al. (1993), Coyle et al. (1993), and Hermanutz et al. (1996) support that dietary exposure to Se, and particularly organic Se, is the critical exposure pathway. In Belews Lake, mean whole-body Se concentrations in centrarchids (e.g., bluegill) ranged from approximately 20 to >100 µg/g dry wt. at water Se concentrations of approximately 10 µg/L (Cumbie and Van Horn 1978). However, Cleveland et al. (1993) exposed bluegill to an aqueous 6:1 selenate:selenite mixture up to approximately 1100 µg/L for 60 days and whole-body Se concentrations only reached approximately 10 µg/g dry wt. In the same study, an aqueous selenite concentration of 640 µg/L was necessary to achieve a whole-body Se concentration of approximately 5 µg/g dry wt. in bluegill. The aqueous Se concentration of 640 µg/L (6:1 selenate:selenite) is higher than some acute selenate and selenite LC50 values for invertebrates (U.S. EPA 2004), which questions the environmental relevance of water-only Se exposures for relating fish tissue residues to toxicity. Coyle et al. (1993) observed that Se accumulation of Se in bluegill ovaries increased with increasing dietary organic Se exposures, but not when bluegill were exposed to an aqueous selenate:selenite mixture of 8.4 µg/L. The results of Hermanutz et al. (1996) also demonstrate the importance of the dietary exposure route for Se bioaccumulation. They conducted three sets of experiments in experimental streams. In the first two experiments, bluegill were exposed to aqueous selenite and dietary Se accumulated in the food items present in the streams, while in the third experiment the aqueous selenite source was turned off and bluegill were only exposed to the residual Se present in the food items in the streams. Se levels in the bluegill from experiment III (diet-only exposure) were nearly as high as Se from the aqueous Se exposures. It is critical, therefore, that the Se exposure route in chronic Se toxicity tests include the dietary exposure pathway.

In the Cleveland et al. (1993) study discussed earlier, 5-month-old juvenile bluegill were exposed to a waterborne 6:1 selenate:selenite mixture and 3-month-old juvenile bluegill to dietary seleno-L-methionine. The relationship between whole-body Se and juvenile mortality was much different in the aqueous exposures than in the dietary exposures. For example, at day 30 mortality was 87.5% in fish exposed to an aqueous Se concentration of 2700 µg/L (for comparison, the surface water Se concentration in Belews Lake was approximately 10 µg/L), which resulted in a whole-body Se concentration of 14.5 µg/g dry wt. In the dietary Se study, fish exposed to the highest dietary Se concentration of 26.0 µg/g dry wt. had 17.5% mortality after 90 days, which was not significantly different from control mortality. These fish had a whole-body Se concentration of 13.5 µg/g dry wt. Accordingly, very similar whole-body Se concentrations were associated with almost complete mortality (when exposed to acutely toxic aqueous Se levels) or low mortality (statistically indistinguishable relative to the controls) when exposed to environmentally relevant dietary Se levels.

Due to these observations, in developing and recommending tissue-based Se thresholds we only considered Se toxicity studies that included dietary exposures to organic Se. DeForest et al. (1999) previously excluded water-only Se exposures in recommending fish tissue-based Se thresholds, as did the U.S. EPA in deriving its draft fish tissue-based Se criterion (U.S. EPA 2004). Ultimately, however, this decision had limited influence on the tissue-based thresholds we recommend since, for reasons discussed later, we considered maternal transfer studies to be the most relevant study type for developing and implementing fish tissue-based Se thresholds. In maternal transfer studies, parent fish are exposed to Se, which is transferred to the ovaries and then the eggs, where, at sufficiently high concentrations, can adversely affect larval development and survival. All of the maternal transfer studies available for deriving egg or ovary Se toxicity thresholds are based on dietary exposures to organic Se (either explicitly in the lab or implicitly in field-exposed fish).

5.3.4 Se Bioaccumulation in Different Tissues

Se concentrations in fish, from both laboratory toxicity studies and monitoring in the field, have been measured in a wide variety of tissues, including whole-body tissue, muscle, liver, ovaries, and eggs. The disparity in tissue types analyzed in toxicity tests has made it difficult to compare the relative sensitivities of different fish species and to develop broadly applicable tissue guidelines or criteria. In deriving its draft whole-body-based Se criterion, the U.S. EPA estimated whole-body Se concentrations from individual tissues when necessary (U.S. EPA 2004). For evaluating the reproductive endpoint of embryo/larval development and mortality, the most appropriate tissues to analyze are the ovaries or eggs, as these tissues are closely linked to the site of toxic action. However, due to practical sampling constraints (e.g., getting adequate tissue mass or seasonal issues), whole-body or muscle Se measurements can also be considered. In general, as discussed later, whole-body Se can provide a reasonable indicator of potential reproductive toxicity in fish. However, relationships between Se in tissues can vary by species, which in turn imparts uncertainty in attempting to develop, and then apply, a whole-body- or muscle-based Se threshold in the field. It is apparent that a whole-body or muscle Se guideline could be either dramatically under-protective for one species or over-protective for another species. The clearest approach for overcoming these species-specific relationships in Se bioaccumulation is to focus on an egg or ovary-based Se threshold. Such a Se threshold could still be related to a site-specific whole-body or muscle Se threshold if a species- and site-specific relationship is derived (deBruyn et al. 2008).

5.4 Se TOXICITY IN FISH

In general, adult fish do not appear to be especially sensitive to Se, and there are numerous studies that found no link between elevated egg Se concentrations and fertilization or hatching success, including Gillespie and Baumann (1986), Coyle et al. (1993), Holm et al. (2005), and Muscatello et al. (2006). The classic pathway of documented Se poisoning in fish is exposure of adult female fish to Se, maternal transfer of the Se to the ovaries and then eggs, and then, if sufficiently high egg Se concentrations are reached, larval deformities and mortality. As discussed in Lemly (2002), excess dietary Se is deposited in the developing eggs of fish, particularly in the yolk. After the eggs hatch, the larval fish rapidly use the yolk as an energy supply and a source of protein for building new body tissues. Lemly (2002) also provides a good overview of Se effects on specific fish organs, as well as Se-induced terata (e.g., spinal, craniofacial, and finfold deformities), which are permanent biomarkers of toxicity that can be lethal to the fish. Edema, or accumulation of fluids in the body cavity and head, is another effect of elevated Se levels in fish (Lemly 2002). In the field, larval terata and edema represent an endpoint that can be used as diagnostic evidence of Se poisoning in the fish. In the laboratory, it has also been shown that juvenile fish exposed to dietary organic Se can experience reduced growth and survivorship. Studies in which fish were exposed to aqueous selenite were not considered for reasons discussed in Section 5.3.3. Table 5.1 lists the toxicity studies in which Se was measured in whole-body fish tissue, but excluded from consideration in recommending a whole-body Se threshold.

Chronic dietary Se toxicity studies with fish, in which Se tissue residues were measured, fall under two general categories: (1) exposure of adults and evaluation of effects on offspring (e.g., larval mortality, teratogenesis, and edema) or (2) direct exposure of juveniles. The studies with adults can be further divided into two additional categories: (1) exposure of adults to dietary Se in the laboratory and (2) natural exposure of adults in the field, stripping of gametes, and rearing of embryos and larvae in the laboratory. This latter technique has typically been used to evaluate Se toxicity in salmonids (e.g., rainbow trout, brook trout, and cutthroat trout) and other rather large species where it is more difficult to rear adults in the laboratory. Further, this latter approach is useful for site-specific toxicity evaluations because the Se exposure history of the fish population at the site is reflected in the toxicity test results.

In DeForest et al. (1999) we provided a study-by-study review of Se toxicity studies with fish, in which Se concentrations in fish tissues were measured during and/or after the Se exposure period. That review summarized the Se toxicity studies available at the time (with the exception of Doroshov et al. [1992] and CP&L [1997], which we were not aware of) and ultimately recommended fish tissue-based Se guidelines by considering only those studies that considered dietary exposures to organic Se forms (typically selenomethionine). The following sections separately summarize

TABLE 5.1
Summary of Toxicity Studies in Which Selenium Was Measured in Whole Body or Ovary Tissue (or Specific Tests Within a Study), but Were Excluded from This Evaluation

Reference	Species	Tissue Se (µg/g dry wt.)	Summary of Test Conditions and Basis for Exclusion
Beyers and Sodergren 2002; Hamilton et al. 2005c	Razorback sucker	—	Both studies evaluated Se toxicity to larval razorback suckers, but identified widely different Se thresholds. Beyers and Sodergren (2002) identified a no effects whole-body Se threshold of >42 µg/g dry wt., while Hamilton et al. (2005c) reported that whole-body Se concentrations of 5.4–6.1 µg/g dry wt. adversely affected survival. Given the disparity in the results and because the toxicity response for each source of larvae in Hamilton et al. (2005c) was not clearly related to either the Se exposure concentrations or the internal tissue Se concentrations (see text), we did not attempt to identify a Se threshold for razorback sucker in our evaluation.
Cleveland et al. 1993	Bluegill sunfish	WB NOEC/LOEC = 3.8/5.0	Juvenile bluegill were exposed to aqueous Se (6:1 selenate:selenite) concentrations ranging from 170 to 2700 µg/L for 60 days. The NOEC and LOEC for mortality were 340 and 680 µg/L, which were associated with whole-body Se concentrations of approximately 3.8 and 5.0 µg/g dry wt., respectively. This study was excluded from our Se threshold evaluation because water-only exposures to inorganic Se were not considered environmentally relevant (see Section 5.3.3).
GEI Consultants 2008	Fathead minnow	—	Fish were collected from streams across a gradient of Se concentrations and spawned in the laboratory. However, an effect level could not be estimated due to high variation in the endpoint showing the greatest response to Se (graduated severity index of larval malformations).
Hamilton and Wiedmeyer 1990	Chinook salmon	WW WB NOEC/LOEC = 10.4/19.5 FW WB NOEC/LOEC = 8.2/14.4 BW WB NOEC = >5.3	Fish were exposed to a mixture of aqueous selenate, selenite, boron, and molybdenum in well water (WW), freshwater (FW), and brackish water (BW) for 60–90 days. This study was excluded from our Se threshold evaluation because fish were exposed to a mixture of elevated elements and because water-only exposures to inorganic Se were not considered environmentally relevant (see Section 5.3.3).
Hamilton et al. 1986	Chinook salmon	WB NOEC/LOEC = 9.6/13	Parr were fed Se-amended mosquitofish collected from areas with potential contamination from other chemicals.
Hamilton et al. 1990	Chinook salmon	WB LOEC: <3.3	In one test from this study juvenile fish were fed Se-amended mosquitofish collected from areas with potential contamination from other chemicals (this study included another test in which fish were fed a diet spiked with seleno-DL-methionine—this test was included in our evaluation [Table 5.3]).

continued

TABLE 5.1 (continued)
Summary of Toxicity Studies in Which Selenium Was Measured in Whole Body or Ovary Tissue (or Specific Tests Within a Study), but Were Excluded from This Evaluation

Reference	Species	Tissue Se (µg/g dry wt.)	Summary of Test Conditions and Basis for Exclusion
Hodson et al. 1980	Rainbow trout	WB NOEC: >1.8	Fish were exposed to five aqueous selenite concentrations ranging from 5.5 to 53 µg/L from the newly fertilized egg stage until 44 weeks posthatch, but tissue Se was only measured in the 53 µg Se/L treatment. Negligible effects were observed in the 53 µg Se/L treatment. In addition, this study was excluded from our Se threshold evaluation because water-only exposures to inorganic Se were not considered environmentally relevant (see Section 5.3.3).
Hunn et al. 1987	Rainbow trout	WB NOEC/LOEC = 2.6/4.3	Fry were exposed to five aqueous selenite concentrations ranging from 7.8 to 100 µg/L for 90 days. The aqueous NOEC and LOEC for mortality were 21.0 and 47.2 µg/L, respectively. This study was excluded from our Se threshold evaluation because water-only exposures to inorganic Se were not considered environmentally relevant (see Section 5.3.3).

WB = whole body; NOEC = no-observed-effect-concentration; LOEC = lowest-observed-effect-concentration.

the maternal transfer and juvenile toxicity studies with Se. The review is then further organized by fish species. Only studies that measured Se concentrations in whole-body tissue, ovaries, or eggs were included in this evaluation. In addition, due to species-specific differences in Se concentration relationships between tissues discussed in Section 5.3.4, this evaluation only considered measured Se concentrations in fish (no attempt was made to estimate Se concentrations in one tissue from concentrations measured in another tissue).

5.4.1 MATERNAL TRANSFER SE EXPOSURES: EFFECTS ON EMBRYO/LARVAL DEVELOPMENT AND SURVIVAL

Studies in which adult fish were exposed to dietary organic Se, and effects on offspring were evaluated, have been conducted with 13 species of fish: bluegills (*L. macrochirus*), brook trout (*Salvelinus fontinalis*), brown trout (*Salmo trutta*), cutthroat trout (*Oncorhynchus clarki*), Dolly Varden (*Salvelinus malma*), fathead minnows (*Pimephales promelas*), largemouth bass (*Micropterus salmoides*), eastern mosquitofish (*Gambusia holbrooki*), western mosquitofish (*Gambusia affinis*), northern pike (*Esox lucius*), rainbow trout (*Oncorhynchus mykiss*), razorback suckers (*Xyrauchen texanus*), and white suckers (*Catostomus commersoni*). The toxicity results are summarized below as no-observed effect concentrations (NOECs) and lowest-observed effect concentrations (LOECs). The NOEC is the highest Se concentration in a study that did not result in statistically significant effects relative to the control, and the LOEC is the lowest Se concentration in a study that did result in statistically significant effects relative to the control. In addition, EC10 and EC20 values (10% and 20% effect concentrations) were identified, which were either reported by the study authors or which we calculated from the concentration-response data reported in the study.

5.4.1.1 Bluegill

Bryson et al. (1984, 1985a, 1985b)—Bryson et al. (1984) collected adult bluegills (*L. macrochirus*) from the Se-contaminated Hyco Reservoir (NC, U.S.) and from a reference lake. All combinations of males and females from Se-contaminated and reference lakes were artificially crossed in different ash pond concentrations (0%, 20%, and 50% ash pond water). Swim-up larvae were fed zooplankton from either Hyco Reservoir or the reference lake. Larvae were observed for 28 days after hatching. All larvae from a Hyco Reservoir female exhibited abnormal development and 100% mortality prior to reaching swim-up. The mean ovary Se concentration in the females from Hyco Reservoir was 49 µg/g dry wt. (Table 5.2).

Bryson et al. (1985a), based on a study similar to Bryson et al. (1984), reported that egg hatchability from female bluegills averaging 9.1 µg/g dry wt. Se in the ovaries was 88%, and that 93% survived through swim-up (Table 5.2). Swim-up was only approximately 5.6% in larvae from females averaging 30 µg Se/g dry wt. in the ovaries (Table 5.2). The results reported in Bryson et al. (1985b) are somewhat difficult to interpret because none of the control fish spawned, but hatching and swim-up success were 83.8–86.6% and 91.1–95.5%, respectively, from parents with a mean ovary Se concentration of 14.8 µg/g dry wt., and 86.0% and 83.3–97.4% from parents with a mean ovary Se concentration of 9.2 µg/g dry wt. (Table 5.2). Overall, the studies by Bryson and colleagues resulted in ovary NOECs of >9.1, >9.2, and >14.8 µg/g dry wt. and ovary LOECs of <30 and <49 µg/g dry wt. (Table 5.2).

Gillespie and Baumann (1986)—Adult bluegills (*L. macrochirus*) were collected from the Se-contaminated Hyco Reservoir and from an uncontaminated reservoir and all possible parent combinations were artificially crossed. Zygotes were reared in uncontaminated water and evaluated for percent fertilization and hatching success. In over 18 crosses during a two year period, no significant differences in percent fertilization or percent hatch were found between all parent combinations. However, all crosses involving a female from the Se-contaminated water body resulted in larvae with gross abnormal morphology (65–100%), and all larvae died before they reached the swim-up stage. The ovary Se concentration in the female producing 65% abnormal larvae was 38.6 µg/g dry wt. (assuming a moisture content of 85%, per Gillespie and Baumann [1986]). Accordingly, the ovary LOEC from this study was <38.6 µg/g dry wt. (Table 5.2).

Doroshov et al. (1992)—Bluegill (*L. macrochirus*) were fed Se concentrations of 5.5, 13.9, and 21.4 µg/g (as seleno-L-methionine). Natural spawning was not successful, so ovulation was hormonally induced and fertilization was conducted *in vitro*. The most sensitive endpoint was larval edema, which was observed during the first 5 days posthatch. Edema was observed in 0%, 5%, and 95.7% of the larvae in the control, 5.5, 13.9, and 21.4 µg/g Se diets, respectively. Because 5% edema was statistically significant ($p \leq .05$) in the 13.9 µg/g Se treatment, 5.5 and 13.9 µg/g were identified as the dietary NOEC and LOEC. The mean egg Se concentrations associated with the NOEC and LOEC were reported as 8.3 and 19.5 µg/g dry wt., and the mean ovary Se concentrations associated with the NOEC and LOEC were reported as 6.34 and 14.1 µg/g dry wt. In terms of identifying a Se toxicity threshold from this study, it is interesting to note that the egg Se concentrations associated with the dietary LOEC of 13.9 µg/g were variable. Edema was measured in six progenies from parents exposed to this dietary Se concentration and the associated egg Se concentrations are reported for three of these progenies in Table 21 of Doroshov et al. (1992). Egg Se concentrations for these three progenies from the 13.9 µg Se/g diet were 8.55, 22.06, and 30.20 µg/g dry wt. Likewise, ovary Se concentrations were 3.94, 21.82, and 20.40 µg/g dry wt. for the 13.9 µg Se/g dietary treatment. Comparison of these Se concentrations to the edema data reported in Appendix 8 of Doroshov et al. (1992) shows that edema was 0% at an egg Se concentration of 8.55 µg/g dry wt. (ovary Se = 3.94 µg/g dry wt.), 10% at 22.06 µg/g dry wt. (ovary Se = 21.82 µg/g dry wt.), and 10% at 30.20 µg/g dry wt. (ovary Se = 20.40 µg/g dry wt.). Thus, based on data reported for three of the six progenies, 10% edema was only observed in progeny associated with egg Se concentrations ranging from 22.06 to 30.20 µg/g dry wt. (geometric mean of 25.81 µg/g dry wt.) or ovary Se concentrations ranging from

TABLE 5.2
Summary of Toxicity Studies That Evaluated Selenium Toxicity to Embryos/Larvae Resulting from Maternal Transfer

Species	Reference	Adult Exposure	Endpoint	Tissue	Se Concentration (μg/g dry wt.)			
					NOEC	LOEC	EC10	EC20
Bluegill	Bryson et al. 1984	Field	Larval mortality	Ovary	—	<49	—	—
	Bryson et al. 1985a	Field	Hatchability/swim-up	Ovary	>9.1	—	—	—
	Bryson et al. 1985b	Field	Hatchability/swim-up	Ovary	—	<30	—	—
		Field	Hatchability/swim-up	Ovary	>14.8	—	—	—
		Field	Hatchability/swim-up	Ovary	>9.2	—	—	—
	Gillespie and Baumann 1986	Field	Larval edema	Ovary	—	<38.6	—	—
	Doroshov et al. 1992	Lab	Larval edema	Ovary	3.94	21.10	16[a]	20[a]
		Lab	Larval edema	Egg	8.55	25.81	21[a]	23[a]
	Coyle et al. 1993	Lab	Larval mortality	WB	7	16	8[a]	8.5[a]
		Lab	Larval mortality	Ovary	20	35	24[a]	27[a]
		Lab	Larval mortality	Egg	22.5	41.3	22[a]	26[a]
	Hermanutz et al. 1996	Mesocosm	Larval edema	WB	4.4	21.8	7.7[a]	9.7[a]
		Mesocosm	Larval edema	Ovary	17.3	69	30[a]	36[a]
Brook trout	Holm et al. 2005	Field	Larval deformities	Egg	17	20	20 (EC06)	—
Brown trout	Formation Environmental 2009	Field	Alevin mortality	Egg	—	—	17.7	21.6
		Field	Larval deformities	Egg	—	—	19.3	21.7

Species	Reference	Study type	Endpoint	Tissue	NOEC	LOEC	EC10	EC20
Cutthroat trout	Kennedy et al. 2000	Field	Larval deformities/ mortality	Egg	>21	—	—	—
	Hardy et al. 2010	Lab	Larval deformities/ mortality	WB	>11.37	—	—	—
		Lab	Larval deformities/ mortality	Egg	>16.04	—	—	—
	Rudolph et al. 2008	Field	Larval deformities	Egg	20.6	46.8	—	—
		Field	Alevin mortality	Egg	—	—	17[a]	23[a]
Dolly Varden	Golder 2009	Field	Larval deformities	Egg	—	—	54	60
Fathead minnow	Ogle and Knight 1989	Lab	Reproduction	WB	>7.5	—	—	—
		Lab	Reproduction	Ovary	>10.92	—	—	—
	Schultz and Hermanutz 1990	Mesocosm	Larval edema/lordosis	Ovary	—	<23.6	—	—
Largemouth bass	CP&L 1997	Lab	Larval mortality	Ovary	—	—	22[a]	24[a]
Northern pike	Muscatello et al. 2006	Field	Larval deformities	Egg	3.80	31.28	20.38	33.55
Rainbow trout	Holm et al. 2005	Field	Larval deformities	Egg	17	25	23[a]	27[a]
Eastern mosquitofish	Staub et al. 2004	Field	Brood size/offspring viability	WB	>11.85	—	—	—
Western mosquitofish	Saiki et al. 2004	Field	Fry mortality and deformities	WB	>17.5	—	—	—
Razorback sucker	Hamilton et al. 2005a, 2005b	Field	Larval deformities	Egg	37.8	46.5	—	—
White sucker	de Rosemond et al. 2005	Field	Larval deformities	Egg	—	—	25.6 (EC13)	—

[a] We calculated these EC10 and EC20 values from the concentration-response data reported in the original studies.

WB = whole body; NOEC = no-observed-effect concentration; LOEC = lowest-observed-effect concentration; EC10 = 10% effect concentration; EC20 = 20% effect concentration.

20.40 to 21.82 µg/g dry wt. (geometric mean of 21.10 µg/g dry wt.). Accordingly, the geometric mean egg and ovary Se concentrations of 25.81 and 21.1 µg/g dry wt., respectively, were identified as the LOECs from this study (Table 5.2). By fitting a probit model to the concentration-response data for edema, we calculated ovary Se EC10 and EC20 values of 16 and 20 µg/g dry wt. and egg Se EC10 and EC20 values of 21 and 23 µg/g dry wt. (Table 5.2).

Coyle et al. (1993)—Bluegill (*L. macrochirus*) were fed diets containing Se concentrations of 0.8, 4.6, 8.5, 16.8, and 33.3 µg/g, and simultaneously exposed to a nominal aqueous Se concentration of 10 µg/L for 140 days. Seleno-L-methionine was used for the dietary exposures and a 6:1 ratio of selenate:selenite was used for aqueous exposures. Spawning frequency, fecundity, and hatching success were monitored during the last 80 days of the test and fry survival was monitored for 30 days after hatch (fry were exposed to the same aqueous Se concentrations as their parents over the 30-day posthatch duration). No effects on spawning frequency, fecundity, or hatching success were observed in any of the treatments. In addition, growth of bluegill fry at either 15 or 30 days posthatch was unaffected by any of the Se treatments tested. However, larvae from females fed the 33.3 µg Se/g diet had greatly reduced survival (7%) relative to other treatments (75–90%) 5–6 days after hatching. All bluegill larvae in this study experienced increased mortality from approximately day-6 through day-30 posthatch, including control fish. Coyle et al. (1993) attributed this increased mortality to the likelihood of starvation resulting from the unsuccessful transition from endogenous to exogenous feeding. The dietary NOEC and LOEC, therefore, was 16.8 and 33.3 µg Se/g, respectively. The corresponding mean ovary Se concentrations at the NOEC and LOEC were 20 and 35 µg/g dry wt., respectively, and the mean egg Se concentrations at the NOEC and LOEC were 22.5 and 41.3 µg/g dry wt. (Table 5.2). On the basis of the whole-body Se concentrations in adults on the day spawning was initiated, the whole-body Se concentrations associated with the NOEC and LOEC were 7 and 16 µg/g dry wt., respectively (Table 5.2). By fitting a probit model to the concentration-response data for larval mortality, we calculated EC10 and EC20 values of 24 and 27 µg/g dry wt. for ovaries, 22 and 26 µg/g dry wt. for eggs, and 7.8 and 9.2 µg/g dry wt. for whole body (Table 5.2).

Hermanutz et al. (1992, 1996); Tao et al. (1999)—Adult bluegill (*L. macrochirus*) were exposed to Se in experimental streams dosed with selenite. The streams contained well-developed assemblages of fish food organisms, so an environmentally relevant dietary Se exposure pathway was simulated. Three Se exposure regimes were tested: (1) bluegills were exposed in streams dosed with 10 and 30 µg/L selenite; (2) bluegills were exposed in streams dosed with 2.5 and 10 µg/L selenite; and (3) bluegills were exposed in streams that had previously been dosed with either 2.5 or 10 µg/L selenite, but the selenite source had been turned off. Of these studies, we focused on the results from the second study that evaluated streams continuously dosed to maintain aqueous Se concentrations of approximately 2.5 and 10 µg/L (plus a control stream). At the end of the exposure period, whole-body Se concentrations in adult bluegill were approximately 1.6, 4.4, and 21.8 µg/g dry wt. (assuming 75% moisture) and ovary Se concentrations were 5.1, 17.3, and 69.0 µg/g dry wt. (assuming 85% moisture, per Gillespie and Baumann [1986]). The most sensitive endpoint was larval edema, which was 0% in the control stream and in the 2.5 µg Se/L stream and 100% in the 10 µg Se/L stream. Thus, the whole-body NOEC and LOEC were 4.4 and 21.8 µg/g dry wt., respectively, and the ovary NOEC and LOEC were 17.3 and 69 µg/g dry wt. (Table 5.2).

5.4.1.2 Brook Trout

Holm (2002); Holm et al. (2003, 2005)—Gametes from adult brook trout (*S. fontinalis*) were collected from a stream with elevated Se (Luscar Creek, downstream from active coal mining), intermediate Se (Gregg River), and a reference stream (Cold Creek) in west-central Alberta, Canada. Eggs were fertilized in the laboratory and deformities and edema in fry were evaluated. The most pronounced effect observed in the trout from Luscar Creek was a 7.9 ± 1.8% (mean ± SE) incidence of craniofacial deformities, compared with a 2.3 ± 0.8 and 2.1 ± 0.6% incidence of craniofacial deformities in fish collected from the Gregg River and the Cold Creek reference stream, respectively.

The incidence of craniofacial deformities was significantly different ($p < .05$) in Cold Creek relative to Luscar Creek, but not in the Gregg River relative Luscar Creek (despite similar levels of craniofacial deformities observed in the Gregg River and Cold Creek). The mean (±SE) egg Se from parent fish collected from Luscar Creek was 7.8 ± 0.7 µg/g wet wt. (20 ± 1.8 µg/g dry wt.), compared to mean egg Se concentrations of 6.6 ± 0.4 µg/g wet wt. (17 ± 1.0 µg/g dry wt.) for Gregg River fish and 1.3 ± 0.1 µg/g wet wt. (3.3 ± 0.26 µg/g dry wt.) for Cold Creek fish (wet-to-dry weight conversions based on an assumed moisture content of 61% [Holm et al. 2005]). Accordingly, egg Se concentrations of 17 and 20 µg/g dry wt. may be considered the NOEC and LOEC from this study, although the LOEC based on Luscar Creek fish is based on just a 6% increase in craniofacial deformities relative to the intermediate Se (Gregg River) and reference site (Cold Creek) (Table 5.2).

5.4.1.3 Brown Trout

Formation Environmental (2009)—Adult brown trout (*S. trutta*) were collected from southeast Idaho streams with varying Se levels. Gametes were stripped and eggs were fertilized in the field. The fertilized eggs were then transported to the laboratory, where they were reared through swim-up. Larvae were observed for mortality and deformities. The resulting egg Se EC10 and EC20 values were 17.7 and 21.6 µg/g dry wt. for alevin mortality and 19.3 and 21.7 µg/g dry wt. for larval deformities (Table 5.2).

5.4.1.4 Cutthroat Trout

Kennedy et al. (2000)—Adult cutthroat trout (*O. clarki lewisi*) were collected from a stream with elevated Se levels (Fording River) and a reference site (Connor Lake) in southeastern British Columbia, Canada. Gametes were stripped and eggs were fertilized and transported to the laboratory for rearing. Larvae and fry were then observed for deformities and mortality. Se concentrations in Fording River eggs were highly variable, ranging from 8.7 to 81.3 µg/g dry wt., with an average of 21.2 µg/g dry wt. No significant ($p < .05$) correlations between Se concentrations in eggs from Fording River and Connor Lake and hatching time, deformities, and mortalities were observed. However, the variability in the data may have masked some of the potential Se-related effects. For example, percent mortalities ranged from 1.5% to 43.7% after yolk absorption, with the correlation between egg Se and mortalities having a *p*-value of 0.15. Because no statistically significant effects could be identified from this study, the mean egg Se concentration of 21.2 µg/g dry wt. was identified as a NOEC. Although there is added uncertainty in this value given the variability in the data and the relationship between egg Se and larval mortality postyolk absorption, the other toxicity studies with cutthroat trout, summarized later, suggest that this egg Se NOEC estimate is not unreasonable.

Hardy et al. (2010)—The sensitivity of southeast Idaho reared cutthroat trout (*O. clarki*) to dietary Se was evaluated in a multiyear laboratory study. Eyed cutthroat trout eggs were obtained from the Henry's Lake hatchery (southeast Idaho) and reared in the laboratory. Once trout reached the swim-up stage, they were fed Se diets of 1.2 (control), 3.8, 6.4, 9.0, 11.5, or 12.0 µg/g through maturation and spawning. After 80 weeks of feeding, control mortality was 19.5%, and then generally decreased with increasing dietary Se concentration (mortality was 3.0% in fish fed the highest Se diet of 12.0 µg/g). The concentration-response relationship for larval deformities was parabolic, with the highest percentage of deformities (15.9–20.2%) being observed at the intermediate exposure levels and the percentage of deformities at the two highest treatments (6.8–7%) were comparable to in the control fish (5.6%). The percentages of total deformities are higher than the percentages of craniofacial deformities. Hardy et al. (2010) noted that for comparison, Lemly (1997) considered deformity rates of <6% for warm water fishes indistinguishable from background and that deformity rates between 6% and 12% the effect of maternal Se intake is deemed slight. Hardy et al. (2010) also noted that this categorization is based on specific deformities of the skull and jaw, not with general deformities often seen in trout and salmon, such as curled tails and cojoined twins. Accordingly, the low frequency of deformities (both total and craniofacial) in the highest dietary Se treatments, coupled with the low incidence (<10%) of craniofacial deformities in all treatments, suggests that none of the dietary Se treatments resulted in significant effects on cutthroat trout in this study. The highest whole-body and egg Se concentrations measured across all dietary Se treatments were 11.37 and 16.04 µg/g dry wt. The study also evaluated Se depuration and found that cutthroat trout fed the highest Se diet excreted Se at the highest rate. Consequently, Hardy et al. (2010) suggested that cutthroat trout have the ability to tolerate relatively high levels of Se exposure by maintaining non-toxic body levels via enhanced excretory mechanism. The whole-body and egg Se NOECs from this study, therefore, were >11.37 and >16.04 µg/g dry wt. (Table 5.2).

Rudolph et al. (2008)—Adult westslope cutthroat trout (*O. clarki lewisi*) were collected from a reference site (O'Rourke Lake) and a site with elevated Se levels (Clode Pond) in southeastern British Columbia, Canada. Similar to Kennedy et al. (2000), gametes were stripped and eggs were fertilized in the field. The fertilized eggs were transported to the laboratory and fry were evaluated for deformities, edema, and mortality. Mortality of Clode Pond larvae was higher than in larvae from the reference site, but no significant differences ($p < .05$) for percent fertilization, percent hatch, or mortality of eggs and alevins were observed. The eggs from four of the Clode Pond fish died before they reached the laboratory—the Se concentration in these eggs ranged from 86.3 to 140 µg/g dry wt. In addition, eggs with Se concentrations >46.8 and <75.4 µg/g dry wt. were successfully fertilized, but no viable fry were produced. The remaining egg Se concentrations in Clode Pond fish ($n = 4$) ranged from 11.8 (estimated from muscle) to 20.6 (measured) µg/g dry wt. (mean egg Se concentration of 16.4 µg/g dry wt.). For comparison, Se concentrations in O'Rourke Lake eggs ranged from 12.3 to 16.7 µg/g dry wt. (mean of 14.1 µg/g dry wt.). All viable fry after yolk absorption at swim-up were assessed for the frequency and severity of skeletal, craniofacial, and finfold deformities, as well as for edema. Deformities and edema were scored form 0 (normal) to 3 (severe). The frequency and severity of skeletal, craniofacial, and finfold deformities observed in Clode Pond offspring were either statistically insignificant or significantly ($p < .05$) lower than the O'Rourke Lake fish. For edema, the frequency was significantly ($p = .007$) greater in Clode Pond offspring than in O'Rourke Lake offspring (87.7% vs. 61.2%) and the severity score was significantly ($p = .007$) greater (0.88 vs. 0.61). However, the authors noted that the edema was not severe (indicative by the total severity score of <1 for edema), as only slight fluid accumulation was observed around the eyes and no edema of the yolk sac or pericardial cavity was observed. However, there did appear to be a clear relationship between egg Se and alevin mortality, which is a Se-relevant endpoint as alevin mortality occurred during yolk sac absorption and, hence, exposure to Se. With one exception, the percentage of dead alevins was ≤15% at egg Se concentrations ranging from 11.8 to 20.6 µg/g dry wt. On the basis of the data plotted in Figure 1 of Rudolph et al. (2008), we estimated egg Se EC10 and EC20 values of 17 and 23 µg/g dry wt. for alevin mortality (Table 5.2).

5.4.1.5 Dolly Varden

Golder (2009)—Adult Dolly Varden (*S. malma*) were targeted for collection from high-Se waters (Upper Waste Rock Creek and Upper Waste Rock Creek ponds), a moderate Se water (Lower Waste Rock Creek), and a low Se water (North Kemess Creek) in northern British Columbia, Canada. However, no mature females were available from the high-Se waters. Gametes were stripped from fish collected from the moderate and low Se sites and eggs were fertilized in an on-site laboratory. The fertilized eggs were then transported to the toxicology laboratory, where they were reared through hatching and the test was terminated when at least 90% of the fry reach the swim-up stage. Fry were evaluated for deformities (skeletal, craniofacial, and finfold defects) and edema, which were scored from 0 (normal) to 3 (severe). Using a probit model, Golder (2009) derived egg Se EC10 and EC20 values of 54 and 60 µg/g dry wt., respectively (Table 5.2).

5.4.1.6 Fathead Minnow

Ogle and Knight (1989)—Approximately 60-day-old fathead minnows (*P. promelas*) were provided a diet spiked with 25% seleno-L-methionine, 25% selenate, and 50% selenite at measured concentrations of 5.2, 10.2, 15.2, 20.3, and 29.5 µg Se/g. Every 2 weeks fish were collected and weighed, and one fish from two of four replicates was removed for Se analysis. Spawning substrates were provided on day 98. On day 105, a male and female were selected and the spawning period was extended for 30 days after the first spawning event in each replicate. Fish were collected for Se analysis at the end of the spawning period. Eggs were collected and examined for fertility and incubated for determination of hatchability. Survival of larval fish after 14 days was determined. After 98 days of exposure, growth was significantly ($p < .05$) reduced (16% relative to the control) in fish fed the diet containing 20.3 µg Se/g. The ecological significance of this level of growth reduction (16%) is questionable given the variability in the responses at which significant effects were observed (1.30 ± 0.22 g in the control and 1.09 ± 0.16 g in fish fed 20.3 µg Se/g) and, that the fish fed the highest Se diet (29.5 µg/g) had no significant ($p > .05$) effects on any reproductive parameter evaluated: number of spawns per pair, number of eggs per spawn, percent hatch, or larval survival to 14 days. For each of these endpoints, the responses were similar to the controls, suggesting that the dietary Se threshold for reproductive effects is greater than the highest dietary concentration measured (29.5 µg Se/g). Correspondingly, no reproductive effects were observed in fish with a mean ovary Se concentration of 10.9 µg/g dry wt. or a whole-body Se concentration of 7.5 µg/g dry wt. (Table 5.2).

Schultz and Hermanutz (1990)—Fathead minnows (*P. promelas*) were exposed to Se in experimental streams continuously dosed with 10 µg/L selenite. The streams contained well-developed assemblages of fish food organisms, so an environmentally relevant dietary exposure pathway was simulated. Submerged spawning platforms were provided and checked daily for embryos. Samples of collected embryos were then reared in incubation cups containing stream water with the same Se concentrations as in the streams from which they were collected. Each sample was observed each posthatch day for edema and lordosis (larvae were not fed, and observations continued until all larvae died). Se concentrations in the ovaries of the adult females were also determined. The incidence of edema and lordosis in larvae from the dosed stream, 24.6% and 23.4%, respectively, was significantly ($p < .05$) greater than in the control stream (0.9% and 5.6%). The mean Se concentration in ovaries of the parental fish was 5.89 µg/g dry wt., or 23.6 µg/g dry wt. assuming 75% moisture in fathead minnow ovaries). Accordingly, the LOEC from this study was identified as <23.6 µg/g dry wt. (Table 5.2).

5.4.1.7 Largemouth Bass

CP&L (1997)—Adult largemouth bass (*M. salmoides*) were provided artificial diets spiked with seleno-DL-methionine for several months prior to spawning. In 1995 bass were provided diets with measured Se concentrations of 2.9, 7.5, and 11.2 µg/g dry wt., and in 1996 a second set of bass were provided diets with measured Se concentrations of 26.7, 53.1, and 78.4 µg/g dry wt. The fish spawned in the laboratory and the resulting larvae were monitored until swim-up for mortality and deformities. The most sensitive endpoint was larval mortality. On the basis of a logistic concentration-response curve developed by the study authors, we calculated ovary Se EC10 and EC20 values of 22 and 24 µg/g dry wt., respectively (Table 5.2).

5.4.1.8 Mosquitofish

Saiki et al. (2004)—Adult western mosquitofish (*G. affinis*) were collected from two Se-contaminated sites (Sand Luis Drain and North Mud Slough) and two reference sites (North Mud Slough upstream of San Luis Drain and a location on the San Joaquin River with no history of receiving seleniferous drainwater) in central California, U.S. Pregnant females were identified (they are a live-bearing fish) and placed in glass beakers. After birth, live and dead fry were counted, then females and fry were killed with an overdose of MS-222 and measured for length and weight, and fry were evaluated for gross evidence of external deformities. Fry survival averaged greater than 96% from all sites and no fry exhibited evidence of teratogenesis. The highest whole-body Se concentration measured in parent fish was 17.5 µg/g dry wt., which we considered a "greater than" NOEC for this study (Table 5.2).

Staub et al. (2004)—Gravid female eastern mosquitofish (*G. holbrooki*) were collected from a coal ash basin system and a reference pond on the Savannah River Site in South Carolina, U.S. After acclimation in the laboratory, fish were transferred into brood chambers, which included a small opening that allowed offspring to avoid cannibalism in an attached bag. Offspring were inspected for mortality and deformities. Brood sizes and offspring viability at birth were not significantly different ($p > .05$) between ash pond and reference site fish. The mean whole-body Se concentration measured in parent females was 11.85 μg/g dry wt., which we considered a "greater than" NOEC from this study (Table 5.2).

5.4.1.9   Northern Pike

Muscatello et al. (2006)—Gametes were obtained from northern pike (*E. lucius*) collected approximately 2 km (high Se), 10 km (intermediate Se), and 15 km (low Se) downstream of a uranium milling operation in northern Saskatchewan, Canada. The mean egg Se concentrations in the high, intermediate, and low Se exposure sites were 48.23, 31.28, and 3.80 μg/g dry wt., respectively. Gametes were also collected from a reference site, which had a mean egg Se concentration of 3.19 μg/g dry wt. Of 19 trace elements analyzed in the egg samples, only Se was present at elevated concentrations in the intermediate and high Se exposure sites. Eggs were fertilized in the field and transported to the laboratory. The endpoints evaluated included fertilization success, larval deformities, and larval edema. A concentration-response relationship was observed between egg Se concentrations and total embryo deformities, and the study authors calculated egg-based Se EC10 and EC20 values of 20.38 and 33.55 μg/g dry wt., respectively (Table 5.2). The NOEC and LOEC were 3.80 and 31.28 μg/g dry wt., respectively (Table 5.2).

5.4.1.10   Rainbow Trout

Holm (2002); Holm et al. (2003, 2005)—Gametes from adult rainbow trout (*O. mykiss*) were collected from a stream with elevated Se (Luscar Creek, downstream from active coal mining), intermediate Se (Gregg River), and reference streams (Deerlick and Wampus creeks) in west-central Alberta, Canada. Eggs were fertilized in the laboratory and embryos and larvae were reared to the swim-up stage. Fry were evaluated for deformities and other indicators of Se toxicity, such as edema. Craniofacial and skeletal deformities in larvae were more prevalent from fish collected in the elevated Se stream being, on average, 33% and 25% at Luscar Creek compared to 7.1–12% and 7.9–9.2% at the reference streams, respectively. Larval edema was also higher at Luscar Creek (34.5%) relative to Deer Creek (6.1%) and Wampus Creek (23.3%), although the difference was only significant ($p < .05$) relative to Deer Creek. The mean ($\pm$SE) egg Se from parent fish collected from Luscar Creek (elevated Se) was 9.9 ± 1.4 μg/g wet wt. (25 ± 3.6 μg/g dry wt.), compared to mean egg Se concentrations of 6.5 ± 1.6 μg/g wet wt. (17 ± 4.1 μg/g dry wt.) for Gregg River fish, and 3.5 ± 0.4 μg/g wet wt. (9.0 ± 1.0 μg/g dry wt.) for both Deer Creek and Wampus Creek fish (wet-to-dry weight conversions based on an assumed moisture content of 61% [Holm et al. 2005]). Holm et al. (2005) estimated there was a 15% probability of observing craniofacial deformities, skeletal deformities, and edema at an egg Se concentration between 8.8 and 10.5 μg/g wet wt. (or 23–27 μg/g dry wt. assuming 61% moisture). Overall, we estimated egg-based NOEC, LOEC, EC10, and EC20 values of 17, 25, 23, and 27 μg/g dry wt., respectively, from this study (Table 5.2).

5.4.1.11   Razorback Sucker

Hamilton et al. (2005a, 2005b)—Adult razorback suckers (*X. texanus*) were exposed for 9 months to a gradient of Se concentrations in ponds and isolated river channels of the Colorado River near Grand Junction, Colorado (U.S.). The water bodies used in the study were: (1) Horsethief ponds, reference site (2.2 μg Se/L); (2) Adobe Creek (3.8 μg Se/L); and (3) North Pond (9.5 μg Se/L). Suckers in Horsethief ponds were fed a commercial diet, while fish in Adobe Creek and North Pond were fed naturally available diets. The Se concentrations in food were 1.1 μg/g in commercial fish food, 4–56 μg/g in Adobe Creek, and 20–81 μg/g in North Pond. After the 9-month-adult exposure, fish were collected and induced to spawn. The eggs were then used in a 9-day study to evaluate egg survival, hatchability, and embryo deformities. The mean egg Se (SE) concentrations (µg/g dry wt.) were 6.5 (0.2) at Horsethief ponds, 46.5 (4.6) at Adobe Creek, and 37.8 (1.4) at North Pond. There were no significant differences between egg sources on viability, survival, hatch, or hatchability, but the percentage of embryos with deformities was significantly ($p < .05$) increased in the Adobe Creek test (26% deformities vs. 18% in Horsethief ponds). Thus, the egg Se concentration of 46.5 µg/g dry wt. from Adobe Creek was associated with significantly increased deformities, while an egg Se concentration of 37.8 µg/g dry wt. was not associated with any adverse effects compared to the reference site (Table 5.2).

5.4.1.12 White Sucker

de Rosemond et al. (2005)—Gametes were obtained from white sucker (*C. commersoni*) collected from Island Lake, which is downstream of the Cluff Lake uranium mine in northern Saskatchewan, Canada. Eggs were fertilized in the field, and embryos were reared in the laboratory. The mean egg Se concentration was 25.6 µg/g dry wt. and mean total developmental deformities were 12.8% (Table 5.2). A limitation of this study is that an insufficient number of suckers could be collected at an uncontaminated reference lake. Nevertheless, the egg Se concentration and level of deformities observed based on gametes collected from the elevated Se site is consistent with the other field-based studies for salmonids and northern pike described earlier.

5.4.2 DIRECT DIETARY SE EXPOSURES: EFFECTS ON LARVAL/JUVENILE GROWTH AND SURVIVAL

Studies in which juvenile fish were exposed to dietary organoselenium and whole-body Se concentrations were measured, and effects on growth and survival were evaluated have been conducted with six species of fish: bluegills (*L. macrochirus*), Chinook salmon (*Oncorhynchus tshawytscha*), fathead minnows (*P. promelas*), rainbow trout (*O. mykiss*), razorback suckers (*X. texanus*), and white sturgeon (*Acipenser transmontanus*). These are summarized below.

5.4.2.1 Bluegill

Lemly (1993a)—Juvenile bluegill (*L. macrochirus*) were exposed to 5.1 µg Se/g (as seleno-L-methionine) in the diet and 4.8 µg Se/L (as 1:1 selenate:selenite) in water for 180 days. One set of fish was exposed to Se under simulated winter conditions, whereby water temperature was gradually reduced from 20°C to 4°C over a period of 8 weeks and photoperiod was simultaneously reduced. Another set of fish was exposed to Se at a water temperature of 20°C over the entire 180-day exposure. Under simulated winter conditions, bluegills reduced activity and feeding. The increased respiratory demand combined with reduced feeding resulted in lipid depletion and subsequent mortality in a significant number of test fish. The whole-body Se concentration at the onset of winter conditions was 5.85 µg/g dry wt., and the whole-body Se concentration after the 120-day exposure to winter conditions was 7.91 µg/g dry wt. Bluegills exposed to the same Se level, but in 20°C water, continued to feed and did not lose sufficient lipids for mortality to occur. We considered the whole-body LOEC from this study, under simulated winter conditions, to be <5.85 µg/g dry wt. (Table 5.3), as this was the whole-body Se concentration at the onset of the final test temperature of 4°C (the higher whole-body Se concentration of 7.91 µg/g dry wt. at test termination simply reflected an increase in the Se concentration due to lipid depletion during the test).

This study was the basis for the U.S. EPA's draft two-part whole-body fish tissue criterion for Se (U.S. EPA 2004). The first part was a trigger concentration of 5.85 µg/g dry wt. that would be applied to whole-body fish Se concentrations measured in summer or fall. If this concentration was exceeded, fish should be sampled in the winter to determine whether the Se concentration exceeded 7.91 µg/g dry wt. Because Lemly (1993a) was the only study to evaluate the combined effects of Se and simulated winter stress in bluegills (or any fish), and was the basis for the U.S. EPA's draft whole-body Se criterion, the study has received an extra level of scrutiny. One limitation of the

TABLE 5.3
Summary of Toxicity Studies That Evaluated Selenium Toxicity to Juveniles from Direct Dietary Exposures (See Table 5.1 for a List of Studies Excluded from This Table)

Species	Reference	Endpoint	Tissue	Se Concentration (μg/g dry wt.)			
				NOEC	LOEC	EC10	EC20
Bluegill	Lemly 1993a	Mortality (winter stress syndrome)	WB	—	<5.85 (4°C)	—	—
	McIntyre et al. 2008	Mortality (winter stress syndrome)	WB	—	—	9.56 (4°C) 13.29 (9°C)	10.16 (4°C) 14.02 (9°C)
	Cleveland et al. 1993	Mortality	WB	—[a]	—[a]	—	—
Chinook salmon	Hamilton et al. 1990	Growth	WB	5.3	10.4	6.4[b]	10.9[b]
		Fingerling mortality following seawater challenge	WB	12.6	23.2	4.3[b]	8.6[b]
Fathead minnow	Bennett et al. 1986	Growth	WB	—	<43.0	—	—
	Bertram and Brooks 1986	Growth	WB	>2.2	—	—	—
	Dobbs et al. 1996	Growth	WB	—	<47.5 to <76.0	—	—
Rainbow trout	Vidal et al. 2005	Larval deformities	WB	—[a]	—[a]	—	—
White sturgeon	Tashjian et al. 2006	Growth	WB	14.7	22.5	15	23

[a] A NOEC and LOEC could not be identified because the concentration-response data were anomalous (see text).

[b] We calculated these EC10 and EC20 values from the concentration-response data reported in the original studies.

WB = whole body; NOEC = no-observed-effect concentration; LOEC = lowest-observed-effect-concentration; EC10 = 10% effect concentration; EC20 = 20% effect concentration.

study in terms of developing a national criterion, is that there was only one Se treatment (4.8 µg Se/L in water and 5.1 µg Se/g in food) and one coldwater temperature regime (4°C). Accordingly, there was interest in evaluating additional Se exposure concentrations to ensure that a concentration-response relationship exists between Se, winter stress, and juvenile mortality. Further, because many water bodies in the country do not achieve winter water concentrations of 4°C, there was interest in whether winter stress syndrome occurred at an intermediate temperature (e.g., 9°C). To address these questions, McIntyre et al. (2008) conducted an additional series of Se toxicity tests with juvenile bluegills, which are summarized below.

McIntyre et al. (2008)—Three Se toxicity tests were conducted with juvenile bluegill. One test was similar to the Lemly (1993a) study, in which juvenile bluegill were exposed to an aqueous Se concentration of approximately 5 µg/L (as 1:1 selenate:selenite) and a dietary Se concentration of approximately 5 µg/g dry wt. (as TetraMin® spiked with seleno-L-methionine) for 182 days. In the two Se replicates, no fish died in one and two fish died in the other. The mean whole-body Se concentration was 10.0 µg/g dry wt., which was therefore identified as a "greater than" NOEC. In the other two toxicity tests, juvenile bluegill were exposed to a series of aqueous and dietary Se concentrations at either 4°C or 9°C. In these studies, the bluegill diet was the oligochaete worm *Lumbriculus variegates*, which had been exposed to various levels of selenized yeast. From these two toxicity tests, McIntyre et al. (2008) calculated whole-body Se EC10 values, based on juvenile mortality, of 9.56 and 13.29 µg/g dry wt. in the 4 and 9°C treatments, respectively (Table 5.3).

The results from McIntyre et al. (2008) did not corroborate those from Lemly (1993a), but rather suggest that juvenile bluegill in this study were no more sensitive than offspring from adults with similar whole-body Se concentrations. At the time this book went to press, the U.S. EPA was revising the draft fish tissue Se criterion.

Cleveland et al. (1993)—Juvenile bluegill were exposed to dietary seleno-L-methionine for 90 days. An anomalous response in bluegill mortality after 90 days was observed, as fish with a whole-body concentration of 4.7 µg/g dry wt. had significantly ($p < .05$) elevated mortality relative to the control fish, while mortality in fish with whole-body concentrations of approximately 7.5–13.5 µg/g dry wt. was not significantly greater. Accordingly, neither a NOEC nor a LOEC could be estimated from this study.

5.4.2.2 Chinook Salmon

Hamilton et al. (1990)—Chinook salmon swim-up larvae (*O. tshawytscha*) were provided diets containing either high-Se mosquitofish collected from the San Luis Drain (California) or mosquitofish collected from a reference site that were subsequently fortified with seleno-DL-methionine. Hamilton et al. (1990) suggested that increased toxicity observed in the San Luis Drain diet was at least partially explained by the presence of other contaminants in the mosquitofish. Accordingly, the following focuses on the study based on the selenomethionine-spiked diet. In the first of two studies, swim-up larvae were exposed to Se-fortified food for 90 days. The larvae were reared in reconstituted water, which simulated a 1:37 dilution from the San Luis Drain (minus trace elements). Se concentrations in the water were always below the detection limits of 1.5–3.1 µg/L. Dietary Se concentrations were 3.2, 5.3, 9.6, 18.2, and 35.4 µg/g. In the second study, fingerlings were exposed to the same dietary Se concentrations for 120 days in brackish water.

Upon termination of the first study at 90 days, survival and growth in Chinook salmon fed a diet containing 18.2 µg Se/g were significantly ($p < .05$) reduced relative to control fish. No significant ($p < .05$) effects on survival or growth were observed in fish fed a dietary Se concentration of 9.6 µg/g. The whole-body Se concentrations at the 90-day NOEC and LOEC were 5.4 and 10.8 µg/g dry wt., respectively. However, the reliability of the 90-day results is questionable given that control survival decreased from 99% on day 60 to 67% on day 90. The high mortality rate in the control fish suggests that some of the mortality observed in the treatment groups may be due to factors other than Se. On the basis of the day 60 data, the whole-body Se NOEC and LOEC for growth were only slightly lower (5.3 and 10.4 µg/g dry wt.) relative to the day 90 data. Based on the concentration-response data provided in the study, we estimated day-60 EC10 and EC20 values of 6.4 and 10.9 µg/g dry wt., respectively (Table 5.3).

In the second study, 70 mm Chinook fingerlings were exposed to the same dietary Se concentrations as above, but in brackish water, for 120 days. After the 120-day exposure, the Chinook were exposed to a 10-day seawater (28 g/L salinity) challenge test and percent mortality was assessed. The whole-body Se NOEC and LOEC after 120 days of exposure and following the 10-day seawater challenge, for both growth and mortality, were 12.6 and 23.2 µg/g dry wt. In terms of the level of effect, mortality from the seawater challenge study was more sensitive than the reduction in growth. We calculated whole-body EC10 and EC20 values of 4.3 and 8.6 µg/g, dry wt. (Table 5.3).

5.4.2.3 Fathead Minnow

Bennett et al. (1986)—Se was followed through a three-tiered food chain (algae-rotifer-fathead minnow larvae) in three laboratory experiments. In the first experiment, 4-day-old larvae were fed Se-contaminated rotifers for 7 days, followed by a control diet for 19 days. In the second experiment, 8-day-old larvae were fed Se-contaminated rotifers for 9 days. In experiment three, 2-day-old larvae were fed Se-contaminated rotifers for 7 days before test termination. Mean Se concentrations in rotifers fed to fathead minnow larvae were >70 and 68 µg/g dry wt. in experiments one and two, respectively. Larval growth in both of these experiments was significantly ($p \leq .05$) reduced relative to control fish. The mean larval Se concentration at the end of the 7-day exposure period was 43.0 µg/g dry wt. in experiment one, and was 51.7 µg/g dry wt. in experiment two. Larval growth in experiment three was significantly reduced at $p \leq .10$. The mean larval Se concentration was 61.1 µg/g dry wt. Accordingly, a LOEC of <43.0 µg/g dry wt. was identified from this study (Table 5.3).

Bertram and Brooks (1986)—Fathead minnows were exposed in the laboratory to either waterborne Se, Se-contaminated food, or a combination of the two for 8 weeks. Tests of fish exposed only to dietary Se were terminated after 11 weeks. No effects on growth were observed in fathead minnows fed a diet containing 7.3 µg/g Se (and 43.5 µg/L Se in the water), the highest dietary concentration tested. The whole-body Se concentration in the highest dietary Se concentration tested was 2.2 µg/g dry wt. (Table 5.3).

Dobbs et al. (1996)—A three trophic level test system consisting of algae, rotifers, and larval fathead minnows was dosed with various concentrations of selenate for 25 days. Fathead minnow larval growth was significantly ($p \leq .05$) reduced relative to the controls in fish fed rotifers with a Se concentration ranging from approximately 33–60 µg/g dry wt. Whole-body Se concentrations in the affected fish ranged from 47.5 to 76.0 µg/g dry wt. (Table 5.3).

5.4.2.4 Rainbow Trout

Vidal et al. (2005)—Juvenile (24 days old at test initiation) rainbow trout (*O. mykiss*) were exposed to dietary seleno-L-methionine concentrations of 4.6, 12, and 18 µg/g dry wt. for 90 days. Fish weight, fork length, and tissue Se concentrations were measured at days 0, 30, 60, and 90. In addition, hepatic lipid peroxidation and reduced glutathione (GSH)-to-oxidized glutathione (GSSG) ratios were measured on days 60 and 90. After 90 days, no significant effects on hepatic lipid peroxidation or GSH-GSSG ratios were observed in the Se treatments; however, fish exposed to dietary Se concentrations of 4.6 and 12 µg/g dry wt. had significantly ($p \leq .05$) reduced body weight and fork length relative to the controls. Growth (weight) of fish exposed to 4.6, 12, and 18 µg/g dietary Se was reduced by 33%, 33%, and 26%, respectively, relative to control fish. The 26% weight reduction in the highest treatment was not statistically different ($p > .05$) from the control. Vidal et al. (2005) identified the lowest treatment (4.6 µg/g) as the dietary LOEC. The whole-body Se concentration in trout exposed to a dietary Se concentration of 4.6 µg/g was 0.58 µg/g wet wt., or 2.3 µg/g dry wt. assuming 75% moisture. However, the authors did not base the whole-body Se LOEC on the concentration corresponding with the dietary treatment of 4.6 µg/g. Rather, they identified 1.20 µg/g wet wt. as the LOEC, or 4.8 µg/g dry wt. assuming 75% moisture, which was associated with the dietary Se concentration of 12 µg/g—presumably because this was the lowest exposure concentration that resulted in a Se tissue level significantly greater than the concentration in the control. This results in a circumstance where exposure to Se resulted in a whole-body Se concentration that was not significantly different than the control, but growth was still significantly reduced. Further, the whole-body-based Se LOEC of 2.3 µg/g dry wt. is also unexpected because it appears to fall within the range of background Se concentrations. As Skorupa (1998b) reported, for example, background concentrations of Se in whole-body fish range from <1 to 4 µg/g dry wt., and are typically <2 µg/g dry wt.

Adding to the uncertainties in terms of identifying a whole-body Se threshold from the Vidal et al. (2005) study, whole-body Se concentrations peaked at day 60 and then decreased by day 90, even in the control fish (whole-body Se concentrations were also lower at day 90 than at day 30). The changes in whole-body Se during the course of the test were substantial enough that by day 60, whole-body Se in control fish was 1.24 µg/g wet wt. (4.96 µg/g dry wt.; 75% moisture), or more than two times the effects-based whole-body LOEC of 0.58 µg/g wet wt. (2.3 µg/g dry wt.; 75% moisture) discussed earlier (Figure 5.1a). Vidal et al. (2005) suggest that the decreases in whole-body Se from day 60 to day 90 likely occurred because of the relative increases in total body mass, which would have decreased overall concentrations. However, the observed reduction in Se concentrations does not appear to be entirely a growth dilution effect because Se burdens (µg Se per fish), calculated based on the whole-body Se concentration and the weight of the fish, also declined from day 60 to 90 (Figure 5.1b). Regardless, temporal variability in whole-body Se concentrations in this study makes it difficult to link whole-body Se concentrations to different levels of growth effects. Although statistically significant effects were not observed until day 90, these effects were a result of internal Se concentrations throughout the 90-day exposure period. Thus, the effects measured at day 90 may actually result from the higher whole-body Se concentrations earlier in the study. Given all of the above, we did not identify a whole-body Se threshold associated with adverse effects in rainbow trout based on this study.

5.4.2.5 Razorback Sucker

Beyers and Sodergren (2002)—This study simulated an algae→rotifer→larval razorback sucker (*X. texanus*) food chain. All three components of the food chain were exposed to a control water or natural waters representing a gradient of Se concentrations. The natural waters were collected from three locations along the Colorado River near Grand Junction, Colorado (U.S.). Aqueous Se concentrations ranged from <1 µg/L in the control to 20.3 µg/L in natural waters, from <0.183 to 3.74 µg/g dry wt. in algae, and from <0.702 to 21.8 µg/g dry wt. in rotifers. Larval razorback suckers (5.0 mg, 10.9 mm) were exposed to the test solutions and diets for 28 days and the toxicity endpoints were survival and growth (weight and length). At the end of the study whole-body Se was measured in the larval suckers. No effects on survival or growth were observed relative to the control fish, with whole-body Se concentrations ranging up to 42.0 µg/g dry wt. in the highest Se treatment. Consequently, the whole-body Se NOEC from this study was identified as >42.0 µg/g dry wt.

Hamilton et al. (2005c)—This study was a continuation of the Hamilton et al. (2005a, b) studies summarized previously. Larval razorback suckers (*X. texanus*) were reared from the Horsethief Canyon State Wildlife Area (HT) (reference area), Adobe Creek (AC), and North Pond (WW), as well as from brood stock from HT. Three spawns were used for each location, with egg Se concentrations (µg/g dry wt.) ranging from 5.8 to 6.6 at HT, 38.0 to 54.5 at AC, 34.3 to 37.2 at WW, and 7.1 for brood stock (a single egg sample was analyzed for the latter). The study was initiated with 5-day-old larvae. Larvae, by source, were exposed to one of the following four treatments: (1) reference food and reference water; (2) reference food and site water; (3) site food and reference water; and (4) site food and site water. The reference food was brine shrimp (2.7 µg Se/g) and the reference water was from the 24-Road Fish Hatchery (<1.6 µg Se/L). For HT and brood stock larvae, the site food source was zooplankton collected from the Horsethief east wetland (HTEW) and averaged

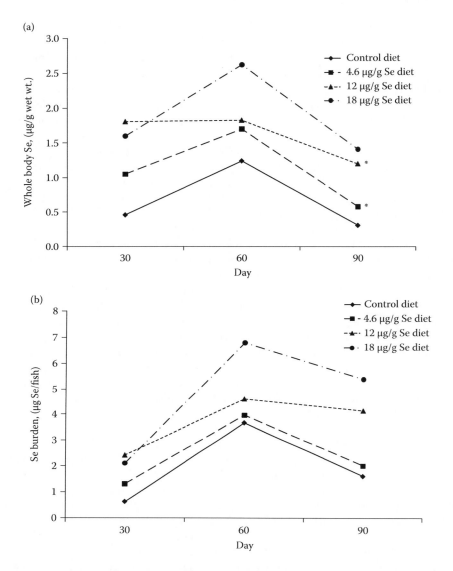

FIGURE 5.1   (a) Whole-body selenium concentrations and (b) selenium body burdens in juvenile rainbow trout following 30, 60, and 90 days. Data from Vidal et al. (2005). Asterisks (*) denote where growth was significantly reduced relative to the control.

5.6 µg Se/g dry wt.; water Se from HT averaged 0.9 µg Se/L. For AC larvae, zooplankton collected from the site averaged 20 µg Se/g dry wt. and water Se averaged 5.5 µg Se/L. For WW larvae, zooplankton collected from the site averaged 39 µg Se/g dry wt. and water Se averaged 10.7 µg/L. Groups of larvae from each fish source were exposed to one of these four possible diet/water combinations for up to 30 days. For the HT, AC, and WW larvae exposed to site diets, whether simultaneously exposed to site water or reference water, 100% mortality was reached in approximately 11, 14, and 7 days, respectively. The associated whole-body Se concentrations (µg/g dry wt.) in the HT, AC, and WW fish were 8.6–8.8 (day 7), 69–70.6 (day 7), and 28.3–44.5 (day 6). For brood stock larvae exposed to site diets, mortality ranged from approximately 40–60% (depending on whether they were exposed to reference or site water) and whole-body Se concentrations were 5.4 and 6.9 µg/g dry wt.

The toxicity response for each source of larvae was not clearly related to either the Se exposure concentrations or the internal tissue Se concentrations. For example, the HT site is a reference area with water and dietary Se concentrations much lower than AC and WW, which resulted in whole-body Se concentrations in HT larvae much lower than those observed in AC and WW larvae. Yet 100% mortality was observed in HT larvae after an exposure period comparable to that required for 100% mortality to be observed in AC and WW larvae. This level of mortality would not be expected due to the low levels of Se and short period of exposure. In addition, the HT larvae and broodstock larvae, which were hatched from eggs with similar Se concentrations and were exposed to the same water and dietary Se source, had different levels of mortality with HT larvae experiencing 100% mortality after about 11 days and brood stock larvae experiencing 40–60% mortality after 30 days. Further, in AC and WW larvae exposed to reference food and reference water, whole-body Se concentrations were 7.7 and 9.7 µg/g dry wt., respectively, which are comparable or higher to the HT larvae and brood stock larvae exposed to site food; however, mortality was much lower in the AC and WW larvae versus HT larvae and brood stock larvae despite similar whole-body Se concentrations. Hamilton et al. (2005c) suggest that this may have resulted from the protective effects of arsenic, which was elevated in the brine shrimp reference food compared to the site zoo-plankton diets. For these reasons, it was not possible to identify a whole-body Se threshold for larval razorback suckers from this study.

Given the inability to derive a whole-body Se threshold for razorback suckers in Hamilton et al. (2005c) and due to differences between Hamilton et al. (2005c) and a similar study by Beyers and Sodergren (2002), that is, NOEC >42 µg/g dry wt., we determined that there was too much uncertainty in deriving a razorback sucker whole-body Se threshold based on the available data.

5.4.2.6 White Sturgeon

Tashjian et al. (2006)—Juvenile white sturgeon (*A. transmontanus*) were provided formulated diets spiked with seleno-L-methionine for 8 weeks. The dietary Se concentrations were 0.4 (control), 9.6, 20.5, 41.7, 89.8, and 191.1 µg/g. Fish were monitored for survival, growth, swimming activity, and histopathological effects. The dietary NOEC and LOEC, based on body weight increase, were 20.5 and 41.7 µg/g, respectively. The corresponding whole-body Se NOEC and LOEC were 14.7 and 22.5 µg/g dry wt., respectively. We derived a whole-body Se EC10 of 15 µg/g dry wt. based on percent body weight increase relative to that in the control fish.

Tashjian et al. (2007)—In a continuation of the juvenile white sturgeon study conducted by Tashjian et al. (2006), fish exposed to dietary organic Se in freshwater were transferred to water with a salinity of either 15 or 20 parts-per-thousand (ppt) in order to evaluate the possible synergistic effects of Se and salinity during seaward migration. Following the 8-week Se exposure described in Tashjian et al. (2006), eight fish from each treatment group were starved for 24 h and transferred to water with a salinity of 15 or 20 ppt for 3 days. Fish survival was defined as those that did not die or lose equilibrium. In fish transferred to 15 ppt salinity water for 3 days and in 20 ppt water for 1 day, the dietary NOEC and LOEC for survival were 89.8 and 191.1 µg/g, respectively. Survival of all fish, including control fish, was low in 20 ppt Se water after 2 days. Overall, therefore, the Se thresholds from Tashjian et al. (2006) based on 8-week Se exposures are protective of juveniles then transferred to 15 ppt salinity for 3 days and 20 ppt salinity water for 1 day.

5.5 RECOMMENDATION FOR A FISH TISSUE Se THRESHOLD

The toxicity studies summarized earlier suggest there are differences in the toxicity data and patterns for maternal transfer toxicity studies versus the juvenile toxicity studies. The following first evaluates whether Se toxicity studies based on maternal transfer or direct toxicity to juveniles is more appropriate for developing a tissue-based Se threshold and then fish tissue-based Se thresholds are recommended.

5.5.1 Life Stage Considerations for a Se Fish Tissue Threshold

The Se toxicity studies reviewed here were categorized as either maternal transfer studies or juvenile direct exposure studies. We recommend that toxicity data from maternal transfer studies, in which egg or ovary Se is linked to larval mortality, deformities, or edema, be used to develop a broadly applicable fish Se tissue threshold. This recommendation is based on (1) the relative sensitivity of maternal transfer and juvenile toxicity studies to Se; (2) consistency of concentration-response relationships within either maternal transfer or juvenile toxicity studies with Se; and (3) consideration of important attributes for identifying a good biomonitor of tissue Se concentrations in the field. Each of these points are discussed further below.

5.5.1.1 Relative Sensitivity of Maternal Transfer and Juvenile Toxicity Studies

The relative sensitivity of two species exposed to a metal or metalloid in the environment is determined by their sensitivity to the external concentration, that is, the substance bioaccessibility (the organism's ability to uptake the substance and transport and sequester it or eliminate the substance). This cannot be inferred simply from relationships between internal tissue concentrations and toxicity under a given set of conditions. The amount of bioavailable metal or metalloid that reaches the internal site of action varies between species due to differences in organism strategies/abilities to detoxify or excrete excess metal and even within species depending on the exposure conditions, all of which influence the bioaccumulation rate and whether toxicity will occur (Luoma and Rainbow 2005). For example, in streams with elevated Se levels, Holm et al. (2005) observed a higher rate of teratogenesis in rainbow trout than in brook trout collected from the same streams. Holm et al. (2005) observed that rainbow trout maternally transferred a larger proportion of their body burden to the eggs than did brook trout. In other words, rainbow trout may not be intrinsically more sensitive to Se than brook trout (e.g., larval development as a function of Se concentration), but may be more sensitive at a given exposure concentration due to species differences in Se toxicokinetics (i.e., the efficiency of maternal transfer of Se to eggs from the muscle during vitellogenesis may be lower in brook trout than in rainbow trout [Holm et al. 2005]).

Following from the above, the relative sensitivity of two species, or the sensitivity of two study types for a given species, is not necessarily based on the level of effects associated with internal (i.e., tissue) Se, but rather the external (i.e., dietary) Se concentration to which they are exposed. For example, suppose a water body contains Species A and Species B. The whole-body tissue Se EC10 values for Species A and Species B are 10 and 12 µg/g, respectively. However, Species A is not necessarily more sensitive than Species B because the Se dietary TTFs (i.e., bioaccumulation potential) for the two species are not necessarily the same. If the dietary TTFs are 1 and 1.5 for Species A and B, respectively, Species B requires a lower dietary Se concentration to achieve the whole-body EC10 than Species A. Thus, the dietary EC10 would be 10 µg/g for Species A and 8 µg/g for Species B and, assuming a similar diet, Species B would therefore be more sensitive than Species A despite having a higher whole-body Se EC10.

The same concept applies when evaluating which fish life stages are more sensitive to Se. Accordingly, when assessing whether the larval development and survival endpoints (resulting from maternal transfer) are more sensitive than the juvenile survival and growth endpoints, one must consider the exposure concentration. For Se, as previously discussed, the most environmentally relevant exposure concentration to consider is dietary organic Se. Relevant toxicity studies for this comparison are limited to laboratory studies with known dietary Se concentrations. The fish species with the most data available for making this comparison is bluegill (*L. macrochirus*). Three laboratory studies exposed adult bluegill to dietary organic Se and evaluated effects in offspring: (1) Woock et al. (1987); (2) Doroshov et al. (1992); and (3) Coyle et al. (1993). The most sensitive endpoints in these three studies were larval mortality in the Woock et al. (1987) and Coyle et al. (1993) studies and larval edema in the Doroshov et al. (1992) study. The concentration-response curves relating dietary Se concentration to the percentage of effect are shown in Figure 5.2. The concentration-response

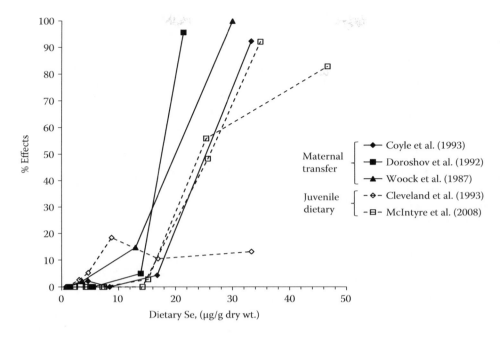

FIGURE 5.2   Selenium effects on bluegill larvae (from maternal transfer) and juveniles (from direct dietary exposures) as a function of dietary selenium.

relationships are similar between the studies, with low-level effects up to a dietary Se concentration of approximately 12–16 µg/g dry wt., and then a rapid increase to a 90–100% effect level.

There are two studies with juvenile bluegills (nonmaternal transfer) from which concentration-response relationships could be derived based on dietary Se: Cleveland et al. (1993) and McIntyre et al. (2008). As shown in Figure 5.2, the concentration-response curve from the Cleveland et al. (1993) study is similar to the maternal transfer studies at low dietary Se concentrations. However, the same rapid increase in adverse effects is not observed at dietary Se concentrations greater than approximately 12 µg/g dry wt. Rather, the mortality response generally reaches a plateau between dietary Se concentrations of 9–33 µg/g dry wt. The concentration-response curves from McIntyre et al. (2008) are similar to those in the maternal transfer studies (Figure 5.2). Significant juvenile mortality did not begin to occur until dietary Se concentrations reach approximately 14–15 µg/g dry wt. and then the concentration-response curves becomes relatively steep. Consequently, the juvenile bluegill in these tests appear to be similarly sensitive to, but no more sensitive than, offspring from adults exposed to similar dietary Se concentrations.

Bryson et al. (1984), as summarized earlier, collected adult female and male bluegill from Hyco Reservoir and a control lake and crossed eggs and sperm from all combinations. The mean ovary Se concentrations were 49 and 3.3 µg/g dry wt. in Hyco Reservoir and control lake bluegill, respectively. All larvae hatched from Hyco Reservoir bluegill eggs experienced 100% mortality, regardless of the source of the male sperm. The surviving larvae (28 days posthatch) from control lake females were then fed zooplankton collected from either Hyco Reservoir (mean Se ± standard deviation = 45 ± 12 µg/g dry wt.) or the control lake (mean Se ± standard deviation = 1.9 ± 2.0 µg/g dry wt.). Mortality was 97.3% in juvenile bluegills fed Hyco zooplankton and 23.7% in juveniles fed control lake zooplankton. Accordingly, Se levels in Hyco Reservoir resulted in almost complete mortality regardless of the life stage and exposure route (100% mortality prior to swim-up when exposed to Se via maternal transfer or 97.3% mortality in juveniles when only exposed to dietary Se). Given the high Se levels in Hyco zooplankton and female ovaries, however, lower effect levels or thresholds could not be identified and used to determine if one life stage was more sensitive than the other.

Overall, the available data suggest that juvenile bluegill are not more sensitive than bluegill larvae exposed to Se via maternal transfer and, in fact, the relative sensitivities of the two life stages appear to be relatively similar based on dietary Se exposures. Although the data available for this comparison are not extensive, the results for bluegill, if translatable across species, suggest the maternal transfer endpoint is more sensitive relative to dietary exposures to juveniles. At this time, similar comparisons cannot be made for other species because data from laboratory adult exposure studies are not available.

5.5.1.2   Consistency of Concentration-Response Relationships

While egg or ovary Se concentrations are strong indicators of whether Se toxicity to fish larvae is likely to occur or not, it is less clear how Se concentration-response relationships for juvenile tissues vary in response to Se exposure concentrations, recognizing that concentrations may vary as a function of age. As discussed earlier, Vidal et al. (2005) observed highly variable whole-body Se concentrations over the 90-day exposure of juvenile rainbow trout to dietary Se, so much so that by day 60, whole-body Se in control fish was 1.24 µg/g wet wt., or more than two times their proposed effects-based whole-body LOEC of 0.58 µg/g wet wt. An observation from this study, which was also noted by the study authors, is that the results may demonstrate the difficulty in using whole-body tissue residues as toxicity thresholds for Se in species that are developmentally immature. However, other studies with juvenile fish, such as Chinook salmon (Hamilton et al. 1990), bluegill (Lemly 1993a, McIntyre et al. 2008), and white sturgeon (Tashjian et al. 2006), did not observe the same fluctuations in tissue Se over time. The Chinook salmon study conducted by Hamilton et al. (1990) resulted in a very different relationship between juvenile whole-body Se and growth effects relative to the Vidal et al. (2005) rainbow trout study even though the two studies have very similar test designs. This variability in basic concentration-response relationships between two closely related species and very similar study designs suggests that the sensitivity of young fish directly exposed to dietary Se is variable, or at least not well understood based on the limited data available.

Teh et al. (2004) provide another example of variability in concentration-response relationships for juvenile fish. Although this study was not included in this review because whole-body Se concentrations were not measured, muscle Se concentrations were measured. Teh et al. (2004) exposed Sacramento splittail (*Pogonichthys macrolepidotus*) to dietary organic Se for 9 months. An anomalous response was observed between dietary Se and larval deformities (the only study that has observed deformities in fish not exposed to egg Se as a developing embryo). Greater than 50% deformities were observed at a dietary Se concentration of 6.6 µg/g dry wt. and then the level of deformities decreased with increasing dietary Se concentration. Growth and survival was not significantly reduced at dietary Se concentrations up to 12.6 µg/g dry wt. The growth data from this study are more consistent with what was observed in Hamilton et al. (1990) for Chinook salmon—weight of fish fed a dietary Se concentration of 12.6 µg/g dry wt. was reduced by 6% relative to the control and by 41% in fish fed 26 µg Se/g. The basis for the anomalous concentration-response data for deformities is unclear. The study is unique due to its longer duration than typical studies with juveniles so there are no results available for direct comparison.

Although Teh et al. (2004) did not measure whole-body Se concentrations during the study, muscle Se concentrations did increase with increasing dietary Se concentrations (and, by inference, whole-body Se would have also increased with increasing dietary Se concentration). As a point of comparison, Lemly (1993c) collected 22 species of fish (representing eight families) from Belews Lake (and two reference reservoirs), which were assessed for deformities and edema and whole-body Se concentrations were measured. Lemly (1993c) observed an increasing relationship between whole-body Se and the prevalence of abnormalities. As shown in Figure 5.3, the relationship between deformities and whole-body or muscle Se is much different between the field data for centrarchids and the laboratory data for the splittail.

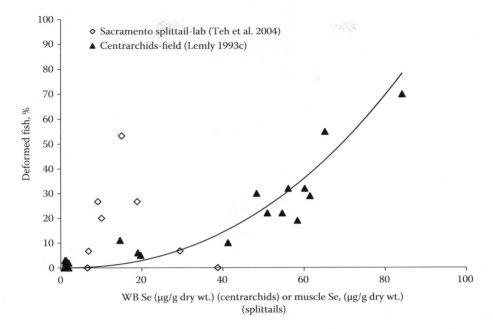

FIGURE 5.3 Relationship between fish deformities and whole-body Se in Belews Lake centrarchids (Lemly 1993c) and muscle Se in Sacramento splittails (Teh et al. 2004).

Overall, there is sufficient variability in concentration-response relationships for Se in young fish such that dietary Se toxicity studies with juveniles may not be appropriate for developing a broadly applicable tissue-based Se threshold for fish. For reasons discussed further below, a Se toxicity threshold based on maternal transfer appears to be most reliable for developing a broadly applicable tissue Se threshold.

5.5.1.3 Considerations for Developing and Applying Tissue-Based Toxicity Thresholds

In developing tissue-based toxicity thresholds for interpreting the concentrations of metals and metalloids in field-collected tissues, an important concept to consider is that there are biomonitors of exposure and biomonitors of effects (Adams et al. 2010). The biomonitor of effects should be a sensitive metric, such as the sensitivity of a species-, population-, or community-based endpoint. However, such a biomonitor of effects may not be readily sampled at a site, particularly across an exposure gradient. As such, a biomonitor of exposure should be identified that is linked to, either directly or by correlation, the biomonitor of effects. As summarized in Adams et al. (2010), a good biomonitor of exposure has the following characteristics: (1) widely distributed; (2) not sensitive to the element of interest; and (3) a strong accumulator of the element such that the internal concentration correlates well with the external exposure concentration.

For Se, maternal transfer studies meet the above characteristics of biomonitors of exposure and biomonitors of effects. Adult fish are relatively insensitive to Se and, therefore, generally amenable to sampling over a range of Se exposure conditions. The Se concentrations measured in the ovaries or eggs collected from adults are then directly linked to the critical endpoint of larval mortality and deformities (Figure 5.4).

5.5.2 Recommendation of a Fish Tissue Se Threshold

In this evaluation, the term "threshold" is intended to represent a Se concentration below which negligible Se toxicity is expected to occur. We selected the EC10 (i.e., 10% effect concentration) for

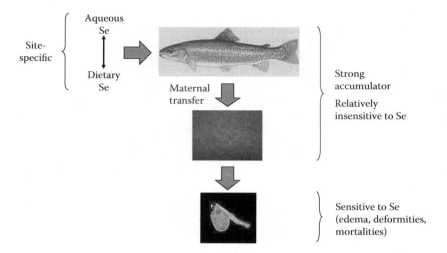

FIGURE 5.4   Schematic representation of selenium maternal transfer studies: Attributes of a good biomonitor.

identifying tissue-based Se thresholds in this evaluation, although it is recognized that lower (e.g., EC05) or higher (e.g., EC20) effect levels are often considered in deriving ecological thresholds, criteria, or guidelines for chemicals. The Se EC10 values identified in this review are often similar to corresponding NOECs from the same study, and always less than the corresponding LOECs. If data were available from studies with similar designs, concentration-response data were pooled for calculation of the EC10. If it was not possible to pool data from separate studies, the lowest EC10 was identified as the threshold for that tissue type. The Se tissue thresholds recommended here are based on maternal transfer studies, for the reasons concluded earlier.

5.5.2.1   Egg Se Threshold

Egg Se-based toxicity data are available for several fish species, including bluegill, brook trout, brown trout, cutthroat trout, Dolly Varden, rainbow trout, northern pike, and white suckers. The minimum EC10 estimated from an individual study was 17 µg/g dry wt., which was based on the relationship between cutthroat trout alevin mortality and egg Se reported in Rudolph et al. (2008) (Table 5.2). This EC10 is based on a simple linear regression between egg Se and alevin mortality data for a reference site and a Se exposed site, and is less than the egg Se NOEC of 20.6 µg/g dry wt. for larval deformities from this same study (Table 5.2). Nevertheless, 17 µg/g dry wt. is assumed to represent a conservative egg Se criterion that should be protective of larval effects. Conveniently and perhaps not surprisingly given that Se is transferred from the ovaries to eggs and that interspecies sensitivity is not highly variable as a function of egg Se, an ovary-based EC10 of 18 µg/g dry wt. is estimated below.

5.5.2.2   Ovary Se Threshold

All of the data relating ovary Se concentrations to adverse effects on fish larvae are for bluegills, largemouth bass, and fathead minnows. The most sensitive endpoints are larval terata, edema, or mortality. The bluegill and fathead minnow data were pooled to estimate an ovary Se EC10 of 18 µg/g dry wt. (Figure 5.5). The raw concentration-response data for largemouth bass were not available for inclusion in Figure 5.5, but the ovary Se EC10 from the logistic regression equation reported in CP&L (1997) was similar, 22 µg/g dry wt. Most of the ovary-based Se effects data are clustered at low effect levels and very high effect levels, which is largely driven by the steepness of the concentration-response relationship. The ovary Se EC10 of 18 µg/g dry wt. falls within the cluster of empirical low effect data. For comparison, there were three bluegill studies with sufficient

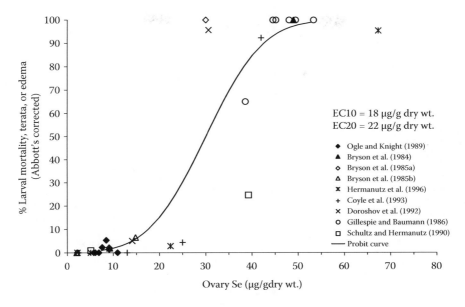

FIGURE 5.5   Relationship between ovary selenium in parent fish versus adverse effects on offspring.

data to estimate EC10 values from the concentration-response relationships between ovary Se and effects on larvae; the EC10 values from these individual studies were 16 µg/g dry wt. (Doroshov et al. 1992), 24 µg/g dry wt. (Coyle et al. 1993), and 30 µg/g dry wt. (Hermanutz et al. 1996). For convenience due to the similarity of the egg Se and ovary Se EC10 values we derived, we recommend that the egg Se threshold of 17 µg/g dry wt. be considered a combined egg or ovary Se threshold.

5.5.2.3   Whole-Body Se Threshold

As previously discussed, we recommend that tissue-based thresholds for Se in fish be based on the Se concentration in the ovaries or eggs, as it is the egg Se concentration that determines whether toxicity in larvae is likely to occur. Because the efficiency with which Se is transferred to the eggs appears to vary between species (e.g., Holm et al. 2005), whole-body-based Se thresholds are expected to be more variable between species. From the earlier review of maternal transfer studies, whole-body-based concentration-response data for bluegill sunfish (Coyle et al. 1993, Hermanutz et al. 1996) and fathead minnow data (Ogle and Knight 1989) were pooled to estimate a whole-body Se EC10 of 8.1 µg/g dry wt., which was associated with adverse effects on larvae (Figure 5.6). As shown in Figure 5.6, whole-body Se concentrations up to 7.5 µg/g dry wt. are associated with ≤5% effects relative to their respective controls, while whole-body Se concentrations ≥16 µg/g dry wt. are associated with >90% effects on larvae. Accordingly, no empirical data are available for whole-body Se concentrations between 7.5 and 16 µg/g dry wt., which adds uncertainty to the concentration-response relationship between this range in whole-body Se concentrations. However, the EC10 of 8.1 µg/g dry wt. appears to be adequately conservative for a threshold, as it falls within the lower end of this range. This EC10 of 8.1 µg/g dry wt. also appears to be protective of maternal transfer toxicity in cutthroat trout, as the NOEC from Hardy et al. (2010) was >11.37 µg/g dry wt. The maternal transfer-based whole-body EC10 of 8.1 µg/g dry wt. is similar to the U.S. EPA's draft whole-body Se criterion of 7.91 µg/g dry wt. (which is being revised when this book went to press), and lower than the EC10 values for juvenile bluegill mortality from the McIntyre et al. (2008) studies and the EC10 value for reduced juvenile white sturgeon growth from Tashjian et al. (2006).

The Chinook salmon juvenile growth and seawater challenge study conducted by Hamilton et al. (1990) indicates that anadromous salmonids may be more sensitive than other fish species tested.

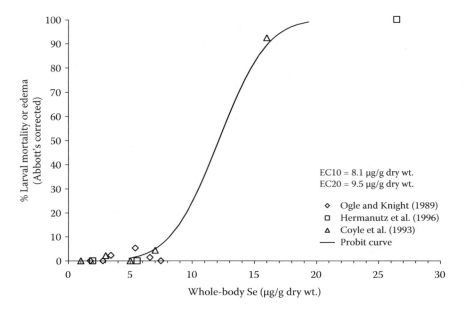

FIGURE 5.6   Relationship between whole-body selenium in parent fish versus adverse effects on offspring.

TABLE 5.4
Comparison of Recommended Selenium Thresholds to the Lowest LOEC and Highest NOEC below the Lowest LOEC

Tissue	Recommended Se Threshold (µg/g dry wt.)	Lowest LOEC (µg/g dry wt.)	Highest NOEC Below Lowest LOEC (µg/g dry wt.)
Egg	17	25	22.5
Ovary	17	21.1	20
Whole body	8	16	>11.85

The EC10 values we derived from Hamilton et al. (1990) were 6.4 µg/g dry wt. for reduced growth and 4.3 µg/g dry wt. for mortality in a 10-day-seawater challenge following a 120-day dietary Se exposure. Thus, a whole-body Se threshold less than 8.1 µg/g dry wt. may be warranted for monitoring Se levels in juvenile anadromous salmon that may be exposed to Se during rearing in fresh water bodies.

5.5.2.4   Summary of Fish Tissue Se Thresholds

From the above review of maternal transfer toxicity studies, we recommend an egg and ovary Se threshold of 17 µg/g dry wt. and a whole-body Se threshold of 8.1 µg/g dry wt. For comparison, and because an EC10 could not be calculated for every study, the lowest whole-body, ovary, and egg Se LOECs identified are 16, 21, and 25 µg/g dry wt., respectively, and the highest whole-body, ovary, and egg Se NOECs that fall below these LOEC values are 11.85, 20, and 22.5 µg/g dry wt., respectively (Table 5.4). Thus, the EC10s recommended as thresholds here are all not only below the lowest LOEC values, but also within the range of NOEC values for most species. Finally, as shown in Figure 5.7, the egg and ovary Se EC10s (or similar effect levels) between most species are not highly variable, with nine of the ten lowest values being within a factor of 1.5. This generally low variability in egg and ovary Se EC10s between species further supports use of egg or ovary Se in deriving a broadly applicable Se threshold.

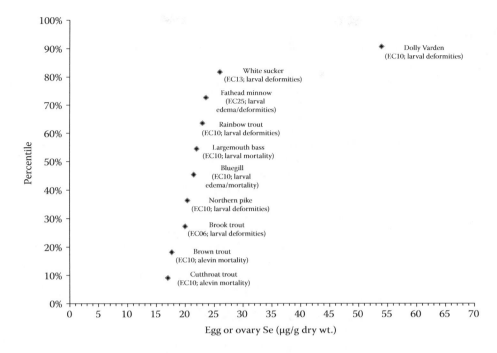

FIGURE 5.7   Distribution of selenium thresholds for eggs or ovaries.

5.5.3 COMPARISONS OF SE FISH TISSUE THRESHOLDS

Although there is a general consensus that monitoring and regulating Se based on concentrations in fish tissue is more useful and appropriate than monitoring Se in water or sediment, there are still a variety of opinions in defining the appropriate fish tissue Se thresholds (e.g., Lemly 1996, DeForest et al. 1999, Hamilton 2002, U.S. EPA 2004, Chapman 2007). The Se toxicity thresholds recommended here are compared to those of Lemly (1996), DeForest et al. (1999), Hamilton (2002), U.S. EPA (2004), and Chapman (2007) in Table 5.5. The egg and ovary Se threshold of 17 µg/g dry wt. recommended in this review is higher than the threshold of 10 µg/g dry wt. recommended in Lemly (1996) and the whole-body Se threshold of 8.1 µg/g dry wt. recommended in this review is higher than the threshold of 4 µg/g dry wt. recommended in Lemly (1996) and Hamilton (2002). The following evaluates the differences between these Se threshold values.

The precise basis for the difference between the recommended Se threshold levels in eggs and ovaries cannot be readily determined because the lowest egg or ovary Se concentration associated with adverse effects identified in Lemly (1996) was an ovary Se concentration of 18 µg/g dry wt., which was associated with reproductive failure in bluegills (Hermanutz et al. 1992). In Hermanutz et al. (1992), a mean ovary Se concentration of 4.4 µg/g wet wt. was associated with significantly ($p \leq .05$) increased mortality, edema, and hemorrhaging in bluegills. Lemly (1996) assumed an ovary moisture content of 75%, which resulted in an estimated dry weight Se concentration of 18 µg/g, while, from Gillespie and Baumann (1986), we assumed a bluegill ovary moisture content of 85% to estimate a dry weight Se concentration of 29 µg/g (we ultimately derived an ovary-based EC10 of 30 µg/g dry wt. based on the updated Hermanutz et al. [1996] report). Therefore, the differences in egg and ovary Se thresholds of 17 µg/g dry wt. from this review and 10 µg/g dry wt. from Lemly (1996) is primarily due to a different percent moisture assumption for bluegill ovaries in the Hermanutz et al. (1992, 1996) studies and an undefined extrapolation from 18 to 10 µg/g dry wt. We believe that the body of evidence, which includes several maternal transfer studies published after Lemly (1996) and DeForest et al. (1999), continues to support that an egg or ovary Se concentration

TABLE 5.5
Comparison of Recommended Thresholds for Selenium in Fish Tissues

Tissue	Threshold (µg/g dry wt.)	Basis	Reference
Egg	17	EC10: Se concentration in fish eggs associated with alevin mortality; data from Rudolph et al. (2008)	This review
	10	Reproductive failure	Lemly 1996
	>16–40	Range in "effects thresholds" for coldwater species	Chapman 2007
Ovary	17	EC10: Se concentration in parent fish ovaries associated with larval mortality and edema in offspring; data from multiple studies with bluegill and fathead minnows (lowered from 18 to 17 µg/g dry wt. for consistency with egg Se threshold)	This review
	10	Reproductive failure	Lemly 1996
	17	EC10: Se concentration in parent fish ovaries associated with larval mortality and edema in offspring; data from multiple studies with bluegill and fathead minnows	DeForest et al. 1999
Whole body	8.1	EC10: Se concentration in parent fish associated with larval mortality and edema in offspring; data from Ogle and Knight (1989), Coyle et al. (1993), and Hermanutz et al. (1996)	This review
	4	Mortality of juveniles and reproductive failure	Lemly 1996
	9	EC10: Se concentration in parent fish associated with larval mortality and edema in offspring; data from Ogle and Knight (1989), Hermanutz et al. (1992), and Coyle et al. (1993)	DeForest et al. 1999
	4	Mortality of juveniles and reproductive failure	Hamilton 2002
	7.91	LOAEC for juvenile bluegill simultaneously exposed to Se and simulated winter conditions, with 5.85 µg/g dry wt. considered a summer/fall trigger for monitoring in winter; data from Lemly (1993a)	U.S. EPA 2004

EC10 = 10% effect concentration; LOAEC = lowest-observed-adverse-effect-concentration.

of 17 µg/g dry wt. is a broadly applicable Se threshold for these tissues (this is the lowest egg or ovary EC10 value that we identified from all available studies).

Although this review and Lemly (1996) both support that eggs and/or ovaries are the most relevant tissues for monitoring the potential for Se toxicity at a site, both studies also provide whole-body-based Se thresholds. The discrepancy in our recommended whole-body Se threshold of 8.1 µg/g dry wt. and the whole-body Se threshold of 4 µg/g dry wt. recommended by Lemly (1996) and Hamilton (2002) is based on the following. First, for reasons discussed earlier, we believe that the variability in relationships between whole-body Se and toxicity in direct toxicity studies with juvenile fish limit the usefulness of these studies for deriving and implementing a broadly applicable tissue-based threshold. Our recommended whole-body Se threshold is based only on maternal transfer studies. Second, we note that the lowest whole-body Se concentrations measured in adversely affected fish were observed in tests where fish were exposed to very high aqueous Se concentrations (Hunn et al. 1987, Cleveland et al. 1993) or exposed to a field-collected diet in which the authors noted that elements in addition to Se may have contributed to the toxicity observed (Hamilton et al. 1990). We identified whole-body chronic values (geometric mean of the NOEC and LOEC) from these studies of 3.3 µg/g dry wt. (rainbow trout; Hunn et al. 1987), 4.4 µg/g dry wt. (bluegill; Cleveland et al. 1993); and <3.3 µg/g dry wt. (Chinook salmon; Hamilton et al. 1990). For perspective, in the study of Se toxicity to juvenile trout conducted by Vidal et al. (2005), the measured whole-body Se concentration in control fish peaked at 4.96 µg/g dry wt. (calculated from a wet weight concentration of 1.24 µg/g and assuming 75% moisture). In addition, under the National Water Quality Assessment (NAWQA) program, the United States Geological Survey (USGS) measured whole-body Se concentrations in fish collected from reference sites in Alaska, Arizona, California, Oregon, Washington, and Wyoming (U.S.) (http://water.usgs.gov/nawqa/). Whole-body Se concentrations in approximately 40% of the reference site fish exceed 4 µg/g dry wt., compared to approximately 5% that exceed 8.1 µg/g dry wt. In our opinion, a whole-body Se threshold of 4 µg/g dry wt. is impractically low for use as a broadly applicable Se threshold. Again, however, we believe that using an egg- or ovary-based Se threshold reduces much of the uncertainty in interpreting Se concentrations measured in field-collected fish.

Summary

Se is a naturally occurring element that typically enters aquatic water bodies, either from natural or anthropogenic sources, as inorganic selenite or selenate. The critical exposure route for fish is dietary exposure to organic Se, with the amount of Se that is ultimately reduced to organic Se compounds and bioavailable to fish being highly dependent on site-specific conditions and uptake by primary consumers. Bioaccumulated Se is maternally transferred to the eggs where, at sufficiently high concentrations, it can cause edema, deformities, or mortality in larval fish during yolk sac absorption. Due to the site-specific factors that influence Se bioavailability and bioaccumulation in fish, the Se concentration in fish tissue, especially in the eggs or ovaries, is a much more reliable indicator of potential toxicity than Se in the water column. The U.S. EPA developed a draft whole-body Se criterion based on the combined effects of Se and simulated winter stress on juvenile bluegill survival, which is being revised at the time this book went to press. Additional recent studies on the toxicity of Se to juvenile bluegills under simulated winter stress suggest that juvenile toxicity studies with Se are not more sensitive than maternal transfer toxicity studies. Maternal transfer studies meet the ideal attributes for deriving and implementing a tissue-based based threshold, as adults are relatively insensitive to Se and amenable for egg or ovary sample collection, with the Se concentration in the eggs or ovaries directly linked to the critical endpoints of larval mortality, deformities, and edema. Further, for development of a broadly applicable tissue-based threshold, egg and ovary Se EC10 values are not highly variable between most freshwater fish species tested to-date. We derived egg and ovary Se EC10 values of 17 and 18 µg/g dry wt., respectively, and recommend an egg and ovary Se threshold of 17 µg/g dry wt. (this is the lowest egg or ovary EC10 value that we identified from all available studies). Although we believe that eggs or ovaries are the more appropriate tissues for developing a broadly applicable fish tissue Se threshold, we also recommend a whole-body Se threshold of 8.1 µg/g dry wt. when egg or ovary Se data are lacking.

ACKNOWLEDGMENT

We thank Jim Meador and an anonymous reviewer for their valuable comments on this chapter.

REFERENCES

Adams, W., et al. 2010. Utility of tissue residues for predicting effects due to metals in aquatic organisms. *Integr. Environ. Assess. Manage.* doi 10.1002/ieam.108.

Barnhart, R. A. 1957. Chemical factors affecting the survival of game fish in a western Colorado reservoir. M.S. Thesis, Colorado State University, Fort Collins, CO.

Bennett, W. N., A. S. Brooks, and M. E. Boraas. 1986. Selenium uptake and transfer in an aquatic food chain and its effects on fathead minnow larvae. *Arch. Environ. Contam. Toxicol.* 15:513–517.

Bertram, P. E., and A. S. Brooks. 1986. Kinetics of accumulation of selenium from food and water by fathead minnows. *Wat. Res.* 20:877–884.

Besser, J. M., T. J. Canfield, and T. W. La Point. 1993. Bioaccumulation of organic and inorganic selenium in a laboratory food chain. *Environ. Toxicol. Chem.* 12:57–72.

Beyers, D. W., and C. Sodergren. 2002. Assessment of exposure of larval razorback sucker to selenium in natural waters. *Arch. Environ. Contam. Toxicol.* 42:53–59.

Brix, K. V., J. E. Toll, L. M. Tear D. K. DeForest, and W. J. Adams. 2005. Setting site-specific water-quality standards by using tissue residue thresholds and bioaccumulation data. Part 2. Calculating site-specific selenium water-quality standards for protecting fish and birds. *Environ. Toxicol. Chem.* 24:231–237.

Bryson, W. T., W. R. Garrett, M. A. Mallin, K. A. MacPherson, W. E. Partin, and S. E. Woock. 1984. 1982 Environmental monitoring studies, Volume II: Hyco Reservoir bioassay studies. Carolina Power and Light Company, New Hill, North Carolina. 65 pp + appendices.

Bryson, W. T., W. R. Garrett, K. A. MacPherson, M. A. Mallin, W. E. Partin, and S. E. Woock. 1985a. Hyco Reservoir 1983 bioassay report. Carolina Power and Light Company, New Hill, North Carolina. 55 pp.

Bryson, W. T., K. A. MacPherson, M. A. Mallin, W. E. Partin, and S. E. Woock. 1985b. Hyco Reservoir 1984 bioassay report. Carolina Power and Light Company, New Hill, North Carolina. 51 pp.

Butler, D. L., R. P. Krueger, B. C. Osmundson, A. L. Thompson, and S. K. McCall. 1991. Reconnaissance investigation of water quality, bottom sediment, and biota associated with irrigation drainage in the Gunnison and Uncompahgre River Basins and at Sweitzer Lake, West-Central Colorado, 1988–89. U.S. Geological Survey. Water-Resources Investigation Report 91–4103. 99 pp.

Carolina Power & Light Company (CP&L). 1997. Largemouth bass selenium bioassay. Roxboro, NC, USA. 14 pp.

Chapman, P. M. 2007. Selenium thresholds for fish from cold freshwaters. *Hum. Ecol. Risk Assess.* 13(1):20–24.

Chapman, P. M., et al. 2009. Ecological assessment of selenium in the aquatic environment: Summary of a SETAC Pellston Workshop. Society of Environmental Toxicology and Chemistry (SETAC). Pensacola, FL, USA.

Cleveland, L., E. E. Little, D. R. Buckler, and R. H. Wiedmeyer. 1993. Toxicity and bioaccumulation of waterborne and dietary selenium in juvenile bluegill (*Lepomis macrochirus*). *Aquat. Toxicol.* 27:265–280.

Coyle, J. J., D. R. Buckler, C. G. Ingersoll, J. F. Fairchild, and T. W. May. 1993. Effect of dietary selenium on the reproductive success of bluegills (*Lepomis macrochirus*). *Environ. Toxicol. Chem.* 12:551–565.

Cumbie, P. M., and S. L. Van Horn. 1978. Selenium accumulation associated with fish mortality and reproductive failure. *Proc. Ann. Conf. S.E. Assoc. Fish Wildl. Agencies* 32:612–624.

de Rosemond, S. C., K. Liber, and A. Rosaasen. 2005. Relationship between embryo selenium concentration and early life stage development in white sucker (*Catostomus commersoni*) from a northern Canadian lake. *Bull. Environ. Contam. Toxicol.* 74:1134–1142.

deBruyn, A. M., A. Hodaly, and P. M. Chapman. 2008. Tissue selection criteria: Selection of tissue types for development of meaningful selenium tissue thresholds in fish. Part 1 of Selenium issue thresholds: Tissue selection criteria, threshold development endpoints, and field application of tissue thresholds. Washington (DC, USA): North America Metals Council-Selenium Working Group.

DeForest, D. K., K. V. Brix, and W. J. Adams. 1999. Critical review of proposed residue-based selenium toxicity thresholds for freshwater fish. *Hum. Ecol. Risk Assess.* 5(6):1187–1228.

DeForest, D. K., K. V. Brix, and W. J. Adams. 2007. Assessing metal bioaccumulation in aquatic environments: The inverse relationship between bioaccumulation factors, trophic transfer factors and exposure concentration. *Aquat. Toxicol.* 84:236–246.

Dobbs, M. G., D. S. Cherry, and J. Cairns, Jr. 1996. Toxicity and bioaccumulation of selenium to a three-trophic level food chain. *Environ. Toxicol. Chem.* 15(3):340–347.

Doroshov, S., et al. 1992. Development of water quality criteria for resident aquatic species of the San Joaquin River. Draft final report to the California State Water Resources Control Board for Contract No. 7–197–250–0. Department of Animal Science, University of California, Davis, CA.

Finley, K. A. 1985. Observations of bluegills fed selenium-contaminated *Hexagenia* nymphs collected from Belews Lake, North Carolina. *Bull. Environ. Contam. Toxicol.* 35:816–825.

Formation Environmental. 2009. Brown trout laboratory reproduction studies conducted in support of development of a site-specific selenium criterion. Unpublished draft report prepared for J.R. Simplot Company, Smoky Canyon Mine, Pocatello, ID, USA.

Garrett, G. P., and C. R. Inman. 1984. Se-induced changes in fish populations of a heated reservoir. *Proc. Ann. Conf. S.E. Assoc. Fish and Wildl. Agencies* 38:291–301.

GEI Consultants. 2008. Maternal transfer of selenium in fathead minnows, with modeling of ovary tissue-to-whole body concentrations. Unpublished report prepared on behalf of Conoco-Phillips. Littleton, CO, USA.

Gillespie, R. B., and P. C. Baumann. 1986. Effects of high tissue concentrations of selenium on reproduction by bluegills. *Trans. Am. Fish Soc.* 115:208–213.

Goettl, Jr., J. P., and P. H. Davies. 1977. Water pollution studies. Job Progress Report, Federal Aid Project F-33-R-12. Department of Natural Resources, Colorado Division of Wildlife, Fort Collins, CO, USA. pp. 39–42.

Golder. 2009. Development of a site-specific selenium toxicity threshold for Dolly Varden char. Prepared for Northgate Minerals Corporation, Smithers, BC, Canada. 17 pp. + tables, figures, and appendices.

Hamilton, S. J. 2002. Rationale for a tissue-based selenium criterion for aquatic life. *Aquat. Toxicol.* 57: 85–100.

Hamilton, S. J., and R. H. Wiedmeyer. 1990. Concentrations of boron, molybdenum, and selenium in chinook salmon. *Trans. Am. Fish Soc.* 119:500–510.

Hamilton, S. J., A. N. Palmisano, G. A. Wedemeyer, and W. T. Yasutake. 1986. Impacts of selenium on early life stages and smoltification of fall chinook salmon. *Trans. 51st N. A. Wildl. Nat. Res. Conf.* 51:343–356.

Hamilton, S. J., K. J. Buhl, N. L. Faerber, R. H. Wiedmeyer, and F. A. Bullard. 1990. Toxicity of organic selenium in the diet to chinook salmon. *Environ. Toxicol. Chem.* 9:347–358.

Hamilton, S. J., K. M. Holley, K. J. Buhl, F. A. Bullard, L. K. Weston, and S. F. McDonald. 2005a. Selenium impacts on razorback sucker, Colorado River, Colorado. I. Adults. *Ecotoxicol. Environ. Saf.* 61:7–31.

Hamilton, S. J., K. M. Holley, K. J. Buhl, and F. A. Bullard. 2005b. Selenium impacts on razorback sucker, Colorado River, Colorado. II. Eggs. *Ecotoxicol. Environ. Saf.* 61:32–43.

Hamilton, S. J., K. M. Holley, K. J. Buhl, and F. A. Bullard. 2005c. Selenium impacts on razorback sucker, Colorado River, Colorado. III. Larvae. *Ecotoxicol. Environ. Saf.* 61:168–169.

Hardy, R. W., L. Oram, and G. Möller. 2010. Effects of dietary selenomethionine on cutthroat trout (*Oncorhynchus clarki bouvieri*) growth and reproductive performance over a life cycle. *Arch. Environ. Contam. Toxicol.* 58(1):237–245.

Hermanutz, R. O., K. N. Allen, T. H. Roush, and S. F. Hedtke. 1992. Effects of elevated selenium concentrations on bluegills (*Lepomis macrochirus*) in outdoor experimental streams. *Environ. Toxicol. Chem.* 11:217–224.

Hermanutz, R. O., K. N. Allen, N. E. Detenbeck, and C. E. Stephan. 1996. Exposure of bluegill (*Lepomis macrochirus*) to selenium in outdoor experimental streams. USEPA Report. Mid-Continent Ecology Division, Duluth, MN.

Hodson, P. V., D. J. Spry, and B. R. Blunt. 1980. Effects on rainbow trout (*Salmo gairdneri*) of a chronic exposure to waterborne selenium. *Can. J. Fish Aquat. Sci.* 37:233–240.

Holm, J. 2002. Sublethal effects of selenium in rainbow trout (*Oncorhynchus mykiss*) and brook trout (*Salvelinus fontinalis*). Masters Thesis; Department of Zoology, University of Manitoba, Winnipeg, Canada.

Holm, J., et al. 2003. An assessment of the development and survival of wild rainbow trout (*Oncorhynchus mykiss*) and brook trout (*Salvelinus fontinalis*) exposed to elevated selenium in an area of active coal mining. The Big Fish Bang. Proceedings of the 26th Annual Larval Fish Conference.

Holm, J., et al. 2005. Developmental effects of bioaccumulated selenium in eggs and larvae of two salmonid species. *Environ. Toxicol. Chem.* 24:2373–2381.

Hunn, J. B., S. J. Hamilton, and D. R. Buckler. 1987. Toxicity of sodium selenite to rainbow trout fry. *Wat. Res.* 21(2):233–238.

Kennedy, C. J., L. E. McDonald, R. Loveridge, and M. M. Strosher. 2000. The effect of bioaccumulated selenium on mortalities and deformities in the eggs, larvae, and fry of a wild population of cutthroat trout (*Oncorhynchus clarki lewisi*). *Arch. Environ. Contam. Toxicol.* 39:46–52.

Lemly, A. D. 1985. Toxicology of selenium in a freshwater reservoir: Implications of environmental hazard evaluation and safety. *Ecotoxicol. Environ. Saf.* 10:314–338.

Lemly, A. D. 1993a. Metabolic stress during winter increases the toxicity of selenium to fish. *Aquat. Toxicol.* 27:133–158.

Lemly, A. D. 1993b. Guidelines for evaluating selenium data from aquatic monitoring and assessment studies. *Environ. Monit. Assess.* 28:83–100.

Lemly, A. D. 1993c. Teratogenic effects of selenium in natural populations of freshwater fish. *Ecotoxicol. Environ. Saf.* 26:181–204.

Lemly, A.D. 1996. Selenium in aquatic organisms. In: *Environmental Contaminants in Wildlife—Interpreting Tissue Concentrations,* eds. W.N. Beyer, G.H. Heinz, and A.W. Redmon-Norwood, pp. 427–445. New York, NY: CRC/Lewis Publishers.

Lemly, A. D. 1997. A teratogenic deformity index for evaluating impacts of selenium on fish populations. *Ecotoxicol. Environ. Saf.* 37:259–66.

Lemly, A. D. 2002. *Selenium assessment in aquatic ecosystems: A guide for hazard evaluation and water quality criteria.* New York, NY: Springer-Verlag.

Lemly, A. D., and G. J. Smith. 1987. Aquatic cycling of selenium: implications for fish and wildlife. U.S. Dept. of the Interior, Fish and Wildlife Service. Leaflet 12. Washington, DC. 10 pp.

Lillebo, H. P, S. Shaner, D. Carlson, N. Richard, and P. DuBowy. 1988. Regulation of agricultural drainage to the San Joaquin River. State Water Resources Control Board. SWRCB Order No. W.Q. 85–1. 151 pp.

Luoma, S. N., and P. S. Rainbow. 2005. Why is metal bioaccumulation so variable? Biodynamics as a unifying concept. *Environ. Sci. Technol.* 39(7):1921–1931.

Luoma, S. N., and T. S. Presser. 2009. Emerging opportunities in management of selenium contamination. *Environ. Sci. Technol.* 43:8483–8487.

Maier, K. J., and A. W. Knight. 1994. Ecotoxicology of selenium in freshwater systems. *Rev. Environ. Contam. Toxicol.* 134:31–48.

Malchow, D. E., A. W. Knight, and K. J. Maier. 1995. Bioaccumulation and toxicity of selenium in *Chironomus decorus* larvae fed a diet of seleniferous *Selenastrum capricornutum. Arch. Environ. Contam. Toxicol.* 29:104–109.

May, T. W., et al. 2001. An evaluation of selenium concentrations in water, sediment, invertebrates, and fish from the Republican River Basin: 1997–1999. *Environ. Monit. Assess.* 72:179–206.

May, T. W., et al. 2008. An evaluation of selenium concentrations in water, sediment, invertebrates, and fish from the Solomon River Basin. *Environ. Monit. Assess.* 137(1–3):213–232.

McIntyre, D. O., M. A. Pacheco, M. W. Garton, D. Wallschläger, and C. G. Delos. 2008. Effect of selenium on juvenile bluegill sunfish at reduced temperature. Health and Ecological Criteria Division, Office of Water, USEPA, Washington, DC. EPA-822-R-08–020.

Muscatello, J. R., P. M. Bennett, K. T. Himbeault, A. M. Belknap, and D. M. Janz. 2006. Larval deformities associated with selenium accumulation in northern pike (*Esox lucius*) exposed to metal mining effluent. *Environ. Sci. Technol.* 40:6506–6512.

Muscatello, J. R., A. M. Belknap, and D. M. Janz. 2008. Accumulation of selenium in aquatic systems downstream of a uranium mining operation in northern Saskatchewan, Canada. *Environ. Pollut.* 156(2):387–393.

Ogle, R. S., and A. W. Knight. 1989. Effects of elevated foodborne selenium on growth and reproduction of the fathead minnow (*Pimephales promelas*). *Arch. Environ. Contam. Toxicol.* 18:795–803.

Ogle, R. S., K. J. Maier, P. Kiffney, M. J. Williams, A. Brasher, L. A. Melton, and A. W. Knight. 1988. Bioaccumulation of selenium in aquatic systems. *Lake Res. Manage.* 4(2):165–173.

Orr, P. L., K. P. Guiguer, and C. K. Russel. 2006. Food chain transfer of selenium in lentic and lotic habitats of a western Canadian watershed. *Ecotoxicol. Environ. Saf.* 63:175–188.

Palace, V. P., J. E. Spallholz, J. Holm, K. Wautier, R. E. Evans, and C.L. Baron. 2004. Metabolism of selenomethionine by rainbow trout (*Oncorhynchus mykiss*) embryos can generate oxidative stress. *Ecotoxicol. Environ. Saf.* 58:17–21.

Riedel, G. F., D. P. Ferrier, and J. G. Sanders. 1991. Uptake of selenium by freshwater phytoplankton. *Water Air Soil Pollut.* 57–58:23–30.

Riedel, G. F., and J. G. Sanders. 1996. The influence of pH and media composition on the uptake of inorganic selenium by *Chlamydomonas reinhardtii. Environ. Toxicol. Chem.* 15(9):1577–1583.

Rudolph, B.-L., I. Andreller, and C. J. Kennedy. 2008. Reproductive success, early life stage development, and survival of westslope cutthroat trout (*Oncorhynchus clarki lewisi*) exposed to elevated selenium in an area of active coal mining. *Environ. Sci. Technol.* 42:3109–3114.

Saiki, M. K., B. A. Martin, and T. M. May. 2004. Reproductive status of western mosquitofish inhabiting selenium-contaminated waters in the grassland water district, Merced County, California. *Arch. Environ. Contam. Toxicol.* 47:363–369.

Sanders, R. W., and C. C. Gilmour. 1994. Accumulation of selenium in a model freshwater microbial food web. *Appl. Environ. Microbiol.* 60(8):2677–2683.

Schultz, R., and R. Hermanutz. 1990. Transfer of toxic concentrations of selenium from parent to progeny in the fathead minnow (*Pimephales promelas*). *Bull. Environ. Contam. Toxicol.* 45:568–573.

Skorupa, J. P. 1998a. Selenium poisoning of fish and wildlife in nature: lessons from twelve real-world examples. In: *Environmental chemistry of selenium,* eds. W. T. Frankenberger and R. A. Engberg, pp. 315–354. New York, NY: Marcel Dekker, Inc.

Skorupa, J. P. 1998b. Selenium. Pages 139–184 in National Irrigation Water Quality Program Information Report No. 3: Guidelines for interpretation of the biological effects of selected constituents in biota, water, and sediment. United States Department of the Interior, November 1998. 198 pp. + appendices.

Staub, B. P., W. A. Hopkins, J. Novak, and J. D. Congdon. 2004. Respiratory and reproductive characteristics of eastern mosquitofish (*Gambusia holbrooki*) inhabiting a coal ash settling basin. *Arch. Environ. Contam. Toxicol.* 46:96–101.

Stewart, R. A., S. N. Luoma, C. E. Schlekat, M. A. Doblin, and K. A. Heib. 2004. Foodweb pathway determine how selenium affects aquatic systems: A San Francisco Bay case study. *Environ. Sci. Technol.* 38:4519–4526.

Tao, J., P. Kellar, and W. Warren-Hicks. 1999. Statistical analysis of selenium toxicity data. Report submitted for US EPA, Health and Ecological Criteria Division. The Cadmus Group, Inc., Durnham, NC.

Tashjian, D. H., S. J. Teh, A. Sogomoyan, and S. S. O. Hung. 2006. Bioaccumulation and chronic toxicity of dietary L-selenomethionine in juvenile white sturgeon (*Acipenser transmontanus*). *Aquat. Toxicol.* 79:401–409.

Tashjian, D., J. J. Cech, Jr., and S. S. O. Hung. 2007. Influence of dietary L-selenomethionine exposure on the survival and osmoregulatory capacity of white sturgeon in fresh and brackish water. *Fish Physiol. Biochem.* 33:109–119.

Teh, S. J., et al. 2004. Chronic effects of dietary selenium on juvenile Sacramento splittail (*Pogonichthys macrolepidotus*). *Environ. Sci. Technol.* 38:6085–6593.

U.S. EPA (United States Environmental Protection Agency). 2004. Draft aquatic life water quality criteria for selenium–2004. Office of Water, Office of Science and Technology, Washington, DC.

Vidal, D., S. M. Bay, and D. Schlenk. 2005. Effects of dietary selenomethionine on larval rainbow trout (*Oncorhynchus mykiss*). *Arch. Environ. Contam. Toxicol.* 49:71–75.

Williams, M. J., R. S. Ogle, A. W. Knight, and R. G. Burau. 1994. Effects of sulfate on selenate uptake and toxicity in the green alga *Selenastrum capricornutum*. *Arch. Environ. Contam. Toxicol.* 17:449–453.

Woock, S. E., W. R. Garrett, W. E. Partin, and W. T. Bryson. 1987. Decreased survival and teratogenesis during laboratory selenium exposures to bluegill, *Lepomis macrochirus*. *Bull. Environ. Contam. Toxicol.* 39:998–1005.

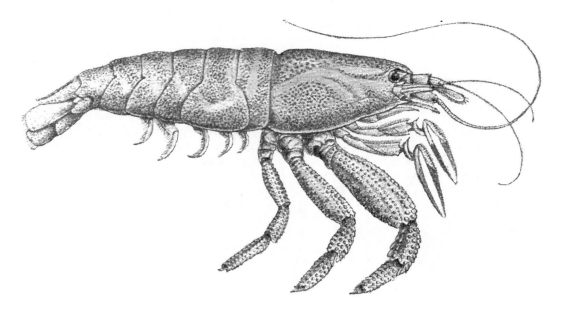

Palaemon serratus

From *The Royal Natural History,* edited by Richard Lydekker, Frederick Warne & Co., London, 1893–94.

6 Trace Metals in Aquatic Invertebrates

Philip S. Rainbow
Samuel N. Luoma

CONTENTS

6.1   Introduction 231
6.2   Accumulation of Trace Metals 232
    6.2.1   Uptake 232
    6.2.2   Excretion 236
    6.2.3   Accumulation 236
        6.2.3.1   Metabolic Requirements for Essential Trace Metals 238
        6.2.3.2   Detoxified Component of Trace Metals 239
    6.2.4   Variation in Accumulation Patterns and Accumulated Concentrations 240
    6.2.5   Biodynamic Modeling 243
    6.2.6   Toxicity and Accumulation 244
        6.2.6.1   Lethal Body Concentrations of Trace Metals 244
        6.2.6.2   Bioaccumulated Metal Guidelines 247
Summary 249
References 249

6.1   INTRODUCTION

This chapter addresses the interpretation of accumulated trace metal concentrations in one component of the biota—in this case aquatic invertebrates. Aquatic invertebrates represent key members of the biota of habitats particularly at risk from toxic metal contamination, such as freshwaters and estuaries receiving metal-laden effluents from mining activities, industrial concerns, and urban and domestic run off. Careful and informed interpretation of the significance of the accumulated metal concentrations of aquatic invertebrates has much to tell us about ambient metal bioavailabilities, and potential ecotoxicological effects. In this chapter, we shall explore general principles to aid such interpretation against a background of the variation in the chemical properties of metals that control much of their biochemistry and physiology in organisms, and the additional variation in how different invertebrates physiologically handle and are sensitive to the toxic metals that they inevitably take up into their bodies.

As in the corresponding chapter in the original version of this book (Rainbow 1996), we interpret the term "trace metal" very widely (see also Luoma and Rainbow 2008), not excluding elements such as arsenic and selenium typically referred to as metalloids. For many of our examples, we have, however, tried to restrict ourselves to relatively few trace metals—cadmium, a metal generally considered nonessential with a known exception in the case of some oceanic phytoplankton

(Cullen et al. 1999, Lane et al. 2005), and two essential metals, zinc and particularly copper. It is worth stating clearly at this stage that all trace metals, whether essential or not, are toxic to organisms above a threshold bioavailability and are all taken up to varying degrees by organisms, including of course aquatic invertebrates.

In the last decade our ability to interpret the significance of trace metal concentrations, particularly in animals, has taken a quantum step forward (Rainbow 2002, 2007, Luoma and Rainbow 2008), not least as a result of the wider application of biodynamic modeling to determine the relative importance of solution and food as metal sources, and to predict accumulated concentrations in different animals in different situations (Wang et al. 1996, Luoma and Rainbow 2005). Furthermore subcellular fractionation of accumulated metals (Wallace et al. 2003) is being used increasingly to understand why the same accumulated total body concentration of a metal in different aquatic invertebrates may or may not be reflected in ecotoxicological effects on that invertebrate (Wang and Rainbow 2006; Luoma and Rainbow 2008).

Unless stated otherwise, concentrations are expressed as µg/g dry weight.

6.2 ACCUMULATION OF TRACE METALS

It is a striking feature of the accumulation of trace metals by aquatic invertebrates that the resulting accumulated concentrations vary hugely, among different invertebrates in the case of one metal, and within one invertebrate for different metals, even in the absence of raised metal bioavailabilities. This is illustrated in Table 6.1, which shows some of the variability within a single taxon, the crustaceans, of accumulated concentrations of just three trace metals—zinc, copper, and cadmium.

Three crustacean taxa are illustrated: barnacles (Cirripedia), and two malacostracan groups—amphipods and caridean decapods, with examples from both metal-contaminated and noncontaminated sites. Barnacles have Zn concentrations an order of magnitude above those of amphipods and caridiens, even when the barnacles are from uncontaminated sites. On the other hand, barnacles from uncontaminated sites typically have body Cu concentrations below those of amphipods and caridiens. Barnacles, however, increase their body Cu concentrations well above those of amphipods and caridiens when at Cu-contaminated sites, for example, Chai Wan Kok, Hong Kong (Phillips and Rainbow 1988) or Dulas Bay, Wales (Walker 1977). Differences between body Cd concentrations in the three crustacean taxa are not so marked.

An accumulated body concentration of trace metal in an aquatic invertebrate is the net result of the processes of uptake of the metal into the body, excretion of the metal from the body and changes in weight of the animal. For example, growth will cause a reduction in accumulated metal concentration (termed growth dilution) if the rate of growth outpaces the rate of accumulation of metal content (a net effect of uptake and excretion), even if net accumulation of metal content is ongoing. Loss of weight, for example, by use of energy reserves or loss of gametes, has the potential to increase accumulated metal concentration with or without changes in accumulated metal content. The same effects can take place at organ or tissue level within an invertebrate.

6.2.1 UPTAKE

Aquatic invertebrates take up trace metals from solution and from the diet. Rates of uptake and the proportional contributions of the two routes vary between metals and/or invertebrate species. These rates and proportional importance also vary intraspecifically with physiological state of the animal (e.g., stage in moult cycle of an arthropod), physicochemical state of the surrounding medium (not only directly via effects on dissolved metal bioavailabilities, not least dissolved metal concentration changes, but also indirectly, such as by salinity changes affecting the animal's osmotic and/or ionic physiology), and by changes in the bioavailability of trace metals in the diet. Here we follow the definition of bioavailability used by Luoma and Rainbow (2008)—a relative measure of that fraction of the total ambient metal that an organism actually takes up when encountering or processing

TABLE 6.1

Variability in Cd, Cu, and Zn Concentrations among Crustaceans

Species	Location	Zinc	Copper	Cadmium	Reference
	Cirripedia				
Tetraclita	Hung Hom, Hong Kong (C)	6963	94.9	2.8	Phillips and Rainbow 1988
squamosa	Tung Chung, Hong Kong	2245	14.9	4.2	Phillips and Rainbow 1988
Balanus	Chai Wan Kok, Hong Kong (C)	9353	3472	7.3	Phillips and Rainbow 1988
amphitrite	Lai Chi Chong, Hong Kong	2726	59.3	5.5	Phillips and Rainbow 1988
Semibalanus	Dulas Bay, Wales (C)	50,280	3750	—	Walker 1977
balanoides	Menai Strait, Wales	19,230	170	—	Rainbow 1987
Elminius	Southend, England	3463	—	23.3	Rainbow and Wang 2001
modestus					
	Malacostraca				
Amphipoda					
Orchestia	Restronguet Creek, England (C)	392	139	9.8	Rainbow et al. 1989, 1999
gammarellus	Dulas Bay, Wales (C)	126	105	9.1	Rainbow et al. 1999
	Millport, Scotland	188	77.5	1.6	Rainbow et al. 1999
Talitrus	Dulas Bay, Wales (C)	235	203	—	Fialkowski et al. 2009
saltator	Kulikovo, Kaliningrad, Russia	126	52.2	1.0	Fialkowski et al. 2009
Talorchestia	St Kilda, Dunedin, New Zealand	481	31.9	17.2	Rainbow et al. 1993
quoyana	Sandfly Bay, Dunedin, New Zealand	133	15.6	8.9	Rainbow et al. 1993
Eucarida					
Decapoda					
Pleocyemata					
Caridea					
Palaemon	Millport, Scotland	80.6	110	0.9	White and Rainbow 1986
elegans					
Pandalus	Firth of Clyde, Scotland	57.5	57.4	—	Nugegoda and Rainbow 1988
montagui					

Note: A selection of body concentrations (µg/g dry wt.) of three trace metals (Zn, Cu, Cd) in a systematic range of crustaceans from clean and metal-contaminated (C) sites. Metal-contaminated sites (C) are contaminated with at least one, but not necessarily all three, of the metals listed.

Source: Partly from Luoma, S. N., and P. S. Rainbow, *Metal Contamination in Aquatic Environments: Science and Lateral Management*, Cambridge University Press, Cambridge, 2008. With permission.

environmental media, summated across all possible sources of metal, including water and food as appropriate.

Even animals buried in sediments take up metals by only these routes, although the sediment can contribute to either or both, dependent on the biology of the infaunal invertebrate. Metals dissolved in interstitial porewater are in steady state with metals adsorbed on the sediment particles, the dissociation constant varying with the metal, the characteristics of the sediment particles (e.g., organic matter content, concentrations of iron and manganese oxides), and the physicochemistry of the porewater, not least its redox potential.

Proportions of dissolved metal uptake from porewater and overlying water column, inevitably of different metal bioavailabilities, will depend on how the organism interacts with its environment.

If porewater bathes a permeable surface of the burrowing invertebrate, such as the external surface of polychaetes, uptake of dissolved metal may occur from this source. However, many burrowing invertebrates derive a respiratory and/or feeding current from the overlying water column, despite living within the sediments. Exposure to porewater will depend on whether the irrigation current mixes with porewater or indeed prevents its access to any body surface. In addition, organisms may migrate in and out of the sediments, with consequent exposure to both porewater and surface water. Thus, even in animals that live in sediments, the dissolved metal in the water column can be, and is commonly, the source of dissolved metal uptake (Warren et al. 1998). Thus, the metal uptake properties of the different exposed epithelia depend on the biology of the burrowing invertebrate. Second, infaunal invertebrates may take up metals from the sediment particles directly by ingestion—many, but not all of them, being deposit feeders. Sediment-ingesting deposit feeders inevitably take up metals from this dietary source, metal uptake depending on the nature of the sediment and its history of contamination and depth in the sediment column, the digestive physiology of the invertebrate and the rate of sediment ingestion. The latter two factors at least depend on the biology of the animal.

Dissolved trace metals are taken up across a surface epithelium via several possible mechanisms. The relative importance of these different mechanisms varies interspecifically and can change intraspecifically, both with the physicochemistry of the surrounding medium and the physiology of the invertebrate (Luoma and Rainbow 2008). Furthermore some animals will swallow water as they feed or as part of osmotic regulation mechanisms, so some metal uptake from solution may take place across the gut epithelium, without these metals strictly being derived from the diet.

The lipid bilayer of the cell membrane represents an impermeable barrier to a dissolved polar entity such as a positively charged metal cation. The metabolism of all cells depends on accessibility to a sufficiency of different metals, be they major ions such as Na^+ or Ca^{2+}, or essential trace metals like copper and zinc playing key biochemical roles in enzymes. Thus, there are mechanisms for trace metals to cross cell membranes; for details refer to Luoma and Rainbow (2008). Such transport mechanisms are typically based on proteins crossing the membrane. Carrier or transporter proteins bind trace metal ions externally, pass them across the lipid bilayer and release them on the intracellular side. This route may well represent the dominant route of uptake for most trace metals from solution by most aquatic invertebrates in the media of most aquatic habitats. It appears unlikely that specific carriers exist for nonessential metals, which may be presumed to be transported via carrier proteins for essential trace metals to which they are chemically similar, as perhaps in the case of cadmium and zinc. Major ions typically enter via channels formed from proteins forming temporary aqueous pores of a specific diameter to allow passage of specific ions. Trace metal ions similar in size and charge to the appropriate major ion will trespass through these major ion channels, as in the case of cadmium (and in some cases zinc) through calcium channels, and silver and copper through sodium channels; the relatively few examples of dissolved metals predominating in anionic form such as vanadate and molybdate may trespass on sulfate channels (Luoma and Rainbow 2008). Another potential mechanism of trace metal uptake (Luoma and Rainbow 2008) can be cotransport with carriers for amino acid or dipeptide uptake (Conrad and Ahearn 2005). This is particularly well represented in the gut, and may be of more importance in uptake from diet than uptake of dissolved metal. Organometallic compounds may cross the membrane without intercession of proteins, and endocytotic uptake of metaliferous colloids or particles (<100 nm) may occur, for example of particulate iron in the gills of bivalves (George et al. 1976). Inevitably this latter route extends beyond what might be considered as dissolved uptake, but illustrates that organisms may be able to take up metals from a continuum of sized materials, from dissolved ions to small molecules to nanosized particles.

Transport of trace metals across membranes through the intercession of proteins, whether in the form of carrier proteins, major ion channels or amino acid transporters, does not typically require energy (Luoma and Rainbow 2008). In these examples, trace metals can cross the membrane by facilitated diffusion, diffusion facilitated by the transmembrane protein, so long as there is a concentration gradient allowing a passive movement of the trace metal from the external medium to the intracellular one. Given the extremely low concentrations of dissolved trace metals in many aquatic habitats, not least the surface waters of oceans (Luoma and Rainbow 2008), and their usually higher intracellular concentrations, it appears counterintuitive that there is a concentration gradient allowing trace metals to enter passively. The explanation lies in the Free Ion Activity Model or FIAM (Sunda and Guillard 1976; Anderson and Morel 1977; Campbell 1995). The model proposes that it is the activity of the free metal ion that is a good predictor of the dissolved form of the metal that binds to the transmembrane protein. In the external medium, trace metals that produce cations (e.g., M^{2+}) do not exist only as this free hydrated ion, but achieve stability by complexing with ligands, both inorganic (e.g., Cl^-, CO_3^{2-}) and organic (dissolved organic matter including fulvic and humic acids). Metal speciation equilibria are set up according to the nature of the medium and complexing anions available, with the free metal ion being available as a percentage of total dissolved metal (Luoma and Rainbow 2008). Any free metal ion binding with the transmembrane protein for subsequent uptake by the epithelial cell can be replenished from the metal speciation equilibrium externally. Intracellularly the free metal ion released from the transmembrane protein binds to one of many high affinity cellular ligands present (perhaps an intracellular protein), maintaining a concentration gradient of the free metal ion from out to in, in spite of the very low external activity of this ion.

The other transmembrane routes above that do not involve a transmembrane protein typically do not follow the FIAM (Campbell 1995). Nevertheless, the FIAM is such a good predictor of the uptake of dissolved trace metals by aquatic invertebrates for many trace metals under most conditions, that uptake via a transmembrane protein (arguably a carrier protein, Luoma and Rainbow 2008) does seem to be the dominant uptake route for dissolved trace metals. As stated earlier, however, different uptake routes coexist and the dominant uptake route, even in one species for one metal, may change for biological or physicochemical reasons.

Aquatic invertebrates will also take up metals from ingested food particles, and the diet can indeed be the predominant route of uptake of a trace metal by an aquatic invertebrate (Wang 2002). Uptake from food is a function of how much food is ingested (the feeding or ingestion rate), the concentration of the metal in the food, and the assimilation efficiency (AE) of that metal from that food (Wang and Fisher 1999; Luoma and Rainbow 2008). The AE varies between metals and food types for a single feeding aquatic invertebrate, and between feeding invertebrates for the same metal and food item (Luoma and Rainbow 2008). AE differences between foods are related to the different chemical nature of the metal accumulated in the food items (Wallace et al. 2003), and AE differences between feeders are related to the differences between the digestive processes of the different feeders (Rainbow et al. 2007).

An initial proposal to explain variability in the AEs of trace metals in copepod crustaceans was that metals bound to soluble components in their phytoplankton food represented that component of dietary metal that was bioavailable to the copepod (Reinfelder and Fisher 1991). As standard operational techniques to define the subcellular compartmentalization of accumulated metals in biota became routine (Wallace et al. 2003), they facilitated further investigation of the role of the chemical binding of metals in food in affecting the subsequent assimilation of metals by a herbivore or predator. Studies of the assimilation of trace metals by aquatic invertebrates spread to further food items and further prey (Wallace and Lopez 1996, 1997; Wallace and Luoma 2003; Ng et al. 2005; Rainbow et al. 2006), and it became appreciated that other (insoluble) components of accumulated metal in the diet could contribute to metal assimilated by the ingesting herbivore or predator. Even metals bound in insoluble metal-rich granules in prey items can yield metal for assimilation in the gut of neogastropod mollusk predators like *Thais clavigera* (Cheung and Wang 2005) or scavengers like *Nassarius festivus* (Rainbow et al. 2007). Thus differences in the chemical nature of the different fractions of dietary metal, differences in the comparative fractionation of accumulated metals between food types, and the different assimilative powers of different feeding animals, combine to produce the different AEs of different metals from different foods.

Biodynamic modeling (Luoma and Rainbow 2005) now represents an important tool in determining the relative importance of dissolved and dietary uptake of different metals to different aquatic invertebrates in different habitats under different degrees of metal contamination.

6.2.2 EXCRETION

The second process affecting the accumulation of trace metals by aquatic invertebrates is that of excretion from the body. The rate of excretion is defined by a species-specific rate constant of loss multiplied by the concentration in the body at any instant. The rate constant can vary widely and is of fundamental importance in the difference in bioaccumulation among species (see later discussion of biodynamics). It is not always the case that significant excretion occurs (rate constant of loss <0.001 of the body concentration per day), if all incoming metal is stored within the body in insoluble forms, for example. But metals can also be excreted very rapidly by some species (e.g., 0.2–0.4 of the body concentration per day), as in the case of some metals in some arthropods. Metal excretion may be in soluble form from organs such as the gills and/or in insoluble form, for example, from the alimentary tract (or organs like a hepatopancreas or digestive gland derived therefrom) or the kidney. Thus amphipod crustaceans excrete copper-rich granules from the ventral caeca of the alimentary tract (Icely and Nott 1980; Nassiri et al. 2000), and mussels excrete zinc granules from the kidney (George and Pirie 1980).

Excretion of a trace metal may be predominantly by a single route or by different routes according to the route of uptake, with many different combinations available according to metal and invertebrate species. For example, a metal taken up from solution may or may not be translocated to a particular organ for temporary storage to be followed by excretion perhaps in insoluble form, or may be lost from the gill with or without intermediate translocation elsewhere in the body. Similarly a metal taken up in the alimentary tract may stay in the gut epithelial cell into which it was assimilated and later be lost from that cell, or may be translocated elsewhere in the body to be excreted from a gill or kidney.

In biodynamic modeling investigations, the efflux of labeled metal can be quantified by live counting the whole organism (with invertebrates, for example) after uptake from solution or diet using a unique label to follow the rate of loss. It also is common experimentally to test, for example, whether the same rate constants (and therefore probably routes) of excretion apply to the different uptake routes. Furthermore curve stripping can be applied to attempt operationally to identify the number of different pools from which metals are being excreted. It seems intuitively obvious that there is a potential to match these dynamic pools to physiological stores of accumulated metal identified separately at organ and/or subcellular level. But little direct evidence is available to tie the two concepts together. It is possible that one or both of these operational tools is insufficiently sensitive, as of yet, to define mechanisms beyond gross rates or gross distributions.

6.2.3 ACCUMULATION

The accumulated concentration of a trace metal in an aquatic invertebrate is the net result of the inevitable uptake of that metal from solution and diet, and any excretion of that metal, be it predominantly via one excretion route or by any combination of excretion routes. A high accumulated concentration can, therefore, result from different combinations of high and low uptake and excretion rates. Furthermore there is typically no absolute definition of what is a high or low accumulated concentration of one particular metal across invertebrates. What is a high concentration in one aquatic invertebrate may be low for another, and vice versa (see Table 6.1).

When a metal crosses the membrane of an epithelial cell and enters an invertebrate, from whatever source, it has the potential to bind with a number of internal ligands, for example proteins, which have a high affinity for that metal (Luoma and Rainbow 2008). In short it is metabolically available. This binding may be key in the metabolism of the organism as in the case of essential trace metals being incorporated into enzymes in which they play a crucial role, or may be the cause of toxic effects, the binding of the wrong metal in the wrong place causing a lack of function for example of a key metabolic enzyme. It is important therefore that the binding of that incoming metal is controlled by temporary or permanent detoxification, whether the metal is essential or not, because all trace metals can elicit toxic effects if left unattended in a cell above a threshold concentration. Such detoxification can include binding to soluble metabolites such as glutathione or detoxificatory proteins like metallothioneins, or ultimately formation of insoluble granules such as metal pyrophosphates or deposits such as tertiary lysosomes (Mason and Jenkins 1995; Marigómez et al. 2002; Amiard et al. 2006; Luoma and Rainbow 2008).

Metals taken up into an epithelial cell by endocytosis, whether into a gut epithelial cell for digestion or into an external epithelium cell such as a gill cell of a lamellibranch bivalve, do not become metabolically available until the metal is solubilized and crosses the vacuole membrane into the cytosol.

Figure 6.1 depicts schematically the accumulated metal content of an aquatic invertebrate such as a decapod crustacean. It is useful conceptually to divide the accumulated metal content into two forms—metabolically available and detoxified, as discussed above. Nevertheless it must be appreciated that these categories simply refer to certain products in a network of interconnected physiological pathways handling the metals at different rates of reaction. Attempts (Wallace et al. 2003) can be made to equate these conceptual divisions to recognizable subcellular forms of bound metal. This is perhaps easier to achieve in the case of detoxified metal, for example metal bound with glutathione, metallothionein or in insoluble granular form, but is more difficult in the case of metabolically available metal.

Adsorbed metal available for physical exchange on the external surface of an aquatic invertebrate such as an arthropod, may represent a significant proportion of the body content of some surface-active metals (iron, manganese) for some species under some environmental conditions, and has been included in Figure 6.1. For most metals, however, the adsorbed fraction is usually quantitatively not significant, and it plays no role in the ecophysiology or ecotoxicology of the metal in the invertebrate for it has not been taken up into the body.

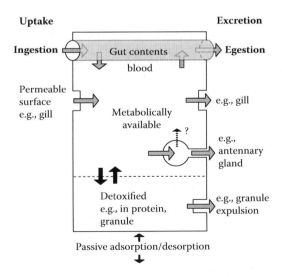

FIGURE 6.1 A schematic representation of the body metal content of an aquatic invertebrate such as a decapod crustacean (after Rainbow 2007). Metal crossing the membrane of an epithelial cell and first entering the body will be metabolically available, before potentially being stored, temporarily or permanently, in detoxified form, probably elsewhere in the body after internal transport. Metals then may or may not be excreted, either from the metabolically available component or from the detoxified store. (From Luoma, S. N., and P. S. Rainbow, *Metal Contamination in Aquatic Environments: Science and Lateral Management*, Cambridge University Press, Cambridge, 2008. With permission.)

6.2.3.1   Metabolic Requirements for Essential Trace Metals

It is possible to make theoretical estimates of the amounts of essential trace metals required to fulfill metabolic needs in an aquatic invertebrate, and thus gain an understanding of the relative size of the metabolically available component of accumulated metal depicted in Figure 6.1 (Rainbow 2007; Luoma and Rainbow 2008). For example in the cases of copper and zinc, theoretical calculations (White and Rainbow 1985) can be based on the number of copper- and zinc-bearing enzymes and their contributions to the total concentrations of enzymes in metabolizing tissues (Table 6.2), albeit with an awareness of the many assumptions being made. There may be other identified requirements for a trace metal beyond the needs of metal-associated enzymes. For example, malacostracan crustaceans (including amphipods and decapods) contain the respiratory protein hemocyanin, which is copper-based, and so estimates of the copper requirements of hemocyanin can be added into any total requirement estimate (White and Rainbow 1985; Rainbow 2007). Table 6.3 therefore shows such a total estimate for copper in an aquatic decapod crustacean, in this case *Pandalus montagui*, taking into account the weight of the metabolically inert exoskeleton in the calculations of enzyme requirements. The total theoretical figure estimated is 38.1 µg/g (Table 6.3). A measured mean concentration of copper in this decapod collected in the Firth of Clyde, Scotland is 57.4 µg/g (Nugegoda and Rainbow 1988), suggesting that such theoretical estimates may have some relevance.

If there is a minimum requirement for a particular trace metal in an invertebrate, are there any examples where invertebrates may be approaching a state of essential metal deficiency? It is appreciated that it would be impossible to identify a true case of absolute metal deficiency given that the invertebrate would be absent and its body concentration not measurable. Perhaps some oceanic decapod crustaceans approach a situation of copper deficiency, for the surface waters of oceans show depleted dissolved copper concentrations (Luoma and Rainbow 2008). Small specimens of the mesopelagic caridean decapod *Systellaspis debilis* have low copper concentrations of about 30 µg/g, only reaching more typical caridean copper concentrations of about 100 µg/g in large adults (White and Rainbow 1987; Rainbow and Abdennour 1989; Ridout et al. 1989). According to similar calculations as laid out in Table 6.3, many juvenile *S. debilis* would only have sufficient body copper to match enzyme needs, whereas larger adults have enough copper for hemocyanin also. In fact small *S. debilis* have been shown to contain little or no hemocyanin while adults contain a more typical caridean decapod quotient (Rainbow and Abdennour 1989). Furthermore juvenile *S. debilis* undergo less distinct vertical migrations than the adults (Roe 1984), perhaps related to a deficiency of copper for hemocyanin and consequently decreased respiratory performance. Two other mesopelagic decapods, the penaeids *Sergia robustus* and *Gennadas valens*, also show a positive relationship between body copper concentration and size, perhaps indicative of the same effect (Ridout et al. 1989).

TABLE 6.2
Essential Metal Requirements in Enzymes of Metabolizing Soft Tissue

	Number of Metal-Associated Enzymes	Percentage of Total Number of Enzymes	Average Number of Metal Atoms per Enzyme Molecule	Estimated Enzyme Metal Requirement in Tissue (µg/g dry wt.)
Copper	30	1.40	2.95	26.3
Zinc	80	3.74	1.41	34.5

Source: From White, S. L., and P. S. Rainbow, *Mar. Ecol. Progr. Ser.* 8:95–101, 1982; Luoma, S. N., and P. S. Rainbow, *Metal Contamination in Aquatic Environments: Science and Lateral Management*, Cambridge University Press, Cambridge, 2008. With permission.

TABLE 6.3
Estimates of the Essential Requirements for Copper
(µg/g dry wt.) in the Caridean Decapod *Pandalus montagui*
(after Luoma and Rainbow 2008) Compared with a
Measured Concentration (Nugegoda and Rainbow 1988)

Pandalus montagui

Percentage distribution dry weight

Exoskeleton	40.0%
Blood	2.1%
Soft tissues	57.9%

Hemocyanin

Blood Cu concentration	44 µg/mL
Blood volume in body	0.52 mL/g dry wt.
Blood Cu concentration in body	22.9 µg/g

Enzyme requirement

Metabolizing soft tissue	26.3 µg/g
Whole body	15.2 µg/g

Total body Cu metabolic requirement

Hemocyanin	22.9 µg/g
Enzymes	15.2 µg/g
Total	38.1 µg/g

Measured body Cu concentration

Mean ± SD	57.4 ± 18.9 µg/g

Note: For further explanation, see text.

6.2.3.2   Detoxified Component of Trace Metals

The second component of accumulated metal depicted in Figure 6.1 is that of detoxified metal, which may be stored permanently or temporarily before excretion. Detoxification is the process by which a metal released intracellularly is bound by a ligand of such high affinity that it in effect prevents that metal from binding to other molecules, in either an essential role or to cause toxicity (Luoma and Rainbow 2008). Detoxification can be in soluble form as in the case of metallothioneins or in insoluble form in one of a variety of insoluble granules or deposits. Detoxification pathways may interrelate. For example a metal bound to a metallothionein may later appear in an insoluble deposit in a tertiary lysosome after autolysis of the metallothionein in a turnover cycle of protein production and breakdown (Mason and Jenkins 1995; Luoma and Rainbow 2008). Copper- and sulfur-rich granules in the talitrid amphipod *Orchestia gammarellus* from copper-contaminated sites probably have this origin (Nassiri et al. 2000). Ferritin, an intracellular storage protein binding iron, is usually present in soluble form, but in stegocephalid amphipod crustaceans it is found as large crystals in the ventral caeca (Moore and Rainbow 1984).

Occasionally essential metals in detoxified form may be released back into metabolically available form under carefully controlled circumstances, as in the release of copper from metallothionein to be incorporated into the respiratory protein hemocyanin in decapod crustaceans (Brouwer et al. 2002; Amiard et al. 2006).

Copper can be used again as an example to illustrate the relative proportions of accumulated metal in metabolically available and detoxified form (Figure 6.1), using the crustaceans listed in Table 6.1.

Unlike malacostracans, barnacles lack hemocyanin and their metabolic requirements for copper may be restricted to enzyme needs. A barnacle body would have a theoretical copper requirement of about 15–26 µg/g (Tables 6.2 and 6.3). Body concentrations of copper in barnacles from noncontaminated sites are as low as 15 µg/g (Table 6.1) suggesting a lack of detoxified storage of copper in these barnacles. Nevertheless barnacle species have the potential to increase accumulated body concentrations of copper hugely (from 59 to 3472 µg/g in *Balanus amphitrite*, and from 170 to 3750 µg/g in *Semibalanus balanoides*—Table 6.1) when exposed to copper contamination, the vast majority of this accumulated copper now being in detoxified form (in fact in copper- and sulfur-rich deposits (Walker 1977), which are probably in tertiary lysosomes derived from the breakdown of copper-bearing metallothionein).

In the case of decapod crustaceans, the metabolic requirements for copper involve both enzymes and hemocyanin, estimated to total 38.1 µg/g for *P. montagui* (Table 6.3). Decapod crustaceans therefore seem to contain enough copper for these needs with some to spare (Table 6.1). Over most copper bioavailabilities the body copper concentrations of decapods is kept relatively constant, apparently in line with metabolic requirements (Rainbow 2002; Luoma and Rainbow 2008). In extremely high copper bioavailabilities, the copper concentration of the decapod *Palaemon elegans* can rise from about 100 µg/g (Table 6.1) to about 600 µg/g (White and Rainbow 1982). In order to survive such circumstances, the organism must hold the extra accumulated copper in detoxified form, again as copper- and sulfur deposits, now in the hepatopancreas, in tertiary lysosomes (Rainbow 1998).

An extreme example of the relative sizes of metabolically available and detoxified components is that of zinc accumulated in barnacles (Rainbow 2002, 2007). Even barnacles from noncontaminated sites have zinc body concentrations of thousands of µg/g with the potential to increase to tens of thousands of µg/g (Table 6.1). Following arguments above for metabolic requirements of a few tens of µg/g (Table 6.2), and the fact that decapods maintain body concentrations of a similar magnitude (50–100 µg/g, Table 6.1), it would appear that the vast percentage of the high body concentration of all barnacles is in detoxified permanently stored form—in this case zinc pyrophosphate granules (Pullen and Rainbow 1991; Rainbow 2002, 2007).

While barnacles store detoxified zinc granules in the general body tissue, in many aquatic invertebrates, particular organs are used for detoxified storage of metals, often the hepatopancreas (or digestive gland, nomenclature varying between invertebrate taxa) or the kidney. On the other hand, oysters actually store detoxified zinc and copper in insoluble forms in blood cells (George et al. 1978).

6.2.4 VARIATION IN ACCUMULATION PATTERNS AND ACCUMULATED CONCENTRATIONS

Aquatic invertebrates show different accumulation patterns for different metals, an understanding of which illuminates how very different accumulated metal concentrations arise in these different invertebrates. Figure 6.2 shows some examples of such accumulation patterns, again using crustaceans and the trace metals zinc, copper, and cadmium as examples (Table 6.1). Figure 6.2a and 6.2b show two extremes of the gradient of metal accumulation patterns—that of regulation (Figure 6.2a) and that of strong net accumulation (Figure 6.2b).

Figure 6.2a shows the trace metal accumulation pattern of an aquatic invertebrate that regulates the total body metal concentration of an essential metal by balancing uptake with excretion, as in the case of Zn in the decapod crustacean *Palaemon elegans* over a wide range of ambient zinc bioavailabilities (Rainbow 1998, 2002, 2007). It is assumed that detoxification of the metal in such species is minimal. All metal appears to be accumulated in the metabolically available component, itself subdivided into the essential metal required for metabolic purposes, and excess metal over and above this metabolic requirement. There is a threshold concentration of metabolically available metal, above which the accumulated metal is toxic. This threshold is reached when the rate of uptake has increased with raised external bioavailability to such an extent that it is no longer matched by the rate of excretion only (in the absence of detoxification). Zinc continues to accumulate in the

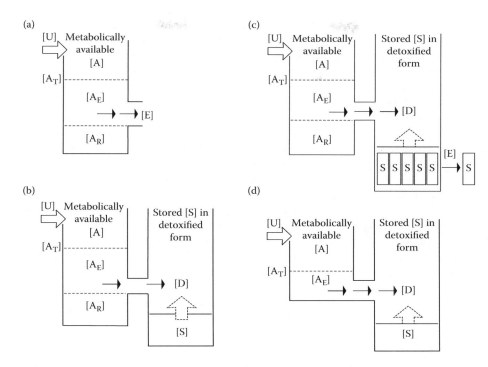

FIGURE 6.2 Trace metal accumulation patterns of aquatic invertebrates, exemplified by crustaceans (after Rainbow, 2002). After uptake [U], metal is accumulated in the metabolically available component [A], itself subdivided into the essential metal required for metabolic purposes [A_R], and excess metal [A_E] over and above this metabolic requirement. There is no [A_R] component in the case of nonessential metals. There is a threshold concentration [A_T] of metabolically available metal, above which the accumulated metal is toxic. Accumulated metal may be moved from the metabolically available component either by excretion [E] or by detoxification [D] to be stored [S], either permanently with no upper concentration limit or temporarily before excretion. (a) Accumulation pattern of an invertebrate that regulates the total body metal concentration of an essential metal by balancing uptake [U] with excretion [E], as in the case of Zn in the decapod crustacean *Palaemon elegans*. (b) Accumulation pattern of an invertebrate that is a net accumulator of an essential metal without significant excretion of metal taken up, as in the case of Zn in barnacles. (c) Accumulation pattern of an invertebrate that shows net accumulation of an essential metal in detoxified form, but excretes some of that accumulated metal in the detoxified component, as in the case of Cu accumulated from the diet by amphipod crustaceans. (d) Accumulation pattern of an invertebrate that shows net accumulation of a nonessential metal in detoxified form with no significant excretion, as in the case of Cd in barnacles. (From Luoma, S. N., and P. S. Rainbow, *Metal Contamination in Aquatic Environments: Science and Lateral Management*, Cambridge University Press, Cambridge, 2008. With permission.)

body without being detoxified, and toxicity ensues. The regulated body concentration of zinc is about 80 µg/g in *P. elegans* (Table 6.1), the lethal body concentration only about 200 µg/g (White and Rainbow 1982), exemplifying the importance of detoxification in the accumulation of raised body concentrations.

Figure 6.2b depicts the trace metal accumulation pattern of an aquatic invertebrate that is a net accumulator of an essential metal without significant excretion of metal taken up, as in the case of Zn in barnacles. Metabolically available metal in excess of requirements is detoxified to be stored as the detoxified component of accumulated metal with no upper concentration limit. Thus all barnacles store accumulated zinc detoxified in zinc pyrophosphate granules, the number of these granules, and hence body zinc concentration, increasing in zinc-contaminated habitats (Table 6.1) (Rainbow 2002, 2007). The fact that barnacles survive with body zinc concentrations in excess of 50,000 µg/g while a body concentration of 200 µg/g is lethal to *P. elegans* again illustrates the importance of detoxification in any appreciation of the significance of an accumulated metal concentration in an aquatic invertebrate. An uptake rate of zinc higher than the rate of detoxification will cause the metabolically available component of accumulated zinc to exceed the toxic threshold, and the barnacle will die. This lethal concentration of metabolically available zinc may well only be of the order of 200 μg/g as seen in *P. elegans*, and bears no relation to the total accumulated concentration, overwhelmingly dominated by a detoxified zinc component built up over time when the rate of detoxification was able to match the rate of uptake, even in relatively high zinc bioavailabilities.

In another example, Figure 6.2c illustrates the conceptual model for the trace metal accumulation pattern of an aquatic invertebrate that shows net accumulation of an essential metal in detoxified form, but excretes some of that accumulated metal from the detoxified component. This is the case for copper accumulated from the diet by amphipod crustaceans as shown in more detail in Figure 6.3 (Luoma and Rainbow 2008). Copper taken up from the diet is incorporated into the epithelial cells of the ventral caeca in the alimentary tract, ultimately as copper- and sulfur-rich insoluble deposits considered to be in tertiary lysosomes and derived from the turnover of metallothionein binding copper (Nassiri et al. 2000). Cells of each ventral caecum originate at the distal end and are progressively moved along the epithelium towards the main alimentary tract where they disintegrate and release all remaining cell contents into the gut lumen (Galay Burgos and Rainbow 1998). Included in the cellular debris are the copper-rich lysosomal residual bodies. Thus at a given concentration of copper in the diet, the cells of the ventral caecum accumulate a certain number of copper-rich deposits, turning over with the cell cycle, and contributing to a steady state caecal (and body) concentration of copper (Figure 6.3). If the copper availability in the diet is raised, more copper is taken up into these cells, more copper-rich deposits are present, and a new steady state body copper concentration is reached (Figure 6.3). Thus, in Cu-contaminated conditions, the body concentration of the amphipod

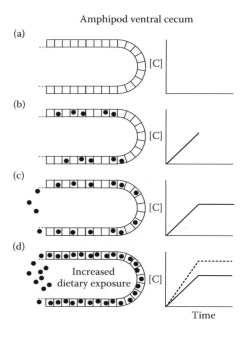

FIGURE 6.3   (a, b, c) The accumulation of detoxified metal (e.g., Cu) in the ventral caecal cells of an amphipod crustacean exposed to a trace metal in the diet, and the corresponding changes in total body metal concentration [C] over time. (d) The effect of increased dietary exposure to the metal. (From Rainbow, P. S., *Environ. Pollut.*, 120, 497–507, 2002; Luoma, S. N., and P. S. Rainbow, *Metal Contamination in Aquatic Environments: Science and Lateral Management*, Cambridge University Press, Cambridge, 2008. With permission.)

increases in response to the elevated bioavailable Cu (Table 6.1). But the increase is not infinite, as the accumulating, detoxified and temporarily stored copper is excreted at increasing rates until its loss rate matches the uptake rate into the organism, and a new steady state has been reached (Figures 6.2 and 6.3). For the amphipod *Talitrus saltator*, 200 µg/g is a high body copper concentration compared to more typical concentrations of 30–75 µg/g (Fialkowski et al. 2009), and much of this increased accumulated copper is in detoxified form. This higher concentration is reached because the rate of detoxification matches the increased rate of excretion upon exposure to the more contaminated diet (in turn later to be matched by the rate of excretion of detoxified copper after an increase in body copper concentration) (Figures 6.2 and 6.3). A very high copper exposure will cause the uptake rate to exceed the maximum detoxification rate, increasing the metabolically available body copper concentration above the threshold, and causing stress or death, even if the total body concentration has not reached a concentration as high as 200 µg/g identifiable in field specimens exposed to high, but not so high, copper exposure over a longer time period.

The final example (Figure 6.2d) is the trace metal accumulation pattern of an aquatic invertebrate that shows net accumulation of a nonessential metal in detoxified form with no significant excretion, as in the case of Cd in barnacles. There is no required component of metabolically available accumulated metal as for essential metals, but still a threshold in this component (however small) above which toxicity occurs.

6.2.5 BIODYNAMIC MODELING

The recent application of biodynamic modeling to trace metal accumulation (Wang et al. 1996; Luoma and Rainbow 2005), added to an understanding of accumulation patterns of different trace metals by different aquatic invertebrates, has led to the ability to predict accumulated concentrations in different invertebrates in different situations. Biodynamic modeling, furthermore, has allowed the determination of the relative importance of solution and food as metal sources, and the comparative significance of different route of excretion.

In this context, biodynamic modeling considers a steady state accumulated metal concentration to be derived from a summation of uptake rate from solution plus uptake rate from food, minus excretion rate and growth rate (Luoma and Rainbow 2005, 2008).

Over an environmentally realistic range of dissolved trace metal concentrations, the rate of metal uptake from solution by an aquatic invertebrate is directly proportional to the dissolved concentration C_w; the uptake rate constant k_u (L/g/d) is the slope of this direct relationship (Luoma and Rainbow 2005). From a measured k_u for a given set of physicochemical variables of a medium (affecting for example the availability of the free metal ion), it is possible to calculate an uptake rate from solution (µg/g/d) as $k_u C_w$. The rate of metal uptake from food is expressed as AE (IR) C_f where IR is the ingestion rate (g/g/d), C_f the concentration of the metal in the food (µg/g), and AE the assimilation efficiency of the metal from that food. AE is measured in the laboratory using radiotracers (Wang and Fisher 1999; Rainbow and Wang 2001). Also measured in the laboratory with radiotracers is the efflux rate constant k_e (per day) (Rainbow and Wang 2001), if appropriate, subdivided into two efflux rate constants if excretion after uptake from water and food are different. Growth rate constants (per day) allow for the effect of growth dilution if necessary. These separate terms are then combined into the biodynamic model to predict the accumulated concentration (C) of a trace metal in an aquatic invertebrate.

$$C = \frac{k_u C_w + AE\,(IR)C_f}{k_e + g}$$

Luoma and Rainbow (2005) made a comparison of accumulated concentrations of trace metals predicted by biodynamic modeling in 15 different studies against independently measured bioaccumulated concentrations in the species of interest from the habitat for which the parameters discussed earlier had been developed. The data covered 6 metals, 13 animal species, and 11 marine, estuarine, and freshwater habitats, and the bioaccumulation parameters used in the predictive modeling had been obtained under environmental conditions broadly similar to those in the field-monitoring studies. Table 6.4 presents a selection of these data. The agreement between predicted and independently measured bioaccumulated concentrations was strong, indicating that the basic assumptions of biodynamic modeling are tenable. Biodynamic modeling allows the prediction and explanation of widely differing accumulated trace metal concentrations in aquatic invertebrates, combining geochemical measurements of environmental metal concentrations with the estimation of key biological parameters.

Figure 6.4 illustrates the explanatory power of biodynamic modeling (Luoma and Rainbow 2005, 2008), presenting accumulated zinc concentrations and physiological coefficients for uptake and loss for a barnacle (*B. amphitrite*) and a mussel (*Perna viridis*) collected simultaneously from the same locations in Hong Kong coastal waters. In fact the barnacles were often growing on the mussel shells. More assimilation from food and much slower excretion explain the consistently higher Zn concentrations in barnacles than in mussels, observed everywhere barnacles and mussels co-occur (metal-contaminated or not) (Luoma and Rainbow 2005, 2008). Both the capability for validation against nature and the explanatory power of biodynamic modeling suggest many uses for this approach in the future. However, to date, most physiological coefficients are more generic than specific to any given circumstance. It is likely that applications to very specific circumstances will require refinement of the coefficients to those conditions.

6.2.6 TOXICITY AND ACCUMULATION

Toxicity occurs when the rate of uptake exceeds the combined rates of excretion and detoxification. Under these circumstances, metal taken up continues to accumulate in the metabolically available component until it reaches a threshold (Figure 6.2), when too much toxic metal is binding to so many vital cellular components (enzymes, etc.) that metabolism is disturbed, ultimately lethally. Both essential and nonessential trace metals follow this scenario, although nonessential metals do not have a required component of metabolically available metal that must be present to meet metabolic needs. According to the accumulation pattern of the particular metal in the particular invertebrate, the metabolically available component may represent nearly all the accumulated metal content (e.g., accumulated zinc and copper in caridean decapod crustaceans regulating the body content with a little to spare, perhaps bound in waiting to the detoxificatory protein metallothionein), or a very tiny percentage of the whole-body content.

6.2.6.1 Lethal Body Concentrations of Trace Metals

Is there, therefore, a lethal body concentration of an accumulated metal that can be recognized as a limiting concentration in field-collected specimens? If the metal is detoxified to be stored permanently or temporarily, then the answer is "No." Thus, a barnacle will die when the local availability of zinc increases for whatever reason so that the rate of zinc uptake exceeds the rate of detoxification, zinc accumulates in the metabolically available fraction and the toxic threshold of metabolically available body zinc is reached. This concept suggests that toxicity can occur independently of how much zinc has been detoxified over the previous life of the barnacle—that is, regardless of the total body zinc concentration.

In the case of a regulator of trace metal body content, for example the caridean decapod *Palaemon elegans* for zinc, the lethal body concentration concept is more valid (*ca*. 200 µg/g, see earlier). This is because most, if not all, of the body zinc appears to remain in metabolically available form with no detoxification. The figure of 200 µg/g was derived from a laboratory experiment (White and Rainbow 1982). In practice, it is highly unlikely to detect or define a lethal body concentration in such a species in a field study. Once regulation has broken down (when uptake rate exceeds

TABLE 6.4
Comparisons between Observed Metal Concentrations (μg/g) Accumulated by Aquatic Invertebrates in Particular Field Situations and Those Predicted by the Biodynamic Model

Species	Ecosystem	Metal	Predicted Concentration			Observed Concentration			Reference
			Minimum	Maximum	Median or Mean	Minimum	Maximum	Median or Mean	
Bivalves									
Mytilus edulis	San Francisco Bay	Cd	2.7	10.1	6.4	4.4	9.4	6.9	Wang et al. 1996
Mytilus edulis	San Francisco Bay	Se	1.0	5.6	3.3	2.5	6.7	4.6	Wang et al. 1996
Mytilus edulis	San Francisco Bay	Zn	54	265	160	54	130	92	Wang et al. 1996
Dreissena polymorpha	Hudson River	Cd			21.7			18.7	Roditi et al. 2000
Macoma balthica	San Francisco Bay	Ag	1.3	21.0	11.2			8.0	Griscom et al. 2002
Macoma balthica	San Francisco Bay	Cd	0.02	0.90	0.46			0.33	Griscom et al. 2002
Macoma balthica	San Francisco Bay	Se	2.2	4.3	3.3			3.0	Luoma et al. 1992
Insect larvae									
Chaoborus punctipenis	Flavrian Lake, Quebec	Cd	0.7	3.0	1.8			1.3	Croteau et al. 2001
Chaoborus punctipenis	La Bruère Lake, Quebec	Cd	0.5	2.2	1.3			1.7	Croteau et al. 2001
Chaoborus punctipenis	Vaudray Lake, Quebec	Cd	1.1	4.4	2.7			2.9	Croteau et al. 2001
Barnacles									
Balanus amphitrite	Hong Kong	Zn	2610	11560	7080	3100	11000	6550	Rainbow et al. 2003
Elminius modestus	English Channel	Cd	7.6	11	9	15	27	21	Rainbow and Wang 2001
Elminius modestus	English Channel	Zn	1500	4400	2950	2470	4730	3600	Rainbow and Wang 2001

Note: Different authors have used mean or median concentrations.
Source: From Luoma, S. N., and P. S. Rainbow, *Environ. Sci. Technol.*, 39, 1921–1931, 2005; Rainbow, P. S., *Environ. Int.* 33, 576–582, 2007. With permission.

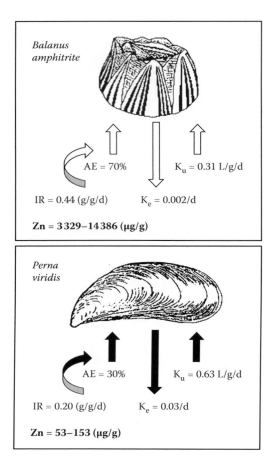

FIGURE 6.4 Zn concentrations and physiological coefficients for uptake and loss for a barnacle (*Balanus amphitrite*) and a mussel (*Perna viridis*) collected simultaneously from the same locations in Hong Kong coastal waters. AE, assimilation efficiency; IR, ingestion rate; K_u, dissolved uptake rate constant; K_e, rate constant of loss. (From Luoma, S. N., and P. S. Rainbow, *Environ. Sci. Technol.*, 39, 1921–1931, 2005; Luoma, S. N., and P. S. Rainbow, *Metal Contamination in Aquatic Environments: Science and Lateral Management*, Cambridge University Press, Cambridge, 2008. With permission.)

excretion rate), the metabolically available (and therefore body) metal concentration will increase from the typically regulated concentration over such a short period that it is highly unlikely that any field biologist will be in the right place at the right time to collect a specimen on the verge of death (the smoking gun in this context).

What about the case for aquatic invertebrates with an intermediate metal accumulation pattern? The same general point holds. The greater the detoxification of accumulated metal, the more difficult it will be to define a lethal body concentration applicable to field collected specimens, especially if exposures are variable. We have seen earlier that talitrid amphipod crustaceans can increase their accumulated copper concentrations (Table 6.1), for example, from 30 to 200 µg/g in the case of *T. saltator* (Fialkowski et al. 2009), when exposed in the field. If such amphipods were to be exposed to very high copper bioavailabilities in the laboratory, for example, in toxicity testing experiments, then so much copper would enter so rapidly that almost all the copper would accumulate in the metabolically available fraction without significant detoxification. With any existing detoxified accumulated metal representing a small and/or reproducible proportion of total accumulated metal, the copper-exposed amphipod would die at a reproducible concentration of metabolically available metal, and hence of total metal. Thus, there will be a lethal body concentration reproduced in laboratory toxicity testing, but this concentration cannot be directly applied to field-exposed amphipods taking up copper at slower rates in the field, even though rates in copper-exposed habitats are raised above normal. The field-exposed amphipods, with an increased proportion of detoxified metal in the body, are likely to accumulate higher concentrations than the artifactual laboratory lethal concentration. Again, in this case, no lethal body concentration of an accumulated metal can be recognized as a limiting concentration in field-collected specimens.

We have argued earlier that if there is no detoxification of accumulated metal, then there is theoretically a lethal body concentration that can be applicable in the field. The same principle would apply if any detoxified component is a fixed percentage of total accumulated metal content.

But are there any exceptions to the general principle that there is no lethal body concentration of a metal applicable to an invertebrate in the field? The answer may well be "Yes" for selenium and certain organometal compounds (Luoma and Rainbow 2008). Total accumulated selenium concentrations can be correlated with toxic effects, at least in fish and birds (Skorupa 1998). There is no known accumulated store of detoxified selenium, analogous to the insoluble concretions formed by other metals, for example. There is net bioaccumulation of Se, however, (it is not regulated); and some species are somewhat slightly more tolerant to an elevated body burden than others. Therefore selenium appears to represent a case of a net accumulator with little detoxification. It is also the element with the narrowest window between essential and toxic concentrations; and this model may be why (Luoma and Rainbow 2008). Tributyltin (TBT) bioaccumulated concentrations can also be correlated with observed toxic effects (Meador 2000; Luoma and Rainbow 2008), because TBT can be metabolized to less toxic forms (i.e., detoxified forms are not bioaccumulated). Its behavior is somewhere between behavior as a traditionally accumulated metal and behavior as an organic compound (Luoma and Rainbow 2008).

If there is no valid lethal total body concentration applicable to a particular species that can be recognized as a limiting concentration in field-collected specimens, can we extend the concept to define a lethal body concentration of metabolically available accumulated metal? The answer is "Yes," as attempted earlier for zinc in *P. elegans*. An alternative approach that is being increasingly attempted is to use subcellular compartmentalization in an attempt to measure this metabolically available fraction (Wallace and Luoma 2003, Wallace et al. 2003, Wang and Rainbow 2006). Such attempts may prove productive, but, against a background of cellular reactions handling metals occurring at different rates, ultimately it may prove impossible to equate directly a subcellular component, operationally separated by techniques such as centrifugation, with the conceptually important metabolically available component of accumulated metal.

6.2.6.2 Bioaccumulated Metal Guidelines

Although, with the exception of the likes of selenium and TBT, it is not possible to define lethal total accumulated body concentrations for field-collected specimens for particular metals, it is possible to recognize what are abnormally high bioaccumulated concentrations of particular metals in particular aquatic invertebrates. Such concentrations have great potential as bioaccumulated metal guidelines providing information on high and potentially ecotoxicological trace metal bioavailabilities in aquatic habitats (Luoma and Rainbow 2008).

Table 6.5 is a compilation of accumulated trace metal concentrations in selected aquatic invertebrates that can be used as biomonitors of trace metal bioavailabilities in coastal waters (Luoma and Rainbow 2008). The table identifies accumulated concentrations in these invertebrates that are indicative of the presence of atypically high local bioavailabilities (Rainbow 2006). The table reinforces earlier conclusions that an absolute concentration that is high in one aquatic invertebrate species may be typical in another with no suggestion of local metal contamination. For example, a very high zinc concentration in the mussel *Mytilus edulis* would be considered a typical zinc concentration in even another bivalve, the tellinid clam *Scrobicularia plana* (Table 6.5).

The high accumulated concentrations recognized here can be used as easily measurable bioaccumulated metal guidelines indicative of atypically high local availabilities, without the continuing need

TABLE 6.5

Accumulated Concentrations of Trace Metals (μg/g dry wt.) in the Soft Tissues of Selected Aquatic Invertebrate Biomonitors of Trace Metals, Categorized Here as "Typical" or "High"

	Ag	As	Cd	Co	Cr	Cu	Ni	Pb	Zn	Reference
Bivalves										
Mytilus edulis (mussel)										
Typical	0.1–1.1	12–31	0.4–2.7	0.3–6.3	0.2–2.5	6–11	0.5–3.6	2–19	45–119	Bryan et al. 1985
	0.1–2.4		0.8–3.4			4–19	0.4–3.7	0.2–3.3	51–150	Goldberg et al. 1983
			0.7–4.6	0.9–6.5		5–13	1.4–3.0	7–25	78–139	Boalch et al. 1981
			0.7–4.7			3–17	1.3–7.1	2–22	32–150	Gault et al. 1983
High	17	5–28	21–65		7.2	262	12	105	198–579	Bryan et al. 1985
			36					58	235–440	Gault et al. 1983
Scrobicularia plana (tellinid clam)										
Typical	0.1–4.6		0.2–9.1	2.0–13.3	0.5–7.1	9–61	1.0–5.8	5–109	256–1514	Bryan et al. 1980, 1985
								14–60		Luoma and Bryan 1978
										Langston 1980
High	259	98–191	31.5–42.7	33.0–106	16.3–23.8	136–752	14.4–22.4	225–3000	2060–4920	Bryan et al. 1980, 1985
Crustaceans										
Balanus amphitrite (barnacle)										
Typical	0.7–9.0	10–71	0.8–10.2	0.1–2.1	1–5	52–415	1.3–8.2	0.4–8.3	2860–8530	Rainbow and Blackmore 2001
					1–8	60–486		1.7–9.2	2730–9350	Phillips and Rainbow 1988
High	22.5	457	2.1–10.1	9.0–11.0	17–38	494–1810	99		14,000–23,300	Rainbow and Blackmore 2001
						1010–3472		37–39	12,000	Phillips and Rainbow 1988
Talitrus saltator (talitrid amphipod)										
Typical			1–5		1–5	30–75			100–250	Fialkowski et al. 2009
High			35		180	200			300	Fialkowski et al. 2009

Note: The "High" concentrations are indicative of atypically raised bioavailability of that metal in the local habitat. After Luoma and Rainbow (2008).

Source: From Luoma, S. N., and P. S. Rainbow, *Metal Contamination in Aquatic Environments: Science and Lateral Management,* Cambridge University Press, Cambridge, 2008. With permission.

to measure the ecological parameters (e.g., species numbers, richness, abundances, etc.) now necessary to recognize ecotoxicological effects at the community level in an aquatic habitat. The concept is well illustrated by the use of accumulated copper concentrations in larvae of species of the caddis fly *Hydropsyche* as bioaccumulated metal guidelines correlating with copper-induced changes in the local mayfly community assemblages of streams in North America affected by mining (Luoma and Rainbow 2008). Such copper concentrations can even be used as a threshold to correlate with the loss of particular mayflies, particularly heptageneiid mayflies (Luoma and Rainbow 2008). Clearly such an approach is particularly applicable in aquatic habitats in which metals are the dominating factor explaining much of any ecotoxicological effect, as in waters affected by mine waste. Nevertheless, an understanding of the significance of such accumulated metal concentrations in local tolerant bioaccumulators will also provide information on the relative importance of these metals as ecotxicological agents in habitats with mixed sources of trace contaminants, or indeed other causes of ecotoxicity.

Summary

We set out in this chapter to explore the general principles involved in the interpretation of the significance of accumulated metal concentrations in aquatic invertebrates, and their use in understanding ambient metal bioavailabilities and potential ecotoxicological effects. As shown earlier, we are now in a position to understand why different aquatic invertebrates accumulate different body concentrations of trace metals and their comparative significance. We know why a bioaccumulated metal concentration is to be considered as high in one invertebrate and low in another. We also know what is a high or low metal concentration in a single invertebrate. We argue that there is no general lethal body concentration of a metal applicable to a particular species that can be recognized as a limiting body concentration in field-collected specimens, if that species carries out any significant storage detoxification of that metal, as is often the case for trace metals in invertebrates. Selenium and organometals do not always meet this latter provision, and recognition of lethal body concentrations is a distinct possibility. Nevertheless, we can use accumulated metal concentrations in hardy biomonitors as bioaccumulated metal guidelines, allowing us to draw environmentally significant conclusions on the ecotoxicological status of particular aquatic habitats.

REFERENCES

Amiard, J.-C., C. Amiard-Triquet, S. Barka, J. Pellerin, and P. S. Rainbow. 2006. Metallothioneins in aquatic invertebrates: their role in metal detoxification and their use as biomarkers. *Aquat. Toxicol.* 76:160–202.

Anderson, D. M., and F. M. M. Morel. 1977. Copper sensitivity of *Gonyaulax tamarensis. Limnol. Oceanogr.* 23:284–293.

Boalch, R., S. Chan, and D. Taylor. 1981. Seasonal variation in the trace metal content of *Mytilus edulis. Mar. Pollut. Bull.* 12:276–280.

Brouwer, M., R. Syring, and T. H. Brouwer. 2002. Role of a copper-specific metallothionein of the blue crab, *Callinectes sapidus*, in copper metabolism associated with degradation and synthesis of hemocyanin. *J. Inorg. Biochem.* 88:228–239.

Bryan, G. W., W. J. Langston, and L. G. Hummerstone. 1980. The use of biological indicators of heavy metal contamination in estuaries. *Occ. Pub. Mar. Biol. Ass. UK* 1:1–73.

Bryan, G. W., W. J. Langston, L. G. Hummerstone, and G. R. Burt. 1985. A guide to the assessment of heavy-metal contamination in estuaries. *Occ. Pub. Mar. Biol. Ass. UK* 4:1–92.

Campbell, P. G. C. 1995. Interaction between trace metals and aquatic organisms: a critique of the free-ion activity model. In: *Metal Speciation and Aquatic Systems*, eds. A. Tessier and D. R. Turner, pp. 45–102. New York: Wiley.

Cheung, M. S., and W.-X. Wang. 2005. Influences of subcellular compartmentalization in different prey on the transfer of metals to a predatory gastropod from different prey. *Mar. Ecol. Progr. Ser.* 286:155–166.

Conrad, E. M., and G. A. Ahearn. 2005. 3H-L-histidine $^{65}Zn^{2+}$ are cotransported by a dipeptide transport system in intestine of lobster *Homarus americanus*. *J. Exptl. Biol.* 208:287–296.

Croteau, M. N., L. Hare, and A. Tessier. 2001. Differences in cadmium bioaccumulation among species of the lake-dwelling biomonitor *Chaoborus*. *Can. J. Fish. Aquat. Sci.* 58:1737–1746.

Cullen, J. T., T. W. Lane, F. M. M. Morel, and R. M. Sherrell. 1999. Modulation of cadmium uptake in phytoplankton by seawater CO_2 concentration. *Nature* (London) 402:165–167.

Fialkowski, W., P. Calosi, S. Dahlke, A. Dietrich, P. G. Moore, S. Olenin, et al. 2009. The sandhopper *Talitrus saltator* (Crustacea: Amphipoda) as a biomonitor of trace metal bioavailabilies in European coastal waters. *Mar. Pollut. Bull.* 58:39–44.

Galay Burgos, M., and P. S. Rainbow. 1998. Uptake, accumulation and excretion by *Corophium volutator* (Crustacea: Amphipoda) of zinc, cadmium and cobalt added to sewage sludge. *Estuar. Coast. Shelf Sci.* 47:603–620.

Gault, N. F. S., E. L. C. Tolland, and J. G. Parker. 1983. Spatial and temporal trends in heavy metal concentrations in mussels from Northern Ireland coastal waters. *Mar. Biol.* 77:307–316.

George, S. G., and B. J. S. Pirie. 1980. Metabolism of zinc in the mussel *Mytilus edulis* (L.): a combined ultrastructural and biochemical study. *J. Mar. Biol. Ass. UK* 60:575–590.

George, S. G., B. J. S. Pirie, A. R. Cheyne, T. L. Coombs, and P. T. Grant. 1978. Detoxication of metals by marine bivalves: an ultrastructural study of the compartmentation of copper and zinc in the oyster *Ostrea edulis*. *Mar. Biol.* 45:147–156.

George, S. G., B. J. S. Pirie, and T. L. Coombs. 1976. The kinetics of accumulation and excretion of ferric hydroxide in *Mytilus edulis* (L.) and its distribution in the tissues. *J. Exp. Mar. Biol. Ecol.* 23:71–84.

Goldberg, E. D., M. Koide, V. Hodge, A. R. Flegal, and J. Martin, 1983. U.S. Mussel Watch: 1977–1978 results on trace metals and radionuclides. *Estuar. Coast. Shelf Sci.* 16:69–93.

Griscom, S. B., N. S. Fisher, and S. N. Luoma. 2002. Kinetic modeling of Ag, Cd and Co bioaccumulation in the clam *Macoma balthica*: quantifying dietary and dissolved sources. *Mar. Ecol. Progr. Ser.* 240:27–41.

Icely, J. D., and Nott, J. A. 1980. Accumulation of copper within the "hepatopancreatic" caeca of *Corophium volutator*. *Mar. Biol.* 57:193–199.

Lane, T. W., M. A. Saito, G. N. George, I. J. Pickering, R. C. Prince, and F. M. M. Morel. 2005. A cadmium enzyme from a marine diatom. *Nature* (London) 435:42.

Langston, W. J. 1980. Arsenic in U. K. estuarine sediments and its availability to benthic organisms. *J. Mar. Biol. Ass. UK* 60:869–881.

Luoma, S. N., and G. W. Bryan. 1978. Factors controlling availability of sediment-bound lead to the estuarine bivalve *Scrobicularia plana*. *J. Mar. Biol. Ass.UK* 58:793–802.

Luoma, S. N., and P. S. Rainbow. 2005. Why is metal bioaccumulation so variable? Biodynamics as a unifying concept. *Environ. Sci. Technol.* 39:1921–1931.

Luoma, S. N., and P. S. Rainbow. 2008. *Metal Contamination in Aquatic Environments: Science and Lateral Management*. Cambridge: Cambridge University Press.

Luoma, S. N., C. Johns, N. S. Fisher, N. A. Steinberg, R. S. Oremland, and J. R. Reinfelder. 1992. Determination of selenium bioavailability to a benthic bivalve from particulate and solute pathways. *Environ. Sci. Technol.* 26:485–491.

Marigómez, I., M. Soto, M. P. Carajaville, E. Angulo, and L. Giamberini. 2002. Cellular and subcellular distribution of metals in molluscs. *Microsc. Res. Techn.* 56:358–392.

Mason, A. Z., and K. D. Jenkins.1995. Metal detoxification in aquatic organisms. In: *Metal Speciation and Aquatic Systems*, eds. A. Tessier and D. R. Turner, pp. 479–608. New York: Wiley.

Meador, J. P. 2000. Predicting the fate and effects of tributyltin in marine systems. *Rev. Environ. Contam. Toxicol.* 166:1–48.

Moore, P. G., and P. S. Rainbow. 1984. Ferritin crystals in the gut caeca of *Stegocephaloides christianiensis* Boeck and other Stegocephalidae (Amphipoda:Gammaridea): a functional interpretation. *Phil. Trans. R. Soc. Lond. B* 306:219–245.

Nassiri, Y., P. S. Rainbow, C. Amiard-Triquet, F. Rainglet, and B. D. Smith. 2000. Trace metal detoxification in the ventral caeca of *Orchestia gammarellus* (Crustacea:Amphipoda). *Mar. Biol.* 136:477–484.

Ng, T. Y.-T., C. Amiard-Triquet, P. S. Rainbow, J.-C. Amiard, and W.-X. Wang. 2005. Physico-chemical form of trace metals accumulated by phytoplankton and their assimilation by filter-feeding invertebrates. *Mar. Ecol. Progr. Ser.* 299:179–191.

Nugegoda, D., and P. S. Rainbow. 1988. Zinc uptake and regulation by the sublittoral prawn *Pandalus montagui* (Crustacea: Decapoda). *Estuar. Coast. Shelf Sci.* 26:619–632.

Phillips, D. J. H., and P. S. Rainbow. 1988. Barnacles and mussels as biomonitors of trace elements: a comparative study. *Mar. Ecol. Progr. Ser.* 49:83–93.

Pullen, J. S. H., and P. S. Rainbow. 1991. The composition of pyrophosphate heavy metal detoxification granules in barnacles. *J. Exp. Mar. Biol. Ecol.* 150:249–266.

Rainbow, P. S. 1987. Heavy metals in barnacles. In: *Barnacle Biology*, ed. A. J. Southward, pp. 405–417. Rotterdam: A. A. Balkema.

Rainbow, P. S. 1996. Heavy metals in aquatic invertebrates. In: *Environmental Contaminants in Wildlife: Interpreting Tissue Concentrations*, eds. W. N. Beyer, G. A. Heinz, and A. W. Redmon-Norwood, pp. 405–425. Boca Raton: Lewis Publishers.

Rainbow, P. S. 1998. Phylogeny of trace metal accumulation in crustaceans. In: *Metal Metabolism in Aquatic Environments*, eds. W. J. Langston, and M. Bebianno, pp. 285–319. London, Chapman and Hall.

Rainbow, P. S. 2002. Trace metal concentrations in aquatic invertebrates: why and so what? *Environ. Pollut.* 120:497–507.

Rainbow, P. S. 2006. Biomonitoring of trace metals in estuarine and marine environments. *Australasian J. Ecotoxicol.* 12:107–122.

Rainbow, P. S. 2007. Trace metal bioaccumulation: models, metabolic availability and toxicity. *Environ. Int.* 33:576–582.

Rainbow, P. S., and C. Abdennour. 1989. Copper and haemocyanin in the mesopelagic decapod crustacean *Systellaspis debilis*. *Oceanologica Acta* 12:91–94.

Rainbow, P. S., and G. Blackmore. 2001. Barnacles as biomonitors of trace metal availabilities in Hong Kong coastal waters: changes in space and time. *Mar. Environ. Res.* 51:441–463.

Rainbow, P. S., and W.-X. Wang. 2001. Comparative assimilation of Cd, Cr, Se, and Zn by the barnacle *Elminius modestus* from phytoplankton and zooplankton diets. *Mar. Ecol. Progr. Ser.* 218:239–248.

Rainbow, P. S., J.-C. Amiard, C. Amiard-Triquet, M.-S. Cheung, L. Zhang, H. Zhong, and W.-X. Wang. 2007. Trophic transfer of trace metals: subcellular compartmentalization in bivalve prey, assimilation by a gastropod predator and *in vitro* digestion simulations. *Mar. Ecol. Progr. Ser.* 348:125–138.

Rainbow, P. S., C. Amiard-Triquet, J.-C. Amiard, B. D. Smith, S. L. Best, Y. Nassiri, and W. J. Langston. 1999. Trace metal uptake rates in crustaceans (amphipods and crabs) from coastal sites in NW Europe differentially enriched with trace metals. *Mar. Ecol. Progr. Ser.* 183:189–203.

Rainbow, P. S., G. Blackmore, and W.-X. Wang. 2003. Effects of previous field exposure history on the uptake of trace metals from water and food by the barnacle *Balanus amphitrite*. *Mar. Ecol. Progr. Ser.* 259:201–213.

Rainbow, P. S., R. H. Emson, B. D. Smith, P. G. Moore, and P. V. Mladenov. 1993. Talitrid amphipods as biomonitors of trace metals near Dunedin, New Zealand. *New Zealand J. Mar. Freshwat. Res.* 27:201–207.

Rainbow, P. S., P. G. Moore, and D. Watson. 1989. Talitrid amphipods as biomonitors for copper and zinc. *Estuar. Coast. Shelf Sci.* 28:567–582.

Rainbow, P. S., L. Poirier, B. D. Smith, K. V. Brix, and S. N. Luoma. 2006. Trophic transfer of trace metals: subcellular compartmentalization in a polychaete and assimilation by a decapod crustacean. *Mar. Ecol. Progr. Ser.* 308:91–100.

Reinfelder, J. R., and N. S. Fisher. 1991. The assimilation of elements ingested by marine copepods. *Science* 251:794–796.

Ridout, P. S., P. S. Rainbow, H. S. J. Roe, and H. R. Jones. 1989. Concentrations of V, Cr, Mn, Fe, Ni, Co, Cu, Zn, As and Cd in mesopelagic crustaceans from the North East Atlantic Ocean. *Mar. Biol.* 100:465–471.

Roditi, H. A., N. S. Fisher, and S. A. Sanudo-Wilhelmy. 2000. Field testing a bioaccumulation model for zebra mussels. *Environ. Sci. Technol.* 34:2817–2825.

Roe, H. S. J. 1984. The diel migration and distribution within a mesopelagic community in the Northeast Atlantic: 2–vertical migrations and feeding of mysids and decapod Crustacea. *Progr. Oceanogr.* 13:269–318.

Skorupa, J. P. 1998. Selenium poisoning of fish and wildlife in nature: lessons from twelve real-world examples. In: *Environmental Chemistry of Selenium*, eds. W. T. Frankenberger Jr. and R. A. Engberg, pp. 315–354. New York: Marcel Dekker Inc.

Sunda, W. G., and R. R. L. Guillard. 1976. The relationship between cupric ion activity and the toxicity of copper to phytoplankton. *J. Mar. Res.* 34:511–529.

Walker, G. 1977. "Copper" granules in the barnacle *Balanus balanoides*. *Mar. Biol.* 39:343–349.

Wallace, W. G., and G. R. Lopez. 1996. Relationship between the subcellular cadmium distribution in prey and cadmium transfer to a predator. *Estuaries* 19:923–930.

Wallace, W. G., and G. R. Lopez. 1997. Bioavailability of biologically sequestered cadmium and the implications of metal detoxification. *Mar. Ecol. Progr. Ser.* 147:149–157.

Wallace, W. G., and S. N. Luoma. 2003. Subcellular compartmentalization of Cd and Zn in two bivalves. II. The significance of trophically available metal (TAM). *Mar. Ecol. Progr. Ser.* 257:125–137.

Wallace, W. G., B. G. Lee, and S. N. Luoma. 2003. Subcellular compartmentalization of Cd and Zn in two bivalves. I. Significance of metal-sensitive fractions (MSF) and biologically detoxified metal (BDM). *Mar. Ecol. Progr. Ser.* 249:183–197.

Wang, W.-X. 2002. Interactions of trace metals and different marine food chains. *Mar. Ecol. Progr. Ser.* 243:295–309.

Wang, W.-X., and N. S. Fisher. 1999. Assimilation efficiencies of chemical contaminants in aquatic invertebrates: a synthesis. *Environ. Toxicol. Chem.* 18:2034–2045.

Wang, W.-X., and P. S. Rainbow. 2006. Subcellular partitioning and the prediction of cadmium toxicity to aquatic organisms. *Environ. Chem.* 3:395–399.

Wang, W.-X., N. S. Fisher, and S. N. Luoma. 1996. Kinetic determinations of trace element bioaccumulation in the mussel *Mytilus edulis. Mar. Ecol. Progr. Ser.* 140:91–113.

Warren, L. A., A. Tessier, and L. Hare. 1998. Modelling cadmium accumulation by benthic invertebrates *in situ*: the relative contributions of sediment and overlying water reservoirs to organism cadmium concentrations. *Limnol. Oceanog.* 43:1442–1454.

White, S. L., and P. S. Rainbow. 1982. Regulation and accumulation of copper, zinc and cadmium by the shrimp *Palaemon elegans. Mar. Ecol. Progr. Ser.* 8:95–101.

White, S. L., and P. S. Rainbow. 1985. On the metabolic requirements for copper and zinc in molluscs and crustaceans. *Mar. Environ. Res.* 16:215–229.

White, S. L., and P. S. Rainbow. 1986. A preliminary study of Cu-, Cd- and Zn-binding components in the hepatopancreas of *Palaemon elegans* (Crustacea: Decapoda). *Comp. Biochem. Physiol.* 83C:111–116.

White, S. L., and P. S. Rainbow. 1987. Heavy metal concentrations and size effects in the mesopelagic decapod crustacean *Systellaspis debilis. Mar. Ecol. Progr. Ser.* 37:147–151.

Fucus-Like Sea Horse

By G. Mutzel, from *The Royal Natural History,* edited by Richard Lydekker, Frederick Warne & Co., London, 1893–94.

7 Organotins in Aquatic Biota
Occurrence in Tissue and Toxicological Significance

James P. Meador

CONTENTS

7.1 Introduction 256
7.2 Background 256
    7.2.1 Legislation 258
    7.2.2 Reporting Concentrations 258
7.3 Environmental Chemistry 259
    7.3.1 Octanol–Water Partition Coefficient 259
    7.3.2 Hydrogen Ion Activity (pH) and the Acid-Dissociation Constant (pKa) 259
    7.3.3 Carbon 260
7.4 Environmental Occurrence 260
    7.4.1 Water and Sediment 260
    7.4.2 Tissue 261
        7.4.2.1 Butyltins 261
        7.4.2.2 Phenyltins 261
        7.4.2.3 Other Organotins 261
    7.4.3 Observations 261
7.5 Organotin Bioaccumulation 263
    7.5.1 Concepts 263
    7.5.2 Bioaccumulation Kinetics 264
        7.5.2.1 Uptake 264
        7.5.2.2 Elimination 265
    7.5.3 Observations 265
7.6 Organotin Toxicity 266
    7.6.1 Toxicity from Ambient Exposure 267
    7.6.2 Responses and Modes and Mechanisms of Toxic Action 269
    7.6.3 Tissue-Residue Toxicity 270
        7.6.3.1 Mortality 270
        7.6.3.2 Growth 272
        7.6.3.3 Immunotoxicity 273
        7.6.3.4 Reproductive 273
        7.6.3.5 Neurological 276
        7.6.3.6 Behavioral 276
        7.6.3.7 Obesogen/Somatogen 277

Summary 277
Acknowledgments 279
References 279

7.1   INTRODUCTION

Organotins are organometallic compounds that exhibit complex environmental chemistry and toxicity. The handbook of Chemistry and Physics (CRC 1975) lists more than 250 organotin compounds. Even though a number of these are specific compounds (e.g., triphenyltin [TPT]) that are listed as various salts (e.g., TPT chloride, sulfide, hydroxide, and bromide), there are dozens of unique compounds. Because a number of organotins will be considered here, Table 7.1 lists the compounds and their abbreviations.

The focus for this chapter is the occurrence of organotin compounds in aquatic organisms and the associated toxic responses. For aquatic organisms, we traditionally define the effective concentration for toxicity based on the ambient-exposure pathway (e.g., water, air, soil/sediment, prey); however, tissue residues reflect the bioavailable and effective target dose more accurately than the conventional "dose." The term "dose" is loosely applied; however, it most accurately defines that bioactive fraction occurring at the site of action. Differences in the inherent toxicity (potency) of compounds within and between mechanisms of toxic action (MeOAs) are more apparent in residue-based dose metrics than exposure-based dose metrics because the influence of many confounding factors can be taken into account and avoided. For example, a high percentage of the range for LC_{50} or EC_{50} values that are based on water or sediment concentrations are due to the variability in the bioavailable fraction and the uptake and elimination rate kinetics that determine bioaccumulation (Meador 2006). When tissue residues are used as the dose metric, differences in bioavailability and bioaccumulation are greatly reduced and we are left with just the potential variability in potency that may occur among species. In many cases, the range in values for a given toxicant among all species can be reduced by 4–5 orders of magnitude for ambient-exposure toxicity metrics to one order of magnitude when tissue-based toxicity metrics are considered (Meador 2006, Meador et al. 2008). The major advantage of this feature is in assessing concentrations of a given toxicant in feral organisms. Hence, we can more accurately determine the likelihood of potential toxic effects for some chemicals measured in field-collected organisms when a low variance is observed among species.

In this chapter I will present an overview for organotins, with some basic information on their environmental chemistry, occurrence, and bioaccumulation. The available data on tissue-residue toxicity for aquatic biota will be presented and discussed. Some of the toxicity information for small mammals will also be shown for comparison and to highlight similarities among diverse taxa. In the summary, section I will provide general conclusions regarding the toxicity of organotin compounds as a function of tissue concentrations.

7.2   BACKGROUND

Organotins have several applications and are primarily used as plasticizers in industrial applications and as biocides to control so-called nuisance organisms. As pesticides they are used as antifoulants, wood preservatives, molluscicides, antihelminthics, and fungicides for textiles and various water systems (Cima et al. 2003, Antizar-Ladislao 2008). Most of the pesticides are triorganotin compounds and of this group tributyltin (TBT) and TPT are the most commonly encountered environmental contaminants because of their biocidal properties and widespread use on boat hulls to prevent the accumulation of fouling organisms. TPT is also applied to some crops as a fungicide and will therefore likely leach into watersheds. Organotins are also used as heat and light stabilizers for

TABLE 7.1
Organotin Properties

Compound	Abbrev	MW	Log$_{10}$ K$_{ow}$	pKa	Predicted BCF	Convert ng Sn/g to ng OT/g	Convert μg OT/g to nmol/g
Tetramethyltin	TeMT	179	−2.2	na	9×10^{-5}	1.50	5.59
Trimethyltin	TMT	164	−2.3	6.6	8×10^{-5}	1.38	6.10
Dimethyltin	DMT	148	−3.1	3.5	1×10^{-5}	1.24	6.76
Monomethyltin	MMT	135	−3.1	2.6	1×10^{-5}	1.13	7.41
Tetraethyltin	TeET	233	—	—	—	1.97	4.29
Triethyltin	TET	205	−1.8	6.8	2×10^{-4}	1.74	4.88
Diethyltin	DET	177	−1.4	3.7	6×10^{-4}	1.49	5.65
Tetrapropyltin	TePrT	291	2.0	na	1.50	2.45	3.44
Tripropyltin	TPrT	248	0.9	6.3	0.12	2.08	4.03
Dipropyltin	DPrT	205	na	na	—	1.72	4.88
Monopropyltin	MPrT	161	na	na	—	1.35	6.21
Tetrabutyltin	TeBT	347	3.9	na	119	2.92	2.88
Tributyltin	TBT	283	4.4	6.5	377	2.44	3.45
Dibutyltin	DBT	233	1.3	3.8	0.30	1.96	4.29
Monobutyltin	MBT	177	0.4	2.0	0.04	1.49	5.65
Tetraphenyltin	TePT	427	4.4	na	377	3.59	2.34
Triphenyltin	TPT	350	3.6	5.2	60	2.94	2.86
Diphenyltin	DPT	273	1.9	2.7	1.2	2.29	3.66
Monophenyltin	MPT	197	1.2	na	0.24	1.66	5.08
Tri-*n*-hexyltin	TnHT	375	3.7	na	75	3.15	2.67
Tri-*c*-hexyltin	TcHT	375	4.1	na	189	3.15	2.67
Azocyclotin	ACT	436	5.4	na	3598	3.66	2.29
Dihexyltin	DHT	298	na	na	—	2.50	3.36
Trioctyltin	TOT	458	na	na	—	3.85	2.18
Dioctyltin	DOT	355	5.8	na	9910	2.98	2.82
Monooctyltin	MOT	252	2.1	na	2.1	2.12	3.97
Inorganic tin	Sn	119	na	na	—	—	—

Many of the log$_{10}$ K$_{ow}$ values were calculated or determined at unspecified pH. Data for tributyltin, triphenyltin, and trihexyltin from Fent (1996), Meador (2000), Arnold et al. (1997), and Tas (1993) were determined at circumneutral pH. Most other data from Wong et al. (1982) and Vighi and Calamari (1985). Convert to ng OT/g is the value to multiply ng Sn/g for the result. Covert to nmol/g is the value to multiply μg OT/g for the result (dividing nmol/g by this factor equals μg OT/g). Predicted BCF based on K$_{ow}$ QSAR for ionization-corrected substituted phenols, BCF = K$_{ow}$ × 0.015 (McCarty 1986). MW is the molecular weight, K$_{ow}$ is the octanol–water partition coefficient, OT is organotin, and na is not available. Note, most organotins are reported as the ionic concentration (i.e., without the anion such as Cl or OH) except for the tetra substituted or neutral forms.

polyvinyl chloride (PVC) plastics and as catalysts for various chemical reactions and account for approximately 70% of the total production (mostly dibutyltins [DBTs]) (Cima et al. 2003). DBT is used as a biocide to treat chickens for tapeworm, and it is also a metabolite of TBT, thus commonly found in tissue after TBT exposure. Also, because this compound is used in PVC production, it will leach into aquatic systems from pipes made of this plastic. Most of the organotin research has been conducted on TBT, although there are several other organotins (e.g., TPT, fenbutatin, azocyclotin, and hexamethylditin) that are widely used, mostly as agricultural pesticides that can end up in aquatic systems.

7.2.1 Legislation

Organotins as antifoulants were introduced early in the 1970s and widely used throughout the 1980s. Late in the 1980s, several countries around the world and states within the United States enacted restrictions on the use of organotins as antifoulants. In 1988, the U.S. government enacted the Organotin Antifouling Paint Control Act of 1988 (OAPAC; U.S. Congress 1988), which restricted the use of tin-based antifoulants on small vessels based on the size of the vessel (<25 m in length) and release rate from the paint surface (4 µg $TBT/cm^2/day$).

An International Convention (treaty) was adopted in 2001 by the International Maritime Organization to prohibit the use of TBT by 2008 (IMO 2001). This treaty came into force in September of 2008 and it requires signatories to prohibit the use of harmful antifoulants (organotins) on ships flagged in their country and to deny entry into their ports for any foreign ship using such antifoulants.

7.2.2 Reporting Concentrations

Many studies report organotin concentrations as ng Sn/g. This is mostly a result of the standard analytical method (flame photometric detection) that quantifies tin concentrations using a tin-selective detector. Reporting organotins as ng Sn/g is misleading because the whole molecule is responsible for the toxic response as a function of its interaction with a receptor. The organotin is the active molecule causing toxicity, not elemental tin (Sn). There are many metal-containing compounds that are not reported in these terms. For example, methyl mercury is not reported as ng Hg/g nor is hemoglobin reported as mg Fe/L. When an organotin is expressed as ng Sn/g the variability among organotins and their toxic potency is masked. In addition, it is generally not appropriate to report the concentration of the various salts or complexes (e.g., tributyltin chloride [TBTCl] or tributyltin hydroxide [TBTOH], bis(tributyltin) oxide [bis-TBTO]) mainly because the standard analytical techniques for quantitation can not distinguish between these species, and it is not known which form (salt or ion) is the toxic species of concern. Once these compounds are introduced to water or tissue, they speciate according to the pH and ionic composition of the receiving water or fluid. Even though TBTOH may be the predominant species that is bioaccumulated, TBT is also found as many different species in plasma and tissue.

In comparing specific compounds, molar concentration is more appropriate because the toxic response is more closely related to the number of molecules interacting with the receptor, not their mass per unit organism weight. Table 7.1 provides conversion factors for mass to molar concentrations and for converting ng Sn/g to ng organotin/g. Organotin concentrations in this review are reported as mass or molar concentrations [e.g., ng or nmol organotin/unit matrix (e.g., g or mL)]. Most organotins are reported as ng organotin/g, except those that are neutral species. Also, all tissue concentrations are reported in terms of wet weight (ww) (unless noted), and body weight is abbreviated as "bw." Many tissue concentrations were originally reported as dry weights, which were converted to wet weights for this review by multiplying the value by 0.2, a standard conversion factor for fish and invertebrate tissue.

The lethal and effective residue designations (LR_p and ER_p) are used in this review to denote the residue (tissue concentration) associated with a response (Meador 1997). The term "effective residue" or "effective concentration" is normally used to characterize a sublethal response and the "p" for each represents the percent or proportion responding. Similar designations are used for the lowest observable effect residue (LOER) and no observable effect residue (NOER). These metrics represent the internal acquired dose, which is generally not equivalent to the externally administered dose. These values (LR_p and ER_p) are distinguished from lethal dose (LD_p) or effective dose (ED_p) values, which are usually based on the administered dose (i.e., dietary as a daily or one-time dose; µg toxicant/g bw/day or µg toxicant/g bw). The administered dose is frequently not the concentration associated with the response, which is our main interest for this review (Meador 2006).

7.3   ENVIRONMENTAL CHEMISTRY

Organotin compounds are a combination of organic moiety and elemental tin (Sn). All organotin compounds contain a metal–carbon bond, and in many cases the organic moiety is an alkyl group or simple ring structure. The most common alkyl moieties are the methyl, ethyl, propyl, butyl, hexyl, and octyl groups and these may occur in series (e.g., methyl, dimethyl, and trimethyl). Tin is also found in coordination with other groups (e.g., phenyl, cyclohexyl, and others) and these can also occur in series (tri-, di-, and monosubstituted). The presence of various organic moieties greatly enhances the compound's hydrophobicity, which increases its bioavailability and toxicity. A list of various organotin compounds and their key physical-chemistry properties can be found in Table 7.1. Some organotin compounds found in the environment are not listed because no data were found. As a result of the increased hydrophobicity, organotins are readily bioaccumulated by organisms and may be more persistent in tissues; however, predicting the bioaccumulated amount is complex.

The most common organotins in aquatic environments occur as triorganotins (e.g., TBT, TPT, trimethyltin [TMT], tripropyltin [TPrT], etc.), diorganotins (DBT, dimethyltin [DMT], diethyltin [DET], etc.), and monoorganotins such as monobutyltin (MBT), and monomethyltin (MMT). There are a very large number of potentially toxic organotins and many of these are found in the environment and are considered significant contaminants. Unfortunately, we know very little about the occurrence, bioaccumulation, and toxicity for most organotins. Organotin environmental chemistry is relatively complex because these compounds are often polar, ionizable, and hydrophobic.

7.3.1   OCTANOL–WATER PARTITION COEFFICIENT

A very useful chemical parameter for predicting the partitioning behavior between water, sediment, and tissue for some organic compounds is the octanol–water partition coefficient (K_{ow}). The K_{ow} is a surrogate measure of the association of organic compounds with lipid or organic carbon. In many cases the K_{ow} is used in quantitative-structure activity relationships (QSARs) to predict sediment-water or water-tissue partitioning. Even though organotins exhibit strong partitioning to lipid and organic carbon (Meador 2000, Brändli et al. 2009) these QSARs do not always predict chemical behavior. In some cases QSARs for sediment-water partitioning are fairly accurate (Meador 2000); however, those for water-tissue partitioning generally are far from predictive. The available K_{ow} values (as $\log_{10}$) for organotin compounds are listed in Table 7.1, and it is important to note the large variability among organotins. The very low (some negative) $\log_{10} K_{ow}$ values indicate that these organotins may not bioaccumulate as much as others; however, bioaccumulation of individual organotins is likely not related to K_{ow} or lipid content (Meador 2000). Surprisingly, many organotins exhibit very low K_{ow} values, which would indicate a low potential to cross biological membranes; however, some exhibit high bioaccumulation factors.

7.3.2   HYDROGEN ION ACTIVITY (pH) AND THE ACID-DISSOCIATION CONSTANT (pKa)

For ionizable organometallic compounds, hydrogen ion activity (pH) and the acid-dissociation constant (pKa) appear to be important chemical controlling factors. Many organotins are ionizable; therefore, pH can have a strong effect on partitioning. Because neutral chemical species, such as TBTOH, are generally more bioavailable for passive diffusion than the ionized form, the proportion of the total compound in solution that is in the neutral form is important for bioaccumulation assessment. Organotins are generally cations and when the pH of a solution containing an ionizable organotin is equal to the pKa, the molecules are equally apportioned between ionized and unionized forms. As the pH increases above the pKa, more of the organotin will be in the neutral form and available for uptake.

As shown by Tsuda et al. (1990) and Arnold et al. (1997) the octanol–water partition coefficient (K_{ow}) for TBT is strongly affected by pH. From a pH of 5.8 to 8.0, the K_{ow} increases from 1600 to

12,000. TPT is also affected by pH and for this same range in pH the K_{ow} increases from 1180 to 3650 (Tsuda et al. 1990). This is an important factor for assessing bioaccumulation because it is the hydrophobic portion that partitions into octanol and this is generally the bioavailable form. As seen in Table 7.1, the pKa for these compounds ranges between 2 and 7, indicating that in most aquatic systems with an alkaline pH, the neutral forms, such as hydroxide species, will predominate.

7.3.3 CARBON

Water-sediment partitioning and bioaccumulation of organotins are known to be affected by the organic carbon and black carbon content in sediment and dissolved organic carbon in the water column (Fent 1996, Meador 2000, Hoch and Schwesig 2004, Veltman et al. 2006, Brändli et al. 2009). Carbon content can affect the amount of free TBT that is available for uptake because of its predominant hydroxide form in aquatic systems that will complex with dissolved or particulate carbon.

7.4   ENVIRONMENTAL OCCURRENCE

7.4.1 WATER AND SEDIMENT

TBT has been reported by several authors in the water column at concentrations commonly ranging from 1 to 200 ng/L in harbors and marinas around the world (Seligman et al. 1989, Fent 1996, Antizar-Ladislao 2008, Harino et al. 2008). In a few cases, aqueous concentrations have been extremely high (500–2000 ng/L) (Clark and Steritt 1988, Antizar-Ladislao 2008); however, observations in this range were not common.

When restrictions on the use of TBT as an antifouling paint were enacted in the late 1980s in many countries, water concentrations declined (Fent 1996); however, sediment concentrations remained relatively high (Krone et al. 1996, Antizar-Ladislao 2008). Even though water concentrations in the United States declined after OAPCA was enacted, levels were still somewhat elevated because it did not impact use on most commercial ships. Due to the large number of small vessels, these restrictions were generally effective causing substantial reductions in aqueous concentrations (Huggett et al. 1992), but less so for sediment. This has also been observed in coastal waters around Japan where the use of TBT and TPT were restricted in 1989 (Harino et al. 2008). The Harino et al. (2008) review on TBT and TPT occurrence in water, sediment, and tissue found large decreases in concentrations shortly after the restrictions were enacted, but only minor reductions were observed after reaching lower levels. These reduced concentrations are still relatively high compared to toxic levels, with water concentrations averaging 11 ng/L (range <1–83 ng/L) and highly variable sediment values (<1–650 ng/g).

TBT appears to be very persistent in sediment and concentrations in the hundreds of parts per billion (ppb) to low parts per million (ppm) range can still be found in harbors and marinas in various countries (de Mora et al. 1989, Fent 1996, Antizar-Ladislao 2008). Although TBT may be quickly degraded to DBT and MBT in the water column (Seligman et al. 1988), degradation appears to be much slower once it is associated with sediment. One study determined half-lives of TBT in sediment ranging from 1 to 2 years in surficial aerobic sediment (Dowson et al. 1996); however, they reported essentially no degradation in anaerobic sediment. These estimates are supported by other studies that observed long half-lives for TBT in sediment (Hwang et al. 1999, Takahashi et al. 1999). As a result of these long half-lives, sediment-associated TBT will likely continue to be a source and lead to elevated water and tissue concentrations.

Fent (1996) provides an excellent review of the environmental concentrations that were found from the early 1980s through the mid-1990s. It is striking to note the extent and frequency of observed sediment concentrations in the low ppm range (i.e., 1–10 µg/g) and water concentrations in the high parts per trillion (pptr) range (0.1–1.0 ng/mL). These elevated concentrations certainly resulted in severe biological effects in many ecosystems considering that the EPA chronic water quality criteria (U.S. EPA 2003) is set to 0.07 ng/mL in freshwater and 0.007 ng/mL in marine ecosystems.

7.4.2 Tissue

7.4.2.1 Butyltins

Most of the available tissue concentration data are for TBT (and its metabolites DBT and MBT) because it is a commonly applied pesticide and is extremely toxic. There are several excellent reviews that provide tables of butyltin concentrations in aquatic species (Tanabe 1999, Maguire 2000, Birchenough et al. 2002, Shim et al. 2005). A recent review article provided an overview of measured tissue concentrations in a variety of fish and invertebrates from American, Asian, and European harbors and marinas (Antizar-Ladislao 2008). As expected the range in tissue levels is very broad, although many of the species exhibit relatively high concentrations (hundreds of ng/g ww) for all butyltin compounds. Shim et al. (2005) presented soft-tissue concentrations of butyltins and TPT in five bivalve species collected worldwide. Most of those samples span from the late 1980s through the 1990s and show very high concentrations for most samples. In this chapter (Table 7.2), we have listed some additional recent tissue concentration data for butyltins. These recent values also indicate very high values for some locations and species.

7.4.2.2 Phenyltins

Phenyltins are also applied as an antifoulant and consequently are commonly found in the tissues of field-collected aquatic organisms. In many cases, TPT is found at elevated and similar concentrations to that observed for TBT (Tolosa et al. 1992, Shim et al. 2005) with concentrations occurring in the range of hundreds of ng/g ww (Table 7.2). A comprehensive review of TPT tissue concentrations in the muscle of wild fish from around the world ($n = 20$ species) found high mean concentrations with many in the 200–600 ng/g ww range (Zhang et al. 2008, Table S3 in supporting information). The review by Shim et al. (2005) also indicates high TPT concentrations (up to 5930 ng/g ww) for many of the bivalve samples from several locations (Korea, Japan, Mediterranean, and The Netherlands). High levels were also reported by Harino et al. (2008) for TPT in mussel tissue (up to 3400 ng/g) in their review of data from Japanese coastal waters.

7.4.2.3 Other Organotins

Essentially all of the other tissue concentration data for organotins consists of values for the TBT and TPT metabolites (DBT, MBT, diphenyltin [DPT], and monophenyltin [MPT]). After extensive searching, very few studies were found that reported tissue concentrations for other organotins in field-collected aquatic animals. One study examined various seafood species for tetramethyltin (TeMT), tetraethyltin (TeET), and TMT (Forsyth and Clerous 1991). No concentrations above the method detection limit (MDL) were observed for TeMT (MDL = 1.2 ng/g) and TeET (MDL = 1.4 ng/g). TMT was found in cockles (1.0 ng/g) and turbot (3.9 ng/g).

7.4.3 Observations

One monitoring study found no decline in TBT over time in mussels from the North Sea even though this antifoulant was banned on small boats there in 1991 (Rüdel et al. 2003). The authors concluded that the absence of decline for tissue residues was likely due to inputs from large vessels that were still using TBT. The recently enforced IMO ban may produce reductions in water concentrations, however, due to the extensive half life of TBT (and likely other organotins) in sediment; these compounds will likely continue to be a concern for many years.

The vast majority of reported tissue concentrations for organotin compounds are for marine species; however, there are studies that examined these compounds in freshwater ecosystems. One

TABLE 7.2

Recent Data on Occurrence of Butyltins and Phenyltins in Aquatic Organisms

Organotin	Spp.	Type	Tissue	D/W	ng ion/g	n	Site	Reference
TBT, DBT, MBT	O.o.	Marine mammal	Liver	W	19 (8), 298 (192), 77 (51)	5	Rausu, Hokkaido, Japan	Harino et al. (2008)
Butyltins								
TBT	C.g.	Oyster	Soft tissue	D	263–10,562	337	Luerman Estuary, Taiwan	Tang and Wang (2008)
TBT	P.v.	Mussel	Soft tissue	D	209–14,000	242	Luerman Estuary, Taiwan	Tang and Wang (2008)
DBT	10 species	Fish	Whole fish	W	<dl–276	27	Several U.S. fw sites	Jones-Lepp et al. (2004)
TBT	Unspec	Mussel	Whole	W	4–381	6 sites	Coastal Japan	Harino et al. (2008)
TBT	15 species	Fish and inverts	Unknown	D	2–240	2 sites	Coastal Japan—deep water	Kono et al. (2008)
TBT, DBT	P.p.	Porpoise	Liver	W	67–266, 88–743	12 sites	Baltic	Ciesielski et al. (2004)
TBT, DBT	11 species	Marine mammal	Liver	W	20–820, 32–2900	10 sites	Worldwide	Kajiwara et al. (2006)
Phenyltins								
TPT	C.g.	Oyster	Soft	D	882 (498)	13	South Korea	Shim et al. (2005)
TPT	M.e.	Mussel	Soft	D	1093 (1071)	5	South Korea	Shim et al. (2005)
TPT	A.p.	Starfish	Soft	D	976 (664)	26	South Korea	Shim et al. (2005)
TPT	15 species	Fish and inverts	Unknown	D	5–460	2 sites	Coastal Japan—deep water	Kono et al. (2008)
TPT	Various*	Fish and inverts	Fish muscle soft tissue inverts	W	1.2–35	48	Bohai Bay, China	Hu et al. (2006)
TPT	11 species	Fish	Muscle	W	25–130	60	Osaka, Japan	Harino et al. (2000)
TPT	10 species	Fish	Whole fish	W	<dl–500	27	Several U.S. fw sites	Jones-Lepp et al. (2004)
TPT, DPT	6 species	Fish	Muscle	W	22–1535, 2–180	32^	Lake system Westeinder, The Netherlands	Stäb et al. (1996)
TPT	Various	Inverts	Whole	W	20–543	22^	Lake system Westeinder, The Netherlands	Stäb et al. (1996)
DPT	Various	Inverts	Whole	W	<dl–736	22^	Lake system Westeinder, The Netherlands	Stäb et al. (1996)
TPT	Unspecified	Mussels	Whole	W	<dl–3400	6 sites	Coastal Japan	Harino et al. (2008)
TPT	15 species	Fish and inverts	Unknown	D	5–460	2 sites	Coastal Japan	Kono et al. (2008)
TPT, DPT, MPT	O.o.	Marine mammal	Liver, lung, blubber, muscle	W	<1–14, <1–17, <1–72	5	Rausu, Hokkaido, Japan	Harino et al. (2008)

Values are mean (standard deviation) for various tissues (tiss). C.g. = *Crassostrea gigas*, A.p. = *Asteria pectinifera*, M.e. = *Mytilus edulis*, P.v. = *Perna viridis*, O.o. = *Orcinus orca*, P.p. = *Phocoena phocoena*. Various* are samples from a number of species including phyto- and zooplankton, benthic invertebrates (inverts), and fish. *n* is the number of individuals for each site, except where number of sites with variable sample sizes indicated. D/W shows dry (D) or wet (W) weight, fw is freshwater, and dl is the detection limit. ^ denotes composite samples. Organotin (OT) abbreviations in Table 7.1.

recent study conducted a survey of DBT and TPT in whole fish from freshwater sites across the United States, which reported values from the detection limit (<1 ng/g) to 276 ng/g for DBT and <1.8–499 ng/g for TPT (Jones-Lepp et al. 2004). Some older studies reported high concentrations of TPT in freshwater organisms. These include bivalves collected in Swiss lakes (Fent 1996) and numerous invertebrates and fish from Dutch lakes (Stäb et al. 1996) (Table 7.2).

Even far offshore in deep (marine) water (100–400 m), elevated concentrations of TBT and TPT have been documented (Kono et al. 2008). These authors found several marine species with concentrations of these two organotins in the range of 2–20 ng/g ww with one value as high as 90 ng/g ww. These authors also reported water concentrations in the range of 0.3–0.8 ng/L for TBT and sediment concentrations up to 16 ng/g dry wt. for this compound and 12 ng/g dry wt. for TPT.

Marine mammals also appear to accumulate relatively high concentrations of organotins. Several recent studies and reviews demonstrate that numerous marine mammal species exhibit high levels in various tissues, including liver, blubber, and muscle. Tanabe (1999) found concentrations of TBT at high concentrations (35–2200 ng/g ww) in several different tissues of finless porpoise (*Neophocaena phocaenoides*) from waters around Japan, with similar high concentrations for DBT and MBT. A review article by Kajiwara et al. (2006) presents data for 11 marine mammals species from various locations (Japan, Great Britain, Mediterranean, United States, Indo-Pacific, and India) showing high concentrations of TBT in liver (mean values 20–820 ng/g ww, maximum = 1200 ng/g). A number of studies examined organotins in killer whales (*Orcinus orca*). Harino et al. (2008) found TBT concentrations in the range of 6–25 ng/g ww and far higher levels of DBT (16–556 ng/g) and MBT (16–152 ng/g) in the liver of this species (Table 7.2). They also report low levels of TPT (<1–58 ng/g) in blubber and liver, which was also noted by Kajiwara et al. (2006) who reported no detectable concentrations of TPT or DPT in killer whales.

7.5 ORGANOTIN BIOACCUMULATION

Assessing bioaccumulation of organotins is complex. Standard QSAR models used for organic compounds are poor predictors of accumulation for these compounds and most organotins do not behave as metals. In general, it appears the pattern of bioaccumulation among organotins is somewhat correlated to K_{ow}; however, using this parameter to predict bioaccumulation for an individual compound across species is not supportable. For example, an analysis of TBT bioaccumulation shows that it does not obey organic-compound QSAR predictions for bioaccumulation or toxicity. The predicted wet-weight bioconcentration factor (BCF) for TBT in species that do not metabolize this compound is approximately 377. This value was determined with the QSAR for ionization-corrected substituted chlorophenols (Saarikoski and Viluksela 1982; also see McCarty 1986). These compounds are known to be uncouplers of oxidative phosphorylation, as are organotins. As shown in Meador (2006), all observed TBT BCFs exceed this predicted value. For those species that exhibit weak biotransformation of this compound, the BCFs are approximately 30–250 fold higher than those predicted using the QSAR.

7.5.1 CONCEPTS

Some of the triorganotins exhibit relatively high K_{ow} values (log_{10} values of 3–4); however, TMT, triethyltin (TET), and TPrT are all very low (Table 7.1). In general, there is a very strong association between the K_{ow} value and the number of carbons in the substituted groups among triorganotins ($r^2 = 0.80$), and based on this relationship the compounds with the highest K_{ow} value and number of carbons would be expected to exhibit the highest BCFs or BAFs. As shown later, this is not the case for some of the organotins, such as TPrT. Unfortunately, we have very few data for other organotins.

Based on the K_{ow} at pH 8.0 from Tsuda et al. (1990) the predicted BCF for TPT at steady state using the same QSAR formula derived for substituted phenols is 60. The observed steady-state BCF for carp (*Cyprinus carpio*) exposed to TPT was determined to be 600, which is 10 times higher.

Because fish are known to extensively metabolize TBT (Lee 1985), the QSAR predicted BCF for this species should be far higher than the observed value. Another study reported the TPT BCFs for two fish species (*Pagrus major* and *Rudarius ercodes*) after 56 days of exposure to be similar at 3200 and 4100, respectively, 53–68 times the predicted value. In general terms, several studies have demonstrated that BCFs for TBT are substantially higher than those for TPT (Tsuda et al. 1988, 1991, Fent 1996), which may be a result of the speciation profile that is controlled by pH, K_{ow}, or some physiologic aspect of bioaccumulation. When tissue-residue toxicity is considered, these differences become less significant because of the increased importance of toxic potency.

Many aquatic invertebrate species are known to have minimal metabolic capacity for organotins (Fent 1996), and they exhibit BCF values that are far higher than those observed for fish. As seen in Meador (2006) the BCFs for TBT in several species are very high ranging from 2000 to 95,000. Similarly, high BCF values have also been observed for invertebrates and other organotins. One study exposed marine snails (*Nucellus lapillus*) to aqueous concentrations of TPrT and reported a BCF value of approximately 15,000 after 30 days exposure (Bryan et al. 1988). Based on the bioaccumulation QSAR for this compound used earlier, the expected steady-state BCF for a species that does not metabolize TPrT is predicted to be 0.12, which is 1.3×10^5 times lower than the observed value for *N. lapillus*. No bioaccumulation is expected for TMT ($\log_{10} K_{ow}$ of −2.3); however, one study reported a BCF value of 75 for the brine shrimp (*Artemia franciscana*) (Hadjispyrou et al. 2001). The same study reported a BCF of 50 for DMT, which exhibits a $\log_{10} K_{ow}$ of −3.1 (essentially 0).

From these data we can conclude that the commonly used bioaccumulation QSAR equations can not be used to predict tissue concentrations for organotins for fish or invertebrates. Therefore, measured toxicokinetic values are the only reliable method for such predictions. This is an important point because a number of guidelines and statutes require chemical screening based on K_{ow} when assessing the potential for chemicals to bioaccumulate and cause harm. As seen earlier, these assumptions on bioaccumulation QSARs are not valid for organotins and possibly other poorly studied contaminants.

Several environmental factors are likely important for determining bioaccumulation for organotins including pH, temperature, redox state, salinity, and organic carbon content in sediment and water. Most of these factors will affect the amount of the bioavailable compound or the rate of uptake. Also, because of the large disparities between actual and predicted bioaccumulation, there must be biological factors (e.g., interspecific differences in rates of uptake, elimination processes, membrane permeability, transport mechanisms, etc.) that are prominent controlling factors for bioaccumulation. When we consider tissue-residue toxicity metrics, these factors are all far less germane because bioavailability and toxicokinetics are accounted for when toxicity is expressed as a tissue concentration (e.g., LR_p or ER_p).

7.5.2 Bioaccumulation Kinetics

A more accurate way to predict bioaccumulation is with uptake and elimination rate constants. As a simple example, the rate of uptake (uptake clearance; k_1) divided by the rate of elimination (k_2) equals the BCF at steady state.

7.5.2.1 Uptake

As mentioned earlier, pH has a large influence on hydrophobic partitioning because it determines the profile of the various organotin species. As pH increases, TBTOH becomes more abundant and the ionic form (TBT⁺) decreases. A few studies have demonstrated that the rate of uptake increases with increasing pH for TBT (Fent 1996) and TPT (Tsuda et al. 1990), which is likely a result of the reduction in the ionic species and an increase in neutral species (e.g., TBTOH). Therefore pH impacts the rate of uptake only because it affects the proportions of the various organotin species in the exposure media. Because of these differences due to pH, marine organisms often exhibit higher BCF values than freshwater species for organotins because the pH of freshwater is often lower than seawater (≈8.1).

We know from several studies that rate of uptake for TBT is highly variable among species. As an example, Meador (1997) observed order of magnitude differences in TBT BCF values and toxicokinetics for two similar amphipods under identical environmental conditions. It is not known if these results are a function of ventilation rate, membrane permeability, or other physiological or morphological differences among species.

7.5.2.2 Elimination

The rate of elimination includes the processes of metabolism, passive diffusion, and excretion. Because total elimination values are calculated, we do not know what portion was metabolized and how much of the parent compound was lost through passive or active processes. Elimination for many organotins is accomplished by metabolic transformation, which occurs via the cytochrome P450 enzyme system that facilitates the degradation of a large number of xenobiotics (Fent 1996). For example, TBT is sequentially debutylated in a series of reactions with the cytochrome P450 system (TBT $\rightarrow$ DBT $\rightarrow$ MBT $\rightarrow$ Sn) (Fent 1996). All these metabolites will be measured in organisms that can metabolize TBT and are exposed to this compound for several days (Meador 1997). TPT is metabolized in a similar fashion (TPT $\rightarrow$ DPT $\rightarrow$ MPT $\rightarrow$ Sn) (Fent 1996). A few studies have compared the elimination rates of TBT and TPT in fish and found that TBT was generally more rapidly eliminated than TPT (Tsuda et al. 1988, 1992) indicating that TPT may persist longer in tissue. A low rate of elimination (k_2) for TPT was also noted by Stoner (1966) for guinea pigs.

For fish species, metabolic rates (k_m) for organotins should be substantial due to the high levels of cytochrome P450. In general, rates of biotransformation via P450 are known to be highly variable among invertebrate taxa (Livingstone 1998). Unfortunately, most studies that examine the elimination of organotins from tissue only report the total loss of the compound over time. It is possible to determine k_m values by quantifying the changes in parent compound and metabolites (e.g., DBT and MBT) over time; however, this is rarely calculated.

7.5.3 OBSERVATIONS

Parental transfer of organotins is an important factor to consider, especially in light of the toxicity information for development. One study reported that maternal transfer of butyltins occurred for the Dall's porpoise; however, the concentrations in fetal tissue were relatively low compared to the adult (Yang and Miyazaki 2006). Concentrations of TBT in the fetus (1.0 kg at 6 months post fertilization) were about 10 times lower than maternal concentrations (15 ng/g vs. 1.4 ng/g, whole-body values determined by summation of organ burdens). Another study (Kajiwara et al. 2006) that examined organotins in stranded killer whales found similar differences (2–20 fold) for TBT concentrations in liver between mature adults and calves that were estimated to be a few months old. Of note were the concentrations of total phenyltins (tri- and di-), which were detected in all three calves and in only one of the five mature females, although the concentrations were low ($\approx$1 ng/g).

The results for marine mammals are in stark contrast to those for fish as demonstrated for viviparous surfperch (*Ditrema temmincki*) (Ohji et al. 2006). The concentration of TBT in fry was 10–16 times higher than values reported for whole-body parental females. The percentage TBT in relation to total butyltins (TBT, DBT, and MBT) was 51% in the females and 81% in the fry indicating a reduced capacity for biotransformation. Due to early life-stage sensitivity of fish to butyltins in tissue, this is an important observation.

TBT bioaccumulation was also observed in algae. Maguire et al. (1984) reported a dry-weight TBT BCF for the green alga *Ankistrodesmus falcatus* of 30,000, which can lead to very high concentrations. The consequences of this high BCF value include the enhancement of dietary uptake by planktivores and direct toxicity to algal species because TBT is known to affect energy production in chloroplasts, heme metabolism, and disrupt ion pumps (Fent 1996).

The reason for the very high bioaccumulation factors for many species can be found in the rates of uptake and elimination. Many QSAR models have been developed for organic compounds that relate K_{ow} with k_1 and k_2 (Connell 1990). These models are generally accurate predictors for passive organic-compound flux in species that exhibit low rates of metabolism. The toxicokinetic model for chlorobenzenes in molluscs was selected as an example (Connell 1990), because low metabolism was expected for this taxa. Based on the K_{ow} for TBT, the predicted values are 280/d for k_1 and 0.96/d for k_2. Using these toxicokinetic values, the predicted BCF is 292, which is very similar to the predicted BCF of 377 that was described earlier using the bioaccumulation QSAR equation for substituted chlorophenols. Measured values for k_1 and k_2 in species that are expected to exhibit low metabolic rates for TBT are generally substantially different than these predicted values. For example Gomez-Arizas et al. (1999) determined the TBT k_2 for clams (*Venerupis decussata*) to be approximately 0.02–0.03/d, which was similar to the values reported by Meador (1997) for an amphipod and Tessier et al. (2007) for a gastropod (*Lymnaea stagnalis*) (each k_2 = 0.04/d). These values are 25–50 fold lower than the expected elimination rates for passive diffusion. The QSAR predicted k_1 value is 280/d, which is approximately 3–27 times less than other reported values (Meador 1997, Gomez-Ariza et al. 1999, Tessier et al. 2007). Given these large differences in predicted and observed toxicokinetic rates, it is not surprising that BCFs are far higher than those predicted with QSARs.

The observed TBT BCFs can be two orders of magnitude or more above predicted levels, which is consistent with the large observed disparity in toxicokinetics. Based on an examination of the limited available data, it appears that the k_2 rate constant exhibits a greater influence than the k_1 value as determinants for the BCF. We can conclude from this that TBT (and likely other organotins) is very slowly eliminated from tissue. Of course, for those species that are able to metabolize TBT, the overall elimination rate will be higher; however, the passive rate of elimination is still an important factor for determining k_2 and bioaccumulation factors. Also noteworthy is that the k_2 values in this range (0.02–0.04/d) indicate that some species will take ≈75–150 days to reach steady-state tissue concentrations.

7.6 ORGANOTIN TOXICITY

In all cases an organotin compound is far more toxic than its individual components. For example, the toxicity of TBT is considerably more toxic than inorganic tin or the component butyl groups. For comparison, this is generally the same pattern for organomercurials, but not for arsenic because methylation reduces toxicity and the inorganic forms tend to be more toxic. Within a series of organotin compounds there are differences in toxicity. When expressed in terms of water exposure, the triorganotins (e.g., TBT, TPT, TMT, and TPrT) are considered more toxic than the mono-, di-, or tetraorganotins (Laughlin et al. 1985, Brüschweiler et al. 1995). These authors proposed that the increased water toxicity for the triorganotin compounds may be a result of several factors such as their higher K_{ow} values, a higher rate of uptake, differences in the MeOA, or the increased propensity for bioaccumulation and persistence.

Many of the factors that should be considered during toxicity assessment from ambient exposure are less important for tissue-residue toxicity. For most contaminants, pH, redox state, organic carbon, and salinity are controlling factors for ambient-toxicity metrics; however, once we consider tissue concentrations these factors are considerably less important. The factors that are important for both ambient-exposure and tissue-residue toxicity metrics include organism health, temperature, and lipid content (Meador et al. 2008). Lipid content is an important parameter for hydrophobic compounds because of internal toxicant partitioning and the relative amount of the active toxicant fraction (i.e., biologically effective dose) (Lassiter and Hallam 1990). There are few data on this subject for organotins; however, one study found that the LR_{50} (lethal tissue residue) for an amphipod (*Rhepoxynius abronius*) was approximately three times lower in individuals containing a reduced lipid content (Meador 1993). When this toxicity metric was normalized to lipid content and expressed on a lipid basis the values for the normal and reduced lipid groups became statistically indistinguishable.

7.6.1 TOXICITY FROM AMBIENT EXPOSURE

It is well known that the toxicity of organotins varies widely among compounds and species when external exposure (e.g., water concentrations) is considered. Water exposure to TBT produces LC_{50} values ranging over two orders of magnitude among aquatic species (~0.5–200 ng/mL) (Figure 7.1) and three orders for most sublethal responses such as growth and reproductive impairment (0.005–5 ng/mL) (Cardwell and Meador 1989, Meador 2000, U.S. EPA 2003).

One comprehensive study examined the aqueous toxicity of seven diorganotins (R_2SnX_2) and eight triorganotins (R_3SnX_2) to crab zoeae (*Rhithropanopeus harrisii*) (Laughlin et al. 1985). The R groups for the triorganotins were methyl, ethyl, propyl, butyl, phenyl, and cyclohexyl. The same list applies for the diorganotins with the addition of benzyl. All of the X groups were oxides, hydroxides, bromides, and chlorides. The X group is essentially unimportant because as soon as the compound is added to water it speciates according to the pH, redox state, and the ionic content of the receiving water. The variability among diorganotins for the day 14 LC_{50} was 250 fold and for the triorganotins was 28 fold. For all compounds the range was four orders of magnitude, which was very similar to that reported by Nagase et al. (1991) who determined the 48 h LC_{50} for killifish (*Oryzias latipes*) exposed to 29 different organotins (Figure 7.1). In the Laughlin et al. (1985) study, a strong linear relationship was found between LC_{50} and the Hansch lipophilicity parameter (π) for both the di- and triorganotins ($r^2 > 0.94$) indicating that the most important factor determining toxicity was the hydrophobic characteristics of each compound. We would expect differences in potency among these compounds when toxicity is based on tissue concentrations; however, when considering ambient-exposure toxicity metrics, the variability due to differences in toxicokinetics

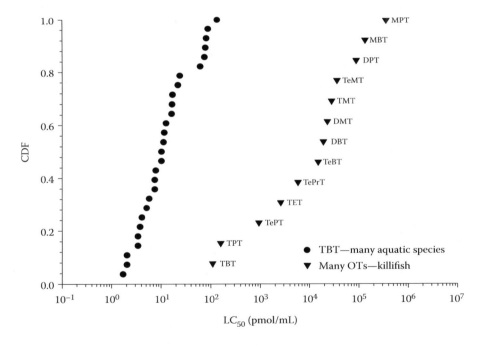

FIGURE 7.1 Circles show 96 h LC_{50} values in pmol/mL for 28 aquatic species (polychaetes, amphipods, copepods, shrimp, and fish) exposed to TBT in water. Data from Cardwell and Meador (1989). Triangles are the 48 h LC_{50} values for one species (killifish, *Oryzias latipes*) exposed to many different organotins (OTs) (Nagase et al. 1991). Of the 29 organotins tested on killifish, 16 were unique (many were salts of one compound). Only three were not shown (*n*-butyltrimethyltin, di-*n*-butyldimethyltin, and tri-*n*-butylmethyltin) all of which exhibited LC_{50} values that fell within the range shown. See Table 7.1 for organotin abbreviations. CDF is cumulative distribution function.

(bioaccumulation) versus that for toxicodynamics (potency) can not be distinguished. As shown later, the differences in potency are relatively minor among all these organotins for lethality.

This observation is supported by an example for DBT and TBT, both of which are known to inhibit adenosine triphosphate (ATP) synthesis. One study found that DBT was more than 50 times less toxic than TBT (Lytle et al. 2003) when based on exposure (water) concentrations; however, another study demonstrated that DBT was only three times less potent than TBT for this MeOA when based on tissue concentrations (Aldridge et al. 1977). These results highlight the differences between aqueous and tissue-residue toxicity metrics among compounds as a function of bioavailability and external toxicokinetics. Based on this information, the diorganotins will likely cause lethality in aquatic species at tissue concentrations that are relatively similar to those determined for TBT lethality.

Another study on organotin toxicity is useful for highlighting the role of bioaccumulation and QSARs. Vighi and Calamari (1985) reported the 24 h acute LC_{50} values for *Daphnia magna* exposed to TMT, TET, TPrT, TBT, and TPT. These values ranged from 13 to 470 ng/mL, which is a factor of 36. This range in toxicity values is relatively minor compared to the approximately six orders of magnitude range for compound K_{ow} values and the expected differences in bioaccumulation as predicted by K_{ow} QSARs. One conclusion from these data is that the predicted BCF values do not accurately reflect the observed values. Based on the highly similar tissue-based mortality data presented here (Tables 7.3 and 7.4) it is

TABLE 7.3
Critical Triorganotin Body Residues for Aquatic Species

Response	Organotin	Species	CBR (nmol/g)	Range (nmol/g)	SD	CV (%)	CBR (ng/g)	Range (ng/g)	n
Mortality	TBT, TPT	Fish and invertebrates	33.4	12–51	12.4	37	9600	3500–14,780	11
Growth impairment	TBT	Fish and invertebrates	2.1	0.56–4.3	1.2	57	640	162–1246	11
Growth stimulation/ obesogen	TBT	Fish	0.04	—	—	—	13	—	1
Behavior	TBT	Fish and invertebrates	0.70	0.35–1.0	—	—	200	100–300	2
Imposex— female sterilization	TBT, TPT	Gastropod snails	0.29	0.05–0.42	0.21	73	85	14–141	11
Imposex— threshold	TBT, TPT	Gastropod snails*	0.10	0.03–0.17	0.06	62	30	10–49	4
Reproductive impairment	TBT, TPT	Invertebrates	0.48	0.006–0.97	0.49	100	140	2–280	3
Reproductive impairment	TBT, TPT	Fish—adult/ juvenile	0.07	0.06–0.08	0.01	20	24	18–29	2
Reproductive impairment	TBT, TPT	Fish—egg	0.28	0.01–0.55	0.23	81	83	5–160	5

Mean, standard deviation (SD), and range in values for critical body residues (CBRs) in nmol/g wet weight and equivalent ng/g value. Mortality CBR is based on LR_{50}. The CBRs for reproductive impairment, growth impairment, and growth stimulation based on LOER and ERp values. The imposex CBR for female sterilization based on definition in Meador et al. (2002). Range shows minimum and maximum values for n studies, most of which are for different species. CV is the coefficient of variation in %. Values are for whole body, except (*), which are whole body or muscle tissue. Last row shows concentrations for eggs. Data from this chapter, Meador (2000), Meador et al. (2002), Meador (in press).

Source: From Meador, J. P., *Rev. Environ. Contam. Toxicol.*, 166, 1–48, 2000; Meador, J. P. et al., *Aquat. Conserv.: Mar. Freshwat. Ecosyst.*, 12, 539–551, 2002. With permission.

TABLE 7.4
Lethal Values for Several Organotins in Small Mammals

Triorganotins	Species	LR_{50} i.p. μg/g	SD	nmol/g
Trimethyltin	Rat	16	—	97.6
Triethyltin	Rabbit	10	—	48.8
Triethyltin	Rat	10	—	48.8
Tributyltin	Rat	10	—	34.6
Triphenyltin	Rat	11.4	(2.0)	32.3
Triphenyltin	Guinea pig	3.7	—	10.7
Triphenyltin	Mouse	7.9	—	37.1
Triphenyltin	Rabbit	16	—	97.6
Tricyclohexyltin	Rat	13	—	34.7
Trioctyltin	Rat	>48	—	>100

Diorganotins	Est. metric	Est. LR_p μg/g		nmol/g
Dimethyltin	LR_{50}	40	—	270.3
Diethyltin	LR_{100}	40	—	226.0
Dipropyltin	LR_{75}	10	—	48.8
Di-isopropyltin	LR_{100}	20	—	97.6
Dibutyltin	LR_{100}	10	—	42.9
Dipentyltin	LR_{100}	20	—	73.5
Diphenyltin	LR_{100}	15.3*	—	56.0
Dihexyltin	LR_{50}	10	—	33.6
Dioctyltin	LR_{100}	10	—	28.2

Calculated LR_{50} values for triorganotins based on intraperitoneal (i.p.) injection. Each diorganotin tested at four doses by intravenous injection (i.v.) to groups of four rats. For diorganotins, LR_p (p for percentage) was estimated by the reported number of mortalities per treatment. The TPT LR_{50} for rat based on four experiments. Mean and standard deviation (SD) for all triorganotin LR_{50} values (except trioctyltin) is 10.9 (3.9) μg/g. * Is i.p. injection. Data from Barnes and Stoner (1958), Stoner (1966), and Kimbrough (1976). See text for details.

likely that the mode or MeOA for mortality is the same for all these organotins, which would reduce the importance of potency as a factor in the observed disparities in LC_{50} values and expected BCFs.

7.6.2 Responses and Modes and Mechanisms of Toxic Action

Toxic response can be considered at three levels. The first level is the organismal response (e.g., mortality, growth, and reproduction), the second level is mode of action (MoOA), and the third is the mechanism of action (MeOA). Organotins are known to cause several adverse effects including growth impairment, growth enhancement, abnormal development, altered behavior, and reproductive effects. The terms "mode and mechanism of toxic action" have distinct meaning (Meador et al. 2008). In general, the mode of action is the higher level of toxicological disturbance to biological function consisting of physico-chemical, physiological, or biochemical pathway alterations resulting from one or more MeOAs. Triorganotins, and some of the disubstituted organotins, are known to act by several MoOAs including inhibition of cellular energy metabolism (Aldridge et al. 1977, Hunziker et al. 2002), endocrine disruption (Matthiessen and Gibbs 1998, Grün et al. 2006), neurotoxicity (Walsh and DeHaven 1988), inhibition of ion pumps (Fent 1996), inhibition of cytochrome

P450 (Fent 1996), inhibition of intracellular enzymes (Walsh and DeHaven 1988), and immune system impairment (Bouchard et al. 1999, De Santiago and Aguilar-Santelises 1999). There is very little toxicity information for the monosubstituted organotins.

Most, if not all, of these MoOAs likely result from multiple MeOAs (the crucial and specific biochemical alteration) or inhibition of specific pathways. Several definitions for MeOA have been proposed; however, in many applications it refers to the biochemical target or specific biochemical pathway affected. More precise definitions have recently been proposed for both mode and MeOA (Borgert et al. 2004, Meador et al. 2008). For example, there are many compounds that are considered uncouplers (a mode of action), but do so by different biochemical mechanisms. This distinction is important, especially when considering the nature of toxicant interactions (e.g., additivity [dose or response] and those that are less or more than additive).

We have some information regarding toxic action for the commonly encountered organotins and limited information for most of the other compounds. There are some similarities among the organotins in the MeOA, especially for the acute mortality response for triorganotins. For the sublethal response, there are a variety of MoOAs and MeOAs for organotins, which are likely a function of dose and the compound's unique stereochemistry and resultant association with biomolecules. Detailed work by a few authors discovered that several of the triorganotins (TMT, TET, TBT, TPT, and Tri-c-hexyltin [TcHT]) and many of the diorganotins (DMT, DET, dipropyltin [DPrT], DBT, DPT, and dihexyltin [DHT]) inhibit respiration (Aldridge 1958, 1976, Aldridge et al. 1977, Connerton and Griffiths 1989). In addition, these diorganotins block the pathway from glutamate to oxoglutarate as part of the Krebs cycle in mitochondria (Aldridge 1976). Tetraorganotins generally do not inhibit respiration (Aldridge 1976); however, once bioaccumulated they are metabolized to triorganotins by many species.

Several recent studies have explored the role of organotins as agonists of nuclear hormone receptors and their role as endocrine disruptors and obesogens (Grün et al. 2006, Grün and Blumberg 2007). One study demonstrated that TBT and TPT were potent activators of the retinoid X receptor (RXR) and peroxisome proliferator-activated receptor (PPARγ) (Grün et al. 2006) at low concentrations (3–20 pmol/g). They also showed that several other butyltins were activators of these receptors but at higher levels; tetrabutyltin (TeBT) (150 pmol/g), DBT (3000 pmol/g), TET (2800 pmol/g), and TMT (>10,000 pmol/g). MBT was not active. These very low potency factors for RXR were confirmed by Hu et al. (2009) for TBT and TPT, which were essentially identical (9.6 and 20 pmol/g, respectively). According to Grün et al. (2006) some organotins (especially TBT and TPT) are potent endocrine disruptors that target adipogenesis by modulating key regulatory transcription factors via RXR and PPARγ.

Specific organotins can act by multiple MoOAs and MeOAs that are likely dose and time dependent. For example, short-term exposure to high doses of TBT leads to mortality and growth inhibition; however, under chronic low-dose exposures endocrine and immunotoxic responses can be observed in a variety of species. For these, and many compounds, it is important to consider critical toxic concentrations among species for a given response, not all responses combined.

7.6.3   Tissue-Residue Toxicity

The following is a brief survey of the various biological responses reported for organotins. For aquatic species, most of the focus has been on TBT and its effects on survival, growth, and reproductive impairment and there are limited data for TPT. In addition, there are toxicity data for several of the tri- and diorganotins for small mammals that are informative. Unfortunately, there are no tissue-residue mortality data for aquatic species exposed to di- or monoorganotins.

7.6.3.1   Mortality

In terms of the tissue-based toxic response, the TBT concentration causing lethality is approximately 100 times less than that for baseline (narcosis) toxicants that has been characterized for a large number of organic compounds (Di Toro et al. 2000). As shown in Table 7.3, sublethal toxic responses can occur at tissue residues that are 700 times lower than those for lethal levels. The lethal concentration for some organotins has been determined in several species and the values are remarkably consistent. One comparative study with guppies (*Poecillia reticulata*) found very similar lethal tissue concentrations for TBT, TPT, and TcHT, although tri-*n*-hexyltin (TnHT) was about 10 times more toxic ($LR_{100} \approx 1$ nmol/g ww) on a tissue-residue basis (Tas 1993). It is important to note here that most of the mortality data are based on short-term exposures (acute). It is possible to observe mortality during chronic exposure, which may be associated with far lower tissue residues. These chronic mortalities are likely a secondary response to the primary effect, such as mortality from a pathogen due to a weakened immune system or starvation from the inhibition of energy producing pathways.

The mortality values in Table 7.4 were determined by intraperitoneal (i.p.) or intravenous (i.v.) injection (as μg organotin injected per gram organism), which were used to estimate the LR_p values. Dosing by injection produces far less variable toxicity metrics than what is generally obtained when these compounds were administered orally (administered dose) to produce an LD_p or ED_p value. The variability in metabolism among species can have a large impact on the LD_{50} when toxicants are introduced orally, which is likely mitigated when using the injection route of exposure. With injection, the toxicant is quickly distributed to the tissues and results in a response when the critical concentration is achieved. Because tissue-residue toxicity metrics are often time independent for many toxicants (Meador 2006), this route of exposure would lead to a reasonable estimate of the tissue (e.g., whole-body) concentration (acquired dose) associated with the biological response (i.e., the LR_p or ER_p). The type of response is also important when considering injection as the route of exposure. These results may be reliable for the acute (short term) lethality response, but less so for sublethal responses that require extended periods of time at relatively constant concentrations to develop and manifest.

The literature on small mammals indicates a remarkable similarity in the LR_{50} values for several triorganotins. Data were found for nine toxicity values from tests with five organotins (TMT, TET, TBT, TPT, and TcHT) and four common laboratory bioassay species. The mean (SD) LR_{50} for all species and triorganotins was 11.0 (3.6) μg/g whole-body ww (Table 7.4). Interestingly, this value is almost identical to the TBT LR_{50} in Table 7.3 for 10 species of fish and invertebrates. Based on the data in Table 7.4 it is likely that these triorganotins act by the same MoOA (and possibly mechanism), which is assumed to be uncoupling of oxidative phosphorylation. Interestingly, the LR_{50} for trioctyltin was far above 48 μg/g via i.p. injection (no response at this concentration) (Table 7.4) (Barnes and Stoner 1958), and this is the one triorganotin that is considered not to be an uncoupler. The low variability for these data also imply that the whole-body tissue distribution for the various routes of exposure (injection, ingestion, and ventilation) result in similar internal tissue partitioning, which may not be the case for other toxicants.

Barnes and Stoner (1958) report lethal toxicity data for rats exposed to several diorganotins (methyl, ethyl, propyl, isopropyl, butyl, pentyl, hexyl, and octyl) that were administered via i.v. injection. Although sample sizes were small (four animals per dose, four doses for each diorganotin), all these diorganotins caused 50–100% mortality within 2–72 h at doses between 10 and 20 μg/g, except for DMT (50% mortality at 40 μg/g) and DET (100% mortality at 40 μg/g). Interestingly, dioctyltin (DOT) was relatively toxic, especially compared to trioctyltin. Because trioctyltin (TOT) is a large and presumably very hydrophobic compound (log K_{ow} for DOT is 5.8 and likely much higher for TOT) it may exhibit steric hindrance for membrane permeability. As mentioned previously, several of the diorganotin compounds are known inhibitors of energy production and because these compounds resulted in mortality at similar concentrations as the triorganotins, it is likely that they also act by the same MoOA or MeOA.

In general, we have very little tissue-residue-based lethality data for organotins other than triorganotins in aquatic species; however, we can estimate the acute toxicity for a methyltin. As described in Meador (2006), multiplying the BCF by the LC_{50} for time-matched values will result in the LR_{50}. A study by Hadjispyrou et al. (2001) provided the 24 h BCF and LC_{50} for the brine shrimp *Artemia franciscana* exposed to DMT. Using this equation, the resulting LR_{50} value is 78 μg/g ww, which is only 1.95 times higher than the value in Table 7.4 for rats. Hadjispyrou et al. (2001) also provide data for TMT; however, the BCF was determined at a concentration far higher than the LC50 and could not be used for this estimation.

Surprisingly, the lethal toxicity for TBT and TPT are very similar among invertebrates, fish, and mammals (Figure 7.2, Table 7.4). As seen in this figure, the LR_{50} for small mammals exposed to TBT via intravenous or intraperitoneal injection is almost identical to the observed LR_{50} for fish and invertebrates exposed via water or diet with most values between 10 and 50 nmol/g. Considering the 28-fold variability in aqueous LC_{50} values reported by Laughlin et al. (1985) and the 1800-fold range by Nagase et al. (1991) for one species exposed to triorganotins, this range of fivefold for tissue-residue-based mortality for five triorganotins in a very wide diversity of taxa is very low. Based on the similarities in the tissue-based mortality metrics for mammals, fish, and invertebrates exposed to triorganotins and the fact that most tri- and diorganotins are considered uncouplers, a reasonable assumption would be that the diorganotins would also result in similar acute lethality toxicity metrics for aquatic species. An important conclusion from these data is that all these compounds (di- and triorganotins) are likely dose additive for this response and all species considered.

7.6.3.2 Growth

Several studies have demonstrated reduced growth in a variety of aquatic species exposed to TBT (Table 7.3). Most of those studies show that growth inhibition occurs at a relatively consistent concentration of approximately 2 nmol/g (0.6 μg/g ww) for whole body (Meador 2000, 2006). A reasonable hypothesis for this response is that inhibition of oxidative phosphorylation causes a reduction in the available energy needed for growth; however, we do not know the actual MoOA

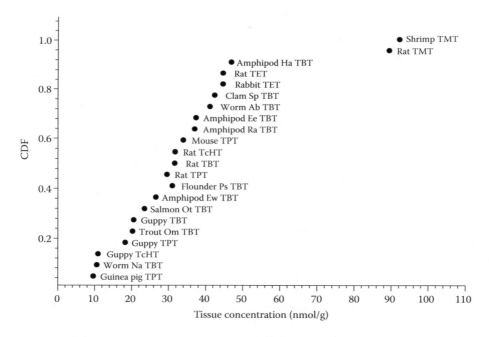

FIGURE 7.2 Values are whole-body lethal concentrations (LR_{50}) for 15 species and five organotins. The small mammal values are equivalent to the intraperitoneal injection concentration. Values range from 10.6 to 100 nmol/g wet weight and species names are centered on the value (e.g., guinea pig TPT is 10.6 nmol/g). Species are shrimp (*Artemia franciscana*), amphipods (*Hyalella azteca*; *Eohaustorius estuarius*; *E. washingtonianus*; *Rhepoxynius abronius*), clam (*Scrobicularia plana*), polychaete worms (*Armandia brevis*; *Neanthes arenaceodentata*), flounder (*Platichthys stellatus*), salmon (*Oncorhynchus tshawytscha*), and trout (*O. mykiss*). See Table 7.1 for organotin abbreviations. CDF is cumulative distribution function.

or MeOA for this response. As discussed later, organotins are known to affect metabolism and adipogenesis and low concentrations are known to affect growth, therefore this response may be due to long-term disruption in the pathway for steroid metabolism or the result of low-level effects on energy production in mitochondria. Widdows and Page (1993) found an impact to the Scope for Growth metric (joules/hour) at a TBT tissue concentration of ≈1 μg/g in mussel (*Mytilus edulis*). This value represents the difference between the energy absorbed from food and the energy expended with respiration.

There are very few studies that demonstrate concordance between lab and field-toxicity data, which has been confirmed for TBT. The bioaccumulation and toxicity data from field studies for TBT are remarkably similar to the values determined in the lab (Salazar and Salazar 1995, 1998). The tissue concentration shown to be associated with impaired growth in mussels from caging studies in contaminated harbors is approximately 0.8 μg/g (wet wt.), which is essentially identical to the value in Table 7.3 that is derived from lab studies.

TPT is also known to affect growth; however, none of the studies examined reported tissue concentrations for this response. Rehage et al. (2002) described reduced growth in larval salamanders exposed to 1 ppb of aqueous TPT, which may have resulted in a relatively low tissue concentration depending on the toxicokinetics for this species. In addition, Stoner (1966) reported a reduction in body mass for guinea pigs fed a relatively low dose of TPT (1 μg/g in diet ≈ 0.1 μg/g bw/day).

Another well-known effect due to TBT exposure is the chambering response in bivalves. Many laboratory (Chagot et al. 1990, Coelho et al. 2006) and field studies (King et al. 1989, Alzieu 2000) have documented that low exposure concentrations of TBT (5–100 ng/L; approximately 10–100 ng/g in tissue) resulted in excessive shell growth (chambering) in oysters and clams. King et al. (1989) found a strong correlation between the number of shell chambers for *Crassostrea gigas* and TBT tissue concentrations with an apparent threshold of ≈100 ng/g ww. This abnormal growth severely impairs the marketability for oysters and may result in reproductive impairment because a large percentage of the available energy is utilized for shell growth leaving less for somatic and reproductive biomass. This response may be considered a growth effect; however, in some cases the shell malformation is observed at concentrations lower than those causing reductions in soft-tissue biomass.

7.6.3.3 Immunotoxicity

This mode of action for organotins may result in several biological changes including atrophy of the thymus, reduction of kidney macrophages, and changes to the spleen (Fent 1996). One of the biochemical mechanisms promoting immunotoxicity is likely related to the disruption of calcium homeostasis (Chow et al. 1992). The human health reference dose for TBT (0.3 ng/g bw/d) is based on immunotoxicity (U.S. EPA 1997), and there is evidence that DBT is also immunotoxic (Fent 1996, O'Halloran et al. 1998). Bouchard et al. (1999) and O'Halloran et al. (1998) concluded that DBT was a more potent immunotoxicant than TBT (based on concentrations in hemolymph and cell cultures), which had important implications for assessing the toxicity of DBT tissue concentrations. When considering organotin tissue residues, DBT is particularly important because of its immunotoxic potency and because high concentrations of DBT are often found in tissue as a result of TBT metabolism. The metabolic conversion of organometallics to other compounds and their often toxic nature provides a strong argument for considering both the parent compound and any metabolites when conducting a toxicity assessment based on tissue residues.

7.6.3.4 Reproductive

7.6.3.4.1 Imposex in Molluscs

Some organotins are also potent endocrine disruptors causing imposex in meso- and neogastropods (Matthiessen and Gibbs 1998), which is the manifestation of secondary male sexual characteristics in female gastropods. The imposex abnormality was the primary driver for the TBT water quality criteria promulgated by the U.S. EPA (2003) because this response occurred at the lowest effect concentrations. For aquatic organisms, this reproductive impairment is one of the most sensitive responses and it also occurs at the lowest tissue concentrations (Meador 2006). The observation of imposex in molluscs at relatively low tissue concentrations is supported by laboratory (Bryan et al. 1988, Horiguchi et al. 1997a, 1997b) and field studies (Horiguchi et al. 1994, Morcillo and Porte 1999, Barreiro et al. 2001, Bech et al. 2002). The threshold for imposex in snails has been reported to occur in the 10–50 ng/g ww range (Table 7.5).

Several theories have been offered over the years on the mechanism for imposex in stenoglossan snails; however, the actual mechanistic response was described recently by Nishikawa et al. (2004). As described in this study, TBT and TPT are potent agonists for the RXRs. In addition to demonstrating that TBT strongly binds RXRs, the authors of this elegant study were able to induce imposex in the rock shell (*Thais clavigera*) within 4 weeks by injecting individuals with 9-cis retinoic acid, the natural ligand for RXR.

TBT is the main focus for imposex studies; however, TPT also causes imposex within the same range of tissue concentrations (Table 7.5; Horiguchi et al. 1997a). Given the similarity in the dose–response relationship for imposex and these two compounds they should be considered additive when assessing imposex. The concentration of TBT and TPT in tissue associated with sterilization in these molluscs occurs at approximately 85 ng/g ww (Meador et al. 2002; Table 7.3). Sterilization for these molluscs is a severe effect and has been linked with adverse population level attributes (Oehlmann et al. 1996, Horiguchi et al. 1997b).

One paper reported the results for six organotins (TBT, DBT, MBT, TPT, DPT, and MPT) and their potential to cause imposex in *Thais clavigera*, a commonly studied snail for this response (Horiguchi et al. 1997a). A sequential series of experiments using high tissue doses (via injection) to screen organotins and lower doses to characterize the degree of increased penis length in female snails found that the triorganotins TBT, TPT, and TPrT were the strongest inducers of imposex. This study also concluded that the di- and monoorganotins (DBT, MBT, DPT, and MPT) did not cause the response. Bryan et al. (1988) also concluded that DBT did not lead to imposex, and they observed mixed results for TPrT (positive results only at high concentrations, i.e., >500 ng/g ww) in the snail *Nucella lapillus*. They also reported no response to TPT when animals were exposed to 590 ng/L in water or injected with 4.4 µg/g bw. TeBT was also tested by Bryan et al. (1988); however, they concluded that the observed positive response may have been caused by TBT contamination, which is commonly found in stock solutions.

A recent study reported a high correlation ($r^2 = 0.94$) between the degree of imposex in snails and the extent of DNA damage as assessed with micronucleus formation in hemocytes (Hagger et al. 2006). This study also found a high correlation between whole-body TBT concentrations and neoplastic proliferations on genital organs of male and female snails.

7.6.3.4.2 Fish Reproduction and Development

At least five studies examined early-life stage effects in fish due to the exposure of eggs to TBT. Three of the studies assessed the effects of maternally transferred TBT using different routes of exposure including dietary (Nakayama et al. 2005, Shimasaki et al. 2006) and aqueous (Zhang et al. 2008). Two other studies exposed the eggs via nanoinjection (Hano et al. 2007, Hu et al. 2009). A number of adverse effects were reported in these studies, and many of those responses occurred when TBT was approximately 5–160 ng/g ww egg (Table 7.5). Based on these five studies, we can conclude that fish embryos are very sensitive to TBT at these low concentrations and that maternal transfer is an important route of exposure.

Reproductive effects were also observed in juvenile and adult fish at very low tissue concentrations. A high rate of sex reversal was observed among genetic female flounder (*Paralichthys olivaceus*) at a whole-body TBT concentration of 18 and 160 ng/g ww (Shimasaki et al. 2003). These authors also reported a statistically significant decrease in growth (body weight and length) at 18 ng/g ww. Other studies on reproductive effects in adult fish include Zhang et al. (2008) who

TABLE 7.5
Tissue Concentrations Associated with Reproductive or Early-Life Stage Responses by Aquatic Species to Tributyltin (TBT) and Triphenyltin (TPT)

	Responses	Species	Type	ng/g	Tissue	nmol/g	Exposure	Reference
			Fish					
TBT	↓Viable hatch, ↓viable larvae, ↓floating egg rate	*Sillago japonica*	Whiting	85–160	Egg	0.30	Maternal lab	Shimasaki et al. (2006)
TBT	↓Fertility, ↓hatchability, ↓swim up rate	*Oryzias latipes*	Medaka	<20–265	Egg	0.07–0.9	Maternal lab	Nakayama et al. (2005)
TBT	↓Swim up rate, ↑mort	*O. latipes*	Medaka	160	Egg	0.55	Nano injection	Hano et al. (2007)
TBT	↓Growth, ↑%male	*Paralichthys olivaceus*	Japanese flounder	18	Whole body	0.06	Dietary lab	Shimasaki et al. (2003)
TPT	↓Protein/egg, ↓hatching success, ↓swim up success, ↓surviving larvae/female/d, ↑hemorrhaging, ↑abnormal ocular development	*O. latipes*	Medaka	4.6	Egg	0.013	Maternal lab	Zhang et al. (2008)
TPT	↑Abnormal ocular development	*Acipenser baerii*	Sturgeon	27	Egg	0.08	Nanoinjection	Hu et al. (2009)
TPT	↓Vitellogenin, ↓spawning frequency, ↓eggs/female/d	*O. latipes*	Medaka	29	Whole body	0.08	Lab	Zhang et al. (2008)
			Invertebrates					
TBT	↑Male repro cells in ovary	*Haliotis gigantea*	Abalone	2.4	Muscle	0.008	Lab	Horiguchi et al. (2002)
TBT	Imposex—100% females	*Bolinus brandaris*	Snail	141	Whole body	0.49	Field	Morcillo and Porte (1999)
TBT	↑Imposex	*Thais distinguenda*	Snail	10	Whole body	0.03	Field	Bech et al. (2002)
TBT	↑Imposex	*Nassarius reticulatus*	Snail	35	Whole body	0.12	Field	Barreiro et al. (2001)
TBT	↑Imposex	*Hydrobia ulvae*	Snail	49	Whole body	0.17	Field	Schulte-Oehlmann et al. (1998)
TBT	↑Imposex	*Thais clavigera*	Snail	20	Whole body	0.07	Lab	Horiguchi et al. (1997b)*
TBT	↑Imposex	*Nucella lapillus*	Snail	330	Whole body	1.1	Lab	Bryan et al. (1988)
TBT	↑Imposex	*T. clavigera*	Snail	221	Whole body	0.76	Lab	Horiguchi et al. (1997a)
TBT	↓Number of young (50% decline; EC$_{50}$)	*Hyalella azteca*	Amphipod	284	Whole body	0.98	Lab	Bartlett et al. (2004)
TPT	↑Imposex	*T. clavigera*	Snail	44	Whole body	0.13	Lab	Horiguchi et al. (1997a)
TPT	↑Male repro cells in ovary	*H. gigantea*	Abalone	126	Muscle	0.44	Lab	Horiguchi et al. (2002)

Values (as wet weights) are from recent laboratory studies or field assessments. * Details for this study in Horiguchi (1993). All responses exhibited statistical p-values ≤ .05 at the stated tissue concentration, which is usually the lowest observed effect residue (LOER).

reported decreased vitellogenin, spawning frequency, and eggs/female/day at very low whole-body concentrations (29 ng/g ww).

Another study on fish reproductive effects due to TBT exposure found significant increases in the sex ratio (males:females) and percentage abnormal sperm in zebrafish (*Danio rerio*) exposed from hatch to day 70 at an aqueous concentration of 0.1 ng/L (McAllister and Kime 2003). They also reported a large and significant decrease in sperm motility at 1 ng/L and the complete absence of flagella in sperm at 10 ng/L. The typical long-term BCF for small fish range from 300 to 5000 (Fent 1996, Zhang et al. 2008). If we use the high BCF estimate of 5000 that was determined for medaka (Zhang et al. 2008), the whole-body tissue concentration in the zebrafish from the McAllister and Kime (2003) study is predicted to be 0.5 ng/g for the 0.1 ng/L exposure concentration and 5 ng/g for the 1 ng/L treatment. This conservative estimate for the BCF results in very low tissue concentrations for these adverse effects that are comparable with measured concentrations and adverse effects for fish reported in Table 7.5.

7.6.3.4.3 Reproductive Effects in Other Species

One study with the starfish (*Leptasteria polaris*) reported large and statistically significant reductions in the diameter of previtellogenic and mature (final-stage) oocytes in females and the thickness of gonadal epithelium for both sexes (Mercier et al. 1994). These responses occurred at TBT concentrations of ≈300 ng/g as measured in the pyloric caeca. Concentrations of TBT were below the detection limit (2.5 ng/g) in the gonads. Throughout this 53-day dietary exposure study the metabolites DBT and MBT continued to increase indicating metabolism of the accumulated TBT. The authors concluded that oocyte development could not be maintained at this tissue concentration due to the thinning gonadal epithelia and consequent lack of available nutrients for the oocytes to develop.

Another study with TBT and copepods found reproductive impairment at exposure concentrations of 10, 50, and 100 ng/L (Johansen and Mohlenberg 1987), which covers that range of aqueous concentrations usually reported for imposex in snails. These results indicate that reproductive effects at these low concentrations are not limited to gastropod snails. Because stenoglossan snails and oysters exhibit the highest BCF values, these copepods are likely exhibiting reproductive effects at similar tissue concentrations or lower to those observed for molluscs.

7.6.3.5 Neurological

Many of the butyltins are neurotoxic, an effect that has been described for small mammals. TET binds myelin with high affinity, and TMT causes cell death in the limbic system, neocortex, and sensory neurons, which is considered a unique pathology (Walsh and DeHaven 1988). Because TMT and TET are potent neurotoxicants, behavioral effects are also common for mammals and likely other vertebrate classes. Unfortunately, very few data exists for aquatic organisms; however, invertebrate neurons do not contain myelin, therefore, this mode of action may be less important for these taxa. TBT and TPT are not generally considered neurotoxic (Fent 1996); however, they do cause behavioral changes. One study examined brain transmitters in rockfish (*Sebastiscus marmoratus*) that were injected intraperitoneally with TBT and TMT. TMT stimulated dose-response increases in the neurotransmitters aspartate and γ-aminobutyric acid (GABA) at all doses injected (10, 100, and 1000 ng/g bw). TBT caused an increase only in GABA at the highest dose. Both organotins affected the N-methyl-D-aspartate receptor (NMDAR) signaling pathway and its components such as calmodulin and calmodulin-dependent kinase II at most doses but in different directions (TBT downregulated and TMT upregulated NMDAR and other genes in this pathway) (Zuo et al. 2009). Changes to the levels of these neurotransmitters can alter neurotransmission, and by extension, cause abnormal neuronal function.

7.6.3.6 Behavioral

Triebskorn et al. (1994) demonstrated alterations in behavior for rainbow trout at whole-body TBT concentrations of approximately 300 ng/g ww. At this concentration fish exhibited hyperactivity by swimming farther and faster for extended periods than control fish and also exhibited more random orientation in tanks. For some organotins, behavioral changes may be due to neurotoxicity; however, for TBT and TPT these responses could also be a result of an energy imbalance and higher than normal metabolism, which may result in lethargy or hyperactivity. Another study reported behavioral effects related to reproduction for medaka at a dietary dose of 1 µg/g bw/d (Nakayama et al. 2004) and concluded that the observed reduction in fertilization success was a result of the behavioral alteration.

Another study on behavioral alterations reported hyperactivity and a complete reversal in phototaxis for *Daphnia magna* exposed to TBT at concentrations between 0.5 and 1.0 ng/mL (Meador 1986). The percentage of individuals that exhibited positive phototaxis increased gradually from 0% to 100% over 6 days at 0.5 ng/mL and at a faster rate for the 0.75 and 1.0 ng/mL treatments. The photopositive animals increased their antennular strokes (an indication of activity and swimming speed) to 5 strokes/sec compared to 2 strokes/sec for the controls. Based on the BCF reported for this species by Fent (1996), the predicted tissue concentration for these responses at the lowest dose (0.5 ng/mL) is approximately 100 ng/g ww. Alteration in behavior is usually important for organism survival in the wild, and these results are especially noteworthy because *Daphnia* spp. and other zooplankton rely on light gradients and contrasts for vertical migration and antipredator behavior.

7.6.3.7 Obesogen/Somatogen

Obesogens are compounds that affect metabolic physiology and can lead to increased body fat, but not necessarily total body mass. Somatogens are compounds that promote increases in body mass (usually muscle) and possibly fat content. These compounds are often endocrine disruptors, and they may act through the same mechanism. Organotins have been implicated as obesogens, via activation of RXR and PPARγ (Grün and Blumberg 2007). These are the same receptors that have been implicated in the imposex response in snails. Because compounds that act as obesogens affect adipogenesis, it appears that they not only affect endocrine systems that control sexual differentiation but also the biochemical pathways that regulate lipid metabolism and growth. This is a new area of research and most of the limited number of studies have been conducted with small mammals. One recent study (Meador et al. in press) with juvenile chinook salmon (10–20 g) found that whole-body TBT concentrations as low as 13 ng/g ww caused significant increases in fish weight, whole-body lipid content, and several physiological parameters measured in plasma (glucose, alkaline phosphatase, lipase, triacylglycerols, and cholesterol). These results were more consistent with a somatogen response rather than the metabolic syndrome associated with the obesogen response. These whole-body tissue concentrations are in the same range as those that are considered threshold values for the imposex response in snails and reproductive effects in fish. Considering the role of RXR for imposex for both TBT and TPT, there is a high probability that TPT will also elicit these metabolic abnormalities at similar tissue concentrations reported for TBT.

The growth response for TBT is a good example of "hormesis," which is characterized by low-dose stimulation and high-dose inhibition (Calabrese and Baldwin 2003). As seen in Table 7.3, there are several studies demonstrating reduced growth in various species at whole-body concentrations of 600 ng/g. In this case we do not know if two separate dose-dependent MeOAs are involved or if these responses are just a continuum for one MeOA that produces different results depending of the degree of receptor interaction and length of time for exposure.

Summary

As shown for many organotins, bioaccumulation values are not predictable using the standard QSAR equations. Because there is no reliable way to predict bioaccumulation, direct observation is the best method for determining tissue concentrations for toxicity assessment. Bioaccumulation can be predicted with toxicokinetic rates; however, chemical concentration data from water, sediment, or diet would be required to estimate tissue concentrations and these are highly variable. Given the highly elevated BCF values observed for TBT and TPT, very high tissue concentrations are expected for low exposure concentrations. Due to the highly toxic nature of these compounds at relatively low tissue concentrations, these organotins warrant the high level of environmental concern they have been given. Many of the other organotins are also very toxic at relatively low tissue concentrations; however, data on their occurrence in a variety of species are lacking therefore precluding a complete assessment of their potential toxicity. Even though BCF values are relatively low for some of these organotins (e.g., TMT and DMT), dietary uptake may be important for some. Many of the organotin compounds have not been detected in field-collected water or sediment and they are rarely analyzed in tissue. Some of the issues involve detection limits and a lack of studies to determine if these compounds occur. The butyltins and phenyltins are extensively studied and several tissue values have been reported for field-collected organisms. Except for the one study on seafood species (Forsyth and Clerous 1991), there are no field data for organotins, other than those mentioned earlier.

Although most of the research has been on TBT and the imposex response, a number of other organotins appear to be very toxic. Most notable is TPT, which exhibits a similar BCF and has been demonstrated to cause imposex at comparable concentrations as TBT. On the basis of the data presented in Table 7.2, it is evident that butyltins and phenyltins can occur in high concentrations in aquatic biota and should therefore always be considered together when assessing this response. When examined in light of the available toxicity information presented in this review, we can conclude that many species of fish and invertebrates exhibit concentrations high enough to result in adverse biological effects when exposed to these two triorganotins.

Surprisingly, the acute lethality values for many of the triorganotins, TMT, TET, TBT, TPT, and TcHT, are very similar among a variety of species from polychaetes to small mammals (Tables 7.3 and 7.4, Figure 7.2). Lethal tissue concentrations for these organotins and species exhibit a relatively tight range of concentrations (10–100 nmol/g; 4–16 ppm) with a low variance. Based on the work of Aldridge (1958), Connerton and Griffiths (1989), and others, most triorganotins (with the exception of trioctyltin) are considered uncouplers of oxidative phosphorylation, which is the likely mode of action responsible for lethality.

Also noteworthy is that many of the diorganotins (DPrT, DBT, DPT, DHT, and DOT) produce lethal responses in rats at very similar concentrations to those observed for triorganotins (Table 7.4). Mortality likely results from respiratory uncoupling and it appears to occur at very similar whole-body concentrations as that reported for triorganotins. As a result of this similarity, these compounds may be dose additive for the mortality response. This information suggests that a lethality-based toxicity assessment for organotins in field-collected species consider the summed concentration of all these di- and triorganotins.

Although not stated precisely as such, the concept put forth by Paracelsus (1493–1541) that the dose makes the poison, is germane here. For TMT we know that the LR_{50} for the brine shrimp is approximately 16.5 µg/g, which is very similar to values for other triorganotins. Due the low K_{ow} (low bioavailability) and high LC_{50} for this compound (220 ng/mL), it may not be an important environmental contaminant unless elevated tissue concentrations occur via dietary uptake. Even though most of the triorganotins lead to mortality at a similar whole-body tissue concentration, it is important to assess those tissue concentrations and their potential to reach adverse levels in feral organisms when concentrations in the environment (water, sediment, and prey) are relatively low.

Because comparable TBT and TPT tissue concentrations lead to similar responses it appears that these compounds may act similarly at the molecular level and bind the same receptors. This hypothesis is supported by the almost identical potency observed for interaction with the RXR, which is linked to the MeOA for endocrine- and metabolic-related responses (reproductive and growth disorders). On the basis of the research presented, comparable tissue concentrations for TBT and TPT will likely result in similar response levels for mortality, growth stimulation and inhibition, and imposex. As a result of these observations, dose additivity would be a reasonable hypothesis for these two compounds, which means their concentrations should be added together when assessing these toxicological impacts (Meador 2006).

Even though imposex in molluscs is considered the most sensitive response, several recent studies have shown that fish respond at similar concentrations of triorganotins as those causing imposex. As more research is conducted, it will likely become evident that these very low tissue concentrations are able to cause adverse effects in a variety of taxa. It is clear from the data that TBT, and likely TPT, are very potent endocrine disruptors and reproductive toxicants for snails as well as other species. Based on these data there is no reason to limit the analysis of reproductive effects only to stenoglossan snails, which has been the intense focus for several years. All species should be considered at risk for reproductive impairment at these low tissue concentrations (10–50 ng/g ww).

The critical body residue (CBR) data for mortality, growth impairment, and population sterility due to imposex (Table 7.3) were used recently to develop tissue and sediment quality guidelines (Meador et al. 2002, Meador 2006). The observed variance for each endpoint-specific mean was relatively low allowing the selection of a mean value that could be used to assess toxic impact in field-collected organisms. Based on the available data presented here, it appears that whole-body tissue concentrations in the low ppb range (10–50 ng/g ww) represent threshold levels for a variety of effects in all aquatic species and in many cases result in serious impairment. Higher concentrations (100–500 ng/g ww) should be considered toxic to all species and likely to cause adverse effects in individuals, and potentially populations, if the exposure is long term. Any tissue concentrations in the low μg/g range (1–10 ppm) should be considered lethal for all species.

For the well-studied organotins (TBT and TPT), the combination of high uptake kinetics and slow rates of elimination coupled with relatively high-potency results in very toxic environmental contaminants. This combination of high bioaccumulation and potency is why organotins are considered one of the most toxic anthropogenic compounds ever released into the environment. Given some of the similarities in bioaccumulation and potency for the other organotins discussed in this review, we can conclude that this class of compounds warrants caution, assessment, and action when found in feral organisms.

ACKNOWLEDGMENTS

I would like to thank Nancy Beckvar, Mike Salazar, Nat Scholz, and Catherine Sloan for conducting a thorough review of this manuscript and providing insightful comments.

REFERENCES

Aldridge, W. N. 1958. The biochemistry of organotin compounds. *Biochem. J.* 69:367–376.
Aldridge, W. N. 1976. The influence of organotin compounds on mitochondrial functions. In *Organotin Compounds: New Chemistry and Applications,* ed. J. J. Zuckerman, pp. 186–196. *Advances in Chemistry Series* 157, Washington, DC: American Chemical Society.
Aldridge, W. N., J. E. Casida, R. H. Fish, E. C. Kimmel, and B. W. Street. 1977. Action on mitochondria and toxicity of metabolites of tri-n-butyltin derivatives. *Biochem. Pharmacol.* 26:1997–2000.
Alzieu, C. 2000. Environmental impact of TBT: the French experience. *Sci. Total Environ.* 258:99–102.
Antizar-Ladislao, B. 2008. Environmental levels, toxicity and human exposure to tributyltin (TBT)-contaminated marine environment. A review. *Environ. Int.* 34:292–308.
Arnold, C. G., A. Weidenhaupt, M. M. David, S. R. Müller, S. B. Haderlein, and R. P. Schwarzenbach. 1997. Aqueous speciation and 1-octanol-water partitioning of tributyl- and triphenyltin: effect of pH and ion composition. *Environ. Sci. Technol.* 31:2596–2602.
Barnes, J. M., and H. B. Stoner. 1958. Toxic properties of some dialkyl and trialkyl tin salts. *Br. J. Ind. Med.* 15:15–22.

Barreiro, B., R. Gonzalez, M. Quintela, and J. M. Ruiz. 2001. Imposex, organotin bioaccumulation and sterility of female *Nassarius reticulatus* in polluted areas of NW Spain. *Mar. Ecol. Prog. Ser.* 218:203–212.

Bartlett, A. J., U. Borgmann, D. G. Dixon, S. P. Batchelor, and R. J. Maguire. 2004. Accumulation of tributyltin in *Hyalella azteca* as an indicator of chronic toxicity: survival, growth, and reproduction. *Environ. Toxicol. Chem.* 23:2878–2888.

Bech, M., J. Strand, and J. A. Jacobsen. 2002. Development of imposex and accumulation of butyltin in the tropical muricid *Thais distinguenda* transplanted to a TBT contaminated site. *Environ. Pollut.* 119:253–260.

Birchenough, A. C., N. Barnes, S. M. Evans, H. Hinz, I. Kronke, and C. Moss. 2002. A review and assessment of tributyltin contamination in the North Sea, based on surveys of butyltin tissue burdens and imposex/intersex in four species of neogastropods. *Mar. Pollut. Bull.* 44:534–543.

Borgert, C. J, T. F. Quill, L. S. McCarty, and A. M. Mason. 2004. Can mode of action predict mixtures toxicity for risk assessment? *Toxicol. Appl. Pharm.* 201:85–96.

Bouchard, N., E. Pelletier, and M. Fournier. 1999. Effects of butyltin compounds on phagocytic activity of hemocytes from three marine bivalves. *Environ. Toxicol. Chem.* 18:519–522.

Brändli, R. C., G. D. Breedveld, and G. Cornelissen. 2009. Tributyltin sorption to marine sedimentary black carbon and to amended activated carbon. *Environ. Toxicol. Chem.* 28:503–508.

Bryan, G. W., P. E. Gibbs, and G. R. Burt. 1988. Comparison of the effectiveness of tri-n-butyltin chloride and five other organotin compounds in promoting the development of imposex in the dog-whelk *Nucella lapillus*. *J. Mar. Biol. Assoc. UK* 68:733–744.

Brüschweiler, B. J, F. E. Würgler, and K. Fent. 1995. Cytotoxicity in vitro of organotin compounds to fish hepatoma cells PLHC-1 (*Poeciliopsis lucida*). *Aquat. Toxicol.* 32:143–160.

Calabrese, E. J., and L. A. Baldwin. 2003. Hormesis: the dose-response revolution. *Annu. Rev. Pharmacol.* 43:175–197.

Cardwell, R. D., and J. P. Meador. 1989. Tributyltin in the environment: an overview and key issues. In *Proceedings of the Organotin Symposium, Oceans 89*. September 18–21, 1989, Seattle, WA, Vol 2: 537–544.

Chagot, D., C. Alzieu, J. Sanjuan, and H. Grizel. 1990. Sublethal and histopathological effects of trace levels of tributyltin fluoride on adult oysters *Crassostrea gigas*. *Aquat. Living Resour.* 3:121–130.

Chow, S. C., et al. 1992. Tributyltin increases cytosolic free Ca^{2+} concentration in thymocytes by mobilizing intracellular Ca^{2+}, activating a Ca^{2+} entry pathway, and inhibiting Ca^{2+} efflux. *Arch. Biochem. Biophys.* 298:143–151.

Ciesielski, T., A. Wasik, I. Kuklik, K. Skora, J. Namiesnik, and P. Szefer. 2004. Organotin compounds in the liver tissue of marine mammals from the Polish Coast of the Baltic Sea. *Environ. Sci. Technol.* 38:1415–1420.

Cima, F., P. J. Craig, and C. Harrington. 2003. Organotin compounds in the environment. In *Organometallic Compounds in the Environment*, ed. P. J. Craig, pp. 101–149. West Sussex: John Wiley and Sons.

Clark, E. A., and R. N. Sterrit. 1988. The fate of tributyltin in the aquatic environment. *Environ. Sci. Technol.* 22:600–603.

Coelho, M. R., W. J. Langston, and M. J. Bebianno. 2006. Effect of TBT on *Ruditapes decussatus* juveniles. *Chemosphere* 63:1499–1505.

Connell, D. W. 1990. Bioconcentration of lipophilic and hydrophobic compounds by aquatic organisms. In *Bioaccumulation of Xenobiotic Compounds*, ed. Connell, D. W., pp. 97–144. Boca Raton, FL: CRC Press.

Connerton, I. F., and D. E. Griffiths. 1989. Organotin compounds as energy-potentiated uncouplers of rat liver mitochondria. *Appl. Organomet. Chem.* 3:545–551.

CRC. 1975. *Handbook of Chemistry and Physics, 56th edition*, ed. R. C. Weast. Cleveland, OH: CRC Press.

de Mora, S. J., N. G. King, and M. C. Miller. 1989. Tributyltin and total tin in marine sediments: profiles and the apparent rate of TBT degradation. *Environ. Technol. Lett.* 10:901–908.

De Santiago, A., and M. Aguilar-Santelises. 1999. Organotin compounds decrease in vitro survival, proliferation and differentiation of normal human B lymphocytes. *Hum. Exp. Toxicol.* 18:619–664.

Di Toro, D. M., J. A. McGrath, and D. J. Hansen. 2000. Technical basis for narcotic chemicals and polycyclic aromatic hydrocarbon criteria. I. water and tissue. *Environ. Toxicol. Chem.* 19:1951–1970.

Dowson, P. H., J. M. Bubb, and J. N. Lester. 1996. Persistence and degradation pathways of tributyltin in freshwater and estuarine sediments. *Estuar. Coast. Shelf Sci.* 42:551–562.

Fent, K. 1996. Ecotoxicology of organotin compounds. *Crit. Rev. Toxicol.* 26:1–117.

Forsyth, D. S., and C. Clerous. 1991. Determination of butyltin, methyltin, and tetraalkyltin in marine food products with gas chromatography-atomic absorption spectrometry. *Talanta* 38:951–957.

Gomez-Ariza, J. L., E. Morales, and I. Giraldez. 1999. Uptake and elimination of tributyltin in clams, *Venerupis decussata*. *Mar. Environ. Res.* 47:399–413.

Grün, F., and B. Blumberg. 2007. Perturbed nuclear receptor signaling by environmental obesogens as emerging factors in the obesity crisis. *Rev. Endocr. Metab. Dis.* 8:161–171.

Grün, F., H. Watanabe, and Z. Zamanian. 2006. Endocrine-disrupting organotin compounds are potent inducers of adipogenesis in vertebrates. *Mol. Endocrinol.* 20:2141–2155.

Hadjispyrou, S., A. Kungolos, and A. Anagnostopoulos. 2001. Toxicity, bioaccumulation, and interactive effects of organotin, cadmium, and chromium on *Artemia franciscana*. *Ecotox. Environ. Safe.* 49:179–186.

Hagger, J. A., M. H. Depledge, J. Oehlmann, S. Jobling, and T. S. Galloway. 2006. Is there a causal association between genotoxicity and the imposex effect? *Environ. Health Persp.* 114:20–26.

Hano, T., et al. 2007. Tributyltin causes abnormal development in embryos of medaka, *Oryzias latipes*. *Chemosphere* 69:927–933.

Harino, H., M. Fukushima, and S. Kawai. 2000. Accumulation of butyltin and phenyltin compounds in various fish species. *Arch. Environ. Contam. Toxicol.* 39:13–19.

Harino, H., M. Ohji, R. L. Brownell, T. Arai, and N. Miyazaki. 2008. Concentrations of organotin compounds in the stranded killer whales from Rausu, Hokkaido, Japan. *Arch. Environ. Contam. Toxicol.* 55:137–142.

Hoch, M., and D. Schwesig. 2004. Parameters controlling the partitioning of tributyltin (TBT) in aquatic systems. *Appl. Geochem.* 19:323–334.

Horiguchi, T. 1993. Imposex induced by organotin compounds in gastropods in Japan. PhD Dissertation, University of Tokyo. (in Japanese).

Horiguchi, T., H. Shiraishi, M. Shimizu, and M. Morita. 1994. Imposex and organotin compounds in *Thais clavigera* and *T. bronni* in Japan. *J. Mar. Biol. Assoc. UK* 74:651–669.

Horiguchi, T., H. Shiraishi, M. Shimizu, and M. Morita. 1997a. Effects of triphenyltin chloride and five other organotin compounds on the development of imposex in the rock shell, *Thais clavigera*. *Environ. Pollut.* 95:85–91.

Horiguchi, T., H. Shiraishi, M. Shimizu, and M. Morita. 1997b. Imposex in sea snails, caused by organotin (tributyltin and triphenyltin) pollutions in Japan: a survey. *Appl. Organomet. Chem.* 11:451–455.

Horiguchi, T., et al. 2002. Tributyltin and triphenyltin induce spermatogenesis in ovary of female abalone. *Haliotis gigantea*. *Mar. Environ. Res.* 54:679–684.

Hu, J., et al. 2006. Trophic magnification of triphenyltin in a marine food web of Bohai Bay, North China: comparison to tributyltin. *Environ. Sci. Technol.* 40:3142–3147.

Hu, J., et al. 2009. Malformations of the endangered Chinese sturgeon, *Acipenser sinensis*, and its causal agent. *P. Nat. Acad. Sci.* 106:9339–9344. plus supporting information.

Huggett, R. J., M. A. Unger, P. F. Seligman, and A. O. Valkirs. 1992. The marine biocide tributyltin. *Environ. Sci. Technol.* 25:232–237.

Hunziker, R. W., B. I. Escher, and R. P. Schwarzenbach. 2002. Acute toxicity of triorganotin compounds: different specific effects on the energy metabolism and role of pH. *Environ. Toxicol. Chem.* 21:1191–1197.

Hwang, H. M., J. R. Oh, S-H. Kahng, and K. W. Lee. 1999. Tributyltin compounds in mussels, oysters and sediments of Chinhae Bay. Korea. *Mar. Environ. Res.* 47:61–70.

IMO. 2001. International Maritime Organization. International Convention on the Control of Harmful Antifouling Systems on Ships. http://www.imo.org (accessed April 20, 2009).

Johansen, K., and F. Mohlenberg. 1987. Impairment of egg production in *Acartia tonsa* exposed to tributyltin oxide. *Ophelia* 27:137–141.

Jones-Lepp, T. L., K. E. Varner, and D. Heggem. 2004. Monitoring dibutyltin and triphenyltin in fresh waters and fish in the United States using micro-lipid chromatography-electrospray/ion trap mass spectrometry. *Arch. Environ. Contam. Toxicol.* 46:90–95.

Kajiwara, N., T. Kunisue, S. Kamikawa, Y. Ochi, S. Yano, and S. Tanabe. 2006. Organohalogen and organotin compounds in killer whales mass-stranded in the Shiretoko Peninsula, Hokkaido, Japan. *Mar. Pollut. Bull.* 52:1066–1076.

Kimbrough, R. D. 1976. Toxicity and health effects of selected organotin compounds: a review. *Environ. Health Persp.* 14:51–56.

King, N., M. Miller, and S. de Mora. 1989. Tributyl tin levels for seawater, sediment, and selected species in coastal Northland and Auckland, New Zealand. *N. Z. J. Mar. Fresh.* 23:287–294.

Kono, K., T. Minami, H. Yamada, H. Tanaka, and J. Koyama. 2008. Bioaccumulation of tributyltin and triphenyltin compounds through the food web in deep offshore water. *Coast. Mar. Sci.* 32:102–107.

Krone, C. A., J. E. Stein, and U. Varanasi. 1996. Butyltin contamination of sediments and benthic fish from the East, Gulf and Pacific coasts of the United States. *Environ. Monit. Assess.* 40:75–89.

Lassiter, R. R., and T. G. Hallam. 1990. Survival of the fattest: implications for acute effects of lipophilic chemicals on aquatic populations. *Environ. Toxicol. Chem.* 9:585–595

Laughlin, R. B. Jr., R. B. Johannesen, W. French, H. Guard, and F. E. Brinckman. 1985. Structure-activity relationships for organotin compounds. *Environ. Toxicol. Chem.* 4:343–351.

Lee, R. F. 1985. Metabolism of tributyltin oxide by crabs, oysters, and fish. *Mar. Environ. Res.* 17:145–148.

Livingstone, D. R. 1998. The fate of organic xenobiotics in aquatic ecosystems: quantitative and qualitative differences in biotransformation by invertebrates and fish. *Comp. Biochem. Physiol.* 120A:43–49.

Lytle, T. F., et al. 2003. Life-cycle toxicity of dibutyltin to the sheepshead minnow (*Cyprinodon variegatus*) and implications of the ubiquitous tributyltin impurity in test material. *Appl. Organomet. Chem.* 17:653–661.

Maguire, R. J. 2000. Review of the persistence, bioaccumulation and toxicity of tributyltin in aquatic environments in relation to Canada's toxic substances management policy. *Water Qual. Res. J. Can.* 35:633–679.

Maguire, R. J., P. T. S. Wong, and J. S. Rhamey. 1984. Accumulation and metabolism of tri-*n*-butyltin cation by green algae, *Ankistrodesmus falcatus*. *Can. J. Fish. Aquat. Sci.* 41:537–540.

Matthiessen, P., and P. E. Gibbs. 1998. Critical appraisal of the evidence for tributyltin-mediated endocrine disruption in mollusks. *Environ. Toxicol. Chem.* 17:37–43.

McAllister, B. G., and D. E. Kime. 2003. Early life exposure to environmental levels of the aromatase inhibitor tributyltin causes masculinisation and irreversible sperm damage in zebrafish (*Danio rerio*). *Aquat. Toxicol.* 65:309–316.

McCarty, L. S. 1986. The relationship between aquatic toxicity QSARs and bioconcentration for some organic chemicals. *Environ. Toxicol. Chem.* 5:1071–1080.

Meador, J. P. 1986. An analysis of photobehavior in *Daphnia magna* exposed to tributyltin. *Oceans 86 Conference, Organotin Symposium*, Vol. 4:1213–1218. Washington DC, Marine Technological Society.

Meador, J. P. 1993. The effect of laboratory holding on the toxicity response of marine infaunal amphipods to cadmium and tributyltin. *J. Exp. Mar. Biol. Ecol.* 174:227–242.

Meador, J. P. 1997. Comparative toxicokinetics of tributyltin in five marine species and its utility in predicting bioaccumulation and toxicity. *Aquat. Toxicol.* 37:307–326.

Meador, J. P. 2000. Predicting the fate and effects of tributyltin in marine systems. *Rev. Environ. Contam. Toxicol.* 166:1–48.

Meador, J. P. 2006. Rationale and procedures for using the tissue-residue approach for toxicity assessment and determination of tissue, water, and sediment quality guidelines for aquatic organisms. *Hum. Ecol. Risk Assess.* 12:1018–1073.

Meador, J. P., T. K. Collier, and J. E. Stein. 2002. Determination of a tissue and sediment threshold for tributyltin to protect prey species for juvenile salmonids listed by the U.S. Endangered Species Act. *Aquat. Conserv.: Mar. Freshwat. Ecosyst.* 12:539–551.

Meador, J. P., L. S. McCarty, B. I. Escher, and W. J. Adams. 2008. The tissue-residue approach for toxicity assessment: concepts, issues, application, and recommendations. *J. Environ. Monitor.* 10:1486–1498.

Meador, J. P., F. C. Sommers, K. Cooper, G. Yanagida. 2011. Tributyltin and the obesogen metabolic syndrome in a salmonid. *Env. Res.* In press.

Mercier, A., E. Pelletier, and J-F. Hamel. 1994. Metabolism and subtle toxic effects of butyltin compounds in starfish. *Aquat. Toxicol.* 28:259–273.

Morcillo, Y., and C. Porte. 1999. Evidence of endocrine disruption in the imposex-affected gastropod *Bolinus brandaris*. *Environ. Res.* 81:349–354.

Nagase, H., T. Hamasaki, T. Sato, H. Kito, Y. Yoshioka, and Y. Ose. 1991. Structure-activity relationships for organotin compounds on the red killifish *Oryzias latipes*. *Appl. Organomet. Chem.* 5:91–97.

Nakayama, K., Y. Oshima, K. Nagafuchi, T. Hano, Y. Shimasaki, and T. Honjo. 2005. Early-life stage toxicity in offspring from exposed parent Medaka, *Oryzias latipes*, to mixtures of tributyltin and polychlorinated biphenyls. *Environ. Toxicol. Chem.* 24:591–596.

Nakayama, K., Y. Oshima, and T. Yamaguchi. 2004. Fertilization success and sexual behavior in male Medaka, *Oryzias latipes*, exposed to tributyltin. *Chemosphere* 55:1331–1337.

Nishikawa, J-I., et al. 2004. Involvement of the retinoid X receptor in the development of imposex caused by organotins in gastropods. *Environ. Sci. Technol.* 38:6271–6276.

Oehlmann, J., P. Fioroni, E. Stroben, and B. Markert. 1996. Tributyltin (TBT) effects on *Ocinebrina aciculata* (Gastropoda: Muricidae): imposex development, sterilization, sex change and population decline. *Sci. Total Environ.* 188:205–223.

O'Halloran, K., J. T. Ahokas, and P. F. A. Wright. 1998. Response of fish immune cells to in vitro organotin exposures. *Aquat. Toxicol.* 40:141–156.

Ohji, M., T. Arai, and N. Miyazaki. 2006. Transfer of tributyltin from parental female offspring in the vivipa-rous surfperch (*Ditrema temmincki*). *Mar. Ecol. Prog. Ser.* 307:307–310.

Rehage, J. S., S. G. Lynn, J. I. Hammond, B. D. Palmer, and A. Sih. 2002. Effects of larval exposure to triph-enyltin on the survival, growth, and behavior of larval and juvenile *Ambystoma barbouri* salamanders. *Environ. Toxicol. Chem.* 21:807–815.

Rüdel, H., P. Lepper, and J. Steinhanses. 2003. Retrospective monitoring of organotin compounds in marine biota from 1985 to 1999: results from the German Environmental Specimen Bank. *Environ. Sci. Technol.* 37:1731–1738.

Saarikoski, J., and M. Viluksela. 1982. Relation between physiochemical properties of phenols and their toxic-ity and accumulation in fish. *Ecotox. Environ. Safe.* 6:501–512.

Salazar, M. H., and S. M. Salazar. 1995. In-situ bioassays using transplanted mussels: I. Estimating chemi-cal exposure and bioeffects with bioaccumulation and growth. In *Environmental Toxicology and Risk Assessment, Third Volume*, eds. G. R. Biddinger, E. Mones, and J. S. Hughes, pp. 216–241. Philadelphia, PA: American Society for Testing and Materials.

Salazar, M. H., and S. M. Salazar. 1998. Using caged bivalves as part of an exposure-dose-response triad to support an integrated risk assessment strategy. In *Proceedings–Ecological Risk Assessment: A meeting of Policy and Science*, eds. A. de Peyster and K. Day, pp. 167–192. Pensacola, FL: SETAC Press.

Schulte-Oehlmann, U., J. Oehlmann, B. Bauer, P. Fioroni, and U-S. Leffler. 1998. Toxico-kinetic and -dynamic aspects of TBT-induced imposex in *Hydrobia ulvae* compared with intersex in *Littorina littorea* (Gastropoda, Prosobranchia). *Hydrobiologia* 378:215–225.

Seligman, P. F., A. O. Valkirs, P. M. Stang, and R. F. Lee. 1988. Evidence for rapid degradation of tributyltin in a marina. *Mar. Pollut. Bull.* 19:531–534.

Seligman, P. F., et al. 1989. Distribution and fate of tributyltin in the United States marine environment. *Appl. Organomet. Chem.* 3:31–47.

Shim, W. J., et al. 2005. Accumulation of butyl- and phenyltin compounds in starfish and bivalves from the coastal environment of Korea. *Environ. Pollut.* 133:489–499.

Shimasaki, Y., T. Kitano, Y. Oshima, S. Inoue, N. Imada, and T. Honjo. 2003. Tributyltin causes masculinization in fish. *Environ. Toxicol. Chem.* 22:141–144.

Shimasaki, Y., et al. 2006. Effect of tributyltin on reproduction in Japanese whiting, *Sillago japonica*. *Mar. Environ. Res.* 62:S245–S248.

Stäb, J. A., T. P. Traas, G. Stroomberg, J. van Kesteren, P. Leonards, and B. van Hattum. 1996. Determination of organotin compounds in the foodweb of a shallow freshwater lake in the Netherlands. *Arch. Environ. Contam. Toxicol.* 31:319–328.

Stoner, H. B. 1966. Toxicity of triphenyltin. *Br. J. Ind. Med.* 23:222–229.

Takahashi, S., S. Tanabe, I. Takeuchi, and N. Miyazaki. 1999. Distribution and specific bioaccumulation of butyltin compounds in a marine ecosystem. *Arch. Environ. Contam. Toxicol.* 37:50–61.

Tanabe, S. 1999. Butyltin contamination in marine mammals—a review. *Mar. Pollut. Bull.* 39:62–72.

Tang, C-H., and W-H. Wang. 2008. Butyltin accumulation in two marine bivalves along a pollution gradient. *Environ. Toxicol. Chem.* 27:2179–2185.

Tas, J. W. 1993. Fate and effects of triorganotins in the aqueous environment: bioconcentration kinetics, lethal body burdens, sorption and physico-chemical properties. PhD Dissertation, University of Utrecht, The Netherlands, 205 pp.

Tessier, E., et al. 2007. (Tri)butyltin biotic degradation rates and pathways in different compartments of a fresh-water model ecosystem. *Sci. Total Environ.* 388:214–233.

Tolosa, I., L. Merlini, N. De Bertrand, J. M. Bayona, and J. Albaiges. 1992. Occurrence and fate of tributyl- and tri-phenyltin compounds in western Mediterranean coastal enclosures. *Environ. Toxicol. Chem.* 11:145–155.

Triebskorn, R., H-R. Kohler, J. Flemming, T. Braunbeck, R-D. Negele, and H. Rahmann. 1994. Evaluation of bis(tri-n-butyltin)oxide (TBTO) neurotoxicity in rainbow trout (*Oncorhynchus mykiss*). 1. Behaviour, weight increase, and tin content. *Aquat. Toxicol.* 30:189–197.

Tsuda, T., S. Aoki, M. Kojima, and T. Fujita. 1992. Accumulation and excretion of tri-n-butyltin chloride and triphenyltin chloride by willow shiner. *Comp. Biochem. Physiol.* 101C:67–70.

Tsuda, T., S. Aoki, M. Kojima, and H. Harada. 1990. The influence of pH on the accumulation of tri-n-butyltin chloride and triphenyltin chloride in carp. *Comp. Biochem. Physiol.* 95C:151–153.

Tsuda, T., S. Aoki, M. Kojima, and H. Harada. 1991. Accumulation of tri-n-butyltin chloride and triphe-nyltin chloride by oral and via gill intake of goldfish (*Carassius auratus*). *Comp. Biochem. Physiol.* 99C:69–72.

Tsuda, T., M. Wada, S. Aoki, and Y. Matsui. 1988. Bioconcentration, excretion and metabolism of bis(tri-*n*-butyltin) oxide and triphenyltin chloride by goldfish. *Toxicol. Environ. Chem.* 18:11–20.

U.S. Congress. 1988. Organotin antifouling paint control act of 1988. Public Law 100-333 (33 USC 2401). 100th Congress of the United States of America, Second Session. Washington, DC.

U.S. EPA. 1997. Toxicological Review. Tributyltin. Integrated Risk Information System (IRIS), Washington, DC.

U.S. EPA. 2003. Ambient Aquatic Life Water Quality Criteria for Tributyltin (TBT)—Final. EPA 822-R-03-031. Environmental Protection Agency, Office of Water, Washington DC.

Veltman, K., M. A. J. Huijbregts, M. J. van den Heuvel-Greve, A. D. Vethaak, and A. J. Hendriks. 2006. Organotin accumulation in an estuarine food chain: comparing field measurements with model estimations. *Mar. Environ. Res.* 61:511–530.

Vighi, M., and D. Calamari. 1985. QSARs for organotin compounds on *Daphnia magna*. *Chemosphere* 14:1925–1932.

Walsh, T. J., and D. L. DeHaven. 1988. Neurotoxicity of the alkyltins. In *Metal Neurotoxicity*, eds. S. C. Bondy, and K. N. Prasad, pp. 87–107. Boca Raton, FL: CRC Press.

Widdows, J., and D. S. Page. 1993. Effects of tributyltin and dibutyltin on the physiological energetics of the mussel *Mytilus edulis*. *Mar. Environ. Res.* 35:233–249.

Wong, P. T. S., Y. K. Chau, O. Kramar, and G. A. Bengert. 1982. Structure-activity relationship of tin compounds on algae. *Can. J. Fish. Aquat. Sci.* 39:483–488.

Yang, J., and N. Miyazaki. 2006. Transplacental transfer of butyltins to fetus of Dall's porpoises (*Phocoenoides dalli*). *Chemosphere* 63:716–721.

Zhang, Z., J. Hu, H. Zhen, X. Wu, and C. Huang. 2008. Reproductive inhibition and transgenerational toxicity of triphenyltin on medaka (*Oryzias latipes*) at environmentally relevant levels. *Environ. Sci. Technol.* 42:8133–8139.

Zou, Z., J. Cai, X. Wang, B. Li, C. Wang, and Y. Chen. 2009. Acute administrations of tributyltin and trimethyltin modulated glutamate and N-methyl-D-aspartate receptor signaling pathway in *Sebastiscus marmoratus*. *Aquat. Toxicol.* 92:44–49.

Great Pipe-Fish and Short-Snouted Sea-Horse

By G. Mutzel, from *The Royal Natural History*, edited by Richard Lydekker, Frederick Warne & Co., London, 1893–94.

8 Active Pharmaceutical Ingredients and Aquatic Organisms

Christian G. Daughton
Bryan W. Brooks

CONTENTS

8.1 Introduction 288
8.2 Exposure 289
    8.2.1 Background..289
    8.2.2 Sources/Origins Leading to Exposure..295
    8.2.3 Exposure Variables 297
    8.2.4 Some General Perspectives and Background Regarding Aquatic Tissue Levels
        of APIs...297
    8.2.5 Predictive Modeling..300
    8.2.6 Overview of Fish Tissue and Other Residue Data for APIs 301
        8.2.6.1 SSRIs/SNRIs (Selective Serotonin and Serotonin-Norepinephrine
            Reuptake Inhibitors)...302
        8.2.6.2 NSAIDs (Nonsteroidal Anti-inflammatory Drugs) 304
        8.2.6.3 Lipid Regulators...305
        8.2.6.4 β-Blockers...305
        8.2.6.5 Fungicides 306
        8.2.6.6 Macrocyclic Lactones 306
        8.2.6.7 Steroids...306
        8.2.6.8 Antibiotics: Informing Environmental Exposure with Data from
            Use of Veterinary Aquaculture Drugs 310
        8.2.6.9 Carbamazepine (CBZ)..311
        8.2.6.10 Triclosan (and Methyl Triclosan) and Triclocarban.............................311
        8.2.6.11 Miscellaneous APIs 313
        8.2.6.12 API Disinfection By-Products (DBPs) and Metabolites.......................314
    8.2.7 Uptake by Aquatic Plants and Aerial Invertebrates 314
    8.2.8 Multianalyte Studies...315
    8.2.9 Summary of Published Data..317
8.3 Factors Influencing Exposure 324
    8.3.1 General Considerations..324

        8.3.2   Select Site-Specific Factors Influencing Exposure 325
                8.3.2.1   Hydrology 325
                8.3.2.2   Wastewater Treatment Technologies... 326
        8.3.3   Site-Specific pH and API pKa 327
        8.3.4   Advancement in Tissue Sampling and Surrogate Monitoring.................... 328
8.4   Models for Predicting Exposure and Potential Effects of APIs 329
        8.4.1   Background and Prioritization... 329
        8.4.2   Physiological-Based Pharmaco(Toxico)Kinetic (PBPK) Models............................ 331
        8.4.3   The Huggett Model 331
Summary 333
Acknowledgments... 334
References.. 337

8.1   INTRODUCTION

The presence of active pharmaceuticals ingredients (APIs) in aquatic systems has led in recent years to a burgeoning literature examining environmental occurrence, fate, effects, risk assessment, and treatability of these compounds. The vast preponderance of studies aimed at identifying and quantifying contaminant residues in aquatic tissues have involved the conventional and legacy pollutants. Comparatively few studies have been targeted at APIs, primarily those that are lipophilic. Although APIs have received much attention as "emerging" contaminants of concern, it is important to recognize that traditional approaches to understand and predict exposure and effects of other environmental organic contaminant classes may or may not be appropriate for APIs. For example, traditional approaches for understanding aquatic effects may not be as useful for some APIs (Brooks et al. 2003), but lessons learned from the study of compounds active at the hypothalamic–pituitary–gonadal axis (endocrine disruptors/modulators) may reduce uncertainties associated with environmental assessments of other APIs (Ankley et al. 2007).

Whereas APIs are often considered as a combined class of environmental contaminants, APIs include diverse groups of chemicals with physiochemical properties ranging in pharmacological potencies, environmental fate profiles, and patient usage patterns. Due to the relatively rudimentary state of knowledge for aquatic exposures to these substances, an understanding of critical body residues (CBRs) necessary to elicit pharmacologically and ecologically relevant responses is not available at this time. Because exposure does not necessarily evoke effects or risk, current challenges include understanding the relationship between exposure and effects within an ecological risk assessment framework. It appears particularly critical to understand whether internal pharmacological doses of APIs in target tissues result from exposures at environmentally relevant or realistic concentrations. Such information can inform ecological risk assessments examining the potential effects of APIs based on their specific mechanism/mode of action (MOA).

Tissue residues of contaminants are commonly used in retrospective ecological risk assessments to support an understanding of environmental exposure (Suter et al. 2000). Bioconcentration factors (BCFs) are useful in both retrospective and prospective assessments of traditional contaminants; BCFs are expressed as the ratios of concentrations in tissues (mass/kg) and the respective concentrations in the surrounding aqueous compartment (mass/L), resulting in units of L/kg. APIs are conceptually no different from these conventional contaminants because tissue residues can provide important information in exposure analysis, particularly when used as indicators of exposure in the field. Whole-body and tissue concentrations are essentially proxies for gauging the actual dose at the site(s) of action, which may or may not be known. When bioconcentration occurs, an obvious advantage of measuring internal concentrations is when the external concentrations are below method detection limits (MDLs).

Compared with many conventional pollutants, APIs are in general comparatively more polar. They therefore tend to not partition to particulates and sediments, but rather remain dissolved in the aqueous phase. For those APIs that have significant tendency to become sediment-bound, little attention has been given to their bioavailability as measured by bioaccumulation. The measure of bioaccumulation of a chemical associated with sediment is the biota-sediment accumulation factor (BSAF). The BSAF is expressed as a chemical's tissue concentration normalized to lipid content relative to its concentration normalized to sediment total organic carbon (Burkhard 2009).

In this chapter we examine relevant information on residues in aquatic organisms, select factors influencing exposure, and available methods to understand relationships between exposure scenarios and effects thresholds. Though APIs are often combined with discussions of personal care products (PCPs) in the literature that has rapidly developed following publication of Daughton and Ternes (1999), we specifically focus on APIs for the purposes of this chapter. A broad literature on PCPs in aquatic organisms continues to develop (e.g., Mottaleb et al. 2009) and was recently summarized by Ramirez (2007).

An insight of particular importance concerns the level of knowledge regarding the linkage between exposure to APIs and adverse effects in aquatic organisms. Regardless of the available published data on exposure of aquatic organisms to APIs (and its size is indeed very limited, especially for real-world scenarios), it rarely intersects a complementary body of data for biological effects. When exposure data do exist, the same conditions, concentrations, and species have rarely been used in toxicity studies; parallel data sets are also usually disconnected temporally. Likewise, a considerable body of API data exists for effects, but exposure conditions (especially the API concentration or route of exposure) are often not environmentally realistic. The ability to routinely connect real-world API occurrence data to documented biological effects is therefore not yet available. Evidence for causality is probably strongest for sex steroids, largely because of their ubiquity and potencies; but even here, evidence can be confusing (e.g., see Vögeli 2008). A major challenge in establishing causality—a lack of correlation between tissue levels and observed effects—could well be the result of delayed effects, such as those only manifesting at later life stages but originating from exposure at earlier developmental stages; this is especially true for organisms with longer life cycles.

8.2 EXPOSURE

8.2.1 BACKGROUND

The published literature on APIs as environmental contaminants is dominated with data on the analysis, occurrence, and fate of these chemicals in the environment, together with evaluation of waste and water treatment technologies. Surprisingly, comparatively little has been published regarding the aquatic toxicology of APIs, especially data relevant to exposure. Little information is available, for example, on the occurrence of APIs in aquatic organisms. This in itself is surprising given that predictive models for bioconcentration in fish are not yet up to the task, and empirical data are needed at least to validate computational approaches. The complexities and limitations of modeling bioconcentration of conventional pollutants (especially the legacy pollutants) in fish are discussed in detail by a number of authors (e.g., Geyer et al. 2000, Gobas and Morrison 2000, Van der Oost et al. 2003, Nichols et al. 2007).

The bulk of the studies on drug residues in aquatic tissues relate to what is known from aquaculture, where exposure is restricted to a very limited number of drugs (almost all being veterinary drugs) and at concentrations orders of magnitude higher than might occur in the ambient environment. Of the thousands of published studies that have been compiled regarding the many aspects of APIs as environmental pollutants (U.S. EPA 2009), roughly only 50 or so are directly relevant to aquatic exposure and tissue levels of APIs, and the majority of these studies have been published since Brooks et al. (2005) reported fish tissue residues of SSRIs (selective serotonin reuptake inhibitors) from an effluent-dominated stream.

APIs have long been assumed to show little propensity to bioconcentrate, no less biomagnify. This has been based largely on their greater water solubility compared with conventional pollutants such as pesticides and many industrial pollutants, especially the persistent organic pollutants (POPs). But APIs are known to sometimes undergo active transport, so this assumption may not be valid. Little is known for assisting the assessment of the bioaccumulation potential in fish (Cowan-Ellsberry et al. 2008); even less is known regarding the bioconcentration of APIs by fish or other aquatic organisms. Bioaccumulation is deemed possible when the BCF (expressed as L/kg) exceeds a range of 500–5000, depending on the standard being applied (Cowan-Ellsberry et al. 2008).

Even when definitive data have been obtained regarding bioconcentration, whether this can be extrapolated among species is unknown. Owen et al. (2007) emphasize the diversity in biology among the 28,000 species of fish. Tissue levels are governed largely by the pharmacokinetics of the API. These authors also point out that almost nothing is known regarding the pharmacokinetics of APIs in aquatic organisms—the xenoestrogen 17α-ethinyl estradiol (EE2) being one exception. Also, while it might be useful, extrapolations between mammals and fish can be challenging and potentially misleading because of key differences in physiology. Owen et al. (2007) stress that a primary route of uptake in fish is via the gills, from where blood is delivered to various organs before reaching the liver, in contrast to oral exposure in mammals, which leads to first-pass metabolism in the liver. Yet a further complication is the wide disparity that can exist among reported K_{ow} data for a particular API; the K_{ow} is the octanol–water partition coefficient—the ratio of a chemical's equilibrium concentration in octanol versus water at a defined temperature. These values can range over several orders of magnitude, pointing to the need for empirical data with which to validate and develop better predictive models. As discussed later, site-specific pH can influence the ionization state of many APIs, reducing the utility of using K_{ow} to predict exposure in retrospective evaluations; methodologies for estimating the BCFs for organic electrolytes have been assessed by Fu et al. (2009).

Exposure is a term commonly used by toxicologists, modelers, and others involved with environmental science. Exposure is a key component of the risk assessment paradigm that can provide insights for ways to reduce biological effects as well as better understand or predict their potential for occurring. Exposure translates the potential of hazard into the reality of risk (see Figure 8.1). But defining what is actually meant by exposure poses significant challenges. It does not necessarily represent a discrete physical or temporal point in the complex series of events that determine the outcomes from physical contact of an organism with a chemical or other stressor. Rather, the processes involved with exposure are spread across a complex spatiotemporal continuum that links a stressor's source or origin with the eventual effects that might occur within biological systems.

Although exposure includes understanding the duration, frequency, and magnitude to which organisms interact with biologically available contaminants, exposure magnitude is available for a limited number of APIs, and exposure duration and frequency is largely unknown for all APIs. Exposure is usually shown in conceptual diagrams as a standalone part of the hazard-risk continuum, an example being shown in Figure 8.1. External factors may influence the bioavailability, absorption, and uptake of an API, while physiochemical properties of an API influences pharmacological bioavailability, and internal factors (e.g., metabolism) will influence the duration of internal dosimetry. Chemical exposure is often visualized as the physicochemical interaction of a biological receptor with the chemical stressor, as when a ligand binds with a receptor. In practice, however, separating exposure from effects can be arbitrary and at times confusing. This is especially true when discussing biomarkers of exposure and biomarkers of effects.

A biomarker can be defined as pathway- or receptor-specific observations that are chemical-induced responses at the biochemical, physiological, or morphological level of an organism (Committee on Biological Markers of the National Research Council 1987). In ecotoxicology, biomarkers represent critical measures to support an understanding of exposure and potential effects to environmental contaminants. Under prospective and retrospective ecological risk assessment frameworks, it is useful to classify the various sublethal responses organisms may exhibit following

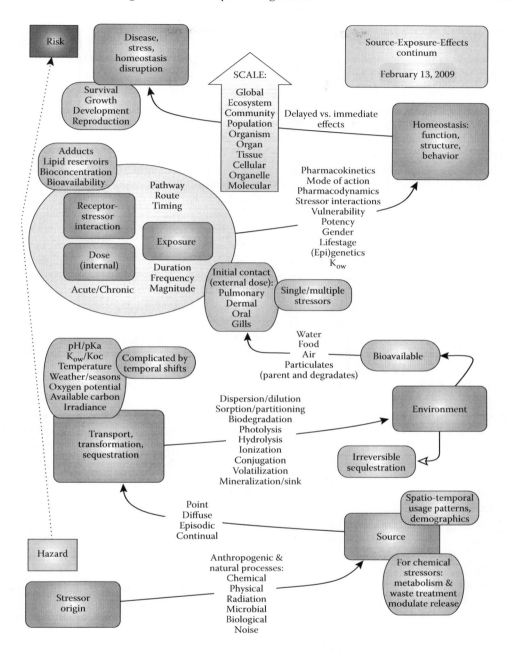

FIGURE 8.1 Source-to-effects continuum generalized for chemical stressors (but all principles also apply to APIs). The rounded rectangular boxes represent major points or processes in time or space along the source-to-effects continuum. The ovalized rectangular boxes and arrow labels represent variables that affect the processes along the continuum. The SCALE arrow represents the level of biological organization that the stressor impacts.

exposure to contaminants as either biomarkers of exposure or biomarkers of effect (Huggett et al. 1992). Biomarkers of exposure differ from biomarkers of effect in that these measures inform whether an organism has been exposed to contaminants (e.g., gene expression) but do not necessarily allow for determination of whether the organism has been adversely impaired. Biomarkers of effect (e.g., egg shell thinning, histopathology), however, are indicators of ecotoxicity, especially when measures of an adverse effect can be linked to a physiologically and ecologically relevant endpoint (e.g., growth) with population-level relevance (Ankley et al. 2007, Brain et al. 2008a). Biomarkers of exposure and effect have been successfully employed in a variety of retrospective ecological risk assessments (Suter et al. 2000), and their use in prospective environmental risk assessment is projected to increase due to a number of factors. Environmental contaminants such as APIs that may exert toxicity through receptor/enzyme interactions are providing an impetus toward "intelligent" ecotoxicity testing, which requires development and interpretation of biomarkers.

Regardless of the difficulty in providing a rigorous definition of exposure, the general concept of exposure is essential for organizing discussions about assessing risk. In this chapter, we use exposure as the framework to better understand the ramifications of APIs in the aquatic environment. After all, in the absence of exposure, hazard cannot translate into risk. Likewise, exposure does not necessarily pose ecologically relevant risk levels—for example, if perturbation of homeostasis does not result.

The prime aspect of exposure that creates the potential for a cascade of events within biological systems is the initial physical association or interaction of chemical stressors with an organism. In the aquatic environment, a prime sign of exposure is the simple occurrence of a stressor within one or more tissues, such as blood, lipid, muscle, bile, liver, ovaries, eggs, or brain. The factors driving the partitioning and distribution of APIs within fish are incompletely understood. For example, fish have a variety of chromatophores, some of which are pigment-containing, such as melanocytes (New World Encyclopedia contributors August 13, 2008). During development, pigmented melanocytes begin appearing within a day of fertilization. The potential significance of these cells to exposure is related to their high binding affinities for a broad spectrum of xenobiotics, including a wide array of APIs known for high affinity to melanin (Testorf et al. 2001, Aubry 2002, Roffey et al. 2007). Since humans only have melanocyte chromatophores, melanin has been most studied. Binding to melanin is known to lead to drug accumulation, one of the routes resulting in the concentration of drugs in hair (Larsson 1993). In fish, binding to melanin (and possibly to other pigments) could lead to API reservoirs within certain tissues.

APIs introduced to the aquatic environment via sewage pose additional challenges in assessing exposure. Because of extreme spatiotemporal fluctuations in concentrations, how accurately can exposure be assessed on the basis of monitoring residues in water? This is especially true for APIs that are released episodically, whether event-driven (such as by the sporadic practice of disposal) or when discharged by diurnal patterns in flushing of sewers or when influenced by seasonality (which can affect waste treatment efficiency as well as the types and quantities of individual APIs that are being used). These factors can result in high but transient concentrations, strongly influencing exposure magnitude, duration, and frequency (Daughton 2007). Traditional sampling using discrete grab samples may not be representative of longer term exposure levels. Passive, integrative samplers can overcome some of these limitations (e.g., Vermeirssen et al. 2008), though the applicability has not been demonstrated for a broad range of APIs, and samplers do not characterize all routes of exposure (e.g., dietary).

A body of published work points to a spectrum of *potential* toxicological consequences following controlled exposure of certain aquatic organisms to a limited number of APIs, but these studies ignore whether real-world exposures actually occur and whether they occur at the requisite realistic concentrations (e.g., Gaworecki and Klaine 2008). Often, effects using traditional endpoints (e.g., survival, growth, and reproduction) in traditional test species (e.g., *Ceriodaphnia dubia*) are found only at levels orders of magnitude higher than known in the environment (e.g., Henry and Black 2008). Such observations are generally supported by several recent papers that reviewed information on aquatic toxicity of APIs (Crane et al. 2006, Fent et al. 2006, Farré et al. 2008). Rather than repeating such efforts here, it is instead important to keep in mind that the effects level is a function of the sensitivity of the targeted endpoint in the test species selected for study; other endpoints—especially those still not recognized—may have lower effective concentrations (below current NOAELs—no observable adverse effect levels), particularly from standardized ecotoxicity assays.

Determining whether a more sensitive endpoint represents an ecologically relevant measure of effect amenable to inclusion in effects analysis of an ecological risk assessment must be considered for APIs (Ankley et al. 2007). For example, the ecologically relevant fathead minnow reproduction and feeding behavior endpoints are more sensitive to EE2 (Ankley et al. 2001) and fluoxetine (Stanley et al. 2007), respectively, than 7-day juvenile fathead minnow growth, which is routinely employed prospectively to assess sublethal effects of individual chemicals, ambient toxicity in surface waters, and whole-effluent toxicity. Other endpoints, which may represent ecologically relevant biomarkers of effect, should be mechanistically linked to ecologically relevant endpoints (Ankley et al. 2007) that can be integrated in ecological risk assessments (Brooks et al. 2009b). One recent example of an ecologically relevant and more sensitive endpoint than traditional morphometric responses (e.g., growth) was presented by Brain et al. (2008b). Disruption of the folate biosynthetic pathway in aquatic plants was demonstrated following sublethal exposure to sulfamethoxazole at concentrations more than an order of magnitude lower than for growth impairment (Brain et al. 2008b).

Why should we care about exposure? Better understanding of exposure provides the opportunity to look both backward and forward in the hazard-risk continuum—ranging from sources to effects. Measurement of exposure can be used in a retrospective ecological risk assessment, for example, to reveal sources of contamination, measure site-remediation success, as a surrogate measure of contaminant concentrations, or to reconstruct dose, or in a prospective ecological risk assessment to estimate the possibility of disease and its time of onset. Some of the major reasons for understanding and measuring exposure in the aquatic environment are captured in Table 8.1.

TABLE 8.1
The Importance of Understanding or Measuring Aquatic Exposure

Aquatic Exposure Data Can Be Used For	Rationale
Advanced warning of effects	Measure potential for adverse population-level effects or individual effects (in the case of threatened and endangered species), should the level of exposure be sustained or rise, or should the onset of effects be delayed.
Prognosis or vulnerability	Predict the likelihood of, or vulnerability for, future disease onset should exposure occur.
Reveal potential for subtle effects	Provide insights on the potential for cumulative subtle effects that might go unnoticed until a level of irreversible harm has been reached.
Prioritization of APIs	Exposure studies can inform, direct, and guide the selection of APIs for more in-depth toxicological study; likewise, occurrence studies (such as from water monitoring), can be used to guide exposure studies.
Reveal inadequacies/flaws in models	Bioconcentrated/bioaccumulated body burdens often do not comport with predictive models (e.g., Maunder et al. 2007).
Reveal potential for food chain disturbances	When bioconcentration and bioaccumulation resist predictive modeling, analysis of tissue levels are therefore required.
Sentinel for terrestrial organism exposure	Being more susceptible to exposure than terrestrial organisms and for development of subtle effects, aquatic exposure can serve as a sentinel model for terrestrial exposures (such as might occur from the recycling of treated sewage).
Potential for human exposure	Can provide information on levels of potential stressors that humans might consume via the food chain (e.g., contaminated fish in the diet).
Corroborate representativeness or need for toxicity testing	Establish whether concentrations/doses used in toxicity testing actually occur in the wild.

continued

TABLE 8.1 (continued)
The Importance of Understanding or Measuring Aquatic Exposure

Aquatic Exposure Data Can Be Used For	Rationale
Stressor sources/origins/distributions	Tissue concentrations can help to reveal the locations, distributions, frequencies, or durations of stressor presence in the ambient environment.
Pollution prevention/source control	Guide actions to prevent or control the source or origin of the stressors, thereby reducing the potential for exposure.
Gauge success/progress of cleanup or remediation efforts, or of new waste treatment technologies	Tissue levels of chemical stressors used to measure ambient concentrations after cleanup actions or remediation, or after implementing new control measures at sewage treatment facilities.
Surrogate measure of pollutant concentrations	Tissue levels of chemical stressors used as surrogate measures of ambient concentrations of APIs in water.
Inadequacy in water monitoring for predicting exposure	Extreme spatiotemporal fluctuations in water concentrations of APIs obscure or complicate evaluation of actual exposure. This is especially true for APIs that are: (i) released episodically (event-driven, such as by the sporadic practice of disposal or events causing large changes in usage rates or in the types of medication being used), (ii) discharged by diurnal patterns in flushing, (iii) influenced by seasonality. Tissue levels may serve to integrate exposure over time.
Magnify levels of stressors not ordinarily toxic	Uptake and bioconcentration of chemical stressors at concentrations below no-effect levels in water can eventually achieve body-burden levels that exceed effects levels.
Magnify levels of stressors not ordinarily detectable	When stressors can be bioconcentrated, exposed organisms serve as sentinels for sources of pollution or ambient levels that might otherwise be too small to directly detect.
Exposure times extended in absence of stressor in ambient environment	Biomagnified residues extend the time during which exposure can occur, even after the API is no longer present in the external environment.
Dose reconstruction	Levels of exposure can be used to reconstruct the original dose.
Correlate ambient, external concentrations with exposure	If actual exposure can be calibrated to external exposure levels, then simple chemical monitoring in water could possibly be used instead of more resource-intensive tissue monitoring for predicting exposure.
Deconvolute contributions to biological effects originating from natural (ambient) background	Natural incidence can play a role in certain biological conditions that also result from exposure to stressors. Natural background can be confused with effect. Understanding exposure helps to distinguish the two causes (e.g., Grim et al. 2007).
Detecting newly emerging stressors	Identifying previously unknown chemicals concentrated in tissues can reveal newly emerging pollutants before any linkage with effects might be suspected; this is particularly germane to new drug entities recently introduced to commerce.

Remarkably, given the voluminous literature on APIs in the aquatic environment, little data exist relevant to the occurrence of APIs as residues in the tissues of aquatic organisms. These limited data come primarily from controlled studies using APIs postulated or known to occur in the environment. These studies involve exposure experiments under controlled conditions (including fish caged in the wild) and especially from studies of veterinary drug residues in aquacultured fish. Another source of data exists in the form of calculated BCFs from predictive models. One of the ramifications of this is that except for the expected residues of certain veterinary APIs used in aquaculture, it is essentially unknown what levels of the numerous APIs occurring in the ambient environment might lead to human exposure via consumption of fish or other aquatic life. No human exposure studies have been done on API residues in wild fish. It is known, however, that tissue residues of certain APIs in aquacultured fish can resist degradation during cooking and can migrate from one tissue to another during cooking, as shown for the antibiotics oxolinic acid (OA) and flumequine (Steffenak et al. 1994).

8.2.2 Sources/Origins Leading to Exposure

Thousands of distinct chemicals are formulated into tens of thousands of commercial products used worldwide in the practice of human and veterinary medicine or for personal care. Those of most interest in the aquatic environment include the APIs that: (i) are most frequently used or used in the greatest quantities (e.g., NSAIDs: nonsteroidal anti-inflammatory drugs) or which are commonly disposed by flushing into sewers, (ii) are the most potent (e.g., synthetic hormones such as EE2) or that have clear potential for environmental effects (such as antibiotics, serotonin regulators, or oncolytics), (iii) are excreted largely unchanged, (iv) share like-modes or mechanisms of action and which could therefore act by concentration addition (e.g., SSRI antidepressants), and (v) can bioconcentrate.

The occurrence and disposition of APIs in the environment is best viewed in terms of their origins or sources, which comprise the locations from which they enter the environment and the pathways by which they enter (and are initially distributed within) the environment. These are many and varied, and even a simplified network flowchart can seem complex (see Figure 1 in Daughton 2008). The major sources that have the potential for impacting the aquatic environment are summarized in Table 8.2; an overview of the many sources and origins of APIs in the environment is

TABLE 8.2
Sources/Origins of APIs Significant to Aquatic Exposure

API Source/Origin	Description/Explanation of API Source
Human and veterinary medications	Human medications (over-the-counter; prescription only or legend drugs) include thousands of distinct APIs comprising numerous categories used in diagnostics, prophylaxis (including vaccines), cosmetics, lifestyle, and therapy. Veterinary medications (including those used in aquaculture) primarily comprise anabolic steroids, parasiticides, and antibiotics, though other medicines are also used for companion animals (e.g., fluoxetine).
Dual-use	Some APIs also have uses outside the practice of medicine. Some have dual use as pesticides (e.g., lindane, pyrethrins, avermectins, azole fungicides, and warfarin); another example is malachite green, a chemical used illegally in aquaculture but where aquatic exposure has been documented as a result of other uses (Schuetze et al. 2008). Some veterinary medicines have even been evaluated for nonaquaculture pesticidal use directly in the aquatic environment; for example, the sedative medetomidine has been proven effective as an antifouling agent against barnacles (Hilvarsson et al. 2007).
Chemical categories	Lower and higher MW synthetics; biologics (derived from organisms; e.g., antibodies, vaccines, and interleukins); natural products; nanomedicines; halogenated APIs are common (especially those containing fluorine).
DBPs and transformation products	Little is known regarding the halogenated disinfection by-products (DBPs) that might be created from any number of APIs (e.g., Flores and Hill 2008); many metabolites or environmental transformation products can be bioactive.
Manufacturing	Release in effluent from pretreated and untreated API manufacturing; perhaps a minor, localized source in the United States, but possibly of greater significance in less developed countries (e.g., Larsson et al. 2007; Fick et al. 2009).
Excretion	Pharmacokinetics for humans differ radically among APIs, resulting in wide spans in excretion efficiency of unmetabolized parent APIs or in excretion of biologically active metabolites; conjugates can act as reservoirs of parent APIs once hydrolyzed during sewage treatment or in the environment. Excretion apportioned between the feces and urine varies among APIs. Maternal-fetal transport of APIs in fish is not understood.

continued

TABLE 8.2 (continued)
Sources/Origins of APIs Significant to Aquatic Exposure

API Source/Origin	Description/Explanation of API Source
Bathing	Bathing and swimming release APIs directly to sewers or surface waters by two mechanisms, both involving dissolution of APIs: (i) from medications applied directly to the skin, and (ii) after excretion to the skin via sweat (Daughton and Ruhoy 2009).
Sewerage	Only the contributions from distributed sewer systems can directly impact the aquatic environment; septic and advanced onsite systems generally do not serve as direct sources, but can be important in extreme conditions (e.g., improperly installed leach fields, direct discharge from advanced onsite systems). Straightpiping of untreated sewage continues to be practiced in some locales. Wet-weather events can cause discharge of untreated sewage.
Treated sewage effluent	The removal efficiency of APIs from the influents to WWTPs varies greatly—from 0 to essentially 100%, as a function of the properties of the individual API and the type of treatment being used. Removal includes "destruction" (alteration of chemical structure) as well as physical sequestration, such as by filtration or sorption to sludge (and subsequent creation of "biosolids").
Biosolids, manure, pet excrement	One disposal option for biosolids (treated sewage sludges), which can contain sorbed or occluded APIs, is amendment of agricultural soils; wet-weather runoff holds the potential for transporting these APIs to surface water via runoff; manure from farm animals and excrement from medicated pets can also serve as a source (via wet-weather runoff).
Runoff	Primarily relevant to veterinary drugs from CAFOs (e.g., wet-weather flooding of retention ponds) or leaching of manure or feces from domestic animals and pets; can also involve human medications, which are experiencing growing usage with pets.
Wet-weather overflows and straightpiping	Flows that exceed the capacity of sewage treatment plants and rural areas where sewers still discharge directly to surface waters represent worst-case sources (e.g., combined sewer overflows), as the concentrations of APIs in these waste streams would not be reduced by treatment.
Locations where unused medications accumulate	Leftover drugs tend to accumulate (and eventually require disposal from) a wide expanse of locations, extending far beyond the domestic medicine cabinet (Ruhoy and Daughton 2008).
Disposal to sewerage	Unwanted medications are often disposed by flushing down sewer drains; this can involve leftover, unused medications and also the residuals contained in used containers or delivery devices (Ruhoy and Daughton 2008).
Disposal to domestic trash	Instead of disposal to sewerage, leftover medications are also disposed to domestic trash (this is currently the preferred method recommended by the U.S. federal government, when alternative collection options are not available) (U.S. FDA 2010; ONDCP 2009 [updated October]). The leachate collected from trash disposed in engineered landfills is sometimes returned to sewage treatment facilities, possibly serving only as a minor and indirect route to the aquatic environment.
Commercial use	Agriculture (antibiotics) and aquaculture (medicated feeds) serve as sources in localized environments; these sources can involve significant quantities of APIs, especially uneaten medicated feeds (Rigos et al. 2004).
Vaporization/ aerosolization	The vapor pressure (volatility) for the vast preponderance of APIs is insufficient to serve as a source (or route of loss) for the aquatic environment; vaporization might, however, serve as a source for certain PCPs, such as synthetic musks (Aschmann et al. 2001; Peck and Hornbuckle 2004, 2006) and cyclic methylsiloxanes (Xu and Chandra 1999); indeed, Galaxolide has been detected in the blubber from Antarctic fur seal (*Arctocephalus gazella*) from the South Shetland Islands (Corsolini 2009). APIs can also get entrained on particulates that can then become airborne (Hamscher et al. 2003; Cecinato and Balducci 2007).

Note: An overview of sources and origins of APIs in the environment is also provided by Daughton (2007, 2008).

provided by Daughton (2007, 2008). Sources and origins are important to understand not just for exposure. A comprehensive inventory of sources and their individual significance relative to each other with respect to contributions to environmental loadings can facilitate the prioritization of actions needed to reduce, minimize, or eliminate the potential for exposure via source reduction or pollution prevention. A complex array of processes acts upon APIs released into sewage to diminish their ultimate concentrations in the aquatic environment. One example situation using the fluoroquinolones is provided by Golet et al. (2002).

8.2.3 EXPOSURE VARIABLES

A vast and complex array of variables and their interactions (involving both the stressor and the target organism) dictate how and to what degree exposure occurs. These include the routes and processes by which exposure occurs. Few of these variables are unique to aquatic exposure involving APIs, and all can play roles, either in concert or in sequence. Not all of the variables, however, have been examined with respect to APIs. Many are relevant to controlled exposure studies and therefore only have hypothetical applicability to exposure in the wild. Table 8.3 summarizes some of the many aspects relevant to aquatic exposure and provides examples from the literature. Some of these factors are incorporated in the conceptualized diagram of the "4Ts": toxicant, totality, tolerance, and trajectory (Daughton 2005). It is also important to recognize that just because a chemical stressor might not be detectable in any tissue does not mean that exposure has not occurred. It could be that the analytical methodology cannot detect the stressor at a sufficiently low level (the MDL is too high). After all, some concentration levels known to have effects are extraordinarily low. For example, effects from EE2 have been documented at the sub-ppt level in surrounding water (i.e., 0.05 ng/L) (Larsen et al. 2008); diclofenac is purported to have pro-inflammatory effects at concentrations as low as 10^{-14} M (~3 pg/L) (Schirmer and Schirmer 2008). Moreover, the kinetics involved with the processes governing exposure might be sufficiently fast that uptake by an organism results in immediate formation of irreversible products of exposure, such as adducts, or in metabolic transformation to bioactive products. These products might merely be biomarkers of exposure or they might alternatively be biomarkers of effect. Regardless, in these situations monitoring for tissue residues of the parent API may well yield negative data, and therefore exposure can be overlooked or underestimated.

The internal dose is a function of the external concentration, chemical state of the API (e.g., ionization), uptake by and distribution within the organism, and bioavailability. Since the amount of stressor that actually interacts with receptors of toxicological consequence cannot be easily determined, surrogate measures are usually used to estimate the actual biologically effective dose. An added difficulty arises in translating "concentrations" associated with sediments and particulates to freely available portions. Studies of uptake from sediments are not common, one example being Higgins et al. (2009), who report on the uptake by an oligochaete of triclocarban.

8.2.4 SOME GENERAL PERSPECTIVES AND BACKGROUND REGARDING AQUATIC TISSUE LEVELS OF APIs

Surprisingly few studies have been published that examine the occurrence of APIs not just in fish but in any aquatic animal or plant. Moreover, most of these data have been obtained for veterinary APIs (primarily antibiotics and estrogenic/androgenic steroids), which are used in aquaculture at levels far exceeding those found in the ambient environment. Most aquatic biomonitoring studies designed to emulate exposure under ambient environmental conditions are performed under controlled conditions, often with exposure concentrations still exceeding those that would be found in the ambient environment. Controlled exposure studies often use exposure concentrations that are one or more orders of magnitude higher than those that exist in the ambient environment, usually to

TABLE 8.3
Variables Involved with Aquatic Exposure

Variable Affecting Exposure	Example
Exposure under controlled conditions	Baths (immersion in static, replacement, or flow-through systems, sometimes using whole effluent from sewage treatment or water collected from native environments), oral (bolus), injection (IP), feeding, caged in wild, whole-effluent toxicity testing.
Route of exposure (free in wild)	Gill (brachial) transport, dermal absorption, oral/gut, olfactory (e.g., transport of nanoparticles via olfactory neurons), lateral-line sensory organ (e.g., Chiu et al. 2008).
Type of organism	Teleost and cartilaginous fish, macroinvertebrates (invertebrates such as insects, crustaceans, mollusks, and worms), periphyton (e.g., Aufwuchs), plants, amphibians, and reptiles, waterfowl, and mammals (e.g., otters).
Environmental location	River, wetland, lake, reservoir, estuary, marine; benthic, pelagic, interfacial monomolecular monolayer; proximity to effluent from treated effluent discharge or raw sewage.
Stressor physical status	Dissolved in water column, particulate-bound (including incorporation into feed), sediment-bound; uptake from water can differ dramatically from dietary uptake.
Stressor properties	Water solubility, K_{ow}, pKa, log D, log K_{lipw}, MW, molecular size or cross section, molecular conformation and steric factors, environmental/metabolic half-life, chirality, vapor pressure.
Ambient environment	pH, temperature (or season), salinity, total dissolved solids, natural organic matter (Galvez et al. 2008), dissolved oxygen or hypoxia, solar irradiance (photolysis), nutrient levels and turbidity (Gordon et al. 2006), geographic locale, and dilution (e.g., effluent-dominated streams yield maximum concentrations). Season can affect performance of sewage treatment (e.g., via temperature) and irradiance (e.g., for photolysis or growth of autotrophs). Season and locale can also affect the types of APIs present in sewage because different medications are used in different seasons and in different locales.
Engineering controls	End-of-pipe treatment can reduce concentrations of some but not all APIs; it can also lead to the formation of biologically active products and API-based DBPs; efficiency is a function of the types of engineered treatments used by manufacturers and municipalities (e.g., Snyder et al. 2007). Types and efficiencies of treatment can vary considerably from country to country (e.g., see Larsson et al. 2007).
Fate	The interactions of the stressor with the ambient environment dictate its fate; key processes are microbial degradation, photolysis, and sequestration via sorption to sediments. But even short environmental half-lives do not preclude a continual presence of those stressors that are continually introduced, such as via treated sewage; this has been termed "pseudopersistence" (Daughton 2002).
Organism status	Food supply, nutrient levels, health, prior exposure history, growth state, lipid content, injuries, disease, age, gender, condition index/factor, body length, over-expression/inhibition of efflux pumps, species/strain. For example, nutrient concentrations and stoichiometry influence lower trophic-level responses to triclosan (Fulton et al. 2009).
Organism behavior and niche	Behavior involving feeding (free-feeding, sessile, filter-feeding), niche, trophic level, swimming/migratory behavior, and attraction/avoidance, and how these influence proximity to source (e.g., sewage outfalls) and aggregation; aquatic, marine, estuarine.
Timing of exposure	Windows of vulnerability, developmental life stage/reproductive status (e.g., embryo, egg, larval, hatchling, fry, juvenile, adult, spawning) (Van Aerle et al. 2002). Simultaneous vs. sequential exposure when multiple stressors are involved.
Magnitude of exposure	Aqueous concentration is a major determinant of actual internal exposure dose for aquatic organisms. For APIs, concentration is partly dictated by the degree of dilution in waters receiving treated or raw sewage effluent (with effluent-dominated streams and streams receiving raw sewage representing worst cases); other determinants include population served, sewage flow rate, and treatment technology. High stressor concentrations in controlled laboratory studies are often not relevant to real-world exposure. Newer exposure studies, however, are exploring lower and lower concentrations; need to ensure that studies are relevant to the real world (Hinton et al. 2005). BCFs are often higher at lower concentrations. Lack of exposure or effects at higher concentrations cannot be used to rule out the importance of lower concentrations because of multiphasic and nonmonotonic dose-response.

TABLE 8.3 (continued)
Variables Involved with Aquatic Exposure

Variable Affecting Exposure	Example
Duration/variability/ frequency of exposure	Constant, pulsed, discrete, episodic, acute/chronic, life cycle, and multigenerational.
Multiple stressors	Simultaneous versus sequential exposure to multiple stressors. Multistressor interaction effects (e.g., additive, potentiated, antagonistic, and synergistic) may follow concentration addition or independent action models; competition for—or facilitation of—transport across membranes.
Exposure history	Prior or simultaneous exposure to other stressors—both chemical and physical (e.g., temperature, salinity, stress from prey, etc.).
Pharmacokinetics	ADME: absorption, distribution, metabolism (e.g., phase I and II), excretion/elimination—all affecting half-life and disposition within organism (including depuration); reactivity within organism (e.g., formation of adducts vs. bioconcentration within lipid).
Pharmacological bioavailability	Residues that are bound versus those that are free (e.g., particulate-bound APIs may not transport across gills); sequestration and/or bioconcentration of body residues, e.g., within lipids, binding with melanins (Larsson 1993; Testorf et al. 2001; Aubry 2002; Roffey et al. 2007) and as other adducts; disposition in blood/plasma (primarily bound to proteins), muscle, bile, liver, brain, gonads, eggs, skin, bone, etc.; these factors also have direct relevance to subsequent human exposure to APIs via the food chain (function of edible tissue distribution).
Aquatic bioavailability	pH influence on ionization state and lipophilicity of APIs. Influence of sorption to suspended particulate matter. Concentration enrichment within sediments. For example, higher K_D values were observed for ciprofloxacin with fine particulate organic matter (FPOM) than coarse particulate organic matter, resulting in a higher magnitude of exposure to benthic macroinvertebrates consuming FPOM (e.g., collector-gatherers) (Belden et al. 2007).
Transformation products	Metabolites and environmental transformation products (e.g., degradates from photolysis or halogenation during disinfection) may prove more important that the parent API.
Resiliency/ vulnerability	Biochemical, physiological, and behavioral repertoire of an organism that determines its ability to avoid exposure or reduce its magnitude or duration; ability to maintain internal homeostasis and adapt to stressors (Clubbs and Brooks 2007).
Biological modulators of exposure	Various biological processes can enhance, facilitate, or reduce exposure. Those that are evolutionarily conserved play key roles. Among the most important considerations include variations of API metabolism among aquatic organisms. Exposure defenses include p-GP efflux pumps, which can also be induced/repressed by other stressors, especially APIs (see: Tan 2007); cellular stress response (e.g., induction of heat shock proteins); facilitation might occur via active transport (e.g., for APIs of low K_{ow} or high MW).

Note: Some of these factors are discussed in detail by Geyer et al. (2000) and van der Oost (2003).

maximize the chances of detecting and quantifying any amount that might accumulate. Most of the aquatic studies involving human APIs have been published only in the last few years.

Some studies examine the presence of APIs (primarily estrogens) indirectly, by way of activity assays. For example, Houtman et al. (2004) examined fish bile for estrogenic activity; but fractionation of the sample is required for assigning activity to a particular API. It is also important to note that in many studies, especially those examining bile, the APIs can be present as metabolically reversible conjugates, which may or may not have been cleaved prior to analysis; so the API is not really present in its parent form, although it can be reabsorbed once the conjugate is excreted into the intestine. The interpretation of tissue levels of many APIs, especially the steroids, is greatly complicated by the relative portions that are free versus conjugated.

The analysis of bile for free and especially conjugated APIs has been well-established for over 30 years. In a summary of data acquired in the 1970s for over a dozen APIs, concentration in the bile of the dogfish shark was already known (Guarino and Lech 1986). Concentrations in the bile versus plasma were known to range up to factors of hundreds (e.g., warfarin and diethylstilbestrol) or thousands (e.g., methotrexate). The utility of bile in exposure monitoring is discussed by Adolfsson-Erici (2005) and Pettersson (2006).

Almost no tissue-monitoring study has examined the optical isomer ratios of racemic APIs; indeed, even the aquatic toxicity of chiral APIs has been little studied, with Stanley et al. (2006) publishing one of the first studies (with propranolol). Stanley et al. (2007) further examined enantiomer-specific sublethal effects of the antidepressant fluoxetine on traditional endpoints (survival and growth) and the ecologically relevant behavioral response of feeding behavior. Both studies by Stanley et al. (2006, 2007) indicated that the more pharmacologically active enantiomer was more toxic to fish when sublethal rather than lethal responses were examined. Clearly enantiomer-specific bioaccumulation and effects of chiral compounds require additional attention and may necessitate *a priori* considerations in ecological risk assessments (Stanley and Brooks 2009).

Although the significance of bioconcentration is largely one of establishing potential internal dose, it is important to note that internal exposure is not necessary for an effect to occur. Certain effects can occur when the target organ is external. This is the case, for example, with exposure of the lateral-line sensory organ (e.g., Chiu et al. 2008) and for olfactory and taste exposure. Theoretically, antibiotics could alter the natural community structure of microorganisms that reside on the external surfaces of any aquatic organism.

An important perspective regarding the range of APIs and metabolites that have been and will continue to be detected in aquatic tissues is that of "self-biasing detectability." Those APIs with the highest probability of being detected are those that (i) can be taken up from water or food, (ii) are present in the highest concentrations, (iii) have the lowest MDLs, and (iv) have available appropriate analytical reference standards. Regarding the second point, those present in the highest concentrations tend to be those with the higher required doses, and which therefore have lower potency (the human acceptable daily intakes [ADIs] for residues of veterinary APIs in food therefore tend to also be higher). The third point is rarely pointed out. MDLs among APIs in a particular tissue can vary by more than 2 orders of magnitude. This means that APIs commonly present in tissues but having high MDLs might not be detected. As an example, Ramirez (2007) targeted 25 APIs/metabolites in fish in effluent-dominated streams and rivers. Of the seven that were detected, three had the lowest MDLs. Of the seven APIs with the highest MDLs, only one (gemfibrozil) was detected. Tissue concentrations below 0.1–1 µg/kg are rarely reported because this usually falls below the current method detection capability, primarily because of matrix interferences. But APIs could nonetheless be present at these levels and will therefore be overlooked or self-censored. In the case of API metabolites, very few deuterated standard compounds are commercially available, which presumably has precluded their analyses in tissues compared with parent drugs.

8.2.5 PREDICTIVE MODELING

Empirical data for the uptake and bioconcentration of APIs by aquatic organisms is very limited. Such data are primarily focused on the legacy POPs. The empirical data that do exist are fraught with quality issues and uncertainty; extrapolations across chemical classes or aquatic species are notoriously unreliable. Acquiring empirical data is met with a number of hurdles, not the least of which are cost and animal welfare concerns. Data from real-world field conditions are even more limited than data obtained under controlled laboratory conditions, and these data are poorly covered in the available electronic databases. The bioaccumulation databases for fish have been summarized by Weisbrod et al. (2007); *in vitro* methods for measuring bioavailability in fish have been reviewed by Weisbrod et al. (2009).

A comprehensive review of the literature on uptake and bioconcentration of organic chemicals by aquatic organisms revealed thousands of BCFs and bioaccumulation factors (BAFs) for 842 organic chemicals in 219 aquatic species (Arnot and Gobas 2006). But nearly half of the BCFs had major sources of uncertainty, and predicted values that usually underestimated empirical values, which were sparse. BAFs under ambient field conditions were generally higher than BCFs obtained under controlled conditions. None of these data, however, included APIs.

Ankley et al. (2005) point out that routine aquatic bioconcentration testing is not common, as conventional APIs tend to be water soluble, with K_{ow} values below 3. Even then, higher molecular weights (MWs) or a propensity for facile transformation (e.g., via hydrolysis) often preclude bioconcentration. On the other hand, given the fact that APIs often rely on active transport for uptake (e.g., Van Bambeke et al. 2000), even low K_{ow} values sometimes may not preclude bioconcentration. A significant aspect of APIs to note is that they are designed to minimize accumulation in the body during intended use, so any build-up that might occur in aquatic tissue could prove toxicologically significant. Another factor that sets APIs apart from conventional POPs is their metabolism—sometimes yielding products that themselves can bioconcentrate. This creates the need to calculate "pseudo" BCFs—the ratio of tissue concentration of a metabolite and the aqueous concentration of the parent API (such as for norfluoxetine).

Van der Oost et al. (2003) also stress that predicting bioaccumulation in fish using simple models (e.g., relying on K_{ow}) is "virtually impossible" and extremely prone to error even with sophisticated models. The dynamics imposed on APIs by pharmacokinetics in particular make prediction extremely difficult to model.

In the absence of empirical data, computed BCFs (e.g., using Quantitative Structure Activity Relationship—QSAR—calculations) are often relied upon—at least to try and inform which APIs might be of concern with regard to bioconcentration. But this approach has major unknowns with respect to pharmaceuticals (Walker et al. 2004a, 2004b). Other approaches for prioritizing which APIs might be of highest exposure concern include those that rely on informatics (e.g., Gunnarsson et al. 2008, Kostich and Lazorchak 2008) or water/sediment monitoring (e.g., Lissemore et al. 2006).

Cunningham et al. (2009) reported on calculated BCFs in fish for 43 APIs. They range up to highs of 353 L/kg (atovaquone), 190 (dutasteride), 64 (beclomethasone), and 51 (nabumetone), but nearly all of the remainder were less than 4 L/kg. In a major analysis of the factors involved with bioconcentration, Geyer et al. (2000) provide calculated K_{ow} values or BCFs for a number of natural and synthetic estrogens and estrogenic chemicals, androgenic steroids, and nonsteroidal antiandrogenic chemicals. Quite a number of computed BCFs are mentioned by Grung et al. (2007), and a number of computed BCFs for lipid-regulators are provided by Hernando et al. (2007). Other predictive modeling approaches are explored in Section 8.4.

The many variables and pitfalls in determining BCFs and their use for predicting BAFs are discussed by Parkerton et al. (2008). These issues, which surround the soundness or quality of BCF/BAF data, were not evaluated in the summary of published data reported in this chapter.

8.2.6 Overview of Fish Tissue and Other Residue Data for APIs

The study of the exposure of fish to APIs is dominated by endocrine-disrupting compounds [EDCs; predominately by the natural and synthetic estrogenic sex steroids—primarily 17β-estradiol (E2) and EE2; comparatively less focus is directed at androgens] and by antibiotics and biocides (such as triclosan). Comparatively few data exist for other drug classes. Most studies regarding EDCs obviously concern reproductive and other direct endocrine effects. Little exists on exposure to APIs having the potential to lead to subtle, difficult-to-detect effects such as alteration of behavior or perturbations of the immune system (e.g., Hoeger et al. 2005, Salo et al. 2007).

Exposure studies regarding endocrine disruption have been dominated by the numerous studies demonstrating that fish are impacted by exposures to treated and untreated sewage. The first of these, which linked exposure to sewage with estrogenicity, was published in 1994 (Purdom et al. 1994). Most evidence for exposure is inferential. Few studies have established the possibility of exposure to either individual or combinations of specific API EDCs.

Excluding the data collected on fish uptake of antibiotics as a direct or indirect result of usage in aquaculture, actual empirical data are rare for fish-tissue residues of APIs resulting from exposures in the environment or exposures under controlled conditions emulating ambient concentrations. Determination of empirical BCFs is rarer yet. These data come from two major categories: (1) fish collected from (or caged in) wild, native environments and (2) fish exposed under controlled conditions in the laboratory. Collectively, these data exist only for about 30 different APIs, some of which were the subject for just a single study (Ramirez et al. 2007).

The limited data on residue levels from exposure in the natural environment are extremely limited, primarily comprising: SSRIs (especially paroxetine, fluoxetine, sertraline, and some principal metabolites such as norfluoxetine and norsertraline), NSAIDs (diclofenac, naproxen, ketoprofen, and ibuprofen), steroids (estrone [E1], E2, and EE2), and diphenhydramine, diltiazem, gemfibrozil, and carbamazepine (CBZ). Also available are data for biocides (triclosan, methyl triclosan, and triclocarban) and malachite green and its leuco metabolite.

Tissue residue levels resulting from exposures under controlled conditions are a bit more common, including many of the same ones as detected under environmental monitoring, but also including: β-blockers (propranolol and atenolol), fungicides (the triazoles: bromuconazole, cyproconazole, metconazole, myclobutanil, penconazole, propiconazole, tebuconazole, tetraconazole, and triadimefon), the macrocyclic lactone avermectin B1, steroids (hydroxyestrone, estriol [E3], 17β-dihydroequilenin, and testosterone), and mono- and di-brominated derivatives of EE2.

These data are discussed in more detail in the succeeding sections. While on the one hand it is interesting that most of these APIs have BCFs greater than unity (considering their generally low K_{ow} values), few have BCFs above 1000 and therefore do not have anywhere near the accumulation potential of the POPs. The tissue levels of low-K_{ow} APIs point to uptake mechanisms beyond passive diffusion. The greatest concentrations tend to be in the bile and liver. Actual bioconcentration of diclofenac seems to be higher than for most APIs. The computed BCF for fluvastatin is among one of the highest for any API. But the highest measured BCFs are for the biocide triclosan, which was found to range upward of 10,000 in the intestines of zebra fish (Orvos et al. 2002). The most empirical data exist for steroids and antibiotics. While residues in edible tissues have clear ramifications for human exposure, residues in organs such as the brain (a focus for monitoring targeted at SSRIs) have implications regarding immediate biological effects. All residues have implications with respect to trophic biomagnification, which has not been studied with respect to APIs.

8.2.6.1 SSRIs/SNRIs (Selective Serotonin and Serotonin-Norepinephrine Reuptake Inhibitors)

Fluoxetine and sertraline and their principal metabolites (norfluoxetine and norsertraline) were detected in all fish tissues (brain, liver, and muscle) from three fish species collected from Pecan Creek TX, an effluent-dominated stream: *Lepomis macrochirus* (bluegill), *Ictalurus punctatus* (channel catfish), and *Pomoxis nigromaculatus* (black crappie) (Brooks et al. 2005). All four analytes were detected in all tissues at levels exceeding 0.1 ng/g; no residues were detected in fish from a reference site not receiving effluent discharges. Compared with average tissue levels of fluoxetine and sertraline, the average levels for norfluoxetine and norsertraline were higher in brain, liver, and muscle. The highest concentrations for each API were detected in the brain, generally followed by the liver: norsertraline (15.6 and 12.94 ng/g), norfluoxetine (8.86 and 10.27 ng/g), sertraline (4.27 and 3.59 ng/g), and fluoxetine (1.58 and 1.34 ng/g); the lowest concentrations were in muscle: norfluoxetine (1.07 ng/g), norsertraline (0.69 ng/g), sertraline (0.34 ng/g), and fluoxetine (0.11 ng/g). These levels are roughly 0–4 orders of magnitude higher than reported in wastewater and effluent-dominated streams, reflecting bioconcentration as well as the possibility of active transport. Although hypothetical human dietary exposure from these fish would yield daily intakes roughly 6 orders of magnitude below therapeutic doses, note that ADIs exist only for veterinary drugs.

Schultz et al. (2010) recently reported extensive data from a monitoring study involving the largest suite of APIs yet targeted from any single therapeutic class (SSRIs/SNRIs); they also collected concomitant data from three matrices (water, bed sediment, and brain tissue of white sucker, *Catostomus commersoni*) from two effluent-impacted streams in Iowa and Colorado. They targeted 10 antidepressants (including two metabolic/transformation products): bupropion, citalopram, duloxetine, fluoxetine (and norfluoxetine), fluvoxamine, paroxetine, sertraline (and norsertraline), and venlafaxine, an SNRI. All but two (duloxetine and fluvoxamine) were found in brain tissue. Of particular interest was the distinct lack of correlation between the types and quantities of these APIs measured in the stream waters versus those in the brain tissues. Venlafaxine was found in most of the stream samples at concentrations consistently higher than the next two most prevalent (bupropion and citalopram), sometimes at levels over an order of magnitude higher. The levels of venlafaxine in the streams sometimes exceeded 0.5 µg/L. In contrast, the primary analytes in brain tissue were norsertraline and sertraline, followed by norfluoxetine and fluoxetine. Indeed little venlafaxine, bupropion, or citalopram were found in brain tissue but were prevalent in sediments (although their relative levels were generally the inverse of those in brain tissue). These data point to the possible involvement of selective uptake of these chemicals into brain tissue. The maximum and the range of mean concentrations (ng/g) in brain tissue were: norsertraline (28.9; 0.01–3), sertraline (4.24; 0.005–1.8), norfluoxetine (3.57; 0.07–0.9), and fluoxetine (1.65; 0.02–0.6). The maximum level reported for norsertraline (28.9 ng/g) is the highest yet reported for an API in brain tissue.

In a more recent study, Brooks et al. (in preparation) observed sertraline, norsertraline, fluoxetine, and norfluoxetine at low ng/g levels in periphyton and three taxa of benthic macroinvertebrates (*Corbicula fluminea*, *Argia* sp., hydropsychidae) collected from Pecan Creek TX, indicating that dietary exposure to fish from these SSRIs and potentially other APIs deserve further study. As noted elsewhere in this document, extraction and analysis approaches to account for matrix differences among aquatic organisms require further development. For example, a recent study advanced extraction techniques for identification of APIs in mollusks (Cueva-Mestanza et al. 2008).

Fish from a Lake Ontario harbor receiving sewage effluent were analyzed for SSRIs by Chu and Metcalfe (2007). Seven fish were collected from three species [three brownbullhead (*Ameiurus nebulosus*), three gizzard shad (*Dorosoma cepedianum*), and one white perch (*Morone americana*)] and analyzed for paroxetine, fluoxetine, and norfluoxetine. Concentrations on the basis of whole wet weight ranged up to 1 µg/kg: paroxetine (0.48–0.58 µg/kg; 3 of 7 samples), fluoxetine (0.14–1.02 µg/kg; 6 of 7 samples), and norfluoxetine (0.15–1.08 µg/kg; 4 of 7 samples); neither fluoxetine nor norfluoxetine was detected in white perch. This occurrence of paroxetine is the first reported in the literature.

In a 2-year study in the Caloosahatchee River, Florida, water samples and the plasma of juvenile bull sharks (*Carcharhinus leucas*) were analyzed for four SSRIs and a metabolite (citalopram, fluoxetine/norfluoxetine, paroxetine, and sertraline) and an SNRI (venlafaxine) (Gelsleichter 2009). Only citalopram, sertraline, and venlafaxine were detected in wastewater or river-water samples. All analytes except for fluoxetine, however, could be detected at very low quantifiable levels in at least one plasma sample from at least one of the 2 years. Sertraline was the only analyte detected in all samples in 2006, while only venlafaxine and citalopram were detectable in 2007.

The bioconcentration of fluoxetine at an exposure concentration of 10 µg/L by Japanese medaka (*Oryzias latipes*) was evaluated under controlled conditions, at three pH values below the pK_a (Nakamura et al. 2008). The empirical BCFs were 8.8, 30, and 260 for the body and 330, 580, and 3100 for the liver, at respective pH values of 7, 8, and 9; the BCF for fluoxetine summed over the body and liver was 11 at pH 7.2. The BCFs increased with increasing pH since fluoxetine is a weak base and uptake of the nonionized molecule is facilitated by diffusion. Not unexpectedly, the N-demethylated metabolite norfluoxetine was similarly recovered. BCFs predicted from liposome/water equilibration did not increase as much with increasing pH as did BCFs predicted by octanol/water partitioning (with control of ionic strength). Bioconcentration of an API such as fluoxetine, while made complicated by pH influencing ionization, differs dramatically from the bioconcentration of a conventional, long-lived POP. It is further complicated when the API (such as fluoxetine) can be metabolized to another chemical species (in this case, norfluoxetine) that is also subject to bioconcentration. Since S-norfluoxetine binds to the serotonin reuptake transporter with affinity similar to that of R-fluoxetine and S-fluoxetine (Wong et al. 1993), the combined sum of the parent compound and active metabolite should be considered in determining bioconcentration and effects in ecological risk assessment.

Using caged fathead minnows in the outfall from sewage treatment plants, Metcalfe et al. (2010) detected venlafaxine, citalopram, sertraline, and a demethylated metabolite from each (as well as norfluoxetine) at levels roughly ranging from 1 to 4 ng/g (whole-body wet weight). Only sertraline, however, was detected in more than one of the three sites examined.

Research on the pharmacokinetics of APIs in fish is very sparse. Schultz et al. (2001) published perhaps the first study—on 17α-ethinylestradiol. Paterson and Metcalfe (2008) published an initial examination of the uptake and elimination of fluoxetine. At the outset of exposure to 0.55 µg/L in water over 7 days, accumulation by medaka was noted within the first 5 h; norfluoxetine was also noted at this time at about 40% of the fluoxetine level. A peak tissue concentration of 49 µg/kg was recorded after 3 days for fluoxetine. After 6 days, norfluoxetine exceeded fluoxetine: 64 µg/kg versus 40 µg/kg. This yielded a BCF of 74 for fluoxetine and a pseudo-BCF of 117 for norfluoxetine.

These studies (Brooks et al. 2005, Chu and Metcalfe 2007, Nakamura et al. 2008, Paterson and Metcalfe 2008, Schultz et al. 2010) collectively show that an API metabolite (such as norfluoxetine) can accumulate to equal or greater tissue concentrations than its parent (e.g., fluoxetine), which is not surprising since norfluoxetine is more nonpolar than fluoxetine. This is important when the metabolite (such as norfluoxetine) is bioactive. Metabolic conversion of the parent API means that its measured BCF will be lower than its actual BCF. Presence of an API in tissue might serve as an indicator of higher exposure to a metabolite of similar or higher biological activity. Whether these residues are bioavailable, however, is a key question, as shown by Zhou et al. (2008) who reported that the BCFs for "free" (unbound and directly bioavailable) fluoxetine were less than unity, while those for total fluoxetine were in the range reported by these previous studies. It is also not always clear whether API metabolites in tissues result from endogenous metabolism or from uptake of pre-existing metabolites from water.

Clearly, the relationship between coexposure to SSRIs/SNRIs, tissue residues, and potential biological effects requires further study. This was highlighted by the recent work of Painter et al. (2009), which identified environmentally relevant concentrations of a mixture of SSRIs/SNRIs to adversely affect fathead minnow predator escape behavior.

8.2.6.2 NSAIDs (Nonsteroidal Anti-inflammatory Drugs)

Given the widespread usage of NSAIDs and the published data on their environmental occurrence and ecotoxicology, surprisingly few studies have monitored for any of the NSAIDs in aquatic organisms.

Brown et al. (2007) exposed caged juvenile rainbow trout to sewage effluent at three sites and measured plasma levels for various NSAIDs (and gemfibrozil). This was the first reported measurement of fish plasma levels of diclofenac, naproxen, and ketoprofen (as well as gemfibrozil) after exposure in the field. All except ketoprofen showed a propensity to bioconcentrate in plasma. Plasma concentrations ranged from tens to several thousand ng/mL, with the highest being for gemfibrozil. Of particular significance was the wide range of BCFs for any particular API across the study sites. The wide variance in BCFs did not seem to be a function of API concentration in the water, temperature, pH, or exposure time; the authors concluded that some other chemical characteristic of the effluents governed uptake, possibly the presence of colloids or surfactants. Compared with predicted BCFs, all of the APIs, with the exception of ibuprofen, had BCFs that ranged from unity to considerably lower.

Under static exposure to high nominal concentrations (490–1000 µg/L), plasma levels ranged from: 60 ng/mL (ketoprofen), 3440 (diclofenac), and 3640 (naproxen), to 4680 (ibuprofen); under the same conditions, the level for gemfibrozil was 21,500. These gave empirical BCFs ranging from: 0.1 (ketoprofen), 4 (naproxen), 7 (diclofenac), to 9 (ibuprofen) [and 63 for gemfibrozil].

During the caged study, the exposure levels at the three sites ranged from 4.5 ng/L (ibuprofen) to 2320 ng/L (diclofenac). Resulting plasma concentrations ranged from undetectable (ketoprofen), 12 ng/mL (diclofenac), 14 (naproxen), and 84 (ibuprofen) (and 210 for gemfibrozil). These gave BCFs ranging from: <11 (ketoprofen), 5 (diclofenac), 56 (naproxen), to 18,667 (ibuprofen) (and 199 gemfibrozil).

With juvenile rainbow trout exposed under continuous flow for 96 h to 920 µg/L ibuprofen, after the first 24 h of exposure, the plasma concentrations of ibuprofen increased, beginning at about 7 µg/mL and ending at about 10.6 µg/mL (Huggett et al. 2004), in rough agreement with the data from Brown et al. (2007).

The bioconcentration of diclofenac by fish was reported for the first time by Schwaiger et al. (2004). Rainbow trout (*Oncorhynchus mykiss*) were exposed for 28 days to concentrations ranging from 1 to 500 µg/L. Concentrations in the liver were about 40-fold greater than in muscle. With exposure to 1 µg/L, tissue residue concentrations were about: 2882 ng/g (liver), 1025 ng/g (kidney), 805 ng/g (gills), and 73 ng/g (muscle), yielding BCFs (L/kg) of 2732 (liver), 971 (kidney), 763 (gills), and 69 (muscle); plasma levels were not reported. Tissue concentrations increased linearly with increasing dose, up to 500 µg/L, which yielded tissue concentrations roughly twice those from the 1-µg/L dose. The lower tissue concentrations with respect to dose are the reason the calculated BCFs decreased with increasing concentrations.

Mehinto (2010) reported bile BCFs in the narrow range 509–657 after 21-day exposures of trout to diclofenac at 0.5, 5.0, and 25 ng/L.

8.2.6.3 Lipid Regulators

Gemfibrozil was shown to bioconcentrate in the plasma of goldfish (*Carassius auratus*) after laboratory exposure to an environmentally relevant concentration of 0.34 µg/L (nominal 1.5), as well as a higher concentration of 852 µg/L (nominal 1500) (Mimeault et al. 2005). After 14 days of exposure, plasma BCFs were 500 and 92, respectively, resulting from respective plasma concentrations of roughly 170 and 78,000 µg/L. Uptake was concluded to occur across the gill membrane but passive diffusion or active transport could not be distinguished.

In the same study with four NSAIDs, gemfibrozil was measured in the plasma from rainbow trout caged in effluent-dominated streams and during a controlled static exposure experiment (Brown et al. 2007). This was the first measurement of a fish plasma level of gemfibrozil after ambient exposure in the field. Of the five APIs, the resulting plasma levels were highest for gemfibrozil. Static exposure to a high level of 510 µg/L gave a plasma level of 21,500 ng/mL, yielding a BCF of 63. In the caged field study, the plasma level reached 210 ng/mL, yielding a BCF of 199.

A compendium of calculated K_{ow} and BCF values for a number of fibrates and statins is provided by Hernando et al. (2007). While the $\log K_{ow}$ values all indicate relatively high lipophilicity (most exceeding 4), computational approaches for estimating BCFs showed relatively low values (3.2) for bezafibrate, gemfibrozil, clofibric acid, fenofibric acid, and pravastatin, and 56 for atorvastatin; values ranged from 120 to 380 for clofibrate, fenofibrate, lovastatin, and mevastatin. Higher computed values were found only for simvastatin (800) and fluvastatin (2000). These values might be useful for targeting these latter two statins for biomonitoring to determine actual empirical BCFs. Note, however, that the low computed BCF for gemfibrozil is not predictive of the empirical BCFs, being 1–2 orders of magnitude lower than those measured by Mimeault et al. (2005) and Brown et al. (2007).

8.2.6.4 *β*-Blockers

Owen et al. (2007; also unpublished data; Owen et al. 2009) exposed juvenile rainbow trout to relatively high levels of propranolol: 10 mg/L for 10 days. Plasma levels of about 5 µg/mL were reached; concentrations were often 40–80% of the water levels after 40 days.

Winter et al. (2008) exposed fathead minnows to relatively high concentrations (0.1–10 mg/L) of atenolol. Compared with water concentrations, plasma concentrations ranged between 1.8% and 6.2% (for males) and 0% and 12.2% (females). The male fish plasma concentration for atenolol corresponding to the exposure concentration for the LOEC (Lowest observed effect concentration) condition index (3.2 mg/L) was 0.0518 mg/L (51 µg/mL).

Cleuvers (2005) reported calculated BCFs of 4.47 and 0.89 for propranolol and metoprolol, and a value for atenolol too low to calculate. But β-blockers were not detected in fish during a study by Brooks et al. (2005, unpublished data).

8.2.6.5 Fungicides

Juvenile rainbow trout (*O. mykiss*) were exposed to nine triazole fungicides (dual-use pesticides and pharmaceuticals) in feed containing each at 23–35 µg/g (wet weight) (Konwick et al. 2006). These triazoles had log K_{ow}'s ranging from 2.9 to 3.9: bromuconazole, cyproconazole, metconazole, myclobutanil, penconazole, propiconazole, tebuconazole, tetraconazole, and triadimefon. Each compound was taken up quickly, reaching steady state after a day of exposure. They quickly reached concentrations in body lipids ranging from roughly 0.5–1 µg/g lipid, yielding biomagnification factors (on the basis of lipid content of fish vs. lipid content of food) ranging from 0.006 (propiconazole) to 0.012 (triadimefon and tebuconazole).

8.2.6.6 Macrocyclic Lactones

The macro-lactone dual-use parasiticides (used in veterinary medicine) are best known as the avermectins. Examples of these large, polycyclic lactones include: abamectin, albendazole, doramectin, emamectin, eprinomectin, ivermectin, morantel, moxidectin, milbemycin, and selamectin. Residues are well-established as occurring primarily in liver and lipid tissues, followed by kidney and muscle. An overview of this chemical class is provided by Danaher et al. (2006). Escher et al. (2007) point out that uptake kinetics and calculated BCFs are lower than predicted based on hydrophobicity. This is hypothesized to result from membrane exclusion because of the large molecular cross section.

Avermectin B1 (abamectin) was shown to resist uptake by sturgeon (into muscle) (Shen et al. 2005); biomagnification therefore would not occur. After a 22-day exposure to 0.2 and 1 ng/mL in water, concentrations in muscle reached steady state in about 2 weeks, giving muscle concentrations of 7.75 and 38.29 ng/g, respectively, yielding BCFs of 41–42.

Exposure for 28 days of bluegill sunfish (*L. macrochirus*) to an aqueous concentration of 0.099 µg/L avermectin B1a gave tissue concentrations of 6.8, 3.0, and 11 µg/kg, in whole fish, fillet, and viscera, respectively, yielding BCFs of 56, 28, and 84 (Van den Heuvel et al. 1996). As with other studies, it was concluded that abamectin does not strongly bioconcentrate and would therefore not be expected to biomagnify.

8.2.6.7 Steroids

Determining the uptake and bioconcentration of steroids is complicated by the fact that many of them have multiple origins. Several of the estrogens, for example, are endogenous to all fish (and some invertebrates but not plankton). Endogenous production can be further complicated by substances that induce synthesis (e.g., via aromatization) or inhibit excretion (e.g., repression of efflux pumps). But they also have at least two other origins. The first is the subject of this chapter—many have origins from the pharmaceutical preparations in which they are used; even β-estradiol is used in certain hormone preparations. The second origin is from other fish, which excrete a variety of steroids, for example as pheromones (Scott and Ellis 2007). These origins become intermingled with that resulting from endogenous synthesis; for estrogens and androgens, this becomes problematic for modeling female and male fish, respectively.

The steroids are also intimately involved in a metabolic cascade that involves interconversion, such as via aromatization, and conjugation. Uptake from surrounding media continually adds to the pool involved with natural metabolic processes. These factors greatly complicate the modeling of uptake and bioconcentration. Given the dynamic state of uptake, interconversion, and excretion, steady-state concentrations are probably rarely reached in laboratory studies; life-cycle studies are rare. This leads to very wide ranges in both predicted and measured tissue concentrations and BCFs—measured both under controlled laboratory conditions and in the wild. An overview of the environmental occurrence and consequences of exposure of fish to natural and synthetic estrogenic chemicals (of which only a small portion are APIs) is provided by Tyler et al. (2008); further discussion on bioaccumulation of E2 and EE2 is provided by Langston et al. (2005). A method using gas chromatography/mass spectrometry was developed for simultaneously quantifying 12 endogenous steroids in plasma and bile, using flounders (*Platichthys flesus*) as the test species (Budzinski et al. 2006); the steroid analytes spanned the estrogen/androgen metabolic cascades from pregnenolone and progesterone to E2 and 11-ketotestosterone.

One major but very limited source of data on uptake and tissue levels of estrogenic and androgenic steroids is from the aquaculture literature. Steroids are used to induce sex reversal in farmed populations. These data are not covered here. The data of Stewart et al. (2001) serve as one example.

Rainbow trout (*O. mykiss*) and roach (*Rutilus rutilus*) exposed to treated sewage effluent in controlled continuous-flow tanks concentrated E1, E2, and EE2 in the bile—at levels beyond endogenous production (Gibson et al. 2005a, 2005b). Most was present as glucuronide conjugates. Bioconcentration was roughly 4000–6000 for EE2 and 10,000–13,000 for E2 and E1 combined. The conjugated equine estrogen (CEE) metabolite, 17β-dihydroequilenin (17β-Eqn), was also detected; while this is perhaps the first report of 17β-Eqn in an aquatic organism (or in any environmental sample), its specific source was unknown (e.g., whether an endogenous metabolic product vs. an ingredient from a conjugated equine hormone preparation).

This work on CEEs (Gibson et al. 2005a) was extended further in a comprehensive examination of treated and untreated sewage and fish exposed under controlled conditions (Tyler et al. 2009). Treated sewage from wastewater treatment plants (WWTPs) in the United Kingdom were examined for six CEEs: equilin (Eq) and equilenin (Eqn), and four metabolites, 17β-dihydroequilin (17β-Eq), 17α-dihydroequilin (17α-Eq), 17β-dihydroequilenin (17β-Eqn), and 17α-dihydroequilenin (17α-Eqn). The bile from two species of fish (rainbow trout and the common carp, *Cyprinus carpio*) exposed to treated sewage effluent was also analyzed. Among these six CEEs, only two (Eqn and its metabolite 17β-Eqn) were detected in wastewaters. Eqn concentrations ranged from 1.32 to 2.59 ng/L (influent) and 0.32–1.32 ng/L (effluent), and 17β-Eqn ranged from <0.2 (LOD) to 0.37 ng/L (influent) and 0.07–0.18 ng/L (effluent), concentrations on par with those of E2 and EE2. The authors pointed out that since these two CEEs occurred in sewage influent, their origin from hormone replacement products was more probable than from exogenous metabolic processes. 17β-Eqn (as in the prior study) and now Eqn (for the first time) were the only two CEEs detected in bile. BCFs for trout exposed to 17β-Eqn were calculated to be 1.5×10^6 and 2.2×10^6 for trout exposed to Eqn.

Notably, the study of Tyler et al. (2009) is one of the very few focusing on aquatic exposure to also extend its findings to the potential for effects from exposure to environmentally realistic concentrations. Concentrations of 17β-Eqn as low as 0.6 ng/L elicited a vitellogenic response in trout, as well as all but the lowest exposure concentration of Eqn (4.2 ng/L); the carp were nearly three orders of magnitude less sensitive. A 17β-Eqn concentration of 0.6 ng/L nearly intersects with its concentrations detected in the treated UK wastewaters, providing a rare linkage between real-world exposure levels and the potential for adverse effects.

The difficulty in tying exposure to effects is demonstrated in another unique study, involving fish showing signs of possible exposure to estrogens. Three separate projects involved male bream (*Abramis brama*) with ovotestis and vitellogenin from two different locations in the Netherlands and whitefish (*Coregonus lavaretus*) with malformed gonads from Lake Thun, Switzerland (Vögeli 2008). In the ovotestis case, while levels of E1, E2, and EE2 in adipose tissue did not differ from controls, levels of E1 and EE2 in bile showed bioaccumulation in the ovotestis fish; E2 did not differ from the control. In contrast, with the case of elevated vitellogenin, levels of E1, E2, and EE2 were higher in the bile of the controls. With the group with malformed gonads, only E1 and E2 were present (above the MDLs) in the bile of all fish but the levels were higher in the fish with normal gonads.

In male Rainbow trout exposed to EE2 at relatively high nominal water levels of 125 ng/L, EE2 was shown to be rapidly absorbed (Skillman et al. 2006). EE2 was detected in plasma upon the first sampling time of 15 min and reached a steady-state range of 60–90 ng/mL within 16 h, yielding a BCF of up to 720. Levels in the liver corresponded with those in the plasma. In the bile, levels of free EE2 were also similar to those of the plasma and liver. In the bile, however, conjugated EE2 continued to increase, until 99% of the total EE2 in the bile comprised conjugated glucuronides. The authors conclude that EE2 in plasma, reaching equilibrium levels several hundred-fold higher than in water, represents a viable means for measuring current environmental levels; bile levels, in contrast, were more representative of cumulative exposure. The study also followed the parallel time course synthesis of vitellogenin and gene expression.

A model developed by Lai et al. (2002b) predicted relative bioconcentration of steroids, ranging from fish at the highest trophic level (1.8 for E3) to fish at the lowest trophic level (332 for EE2). In another study (Lai et al. 2002a) examined the uptake of natural (E1, E2, hydroxyestrone, and E3) and synthetic (EE2 and estradiol valerate) estrogens by the freshwater alga, *Chlorella vulgaris*. Under static conditions, all the estrogens were taken up, but E3, hydroxyestrone, E2, and EE2 could not be detected, because of metabolism. No equilibria could be reached, except for E1, for which a BCF of about 27 was calculated. On the basis of K_{ow}, EE2 would be expected to accumulate more, but did not—possibly because of active transport of the endogenous estrogens or active efflux of EE2.

Juvenile rainbow trout (*O. mykiss*), under controlled conditions, were exposed to sewage under continuous flow before and after treatment by sand filtration (Pettersson et al. 2006). After 28 days, bile was sampled. When exposed to untreated water, levels of E1 were two orders of magnitude higher than in controls (4.0 µg/g vs. 0.04 µg/g). Bile levels were also higher compared with controls for EE2 (0.25 µg/g vs. 0.10 µg/g) and E2 (0.17 µg/g vs. 0.04 µg/g). When exposed to treated water (posts and filtration), the bile concentrations for E1 (0.17 µg/g) and E2 (0.04 µg/g) were reduced considerably. The concentration for EE2, however, was slightly higher (0.38 µg/g).

In a subsequent study, Pettersson et al. (2007) examined the bile of perch (*Perca fluviatilis* L.) from the coastal waters of the Swedish Baltic Sea impacted by sewage for E1, E2, and EE2. Studies of fish from the wild are uncommon. EE2 was never detected, in contrast with E1 and E2, which were almost always present. These levels did not differ significantly from samples obtained from reference sites. These findings, however, corroborated lack of signs of endocrine effects, possibly because of efficient sewage treatment practices.

In another study using fish in the native ambient environment, Vermeirssen et al. (2005) used caged brown trout downstream of sewage effluents at five sampling sites. They measured E2, E1, and EE2 but did not report them separately—only as estradiol equivalents. Houtman et al. (2004) also indirectly measured EE2 via estrogen assay (in the bile of male bream, *A. brama*).

In a study of juvenile bull sharks (*C. leucas*) in the Caloosahatchee River, Florida, EE2 was detected in plasma at levels only up to slightly above the MDL (Gelsleichter 2009). EE2 was detected only during the second of 2 years of sampling (2006–2007), being detected in 7 of the 12 sharks sampled; EE2 was not detected in the Myakka River, a control river that did not receive treated wastewater. Levels of EE2 in the river ranged only up to 0.23 ng/L. Of the seven shark plasma samples with detectable residues, the two that could be quantified ranged up to 3.79 ng/mL plasma.

Perhaps the first bioconcentration study of EE2 in fish was reported by Larsson et al. (1999). Caged juvenile rainbow trout (*O. mykiss*) were exposed to an effluent-dominated stream, and E1, E2, and EE2 were measured in bile. The respective concentrations (conjugated and unconjugated combined) for E1, E2, and EE2 in the bile after 2 and 4 weeks were (approximately): 0.6 and 2.5 µg/g; ND and 1.0 µg/g; and 0.3 and 1.1 µg/g. These bile concentrations of roughly 1 ppm were about 4–6 orders of magnitude higher than the water levels. A separate static study using juvenile rainbow trout exposed for 46 h to 5 µg/L of either E1, E2, or EE2 produced respective bile concentrations of >400, ~200, and ~350 µg/g.

Perhaps the first full life-cycle bioconcentration study of EE2 in fish was reported by Lange et al. (2001). A life-cycle exposure study, using the fathead minnow (*Pimephales promelas*) was done using newly fertilized embryos (24-h old) under continuous flow for 305 days at five concentrations of EE2: 0.2, 1.0, 4.0, 16, and 64 ng/L (Lange et al. 2001); note that the two highest concentrations were toxic. No EE2 could be detected in tissue (<0.38 ng/g) after exposure at 0.2 and 1.0 ng/L test concentrations 192 days posthatch. At 16 ng/L (239 days posthatch) and 64 ng/L (153 days post-hatch), the EE2 tissue levels were 7.3 and 31 ng/g, yielding BCFs of 610 and 660, respectively. The authors concluded that the BCF (L/kg) was likely less than 500 (and probably less than 2400) for healthy fish. A more recent study by Caldwell et al. (2008) provided an HC5 value (hazardous concentration predicted to negatively affect 5% of the population) of 0.343 ng/L for a species sensitivity distribution (SSD) of EE2, highlighting the high potency of this API.

The bioconcentration of steroids is yet further complicated by the possibility that uptake is being augmented by facilitated transport to yield tissue levels far beyond what would be predicted with existing models assuming passive brachial uptake.

A study using the three-spined stickleback (*Gasterosteus aculeatus*) used 6-day static exposure concentrations of 1 µg/L (nominal) of either E2, testosterone (T), or E2 and T combined (Maunder et al. 2007). Plasma levels climbed rapidly within the first 6 h to within the range of 20–90 ng/mL. These bioconcentrated levels were 50-fold (E2) and 200-fold (T) greater than the measured exposure concentrations. The authors postulated that the faster and greater uptake than predicted of E2 and T might be due to the presence of a plasma sex hormone-binding globulin (SHBG). Scott et al. (2005) also postulated that SHBG is responsible for enhanced uptake of many of the steroids. This hypothesis is set forth in more detail by Miguel-Queralt and Hammond (2008).

Miguel-Queralt and Hammond (2008) report that natural and synthetic estrogens and androgens are actively taken up by fish via the gills by way of binding to SHBG in the brachial filaments. This uptake mechanism is extremely fast, with up to 70% of T or EE2 being removed from water in 90 min. A broad range of steroids have a high affinity for fish SHBG, whose ligand specificity varies widely across species. Trace amounts (e.g., 50 pmol) of ligand can be taken up from water within minutes. After uptake, residues are rapidly distributed throughout the body; EE2 was reported to then accumulate in the brain, ovaries/eggs, and muscle. The authors point out that SHBG also has a high affinity for at least two of the more common progestin APIs—levonorgestrel and 19-norethindrone. Since these APIs may be frequently present in sewage-derived waters, sometimes at relatively high concentrations, this points to the possibility of progestins occurring in fish. Progestins, however, have only rarely been targeted in environmental monitoring. Sediments in Puget Sound were analyzed for the synthetic progestogen 19-norethindrone. Levels ranged from 419 to 890 ng/g, but the analysis was done with GC-FID rather than GC-MS (Kimball 2008). 19-Norethindrone was the most frequently detected and abundant (26–224 ng/L) of all the synthetic estrogens/progesterones in sewage influent samples (Fernandez et al. 2007). López de Alda et al. (2002) reported 19-norethindrone as frequently occurring in sediments but at low ng/g-levels. Viglino et al. (2008) reported levonorgestrel and 19-norethindrone concentrations in sewage effluent ranging between 30 and 53 ng/L, respectively.

Others postulate that steroid residues in food may contribute more to bioaccumulation by fish than do the residues at significantly lower concentrations in water. Takahashi et al. (2003) report E2 concentrations ranging from 0.0001 to 0.0076 µg/L in water, compared with 0.09–2.26 µg/kg-wet in the periphytons and less than 0.01–0.22 µg/kg-wet in the benthos. Bioaccumulation factors of E2 were estimated at 64–1200 for the periphyton and 100–160 for the benthos.

It is important to note that even though the BCFs for EE2 do not indicate a propensity for bioaccumulation, the extremely low no-effect levels for this steroid have led a number of investigators to recommend more detailed examinations (e.g., Lyssimachou and Arukwe 2007).

Direct uptake from water of E1 by *Daphnia magna* gave a BCF of 228 (Gomes et al. 2004); biomagnification via feeding on *C. vulgaris* was not as efficient.

In a controlled study using artificial sediment and radiolabeled EE2, a benthic freshwater oligochaete (*Lumbriculus variegatus*) was exposed over 35 days to a nominal concentration of 300 ng/g wet-weight sediment (556 ng/g dry weight) (Liebig et al. 2005). Continuous linear uptake never reached steady state. The BSAF was 75 after the 35 days. A calculated steady state (after 360 days of exposure) would yield a BSAF of 190—higher than predicted by K_{ow}. A study with two invertebrates—a midge (*Chironomus tentans*) and amphipod (*Hyalella azteca*)—followed 21-day EE2 exposures using spiked water and water with sediments (Dussault et al. 2009). The exposure concentrations, however, ranged up to 3.1 ppm, orders of magnitude higher than those found in the ambient environment. At one of the lower, but still high, water-only exposures (20 μg/L), the BCFs were 31 for *C. tentans* and 142 for *H. azteca*; BSAFs were 0.8 and 1.5, respectively.

8.2.6.8 Antibiotics: Informing Environmental Exposure with Data from Use of Veterinary Aquaculture Drugs

APIs are used in aquaculture at levels many orders of magnitude higher than their occurrence in the ambient environment. Exposure data in aquaculture settings are obtained usually to assess if therapeutic or prophylactic doses are reached and to assess subsequent depuration of residues to ensure consumer safety. For this reason, the exposure concentrations are orders of magnitude higher than ambient levels, and the antibiotics studied tend to be restricted to those used in veterinary practice (although use of unapproved, illegal drugs also occurs). An overview of antibiotics used in aquaculture is provided by Sapkota et al. (2008). In a Canadian Total Diet Study focused on residues of 39 different veterinary drugs, levels tended to be in the range of low nanograms per gram (Tittlemier et al. 2007).

Even though exposures emulating those during aquaculture occur at higher ambient levels of APIs, they might be useful as worst-case scenarios to inform the potential for bioconcentration under ambient conditions. As one example, trout raised in aquaculture receiving medicated feed with roughly 0.6% oxytetracycline (OTC), which yielded a maximum water concentration of about 0.8 ppm, reached a maximum muscle-tissue concentration of 1.8 ppm (Bebak-Williams et al. 2002). This maximum level rapidly dissipated once the aqueous concentration dissipated. This shows that at high exposure concentrations, the muscle-tissue level shows very little bioconcentration. The literature on veterinary drug exposure is comparatively large, just two examples being Hou et al. (2003) and Chu et al. (2008), who examined the uptake into muscle and depuration of sulfamethazine and nitrofurans.

The study of aquatic exposure to APIs actually began several decades ago. The study of antibiotics used in aquaculture led to the need for examining aquatic tissue levels to assess therapeutic dose levels while assuring levels were sufficiently low for human consumption via the food supply. But even then, the potential for environmental impacts was also a consideration; the early work of Coats et al. (1976) using model ecosystems is an example.

Early studies on the environmental fate and possible biomagnification of veterinary drugs, particularly parasiticides, antibiotics, and other biocides, began in the 1960s and 1970s. Many of these studies were comprehensive and generated considerable data, as they used traditional radiolabeled materials to try and reach closure around mass balances. For example, 3-day uptake in fish of four veterinary drugs was studied in aquatic model ecosystems, using radiolabeled anthelmintic phenothiazine, the coccidiostat clopidol, the bacteriostat sulfamethazine, and the growth promoter diethylstilbestrol (Coats et al. 1976).

Another route of exposure as a result of aquaculture, however, occurs because 70–80% of the APIs used in medicated feed are released to the ambient environment as a result of excretion or escape by way of feed that is not consumed (Pouliquen et al. 2009). Native fish in the vicinity can then be unintentionally exposed—to levels exceeding ambient background concentrations. Usage of antibiotics in aquaculture, however, is episodic and occurs for very limited number of days, but concentrations in sediments immediately below can exceed the ppm-level (Pouliquen et al. 2009). Samuelson et al. (1992) reported that levels of several antibiotics in aquatic organisms nearby aquaculture exceeded levels considered safe for human consumption; also see Cabello (2006).

Blue mussels (*Mytilus edulis*) were evaluated for their ability to bioconcentrate two veterinary antibiotics: OA and OTC (Le Bris and Pouliquen 2004). Exposure concentrations were intended to emulate unintended exposure by what might be encountered near aquaculture. Exposure concentrations were roughly 0.95 mg/L for OTC and 1.46 mg/L for OA. Uptake was determined for foot, muscle, mantle, viscera, gills, and shell. OTC concentrations were higher in viscera (1.83 mg/kg) than gills (0.37 mg/L), with other parts less than 0.2 mg/kg. OA concentrations were highest in gills (0.79 mg/kg) followed by shell (0.19 mg/kg). BAFs less than 1.0 were expected for these two highly ionized APIs.

Nie et al. (2008) found the bioaccumulation of ciprofloxacin by carp (*Allogynogenetic crucian*) under controlled feeding conditions to vary greatly, depending on several exposure scenarios. Feeding resulted in much higher residues (in visceral and muscle tissues) than via exposure to water. Uptake was fast, with maximum levels being reached within a day. The tissue concentrations (µg/kg) resulting from each type of exposure were: water (muscle: 10; viscera: 42); feeding (muscle: 73; viscera: 645); and dual exposure (muscle: 43; viscera: 368).

8.2.6.9 Carbamazepine (CBZ)

Juvenile rainbow trout were exposed under continuous flow for 96 h to 200 µg/L CBZ (Huggett et al. 2004). After 24 h of exposure, the plasma concentrations of CBZ decreased, beginning at about 2.5 ng/mL and ending at less than 1 ng/mL, showing a low propensity to bioconcentrate. See the results for CBZ published by Ramirez (2007), Ramirez et al. (2007), and Zhou et al. (2008) summarized under the section "Multianalyte Studies." In those studies, CBZ was also shown to poorly bioconcentrate, having a low BCF (<1).

After a 60-day exposure to a high 19-ppm concentration of CBZ, no intracellular accumulation could be detected in the algae *Ankistrodesmus braunii* (Andreozzi et al. 2002). A method developed for determining CBZ in tissues was used to analyze a crustacean (*Thamnocephalus platyurus*) after it fed on algae (*Pseudokirchneriella subcapitata*) that had been previously exposed to CBZ at 250 mg/L (Lajeunesse et al. 2009). The mean CBZ concentration in dried *T. platyurus* was 129 (±57) µg/mg.

8.2.6.10 Triclosan (and Methyl Triclosan) and Triclocarban

With respect to the most heavily used biocides, triclosan [TCS: 5-chloro-2-(2,4-dichlorophenoxy)phenol] has been studied more frequently than triclocarban [TCC: N-(4-chlorophenyl)-N′-(3,4-dichlorophenyl)urea]. In general, the transformation of triclosan to the more lipophilic methyl triclosan (MTCS: 4-chloro-1-(2,4-dichlorophenoxy)-2-methoxybenzene) leads to lower tissue levels of TCS compared with MTCS. Most of the research has been conducted in Europe and Scandinavia, with very recent studies in the United States (e.g., Leiker et al. 2009). Tissue residue levels of MTCS generally exceed those of any API—a result of higher BCFs and higher exposure levels.

MTCS was first identified in fish by Miyazaki et al. (1984). Up to 38 ng/g was detected in the whole bodies of a freshwater fish (*Carassius carassius*) collected from Tama River, Tokyo Bay. Samsøe-Petersen et al. (2003) report on a monitoring study that sampled various aquatic species from 12 locations in Sweden, where concentrations ranged from less than 0.1 to 13 µg/kg (wet weight). Much lower concentrations (in blood plasma) were reported in perhaps the first study from the United States, where Alaee et al. (2003) reported on fish from the Detroit River (Michigan-Ontario) having TCS in the blood plasma of all 13 species surveyed; levels ranged from 0.61 ng/g wet weight (brown bullhead) to 10.4 ng/g (white bass). In contrast, MTCS was detected in the plasma of all 13 species but ranged only from 0.0004 ng/g for common carp to 0.0132 ng/g for largemouth bass; the presence of TCS at 3 orders of magnitude higher concentration than MTCS was ascribed to the higher lipophilicity of MTSC and its possible preferential partitioning to lipid tissue.

Bile was analyzed for TCS in fish subjected to various exposure scenarios involving three WWTPs in Sweden, ranging from caged within effluent-dominated flows, to wild (or directly exposed to sewage under controlled compositions) (Adolfsson-Erici et al. 2002). Concentrations ranged from 0.24 to 4.4 mg/kg bile (for wild fish) to 34–120 mg/kg (for those exposed directly to treated sewage effluent).

Orvos et al. (2002) assessed TCS bioconcentration in zebrafish (*Danio rerio*) using continuous flow with 3 and 30 µg/L for 5 weeks followed by 2 weeks of depuration. During the 5-week exposure, the BCFs ranged between 2000 and less than 3500 for the 39-µg/L exposure and from 3500 to about 5200 for the 3-µg/L exposure, giving average BCFs during the exposure period of 4157 at 3 µg/L and 2532 at 30 µg/L. BCFs for head/scale and fillet ranged from about 1000 to 2000, whereas they ranged from about 8000 to 11,000 in the intestines for the low and high exposures, respectively. After depuration, the BCFs were 30 and 41 for the high and low exposures, respectively, so half-life residence time within the body was short compared with POPs.

Boehmer et al. (2004) performed a rare 10-year retrospective study (1994–2003) of breams (*A. brama*) from representative German rivers. The study revealed that TCS was rarely present in muscle while MTCS was detected in all specimens collected. In general, when present, TCS muscle-tissue concentrations remained relatively constant but low for any given river—less than 1 ng/g wet weight. TCS concentrations were always lower than MTCS, which had excursions above 30 ng/g wet-weight muscle.

Balmer et al. (2004) reported MTCS in fish (white fish, *Coregonus* sp.; roach, *R. rutilus*) from various lakes in Switzerland receiving treated sewage effluents. Concentrations ranged up to 35 ng/g (wet weight) or 365 ng/g (lipid basis) and fell within narrow ranges for a given lake. In another study of Swiss lakes, Balmer et al. (2005) measured lipid levels of MTCS in fish lipids, where levels (ng/g) ranged from undetectable (perch) and 4–233 (roach), to 4–56 (whitefish). Buser et al. (2006) examined the muscle tissue of brown trout (*Salmo trutta fario*), from seven Swiss rivers that receive treated sewage effluent, for MTCS. All concentrations were higher than those reported by Balmer et al. (2005) for lake fish (white fish, *Coregonus* sp. and roach, *R. rutilus*). Concentrations ranged from 130 to 2100 ng/g, compared with the previous lake fish study of 4–370 ng/g. The concentrations for river fish had considerable variation, possibly due to a more fluctuating input from sewage; river fish had higher concentrations probably because the exposure levels were higher.

In a recent study, a survey of common carp from Las Vegas Bay revealed MTCS (but not TCS) in all 29 male common carp at a mean whole-body concentration of 600 µg/kg wet weight (7400 µg/kg on the basis of lipid, giving a BCF of 1.8×10^5) (Leiker et al. 2009). Three chlorinated analogs (3- and 5-chloromethoxy triclosan and 3,5-dichloromethoxy triclosan) were also present but less often and at lower concentrations, ranging from 0.5 to 13 µg/kg in 21–76% of the samples; the brominated analog (bromomethoxy triclosan) was detected but not quantified.

In contrast with MTCS, reported TCS levels are not as common and almost always lower (with the exception of blood plasma). The average TCS accumulation factor for zebrafish over a 5-week test period was 4157 at 3 mg/L and 2532 at 30 mg/L (Orvos et al. 2002). TCS concentrations were highest in the digestive tract; head and muscle concentrations were similar. Following a 2-week depuration, the average BAF was 41 for 3 mg/L exposure and 32 for 30 mg/L exposure. The BCF was predicted to be roughly 2500.

Houtman et al. (2004) identified TCS at ppm levels in the nonpolar residual fraction of bile from wild fish in the Netherlands. Bile concentrations were about 14 µg/mL for fish from the North Sea Canal and 80 µg/mL for fish from the River Dommel.

In a rare cross-species survey, TCS and MTCS were measured in the blood plasma of 13 species of fish (both benthic and pelagic) from a stretch of the "highly contaminated" Detroit River (Valters et al. 2005). TCS levels ranged from 0.750 to 10.0 ng/g, while MTCS was present at 0.4–13.4 pg/L, 3 orders of magnitude lower. TCS in the estuarine water samples averaged 7.5 ng/L, although the tissue and water sampling were temporally disconnected. Another feature of this study was the parallel analyses for a spectrum of brominated diphenyl ethers.

Algae were shown to bioconcentrate TCS, MTCS, and TCC (Coogan et al. 2007) by roughly 3 orders of magnitude when collected from Pecan Creek, the same effluent-dominated stream in north Texas, U.S., previously studied by Brooks et al. (2005) and Ramirez et al. (2007). This may be the first report of the bioconcentration of any of these three chemicals in algae; it may also be the first report of the bioconcentration of TCC by any organism. Dissolved concentration ranges (and algal wet-weight bioconcentration ranges and BAFs, L/kg) from four sampling sites for each of the three analytes were: TCS levels of <10–120 ng/L (<10–146 µg/L; BAFs nil-2100); MTCS levels of <5–80 ng/L (<5–89 µg/L; BAFs nil-1500); and TCC levels of <15–190 ng/L (<10–401 µg/L; BAFs nil-2700).

Coogan and La Point (2008) extended these initial algal bioconcentration studies to examine snail (*Helisoma trivolvis*) bioaccumulation of TCS, MTCS, and TCC from the effluent outfall to Pecan Creek. Dissolved concentration ranges (and snail wet-weight bioconcentration level and BAFs) were: TCS level of 112 ng/L (58.7 µg/L; BAF 500); MTCS level of 41 ng/L (49.8 µg/L; 1200); and TCC 191 ng/L (299 µg/L; 1600). Bioaccumulation of antimicrobials has been observed in other macroinvertebrates; adult grass shrimp accumulated MTCS after a 14-day exposure to 100 µg/L TCS (Delorenzo et al. 2008).

More recently, Mottaleb et al. (2009) reported mean ($n = 11$, ±SD) TCS levels at 21 ng/g (±4) in *L. macrochirus* (bluegill) from Pecan Creek, Texas. Although TCS was not detected in fish (Sonora sucker) collected from a relatively pristine location in the East Fork Gila River, New Mexico, TCS was detected at 12 ng/g in bluegill from Clear Creek, Texas, a regional reference site studied by Brooks et al. (2005) and Ramirez et al. (2007). This site does not receive point-source munici-pal effluent, but may be influenced by onsite wastewater. In the Mottaleb et al. (2009) study, fish samples examined from Pecan Creek were the same organisms analyzed previously for target APIs by Ramirez et al. (2007).

The most in-depth controlled study of TCC involved its uptake from sediments by the fresh-water oligochaete *L. variegatus* (Higgins et al. 2009); depuration was also studied. TCC BSAFs were calculated and determined empirically during a 56-day study. Sediment spiked with TCC maintained a constant (and environmentally relevant) concentration over 56 days (22.4 ± 7.6 µg/g dry weight); the TCC concentration in the surrounding water also maintained constant, at 820 ± 220 ng/L. Uptake by *L. variegatus* was rapid, reaching a maximum of 1310 ± 60 µg/g lipid or 42 ± 2 µg/g wet weight at 5 days, after which levels began to decline. Bioaccumulation comported with predic-tions from conventional models. Depuration was rapid. After 21 days in clean sediment, the TCC concentration in *L. variegatus* had declined to 9.6 ± 0.3 µg/g lipid (0.31 ± 0.01 µg/g wet weight). The BSAF ([mass of sediment organic carbon]/[mass of tissue lipid organic carbon]) after 56 days was calculated as 1.6 ± 0.6.

In a very rare study of higher-tropic-level aquatic wildlife, triclosan was measured for the first time in a marine mammal—bottlenose dolphins (*Tursiops truncatus*) from two estuarine sites (Charleston, South Carolina, and Indian River Lagoon, Florida) (Fair et al. 2009). Both sites are influenced by discharged treated wastewaters. Blood-plasma levels of TCS for one site ranged from 0.12 to 0.27 ng/g (with 4 of the 13 having levels exceeding the MDL of 0.033 ng/g), and for the other site ranged from 0.085 to 0.106 (with 3 of 13 having detectable levels). These are possibly the highest plasma levels yet reported for any aquatic organism. TCS levels in the respective waters for the two sites averaged 7.5 ng/L, with a maximum of 13.7 ng/L.

8.2.6.11 Miscellaneous APIs

In perhaps its first reported occurrence in fish from the wild, diazepam was quantified in liver samples from 10 hornyhead turbot (*Pleuronichthys verticalis*) collected near MWTP ocean dis-charges in southern California (Kwon et al. 2009). The levels in five females ranged from 23 to 45 ng/g (wet weight) and in five males from 58 to 110 ng/g (wet weight); EE2, CBZ, simvastatin, and oxybenzone were also targeted but not detected.

Malachite green is a multiuse chemical. Although it has useful properties in aquaculture, its use in food is prohibited worldwide (see Sudova et al. 2007); nonetheless, it still experiences clandes-tine use in aquaculture and can be used legally for ornamental fish. Because it is a chromophore, it also has a variety of other commercial uses unrelated to veterinary medicine—particularly as a dye.

Malachite green bioconcentrates readily in the lipid of aquatic organisms, primarily as its metabolite leuco malachite green, which occurs at a ratio of 5–7:1. It persists in tissues, being found in the highest concentration in the liver. Schuetze et al. (2008) documented the occurrence of malachite green in the European eel (*Anguilla anguilla* L.) from lakes, rivers, and a canal in Berlin, Germany. Total concentrations of malachite green and the leuco form ranged up to 0.765 µg/kg (wet weight) in 25 of the 45 eels collected. Exposure was concluded to result from treated sewage. Although some of the bioconcentrated residue may have come from the use of malachite green for illegal and legal treatment of fish, an unknown but possibly large portion undoubtedly resulted from other commercial uses, such as dyed textiles.

8.2.6.12 API Disinfection By-Products (DBPs) and Metabolites

Chlorination of either drinking water or wastewater containing steroids is known to produce mono- and di-chlorinated products of varying estrogenic activity. Little has been published on DBPs from APIs. In the presence of bromide, which often occurs in surface waters and wastewaters, multiply-brominated analogs can be formed (Lu and Korshin 2008). In particular, Lu and Korshin (2008) demonstrated the formation of stable dibromo-EE2. Buth et al. (2007) identified a number of products from the reaction of cimetidine with chlorine; Dodd and Huang (2004) identified products from sulfamethoxazole; and DellaGreca et al. (2009) identified various chlorinated and nonchlorinated products from atenolol. Similarly, Nakamura et al. (2007) identified a number of chlorinated estrones. The bioaccumulative potential for these reaction products is unknown. Similar issues surround the complex array of potential metabolites and other transformation products from parent APIs; many examples are reviewed by Farré et al. (2008) and by Kosjek and Heath (2008). Little work has been published on the possible metabolites from aquatic organisms. The recent work of Mehinto et al. (2010) revealed some possible metabolites from diclofenac in fish.

Despite an increasing number of studies on API DBPs and other transformation products, there are very few studies regarding their uptake by aquatic organisms. In one of the only such studies, roach (*R. rutilus*) were exposed for 5 days in an aquarium filled with drinking water and spiked with EE2 at a nominal concentration of 30 ng/L; the measurable concentration in the test situation, however, was below the limit of detection (0.6 ng/L) (Flores and Hill 2008). EE2 was found to be rapidly brominated (yielding mono- and di-brominated EE2). Di-brominated EE2 (but no detectable mono-brominated EE2) accumulated in the ovaries and liver to levels 18- to 67-fold greater than the parent EE2. Concentrations (ng/g wet weight) of EE2 and dibromo-EE2 detected were: liver (EE2: 2.7 and dibromo-EE2: 92.3) and ovaries (EE2: 0.2 and dibromo-EE2: 2.3), yielding a BCF for the ovaries of 130 and for the liver of 7894.

8.2.7 Uptake by Aquatic Plants and Aerial Invertebrates

The uptake of APIs by plants and algae, which compose an important part of the aquatic food chain, might prove a significant part of dietary exposure. Indeed, uptake of APIs by certain plants is so efficient that they have been evaluated for *in-situ* phytoremediation of contaminated waters and sediments (e.g., Forni et al. 2002). Plant uptake has been particularly germane to aquaculture sites.

A discrete body of work has been published on the uptake of APIs by aquatic plants. The aquatic bryophyte *Fontinalis antipyretica* is known to bioconcentrate metals, pesticides, and PAHs and has therefore been used *in situ* as a bioindicator for integrative monitoring. A study of the uptake of three antibiotics widely used in aquaculture (OA, flumequine, and OTC) showed BCFs ranging from 75 (flumequine) to 450 (OTC) (Delepee et al. 2004). These antibiotics had mean tissue residence times of 18 and 59 days. The study was conducted at relatively high concentrations of 100 and 1000 ppb. BCFs were higher at the lower concentrations and were an inverse function of K_{ow}—increasing according to ionization instead of lipophilicity.

In a study of transpiration stream concentration factors (TSCFs) versus polarity, Dettenmaier et al. (2009) showed that polar but nonionizable, highly water soluble organic compounds can be easily taken up by plant roots and translocated to shoot tissue. Studies on uptake of APIs by plants (primarily bryophyte) have generally revealed rather high levels, and sometimes the source was not necessarily related to aquaculture, as upstream samples have at times shown similar levels. Pouliquen et al. (2009) examined bryophytes as biomonitors downstream of aquaculture and sewage. They reported maximum tissue concentrations (ng/g) for OA (47), flumequine (~600), OTC (1200), and florfenicol (513).

Migliore et al. (2000) exposed an aquatic weed (*Lythrum salicaria* L.) to flumequine. After 35 days, the dry-weight tissue concentrations were in the ppm range: 13.3, 8.7, 0.7, 0.3, and 0.2 µg/g at flumequine aqueous concentrations of 5000, 1000, 500, 100, and 50 µg/L, respectively. Exposure of an aquatic fern (*Azolla filiculoides* Lam.) to sulfadimethoxine for 5 weeks at concentrations of 50, 150, 300, and 450 mg/L resulted in uptake at the mg/g dry-weight level (1000 ppm) (Forni et al. 2002). Typha was shown to rapidly absorb clofibric acid at 20 µg/L, removing more than 50% within 48 h (Dordio et al. 2008).

Redshaw et al. (2008) recently used a *Brassicaceae* (cauliflower) model to examine fluoxetine uptake by plants. Following a 12-week exposure to 280 µg/L fluoxetine in growth media, fluoxetine concentrations were higher in the stems (0.49 µg/g wet weight) than in leaf tissues (0.26 µg/g wet weight) of *Brassicaceae*. This study did not examine steady-state tissue levels of fluoxetine; this is important because fluoxetine is photolabile and should have degraded over the 12-week study period (Redshaw et al. 2008). However, presence of low µg/g-levels of fluoxetine suggest bioconcentration, which did not correlate with lipid content in leaf and stem tissues of *Brassicaceae*. Although cauliflowers are terrestrial plants and this study was specifically interested in estimating potential fluoxetine uptake in terrestrial plants exposed to biosolid-amended soils, it suggests that fluoxetine accumulation by nonrooted aquatic macrophytes such as *Lemna* sp. should be considered (Redshaw et al. 2008).

Coexposure to APIs will often occur with varying nutrient ratios and stoichiometries. Nutrient enrichment was previously demonstrated to influence the magnitude of triclosan toxicity to *L. gibba*, for both traditional morphometric endpoints (Fulton et al. 2009) and nontraditional responses, such as internal C:N:P and nitrate uptake kinetics (Fulton et al. 2010). Because nutrient stoichiometry can also influence internal lipid metabolism and concentrations in plants and algae, site-specific nutrient enrichment differences may result in differential bioconcentration of APIs (Fulton et al. 2010).

The potential for trophic transfer of APIs out of the aquatic realm was recently shown by Park et al. (2009). EE2 was determined in aerial invertebrates (primarily Diptera) whose larval stages develop in STP percolating filter beds. EE2 concentrations in insects captured near STPs were significantly higher than in those over 2 km away. The median EE2 tissue concentration was 42 ng/g (with the 75th percentile 140 ng/g) from insects near the STPs, compared with a median level of less than 3 ng/g (and 9 ng/g 75th percentile level) detected in the insects more distant from the STPs. Further transfer to insectivorous bats and birds was postulated. Rough calculations estimated that daily exposure to EE2 for bats feeding on insects near the STPs could range from 9 to 159 ng/g.

8.2.8 MULTIANALYTE STUDIES

Studies that target multiple APIs to gauge ambient exposure are indeed rare. The recent study of Schultz et al. (2010) targeting 10 antidepressants (including two metabolic/transformation products) was discussed earlier. The first and most comprehensive multianalyte study to date on fish tissue was by Ramirez et al. (2007). From a target list of 23 APIs and 2 metabolites, only four were reported as being detected. Fish (*Lepomis* sp.) were sampled in Texas from an effluent-dominated stream ($n = 11$) and from another creek ($n = 20$) that served as reference. The four APIs were detected in muscle from all samples in the study site. The range (and mean; ng/g wet weight) were: diphenhydramine [0.66–1.32 (0.96)]; diltiazem [0.11–0.27 (0.21)]; CBZ [0.83–1.44 (1.16)]; and norfluoxetine [3.49–5.14 (4.37)]. This is the first report of diphenhydramine, diltiazem, and CBZ in wild fish.

With impetus provided by the Brooks et al. (2005) study, the U.S. EPA initiated the National Pilot Project of pharmaceuticals and PCPs (PPCPs) in Fish Tissue (U.S. EPA 2008a), which represents the first national-scale reconnaissance study of PPCPs in fish tissue (Ramirez 2007, Ramirez et al. 2009). Sample collection and processing procedures followed approaches previously used during the U.S. EPA's National Study of Chemical Residues in Lake Fish Tissue. Analytical methods for PPCPs in the National Pilot Project employed approaches previously developed by Ramirez et al. (2007) and Mottaleb et al. (2009). Because effluent-dominated and effluent-dependent ecosystems represent worst-case scenarios for API exposure (Brooks et al. 2006), five effluent-dominated river systems were selected for study: Phoenix, AZ; Orlando, FL; Chicago, IL; West Chester, PA; and Dallas, TX. The Gila River, NM was selected as a reference site for this study. Ramirez et al. (2009) targeted 25 APIs/metabolites in the fillets and livers from wild-caught fish: acetaminophen, atenolol, caffeine, *CBZ**, cimetidine, clofibric acid, codeine, *diltiazem**, 1,7-dimethylxanthine, *diphenhydramine**, erythromycin, fluoxetine*, gemfibrozil*, ibuprofen, lincomycin, metoprolol, miconazole, norfluoxetine*, propranolol, sertraline*, sulfamethoxazole, thiabendazole, trimethoprim, tylosin, *warfarin*. The four shown in italics had MDLs below 1 ng/g. The seven with asterisks were detected in multiple fish from multiple locations. Of these seven that were detected, three had the lowest MDLs (CBZ, diltiazem, and diphenhydramine) in both fillet (less than 1 ng/g) and liver (less than 2 ng/g), while only two (fluoxetine and gemfibrozil) were among those with the highest MDLs (greater than 6 ng/g in fillet and 12 ng/g in liver). Of the seven APIs with the highest MDLs, only one (gemfibrozil) was detected.

At only one of the five sites, receiving effluent from a sewage treatment facility using tertiary treatment, none of the target APIs was detected in the fillet from any fish; also no API was detected in any fish from a nonimpacted reference site. All of the seven APIs detected among the 25 targeted APIs were detected in fish from only one site, which received effluent from secondary treatment. Mean concentrations in fillet for all the detected APIs were generally less than 3 ng/g and ranged from 0.04 to 11 ng/g. The majority of the mean concentrations in liver for all the detected APIs were generally greater than 6 ng/g and ranged from 0.03 to a high of 380 ng/g (sertraline). Except for one site where fluoxetine was found in fillet but not in liver (and where the liver also contained substantially more norfluoxetine), the API concentrations in livers were always larger—by several fold or by over one order of magnitude. Of significance, API concentrations did not correlate with lipid content—a finding shared with other published studies.

Another study targeted five APIs during the course of ground truthing a new *in vivo* tissue sampling method using implanted solid-phase microextraction (SPME) fibers (Zhou et al. 2008). The targeted APIs were the ones previously reported by Ramirez et al. (2007): diltiazem, diphenhydramine, CBZ, and norfluoxetine. Under controlled exposure conditions, rainbow trout (*O. mykiss*) gave BCFs for CBZ in muscle after 7- and 14-day exposures of 0.44 and 0.22, respectively. Significantly, free and total tissue levels after 14 days were lower than those after 7 days. The authors postulated that CBZ metabolism was upregulated during the exposure time. The bioconcentration of fluoxetine differed markedly. While the respective free concentrations in muscle after 7 and 14 days of exposure were only 0.30 and 0.65 times those in the aqueous media, the BCFs for total fluoxetine in muscle were 62 and 84, respectively. The same approach was used for determining muscle levels of free API in wild fish captured from streams that received treated sewage. These are the only reports of "free" APIs in aquatic tissues. In the wild fish, only diltiazem and diphenhydramine were detected. Free diltiazem muscle concentrations were 2.04 and 5.69 pg/g in the white sucker (*C. commersoni*) and Johnny darter (*Etheostoma nigrum*), respectively. Free diphenhydramine concentrations were 32.0 and 81.6 pg/g for white sucker and Johnny darter, respectively. These concentrations are several orders of magnitude lower than the conventional "total" levels reported by all previous investigations.

A more recent study (Fick et al. 2010), which was built on previous efforts by Brown et al. (2007), exposed rainbow trout to final treated effluent for 14 d at three different sites in Sweden (Umeå, Stockholm, and Gothenburg). Of the 25 API analytes targeted, 16 were detected in fish plasma (sampled from at least one study location) at levels exceeding one-thousandth of their respective human plasma levels associated with therapeutic dose (C_{max}). One of these 16 APIs of particular significance was the synthetic progestin levonorgestrel—at plasma levels of 8.5 and 12 ng/mL, a level 4-fold greater than human plasma C_{max}. This plasma level of levonorgestrel represented an empirical BCF of 12,000, which was 200-fold higher than the predicted BCF. Such approaches for relating internal-dose API exposures to potential effect thresholds are explored further below.

Finally, the study of Kwon et al. (2009), as discussed earlier, targeted five APIs: EE2, diazepam, CBZ, simvastatin, and oxybenzone. Only the first two, however, were detected in liver samples from hornyhead turbot (*P. verticalis*) from southern California.

8.2.9 SUMMARY OF PUBLISHED DATA

Much of the data for APIs/metabolites and related DBPs in this chapter on aquatic tissue-levels and BCFs/BAFs compiled from the published literature is summarized in Table 8.4. Included in the table is an indication of historical precedence—whether the data were the first to be reported; most of the data are "firsts," revealing that the depth of the published data in terms of repeated measurements is very shallow.

TABLE 8.4
Summary of Bioconcentration Data for APIs in Aquatic Tissues

APIs Studied in Wild Specimens (Controlled *In Situ* Studies Indicated by Asterisk*)	Maximum Concentration in Wild Specimens (µg/kg) (Controlled Studies Indicated by Asterisk*)	Historical Precedence in Literature	Notes	Reference
Antidepressants				
Bupropion	0.013–0.07 brain	Probably 1st report	Detected in samples from 5 of 8 streams; upper range = 0.348 ng/g. Water = 20–50 ng/L.	Schultz et al. (2010)
Citalopram	0.57 plasma	Possibly 1st report	Sharks.	Gelsleichter (2009)
	0.01–0.07 brain	Probably 1st report	Detected in samples from 4 of 8 streams; upper range = 0.212 ng/g. Water = 4.5–70 ng/L.	Schultz et al. (2010)
Fluoxetine	1.58 brain 1.34 liver	Probably 1st reports	Lowest concentrations in muscle. Empirical BCFs up to 260 for body and 3100 for liver (Nakamura et al. 2008); but BCF for "free" fluoxetine less than unity (Zhou et al. 2008). Controlled exposure to 0.55 g/L gave peak concentration of 49 µg/kg tissue (Paterson and Metcalfe 2008).	Brooks et al. (2005) [also Chu and Metcalfe (2007)]
	0.02–0.6 brain		Detected in samples from 6 of 8 streams; upper range = 1.6 ng/g. Water = 1–9 ng/L.	Schultz et al. (2010)

continued

TABLE 8.4 (continued)
Summary of Bioconcentration Data for APIs in Aquatic Tissues

APIs Studied in Wild Specimens (Controlled *In Situ* Studies Indicated by Asterisk*)	Maximum Concentration in Wild Specimens (µg/kg) (Controlled Studies Indicated by Asterisk*)	Historical Precedence in Literature	Notes	Reference
Norfluoxetine	10.27 liver 8.86 brain	Probably 1st reports	Lowest concentrations in muscle. Controlled exposure to 0.55 ng/L of fluoxetine gave peak concentration of 64 µg/kg tissue (Paterson and Metcalfe 2008).	Brooks et al. (2005) [also Chu and Metcalfe (2007)]
	0.07–0.9 brain		Detected in samples from 5 of 8 streams; upper range = 3.6 ng/g. Water = 0.9–4 ng/L.	Schultz et al. (2010)
Paroxetine	0.58 whole body	Probably 1st report		Chu and Metcalfe (2007)
	0.005–1.8 brain		Detected in samples from 6 of 8 streams; upper range = 4.2 ng/g. Water = 2–4 ng/L.	Schultz et al. (2010)
Sertraline	4.27 brain 3.59 liver	Probably 1st reports	Lowest concentrations in muscle.	Brooks et al. (2005)
	1.1–1.2 plasma		Fish exposed to treated sewage effluent; 2 of 3 sites in Sweden; BCF > 138–240 (predicted = 959).	Fick et al. (2010)
Norsertraline	15.6 brain 12.94 liver	Probably 1st reports	Lowest concentrations in muscle.	Brooks et al. (2005)
	0.01–3 brain		Detected in samples from 7 of 8 streams; upper range = 28.9 ng/g. Water = 1.1–6 ng/L.	Schultz et al. (2010)
	0.01–0.02 brain		Detected in samples from 3 of 8 streams; upper range = 0.113 ng/g. Water = 0.8–4 ng/L.	Schultz et al. (2010)
Venlafaxine	0.32 plasma	Possibly 1st report	Sharks.	Gelsleichter (2009)
	0.02–0.1 brain		Detected in samples from 2 of 8 streams; upper range = 1.12 ng/g. Water = 102–220 ng/L.	Schultz et al. (2010)
NSAIDs				
Diclofenac	12 plasma	Probably 1st report	BCF = 5. First report of bioconcentration (under controlled conditions) gave mg/kg concentrations in liver and kidney, with BCFs of nearly 3000 [Schwaiger et al. (2004)].	Brown et al. (2007)
	2.2–20 plasma		Fish exposed to treated sewage effluent; 3 of 3 sites in Sweden; BCF = 2.5–29 (predicted = 93).	Fick et al. (2010)
	328 bile	Probably 1st report in bile	21-day exposure of trout to 0.5 ng/mL; BCF = 657.	Mehinto et al. (2010)

TABLE 8.4 (continued)
Summary of Bioconcentration Data for APIs in Aquatic Tissues

APIs Studied in Wild Specimens (Controlled *In Situ* Studies Indicated by Asterisk*)	Maximum Concentration in Wild Specimens (µg/kg) (Controlled Studies Indicated by Asterisk*)	Historical Precedence in Literature	Notes	Reference
Ibuprofen	84 plasma	Possibly 1st report	BCF = 18,667.	Brown et al. (2007)
	5.5–102 plasma		Fish exposed to treated sewage effluent; 3 of 3 sites in Sweden; BCF = 21–58 (predicted = 77).	Fick et al. (2010)
Ketoprofen	Undetected	Probably 1st attempted analysis	Did not bioconcentrate.	Brown et al. (2007)
	15–107 plasma	Probably 1st report	Fish exposed to treated sewage effluent; 3 of 3 sites in Sweden; BCF = 3.5–48 (predicted = 20).	Fick et al. (2010)
Naproxen	14 plasma	Probably 1st report	BCF = 56.	Brown et al. (2007)
	33–46 plasma		Fish exposed to treated sewage effluent; 3 of 3 sites in Sweden; BCF = 22–28 (predicted = 24).	Fick et al. (2010)
Lipid Regulators	Calculated BCFs for a number of fibrates and statins are provided by Hernando et al. (2007); values ranged 120–380 for clofibrate, fenofibrate, lovastatin, and mevastatin, but higher for simvastatin (800) and fluvastatin (2000).			
Gemfibrozil	210 plasma	Probably 1st report	BCF = 199. Plasma concentration of 170 µg/L yielded a BCF of 500 after controlled exposure to 0.34 µg/L (Mimeault et al. 2005).	Brown et al. (2007)
β-Blockers			Not detected in studies of Brooks et al. (2005) and Brooks (unpublished data).	
Atenolol*	51 plasma*	Probably 1st report (for controlled exposure)	Controlled exposure to 3.2 mg/L. Calculated BCF diminishingly low (Cleuvers 2005).	Winter et al. (2008)
Propranolol*	5 plasma*	Probably 1st report (for controlled exposure)	Controlled exposure to 10 mg/L for 10 days.	Owen et al. (2007; also unpublished data)
Fungicides				
Nine triazoles*	500–1000 body lipid*	Probably 1st report (for controlled exposure)	Exposed via feed at concentrations of 23–35 mg/kg w/w.	Konwick et al. (2006)

continued

TABLE 8.4 (continued)
Summary of Bioconcentration Data for APIs in Aquatic Tissues

APIs Studied in Wild Specimens (Controlled *In Situ* Studies Indicated by Asterisk*)	Maximum Concentration in Wild Specimens (µg/kg) (Controlled Studies Indicated by Asterisk*)	Historical Precedence in Literature	Notes	Reference
Macrocyclic Lactones (avermectins)			Residues well-established as occurring in liver and lipids.	See overview: Danaher et al. (2006).
Abamectin*	38.29 muscle*		Controlled 22-d exposure to 1 µg/L; BCF = 42.	Shen et al. (2005)
Avermectin B1a*	6.8, 3.0, and 11* in whole fish, fillet, and viscera	One of earliest reports for controlled exposure	Controlled 28-d exposure to 1 µg/L; BCFs of 56, 28, and 84.	Van den Heuvel et al. (1996)
Steroids			Considerable data exist on uptake of endogenous estrogens under controlled *in situ* conditions (not summarized here).	See overview: Tyler et al. (2008).
EE2	1100 bile	Perhaps 1st bioconcentration study	4-week exposure in effluent-dominated stream; controlled 46-h exposure to 5 µg/L gave 350,000 µg/L in bile; bioconcentration is in the range of 4–6 orders of magnitude.	Larsson et al. (1999)
	3.79 plasma		Juvenile sharks in wild.	Gelsleichter (2009)
EE2*	31 tissue*	Perhaps the first full life-cycle bioconcentration study of EE2	Lifecycle/posthatch exposure to 64 ng/L. BCFs probably less than 500–2400. No detectable residues after exposure to 1 ng/L.	Lange et al. (2001)
	32–40 (E2eq) bile*		Also detected in testes and ovaries.	Gibson et al. (2005a)
EE2 dibrominated*	92.3 liver* 2.3 ovaries*	Perhaps 1st study (controlled) targeted at API DBPs	Roach exposed for 5 days to drinking water with measurable EE2 of 0.6 ng/L. Accumulated concentrations 18–67 greater than those measured for EE2. BCFs 7894 (liver) and 130 (ovaries).	Flores and Hill (2008)
Levonorgestrel	8.5–12 plasma	Probably 1st report	Fish exposed to treated sewage effluent; 2 of 3 sites in Sweden; BCF = 12,000 (predicted = 46).	Fick et al. (2010)
Testosterone*	80 plasma*		6-day exposure to 1 µg/L; plasma levels dropped quickly upon cessation of exposure.	Maunder et al. (2007)

TABLE 8.4 (continued)
Summary of Bioconcentration Data for APIs in Aquatic Tissues

APIs Studied in Wild Specimens (Controlled *In Situ* Studies Indicated by Asterisk*)	Maximum Concentration in Wild Specimens (µg/kg) (Controlled Studies Indicated by Asterisk*)	Historical Precedence in Literature	Notes	Reference
Equilenin*		Probably 1st report (for exposure to treated sewage)	$BCF = 2.2 \times 10^6$	Tyler et al. (2009)
17β-Dihydroequilenin*	30–40 (E2eq) bile*	Probably 1st report (for exposure to treated sewage)	$BCF = 1.5 \times 10^6$	Gibson et al. (2005a, 2005b)
Antibiotics	Considerable data exist for tissue levels resulting from the high exposures used in aquaculture. Some data exists from controlled studies that simulate the indirect exposure that might occur for organisms in the vicinity of an aquaculture operation. See text for discussion.			
Miscellaneous				
Carbamazepine (CBZ)	2.5 plasma*	Perhaps 1st report (for controlled exposure)	24-h exposure to 200 µg/L; BCF < 1.	Huggett et al. (2004)
	0.3–1.0 plasma		Fish exposed to treated sewage effluent; 3 of 3 sites in Sweden; BCF = 0.8–4.2 (predicted = 6).	Fick et al. (2010)
	0.83–1.44 muscle	1st report in wild fish	11 specimens from effluent-dominated stream.	Ramirez et al. (2007)
	129,000* crustacean	Trophic-level transfer	Concentration on basis of dried weight; fed algae that had been exposed to CBZ.	Lajeunesse et al. (2009)
	None detected in algae		Exposed to 19 ppm.	Andreozzi et al. (2002)
Cilazapril	0.1–0.7 plasma	Possibly 1st report	Fish exposed to treated sewage effluent; 2 of 3 sites in Sweden; BCF > 100–700 (predicted = 6).	Fick et al. (2010)
Diazepam	23–110 liver	Possibly 1st report	CBZ and simvastatin also targeted but not detected.	Kwon et al. (2009)
Diltiazem	0.11–0.27 muscle	1st report in wild fish	11 specimens from effluent-dominated stream.	Ramirez et al. (2007)
	0.002–0.0056 muscle	1st report of "free" concentrations in wild		Zhou et al. (2008)
	0.9 plasma		Fish exposed to treated sewage effluent; 1 of 3 sites in Sweden; BCF = 24–139 (predicted = 14).	Fick et al. (2010)

continued

TABLE 8.4 (continued)
Summary of Bioconcentration Data for APIs in Aquatic Tissues

APIs Studied in Wild Specimens (Controlled *In Situ* Studies Indicated by Asterisk*)	Maximum Concentration in Wild Specimens (µg/kg) (Controlled Studies Indicated by Asterisk*)	Historical Precedence in Literature	Notes	Reference
Diphenhydramine	0.66–1.32 muscle	1st report in wild fish	11 specimens from effluent-dominated stream.	Ramirez et al. (2007)
	0.032–0.082 muscle	1st report of "free" concentrations in wild		Zhou et al. (2008)
Haloperidol	1.2 plasma	Possibly 1st report	Fish exposed to treated sewage effluent; 1 of 3 sites in Sweden; BCF = 3.2 (predicted = 153).	Fick et al. (2010)
Meclozine (meclizine)	0.1–0.7 plasma	Possibly 1st report	Fish exposed to treated sewage effluent; 2 of 3 sites in Sweden; BCF > 200–1400 (predicted = 2521).	Fick et al. (2010)
Memantine	2.3 plasma	Probably 1st report	Fish exposed to treated sewage effluent; 1 of 3 sites in Sweden; BCF < 50–164 (predicted = 36).	Fick et al. (2010)
Orphenadrine	0.9 plasma	Probably 1st report	Fish exposed to treated sewage effluent; 1 of 3 sites in Sweden; BCF < 63–100 (predicted = 61).	Fick et al. (2010)
Oxazepam	0.2–0.7 plasma	Probably 1st report	Fish exposed to treated sewage effluent; 3 of 3 sites in Sweden; BCF = 0.7–3.6 (predicted = 7).	Fick et al. (2010)
Risperidone	0.3–2.4 plasma	Probably 1st report	Fish exposed to treated sewage effluent; 3 of 3 sites in Sweden; BCF > 60–480 (predicted = 47).	Fick et al. (2010)
Tramadol	1.1–1.9 plasma	Probably 1st report	Fish exposed to treated sewage effluent; 3 of 3 sites in Sweden; BCF = 2.3–3.3 (predicted = 20).	Fick et al. (2010)
Verapamil	0.7 plasma	Probably 1st report	Fish exposed to treated sewage effluent; 1 of 3 sites in Sweden; BCF < 33–175 (predicted = 40).	Fick et al. (2010)
Triclosan, Methyl triclosan, Triclocarban	TCS: 5-chloro-2-(2,4-dichlorophenoxy)phenol MTCS: 4-chloro-1-(2,4-dichlorophenoxy)-2-methoxybenzene TCC: N-(4-chlorophenyl)-N′-(3,4-dichlorophenyl)urea			
TCS	0.61–10.4 plasma	1st report in U.S.	Detected in all 13 species surveyed.	Alaee et al. (2003)

TABLE 8.4 (continued)
Summary of Bioconcentration Data for APIs in Aquatic Tissues

APIs Studied in Wild Specimens (Controlled In Situ Studies Indicated by Asterisk*)	Maximum Concentration in Wild Specimens (µg/kg) (Controlled Studies Indicated by Asterisk*)	Historical Precedence in Literature	Notes	Reference
	0.75–10.0 plasma		13 species from Detroit River. MCTS levels 0.0004–0.013 µg/kg, 3 orders of magnitude lower.	Valters et al. (2005)
	85–270 plasma	1st report in marine mammal	Detected in 7 of 26 dolphins. Possibly highest plasma level reported for any aquatic organism.	Fair et al. (2009)
	240 to 4400 bile	Possibly 1st report (in bile of wild fish)		Adolfsson-Erici et al. (2002)
	14,000–80,000 bile		The Netherlands	Houtman et al. (2004)
	35 whole body; 365 lipid		Swiss lakes	Balmer et al. (2004)
	21 whole body			Mottaleb et al. (2009)
	146 algae whole body	1st report in snail	Algal BAF: <2100 Snail BAF: 500	Coogan et al. (2007)
	58.7 snail whole body			Coogan and La Point (2008)
TCS—halogenated analogs	0.5–13 whole body	1st report of halogenated products	3- and 5-chloromethoxy TCS; 3,5-dichloromethoxy TCS; unidentified bromomethoxy TCS; detected in 21–76% of 29 carp from Las Vegas Bay.	Leiker et al. (2009)
MTCS	38 whole bodies	1st report		Miyazaki et al. (1984)
	0.1–13 whole bodies		Monitoring study of multiple species at 12 locations in Sweden.	Samsøe-Petersen et al. (2003)
	600 whole body; 7000 lipid		Detected in all 29 carp from Las Vegas Bay. TCS not detected. BCF = 1.8×10^5.	Leiker et al. (2009)
	0.0004–0.0132 plasma	1st report in U.S.	Detected in all 13 species surveyed.	Alaee et al. (2003)
	30 muscle		10-year retrospective study of breams. MTCS was always detected, but TCS was rarely present.	Boehmer et al. (2004)
	130–2100 muscle		Seven Swiss rivers	Buser et al. (2006)

continued

TABLE 8.4 (continued)
Summary of Bioconcentration Data for APIs in Aquatic Tissues

APIs Studied in Wild Specimens (Controlled *In Situ* Studies Indicated by Asterisk*)	Maximum Concentration in Wild Specimens (μg/kg) (Controlled Studies Indicated by Asterisk*)	Historical Precedence in Literature	Notes	Reference
	4–233 lipid		Swiss lakes	Balmer et al. (2005)
	89 algae whole body	1st report in snail	Algal BAF: <1500	Coogan et al. (2007)
	49.8 snail whole body		Snail BAF: 1200	Coogan and La Point (2008)
TCC	401 algae whole body	1st report in any organism	Algal BAF: <2700	Coogan et al. (2007)
	299 snail whole body	1st report of BSAF	Snail BAF: 1600	Coogan and La Point (2008)
	42,000 worm whole body*		Worm BSAF: ca 2	Higgins et al. (2009)

Considering the data for all APIs (excluding triclosan and its derivatives) on aquatic tissue levels or bioconcentration, the following can be noted. Only a handful of studies predate 2003. The concentrations for the majority of APIs range from 1 to 100 μg/kg regardless of tissue type. Those APIs showing higher concentrations include gemfibrozil and triazole fungicides. Most data are for controlled *in situ* exposures rather than for organisms sampled in the wild. Data for tissue levels in wild samples exist for roughly 21 APIs and metabolites. Data for controlled studies exist for about 9 APIs; the study of Fick et al. (2010) exposed fish to treated sewage effluent and quantified an additional 10 unique APIs. Data for tissue levels exist for roughly 40 human APIs/metabolites (excluding antibiotics) but many are from single studies. Steroids are commonly quantified as total (conjugates combined with free). Of the existing calculated empirical BCFs, except those for steroids, nearly all are lower than several thousand, most being lower than 100. More data exist for estrogens (especially endogenous estrogens) and triclosan (including MTCS and other derivatives) than for any other class of APIs; surprisingly, despite its high usage (similar to triclosan), very little data exists for triclocarban. MTCS, unlike TCS, does not concentrate in plasma. Tissue concentrations for both TCS and MTCS can exceed tens of thousands μg/kg, with BCFs up to the range of several million.

8.3 FACTORS INFLUENCING EXPOSURE

8.3.1 General Considerations

It is important to keep in mind the difficulty in comparing BCFs between APIs (or even for a given API) or between species of fish and other aquatic organisms. The wide range of variables in Table 8.3 can add tremendous variability to these values. But moreover, BCFs are reported on different bases, not just whole body; these include different tissues or on a wet-weight basis or on the basis of lipid content. They can also use empirical data generated by static (steady-state equilibration and nonsteady-state) or kinetic uptake measurements, as well as nominal exposure levels that span one or more orders of magnitude (sometimes exceeding the solubility, and other times the uptake rate is the limiting factor, resulting in lower BCFs at higher exposures). These factors make it difficult to distill existing data into succinct generalizations.

Various models have been developed in attempting to link aquatic tissue residues with biological effects. As an example, attempting to establish a more realistic measurable linkage of exposure with effects, the CBR concept holds that the whole-body concentrations across species does not vary wildly among chemical stressors sharing the same MOA for a given biological endpoint. The CBR is supposedly relatively consistent for a given endpoint, whether acute or chronic. Its appeal derives from the assumption that levels of chemical stressors internal to an organism more directly dictate receptor interaction than doses calculated from surrounding ambient concentrations. By the nature of its definition, CBR should be relatively independent of the stressor's ambient concentration in the immediate aqueous environment. The CBR concept supposedly accounts for a measure that is more closely associated with the level of stressor that would actually interact with the receptor. However, in a critical examination of CBR by Barron et al. (2002), published data were not found to support the CBR concept among members from groups of chemicals sharing the same MOA; variability in correlation with effects was found to be as great as other measures such as ambient concentration. Many variables may be at work here. For example, it is not known whether bioaccumulated residues are readily bioavailable, or if rather, only the free residues are (e.g., Zhou et al. 2008).

These issues, together with the many terms used in aquatic exposure (e.g., bioconcentration, bio-accumulation, biomagnification, bioavailability, and biomarkers) and exposure's role in assessing aquatic health, are discussed in the comprehensive work of Geyer et al. (2000), Gobas and Morrison (2000), and van der Oost et al. (2003). What measure of stressor level experienced by an organism serves as the best surrogate for true dose remains elusive. Below we examine several important variables that may be critical for ecological risk assessments of APIs.

8.3.2 Select Site-Specific Factors Influencing Exposure

8.3.2.1 Hydrology

Effluent-dominated ecosystems may be defined as receiving systems in which more than 50% of the in-stream flow results from effluent discharges. Effluent-dependent conditions result seasonally when the in-stream flow of these receiving systems is entirely dependent on effluent discharges. In more arid or semiarid regions experiencing rapid urbanization, effluent-dominated or dependent conditions are common (Brooks et al. 2006). Examples of effluent-dominated large river systems include the Trinity River in Texas and the South Platt River in Colorado (Brooks et al. 2006). Prospective environmental assessments of APIs often include a default in-stream dilution factor of 10 when predicting expected environmental concentrations (Brooks et al. 2003), which are not representative or protective of effluent-dominated or dependent ecosystems.

In an attempt to estimate effluent-dominated conditions in the United States, Brooks et al. (2006) examined information from the U.S. EPA on receiving system critical dilution limits included in the National Pollutant Discharge Elimination System (NPDES) program (U.S. EPA 1991). Under annual mean flow, it was estimated that less than 20% of discharges entered receiving systems with less than 10-fold dilution, but this value increased three fold to approximately 60% of in-stream dilution occurring at less than 10 fold during low flow conditions (e.g., 7Q10, the seven consecutive-day lowest flow with a 10-year recurrence interval) (Brooks et al. 2006).

The NPDES data summarized above was quite dated. New discharges or increased treatment demands on existing dischargers frequently result from increasing population growth. Thus, Brooks et al. (2006) examined a representative sample of NPDES permits (582) in U.S. EPA Region 6, which comprises the states of Arkansas, Louisiana, Oklahoma, New Mexico, Texas, and a number of Tribes. The minimum (or critical) dilution limit for a wastewater stream is the smallest degree of dilution that can avoid reasonable potential to exceed water quality criteria. Of the permits examined during the late 1990s and early 2000s by U.S. EPA Region 6 staff, 58% included critical dilution limits of >50%, suggesting effluent-dominated or dependent conditions under low flows. Critical dilution limits of 100%, indicating effluent-dependent conditions, were observed in 37% of permits evaluated (Brooks et al. 2006).

As noted earlier, in-stream hydrology is an important consideration because effluent-dominated conditions present worst-case locations for API exposures in developed countries. Daughton (2002) proposed the term "pseudopersistent" to describe the unique exposure scenarios to APIs in these ecosystems. Although APIs are designed to be stable enough to ensure parent stability through the manufacturing-distribution-prescription-treatment continuum, APIs are generally considered to have lower environmental persistence than conventional priority pollutants. However, human APIs (and ingredients from PCPs) may be unique compared with conventional contaminants because they can be continuously introduced via effluent to a receiving system (Daughton 2002). Under these conditions the half-lives of the compounds may exceed in-stream hydrologic retention times, increasing the effective exposure duration experienced by organisms residing in the receiving system (Ankley et al. 2007). Of course, increased effective exposure duration could also apply to other effluent contaminants in these scenarios. Unfortunately very little information is available for in-stream magnitude, frequency, and duration of exposure to APIs originated from any of numerous sources, so the influence of hydrology on "pseudopersistence" of APIs requires more study.

8.3.2.2 Wastewater Treatment Technologies

Though effluent-dominated or effluent-dependent conditions described earlier deserve particular attention for API exposures in developed countries, treatment capabilities of WWTPs discharging to these systems are likely to be relatively high because effluent dilution limits are generally more stringent to meet effluent quality goals (e.g., water quality criteria, whole effluent toxicity). An understanding of treatment capabilities for APIs has grown in recent years, though an understanding of site-specific loading of APIs will be influenced by a number of factors. The most comprehensive study to date was commissioned by American Water Works Association Research Foundation (Snyder et al. 2007). During this study, various treatment technologies were evaluated singularly and in combination for their efficiencies in removing select APIs, PCPs, and endocrine-disrupting compounds.

Snyder et al. (2007) concluded that conventional processes for coagulation, flocculation, sedimentation, and ultraviolet radiation (for disinfection) were largely ineffective for many of the target analytes examined, including a number of APIs. More advanced treatment technologies such as reverse osmosis, activated carbon, advanced oxidation processes, and nanofiltration were considered relatively highly effective for target analytes, though API structural properties influenced treatability among tested technologies (Snyder et al. 2007). This study highlighted the importance of understanding ecological risk from specific APIs prior to making risk-based management decisions (U.S. EPA 1999), because risk mitigation technologies such as advanced treatment processes for APIs may be cost-prohibitive for municipal dischargers.

In developing countries, however, advanced WWTP technologies might not be employed, regulatory guidelines not be developed, or enforcement of regulations may not be as prevalent as in the developed world. A recent study by Larsson et al. (2007) examined select APIs in effluent from a WWTP in Patancheru, India. This WWTP was reported to primarily receive influent wastewater from approximately 90 pharmaceutical manufacturers. Although isotope dilution was not employed and extraction efficiencies were not reported in this screening study, high levels of several APIs were reported in grab samples collected on two consecutive days, ranging from 90 (ranitidine) to 31,000 (ciprofloxacin) µg/L. Further, 21 of 59 target pharmaceuticals were reported earlier to be 1 µg/L (Larsson et al. 2007). Levels of most of these APIs represent the highest concentrations reported in the peer-reviewed literature, highlighting the importance of understanding site-specific ecological exposure and risks in less developed countries. A follow-up study (Fick et al. 2009) revealed concentrations of APIs surface- and well-water levels that may be the highest yet reported in the ambient environment—above the ppm (mg/L) level. Lakes receiving treated wastewater effluent contained ppm levels of three fluoroquinolone antibiotics (ciprofloxacin, norfloxacin, and enoxacin) and cetirizine, an antihistamine. These studies show that in special cases, aquatic exposure levels have the potential to reach concentrations that exceed human plasma levels achieved during therapeutic treatment.

8.3.3 SITE-SPECIFIC pH AND API pKA

Many APIs are weak acids or weak bases. Because these compounds are ionizable, their pKa and the pH of the medium influence the proportion of the molecules present in a nonionized form. The nonionized/ionized ratio of an API in a matrix (e.g., body compartment) is an important consideration in pharmacology and toxicology—influencing absorption and disposition profiles of APIs following dosage in target organisms (Klaassen and Watkins 2003). As noted earlier, additional uptake mechanisms are possible for APIs, but the nonionized form of a drug is more nonpolar and thus considered to passively cross membranes more readily than the ionized form of an API (Kah and Brown 2008). Such observations for APIs are included in physiological-based pharmacokinetic models, which are discussed in greater detail later.

For conventional contaminants, such as pentachlorophenol and ammonia, the more nonionized form is believed to be more bioavailable and toxic to aquatic life. Subsequently, the U.S. EPA developed National Ambient Water Quality Criteria for ammonia (U.S. EPA 1985) and pentachlorophenol (U.S. EPA 1986) that incorporate adjustment factors for site-specific differences in pH. Similarly, the nonionized forms of APIs are likely more bioavailable and potentially more toxic to aquatic life residing in receiving systems (Kah and Brown 2008). An example is provided in Figure 8.2 for the SSRIs fluoxetine (Figure 8.2a) and sertraline (Figure 8.2b), which were reported in three fish species in a receiving system with in-stream pH commonly >8.0 (Brooks et al. 2005).

For chemicals that can ionize, distribution into lipid is a function of the pH. For these dissociative systems, a "distribution" coefficient "D" (as opposed to partition coefficient) is calculated; D can be viewed as an "apparent" partition coefficient—one that depends on pH and the degree of ionization. Both fluoxetine (pKa = 10.05 ± 0.10) and sertraline (pKa = 9.47 ± 0.40) are weak bases with log D values and associated BCFs that are predicted to increase over environmentally relevant pH ranges (Figure 8.2); however, the liposome-water distribution coefficient (log D_{lipw}) may be more useful than log D for predicting accumulation of ionizable compounds (Escher et al. 2000).

As presented previously, Nakamura et al. (2008) observed fluoxetine toxicity for, and BCFs in, Japanese medaka to increase with increasing pH in laboratory studies. Valenti et al. (2009) reported similar toxicity-pH observations with juvenile fathead minnows exposed to sertraline. Further, Valenti et al. (2009) performed a time-to-death fathead minnow study with 500 μg/L of sertraline, and estimated associated LT50 values of >48, 31.9, and 4.9 h at pH treatment levels of 6.5, 7.5, and 8.5, respectively. Such observations support the findings of Nakamura et al. (2008) because if more nonionized sertraline exists at higher pH treatment levels, then sertraline should be more bioavailable and more readily absorbed by juvenile fathead minnows, resulting in the observed more rapid onset of mortality at increasingly higher pHs.

Nakamura et al. (2008) further used pH and the BCF values calculated in their study to predict aqueous fluoxetine levels in Pecan Creek, TX that would result in reported levels of accumulation of fluoxetine in fish (Brooks et al. 2005). Interestingly Nakamura et al. (2008) imputed that the fluoxetine concentrations in Pecan Creek, TX should be ~11 ng/L, which is representative of fluoxetine levels routinely observed in Pecan Creek over the past few years (Brooks unpublished data). However, these estimates do not account for other routes of exposure such as diet. Brooks et al. (in preparation) have quantitated levels of sertraline, norsertraline, fluoxetine, and norfluoxetine in periphyton and benthic macroinvertebrates from Pecan Creek, suggesting that future studies should understand the relative contribution of bioconcentration to bioaccumulation of these and other APIs in aquatic life.

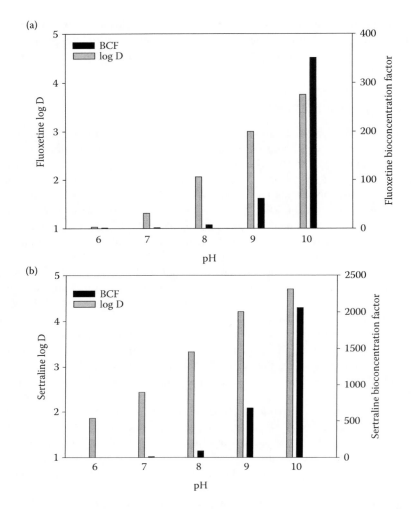

FIGURE 8.2 Bioconcentration factors and log D values for the selective serotonin reuptake inhibitors fluoxetine (a) and sertraline (b) across environmentally relevant pH ranges. Values calculated using Advanced Chemistry Development (ACD/Labs) Software V8.14.

8.3.4 ADVANCEMENT IN TISSUE SAMPLING AND SURROGATE MONITORING

Site-specific exposure may be estimated *in situ* using surrogate measures of API bioavailability. Passive sampling devices have long been used for obtaining estimates of aquatic uptake; see overviews in Greenwood et al. (2007). A variety of devices have been developed for field-deployment to emulate the uptake of xenobiotics by fish via diffusion into lipid. These include semipermeable membrane devices (SPMD) (Barber et al. 2006) and the polar organic chemical integrative sampler (POCIS) (Vermeirssen et al. 2005). Bayen et al. (2009) discuss the variables in the use of passive sampling devices for predicting uptake of hydrophobic chemicals, which would be applicable to only a portion of APIs. Key characteristics for the device and organism are surface-to-volume/weight ratios.

Nonlethal sampling via biopsy has been used for other pollutants but not yet for APIs. One example is the acquisition of tissue samples from fins (Rolfhus et al. 2008).

Conventional sampling devices and tissue sampling approaches suffer from cost for the devices and expenses associated with sample preparation, including organic solvents (and disposal) and analyst time. More significantly, with respect to exposure studies, tissue extraction usually only measures total levels of the analyte as opposed to free, unbound residues that are more readily bio-available. A recently developed approach uses *in-vivo* sampling by way of implanted fibers coated with the sorbent poly(dimethylsiloxane) in a SPME format (Zhou et al. 2008); another approach for establishing chemical activity in tissues is the silicone membrane equilibrator developed by Mayer et al. (2009). SPME avoids many of these limitations and serves to collect only free resi-dues. However, an understanding of the utility of various SPMDs, SPMEs, and POCIS technologies across API classes ranging in physiochemical properties under varied environmentally relevant pH ranges is not available at this time.

8.4 MODELS FOR PREDICTING EXPOSURE AND POTENTIAL EFFECTS OF APIs

8.4.1 BACKGROUND AND PRIORITIZATION

Although several recent book chapters reviewed approaches for predicting human API (Versteeg et al. 2005) and veterinary API (Metcalfe et al. 2008) concentrations in aquatic systems, limited approaches are available for predicting exposure within an organism and linking exposure to poten-tial ecological effects. Prospective assessments often include trigger values for further testing based on predicted environmental concentrations (e.g., 1 µg/L for human APIs in the United States). These predicted concentrations are often driven by production volumes and associated patient uses, and do not consider API potency. Ankley et al. (2007) reviewed assumptions associated with API trigger values based on usage, noting that a trigger value of 1 µg/L equates to a production volume in the United States of 44,000 kg/year, but that this approach is not appropriate for highly potent APIs such as EE2. Despite its very low production volume, it is highly potent (C_{max} is less than 100 pg/mL, where C_{max} is the maximum plasma level reached during therapeutic dosing) and lipophilic (log P ~ 4).

Because APIs represent compounds with a wide range of potencies and physiochemical proper-ties (log P or D, pKa), screening approaches that examine similar properties for the large expanse of thousands of APIs may be useful for prioritizing substances for further bioaccumulation or eco-toxicity studies. Although risk-based prioritization approaches have been developed for veterinary APIs (Boxall et al. 2003, Capleton et al. 2006, Kools et al. 2008) and pesticide transformation products (Sinclair et al. 2006), few approaches have been published for prioritizing human APIs (see Gunnarsson et al. 2008, Kostich and Lazorchak 2008). A powerful tool for such studies may be derived from probabilistic hazard/risk assessment. Chemical toxicity distributions (CTDs) represent robust probabilistic approaches for predicting a specific toxicological response in a model organism (e.g., fathead minnow reproduction) associated with the universe of chemicals that share a com-mon MOA. CTDs are derived by plotting toxicity property data (e.g., NOAELs for fathead minnow reproduction) for a number of chemicals against a probability scale. This represents an approach conceptually similar to SSDs, which plot a distribution of toxicity benchmarks for various species exposed to a common chemical. Much like SSDs, which allow an assessor to estimate the concen-tration below which a certain percentage of aquatic species would respond to a chemical (e.g., an HC5 or 5th centile value), a CTD allows for predictions of the concentration below which a specific percentage of chemicals with a common MOA (or theoretically any other common data property) will still elicit a specific response (e.g., below the NOAEL for fathead minnow reproduction). For example, CTDs were previously demonstrated to predict toxicity of carcinogens (Munro 1990), anti-biotics (Brain et al. 2006), estrogen agonists (Dobbins et al. 2008), and the antimicrobial parabens (Dobbins et al. 2009). This approach is particularly useful for environmental contaminants such as APIs that have limited environmental exposure information (Brain et al. 2006, Dobbins et al. 2009). CTDs were further demonstrated to exhibit diagnostic capabilities to predict differences in sensitivities among common *in vitro* and *in vivo* models of estrogen agonist activity (Dobbins et al. 2008). CTDs are conceptually similar to but provide a more quantitative approach than Threshold of Toxicological Concern methodologies (Gross et al. 2010) previously used in human health risk assessment (Brooks et al. 2009a).

We explored the utility of using probabilistic therapeutic distributions (PTDs), which are identical to CTDs with the exception being that therapeutic plasma data (C_{max}) are examined, to represent the full spectrum of API potencies. Figure 8.3 presents a PTD of C_{max} values for 275 human APIs, and Table 8.5

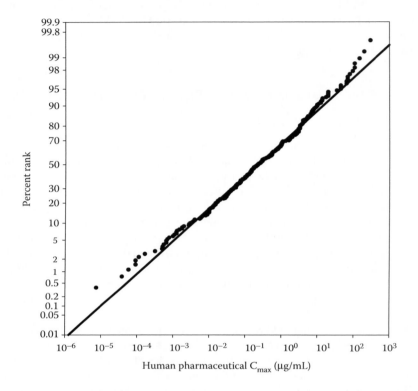

FIGURE 8.3 Probabilistic therapeutic distribution (PTD) of human plasma C_{max} values for 275 Active Pharmaceutical Ingredients (APIs; $r^2 = 0.99$). For APIs with multiple C_{max} values, distribution values are C_{max} concentrations associated with the most common dosage; also see Table 8.5.

TABLE 8.5
Probabilistic Therapeutic Distribution Centiles and Predicted C_{max} Values Derived from C_{max} Plasma Concentrations for 275 Human Pharmaceuticals; also See Figure 8.3

Centile Value	C_{max} (µg/mL)
1%	0.0001169
5%	0.00107
10%	0.00347
25%	0.0248
50%	0.221
75%	1.97
90%	14.15
95%	45.97
99%	419.32

identifies concentrations associated with specific centiles (e.g., 5th, 95th centile) of the distribution. Using this PTD approach $\leq 25\%$ of human APIs are predicted to have C_{max} values less than 0.0248 or greater than 1.97 µg/mL (Table 8.5). Chemical classes with relatively high potencies include endocrine active substances ($n = 12$; range: 0.0000922–0.0595 µg/mL), whereas NSAIDs ($n = 11$; range: 0.705–110 µg/mL) have relatively lower potencies.

Although the PTD approach presented here only examined C_{max} values to compare relative potencies among a wide range of APIs, PTDs could be developed for other API property data useful for predicting accumulation (e.g., BCF). For example, maximum log D (or log K_{lipw}) values for weak acids and weak bases could be examined over an environmentally relevant pH range (e.g., pH 6–9). Maximum log D PTDs could be developed for weak acids and weak bases (e.g., pH 9 used for a weak base with a pKa > 9) to predict the proportion of APIs that may be expected to have log D values greater than some screening threshold (e.g., ~3) at environmentally relevant conditions. Such approaches could provide useful rankings of relative therapeutic property data and for predicting potential property data of new medicines within API classes (Brooks et al. 2009a). For example, Berninger and Brooks (2010) provide a more extensive examination of the utility of PPDs. Specifically PPDs to prioritize pharmaceutical classes for further study, based on a statistically significant relationship between a mammalian margin of safety corollary and fish acute-to-chronic ratios, when fish chronic responses were plausibly linked to therapeutic MOA (Berninger and Brooks 2010).

8.4.2 Physiological-Based Pharmaco(Toxico)Kinetic (PBPK) Models

Pharmacokinetics often utilizes one- and two-compartment models to examine potential systemic effects following exposure. These relatively simple approaches model the distribution of a contaminant (or therapeutic) in a whole body or plasma compartment (one compartment). In a two-compartment model, disposition in a whole body or plasma compartment is coupled with a second compartment, which represents movement to storage depots (e.g., fat) or metabolism. Although these models are useful for deriving parameters such as clearance rates, multicompartment PBPK models are also useful tools for predicting uptake and disposition of environmental contaminants. These more advanced models can: incorporate physiological processes to predict distribution of a compound among various tissues; extrapolate among organisms, exposure routes and ages; and estimate internal dose (Andersen and Dennison 2002, Barton et al. 2007). Subsequently, PBPK models are routinely used in human health risk assessments (U.S. EPA 2006) and increasingly developed for ecotoxicological applications in fish models (see other chapters in this book).

Whereas a number of investigators have examined the utility of physiological models for predicting environmental contaminant uptake and distribution in fish (Erickson et al. 2006a, 2006b, 2008), Erickson et al. (2006a), recently developed a model in trout for describing uptake and elimination of ionizable organic chemicals (chlorophenols) at fish gills. Erickson et al. (2006b) further applied this model to several weak acids with pKa values ranging from 4.74 to 8.62 and log K_{ow} values ranging from 2.75 to 5.12. This model was found to predict uptake of ionizable chemicals based on physiochemical properties under the exposure conditions evaluated with trout (Erickson et al. 2006b). Similar approaches with APIs would be useful for predicting uptake. Although multicompartment PBPK models have not been developed for ionizable APIs and fish, such efforts could be critical to estimate internal dose of APIs to target tissues under environmentally realistic API exposures and pH gradients. This area deserves additional study to characterize API exposure in various tissues of aquatic organisms where therapeutic targets are present.

8.4.3 The Huggett Model

In addition to exhibiting wide ranges in potency, lipophilicity of APIs demonstrates marked variability along a polar-nonpolar continuum. Accounting for such differences in lipophilicity and potency, Huggett et al. (2003) proposed the following model to prioritize human APIs for additional chronic testing (Equation 8.1):

$$F_{SS}PC = EC \times (P_{Blood:Water}) \tag{8.1}$$

where $F_{SS}PC$ is predicted fish steady-state plasma concentration, EC is the aqueous exposure concentration, and $P_{Blood:Water}$ is the predicted partition coefficient in blood from aqueous exposure medium. Fish were selected for model development because more information is available for these organisms, and fish appear to contain relatively high evolutionary conservation of API targets (Huggett et al. 2003, Gunnarsson et al. 2008). The Huggett Model (Huggett "mammalian-fish leverage model") simply proposes that the higher an API's predicted plasma concentration in fish (FSSPC) compared with that of a mammal (e.g., human therapeutic plasma concentration [HTPC] or a C_{max} value) the higher the likelihood of chronic adverse effects (Huggett et al. 2003). As the effect ratio (ER) (Equation 8.2) inflates, the likelihood of an API causing chronic effects drops. As the ER drops, and especially when it becomes less than 1, adverse effects become more probable.

$$ER = H_{T}PC/F_{SS}PC \tag{8.2}$$

The core calculation of this model ($P_{Blood:Water}$) employed an empirical relationship between log K_{ow} and plasma concentrations in trout and *in vitro* partitioning data, which was developed for hydrophobic compounds by Fitzsimmons et al. (2001; Equation 8.3):

$$\log P_{Blood:Water} = 0.73 \times \log K_{ow} - 0.88 \tag{8.3}$$

Though Huggett et al. (2003) used Equation 8.3 for development of Equation 8.2, another relationship (Equation 8.4) reported by Fitzsimmons et al. (2001) appears even more important for APIs with apparent log P values lower than 2:

$$\log P_{BW} = \log [(10^{0.73\log K_{ow}} \times 0.16) + 0.84] \tag{8.4}$$

To derive Equations 8.3 and 8.4, Fitzsimmons et al. (2001) coupled *in vivo* log $P_{Blood:Water}$ values for compounds having log K_{ow} values ranging from 3.1 to 8.2 with previously published *in vitro* data for compounds with lower log K_{ow}'s (Bertelsen et al. 1998).

Brooks et al. (2009a) extended this approach for fish exposed to veterinary medicines (Equation 8.5):

$$ER = EIC_{Plasma}/C_{max} \tag{8.5}$$

where EIC_{Plasma} is the concentration in fish plasma resulting from environmental exposure. Similar to the Huggett Model, the EIC_{Plasma} value proposed by Brooks et al. (2008) would include an uptake prediction from aqueous exposure, and C_{max} would be derived from animal efficacy studies (e.g., in livestock). As previously noted, the study by Fick et al. (2010) identified accumulation of select pharmaceuticals approaching or exceeding human therapeutic levels in plasma of caged fish below effluent discharges.

It is important to note, however, that the models of Huggett et al. (2003) and Brooks et al. (2009b) considered neither bioaccumulation through dietary sources nor the metabolism potential once an API is absorbed into the fish. They also did not include log D in model derivation. Despite its limitations, the Huggett Model appears to provide a reasonable screening approach that is amenable to further refinement. Remaining to be determined is whether: (i) a relationship similar to the equation of Fitzsimmons et al. (2001) (Equation 8.3) would be appropriate for ionizable APIs if, for example, log K_{ow} were substituted with log D; (ii) dietary exposure is a concern for specific APIs; (iii) clearance rates could be predicted in fish using mammalian or target organism information and allometric scaling approaches (if the metabolic pathway for eliminating an API [e.g., CYP450 isoenzyme] is present in a study species); or (iv) API target densities and functional responses would be different between mammals and fish. Similar approaches have not been developed for other organisms (e.g., invertebrates).

Summary

The preponderance of studies published on APIs as contaminants in the aquatic environment have focused on establishing the presence of APIs in the abiotic environment—primarily levels in water and sediment. Comparatively few studies document tissue concentrations. Even fewer studies examine bioaccumulation from sediments. This is a major limitation in being able to establish correlations between biological effects observed in the field with exposure, especially because exposure usually involves multiple chemical (and other) stressors acting in unison or in sequence. This is a critical step in being able to establish cause and effect. Few data are available on tissue levels from free-ranging, migratory fish in locales not directly impacted by sewage effluent. Even fewer controlled exposure studies follow the emergence of any type of biological effect. As one example, a range of effects have been reported for trout chronically exposed to a minimum of 500 ng/L of diclofenac for 21 days (Mehinto et al. 2010).

As part of the EPA's *National Rivers and Streams Assessment (NRSA)* study (U.S. EPA 2008b), plans include analysis of water and fish fillets from 183 urban rivers in the United States for 54 APIs targeted from a range of therapeutic classes, as well as for four synthetic musks and two of their metabolites (Blocksom et al. 2009). The NRSA will attempt to provide the largest dataset yet on the occurrence of multiple APIs in fish tissue. Sample collection began in 2008, and plans are to report on the data by 2011.

A concerted effort is needed to synthesize the data and knowledge that have already been published, especially in the non-English literature not covered in this review (Daughton 2009). This knowledge basically languishes in the published literature, reducing the ability to rationally prioritize and design the most needed research. As such, it is also unable to prevent duplication of effort. While actual empirical data on BCFs or tissue levels of APIs are extremely scarce, a range of modeling techniques can be used to prioritize APIs according to predicted BCF or known tendency for active transport. A limited list of APIs could then be targeted for in-field monitoring to corroborate predictions. Computed BCFs from models are more available than empirical BCFs, but these have never been compiled in any database. Further, BAFs (and BSAFs) for APIs should be further considered to account for site-specific pH influences on ionization and partitioning, dietary exposures, and potential trophic transfer. The literature focusing on veterinary drug use in aquaculture, although limited primarily to antibiotics and steroids, may be of use in extrapolating to the ambient environment.

Predictive models and studies under controlled laboratory conditions directed at APIs that have yet to be identified in aquatic tissue from the wild could be more widely used to better inform the prioritization of APIs for field biomonitoring. For example, the fact that some synthetic progestins (i.e., levonorgestrel and 19-norethindrone) have a high affinity for fish sex hormone-binding globulin, points to the possibility that their enhanced uptake could lead to their being found in tissues. Observations by Fick et al. (2010) for levonorgestrel appear to confirm this perspective. This is critical because synthetic progestins have been identified to adversely affect fish reproduction at low or even sub-part-per-trillion concentrations (Zeilinger et al. 2009).

More study is needed on the bioaccumulation of metabolites, especially those that are bioactive themselves. Likewise, more needs to be known about halogenated DBPs, which might very well have higher BCFs than their parent APIs. Such data are important with regard to trophic chain transfer and accumulation and with respect to human consumption, particularly in effluent-dominated or dependent ecosystems. Given the probably significant role in medicine to be played by nanomaterials (especially in better targeting drug delivery), studies are needed on the uptake of engineered nanoparticles; see overview by Baun et al. (2008).

Exposure studies under controlled laboratory conditions need to use concentrations that have relevance to the environment. Earlier studies (as late as the 1990s) focused on acute toxicity studies, yielding results using concentrations that far exceeded realistic ambient levels. Concentrations that occur in the aquatic environment are often orders of magnitude lower than those used in many controlled studies. Such low concentrations are capable of eliciting sublethal effects that are much more difficult to detect, especially those leading to subtle behavioral effects or changes that are of delayed onset. Further studies relating tissue-specific internal exposures to sublethal responses predicted by API MOAs are warranted.

Perhaps the gold standard for the study of exposure is the "lossless" model using radiolabeled APIs. This permits attempts at achieving mass balance closures for residues distribution over the whole body (e.g., Junker et al. 2006) and in target tissues where therapeutic targets for many APIs are present at highest densities. Even with radiolabeled materials, however, it is rare that mass balances can be fully achieved (Roffey et al. 2007). Unfortunately, radiolabeled APIs are expensive and not widely available, presumably precluding such studies to date.

A major question is whether measurements of whole body or of certain tissues or compartments accurately reflect the internal dose at target organs better than exposure measurements imputed from aqueous concentrations. While assumed to better represent exposure than do concentrations external to the organism, the actual bioavailability of these residues is essentially unknown. Given the exposure continuum that ultimately leads to stressor-receptor interaction, stressor residues residing in tissues such as fat depots, cartilage, bone/scale, gonads, and blood may not yet have reached their ultimate destinations where maximal biological activity can be affected. New approaches using biopsy or *in vivo* equilibrative extraction of target tissues could illuminate this unknown. As an example, SPME has been used to examine *in vivo* the body/tissue concentrations of free stressors (Wen et al. 2006, Zhou et al. 2008). Other new approaches for measuring tissue concentrations include caudal fin biopsy (e.g., Rolfhus et al. 2008).

The difficulty in assessing tissue concentrations is the driving force behind the development of physical model systems that can emulate biological exposure. A number of approaches have evolved, primarily aqueous sampling devices that act as surrogates for whole-body exposure (to estimate concentrations within the body). These devices include SPMD, SPME, and POCIS. However, the utility of such technologies to be predictive of accumulation in organ systems within an organism where API targets are present is not understood. Coupling radiolabeled API studies with development of PBPK fish models presents a useful approach for future efforts, particularly for understanding uptake of ionizable APIs across environmentally relevant pH gradients.

Perhaps the ultimate question is whether the assessment of exposure is truly meaningful if the "totality" of exposure is not considered—that is, all the stressors to which an organism is exposed, both anthropogenic and naturally occurring. This includes the universe of chemical stressors and the numerous other stressors, including those that are physical, electromagnetic, radiological, or biological. Relative risks of APIs for aquatic organisms in the field, compared with other stressors, remains a significant research need. Perhaps the major question with regard to APIs in the aquatic environment is the importance of long-term, low-level, multistressor exposure.

A range of other gaps and limitations with respect to better understanding APIs in aquatic organisms is summarized in Table 8.6.

The United States Environmental Protection Agency through its Office of Research and Development partially funded and collaborated in the research described here. It has been subjected to Agency review and approved for publication.

ACKNOWLEDGMENTS

We thank Jason Berninger for assistance developing Figure 8.3 and Table 8.5. BWB was supported by grants from the U.S. Environmental Protection Agency and WateReuse Foundation during development of this manuscript.

TABLE 8.6
Tissue Levels of APIs in Fish: Limitations, Gaps, and Unique Aspects

Limitation of Data	Example/Explanation	Additional Details
Limited data on tissue concentrations	Of the thousands of publications covering the many aspects of APIs as environmental contaminants, only about 50 published studies have reported API concentrations in fish tissues.	Only a subset of these studies has reported residues in fish exposed in wild, native environments.
Ecotoxicity studies rarely report tissue occurrence of APIs	Ecotoxicity studies of APIs have been conducted to assess effects endpoints. They generally provide little data or insights regarding either the parameters involved with internal exposure or tissue concentrations.	
Linkage between tissue residues and biological effects	Do concentrations in aquatic tissues better represent the potential for toxicity than concentrations free in the surrounding aquatic environment? Do they better reflect the actual dose? Tissue residues in target systems could reflect relevant internal dose compared to whole organism CBR approaches.	Very few studies have examined internal dose (in plasma) following exposure in the field.
Tissue data almost always report total concentrations	The portion of the total tissue-level of an API that is bioavailable is almost always unknown; unknown portions can be sequestered as adducts. Free versus bound residues are rarely distinguished, likely because radiolabeled APIs are scarce.	Total residue levels often involve conjugates, which do not pose a readily bioavailable source.
Limited scope of targeted analytes	Extremely few APIs have been targeted for tissue analysis. Empirical tissue levels in nonaquacultured fish have been published for roughly only 30 different APIs, excluding antibiotics (roughly only 20 in samples from native environments); the vast majority of all possible APIs identified in the aquatic environment (and those not yet identified) have never been targeted for tissue analysis.	The limited published data for tissue residues are focused primarily on antibiotics/biocides and natural and synthetic sex steroids; very few studies have targeted the simultaneous presence of multiple APIs.
Exposure concentration verification in laboratory studies	With aqueous exposure under controlled conditions, the actual dosage needs to be measured; the nominal added dosage is likely to be different than the measured dosage (as a result of sorption to container walls and other solids, degradation, etc).	Inaccurate exposure concentrations can lead to calculated BCFs that are too high or too low by one or more orders of magnitude.
Exposure duration and frequency	Little is known regarding the life-cycle body burden of APIs. Multigenerational studies are extremely rare.	Multiyear monitoring studies are even rarer (e.g., Boehmer et al. 2004).
Self-biasing	APIs with the lowest MDLs have the best chance of being detected. Those with the highest MDLs have the lowest chance of being detected. API MDLs for a particular tissue can vary by more than 2 orders of magnitude. This means that APIs commonly present in tissues but having high MDLs might not be detected.	Tissue concentrations below 0.1–1 µg/kg are rarely reported because this usually is below the method limit of detection—because of matrix interferences. So an unknown number of APIs could be present at these levels.
Tissue and measurement basis	API residues have been reported in a wide range of tissues. They are also normalized on different bases (e.g., wet weight, dry weight, lipid content, etc.).	This makes intercomparisons between studies very difficult.

continued

TABLE 8.6 (continued)
Tissue Levels of APIs in Fish: Limitations, Gaps, and Unique Aspects

Limitation of Data	Example/Explanation	Additional Details
Mass balance	Few studies attempt to achieve mass closure around the total body burden of APIs and their distribution across all tissues.	Rigorous closure studies usually require the use of radiolabeled APIs, but the commercial availability of radiolabeled compounds is extremely limited.
Endogenous contributions	For the endogenous sex steroids, levels metabolically synthesized are augmented by unknown levels contributed by uptake of exogenous residues in water and sorbed to food.	Exogenous contributions, even for endogenous steroids, can have origins from pharmaceuticals.
Veterinary medicine exposure studies	The vast preponderance of bioconcentration studies for fish has been conducted because of the concern for food residues resulting from API usage in aquaculture. Knowledge gained from these studies could possibly be evaluated for relevance to exposure in the ambient environment.	Aquaculture studies comprise two major scenarios: (1) direct incorporation of API in fish being treated with high levels during aquaculture, and (2) indirect incorporation of API in wild fish that become exposed to aquaculture residues leftover from uneaten feed and excreted residues.
Controlled exposure studies versus real-world settings	Tissue data are primarily obtained under controlled exposure studies. Of the very limited studies on tissue concentrations, only about two-thirds have been obtained under ambient conditions with fish in their native environments.	Even in native environments, exposure studies are often controlled by using caged wild fish. More studies are needed using fish captured from waters less dominated by sewage effluents.
Real-world exposure scenarios and exposure concentrations	Tissue data obtained under controlled conditions are often derived from exposure concentrations far higher than those that occur in native environments. Controlled exposure studies often use exposure concentrations that are one or more orders of magnitude higher than those that exist in the ambient environment, to maximize the chances of detecting and quantifying any amounts that have been accumulated.	BCFs usually drop off as exposure concentrations increase, leading to gross underestimates for real-world settings at much lower concentrations.
Inadequate uptake models	Lipid-solubility (e.g., as modeled by octanol–water coefficient) is an inadequate predictor of API uptake. K_{ow} coupled with pK_a (log D, log K_{lipw}) provides a more realistic model.	Other mechanisms may be involved, especially those involving active transport, facilitating the uptake of polar (ionizable) APIs.
K_{ow} data unreliable	Published K_{ow} data are extremely variable, probably a result of ionization and localized charges and because of interaction with other ions.	Models employed for predictions of physicochemical properties inherently vary.
Models for predicting BAFs	Bioaccumulation of APIs cannot currently be modeled.	Too many variables compared with the conventional nonpolar pollutants (e.g., legacy POPs), whose accumulation is dictated primarily by lipid-solubility and metabolism is minimal.
Metabolites	The metabolism of APIs creates the potential for bioconcentration of metabolites, some of which are bioactive themselves.	Little is known regarding pharmacokinetics of APIs in fish or other aquatic organisms.

TABLE 8.6 (continued)
Tissue Levels of APIs in Fish: Limitations, Gaps, and Unique Aspects

Limitation of Data	Example/Explanation	Additional Details
Tissue depots/reservoirs	It is poorly understood how APIs are stored and accumulate in aquatic organisms.	Mechanisms could include adducts with DNA, pigments such as melanin, and a wide array of endogenous proteins (especially those in plasma).
Tissue levels are not necessary for adverse effects	Important to note that internal exposure is not necessary for an effect when the target organ is external.	This is the case, for example, with exposure of the lateral-line sensory organ.
Significant BCFs are not necessary for adverse effects	APIs are designed to minimize accumulation during therapy in the body. Any build-up in aquatic tissue could be important, regardless of how low the BCF might be.	
Interspecies extrapolations	Variance in uptake and pharmacokinetics of fish species makes usefulness of extrapolations questionable, requiring further investigation.	Comparative metabolism and other potential applications of pharmacological "read-across" are largely unknown.
Surrogates for bioconcentration	Sampling devices based on partitioning of analytes from aqueous media (e.g., semipermeable membranes) have not been calibrated for fidelity to bioconcentration under varying field conditions and do not account for bioaccumulation through dietary routes of exposure.	

REFERENCES

Adolfsson-Erici, M. 2005. Fish bile in environmental analysis. Doctoral diss., Department of Applied Environmental Science, Stockholm University, Stockholm, Sweden. http://www.diva-portal.org/su/abstract.xsql?dbid=494 (accessed February 3, 2010).

Adolfsson-Erici, M., M. Pettersson, J. Parkkonen, and J. Sturve. 2002. Triclosan, a commonly used bactericide found in human milk and in the aquatic environment in Sweden. *Chemosphere* 46(9–10):1485–1489.

Alaee, M., I. D'Sa, E. Bennett, and R. Letcher. 2003. Levels of triclosan and methyl triclosan in the plasma of fish from the Detroit river. *Organohalogen Compd.* 62:136–139.

Andersen, M., and J. Dennison. 2002. Toxicokinetics models: where we've been and where we need to go! *Hum. Ecol. Risk Assess.* 8(6):1375–1396.

Andreozzi, R., R. Marotta, G. Pinto, and A. Pollio. 2002. Carbamazepine in water: persistence in the environment, ozonation treatment and preliminary assessment on algal toxicity. *Water Res.* 36(11):2869–2877.

Ankley, G. T., M. C. Black, J. Garric, T. H. Hutchinson, and T. Iguchi. 2005. A framework for assessing the hazard of pharmaceutical materials to aquatic species. In *Human Pharmaceuticals: Assessing the Impacts on Aquatic Ecosystems*, ed. R. T. Williams, pp. 183–237. Pensacola, FL: SETAC Press.

Ankley, G. T., B. W. Brooks, D. B. Huggett, and J. P. Sumpter. 2007. Repeating history: pharmaceuticals in the environment. *Environ. Sci. Technol.* 41(24):8211–8217.

Ankley, G. T., K. M. Jensen, M. D. Kahl, J. J. Korte, and E. A. Makynen. 2001. Description and evaluation of a short-term reproduction test with the fathead minnow (*Pimephales promelas*). *Environ. Toxicol. Chem.* 20(6):1276–1290.

Arnot, J. A., and F. A. Gobas. 2006. A review of bioconcentration factor (BCF) and bioaccumulation factor (BAF) assessments for organic chemicals in aquatic organisms. *Environ. Rev.* 14(4):257–297.

Aschmann, S. M., J. Arey, R. Atkinson, and S. L. Simonich. 2001. Atmospheric lifetimes and fates of selected fragrance materials and volatile model compounds. *Environ. Sci. Technol.* 35(18):3595–3600.

Aubry, A.-F. 2002. Applications of affinity chromatography to the study of drug-melanin binding interactions. *J. Chromatogr. B* 768 (1):67–74.

Balmer, M. E., H.-R. Buser, M. D. Müller, and T. Poiger. 2005. Occurrence of some organic UV filters in wastewater, in surface waters, and in fish from Swiss lakes. *Environ. Sci. Technol.* 39(4):953–962.

Balmer, M. E., et al. 2004. Occurrence of methyl triclosan, a transformation product of the bactericide triclosan, in fish and from various lakes in Switzerland. *Environ. Sci. Technol.* 38(2):390–395.

Barber, L. B., S. H. Keefe, R. C. Antweiler, H. E. Taylor, and R. D. Wass. 2006. Accumulation of contaminants in fish from wastewater treatment wetlands. *Environ. Sci. Technol.* 40(2):603–611.

Barron, M. G., J. A. Hansen, and J. Lipton. 2002. Association between contaminant tissue residues and effects in aquatic organisms. *Rev. Environ. Contam. Toxicol.* 173:1–37.

Barton, H., et al. 2007. Characterizing uncertainty and variability in physiologically based pharmacokinetic models: state of the science and needs for research and implementation. *Toxicol. Sci.* 99(2):395–402.

Baun, A., N. Hartmann, K. Grieger, and K. Kusk. 2008. Ecotoxicity of engineered nanoparticles to aquatic invertebrates: a brief review and recommendations for future toxicity testing. *Ecotoxicology* 17(5):387–395.

Bayen, S., T. L. ter Laak, J. Buffle, and J. L. M. Hermens. 2009. Dynamic exposure of organisms and passive samplers to hydrophobic chemicals. *Environ. Sci. Technol.* 43(7):2206–2215.

Bebak-Williams, J., G. Bullock, and M. C. Carson. 2002. Oxytetracycline residues in a freshwater recirculating system. *Aquaculture* 205(3–4):221–230.

Belden, J. B., J. D. Maul, and M. J. Lydy. 2007. Partitioning and photo degradation of ciprofloxacin in aqueous systems in the presence of organic matter. *Chemosphere* 66(8):1390–1395.

Berninger, J. P., and B. W. Brooks. 2010. Leveraging mammalian pharmaceutical toxicology and pharmacology data to predict chronic fish responses to pharmaceuticals. *Toxicol. Lett.* 193(1):69–78.

Bertelsen, S. L., A. D. Hoffman, C. A. Gallinat, C. M. Elonen, and J. W. Nichols. 1998. Evaluation of log kow and tissue lipid content as predictors of chemical partitioning to fish tissues. *Environ. Toxicol. Chem.* 17(8):1447–1455.

Blocksom, K., Batt, J. Lazorchak, et al. 2009. National rivers and streams assessment: fish tissue contaminants. In *8th Annual Surface Water Monitoring and Standards (SWiMS) Meeting* (http://www.epa.gov/region5/water/wqb/swims.htm). Chicago, IL: US EPA.

Boehmer, W., H. Ruedel, A. Wenzel, and C. Schroeter-Kermani. 2004. Retrospective monitoring of triclosan and methyl-triclosan in fish: results from the German environmental specimen bank. *Organohalogen Compd.* 66:1516–1521.

Boxall, A. B. A., L. A. Fogg, P. Kay, P. A. Blackwell, E. J. Pemberton, and A. Croxford. 2003. Prioritisation of veterinary medicines in the UK environment. *Toxicol. Lett.* 142(3):207–218.

Brain, R. A., M. L. Hanson, K. R. Solomon, and B. W. Brooks. 2008a. Aquatic plants exposed to pharmaceuticals: effects and risks. *Rev. Environ. Contam. Toxicol.* 192:67–115.

Brain, R. A., A. J. Ramirez, B. A. Fulton, C. K. Chambliss, and B. W. Brooks. 2008b. Herbicidal effects of sulfamethoxazole in lemna gibba: using p-aminobenzoic acid as a biomarker of effect. *Environ. Sci. Technol.* 42(23):8965–8970.

Brain, R. A., H. Sanderson, P. K. Sibley, and K. R. Solomon. 2006. Probabilistic ecological hazard assessment: evaluating pharmaceutical effects on aquatic higher plants as an example. *Ecotoxicol. Environ. Saf.* 64(2):128–135.

Brooks, B., et al. [in preparation]. Determination of selective serotonin reuptake inhibitors in macroinvertebrates and periphyton from an effluent-dominated stream.

Brooks, B. W., G. T. Ankley, J. F. Hobson, J. M. Lazorchak, R. D. Meyerhoff, and K. R. Solomon. 2008. Assessing the Aquatic Hazards of Veterinary Medicines. In *Veterinary Medicines in the Environment*, eds. M. Crane, A. Boxall, and K. Barrett, pp. 97–153. Pensacola, FL: SETAC Press.

Brooks, B. W., et al. 2005. Determination of select antidepressants in fish from an effluent-dominated stream. *Environ. Toxicol. Chem.* 24(2):464–469.

Brooks, B. W., et al. 2003. Aquatic ecotoxicology of fluoxetine. *Toxicol. Lett.* 142(3):169–183.

Brooks, B. W., D. Huggett, and A. Boxall. 2009a. Pharmaceuticals and personal care products: research needs for the next decade. *Environ. Toxicol. Chem.* 28(12):2469–2472.

Brooks, B. W., D. B. Huggett, R. A. Brain, and G. Ankley. 2009b. Risk asessment considerations for veterinary medicines in aquatic systems. In *Veterinary Pharmaceuticals in the Environment*, eds. K. Henderson and J. Coats, pp. 205–223. Washington, DC: American Chemical Society.

Brooks, B. W., T. M. Riley, and R. D. Taylor. 2006. Water quality of effluent-dominated ecosystems: ecotoxicological, hydrological, and management considerations. *Hydrobiologia* 556(1):365–379.

Brown, J. N., N. Paxeus, L. Forlin, and D. G. J. Larsson. 2007. Variations in bioconcentration of human pharmaceuticals from sewage effluents into fish blood plasma. *Environ. Toxicol. Pharmacol.* 24(3): 267–274.

Budzinski, H., M. H. Devier, P. Labadie, and A. Togola. 2006. Analysis of hormonal steroids in fish plasma and bile by coupling solid-phase extraction to GC/MS. *Anal. Bioanal. Chem.* 386(5):1429–1439.

Burkhard, L. 2009. Estimation of biota sediment accumulation factor (BSAF) from paired observations of chemical concentrations in biota and sediment EPA/600/R-06/047. Cincinnati, OH: U.S. Environmental Protection Agency, Ecological Risk Assessment Support Center, 35 pp. http://oaspub.epa.gov/eims/eim-scomm.getfile?p_download_id=488260 (accessed February 3, 2010).

Buser, H.-R., M. E. Balmer, P. Schmid, and M. Kohler. 2006. Occurrence of UV filters 4-methylbenzylidene camphor and octocrylene in fish from various Swiss rivers with inputs from wastewater treatment plants. *Environ. Sci. Technol.* 40(5):1427–1431.

Buth, J. M., W. A. Arnold, and K. McNeill. 2007. Unexpected products and reaction mechanisms of the aqueous chlorination of cimetidine. *Environ. Sci. Technol.* 41(17):6228–6233.

Cabello, F. C. 2006. Heavy use of prophylactic antibiotics in aquaculture: a growing problem for human and animal health and for the environment. *Environ. Microbiol.* 8(7):1137–1144.

Caldwell, D. J., et al. 2008. Derivation of an aquatic predicted no-effect concentration for the synthetic hormone, 17a-ethinyl estradiol. *Environ. Sci. Technol.* 42(19):7046–7054.

Capleton, A. C.,et al. 2006. Prioritising veterinary medicines according to their potential indirect human exposure and toxicity profile. *Toxicol. Lett.* 163(3):213–223.

Cecinato, A., and C. Balducci. 2007. Detection of cocaine in the airborne particles of the Italian cities Rome and Taranto. *J. Sep. Sci.* 30(12):1930–1935.

Chiu, L., L. Cunningham, D. Raible, E. Rubel, and H. Ou. 2008. Using the zebrafish lateral line to screen for ototoxicity. *JARO J. Assoc. Res. Otolaryngol.* 9(2):178–190.

Chu, P.-S., M. I. Lopez, A. Abraham, K. R. El Said, and S. M. Plakas. 2008. Residue depletion of nitrofuran drugs and their tissue-bound metabolites in channel catfish (ictalurus punctatus) after oral dosing. *J. Agric. Food Chem.* 56(17):8030–8034.

Chu, S., and C. D. Metcalfe. 2007. Analysis of paroxetine, fluoxetine and norfluoxetine in fish tissues using pressurized liquid extraction, mixed mode solid phase extraction cleanup and liquid chromatography-tandem mass spectrometry. *J. Chromatogr.* 1163(1–2):112–118.

Cleuvers, M. 2005. Initial risk assessment for three β-blockers found in the aquatic environment. *Chemosphere* 59(2):199–205.

Clubbs, R. L., and B. W. Brooks. 2007. *Daphnia magna* responses to a vertebrate estrogen receptor agonist and an antagonist: a multigenerational study. *Ecotoxicol. Environ. Saf.* 67(3):385–398.

Coats, J. R., R. L. Metcalf, P. Y. Lu, D. D. Brown, J. F. Williams, and L. G. Hansen. 1976. Model ecosystem evaluation of the environmental impacts of the veterinary drugs phenothiazine, sulfamethazine, clopidol, and diethylstilbestrol. *Environ. Health Perspect.* 18:167–179.

Committee on Biological Markers of the National Research Council. 1987. Biological markers in environmental health research. *Environ. Health Perspect.* 74:3–9.

Coogan, M. A., R. E. Edziyie, T. W. La Point, and B. J. Venables. 2007. Algal bioaccumulation of triclocarban, triclosan, and methyl-triclosan in a North Texas wastewater treatment plant receiving stream. *Chemosphere* 67(10):1911–1918.

Coogan, M. A., and T. W. La Point. 2008. Snail bioaccumulation of triclocarban, triclosan, and methyltriclosan in a North Texas, USA, stream affected by wastewater treatment plant runoff. *Environ. Toxicol. Chem.* 27(8):1788–1793.

Corsolini, S. 2009. Industrial contaminants in Antarctic biota. *J. Chromatogr.* 1216(3):598–612.

Cowan-Ellsberry, C. E., et al. 2008. Approach for extrapolating *in vitro* metabolism data to refine bioconcentration factor estimates. *Chemosphere* 70(10):1804–1817.

Crane, M., C. Watts, and T. Boucard. 2006. Chronic aquatic environmental risks from exposure to human pharmaceuticals. *Sci. Total Environ.* 367(1):23–41.

Cueva-Mestanza, R., Z. Sosa-Ferrera, M. E. Torres-Padrón, and J. J. Santana-Rodríguez. 2008. Preconcentration of pharmaceuticals residues in sediment samples using microwave assisted micellar extraction coupled with solid phase extraction and their determination by HPLC-UV. *J. Chromatogr. B* 863(1): 150–157.

Cunningham, V. L., S. P. Binks, and M. J. Olson. 2009. Human health risk assessment from the presence of human pharmaceuticals in the aquatic environment. *Regul. Toxicol. Pharmacol.* 53(1):39–45.

Danaher, M., L. C. Howells, S. R. H. Crooks, V. Cerkvenik-Flajs, and M. O'Keeffe. 2006. Review of methodology for the determination of macrocyclic lactone residues in biological matrices. *J. Chromatogr. B* 844(2):175–203.

Daughton, C. G. 2002. Environmental stewardship and drugs as pollutants. *The Lancet* 360(9339): 1035–1036.

Daughton, C. G. 2005. Emerging chemicals as pollutants in the environment: a 21st century perspective. *Renewable Resources Journal* 23(4):6–23.

Daughton, C. G. 2007. Pharmaceuticals in the environment: sources and their management. In *Analysis, Fate and Removal of Pharmaceuticals in the Water Cycle*, eds. M. Petrovic and D. Barcelo, pp. 1–58. Amsterdam, The Netherlands: Elsevier Science.

Daughton, C. G. 2008. Pharmaceuticals as environmental pollutants: the ramifications for human exposure. In *International Encyclopedia of Public Health*, eds. K. Heggenhougen and S. Quah, pp. 66–102. Oxford, England: Academic Press.

Daughton, C. G. 2009. Chemicals from the practice of healthcare: challenges and unknowns posed by residues in the environment. *Environ. Toxicol. Chem.* 28(12):2490–2494.

Daughton, C. G., and I. S. Ruhoy. 2009. Environmental footprint of pharmaceuticals—the significance of factors beyond direct excretion to sewers. *Environ. Toxicol. Chem.* 28(12):2495–2521.

Daughton, C. G., and T. A. Ternes. 1999. Pharmaceuticals and personal care products in the environment: agents of subtle change? *Environ. Health Persp.* 107(Suppl 6):907–938.

Delepee, R., H. Pouliquen, and H. Le Bris. 2004. The bryophyte *Fontinalis antipyretica* Hedw. bioaccumulates oxytetracycline, flumequine and oxolinic acid in the freshwater environment. *Sci. Total Environ.* 322(1–3):243–253.

DellaGreca, M., M. R. Iesce, P. Pistillo, L. Previtera, and F. Temussi. 2009. Unusual products of the aqueous chlorination of atenolol. *Chemosphere* 74(5):730–734.

Delorenzo, M. E., J. M. Keller, C. D. Arthur, M. C. Finnegan, H. E. Harper, V. L. Winder, and D. L. Zdankiewicz. 2008. Toxicity of the antimicrobial compound triclosan and formation of the metabolite methyl-triclosan in estuarine systems. *Environ. Toxicol.* 23(2):224–232.

Dettenmaier, E. M., W. J. Doucette, and B. Bugbee. 2009. Chemical hydrophobicity and uptake by plant roots. *Environ. Sci. Technol.* 43(2):324–329.

Dobbins, L., R. Brain, and B. Brooks. 2008. Comparison of the sensitivities of common *in vitro* and *in vivo* assays of estrogenic activity: application of chemical toxicity distributions. *Environ. Toxicol. Chem.* 27(12):2608–2616.

Dobbins, L., S. Usenko, R. Brain, and B. Brooks. 2009. Probabilistic ecological hazard assessment of parabens using *Daphnia magna* and *Pimephales promelas*. *Environ. Toxicol. Chem.* 28(12):2744–2753.

Dodd, M. C., and C. H. Huang. 2004. Transformation of the antibacterial agent sulfamethoxazole in reactions with chlorine: kinetics mechanisms, and pathways. *Environ. Sci. Technol.* 38(21):5607–5615.

Dordio, A. V., C. Duarte, M. Barreiros, A. J. P. Carvalho, A. P. Pinto, and C. T. da Costa. 2008. Toxicity and removal efficiency of pharmaceutical metabolite clofibric acid by Typha spp.—potential use for phytoremediation? *Bioresour. Technol.* 100(3):1156–1161.

Dussault, È. B., V. K. Balakrishnan, U. Borgmann, K. R. Solomon, and P. K. Sibley. 2009. Bioaccumulation of the synthetic hormone 17α-ethinylestradiol in the benthic invertebrates *Chironomus tentans* and Hyalella azteca. *Ecotoxicol. Environ. Saf.* 72(6):1635–1641.

Erickson, R. J., J. M. McKim, G. J. Lien, A. D. Hoffman, and S. L. Batterman. 2006a. Uptake and elimination of ionizable organic chemicals at fish gills: I. Model formulation, parameterization, and behavior. *Environ. Toxicol. Chem.* 25(6):1512–1521.

Erickson, R. J., J. M. McKim, G. J. Lien, A. D. Hoffman, and S. L. Batterman. 2006b. Uptake and elimination of ionizable organic chemicals at fish gills: II. Observed and predicted effects of pH, alkalinity, and chemical properties. *Environ. Toxicol. Chem.* 25(6):1522–1532.

Erickson, R., J. Nichols, P. Cook, and G. Ankley. 2008. Bioavailability of chemical contaminants in aquatic systems. In *The Toxicology of Fishes*, eds.R. T. Di Giulio and D. E. Hinton, pp. 9–54. Boca Raton, FL: CRC Press.

Escher, B. I., et al. 2007. Membrane-water partitioning, membrane permeability and baseline toxicity of the parasiticides ivermectin, albendazole, and morantel. *Environ. Toxicol. Chem.* 27(4):909–918.

Escher, B., R. Schwarzenbach, and J. Westall. 2000. Evaluation of liposome-water partitioning of organic acids and bases. 2. Comparison of experimental determination methods. *Environ. Sci. Technol.* 34(18):3962–3968.

Fair, P. A.,et al. 2009. Occurrence of triclosan in plasma of wild Atlantic bottlenose dolphins (*Tursiops truncatus*) and in their environment. *Environ. Pollut.* 157(8–9):2248–2254.

Farré, M. L., S. Pérez, L. Kantiani, and D. Barceló. 2008. Fate and toxicity of emerging pollutants, their metabolites and transformation products in the aquatic environment. *TrAC, Trends Anal. Chem.* 27(11):991–1007.

Fent, K., A. A. Weston, and D. Caminada. 2006. Ecotoxicology of human pharmaceuticals. *Aquat. Toxicol.* 76(2):122–159.

Fernandez, M. P., M. G. Ikonomou, and I. Buchanan. 2007. An assessment of estrogenic organic contaminants in Canadian wastewaters. *Sci. Total Environ.* 373(1):250–269.

Fick, J., R. H. Lindberg, J. Parkkonen, B. Arvidsson, M. Tysklind, and D. G. J. Larsson. 2010. In press. Therapeutic levels of Levonorgestral detected in blood plasma of fish: results from screening rainbow trout exposed to treated sewage effluents. *Environ. Sci. Technol.* 44(7): 2661–2666.

Fick, J., H. Söderström, R. H. Lindberg, C. Phan, M. Tysklind, and D. G. J. Larsson. 2009. Contamination of surface, ground, and drinking water from pharmaceutical production. *Environ. Toxicol. Chem.* 28(12):2522–2527.

Fitzsimmons, P. N., J. D. Fernandez, A. D. Hoffman, B. C. Butterworth, and J. W. Nichols. 2001. Branchial elimination of superhydrophobic organic compounds by rainbow trout (*Oncorhynchus mykiss*). *Aquat. Toxicol.* 55(1–2):23–34.

Flores, A., and E. M. Hill. 2008. Formation of estrogenic brominated ethinylestradiol in drinking water: implications for aquatic toxicity testing. *Chemosphere* 73(7):1115–1120.

Forni, C., A. Cascone, M. Fiori, and L. Migliore. 2002. Sulphadimethoxine and *Azolla filiculoides* Lam.: a model for drug remediation. *Water Res.* 36(13):3398–3403.

Fu, W., A. Franco, and S. Trapp. 2009. Methods for estimating the bioconcentration factor of ionizable organic chemicals. *Environ. Toxicol. Chem.* 28(7):1372–1379.

Fulton, B., R. Brain, S. Usenko, J. Back, R. King, and B. Brooks. 2009. Influence of nitrogen and phosphorus concentrations and ratios on Lemna gibba growth responses to triclosan in laboratory and stream mesocosm experiments. *Environ. Toxicol. Chem.* 28(12):2610–2621.

Fulton, B. A., R. A. Brain, S. Usenko, J. A. Back, and B. W. Brooks. 2010. Exploring Lemna gibba thresholds to nutrient and chemical stressors: differential effects of triclosan on internal stoichiometry and nitrate uptake across a N:P gradient. *Environ. Toxicol. Chem.* In press. 10.1002/etc.289.

Galvez, F., A. Donini, R. C. Playle, D. S. Smith, M. J. O'Donnell, and C. M. Wood. 2008. A matter of potential concern: natural organic matter alters the electrical properties of fish gills. *Environ. Sci. Technol.* 42(24):9385–9390.

Gaworecki, K. M., and S. J. Klaine. 2008. Behavioral and biochemical responses of hybrid striped bass during and after fluoxetine exposure. *Aquat. Toxicol.* 88(4):207–213.

Gelsleichter, J. 2009. Project profile: exposure of freshwater sharks to human pharmaceuticals. Evaluating the risks that pharmaceutical-related pollutants pose to Caloosahatchee River wildlife: observations on the bull shark, *Carcharhinus leucas*. In *Final Report: Charlotte Harbor National Estuary Program*, CAC 2–18-09. Sarasota, FL: Mote Marine Laboratory, 18 pp. http://www.chnep.org/NEP/agendas-2009/CAC/CAC2–18-09_3c_pharmaceuticals.pdf (accessed February 3, 2010).

Geyer, H., G. Rimkus, I. Scheunert, A. Kaune, K.-W. Schramm, A. Kettrup, M. Zeeman, D. Muir, L. Hansen, and D. Mackay. 2000. Bioaccumulation and occurrence of endocrine-disrupting chemicals (EDCs), persistent organic pollutants (POPs), and other organic compounds in fish and other organisms including humans. In *Bioaccumulation—New Aspects and Developments*, ed. B. Beek, pp. 1–166. Heidelberg: Springer Berlin.

Gibson, R., M. D. Smith, C. J. Spary, C. R. Tyler, and E. M. Hill. 2005a. Mixtures of estrogenic contaminants in bile of fish exposed to wastewater treatment works effluents. *Environ. Sci. Technol.* 39(8):2461–2471.

Gibson, R., C. R. Tyler, and E. M. Hill. 2005b. Analytical methodology for the identification of estrogenic contaminants in fish bile. *J. Chromatogr.* 1066(1–2):33–40.

Gobas, F. A. P. C., and H. A. Morrison. 2000. Bioconcentration and biomagnification in the aquatic environment. In *Handbook of Property Estimation Methods for Chemicals*, eds. D. Mackay and R. S. Boethling, pp. 189–231. Boca Raton, FL: CRC Press.

Golet, E. M., A. C. Alder, and W. Giger. 2002. Environmental exposure and risk assessment of fluoroquinolone antibacterial agents in wastewater and river water of the Glatt Valley Watershed, Switzerland. *Environ. Sci. Technol.* 36(17):3645–3651.

Gomes, R. L., H. E. Deacon, K. M. Lai, J. W. Birkett, M. D. Scrimshaw, and J. N. Lester. 2004. An assessment of the bioaccumulation of estrone in *Daphnia magna*. *Environ. Toxicol. Chem.* 23(1):105–108.

Gordon, D. A., et al. 2006. Effects of eutrophication on vitellogenin gene expression in male fathead minnows (*Pimephales promelas*) exposed to 17a-ethynylestradiol in field mesocosms. *Environ. Pollut.* 142(3):559–566.

Greenwood, R., B. Vrana, and G. Mills, eds. 2007. *Passive Sampling Techniques in Environmental Monitoring*. Wilson & Wilson's Comprehensive Analytical Chemistry series ed. D. Barcelo. Amsterdam, The Netherlands: Elsevier.

Grim, K. C., M. Wolfe, W. Hawkins, R. Johnson, and J. Wolf. 2007. Intersex in japanese medaka (oryzias latipes) used as negative controls in toxicologic bioassays: a review of 54 cases from 41 studies. *Environ. Toxicol. Chem.* 26 (8):1636–1643.

Gross, M., et al. 2010. Thresholds of toxicological concern for endocrine active substances in the aquatic environment. *Integr. Environ. Assess. Manage.* 6 (1):2–11.

Grung, M., et al. 2007. Human and veterinary pharmaceuticals, narcotics, and personal care products in the environment: current state of knowledge and monitoring requirements, TA-2325/2007. Oslo, Norway: Swedish Environmental Research Institute, 98 pp. http://www.sft.no/publikasjoner/2325/ta2325.pdf (accessed February 3, 2010).

Guarino, A. M., and J. J. Lech. 1986. Metabolism, disposition, and toxicity of drugs and other xenobiotics in aquatic species. *Vet. Hum. Toxicol.* 28 (Suppl. 1):38–44.

Gunnarsson, L., A. Jauhiainen, E. Kristiansson, O. Nerman, and D. G. J. Larsson. 2008. Evolutionary conservation of human drug targets in organisms used for environmental risk assessments. *Environ. Sci. Technol.* 42(15):5807–5813.

Hamscher, G., H. T. Pawelzick, S. Sczesny, H. Nau, and J. Hartung. 2003. Antibiotics in dust originating from a pig-fattening farm: a new source of health hazard for farmers? *Environ. Health Perspect.* 111(13):1590–1594.

Henry, T., and M. Black. 2008. Acute and chronic toxicity of fluoxetine (selective serotonin reuptake inhibitor) in western mosquitofish. *Arch. Environ. Contam. Toxicol.* 54(2):325–330.

Hernando, M., A. Agüera, and A. Fernández-Alba. 2007. LC-MS analysis and environmental risk of lipid regulators. *Anal. Bioanal. Chem.* 387(4):1269–1285.

Higgins, C. P., Z. J. Paesani, T. E. A. Chalew, and R. U. Halden. 2009. Bioaccumulation of triclocarban in *Lumbriculus variegatus*. *Environ. Toxicol. Chem.* 28(12):2580–2586.

Hilvarsson, A., H. P. Halldorsson, and A. Granmo. 2007. Medetomidine as a candidate antifoulant: sublethal effects on juvenile turbot (Psetta maxima L.). *Aquat. Toxicol.* 83(3):238–246.

Hinton, D. E., et al. 2005. Resolving mechanisms of toxicity while pursuing ecotoxicological relevance? *Mar. Pollut. Bull.* 51(8–12):635–648.

Hoeger, B., B. Hitzfeld, B. Kollner, D. R. Dietrich, and M. R. van den Heuvel. 2005. Sex and low-level sampling stress modify the impacts of sewage effluent on the rainbow trout (*Oncorhynchus mykiss*) immune system. *Aquat. Toxicol.* 73(1):79–90.

Hou, X., J. Shen, S. Zhang, H. Jiang, and J. R. Coats. 2003. Bioconcentration and elimination of sulfamethazine and its main metabolite in sturgeon (Acipenser schrenkii). *J. Agric. Food Chem.* 51(26):7725–7729.

Houtman, C. J., A. M. Van Oostveen, A. Brouwer, M. H. Lamoree, and J. Legler. 2004. Identification of estrogenic compounds in fish bile using bioassay-directed fractionation. *Environ. Sci. Technol.* 38(23):6415–6423.

Huggett, D. B., J. C. Cook, J. F. Ericson, and R. T. Williams. 2003. A theoretical model for utilizing mammalian pharmacology and safety data to prioritize potential impacts of human pharmaceuticals to fish. *Hum. Ecol. Risk Assess.* 9(7):1789–1799.

Huggett, D. B., J. F. Ericson, J. C. Cook, and R. T. Williams. 2004. Plasma concentrations of human pharmaceuticals as predictors of pharmacological responses in fish. In *Pharmaceuticals in the Environment—Sources, Fate, Effects and Risks*, ed. K. Kümmerer, pp. 373–386. Berlin, Germany: Springer.

Huggett, R., R. Kimerle, P. Mehrle, and H. Bergman, eds. 1992. *Biomarkers: Biochemical, Physiological, and Histological Markers of Anthropogenic Stress*. Boca Raton, FL: Lewis Publishers.

Junker, T., R. Alexy, T. Knacker, and K. Kummerer. 2006. Biodegradability of 14C-labeled antibiotics in a modified laboratory scale sewage treatment plant at environmentally relevant concentrations. *Environ. Sci. Technol.* 40(1):318–324.

Kah, M., and C. D. Brown. 2008. Log D: Lipophilicity for ionisable compounds. *Chemosphere* 72(10): 1401–1408.

Kimball, B. 2008. Endocrine disrupting chemical (EDC) accumulation in Puget Sound sediments and the implications for native fish populations. Senior Thesis diss., School of Oceanography, University of Washington. https://digital.lib.washington.edu/xmlui/handle/1773/3770 (accessed February 3, 2010).

Klaassen, C., and J. Watkins. 2003. *Casarett & Doull's Essentials of Toxicology*. New York, NY: McGraw-Hill.

Konwick, B. J., A. W. Garrison, J. K. Avants, and A. T. Fisk. 2006. Bioaccumulation and biotransformation of chiral triazole fungicides in rainbow trout (*Oncorhynchus mykiss*). *Aquat. Toxicol.* 80(4):372–381.

Kools, S. A., A. Boxall, J. F. Moltmann, G. Bryning, J. Koschorreck, and T. Knacker. 2008. A ranking of European veterinary medicines based on environmental risks. *Integr. Environ. Assess. Manage.* 4(4):399–408.

Kosjek, T., and E. Heath. 2008. Applications of mass spectrometry to identifying pharmaceutical-transformation products in water treatment. *TrAC, Trends Anal. Chem.* 27(10):807–820.

Kostich, M. S., and J. M. Lazorchak. 2008. Risks to aquatic organisms posed by human pharmaceutical use. *Sci. Total Environ.* 389(2–3):329–339.

Kwon, J. W., K. L. Armbrust, D. Vidal-Dorsch, and S. M. Bay. 2009. Determination of 17α-ethynylestradiol, carbamazepine, diazepam, simvastatin, and oxybenzone in fish livers. *J. AOAC Int.* 92(1):359–69.

Lai, K. M., M. D. Scrimshaw, and J. N. Lester. 2002a. Biotransformation and bioconcentration of steroid estrogens by *Chlorella vulgaris*. *Appl. Environ. Microbiol.* 68(2):859–64.

Lai, K. M., M. D. Scrimshaw, and J. N. Lester. 2002b. Prediction of the bioaccumulation factors and body burden of natural and synthetic estrogens in aquatic organisms in the river systems. *Sci. Total Environ.* 289(1–3):159–168.

Lajeunesse, A., G. Vernouillet, P. Eullaffroy, C. Gagnon, P. Juneau, and S. Sauvé. 2009. Determination of carbamazepine in aquatic organisms by liquid–liquid extraction and liquid chromatography-tandem mass spectrometry. *J. Environ. Monit.* 11:723–725.

Lange, R., et al. 2001. Effects of the synthetic estrogen 17α-ethinylestradiol on the life-cycle of the fathead minnow (*Pimephales promelas*). *Environ. Toxicol. Chem.* 20(6):1216–1227.

Langston, W. J., G. R. Burt, B. S. Chesman, and C. H. Vane. 2005. Partitioning, bioavailability and effects of oestrogens and xeno-oestrogens in the aquatic environment. *J. Mar. Biol. Assoc. U.K.* 85(1):1–31.

Larsen, M., K. Hansen, P. Henriksen, and E. Baatrup. 2008. Male zebrafish (*Danio rerio*) courtship behaviour resists the feminising effects of 17α-ethinyloestradiol—morphological sexual characteristics do not. *Aquat. Toxicol.* 87(4):234–244.

Larsson, B. S. 1993. Interaction between chemicals and melanin. *Pigment Cell Res.* 6(3):127–133.

Larsson, D. G. J., C. de Pedro, and N. Paxeus. 2007. Effluent from drug manufactures contains extremely high levels of pharmaceuticals. *J. Hazard. Mater.* 148(3):751–755.

Larsson, D. G. J., et al. 1999. Ethinyloestradiol—an undesired fish contraceptive? *Aquat. Toxicol.* 45(2–3):91–97.

Le Bris, H., and H. Pouliquen. 2004. Experimental study on the bioaccumulation of oxytetracycline and oxolinic acid by the blue mussel (*Mytilus edulis*). An evaluation of its ability to bio-monitor antibiotics in the marine environment. *Mar. Pollut. Bull.* 48(5–6):434–440.

Leiker, T. J., S. R. Abney, S. L. Goodbred, and M. R. Rosen. 2009. Identification of methyl triclosan and halogenated analogues in male common carp (Cyprinus carpio) from Las Vegas Bay and semipermeable membrane devices from Las Vegas Wash, Nevada. *Sci. Total Environ.* 407(6):2102–2114.

Liebig, M., P. Egeler, J. Oehlmann, and T. Knacker. 2005. Bioaccumulation of ^{14}C-17a-ethinylestradiol by the aquatic oligochaete *Lumbriculus variegatus* in spiked artificial sediment. *Chemosphere* 59(2):271–280.

Lissemore, L., C. Hao, P. Yang, P. K. Sibley, S. Mabury, and K. R. Solomon. 2006. An exposure assessment for selected pharmaceuticals within a watershed in Southern Ontario. *Chemosphere* 64(5):717–729.

López de Alda, M. J., A. Gil, E. Paz, and D. Barceló. 2002. Occurrence and analysis of estrogens and progestogens in river sediments by liquid chromatography-electrospray-mass spectrometry. *Analyst* 127(10):1299–1304.

Lu, J., and G. V. Korshin. 2008. A spectroscopic study of the bromination of the endocrine disruptor ethynylestradiol. *Chemosphere* 72(3):504–508.

Lyssimachou, A., and A. Arukwe. 2007. Alteration of Brain and Interrenal StAR Protein, P450scc, and Cyp11β mRNA Levels in Atlantic Salmon after Nominal Waterborne Exposure to the Synthetic Pharmaceutical Estrogen Ethynylestradiol. *J. Toxicol. Environ. Health, A* 70(7):606–613.

Maunder, R. J., P. Matthiessen, J. P. Sumpter, and T. G. Pottinger. 2007. Rapid bioconcentration of steroids in the plasma of three-spined stickleback *Gasterosteus aculeatus* exposed to waterborne testosterone and 17ß-oestradiol. *J. Fish Biol.* 70(3):678–690.

Mayer, P., L. Toraäg, N. Glaesner, and J. Å. Jönsson. 2009. Silicone membrane equilibrator: measuring chemical activity of nonpolar chemicals with poly(dimethylsiloxane) microtubes immersed directly in tissue and lipids. *Anal. Chem.* 81(4):1536–1542.

Mehinto, A. C., E. M. Hill, and C. R. Tyler. 2010. Uptake and biological effects of environmentally relevant concentrations of the nonsteroidal anti-inflammatory pharmaceutical diclofenac in rainbow trout (*Oncorhynchus mykiss*). *Environ. Sci. Technol.* 44(6):2176–2182.

Metcalfe, C. D., A. Boxall, K. Fenner, D. Kolpin, M. Servos, E. Silberhorn, and J. Staveley. 2008. Exposure assessment of veterinary medicines in aquatic systems. In *Veterinary Medicines in the Environment*, eds. M. Crane, K. Barrett and A. B. A. Boxall, pp. 57–96. Pensacola, FL: SETAC Press.

Metcalfe, C. D., S. Chu, C. Judt, H. Li, K. D. Oakes, M. R. Servos, and D. M. Andrews. 2010. Antidepressants and their metabolites in municipal wastewater, and downstream exposure in an urban watershed. *Environ. Toxicol. Chem.* 29(1):79–89.

Migliore, L., S. Cozzolino, and M. Fiori. 2000. Phytotoxicity to and uptake of flumequine used in intensive aquaculture on the aquatic weed, Lythrum salicaria L. *Chemosphere* 40(7):741–750.

Miguel-Queralt, S., and G. L. Hammond. 2008. Sex hormone-binding globulin in fish gills is a portal for sex steroids breached by xenobiotics. *Endocrinology* 149(9):4269–4275.

Mimeault, C., A. J. Woodhouse, X. S. Miao, C. D. Metcalfe, T. W. Moon, and V. L. Trudeau. 2005. The human lipid regulator, gemfibrozil bioconcentrates and reduces testosterone in the goldfish, *Carassius auratus. Aquat. Toxicol.* 73(1):44–54.

Miyazaki, T., T. Yamagishi, and M. Matsumoto. 1984. Residues of 4-chloro-1-(2,4-dichlorophenoxy)-2-methoxybenzene(triclosan methyl) in aquatic biota. *Bull. Environ. Contam. Toxicol.* 32(2):227–32.

Mottaleb, M. A., S. Usenko, J. G. O'Donnell, A. J. Ramirez, B. W. Brooks, and C. K. Chambliss. 2009. Gas chromatography-mass spectrometry screening methods for select UV filters, synthetic musks, alkylphenols, an antimicrobial agent, and an insect repellent in fish. *J. Chromatogr.* 1216(5):815–823.

Munro, I. 1990. Safety assessment procedures for indirect food additives: an overview. Report of a workshop. *Regul. Toxicol. Pharmacol.* 12(1):2–12.

Nakamura, H., R. Kuruto-Niwa, M. Uchida, and Y. Terao. 2007. Formation of chlorinated estrones via hypochlorous disinfection of wastewater effluent containing estrone. *Chemosphere* 66(8):1441–1448.

Nakamura, Y., H. Yamamoto, J. Sekizawa, T. Kondo, N. Hirai, and N. Tatarazako. 2008. The effects of pH on fluoxetine in Japanese medaka (Oryzias latipes): acute toxicity in fish larvae and bioaccumulation in juvenile fish. *Chemosphere* 70(5):865–873.

New World Encyclopedia contributors. 13 August 2008. Chromatophore. In *New World Encyclopedia.*

Nichols, J. W., P. N. Fitzsimmons, and L. P. Burkhard. 2007. *In vitro-in vivo* extrapolation of quantitative hepatic biotransformation data for fish. II. Modeled effects on chemical bioaccumulation. *Environ. Toxicol. Chem.* 26(6):1304–1319.

Nie, X.-P., J.-F. Chen, X. Wang, X.-Z. Zhou, J.-Y. Lu, and Y.-F. Yang. 2008. Bioaccumulation of Ciprofloxacin in *Allogynogenetic crucian* carp and its toxic effects (in Chinese). *Acta Ecologica Sinica* 28(1):246–252.

ONDCP. 2009 [updated October]. Proper disposal of prescription drugs: Federal Guidelines. Washington, DC: White House Office of National Drug Control Policy, http://www.whitehousedrugpolicy.gov/publications/pdf/prescrip_disposal.pdf (accessed February 3, 2010).

Orvos, D. R., D. J. Versteeg, J. Inauen, M. Capdevielle, A. Rothenstein, and V. Cunningham. 2002. Aquatic toxicity of triclosan. *Environ. Toxicol. Chem.* 21(7):1338–1349.

Owen, S. F., et al. 2007. Comparative physiology, pharmacology and toxicology of ß-blockers: mammals versus fish. *Aquat. Toxicol.* 82(3):145–162.

Owen, S. F., et al. 2009. Uptake of Propranolol, a cardiovascular pharmaceutical, from water into fish plasma and its effects on growth and organ biometry. *Aquat. Toxicol.* 93(4):217–224.

Painter, M. M., et al. 2009. Antidepressants at environmentally relevant concentrations affect predator avoidance behavior of larval fathead minnows (*Pimephales promelas*). *Environ. Toxicol. Chem.* 28(12): 2677–2684.

Park, K. J., C. T. Müller, S. Markman, O. Swinscow-Hall, D. Pascoe, and K. L. Buchanan. 2009. Detection of endocrine disrupting chemicals in aerial invertebrates at sewage treatment works. *Chemosphere* 77(11):1459–1464.

Parkerton, T. F., et al. 2008. Guidance for evaluating *in vivo* fish bioaccumulation data. *Integr. Environ. Assess. Manage.* 4(2):139–155.

Paterson, G., and C. D. Metcalfe. 2008. Uptake and depuration of the anti-depressant fluoxetine by the Japanese medaka (Oryzias latipes). *Chemosphere* 74(1):125–130.

Peck, A. M., and K. C. Hornbuckle. 2004. Synthetic musk fragrances in Lake Michigan. *Environ. Sci. Technol.* 38(2):367–72.

Peck, A. M., and K. C. Hornbuckle. 2006. Synthetic musk fragrances in urban and rural air of Iowa and the Great Lakes. *Atmos. Environ.* 40(32):6101–6111.

Pettersson, M. 2006. Endocrine disrupting compounds in effluent waters: Chemical analysis to evaluate exposure of fish. Doctoral diss., Department of Applied Environmental Science (ITM), Stockholm University, Stockholm. http://www.diva-portal.org/su/theses/abstract.xsql?dbid=1373 (accessed February 3, 2010).

Pettersson, M., M. Adolfsson-Erici, J. Parkkonen, L. Forlin, and L. Asplund. 2006. Fish bile used to detect estrogenic substances in treated sewage water. *Sci. Total Environ.* 366(1):174–186.

Pettersson, M., E. Hahlbeck, I. Katsiadaki, L. Asplund, and B.-E. Bengtsson. 2007. Survey of estrogenic and androgenic disruption in Swedish coastal waters by the analysis of bile fluid from perch and biomarkers in the three-spined stickleback. *Mar. Pollut. Bull.* 54(12):1868–1880.

Pouliquen, H., et al. 2009. Comparison of water, sediment and plants for the monitoring of antibiotics: a case study on a river dedicated to fish farming. *Environ. Toxicol. Chem.* 28(3):496–502.

Purdom, C. E., P. A. Hardiman, V. V. J. Bye, N. C. Eno, C. R. Tyler, and J. P. Sumpter. 1994. Estrogenic effects of effluents from sewage treatment works. *Chem. Ecol.* 8(4):275–285.

Ramirez, A., et al. 2009. Occurrence of pharmaceuticals and personal care products (PPCPs) in fish tissues: results of a national pilot study in the U.S. *Environ. Toxicol. Chem.* 28(12):2587–2597.

Ramirez, A. J. 2007. Determination of Pharmaceuticals and Personal Care Products in Fish Using High Performance Liquid Chromatography-Tandem Mass Spectrometry and Gas Chromatography-Mass Spectrometry. PhD diss., Department of Chemistry and Biochemistry, Baylor University, Waco, TX. https://beardocs.baylor.edu/handle/2104/5119 (accessed February 3, 2010).

Ramirez, A. J., M. A. Mottaleb, B. W. Brooks, and C. K. Chambliss. 2007. Analysis of pharmaceuticals in fish using liquid chromatography-tandem mass spectrometry. *Anal. Chem.* 79(8):3155–3163.

Redshaw, C. H., V. G. Wootton, and S. J. Rowland. 2008. Uptake of the pharmaceutical Fluoxetine Hydrochloride from growth medium by Brassicaceae. *Phytochemistry* 69(13):2510–2516.

Rigos, G., I. Nengas, M. Alexis, and G. M. Troisi. 2004. Potential drug (oxytetracycline and oxolinic acid) pollution from Mediterranean sparid fish farms. *Aquat. Toxicol.* 69(3):281–288.

Roffey, S. J., R. S. Obach, J. I. Gedge, and D. A. Smith. 2007. What is the objective of the mass balance study? A retrospective analysis of data in animal and human excretion studies employing radiolabeled drugs. *Drug Metab. Rev.* 39(1):17–43.

Rolfhus, K. R., M. B. Sandheinrich, J. G. Wiener, S. W. Bailey, K. A. Thoreson, and C. R. Hammerschmidt. 2008. Analysis of fin clips as a nonlethal method for monitoring mercury in fish. *Environ. Sci. Technol.* 42(3):871–877.

Ruhoy, I. S., and C. G. Daughton. 2008. Beyond the medicine cabinet: an analysis of where and why medications accumulate. *Environ. Int.* 34(8):1157–1169.

Salo, H. M., N. Hébert, C. Dautremepuits, P. Cejka, D. G. Cyr, and M. Fournier. 2007. Effects of montreal municipal sewage effluents on immune responses of juvenile female rainbow trout (*Oncorhynchus mykiss*). *Aquat. Toxicol.* 84(4):406–414.

Samsøe-Petersen, L., M. Winther-Nielsen, T. Madsen, and DHI Water and Environment. 2003. Fate and effects of Triclosan, Environmental Project No. 861 2003: Danish EPA, 47 pp. http://www.miljoestyrelsen.dk/udgiv/publications/2003/87–7972–984-3/pdf/87–7972–985-1.pdf (accessed February 3, 2010).

Samuelsen, O. B., B. T. Lunestad, B. Husevag, T. Holleland, and A. Ervik. 1992. Residues of oxolinic acid in wild fauna following medication in fish farms. *Dis. Aquat. Org.* 12:111–119.

Sapkota, A., et al. 2008. Aquaculture practices and potential human health risks: current knowledge and future priorities. *Environ. Int.* 34(8):1215–1226.

Schirmer, K., and M. Schirmer. 2008. Who is chasing whom? A call for a more integrated approach to reduce the load of micro-pollutants in the environment. *Water Sci. Technol.* 57(1):145–150.

Schuetze, A., T. Heberer, and S. Juergensen. 2008. Occurrence of residues of the veterinary drug malachite green in eels caught downstream from municipal sewage treatment plants. *Chemosphere* 72(11):1664–1670.

Schultz, I. R., G. Orner, J. L. Merdink, and A. Skillman. 2001. Dose-response relationships and pharmacokinetics of vitellogenin in rainbow trout after intravascular administration of 17α-ethynylestradiol. *Aquat. Toxicol.* 51(3):305–318.

Schultz, M. M., et al. 2010. Antidepressant pharmaceuticals in two U.S. effluent-impacted streams: occurrence and fate in water and sediment, and selective uptake in fish neural tissue. *Environ. Sci. Technol.* 44(6):1918–1925.

Schwaiger, J., H. Ferling, U. Mallow, H. Wintermayr, and R. D. Negele. 2004. Toxic effects of the non-steroidal anti-inflammatory drug diclofenac: Part I: histopathological alterations and bioaccumulation in rainbow trout. *Aquat. Toxicol.* 68(2):141–150.

Scott, A., M. Pinillos, and M. Huertas. 2005. The rate of uptake of sex steroids from water by *Tinca tinca* is influenced by their affinity for sex steroid binding protein in plasma. *J. Fish Biol.* 67(1):182–200.

Scott, A. P., and T. Ellis. 2007. Measurement of fish steroids in water—a review. *Gen. Comp. Endocrinol.* 153 (1–3):392–400.

Shen, J., Q. Zhang, S. Ding, S. Zhang, and J. R. Coats. 2005. Bioconcentration and elimination of avermectin B1 in sturgeon. *Environ. Toxicol. Chem.* 24(2):396–399.

Sinclair, C. J., A. B. A. Boxall, S. A. Parsons, and M. R. Thomas. 2006. Prioritization of pesticide environmental transformation products in drinking water supplies. *Environ. Sci. Technol.* 40(23):7283–7289.

Skillman, A. D., J. J. Nagler, S. E. Hook, J. A. Small, and I. R. Schultz. 2006. Dynamics of 17a-ethynylestradiol exposure in rainbow trout (*Oncorhynchus mykiss*): absorption, tissue distribution, and hepatic gene expression pattern. *Environ. Toxicol. Chem.* 25(11):2997–3005.

Snyder, S. A., E. C. Wert, H. Lei, P. Westerhoff, Y. Yoon, and AWWA Research Foundation. 2007. Removal of EDCs and Pharmaceuticals in Drinking and Reuse Treatment Processes [Project #2758] Denver, CO: Awwa Research Foundation, 331 pp. http://ualweb.library.ualberta.ca/uhtbin/cgisirsi/x/0/0/57/5?user_id =WUAARCHIVE&searchdata1=4198864 (accessed February 3, 2010).

Stanley, J. K., and B. W. Brooks. 2009. Perspectives on ecological risk assessment of chiral compounds. *Integr. Environ. Assess. Manage.* 5(3):364–373.

Stanley, J. K., A. J. Ramirez, C. K. Chambliss, and B. W. Brooks. 2007. Enantiospecific sublethal effects of the antidepressant fluoxetine to a model aquatic vertebrate and invertebrate. *Chemosphere* 69(1):9–16.

Stanley, J. K., A. J. Ramirez, M. Mottaleb, C. K. Chambliss, and B. W. Brooks. 2006. Enantiospecific toxicity of the ß-blocker propranolol to *Daphnia magna* and *Pimephales promelas. Environ. Toxicol. Chem.* 25(7):1780–1786.

Steffenak, I., V. Hormazabal, and M. Yndestad. 1994. Effect of cooking on residues of the quinolones oxolinic acid and flumequine in fish. *Acta Vet. Scand.* 35(3):299–301.

Stewart, A. B., A. V. Spicer, E. K. Inskeep, and R. A. Dailey. 2001. Steroid hormone enrichment of Artemia nauplii. *Aquaculture* 202(1–2):177–181.

Sudova, E., J. Machova, Z. Svobodova, and T. Vesely. 2007. Negative effects of malachite green and possibilities of its replacement in the treatment of fish eggs and fish: a review. *Vet. Med. (Praha).* 52(12):527–539.

Suter, G. W., R. A. Efroymson, B. E. Sample, and D. S. Jones. 2000. *Ecological risk assessment for contaminated sites.* Boca Raton, FL: Lewis Publishers.

Takahashi, A., T. Higashitani, Y. Yakou, M. Saitou, H. Tamamoto, and H. Tanaka. 2003. Evaluating bioaccumulation of suspected endocrine disruptors into periphytons and benthos in the Tama River. *Water Sci. Technol.* 47(9):71–76.

Tan, X. 2007. P-glycoprotein and Membrane Permeability as Determinants for Xenobiotic Bioavailability and Bioaccumulation PhD diss., Comparative Biomedical Sciences Shandong Medical University Shandong, China. http://etd.lsu.edu/docs/available/etd-11152007–101053/ (accessed 3 February 2010).

Testorf, M. F., R. Kronstrand, S. P. S. Svensson, I. Lundström, and J. Ahlner. 2001. Characterization of [3H] flunitrazepam binding to melanin. *Anal. Biochem.* 298(2):259–264.

Tittlemier, S. A., et al. 2007. Analysis of veterinary drug residues in fish and shrimp composites collected during the Canadian Total Diet Study, 1993–2004. *Food Addit. Contam.* 24(1):14–20.

Tyler, C., E. Routledge, and R. van Aerle. 2008. Estrogenic effects of treated sewage effluent on fish: steroids and surfactants in English rivers. In *The Toxicology of Fishes*, eds. R. T. DiGiulio and D. E. Hinton, pp. 971–1002. Boca Raton, FL: CRC Press.

Tyler, C. R., et al. 2009. Environmental health impacts of equine estrogens derived from hormone replacement therapy. *Environ. Sci. Technol.* 43(10):3897–3904.

U.S. EPA. 1985. Ambient water quality criteria for ammonia—1984, EPA-440–5-85–001. Washington, DC: Office of Regulations and Standards, Criteria and Standards Division, http://www.epa.gov/waterscience/criteria/library/ambientwqc/ammonia1984.pdf (accessed February 3, 2010).

U.S. EPA. 1986. Ambient water quality criteria for pentachlorophenol-1986, EPA-440–5-86–009. Washington, DC: Office of Regulations and Standards, Criteria and Standards Division.

U.S. EPA. 1991. Technical Support Document for Water Quality-Based Toxics Control, EPA/505/2–90-001, 335 pp. http://www.epa.gov/npdes/pubs/owm0264.pdf (accessed February 3, 2010).

U.S. EPA. 1999. Guidelines for Ecological Risk Assessment, April 1998, Risk Assessment Forum, EPA/630/R-95/002F. Washington, DC, http://www.epa.gov/ncea/raf/pdfs/ecotxtbx.pdf (accessed February 3, 2010).

U.S. EPA. 2006. Approaches for the Application of Physiologically Based Pharmacokinetic (PBPK) Models and Supporting Data in Risk Assessment (Final Report), EPA/600/R-05/043F, http://cfpub.epa.gov/ncea/cfm/recordisplay.cfm?deid=157668 (accessed February 3, 2010).

U.S. EPA. 2008a. *EPA Pilot study of pharmaceuticals in fish tissue: Office of Water*, http://www.epa.gov/waterscience/ppcp/files/fish-pilot.pdf (accessed February 3, 2010).

U.S. EPA. 2008b. *National Rivers and Streams Assessment (NRSA)*. US Environmental Protection Agency. Washington, DC. http://www.epa.gov/owow/riverssurvey (accessed February 3, 2010).

U.S. EPA. 2009. *Pharmaceuticals and Personal Care Products (PPCPs): Relevant Literature*. US Environmental Protection Agency. http://www.epa.gov/ppcp/lit.html (accessed March 10, 2009).

U.S. FDA. 2010. *How to Dispose of Unused Medicines, Consumer Health Information: US Food and Drug Administration*, http://www.fda.gov/ForConsumers/ConsumerUpdates/ucm101653.htm (accessed February 3, 2010).

Valenti Jr, T. W., P. Perez Hurtado, C. K. Chambliss, and B. W. Brooks. 2009. Aquatic toxicity of sertraline to *Pimephales promelas* at environmentally relevant surface water pH. *Environ. Toxicol. Chem.* 28(12):2685–2694.

Valters, K., et al. 2005. Polybrominated diphenyl ethers and hydroxylated and methoxylated brominated and chlorinated analogues in the plasma of fish from the Detroit River. *Environ. Sci. Technol.* 39(15):5612–5619.

Van Aerle, R., N. Pounds, T. H. Hutchinson, S. Maddix, and C. R. Tyler. 2002. Window of sensitivity for the estrogenic effects of ethinylestradiol in early life-stages of fathead minnow, *Pimephales promelas*. *Ecotoxicology* 11(6):423–434.

Van Bambeke, F., E. Balzi, and P. M. Tulkens. 2000. Antibiotic efflux pumps. *Biochem. Pharmacol.* 60(4):457–470.

Van den Heuvel, W. J. A., A. D. Forbis, B. A. Halley, C. C. Ku, T. A. Jacob, and P. G. Wislocki. 1996. Bioconcentration and depuration of avermectin B1a in the bluegill sunfish. *Environ. Toxicol. Chem.* 15(12):2263–2266.

Van der Oost, R., J. Beyer, and N. P. E. Vermeulen. 2003. Fish bioaccumulation and biomarkers in environmental risk assessment: a review. *Environ. Toxicol. Pharmacol.* 13(2):57–149.

Vermeirssen, E., R. Eggen, B. Escher, and M. Suter. 2008. Estrogens in Swiss rivers and effluents—sampling matters. *Chimia* 62(5):389–394.

Vermeirssen, E. L. M., O. Körner, R. Schönenberger, M. J.-F. Suter, and P. Burkhardt-Holm. 2005. Characterization of environmental estrogens in river water using a three pronged approach: active and passive water sampling and the analysis of accumulated estrogens in the bile of caged fish. *Environ. Sci. Technol.* 39(21):8191–8198.

Versteeg, D. J., A. C. Alder, V. L. Cunningham, D. W. Kolpin, R. Murray-Smith, and T. Ternes. 2005. Environmental exposure modeling and monitoring of human pharmaceutical concentrations in the environment. In *Human Pharmaceuticals: Assessing the Impacts on Aquatic Ecosystems*, ed. R. T. Willams, pp. 71–110. Pensacola, FL: SETAC Press.

Viglino, L., K. Aboulfadl, M. Prévost, and S. Sauvé. 2008. Analysis of natural and synthetic estrogenic endocrine disruptors in environmental waters using online preconcentration coupled with LC-APPI-MS/MS. *Talanta* 76(5):1088–1096.

Vögeli, A. C. 2008. Endocrine disrupting chemicals—linking internal exposure to effects in wild fish. Doctoral diss., ETH Zurich, Zurich, Switzerland. http://e-collection.ethbib.ethz.ch/eserv/eth:30864/eth-30864-02.pdf (accessed February 3, 2010).

Walker, J., D. Dnaebel, K. Mayo, J. Tunkel, and D. Gray. 2004a. Use of QSARs to promote more cost-effective use of chemical monitoring resources. 1. Screening industrial chemicals and pesticides, direct food additives, indirect food additives and pharmaceuticals for biodegradation, bioconcentration and aquatic toxicity potential. *Wat. Qual. Res. J. Canada* 39(1):35–39.

Walker, J. D., D. Plewak, O. Mekenyan, S. Dimitrov, and N. Dimitrova. 2004b. Use of QSARs to promote more cost-effective use of chemical monitoring resources. 2. Screening chemicals for hydrolysis half-lives, Henry's law constants, ultimate biodegradation potential, modes of toxic action and bioavailability. *Wat. Qual. Res. J. Canada* 39(1):40–49.

Weisbrod, A. V., et al. 2007. Workgroup report: review of fish bioaccumulation databases used to identify persistent, bioaccumulative, toxic substances. *Environ. Health Perspect.* 115(2):255–61.

Weisbrod, A. V., et al. 2009. The state of *in vitro* science for use in bioaccumulation assessments for fish. *Environ. Toxicol. Chem.* 28(1):86–96.

Wen, Y., Y. Wang, and Y.-Q. Feng. 2006. Simultaneous residue monitoring of four tetracycline antibiotics in fish muscle by in-tube solid-phase microextraction coupled with high-performance liquid chromatography. *Talanta* 70(1):153–159.

Winter, M. J., et al. 2008. Defining the chronic impacts of atenolol on embryo-larval development and reproduction in the fathead minnow (*Pimephales promelas*). *Aquat. Toxicol.* 86(3):361–369.

Wong, D. T., F. P. Bymaster, L. R. Reid, D. A. Mayle, J. H. Krushinski, and D. W. Robertson. 1993. Norfluoxetine enantiomers as inhibitors of serotonin uptake in rat brain. *Neuropsychopharmacology* 8(4):337–344.

Xu, S. H., and G. Chandra. 1999. Fate of cyclic methylsiloxanes in soils. 2. Rates of degradation and volatilization. *Environ. Sci. Technol.* 33(22):4034–4039.

Zeilinger, J., T. Steger-Hartmann, E. Maser, S. Goller, R. Vonk, and R. Länge. 2009. Effects of synthetic gestagens on fish reproduction. *Environ. Toxicol. Chem.* 28(12):2663–2670.

Zhou, S. N., K. D. Oakes, M. R. Servos, and J. Pawliszyn. 2008. Application of solid-phase microextraction for *in vivo* laboratory and field sampling of pharmaceuticals in fish. *Environ. Sci. Technol.* 42(16):6073–6079.

Common Seal

By T. W. Wood, from *The Royal Natural History*, edited by Richard Lydekker, Frederick Warne & Co., London, 1893–94.

9 Organic Contaminants in Marine Mammals

Concepts in Exposure, Toxicity, and Management

Lisa L. Loseto
Peter S. Ross

CONTENTS

9.1 Introduction 350
9.2 Exposure Routes and Vulnerability 351
    9.2.1 Marine Mammal Lifestyles 352
    9.2.2 Spatial Variability of PBTs in True Seals 355
9.3 Toxic Effects of Organic Contaminants 356
    9.3.1 Persistent Bioaccumulative Toxic Contaminants................................. 356
        9.3.1.1 Legacy Organic Contaminants 356
        9.3.1.2 Chemicals of Emerging Concern... 357
        9.3.1.3 Metabolism and Toxic Action 358
    9.3.2 Nonpersistent Bioaccumulative Toxic Contaminants 359
        9.3.2.1 Polycyclic Aromatic Hydrocarbons 359
        9.3.2.2 Pesticides, Pharmaceuticals, Personal Care Products, and Plastics 359
    9.3.3 Approaches in Marine Mammal Toxicology: Study Design 360
        9.3.3.1 Associative Studies 360
        9.3.3.2 Correlative Studies 362
        9.3.3.3 Captive Studies 363
        9.3.3.4 *In Vitro* Studies 363
9.4 Concepts in Risk Assessment and Management 363
    9.4.1 The Challenge of Complex Mixtures 363
    9.4.2 Weight of Evidence in Marine Mammal Toxicology 364
    9.4.3 International Treaties, National Regulations, and Local Application 364
    9.4.4 Regulatory Guidelines for Environmental Media 365
Summary 367
References... 367

9.1 INTRODUCTION

Marine mammals live partly or entirely in the aquatic environment, which is at the receiving end of deliberate or accidental discharges of thousands of industrial, agricultural, and urban pollutants. As a consequence, marine mammals are exposed to a large array of contaminants throughout their lives. Characterizing the resultant contaminant risks to the health of marine mammal populations is complicated in part by the large number of contaminants in their environment, and the variations in the properties of these contaminants. These properties include environmental half-lives under different conditions, water solubility, vulnerability to abiotic breakdown, vulnerability to metabolic breakdown in different biota, and propensity to travel long distances through atmospheric or oceanic transport. Although these challenges can at times frustrate the assessment of particular risks associated with a given chemical, it is possible to categorize contaminants in support of a basic understanding of the threat that they may pose to marine mammals. Such an approach entails a combination of a chemocentric approach, coupled with an understanding of marine mammal vulnerabilities, and can support a set of principles by which managers and conservationists can make remediation decisions regarding a particular marine mammal species, or a contaminated site.

Because many marine mammals are long lived, often occupy high trophic levels and have large stores of fatty tissue, there has been longstanding concern about contaminants possessing persistent, bioaccumulative, and toxic (PBT) properties; namely those contaminants deemed PBT. These chemicals are as follows:

- *Persist (P) in the environment and do not readily breakdown;*
- *Bioaccumulate (B) in organisms over time due to the inability to metabolize and eliminate the contaminant, resulting food web magnification;*
- *Cause Toxic (T) injury or harm.*

Organic contaminants that do not exhibit PBT characteristics (non-PBTs) may nonetheless remain a concern for some marine mammals, as they may mediate injury via different exposure and mechanistic pathways. They may cause direct harm to marine mammals or they may cause indirect harm by impacting prey quality and quantity. Since marine mammals are likely to be exposed to both PBTs and non-PBTs in their environment, it becomes important to categorize and prioritize the exposure routes and means of toxic injury that may act synergistically or antagonistically.

Healthy marine mammal populations reflect a healthy aquatic environment, thus marine mammals can serve as integrative indicators of aquatic food web quality (Ross 2000, Wells et al. 2004). By establishing regulations, guidelines, best practices, and principles that protect marine mammals from different types of contaminants, one may achieve a broader objective, namely the protection of marine ecosystem health. To develop defensible and effective practices that protect marine mammals from contaminants, it is also important to recognize some of the other threats that have affected marine mammal populations for decades, if not centuries. These include the historical and current impacts of commercial and scientific harvesting, fisheries by-catch, habitat degradation, harmful algal blooms, competition with fisheries for resources, and altered ocean productivity associated with climate change (Lewison et al. 2004, Halpern et al. 2008).

Numerous approaches to characterizing contaminant risks to an organism exist, and these may vary by site, species, management objective or jurisdiction. Risks can be characterized through the delineation of exposure routes and associated effects (dose). Before pursuing a chemical-specific assessment of risk associated with a given dose in a marine mammal, it is important to consider the diversity of life history features among the different marine mammal species, and the complexity of contaminants to which they are exposed.

9.2 EXPOSURE ROUTES AND VULNERABILITY

Exposure routes are in part defined by physicochemical properties of the contaminant of concern, coupled with a consideration of marine mammal habitat use, life history, trophic level, foraging ecology and physiology. Exposure will vary with contaminant type and vulnerability will differ with the species of marine mammal; together these will drive the risk of toxic effect at the individual or population level. There are three major exposure routes by which organic contaminants may affect marine mammal health, these being related to the class of contaminant of concern (Figure 9.1):

1. *Dietary exposure* represents the primary means by which marine mammals are exposed to environmental contaminants of concern, namely those with PBT properties, through the ingestion of prey. The importance of this route is evidenced by the abundant literature available on various PBT chemicals in different marine mammals and associations with adverse health effects.

The degree to which marine mammals become contaminated with PBT contaminants varies as a direct function of feeding preferences and the proximity to a contaminant source. Because of their low water solubility or high octanol/water partitioning coefficients (high log $K_{o/w}$), PBT contaminants can undergo long-range atmospheric transport away from point sources to be deposited into remote regions of the world (Wania and Mackay 1996). Once deposited in aquatic environments, PBT contaminants partition into lipids at the bottom of food webs where they accumulate over time and magnify up food chains, often reaching very high concentrations in top predators (Borga et al. 2001, Hoekstra et al. 2003). The high affinity of PBT contaminants for fatty tissues presents a significant concern for marine mammals where lipids represent the currency of energy. The bioaccumulative and biomagnificative properties of these contaminants highlight the importance of

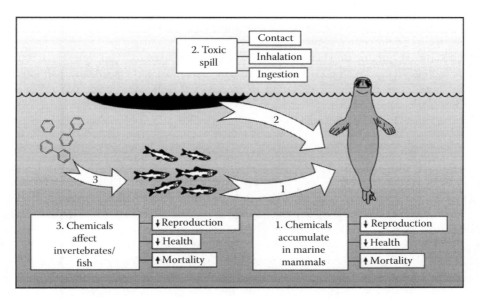

FIGURE 9.1 Marine mammals may be affected by organic contaminants in three general ways: (1) through dietary exposure to chemicals deemed to be persistent, bioaccumulative and toxic, resulting in risk of developmental, neurological, reproductive and/or immunological dysfunction; (2) through contact, inhalation and/or ingestion when a toxic slick, such as an oil spill, contaminates the water's surface; and (3) through an indirect effect, where the prey base of marine mammals may be impacted by chemicals which might not otherwise be considered as a concern to marine mammals because they do not persist or biomagnify in food webs (From Garrett, C., and P. S. Ross, *Can. Tech. Rep Fish. Aquat. Sci.*, pp. 243, 2010. With permission.).

considering the trophic level of the marine mammal species of concern. For example, mammal-eating polar bears and killer whales are exposed to concentrations of PBTs 10 to 100 fold greater than piscivorous marine mammals (Letcher et al. 1995, Ross et al. 2000).

2. *Direct contact* to a given chemical, where exposure in surrounding media (water or air) may cause direct and often acute toxic injury to eyes, oral cavity, skin, airway and lungs, gastro-intestinal tract, liver, and/or the nervous system. Contaminants that elicit harm through direct exposure, including inhalation, ingestion, and/or dermal/fur contact are typically non-PBT contaminants such as petro-chemicals. For example, crude oil spills can introduce hundreds of chemicals to marine habitat. Crude oil contaminants can be grouped into four classes based on solubility: saturates, resins, asphaltenes and aromatics; the latter category includes the toxic polycyclic aromatic hydrocarbons (PAHs) (Leahy and Colwell 1990).

Despite the constraints to an accumulation of PAHs as a result of relatively efficient metabolism in members of aquatic food webs, some marine mammals may nonetheless be heavily exposed through diet because of their reliance on lower trophic level prey items and/or their physiology. Baleen whales, shellfish-eating pinnipeds, and sea otters may be at risk for elevated PAH exposure as a consequence of their diet of invertebrates. For any marine mammal, risk of PAH toxicity is related to their proximity to, and duration spent in, an affected area.

3. *Indirect ecosystem impacts* may take place, where food webs may be affected by pollu-tion, thereby diminishing food availability or food quality for marine mammals. Such indirect effects typically vary directly as a function of proximity to a pollutant source. Over the past few decades, the regulatory and chemical engineering paradigm has increas-ing favored less persistent chemical properties (i.e., "non-PBT" properties), increasing the number of contaminants that have poorly described metabolic or breakdown products in the environment. Non-PBTs may elicit direct impacts to lower trophic level marine mam-mals; however, the indirect ecosystem risks remain extremely difficult to measure or pre-dict. Reduced prey quality and/or quantity have been clearly linked to reduced marine mammal population health. For example, El Niño events lead to reduced ocean productiv-ity and the subsequent nutritional stress and population decline of the South American sea lions (*Otaria flavescens*) (Soto et al. 2004).

9.2.1 Marine Mammal Lifestyles

An understanding of the rich diversity in the biology, ecology, behavior and life histories of marine mammals is an important, yet all-too-often overlooked, component of characterizing contaminant risks. There are over 120 marine mammal species across four phylogenetic Orders, including the toothed and baleen whales (Order Cetacea), the manatees and dugongs (Order Sirenia), true seals, eared seals, and the walrus (Order Pinnipedia), sea otters, and the polar bear (Order Carnivora). Although our chapter considers marine mammals, it is worth noting that several nonmarine species or populations of seals exist, including the freshwater Baikal seals (*Phoca sibirica*), Caspian seals (*Phoca caspica*), and Ungava harbor seal subspecies (*Phoca vitulina mellonae*).

As an initial basis for this chapter, we describe marine mammals as a function of their feeding ecology, thereby providing a framework for characterizing risks that are related to dietary exposure (Figure 9.2). This schematic depicts the divergent risks between high trophic level marine mam-mals (most vulnerable to PBT contaminant accumulation) and low trophic level marine mammals (less vulnerable to PBT contaminant accumulation but more vulnerable to non-PBT contaminants). As one starts at the bottom of the aquatic food web and then proceeds to climb up toward the apex predators, this divergence of risks associated with these two categories of contaminants, the PBT and the non-PBT contaminants, become evident.

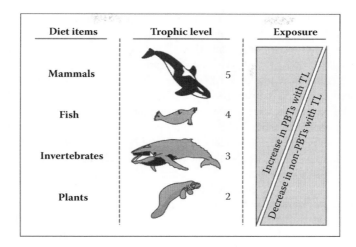

FIGURE 9.2 The trophic level of different marine mammals presents a functional means of characterizing risks associated with dietary exposure. In this way, chemicals deemed to be Persistent, Bioaccumulative, and Toxic (PBT) present the highest risk to marine mammals occupying high trophic positions. Conversely, PBT chemicals present relatively low risks to marine mammals occupying low trophic positions. Those chemicals with non-PBT properties, however, may present an elevated risk to low trophic level marine mammals, as much shorter food webs will inevitably reduce the protective actions of metabolism and degradation before ingestion. In all cases, proximity to source remains an important consideration.

The fully aquatic manatee and dugong are the only herbivorous marine mammals. Feeding on plants and algae, they are not exposed to high concentrations of contaminants that bioaccumulate and biomagnify up food webs. Typically, they inhabit warm environments and lack a thick blubber layer for thermoregulation. The manatee relies on coastal and estuarine habitats that are often in close proximity to industrial and urban areas, consequently exposing these marine mammals to a complex mixture of non-PBT contaminants associated with sewage and other effluents (O'Shea 2003, Bonde et al. 2004).

At the next trophic level are the baleen whales that graze the ocean for plankton, and the walrus, a benthic feeder that favors a diet of filter feeding bivalves. Baleen whales are wholly aquatic and undergo some of the longest mammalian migrations, typically in offshore marine regions. On the other hand, the walrus uses sea ice and shallow Arctic shelves for habitat. Both the baleen whales and the walrus lack fur and have thick blubber layers for energy storage, thermoregulation, and buoyancy. In general terms, these species can be found in habitats far removed from industrialized regions. This, coupled with their low trophic level, limits the exposure risk for both PBT and non-PBT contaminants for these species (O'Shea and Brownell 1994, Hoekstra et al. 2003).

The true (phocid) and eared (otariid) seal species, and the sea otter, are essentially omnivores that feed on low- to mid- trophic level fish, shellfish and/or crustaceans. Many of the toothed whales also feed on a mix of prey at mid-trophic levels, including fish and cephalopods. Thus, these marine mammals can be exposed to higher concentrations of PBT contaminants that biomagnify in aquatic food webs. The true seals are typically temperate and polar species that use both fur and blubber for thermoregulation, whereas eared seals and sea otters have little blubber and rely heavily on fur for insulation. With over 70 species, the toothed whales have member species that have variable sizes and morphologies, as well as habitat types. Habitat use among toothed whales ranges from small home ranges in coastal estuarine habitats (e.g., Amazon river dolphins [*Inia geoffrensis*] (McGuire and Winemiller 1998) to large home ranges in the offshore marine environment (e.g., sperm whales [*Physeter macrocephalus*] [Whitehead 2000]). The pinnipeds and sea otters are partially aquatic, spending variable amounts of time on land or on sea ice platforms, which may also render them vulnerable to contaminants that are found in nearshore or intertidal areas.

The apex predators among the marine mammals are represented by three mammal-eating species, including the polar bear, several populations of killer whales (Ford et al. 1998) and interestingly, some seal-eating walruses (Muir et al. 1995). Due to their top trophic position, these marine mammals are exposed to some of the highest concentrations of PBT contaminants of any vertebrate species in the world (Letcher et al. 1995, Ross et al. 2000). Although these marine mammals may share a similar trophic position, their morphologies, life histories, and habitat use differ widely. The polar bear roams the Arctic sea ice in search of prey, notably the blubber-rich ringed seal (*Phoca hispida*). Polar bears are fur-bearing and have an adipose layer that is easily mobilized for energy. The marine mammal-eating transient killer whales of the NE Pacific Ocean have some of the highest PCB levels among marine mammals (Ross et al. 2000). Although walrus typically feed on low trophic level benthic mollusks, some have demonstrated high PBT levels along with high δ^{15}N that supports feeding on higher trophic level species such as ringed seals (Muir et al. 1995).

While some marine mammals are specialist feeders and may have physiological or behavioral adaptations to feeding on a particular species (e.g., the consumption of squid by sperm whales), others are generalists that can feed on several species (e.g., the consumption of various species of fish and invertebrate by beluga whales). Specialists may be prone to higher exposure risks if their preferred prey become highly contaminated. While dietary diversity and trophic level drive, in large, the degree to which a given marine mammal will be contaminated, there is often a large spatial component to marine mammal foraging. Many marine mammals have large home ranges, can undergo extensive migrations and may seasonally feed in different ecosystems, complicating the interpretation of exposure and associated health risks.

Due to the difficulties in observing marine mammal feeding, research has pursued a variety of approaches to characterize feeding ecology, such as the collection and identification of stomach and/or scat contents (Harvey 1994), the use of tissue biochemical markers such as stable isotopes (e.g., Hobson and Welch 1992, Kurle and Worthy 2002) and fatty acids (e.g., Iverson 1993, Iverson et al. 2004, Loseto et al. 2009), as well as satellite and/or radio tags equipped with video recorders (Bowen et al. 2002). Stable isotopes have provided substantial information about the trophic levels of marine mammals (Pauly et al. 1998). A clear understanding of dietary preferences is the basis for the characterization of PBT exposure-related risks, as prey represents the route of exposure to contaminants and hence the "dose" to which they are exposed.

In addition to the factors that influence dietary exposure among different marine mammals, a number of factors influence exposure and accumulation within a species. The age, sex, size, and reproductive status of a marine mammal shapes contaminant burdens, reflecting differences in life history and dietary intakes within or among different species (Ross et al. 2000). Concentrations of PBT contaminants in adult females and males differ due to the ability of reproductively mature females to transfer fat-soluble PBTs to offspring, thereby lowering their body burden. Males, on the other hand, do not have such an opportunity, and their body burden of PBTs reflect a life time accumulation.

As a result of females offloading PBTs to offspring, marine mammals at the pre and post-natal life history stages are exposed to some of the highest levels of PBTs they will encounter during their lifetime (Addison and Brodie 1987, Hickie et al. 2005). At the pre-natal stage, the fetus is exposed to contaminants passed through the placenta via blood and fat that is mobilized from the blubber of the mother (Duinker and Hillebrand 1979, Tanabe et al. 1982). While transplacental transfer does deliver a notable fraction of the mother's burden to her fetus, lactation accounts for more than 80% of the total transfer to her offspring (Subramanian et al. 1988, Borrell et al. 1995, Salata et al. 1995, Wolkers et al. 2004a). Lactational exposure levels are typically an order of magnitude higher than levels in the prey of the mother. First born calves or pups will receive the largest amount relative to subsequent young, as females offload PBT contaminants they have accumulated until they attain sexual maturity (Aguilar and Borrell 1994, Beckmen et al. 1999). The concentration of PCBs as well as the congener composition varies during the lactation period in relation to selective mobilization and metabolism of different congeners (Jenssen et al. 1996, Debier et al. 2003a, 2003b). The duration of lactation varies among marine mammals, thus extending or shortening the period of high PBT exposure to neonates.

9.2.2 Spatial Variability of PBTs in True Seals

Given the wide range in diet, foraging behavior, physiology and life history among marine mammals, combined with the different physicochemical properties of the various PBT contaminants (which control long-range transport, half-lives, and accumulation potential), it is difficult to draw direct comparisons for PBTs in marine mammals across geographic regions. Such a comparison is constrained by the variability in sampling techniques and study design (i.e., biopsies, strandings, and tissue type), or analytical methods and capabilities (i.e., number of PCBs congeners measured) that change over time and across laboratories. Overlaying these complexities with among-species or among-population variability related to sex, size, age, and reproductive status reduces the ability to adequately assess global trends in marine mammal contaminant levels.

However, a basic comparative approach can be gleaned from recent studies of like-species and similar age categories in the northern hemisphere (Table 9.1). The limited range in PCB concentration in true seals across the northern hemisphere is notable, and underscores the importance

---

TABLE 9.1
Polychlorinated Biphenyl (PCB) Concentrations Vary among and within Marine Mammal Species

Region	Species	ΣPCBs	Reference
Russia			
White Sea, Russia	Ringed Seal	1.7	Muir et al. 2003
East Ice, Russia	Harp seal	1.7	Kleivane et al. 1997
Europe			
Jarfjord, Norway	Ringed seal	1.6	Muir et al. 2000
Svalbard, Norway	Ringed seal	0.5	Wolkers et al. 1998
Norway	Gray seal	1.5	Sormo et al. 2005
Baltic Sea	Gray seal	5.4	Sormo et al. 2005
Greenland			
Qeqertarsaug, W. Greenland	Ringed seal	0.3	Muir et al. 2000
Nanortalik, W. Greenland	Ringed seal	0.6	Muir et al. 2000
E. Greenland	Harp seal	0.6	Oehme et al. 1995
North America			
Southern Labrador, Canada	Harp seal	0.9	Zitko et al. 1998
Sable Island, Canada	Gray seal	7.0	Addison and Stobo 2001
St. Lawrence Estuary, Canada	Gray seal	4.9	Bernt et al. 1999
St. Lawrence Estuary, Canada	Harbor seal	13.3	Bernt et al. 1999
Queen Charlotte Strait, Canada	Harbor seal	1.1	Ross et al. 2004
Strait of Georgia, Canada	Harbor seal	2.5	Ross et al. 2004
Puget Sound, U.S.	Harbor seal	18.0	Ross et al. 2004
PWS AK, U.S.	Harbor seal	0.6	Krahn et al. 1997
PWS AK, U.S.	Ringed seal	0.3	Krahn et al. 1997
AK, U.S.	Ribbon seal	0.5	Quakenbush and Sheffield 2007
Asia			
Hokkaido, Japan	Ribbon seal	0.7	Chiba et al. 2001

Note: This table represents recent PCB concentrations (mg/kg lw) in blubber from small true seals (subfamily Phocinae) of a similar age range (pups and juveniles), thereby reducing the confounding effects of age and lifestyle, and enabling a more accurate overview of geographical variation within the Northern Hemisphere.

of long-range transport in distributing PBT contaminants globally (Table 9.1). Despite the lack of large-scale variation among these seals, there are nonetheless indications of localized contamination. For example, the most contaminated seals are found near local industrial and urban sources, in particular the St Lawrence Estuary (Canada), Puget Sound (U.S.), and the Baltic Sea (northern Europe) (Table 9.1). On the other hand, ringed seals from Arctic regions had the lowest PCB levels observed, reflecting the limited contribution of local sources. The difference between two seal species in the St. Lawrence Estuary, the harbor and the gray seal, likely reflect the differences in feeding ecology (Bernt et al. 1999), whereas the harbor and ringed seals from Alaska had similar PCB levels (Krahn et al. 1997).

Since these phocids represent an important food source for top predators, it is interesting to note that the average PCB concentrations in all of these pinnipeds exceed the recommended dietary intake of 0.05 mg/kg PCBs required to keep killer whales below the health effects threshold of 17 mg/kg in their tissues (Hickie et al. 2007).

9.3 TOXIC EFFECTS OF ORGANIC CONTAMINANTS

9.3.1 PERSISTENT BIOACCUMULATIVE TOXIC CONTAMINANTS

9.3.1.1 Legacy Organic Contaminants

Several PBT contaminants have been banned under the terms of the international Stockholm Convention (UNEP 2004). Polychlorinated biphenyls (PCBs) were once used as heat-resistant oils in transformers and industrial applications, but were also used in plastics, adhesives, and paints. The chemical structure of PCBs consists of two chlorinated benzene rings, and there exist 209 theoretical congeners. The structure of PCBs dictates toxicity through (a) the number and position of chlorine atoms on the benzene rings and (b) how close to planarity (benzene rings on the same plane) the congener is. The toxicity of particular PCB congeners can be described in terms of their structural similarity to the planar dioxin molecule (2,3,7,8-tetrachlorodibenzo-p-dioxin or TCDD; dioxin-like toxicity) that elicits toxicity via the Aryl hydrocarbon Receptor (AhR); or the more "globular" structures (nondioxin like) that are carcinogenic, neurotoxic, and endocrine-disrupting (Giesy and Kannan 1998).

Polychlorinated dibenzo-p-dioxins (PCDDs) and polychlorinated dibenzofurans (PCDFs) were never intentionally produced, and represent by-products resulting from a number of processes. These include incomplete combustion, low temperature incineration (<800°C), manufacturing of pesticides, the use of elemental chlorine in the pulp and paper bleaching process, and the use of pentachlorophenol (PCP) for wood chip preservation. PCDDs and PCDFs are structurally related to PCBs, both having chlorinated phenyl rings. Among the 75 PCDD congeners, 2,3,7,8-TCDD is the most toxic, based on acute toxicity studies on cells and organisms (Safe 1992). Dioxin toxicity is mediated through the AhR, a cytosolic receptor found in all vertebrates (Hahn 1998). The toxic potency of dioxin-like PCB, PCDD, and PCDF congeners in complex mixtures is established on the basis of estimated Toxic Equivalency Factors (TEFs) (Safe 1992, Van den Berg et al. 1998, 2006). The TEF concept represents an important means of characterizing the overall risk of a relatively complex mixture of contaminants, namely for dioxin-like effects for the theoretically possible 419 PCB, PCDD, and PCDF congeners.

Although the TEF approach is widely accepted, it has limitations to its application. TEFs assume that the relative potencies remain the same across species and affect the same endpoints; however, both potencies and effects vary among species (Giesy and Kannan 1998). Second, TEFs do not address toxicity associated with nondioxin-like structures that are globular (nonplanar) and do not induce AhR, but exhibit other effects such as endocrine disruption (Brouwer 1991, Brouwer et al. 1999), neurotoxicity (Tilson et al. 1998), and carcinogenicity (Ahlborg et al. 1995).

Dichloro diphenyl trichloroethane (DDT) was introduced as a pesticide for agricultural practices as well as for antimalarial applications. DDT and its metabolites (breakdown products) are well known for effects on endocrine and reproductive systems in wildlife brought to attention by Rachel Carson's Silent Spring (Carson 1962). DDT effects on egg shell thinning resulted in the near extinction of fish eating birds (Hickey and Anderson 1968). The use of DDT continues for mosquito control and the prevention of malaria in developing countries.

There are numerous other organochlorine pesticides that are considered to be PBT contaminants, many of which have been targeted for elimination under the terms of the Stockholm Convention. While typically of less concern to marine mammals than the PCBs, dioxins, furans and DDT, they may nonetheless contribute to the toxicity of the complex mixture to which marine mammals are exposed (Table 9.2).

9.3.1.2 Chemicals of Emerging Concern

A gradual shift in chemical design and management paradigms away from compounds with PBT characteristics has contributed to a reduction of contaminant-associated risks in marine mammals over the last four decades. However, some PBT chemicals remain in use, in part due to grandfather clauses for chemicals that entered the market prior to new legislative actions. Polybrominated diphenyl ethers (PBDEs) and perfluoroalkylated surfactants (PFAS) are PBT chemicals of emerging concern that are widely used in consumer products, and have been measured in marine mammal food webs around the world (e.g., Giesy and Kannan 2001, AMAP 2004, Wolkers et al. 2004b, Tomy et al. 2009).

PBDEs are a type of brominated flame retardant (BFR) widely used in textiles, appliances, plastics and electronic circuitry. They are similar in structure to the PCBs, with two benzene rings that are brominated rather than chlorinated. Similar to PCBs, PBDEs have 209 possible congeners, are lipophilic, appear to be metabolized through similar enzyme pathways (McKinney et al. 2006), and are endocrine disrupting (Hallgren et al. 2001, Zhou et al. 2002). Like PCBs, PBDEs can be found in marine mammals as far as the remote Arctic (Wolkers et al. 2004b, Tomy et al. 2009). PBDEs face increasing regulatory scrutiny globally such that certain formulations have been added to the list of candidates for elimination under the terms of the Stockholm Convention (Ross et al. 2009).

PFAS are used as refrigerants, polymers, components of pharmaceuticals, lubricants and impregnation agents for textiles. Of particular concern are the perfluorooctane sulfonic acids (PFOS) and

---

TABLE 9.2
The "Dirty Dozen" Chemicals Banned by the United Nations Environmental Program (UNEP) at the Stockholm Convention Include Those Deemed to Be Persistent in the Environment, Bioaccumulative, and Toxic (PBT), as Well as Subject to Long-Range Transport

Chemical	Use
Aldrin	Insecticide
Chlordane	Insecticide
DDT[a]	Insecticide
Dieldrin	Insecticide
Dioxins	By-product of combustion of chlorine products and paper bleaching
Endrin	Rodenticide and insecticide
Furans	By-product of PCB manufacturing
Heptachlor	Insecticide
Hexachlorobenzene	Fungicide, by-product of pesticide manufacturing
Mirex	Insecticide, flame retardant
PCBs[b]	Coolants, lubricants for transformers, weatherproofing, adhesives
Toxaphene	Insecticide

[a] Dichlorodiphenyltrichloroethane.
[b] Polychlorinated biphenyls.

perfluorooctanoic acid (PFOA). These contaminants have received less attention than chlorinated and brominated compounds, partly due to the difficulties in their analytical measurement, and partly due to their high molecular weights, low volatility and insolubility in water as well as fatty tissues (Jensen and Leffers 2008). As a result, PFOS do not concentrate blubber, but rather bind to proteins in blood liver, kidney, and the brain of mammals (Jones et al. 2003). The persistence and widespread occurrence in the environment and in wildlife highlights the emergence of PFAS as a global environmental concern (Giesy and Kannan 2001).

9.3.1.3   Metabolism and Toxic Action

Adverse effects associated with PBT contaminants are largely mediated by the disruption of physiological functions, with the endocrine, immune, neurological, and reproductive systems being particularly vulnerable. The similar structures of many hormones and PBT contaminants explain, in part, the endocrine disrupting nature of PBT contaminants. Dioxin and dioxin-like chemicals may bind to thyroid hormone receptors due to their similar structure to thyroxine (McKinney 1989). The estrogenic properties of DDT and its metabolites are thought to have resulted in the thinning of eggshells and associated reproductive failure in fish-eating birds (Hickey and Anderson 1968), feminization in male gull embryos (Fry and Toone 1981), synthesis of vitellogenesis in fish (Donohoe and Curtis 1996), and alteration of steroid metabolism in gray seals (Lund 1994).

PBTs are highly resistant to metabolic removal, and yet marine mammals, along with most vertebrates, do have some limited ability to metabolize PBTs. The elimination of contaminants such as PCBs or DDT (and PAHs) occurs through the biotransformation and metabolism of lipophilic parent chemicals into more hydrophilic products that can be more readily excreted. Although PBT metabolism reduces the body burden of these compounds, the metabolite breakdown products are also associated with toxic injury (Letcher et al. 2000). PCBs metabolize to hydroxylated (OH-PCBs) and methyl sulfone (MeSO2) PCBs that have been associated with a disruption of the thyroid hormone system (Brouwer et al. 1998, Letcher et al. 2000), while DDT metabolites (e.g., o,p' DDE, p,p' DDD) are known as estrogen agonists (Klotz et al. 1996).

Metabolism is carried out by two major enzyme groups in vertebrates: (a) Phase I, or oxidation enzymes, including cytochrome (CYP) P450s and other oxidation-reduction enzymes that assist with metabolism by the insertion of a polar group or removal of chlorine molecule (dehalogenation); (b) Phase II, or conjugation enzymes, include glutathione S-transferases, among others, that increase the hydrophilicity of the contaminant (Parkinson 1996). Expression of phase I and II enzymes occurs when contaminants interact as ligands with the AhR. The AhR then dimerizes with the AhR nuclear transporter (ARNT) and binds to xenobiotic response elements (XRE) that regulate the transcription of contaminant inducible genes (Whitlock 1999).

Cytochrome P450 enzymes (also known as the microsomal mixed-function monoxygenase enzymes) are essential in the metabolism of steroid hormones and fatty acids. Thus, the induction of P450 by PBTs may also result in the alteration of hormone and fatty acid metabolism, disrupting normal physiological functions. CYP1A1 (cytochrome P450 gene family 1, subfamily A, gene 1) is the gene most commonly associated with planar PCB and 2,3,7,8-TCDD metabolism. CYP1A activities have been associated with PBTs in cetaceans and pinnipeds including beluga, bottlenose dolphins, ringed and gray seal, harbor seal, and polar bears (White et al. 1994, Letcher et al. 1996, Nyman et al. 2003, Miller et al. 2005, Assuncao et al. 2007, Montie et al. 2008).

Non dioxin-like PBTs that are globular or nonplanar in structure may not induce the AhR, yet may cause toxic injury by various mechanisms that can disrupt neurological, behavioral, and endocrine systems (summarized in Giesy and Kannan 1998). Toxic effects of nondioxin-like PBTs include disruption of vitamin A and thyroid hormones (Brouwer 1991). As a result vitamin A and thyroid hormones, like CYP1A, have been used as biomarkers of toxic effects in marine mammals in correlative and associative studies (e.g., Jensen et al. 2003, Braathen et al. 2004, Hall and Thomas 2007, Mos et al. 2007).

9.3.2 Nonpersistent Bioaccumulative Toxic Contaminants

9.3.2.1 Polycyclic Aromatic Hydrocarbons

While natural geological seeps release PAHs into many marine areas, anthropogenic oil spills from ships, pipelines, and storage facilities can deliver sudden and catastrophic quantities to the ocean surface where marine mammals live, breathe, and reproduce. The final sink for PAHs, after volatilization and mixing in the water column is the ocean sediments. These sediments, in turn, become a source to benthic invertebrates and their predators (Latimer and Zhang 2003). Although the ability of most marine invertebrates to metabolize these chemicals is limited, marine mammals and other vertebrates are reasonably able of metabolizing PAHs (Meador et al. 1995). Metabolism by vertebrates removes the opportunity for food web biomagnification, and hence, reduces the risk of dietary exposure and subsequent accumulation in marine mammals (Nakata 2003). Nevertheless, direct contact has been associated with large-scale impacts on marine mammals.

Marine mammal populations were seriously affected by the Exxon Valdez Oil Spill (EVOS) in Prince William Sound, Alaska (U.S.) in 1989. This accident released 40 million litres of crude oil, and was implicated in the deaths of killer whales (*Orcinus orca*), harbor seals (*Phoca vitulina*), and sea otters (*Enhydra lutris*) (Loughlin 1994). The estimated acute mortality of 302 harbor seals (Loughlin and York 2000) was attributed to inhalation and/or ingestion that likely resulted in brain lesions and disorientation causing death (Spraker et al. 1994). Effects from inhalation range from mild irritation to respiratory stress, narcosis and drowning, lung congestion leading to pneumonia, liver damage and death in marine mammals (Geraci 1990, Neff 1990). Killer whales, like other cetaceans, lack an olfactory system making it difficult for them to detect and avoid spills (Matkin et al. 2008). The EVOS-associated oiling of sea otter fur caused matting and a loss of insulative capacity leading to death by hypothermia, smothering, or drowning with mortalities estimated at more than 1000 individuals (Garrott et al. 1993).

Similarly, eared seals (e.g., sea lions *Zalophus californianus*) rely in part on their fur for thermoregulation, whereas the harbor seals have a thick blubber layer in addition to fur, making them less susceptible to hypothermia (St. Aubin 1990). Nonfur-bearing marine mammals such as bottlenose dolphins (*Tursiops truncatus*) have shown little dermal toxicity in captive experiments due to a thick and dense epidermis (Geraci and St. Aubin 1985).

Chronic exposure at sublethal levels may affect wildlife and recovery for protracted periods of time (Peterson et al. 2003). For example, killer whales that fed on oil-covered harbor seals may have been more vulnerable to oil-related impacts as a consequence of existing exposure to PBT contaminants (Matkin et al. 2008). And while metabolic elimination reduces the scope for food web related amplification, PAH breakdown products are highly reactive and may be harmful to marine mammals over the long term (Hawkins et al. 2002).

9.3.2.2 Pesticides, Pharmaceuticals, Personal Care Products, and Plastics

By virtue of their design, currently used pesticides have shorter half-lives, higher water solubilities, and undergo relatively rapid degradation or metabolism. Although such chemicals are designed to target specific pests through acute toxicity, some have been found to impact on nontarget species such as salmon (Scholz et al. 2000, Tierney et al. 2008), a key component of killer whale diet in the NE Pacific Ocean (Ford and Ellis 2006). Reduced returns of Atlantic salmon in eastern Canada were associated with the presence of 4-nonylphenol as an adjuvant in the aerial application of the forestry pesticide Matacil 1.8D (Brown and Fairchild 2003). Even low-level applications of pesticides may alter community structure in ways that ultimately reduce food availability for predators (Relyea 2009, Vonesh and Kraus 2009).

Unlike the target (pest)-specific chemicals that are applied in the environment, many other products found in aquatic ecosystems have been inadvertently released following household or industrial use. These include pharmaceuticals and personal care products (PPCPs) and plastics. PPCPs encompass a large variety of chemicals that have been measured in marine and freshwater environments (Kolpin et al. 2002). The detection of PPCPs in the environment has been constrained by analytical challenges related to high detection limits, transformation, and degradation products. Concentrations and half-lives of these chemicals are typically low, yet their continual discharge in effluent has resulted in "pseudopersistence" (Ankley et al. 2007).

Plasticizers are softeners used in products made with polyvinyl chloride (PVC) to make the product flexible (e.g., plastic bags, toys, clothing, flooring, and medical supplies). The most commonly used plasticizer is di(2-ethylhexyl) phthalate (DEHP). Bisphenol A (BPA) is used as a hardening agent in plastics and is one of the highest-volume chemicals produced worldwide, with 6–10% growth in demand expected each year (Burridge 2003 in [vom Saal and Hughes 2005]). BPA has been shown to reduce antioxidants and cause an early onset of sexual maturation in females at low doses (reviewed in [vom Saal and Hughes 2005]). Given the size of the global plastics industry, and the presence of plastic constituents in the aquatic environment, there may increasing impact on invertebrate and fish communities.

Like PAHs, BPA and phthalates can also undergo metabolic breakdown, and the metabolites may cause toxic injury (Ike et al. 2002). BPA and phthalates are known xenoestrogens and are acutely toxic (vom Saal and Hughes 2005), under chronic exposure scenarios may lead to increased risk of cancer (Parkinson 1996). The long-term sublethal effects of low-level PPCPs may be of more concern than the short acute exposure due to impacts on critical physiological functions. For example, altered hormone profiles and reduced reproductive success were caused by the feminization of male fish exposed to estrogenic steroid hormones in effluent (Williams et al. 2009).

9.3.3 Approaches in Marine Mammal Toxicology: Study Design

The long lives, limited metabolic capacity to eliminate contaminants, large habitat needs, and often high trophic level, underlie the special vulnerability of the marine mammals to organic contaminants, notably those with PBT properties. Despite the extensive literature documenting the extent to which many marine mammals have become heavily contaminated by PBT contaminants, it remains very difficult to establish cause-and-effect evidence of adverse health effects associated with a given contaminant. This is due to three fundamental challenges: (1) the highly complex mixtures of contaminants, to which free-ranging marine mammals are exposed; (2) the multitude of confounding factors other than contaminants that affect the health of free-ranging marine mammals, and may confound contaminant-oriented studies (e.g., age, sex, disease, nutritional status and stress, as well as anthropogenic influences such as noise disturbance, climate change, and habitat destruction); and (3) the legal, ethical, and technical challenges that typically preclude captive dosing studies from being carried out (Ross et al. 2003).

Environmental contaminants have been associated with impacts on several health endpoints in captive and free-ranging marine mammals (Table 9.3). Lines of evidence have been derived from a variety of approaches, including associative, correlative, captive feeding, laboratory animal, and *in vitro* study designs (Ross 2000, Ross et al. 2003).

9.3.3.1 Associative Studies

Associative studies include observations of mass mortality, infectious disease outbreaks, physiological abnormalities, and/or low recruitment in free-ranging marine mammals inhabiting contaminated areas. Early examples of associative findings include pathogen-induced abortions in California sea lions (*Z. californianus*) (DeLong et al. 1973). More recently, the same population is experiencing a very high prevalence (18%) of neoplasms among stranded adult pinnipeds (Gulland et al. 1996). Death via carcinoma, examined with a logistic model, revealed animals with carcinoma had higher mean concentrations of PCBs and DDTs in blubber (Ylitalo et al. 2005).

The mass mortality of harbor seals attributed to phocine distemper virus (PDV) provided some support for contaminant induced immunosuppression because individuals who survived the outbreak had significantly lower organochlorine levels (Hall et al. 1992). Associative studies paired with

TABLE 9.3
Health Effects Related to Elevated Exposure to Chemicals Deemed to Be Persistent, Bioaccumulative, and Toxic (PBT) Have Been Observed in Marine Mammals, Using a Combination of Different Study Designs

Effect/Biomarker	∑PCB	Tissue	Species	Reference
Captive Feeding				
↓ Implantation, vitamin A, thyroid hormones	26.00	Blood	Harbor seal	Brouwer et al. 1989, Reijnders 1986
↓ Immune system	0.02	Blubber	Harbor seal	de Swart et al. 1996, Ross et al. 1996b
Correlative				
↑ CYP activity	0.42	Blubber	Beluga whale	White et al. 1994
↑ CYP activity	5.00[a]	Skin	Fin whale	Marsili et al. 1998
↑ CYP activity	28.00	Liver	Gray seal	Nyman et al. 2003
↑ CYP activity	66.00	Liver	Ringed seal	Nyman et al. 2003
↓ Testosterone	9.00[b]	Blubber	Dalls Porpoise	Subramanian et al. 1987
↓ Thyroid hormones	7.02	Plasma	Polar bear	Braathen et al. 2004
↓ Thyroid hormones	0.74	Blubber	Ribbon seal	Chiba et al. 2001
↓ Thyroid hormones	3.20	Blubber	Spotted seal	Chiba et al. 2001
↓ Thyroid hormones	2.50	Blubber	Harbor seal	Tabuchi et al. 2006
↑ Thyroid hormones	8.20	Blubber	Harbor seal	Hall and Thomas 2007
↓ Thyroid hormones	5.40	Blubber	Gray seal	Sormo et al. 2005
↓ Thyroid hormones	2.80	Plasma	Gray seal	Jenssen et al. 1994
↑ Vitamin A	15.40	Blubber	Harbor seal	Simms et al. 2000
↓ Vitamin A	2.50	Blubber	Harbor seal	Mos et al. 2007
↓ Vitamin A	0.80	Blood	Gray seal	Jensen et al. 2003
↓ Bone mineral density	8.60	Fat	Polar bear	Sonne et al. 2004
↓ Lymphocyte/Immune	2.52	Blubber	Harbor seal	Levin et al. 2005
↑ IgG, antibodies	0.05	Plasma	Polar bear	Lie et al. 2004
↓ Lymphocyte/Immune	3.00	Blubber	Harbor seal	Mos et al. 2006

Note: These include both captive feeding studies where controlled conditions enable a high degree of confidence on the effects of PBT contaminants, as well as correlative studies in free-ranging populations of marine mammals where less control is afforded, but the context is representative of a population in the "real world." Most literature points to the PCBs as the primary PBT contaminant underlying these effects, although a contribution of other PBT contaminants cannot be ruled out. It should be noted that the ∑PCB values presented here represent mean values taken from the studies, rather than a threshold associated with the effect observed.

[a] Dry weight.
[b] Wet weight.

controlled feeding studies can assist in determining contaminant thresholds. For example, Jepson et al. (2005) found the previously identified PCB threshold of 17 mg/kg in blubber for adverse health effects including immunotoxicity (Ross et al. 1995), effectively delineated the stranded harbor porpoises into an infectious disease group and a physical trauma stranding group.

Similar to observations in the California sea lions, the St. Lawrence Estuary (SLE) beluga population has exhibited numerous cancer tumors, low recruitment, and the first case of true hermaphroditism in a cetacean (De Guise et al. 1994a, 1994b, 1995, Martineau et al. 1994). The proximity of SLE belugas to industrial effluent provided a basis to consider the above effects as being in part due to their high contaminant exposure. More recently, the SLE beluga population has been used as an indicator for human health, as residents living near the SLE are experiencing elevated incidence of cancer (Martineau et al. 2002). Limited opportunities to obtain samples make it difficult to demonstrate cause and effect between specific contaminants. Therefore, associative field studies provide indications of a marine mammal population that show signs of contaminant related effects, with limited mechanistic evidence.

9.3.3.2 Correlative Studies

Correlative studies use documented statistical relationships between contaminant levels and physiological endpoints known to be sensitive to disruption by contaminants such as the endocrine, reproductive, and immune systems. Toxicological endpoints or biomarkers for endocrine disruption and reproductive impairment in wildlife have been reviewed by (Rolland 2000, Simms and Ross 2001, Hahn 2002, Peakall et al. 2002).

CYP1A has been used as a biomarker of PBT exposure and the onset of physiological effects in fish and marine mammals (Hahn 2002) (Table 9.3). Beluga whales from the St. Lawrence estuary were initially thought to have higher CYP1A activity relative to Arctic belugas in response to higher PCB exposure (Muir et al. 1996), but a later study did not support those findings (Wilson et al. 2005). This may be explained in part by differences in sampling techniques. A recent study of bottlenose dolphins revealed CYP1A1 expression in relation to 2,3,7,8 TCDD and OH-PCBs to be significant only in the deep blubber layer (Montie et al. 2008). This was the first study to find stratification of CYP1A1 expression in blubber, demonstrating the importance of understanding the biomarker selected to study effects. Overall, PBT induced CYP1A activity appears to be a reliable indicator of dioxin-like responses in pinnipeds (Nyman et al. 2003) and cetaceans (e.g., Montie et al. 2008). However, effort is needed to ensure better comparability among samples and study designs that incorporate and/or address the confounding effects of age, sex, condition, and sampling techniques.

Both steroid hormones and thyroid hormones have been used as biomarkers of toxic effects (Table 9.3). Altered circulating hormone concentrations with an increase in contaminant levels such as PCBs are thought to result from (a) steroid breakdown via PCB and/or PCB metabolite-induced enzyme activity (P450 induction) and/or (b) a competitive displacement by PCBs and/or their metabolites of hormones from their receptors. A decrease in thyroid hormone concentrations with increasing PBT contaminant levels was observed in spotted seals (*Phoca largha*), ribbon seals (*Phoca fasciata*), harbor seals, and polar bears (Chiba et al. 2001, Braathen et al. 2004, Sormo et al. 2005, Tabuchi et al. 2006) (Table 9.3). In addition to a reduction in circulating thyroid hormones with increased PCB levels in free-ranging harbor seals, Tabuchi et al. (2006) revealed alterations in thyroid hormone gene expression (*TH α*) in blubber. Genotoxic endocrine disruption in blubber suggests a potential consequence on blubber functions such as lipid metabolism and thermoregulation (Tabuchi et al. 2006).

Vitamin A is an essential dietary component required for growth, development, and maintenance of endocrine, reproductive, and immune systems (Rolland 2000, Simms and Ross 2000). Levels of vitamin A in free-ranging marine mammals are thought to have been disrupted by PCBs through several modes of toxic action (Rolland 2000, Jensen et al. 2003, Mos et al. 2007). As with all biomarkers, an understanding of their baseline physiology, along with the roles of potentially confounding factors such as nutritional condition, size, sex, and reproductive status, is a critical part of study design and interpretation (Simms and Ross 2000, 2001). Selecting an appropriate tissue was demonstrated in gray seal pups where plasma levels related more significantly with PCB levels than whole blood media (Jensen et al. 2003).

While serious outbreaks of disease have underscored the possible contribution of contaminant-related immunotoxicity among marine mammals, measuring immune function in free-ranging marine mammals is fraught with challenges. These include the difficulties in obtaining fresh blood samples from healthy individuals, the confounding influences of age, sex, disease, condition, and stress on immunological endpoints, and the need to conduct functional evaluations of fresh blood cells at the same time to ensure interindividual comparability. Endpoints to evaluate immune function include aspects of nonspecific immunity such as natural killer cell activity, and specific immunity such as lymphocyte function, or the determination of cytokines, proteins, enzymes, or antibodies (reviewed in Ross 2002). The complexity of the immune system has limited the number of studies on free-ranging marine mammals, with most evidence emanating from the well-studied harbor seal (Hall et al. 1992, 2006, Lahvis et al. 1995, Levin et al. 2005, Mos et al. 2006).

9.3.3.3 Captive Studies

Captive studies offer a means of studying the effects of contaminants in a controlled setting that can help to elucidate causal linkages between contaminants and toxicity endpoints in marine mammals. Elevated PBT concentrations were implicated in reduced reproductive success in one group of captive harbor seals fed fish from the contaminated Wadden Sea compared to a control group (Reijnders 1986) (Table 9.3). Those seals fed contaminated fish experienced altered vitamin A and circulating thyroid hormone levels (Brouwer et al. 1989), an effect later attributed to PCB metabolite-associated displacement of the thyroid hormone-carrier complex in circulation (Brouwer et al. 1999). In a later captive-feeding study, reduced function of natural killer cells and T-cells, and impaired delayed-type hypersensitivity, were observed in harbor seals fed herring from the contaminated Baltic Sea compared to seals fed herring from the North Sea (reviewed in de Swart et al. 1996, Ross et al. 1996b).

The cumulative evidence from these major studies provided support for the notion that some of the virus-associated mass mortalities in Europe may have been exacerbated by PBT contaminants, as many affected populations had PCB concentrations above those which elicited significant changes in immune function in the captive seals (Ross et al. 1996a). Captive dolphins have been used to study contaminant pharmacokinetics, including partitioning between blood and blubber, transfer during different stages of lactation, and relationship with calving success (Ridgway and Reddy 1995, Reddy et al. 1998, 2001).

9.3.3.4 *In Vitro* Studies

In vitro research offers additional opportunities to explore the effects of contaminants on the health of marine mammal cells or tissues, with the advantage of being noninvasive. Such approaches allow for a careful control of contaminant exposures in the context of a strictly defined response. These studies can provide information of cause and effect linkages between individual contaminants and an effect, or the effects of a complex mixture. Several *in vitro* studies have been carried out using cells isolated from beluga whales (De Guise et al. 1996, 1997, 1998, Levin et al. 2004, Mori et al. 2006). These were aimed at characterizing possible mechanisms of action for immunotoxicity by incubating peripheral blood leukocytes with a series of single/multiple chemical exposures. The ultimate challenge of *in vitro* research results is to extrapolate the linkage to the organism and population level. At an interspecies level, an *in vitro* PCB exposure in pinnipeds revealed reduced immune function in harbor seals and no effect on sympatric gray seals, suggesting the harbor seal to be more vulnerable to immunotoxicity and possibly disease (Hammond et al. 2005).

9.4 CONCEPTS IN RISK ASSESSMENT AND MANAGEMENT

9.4.1 THE CHALLENGE OF COMPLEX MIXTURES

Concentrations of PCBs or DDT are typically an order of magnitude higher than new compounds such as PBDEs and PFOS in biota. This increasing complexity of contaminants found in marine mammals renders it difficult to elucidate the effects of emerging chemicals in free-ranging populations. Controlled laboratory studies assist in determining modes of action of new compounds, but do not address the additive, synergistic, or antagonistic effects of the complex mixtures in the marine environment.

PBDE-exposed rodents revealed a disruption in thyroid hormone circulation (Hallgren et al. 2001, Zhou et al. 2002) as well as developmental neurotoxic and behavioral effects (Eriksson et al. 2001, Viberg et al. 2003a, 2003b). One PBDE field study has been conducted to evaluate possible PBDE-mediated effects in postweaned gray seal pups and juveniles (Hall et al. 2003). Results demonstrated a relationship between circulating thyroid hormones, transport proteins and PBDE uptake (0.2; 0.5 mg/kg lw in pups and juveniles). Further research is needed to better characterize the relative contributions of PBDEs to the possibly pre-existing effects of the other PBT contaminants that are typically found at much higher concentrations in marine mammals (Law et al. 2002).

There have been few studies of the effects of PFOS on wildlife, but laboratory studies have demonstrated the potential for endocrine disruption (Lau et al. 2004, 2007). High exposure resulted in decreased testosterone, and gene expression involved in steroidogenesis, and increased oestriadiol (Biegel et al. 1995, Shi et al. 2007). In addition, reduced body weight, increased liver weight, and hepatocellular adenomas have been described in laboratory mammal studies (Seacat et al. 2002, 2003). To date there has only been one field study linking high levels of PFOA (0.06 ± 0.03 mg/kg ww) and PFOS (0.05 ± 0.01 mg/kg ww) in sea otters to infectious disease prevalence (Kannan et al. 2006). However, these same individuals also had high levels of PCBs, DDT and heavy metals (Kannan et al. 1998, 2004, Nakata et al. 1998), confounding a mechanistic understanding of this observation.

9.4.2 Weight of Evidence in Marine Mammal Toxicology

No single line of research provides sufficient evidence of causality at the marine mammal population level for the effects of a given contaminant, underscoring the need for a "weight of evidence" approach (Ross 2000). This approach is akin to that used in human health research (Ross and Birnbaum 2003). The integration of human and ecological approaches to toxicology offers benefits to furthering an understanding of PBT health risks at the ecosystem level (Munns et al. 2003).

Advantages of *associative* or *correlative* field research include the "real world" nature of the study, and the relevance to conservation of management questions. Disadvantages to these field-based approaches include the near-impossible task of identifying a single chemical involved in observed effects, and the confounding influences of multiple factors that affect health in the wild. Advantages of *captive studies* include the ability to obtain high quality samples from known animals, the ability to minimize confounding factors, and the ability to adequately characterize contaminant mixtures to which they are exposed in their diet; disadvantages include the ethical constraints associated with marine mammal research, a still complex contaminant mixture in their diet, and the need to ultimately link the results with the real world. Advantages of *in vitro* studies include the ability to work with carefully defined chemicals alone or in combination, the control over simple physiological endpoints or systems with the aim of mechanistic evaluation, and the ability to work with little or no impact on marine mammals; disadvantages include the lack of organismal context and the need to relate back to individual and population health.

Ultimately, however, each and every one of these lines of research has helped to deliver a "weight of evidence" which has clarified the putative roles of PBT contaminants, notably the PCBs, dioxin-like mixtures, and PCB metabolites, in reduced immunological fitness, reproductive impairment, skeletal malformations, tumors, and endocrine disruption in marine mammals. Future toxicological studies need to feed results into a broader context for risk assessment managers to better decide on appropriate mitigation or management actions in support of the conservation of healthy marine mammal populations.

9.4.3 International Treaties, National Regulations, and Local Application

In response to concern for the impacts of PBT contaminants on human and environmental health on a global scale, the Stockholm Convention on Persistent Organic Pollutants (POPs) was adopted as an international treaty in 2004. The convention requires signatories to eliminate or reduce the release of chemicals that are persistent and capable of long-range transport, bioaccumulation, and toxic injury (PBTs). The global treaty administered the removal and regulation of chemicals referred to as the "Dirty Dozen," and included PCBs and DDT (Table 9.2).

Most countries have adopted sets of regulations to implement the terms of the Stockholm Convention. Guidelines on chemical regulations are complex as management and mitigation strategies differ among nations and within nations. For example, in the United States, chemical products are regulated by the United States Environmental Protection Agency (U.S. EPA). Within the U.S. EPA, the Toxic Substances and Control Act provides guidance on regulating the production, importation, use and disposal of chemicals such as PCBs, whereas other chemical substances such as pharmaceuticals, cosmetics, food, and pesticides are regulated under other acts. Typically these regulations require chemical manufactures to go through a "gatekeeper" screening process.

In Canada, the Canadian Environmental Protection Act (CEPA), created in 1999, manages toxic chemical substances, but as many as 25 laws govern chemical substances under food, drug, and pesticide components. Regulations for such products require that all new chemicals that are manufactured or imported be assessed against specific criteria.

The European Union (EU) has developed a comprehensive set of regulations for chemicals, known as the Regulation on Registration, Evaluation, Authorization and Restriction of Chemicals (REACH). This became law in the EU in 2007, and consists of an intensive "cradle to grave" program to ensure that manufacturers comply with chemical regulations, as well as take on the responsibility of managing risks of substances entering the market.

9.4.4 REGULATORY GUIDELINES FOR ENVIRONMENTAL MEDIA

In addition to providing regulatory oversight for chemical design and management, international treaties and national regulations translate into state/provincial and municipal jurisdictions to support information needs for management and decisions for the protection of wildlife and human health. Environmental quality guidelines for chemicals in media such as water and sediments rely on benchmarks, reference points or thresholds to protect wildlife by providing indicators and a method of interpreting progress when mitigating site-specific impacts or preventing further damage following the use, release or disposal of a chemical. Guidelines have been developed by several programs that include but are not limited to the following: Canadian Council of Ministers of the Environment (CCME), the Agency for Toxic Substances and Disease Registry (ATSDR), New York Department of Environmental Conservation, United States National Academy of Sciences, and the European Chemical Agency.

Guidelines for mitigation, source control, and conservation have been developed to protect site- or species-specific effects of contaminants. These typically include water, sediment, and prey tissue compartments. Since PBTs in water do not represent a significant mode of exposure for marine mammals, guidelines for PBTs in water are not relevant to marine mammals. PBT contaminants do partition into sediments, such that sediment contaminants are relevant when contaminants are mobilized through biotic and abiotic processes, and taken up into aquatic food webs. Thus, biota-sediment accumulation factor (BSAF) models have been used to predict bioaccumulation of PBT chemicals in fish and aquatic biota using sediment concentrations and equilibrium partitioning (Gobas and Morrison 2000, Wong et al. 2001).

Because diet represents the primary means of PBT exposure in marine mammals, prey tissue guidelines provide the most relevant means of assessing risk and designing mitigative strategies. Tissue Residue Guidelines (TRGs) have been used to characterize risk to mammals and birds (CCME 1998). A TRG is a type of toxicity reference value (TRV) that is determined for a chemical in an exposure medium (Sample et al. 1996). Methods used to select the appropriate TRG for a species and contaminant will vary among agencies. The protocol for TRGs for the CCME (CCME 1998) is partially based on methods developed by the New York State Department of Environmental

Conservation (Newell et al. 1987). The latter relies on the tolerable daily intake (TDI) (calculated in milligrams per kilogram of body weight per day [mg/kg bw/d]), such that a guideline is anticipated to not result in any adverse health effects following exposure. TDIs are calculated with the lowest observed adverse effect level (LOAEL) and the no observed adverse effect level (NOAEL) from toxicological studies, with the incorporation of an appropriate uncertainty factor (UF):

$$TDI = \frac{(LOAEL \cdot NOAEL)^{0.5}}{UF}$$

If the NOAEL is unknown, it can be extrapolated from dose–response curves. Using the TDI, reference concentrations (RC) in mg/kg can be calculated for a specific species using daily food ingestion (FI) rates in kg/d and body weights (bw).

$$RC = \frac{TDI}{FI/bw}$$

Data to run models and determine TRGs are largely derived from laboratory species such as rats or mice that are fed a diet of a given chemical to mimic environmental levels of concern. There are strict guidelines on the toxicological data deemed acceptable prior to extrapolating, and uncertainty surrounding sensitivity variability among species are considered and accounted for. Despite these strict guidelines, there are limitations to this approach. First, the use of laboratory animals constrains the relevance and limits the environmental variability when resultant TRGs are used for the protection of wildlife such as marine mammals. Extrapolation from a controlled laboratory study to a free-ranging animal does not adequately account for (a) environmental interactions; (b) variability in exposure/ feed (how the contaminant was added to food material), ingestion rate; (c) differences in uptake, metabolism, absorption, and excretion of different compounds; and (d) body mass, that together present challenges to size scaling or extrapolation among species (Sample and Arenal 1999).

Methods to overcome such interspecies challenges do exist and provide a means to extrapolate from lab to field studies using mode of toxicity classification groups (Raimondo et al. 2007). This approach, as well as the majority of laboratory effects studies, typically focus on acute effects, rather than chronic approaches that may be more environmentally relevant. Chronic studies are generally much more costly, difficult to carry out, and involve low-level effects that may not be valued under current risk assessment paradigms.

TRG- and TDI-based approaches, are largely based on dose–response relationships from laboratory rodent feeding studies. This is very different from a field-based approach (i.e., correlative) that evaluates contaminant concentrations in free-ranging wildlife and associates concentrations with disrupted physiological endpoints. Moreover, the physiology and energetics of marine mammals typically involves seasonal foraging on lipid rich prey followed by living off of blubber energy stores (Costa 2008). This may predispose marine mammals to a different set of risk scenarios when compared to the physiological energetics of laboratory species where diets are largely derived from carbohydrates and proteins (Young 1976). A wider acceptance of field-study data by risk assessors may help to improve the relevance of ecotoxicological studies and improve society's ability to better conserve and protect marine mammals at the population level.

Marine mammals may offer some guidance on how to improve our ability to design and regulate chemicals, and to establish site- or specific-specific protections. For example, while the implementation of PCB regulations in the mid-1970s helped to reduce environmental concentrations of this chemical, it is increasingly clear that current benchmarks and guidelines do not offer adequate protection for marine mammals. In a case study, it was demonstrated that the fish-eating killer whales of the NE Pacific Ocean will not be "protected" from PCB-related health risks for another 30–70 years, and that if these cetaceans were to consume prey containing a "safe" level of PCBs (0.050 mg/kg wet weight), fully 95% of the population would be at risk for adverse health effects (Hickie et al. 2007).

This disconnect between toxicological research and risk assessment paradigms underscores the need for a more encompassing consideration of field data and life history features by risk assessors. A "weight of evidence" offers a basic start in this regard. On the other hand, field-based scientists should be made aware of the needs of risk assessors, such that the resultant research will be of greater relevance at the regulatory level. Captive feeding studies has been suggested as a means to bridge the field and laboratory approach, and can offer accurate toxicity data for acute and chronic injury (O'Shea et al. 1999). However, the logistical and ethical associated with such studies constrains the opportunities in this regard.

Summary

Based on an accumulating weight of evidence, adverse health effects in different marine mammal populations have been linked to two principal types of contaminants. First, elevated PBT contaminants have been implicated in endocrine disruption, immunotoxicity, increased vulnerability to disease, and reproductive impacts in the case of mid- to high trophic level species. Among the complex mixture of environmental PBTs, PCBs represent the prime culprits in these observed health impacts in high trophic level marine mammals in the northern hemisphere. Based on the commonly used effect threshold of 17 ppm for PCBs in relation to the lower concentrations with observed effects presented in Table 9.3, highlights the need for more effects research on free-ranging marine mammals.

In addition to effects of PBTs, hydrocarbons related to oil spills have caused acute impacts in the case of several species, notably those that are furred and frequent nearshore and intertidal areas. Evidence is more scant for the effects of a wide range of metals, pesticides, and PPCPs, but exposures have been documented, particularly among lower trophic level marine mammals. Risks for these contaminants will vary as a function of proximity to source and dose, as well as the life history features of the marine mammal in question.

A causal understanding of health effects in marine mammals exposed to highly complex mixtures remains a challenge due to the complex mixture to which marine mammals are exposed in their habitat, the confounding factors facing researchers such as age, sex, disease, or nutritional status, and the ethical, logistical, and legal constraints associated with conducting research on free-ranging marine mammals. Further research and management action is needed to better protect marine mammals from contaminant-related effects at a population level. Areas of interest would include risk assessment, chemical regulation, chemical engineering, source control, remediation, and conservation measures; each of these areas should better take into account the unique vulnerabilities of different marine mammals. At the end of the day, however, the exposure of marine mammals to high levels of endocrine disrupting contaminants signals a failure at the regulatory level. A better consideration of such vulnerable species, as well as a more all-encompassing inclusion of the precautionary principle, would go a long way in protecting the world's marine mammals.

REFERENCES

Addison, R. F., and P. F. Brodie. 1987. Transfer of organochlorine residues from blubber through the circulatory system to milk in the lactating gray seal (*Halichoerus grypus*). *Can. J. Fish. Aquat. Sci.* 44:782–786.

Addison, R. F., and W. T. Stobo. 2001. Trends in organochlorine residue concentrations and burdens in grey seals (*Halichoerus grypus*) from Sable Is., NS, Canada, between 1974 and 1994. *Environ. Pollut.* 112:505–513.

Aguilar, A., and A. Borrell. 1994. Reproductive transfer and variation of body load of organochlorine pollutants with age in fin whales (*Balaenoptera physalus*). *Arch. Environ. Contam. Toxicol.* 27:546–554.

Ahlborg, U. G., et al. 1995. Organochlorine compounds in relation to breast cancer, endometrial cancer and endometriosis: an assessment of the biological and epidemiological evidence. *Crit. Rev. Toxicol.* 25:463–531.

AMAP. 2004. AMAP Assessment 2002: Persistent Organic Pollutants (POPs) in the Arctic. Arctic Monitoring and Assessment Programme, Oslo, Norway, pp. xvi–310 pp.

Ankley, G. T., B. W. Brooks, D. B. Huggett, and J. P. Sumpter. 2007. Repeating history: pharmaceuticals in the environment. *Environ. Sci. Technol.* 41:8211–8217.

Assuncao, M. G. L., K. A. Miller, N. J. Dangerfield, S. M. Bandiera, and P. S. Ross. 2007. Cytochrome P4501A expression and organochlorine contaminants in harbour seals (*Phoca vitulina*): evaluating a biopsy approach. *Comp. Biochem. Physiol. C* 145:256–264.

Beckmen, K. B., G. M. Ylitalo, R. G. Towell, M. M. Krahn, S. O'Hara, and J. E. Blake. 1999. Factors affecting organochlorine contaminant concentrations in milk and blood of northern fur seal (*Callorhinus ursinus*) dams and pups from St. George Island, Alaska. *Sci. Total Environ.* 231:183–200.

Bernt, K. E., M. O. Hammill, M. Lebeuf, and K. M. Kovacs. 1999. Levels and patterns of PCBs and OC pesticides in harbour and grey seals from St Lawrence Estuary, Canada. *Sci. Total Environ.* 243/244:243–262.

Biegel, L. B., R. C. Liu, M. E. Hurtt, and J. C. Cook. 1995. Effects of ammonium perfluorooctanoate on Leydig cell function: *in vitro, in vivo* and ex vivo studies. *Toxicol. Appl. Pharmacol.* 134:18–25.

Bonde, R. K., A. Alonso, and J. Powell. 2004. Manatees as sentinels of marine ecosystem health: are they the 2000-pound canaries? *Ecohealth.* 1:255–262.

Borga, K., G. W. Gabrielsen, and J. U. Skaare. 2001. Biomagnification of organochlorines along a Barents Sea food chain. *Environ. Pollut.* 113:187–198.

Borrell, A., D. Bloch, and G. Desportes. 1995. Age trends and reproductive transfer of organochlorine compounds in long-finned pilot whales from the Faroe Island. *Environ. Pollut.* 88:283–292.

Bowen, W. D., D. Tully, D. J. Boness, B. M. Bulheier, and G. J. Marshall. 2002. Prey-dependent foraging tactics and prey profitability in a marine mammal. *Mar. Ecol. Prog. Ser.* 244:235–245.

Braathen, M., A. E. Derocher, O. Wiig, E. G. Sormo, E. Lie, and J. U. Skaare. 2004. Relationships between PCBs and thyroid hormones and retinol in female and male polar bears. *Environ. Health Persp.* 112:826–833.

Brouwer, A. 1991. Role of biotransformation in PCB-induced alterations in vitamin A and thyroid hormone metabolism in laboratory and wildlife species. *Biochem. Soc. T.* 19:731–737.

Brouwer, A., P. J. H. Reijnders, and J. H. Koeman. 1989. Polychlorinated biphenal (PCB)-contaminated fish induces vitamin A and thyroid hormone deficiency in the common seal (*Phoca vitulina*). *Aquat. Toxicol.* 15:99–106.

Brouwer, A., et al. 1998. Interactions of persistent environmental organohalogens with the thyroid hormone system: mechanisms and possible consequences for animal and human health. *Toxicol. Ind. Health.* 14:59–84.

Brouwer, A., et al. 1999. Characterization of potential endocrine-related health effects at low-dose levels of exposure to PCBs. *Environ. Health Persp.* 107:639–649.

Brown, S. B., and J. F. Fairchild. 2003. Evidence for a causal link between exposure to an insecticide formulation and declines in catch of Atlantic Salmon. *Hum. and Ecol. Risk Assess.* 9:137–148.

Carson, R. 1962. *Silent spring.* New York, NY: Houghton Mifflin.

CCME. 1998. Protocol for the derivation of Canadian tissue residue guidelines for the protection of wildlife that consume aquatic biota. Canadian Council of Ministers of the Environment. Government of Canada.

Chiba, I., et al. 2001. Negative correlations between plasma thyroid hormone levels and chlorinated hydrocarbon levels accumulated in seals from the coast of Hokkaido, Japan. *Environ. Toxicol. Chem.* 20:1092–1097.

Costa, D. P. 2008. Energetics. In *Encyclopedia of marine mammals*, second edition, ed. W. F. Perrin. San Diego, CA: Academic Press.

De Guise, S., A. Lagace, and P. Beland. 1994a. Tumors in St. Lawrence beluga whales (*Delphinapterus leucas*). *Vet Pathol* 31:444–449.

De Guise, S., A. Lagace, and P. Beland. 1994b. True Hemaphroditism in a St. Lawrence Beluga Whale. *J. Wildlife Dis.* 30:287–290.

De Guise, S., D. Martineau, P. Beland, and M. Fournier. 1998. Effects of *in vitro* exposure of beluga whale leukocytes to selected organochlorines. *J. Toxicol. Env. Health, Part A* 55:479–493.

De Guise, S., A. Lagace, P. Beland, C. Girard, and R. Higgins, 1995. Non-neoplastic lesions in beluga whales (*Delphinapterus leucas*) and other marine mammals from the St. Lawrence Estuary. *J. Comp. Pathol.* 112:257–271.

De Guise, S., J. Bernier, D. Martineau, P. Beland, and M. Fournier. 1996. Effects of *in vitro* exposure of beluga whale splenocytes and thymocytes to heavy metals. *Environ. Toxicol. Chem.* 15:1357–1364.

De Guise, S., P. S. Ross, A. D. M. E. Osterhaus, D. Martineau, P. Beland, and M. Fournier, 1997. Immune functions in beluga whales (*Delphinapterus leucas*): evaluation of natural killer cell activity. *Vet. Immunol. Immunop.* 58:345–354.

de Swart, R., P. S. Ross, D. J. Vos, and A. D. M. E. Osterhaus. 1996. Impaired immunity in harbour seals (*Phoca vitulina*) exposed to bioaccumulated environmental contaminants: review of a long-term feeding study. *Environ. Health Persp.* 104:823–828.

Debier, C., et al. 2003a. Dynamics of PCB Transfer from other to pup during lactation in UK grey seals *Halichoerus grypus*: differences in PCB profile between compartments of transfer and changes during the lactation period. *Mar. Ecol. Prog. Ser.* 247:249–256.

Debier, C., et al. 2003b. Quantitative dynamics of PCB transfer from mother to pup during lactation in UK grey seals *Halichoerus grypus*. *Mar. Ecol. Prog. Ser.* 247:237–248.

DeLong, R. L., W. G. Gilmartin, and J. G. Simpson. 1973. Premature births in California sea lions: association with high organochlorine pollutant residue levels. *Science* 181:1168–1170.

Donohoe, R. M., and L. R. Curtis. 1996. Estrogenic activity of chlordecone, *o,p'*-DDT and *o,p'*-DDE in juvenile rainbow trout: induction of vitellogenesis and interaction with hepatic estrogen binding sites. *Aquat. Toxicol.* 36:31–52.

Duinker, J. C., and M. T. J. Hillebrand. 1979. Mobilization of organochlorines from female lipid tissue and transplacental transfer to fetus in a harbour porpoise (*Phocoena phocoena*) in a contaminated area. *Bull. Environ. Contam. Toxicol.* 23:728–732.

Eriksson, P., E. Jakobsson, and A. Fredrikson. 2001. Brominated flame retardants: a novel class of developmental neurotoxicants in our environment? *Environ. Health Persp.* 109:903–908.

Ford, J. K. B., and G. M. Ellis. 2006. Selective foraging by fish-eating killer whales Orcinus orca in British Columbia. *Mar. Ecol. Prog. Ser.* 316:185–199.

Ford, J. K. B., G. M. Ellis, L. Barrett-Lennard, A. B. Morton, R. S. Palm, and K. C. Balcomb. 1998. Dietary specialization in two sympatric populations of killer whales (*Orcinus orca*) in coastal British Columbia and adjacent waters. *Can. J. Zool.* 76:1456–1471.

Fry, D. M., and C. K. Toone. 1981. DDT-induced feminization of gull embryos. *Science* 213:922–924.

Garrett, C., and P. S. Ross. 2010. Recovering resident killer whales: a guide to contaminant sources, mitigation, and regulations in British Columbia. In: Sci., C.T.R.F.A. (Ed.), *Can. Tech. Rep Fish. Aquat. Sci.* 2894: 224 + xiii pp.

Garrott, R. A., L. L. Ebehardt, and D. M. Burn. 1993. Mortality of sea otters in Prince William Sound following the Exxon Valdez oil spill. *Mar. Mamm. Sci.* 9:343–359.

Geraci, J. R. 1990. Physiologic and toxic effects on cetaceans. In *Sea Mammals and oil: confronting the risks*, eds. J. R. Geraci, and D. St. Aubin. New York, NY: Academic Press.

Geraci, J. R., and D. J. St. Aubin. 1985. Expanded studies of the effects of oil on Cetaceans. Final report (Part l) to U.S. Department of Interior, BLM contract 14-12-0001-29169.

Giesy, J. P., and K. Kannan. 1998. Dioxin-like and non-dioxin-like toxic effects of polychlorinated biphenyls (PCBs): implications for risk assessment. *Crit. Rev. Toxicol.* 28:511–569.

Giesy, J. P., and K. Kannan. 2001. Global distribution of perfluorooctane sulfonate in wildlife. *Environ. Sci. Technol.* 35:1339–1342.

Gobas, F. A. P. C., and H. A. Morrison. 2000. Bioconcentration and biomagnification in the aquatic environment. In *Handbook of property estimation methods for chemicals: Environmental and health sciences*, eds. R. S. Boethling and D. Mackay. Boca Raton, FL: Lewis Publishers.

Gulland, F. M. D., J. G. Trupkiewicz, T. R. Spraker, and L. J. Lowenstine. 1996. Metastatic carcinoma of probable transitional cell origin in 66 free-living California sea lions (*Zalophus californianus*), 1979 to 1994. *J. Wildlife Dis.* 32:250–258.

Hahn, M. E. 1998. The aryl hydrocarbon receptor: a comparative perspective. *Comp. Biochem. Phys. C* 121:23–53.

Hahn, M. E. 2002. Biomarkers and bioassays for detecting dioxin-like compounds in the marine environment. *Sci. Total Environ.* 289:49–69.

Hall, A. J., and G. O. Thomas. 2007. Polychlorinated biphenyls, DDT, Polybrominated diphenyl ethers, and organic pesticides in United Kingdom harbor seals (*phoca vitulina*)—mixed exposures and thyroid homeostasis. *Environ. Toxicol. Chem.* 26:851–861.

Hall, A. J., K. Hugunin, R. Deaville, R. J. Law, C. R. Allchin, and P. D. Jepson. 2006. The risk of infection from polychlorinated biphenyl exposure in the harbor porpoise (*Phocoena phocoena*): a case-control approach. *Environ. Health Persp.* 114:704–711.

Hall, A. J., O. I. Kalantzi, and G. O. Thomas. 2003. Polybrominated diphenyl ethers (PBDEs) in grey seals during their first year of life-are they thyroid hormone endocrine distrupters? *Environ. Pollut.* 126:29–37.

Hall, A. J., et al. 1992. Organochlorine levels in common seals (*Phoca vitulina*) which were victims and survivors of the 1988 phocine distemper epizootic. *Sci. Total Environ.* 115:145–162.

Hallgren, S., T. Sinjari, H. Hakasson, and P.O. Darnerud. 2001. Effects of polybrominated diphenyl ethers (PBDEs) and polychlorinated biphenyls (PCBs) on thyroid hormone and vitamin A levels in rats and mice. *Arch. Toxicol.* 75:200–208.

Halpern, B. S., et al. 2008. A global map of human impact on marine ecosystems. *Science* 319:948–952.

Hammond, P. S., A. J. Hall, and E. A. Dyrynda. 2005. Comparison of polychlorinated biphenyl (PCB) induced effects on innate immune functions in harbour and grey seals. *Aquat. Toxicol.* 74:126–138.

Harvey, J. T. 1994. Biases associated with non-lethal methods of determining the diet of northern elephant seals. *Mar. Mamm. Sci.* 10:178–187.

Hawkins, S., S. M. Billiard, S. P. Tabash, S. R. Brown, and P. V. Hodson. 2002. Altering cytochrome P4501A activity affects polycyclic aromatic hydrocarbon metabolism and toxicity in Rainbow Trout (*Oncorhynchus mykiss*). *Environ. Toxicol. Chem.* 21:1845–1853.

Hickey, J. J., and D. W. Anderson. 1968. Chlorinated hydrocarbons and eggshell changes in raptorial and fish-eating birds. *Science* 162:271–273.

Hickie, B. E., D. C. G. Muir, R. F. Addison, and P. F. Hoekstra. 2005. Development and application of bioaccumulation models to assess persistent organic pollutant temporal trends in arctic ringed seal (*Phoca hispida*) populations. *Sci. Total Environ.* 351–352:413–426.

Hickie, B. E., P. S. Ross, R. W. Macdonald, and J. Ford. 2007. Killer whales (*Orcinus orca*) face protracted health risks associated with lifetime exposure to PCBs. *Environ. Sci. Technol.* 41:6613–6619.

Hobson, K. A., and H. E. Welch. 1992. Determination of trophic relationships within a high Arctic marine food web using del carbon and del nitrogen analysis. *Mar. Ecol. Prog. Ser.* 84:9–18.

Hoekstra, P. F., T. M. O'Hara, A. T. Fisk, K. Borga, K. R. Solomon, and D. C. G. Muir. 2003. Trophic transfer of persistent organochlorine contaminants (OCs) within an Arctic marine food web from the southern Beaufort-Chukchi Seas. *Environ. Pollut.* 124:509–522.

Ike, M., M. Y. Chen, C. S. Jin, and M. Fujita. 2002. Acute toxicity, mutagenicity, and estrogenicity of biodegradation products of bisphenol-A. *Environ. Toxicol.* 17:457–461.

Iverson, S. J. 1993. Milk secretion in marine mammals in relation to foraging: can milk fatty acids predict diet? *Sym. Zool. S.* 66:509–516.

Iverson, S. J., C. Field, W. D. Bowen, and W. Blanchard. 2004. Quantitative fatty acid signature analysis: a new method of estimating predator diets. *Ecol. Monogr.* 74:211–235.

Jensen, A. A., and H. Leffers. 2008. Emerging endocrine disrupters: perfluoroalkylated substances. *Int. J. Androl.* 31:161–169.

Jensen, B. M., O. Haugen, E. G. Sormo, and J. U. Skaare. 2003. Negative relationship between PCBs and plasma retinol in low-contaminated free-ranging gray seal pups (*Halichoerus grypus*). *Environ. Res.* 93:79–87.

Jenssen, B. M., J. U. Skaare, and M. Ekker. 1994. Blood sampling as a non-destructive method for monitoring levels and effects of organochlorines (PCB and DDT) in seals. *Chemosphere* 28:3–10.

Jenssen, B. M., J. U. Skaare, M. Ekker, D. Vongraven, and S. H. Lorentsen. 1996. Organochlorine compounds in blubber, liver and brain in neonatal grey seal pups. *Chemosphere* 32:2115–2125.

Jepson, P. D., P. M. Bennett, R. Deaville, C. R. Allchin, J. R. Baker, and R. J. Law. 2005. Relationships between polychlorinated biphenyls and health status in harbor porpoises (phocoena phocoena) stranded in the United Kingdom. *Environ. Toxicol. Chem.* 24:238–248.

Jones, P. D., W. Hu, W. De Coen, J. Newsted, and J. P. Giesy. 2003. Binding of perfluorinated fatty acids to serum proteins. *Environ. Toxicol. Chem.* 22:2639–2649.

Kannan, K., K. S. Guruge, N. J. Thomas, S. Tanabe, and J. P. Giesy. 1998. Butyltin residues in southern sea otters (*Enhydra lutris nereis*) found dead along California coastal waters. *Environ. Sci. Technol.* 32:1169–1175.

Kannan, K., E. Perrotta, and N. J. Thomas. 2006. Association between perfluorinated compounds and pathological conditions in southern sea otters. *Environ. Sci. Technol.* 40:4943–4948.

Kannan, K., et al. 2004. Profiles of polychlorinated biphenyl congeners, organochloine pesticides, and butyltins in southern sea otters and their prey. *Environ. Toxicol. Chem.* 23:49–56.

Kleivane, L., O. Espeland, K. A. Fagerheim, K. Hylland, A. Polder, and J. U. Skaare. 1997. Organochlorine pesticides and PCBs in the east ice harp seal (*Phoca groenlandica*) population. *Mar. Environ. Res.* 43:117–130.

Klotz, D. M., B. S. Beckman, S. M. Hill, J. A. McLachlan, M. R. Walters, and S. F. Arnold. 1996. Identification of environmental chemicals with estrogenic activity using a combination of *in vitro* assays. *Environ. Health Persp.* 104:1084–1089.

Kolpin, D. W., et al. 2002. Pharmaceuticals, hormones and other organic wastewater contaminants in U.S. streams, 1999–2000: a national reconnaissance. *Environ. Sci. Technol.* 36:1202–1211.

Krahn, M. M., P. R. Becker, K. L. Tilbury, and J. E. Stein. 1997. Organochlorine contaminants in blubber of four seal species: integrating biomonitoring and specimen banking. *Chemosphere* 34:2109–2121.

Kurle, C. M., and G. A. J. Worthy. 2002. Stable nitrogen and carbon isotope ratios in multiple tissues of the northern fur seal *Callorhinus ursinus*: implications for dietary and migratory reconstructions. *Mar. Ecol. Prog. Ser.* 236:289–300.

Lahvis, G. P., R. S. Wells, D. W. Kuehl, J. L. Stewart, H. L. Rhinehart, and C. S. Via. 1995. Decreased lymphocyte responses in free-ranging bottlenose dolphins (*Tursiops truncatus*) are associated with increased concentrations of PCBs and DDT in peripheral blood. *Environ. Health Persp.* 103:67–72.

Latimer, J. S., and J. Zhang. 2003. The sources, transport, and fate of PAHs in the marine environment. In *PAHs: An ecotoxicological perspective, ecological and environmental toxicology series,* ed. P.E.T. Douben. West Sussex, Eng.: John Wiley and Sons, Ltd.

Lau, C., K. Anitole, C. Hodes, D. Lai, A. Pfahles-Hutchens, and J. Seed. 2007. Perfluoroalkyl acids: a review of monitory and toxicological findings. *Toxicol. Sci.* 99:366–394.

Lau, C., J. L. Butenhoff, and J. M. Rogers. 2004. The developmental toxicity of perfluoroalkyl acids and their derivatives. *Toxicol. Appl. Pharmacol.* 15:231–241.

Law, R. J., C. R. Allchin, M. E. Bennett, S. Morris, and E. Rogan. 2002. Polybrominated diphenyl ethers in two species of marine top predators from England and Wales. *Chemosphere* 46:673–681.

Leahy, J. G., and R. R. Colwell. 1990. Microbial degradation of hydrocarbons in the environment. *Microbiol. Rev.* 54:305–315.

Letcher, R., R. J. Norstrom, and A. Bergman. 1995. Geographical distribution and identification of methyl sulphone PCB and DDE metabolites in pooled polar bear (*Ursus maritimus*) adipose tissue from western hemisphere Arctic and Subarctic regions. *Sci. Total Environ.* 160–161:409–420.

Letcher, R., R. J. Norstrom, S. Lin, M. A. Ramsay, and S. M. Bandiera. 1996. Immunoquantitation and microsomal monooxygenase activities of hepatic cytochromes P4501A and P4502B and chlorinated hydrocarbon contaminant levels in polar bear (*Ursus maritimus*). *Toxicol. Appl. Pharmacol.* 137:127–140.

Letcher, R. J., E. Klasson-Wehler, and A. Bergman. 2000. Methyl sulfone and hydroxylated metabolites of polychlorinated biphenyls. In *The handbook of environmental chemistry,* ed. J., Paasivirta, pp. 315–360. Berlin: Springer-Verlag, Berlin.

Levin, M., S. De Guise, and P. S. Ross. 2005. Association between lymphocyte proliferation and polychlorinated biphenyls in free-ranging harbor seals (*Phoca vitulina*) pups from British Columbia, Canada. *Environ. Toxicol. Chem.* 24:1247–1252.

Levin, M., B. Morsey, C. Mori, and S. De Guise. 2004. Specific non-coplanar PCB-Mediated modulation of bottlenose dolphin and beluga whale phagocytosis upon *in vitro* exposure. *J. Toxicol. Env. Health, Part A* 67:1517–1537.

Lewison, R. L., L. B. Crowder, A. J. Read, and S. A. Freeman. 2004. Understanding impacts of fisheries bycatch on marine megafauna. *Trends Ecol. Evol.* 19:598–604.

Lie, E., et al. 2004. Does high organochlorine (OC) exposure impair the resistance to infection in polar bears (*Ursus maritimus*)? Part I: effect of OCs on the humoral immunity. *J. Toxicol. Env. Health, Part A* 67:555–582.

Loseto, L. L., G. A. Stern, T. A. Connelly, D. Deibel, B. Gemmill, A. Prokopowicz, L. Fortier, and S. H. Ferguson. 2009. Summer diet of beluga whales inferred by fatty acids analysis of the eastern Beaufort Sea food web. *J. Exp. Mar. Bio. Ecol.* 374:12–28.

Loughlin, T. R. 1994. *Marine mammals and exxon valdez.* San Diego, CA: Academic Press.

Loughlin, T. R., and A. E. York. 2000. An accounting of the sources of Steller Sea Lion, *Eumetopias jubatus*, mortality. *Mar. Fish. Rev.* 62:40–44.

Lund, B. O. 1994. *In vitro* adrenal bioactivation and effects on steriod metabolism of DDT, PCBs and their metabolites in the grey seal (*Halichoerus grypus*). *Environ. Toxicol. Chem.* 13:911–917.

Marsili, L., et al. 1998. Relationship between organochlorine contaminants and mixed function oxidase activity in skin biopsy specimens of Mediterranean fin whales (*Balaenoptera physalus*). *Chemosphere* 37:1501–1510.

Martineau, D., et al. 1994. Pathology and toxicology of beluga whales from the St. Lawrence Estuary, Quebec, Canada. Past, present and future. *Sci. Total Environ.* 154:201–215.

Martineau, D., et al. 2002. Cancer in wildlife, a case study: beluga from the St. Lawrence Estuary, Quebec, Canada. *Environ. Health Perspect.* 110:285–292.

Matkin, C. O., E. L. Saulitis, G. M. Ellis, P. Olesiuk, and S. D. Rice. 2008. Ongoing population-level impacts on killer whales *Orcinus orca* following the "Exxon Valdez" oil spill in Prince William Sound, Alaska. *Mar. Ecol. Prog. Ser.* 356:269–281.

McGuire, T. L., and K. O. Winemiller. 1998. Occurrence patterns, habitat associations, and potential prey of the River Dolphin, *Inia geoffrensis*, in the Cinaruco River, Venezuela. *Biotropica* 30:625–638.

McKinney, J. D. 1989. Multifunctional receptor model for dioxin and related compound toxic action: possible thyroid hormone-responsive effector-linked site. *Environ. Health Perspect.* 82:323–336.

McKinney, M. A., S. De Guise, D. Martineau, P. Beland, A. Arukwe, and R. Letcher. 2006. Biotransformation of polybrominated diphenyl ethers and polychlorinated biphenyls in beluga whale (*Delphinapterus leucas*) and rat mammalian model using an *in vitro* hepatic microsomal assay. *Aquat. Toxicol.* 77:87–97.

Meador, J. P., J. E. Stein, W. L. Reichert, and U. Varanasi. 1995. Bioaccumulation of polycyclic aromatic hrydrocarbons by marine organisms. *Rev. Environ. Contam. T.* 143:79–165.

Miller, K. A., M. G. L. Assuncao, N. J. Dangerfield, S. M. Bandiera, and P. S. Ross. 2005. Assessment of cytochrome P450 1A in harbour seals (*Phoca vitulina*) using a minimally-invasive biopsy approach. *Mar. Environ. Res.* 60:153–169.

Montie, E. W., et al. 2008. Cytochrome P4501A1 expression, polychlorinated biphenyls and hydroxylated metabolites, and adipocyte size of bottlenose dolphins from the Southeast United States. *Aquat. Toxicol.* 86:397–412.

Mori, C., B. Morsey, M. Levin, P. R. Nambiar, and S. De Guise. 2006. Immunomodulatory effects of *in vitro* exposure to organochlorines on T-cell proliferation in marine mammals and mice. *J. Toxicol. Environ. Health, Part A* 69:283–302.

Mos, L., M. Tabuchi, N. J. Dangerfield, S. J. Jefferies, B. F. Koop, and P. S. Ross. 2007. Contaminant-associated disruption of vitamin A its receptor (retinoic acid receptor a) in free-ranging harbour seals (*Phoca vitulina*). *Aquat. Toxicol.* 81:319–238.

Mos, L., et al. 2006. Chemical and biological pollution contribute to the immunological profiles of free-ranging harbor seals. *Environ. Toxicol. Chem.* 25:3110–3117.

Muir, D. C. G., C. A. Ford, B. Rosenberg, R. J. Norstrom, M. Simon, and P. Beland. 1996. Persistent organochlorines in beluga whales (*Delphinapterus leucas*) from the St. Lawrence River Estuary. 1. Concentrations and patterns of specific PCBs chlorinated pesticides and polychlorinated dibenzo-p-dioxins and dibenzofurans. *Environ. Pollut.* 93:219–234.

Muir, D., T. Savinova, V. Savinova, L. Alexeeva, V. Potelov, and V. Svetochev. 2003. Bioaccumulation of PCBs and chlorinated pesticides in seals, fishes and invertebrates from the White Sea, Russia. *Sci. Total Environ.* 306:111–131.

Muir, D., M. D. Segstro, K. A. Hobson, C. A. Ford, R. E. A. Stewar, and S. Olpinsk. 1995. Can seal eating explain elevated levels of PCBs and organochlorine pesticides in walrus blubber from eastern Hudson Bay (Canada)? *Environ. Pollut.* 90:335–348.

Muir, D., et al. 2000. Circumpolar trends of PCBs and organochlorine pesticides in the arctic marine environment inferred from levels in Ringed Seals. *Environ. Sci. Technol.* 34:2431–2438.

Munns, W. R., G. W. Suter, T. Damstra, R. Kroes, L. W. Reiter, and E. Marafante. 2003. Integrated risk assessment—results from an international workshop. *Hum. Ecol. Risk Assess.* 9:379–386.

Nakata, H. 2003. Bioaccumulation and toxic potencies of polychlorinated biphenyls and polycyclic aromatic hydrocarbons in tidal flat and coastal ecosystems of the Ariake Sea, Japan. *Environ. Sci. Technol.* 37:3513–3521.

Nakata, H., K. Kannan, L. Jing, N. J. Thomas, S. Tanabe, and J. P. Giesy. 1998. Accumulation patterns of organochlorine pesticides and polychlorinated biphenyls in southern sea otters (*Enhydra lutris nereis*) found stranded along coastal California, USA. *Environ. Pollut.* 103:45–53.

Neff, J. M. 1990. Composition and fate of petroleum and spill-treating agents in the marine environment. In *Sea mammals and oil: Confronting the risks*, eds. J. R. Geraci, and D. St. Aubin, pp. 1–33. New York, NY: Academic Press.

Newell, A. J., D. W. Johnson, and L. K. Allen. 1987. Niagara River biota contamination project: fish flesh criteria for piscivorous wildlife, Technical Report 87–3. New York State Department of Environmental Conservation, Albany, NY.

Nyman, M., et al. 2003. Contaminant exposure and effects in Baltic ringed and grey seals as assessed by biomarkers. *Mar. Environ. Res.* 55:73–99.

O'Shea, T. J. 2003. Toxicology of sirenians. In *Toxicology of marine mammals*, ed. D. J. Vos. London: Taylor and Francis.

O'Shea, T. J., and R. J. Brownell. 1994. Organochlorine and metal contaminants in baleen whales: a review and evaluation of conservation implications. *Sci. Total Environ.* 154:179–200.

O'Shea, T. J., R. R. Reeves, and A. K. Long. 1999. Marine mammals and persistent ocean contaminants: proceedings of the marine mammal commission workshop keystone, Colorado, October 12–15, 1998. Marine Mammal Commission, Bethesda, Maryland, USA, pp. 151.

Oehme, M., M. Schlabach, K. Hummert, B. Luckas, and E. S. Nordoy. 1995. Determination of levels of polychlorinated dibenzo-p-dioxins, dibenzofurans, biphenyls and pesticides in harp seals from the Greenland Sea. *Sci. Total Environ.* 162:75–91.

Parkinson, A. 1996. Biotransformation of xenobiotics. In *Casarett and Doull's toxicology: The basic science of Poisons,* ed. C. D. Klaassen, pp. 113–186. New York, NY: McGraw-Hill.

Pauly, D., A. W. Trites, E. Capuli, and V. Christensen. 1998. Diet composition and trophic levels of marine mammals. *ICES J. Mar. Sci.* 55:467–481.

Peakall, D. B., H. Thomson, and E. Baatrup. 2002. Relationship between behaviour and the biochemical/physiological biomarkers of exposure to environmental pollutants. In *Behavioural ecotoxicology,* ed. G. Dell'Ommo. Chichester, UK: John Wiley and Sons Ltd.

Peterson, C. H., et al. 2003. Long-term ecosystem response to the exxon valdez oil spill. *Science* 302:2082–2086.

Quakenbush, L., and G. G. Sheffield. 2007. Ice seal bio-monitoring in the Bering-Chukchi Sea region. North Pacific Research Board Final Report, 46 p.

Raimondo, S., P. Mineau, and M. G. Barron. 2007. Estimation of chemical toxicity to wildlife species using interspecies correlation models. *Environ. Sci. Technol.* 41:5888–5894.

Reddy, M., S. Echols, B. Finklea, D. Busbee, J. Rief, and S. H. Ridgway. 1998. PCBs and chlorinated pesticides in clinically healthy *Tursiops truncatus*: relationships between levels in blubber and blood. *Mar. Pollut. Bull.* 36:892–903.

Reddy, M., J. S. Reif, A. Bachand, and S. H. Ridgway. 2001. Opportunities for using Navy marine mammals to explore associations between organochlorine contaminants and unfavorable effects on reproduction. *Sci. Total Environ.* 274:171–182.

Reijnders, P. J. H. 1986. Reproductive failure in common seals feeding on fish from polluted coastal waters. *Nature* 324:456–457.

Relyea, R. A. 2009. A cocktail of contaminants: how mixtures of pesticides at low concentrations affect aquatic communities. *Community Ecol.* 159:363–376.

Ridgway, S. H., and M. Reddy. 1995. Residue levels of several organochlorines in *Tursiops truncatus* milk collected at varied stages of lactation. *Mar. Pollut. Bull.* 30:609–614.

Rolland, R. M. 2000. A review of chemically-induced alterations in thyroid and vitamin A status from field studies of wildlife and fish. *J. Wildlife Dis.* 36:615–635.

Ross, P. S. 2000. Marine mammals as sentinels in ecological risk assessment. *Hum. Ecol. Risk Assess.* 6:29–46.

Ross, P. S. 2002. The role of immunotoxic environmental contaminants in facilitating the emergence of infectious diseases in marine mammals. *Hum. Ecol. Risk Assess.* 8:277–292.

Ross, P. S., and L. S. Birnbaum. 2003. Integrated human and ecological risk assessment: a case study of persistent organic pollutants (POPs) in humans and wildlife. *Hum. Ecol. Risk Assess.* 9:303–324.

Ross, P. S., K. B. Beckmen, and S. Pillet. 2003. Immunotoxicology of free-ranging pinnipeds: approaches to study design. In *Toxicology of marine mammals,* eds. D. J. Vos, et al. New York, NY: Taylor and Francis.

Ross, P. S., R. L. De Swart, P. J. H. Reijnders, H. Van Loveren, J. G. Vos, and A. D. M. E. Osterhaus. 1995. Contaminant-related suppression of delayed-type hypersensitivity and antibody responses in harbor seals fed herring from the Baltic Sea. *Environ. Pollut.* 103:162–167.

Ross, P. S., R. De Swart, H. Van Loveren, A. Osterhaus, and J. G. Vos. 1996a. The immunotoxicity of environmental contaminants to marine wildlife: a review. *Ann. Rev. Fish Dis.* 6:151–165.

Ross, P. S., G. M. Ellis, M. G. Ikonomou, L. Barrett-Lennard, and R. F. Addison. 2000. High PCB concentrations in free-ranging pacific killer whales, *Orcinus orca*: effects of age, sex and dietary preference. *Mar. Pollut. Bull.* 40:504–515.

Ross, P. S., S. J. Jefferies, M. B. Yunker, R. F. Addison, M. G. Ikonomou, and J. C. Calambokidis. 2004. Harbor seals (*Phoca vitulina*) in British Columbia, Canada, and Washington State, USA, Reveal A combination of local and global polychlorinated biphenyl, dioxin and furan signals. *Environ. Toxicol. Chem.* 23:157–165.

Ross, P. S., et al. 1996b. Suppression of natural killer cell activity in harbour seals (*Phoca vitulina*) fed Baltic Sea herring. *Aquat. Toxicol.* 34:71–84.

Ross, P. S., et al. 2009. Large and growing environmental reservoirs of Deca-BDE present an emerging health risk for fish and marine mammals. *Mar. Pollut. Bull.* 58:7–10.

Safe, S. 1992. Development, validation and limitations of toxic equivalency factors. *Chemosphere* 25:61–64.

Salata, G. G., T. L. Wade, J. L. Sericano, J. W. Davis, and J. M. Brooks. 1995. Analysis of Gulf of Mexico bottlenose dolphins for organochlorine pesticides and PCBs. *Environ. Pollut.* 88:167–175.

Sample, B. E., and C. A. Arenal. 1999. Allometric models for interspecies extrapolation of wildlife toxicity data. *Bull. Environ. Contam. Toxicol.* 62:653–663.

Sample, B. E., D. M. Opresko, and G. W. Suter. 1996. Toxicological benchmarks for wildlife: 1996 revision, ES/ER/TM-86/R3. Oak Ridge, Tennessee, Risk Assessment Program, Health Sciences Research Division.

Scholz, N. L., et al. 2000. Diazinon disrupts antipredator and homing behaviors in chinook salmon (*Oncorhynchus tshawytscha*). *Can. J. Fish. Aquat. Sci.* 57:1911–1918.

Seacat, A. M., P. J. Thomford, K. J. Hansen, G. W. Olsen, M. T. Case, and J. L. Butenhoff. 2002. Subchronic toxicity studies on perfluorooctanesulfonate potassium salt in Cynomolgus monkeys. *Toxicol. Sci.* 68:249–264.

Seacat, A. M., et al. 2003. Sub-chronic dietary toxicity of potassium perfluorooctanesulfonate in rats. *Toxicology* 183:117–131.

Shi, Z., H. Zhang, Y. Liu, M. Xu, and J. Dai. 2007. Alterations in gene expression and testosterone synthesis in the testes of male rats exposed to perfluorododecanoic acid. *Toxicol. Sci.* 98:206–215.

Simms, W., and P. S. Ross. 2000. Developmental changes in circulatory vitamin A (retinol) and its transport proteins in free-ranging harbour seal (*phoca vitulina*) pups. *Can. J. Zool.* 78:1862–1868.

Simms, W., and P. S. Ross. 2001. Vitamin A physiology and its application as a biomarker of contaminant-related toxicity in marine mammals: a review. *Toxicol. Ind. Health* 16:291–302.

Simms, W., S. J. Jefferies, M. G. Ikonomou, and P. S. Ross. 2000. Contaminant-related disruption of vitamin A dynamics in free-ranging harbor seal (*phoca vitulina*) pups from British Columbia, Canada, and Washington State, USA. *Environ. Toxicol. Chem.* 19:2844–2849.

Sonne, C., et al. 2004. Is bone mineral composition disrupted by organochlorines in East Greenland polar bears (*Ursus maritums*)? *Environ. Health Perspect.* 112:1711–1716.

Sormo, E. G., I. Jussi, M. Jussi, M. Braathen, J. U. Skaare, and B. M. Jenssen. 2005. Thyroid hormone status in gray seal (*Halichoerus Grypus*) pups from the Baltic Sea and the Atlantic Ocean in relation organochlorine pollutants. *Environ. Toxicol. Chem.* 24:610–616.

Soto, K. H., A. W. Trites, and M. Arias-Schreiber. 2004. The effects of prey availability on pup mortality and the timing of birth of South American sea lions (*Otaria flavescens*) in Peru. *J. Zool., London.* 264:419–428.

Spraker, T. R., L. F. Lowry, and K. J. Frost. 1994. Gross necropsy and histopathological lesions found in harbour seals. In *Marine mammals and the "Exxon Valdez,"* ed. T. R. Loughlin. San Diego, CA: Academic Press.

St. Aubin, D. 1990. Physiological and toxic effects on pinnipeds. In *Sea mammals and oil: Confronting the risks,* eds. J. R. Geraci, and D. St. Aubin, pp. 103–127. San Diego, CA: Academic Press Inc.

Subramanian, A., S. Tanabe, and R. Tatsukawa. 1988. Use of organochlorines as chemical tracers in determining some reproductive parameters in dalli-type Dall's porpoise *Phocoenoides dalli. Mar. Environ. Res.* 25:161–174.

Subramanian, A., S. Tanabe, R. Tatsukawa, S. Saito, and N. Miyazak. 1987. Reduction in the testosterone levels by PCBs and DDE in Dall's porpoises of Northwestern North Pacific. *Mar. Pollut. Bull.* 18:643–646.

Tabuchi, M., N. Veldhoen, N. J. Dangerfield, S. J. Jefferies, C. C. Helbing, and P. S. Ross. 2006. PCB-related alteration of thyroid hormones and thyroid hormone receptor gene expression in free-ranging harbor seals (*Phoca vitulina*). *Environ. Health Perspect.* 114:1024–1030.

Tanabe, S., R. Tatsukawa, K. Maruyama, and N. Miyazak. 1982. Transplacental transfer of PCBs and chlorinated hydrocarbon pesticides from the Pregnant Striped Dolphin (*Stenella coeruleoalba*) to her fetus. *Agric. Biol. Chem.* 46:1249–1254.

Tierney, K. B., J. L. Sampson, P. S. Ross, J. Sekela, and C. J. Kennedy. 2008. Salmon olfaction is impaired by an environmentally realistic pesticide mixture. *Environ. Sci. Technol.* 42:4996–5001.

Tilson, H. A., P. R. Kodavanti, W. R. Mundy, and P. J. Bushnell. 1998. Neurotoxicity of environmental chemicals and their mechanism of action. *Toxicol. Lett.* 102–103:631–635.

Tomy, G. T., et al. 2009. Trophodynamics of some PFCs and BFRs in a Western Canadian Arctic marine food web. *Environ. Sci. Technol.* 43(11):4076–4081.

UNEP. 2004. www.unep.org

Van den Berg, M., et al. 1998. Toxic equivalency factors (TEFs) for PCBs, PCDDs, PCDFs for humans and wildlife. *Environ. Health Perspect.* 106:775–792.

Van den Berg, M., et al. 2006. The 2005 World Health Organization reevaluation of human and mammalian toxic equivalency factors for dioxins and dioxin-like compounds. *Toxicol. Sci.* 93:223–241.

Viberg, H., A. Fredrikson, and P. Eriksson. 2003a. Neonatal exposure to polybrominated diphenyl ether (PBDE 153) disrupts spontaneous behaviour, impairs learning and memory, and decreases hippocampal cholinergic receptors in adult mice. *Toxicol. Appl. Pharmacol.* 192:95–106.

Viberg, H., A. Fredrikson, E. Jakobsson, U. Orn, and P. Eriksson. 2003b. Neurobehavioral derangements in adult mice receiving decabrominated diphenyl ether (PBDE 209) during a defined period of neonatal brain development. *Toxicol. Sci.* 76:112–120.

vom Saal, F. S., and C. Hughes. 2005. An extensive new literature concerning low-dose effects of bisphenol A shows the need for a new risk assessment. *Environ. Health Perspect* 113:926–933.

Vonesh, J. R., and J. M. Kraus. 2009. Pesticide alters habitat selection and aquatic community composition. *Conserv. Biol.* 160:379–385.

Wania, F., and D. Mackay. 1996. Tracking the distribution of persistent organic pollutants. *Environ. Sci. Technol.* 30:390A–396A.

Wells, R. S., et al. 2004. Bottlenose dolphins as marine ecosystem sentinels: developing a health monitoring system. *Ecohealth* 1:246–254.

White, R. D., M. E. Hahn, L. Lockhart, and J. J. Stegeman. 1994. Catalytic and immunochemical characterization of hepatic microsomal cytochromes P450 in beluga whale (*Delphinapterus leucas*). *Toxicol. Appl. Pharm.* 126:45–57.

Whitehead, H. 2000. Density-dependent habitat selection and the modeling of sperm whale (*Physeter macrocephalus*) exploitation. *Can. J. Fish. Aquat. Sci.* 57:223–230.

Whitlock, J. J. 1999. Induction of cytochrome P4501A1. *Annu. Rev. Pharmacol. Toxicol.* 39:103–125.

Williams, R. J., et al. 2009. A national risk assessment for intersex in fish arising from steriod estrogens. *Environ. Toxicol. Chem.* 28:220–230.

Wilson, J. Y., et al. 2005. Systemic effects of arctic pollutants in beluga whales indicated by CYP1A1 expression. *Environ. Health Perspect.* 113:1594–1599.

Wolkers, H., C. Lydersen, and K. M. Kovacs. 2004a. Accumulation and lactational transfer of PCBs and pesticides in harbor seals (*Phoca vitulina*) from Svalbard, Norway. *Sci. Total Environ.* 319:137–146.

Wolkers, H., B. Van Bavel, A. E. Derocher, O. Wiig, K. M. Kovacs, and C. G. L. Lydersen. 2004b. Congener-specific accumulation and food chain transfer of polybrominated diphenyl ethers in two arctic food chains. *Environ. Sci. Tech.* 38:1667–1674.

Wolkers, J., I. C. Burkow, C. Lydersen, S. Dahle, M. Monshouwer, and R. F. Witkamp. 1998. Congener specific PCB and polychlorinated camphene (toxaphene) levels in Svalbard ringed seals (*Phoca hispida*) in relation to sex, age, condition and cytochrome P450 enzyme activity. *Sci. Total Environ.* 216:1–11.

Wong, C. S., P. D. Capel, and L. H. Nowell. 2001. National-scale, field-based evaluation of the biota-sediment accumulation factor model. *Environ. Sci. Technol.* 35:1709–1715.

Ylitalo, G. M., et al. 2005. The role of organochlorines in cancer-associated mortality in California sea lions (*Zalophus californianus*). *Mar. Pollut. Bull.* 50:30–39.

Young, R. A. 1976. Fat, energy and mammalian survival. *Integr. Comp. Biol.* 16:699–710.

Zhou, T., M. M. Taylor, M. J. DeVito, and K. A. Crofton. 2002. Developmental exposure to brominated diphenyl ethers results in thyroid hormone disruption. *Toxicol. Sci.* 66:105–116.

Zitko, V., G. Stenson, and J. Hellou. 1998. Levels of organochlorine and polycyclic aromatic compounds in harp seal beaters (*Phoca groenlandica*). *Sci. Total Environ.* 221:11–29.

School of Narwhals

By G. Mutzel, from *The Royal Natural History*, edited by Richard Lydekker, Frederick Warne & Co., London, 1893–94.

10 Select Elements and Potential Adverse Effects in Cetaceans and Pinnipeds

Todd O'Hara
Takashi Kunito
Victoria Woshner
Shinsuke Tanabe

CONTENTS

10.1 Introduction 378
    10.1.1 Basic Approach and Philosophy of Chapter 378
    10.1.2 Definition of Metals 379
    10.1.3 Effects Assessment (Toxicity)...380
        10.1.3.1 *In Vitro* Approaches...381
        10.1.3.2 Genomic Tools 381
        10.1.3.3 Oral Intake/Feeding Ecology Approaches 381
10.2 Mercury 382
    10.2.1 Cetaceans 382
        10.2.1.1 General...382
        10.2.1.2 Odontocetes (Case Studies or Species of Focus) 383
        10.2.1.3 Mysticetes 388
    10.2.2 Pinnipeds 389
10.3 Hg and Nutrient Interactions 391
    10.3.1 Hg and Se..391
10.4 Cadmium 393
    10.4.1 General...393
    10.4.2 Cetaceans 394
        10.4.2.1 Odontocetes 394
        10.4.2.2 Mysticetes 395
Summary 396
References..397

10.1 INTRODUCTION

As stated in Woshner et al. (2002):

> The propensity for various marine mammal species to accumulate some inorganic elements, particularly mercury (Hg), in tissues to concentrations associated with toxicosis in domestic species has provoked conjecture as to the potential adverse effects of these residues.

Relatively moderate to high concentrations of many classes of contaminants have been observed in high trophic level marine mammals raising concerns about possible adverse health effects. Some scientists have viewed these results with alarm, while others have marveled at the apparent physiological capacity of these animals to tolerate such concentrations absent acute health impacts. The most appropriate investigative approach should strike a balance between these two extremes, neither assuming that elevated elemental concentrations in marine mammals necessarily precipitate the harmful effects witnessed in some terrestrial species, nor trusting that marine mammals are impervious to such adverse effects.

Temporal trends indicate that certain elements are increasing globally and/or locally (Riget and Dietz 2000, Riget et al. 2000, 2004). For example, mercury (Hg) levels in the Beaufort Sea ecosystem are rising due to inputs from the Mackenzie River discharge, climate change, and/or industrial emissions (Macdonald 2005, Leitch et al. 2007). Concerns with regard to Hg, as well as other contaminant classes, include: neurotoxicity, reproductive impairment, reduced immunocompetence with consequent increased susceptibility to infectious agent diseases, endocrine disruption, and altered growth and development. Some of these health effects are perceptible at the population level, including decreased abundance, increased rates of mortality, and decreased reproductive output. Hg has been shown to interfere with endocrine processes in marine mammals (Freeman et al. 1975) as have some organic compounds (Ross et al. 1996a, 1996b, Simms and Ross 2001, Rolland 2000). Metal-induced disruption of natural hormones can have serious implications for growth, metabolism, development, reproduction, and immune function in certain marine mammals. For example, Kakuschke et al. reported immune dysfunction among both captive (2008) and wild (2005) harbor seals (*Phoca vitulina*) from the North Sea including hypersensitivity reactions (2006).

10.1.1 Basic Approach and Philosophy of Chapter

Basic principles of comparative toxicology dictate that we evaluate the toxicity of some elements (often referred to as "metals") in marine mammals with caution in that extrapolation of studies of laboratory animals and humans are fraught with assumptions (many unwarranted even among mammals). Even among marine mammals extrapolation must be done with care given their evolutionary and taxonomic diversity. Underlying differences in ecology, anatomy, physiology, and biochemistry of individual species must be taken into account. Whenever possible, direct studies on given taxa should be conducted to elucidate potential effects of individual elements. For example, it is no more appropriate to infer toxic effects of a particular element in beluga whales (*Delphinapterus leucas*) from those in dugongs than it is to extrapolate results from cats to cattle. Thus, we focus on cetaceans and pinnipeds in this chapter.

It is not the intent of this chapter to review sources and pathways of elements to marine mammals. We refer readers to documents that review and/or determine fate and transport of these elements (Dietz et al. 1990, 1995, 1996, 1998a, 1998b, 2006, Pacyna and Winchester 1990, Joiris et al. 1991, Pacyna and Keeler 1995, AMAP 1997, 1998, Joiris 1997, UNEP 1999, 2002). However, the marine environment and associated marine mammal food web is intimately linked to both natural and anthropogenic sources (Booth and Zeller 2005). It is important to remember that nonessential elements may mimic absorptive and physiologic pathways of essential elements and this often relates to their toxicity.

When considering elements, we need to recognize that many are essential for life. These are generally classed as (1) macroelements (macronutrients such as the electrolytes sodium [Na], potassium [K], chlorine [Cl], etc.), or as (2) trace elements (copper [Cu], selenium [Se], zinc [Zn], iron [Fe], etc.) that are critical for numerous biochemical functions. Some essential elements can be toxic at high concentrations (e.g., Se and Fe). Other elements, including Hg, Cd, and lead (Pb), have no known physiologic function and thus are categorized as "nonessential" or "toxic." The essentiality of certain elements is controversial, such as arsenic (As) (Uthus 1992, Neff 1997). With respect to marine mammals, essentiality of specific elements has been presumed based on other homeothermic vertebrates, because studies have not been conducted to directly assess the need for specific elements in marine mammals. As preposterous as it may seem, one could propose that Hg is "essential" for the detoxification of Se in marine mammals because we know that these two elements covalently link to yield the insoluble compound tiemannite. Scientists interpret this interaction as a detoxification *of* Hg *by* Se; however, a quick scan of the veterinary agricultural literature will show that hepatic Se concentrations typical for many marine mammals is quite high and potentially lethal for other mammals (e.g., Puls [1994] cites toxic Se concentrations in liver of cattle as 1.25–7.0 ppm wet weight [ww]). So, is Hg "essential" for prevention of Se toxicity in some pinnipeds and cetaceans? Clearly this is exaggerative speculation for these mammals, but the fact remains that one cannot underestimate the importance of elemental interactions in the expression of toxicity. In 1974, it was reported in an avian model that Hg could alleviate selenosis; thus, the phenomenon has been demonstrated (Hill 1974). Similarly, the relative balance between particular elements may be crucial to manifestation of appropriate function. For example, inadequate concentrations of magnesium (Mg) can be exhibited as functional calcium toxicity. Appraisals of element essentiality are sorely needed in marine mammal species and could be an area of research for captive and live-capture operations.

This chapter is intended to represent a global, or multiocean, perspective and thus addresses elements of general concern. This does not diminish the need to better understand and manage locally relevant impacts of point source metal contamination on particular populations of marine mammals. This chapter mostly references peer-reviewed published research and for the most part omits many agency or organizational reports, so-called gray literature. Because of our focus on globally germane metals in the published literature, information about emerging metals or metals not typically evaluated will not be represented in this chapter. Other efforts have provided expansive tables of element concentrations in marine mammals and we suggest readers refer to these sources (see tabulations for pinnipeds, cetaceans and polar bears compiled by O'Shea 1999 and updated information in O'Hara and O'Shea 2001, 2005, O'Shea and Aguilar 2001, O'Hara et al. 2003, O'Shea 2003, O'Shea and Tanabe 2003).

Some regions and species of animals will be better represented due to the authors' experience and awareness, the focus of intense research programs, commercial and political necessity, relative wealth of certain countries and regions, our limitations to access all possible resources of information, intensively managed stranding networks or sanctioned hunts, and so on. This chapter focuses on pinnipeds and cetaceans; and does not include assessments of polar bears, sea otters, or sirenians because of page limitations.

10.1.2 Definition of Metals

"Heavy metal" is an ambiguous term that does not signify a common set of properties (such as high toxicity, high atomic weight, etc.); rather, it basically refers to aspects of density (>5 g/mL) and other characteristics but is ill-defined (Duffus 2002). In common scientific parlance and as delineated by the periodic table, "heavy metal" implies to the chemical and physical properties of a metallic substance at room temperature and includes a number of higher atomic weight elements. Other nonmetallic elements share some characteristics with metals and consequently have been dubbed "metalloids" (e.g., As, Se). To scientists in certain disciplines, "heavy metal" denotes a subset of elements that possess potent toxicity, such as Hg, Cd, Pb, and As. Throughout this chapter we refer to selected essential and nonessential elements as "metals." Essentiality is based on the scientific documentation that an element is needed for a critical function in a vertebrate (usually mammal) and includes Se (a metalloid chemically similar to sulfur [S]). However, this chapter will focus mostly on Hg and Se.

10.1.3 Effects Assessment (Toxicity)

It is not the intent of this chapter to provide a general review of metal toxicology. We refer readers to other resources for such general information (e.g., Casarett and Doull's Toxicology: The Basic Science of Poisons [Klaassen 2007], Zalups and Koropatnick 2000). It is important to recognize that metal toxicity is an expression of an interaction between a given element and the individual exposed. As such, it is influenced by a vast array of parameters (e.g., valence, associated organic and inorganic components, chemical form of the element, acute or chronic nature of the exposure, uptake and physiologic interactions with other elements, etc.). This panoply of factors exceeds the limited scope of this chapter, which will focus on selected elements and species of marine mammals.

A propensity to report tissue concentrations of elements with little or no meaningful data pertaining to measures of health outcome and/or biology has continued to hamper development of meaningful tissue residue criteria for prediction of toxicological risk. Quite frankly, the continued publication of well-documented differences in element concentrations between sexes, or among age groups or tissue types in marine mammals is neither novel nor enlightening with respect to evaluation of adverse effects. We do recognize that such data may have relevance to spatial and temporal assessments, bioaccumulation, biomagnification, and human food safety and quality determination. Some frequently reported measures such as metallothionein (MTH) induction do represent a biological response to an element, but are not adverse or toxic responses per se, and are very difficult to interpret at the population level (epidemiology). This current situation is understandable considering the compulsory reliance on opportunistic examination and sample acquisition from stranded (live and dead, Bouquegneau et al. 1997), hunter-killed and other less than ideal research candidates. As stated in other reviews on this topic (e.g., O'Shea et al. 1998, O'Shea 1999), controlled studies with appropriate measures (e.g., pathologic, physiologic, biochemical, and genetic) and time series of relevant exposures are required to relate an adverse effect to a specific toxicant or to understand an element's toxicodistribution and toxicokinetics (e.g., Nigro and Leonzio 1996, Nigro et al. 2002). Without these data we struggle to develop criteria for toxicity based solely on tissue concentrations in cetaceans and pinnipeds.

Among the difficulties inherent in the assessment of "effects" are that "effects" can be defined broadly or narrowly, and qualitatively and/or quantitatively determined. Some effects may be overtly negative (e.g., death, decreases in abundance) while others are of debatable consequence (e.g., induction of enzyme or binding protein). Condition or nutritional status (including feeding ecology) may be assessed through measurement of various endpoints such as total lipid, lipid classes, lipid peroxidation, fatty acid signatures, and carbon (C), and nitrogen (N) stable isotopes of specific organic acids. Estimation of such endpoints in conjunction with complementary health evaluations, modeling efforts, and risk characterizations can assist our interpretations of effects if *appropriate* data and suitable epidemiologic tools are used. For example, the well-known positive correlation between renal cadmium (Cd) concentration and age and/or body size is not noteworthy, whereas an association with a biomarker of renal function might be (e.g., blood urea nitrogen, creatinine, specific peptide markers, and specific histologic changes). An ancillary area of effects assessment is the evaluation of contaminant concentrations in tissue versus that observed in prey (food, oral intake measures), in conjunction with consumption guidelines published for particular species (e.g., humans) or an ecological context (Ross 2000). Organizations engaged in work with captive marine mammals where diet is routinely controlled could make valuable contributions to the field of marine mammal toxicology through the establishment of consumption criteria that specify the amount of a given toxicant that may be safely consumed over a specific period (Nielsen 1995, Kakuschke et al. 2008). The authors strongly encourage increased efforts to access oral intake of elements with respect to toxicology (measures of health status and appropriate biomarkers) via enhanced collaborative efforts of scientists and establish linkages with corresponding tissue residues.

10.1.3.1 *In Vitro* Approaches

Controlled toxicity experiments using viable tissues or cells cultured from marine mammals (De Guise et al. 1996, Gauthier et al. 1998, Goodwin et al. 2000, Pillet et al. 2000, Wang et al. 2001, Godard et al. 2006, Wise et al. 2008) provide insight to toxicity but may not directly extrapolate to the whole animal. Nevertheless, such *in vitro* techniques serve as valuable tools for the toxicologist, not only for estimating toxicity, but for elucidating mechanisms of both the toxicants and the physiologic defenses of marine mammals. Such work will improve our ability to assess adverse effects on whole animals and ultimately on populations (e.g., develop specific biomarkers). Development or application of existing genomic, endocrine, and physiologic tools to characterize the health of marine mammals with respect to essential and nonessential elements could allow investigators to relate contaminant concentrations to health endpoints, prey, and international consumption guidelines, thereby facilitating critical evaluation of health risks associated with exposure and accumulation. Ideally we could develop models to explore contaminant-related health risks over time and/or design and conduct laboratory-based toxicity experiments to evaluate effects of elements on marine mammals. However, the limited funding and time-frame (e.g., 1–3 years) associated with most awards (projects) present obstacles to designing studies with such a long-term (decadal) view.

10.1.3.2 Genomic Tools

Absent knowledge of exposures in the wild, which by definition comprises uncontrolled environments that generally contain complex contaminant mixtures rather than individual compounds, researchers frequently have relied on correlative analyses to associate tissue contaminant concentrations with health endpoints. Other impediments to judging contaminant impacts on marine mammal health include acquisition of suitable samples, confounding variables of age and sex, and the paucity of species-specific reagents and techniques for many marine mammals. Some species-specific assays are under development; using toxicogenomics (e.g., Mollenhauer et al. 2009) as an example, relevant efforts include: bottlenose dolphin arrays (e.g., Romano and Warr 2004), aquaSCOPE (Aquatic Sentinel species Comparative Omics for the Environment as currently implemented in British Columbia, Canada), and MAGEX (Multispecies Analysis of Gene Expression, e.g., Helbing et al. 2003). A strategic approach would focus on detoxification (biotransformation) pathways, immune system processes, transcriptional regulatory factors, and modulators of energy and nutrient storage and mobilization. Reliable data generated through these and other *in vitro* techniques would lend themselves to rigorous statistical analyses and ultimately to more defensible determinations of cause and effect than may be had by correlations between tissue contaminant concentrations and various health indices. Resources and expertise need to be focused on critical answerable questions to improve understanding and management of cetaceans and pinnipeds, especially with respect to the role of elements in health and disease.

10.1.3.3 Oral Intake/Feeding Ecology Approaches

Evaluation of marine mammal diets has the potential to yield information on exposure as a consumption rate (e.g., mg of element/kg of body weight/unit time). For each marine mammal species of concern, this would require comprehensive knowledge of both their dietary components and the concentrations of the elements (including the specific chemical forms) therein. This degree of knowledge is generally impracticable, but captive marine mammal situations have permitted calculation of dietary intake for a few cetaceans (Reddy et al. 2001, Nigro et al. 2002) and pinnipeds (Reijnders 1986, Ross et al. 1996a, 1996b) for some contaminants. Captive environments offer opportunities for controlled time series experiments as well as measures of subtle responses, especially if conducted pursuant to the cutting-edge *in vitro* techniques to which we alluded in this chapter's introduction. Knowledge of dietary exposure via these studies and resultant tissue element concentrations of the consumers would be an important link to understand (e.g., amount ingested per kg of body mass results in X mg/kg of various compartments).

Determinations of stable isotopes of carbon and nitrogen in marine mammals tissues can serve as a basic tool to characterize feeding ecology (Hobson and Welch 1992) as well as nutritional status and have been used extensively by one of this chapter's authors (O'Hara). The isotopic signature of carbon can convey spatial origins of dietary constituents (i.e., nearshore vs. offshore, benthic/littoral vs. pelagic), while that of nitrogen relates to the trophic level of prey items (DeNiro and Epstein 1981). These tools have been used to assess Hg dynamics in harbor seals (Brookens et al. 2007) and marine ecosystems (Atwell et al. 1998). The biochemical processes that govern selective accumulation of specific isotopes are also influenced by nutritional stress (Hobson et al. 1993), such that during starvation, heavier stable isotopic forms become enriched. Analysis of stable isotopes in specific biologic molecules (such as specific fatty acids or amino acids) can provide data regarding specific nutrient interactions and sources (Thiemann et al. 2007, 2008, Budge et al. 2008a, 2008b). Such information, in turn, may assist in the interpretation of concentrations of other elements of interest (e.g., Hg, Se, etc.), including their potential sources and effects. Ecological or spatial factors that influence Hg exposure can be reflected through a variety of measurements and techniques (e.g., telemetry, individual animal identifiers) but do not elucidate physiologic impacts. These ecological/biological measures can include stable isotopes of C and N, morphological measures, or life history parameters (including season, sex, and age) and should be integrated with epidemiological approaches (Woshner et al. 2008). It is important to not confuse associations (correlations) with documented impacts.

10.2 MERCURY

Hg is a toxic nonessential element that both bioaccumulates and biomagnifies in most piscivorous marine mammals (Law et al. 1991, Paludan-Müller et al. 1993), especially in areas where Hg of geologic origin is high. Point sources of Hg associated with anthropogenic emissions may also occur, such as mining areas near Amazon River dolphin (*Inia geoffrensis*) habitat (Rosas and Lehti 1996). It is now well established that Hg in marine mammal liver increases with age (see tabulations for pinnipeds, cetaceans, and polar bears compiled by O'Shea 1999 and updated information in O'Hara and O'Shea 2001, 2005, O'Shea and Aguilar 2001, O'Hara et al. 2003, O'Shea 2003, O'Shea and Tanabe 2003). While researchers initially found "high" hepatic Hg concentrations in marine mammals shocking (e.g., >13,000 ppm dry weight (dw) in bottlenose dolphins (*Tursiops truncatus*) from the Mediterranean Sea (Leonzio et al. 1992), this should no longer be the case. We will discuss Hg detoxification with respect to Se in a separate section. It is well recognized that methylmercury can be sequestered directly into keratinized structures and eliminated by hair loss or epidermal sloughing (Scheuhammer 1991, Clarkson 1994, Gaggi et al. 1996, Brookens et al. 2008) and should be considered in assessments of Hg and Se.

10.2.1 CETACEANS

10.2.1.1 General

Cetaceans include the large whales, dolphins, and porpoises. Body size varies tremendously among cetaceans along with many other features of behavior, physiology, anatomy, and life history. Cetaceans are divided into two major taxonomic groups: odontocetes (toothed whales) and mysticetes (baleen whales). Cetacean species tend to be relatively long-lived, with dolphins easily living 40 years or more (Wells et al. 2004, 2005), while the longevous bowhead whale (*Balaena mysticetus*) may achieve 150–200 years (George et al. 1999). Various tissues demonstrate metal accumulation with age, with some evidence that it may plateau or slightly decrease in tissues of the oldest whales

(Dehn et al. 2006). Examples of this accumulation pattern include Cd and Hg in liver and kidney. This correlation between age and metal concentration presents profound difficulties for differentiating tissue changes directly caused by metal accumulation from normal age-associated degeneration, particularly since neither etiology induces distinctive (i.e., pathognomonic) lesions. Rather, both chronic metal exposure and nonspecific stressors associated with aging metabolism can induce tissue changes such as fibrosis or pigment accumulation, as well as alteration in other biochemical or physiologic indices, such as increases in MTH. For large whales, tissue changes such as fibrosis may be adaptations to large body mass and/or diving that are coincidentally age-related (i.e., older animals tend to be bigger) as reported in Rosa et al. (2008). These "lesions" could include some adaptations to potentially toxic elements (Dietz et al. 1998a). Since most studies depend on correlative analyses it is very difficult to infer a cause and effect relationship given that, for many elements in various tissues, age and metal concentration are positively associated.

10.2.1.2 Odontocetes (Case Studies or Species of Focus)

10.2.1.2.1 Background (Odontocete Examples)

Various regions report Hg concentrations for bottlenose dolphins (mostly stranded) including the North American coastline (Rawson et al. 1993, Kuehl et al. 1994, Kuehl and Haebler 1995, Beck et al. 1997, Meador et al. 1999, Carvalho et al. 2002, Woshner et al. 2006, 2008, Stavros et al. 2007, 2008); European waters (Rudneva and Pronin 1996, Holsbeek et al. 1998, 1999, Frodello and Marchand 2001, Roditi-Elasar et al. 2003, Decataldo et al. 2004) and Asia (Endo et al. 2005, 2006). Because many of these reports do not offer insights into adverse health effects and space is limited, we will focus on the more recent reports, a few of which have been evaluated within or across stock variability in Tursiops (Stavros et al. 2007, 2008, Woshner et al. 2008). Because the coastal bottlenose dolphin is common, widely distributed, and demonstrates site fidelity (Das et al. 2000, 2003, Fair and Becker 2000, Wells et al. 2004, Bossart 2006, Stavros et al. 2007) this species has been proposed as a sentinel for monitoring spatial and temporal trends of metals. While these are compelling considerations, dolphin movement patterns need closer assessment, and the geographic range of their prey base must also be taken into account. Thus, we should not just consider the range of the cetacean consumer (the matrix intended for Hg analysis) but the range of the consumer along with that of its prey, together with long-range Hg deposition (including abiotic processes). Human affinity for bottlenose dolphins is also a driving force for their study (the "*Flipper* factor"), and in some regions *Tursiops* are a human food source so that Hg presents a controversial food safety issue (Endo et al. 2005).

The relatively high trophic level and piscivory of beluga whale feeding predisposes this odontocete to accumulation of Hg in many tissues, as well as other contaminants (Wagemann et al. 1990, 1996, Becker et al. 2000, Woshner et al. 2001b, Lockhart et al. 2005, Wagemann and Kozlowska 2005, Dehn et al. 2006, Loseto et al. 2007). Numerous reviews and studies address OHs and/or PAHs in beluga whales, especially for the Saint Lawrence Estuary (SLE) beluga whales (a small, endangered population between the United States and Canada). Similar impacts on the immune system, reproduction, neoplasia, and general morbidity and mortality can result from some heavy metals as well (different mechanisms but the same pathologic effect and/or target). For example, adverse health effects of methylmercury occur in the reproductive (Koller 1979), immune (Koller 1979), and central nervous systems (CNS) (Scheuhammer 1991, Clarkson 1994). Thus, we suggest caution in implicating one contaminant over another with respect to observed adverse health effects and also admonish that other stressors (infectious agents, nutritional deficiencies) can act alone or in concert to produce similar effects (e.g., neoplasia in general and gastric papillomas in beluga whales as reported by Martineau et al. 1994, 2002, and De Guise et al. 1994, respectively). This applies to all marine mammals, but considering the highly stressed and generally contaminated SLE beluga whales we are hesitant to draw conclusions about impacts of specific elements alone on the occurrence of disease for these animals. Thus, we limit our discussion of the SLE beluga population in this review, but this does not indicate a lack of concern or urgency for the SLE region. It is simply the chemical milieu of the SLE that makes teasing out specific effects of certain toxicants (e.g., nonessential elements) more difficult than for other already complicated scenarios even in relatively "pristine" cetacean environments, for example, Arctic and Antarctic.

10.2.1.2.2 In Vivo (Odontocetes)

In whole blood of bottlenose dolphins from Sarasota Bay, Florida (U.S.), mean total Hg exceeded 500 µg/L (Woshner et al. 2006, 2008). Such concentrations are about 9 times the 58 µg/L benchmark dose level for human cord blood and 100 times the level at which adverse neurodevelopmental effects may occur (5.8 µg/L in human cord blood; Trasande et al. 2005). However, this must be interpreted carefully since the regulatory levels are very conservative and derived for human risk assessments rather than for cetaceans or marine mammals. Due to the interaction between Hg and Se, any assessment of adverse Hg effects in dolphins should include indices of Se status (direct or indirect), including measures of thyroid hormones, blood glutathione peroxidase activity, routine hematology, and serum chemistry panels as in Woshner et al. (2008). In addition, *in vitro* immune function assays using a range of environmentally relevant blood Hg concentrations will improve our understanding of the potential effects of Hg on the immune system of bottlenose dolphins (Pellisso et al. 2008).

Hg in blood and epidermis is mostly methylmercury (Stavros et al. 2007, Woshner et al. 2008), whereas in liver and kidney there is an age-associated increase of total Hg but negligible change in methylmercury. Concentrations of Hg in epidermis correlate with those in blood, with blood concentrations being much lower (Woshner et al. 2008). Hg in blood correlates strongly with serum Se concentration, the heavy isotope of N (^{15}N or increased trophic level), and age. These ecological and elemental interactions are important considerations for interpreting observed Hg concentrations (but still do not indicate impact) and import of exposure pathways. Based on hematology and clinical chemistries, Woshner et al. (2008) concluded that dolphins evidently are subject to seasonal exacerbation of oxidative stress, which might render them more vulnerable to toxic effects of Hg based on Se status and other antioxidant processes. Thus, Hg concentrations and associated risk of toxicity may be seasonally and/or nutrient dependent. Although this remains speculative, the Hg and Se interaction is more than simple complexation (Hg–Se formation) as Se is needed for numerous functions including protection from oxidants.

10.2.1.2.3 Elemental and MTH Interactions

Since upper trophic level organisms have increased methylmercury exposure (Das et al. 2000a, 2000c) and Se antagonizes Hg toxicosis we must consider nutrient interactions, which will be discussed in Section 10.3. Chelation (binding) of metals with MTH and MTH-like proteins is very important to consider with respect to uptake, storage, and toxicity of metals (Das et al. 2000b, 2001, 2004, 2006, Decataldo et al. 2004). Hepatic and renal Hg, Cd, and Zn were determined from *Stenella coeruleoalba* and *Tursiops truncatus* stranded along the southern coast of Italy. Decataldo et al. (2004) confirmed that in odontocetes: (1) Hg is mainly found in the insoluble fractions in liver and kidney (thus MTHs seem to have no or a minimal role in end stage Hg detoxification); (2) the hepatic molar ratio between Se and inorganic Hg approximates 1:1 suggesting that the end-stage product of methylmercury detoxification is likely mercuric selenide (HgSe) (tiemannite); and (3) Cd and Zn exist in the purified heat-stable fraction (including MTH-like proteins) of most samples. Nevertheless, the uptake and initial biotransformation of Hg is a dynamic process and despite these results (Decataldo et al. 2004) the possibility remains that MTH may be involved in the early stages of Hg binding, producing a short-lived intermediate that is subsequently processed into the end-stage Hg–Se form. The study design of Decataldo et al. (2004) does not entertain the early phase of Hg toxicodistribution in the odontocetes. Thus, it remains to be seen whether MTH may play a role in the initial phases of Hg metabolism, which ultimately results in the MTH-free, insoluble Hg–Se complex.

Decataldo et al. (2004) state: "The remarkable tolerance of marine mammals to heavy metals is well known. It has been reported that they can concentrate Hg from fish and shellfish, and, at the same time, be able to tolerate it with no apparent symptoms of poisoning. This tolerance could be indicative of detoxification processes, based on the capability of these organisms to store contaminants in a nontoxic physicochemical form. The processes of intracellular storage tend to increase concentrations of metals in certain organs, but, at the same time, reduce the toxicological risk."

Information available to date appears to support this view. However, one might question whether dolphins display Hg toxicity differently from other mammals or if parallel consumption of critical dietary components (e.g., Se and specific fatty acids) is required to prevent Hg toxicosis. Even though intracellular storage of tiemannite granules may serve as an effective detoxification mechanism, there is surely some upper limit of Hg that might overwhelm this process, either in the production of the granules, or the cellular storage capacity for them. We are unable to estimate this upper bound of exposure with enough certainty for this chapter. Moreover, researchers may not even be monitoring the appropriate organ systems to reflect Hg toxicosis. We certainly lack the capacity to evaluate potential CNS effects of Hg on free-ranging animals, but with new technologies being developed to assess live-captured and captive dolphins, we may soon be able to relate these subtle physiologic measures to toxicant exposure, including Hg (such as is occurring for some human populations of the Seychelles Islands and Faroe Islands).

Stavros et al. (2007) reported concentrations of many trace elements in "skin" (likely epidermal) samples from free-ranging bottlenose dolphins near Charleston, South Carolina and Indian River Lagoon, Florida (U.S.). As is classically done, distributions of trace element concentrations by age, gender, and study area were examined. Their paper, along with many others, reports significant site differences while considering correlations with age. In many cases, the critical reader should evaluate the biological significance of these differences (actual ppm difference and some relevance to dose and response) and not rely solely on a statistical determination of difference. To have biological relevance, the actual difference in the concentrations should be interpreted in a toxicological context that utilizes the basic principles of the dose–response curve and how it converts to a difference in magnitude of adverse response. Again, the lack of data relevant to effects limits our ability to determine "safe" upper bound exposure indices.

10.2.1.2.4 Liver and Kidney Mercury

Postmortem examination of stranded Atlantic bottlenose dolphins (Florida) revealed suspect age-correlated hepatic (portal) and renal proximal tubular lipofuscinosis (Rawson et al. 1993). Rawson et al. (1993, 1995) noted that lipofuscin-like pigment granules and other findings (fat globules, central necrosis, and lymphocytic infiltrates) were associated with relatively high concentrations of hepatic total Hg (maximum 443 ppm ww). Pigment granules were found in all dolphins with 61 ppm or more total Hg (older animals). Energy dispersive X-ray analysis (EDAX) electron microscopy confirmed Hg aggregates in hepatic lysosomes. Because dolphins with hepatic lipofuscinosis had higher hepatic total Hg concentrations than animals without pigment, Rawson et al. (1993) suggested a causal relationship between hepatic Hg concentrations, lipofuscin accumulation, and hepatic disease. Woshner et al. (2002) and Woshner (2000) also found that Hg concentration and lipofuscin were correlated in livers of beluga whales and bowhead whales from Alaska (U.S.). Nevertheless they questioned Rawson's assertion that the lipofuscin pigment itself was a direct result of hepatic Hg deposition, because in the arctic cetaceans the pigment occurs in young belugas with comparatively low hepatic Hg concentrations, as well as in the lower trophic feeding bowhead whales, which have overall low Hg exposure. Lipofuscin may result from Hg intoxication but this pigment is not a specific indicator of Hg exposure. Rather, lipofuscin can be found in many species across widely varying taxonomic groups, including marine mammals, and is known to be generally associated with aging and oxidative stress (Gray and Woulfe 2005). Siebert et al. (1999) determined total Hg concentrations and gross and microscopic lesions in tissues of harbor porpoises (*Phocoena phocoena*) and white-beaked dolphins (*Lagenorhynchus albirostris*) from the Baltic and North Seas (e.g., Das et al. 2004). No lesions attributable to Hg intoxication were noted with hepatic concentrations of total Hg up to 449 ppm dw and of methylmercury up to 26 ppm dw. AMAP (2005) used a very uncertain threshold for marine mammal liver toxicity of 60 ppm (ww) as cited from Law (1996) and Thompson (1996) for terrestrial mammals at 30 ppm (ww). This emphasizes what was stated in AMAP (2005):

> Metal concentrations in tissues can be determined routinely and with high precision. However, tissue concentrations alone do not provide accurate predictions of biological effects.

Thus, we do not prescribe threshold levels for marine mammals in this effort and encourage direct assessment of toxicity (clinical signs, biochemical and anatomic lesions, etc.). In fact, some published work has used domestic animal toxicity criteria (e.g., Se) and without hesitation extrapolated to tissue residues determined in marine mammals (limited sampling effort) and have made unfortunate conclusions. This occurred for measures of Se in California sea lions (Edwards et al. 1989) that concluded selenosis. These concentrations of Se are now considered routine findings.

The bottlenose dolphin provides an opportunity to better understand metabolism and effects of Hg in cetaceans. We require a basic understanding of absorption, distribution, metabolism, and excretion (ADME) of Hg and the various systems that could be impacted. Current exposures to Hg are of interest and these opportunities should be exploited, particularly through the physiologic and pathologic assessment of live-captured and captive animals. Continued development of *in vitro* methods should be aggressively pursued as well.

The gross distribution of Hg within different parts (lobes) of the liver is basically homogenous. However, the accumulation of Hg does not appear to be homogenous at a fine tissular scale. Rather, Hg and Se occur as dense intracellular granules (presumably bound by lysosomes) collecting primarily in macrophages within the liver, but also in spleen, bone marrow, and lungs (Martoja and Berry 1980, Nigro and Leonzio 1996, Woshner et al. 2002). Macrophages may accumulate Hg through phagocytosis of erythrocytes (Nigro and Leonzio 1996). Woshner et al. (2002) characterized inorganic Hg deposition in kidney and liver of free-ranging, hunter-killed beluga and bowhead whales through autometallography (AMG). This technique localizes and amplifies the intracellular Hg so that it may be visualized microscopically (Danscher et al. 1994); thus, the technique is considered semiquantitative. As expected, AMG granules were not evident in bowhead tissues, confirming low Hg concentrations (mean total Hg 0.06 µg/g ww). In belugas, mean total Hg was 23.29 and 4.97 µg/g ww in liver and kidney, respectively, and granules were easily visible. In kidney, the AMG granules were in cortical tubular epithelial cytoplasm in belugas with lower tissue Hg concentrations; whereas, whales with higher concentrations displayed AMG granules throughout the uriniferous tubular epithelium. Hepatic AMG granular densities differed between lobular zones, concentrating in stellate macrophages and bile canalicular regions of hepatocytes. AMG granules were aggregated in periportal regions in belugas with lower hepatic Hg concentrations, yet among whales with higher Hg the AMG granule deposition extended to pericentral and midzonal regions of liver lobules. In beluga livers, AMG staining density was not associated with lipofuscin quantity (an index of oxidative damage) as estimated by digital image analysis. Further, although AMG and lipofuscin granules were occasionally co-localized, more often they were not, implying that Hg did not figure prominently in hepatic lipofuscin deposition in belugas (Woshner 2000).

Considering that sample collection from marine mammals is often opportunistic and may involve substantial logistical constraints, every effort should be made to routinely collect samples for standard histological examination to critically assess the role of toxicants in observed tissue changes as well as to rule out other causes of disease and better characterize overall health status. Focusing on one or a few key biomarkers does not allow for an overall assessment of animal condition that is critical for putting any findings into the proper context, whereas histological examination of tissues provides both general and specific indications of health and disease status. Moreover, results of chemical analyses may indicate avenues of inquiry that could be addressed through special histologic stains, immunohistochemistry, or electron microscopy.

10.2.1.2.5 *In Vitro (Odontocetes)*

Beluga whale investigators have applied a diverse set of *in vitro* and *in vivo* tools to assess contaminant impacts on a species that is difficult to study. These *in vitro* studies may assist researchers in discriminating prospective effects of individual components in mixed contamination conditions (e.g., chemical milieu) such as those described above for the SLE whales. One can isolate certain viable cell types for exposure to very specific toxicants to ascertain a dose (concentration) response relationship under controlled conditions, including chemical mixtures of interest. For example, De Guise et al. (1996) exposed beluga whale splenocytes and thymocytes to different concentrations of Hg, Cd and Pb (salts of Cl). Increased cell death was observed in Con-A-stimulated (mitogen) thymocytes for Hg, while splenocytes were not killed by these elements at the concentrations employed. However, decreased splenocyte and thymocyte proliferation (decreased function) occurred with the highest concentration of Hg and Cd (10^{-5} M), while lower concentrations (10^{-6} and 10^{-7} M of Hg and Cd) and all concentrations of Pb did not alter cell proliferation. Micronuclei were induced by Hg in fibroblasts from epidermal samples of beluga whales; methylmercury inhibited cell proliferation and induced micronuclei (Gauthier et al. 1998). Ross et al. (personal communication) have proposed exposing various types of beluga whale cells *in vitro* to concentration gradients of Hg and methylmercury in combination with Se relevant to the Hg:Se molar concentrations reported *in vivo* (*in situ*) for liver, kidney, and epidermis. Such studies will certainly promote a better understanding of the mechanisms whereby elements interact, and further cross-species comparisons (De Guise et al. 1996, Gauthier et al. 1998, Taddei et al. 2001, Wang et al. 2001). It is important to recognize that excessive Se can be toxic as well, so that Hg could conceivably even play a protective role with respect to Se in beluga whales. Betti and Nigro (1996) found that cultured lymphocytes of bottlenose dolphins have greater resistance to cytotoxic and genotoxic (single-strand DNA breakage) effects of methylmercury than cells of rats or humans.

10.2.1.2.6 *Hg Toxicodistribution (Odontocetes)*

Dall's porpoise (*Phocoenoides dalli*) is a species of small toothed whale that inhabits the North Pacific Ocean and adjacent seas. For this species, we will discuss the importance of tissular distribution and intracellular distribution and binding as an example of cetacean toxicodistribution. Dall's porpoises showed the highest mean Hg concentration in liver (26 µg/g dw) and kidney (14 µg/g dw), and the lowest in bone (0.2 µg/g dw) and blubber (0.2 µg/g dw) among 14 tissues examined (Yang et al. 2002). In a mature male from which 14 tissues were analyzed, the whole body burden of 0.19 g Hg was distributed primarily in muscle (74.5%), followed by liver (12.8%) (Yang et al. 2006). Ratio of organic Hg to total Hg was highest in skin (0.70) and muscle (0.67), and lowest in lung (0.11), kidney (0.12), liver (0.13) and spleen (0.14) among 14 tissues of a mature female (Yang et al. 2002). Inorganic Hg predominated in liver, while organic Hg was predominant in muscle with increasing total Hg in each tissue (Yang et al. 2003). In a maternal-fetal pair, the hepatic Hg level of the mother (3.2 µg/g ww) was four times that of her fetus (0.82 µg/g ww) (Yang et al. 2004). In cells of both the mother and fetus, about half of the Hg was distributed in cytosol, in which 60–70% was bound to high molecular weight substances, while 16% and 40% was bound to MTH in mother and fetus, respectively (Yang et al. 2004). It has been suggested that Hg forms a complex with Se and then binds to a high molecular weight substance in cytosol (Ikemoto et al. 2004a). In the nuclear and mitochondrial fractions, more than 80% of the Hg was extractable with a 2-mercaptoethanol + guanidinium thiocyanate solution; a weak extractant, 2% SDS solution, was able to extract about half of the Hg (Ikemoto et al. 2004b). These results suggest that most of the Hg was present as labile forms in the liver of the specimens examined, probably due to their low mean total Hg level (3.8 µg/g ww). In kidney, maternal Hg concentration (2.7 µg/g ww) was 12 times higher than that of her fetus (0.23 µg/g ww) (Yang et al. 2008). In both maternal and fetal renal cells, about 50–60% of the Hg was present in cytosol. In maternal kidney, less Hg was distributed in the fraction containing high molecular weight substances (48%) than in the liver (70%), and 41% of renal Hg was bound to

MTH (Yang et al. 2008). Intracellular distribution was not determined for fetal tissue because of the low Hg concentration.

In Dall's porpoises, equimolar accumulation of Hg and Se occurred above the threshold Hg concentration of 20–30 µg/g dw (Yang et al. 2007), which was lower than the threshold for several other species of marine mammals (40–60 µg/g dw; S. Tanabe unpublished results) and northern fur seals (*Callorhinus ursinus*) (150–200 µg/g dw; Ikemoto 2004). Across species, a lower threshold concentration of hepatic Hg for equimolar accumulation of Hg and Se might indicate a comparatively lower ability to detoxify further Hg via formation of HgSe (Yang et al. 2007).

Very little information is available for metals in short-finned pilot whales (*Globicephala macrorhynchus*) from the Pacific Ocean (or for that matter, other oceans). However, the mean concentration of Hg in short-finned pilot whales from offshore water of Japan deserves comment since it was 534 µg/g dw in liver, 62.9 in kidney, and 19.6 in muscle (Tanabe et al. unpublished results). Specimens from New Caledonia exhibited 1411 and 1452 µg/g dw in liver, 32.8 and 27.3 µg/g dw in muscle, and 11.0 and 3.20 µg/g dw in blubber (Bustamante et al. 2003). Mean Hg concentration was 231 µg/g ww in liver, 24.3 in kidney, and 1.02 in blubber of short-finned pilot whales stranded at Cumberland Island, U.S. (Stoneburner 1978). Rawson et al. (1995) reported the presence of HgSe in liver, lungs, and hilar lymph nodes of short-finned pilot whales from Florida, U.S. For long-finned pilot whales (*Globicephala melaena*) from Cape Cod, high concentration of Hg was reported in some specimens: 4.4 µg/g dw in brain, 626 in liver, 49.7 in kidney, and 173 in ovary (Meador et al. 1993). Pilot whales are likely an odontocete worthy of further investigation with respect to the biology and physiology of Hg and Se.

Compared to most other marine mammal species, potential mechanisms of Hg detoxification have been studied extensively for striped dolphins (*Stenella coeruleoalba*) from Japanese coastal waters. Among 14 tissues examined, liver showed by far the highest mean Hg level (205 µg/g ww), followed by spleen (22.6 µg/g ww) (Itano et al. 1984). The hepatic Hg concentration is higher in older animals, whereas in kidney and muscle the concentration after about 18 years of age is lower (Honda et al. 1983). In kidney homogenates, only 17% of Hg was present in supernatant, and 41% of this Hg was bound to MTH (Kwohn et al. 1986), the amino acid sequence of which showed 93% homogeneity with human MTH II (Kwohn et al. 1988). The Hg:Se molar ratio was larger than unity in liver of individuals with hepatic Hg levels greater than 400–600 µg/g dw (Agusa et al. 2008). In these dolphins, S, as well as Se, might be involved in the detoxification of Hg, in a similar scenario to that initially proposed by Pelletier (1985) and outlined in Section 3.1 of this chapter. Ng et al. (2001) identified a solid solution Hg(Se, S) in liver of striped dolphins. HgSe occasionally has been found in livers with the highest Hg levels. However, even in muscle, the molar ratio of Hg:Se was near 1:1 in the salt-insoluble fraction for striped dolphins (Itano et al. 1985a, 1985b). Recently, HgSe was identified in various tissues other than liver, including muscle, kidney, lung, spleen, pancreas, and brain for a striped dolphin (Nakazawa et al. unpublished data). A water-soluble low-molecular-weight compound (molecular weight < 1000) containing both Hg and Se was also purified from a liver sample (Ping et al. 1986). This compound was shown to contain amino-groups, as well as Hg and Se at a 1:1 molar ratio, but has not yet been identified. Again, this emphasizes the piscivorous odontocete as a critical model for understanding the Hg and Se toxicant-nutrient interaction between these two elements and other cellular constituents.

10.2.1.3 Mysticetes

Many scientists believe that Hg and other metals are of minimal toxicologic importance to the baleen whales (e.g., O'Shea and Brownell 1994). In general, this is a likely scenario based on their low trophic level feeding habits (Hoekstra et al. 2002, Dehn et al. 2006). Highly specialized baleen whales (such as gray, bowhead, and North Atlantic right whales) that feed almost exclusively on invertebrates, or baleen whales that consistently eat a mixture of fish and invertebrates (e.g., humpbacks, *Megaptera novaeangliae*) likely have somewhat constant and predictable exposures to Hg and Cd. However, we need to recognize that some mysticetes (e.g., common minke whales,

Balaenoptera acutorostrata) can make abrupt changes in diet that will elicit sudden alterations in exposure to Hg and Cd. Such shifts from invertebrate prey (associated with Cd) to vertebrate prey (fish containing methylmercury) are important to consider (e.g., Hansen et al. 1990).

Southern minke (*Balaenoptera bonaerensis*) whales are distributed in the Antarctic Ocean. They undertake substantial migrations from summer feeding grounds in the Antarctic to tropical or sub-tropical winter breeding grounds. Hg concentrations in southern minke whales are lower than those of many marine mammals from other regions (Kunito et al. 2002). Hg concentration in southern minke whales was 0.004–0.062 µg/g ww in muscle, 0.026–0.163 in liver, and 0.005–0.132 in kidney (Honda et al. 1987b). The low Hg concentration reflects that in their prey, antarctic krill. In general, relatively low Hg levels were observed in other marine organisms from the Antarctic Ocean. Honda et al. (1987b) demonstrated that the low Hg levels characteristic of antarctic marine organisms were associated with concentrations in seawater, in conjunction with the simple and short food webs of the antarctic marine ecosystem. Thus, Hg exposure seems to be of minor importance in southern minke whales. Interestingly, Hg concentrations in tissues of southern minke whales apparently fluctuate in response to competition for food among baleen whales in the Antarctic Ocean and to differences in prey availability caused by the annual changes in sea-ice extent (Honda et al. 1987a, 2006). Hg and Se concentrations are quite low in bowhead whales (Woshner et al. 2001b) and their occasional arctic cohabitants the gray whales (*Eschrichtius robustus*; Dehn et al. 2006). However, Cd is an element of interest for bowhead whales that will be discussed in Section 10.4.2.2.

10.2.2 PINNIPEDS

Historically we note a few experimentally based (i.e., direct exposure) investigations and *in vitro* studies of Hg in pinnipeds. Among the general findings of these studies were: the demethylation of methylmercury, recognition of a role for Se in ameliorating toxicoses due to Hg, and observations of induced acute toxicosis. Freeman et al. (1975) dosed harp seals (*Phoca groenlandica*; syn. *Pagophilus groenlandicus*) with 0.25 mg/kg methylmercury daily for 2 months and found substantial demethylation (70%). Administration of methylmercury to gray seals (*Halichoerus grypus*) elicited increased hepatic Hg and Se, but only increased Hg (not Se) concentrations in other tissues (van de Ven et al. 1979). Tillander et al. (1970) calculated rates of two excretion phases (55% with a half-life of 3 weeks, the balance 500 days) for radioactively labeled methylmercury-dosed ringed seal (*Phoca hispida*). Harp seals administered 25 mg/kg/day methylmercury over 20–26 days developed renal failure, uremia, toxic hepatitis and died. While this shows that there is a limit to Hg tolerance, the role of Se is elusive in these studies since it may or may not have been provided in relevant amounts, temporal contexts, and/or chemical forms (e.g., Se concentrations and forms in a fish diet). No obvious pathologic changes resulted from a single dose of 0.25 mg/kg (Holden 1978). Harp seals dosed with methylmercury also showed nonspecific, low level damage to sensory cells of the organ of Corti which included missing or damaged stereocilia, reticular scars, and collapsed sensory cells (Ramprashad and Ronald 1977). Although the relevance of this damage is unclear without some measure of auditory performance, because damage to hair cells is permanent, one can conjecture that animals confronted with chronic, lifelong exposure to methylmercury might develop auditory or other sensory deficits that could severely compromise their ability to survive in the wild. Alterations of *in vitro* gonadal and adrenal steroid synthesis were reported in harp and gray seals administered 0.25 mg/kg methylmercury (Freeman et al. 1975). Most of these investigations were done 30 years ago, but are still relevant, particularly since it has become very difficult to conduct standard toxicity tests on marine mammals.

Basu et al. (2006) evaluated the effect of Hg and various organochlorines on muscarinic acetylcholine receptors in isolated cellular membrane preparations from ringed seal cerebrum. Of all neurotoxicants tested, only Hg inhibited binding of the radiolabeled acetylcholine receptor agonist [3H]quinuclidinyl benzilate by more than 50%. Inhibition elicited by divalent inorganic Hg was significantly greater than that instigated by methylmercury, but this difference disappeared when data were normalized for calculation of inhibition constants. Hg inhibition of acetylcholine receptors was saturable, and apparently mediated by competitive binding (Basu et al. 2006).

Piscivorous harbor seals accumulate significant amounts of Hg (Koeman and van de Ven 1975, Smith and Armstrong 1978, Reijnders 1980, Himeno et al. 1989, Brookens et al. 2007, 2008). Harbor seals may be useful mammalian integrators for biomonitoring Hg in coastal ecosystems (Griesel et al. 2008), and in the context of this chapter applicable as an apt pinniped model for the evaluation of adverse health effects of toxicants (including Hg). The proposal that harbor seals serve as a prototypical pinniped for metal toxicology is akin to the use of the harbor seal in studies of organohalogens. The harbor seal is relatively well studied as compared to other pinnipeds, especially with respect to abundance, stock structure, ecology, physiology, and some contaminants. This species was nominated as a model for pinniped toxicology by O'Shea (1999). For the sake of brevity we will consider the harbor seal as our model pinniped and provide limited additional information on other pinnipeds. The ringed seal has been selected as a key indicator species for circumpolar arctic assessments (AMAP 1997, 1998, 2002, 2005, http://www.amap.no/). A recent review of ecosystem responses to Hg contamination suggested that some marine mammals, such as ringed seals and harbor porpoises would be potentially good indicators of changing Hg loads in the coastal environments (Wolfe et al. 2007). These species were selected as indicators based on a combination of criteria that included: well-characterized life history; capacity for bioconcentrating and accumulating contaminants of concern; common species in the environment; geographically widespread; sensitive and hence indicative of change; easily collected and measured; adequate size to permit resampling of tissue; occurrence in both polluted and unpolluted areas; display correlation with environmental levels of contaminants; and have background data on the natural condition (Wolfe et al. 2007).

Brookens et al. (2007) reported tissue concentrations of total Hg, methylmercury, Se, and Pb for harbor seals of central and northern California. Hepatic total Hg concentrations increased with age (as had been noted previously by Himeno et al. 1989), and trophic level (Brookens et al. 2007). Methylmercury concentrations in liver increased exponentially until approximately 5 years of age with an asymptote at 1.3 µg/g ww. However, methylmercury expressed as a percentage of total Hg decreased to a minimum (≤5%) at about 4 years of age in what many consider a classic relationship between methylmercury and total Hg in marine mammals (i.e., total Hg increases with age while methylmercury stays constant and relatively low). Hepatic Se also increased with age and was in equimolar ratio with total Hg in adults, whereas, the molar ratio of Se:total Hg in pups deviated from 1:1. These age-dependent interactions are very likely important for detoxification and failure of demethylation and/or Se-dependent protection, either as an antioxidant or as a direct chelator, could result in compromised animals prior to reproductive maturity. The Hg–Se interaction is discussed later in more detail. Metabolic (Hg biotransformation) functions of fetuses and neonates are in a dynamic state of development concurrent with maturing reproductive, immune and nervous systems (Maramba et al. 2006). Because fetal and neonatal seals act as a Hg "sink" for transplacental and transmammary methylmercury transfer (Jones et al. 1976, Reijnders 1980, Chang and Reuhl 1983, Wagemann et al. 1988) they have greater percentage methylmercury than adults, and are thus considered the cohort of concern for exposure to and effects of Hg, as is the case for most mammals (including humans). However, such effects are not merely dependent on Hg concentration but on the distribution and overall burden of Hg throughout the body as it relates to target organs or depots of limited clinical significance (e.g., sequestration of Hg in hair and epidermis). These depots form *in utero* and may be of questionable benefit in early development. Periods of exposure during gestation and lactation require further investigation.

In response to concerns about run off from old mining sites in the San Francisco Bay area (California), Brookens et al. (2008) recently reported Hg tissue burdens of fetal and neonatal harbor seals in relation to feeding ecology. They considered the toxicodistribution of Hg in this cohort with a concurrent assessment of tissue and body burdens. Tissues were weighed and analyzed for total Hg concentration (µg/g fresh weight) to calculate burdens (mg) in each tissue compartment.

Total Hg concentrations were related as follows: hair >> liver = kidney = pelt > muscle > other = heart > brain > blubber > bone. Total Hg burden (mass of Hg in the specified compartment), however, was related as: pelt = muscle > liver = other > kidney = blubber > brain = heart > bone. It is known methylmercury can be sequestered directly into keratinized structures and eliminated by hair/feather loss or epidermal sloughing (Scheuhammer 1991, Clarkson 1994, Gaggi et al. 1996, Wagemann and Kozlowska 2005). This excretory mechanism, whereby the toxic methylmercury is rendered biologically unavailable appears to be especially important for the fetus/neonate via molt of the lanugo.

In humans, methylmercury and inorganic Hg are retained by the liver and kidney, recirculated throughout the body, and eventually excreted slowly in the feces and urine as part of a complicated distribution and excretion process (Morton et al. 2004). Inorganic Hg acts as a free radical and induces various forms of molecular injury due to oxidative stress (Sarafian 1999). The percentage of methylmercury in the livers of harbor seal pups (22–68%; Brookens et al. 2007) versus that in livers of mature pinnipeds (5–15%; Gaskin et al. 1973, Reijnders 1980, Woshner et al. 2001a, Dehn et al. 2005, Brookens et al. 2007) indicates that the demethylation process likely is not well developed until adulthood (Wagemann et al. 1988, Caurant et al. 1996).

Although data are sparse, Hg in Antarctic Weddell seals (*Leptonychotes weddellii*) appears low and thus unlikely to be of concern (Noda et al. 1993). Yamamoto et al. (1987) reported mean concentrations of Hg in adult Weddell seals of 5.8 μg/g ww in liver, 0.67 in kidney, and 0.14 in muscle ($n = 2$). Other studies found 38.0 μg/g dw in liver, 12.8 in kidney, and 2.21 in muscle ($n = 2–3$; Szefer et al. 1993, Bargagli et al. 1998).

10.3 Hg AND NUTRIENT INTERACTIONS

10.3.1 Hg AND Se

Researchers have proposed a role for Se in lessening the toxicity of Hg (and some other elements) for marine mammals. This is mostly based on the correlative analysis of increasing Se and Hg concentrations with age in the liver of many marine mammals, juxtaposed with well-established laboratory animal research (*in vivo* and *in vitro*) showing decreased toxicity of Hg in the presence of higher Se levels (e.g., Christensen et al. 1989). Consequently we dedicate a section of this chapter to this issue.

Vertebrates have evolved for millions of years in the presence of Hg, particularly in marine environments where deep sea vents periodically spew volcanic material. It is intuitive to reason that mechanisms for prevention of toxicity exist for the "natural" forms of Hg. In seminal papers, Koeman and van de Ven (1975) and Koeman et al. (1973) first reported that marine mammals exhibited hepatic Hg concentrations that would be considered toxic in other species, and that the Hg appeared to exist in a 1:1 molar ratio with Se in liver (Koeman et al. 1973 and many others, Koeman and van de Ven 1975). Because Se antagonizes Hg toxicosis in some laboratory studies, the coincident increase of Se and Hg in tissues of odontocetes implicates a protective role for Se (Parizek and Ostadalova 1967, Martoja and Viale 1977, Cuvin-Aralar and Furness 1991, Caurant et al. 1996, Nigro and Leonzio 1996, Wang et al. 2001, Woshner et al. 2001b, see summary in O'Shea 1999). Se potentially binds to Hg in equimolar ratios in the liver of marine mammals (Koeman and van de Ven 1975, Smith and Armstrong 1978); however, some researchers have determined molar ratios deviating from 1, generally in organs other than liver, or in hepatic tissue of younger animals (Wagemann et al. 1988, Himeno et al. 1989, Woshner et al. 2001a, Dehn et al. 2005). The 1:1 molar relationship representing equimolar binding of Hg and Se may serve as a protective mechanism against Hg toxicity by biotransformation of ingested methylmercury into a less toxic chemical form. X-ray microanalytic evidence has provided support for the theory that the hepatic Hg to Se molar ratio of 1:1 first noted by Koeman et al. (1973) in marine mammals results from deposition of tiemannite HgSe crystals as an end product of methylmercury biotransformation

(Martoja and Berry 1980, Nigro and Leonzio 1996). Analysis of liver and kidney from northern fur seal (*Callorhinus ursinus*) and Risso's dolphin (*Grampus griseus*) by extended X-ray absorption fine structure spectroscopy confirmed that the Hg accumulated in marine mammals exists in the form of HgSe (tiemannite; Arai et al. 2004). Depending on their direction, deviations from the 1:1 molar ratio could indicate that Se reservoirs exceed Hg challenge (Agusa et al. 2008), or conversely, that the mechanism for Hg detoxification is overwhelmed or is not completely Se-dependent (Woshner et al. 2001a, 2008); the role of compounds containing sulfhydryl groups could also be important (MTH and MTH-independent), and other nutrients such as fatty acids may be involved.

In liver, the Hg:Se molar ratio is positively correlated with Hg concentrations below 100 ppm, at which point it approximates 1:1 (Krone et al. 1999). The relationship is not consistent among vertebrates and is different in fish (Koeman et al. 1973, Koeman and deVen 1975, Nigro and Leonzio 1996). The biochemistry of demethylation and the likely protective effect of Se are not completely understood. Pelletier (1985) cited the probable form of the reactions whereby glutathione (GSH) could reduce inorganic Se to hydrogen selenide, which subsequently participates in the demethylation of methylmercury to produce HgSe:

$$\text{GSH} \quad \text{GSH}$$

$$\text{SeO}_3^{2+} \rightarrow \text{GSSeSG} \rightarrow \text{H}_2\text{Se}$$

$$\text{H}_2\text{Se} + 2\text{CH}_3\text{Hg}^+ \rightarrow (\text{CH}_3\text{Hg})_2\text{Se} \rightarrow (\text{CH}_3)_2\text{Hg} + \text{HgSe(s)}$$

Alternatively, Pelletier (1985) suggested the following reaction in which the hypothetical intermediate complex formed through oxidation of GSH by Hg^{2+} would yield oxidized GSH and insoluble HgSe:

$$\text{GSH} \; \text{Hg}^{2+}$$

$$\text{SeO}_3^{2-} \rightarrow \text{GSSeSG} \rightarrow [\text{GSHgSeSG}] \rightarrow \text{GSSG} + \text{HgSe(s)}$$

Defense against Hg toxicosis in the kidney, CNS, and other sensitive organs may consist chiefly of this binding of Hg to Se, resulting in formation of an insoluble HgSe complex, with ancillary protection provided by sequestration of methylmercury and other elements in hair, muscle, epidermis, etc. in an apparently non-Se dependent manner (Wagemann et al. 1983, 1990, Wagemann and Kozlowska 2005, Andrade et al. 2007, Brookens et al. 2008). In the presence of low Se concentrations, Hg may be detoxified by binding to MTH, but how effective this may be in marine mammals is unknown.

Shibata et al. (1992) reviewed the chemical forms of Se and some associated biological functions; we emphasize that simply referring to "Se" is not as meaningful as measures of specific forms of Se (e.g., valence, organic moieties). Some analytical methods allow speciation of Se, but this is rarely conducted and is rather difficult (but should be considered). The Hg–Se complex in an odontocete liver has been described as a cationic, water-soluble, low molecular-weight compound containing Se and Hg in a 1:1 molar ratio that is different from a known Hg–Se complex, bis(methylmercuric) selenide. Other studies have addressed the insoluble, rock-like nature of Hg–Se, which implies that the elements are not in a "functional" form. This is an important issue with respect to the bioavailability, toxicodistribution, and toxicokinetics of these two elements. However, the plethora of correlative studies conducted to date provides little physiologic insight. While researchers generally maintain that Se most likely ameliorates Hg toxicity in marine mammals, one could also argue that toxicity of Se is being prevented by the Hg, or that formation of Hg–Se complexes promotes Se deficiency, as well as a number of other scenarios. The dearth of data from controlled experiments coupled with correlative analysis of observed chemical data allow any number of interpretations that may be largely determined by researcher bias.

Varying tissular and subcellular distributions will ultimately determine whether a given element will reach its target cells and organs at concentrations sufficient for expression of toxicity. How marine and terrestrial mammals may differ in this respect is not known. Ikemoto et al. (2004b) characterized the properties of Se, Hg, Ag, Cu, Zn, and Cd in liver of northern fur seals and Dall's porpoises through the use of differential extractions including: (1) sodium dodecylsulfate (SDS); (2) 2-mercaptoethanol; (3) 2-mercaptoethanol + guanidinium thiocyanate; and (4) copper sulfate ($CuSO_4$). Each differential extraction was employed over several concentrations of nuclear, lysosomal, and mitochondrial fractions. With increasing hepatic concentrations, Hg, Se, and Ag preferentially accumulated in nuclear, lysosomal, and mitochondrial fractions, whereas Cu, Zn, and Cd accumulated mainly in cytosol. As Hg concentration in the extraction increased, the molar ratio of Se to Hg approached unity in the nonextractable fraction of 2-mercaptoethanol + guanidinium thiocyanate, suggesting formation of HgSe with increasing hepatic Hg. It is notable that the state of Ag was similar to that of Hg as judged by their subcellular distribution and the extraction tests, suggesting that Ag also interacted with Se in the liver of marine animals in this study.

Many of the studies discussed earlier address the direct interaction of Hg and Se, for example, as HgSe. As an essential element and a component of the amino acid selenocysteine, the concentration of Se is physiologically regulated. Se is a constituent of numerous proteins, including the Se-dependent glutathione peroxidases, which have antioxidant properties (Brigelius-Flohé 1999). Other selenoproteins in blood include type I iodothyronine deiodinase, which converts thyroxine (T4) to the more active 5-triiodothyronine (T3); this conversion and its consequent impact on metabolic rate have been linked to Se availability (Hawkes and Keim 2003). In addition to a strong affinity for Se, Hg binds readily to sulfhydryls; thus, Hg concentration may relate to numerous blood parameters reflecting oxidative and metabolic status (Woshner et al. 2008). We suggest that more attention be given to the physiologic functions of Se in pinnipeds and cetaceans with respect to antioxidant action, Se-dependent proteins (including enzymes), as well as its role in protection from Hg toxicity through direct binding.

10.4 CADMIUM

10.4.1 GENERAL

Like Hg, Cd is toxic to a variety of tissues, depending on route of exposure, level and duration of intake, and tissue concentrations achieved. As for Hg, toxic effects of Cd at the cellular and subcellular level result from generation of reactive oxygen species (Thévenod et al. 2000). The kidney is considered the primary target organ of chronic Cd toxicosis, which causes proximal tubular necrosis progressing to interstitial fibrosis with prolonged exposure (Dorian et al. 1992). However, in rats with chronic exposure to low Cd levels, hepatic injury preceded nephrotoxicity (Dudley et al. 1985). With long-term exposure, bone undergoes increased osteoclastic activity producing osteoporosis because Cd deposition promotes a relative Ca deficiency (Ando et al. 1978, Wang et al. 1994, Brzóska and Moniuszko-Jakoniuk 2004). Cadmium is also a known carcinogen (Goyer et al. 2004).

MTH is an inducible, low molecular-weight protein with a strong affinity for Cd that is found in a variety of tissue types and is particularly abundant in the liver. Binding to MTH is generally considered a mode of Cd detoxification (Klaassen and Liu 1998). Cadmium bound by MTH in the liver is subsequently released into the blood as a Cd-MTH complex (Webb and Cain 1982). The Cd-MTH is filtered by the kidneys and resorbed at the renal proximal tubules where Cd-induced damage transpires, with experimental evidence suggesting that Cd-MTH is the entity that mediates nephrotoxicity (Cherian et al. 1976, Dorian et al. 1992). Moreover, the form in which Cd is encountered markedly affects its tissue disposition. Cadmium ingested as $CdCl_2$ distributes preferentially to the liver (>73%), whereas the overwhelming amount of Cd taken in as Cd-MTH accumulates in the kidney (~70%; Cherian 1983). It is important to note that most Cd present in edible tissues is believed to exist in the MTH-bound form (Cherian 1983).

Arai et al. (2004) used extended X-ray absorption fine structure spectroscopy to evaluate the detoxification mechanism of Cd in liver and kidney of northern fur seals and Risso's dolphins (*Grampus griseus*). Cd concentrations in kidney were higher than those in liver for both species, which is typical for most marine mammals. Arai et al. (2004) found that Cd was bound to sulfur, which was probably derived from MTH and also observed the Cd–oxygen bond in the tissues of northern fur seal. The latter phenomenon is probably due to the presence of Cd-phosphoryl or Cd-carboxyl complexes in northern fur seals, suggesting that Cd is bound to phosphate and/or carboxyl groups. It may be that the Cd-oxygen moieties serve as intermediates in a metabolic pathway that culminates in formation of Cd-MTH or Cd–Se complexes, similar to the scenario proposed for Hg. Although speculative, these are important considerations as we try to learn more about marine mammal adaptations to the presence of Cd in their prey.

Gallien et al. (2001) documented electron dense concretions in basement membranes of renal proximal tubules from two of three adult white-sided dolphins in which the range of renal Cd concentrations was 22.7–31.1 µg/g ww. Energy dispersive X-ray microanalysis revealed that the granules consisted of clusters of spheres comprising calcium, Cd, and phosphorous, with the former two elements present in a 10:1 molar ratio. Light microscopic changes observed in proximal tubule cells included cytoplasmic vacuolation of tubular cells, thickened basement membranes, and swollen capillary loops. While Gallien et al. (2001) suggested that Cd in proximal tubular cells of sampled dolphins could reasonably have achieved concentrations in excess of the 50 µg/g ww cited as a critical level; they also noted that histopathological changes were nonspecific, and may have been attributable to postmortem autolysis. Even if the microscopic lesions could have been definitively ascribed to Cd, it was unclear whether renal damage would have been sufficient to affect kidney function (Gallien et al. 2001).

10.4.2 CETACEANS

10.4.2.1 Odontocetes

In a mature male Dall's porpoise, Cd concentration was highest in kidney (30.2 µg/g ww), followed by liver (10.3 µg/g ww), which accounted for about 41.9% and 29.7%, respectively, of the whole body burden of 0.17 g Cd (Yang et al. 2006). Cd was very low in fetal Dall's porpoises (Fujise et al. 1988), suggesting limited placental Cd transfer. The Cd concentration in liver of one female (4.3 µg/g ww) was 80 times that of her fetus (0.052 µg/g ww; Yang et al. 2004). In both mother and fetus, greater than 80% of the Cd present was in cytosol, and 90% of the maternal cytosolic Cd was bound to MTH (because fetal Cd was negligible, cytosolic binding was not determined; Yang et al. 2004). The high distribution of hepatic Cd in the MTH-containing subcellular fraction was also observed in other specimens of Dall's porpoise (Ikemoto et al. 2004a). In nuclear and mitochondrial fractions, over 70% of the Cd was extractable with 2-mercaptoethanol + guanidinium thiocyanate, while more than half was extracted with the less potent 2% SDS solution (Ikemoto et al. 2004b). These results suggest that Cd was not present in a form with stability comparable to the HgSe complex. The maternal renal Cd concentration (58 µg/g ww) was >2000 times that of her fetus (0.028 µg/g ww; Yang et al. 2008). Subcellular Cd distribution was similar in kidney and liver, with about 80% of the Cd in the renal cytosol of both mother and fetus, and 91% of the maternal cytosolic Cd bound to MTH (Yang et al. 2008). Thus, MTH is involved in Cd detoxification in Dall's porpoises and transplacental transfer appears insignificant, in direct contrast to the scenario for methylmercury, which readily crosses the placenta and appears to be ultimately detoxified with little to no MTH involvement.

Cadmium and silver (Ag) accumulation in beluga whales is worthy of some mention. Cd increases with age in liver and kidney and could induce some changes in tissue (Wagemann et al. 1990, 1996, Woshner et al. 2001b, Lockhart et al. 2005, Dehn et al. 2006); for the sake of space we will discuss

Cd and possible renal impacts with respect to work done on bowhead whales. Ag has been shown to accumulate in beluga whales (Becker et al. 1995 reports mean hepatic Ag in Alaskan belugas of up to 24.3 and 46.6 ppm ww in males and females, respectively), but little is known as to whether this is a unique essential element or coincident accumulation with other elements such as Hg and/or Se. Ag bioaccumulation may also reflect a unique food source exploited by beluga whales (Dehn et al. 2006, Rouleau et al. 2000).

Mean concentrations of Cd in short-finned pilot whales from offshore waters of Japan were 71.0, 296, and 0.800 µg/g dw in liver, kidney, and muscle, respectively (Tanabe et al. unpublished results). Specimens from New Caledonia exhibited 225 and 464 µg/g dw in liver, 0.79 and 1.48 in muscle, and 0.95 and 0.84 in blubber (Bustamante et al. 2003). For short-finned pilot whales stranded at Cumberland Island, U.S., had mean Cd concentrations of 13.9, 31.4, and 0.54 µg/g ww in liver, kidney, and blubber, respectively (Stoneburner 1978).

10.4.2.2 Mysticetes

Concentrations of Cd ranged from 0.01 to 0.20 µg/g ww in muscle, 2.32–41.4 in liver, and 3.50–85.0 in kidney of southern minke whales ($n = 37$; Honda et al. 1987b). Cd levels in southern minke whales are an order of magnitude higher than those measured in common minke whales, which are their counterpart in the northern hemisphere (Bargagli 2001). The Cd concentrations of southern minke whales are considered to be unusually high (O'Shea 1999), reflecting Cd levels in antarctic krill, upon which they feed almost exclusively (Ichii and Kato 1991), whereas Hg concentrations are lower than those of marine mammals from other regions as described earlier. The generally high Cd levels observed for the antarctic marine organisms have been attributed to the upwelling of Cd-rich deep water (Honda et al. 1987b, Sanchez-Herandez 2000).

The high Cd level might pose a potential threat to the southern minke whales, but a detailed evaluation has not been performed. About 70% of hepatic Cd appears to be distributed in the cytosol in southern minke whales, and almost all of this cytosolic Cd is bound to MTH (Tanabe et al. unpublished results). Because southern minke whales showed marked accumulation of both Cd and Se in tissues, with concentrations of these elements positively correlated, Kunito et al. (2002) assumed the presence of insoluble Cd selenide (CdSe). However, a Cd–S rather than a Cd–Se interaction was implicated by X-ray absorption fine structure spectroscopy of southern minke whale tissues (Tanabe et al. unpublished results).

The bowhead whale is an endangered species of great cultural and subsistence value (O'Hara et al. 2006) to arctic residents of eastern Russia, Canada, and northern Alaska. Most research on bowhead whales focuses on the Bering-Chukchi-Beaufort Sea stock. Woshner et al. (2001b) assessed numerous elements (As, Cd, Co, Cu, Pb, Mg, Mn, Hg, Mo, Se, Ag, and Zn) in tissues of bowheads from samples collected between 1983 and 1997 (building on efforts of Bratton et al. 1997). In some tissues, Cd concentrations associated with toxicoses in domestic species were found. Nevertheless, tissue levels of all elements were within ranges reported previously in marine mammals. Significant associations included: Cd with age, Zn, or Cu. Using data from Woshner et al. (2001b) with samples from additional years, Dehn et al. (2006) examined metal concentrations in both bowhead and gray whales in relation to chemical feeding ecology, via analysis of C and N stable isotopes in muscle. Gray whales had lower Cd concentrations in liver and kidney than bowhead whales and a sigmoid correlation of Cd with length (surrogate for age) was noted. This indicates that renal Cd concentrations plateau in older animals and may actually decrease in older or "geriatric" whales. A plausible physiologic/anatomic explanation for this observation is that renal tissue of older animals tends to be more fibrotic and so has less Cd binding and storage capacity; also, the tubular epithelium, where Cd aggregates, tends to shed in geriatric kidneys. However, both renal fibrosis and tubular sloughing could ensue from chronic Cd toxicosis; thus, we are back to the dilemma posed by a number of metals that exert toxicity via oxidative damage—namely, that histologic changes observed with chronic metal toxicosis may be indistinguishable from those incurred by aging oxidative metabolism.

To more critically evaluate the potential effects of Cd on renal tissue, Goodwin et al. (2000) developed a bowhead whale cell culture model (two- and three-dimensional renal proximal tubular cell cultures) and characterized renal MTH (GenBank accession number AF022117). Woshner (2000) and Woshner et al. (2002) histologically assessed hepatic and renal tissues of bowheads and highlighted that fibrosis was relatively extensive in renal tissue of older whales with the highest concentrations of Cd. These findings resulted in an effort to increase sample size and more critically evaluate the role of aging and/or Cd concentrations in renal fibrosis (Rosa et al. 2008). Rosa et al. (2008) evaluated essential and nonessential elements in liver and kidney of bowhead whales by expanding efforts of Woshner et al. (2001b) to include samples from additional years (up to 2001). These data were interpreted using improved aging techniques (aspartic acid racemization of the eye lens and baleen stable isotopes of carbon as in George et al. 1999). Histological evaluation revealed the degree of lung fibromuscular hyperplasia, and renal fibrosis was associated with hepatic and renal Cd concentrations. A significant age effect was found for both pulmonary fibromuscular hyperplasia and renal fibrosis as well, making age a confounding factor (Rosa et al. 2008). Improvements in aging techniques, along with more specific histological indices, may help to clarify how concentrations of elements are related to each other, to life history parameters, and to health. To place the above in context, most of these sampled bowhead whales were deemed to be in very good condition by scientists and hunters. From current information we can conclude that the observed fibrosis is mostly related to age and size of the whales. However, we require a more specific evaluation of the role Cd may play (e.g., biochemical, physiologic, and molecular tools).

Summary

For decades, researchers have documented the comparatively elevated concentrations of potentially toxic metals (particularly Hg and Cd) in marine mammals and debated about the significance of these levels to individual and population health. There is a substantial body of observational data that shows Se correlates strongly with Hg in marine mammals and serves an important role in Hg detoxification in several tissues. Nevertheless, it is evident from a few studies—dosing whole animals or *in vitro*—that there is a clear limit to the marine mammalian capacity for metal tolerance. Moreover, at least one extreme case of environmental pollution involving Hg (among other contaminants) has clearly adversely affected marine mammal health—that is, the St. Lawrence Estuary beluga population.

Given the extensive history of research on metal toxicants in marine mammals, one would think there would be more to show for it. Years of study have culminated in repetitive monitoring reports and correlative analyses with little substantive assessment of whether the toxicant levels measured are impinging upon marine mammal health. As previously acknowledged, monitoring studies can help pinpoint locales where contaminants may be of particular concern, as well as track temporal or geographic trends, but they tell us little about the toxicologic import of metals in marine mammals. As it stands, evidence that contaminants are adversely affecting marine mammals is meager, because such data cannot be gathered in a context conducive to defensible conclusions. So, in a bizarre catch-22, "absence of data" may be construed as "data of absence (of effects)" and in this roundabout way the very restrictions meant to defend marine mammals may hasten their decline.

Fortunately, in the meantime, some researchers have begun to apply tools from assorted scientific disciplines, including molecular biology, genetics, epidemiology, computer modeling, and other fields to the question of contaminants in marine mammals. Marine mammal toxicology must incorporate a plethora of techniques and embrace collaborations among scientists whose expertise resides in complementary spheres if we are to realize significant scientific advances with meaningful ecological and management applications.

REFERENCES

Agusa, T., et al. 2008. Interelement relationships and age-related variation of trace element concentrations in liver of striped dolphins (*Stenella coeruleoalba*) from Japanese coastal waters. *Mar. Pollut. Bull.* 57:807–815.

Arctic Monitoring and Assessment Programme (AMAP). 1997. AMAP Assessment Report: *Arctic pollution issues: A state of the Arctic environmental report.* Arctic Monitoring and Assessment Programme, Oslo, Norway. xii+859 pages.

Arctic Monitoring and Assessment Programme (AMAP). 1998. AMAP Assessment Report (JC Hansen, A Gilman, V Klopov): Arctic Pollution Issues. Arctic Monitoring and Assessment Programme, Oslo, Norway. Xii+859pp.

Arctic Monitoring and Assessment Programme (AMAP) 2002. Arctic Pollution 2002. http://www.amap.no/ 112 pages.

Arctic Monitoring and Assessment Programme (AMAP). 2005. AMAP Assessment 2002: Heavy Metals in the Arctic. Arctic Monitoring and Assessment Programme (AMAP), Oslo, Norway. xvi+265 pp.

Ando, M., Y. Sayato, and T. Osawa. 1978. Studies on the disposition of calcium in bones of rats after continuous oral administration of cadmium. *Toxicol. Appl. Pharmacol.* 46:625–632.

Andrade, S., A. R. Carlini, C. Vodopivez, and S. Poljak. 2007. Heavy metals in molted fur of the southern elephant seal Mirounga leonine. *Mar. Pollut. Bull.* 54:602–625.

Arai, T., et al. 2004. Chemical forms of mercury and cadmium accumulated in marine mammals and seabirds as determined by XAFS analysis. *Environ. Sci. Technol.* 38(24):6468–6474.

Atwell, L., K. A. Hobson, and H. E. Welch. 1998. Biomagnification and bioaccumulation of mercury in an arctic marine food web: insights from stable nitrogen isotope analysis. *Can. J. Fish. Aquat. Sci.* 55:1114–1121.

Bargagli, R. 2001. Trace metals in Antarctic organisms and the development of circumpolar biomonitoring networks. *Rev. Environ. Contam. Toxicol.* 171:53–110.

Bargagli, R., F. Monaci, J. C. Sanchez-Hernandez, and D. Cateni. 1998. Biomagnification of mercury in an Antarctic marine coastal food web. *Mar. Ecol. Prog. Ser.* 169:65–76.

Basu, N., M. Kwan, and H. M. Chan. 2006. Mercury but not organochlorines inhibits muscarinic cholinergic receptor binding in the cerebrum of ringed seals (*Phoca hispida*). *J. Toxicol. Environ. Health, Part A* 69:1133–1143.

Beck, K. M., P. A. Fair, W. McFee, and D. Wolf. 1997. Heavy metals concentrations in livers of bottlenose dolphins stranded along the South Carolina coast. *Mar. Pollut. Bull.* 34:734–739.

Becker, P. R., et al. 1995. Relationship of silver with selenium and mercury in liver of two species of toothed whales (Odontocetes). *Mar. Pollut. Bull.* 30(4):262–271.

Becker, P. R., et al. 2000. Concentrations of polychlorinated biphenyls (PCBs), chlorinated pesticides, and heavy metals and other elements in tissues of beluga whales (Delphinapterus leucas) from Cook Inlet, Alaska. *Mar. Fish. Rev.* 62(3):81–98.

Betti, C., and M. Nigro. 1996. The Comet assay for the evaluation of the genetic hazard of pollutants in cetaceans: preliminary results on the genotoxic effects of methylmercury on the bottle-nosed dolphin lymphocytes *in vitro*. *Mar. Pollut. Bull.* 32: 545–548.

Booth, S., and D. Zeller. 2005. Mercury, food webs, and marine mammals: implications of diet and climate change for human health. *Environ. Health Perspect.* 113:521–526.

Bossart, G. 2006. Marine mammals as sentinel species for oceans and human health. *Oceanography* 19(2):134–138.

Bouquegneau J. M., V. Debacker, S. Gobert, and J. P. Nellissen. 1997. Toxicological investigations on four sperm whales stranded on the Belgian coast: inorganic contaminants. *Bull. Soc. R. Sci.Liège* 67(suppl.):75–78.

Bratton, G. R., W. Flory, C. B. Spainhour, and E. M. Haubold. 1997. Assessment of selected heavy metals in liver, kidney, muscle, blubber, and visceral fat of Eskimo harvested bowhead whales *Balaena mysticetus* from Alaska's north coast. Final report submitted to the North Slope Borough Dept. of Wildlife Management, Barrow, Alaska, 233 pp.

Brigelius-Flohé, R. 1999. Tissue-specific functions of individual glutathione peroxidases. *Free Radic. Biol. Med.* 27:951–965.

Brookens, T. J., J. T. Harvey, and T. M. O'Hara. 2007. Trace element concentrations in the Pacific harbor seal (*Phoca vitulina richardii*) in central and northern California. *Sci. Total Environ.* 372:676–692.

Brookens, T. J., T. M. O'Hara, R. J. Taylor, G. R. Bratton, and J. T. Harvey. 2008. Total mercury body burden in Pacific harbor seal, *Phoca vitulina richardii*, pups from central California. *Mar. Pollut. Bull.* 56(1): 27–41.

Brzóska, M. M., and J. Moniuszko-Jakoniuk. 2004. Low-level exposure to cadmium during the lifetime increases the risk of osteoporosis and fractures of the lumbar spine in the elderly: studies on a rat model of human environmental exposure. *Toxicol. Sci.* 82:468–477.

Budge, S. M., A. M. Springer, S. J. Iverson, G. Sheffield, and C. Rosa. 2008a. Blubber fatty acid composition of bowhead whales, *Balaena mysticetus*: Implications for diet assessment and ecosystem monitoring. *J. Exp. Mar. Biol. Ecol.* 359:40–46.

Budge, S. M., M. J. Wooller, A. M. Springer, S. J. Iverson, C. P. McRoy, and G. J. Divoky. 2008b. Tracing carbon flow in an Arctic marine food web using fatty acid-stable isotope analysis. *Oecologia*. doi: 10.1007/s00442-008-1053-7.

Bustamante, P., et al. 2003. Trace elements in two odontocete species (*Kogia breviceps* and *Globicephala macrorhynchus*) stranded in New Caledonia (South Pacific). *Environ. Pollut.* 124:263–271.

Carvalho, M., R. Pereira, and J. Brito. 2002. Heavy metals in soft tissues of *Tursiops truncatus* and *Delphinus delphis* from west Atlantic Ocean by X-ray spectrometry. *Sci. Total Environ.* 292:247–254.

Caurant, F., M. Navarro, and J. C. Amiard. 1996. Mercury in pilot whales: possible limits to the detoxification process. *Sci. Total Environ.* 186:95–104.

Chang, L. W., and K. R. Reuhl. 1983. Mercury in human and animal health. In; *Trace elements in health: A review of current issues*, ed. J. Rose, pp. 123–154. London: Butterworths.

Cherian, M. G. 1983. Absorption and tissue distribution of cadmium in mice after chronic feeding with cadmium chloride and cadmium-metallothionein. *Bull. Environ. Contam. Toxicol.* 30:33–36.

Cherian, M. G., R. A. Goyer, and L. Delaquerriere-Richardson. 1976. Cadmium-metallothionein-induced nephropathy. *Toxicol. Appl. Pharmacol.* 38:399–408.

Christensen, M., J. Rungby, and S. Mogensen. 1989. Effects of selenium on toxicity and ultrastructural localization of mercury in cultured murine macrophages. *Toxicol. Lett.* 47:259–270.

Clarkson, T. W. 1994. The toxicology of mercury and its compounds. In *Mercury pollution: Integration and synthesis*, eds. C. J. Watras and J. W. Huckabee, pp. 631–642. Boca Raton, FL: CRC Press.

Cuvin-Aralar, M. L. A., and R. W. Furness. 1991. Mercury and selenium interaction: a review. *Ecotoxicol. Environ. Saf.* 21:348–364.

Danscher, G., M. Stoltenberg, and S. Juhl. 1994. How to detect gold, silver and mercury in human brain and other tissues by autometallographic silver amplification. *Neuropathol. Appl. Neurobiol.* 20:454–457.

Das, K., A. De Groof, T. Jauniaux, and J.-M. Bouquegneau. 2006. Zn, Cu, Cd and Hg binding to metallothioneins in harbour porpoises *Phocoena phocoena* from the southern North Sea. *BMC Ecology*, 6:2 doi:10.1186/1472-6785-6-2.

Das, K., V. Debacker, and J. M. Bouquegneau. 2000b. Metallothioneins in marine mammals. *Cell. Mol. Biol.* 46:283–294.

Das, K., V. Jacob, and J. M. Bouquegneau. 2001 White-sided dolphin metallothionein: purification, characterisation and potential role. *Comp. Biochem. Physiol. Part C* 131(3):245–251.

Das, K., G. Lepoint, V. Loizeau, V. Debacker, P. Dauby, and J. M. Bouquegneau. 2000c. Tuna and dolphin associations in the North-East Atlantic: evidence of different ecological niches from stable isotope and heavy metal measurements. *Mar. Pollut. Bull.* 40:102–109.

Das, K., U. Siebert, M. Fontaine, T. Jauniaux, L. Holsbeek, and J. M. Bouquegneau. 2004. Ecological and pathological factors related to trace metal concentrations in harbour porpoises (*Phocoena phocoena*) from the North Sea and adjacent areas. *Mar. Ecol. Prog. Ser.* 281:283–295.

Das, K., et al. 2000a. Marine mammals stranded on the Belgian and Dutch coasts: Approach of their feeding ecology by stable isotope and heavy metal measurements, pp. 219–222 in: *European Research on Cetaceans*. Proceedings of the Fourteenth Annual Conference of the European Cetacean Society, Cork, Ireland, April 2–5, 2000 (Eds. P.G.H. Evans, R. Pitt- Aiken, E. Rogan) 384 pp.

Das, K., et al. 2003. Marine mammals from northeast Atlantic: relationship between their trophic status as determined by d13C and d15N measurements and their trace metal concentrations. *Mar. Environ. Res.* 56:349–365.

Decataldo, A., A. Di Leo, S. Giandomenico, and N. Cardellicchio. 2004. Association of metals (mercury, cadmium and zinc) with metallothionein-like proteins in storage organs of stranded dolphins from the Mediterranean Sea (Southern Italy). *J. Environ. Monit.* 6:361–367.

De Guise, S., A. Lagacé, and P. Béland. 1994. Gastric papillomas in eight St. Lawrence beluga whales (*Delphinapterus leucas*). *J. Vet. Diagn. Invest.* 6:385–388.

De Guise, S., J. Bernier, D. Martineau, P. Béland, and M. Fournier. 1996. Effects of *in vitro* exposure of Beluga whale splenocytes and thymocytes to heavy metals. *Environ. Toxicol. Chem.* 15:1357–1364.

Decataldo, A., A. Di Leo, S. Giandomenico, and N. Cardellicchio. 2004. Association of metals (mercury, cadmium and zinc) with metallothionein-like proteins in storage organs of stranded dolphins from the Mediterranean Sea (Southern Italy). *J. Environ. Monit.* 6:361–367.

Dehn, L.-A., et al. 2005. Trace elements in tissues of phocid seals harvested in the Alaskan and Canadian Arctic—influence of age and feeding ecology. *Can. J. Zool.* 83(5):726–746.

Dehn, L.-A., et al. 2006. Stable isotope and trace element status of subsistence hunted bowhead (*Balaena mysticetus*) and beluga whales (Delphinapterus leucas) in Alaska and gray whales (*Eschrichtius robustus*) in Chukotka. *Mar. Pollut. Bull.* 52:301–319.

DeNiro, M. J., and S. Epstein. 1981. Influence of diet on the distribution of nitrogen isotopes in animals. *Geochimica et Cosmoshimica Acta* 45:341–351.

Dietz, R., E. W. Born, C. T. Agger, and C. O. Nielsen. 1995. Zinc, cadmium, mercury, and selenium in polar bears (*Ursus maritimus*) from East Greenland. *Polar Biol.* 15:175–185.

Dietz, R., C. O. Nielsen, M. M. Hansen, and C. T. Hansen. 1990. Organic mercury in Greenland birds and mammals. *Sci. Total Environ.* 95:41–51.

Dietz, R., J. Nørgaard, and J. C. Hansen. 1998a. Have Arctic marine mammals adapted to high cadmium concentrations? *Mar. Pollut. Bull.* 36(6):490–492.

Dietz, R., P. Paludan-Müller, C. T. Agger, and C. O. Nielsen. 1998b. Cadmium, mercury, zinc and selenium in ringed seals (*Phoca hispida*) from Greenland and Svalbard. *NAMMCO Scientific Contributions* 1:242–273.

Dietz, R., F. Riget, and P. Johansen. 1996. Lead, cadmium, mercury and selenium in Greenland marine animals. *Sci. Total Environ.* 186:67–93.

Dietz, R., et al. 2006. Trends in mercury in hair of Greenlandic polar bears (*Ursus maritimus*) during 1892–2001. *Environ. Sci. Technol.* 40:1120–1125.

Dorian, C., V. H. Gattone, and C. D. Klaassen. 1992. Renal cadmium deposition and injury as a result of accumulation of cadmium-metallothionein (CdMT) by the proximal convoluted tubules—a light microscopic autoradiography study with [109]CdMT. *Toxicol. Appl. Pharmacol.* 114:173–181.

Dudley, R. E., L. M. Gammal, and C. D. Klaassen. 1985. Cadmium-induced hepatic and renal injury in chronically exposed rats: likely role of hepatic cadmium metallothionein in nephrotoxicity. *Toxicol. Appl. Pharmacol.* 77:414–426.

Duffus, J. H. 2002. "'Heavy metals' a meaningless term? (IUPAC Technical Report)" *Pure Appl. Chem.* 74:793–807.

Edwards, W. C., Whitenack, D. L., Alexander, J. W., and Solangi, M. A. 1989. Selenium toxicosis in three California sea lions (*Zalophus californianus*). *Vet. Hum. Toxicol.* Dec, 31(6):568–570.

Endo, T., et al. 2005. Total mercury, methyl mercury, and selenium levels in the red meat of small cetaceans sold for human consumption in Japan. *Environ. Sci. Technol.* 39:5703–5708.

Endo, T., et al. 2006. Distribution of total mercury, methyl mercury and selenium in pod of killer whales (*Orcinus orca*) stranded in the northern area of Japan: comparison of mature females with calves. *Environ. Pollut.* 144:145–150.

Fair, P. A., and P. R. Becker. 2000. Review of stress in marine mammals. *J. Aquatic Ecosystem Stress Recovery* 7:335–354.

Freeman, H. C., G. Sanglang, J. F. Uthe, and K. Ronald. 1975. Steroidogensis *in vitro* in the harp seal without and with methyl mercury treatment *in vivo*. *Environ. Physiol. Biochem.* 5:428–439.

Frodello, J. P., and B. Marchand. 2001. Cadmium, copper, lead and zinc in five toothed whales species of the Mediterranean Sea. *Int. J. Toxicol.* 20:339–343.

Fujise, Y., K. Honda, R. Tatsukawa, and S. Mishima. 1988. Tissue distribution of heavy metals in Dall's porpoise in the Northwestern Pacific. *Mar. Pollut. Bull.* 19:226–230.

Gaggi, C., F. Zino, M. Duccini, and A. Renzoni. 1996. Levels of mercury in scalp hair of fisherman and their families from Camara de Lobos-Madeira (Portugal): a preliminary study. *Bull. Environ. Contam. Toxicol.* 56:860–865.

Gallien, I., et al. 2001. Cadmium-containing granules in kidney tissue of the Atlantic white-sided dolphin (*Lagenorhyncus acutus*) off the Faroe Islands. *Comp. Biochem. Physiol. Part C* 130:389–395.

Gaskin, D. E., R. Frank., M. Holdrinet, K. Ishida, C. J. Walton, and M. Smith. 1973. Mercury, DDT, and PCB in harbour seals (*Phoca vitulina*) from the Bay of Fundy and Gulf of Maine. *J. Fish. Res. Board Can.* 30:471–475.

Gauthier, J. M., H. Dubeau, and E. Rassart. 1998. Mercury-induced micronuclei in skin fibroblasts of beluga whales. *Environ. Toxicol. Chem.* 17:2487–2493.

George, J. C., et al. 1999. Age and growth estimates of bowhead whales (*Balaena mysticetus*) using aspartic acid racemization. *Can. J. Zool.* 77:571–580.

Godard, C. A. J., et al. 2006. Benzo[a]pyrene cytotoxicity in right whale (*Eubalaena glacialis*) skin, testis and lung cell lines. *Environ. Res.* 62:S20–S24.

Goodwin, T. J., L. Coate-Li, R. M. Linnehan, and T. G. Hammond. 2000. Selected contribution: a three-dimensional model for assessment of *in vitro* toxicity in *Balaena mysticetus* renal tissue. *J. Appl. Physiol.* 89 [Cellular Responses to Mechanical Stress]:2508–2517.

Goyer, R. A., J. Liu, and M. P.Waalkes. 2004. Cadmium and cancer of prostate and testis. *Biometals* 17(5):555–558.

Gray, D. A., and J. Woulfe. 2005. Lipofuscin and aging: a matter of toxic waste. *Sci. Aging Knowl. Environ.* 2005 (5), re1. http://sageke.sciencemag.org/cgi/content/full/2005/5/re1

Griesel, S., A. Kakuschke, U. Siebert, and A. Prange. 2008. Trace element concentrations in blood of harbor seals (*Phoca vitulina*) from the Wadden Sea. *Sci. Total Environ.* 392:313–323.

Hansen, C. T., C. O. Nielsen, R. Dietz, and M. M. Hansen. 1990. Zinc, cadmium, mercury and selenium in minke whales, belugas, and narwhals from West Greenland. *Polar Biol.* 10:529–539.

Hawkes, W. C., and N. C. Keim. 2003. Dietary selenium intake modulates thyroid hormone and energy metabolism in men. *J. Nutr.* 133:3443–3448.

Helbing, C. C., K. Werry, D. Crump, D. Domanski, N. Veldhoen, and C. M. Bailey. 2003. Expression profiles of novel thyroid hormone-responsive gene and proteins in the tail of *Xenopus laevis* tadpoles undergoing precocious metamorphosis. *Mol. Endocrinol.* First published April 10, 2003 as doi:10.1210/me.2002–0274.

Hill, C. H. 1974. Reversal of selenium toxicity in chicks by mercury, copper, and cadmium. *J. Nutr.* 104:593–598.

Himeno, S., C. Watanabe, T. Hongo, T. Suzuki, A. Naganuma, and N. Imura. 1989. Body size and organ accumulation of mercury and selenium in young harbor seals (*Phoca vitulina*). *Bull. Environ. Contam. Toxicol.* 42(4):503–509.

Hobson, K. A., R. T. Alisauskas, and R. G. Clark. 1993. Stable-nitrogen isotope enrichment in avian tissues due to fasting and nutritional stress: Implications for isotopic analyses of diet. *Condor* 95:388–394.

Hobson, K. A., and H. E. Welch. 1992. Determination of trophic relationships within a high Arctic marine food web using d13C and d15N analysis. *Mar. Ecol. Prog. Ser.* 84:9–18.

Hoekstra, P. F., L.-A. Dehn, J. C. George, D. C. G. Muir, and T. M. O'Hara. 2002. Trophic ecology of bowhead whales (*Balaena mysticetus*) compared to other arctic marine biota as interpreted from C, N, and S isotope signatures. *Can. J. Zool.* 80(2):223–231.

Holden, A. V. 1978. Pollutants and seals: a review. *Mammal. Rev.* 8:53–66.

Holsbeek, L., U. Siebert, and C. Joiris. 1998. Heavy metals in dolphins stranded of the French Atlantic coast. *Sci. Total Environ.* 217:241–249.

Holsbeek, L., et al. 1999. Heavy metals, organochlorines and polycyclic aromatic hydrocarbons in sperm whales stranded in the southern North Sea during the 1994/1995 winter. *Mar. Pollut. Bull.* 38:304–313.

Honda, K., M. Aoki, and Y. Fujise. 2006. Ecochemical approach using mercury accumulation of Antarctic minke whale, *Balaenoptera bonaerensis*, as tracer of historical change of Antarctic marine ecosystem during 1980–1999. *Bull. Environ. Contam. Toxicol.* 76:140–147.

Honda, K., R. Tatsukawa, K. Itano, N. Miyazaki, and T. Fujiyama. 1983. Heavy metal concentrations in muscle, liver and kidney tissue of striped dolphin, *Stenella coeruleoalba*, and their variations with body length, weight, age and sex. *Agric. Biol. Chem.* 47:1219–1228.

Honda, K., Y. Yamamoto, H. Kato, and R. Tatsukawa. 1987a. Heavy metal accumulations and their recent changes in southern minke whales *Balaenoptera acutorostrata*. *Arch. Environ. Contam. Toxicol.* 16:209–216.

Honda, K., Y. Yamamoto, and R. Tatsukawa. 1987b. Distribution of heavy metals in Antarctic marine ecosystem. *Proc. NIPR Symp. Polar Biol.* 1:184–197.

Ichii, T., and H. Kato. 1991. Food and daily food consumption of southern minke whales in the Antarctic. *Polar Biol.* 11:479–487.

Ikemoto, T. 2004. Detoxification mechanism of mercury in high trophic level marine animals. PhD diss., Ehime University, Matsuyama, Japan (in Japanese).

Ikemoto, T., T. Kunito, H. Tanaka, N. Baba, N. Miyazaki, and S. Tanabe. 2004b. Detoxification mechanism of heavy metals in marine mammals and seabirds: interaction of selenium with mercury, silver, copper, zinc, and cadmium in liver. *Arch. Environ. Contam. Toxicol.* 47:402–413.

Ikemoto, T., et al. 2004a. Association of heavy metals with metallothionein and other proteins in hepatic cytosol of marine mammals and seabirds. *Environ. Toxicol. Chem.* 23:2008–2016.

Itano, K., S. Kawai, N. Miyazaki, R. Tatsukawa, and T. Fujiyama. 1984. Mercury and selenium levels in striped dolphins caught off the Pacific coast of Japan. *Agric. Biol. Chem.* 48:1109–1116.

Itano, K., S. Kawai, and R. Tatsukawa. 1985a. Properties of mercury and selenium in salt-insoluble fraction of muscles in striped dolphin. *Bull. Jpn. Soc. Sci. Fish.* 51:1129–1131.

Itano, K., S. Kawai, and R. Tatsukawa. 1985b. Distribution of mercury and selenium in muscle of striped dolphins. *Agric. Biol. Chem.* 49:515–517.

Joiris, C. R. 1997. Ecotoxicology of stable pollutants: organochlorines and heavy metals in seabirds and marine mammals. *Bull. Soc. R. Sci. Liege* 66(1–3):51–59.

Joiris, C. R., L. Holsbeek, J. M. Bouquegneau, and M. Bossicart. 1991. Mercury contamination of the harbour porpoise *Phocoena phocoena* and other cetaceans from the North Sea and the Kattegat. *Water, Air Soil Pollut.* 56:283–293.

Jones, D., K. Ronald, D. M. Lavigne, R. Frank, M. Holdrinet, and J. F. Uthe. 1976. Organochlorine and mercury residues in the harp seal (*Pagophilus groenlandicus*). *Sci. Total Environ.* 5:181–195.

Kakuschke, A., E. Valentine-Thon, S. Fonfara, S. Griesel, U.Siebert, and A. Prange. 2006. Metal sensitivity of marine mammals: a case study of a gray seal (*Halichoerus grypus*). *Mar. Mamm. Sci.* 22(4):985–996.

Kakuschke, A., E. Valentine-Thon, S. Griesel, S. Fonfara, U. Siebert, and A. Prange. 2005. Immunological impact of metals in harbor seals (*Phoca vitulina*) of the North Sea. *Environ. Sci. Technol.* 39:7568–7575.

Kakuschke, A., et al. 2008. Blood metal levels and metal-influenced immune functions of harbour seals in captivity. *Mar. Pollut. Bull.* 56:764–769.

Klaassen, C. D. 2007. *Casarett and Doull's toxicology: The basic science of poisons*, 7th ed. New York, NY: McGraw-Hill, 1280pp.

Klaassen, C. D., and J. Liu. 1998. Induction of metallothionein as an adaptive mechanism affecting the magnitude and progression of toxicological injury. *Environ. Health Perspect.* 106(Suppl. 1):297–300.

Koeman, J., W. Peeters, C. Koudstaal-Hol, P. Tijoe, and J. De Goeij. 1973. Mercury-selenium correlations in marine mammals. *Nature* 245:385–386.

Koeman, J., and W. van deVen. 1975. Mercury and selenium in marine mammals and birds. *Sci. Total Environ.* 3:279–287.

Koller, L. D. 1979. Effects of environmental contaminants on the immune system. *Adv. Vet. Sci. Comp. Med.* 23:267–295.

Krone, C. A., et al. 1999. Heavy metals and other elements in liver tissues of bowhead whales, *Balaena mysticetus. Mar. Mamm. Sci.* 15(1):123–142.

Kuehl D. W., and R. Haebler. 1995. Organochlorine, organobromine, metal, and selenium residues in bottlenose dolphins (*Tursiops truncatus*) collected during an unusual mortality event in the Gulf of Mexico, 1990. *Arch. Environ. Contam. Toxicol.* 28:494–499.

Kuehl D. W., R. Haebler, and C. Potter. 1994. Coplanar PCB and metal residues in dolphins from the U. S. Atlantic coast including Atlantic bottlenose obtained during the 1987/1988 mass mortality. *Chemosphere* 128:1245–1253.

Kunito, T., I. Watanabe, G. Yasunaga, Y. Fujise, and S. Tanabe. 2002. Using trace elements in skin to discriminate the populations of minke whales in southern hemisphere. *Mar. Environ. Res.* 53:175–197.

Kwohn, Y.-T., A. Okubo, H. Hirano, H. Kagawa, S. Yamazaki, and S. Toda. 1988. Primary structure of striped dolphin renal metallothionein II. *Agric. Biol. Chem.* 52:837–841.

Kwohn, Y.-T., S. Yamazaki, A. Okubo, E. Yoshimura, R. Tatsukawa, and S. Toda. 1986. Isolation and characterization of metallothionein from kidney of striped dolphin, *Stenella coeruleoalba. Agric. Biol. Chem.* 50:2881–2885.

Law, R. J. 1996. Metals in marine mammals. In *Environmental contaminants in wildlife: Interpreting tissue concentrations,* eds. W. N. Beyer, G. H. Heinz, and A. W. Redmon-Norwood, pp. 357–376. Boca Raton, FL: CRC Press.

Law, R. J., et al. 1991. Concentrations of trace metals in the livers of marine mammals (seals, porpoises and dolphins) from waters around the British Isles. *Mar. Pollut. Bull.* 22(4):183–191.

Leitch, D. R., J. Carrie, D. Lean, R. W. Macdonald, G. A. Stern, and F. Wang. 2007. The delivery of mercury to the Beaufort Sea of the Arctic Ocean by the Mackenzie River. *Sci. Total Environ.* 373(1):178–195.

Leonzio, C., S. Focardi, and C. Fossi. 1992. Heavy metals and selenium in stranded dolphins of the northern Tyrrhenian (NW Mediterranean). *Sci. Total Environ.* 119:77–84.

Lockhart, W. L., et al. 2005. Concentrations of mercury in tissues of beluga whales (*Delphinapterus leucas*) from several communities in the Canadian Arctic from 1981 to 2002. *Sci. Total Environ.* 351/352:391–412.

Loseto, L. L., et al. 2007. Linking mercury exposure to habitat and feeding behaviour in Beaufort Sea beluga whales. *J. Mar. Syst.* 74(3–4):1012–1024.

Macdonald, R. W. 2005. Climate change, risks and contaminants: a perspective from studying the Arctic. *HERA* 11:1099–1104.

Maramba, N. P. C., et al. 2006. Environmental and human exposure assessment monitoring of communities near an abandoned mercury mine in the Phillipines: a toxic legacy. *J. Environ. Manage.* 81:135–145.

Martineau, D., et al. 1994. Pathology and toxicology of beluga whales from the St. Lawrence estuary, Quebec, Canada—past, present and future. *Sci. Total Environ.* 154:201–215.

Martineau, D., et al. 2002. Cancer in wildlife, a case study: Beluga from the St. Lawrence Estuary, Québec, Canada. *Environ. Health Perspest* 110(3):285–292.

Martoja, R., and J. Berry. 1980. Identification of tiemannite as a probable product of demethylation of mercury by selenium in cetaceans. A complement to the scheme of the biological cycle of mercury. *Vie Millieu* 30:7–10.

Meador, J. P., U. Varanasi, P. A. Robisch and S.-L. Chan. 1993. Toxic metals in pilot whales (*Globicephala melaena*) from strandings in 1986 and 1990 on Cape Cod, MA. *Can. J. Fish. Aquat. Sci.* 50:2698–2706.

Meador, J. P., et al. 1999. Comparison of elements in bottlenose dolphins stranded on the beaches of Texas and Florida in the Gulf of Mexico over a one-year period. *Arch. Environ. Contam. Toxicol.* 36:87–98.

Mollenhauer, M. A., B. J. Carter, M. M. Peden-Adams, G. D. Bossart, and P. A. Fair. 2009. Gene expression changes in bottlenose dolphin, *Tursiops truncatus*, skin cells following exposure to methylmercury (MeHg)or perfluorooctane sulfonate (PFOS). *Aquatic Toxicol.* 91:10–18.

Morton, J., H. J. Mason, K. A. Ritchie, and M. White. 2004. Comparison of hair, nails and urine for biological monitoring of low level inorganic mercury exposure in dental workers. *Biomarkers* 9:47–55.

Neff, J. M. 1997. Ecotoxicology of arsenic in the marine environment—review. *Environ. Toxicol. Chem.* 16:917–927.

Ng, P.-S., et al. 2001. Striped dolphin detoxificates mercury as insoluble Hg(S, Se) in the liver. *Proc. Jpn. Acad.* 77(Ser. B):178–183.

Nielsen, J. 1995. Immunological and hematological parameters in captive harbor seals (*Phoca vitulina*). *Mar. Mamm. Sci.* 11(3):314–323.

Nigro, M., A. Campana, E. Lanzillotta, and R. Ferrara. 2002. Mercury exposure and elimination rates in captive bottlenose dolphins. *Mar. Pollut. Bull.* 44(10):1071–1075.

Nigro, M., and C. Leonzio. 1996. Intracellular storage of mercury and selenium in different marine vertebrates. *Mar. Ecol. Prog. Ser.* 135:137–143.

Noda, K., T. Kuramochi, N. Miyazaki, H. Ichihashi, and R. Tatsukawa. 1993. Heavy metal distribution in Weddell seals (*Leptonychotes weddellii*) from the Antarctic during JARE-32. *Proc. NIPR Symp. Polar Biol.* 6:76–83.

O'Hara, T. M., C. Hanns, G. Bratton, R. Taylor, and V. M. Woshner. 2006. Essential and non-essential elements in eight tissue types from subsistence hunted bowhead whale: nutritional and toxicological assessment. *Int. J. Circumpolar Health* 65(3):228–242.

O'Hara, T. M., and T. J. O'Shea. 2001. Toxicology. In *Marine mammal medicine: Health, disease and rehabilitation*, 2nd ed., eds. L. Dierauf and F. Gulland, pp. 471–520. Boca Raton, FL: CRC Press.

O'Hara, T. M., and T. J. O'Shea. 2005. Assessing impacts of environmental contaminants. In *Marine mammal research: conservation beyond crisis*, eds. J. E. Reynolds, III, W. F. Perrin, R. R. Reeves, S. Montgomery, and T. J. Ragen, pp. 63–83. Baltimore, MD: Johns Hopkins University Press.

O'Hara, T. M., V. Woshner, and G. Bratton. 2003. Inorganic pollutants in Arctic marine mammals. In *Toxicology of marine mammals*, eds. J. G. Vos, G. D. Bossart, M. Fournier, and T. J. O'Shea, pp. 207–246. New York: Taylor & Francis.

O'Shea, T. J. 1999. Environmental contaminants and marine mammals. In *Biology of marine mammals*, eds. J. E. Reynolds, and S. A. Rommel, pp. 485–536. Washington, DC: Smithsonian Institution Press.

O'Shea, T. J. 2003. Toxicology of sirenians. In *Toxicology of marine mammals*, eds. J. G. Vos, G. D. Bossart, M. Fournier, and T. J. O'Shea, pp. 270–287. New York, NY: Taylor & Francis.

O'Shea, T. J., and A. Aguilar. 2001. Cetacea and sirenia. In *Ecotoxicology of wild mammals*, eds. R. F. Shore, and B. A. Rattner, pp. 427–496. West Sussex, UK: John Wiley & Sons, Ltd.

O'Shea, T. J, and R. L. Brownell. 1994. Organochlorine and metal contaminants in baleen whales: a review and evaluation of conservation implications. *Sci. Total Environ.* 154:179–200.

O'Shea, T. J., R. R. Reeves, and A. K. Long (eds). 1998. Marine mammals and persistent ocean contaminants: Proceedings of the Marine Mammal Commission workshop, pp. 104–109. Keystone, Colorado, 12–15 October, 1998. Bethesda: Marine Mammal Commission, 150 pp. + vii.

O'Shea, T. J., and S. Tanabe. 2003. Persistent ocean contaminants and marine mammals: a retrospective overview. In *Toxicology of marine mammals*, eds. J. G. Vos, G. D. Bossart, M. Fournier, and T. J. O'Shea, pp. 99–134. New York, NY: Taylor and Francis.

Pacyna, J. M., and G. J. Keeler. 1995. Sources of mercury in the Arctic. *Water Air Soil Pollut.* 80:621–632.

Pacyna, J. M., and J. W. Winchester. 1990. Contamination of the environment as observed in the Arctic. *Palaeogeography, Palaeoclimatology, Palaeoecology* (Global and Planetary Change Section) 82:149–157.

Paludan-Müller, P., C. T. Agger, R. Dietz, and C. C. Kinze. 1993. Cadmium, mercury, selenium, copper and zinc in harbour porpoise (*Phocoena phocoena*) from west Greenland. *Polar Biol.* 13:311–320.

Parizek, J., and I. Ostadalova. 1967. The protective effect of small amounts of selenite in sublimate intoxication. *Experientia* 23:142–143.

Pelletier, E. 1985. Mercury-selenium interactions in aquatic organisms: a review. *Mar. Environ. Res.* 18:111–132.

Pellisso, S. C., M. J. Munoz, M. Carballo, and J. M Sanchez-Vizcaino. 2008. Determination of the immunotoxic potential of heavy metals on the functional activity of bottlenose dolphin leukocytes *in vitro. Vet. Immunol. Immunopathol.* 121(3–4):189–198.

Pillet, S., V. Lesage, M. Hammil, D. G. Cyr, J. M. Bouquegneau, and M. Fournier. 2000. *In vitro* exposure of seal peripheral blood leukocytes to different metals reveals a sex-dependant effect of zinc on phagocytic activity. *Mar. Pollut. Bull.* 40(11):921–927.

Ping, L., H. Nagasawa, K. Matsumoto, A. Suzuki, and K. Fuwa. 1986. Extraction and purification of a new compound containing selenium and mercury accumulated in dolphin liver. *Biol. Trace Elem. Res.* 11:185–199.

Puls, R. 1994. *Mineral levels in animal health: Diagnostic data*, 2nd ed., Clearbrook, BC: Sherpa International.

Ramprashad, F., and K. A. Ronald. 1977. Surface preparation study on the effect of methyl mercury on the sensory hair cell population in the cochlea of the harp seal (*Pagophilus groenlandicus* Erxleben, 1777). *Can. J. Zool.* 55:223–230.

Rawson, A., G. Patoon, S. Hofmann, G. Pietra, and L. Johns. 1993. Liver abnormalities associated with chronic mercury accumulation in stranded Atlantic bottlenose dolphins. *Ecotoxicol. Environ. Saf.* 25:41–47.

Rawson, A. J., J. P. Bradley, A. Teetsov, S. B. Rice, E. M. Haller, and G. W. Patton. 1995. A role for airborne particulates in high mercury levels of some cetaceans. *Ecotoxicol. Environ. Saf.* 30:309–314.

Reddy, M. L., J. S. Reif, A. Bachand, and S. H. Ridgway. 2001. Opportunities for using Navy marine mammals to explore associations between organochlorine contaminants and unfavorable effects on reproduction. *Sci. Total Environ.* 274(1–3):171–182.

Reijnders, P. J. H. 1980. Organochlorine and heavy metal residues in harbour seals from the Wadden Sea and their possible effects on reproduction. *Neth. J. Sea Res.* 14:30–65.

Reijnders, P. J. H. 1986. Reproductive failure in common seals feeding on fish from polluted coastal waters. *Nature* 324:456.

Riget, F., and R. Dietz. 2000. Temporal trends of cadmium and mercury in Greenland marine biota. *Sci. Total Environ.* 245:29–60.

Riget, F., R. Dietz, P. Johansen, and G. Asmund. 2000. Lead, cadmium, mercury and selenium in Greenland marine biota and sediments during AMAP phase 1. *Sci. Total Environ.* 245:3–14.

Riget, F., R. Dietz, K. Vorkamp, P. Johansen, and D. Muir. 2004. Spatial and temporal trends of contaminants in Greenland terrestrial, fresh water and marine biota: an update. *Sci. Total Environ.* 331(1–3):29–52.

Roditi-Elasar, M., et al. 2003. Heavy metal levels in bottlenose and striped dolphins off the Mediterranean coast of Israel. *Mar. Pollut. Bull.* 46:491–512.

Rolland, R. M. 2000. A review of chemically-induced alterations in thyroid and vitamin A status from field studies of wildlife and fish. *J. Wildl. Dis.* 36(4):615–635.

Romano, T. A., and G. Warr. 2004. A functional genomics approach to understanding and evaluating health in Navy dolphins: Final rept. (U.S. Department of Defense Report Number: A109524), 6 pp.

Rosa, C., J. E. Blake., G. R. Bratton., L.-A. Dehn, M. J. Gray, and T. M. O'Hara. 2008. Heavy metal and mineral concentrations and their relationship to histopathological findings in the bowhead whale. *Sci. Total Environ.* 399:165–178.

Rosas, F. C., and K. K. Lehti. 1996. Nutritional and mercury content of milk of the Amazon river dolphin, *Inia geoffrensis. Comp. Biochem. Physiol.* 115A:117–119.

Ross, P. S. 2000. Marine mammals as sentinels in ecological risk assessment. *HERA* 6:29–46.

Ross, P. S., R. L. De Swart, R. F. Addison, H. Van Loveren, J. G. Vos, and A. D. M. E. Osterhaus. 1996a. Contaminant-induced immunotoxicity in harbour seals: wildlife at risk? *Toxicology* 112(2):157–169.

Ross, P. S., R. L. De Swart, H. Van Loveren, A. D. M. E. Osterhaus, and J. G. Vos. 1996b. The immunotoxicity of environmental contaminants to marine wildlife: a review. *Ann. Rev. Fish Dis.* 6:151–165.

Rouleau, C., C. Gobeil, and H. Tjälve. 2000. Accumulation of silver from the diet in two marine benthic predators: the snow crab (*Chionoectetes opilio*) and American plaice (*Hippoglossoidesplatessoides*). *Environ. Toxicol. Chem.* 19:631–637.

Rudneva, N. A., and N. M. Pronin. 1996. Microelement composition in seal organs. *Ekologiia (Moscow)* 4:313–315.

Sanchez-Herandez, J. C. 2000. Trace element contamination in antarctic ecosystems. *Rev. Environ. Contam. Toxicol.* 166:83–127.

Sarafian, T. A. 1999. Methyl mercury induced generation of free radicals: biological implications. *Met. Ions Biol. Syst.* 36:415–444.

Scheuhammer, A. M. 1991. Effects of acidification on the availability of toxic metals and calcium to wild birds and mammals. *Environ. Pollut.* 71:329–375.

Shibata, Y., M. Morita, and K. Fuwa 1992. Selenium and arsenic in biology: their chemical forms and biological functions. *Adv. Biophysics* 28:31–80.

Siebert, U., et al. 1999. Potential relation between mercury concentrations and necropsy findings in cetaceans from German waters of the North and Baltic Seas. *Mar. Pollut. Bull.* 38(4):285–295.

Simms, W., and P. S. Ross. 2001. Vitamin A physiology and its application as a biomarker of contaminant-related toxicity in marine mammals: a review. *Toxicol. Ind. Health* 16:291–302.

Smith, T. G., and F. A. J. Armstrong. 1978. Mercury and selenium in ringed and bearded seal tissues from arctic Canada. *Arctic* 31:75–84.

Stavros, H. C., G. D. Bossart, T. C. Hulsey, and P. A. Fair. 2007. Trace element concentrations in skin of free-ranging bottlenose dolphins (*Tursiops truncatus*) from the southeast Atlantic coast. *Sci. Total Environ.* 388(1–3):300–315.

Stavros, H. C., G. D. Bossart, T. C. Hulsey, and P. A. Fair. 2008. Trace element concentrations in blood of free-ranging bottlenose dolphins (*Tursiops truncatus*): influence of age, sex and location. *Mar. Pollut. Bull.* 56:371–379.

Stoneburner, D. L. 1978. Heavy metals in tissues of stranded short-finned pilot whales. *Sci. Total Environ.* 9:293–297.

Szefer, P., W. Czarnowski, J. Pempkowiak, and E. Holm. 1993. Mercury and major essential elements in seals, penguins, and other representative fauna of the Antarctic. *Arch. Environ. Contam. Toxicol.* 25:422–427.

Taddei, F., V. Scarcelli, G. Frenzilli, and M. Nigro. 2001. Genotoxic hazard of pollutants in cetaceans: DNA damage and repair evaluated in the bottlenose dolphin (*Tursiops truncatus*) by the Comet Assay. *Mar. Pollut. Bull.* 42:324–328.

Thévenod, F., J. M. Friedmann, A. D. Katsen, and I. A. Hauseri. 2000. Up-regulation of multidrug resistance P-glycoprotein via Nuclear Factor-kB activation protects kidney proximal tubule cells from cadmium- and reactive oxygen species-induced apoptosis. *J. Biol. Chem.* 275(3):1887–1896.

Thiemann, G. W., S. J. Iverson, and I. Stirling. 2007. Variability in the blubber fatty acid composition of ringed seals across the Canadian Arctic. *Mar. Mamm. Sci.* 23(2):241–261.

Thiemann, G. W., S. J. Iverson, and I. Stirling. 2008. Variation in blubber fatty acid composition among marine mammals in the Canadian Arctic. *Mar. Mamm. Sci.* 24(1):91–111.

Thompson, D. R. 1996. Mercury in birds and terrestrial mammals. In *Environmental contaminants in wildlife: Interpreting tissue concentrations*, eds. W. N. Beyer, G. H. Heinz, and A. W. Redmon-Norwood, pp. 341–356. Boca Raton, FL: CRC Press.

Tillander, M., J. K. Miettinen, and I. Koivisto. 1970. Excretion rate of methyl mercury in the seal (*Pusa hispida*). In *Marine pollution and sea life*, ed. M. Ruivo, pp. 303–305. Surrey, England: Fishing News (Books) Ltd. F.A.O. Pap., FIR-MP MP70E67, Fish. Rep., N.99/159.

Trasande, L., P. J. Landrigan, and C. Schechter. 2005. Public health and economic consequences of methyl mercury toxicity to the developing brain. *Environ. Health Perspect* 113:590–996.

United Nations Environment Programme (UNEP). 1999. *Inventory of information sources on chemicals persistent organic pollutants*. Geneva: UNEP Chemicals, 148 pp.

United Nations Environment Programme (UNEP) Chemicals. 2002. *MASTER list of actions on the reduction and/or elimination of the releases of persistent organic pollutants*, 3rd ed. Geneva: UNEP Chemicals.

Uthus, E. O. 1992. Evidence for arsenic essentiality. *Environ. Geochem. Health* 14:55–58.

van de Ven, W. S. M., J. H. Koeman, and A. Svenson. 1979. Mercury and selenium in wild and experimental seals. *Chemosphere* 8:539–555.

Wagemann, R., S. Innes, and P. Richard. 1996. Overview and regional and temporal differences of heavy metals in Arctic whales and ringed seals in the Canadian Arctic. *Sci. Total Environ.* 186:41–66.

Wagemann, R., and H. Kozlowska. 2005. Mercury distribution in the skin of beluga (*Delphinapterus leucas*) and narwhal (*Monodon monoceros*) from the Canadian Arctic and mercury burdens and excretion by moulting. *Sci. Total Environ.* 351/352:333–343.

Wagemann, R., N. Snow, A. Lutz, and D. Scott. 1983. Heavy metals in tissues and organs of the narwhal (*Monodon monoceros*). *Can. J. Fish. Aquatic Sci.* 40(Suppl. 2):206–214.

Wagemann, R., R. Stewart, P. Béland, and C. Desjardins. 1990. Heavy metals and selenium in tissues of beluga whales, *Delphinapterus leucas*, from the Canadian Arctic and the St. Lawrence Estuary. *Can. Bull. Fish. Aquatic Sci.* 224:191–206.

Wagemann, R., R. E. A. Stewart, W. L. Lockhart, B. E. Stewart, and M. Povoledo. 1988. Trace metals and methyl mercury: Associations and transfer in harp seal (*Phoca groenlandica*) mothers and their pups. *Mar. Mamm. Sci.* 4:339–355.

Wang, A., D. Barber, and C. J. Pfeiffer. 2001. Protective effects of selenium against mercury toxicity in cultured Atlantic spotted dolphin (*Stenella plagiodon*) renal cells. *Arch. Environ. Contam. Toxicol.* 41:403–409.

Wang, C., S. Brown, and M. H. Bhattacharyya. 1994. Effect of cadmium on bone calcium and ^{45}Ca in mouse dams on a calcium-deficient diet: evidence of Itai-Itai-like syndrome. *Toxicol. Appl. Pharmacol.* 127:320–330.

Webb, M., and K. Cain. 1982. Functions of metallothionein. *Biochem. Pharmacol.* 31(2):137–142.

Wells, R. S., et al. 2004. Bottlenose dolphins as marine ecosystem sentinels: developing a health monitoring system. *EcoHealth* 1:246–254.

Wells, R. S., et al. 2005. Integrating life-history and reproductive success data to examine potential relationships with organochlorine compounds for bottlenose dolphins (*Tursiops truncatus*) in Sarasota Bay, Florida. *Sci. Total Environ.* 349:106–119.

Wise, J. P. Sr., et al. 2008. Hexavalent chromium is cytotoxic and genotoxic to the North Atlantic Right whale (*Eubalaena glacialis*) lung and testes fibroblasts. *Mutat. Res.* 650:30–38.

Wolfe, M. F., et al. 2007. Wildlife indicators. In *Ecosystem responses to mercury contamination: Indicators of change*, eds. R. Harris, D. P. Krabbenhoft, R. Mason, M. W. Murray, R. Reash, and T. Saltman, pp. 123–189. Boca Raton, FL: SETAC/CRC Press.

Woshner, V. M. 2000. Concentrations and interactions of selected elements in tissues of four marine mammal species harvested by Inuit hunters in arctic Alaska, with an intensive histologic assessment, emphasizing the beluga whale. PhD dissertation, University of Illinois.

Woshner, V., K. Knott, R. Wells, C. Willetto, R. Swor, and T. O'Hara. 2008. Mercury and selenium in blood and epidermis of bottlenose dolphins (*Tursiops truncatus*) from Sarasota Bay, FL: Interaction and relevance to life history and hematologic parameters. *EcoHealth* 5(3):360–370.

Woshner V., K. Knott, R. Wells, C. Willetto, R. Swor, and T. O'Hara. 2006. Mercury and selenium in blood of bottlenose dolphins (*Tursiops truncatus*): interaction and reference to life history and hematologic parameters. Paper SC/58/E24 presented to the IWC Scientific Committee, June 2006, St. Kitts and Nevis, WI, 9 pp.

Woshner, V. M., T. M. O'Hara, G. R. Bratton, and V. R. Beasley. 2001a. Concentrations and interactions of selected essential and non-essential elements in ringed seals and polar bears of arctic Alaska. *J. Wildl. Dis.* 37:711–721.

Woshner, V. M., T. M. O'Hara, G. R. Bratton, R. S. Suydam, and V. R. Beasley. 2001b. Concentrations and interactions of selected essential and non-essential elements in bowhead and beluga whales of arctic Alaska. *J. Wildl. Dis.* 37:693–710.

Woshner, V. M., et al. 2002. Distribution of inorganic mercury in liver and kidney of beluga whales, compared to bowhead whales, through autometallographic development of light microscopic tissue sections. *Toxicol. Pathol.* 30(2):209–215.

Yamamoto, Y., K. Honda, H. Hidaka, and R. Tatsukawa. 1987. Tissue distribution of heavy metals in Weddell seals (*Leptonychotes weddellii*). *Mar. Pollut. Bull.* 18:164–169.

Yang, J., T. Kunito, Y. Anan, S. Tanabe, and N. Miyazaki. 2004. Total and subcellular distribution of trace elements in the liver of a mother-fetus pair of Dall's porpoises (*Phocoenoides dalli*). *Mar. Pollut. Bull.* 48:1122–1129.

Yang, J., T. Kunito, Y. Anan, S. Tanabe, and N. Miyazaki. 2008. Subcellular distribution of trace elements in kidney of a mother-fetus pair of Dall's porpoises (*Phocoenoides dalli*). *Chemosphere* 70:1203–1210.

Yang, J., T. Kunito, S. Tanabe, and N. Miyazaki. 2002. Mercury in tissues of Dall's porpoise (*Phocoenoides dalli*) collected off Sanriku coast of Japan. *Fish. Sci.* 68(Suppl. 1):256–259.

Yang, J., T. Kunito, S. Tanabe, and N. Miyazaki. 2003. Comparative study of mercury in liver and muscle of Dall's porpoise (*Phocoenoides dalli*) off Sanriku coast of Japan. *J. Phys. IV* 107:1393–1398.

Yang, J., T. Kunito, S. Tanabe, and N. Miyazaki. 2007. Mercury and its relation with selenium in the liver of Dall's porpoises (*Phocoenoides dalli*) off the Sanriku coast of Japan. *Environ. Pollut.* 148:669–673.

Yang, J., N. Miyazaki, T. Kunito, and S. Tanabe. 2006. Trace elements and butyltins in a Dall's porpoise (*Phocoenoides dalli*) from the Sanriku coast of Japan. *Chemosphere* 63:449–457.

Zalups, R., and Koropatnick (eds.). 2000. *Molecular biology and toxicology of metals*. New York, NY: Taylor & Francis.

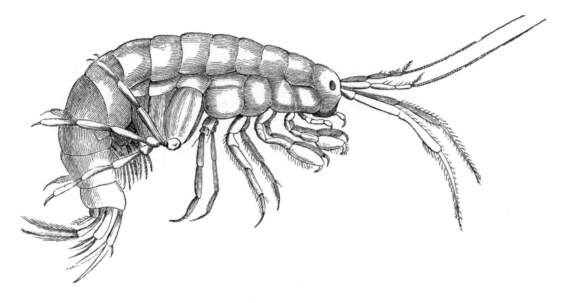

Fresh water shrimp

From *The Royal Natural History*, edited by Richard Lydekker, Frederick Warne & Co.,
London, 1893–94.

11 Toxicological Significance of Pesticide Residues in Aquatic Animals

Michael J. Lydy
Jason B. Belden
Jing You
Amanda D. Harwood

CONTENTS

11.1 Introduction 409
11.2 Background...409
11.3 What is the Toxicological Significance of Pesticide Bioaccumulation? 411
11.4 Relationship between Body Residues and Toxicity 411
11.5 Biotransformation Processes in Aquatic Animals..413
11.6 How Does Biotransformation of Pesticides Impact Toxicological Endpoints? 416
11.7 Environmental and Physiological Factors That Influence Bioaccumulation and Toxicity........417
Summary 419
References..419

11.1 INTRODUCTION

Pesticides are intentionally released into the environment to protect crops, homes, and livestock from undesirable insects, weeds, and pathogens. With designed biological activities, pesticides are toxic to living organisms, and are generally classified as insecticides, herbicides, and fungicides. Extensive usage of pesticides has led to nearly ubiquitous contamination in surface water in many places in the world, including most of the United States (Gilliom 2007). Therefore, aquatic animals are at risk of pesticides entering their body, accumulating, and potentially causing toxicity. The objectives of this chapter are to describe (1) the significance of pesticide bioaccumulation; (2) the relationship between body residues and toxicity; (3) biotransformation processes in aquatic animals; (4) the impact of biotransformation on toxicological endpoints; and (5) the environmental and physiological factors that can influence bioaccumulation and toxicity.

11.2 BACKGROUND

Pesticides are used worldwide and their use has increased dramatically over the years, exceeding 23 million kg being applied worldwide and over 5.4 million kg being used in the United States in 2001 (Kiely and Donaldson 2004). Although the application of pesticides highly benefits crop production, the unintended loss of pesticides by incidental spill, spray drift, and surface runoff has caused contamination to aquatic ecosystems. A recent assessment conducted by the U.S. Geological Survey's (USGS) National Water-Quality Assessment (NAWQA) Program reported frequent occurrences of pesticides in surface water in the United States, with 97% of streams in agricultural and urban areas having one or more detections of the 83 monitored pesticides and degradates. The study also indicated that 56% and 83% of the streams in agricultural and urban areas, respectively had pesticides detected at concentrations that may cause adverse effects to aquatic life and fish-eating wildlife. The detected pesticides included not only the current-use pesticides, such as organophosphate (OP) and carbamate insecticides and multiple classes of herbicides, but also organochlorine (OC) insecticides, some of which were banned years ago (Gilliom 2007).

Although herbicides were the most frequently detected pesticides and were found at the highest concentrations in the streams surveyed by the NAWQA program (Gilliom 2007), insecticides were also commonly found, especially in urban areas. Insecticides were of the greatest concern due to their increased potential to harm nontarget animals. Thus, this chapter will focus on the impacts of insecticide toxicity on aquatic animals, even though the general concepts covered are applicable to all pesticides.

Most current-use insecticides, including OPs, carbamates, and pyrethroids, are neurotoxicants, and can elicit acute toxic effects to nontarget aquatic vertebrates and invertebrates. Two commonly used OPs, chloropyrifos and diazinon, and a carbamate, carbaryl were the most frequently detected insecticides in urban streams in the USGS' NAWQA study (Gilliom 2007), and the application of OPs has been linked to surface-water toxicity (Kuivila and Foe 1995). Pyrethroids are extremely toxic to fish and aquatic invertebrates (Bradbury and Coats 1989), and they were identified as the major contributors to the acute sediment toxicity to benthic invertebrates, *Hyallela azteca* and *Chironomus dilutus* in both agricultural and urban areas of California (Weston et al. 2004, 2005, Amweg et al. 2006) and in urban areas of Texas (Hintzen et al. 2009).

Most OC insecticides are no longer in use; however, they are still frequently detected in sediment and fish tissues due to their persistence in the environment (Gilliom 2007). Compared to OPs and pyrethroids, some OCs, such as DDT and its metabolites (DDD and DDE) have little acute toxicity, but can cause a variety of sublethal effects, including reduced growth and reproduction (Lotufo et al. 2001a). Because of the structural similarity to natural hormones, OCs have been reported to induce endocrine-disrupting effects by binding to androgen and/or estrogen receptors (van den Berg et al. 2003). In addition, OCs can cause adverse effects on organisms that were not directly exposed, since OCs are well known for their potential to bioaccumulate and biomagnify through tropic transfer, which poses a greater threat to consumers at the higher tropic levels (Fisk et al. 1998, Lotufo et al. 2001b, Kwong et al. 2008).

Body residues, bioaccumulation factors (BAFs), bioconcentration factors (BCFs) and biomagnification factors (BMFs) have been introduced as bioaccumulation parameters. Body residue is simply the concentration of the pesticide within the organism. BAF is the ratio of the concentration of pesticide within the organism divided by the concentration in the environment. BAF makes no assumptions as to the uptake route of the pesticide or which matrix accounts for most of the exposure (i.e., sediment or water). BCF is the ratio of the concentration of the pesticide within the organism divided by the pesticide concentration of the water the organism is living in. BCF assumes uptake and loss of the pesticide solely related to aqueous exposure. Finally, BMF is based on the potential for higher trophic levels to have higher body residues of pesticides. The BMF is calculated by dividing the body residue of the animal from the higher trophic level by the body residue of the animal from the lower trophic level. Each of the values can be calculated empirically using experimental method designs that are well established (Landrum et al. 1992). In addition, estimates for these values, especially BCF, are commonly reported using quantitative structure-activity relationships (QSAR) using octanol/water partition coefficients (K_{ow}) for both freshwater and marine species (Zaroogian et al. 1985, Fisk et al. 1998). However, application of QSAR modeling to estimate bioaccumulation and adverse effects caused by pesticides may be restricted due to biotransformation in the animal (De Bruijn and Hermens 1991).

11.3 WHAT IS THE TOXICOLOGICAL SIGNIFICANCE OF PESTICIDE BIOACCUMULATION?

The occurrence and magnitude of bioaccumulation provides insight into the ecotoxicology of pesticides in three ways including trophic transfer, markers of exposure, and critical body residues. First, and most traditionally, bioaccumulation provides a measure of exposure to higher trophic levels. For example, OC insecticides such as chlordane and DDT are not easily biotransformed or depurated from most aquatic species and thus can biomagnify. At higher trophic levels, uptake through ingestion of organisms from lower trophic levels becomes the major route of exposure. For example, marine copepods receive greater than 50% of DDT exposure from their phytoplankton food source (Wang and Wang 2005). Organisms at higher trophic levels typically have even greater contributions from their food. Harbor seals and Gray seals in northern Norway have 10–30 times more body burden of DDT as compared to the fish they use as food even though their exposure is nearly entirely through ingestion of fish (Ruus et al. 1999).

Second, detection of a specific pesticide in an animal indicates that exposure has occurred, which is especially important when investigating an environmental incident such as a fish kill. Basic evidence of exposure can also be important if working with animals that have a large range that overlaps possible areas of contamination or for systems where the bioavailability of the pesticide is in question. For instance, in Sparling et al. (2001), the researchers were investigating if pesticides from coastal areas with high intensity pesticide usage (Central Valley of California, U.S.) are transported to a remote mountainous area where there is a documented amphibian decline (Sierra Nevada). The presence of OP and OC insecticide body residues in the Pacific Tree frog, *Hyla regilla,* indicated that exposure to pesticides was occurring and could be relevant in determining the causes of the decline.

Finally, in order for toxicants to have an effect they must enter an animal and be present at high enough concentrations at the site of toxic action to result in a toxic response. The toxicological process includes the movement of the pesticide to the site of toxic action (toxicokinetics) and the action of the pesticide at the toxicological receptor site (toxicodynamics). Although factors may influence toxicodynamics, changes in toxicokinetics likely account for most of the variability in the toxicological response. Therefore, knowing the concentration of the pesticide in the animal should provide a much better estimate of the toxicity that may occur compared to only knowing the external environmental concentration.

11.4 RELATIONSHIP BETWEEN BODY RESIDUES AND TOXICITY

Since dose is typically not known for aqueous exposures, the concentration of the water is typically used as a measure of the dose or "dose metric." However, the relationship between the pesticide concentration at the site of toxic action within the animal and the concentration in the water around the animal can be variable due to a variety of factors that are discussed later. Since, the resulting variability can lead to uncertainty in regard to the toxicity of the pesticide; a better dose metric is sometimes needed. Ideally, measurement of the pesticide concentration at the site of toxic action would be an ideal dose metric, but it is very difficult to measure. An alternative is to measure the pesticide concentration that is present in the whole body or a specific tissue. Although variability still exists within the relationship of the whole body residue to the concentration at the site of toxic action, whole body residue should theoretically provide a better dose metric than using the external concentration. The whole body residue should be presented in molar terms, as toxicity is related to the number of pesticide molecules and not the mass of the pesticide. The whole body residue that can be linked to causing a biological effect is termed critical body residue (CBR).

The CBR measurement has been determined for a variety of chemical classes including some types of pesticides; however, the approach generally works better for chemicals with nonspecific modes of action (e.g., narcosis) as compared to specific modes of action (e.g., acetycholinesterase inhibition) (review available from Barron et al. 2002). Most pesticides have specific modes of action, at least for organisms that are related to the pesticides target. For example, many OP insecticides inhibit acetycholinesterase through the formation of a metabolite, which is further chemically altered during inhibition of the enzyme. Thus, molecules that have found their target are no longer measurable as the pesticide itself. This increases the chance that factors such as length of exposure and pulsed exposure could drastically change the measured CBR based on the parent pesticide. Several studies have measured and reported CBRs for OP insecticides; however, the consistency of the values may be problematic due to the importance of biotransformation in the toxic action. For example, a study that calculated CBR values for short-term high concentration and long-term lower concentration exposure regimes found that the CBR values fell within a factor of two for a few OPs, such as diazinon. However, the CBR values for the different exposure regimes differed by greater than a factor of 10 for other pesticides such as azinophos-methyl and parathion (Deneer et al. 1999, Beckvar et al. 2005; Table 11.1). For other pesticides that are not inactivated during receptor site binding, CBRs may be consistent and thus predictive of toxicity. For example, pentachlorophenol (PCP), an uncoupler of oxidative phosphorylation, has been well studied and tends to provide a stable, predictive CBR that may have some utility among different organisms and exposure regimes (Kukkenon et al. 2002; Table 11.1).

TABLE 11.1
Critical Body Residues (µmol/g) in Fish and Invertebrate Species for Various Toxicological Endpoints

Compound	Species	Residue (µmol/g)	Effect	Time (d)	Reference
DDT	Goldfish	0.00465	Behavior[o]	ukn	Davy et al. 1972
DDT	Pinfish	0.00155	Lethality[o]	ukn	Butler 1969
DDT	Lake trout	0.000818	Lethality[o]	ukn	Berlin et al. 1981
DDT	Brook trout	0.0316	Growth[o]	ukn	Macek 1968a
DDT	Chinook salmon	0.0103	Lethality[o]	ukn	Buhler et al. 1969
DDT	Cutthroat salmon	0.0031	Lethality[o]	ukn	Allison et al. 1963, 1964
DDT	Brook trout	0.0214	Reproduction[o]	ukn	Macek 1968b
DDT	Fathead minnow	0.318	Lethality[o]	ukn	Jarvin et al. 1976, 1977
DDT	Coho salmon	0.0953	Lethality[o]	ukn	Buhler et al. 1969
Azinphos-ethyl	Guppy	0.1 ± 0.1	Lethality	ukn	de Bruijn et al. 1991
Bromophos-methyl	Guppy	5.0 ± 1.1	Lethality	ukn	Ohayo-Mitoko et al. 1993
Chlorothion	Guppy	0.4 ± 0.1	Lethality	ukn	Ohayo-Mitoko et al. 1993
Chlorpyrifos	Guppy	8.0 ± 3.0	Lethality	ukn	de Bruijn et al. 1991
Diazinon	Guppy	8.0 ± 2.0	Lethality	ukn	de Bruijn et al. 1991
Fenthion	Guppy	10.2 ± 4.3	Lethality	ukn	Ohayo-Mitoko et al. 1993
Methidathion	Guppy	0.003 ± 0.002	Lethality	ukn	Ohayo-Mitoko et al. 1993
Parathion	Guppy	0.1 ± 0.1	Lethality	ukn	Deneer et al. 1999
Azinphos-ethyl	Guppy	0.011 ± 0.005	Lethality	11	Deneer et al. 1999
Azinphos-methyl	Guppy	0.09 ± 0.06	Lethality	6	Deneer et al. 1999
Carbophenothion	Guppy	0.41 ± 0.36	Lethality	5	Deneer et al. 1999
Chlorpyrifos	Guppy	0.54 ± 0.38	Lethality	11	Deneer et al. 1999

TABLE 11.1 (continued)
Critical Body Residues (µmol/g) in Fish and Invertebrate Species for Various Toxicological Endpoints

Compound	Species	Residue (µmol/g)	Effect	Time (d)	Reference
Chlorpyrifos-methyl	Guppy	9.9 ± 5.6	Lethality	7	Deneer et al. 1999
Diazinon	Guppy	1.8 ± 0.7	Lethality	7	Deneer et al. 1999
Dichlofenthion	Guppy	29 ± 17	Lethality	11	Deneer et al. 1999
Fenchlorphos	Guppy	18 ± 13	Lethality	11	Deneer et al. 1999
Fenthion	Guppy	13.6 ± 5.3	Lethality	5	Deneer et al. 1999
Fonofos	Guppy	0.02 ± 0.01	Lethality	5	Deneer et al. 1999
Methidathion	Guppy	0.013 ± 0.005	Lethality	6	Deneer et al. 1999
Parathion	Guppy	0.41 ± 1.8	Lethality	15	Deneer et al. 1999
Parathion-methyl	Guppy	7.9 ± 1.8	Lethality	2	Deneer et al. 1999
26 DCP	*Lumbriculus variegatus*	1.72 (1.68–1.76)	Lethality	1	Deneer et al. 1999
26 DCP	*L. variegatus*	1.24 (1.22–1.27)	Lethality	2	Deneer et al. 1999
245 TCP	*L. variegatus*	0.63 (0.60–0.66)	Lethality	1	Kukkonen et al. 2002
245 TCP	*L. variegatus*	0.54 (0.53–0.55)	Lethality	2	Kukkonen et al. 2002
246 TCP	*L. variegatus*	0.81 (0.80–0.83)	Lethality	1	Kukkonen et al. 2002
246 TCP	*L. variegatus*	0.66 (0.64–0.67)	Lethality	2	Kukkonen et al. 2002
2346 TeCP	*L. variegatus*	0.95 (0.87–1.15)	Lethality	1	Kukkonen et al. 2002
2346 TeCP	*L. variegatus*	0.91 (0.85–0.96)	Lethality	2	Kukkonen et al. 2002
PCP	*L. variegatus*	0.7 (0.67–0.74)	Lethality	1	Kukkonen et al. 2002
PCP	*L. variegatus*	0.45 (0.44–0.47)	Lethality	2	Kukkonen et al. 2002
PCP	*Chironomus riparius*	0.22 (0.20–0.24)	Lethality	1	Kukkonen et al. 2002
PCP	*C. riparius*	0.15 (0.14–0.16)	Lethality	2	Kukkonen et al. 2002

DDT = dichlorodiphenyltrichloroethane, DCP = dichlorophenol, TCP = trichlorophenol, TeCP = tetrachlorophenol, PCP = pentachlorophenol, °low effects residue, ukn = unknown, ± standard deviation, parentheses = 95% confidence intervals.

CBR values can also be determined empirically by measuring the body residue of the pesticide during toxicity testing. The toxicological endpoint measured can then be compared to the body residue to obtain a CBR value. CBR values can also be calculated using toxicological endpoints and bioconcentration factors (BCF). Table 11.2 provides examples of this technique. However, these values should be matched based on species and the exposure conditions should be similar. BCF values should either be time matched with the toxicity test or the duration of the toxicity test should be long enough to allow the body residue concentration within the organisms to obtain steady state. In addition, the body residue measurement should be made using an accurate technique that can specify pesticide or pesticide metabolite of interest. If measurements include both parent and metabolites in a single value or are otherwise not specific to the chemical of interest, the CBR may not be useful to future investigators, especially if other measurement techniques are used later. For example, care should be taken when using radiolabeled pesticide for measurement of CBR to insure that only the chemical of interest is being measured and not transformation products.

11.5 BIOTRANSFORMATION PROCESSES IN AQUATIC ANIMALS

Biotransformation plays an important role on pesticide disposition, bioaccumulation, biomagnification, and toxic effects. This process, however, is highly variable as the capacity of aquatic animals to biotransform insecticides is dependant on both the enzymatic capabilities of the specific animal and

TABLE 11.2
Critical Body Residues (CBR, µmol/g) in *Chironomus dilutus* (Aquatic Insect) for Lethality

Pesticide	Temperature (°C)	LC_{50} (µg/L)	BCF (mL/g)	CBR (µg/kg)	CBR (µmol/g)
Chlorpyrifos	13	0.825	831	686	0.00196
	23	0.459	338	155	0.000442
DDT	13	3.88	48,800	189,000	0.533
	23	6.26	30,000	188,000	0.530
Lambda-cyhalothrin	13	0.0124	170	2.11	4.69 E-06
	23	0.0379	92.5	3.51	7.80 E-06
Permethrin	13	0.0585	315	18.4	4.70 E-05
	23	0.189	87.2	16.4	4.19 E-05

Note: All values were calculated by multiplying the LC_{50} and the BCF values reported in Harwood et al. (2009) to obtain CBR values.

properties of the insecticide. Biotransformation is the enzyme-catalyzed processes, which transforms xenobiotics to more water-soluble metabolites, enhancing polarity and subsequent excretion, and it is generally divided into two components. Phase I biotransformation involves hydrolysis, reduction and oxidation reactions, whereas the reactions of phase II (conjugation) reduce toxicity by addition of a variety of more polar moiety to contaminants or their phase I products. Enzymes generally involved in phase I biotransformation of pesticides by aquatic animals are cytochrome P450 monoxygenases (CYP450), flavin-containing monooxygenases (FMOs), esterases, and other hydrolases, while glutathione-S-transferase (GST), glucuronosyltransferases, sulfotransferase, and acetyl transferase may involve in phase II reactions (Lech and Vodicinik 1985).

Many enzymes involved in biotransformation of pesticides by aquatic animals are similar to those in mammals; however, specific biotransformation features and pathways exist in aquatic animals and are often less advanced than in mammals. For instance, the pyrethroid insecticide permethrin was not only biotransformed at a slower rate by rainbow trout than mice, but also the dominate detoxification pathways differed for the two species (Bradbury and Coats 1989). In addition, aquatic invertebrates and fish are poikotherms, so their biotransformation rates are significantly affected by changes in temperature (Harwood et al. 2009). The differential abilities to biotransform pesticides were demonstrated among aquatic animals as well. Livingstone (1998) has reported an approximately four times higher biotransformation rate for the biocide PCP by fish (*Carassius auratus*) compared to aquatic invertebrates (Molluscan species), and the difference in biotransformation rates were explained by the difference in enzymatic activity.

Fish are capable of biotransforming multiple classes of pesticides, as noted by the induction of enzymatic activity in fish after exposure to pesticides. Other evidence, such as measurements of body residue and bioaccumulation also indicated fish readily biotransform different classes of pesticides. A continuous decline in BMF values of chloropyrifos in the fish *Aphaius iberus* which was fed previously exposed crustacean *Artemia* indicated capability of *A. iberus* to biotransform this OP insecticide (Varo et al. 2002). A wide variation in lethal body residues was observed for four OPs, bromophos, fenthion, chlorothion, and methidathion in guppy, and it was attributed to the differential biotransformation in oxidative and glutathione-mediated *in-vitro* systems (De Bruijn and Hermens 1991, De Bruijn et al. 1991). Extensive biotransformation of triazole fungicides has been reported and it accounted for 60–90% of the elimination of the fungicides by juvenile rainbow trout (Konwick et al. 2006a). The phenylpyrazole insecticide fipronil was rapidly biotransformed by rainbow trout and 88% of its elimination was due to the biotransformation (Konwick et al. 2006b). Fish are also well equipped to biotransform OCs, and different biotransformation pathways have been reported for DDT in different organs in marine black sea bream (*Acanthopagrus schlegeli*) (Kwong et al. 2008).

Interspecies variation in biotransformation potential was also shown among aquatic inverte-brates. Both the endogenous enzymatic activity and the different sensitivity in induction of the enzymes could influence the biotransformation potential of the test species. Guerrero et al. (2002) reported significant differences in biotransformation potential of PCP and two herbicides, fenpropi-din and trifluralin, by three freshwater invertebrates including an annelid *Lumbriculus variegatus*, a bivalve *Sphaerium corneum*, and an arthropod *Chironomus riparius*. Although C. *riparius* exten-sively biotransformed the pesticides, *L. variegatus* and *S. corneum* showed limited biotransforma-tion ability. Ankley and Collyard (1995) suggested that the arthropods, *H. azteca* and *C. riparius,* biotransformed OPs through CYP450 catalyzed desulfuration reactions, while the oligochaete *L. variegatus* was incapable of this bioactivation step. Because of its low biotransformation poten-tial, *L. variegatus* was selected as the standard organism for sediment bioaccumulation testing (U.S. EPA 2000). However, recent studies demonstrated the biotransformation potential of *L. variegatus*. Mäenpää et al. (2008) reported extensive biotransformation of PCP by *L. variegatus* with less than 3% of parent compound being left after a 28-day sediment exposure. With a polar–OH group, PCP may be directly biotransformed through phase II glucosidation (Livingstone 1998). Our recent study (You et al. 2009) also showed that *L. variegatus* was capable of biotransforming pyrethroids, with nearly half of the bifenthrin and permethrin being biotransformed to more hydrophilic metabolites after the 14-day sediment exposure. Therefore, phase II glucosidation and/or hydrolysis reactions catalyzed by carboxylesterases may be the major pathways for the biotransformation of pesticides in *L. variegatus* due to the possible lack of oxidative activity (Ankley and Collyard 1995).

Arthropods showed relatively greater biotransformation potential than other phyla of aquatic invertebrates (Ankley and Collyard 1995, Guerrero et al. 2002). Lotufo et al. (2000, 2001b) reported species-specific biotransformation of DDT in two freshwater amphipods, *H. azteca* and *Diporeia* spp. DDT was mainly biotransformed to DDE in *H. azteca,* while DDD was formed in *Diporeia* spp. and the greater biotransformation and elimination in *H. azteca* resulted in lower toxicity. *H. azteca* was capable of rapidly biotransforming PCP and the OP methyl parathion as well, which may involve multiple types of enzymes, including phase I esterases, hydrolases, and oxidase and the phase II enzyme GST (Nuutinen et al. 2003). *Chironomus* spp. was another model arthropod for biotransformation studies (Buchwalter et al. 2004).

Crustacean species had similar CYP450 and GST activity as fish and generally had higher CYP450 activity compared with other invertebrates suggesting a possible higher biotransformation potential (Livingstone 1998). In an *in-vitro* study (Escartin and Porte 1996) using hepatopancreas microsomes of red swamp crayfish (*Procambarus clarkii*), the microsomes biotransformed the OP fenitrothion to its metabolites fenitrooxon and 3-methyl-4-nitrophenol, indicating two major path-ways, suggesting detoxification by GST and bioactivation by CYP450 enzymes. The EROD activity indicated the existence of CYP4501A enzymes in the crayfish, and significant induction of EROD occurred in field-collected crayfish (Porte and Escartin 1998). Therefore, CYP450 plays an impor-tant role in pesticide biotransformation by crustaceans. Compared to fish and crustaceans, mollusks have 4–5 times lower CYP450 activity, two times higher FMO activity and 4–6 times higher GST activity (Livingstone 1998), therefore, biotransformation pathways may vary significantly among species. Boutet et al. (2004) identified two non-P450 phase I enzymes, FMO-2 and monoamine oxidase A involved in biotransformation of the herbicides, glyphosate, atrazine, diuron, and isopro-turon by the Pacific oyster, *Crassostrea gigas*.

In summary, aquatic animals are capable of biotransforming pesticides, and typically in the order of fish, arthropods, mussels, and annelids. The CYP450 enzymes play an important role in phase I biotransformation for fish and arthropods, while oxidized biotransformation reactions were still uncertain in mussels and annelids. Glucosidation was the major phase II biotransformation pathway for aquatic invertebrates (Livingstone 1998). It is desirable to obtain the information on the biotransformation rates, pathways, and enzymes activity to better understand biotransformation processes *in vivo*. However, there are limited studies on enzymes involved in biotransformation for aquatic animals, especially for aquatic invertebrates.

11.6   HOW DOES BIOTRANSFORMATION OF PESTICIDES IMPACT TOXICOLOGICAL ENDPOINTS?

There are three possible toxicological responses to biotransformation: (1) an increase in toxicity, (2) a decrease in toxicity, or (3) a lack of biotransformation, which could lead to bioaccumulation. Each of these outcomes may influence CBR. Metabolic activation, or increased toxicity with biotransformation, is relatively rare; one notable exception is the OPs (Ankley and Collyard 1995, Bailey et al. 1996, Escartin and Porte 1996, Carr et al. 1997). Chlorpyrifos, diazinon, malathion, methyl parathion, and parathion are commonly used metabolically activated insecticides. Most OP insecticides are metabolically activated, with a few exceptions including dichlorvos. Nonmetabolically activated OPs can be identified by the presence of terminal oxygen in the phosphate group, whereas metabolically activated OPs generally contain terminal sulfur (phosphoryl vs. thiophosphoryl). The replacement of this sulfur with oxygen creates the toxic form of the insecticide. The biotransformation requirement for most OP insecticides causes delayed toxicity as compared to other cholinesterase inhibitors such as the carbamate insecticides. Carbamates can undergo different types of bioactivation processes. For example, Schlenk (1995) suggested that FMO was responsible for bioactivating the carbamate insecticide, aldicarb, in fish. The low sensitivity of channel catfish compared to rainbow trout and bluegill to aldicarb toxicity was the result of a lower level of bioactivation of aldicarb to aldicarb sulfoxide, which is a more potent AChE inhibitor (Perkins and Schlenk 2000). *In-vitro* studies provided valuable information on enzymes involved in the biotransformation reactions. Perkins et al. (1999) demonstrated that the biotransformation of aldicarb to more toxic sulfoxide and less toxic sulfone not only involved CYP450 and FMO enzymes, but also other unidentified CYP isoforms. Therefore, the consistency of CBR values determined for pesticides that are transformed to more toxic products may be low if the body residue of the parent is used as the CBR.

For the majority of the insecticides, however, toxicity is reduced with biotransformation. This includes the carbamates, a few organophosphates, phenylpyrazoles, and pyrethroids that are biotransformed into less toxic degradates (Lydy et al. 2004). Carbamates undergo one of several reactions depending on insecticide and organism and these include ester hydrolysis, oxidation, and conjugation. Pyrethroids, another class of compounds that become less toxic as they are biotransformed, are typically biotransformed via ester hydrolysis. The OC insecticides are unique due to the diversity of compounds; therefore, OCs could fit in several of the aforementioned categories. Toxicity, biotransformation, and the results of biotransformation are highly variable by compound, species, and a variety of environmental factors. Compounds such as endrin, lindane, and methoxychlor have short half-lives, whereas chlordane, DDT, dicofol, and mirex bioaccumulate (Cole et al. 1970, Wolfe and Norment 1973, Brooks 1974, Kawano et al. 1988, Colombo et al. 1995, Lotufo et al. 2000, 2001a, 2001b, Stansley et al. 2001). CBR values for biotransformed pesticides, where the parent is the primary source of toxicity, may provide a more consistent dose metric since differential biotransformation of the pesticide would be accounted for in the body residue measurement, but not in the measurement of the external media.

The aforementioned classes of compounds are often acutely toxic. Hence, the rate of biotransformation can directly impact an insecticide's potential for eliciting toxicity. For example, the potential for toxicity is reduced if an acutely toxic compound is rapidly biotransformed into a less toxic byproduct. This may explain the differences in relative susceptibilities among species, since some organisms have greater biotransformation capabilities. There appears to be a trade-off between insecticides that are relatively acutely toxic to aquatic species, that is, the carbamates, OPs, and pyrethroids that are rapidly biotransformed and those that are relatively less toxic, with slower biotransformation and subsequent bioaccumulation such as some OCs. These properties seem to be present not only in how organisms as individuals process these insecticides, but also in how an ecosystem processes them. Those compounds not readily biotransformed by organisms also tend to be environmentally persistent.

The majority of current-use insecticides have been designed to degrade more rapidly than their bioaccumulating predecessors to prevent long-term contamination. These properties subsequently make the compound easier for organisms to process, particularly birds and mammals, thus decreasing the toxicity to these species. Since many invertebrates, especially insects, lack certain metabolic pathways they are more susceptible to toxicity. This design utilizes the species differences to be insect specific. Fipronil for example, is much more selective to insect receptors decreasing its potential to harm fish and mammal communities directly. However, it is still acutely toxic to nontarget species. Since most commonly used insecticides are biotransformed by most species, estimating the bioaccumulation is difficult. High rates of biotransformation also reduce the ability to accurately determine residues in organisms as the concentration of parent compound may be rapidly changing. Furthermore, previously established predictors of toxicity, such as hydrophobicity are no longer applicable because as the compounds constantly change form the concentration of the toxic form within the animal is altered. This dynamic internal concentration makes it difficult to establish relationships that can be developed in slowly biotransformed compounds or in animals that do not readily biotransform these insecticides. The difficulty in establishing these predictors makes the hazard assessments for these compounds more complex.

Traditional QSAR models for predicting toxicity and bioaccumulation potential of compounds have used the octanol–water partition coefficient (K_{ow}) and this predictor worked well for hydrophobic pesticides that are not biotransformed (Lydy et al. 1990, 2004). However, as stated earlier, many pesticides are readily biotransformed to more or less toxic metabolites and bioaccumulation is influenced by the formation of more polar or more easily excreted metabolites. Therefore, other predictors other than simple water/lipid partitioning are needed. Fisher et al. (1993) examined five different molecular descriptors, including molecular volume, Henry's law constant, K_{ow}, molecular connectivity and linear solvation models, as potential predictors of pesticide toxicity. They found that when sediment was present in the system, sorptive interactions between the pesticides and sediment dominated the processes affecting toxicity. Conversely, in the absence of sediment, the molecular structure of the pesticides was more important than simply partitioning for determining toxicity and the molecular connectivity and linear solvation modeling successfully described the toxicity with up to 96% of the variability in toxicity being described by these models (Fisher et al. 1993).

11.7 ENVIRONMENTAL AND PHYSIOLOGICAL FACTORS THAT INFLUENCE BIOACCUMULATION AND TOXICITY

Bioaccumulation studies tend to focus broadly on individual species or higher organizational levels. As a result, BCF and CBR values are typically provided for a contaminant and a species with limited discussion of how associated factors such as age of the organism and water chemistry may influence pesticide uptake and depuration rates. However, many environmental and physiological factors may influence the extent that an individual or population bioaccumulates a pesticide. Three broad areas that should be considered when measuring bioaccumulation include: (1) the physical/chemical properties of the environment, (2) the physiological status of the organism, and (3) the presence of other chemical stressors.

Several physical and chemical properties can greatly influence bioaccumulation. For example, if a chemical is ionizable and the pK_a is near environmentally relevant pHs, the pH of the water can cause considerable changes in uptake of the pesticide. For example, PCP (pK_a = 4.74) has BCF values of 1066 at pH 6.5 and 281 at pH 8.5. The relationship between pH and log BCF is linear throughout the pH range tested (6.5–8.5) indicating that the decrease in bioaccumulation is caused by a lower uptake rate of the ionized form of the molecule (Spehar et al. 1985). In this case, a CBR approach will improve assessment of toxicity, as changes in uptake would correspond to changes in body residue. Changes in temperature can also change the toxicokinetics of pesticides. In general, increased temperature increases uptake rate (Lydy et al. 1999, Harwood et al. 2009), which is likely the mechanism behind the increase in toxicity reported for most chemicals (Mayer and Ellersieck 1988) and will by itself increase the bioaccumulation of the compound. However, increased temperature also increases the rate of pesticide biotransformation and depuration. For example, despite an increased uptake rate, the OP insecticides chlorpyrifos and methyl parathion have lower bioaccumulation rates for overall accumulation (parent and metabolites) and parent pesticide at higher temperatures (Lydy et al. 1999). In this case, the CBR values also change because as temperature increases, accumulation drops and toxicity increases. Since OP insecticides are pro-insecticides and their oxon metabolite is more toxic, an increase in biotransformation rate could result in a greater occurrence of receptor site binding that is not reversible within a biological time scale resulting in increased toxicity. However, for the majority of highly biotransformed insecticides, where the biotransformation product is less toxic, this increase in biotransformation subsequently decreases toxicity. As shown in Table 11.2, the CBR values likely differ greatly for chlorpyrifos across temperatures due to changes in toxicokinetics, since it is a pro-insecticide. In contrast, the CBRs for the pyrethroids and OC insecticides are expected to be more stable.

The chemical and physical properties of the external matrix containing the pesticide (e.g., water, sediment) may also influence the toxicity. For example, Maul et al. (2008) found that bifenthrin toxicity to *H. azteca* was greater in systems with either leaf material or sediment alone compared to mixed media systems. In addition, partitioning of bifenthrin was greatest within sediment and lowest in coarse leaf material. Therefore, greater partitioning onto the sediment and preferential use of leaf substrates may drive *H. azteca* survival. Utility of a CBR approach would likely improve assessment of toxicity in cases, such as this, where the bioavailability and uptake routes are not well understood and thus the concentration in the external media is not as good of a dose metric.

The physiological status of the organism can also influence the rate of bioaccumulation. Although a large number of factors change the "physiology" of an organism to a degree that bioaccumulation would likely change, very few of the factors have ever been well tested. Examples of factors include: age, sex, developmental characteristics (presence of gills or an oil gland, active metamorphosis), and reproductive status. Several of these factors were studied in relationship to DDT accumulation in the polychaete *Neanthes arenaceodentata*. Sex of the worms proved to be one of the biggest factors as elimination rates were nearly five times faster in males and accumulation was 2.5 times lower (lipid based; Lotufo et al. 2000).

Several studies have evaluated the effect of secondary pesticides on the bioaccumulation of the pesticide of interest with mixed results. For example, Tsuda et al. (1995) determined BCF values for OP insecticides fenthion, fenitrothion, and diazinon singly and in combination. In this case, minimal differences were noted between BCF values determined after exposure to a single pesticide as compared to the BCF of the same pesticide when jointly exposed to the other pesticides as well. However, other studies have demonstrated that pesticides can change the toxicokinetics of other pesticides. The classic example is the synergist piperonyl butoxide, which inhibits mixed function oxygenases that biotransform many pesticides. As a result, the biological half-life of many pesticides, especially the pyrethroids insecticides, is increased resulting in greater accumulation and toxicity (Casida et al. 1970). Other than the known synergists, other changes in toxicokinetics have been demonstrated. For example, several OP insecticides inhibit esterase activity. Through toxicity testing, they have been shown to increase the toxicity, and likely the bioaccumulation factor, of pyrethroid insecticides (Denton et al. 2003, Belden and Lydy 2006). In another example, the herbicide atrazine has been shown to change the toxicokinetics of several OP insecticides including chlorpyrifos. Although bioaccumulation of parent compound does not increase due to the joint exposure to atrazine, bioaccumulation of chlorpyrifos metabolites does increase, as does the observed toxicity (Belden and Lydy 2000). Similarly to the scenario discussed for OP insecticides and temperature, the increase rate of biotransformation leads to an increase in formation of the active metabolite (the oxon form). Atrazine has been shown to increase the specific activity of microsomal enzymes involved in this transformation (Belden and Lydy 2000) and the enzyme responsible and an increase in expression has been noted (Miota et al. 2000).

Similarly as discussed in the section of biotransformation, using body residues as a dose metric will usually result in less variability for most pesticides that have variable toxicity due to changing environmental and physiological conditions. Many of these conditions change the toxicokinetics of the pesticide, which is at least partially considered in CBR measurements. For example, in Table 11.2, the estimated CBRs for both permethrin and lambda-cyhalothrin differed by less than 40% between the two temperatures evaluated; however, the LC_{50} values were nearly 300% different. In contrast, for the pro-insecticide chlorpyrifos, the difference between the estimated CBR values was greater than the difference between the LC_{50} values.

Summary

In this chapter we have attempted to summarize the toxicological significance of pesticide accumulation to aquatic animals. Toxicological significance of bioaccumulation includes sublethal and lethal effects and exposure to higher trophic levels. In general, greater bioaccumulation occurs for pesticides that are not readily biotransformed and toxicity and bioaccumulation potential are predictable from simple lipid-water partitioning (e.g., K_{ow}, Fisher et al. 1993). For these compounds, CBR has been proven as a good dose metric and is better than using external media concentrations. However, many pesticides are biotransformed by aquatic animals and biotransformation can complicate the predictive process. This is especially true when the transformation products either bioaccumulate or contribute to the toxicity. The question then arises whether parent, metabolite or both should be used as the true dose metric. This assessment is further complicated by the fact that many of the biotransformation products are difficult to measure analytically due to their presence at very low concentrations or because of a lack of appropriate analytical methods. In the near future, greater availability of mass spectrometry and applications such as biomimetic devices will increase our ability to measure CBRs for pesticides. In addition, other environmental and physiological factors like the physical/chemical properties of the environment; the physiological status of the animal, and the presence of other chemical stressors can complicate the estimation. Future research should target techniques that will help us address these challenges and provide a more accurate framework for predicting toxicity of pesticides.

REFERENCES

Allison, D., B. J. Kallman, O. B. Cope, and C. Van Valin. 1963. Insecticides: effects on cutthroat trout of repeated exposure to DDT. *Science (Wash DC)*. 142:958–961.

Allison, D., B. J. Kallman, O. B. Cope, and C. Van Valin. 1964. Some chronic effects of DDT on cutthroat trout. Research Report 64. U.S. Department of the Interior, U.S. Fish and Wildlife Service, Bureau of Sport Fisheries and Wildlife, Washington, DC.

Amweg, E. L., D. P. Weston, J. You, and M. J. Lydy. 2006. Pyrethroid insecticides and sediment toxicity in urban creeks from California and Tennessee. *Environ. Sci. Technol.* 40:1700–1706.

Ankley, G. T., and S. A. Collyard. 1995. Influence of piperonyl butoxide on the toxicity of organophosphate insecticides to three species of freshwater benthic invertebrates. *Comp. Biochem. Physiol.* 110C:149–155.

Bailey, H. C., C. DiGiorgio, K. Kroll, J. L. Miller, D. E. Hinton, and G. Starrett. 1996. Development of procedures for identifying pesticide toxicity in ambient waters: carbofuran, diazinon, chlorpyrifos. *Environ. Toxicol. Chem.* 15:837–845.

Barron, M. G., J. A. Hansen, and J. Lipton. 2002. Association between contaminant tissue residues and effects in aquatic organisms. *Rev. Environ. Contam. Toxicol.* 173:1–37.

Beckvar, N., T. M. Dillon, and L. B. Read. 2005. Approaches for linking whole-body fish tissue residues of mercury or DDT to biological effects thresholds. *Environ. Toxicol. Chem.* 24:2094–2105.

Belden, J. B., and M. J. Lydy. 2000. Impact of atrazine on organophosphate insecticide toxicity. *Environ. Toxicol. Chem.* 19:2266–2274.

Belden, J. B., and M. J. Lydy. 2006. Joint toxicity of chlorpyrifos and esfenvalerate to fathead minnows and midge larvae. *Environ. Toxicol. Chem.* 25:623–629.

Berlin, W. H., R. J. Hasselberg, and M. J. Mac. 1981. Growth and mortality of fry of Lake Michigan lake trout during chronic exposure to PCBs and DDE. Technical Paper 105. U.S. Fish and Wildlife Service, Ann Arbor, MI, pp. 11–22.

Boutet, I., A. Tanguy, and D. Moraga. 2004. Molecular identification and expression of two non-P450 enzymes, monoamine oxidase A and flavin-containing monooxygenase 2, involved in phase I of xenobiotic biotransformation in the Pacific oyster, *Crassostrea gigas*. *BBA—Gene Struct. Expr.* 1679:29–36.

Bradbury, S. P., and J. R. Coats. 1989. Toxicokinetics and toxicodynamics of pyrethroid insecticides in fish. *Environ. Toxicol. Chem.* 8:373–380.

Brooks, G. T. 1974. *Chlorinated insecticides*. Vol. 1. Biological and environmental aspects. Cleveland, OH: CRC Press.

Buchwalter, D. B., J. F. Sandahl, J. J. Jenkins, and L. R. Curtis. 2004. Roles of uptake, biotransformation, and target site sensitivity in determining the differential toxicity of chlorpyrifos to second to fourth instar *Chironomus riparius* (Meigen). *Aqua. Toxicol.* 66:149–157.

Buhler, D. R., M. E. Rasmusson, and W. E. Shanks. 1969. Chronic oral DDT toxicity in juvenile Coho and Chinook salmon. *Toxicol. Appl. Pharmacol.* 14:535–555.

Butler, P. A. 1969. The significance of DDT residues in estuarine fauna. In *Chemical fallout: Current research on persistent pesticides*, eds. M. W. Miller and G. G. Berg, pp. 205–220. Springfield, IL: C.C. Thomas.

Carr, R. L., L. L. Ho, and J. E. Chambers. 1997. Selective toxicity of chlorpyrifos to several species of fish during an environmental exposure: biochemical mechanisms. *Environ. Toxicol. Chem.* 16:2369–2374.

Casida, J. E. 1970. Mixed function oxidase involvement in the biochemistry of insecticide synergists. *J. Agric. Food Chem.* 18:753–772.

Cole, J. F., L. M. Klevay, and M. R. Zavon. 1970. Endrin and dieldrin: a comparison of hepatic excretion in the rat. *Toxicol. Appl. Pharmacol.* 16:574.

Colombo, J. C., C. Bilos, M. Campanaro, M. J. Rodriguez Presa, and J. A. Catoggio. 1995. Bioaccumulation of polychlorinated biphenyls and chlorinated pesticides by the Asiatic clam *Corbicula fluminea*: its use as sentinel organism in the Rio de La Plata Estuary, Argentina. *Environ. Sci. Technol.* 29:914–927.

Davy, F. B., H. Kleerekoper, and P. Gensler. 1972. Effects of exposure to sublethal DDT on the locomotor behavior of the goldfish (*Carassius auratus*). *J. Fish Res. Board Can.* 29:1333–1336.

De Bruijn, J., and J. Hermens. 1991. Qualitative and quantitative modeling of toxic effects of organophosphorus compounds to fish. *Sci. Total Environ.* 109–110:441–455.

De Bruijn, J., E. Yedema., W. J. Seinen, and J. Hermens.1991. Lethal body burdens of four organophosphorous pesticides in the guppy (*Poecilia reticulata*). *Aqua. Toxicol.* 20:111–122.

Deneer, J. W., B. J. Budde, and A. Weijers. 1999. Variations in the lethal body burdens of organophosphorus compounds in the guppy. *Chemosphere* 38:1671–1683.

Denton, D. L., C. E. Wheelock, S. A. Murray, L. A. Deanovic, B. D. Hammock, and D. E. Hinton. 2003. Joint acute toxicity of esfenvalerate and diazinon to larval fathead minnows (*Pimephales promelas*). *Environ. Toxicol. Chem.* 22:336–341.

Escartin, E., and C. Porte. 1996. Bioaccumulation, metabolism and biochemical effects of the organophosphorus pesticide fenitrothion in *Procambarus clarkii*. *Environ. Toxicol. Chem.* 15:915–920.

Fisher, S. W., M. J. Lydy, J. Barger, and P. F. Landrum. 1993. Quantitative structure-activity relationships for predicting the toxicity of pesticides in aquatic systems with sediment. *Environ. Toxicol. Chem.* 12:1307–1318.

Fisk, A. T., R. J. Norstrom, C. D. Cymbalisty, and D. C. G. Muir. 1998. Dietary accumulation and depuration of hydrophobic organochlorines: bioaccumulation parameters and their relationship with the octanol/water partition coefficient. *Environ. Toxicol. Chem.* 12:951–961.

Gilliom, R. J. 2007. Pesticides in U.S. streams and groundwater. *Environ. Sci. Technol.* 41:3408–3414.

Guerrero, N. R. V., et al. 2002. Evidence of differences in the biotransformation of organic contaminants in three species of freshwater invertebrates. *Environ. Pollut.* 117:523–530.

Harwood, A. D., J. You, and M. J. Lydy. 2009. Temperature as a toxicity identification evaluation for pyrethroid insecticides: toxicokinetic confirmation. *Environ. Toxicol. Chem.* 28(5):1051–1058.

Hintzen, E. P., M. J. Lydy, and J. B. Belden. 2009. Occurrence and potential toxicity of pyrethroids and other insecticides in bed sediments of urban streams in central Texas. *Environ. Pollut.* 157:110–116.

Jarvin, A. W., M. J. Hoffman, and T. W. Thorslund. 1976. Toxicity of DDT food and water exposure to fathead minnows. EPA-600/3-76/114. U.S. Environmental Protection Agency, Office of Research and Development, Duluth, M.N.

Jarvin, A. W., M. J. Hoffman, and T. W. Thorslund. 1977. Long-term toxic effects of DDT food and water exposure on fathead minnows. *J. Fish Res. Board Can.* 34:2089–2103.

Kawano, M., T. Inoue, T. Wada, H. Hidaka, and R. Tatsukawa.1988. Bioconcentration and residue patterns of chlordane compounds in marine animals: invertebrates, fish, mammals, and seabirds. *Environ. Sci. Technol.* 22:792–797.

Kiely, T., and D. Donaldson. 2004. Pesticides industry sales and usage 2000 and 2001 market estimates, biological and economic analysis. 2004. Division Office of Pesticide Programs Office of Prevention, Pesticides, and Toxic Substances U.S. Environmental Protection Agency, Washington, DC.

Konwick, B. J., A. W. Garrison, J. K. Avants, and A. T. Fisk. 2006a. Bioaccumulation and biotransformation of chiral triazole fungicides in rainbow trout (*Oncorhynchus mykiss*). *Aqua. Toxicol.* 80:372–381.

Konwick, B. J., A. W. Garrison, M. C. Black, J. K. Avants, and A. T. Fisk. 2006b. Bioaccumulation, biotransformation and metabolite formation of fipronil and chiral legacy pesticides in Rainbow Trout. *Environ. Sci. Technol.* 40:2930–2936.

Kuivila, K. M., and C. G. Foe 1995. Concentration, transport, and biological effects of dormant spray pesticides in the San Francisco Estuary, California. *Environ. Toxicol. Chem.* 14:1141–1150.

Kukkenon, J. V. K. 2002. Lethal body residue of chlorophenols and mixtures of chlorophenols in benthic organisms. *Arch. Environ. Contam. Toxicol.* 43:214–220.

Kwong, R. W. M., P. K. N. Yu, P. K. S. Lam, and W. X. Wang. 2008. Uptake, elimination and biotransformation of aqueous and dietary DDT in marine fish. *Environ. Toxicol. Chem.* 27:2053–2063.

Landrum, P. F., H. Lee II, and M. J. Lydy. 1992. Toxicokinetics in aquatic systems: model comparisons and use in hazard assessment. *Environ. Toxicol. Chem.* 11:1709–1725.

Lech, J. J., and M. J. Vodicnik. 1985. Biotransformation. In *Fundamentals of aquatic toxicology*, eds. G. M. Rand and S. R. Petrocelli, pp. 526–557. Washington: Hemisphere Publishing Co.

Livingstone, D. R. 1998. The fate of organic xenobiotics in aquatic ecosystems: quantitative and qualitative differences in biotransformation by invertebrates and fish. *Comp. Biochem. Physiol. Part A* 120:43–49.

Lotufo, G. R., P. F. Landrum, M. L. Gedeon, E. A. Tigue, and L. R. Herche. 2000. Comparative toxicity and toxicokinetics of DDT and its major metabolites in freshwater amphipods. *Environ. Toxicol. Chem.* 13:368–379.

Lotufo, G. R., J. D. Farrar, B. M. Duke, and T. S. Bridges. 2001a. DDT toxicity and critical body residue in the amphipod *Leptocherius plumulosus* in exposures to spiked sediment. *Arch. Environ. Contam. Toxicol.* 41:142–150.

Lotufo, G. R., P. F. Landrum, and M. L. Gedeon. 2001b. Toxicity and bioaccumulation of DDT in freshwater amphipods in exposures to spiked sediments. *Environ. Toxicol. Chem.* 20:810–825.

Lydy, M. J., K. A. Bruner, D. M. Fry, and S. W. Fisher. 1990. Effects of sediment and the route of exposure on the toxicity and accumulation of neutral lipophilic and moderately water-soluble metabolizable compounds in the midge, *Chironomus riparius*. In *Aquatic toxicology and risk assessment, Vol, 12. STP 1096*, eds. W. G. Landis and W. H. Van der Schalie, pp. 140–164. Philadelphia, PA: American Society for Testing and Materials.

Lydy, M. J., J. B. Belden, and M. A. Ternes. 1999. Effects of temperature on the toxicity of m-parathion, chlorpyrifos, and pentachlorobenzene to *Chironomus tentans*. *Arch. Environ. Contam. Toxicol.* 37:542–547.

Lydy, M. J., J. B. Belden, C. E. Wheelock, B. D. Hammock, and D. L. Denton. 2004. Challenges in regulating pesticide mixtures. *Ecology Society.* 9:1–15.

Macek, K. J. 1968a. Growth and resistance to stress in brook trout fed sublethal levels of DDT. *J. Fish Res. Board Can.* 25:2443–2451.

Macek, K. J. 1968b. Reproduction in brook trout fed sublethal concentrations of DDT. *J. Fish Res. Board Can.* 25:1787–1796.

Mäenpää, K., K. Sorsa, M. Lyytikäinen, M. T. Leppänen, and J. V. K. Kukkonen. 2008. Bioaccumulation, sublethal toxicity, and biotransformation of sediment-associated pentachlorophenol in *Lumbriculus variegatus* (Oligochaeta). *Ecotox. Environ. Safe.* 69:121–129.

Maul, J. D., A. T. Trimble, and M. J. Lydy. 2008. Partitioning and matrix-specific toxicity of bifenthrin among sediments and leaf-sourced organic matter. *Environ. Toxicol. Chem.* 27:945–952.

Mayer, F. L. Jr., and M. R. Ellersieck. 1988. Experiences with single-species tests for acute toxic effects in freshwater animals. *Ambio* 17:367–375.

Miota, F., B. D. Siegfried, M. F. Scharf, and M. J. Lydy. 2000. Atrazine induction of cytochrome P450 in *Chironomus tentans* larvae. *Chemosphere* 40:285–291.

Nuutinen, S., P. F. Landrum, L. J. Schuler, J. V. K. Kukkonen, and M. J. Lydy. 2003. Toxicokinetics of organic contaminants in *Hyalella azteca*. *Arch. Environ. Contam. Toxicol.* 44:467–475.

Ohayo-Mitoko, G. J. A., and J. W. Deneer. 1993. Lethal body burdens of four organophosphorus pesticides in the guppy (*Poecilia reticulata*). In *Proceedings of the 2nd European Conference on Ecotoxicology*: *Recent Advances in Ecotoxicology*, eds. W. Slooff and H. de Kruijf, pp. 559–565. *Sci. Total Environ.*

Perkins, E. J., A. El-Alfy, and D. Schlenk. 1999. *In vitro* sulfoxidation of aldicarb by hepatic microsomes of channel catfish, *Ictalurus punctatus. Toxicol. Sci.* 48:67–73.

Perkins, E. J. Jr., and D. Schlenk. 2000. *In vivo* acetycholinesterase inhibition, metabolism, and toxicokinetics of aldicarb in channel catfish: role of biotransformation in acute toxicity. *Toxicol. Sci.* 53:308–315.

Porte, C., and E. Escartin. 1998. Cytochrome P450 system in the hepatopancreas of the red swamp crayfish *Procambarus clarkii*: a field study. *Comp. Biochem. Physiol. Part C* 121:333–338.

Ruus, A., K. I. Ugland, O. Espeland, and J. U. Skaare. 1999. Organochlorine contaminants in a local marine food chain from Jarfjord, Northern Norway. *Marine Environ. Res.* 48:131–146.

Schlenk, D. 1995. Use of aquatic organisms as models to determine the *in vivo* contribution of flavin-containing monooxygenases in xenobiotic biotransformation. *Mol. Marine Bio. Biotechnol.* 4:323–330.

Sparling, D. W., G. M. Fellers, and L. L. McConnells. 2001. Pesticides and amphibian population declines in California, USA. *Environ. Toxicol. Chem.* 20:1591–1595.

Spehar, R. L., H. P. Nelson, M. J. Swanson, and J. W. Renoos. 1985. Pentachlorophenol toxicity to amphipods and fathead minnows at different test pH values. *Environ. Toxicol. Chem.* 4:389–397.

Stansley, W., D. E. Roscoe, E. Hawthorne, and R. Myer. 2001. Food chain aspects of chlordane poisoning in birds and bats. *Arch. Environ. Contam. Toxicol.* 40:285–291.

Tsuda, T., S. Aoki, T. Inoue, and M. Kojima. 1995 Accumulation and excretion of diazinon, fenthion and fenitrothion by killifish: comparison of individual and mixed pesticides. *Water Res.* 29:55–458.

United States Environmental Protection Agency (U.S. EPA). 2000. *Methods for measuring the toxicity and bioaccumulation of sediment-associated contaminants with freshwater invertebrates*, 2nd ed. EPA 600/R-99/064. Duluth MN.

van den Berg, M., T. Sanderson, N. Kurihara, and A. Katayama. 2003. Role of metabolism in the endocrine-disrupting effects of chemicals in aquatic and terrestrial systems. *Pure Appl. Chem.* 75:1917–1932.

Varo, I., R. Serrano, E. Pitarch, F. Amat, F. J. Lopez, and J. Navarro. 2002. Bioaccumulation of chlorpyrifos through and experimental food chain: study of protein HSP70 as biomarker of sublethal stress in fish. *Arch. Environ. Contam. Toxicol.* 42:229–235.

Wang, X., and W. X. Wang. 2005. Uptake, absorption efficiency and elimination of DDT in marine phytoplankton, copepods and fish. *Environ. Pollut.* 136:453–464.

Weston, D. P., J. You, and M. J. Lydy. 2004. Distribution and toxicity of sediment-associated pesticides in agriculture-dominated water bodies of California's Central Valley. *Environ. Sci. Technol.* 38:2752–2759.

Weston, D. P., R. W. Holmes, J. You, and M. J. Lydy. 2005. Aquatic toxicity due to residential use of pyrethroid insecticides. *Environ. Sci. Technol.* 39:9778–9784.

Wolfe, J. L., and B. R. Norment. 1973. Accumulation of mirex residues in selected organisms after an aerial treatment, Mississippi, 1971–1972. *Pest Monit.* 7:112–116.

You, J., A. A. Brennan, and M. J. Lydy. 2009. Bioavailability and biotransformation of sediment-associated pyrethroid insecticides in *Lumbriculus variegatus. Chemosphere* 75:1477–1482.

Zaroogian, G. E., J. F. Heltshe, and M. Johnson. 1985. Estimation of bioconcentration in marine species using structure-activity models. *Environ. Toxicol. Chem.* 4:3–12.

Osprey and Young

By G. Mutzel, from *The Royal Natural History*, edited by Richard Lydekker, Frederick
Warne & Co., London, 1893–94.

12 DDT, DDD, and DDE in Birds

Lawrence J. Blus

CONTENTS

12.1 Introduction 425
12.2 Interpreting Lethal Residues 426
    12.2.1 Brain 426
    12.2.2 Liver 429
    12.2.3 Other Tissues 431
12.3 Interpreting Sublethal Residues 431
    12.3.1 Eggs 431
        12.3.1.1 Eggshell Thinning..431
        12.3.1.2 Eggshell Strength 434
        12.3.1.3 Productivity..435
12.4 Food 439
12.5 Other Tissues 440
Summary 440
References 440

12.1 INTRODUCTION

The organochlorine compound known as dichlorodiphenyltrichloroethane (DDT) was synthesized in 1874. Paul Müller discovered its insecticidal activity in 1939 and subsequently received the Nobel Prize for this discovery (Carson 1962). DDT was used extensively in human health operations during World War II. Agricultural applications started immediately after the war, and the amounts used increased exponentially after that time (Hayes 1991).

Concern about the effects on wildlife began almost immediately (Cottam and Higgins 1946). Early field studies with DDT were concerned with the short-term effects after heavy rates of application; for example, 5.6 kg of DDT per hectare (ha) resulted in immediate reductions in the populations of songbirds and invertebrates in an upland hardwood forest (Hotchkiss and Pough 1946). In contrast, 5.6 kg/ha had no effect on the eggs or nestlings of forest birds, but the DDT spray was limited to an area of only 0.09 m^2 around each nest (Mitchell 1946). DDT applied to a bottom land hardwood forest at a rate of 2.2 kg/ha had no effect on bird populations when applied for only 1 year (Stewart et al. 1946). We know now that the approximate cause-and-effect relation of DDT for mortality may occur immediately after application as well as many months or years after application.

The first important step in uncovering long-term relations was dependent on the development of precise and accurate analytical techniques that could detect DDT and its principal metabolites—DDD and DDE—in environmental samples. Heyroth (1950) summarized the early efforts to develop analytical methodology for detecting residues of DDT. Some of the early work was of limited value, because DDT and one or more metabolites were lumped together or DDE was not measured. Residue

Reprinted from the first edition of this book, Environmental Contaminants in Wildlife: Interpreting Tissue Concentrations, published by Lewis Publishers in 1996.

analysis improved with time, vastly progressing with the development of electron-capture gas chromatography, and was essentially perfected with the development of mass spectrometry. With these advances, the residues of DDT and its metabolites in wildlife could be related to lethal as well as sublethal effects, especially eggshell thinning and reduced reproductive success. Technical DDT, the insecticidal formulation applied in the field, consists of several compounds that may be changed or broken down by a number of physical or biological factors in the environment. Of these compounds, only p,p'-DDT (DDT), p,p'-DDD (DDD), and p,p'-DDE (DDE) have been related to adverse environmental effects. The residue data reported here are on a wet-weight basis unless otherwise indicated.

The purposes of this chapter are to summarize the residue levels of these three compounds in birds that are diagnostic for or are associated with mortality and important sublethal effects and to suggest improvements in the design of contemporary field studies that will result in maximum usefulness in interpreting residue data.

12.2 INTERPRETING LETHAL RESIDUES

12.2.1 BRAIN

The first experimental attempts to measure lethal levels in animals fed DDT-contaminated diets included analyses of several tissues, including the brain. With the limitations of analytical methodology, some of the studies combined DDT with DDD, or DDE was not detected. Considering DDT and DDD combined, Bernard (1963), Stickel et al. (1966), and Stickel and Stickel (1969) concluded that 30 µg/g in the brain were a useful approximation of the lower level representing serious danger and possible death. Most measurements of DDT + DDD in the brains of birds dying from DDT were above this level, but lethal levels were as low as 25 µg/g in house sparrows used in experiments (*Passer domesticus*; Bernard 1963) and 17 µg/g in wild American robins (*Turdus migratorius*) dying with tremors (Hunt 1968). Stickel et al. (1966) concluded that the relative importance of DDT and DDD was not apparent from these data.

With improvements in analytical methodology, residues of DDT, DDD, and DDE were determined in the brains of experimental and wild animals killed by DDT, and more definitive evaluations of the contribution of the individual compounds to lethality were established. Weighting was necessary, because residues in the brains at death ranged from nearly all DDD to nearly all DDT (Stickel et al. 1970). It was also necessary to evaluate residues in the brains of apparently normal animals exposed to DDT and euthanized at periods when others were dying from accumulated dosage and to evaluate the effects of exposure routes, time to death, age, sex, and various stresses on lethal levels in the brain. It was concluded by Stickel et al. (1970) that there is little or no postmortem breakdown of DDT to DDD. Measurements of the lethal levels of DDE and DDD in animals exposed to these individual compounds also helped in evaluating the relative contributions of each toxicant. By considering the levels of DDT and its major metabolites in the brains of animals on DDT dosage that either died or were euthanized, an excellent relation of residues to lethality was established (Table 12.1). This separation was possible because residues in the brain increase rapidly shortly before death, and concentrations uniformly meet or exceed the lower lethal limit in relation to exposure routes, time to death, and the other variables mentioned above (Stickel et al. 1970). There was variation in the means of DDT and the major metabolites in the brain at death; for example, DDT ranged from 15 µg/g in American robins (Wurster et al. 1965) to 40 µg/g in brown-headed cowbirds (*Molothrus ater;* Stickel and Stickel 1969), and DDD ranged from 2 µg/g in northern bobwhite quail (*Colinus virginianus*; Hill et al. 1971) to 99 µg/g in brown-headed cowbirds (Stickel and Stickel 1969). Because of these variations, Stickel et al. (1970) developed the concept of a DDT equivalent, wherein 1 µg/g of DDT equals 5 µg/g of DDD or 15 µg/g of DDE. Using this weighting system, Stickel et al. (1970) indicated that 10 DDT equivalents in the brain constitute an approximate lower lethal limit. American robins that died in tremors had as little as 10 DDT equivalents in their brains. In Table 12.1, all the mean DDT equivalents in the brains of animals that died were

TABLE 12.1

Residues of DDT, DDD, and DDE in Brains of Birds that Died or Were Euthanized While on Experimental DDT Dietary Dosage and in Brains of Wild American Robins that Died in Tremors

Species	Sex/Fate[a]	DDT		DDD		DDE		Mean DDT Equivalent	Reference
		μg/g ww							
		Mean	Range	Mean	Range	Mean	Range		
Brown-headed cowbird	M/D	39	27–90	59	29–99	7	5–12	51	Stickel and Stickel 1969
	M/E	7	3–19	9	4–17	1	<1–2	9	Stickel and Stickel 1969
	F/D	40	27–77	50	27–71	8	6–10	51	Stickel and Stickel 1969
	F/E	9	6–21	9	6–17	1	<1–3	11	Stickel and Stickel 1969
House sparrow	B/D	28	18–38	16	8–29	9	5–18	35	Hill et al. 1971
Northern bobwhite									
Wild	B/D	23	17–29	8	6–14	11	9–13	25	Hill et al. 1971
Game farm	B/D	25	19–32	3	2–4	9	8–11	26	Hill et al. 1971
Northern cardinal (*Cardinalis cardinalis*)	B/D	19	17–24	8	6–10	3	2–3	21	Hill et al. 1971
Blue jay (*Cyanocitta cristata*)	B/D	16	12–20	7	6–9	3	2–4	18	Hill et al. 1971
American robin	B/D	15	NL	39	NL	57	NL	27	Wurster et al. 1965
Clapper rail	M/D	25	19–31	18	13–27	4	3–7	29	Van Velzen and Kreitzer 1975
(*Rallus*	F/D	26	20–31	19	15–23	4	3–5	30	Van Velzen and Kreitzer 1975
longirostris)	M/E	6	2–10	8	4–13	1	<1–2	8	Van Velzen and Kreitzer 1975

[a] B, both sexes; D, died; E, euthanized, appeared normal.
NL, not listed.

≥18, and the mean DDT equivalents in the brains of brown-headed cowbirds that were euthanized on DDT dosage ranged from 8 to 11. Stickel et al. (1970) indicated that the DDT equivalent system was approximate; a 50% margin of error was estimated when all series of data were included. The DDT equivalent weighting system remains a valuable interpretive tool, although the equivalent for DDE probably should be raised to 20 or 25 to reflect the lethal level of DDE alone.

With two notable exceptions, DDE residues in the brains of animals dying from DDT ranged from <1 to 28 µg/g (Table 12.1). Wurster et al. (1965) reported 57 µg/g of DDE in the brains of wild American robins that died from DDT sprayed for Dutch elm disease; DDE exceeded DDT and DDD in these birds. Another more striking exception was the high mean levels of DDE in the brains of cockerels (*Gallus gallus*) that died after being fed DDT; DDE equaled or exceeded the levels of DDT + DDD in these birds with a high of 227 µg/g in a series of birds fed 250 µg of DDT per day (Ecobichon and Saschenbrecker 1968). In comparison with the results of other studies, these high levels of DDE seem anomalous, possibly because of problems in analytical methodology or species differences in the metabolism of DDT.

In animals given diets containing DDE during experiments, residues of only DDE were detected in their brains (Table 12.2). The mean lethal residue of DDE was 499 µg/g in four species of passerine birds, with the lowest individual level of 250 µg/g in a brown-headed cowbird. Stickel et al. (1984) concluded that, for all species tested, residues of DDE in the brain were clearly diagnostic; there was a strong likelihood for death with residues ≥300 µg/g. DDE residues ranged from 52 to 400 µg/g in the brains of birds that were euthanized while receiving dietary levels of DDE that were lethal to other birds. The DDE residue level in the brain of only one euthanized bird overlapped the levels in birds that died (Stickel et al. 1970, 1984).

Regarding DDT equivalents in the brains of birds receiving DDE dosage, the means ranged from 35 to 39 for those that died and from 9 to 10 for those that were euthanized (Table 12.2). Few possible cases of lethal levels of DDE in the brains of wild birds in the United States exist; these include a bald eagle (*Haliaeetus leucocephalus*) with 385 µg/g (Belisle et al. 1972), a great blue heron (*Ardea herodias*) with 246 µg/g (Call et al. 1976), and a black-crowned night heron (*Nycticorax nycticorax*) with 230 µg/g (Ohlendorf et al. 1981). Also, two experimental American kestrels (*Falco sparverius*) that died after a long period on a low dietary dosage of DDE (2.8 ppm) had 213 and 301 µg/g of DDE in their brains (Porter and Wiemeyer 1972). Three other American kestrels died several days

TABLE 12.2

Residues of DDE and DDD in Brains of Birds that Died or Were Euthanized after Experimental Dietary Exposure to DDE or DDD

Species	Sex/Fate[a]	µg/g ww		Mean DDT Equivalent	Reference
		Mean	Range		
DDE					
Brown-headed cowbird	M/D	499	250–660	39	Stickel et al. 1970
	M/E	152	67–400	10	Stickel et al. 1970
Passerines[b] (4 species)	B/D	499	305–694	39	Stickel et al. 1984
	B/E	137	52–219	9	Stickel et al. 1984
DDD					
Brown-headed cowbird	M/D	172	86–358	34	Stickel et al. 1970
	M/E	42	19–105	8	Stickel et al. 1970

[a] B, both sexes; D, died; E, euthanized, appeared normal.

[b] Combined data for brown-headed cowbird, common grackle (*Quiscalus quiscula*), red-winged blackbird (*Agelaius phoeniceus*), and European starling (*Sturnus vulgaris*).

after receiving diets containing 160–250 ppm of DDE; their brains contained from 230 to 280 µg/g of DDE (Henny and Meeker 1981). Although the lower lethal limit of 300 µg/g seems to provide a reliable criterion, there is some evidence that lower levels occasionally prove lethal.

Concerning lethal residues of DDD in the brain, birds dying on DDT dosage had mean levels that varied from 3 to 59 µg/g, with individual levels as high as 151 µg/g (Table 12.1). The brains of birds on experimental dietary dosages of DDD contained an average of 172 µg/g of DDD at death, with an individual lower level of 86 µg/g (Table 12.2). Birds on DDD dosage that were euthanized had 19–105 µg/g in their brains. Stickel et al. (1970) concluded that brain concentrations ≥65 µg/g indicate an increasing likelihood that death was due to poisoning from DDD. DDT equivalents were 34 in those dying on DDD dosage and 8 in those euthanized on that same dosage (Table 12.2). The only confirmed instance of DDD poisoning in a wild animal was that of a common loon (*Gavia immer*). Its brain contained 200 µg/g of DDD, 130 µg/g of DDE, and 2 µg/g of DDT; DDT equivalents totaled 41 (Prouty et al. 1975). The most vivid example of the effects of DDD occurred when Clear Lake in California was treated with DDD for several years. Although DDD almost certainly adversely affected western grebe (*Aechmophorus occidentalis*) survival and reproductive success, the only two brains analyzed (both females found moribund and euthanized) had DDD residues of 46 and 48 µg/g (Rudd and Herman 1972); these levels were less than the lowest individual residues of birds dying on DDD dosage but were slightly greater than the mean level found in birds euthanized (Table 12.2).

One problem with interpreting residues of the DDT group is that other contaminants including organochlorines are frequently present in the eggs or tissues of wild birds. Regarding organochlorines in the brain, Sileo et al. (1977) assumed straightforward additivity of the toxic effects and developed an "organochlorine index" based on the addition of the proportions of lower lethal levels for all compounds; for example, one half of a lower lethal level contributes 0.5 to the index on the basis of 1 indicating lethality. This index has received little use, one reason being that the lower lethal level of 150 µg/g of DDE ascribed by Sileo et al. (1977) is nearly 100 µg/g less than the accepted lower lethal limit (Table 12.2). Another reason is that the lethal limits of individual organochlorines in wild birds are usually distinct from one another. Finally, the assumed additivity of organochlorines related to lethality in birds has received little verification from experimental studies, although additivity seems the most common joint action (Smyth et al. 1969).

12.2.2 LIVER

Residues in the livers of animals dying during experiments while receiving dietary dosages of DDT differ from residues in their brains in that DDD constitutes the bulk of the residues (Table 12.3). Mean levels of DDT ranged from 1 to 35 µg/g, with a range in individual values from <1 to 254 µg/g. Residues of DDE were relatively low in livers except in American robins, where the residues exceeded those of DDD (Wurster et al. 1965); this is the same series that had exceptionally high residues of DDE in their brains. Cockerels dying while receiving a DDT dietary dosage also had levels of DDE in their livers that exceeded the levels of DDD and DDT combined; the same relation held for residues in their brains as previously mentioned (Ecobichon and Saschenbrecker 1968).

Mean residues of DDE in the livers of brown-headed cowbirds receiving dietary DDE (Table 12.4) were 3883 µg/g (range, 460–11725 µg/g) in those that died and 523 µg/g (range, 266–1560 µg/g) in those that were euthanized (Stickel et al. 1970). In Great Britain, Newton et al. (1992) reported that 23 Eurasian kestrels (*Falco tinnunculus*) and 10 Eurasian sparrowhawks (*Accipiter nisus*) died with lethal levels of DDE in their livers, but the lower lethal limit of 100 µg/g was based on correlative field evidence related to DDT and metabolites (Cooke et al. 1982). According to experimentally derived lethal levels (Stickel et al. 1970), DDE residues in Eurasian sparrowhawk livers (140–254 µg/g) were too low to ascribe lethality to DDE, but DDE residues in the livers of at least three Eurasian kestrels (812, 1474, and 1500 µg/g) were within the lethal range. In the United States, the

TABLE 12.3

Residues of DDT, DDD, and DDE in Livers of Birds that Died or Were Euthanized after Experimental Dietary Exposure to DDT and in Wild American Robins that Died in Tremors

Species	Sex/Fate[a]	μg/g ww						Reference
		DDT		DDD		DDE		
		Mean	Range	Mean	Range	Mean	Range	
Brown-headed cowbird	M/D	34	3–254	768	215–1,640	55	25–104	Stickel and Stickel 1969
	M/E	5	1–20	58	30–115	3	2–6	Stickel and Stickel 1969
	F/D	35	9–161	552	292–1,063	53	32–88	Stickel and Stickel 1969
	F/E	8	4–16	72	61–107	4	2–8	Stickel and Stickel 1969
American robin	B/D	1	NL	139	NL	165	NL	Wurster et al. 1965
Clapper rail	M/D	3	<1–5	308	75–938	24	7–47	Van Velzen and Kreitzer 1975
	F/D	4	<1–10	229	130–337	19	13–27	Van Velzen and Kreitzer 1975
	M/E	3	<1–7	157	38–352	11	2–36	Van Velzen and Kreitzer 1975

[a] B, both sexes; D, died; E, euthanized, appeared normal.

NL, not listed.

TABLE 12.4

Residues of DDE and DDD in Livers of Male Brown-Headed Cowbirds that Died or Were Euthanized after Exposure to DDE or DDD in Their Diets (after Stickel et al. 1970)

Fate[a]	μg/g ww	
	Mean	Range
DDE		
D	3,883	460–11,725
E	523	266–1,560
DDD		
D	1,219	79–5,300
E	521	104–2,854

[a] D, died; E, euthanized, appeared normal.

only liver analysis from a wild bird that apparently died from DDE was that of a great blue heron that had DDE residues of 246 μg/g in the brain and 570 μg/g in the liver (Call et al. 1976).

The mean residue levels of DDD in the livers of birds receiving dietary dosages of DDD were 1219 μg/g (range, 79–5300 μg/g) in those that died and 521 μg/g (range, 104–2854 μg/g) in those that were euthanized (Stickel et al. 1970).

To interpret the lethal levels of DDT and its metabolites in the liver, researchers should devise a weighting system, such as that developed for the brain, to determine the relative contribution of each of the compounds. Also, Stickel et al. (1970) indicated that, for birds killed by DDT, residue levels of

DDT and its metabolites in the livers of wild birds were lower than those of laboratory birds. Cooke et al. (1982) indicated that starvation of birds dying from organochlorine pesticides complicated the interpretation of lethal levels in the liver. Bernard (1963) concluded that residues in the brain were more consistent than those in the liver with regard to the interpretation of lethal residues.

12.2.3 OTHER TISSUES

A number of tissues including blood plasma, carcass remainder, kidney, heart, breast muscle, intestinal tract, skin, and fat have been analyzed in birds that were dying or euthanized while receiving a dosage of DDT (Ecobichon and Saschenbrecker 1968, Stickel et al. 1970). Although residues in these tissues have not received the same scrutiny as those in the brain during the assessment of diagnostic lethal levels, Stickel et al. (1970) concluded that residue levels of DDD + DDT in the carcasses of birds dying from DDT increased with the time on dietary dosage and that residue levels in those that were euthanized were essentially indistinguishable from those that died on dosage. In contrast, DDE residues, expressed on a lipid basis, in the carcasses of brown-headed cowbirds sacrificed while receiving DDE dietary dosage differed markedly from those that died on that dosage (Stickel et al. 1984). The same authors concluded that residues in carcass lipids accurately predicted lethal brain residues.

12.3 INTERPRETING SUBLETHAL RESIDUES

12.3.1 EGGS

12.3.1.1 Eggshell Thinning

The classic paper by Ratcliffe (1967a) described eggshell thinning in eggs of peregrine falcons (*Falco peregrinus*) and Eurasian sparrowhawks in Great Britain that occurred following the introduction of DDT.

Soon thereafter in the United States, eggshell thinning was documented in several species of raptorial and fish-eating birds, and the inverse relation between DDE residues in eggs and shell thickness was first established (Hickey and Anderson 1968). Also, Hickey and Anderson (1968) were the first to document decreases in the mean eggshell thickness over a period of years in relation to population declines. Heath et al. (1969) first documented eggshell thinning and associated lowered reproductive success of experimental birds on DDE diets. Subsequently, there have been a substantial number of experimental and field studies that document eggshell thinning and a smaller number that relate residues in eggs to thinning (Tables 12.5 and 12.6).

Birds in experiments on dietary dosages of DDE laid eggs that had considerably thinner shells than did birds on "clean" diets without the compound (Table 12.5). Barn owls (*Tyto alba*) exhibited

TABLE 12.5
Relation of DDE Residues in Eggs to Eggshell Thinning in Birds on Experimental Dietary Dosages of DDE

| Species | Thinning (%) | | Residues (µg/g ww) | | Reference |
	Mean	Range	Mean	Range	
Black duck	18	12–29	46	34–63	Longcore et al. 1971
	24	12–32	144	96–219	Longcore et al. 1971
American kestrel	10	1–18	32	17–44	Wiemeyer and Porter 1970
Barn owl	20	NL	12	NL	Mendenhall et al. 1983
	28	NL	41	NL	Mendenhall et al. 1983

NL, not listed.

TABLE 12.6
DDE Residues in Eggs Associated with Eggshell Thinning in Wild Birds

Species	Area[a]	Mean		Reference
		µg/g ww	% Thinning	
Brown pelican	CA	59[b]	44	Risebrough 1972
	BC	66[b]	46	Jehl 1973
	BC	25[b,c]	47	Jehl 1973
	BC	8[b]	26	Jehl 1973
	BC	3[b]	18	Jehl 1973
	SC	5	17	Blus et al. 1974, 1979
	SC	3	16	Blus et al. 1974, 1979
	SC	1	10	Blus et al. 1974, 1979
	FL	1	5	Blus et al. 1974, 1979
	FL	2	11	Blus et al. 1974, 1979
	TX	3	11	King et al. 1977
American white pelican	CA	2	15	Boellstorff et al. 1985
(*Pelecanus erythrorhynchos*)	CA	2	10	Boellstorff et al. 1985
Western grebe	CA	1	1	Boellstorff et al. 1985
Great blue heron	WA	4	10	Fitzner et al. 1988
	WA	5	13	Fitzner et al. 1988
Peregrine falcon	AK	2	3[d]	White 1973
	AK	7	8[d]	White 1973
	AK	4	7[d]	White 1973
	AK	44	22[d]	Cade et al. 1971
	AK	34	17[d]	Cade et al. 1971
	AK	8	8[d]	Cade et al. 1971
	AU	18	20[d]	Pruett-Jones et al. 1988
Northern gannet (*Sula bassanus*)	QU	19	17	Elliott et al. 1988
Double-crested cormorant	CA[b,c]	32	11	Gress et al. 1973
	BC[b,c]	24	30	Gress et al. 1973
	ON	24	15	Weseloh et al. 1983
Snowy egret	NV	2	12	Henny et al. 1985
	NV	1	3	Henny et al. 1985
White-faced ibis	NV	2	12	Henny et al. 1985
	NV	1	8	Henny et al. 1985
	TX	1	3	King et al. 1980
	TX	3	14	King et al. 1980
White-tailed eagle (*Haliaeetus albicilla*)	FI	30	15	Koivusaari et al. 1980
Black-crowned night heron	CO, WY	4	9	McEwen et al. 1984
	QU	2	<1[d]	Tremblay and Ellison 1980
	MA	4	4	Custer et al. 1983
	MA	2	0	Custer et al. 1983
	RI	4	6	Custer et al. 1983
	RI	1	0	Custer et al. 1983
Gray heron	GB	6	12[d]	Cooke et al. 1976
(*Ardea cinerea*)	GB	3	19[d]	Cooke et al. 1976
Eurasian sparrowhawk	GB	7	18[d]	Newton and Bogan 1974
Black skimmer	TX	12	12	White et al. 1984
(*Rynchops niger*)	TX	3	0	White et al. 1984
Osprey	CT	9	15	Wiemeyer et al. 1975

TABLE 12.6 (continued)
DDE Residues in Eggs Associated with Eggshell Thinning in Wild Birds

Species	Area[a]	Mean μg/g ww	% Thinning	Reference
	MD	2	12	Wiemeyer et al. 1975
Bald eagle	OR, WA	10	10	Anthony et al. 1993
Golden eagle	GB	0.1	7[d]	Newton and Galbraith 1991
(*Aquila chrysaetos*)	GB	0.1	1[d]	Newton and Galbraith 1991
	GB	0.2	3[d]	Newton and Galbraith 1991
	GB	0.3	4[d]	Newton and Galbraith 1991
	GB	0.3	5[d]	Newton and Galbraith 1991

[a] CA, California; BC, Baja California; SC, South Carolina; FL, Florida; TX, Texas; WA, Washington; AK, Alaska; AU, Australia; QU, Quebec; ON, Ontario; NV, Nevada; FI, Finland; CO, Colorado; WY, Wyoming; MA, Massachusetts; RI, Rhode Islands; GB, Great Britain; CT, Connecticut; MD, Maryland; OR, Oregon.

[b] Approximate value converted from lipid basis.

[c] Authors suspected residues were too low because of analytical errors.

[d] Percentage of thinning based on thickness index (Ratcliffe 1967a); all others based on eggshell thickness.

[e] Intact eggs used only for eggshell thickness and residue analysis; the mean eggshell thinning of both crushed and intact eggs was 29% in CA and 38% in BC.

20% eggshell thinning when eggs contained 12 μg/g of DDE (Mendenhall et al. 1983), and black ducks (*Anas rubripes*) exhibited 18% thinning when eggs contained 46 μg/g (Longcore et al. 1971). Although there were several studies of eggshell thinning of birds that were given diets containing technical DDT, most did not list residues in the eggs, and DDE—not DDT—comprises most of the dietary exposure of wild birds with significant eggshell thinning (Stickel 1973). Eggs and tissues of ring-necked pheasants (*Phasianus colchicus*) accumulated high levels of DDT, about 5 times greater than DDE, in areas of intense application of technical DDT (Hunt and Keith 1963). Domestic chickens given diets containing 300 μg/g of technical DDT showed no effects on eggshell thickness compared with controls, even though the eggs of dosed birds contained mean levels of 10 μg/g of DDE and 87 μg/g of DDT (approximate conversion from egg yolk basis at 14 days of incubation; Waibel et al. 1972). Results of the various studies of DDE relations to eggshell thickness in wild birds indicated extreme species differences in sensitivity. Brown pelicans (*Pelecanus occidentalis*) in California and Baja California (Risebrough 1972, Jehl 1973) displayed extreme eggshell thinning and high residues of DDE in 1969 and the early 1970s, with nearly all eggs breaking in the most heavily contaminated colonies. In South Carolina, Florida, and Texas, much lower residues still resulted in mean eggshell thinning of 5–17% (Blus et al. 1974, 1979, King et al. 1977). While there was a statistically significant relation between DDE and eggshell thickness or the thickness index, there were some marked intraspecific differences in response. For example, when considering means (Table 12.6), the peregrine falcon in Alaska showed 22% thinning at 44 μg/g of DDE (Cade et al. 1971) compared with 20% thinning at 18 μg/g of DDE in Australia (Pruett-Jones et al. 1980). Intact eggs of the double-crested cormorant (*Phalacrocorax auritus*) in California exhibited 11% shell thinning at 32 μg/g compared with 24% thinning at 30 μg/g in Baja California (Gress et al. 1973). Many of the cormorant eggs were crushed in both colonies; overall thinning in collections containing both crushed and intact eggs reached 29% in California and 38% in Baja California.

Eggshell thinning is based on either eggshell thickness or the thickness index. In comparisons of the thickness index with eggshell thickness using museum specimens, the index indicated ≥% thinning 76% of the time (Anderson and Hickey 1972) with extreme differences of 10% for each measurement. Thus, it is obvious that either of these measurements represents an accurate indication of eggshell thinning, but thickness is probably the measurement of choice in most instances, particularly when shells are cut because of loss of fragments.

The rate of thinning per microgram per gram of DDE is much greater at lower residues. Using the brown pelican as an example, there is 5–10% thinning at 1 µg/g of DDE compared with 44% at 59 µg/g (Table 12.6). While there is evidence that certain other contaminants and physiological conditions may induce eggshell thinning, the burden of proof overwhelmingly indicates that DDE is the major cause of the eggshell thinning syndrome. There have been attempts to relate DDE residues in the egg to a level of eggshell thinning that is associated with population decline if such thinning persists over a period of years. Initially, Hickey and Anderson (1968) concluded that ≥18% thinning was associated with declining populations; Anderson and Hickey (1972) modified this to "above 15–20% for a period of years." Thus, some have taken 15% as an effect level; however, with few exceptions, 18% is probably a more accurate indicator.

One notable exception was that of a declining Eurasian sparrowhawk population in the Netherlands that had poor production, 18% eggshell thinning, and mean DDE residues of 25 µg/g. Even though there was no significant relation between DDE and eggshell thickness, Koeman et al. (1972) suggested that DDE was responsible for the reproductive problems of Eurasian sparrowhawks. Also, Wiemeyer et al. (1972) found a poor correlation between eggshell thinning and DDE concentrations in bald eagle eggs, but there was a significant relation established when additional data were accumulated (Wiemeyer et al. 1988). One of the problems in these relations is that a greater effect per microgram per gram of DDE occurs at lower levels and, if residues are clumped, particularly on the high side, statistical relations are more difficult to establish. Ideally, a wide spread in residues is optimal for detecting effects on eggshell thickness.

When regression analysis was used to relate DDE levels to 20% eggshell thinning (Table 12.7), the critical estimates have ranged from 5 µg/g for the California condor (*Gymnogyps californianus*; Kiff et al. 1979) to 60 µg/g (fresh eggs) to 110 µg/g (failed eggs) for the bald eagle (Wiemeyer et al. 1993). Estimates in Table 12.7 are of value, but they must be interpreted with some caution. For example, the regression equation listed by Cade et al. (1971) indicated that 20% eggshell thinning was associated with 22 µg/g of DDE, whereas their tabular data indicated 17% thinning at 34 µg/g and 22% thinning at 44 µg/g (Tables 12.6 and 12.7). Blus (1984) reported that most of the error was related to extending the regression line beyond the range of the data. Thus, the critical level of 19 µg/g for the great blue heron is much too low, and that of 54 µg/g for the black-crowned night heron is much too high. There are wide disparities in the estimates for the common loon, osprey (*Pandion haliaetus*), and the peregrine falcon. While the California condor is listed as the most sensitive to DDE-induced thinning, the estimated critical level is based on the measurement of eggshell fragments and extraction of DDE from eggshell membranes and then the calculation of residues in the entire egg from these measurements. From work on the extraction of DDE from membranes of intact peregrine falcon eggshells in museums, lower residues are associated with a far greater degree of thinning than are those of intact eggs collected from the field (Peakall and Kiff 1979). Therefore, the accuracy of this technique requires experimental verification. Although Fox (1979) indicated that the measurement of eggshell thickness was a reliable indicator of the DDE content of the egg in some populations, Blus (1984) concluded that the DDE-thickness relation is not tight enough to do this for individual eggs and that residue analysis is essential for interpretation.

The calculated no-effect level for DDE in eggs related to the effects on eggshell thickness ranged from 0.1 µg/g for the brown pelican (Blus 1984) to 2 µg/g for the peregrine falcon (Cade et al. 1971). An earlier estimate for the brown pelican was 0.5 µg/g (Blus et al. 1974), but the sample size and range in residues were much smaller than in the subsequent study.

12.3.1.2 Eggshell Strength

The strength of eggshells, as determined by various mechanical devices, is related to eggshell thickness and, therefore, to DDE residues in the egg. Shell strength decreased more than eggshell thickness per unit of DDE; for example, 8–16 µg/g in sample eggs of the white-faced ibis were associated

TABLE 12.7
Estimated Residues of DDE in Eggs Associated with Eggshell Thinning of 20% in Wild Birds

Species	DDE (µg/g ww)	Reference
Common loon	14	Price 1977
	47	Fox et al. 1980
California condor	5[a]	Kiff et al. 1979
Peregrine falcon	15–20[b]	Peakall et al. 1975
	22[b]	Cade et al. 1971
	18[b]	Pruett-Jones et al. 1980
	20	Enderson and Wrege 1973
Brown pelican	8	Blus 1984
Black-crowned night heron	54	Blus 1984, Henny et al. 1984
Prairie falcon	7	Enderson and Wrege 1973
Great blue heron	19	Blus 1984
Osprey	9	Wiemeyer et al. 1988
	29[c]	Wiemeyer et al. 1988
	41[c]	Spitzer et al. 1978
Eurasian sparrowhawk	10[c]	Newton et al. 1986
Merlin	16[c]	Newton et al. 1982
White-faced ibis	7	Henny and Herron 1989
Bald eagle	60–110	Wiemeyer et al. 1993

[a] Based on thickness of eggshell fragments and DDE content of shell membranes.

[b] Percentage of thinning based on thickness index (Ratcliffe 1967a); all others based on eggshell thickness.

[c] Eggs collected after nest failure; all other studies except that of Kiff et al. (1979) included at least some eggs collected while nests were active.

with a decrease of 16% in thickness and 37% in strength. In addition, productivity of the young was related to DDE, shell thickness, and shell strength. Although shell strength may provide a more sensitive indicator of potential egg failure due to DDE, simple thickness measurements have served very well in that regard, and there are fewer logistical and financial constraints than are required to measure strength.

12.3.1.3 Productivity

DDT, primarily through its major metabolite DDE, also affects the reproductive success of birds. Eggshell thinning is an important, but not exclusive, factor related to reproductive problems (Blus 1984). Unfortunately, most experimental studies of the reproductive effects of DDE or DDT did not present residues. This was a loss for interpreting the effects of residues on reproductive success in field studies.

There are several methods of expressing reproductive success relative to residues in birds. One method relates the overall reproductive success of a colony or other breeding group, such as a pen of experimental birds, to the mean residue content in their eggs (Table 12.8). Black ducks (*A. rubripes*) receiving dietary dosages of 10 or 30 ppm of DDE had a significantly reduced survival of embryonated eggs or hatchlings to 3 weeks posthatch in relation to controls; DDE averaged 46 and 144 µg/g in eggs of treated birds (Longcore et al. 1971). Barn owls on a diet containing 3 ppm of DDE had hatchling and fledging rates that were reduced about 75% from control values over 2 years when eggs contained an average of 12 µg/g the first year and 41 µg/g the second year (Mendenhall et al. 1983). The response was approximately the same each year, even though residues in eggs of the barn owls were much higher the second year. Although these studies document the effects from DDE,

TABLE 12.8
Residues of DDE in Eggs Related to the Reproductive Success of Birds on Experimental Dietary Dosages of DDE

Species	Reference	Treatment	Residues (µg/g ww)		Reproductive Success (% of Survival to 3 Weeks Posthatch)	
			Mean	Range	Hatchlings	Embryonated Eggs
Black duck	Longcore et al. 1971	Control	0.28	0.14–0.67	91	38
		10 ppm	46	34–63	64[a]	23[a]
		30 ppm	144	96–219	50[a]	9[a]

					Mean per Pair	
					Eggs Hatched	Young Fledged
Barn owl	Mendenhall et al. 1983	Control (1st year)	0.25	NL	3.2	2.9
		Control (2nd year)	0.40	NL	3.7	3.1
		3 ppm (1st year)	12	NL	1.1[a]	0.7[a]
		3 ppm (2nd year)	41	NL	0.9[a]	0.7[a]

[a] Significantly different ($p \leq .05$) from controls.
NL, not listed.

TABLE 12.9
Mean Residues of DDE in Eggs Related to Mean Reproductive Success of Wild Birds

Species	Area[a]	Residues (µg/g ww)	Young Produced per Active Nest	Reference
Double-crested cormorant	CA	32[b]	0.0	Gress et al. 1973
	BC	24[b]	<0.1	Gress et al. 1973
	ON	14–16	0–0.1[c]	Weseloh et al. 1983
	ON	5	0.3	Weseloh et al. 1983
Black-crowned night heron	QU	2	2.4	Tremblay and Ellison 1980
Common loon	SA	6	0.7	Fox et al. 1980
Brown pelican	CA	59[d]	<0.1	Risebrough 1972
	BC	66[d]	0.0	Jehl 1973
	BC	25[d,e]	<0.1	Jehl 1973
	BC	8[d]	0.1	Jehl 1973
	BC	3[d]	−0.8	Jehl 1973
Bald eagle	OR, WA	10	0.6	Anthony et al. 1993

[a] CA, California; BC, Baja California; ON, Ontario; QU, Quebec; SA, Saskatchewan; OR, Oregon; WA, Washington.
[b] Approximate conversion from dry weight basis.
[c] Five colonies.
[d] Approximate conversion from lipid weight basis.
[e] Authors suspected residues were too low because of analytical errors.

a narrow part of the relation is presented; for example, there is no indication of the dietary levels or residues in eggs at which problems first appear or where they initially become serious.

Some of the field data on reproductive success also follow the average residue-average effect design (Table 12.9). A method that gives more insight into the effects of residues on reproductive success is the sample egg technique (Table 12.10), whereby one egg is taken from a nest and analyzed, the nest is marked, its fate is monitored through periodic visits, and the residues in eggs are related to nest success (Blus 1984). This is particularly valuable in the field, where many factors may influence reproductive outcome. Blus (1984) lists advantages and disadvantages of this method. Where nest predation is a problem and where clutch size permits, one egg may be collected for residue analysis and another taken and placed in an incubator.

There are several variations to the sample egg technique as shown in Table 12.10. One involves work with threatened or endangered species or other special situations where the sample egg is not collected until the fate of the marked nest is determined (Spitzer et al. 1978). The major bias with collecting eggs after the fact is that those with thin eggshells and high DDE residues have a greater chance of being crushed or cracked and therefore lost from the population. Other variations relate to statistical analysis of the data, regardless of the time of egg collections. One method ranks young fledged versus residues (Fyfe et al. 1976, Spitzer et al. 1978), and another method ranks residues or a range of residues versus young fledged (Blus et al. 1980, Blus 1982, Henny et al. 1984, 1985, Henny and Herron 1989, Ambrose et al. 1988, Wiemeyer et al. 1993). Of these two methods, I recommend the second, because

TABLE 12.10
Residues of DDE in Sample Eggs of Wild Birds Related to Reproductive Success

Species	Area[a]	Residues (µg/g ww)	Young Produced per Active Nest[b]	Reference
Osprey	CT, NY	23[c,d,e]	0.0	Spitzer et al. 1978
	CT, NY	12[c,d,e]	1.0	Spitzer et al. 1978
	CT, NY	6[c,d,e]	2.1	Spitzer et al. 1978
	ID	14	0.0	Johnson et al. 1975
	ID	6	1.6	Johnson et al. 1975
Peregrine falcon	AK	≤15	1.8	Ambrose et al. 1988
	AK	15–30	2.0	Ambrose et al. 1988
	AK	>30	1.0	Ambrose et al. 1988
Snowy egret	NV	≤1	2.2	Henny et al. 1985
	NV	1–5	2.4	Henny et al. 1985
	NV	5–10	1.0	Henny et al. 1985
	NV	10–20	1.0	Henny et al. 1985
Prairie falcon	AB	2[c]	0.0	Fyfe et al. 1976
	AB	2[c]	1.0	Fyfe et al. 1976
	AB	2[c]	2.0	Fyfe et al. 1976
	AB	2[c]	3.0	Fyfe et al. 1976
	AB	1[c]	4.0	Fyfe et al. 1976
Merlin	AB	11[c]	0.0	Fyfe et al. 1976
	AB	11[c]	1.0	Fyfe et al. 1976
	AB	6[c]	2.0	Fyfe et al. 1976
	AB	5[c]	3.0	Fyfe et al. 1976
	AB	6[c]	4.0	Fyfe et al. 1976
Brown pelican	SC[f]	≤1.5	0.6 (FL),[g] 0.8 (EM)	Blus 1982
	SC[f]	1.5–3	0.6 (FL), 0.8 (EM)	Blus 1982
	SC[f]	>3	0.0 (FL), 0.6 (EM)	Blus 1982
Bald eagle	U.S.[e]	<2.2	1.0	Wiemeyer et al. 1993
	U.S.[e]	2.2–3.5	1.0	Wiemeyer et al. 1993
	U.S.[e]	3.6–6.2	0.5	Wiemeyer et al. 1993
	U.S.[e]	6.3–11.9	0.3	Wiemeyer et al. 1993
	U.S.[e]	≥12	0.2	Wiemeyer et al. 1993

continued

TABLE 12.10 (continued)
Residues of DDE in Sample Eggs of Wild Birds Related to Reproductive Success

Species	Area[a]	Residues (µg/g ww)	Young Produced per Active Nest[b]	Reference
Black-crowned	U.S.	≤1	2.0	Henny et al. 1984
night heron	U.S.	1–4	1.7	Henny et al. 1984
	U.S.	4–8	1.5	Henny et al. 1984
	U.S.	8–12	1.1	Henny et al. 1984
	U.S.	12–16	1.0	Henny et al. 1984
	U.S.	16–25	0.8	Henny et al. 1984
	U.S.	25–50	0.4	Henny et al. 1984
White-faced Ibis	NV	≤1	1.8	Henny and Herron 1989
	NV	1–4	1.8	Henny and Herron 1989
	NV	4–8	1.3	Henny and Herron 1989
	NV	8–16	0.8	Henny and Herron 1989
	NV	>16	0.6	Henny and Herron 1989
Great blue heron	OR, WA	3	1.7–2.0	Blus et al. 1980

[a] CT, Connecticut; NY, New York; ID, Idaho; AK, Alaska; NV, Nevada; AB, Alberta and nearby areas; SC, South Carolina;
   U.S., various locations within the U.S.
[b] Young produced not adjusted for sample egg collected.
[c] Approximate adjustment from dry weight basis.
[d] Elevated levels (17–29 µg/g) of polychlorinated biphenyls also present.
[e] All or most eggs were collected after the fate of marked nests was determined. Production of young at each nest is based
   on a 5-year mean.
[f] Sample egg either freshly laid or embryonated when collected.
[g] FL, freshly laid; EM, embryonated.

it seems to more closely approximate the dependent variable-independent variable relation; however, it should be recognized that these methods have not been subjected to rigid statistical testing.

Problems with comparing the young fledged per nest with the residue content of sample eggs seemed evident in a study of prairie falcons (*Falco mexicanus*) in Alberta, Canada, and surrounding areas, where the mean DDE levels of 1–2 µg/g were said to adversely affect fledging success (Table 12.10; Fyfe et al. 1976). However, differences in the mean residue content were not statistically different, and the egg with the highest level of 11 µg/g was from a nest that fledged five young (Fyfe et al. 1976). In merlins (*Falco columbarius*), fledging success was significantly related to DDE residues in sample eggs with an effect level of near 10 µg/g but, again, high levels of about 31 and 26 µg/g were found in eggs from nests that fledged one and five young, respectively (Fyfe et al. 1976). On the basis of addled or deserted eggs collected from merlin nests in Great Britain, Newton et al. (1982) indicated a positive correlation between fledging success and DDE residues, with zero young fledged when eggs contained 5 µg/g increasing to four young fledged at 8 µg/g. Although the lower critical level of DDE that adversely affects the reproductive success of peregrine falcons was considered to be 15–20 µg/g (Peakall 1976), nest success in Great Britain seemed unaffected by DDE, with the highest residues of 25 and 31 µg/g being detected in sample eggs from successful nests (Ratcliffe 1967b). More recent, albeit limited, evidence from peregrine falcons in Alaska indicated that the effects on nest success occur only when residues exceed 30 µg/g (Ambrose et al. 1988).

Considering DDE residues versus the young produced per nest, declines in the productivity of brown pelicans (Blus 1982), bald eagles (Wiemeyer et al. 1993), black-crowned night herons (Henny et al. 1984), snowy egrets (*Egretta thula*; Henny et al. 1985), and the white-faced ibis (*Plegadis chihi*; Henny and Herron 1989) are obvious (Table 12.10). One can determine the level at which residues first begin having an adverse effect on the number of young produced, for example, 4–8 μg/g in the white-faced ibis (Henny and Herron 1989), and at which few or no young are produced, for example, ≥15 μg/g in the bald eagle. Black-crowned night herons demonstrate an impressive gradual decline in productivity with an increase in residues; however, a few young are produced even at levels >25 μg/g (Henny et al. 1984). In the brown pelican, Blus (1982) indicated a dramatic effect above 3 μg/g, with no young produced in sample eggs that were freshly laid when collected and a 25% reduction in those that were embryonated when collected (Table 12.10). The brown pelican is apparently the most sensitive avian species to DDE, with reproductive failure when residues in eggs exceed 3.7 μg/g (Blus 1982). In South Carolina, a combination of effects from DDE, including eggshell thinning, seemed to adversely affect reproductive success, whereas in California and Baja California nearly every egg collapsed from extreme eggshell thinning in 1969 and several subsequent years (Risebrough 1972, Jehl 1973). To interpret what a DDE-related reduction in the number of young fledged means in terms of population reduction, one has to know a great deal about the population, for example, an approximate recruitment standard (number of young that must be fledged per pair of breeding age to maintain a stable population) and adult mortality compensating mechanisms, including renesting and other factors. Recruitment standards are 1.2–1.5 young for the brown pelican (Henny 1972), 1.0 young for the bald eagle (Wiemeyer et al. 1993), 1–1.3 young (Henny and Wight 1969) and 0.8 young (Spitzer et al. 1983) for the osprey, 1.9 young for the great blue heron (Henny 1972), and 2–2.1 young for the black-crowned night heron (Henny 1972). Variations in the recruitment standard for the osprey probably result from whether active nests (Spitzer et al. 1983) or pairs of breeding age (Henny and Wight 1969) are used in the calculations.

In Table 12.10, no compensation in the young produced was made for collection of the sample egg. Therefore, most of these productivity data are probably biased low, compared with nests without an egg collected. To measure this bias, Henny and Herron (1989) compared production in white-faced ibis nests with an egg collected to that without an egg collected; sample egg collection was associated with a 30% reduction in the young produced per active nest. The percentage of the reduction, of course, would be influenced by clutch size in the species of interest.

Although DDE was responsible for most reproductive failure in birds, very high levels of DDT in ring-necked pheasant eggs in California may have caused reproductive problems, such as crippling and mortality of young; however, the link between them was never clearly established (Hunt and Keith 1963).

Domestic chickens given diets containing 300 μg/g of technical DDT showed no significant effects on reproduction compared with controls, even though their eggs contained mean levels of 10 μg/g of DDE and 87 μg/g of DDT (approximate conversion from egg yolk basis at 14 days of incubation; Waibel et al. 1972).

12.4 FOOD

The lowest dietary concentration of DDE that resulted in critical eggshell thinning and decreased production in the peregrine falcon was estimated at 1 μg/g (Enderson et al. 1982). A more recent study used 3 μg/g as a critical level, but this was based on dietary levels given to experimental raptors that experienced serious reproductive problems and even adult mortality (DeWeese et al. 1986). For the brown pelican, the lower critical dietary level of DDE was estimated at about 0.1 μg/g on the basis of 31 × biomagnification from fish to pelican egg; however, because the chief prey fish also contained DDT at one-half the amount of DDE, the 0.1-μg/g level probably should be raised slightly to account for metabolism from DDT to DDE (Blus et al. 1977). These examples included a highly sensitive species and a moderately sensitive species, so higher estimates for less sensitive species are expected based on experimental studies of the domestic chicken, one of the least sensitive species (Waibel et al. 1972); but lower estimates are unlikely. Because of the lipophilicity and bioaccumulativeness of all three compounds, the highest residues and, depending on species sensitivity, the most extreme effects are found in species at the highest trophic levels, as is evident in most of the studies summarized in this review.

12.5 OTHER TISSUES

There are other measurements of the sublethal effects related to residues of DDT and its metabolites in other tissues of animals used in experiments, but most of these data are fragmentary, and few of these measurements have proven useful in field studies.

Summary

Although technical DDT was initially hailed as a tremendous tool in pest control, the environmental problems soon outweighed the positive aspects. As a result, this compound was banned over much of the world; however, use of DDT continues in some countries, especially for control of insects that are disease vectors.

One of the first findings related to the use of technical DDT was the mortality of wildlife after heavy applications, but suitable analytical techniques were required to detect residues because many adverse effects occurred sometime after exposure. Residues in tissues, particularly the brain, have proven to be diagnostic of lethality in animals on dietary dosages of DDT, DDD, and DDE in experiments. When used in field investigations, this technique made possible the interpretation of lethality when DDT equivalents (weighting system where an equivalent equals 1 µg/g of DDT, 5 µg/g of DDD, or 15 µg/g of DDE) are as low as 10 in brains; however, most birds or mammals that die from DDT have DDT equivalents >20. Few dead wild birds have been found with lethal levels of DDE or DDD in their brains. Residues in livers also have been used to establish the lethality of DDT in wild animals, but a system for the weighting of the three compounds has not been developed.

Residues in the eggs of birds are a reliable indicator of eggshell thinning and reproductive success. Of the three compounds reviewed in this chapter, evidence overwhelmingly indicates that DDE is responsible for most eggshell thinning, reproductive problems, and population reductions. There is a tremendous variation in species sensitivity to these compounds. The brown pelican is the most sensitive, with eggshell thinning and depressed productivity occurring at 3.0 µg/g of DDE in the egg and total reproductive failure when residues exceed 3.7 µg/g. In contrast, adverse effects on the reproductive success of peregrine falcons first occur when DDE residues in the egg are about 10-fold higher, that is, 30 µg/g. Black-crowned night herons demonstrate a different pattern involving a gradual decline in productivity with increasing residues. A few young are still produced at levels >25 µg/g. The domestic chicken is very tolerant of high dietary exposure to technical DDT. By efficient use of the sample egg technique, the effects induced by DDE residues within one colony or breeding area, or compared with a reference colony or area where residues are low, can be quantified and related to the adverse effects on the individual and the population. Techniques for quantifying the relation between residues of DDT and its metabolites and the effects on the biota have been successful, but the process required much time, effort, and financial outlay. Many contemporary field studies are designed inefficiently with regard to quantifying residues, with little or no consideration given to establishing the effects, or less commonly, establishing the effects without evidence from residues. In addition, few experimental studies are directly applicable to the field. Results of experimental and field studies could be made more pertinent to interpretation of field data by changes in the experimental design, so that efficient use can be made of the establishing and measuring of effects induced by residues.

REFERENCES

Ambrose, R. E., C. J. Henny, R. E. Hunter, and J. A. Crawford. 1988. Organochlorines in Alaskan peregrine falcon eggs and their current impact on productivity. In *Peregrine falcon populations: Their management and recovery,* eds. T. J. Cade, J. H. Enderson, C. G. Thelander, and C. M. White, pp. 385–393. Boise: The Peregrine Fund, Inc.

Anderson, D. W., and J. J. Hickey. 1972. Eggshell changes in certain North American birds. *Proc. Int. Ornithol. Congr.* 15:514–540.

Anthony, R. G., M. G. Garrett, and C. A. Schuler. 1993. Environmental contaminants in bald eagles in the Columbia River estuary. *J. Wildl. Manage.* 57:10–19.

Belisle, A. A., et al. 1972. Residues of organochlorine pesticides, polychlorinated biphenyls, and mercury, and autopsy data for bald eagles, 1969 and 1970. *Pestic. Monit. J.* 6:133–138.

Bernard, R. F. 1963. Studies of the effects of DDT on birds. *Mich. State Univ. Mus. Publ. Biol. Serv.* 2:155–192.

Blus, L. J. 1982. Further interpretation of the relation of organochlorine residues in brown pelican eggs to reproductive success. *Environ. Pollut.* 28:15–33.

Blus, L. J. 1984. DDE in birds' eggs: comparison of two methods for estimating critical levels. *Wilson Bull.* 96:268–276.

Blus, L. J., A. A. Belisle, and R. M. Prouty. 1974. Relations of the brown pelican to certain environmental pollutants. *Pestic. Monit. J.* 7:181–194.

Blus, L. J., C. J. Henny, and T. E. Kaiser. 1980. Pollution ecology of breeding great blue herons in the Columbia Basin, Oregon and Washington. *Murrelet* 61:63–71.

Blus, L. J., T. G. Lamont, and B. S. Neely, Jr. 1979. Effects of organochlorine residues on eggshell thickness, reproduction, and population status of brown pelicans (*Pelecanus occidentalis*) in South Carolina and Florida, 1969–76. *Pestic. Monit. J.* 12:172–184.

Blus, L. J., B. S. Neely, Jr., T. G. Lamont, and B. M. Mulhern. 1977. Residues of organochlorines and heavy metals in tissues and eggs of brown pelicans, 1969–73. *Pestic. Monit. J.* 11:40–53.

Boellstorff, D. E., H. M. Ohlendorf, D. W. Anderson, E. J. O'Neill, J. O. Keith, and R. M. Prouty. 1985. Organochlorine chemical residues in white pelicans and western grebes from the Klamath Basin, California. *Arch. Environ. Contam. Toxicol.* 14:485–493.

Cade, T. J., J. L. Lincer, C. M. White, D. G. Roseneau, and L. G. Swartz. 1971. DDE residues and eggshell changes in Alaskan falcons and hawks. *Science* 172:955–957.

Call, D. J., H. J. Shave, H. C. Binger, M. E. Bergeland, B. D. Ammann, and J. J. Worman. 1976. DDE poisoning in wild great blue heron. *Bull. Environ. Contam. Toxicol.* 16:310–313.

Carson, R. 1962. *Silent spring*. Boston: Houghton Mifflin Co.

Cooke, A. S., A. A. Bell, and M. B. Haas. 1982. *Predatory birds, pesticides, and pollution*. Cambridge: Institute of Terrestrial Ecology.

Cooke, A. S., A. A. Bell, and I. Prestt. 1976. Egg shell characteristics and incidence of shell breakage for grey herons (*Ardea cinerea*) exposed to environmental pollutants. *Environ. Pollut.* 11:59–84.

Cottam, C., and E. Higgins. 1946. DDT: its effects on fish and wildlife. *U.S. Fish Wildl. Serv. Circ.* 11, 14 pp.

Custer, T. W., C. M. Bunck, and T. E. Kaiser. 1983. Organochlorine residues in Atlantic Coast black-crowned night-heron eggs, 1979. *Colonial Waterbirds* 6:160–167.

DeWeese, L. R., L. C. McEwen, G. L. Hensler, and B. E. Peterson. 1986. Organochlorine contaminants in Passeriformes and other avian prey of the peregrine falcon in the western United States. *Environ. Toxicol. Chem.* 5:675–693.

Ecobichon, D. J., and P. W. Saschenbrecker. 1968. Pharmacodynamic study of DDT in cockerels. *Can. J. Physiol. Pharmacol.* 46:785–794.

Elliott, J. E., R. J. Norstrom, and J. A. Keith. 1988. Organochlorines and eggshell thinning in northern gannets (*Sula bassanus*) from eastern Canada, 1968–1984. *Environ. Pollut.* 52:81–102.

Enderson, J. H., G. R. Craig, W. A. Burnham, and D. D. Berger. 1982. Eggshell thinning and organochlorine residues in Rocky Mountain peregrines, *Falco peregrinus*, and their prey. *Can. Field-Nat.* 96:255–264.

Enderson, J. H., and P. H. Wrege. 1973. DDE residues and eggshell thickness in prairie falcons. *J. Wildl. Manage.* 37(4):476–478.

Fitzner, R. E., L. J. Blus, C. J. Henny, and D. W. Carlile. 1988. Organochlorine residues in great blue herons from the northwestern United States. *Colonial Waterbirds* 11:293–300.

Fox, G. A. 1979. A simple method of predicting DDE contamination and reproductive success of populations of DDE-sensitive species. *J. Appl. Ecol.* 16:737–741.

Fox, G. A., K. S. Yonge, and S. G. Sealy. 1980. Breeding performance, pollutant burden, and eggshell thinning in common loons (*Gavia immer*) nesting on a boreal forest lake. *Ornis Scand.* 11:243–248.

Fyfe, R. W., R. W. Risebrough, and W. Walker II. 1976. Pollutant effects on the reproduction of the prairie falcons and merlins of the Canadian prairies. *Can. Field-Nat.* 90:346–355.

Gress, F., R. W. Risebrough, D. W. Anderson, L. F. Kiff, and J. R. Jehl, Jr. 1973. Reproductive failures of double-crested cormorants in southern California and Baja California. *Wilson Bull.* 85:197–208.

Hayes, W. J., Jr. 1991. Introduction. In *Handbook of pesticide toxicology*, eds. W. J. Hayes, Jr., and E. R. Laws, Jr., pp. 1–37. San Diego, CA: Academic Press.

Heath, R. G., J. W. Spann, and J. F. Kreitzer. 1969. Marked DDE impairment of mallard reproduction in controlled studies. *Nature (Lond.)*. 224:47–48.

Henny, C. J. 1972. An analysis of the population dynamics of selected avian species—with special reference to changes during the modern pesticide era. *U.S. Fish Wildl. Serv. Wildl. Res. Rep.* No. 1, 99 pp.

Henny, C. J., L. J. Blus, and C. S. Hulse. 1985. Trends and effects of organochlorine residues on Oregon and Nevada wading birds, 1979–1983. *Colonial Waterbirds* 8:117–128.

Henny, C. J., L. J. Blus, A. J. Krynitsky, and C. M. Bunck. 1984. Current impact of DDE on black-crowned night-herons in the intermountain West. *J. Wildl. Manage.* 48:1–13.

Henny, C. J., and G. B. Herron. 1989. DDE, selenium, mercury, and white-faced ibis reproduction at Carson Lake, Nevada. *J. Wildl. Manage.* 53:1032–1045.

Henny, C. J., and D. L. Meeker. 1981. An evaluation of blood plasma for monitoring DDE in birds of prey. *Environ. Pollut. (Ser. A).* 25:291–304.

Henny, C. J., and H. M. Wight. 1969. An endangered osprey population: estimates of mortality and production. *Auk.* 86:188–198.

Heyroth, F. F. 1950. The toxicity of DDT. Part II. A survey of the literature, pp. 72–233. Kettering Laboratory, College of Medicine, Univ. Cincinnati, Cincinnati, OH.

Hickey, J. J., and D. W. Anderson. 1968. Chlorinated hydrocarbons and eggshell changes in raptorial and fish-eating birds. *Science* 162:271–273.

Hill, E. F., W. E. Dale, and J. W. Miles. 1971. DDT intoxication in birds: subchronic effects and brain residues. *Toxicol. Appl. Pharmacol.* 20:502–514.

Hotchkiss, N., and R. H. Pough. 1946. Effect on forest birds of DDT used for gypsy moth control in Pennsylvania. *J. Wildl. Manage.* 10:202–207.

Hunt, E. G., and J. O. Keith. 1963. Pesticide-wildlife investigations in California, 1962. Proceeding of 2nd Annual Conference on the Use of Agricultural Chemicals in California, Davis, CA, 29 pp.

Hunt, L. B. 1968. Songbirds and insecticides in a suburban elm environment. PhD Thesis, University of Wisconsin, Madison, WI.

Jehl, J. R., Jr. 1973. Studies of a declining population of brown pelicans in Northwestern Baja California. *Condor* 75:69–79.

Johnson, D. R., W. E. Melquist, and G. J. Schroeder. 1975. DDT and PCB levels in Lake Coeur d'Alene, Idaho, osprey eggs. *Bull. Environ. Contam. Toxicol.* 13:401–405.

Kiff, L. F., D. B. Peakall, and S. R. Wilbur. 1979. Recent changes in California condor eggshells. *Condor* 81:166–172.

King, K. A., E. L. Flickinger, and H. H. Hildebrand. 1977. The decline of brown pelicans on the Louisiana and Texas Gulf Coast. *Southwest. Nat.* 21:417–431.

King, K. A., D. L. Meeker and D. M. Swineford. 1980. White-faced ibis populations and pollutants in Texas, 1969–1976. *Southwest. Nat.* 25:225–240.

Koeman, J. H., C. F. Van Beusekom, and J. J. M. De Goeij. 1972. Eggshell and population changes in the sparrow-hawk *(Accipiter nisus)*. *TNO-Nieuws.* 27:542–550.

Koivusaari, J., I. Nuuja, R. Palokangas, and M. Finnlund. 1980. Relationships between productivity, eggshell thickness, and pollutant contents of addled eggs in the population of white-tailed eagles *Haliaetus albicilla* L. in Finland during 1969–1978. *Environ. Pollut. Ser. A Ecol. Biol.* 23:41–52.

Longcore, J. R., F. B. Samson, and T. W. Whittendale, Jr. 1971. DDE thins eggshells and lowers reproductive success of captive black ducks. *Bull. Environ. Contam. Toxicol.* 6:485–490.

McEwen, L. C., C. J. Stafford, and G. L. Hensler. 1984. Organochlorine residues in eggs of black-crowned night-herons from Colorado and Wyoming. *Environ. Toxicol. Chem.* 3:367–376.

Mendenhall, V. M., E. E. Klass, and M. A. R. McLane. 1983. Breeding success of barn owls *(Tyto alba)* fed low levels of DDE and dieldrin. *Arch. Environ. Contam. Toxicol.* 12:235–240.

Mitchell, R. T. 1946. Effects of DDT spray on eggs and nestlings of birds. *J. Wildl. Manage.* 10:192–194.

Newton, I., and J. Bogan. 1974. Organochlorine residues, eggshell thinning, and hatching success in British sparrowhawks. *Nature (Lond.).* 249:582–583.

Newton, I., J. Bogan, E. Meek, and B. Little. 1982. Organochlorine compounds and shell-thinning in British merlins *(Falco columbarius)*. *Ibis* 124:328–335.

Newton, I., J. A. Bogan, and P. Rothery. 1986. Trends and effects of organochlorine compounds in sparrowhawk eggs. *J. Appl. Ecol.* 23:461–478.

Newton, I., and E. A. Galbraith. 1991. Organochlorines and mercury in the eggs of golden eagles *Aquila chrysaetos* from Scotland. *Ibis* 133:115–120.

Newton, I., I. Wyllie, and A. Asher. 1992. Mortality from the pesticides aldrin and dieldrin in British sparrowhawks and kestrels. *Ecotoxicology* 1:31–44.

Ohlendorf, H. M., D. M. Swineford, and L. N. Locke. 1981. Organochlorine residues and mortality of herons. *Pestic. Monit. J.* 14:125–135.

Peakall, D. B. 1976. The peregrine falcon (*Falco peregrinus*) and pesticides. *Can. Field. Nat.* 90:301–307.

Peakall, D. B., T. J. Cade, C. M. White, and J. R. Haugh. 1975. Organochlorine residues in Alaskan peregrines. *Can. Field. Nat.* 104:244–254.

Peakall, D. B., and L. F. Kiff. 1979. Eggshell thinning and DDE residue levels among peregrine falcons (*Falco peregrinus*): a global perspective. *Ibis* 121:200–204.

Porter, R. D., and S. N. Wiemeyer. 1972. DDE at low dietary levels kills captive American kestrels. *Bull. Environ. Contam. Toxicol.* 8:193–199.

Price, I. M. 1977. Environmental contaminants in relation to Canadian wildlife. *Trans. N. Am. Wildl. Nat. Resour. Conf.* 42:382–396.

Prouty, R. M., J. E. Peterson, L. N. Locke, and B. M. Mulhern. 1975. DDD poisoning in a loon and the identification of the hydroxylated form of DDD. *Bull. Environ. Contam. Toxicol.* 14:385–388.

Pruett-Jones, S. G., C. M. White, and W. B. Emison. 1980. Eggshell thinning and organochlorine residues in eggs and prey of peregrine falcons from Victoria, Australia. *Emu* 80:281–287.

Ratcliffe, D. A. 1967a. Decrease in eggshell weight in certain birds of prey. *Nature (Lond.).* 215:208–210.

Ratcliffe, D. A. 1967b. The peregrine situation in Great Britain—1965–1966. *Bird Study* 14:238–246.

Risebrough, R. W. 1972. Effects of environmental pollutants upon animals other than man. In Proceedings of 6th Berkeley Symposium on Mathematical Statistics and Probability, pp. 443–463. Berkeley: University of California Press.

Rudd, R. L., and S. G. Herman. 1972. Ecosystemic transferral of pesticides in an aquatic environment. In *Environmental toxicology of pesticides,* eds. F. Matsumura, G. Boush, and T. Misato, pp. 471–485. New York, NY: Academic Press.

Sileo, L., L. Karstad, R. Frank, M. V. H. Holdrinet, E. Addison, and H. E. Braun. 1977. Organochlorine poisoning of ring-billed gulls in southern Ontario. *J. Wildl. Dis.* 13:313–322.

Smyth, H. F., Jr., C. S. Weil, J. S. West, and C. P. Carpenter. 1969. An exploration of joint toxic action: twenty-seven industrial chemicals intubated in rats in all possible pairs. *Toxicol. Appl. Pharmacol.* 14:340–347.

Spitzer, P. R., A. F. Poole, and M. Scheibel. 1983. Initial population recovery of breeding ospreys in the region between New York and Boston. In *Biology and management of bald eagles and ospreys,* ed. D. M. Bird, pp. 231–241. Quebec, Can: Harpell Press.

Spitzer, P. R., R. W. Risebrough, W. Walker II, R. Hernandez, A. Poole, D. Puleston, and I. C. T. Nisbet. 1978. Productivity of ospreys in Connecticut-Long Island increases as DDE residues decline. *Science* 202:333–335.

Stewart, R. E., J. B. Cope, C. S. Robbins, and J. W. Brainerd. 1946. Effects of DDT on birds at the Patuxent Research Refuge. *J. Wildl. Manage.* 10:195–201.

Stickel, L. F. 1973. Pesticide residues in birds and mammals. In *Environmental pollution by pesticides,* ed. C. A. Edwards, pp. 254–312. London, UK: Plenum Press.

Stickel, L. F., and W. H. Stickel. 1969. Distribution of DDT residues in tissues of birds in relation to mortality, body condition, and time. *Ind. Med. Surg.* 38:44–53.

Stickel, L. F., W. H. Stickel, and R. Christensen. 1966. Residues of DDT in brains and bodies of birds that died on dosage and in survivors. *Science* 151:1549–1551.

Stickel, W. H., L. F. Stickel, and F. B. Coon. 1970. DDE and DDD residues correlated with mortality of experimental birds. In *Pesticides symposia. Seventh Int. Am. Conf. Toxicol. Occup. Med.,* ed. W. B. Deichmann, pp. 287–294. Miami, FL: Helios and Association.

Stickel, W. H., L. F. Stickel, R. A. Dyrland, and D. L. Hughes. 1984. DDE in birds: lethal residues and loss rates. *Arch. Environ. Contam. Toxicol.* 13:1–6.

Tremblay, J., and L. N. Ellison. 1980. Breeding success of the black-crowned night heron in the St. Lawrence estuary. *Can. J. Zool.* 58:1259–1263.

Van Velzen, A., and J. F. Kreitzer. 1975. The toxicity of *p,p'*-DDT to the clapper rail. *J. Wildl. Manage.* 39:305–309.

Waibel, G. P., G. M. Speers, and P. E. Waibel. 1972. Effects of DDT and charcoal on performance of white leghorn hens. *Poult. Sci.* 51:1963–1967.

Weseloh, D. V., S. M. Teeple, and M. Gilbertson. 1983. Double-crested cormorants of the Great Lakes: egg-laying parameters, reproductive failure, and contaminant residues in eggs, Lake Huron 1972–1973. *Can. J. Zool.* 61:427–436.

White, C. M., W. B. Emison, and F. S. L. Williamson. 1973. DDE in a resident Aleutian Island peregrine population. *Condor* 75:306–311.

White, D. H., C. A. Mitchell, and D. M. Swineford. 1984. Reproductive success of black skimmers in Texas relative to environmental pollutants. *J. Field Ornithol.* 55:18–30.

Wiemeyer, S. N., C. M. Bunck, and A. J. Krynitsky. 1988. Organochlorine pesticides, polychlorinated biphenyls, and mercury in osprey eggs—1970–1979—and their relationships to shell thinning and productivity. *Arch. Environ. Contam. Toxicol.* 17:767–787.

Wiemeyer, S. N., C. M. Bunck, and C. J. Stafford. 1993. Environmental contaminants in bald eagle eggs—1980–84—and further interpretations of relationships to productivity and shell thickness. *Arch. Environ. Contam. Toxicol.* 24:213–227.

Wiemeyer, S. N., et al. 1972. Residues of organochlorine pesticides, polychlorinated biphenyls, and mercury in bald eagle eggs and changes in shell thickness, 1969 and 1970. *Pestic. Monit. J.* 6:50–55.

Wiemeyer, S. N., and R. D. Porter. 1970. DDE thins eggshells of captive American kestrels. *Nature (Lond.).* 227:737–738.

Wiemeyer, S. N., P. R. Spitzer, W. C. Krantz, T. G. Lamont, and E. Cromartie. 1975. Effects of environmental pollutants on Connecticut and Maryland ospreys. *J. Wildl. Manage.* 39:124–139.

Wurster, D. H., C. F. Wurster, Jr., and W. N. Strickland. 1965. Bird mortality following DDT spray for Dutch elm disease. *Ecology* 46:488–499.

Little Owl

By G. Mutzel, from *The Royal Natural History,* edited by Richard Lydekker, Frederick Warne & Co., London, 1893–94.

13 Cyclodiene and Other Organochlorine Pesticides in Birds

John E. Elliott
Christine A. Bishop

CONTENTS

13.1 Introduction 447
13.2 Cyclodienes...449
    13.2.1 Dieldrin...450
            13.2.1.1 Acute Effects...451
            13.2.1.2 Sublethal Effects...451
    13.2.2 Endrin 460
    13.2.3 Chlordane...460
    13.2.4 Heptachlor..463
13.3 Other OC Pesticides in Birds...463
    13.3.1 Endosulfan...464
    13.3.2 Hexachlorobenzene..464
    13.3.3 Hexachlorocyclohexane 465
    13.3.4 Methoxychlor...465
    13.3.5 Mirex 466
    13.3.6 Toxaphene..466
13.4 An Additive Model for Toxicity of OC Mixtures...466
13.5 Extrapolation of Dieldrin Equivalents to Eggs..468
Summary..468
Acknowledgments..469
References..469

13.1 INTRODUCTION

Based on chemical structure, the organochlorine (OC) insecticides can be placed into five groups: DDT and related compounds (reviewed separately by Blus, this volume), hexachlorocyclohexane (HCH), cyclodienes and related compounds, mirex (and chlordecone), and toxaphene. This chapter focuses mainly on the cyclodiene group of OC pesticides, which includes aldrin, dieldrin, and endrin (the "drins"), as well as heptachlor and chlordane. We also briefly consider several other OC pesticides such as endosulfan, hexachlorocyclohexane, and methoxychlor, which continue to be widely used, and hexachlorobenzene, mirex, and toxaphene, for which detectable residues continue to be reported regularly in wild birds. Since the reviews by Peakall and Wiemeyer in the first edition of this book, there has been a paucity of new data published on acute lethal effects of OC pesticides in birds. As a result, we have leaned heavily on those earlier efforts and added new information where available. We have, however, extended the assessments of sublethal effects, particularly for anorexic effects of dieldrin and consequences for breeding birds and migrants.

Acute toxicity of the OC pesticides varies considerably by chemical and among species for the same chemical (Table 13.1). Those data are useful for predicting relative capacity to kill wild birds in acute exposure scenarios, such as following application events or in highly contaminated environments. However, as is well known, chronic exposure to some OC pesticides can significantly alter other endpoints, not readily predicted from structure or acute toxicity. Those include the effects of the DDT metabolite, DDE, on eggshell quality and embryotoxicity (Blus 2003), or the endocrine-disrupting effects of methoxychlor (Ottinger et al. 2005). Where possible, we also examine the literature on chronic effects of other OC pesticides in birds, and try to suggest criteria.

Tissues of wild birds rarely contain quantifiable residues of only a single OC compound. Analysis and interpretation of data on exposure to multiple compounds has always posed a challenge. For some structurally and mechanistically linked chemicals, such as the dioxin-like compounds, toxic equivalency schemes have been developed (Safe 1990). The structural and functional similarity of the cyclodiene compounds does provide some basis for development of a toxicity index. We suggest a simple model of "Dieldrin Toxic Equivalents" for assessing lethal concentrations of a mix of cyclodienes and also of DDTs in brain of birds. We also suggest that for the cyclodienes, but not the DDTs, that index could be extended to interpreting residue concentrations of mixtures commonly determined in eggs.

TABLE 13.1
Acute Toxicity of Some Organochlorine Pesticides to Birds (Hudson et al. 1984)

Chemical	Test Species	N	Sex	Age (months)	Acute Oral (LD$_{50}$ mg/kg)
Endrin	Mallard (*Anas platyrhynchos*)	12	F	12	5.6
	California Quail (*Callipepla californica*)	12	F	9–10	1.2
	Pheasant (*Phasianus colchicus*)	12	M	3–4	1.8
	Rock dove (*Columba livia*)	16	M,F	—	2.0–5.0
Aldrin	Mallard	16	F	3–4	530
	Bobwhite quail (*Colinus virginianus*)	12	F	3–4	6.59
	Pheasant	12	F	3–4	16.8
Dieldrin	Mallard	12	F	6–7	381
	California Quail	12	M	7	8.8
	Pheasant	9	M	10–23	79.0
	Rock dove	15	M,F	—	26.6
DDT	Mallard	8	F	3	>2,240
	California Quail	12	M	6	595
	Pheasant	15	F	3–4	1,334
Beta-HCH	Mallard	15	F	3	>1,414
	Pheasant	12	F	3–4	118
Gamma-HCH (Lindane)	Mallard	12	M	3–4	>2,000
Chlordane	Mallard	12	F	4–5	1,200
	California Quail	12	M	12	14.1
	Pheasant	4	F	3	24.0–72.0
Endosulfan	Mallard	20	F	12	31.2
	Pheasant	12	F	3–4	190

13.2 CYCLODIENES

As a group, the cyclodienes are characterized as the most acutely toxic of the persistent OC pesticides that were registered and widely used (Table 13.1). Although very similar in structure (Figure 13.1), there is some variation in toxicity of each compound among species. That variation is likely due in part to differences in capacity to metabolize those xenobiotics (Ronis and Walker 1989).

The cyclodienes were used to control a wide range of pests in many different jurisdictions across the globe. Usage in North America probably peaked by the early to mid-1960s for aldrin/dieldrin, while chlordane was heavily used during the early 1970s as a substitute for other restricted OC pesticides (NRCC 1974). Aldrin is rapidly converted to dieldrin in environmental and biological media, and if detected, is usually at much lower concentrations than dieldrin. Endrin is another chemical isomer of dieldrin, and is about an order of magnitude more toxic than dieldrin to most vertebrates (Table 13.1). Chlordane and heptachlor are readily metabolized and multiple compounds with varying persistence and toxicity are often reported, or they are combined and reported as total chlordane.

All of the cyclodienes act through a similar central nervous system (CNS) mechanism (Hayes and Laws 1991). The primary mechanism involves inhibition of the CNS neurotransmitter, gamma-aminobutyric acid (GABA, Figure 13.2). GABA is secreted at the terminal ends of neurons in the spinal chord, cerebellum, basal ganglia, retina, and parts of the cortex. It is an important neurotransmitter for the functioning of inhibitory synapses and is thought to act by inhibiting the inflow of chloride ions, which apparently results in only partial repolarization of neurons leading to a state of continued excitation (Matsumura 1975, Matsumura and Giasudding 1983, Bloomquist and Soderlund 1985, Eldefrawi et al. 1985, Coats 1990, Guyton 1991). Cyclodienes also inhibit neuronal Na^+–K^+–ATPase and Ca^+–Mg^+–ATPase, with the latter enzyme being crucial for control of free Ca^+ at the synaptic membrane of the nerve terminus. Elevated $[Ca^+]$ at the terminus region induces release of neuro-transmitters from storage vesicles and the subsequent stimulation of adjacent neurons and increased CNS stimulation (Matsumura 1975, Wafford et al. 1989).

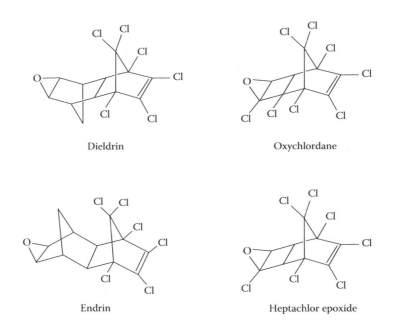

FIGURE 13.1 Molecular structures for some representative cyclodienes; formulas are provided in the text. Cyclodienes all share a characteristic endomethylene bridge structure. (From Matsumura, F., *Toxicology of Insecticides*, Plenum Press, New York, NY, 1975. With permission.)

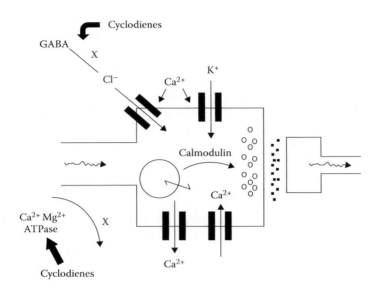

FIGURE 13.2 Proposed mechanism of action of cyclodiene insecticides such as dieldrin. Two primary processes are involved: inhibition of the neurotransmitter, GABA, thus affecting chloride ion transport; and inhibition of Na^+-K^+-ATPase and Ca^+-Mg^+-ATPase. (From Matsumura, F., *Toxicology of Insecticides*, Plenum Press, New York, NY, 1975; Ecobichon, D. J., 1991. *Casarett and Doull's Toxicology*, pp. 565–622, Pergamon, New York, 1991. With permission.)

The pharmacokinetics and dynamics of some cyclodienes, particularly dieldrin, have been examined. Dietary exposure is considered the most important route, and, therefore, has been given the most attention, although dieldrin is readily absorbed through skin (WHO 1989). Chronic dosing with low levels of dieldrin produces a steady accumulation of the compound in fat depots of mammals and birds. Robinson et al. (1967) fed rock doves (*Columbia livia*) dieldrin at 50 ppm in diet for 6 months. Birds were then fed a clean diet and sacrificed periodically. Dieldrin concentrations decreased steadily with a calculated half-life of 47 days. They also reported that ratios of dieldrin in fat, liver, muscle, and brain remained constant during the decline phase, indicating a dynamic equilibrium of the dieldrin body burden, the result of circulation of dieldrin in blood and high perfusion rates of those tissues. A longer half-life of 82 days was calculated for wild adult female herring gulls (*Larus argentatus*) (Clark et al. 1987).

Half-lives for the chlordane metabolite, oxychlordane, have been reported as 66 days in blackbirds (*Agelaius phoeniceus*) (Stickel et al. 1979c), with a much longer half-life of 215 days reported for the herring gull (Clark et al. 1987).

13.2.1 Dieldrin

Dieldrin (1,2,3,4,10,10-hexachloro-6,7-epoxy-1,4,4a,5,6,7,8,8a, octahydro-1,4,5,8-dimethanonaphthalene) has poisoned large numbers of birds, mammals, and other wildlife. The introduction in 1956 of aldrin and dieldrin in Britain as seed treatments resulted in immediate poisonings of birds, particularly wood pigeons (*Columba palumbus*) and pheasants (*Phasianus colchicus*). Those poisonings involved hundreds of incidents a year and thousands of individual birds and continued unabated until use of dieldrin as a wheat seed treatment was discontinued in 1975 (Stanley and Bunyan 1979, Stanley and Fletcher 1981, Sheail 1985).

Similarly, during the 1960s in the United States there was extensive poisoning of pheasants and other wildlife from use of dieldrin to control Japanese beetle (*Popillia japonica*) (Blus 2003). Aldrin-treated rice seed caused poisoning of numerous waterfowl and other birds during the 1960s and

1970s in Texas, and apparently caused the fulvous whistling duck population to decline (Flickinger and King 1972), an effect which persisted into the 1980s (Flickinger et al. 1986).

13.2.1.1　Acute Effects

There have been numerous experimental studies of dieldrin toxicity (Table 13.2). Two commonly cited sources for data on toxicity and lethal tissue levels of dieldrin are the experimental work by Robinson et al. (1967) and Stickel et al. (1969), both in Japanese quail (*Coturnix japonica*). There were differences in some factors such as time to death and possibly greater sensitivity of males in the studies of Stickel et al.; however, both groups concluded that lethal brain concentrations were independent of dose. Robinson et al. (1967) calculated a critical concentration of dieldrin in the brain as 10 µg/g, whereas due to greater variation in their data, Stickel et al. (1969) recommended a threshold concentration of 4–5 µg/g, the lower end of their 95% confidence interval. Linder et al. concurred as they reported a mean brain concentration of 5.8 µg/g dieldrin in pheasants that died after being fed varying doses, and concluded that 3–4 µg/g in avian brain was diagnostic of toxicity. Other peer-reviewed studies report a range of potential lethal concentrations, including up to 20 µg/g in rock dove (Robinson et al. 1967).

Those laboratory studies are important for determining dose–response relationships and diagnostic concentrations in target organs, under controlled conditions of constant light and temperature and adequate food. However, under wild conditions, an animal is subject to multiple stresses including inclement weather, food shortages, predation pressure, disease, and other contaminants. Table 13.3 reports the results of some of the many field studies of avian mortality ascribed to dieldrin. In some reported mortality incidents for the American robin (*Turdus migratorius*), the snow goose (*Chen caeruluescens*), the peregrine falcon (*Falco peregrinus*), the sparrow hawk (*Accipter nisus.*), barn owl (*Tyto alba*), and the sandwich tern (*Sterna sandivicensis*), the lower range of concentrations of dieldrin in brain is approximately 5–6 µg/g, in some cases less. Stickel et al. (1969) argued that prevalence of lower brain residue concentrations in field samples may be partly the result of more susceptible individuals succumbing first; they later suggested that rapid mobilization of a chemical caused, for example, by stress events may kill an animal at exposure levels which normally would be tolerated (Stickel et al. 1973). Thus, they argued there is reasonable support for use of the value of 4 µg/g dieldrin in brain as a critical level. In our opinion this value is a more protective benchmark of avian mortality caused by exposure to dieldrin.

13.2.1.2　Sublethal Effects

There has been some debate about the potential of dieldrin, and possibly other cyclodienes, to affect wild birds at exposures lower than those associated with lethality. While a number of such studies have reported significant effects of dieldrin, they have proved of limited use for inferring critical tissue concentrations. For example, Friend and Trainer (1974) reported immune system effects of dieldrin in mallard (*Anas platyrhincos*) ducklings using challenge with duck hepatitis virus. Baxter et al. (1969) reported that ring-necked pheasant chicks (*Phasianus colichicus*) produced by dieldrin-dosed hens showed reduced feeding efficiency by choosing the deep side of a visual cliff, while controls tended to choose the shallow side. Loggerhead shrikes (*Lanius ludovicianus*) fed 1 µg/g dieldrin per day showed significant effects on their capability to capture mice (Busbee, 1977). Besides convulsions, another sign of impending death in the shrikes was a weight loss beginning a few days prior to death. Sharma et al. (1976) further investigated the effects of dieldrin on biogenic amines and on behavior. When fed 10–30 ppm dieldrin, encounters between mallard drakes indicated a significant decline in the tendency for dieldrin exposed drakes to take the initiative and establish rights of access in approach confrontations (Sharma et al. 1976). Based on such evidence of neurological and behavioral effects of dieldrin on birds, Walker and Newton (1999) concluded there was potential for sublethal dieldrin exposure to impact hunting behavior of raptors, and to have played a role in population declines.

TABLE 13.2
Experimental Studies Determining Lethal Levels of Dieldrin—Birds

Species	Sex	Sample Size No.	Organ	Residue Level (µg/g ww) Mean	Range	Reference
Bobwhite quail (Colinus virginianus)	—	—	Brain	12–14	—	Rudd and Genelly 1956
	M	11	Brain	12.4	8.5–17.8	Gesell and Robel 1979
Japanese quail (Coturnix coturnix japonica)						
Died	M	30	Brain	17.4	15.7–19.3[a]	Robinson et al. 1967
Died	F	35	Brain	17.3	15–20[a]	Robinson et al. 1967
Survived	M,F	12	Brain	6.9	3.1–15.0	Robinson et al. 1967
	M	17[b]	Liver	32.3–41.2	—	Robinson et al. 1967
	F	19[b]	Liver	37.5–57.7	—	Robinson et al. 1967
Died	M	10	Brain	14.8	4.9–44.3[a]	Stickel et al. 1969
Died	F	7	Brain	21.7	10.4–45.4[a]	Stickel et al. 1969
Survived	M	8	Brain	2.6	0.8–8.8[a]	Stickel et al. 1969
Survived	F	8	Brain	4.1	0.1–18.4[a]	Stickel et al. 1969
Pheasant (Phasianus colchicus)						
Died	M,F	39	Brain	5.8	1.2–27	Linder et al. 1970
Survived	F	10	Brain	0.96	0.2–2.2	Linder et al. 1970
Sharp-tailed grouse (Tympanuchus phasianellus)	M	12	Brain	6.9	4.4–10.9	McEwen and Brown 1966
Rock dove (Columba livia)						
Died	M,F	19	Brain	20	18.1–22.0[a]	Robinson et al. 1967
Died	M,F	20	Liver	45.6	37.5–55.5[a]	Robinson et al. 1967
Survived	M	11	Brain	7.7	2.1–8.5	Robinson et al. 1967
Died	M,F	2	Brain	27.7	19.2 and 36.2	Jefferies and French 1972
			Liver	77.2	51.9 and 102.6	Jefferies and French 1972
Survived	M,F	4	Brain	n.g.	5.4–9.5	Jefferies and French 1972
Cowbird (Molothrus ater)						
Died	M	21	Brain	16.3	9.8–23.5	Heinz and Johnson 1981
Ceased feeding		17		6.8	1.5–11.7	Heinz and Johnson 1981
Red-winged blackbird (Agelaius phoeniceus)						
Unstressed	M	30	Brain	19.8	(1.7)[c]	Clark 1975
Food stressed	M	30	Brain	22.2	(0.8)[c]	Clark 1975

[a] 95% confidence intervals.

[b] Three groups of birds on different dosages. Values are geometric means.

[c] Numbers in parentheses, standard deviation.

TABLE 13.3
Field Studies of Dieldrin Mortality of Birds

Species	Sample Size (No.)	Organ	Residue Level (µg/g ww) Mean	Residue Level (µg/g ww) Range	Reference
Sarus crane (*Grus antigone*)	5	Brain	19.3	3.6–43.5	Muralidharan 1993
	5	Liver	56.8	10.5–92.6	Muralidharan 1993
Pink-footed goose (*Anser brachyrhynchus*)	6	Liver	31	15–48	Stanley and Bunyan 1979
Snow goose (*Chen caerulescens*)	5/157[a]	Brain	17.3	13.0–24.5	Babcock and Flickinger 1977
	8/112[a]	Brain	8.2	4.9–14	Flickinger 1979
Lesser scaup (*Aythya affinis*)	4	Brain	11.9	7.7–16	Sheldon et al. 1963 cited in Stickel et al. 1969
Buzzard (*Buteo buteo*)	14	Liver	19.7[b]	7.8–31.2	Fuchs 1967
Lanner falcon (*Falco biarmicus*)	2	Brain	2.0 and 3.3		Jefferies and Prestt 1966
Peregrine (*Falco peregrinus*)	2	Brain		6.8 and 16.4	Bogan and Mitchell 1973
		Liver		17.3 and 53.9	Bogan and Mitchell 1973
	1	Brain		5.4[c]	Reichel et al. 1974
	2	Brain		3.5 and 7.8[d]	Jefferies and Prestt 1966
		Liver		4.0 and 9.3[e]	Jefferies and Prestt 1966
European kestrel (*Falco tinnunculus*)	84	Liver		6–30[f]	Newton et al. 1992
Sparrow hawk (*Accipter nisus*)	25	Liver		5–21[g]	Newton et al. 1992
Owls[h]	10	Brain	15	11–25	Jones et al. 1978
Barn owl (*Tyto alba*)	51	Liver	14	6–44	Newton et al. 1991
Sandwich tern (*Sterna sandvicensis*)					
Chick	6	Liver	5.6	2.4–12[i]	Koeman et al. 1967
Juvenile	8		4.6	1.9–6.6[j]	Koeman et al. 1967
Adult	5		5.5	4.7–7.2[k]	Koeman et al. 1967
Meadowlark (*Sturnella magna*)	5	Brain	9.3	8.6–12.1	Stickel et al. 1969
		Liver	13.1	7.9–15.9	Stickel et al. 1969
American robin (*Turdus migratorius*)	7	Brain	9.6	5.0–17.0	Stickel et al. 1969
Pigeon (*Columba livia*)	4	Liver		15–24	Turtle et al. 1963 cited in Koeman et al. 1967
Collared dove (*Steptopelia decaocto*)	3	Brain	15.2	7.4–20.7	Muralidharan 1993
	3	Liver	36	20–66.2	Muralidharan 1993

[a] Number at each site presumed dead from dieldrin over total numbers.

[b] Wet weight assumed.

[c] Also 34 µg/g of DDE and 55 µg/g of PCBs.

[d] Also 45 µg/g and 44 µg/g of DDE.

[e] Also 70 µg/g and 60 µg/g of DDE.

[f] Measurement given as µg/g; a few outliers as high as 99 µg/g.

[g] Measurement given as µg/g; a few outliners as high as 85 µg/g.

[h] A number of different species maintained at the London Zoo.

[i] Also 2.3 µg/g of telodrin and 0.47 µg/g of endrin.

[j] Also 0.86 µg/g of telodrin and 0.43 µg/g of endrin.

[k] Also 1.0 µg/g of telodrin and 0.67 µg/g of endrin.

13.2.1.2.1 Suppression of Feeding

A number of investigators have noted that sublethal brain concentrations of dieldrin appear to affect appetite and feeding behavior. We thought those findings warranted more consideration, given the particular vulnerability to food shortage in many bird populations during reproduction and migration.

Heinz and Johnson (1981) investigated dieldrin effects on feeding and toxicity in the brown-headed cowbird (*Molothrus ater*). In a pilot study, they found that if dieldrin-dosed birds made less than four feeding attempts within a 5-h period then they rarely resumed eating, and eventually died. Thus, they set a "5 hour-stop feeding" criterion as the indicator for cessation of eating. In a comprehensive study, they fed 20 ppm dieldrin to male cowbirds and sacrificed them as soon as they stopped feeding for 5 h. Brain residues in the "stopped feeding" group were compared to results from birds which continued to feed on a 20 ppm dieldrin diet until death.

In both dieldrin treatment groups, those that ceased feeding and those that continued feeding till death, all but four individual birds displayed substantial weight loss. All of the control birds gained weight. Brain dieldrin residues in birds that stopped feeding averaged 4.4 mg/kg for a subgroup which fed in the morning, then ceased feeding and were sacrificed later in the day. Among those was a bird containing 1.5 mg/kg dieldrin in brain, leading Heinz and Johnson to speculate that 1 mg/kg dieldrin residues in brain could be sufficient to induce anorexic behavior. Consistent with that suggestion, a pharmacokinetic study of rats demonstrated that following a single oral dose of 10 µg/g, dieldrin partitioned rapidly to brain and remained at 1 µg/g for 2 days (Hayes 1974). Those findings suggest that even a single exposure, as potentially experienced by a hyperphagic songbird, could induce cessation of feeding. This is consistent with the feeding patterns seen in some of the earlier dieldrin exposure studies (Table 13.4).

TABLE 13.4
A Summary of Studies that Examined the Effect of Dieldrin on Feeding of Birds

Species	Dose	Effect	Reference
Ring-necked pheasant (*Phasianus colchicus*)	0, 25, 50 µg/g dieldrin in diet to breeding males and females through courtship, breeding and egg laying (no time period given)	Mean daily consumption by bird/day reduced from 44.6 g (25 ppm) to 36.6 g (50ppm) compared to 51.6 g in control group; mean live weight change of females (g) increased by 64 g in controls and decreased by 14 g (25 ppm) and 36 g (50 ppm). No statistical analysis reported.	Genelly and Rudd 1956
Game farm pheasants (*Phasianus colchicus*)	6 mg weekly dose for 1 year	Dosed pheasants consumed less food and lost more weight in 1 of 2 years. Hens began breeding season at relative heavy weights and no mortality occurred at 0, 2, 4, 6 mg/week. In the other year of the study, hens began breeding season at relatively low weights and mortality occurred at a higher rate on 4 mg than those on 1 or 2 mg/week.	Atkins and Linder 1967
Homing pigeon (*Columbia livia var.*)	0 (9 weeks only), 2 mg/kg body weight for 2 years 10 months	"Body fat was not found in any of the [dead] birds at autopsy an observation similar to that for the experimental poisoned birds that had been housed in cages (Robinson Brown et al. 1967)." Body weights of males during 110 weeks showed no increase and showed occasional sudden decreases and females showed steady increased in weights during 110 weeks of monitoring (no statistics).	Robinson and Crabtree 1969

TABLE 13.4 (continued)
A Summary of Studies that Examined the Effect of Dieldrin on Feeding of Birds

Species	Dose	Effect	Reference
Pheasant (*Phasianus colchicus*)	6, 8, 12 mg/week	Pheasants "reduced food consumption and subsequently weights" at 6, 8, 12 mg/week as breeding hens in second generation.	Baxter et al. 1969
Japanese quail (*Coturnix coturnix japonica*)	Dietary concentrations of 0, 10, 20, 30, 40 µg/g dieldrin	Weights not reported.	Walker et al. 1969
Fulvous tree duck (*Dendrocygna bicolour*)	Rice was planted (112 kg/ha or 100 lb/acre) with aldrin + 2.5 g/kg seed (4 oz/100 lb) and Ceresan L, a mercurial fungicide (0.75 g/kg or 1.2 oz/100 lb)	Within 3 days of feeding on contaminated dieldrin seed the adult ducks were "markedly underweight and in poorer physical condition than the controls." "The hungry fulvous tree ducks were not at first repelled by the treated seed, but those that survived initial exposure virtually stopped feeding probably as a result of a physiological response in intoxication."	Flickinger and King 1972
Pheasant (*Phasianus colchicus*)	Weekly doses to pheasants; hens received 1, 6, 10 mg and cocks received 10, 4, 6, mg. F1 and F2 were bred	"Treatment did not affect cock or hen weight ($p > .05$) until just prior to death. At that time, birds stopped eating and lost weight rapidly."	Dahlgren and Linder 1974.
Mallard (*Anas platyrynchos*)	0, 4, 10, 30 µg/g dieldrin in diet to parents and offspring and endpoints evaluated in offspring at 60–75 days old and in adult drakes	Increase in brain: body weight ratio at 30 ppm. Dominance assessment by encounter scores were significantly different than controls in adults and juveniles at 4, 10, 30 ppm.	Sharma et al. 1976
Japanese quail (*Coturnix coturnix japonica*)	Expt 1. at 25% egg production quail fed 0, 10, 25 µg/g dieldrin for 6, 28-day periods Expt 2. 0, 5, 25 ppm of dieldrin for 4, 28 day periods	Body weights and food consumption were not significantly affected by any dieldrin treatments.	Reading et al. 1976
Leghorn chicken (*Gallus domesticus*)	0, 10, and 20 µg/g dieldrin in diet for 12 weeks	Survival time after feed removal was significantly lower in 20 ppm; Fat in carcass dry matter % was significantly higher in 10 and 20ppm groups.	Davison et al. 1971
Brown-headed cowbird (*Molothrus ater*)	20 µg/g dieldrin in diet	Of the birds that were 'sacrificed' as soon as they did not eat for 5 h: 14 of 17 birds lost 1–10 g of body mass during the study. Of the birds that were allowed to die of dieldrin exposure in their own time, all birds ($n = 21$) lost weight during the study. While the sample of control birds ($n = 4$) was small, all gained weight during the study despite the availability of food *ad libitum*.	Heinz and Johnson 1981
Loggerhead shrike (*Lanius ludovicianus*)	To test behavioral effects: Doses: 0 or 1 mg/kg body weight per day; To determine mortality: 0, 1, 2, 3, 4 mg/kg/day	At 1 mg/kg/day 7 of 11 behavior endpoints were significantly different than controls. "Weight loss began a few days before death" occurred.	Busbee 1977

Peakall (1996) considered the work of Heinz and Johnson (1981) interesting from a mechanistic perspective, but rejected the results as a basis for interpreting tissue residues. He concluded that acute lethality was the only endpoint sufficiently supported by the literature to recommend critical concentrations. However, in our opinion, given the consequences of anorexic effects on migrating and breeding birds, we have examined the relevant literature in further detail with the aim of suggesting criteria for sublethal effects.

13.2.1.2.2 Implications for Migrants

Migration is a costly strategy, with substantial energy expended by migrants both in sustaining prolonged flight and in rebuilding fuel stores at stopover locations (Moore and Kerlinger 1987, Woodrey and Moore 1997, Wikelski et al. 2003). Cool weather and storm conditions often encountered during migration further increase energetic needs. In assessing the risks and costs, spring and fall migration must be considered separately. Spring migrants are returning to breeding sites, and overall energy costs of migration may be higher in spring than fall migration (Scott et al. 1994). Both body condition and timing of arrival at the breeding grounds are critical factors determining reproductive success and overall fitness (Marra et al. 1998, Verboven and Visser 1998, Both and Visser 2001). However, fall migration includes large numbers of inexperienced juvenile birds and many species moult during fall migration, further increasing energy demands (Woodrey and Moore 1997, Schaub and Jenni 2001).

To undertake and sustain migration, birds must acquire and store considerable amounts of energy, primarily in the form of lipid stores (e.g., Bairlein and Gwinner 1994, Deviche 1995). Deposition of fat stores prior to and at stopover sites during migration is at least partially due to hyperphagia (reviewed in Bairlein and Gwinner 1994), although factors associated with increased efficiency of assimilation and of fat, protein and carbohydrate use, and shifts in food selection also appear to be involved. Although detailed mechanisms are not well understood, there is evidence that GABA plays a role in appetite stimulation (Pu et al. 1999), and thus there is at least a tentative mechanistic link between GABA-inhibiting compounds, such as the cyclodienes, and appetite suppression.

The ultimate factor inducing hyperphagic behavior in both spring and fall is the change in photoperiod; however, there are a variety of proximal factors involved in modifying and attenuating that behavior in different species, even including some evidence that secondary plant metabolites may play a role (Bairlein and Gwinner 1994). Hyperphagic behaviors are also associated with the behavioral processes which induce the inclination of birds to migrate, termed "migratory restlessness." Migratory restlessness can be inhibited following spontaneous or induced body mass loss (Bairlein and Gwinner 1994).

Stopover times and fat deposition rates of migrant warblers arriving at sites along the U.S. gulf coast during spring migration were studied (Moore and Kerlinger 1987), and many factors influenced body condition of migrant birds on their arrival at a stopover site. Those included species differences, location of departure, flight times, weather and habitat at stopover sites as well as innate species variation in body condition. However, in general, the majority of birds arrive at migratory stopover sites with depleted fat stores (Moore and Kerlinger 1987). For example, 41.1% of 627 migrant songbirds crossing the Gulf of Mexico and stopping along the northern coast were lean (fat class = 0). Significant impediments to foraging efficiency or alterations to the physiological processes regulating energy intake and/or fat deposition would reduce body condition. In the extreme, an affected bird might not survive migration, and if it reached the breeding grounds would likely have reduced productivity. Overall fitness of the bird would, therefore, be lower. At sites where contamination by dieldrin and related chemicals persist (e.g., Frank and Lutz 1999, Stansley and Roscoe 1999), migrating birds could experience substantial exposure even during short stays.

Among the cowbirds studied by Heinz and Johnson, those that were heavier before dieldrin treatment took significantly longer to stop feeding. Migrant songbirds often arrive at stop over sites in reduced body condition, and continue to lose weight before locating food (Winker et al. 1992, Scott et al. 1994, Schaub and Jenni 2001). In that situation, a bird would be at risk of an anorexic effect if exposed to dieldrin or another anorexic agent at lower concentrations than would normally cause death.

As reported by Flickinger (1979), migrating birds can be affected by dieldrin contamination at stopover sites and can die at locations some distance from the contaminated area. He reported large numbers of dead and debilitated waterfowl in Texas rice fields treated with aldrin, and no mortality associated with untreated reference fields. A snow goose (*Chen c. caerulescens*) and a blue-winged teal (*Anas discors*) were observed 5.6 km from the nearest rice field displaying symptoms of dieldrin poisoning; both birds eventually died. The dieldrin concentration in the brain of one goose was 14 µg/g, and was 19 µg/g in the brain of a blue-winged teal. The goose was determined to be a migrant because only migrating geese were observed in the area where it was found; the nearest wintering flock of geese was located 57.9 km southeast of the site (Flickinger 1979). They further reported that "dieldrin residues may affect reproduction once geese reached their nesting grounds." Band returns showed that many of the snow geese which spent the winter on the Garwood Prairie and other parts of southeast Texas eventually nested on the McConnell River on the western shore of Hudson Bay, Canada. Mortality of snow geese at the McConnell River nesting area in 1967, 1971, and 1972 coincided with goose mortality on the Garwood Prairie. Four of five reproductively unsuccessful nesting females found dead at McConnell River in 1972, and four of five females from failed nests contained traces of dieldrin in their brains, but only one of ten that nested successfully contained such traces. While mercurial fungicides were also used at the Texas site, mercury residues were almost negligible in those geese (Flickinger 1979).

From this evidence, we conclude that the presence of 1 µg/g of dieldrin, or dieldrin equivalents, in brain or other tissues of migrating birds is cause for concern. We base that determination on the evidence for potential effects on feeding behavior, which at the extreme could reduce overall survival, or affect fitness by delaying arrival on breeding grounds.

13.2.1.2.3 Effects on Reproduction

In contrast to the lethal effects of dieldrin and thus on survival, there is no clear evidence that dieldrin affects the actual reproductive processes of birds. This is in contrast, for example, to DDT where the persistent metabolite, DDE, reduces the nest success of captive and wild birds through effects on shell quality and likely embryotoxic mechanisms (Blus 2003).

Dietary intakes of dieldrin at 10 µg/g in Japanese quail, which resulted in egg yolk dieldrin concentrations of 23 µg/g, had no effects on reproduction (Walker et al. 1969). Similarly, 15 µg/g in crowned guinea-fowl (*Numida meleagris*) (Wiese et al. 1969) did not significantly affect reproductive parameters such as egg production, fertility, or hatchability. However, in the case of the study of Wiese et al. chick survival was significantly reduced in the 15 µg/g group and an attempt to repeat the experiment at this higher dose failed due to mortality of breeding flocks. Experiments with American kestrels (*Falco sparverius*) (Wiemeyer et al. 1986), and a field study of common (*Gallinula chloropus*) and purple gallinules (*Porphyrio porphyrio*) (Fowler et al. 1971) did not find significant effects of dieldrin on reproductive outcome, particularly egg hatchability. Studies with barn owls, which produced average concentrations of 9.4 µg/g dieldrin in carcasses, had no effect on hatching or fledging (Mendenhall et al. 1983). Chronic dosing of chickens (*Gallus gallus*) with concentrations of 0.2–5 ppm dieldrin in feed, which produced more environmentally realistic concentrations in eggs from 0.36 to 4.20 µg/g, had no effects on egg production, hatching, or survival of chicks (Graves et al. 1969). Some chronic dosing studies of the effects of dieldrin on reproduction of captive birds did, however, report that both juvenile and adult birds suffered mortality (Wiese et al. 1969, Mendenhall et al. 1983).

More recently, Frank and Lutz (1999) studied reproduction of a wild population of great horned owls (*Bubo virginianus*) at a contaminated site, the Rocky Mountain Arsenal National Wildlife Refuge (RMA NWR). Great horned owls were radio-tagged from nests within the core contaminated area of the RMA NWR and from nests where adult birds had been observed foraging in the more contaminated areas. They were designated as "contaminated pairs." There were no differences in hatching and fledging success between owls nesting in areas considered contaminated versus uncontaminated. They used a combination of radio-telemetry and blood sampling which showed a significant negative relationship between dieldrin concentrations in plasma of juvenile owls and post fledging survival. Interval survival of juvenile owls in a high dieldrin plasma concentration category, >0.1 µg/g (100 µg/L) was significantly lower ($p < .01$) than that of birds classified as in a low dieldrin plasma concentration category, <0.05 µg/g (50 µg/L). Although based on only a single study, those results suggest an effect of dieldrin on nestling survival when plasma concentrations exceed 0.1 µg/g. Vander Lee and Lutz (2000) administered dieldrin to nestling black-billed magpies (*Pica pica*), and examined plasma and tissue distribution of dieldrin residues. They reported a good relationship between plasma and brain ($y = 0.882x + 0.007$, $p < .001$, $r^2 = 0.959$, $n = 38$). The results of those studies suggest that brain concentrations as low as 0.1 µg/g dieldrin are associated with reduced survival of juvenile birds. While those findings are interesting, they are not sufficient to recommend criteria.

Roy (1997) reported the results of an investigation (1993–1996) of American kestrels breeding in nest boxes placed at the RMA NWR, and at reference sites. There were no significant differences in hatching success among core and peripheral areas of that highly contaminated site, and a reference area. Rather, he detected a significant effect of dieldrin on young fledged per nest attempt and nest success. However, there were confounding relations with DDE as well, despite low mean DDE concentrations, for example 0.37 µg/g in 32 eggs from the core area, being well below a putative effect level of 10 µg/g. He compared nest success at varying intervals of dieldrin in eggs and calculated a dieldrin "Effect Zone" (Henny 1997) of greater than 0.5–1 µg/g in kestrel eggs for nest success, hatching success and number of young fledged. That was based on comparisons, for example, of mean nest success of 66 nests with dieldrin concentrations in eggs less than 0.5 of 75.8% compared to 48.4% for nests with egg dieldrin greater than 0.5 µg/g. The general trend of the relationships seem to hold, although eight nests with greater than 5.0 to 10.0 µg/g dieldrin in eggs had mean nest success of 62.5%, not much different from the 34 nests with less than 0.15 µg/g at 67.7%. They discussed the likely mechanism for reduced nest success, and stated that there did not appear to be any effects of dieldrin on hatchability or embryo survival or any dieldrin-related nestling mortality. Cases of reduced nest success were the result of "entire nest failure" during egg laying or early incubation periods. They concluded that egg concentrations were a barometer of adult exposure during the nesting period. They made a link, therefore, between nest failure and disappeared and presumably dead males. This was supported by evidence of dead or moribund kestrels, mostly males found during the breeding season, some of which had brain dieldrin concentrations diagnostic of lethal poisoning. The interval suggested by Roy (1997) of 0.5–1.0 µg/g in eggs is consistent with the value of 0.7 µg/g dieldrin in addled eggs of peregrines suggested by Newton (1988) as associated with population decline of that species.

A study of golden eagles (*Aquila chrysaetos*) in Scotland reported that nest success increased significantly coincident with a decrease in dieldrin concentrations in eggs, and that no other causal factor could be identified (Lockie et al. 1969); a level of dieldrin causing reproductive effects of 1 µg/g was implied. Later, however, Newton and Galbraith (1991) were less convinced of the possible role of chlorinated hydrocarbon contaminants in impacting golden eagle nesting success during the 1960s in Scotland.

Concentrations of OCs in eggs are directly indicative of female body burden; ratios of dieldrin and other cyclodienes in eggs to liver of female herring gulls were close to one (Braune and Norstrom 1989). Thus, concentrations of dieldrin or dieldrin equivalents (see Section 13.5 below for further details) in eggs greater than 1 µg/g provide an indication that adult body burdens are, and, therefore, potentially brain concentrations are greater than 1 µg/g, and, within the anorexic danger zone. This would have greater consequences for male birds, which are not able to reduce their body burden via deposition to eggs, and in some species are energetically stressed by the extra demands of feeding chicks and mates.

13.2.1.2.4   Population Level Assessments

The causative role of dieldrin in the decline of a variety of British raptors, sparrow hawks, merlins (*Falco columbarius*), and barn owls has been extensively documented (Newton and Bogan

1974, 1978, Newton and Haas 1988, Newton 1986, Newton and Galbraith 1991, Newton et al. 1992, Walker and Newton 1999, Sibly et al. 2000). Despite evidence of relatively high concomitant exposure to DDT, and evidence of effects of DDT on shell quality and reproduction (Newton 1986), the reduction in juvenile and adult survival due to dieldrin poisoning has been determined to be the more significant factor influencing population stability (Sibly et al. 2000).

Nisbet (1998) reviewed and summarized dieldrin mortality incidents during 1940s–1960s in the United States. He was primarily interested in the role of dieldrin in population decline of the peregrine falcon, and contended that the rapid population crash in the 1950s was more consistent with a mechanism of mortality of breeders, rather than reproductive failure due to DDE effects on shell quality. The timing of the introduction and withdrawal of dieldrin appeared to fit the pattern of peregrine decline and recovery better than did DDT. Similarly, Ratcliffe (1973, 1988) and Newton (1988) argued that adult poisoning caused by the cyclodienes dieldrin, aldrin, and also possibly heptachlor had more to do with the decline of British and European peregrines than DDT.

Nisbet (1989) also examined the possible role of dieldrin in the decline of the bald eagle in North America, and argued that dieldrin may have been an important factor, particularly in some regions. However, he also concluded that the effects of DDE on shell quality were dominant, as have others (Wiemeyer et al. 1993, Elliott and Harris 2001). In both the bald eagle and peregrine cases, the direct evidence from chemical analysis of tissues from dead peregrines is very limited. However, North American *accipiter* hawks collected from Ontario, Canada had higher concentrations of dieldrin and other cyclodienes, compared for example to primarily aquatic feeding species such as gulls (e.g., *Larus argentatus*) or ospreys (*Pandion haliaetus*) from the Great Lakes (Elliott and Martin 1994). They compared temporal trends in accipiter populations with sales of dieldrin and DDT, and found the strongest association with dieldrin exposure.

Newton (1988) presented evidence that British sparrowhawk populations did not recover from low numbers caused primarily by dieldrin mortality until liver concentrations were less than a geometric mean of 1 µg/g, and argued for that as a "Level Leading to Population Decline" (LPD) for dieldrin in wild raptors. Newton compared data from mortality incidents of kestrels and suggested there was evidence that raptors were perhaps more sensitive to dieldrin toxicity than laboratory species such as rock doves or Japanese quail. Walker and Newton (1999) further examined the mortality monitoring database as did Sibly et al. (2000) and argued that there was support for sublethal effects of dieldrin, in the range of 3–9 µg/g in liver particularly of sparrowhawks, which use speed and agility to capture prey and thus would be more susceptible to any subtle neurotoxic effects of dieldrin. They also discussed the data reported in Newton et al. (1992) that a number of birds with residues in that range died from trauma after hitting trees or telephone wires, perhaps suggestive of neurotoxic effects.

When combined with the argument that 1 µg/g dieldrin in the brain is potentially associated with threshold value above which at least some proportion of birds cease to feed, we propose the adoption of concentrations of 1 µg/g of dieldrin equivalents in brain, plasma or eggs and 3 µg/g in liver as protective effect levels (Table 13.5). Dieldrin concentrations in brain and plasma of nestling

TABLE 13.5

Concentrations of Dieldrin or Dieldrin Equivalents in Various Tissues Associated with Significant Toxicological or Ecological Outcomes

Endpoint	Tissue	Concentration (µg/g)	Reference
Lethality	Brain	4	Stickel et al. 1969
Anorexia	Brain, egg, plasma	1	Heinz and Johnson 1981, this chapter
Population decline	Egg	0.7	Newton 1988
Population decline	Liver	3	Newton 1988, Walker and Newton 1999, Elliott and Bishop, this chapter

birds were reported to be close to 1:1 (Vander Lee and Lutz 2000). Thus 1 μg/g dieldrin in plasma indicates a similar brain concentration and can be considered a value associated with cessation of feeding (Table 13.5). However, the measurement of an order of magnitude less, 0.1 μg/g dieldrin was associated with reduced survival of great horned owls (Frank and Lutz 1999), suggesting that lower threshold should be considered for nestlings.

13.2.2 ENDRIN

Endrin ((1a,2,2a,3,6,6a,7,7a)-3,4,5,6,9,9-hexachloro-1a,2,2a,3,6,6a,7,7a-octahydro-2,7:3,6-dimethanonaphth [2,3-b]oxirene) is extremely toxic to birds (Table 13.1). In some mammals, but apparently not in birds, endrin is converted metabolically to 12-ketoendrin which for example in rats is 5-times more toxic than the parent compound (Bedford et al. 1975, Stickel et al. 1979a, Blus et al. 1983). In contrast to many of the other OC pesticides, in birds, endrin has a shorter metabolic half-life (Heinz and Johnson 1979). That more rapid clearance likely accounts for endrin rarely being reported in wildlife tissues, when compared for example to dieldrin and HCB for which uses have been restricted or banned for a similar time period.

The toxicity of endrin to terrestrial vertebrate wildlife was reviewed by Blus (2003). A number of studies have reported endrin to be the cause of relatively wide scale mortality of wild birds. For example, Blus et al. (1979) determined that endrin killed a large percentage of brown pelicans (*Pelecanus occidentalis*) that had been re-established to Louisiana estuaries in the 1970s. Pelicans had lethal brain concentrations greater than or equal to 0.8 μg/g. Use of endrin as a rodenticide in Washington State orchards caused mortality of large numbers of birds (Fleming et al. 1982).

Stickel et al. (1979a) experimentally investigated endrin toxicity in mallards, starlings (*Sturnus vulgaris*), cowbirds, and blackbirds (*Icteridae*) and recommended a lethal brain concentration of 1 μg/g.

13.2.3 CHLORDANE

Chordane (1,2,4,5,6,7,8,8-octachloro-2,3,3a,4,7,7a-hexahydro-4,7-methanoindene) was produced as a complex mixture of some 45 compounds, primarily *cis*- and *trans*-chlordane, heptachlor, and *cis*- and *trans*-nonachlor. Oxychlordane is a metabolite of *cis*- and *trans*-chlordane, which along with another oxidative product, heptachlor epoxide, constitutes the major toxicological components (Eisler 1995). Chlordane was widely used in North America, and heaviest use occurred in the early 1970s as a substitute for other restricted OC pesticides (NRCC 1974). However, like the compounds it replaced, chlordane components are particularly persistent and lipophilic, and most uses were soon restricted, although underground use for termite control continued at least until the late 1980s in many North American jurisdictions. Oxychlordane and heptachlor epoxide continue to be detected in wildlife from remote locations (e.g., Elliott et al. 2000, Henricksen et al. 2000). Continuing until at least the late 1990s, there were reports of lethal concentrations of chlordane compounds, along with dieldrin, in brains of birds found dead and moribund at various locations in the northeastern American states (Stone and Okoniewski 1988, Okoniewski and Novesky 1993, Stansley and Roscoe 1999). Cyclodienes were diagnosed as the cause of death in various species of songbirds and in the secondary poisoning of raptors, mainly Cooper's hawks (*Accipiter cooperii*). The food chain uptake of chlordane in particular was subsequently investigated, and it was suggested that the problem was caused by efforts to control scarab beetles, particularly the grub of the oriental beetle (*Anomala orientalis*) and the Japanese beetle (*Popilla japonica*; Stansley et al. 2001). Apparently a key factor was the degree to which those scarabs developed resistance to chlordane use, resulting in increased applications. Data from those studies are summarized in Table 13.6, which also includes our own calculations of dieldrin equivalents, as an example of that approach. As suggested by Stansley et al. (2001) the problem was likely quite widespread, and, given the persistence of chlordane and dieldrin, likely to be continuing at some sites.

TABLE 13.6
Select Field Studies of Cyclodiene Residues in Brains of Birds (µg/g ww, Mean, Range in Brackets) Associated with Mortality

Species	Location	Sample Size	Heptachlor Epoxide (0.5)	Oxychlordane (0.8)	Dieldrin (1)	Toxic Equivalents	Reference
Starling	NJ	6	4.4 (3.5–5.7)	4.6 (2.5–5.7)	0.68 (0.24–1.3)	6.53 (5.16–7.73)	Stansley and Roscoe 1999
Grackle	NJ	17	5.0 (2.5–8.7)	4.5 (1.6–6.4)	1.21 (0.62–2.8)	7.32 (3.15–11.48)	Stansley and Roscoe 1999
Robin	NJ	3	3.8 (3.3–4.5)	4.4 (4.2–4.9)	0.34 (0.15–0.56)	5.77 (5.16–6.73)	Stansley and Roscoe 1999
American crow	NJ	1	4.10	1.80	0.30	6.24	Stansley and Roscoe 1999
Blue jay	NJ	1	3.30	2.90	0.70	6.93	Stansley and Roscoe 1999
House sparrow	NJ	1	5.00	2.80	<0.19	7.99	Stansley and Roscoe 1999
Cooper's hawk	NJ	9	3.9 (2.5–5.3)	2.8 (1.5–5.5)	1.17 (0.41–2.60)	5.36 (3.70–8.25)	Stansley and Roscoe 1999
Sharp-shinned hawk	NJ	1	3.20	5.30	0.65	9.15	Stansley and Roscoe 1999
Red-tailed hawk	NJ	2	4.3 (2.5–6.6)	4.1 (1.5–5.7)	1.31 (0.41–3.10)	7.02 (3.70–9.87)	Stansley and Roscoe 1999
Great horned owl	NJ	1	3.90	6.30	<0.66	10.86	Stansley and Roscoe 1999
Northern flicker	NY	1	1.13	7.64	2.39	11.16	Okoniewski and Novesky 1993
Blue jay	NY	20	2.68 (0.67–5.94)	2.28 (0.56–3.75)	2.86 (0.14–9.78)	7.82 (3.47–12.39)	Okoniewski and Novesky 1993
Blue jay	MD	5	4.01 (2.84–5.07)	5.28 (3.89–6.83)	0.95 (0.44–1.79)	10.33 (8.64–11.89)	Okoniewski and Novesky 1993
Blue jay	NJ	3	2.21 (1.32–3.39)	2.92 (2.57–3.28)	0.82 (0.43–1.57)	5.95 (4.91–7.88)	Okoniewski and Novesky 1993
American crow	NY	4	3.17 (1.94–5.16)	3.12 (2.39–4.26)	0.14 (0.03–0.39)	6.43 (5.08–9.81)	Okoniewski and Novesky 1993
American crow	MD	1	3.91	3.59	0.70	8.20	Okoniewski and Novesky 1993
Eastern bluebird	NY	2	6.41 (3.3–9.52)	5.54 (1.08–10.00)	0.17 (0.05–0.29)	12.12 (10.65–13.59)	Okoniewski and Novesky 1993
American robin	NY	17	4.25 (0.58–9.12)	2.46 (0.39–6.96)	5.34 (0.04–15.7)	12.05 (4.84–26.83)	Okoniewski and Novesky 1993

continued

TABLE 13.6 (continued)
Select Field Studies of Cyclodiene Residues in Brains of Birds (μg/g ww, Mean, Range in Brackets) Associated with Mortality

Species	Location	Sample Size	Heptachlor Epoxide (0.5)	Oxychlordane (0.8)	Dieldrin (1)	Toxic Equivalents	Reference
Gray catbird	NY	1	6.29	5.19	1.96	13.44	Okoniewski and Novesky 1993
Northern mockingbird	MD	1	2.58	4.52	0.30	7.40	Okoniewski and Novesky 1993
Northern mockingbird	NY	3	6.30 (1.57–15.6)	4.14 (1.61–6.75)	3.58 (0.2–7.5)	14.03 (5.83–25.4)	Okoniewski and Novesky 1993
European starling	NJ	1	3.16	3.98	0.52	7.66	Okoniewski and Novesky 1993
European starling	NY	18	1.85 (0.07–5.12)	1.81 (0.11–7.72)	9.14 (0.41–20.5)	12.79 (4.27–25.44)	Okoniewski and Novesky 1993
Song sparrow	NY	1	1.20	5.50	1.10	7.80	Okoniewski and Novesky 1993
Common grackle	MD	1	2.34	4.12	1.82	8.28	Okoniewski and Novesky 1993
Common grackle	NY	9	3.32 (1.19–6.06)	3.72 (0.63–6.34)	5.37 (0.36–13.4)	12.40 (5.87–17.33)	Okoniewski and Novesky 1993
House sparrow	NY	15	3.17 (1.03–6.65)	2.74 (0.58–7.04)	2.38 (0.08–8.17)	8.28 (5.34–14.54)	Okoniewski and Novesky 1993
Sharp-shinned hawk	NY	4	2.91 (1.45–5.03)	6.47 (2.1–12.2)	2.4 (0.87–4.09)	11.78 (4.78–21.32)	Okoniewski and Novesky 1993
Cooper's-hawk	NY	1	1.64	3.90	1.58	7.12	Okoniewski and Novesky 1993
Red-shouldered hawk	MD	3	4.01(1.93–7.36)	2.89(0.04–5.15)	4.34(0.11–11.3)	11.24(9.49–13.27)	Okoniewski and Novesky 1993
Red-tailed hawk	NY	3	2.44 (1.93–3.00)	3.38 (2.24–4.13)	3.42 (0.47–7.31)	9.24 (7.24–11.48)	Okoniewski and Novesky 1993
American kestrel	MD	2	1.39 (0.08–2.69)	2.97 (2.63–3.3)	0.94 (0.75–1.13)	5.29 (4.13–6.45)	Okoniewski and Novesky 1993
American kestrel	NY	3	3.51 (2.3–5.17)	3.12 (1.97–5.15)	1.97 (0.83–3.7)	8.61 (6.7–11.15)	Okoniewski and Novesky 1993

TABLE 13.6 (continued)
Select Field Studies of Cyclodiene Residues in Brains of Birds (µg/g ww, Mean, Range in Brackets) Associated with Mortality

Species	Location	Sample Size	Heptachlor Epoxide (0.5)	Oxychlordane (0.8)	Dieldrin (1)	Toxic Equivalents	Reference
Common barn owl	NY	2	2.04 (1.71–2.36)	3.71 (2.91–4.51)	0.25 (0.09–0.40)	5.99 (5.02–6.96)	Okoniewski and Novesky 1993
Eastern screech owl	NY	4	1.73 (0.71–2.36)	4.21 (3.83–4.51)	0.17 (0.04–0.35)	6.11 (4.58–7.00)	Okoniewski and Novesky 1993
Great horned owl	NY	13	3.22 (0.44–14.6)	3.59 (1.16–5.87)	2.33 (0.01–10.3)	8.97 (5.17–17.31)	Okoniewski and Novesky 1993
Barred owl	MD	1	2.41	3.20	0.39	6.00	Okoniewski and Novesky 1993

Note: Dieldrin equivalents are calculated using the dieldrin equivalence factors suggested in this report and as shown below the compound.

Units = µg/g.

NJ = New Jersey, NY = New York, MD = Maryland.

The toxicity of chlordane compounds to passerine birds including red-winged blackbirds, common grackles (*Quiscalus quiscula*), brown-headed cowbirds, and starlings has been examined in feeding studies. Wiemeyer (1996) examined data from studies with oxychlordane or mixtures containing chlordane that compared brain residues in birds that died with those that survived and were sacrificed, and recommended that a concentration of 5 µg/g oxychlordane in brain could be considered diagnostic of lethality (Stickel et al. 1979c, 1983).

13.2.4 HEPTACHLOR

Heptachlor (1,4,5,6,7,8,8-heptachloro-3a,4,7,7,7a-tetrahydro-4,7-methanoindene), another cyclodiene insecticide, was primarily used to control soil pests, including termites (Fairchild 1976). Most uses of heptachlor were phased out by 1983 in the United States. Heptachlor is readily metabolized to heptachlor epoxide in vertebrates.

Lethal concentrations of heptachlor epoxide were examined in feeding studies with technical heptachlor in passerine birds. Comparisons of brain residues of heptachlor epoxide were compared between birds that died and survivors that were later sacrificed, and a critical value of 8 µg/g heptachlor epoxide has been suggested (Stickel et al. 1979c).

Effects of heptachlor treated seed on wildlife were investigated in field studies during the late 1970s in the Columbia Basin of Oregon. Canada geese (*Branta canadensis*) fed on spilled and dumped seed, and exhibited reduced nest success when heptachlor epoxide residue concentrations exceeded 10 µg/g in eggs (Blus et al. 1984). American kestrels that fed on contaminated prey had reduced nest success when egg resides exceeded 1.5 µg/g.

13.3 OTHER OC PESTICIDES IN BIRDS

Numerous publications continue to report the presence of OC pesticides in tissues of wild birds, and in some cases to report correlative associations with toxicological or ecological parameters (e.g., Henricksen et al. 2000, Bustnes et al. 2005, Helberg et al. 2005). However, in most cases there is relatively limited new information from which to derive critical tissue concentrations for endosulfan, hexachlorocyclohexane, methoxychlor, or toxaphene. Much of the information presented here on the characteristics of each chemical is summarized from Wiemeyer (1996) in the previous edition of this book, with reference to the original sources. Criteria, where discernible, are provided in Table 13.7.

13.3.1 ENDOSULFAN

Endosulfan (6,7,8,9,10,10-hexachloro-1,5,5a,6,9,9a-hexahydro-6,9-methano-2,4,3-benzodioxathiepin-3-oxide) is a member of the cyclodiene group but differs so greatly in properties and effects that it is recommended that it not be considered one of them (Maier-Bode 1968). The technical grade includes two isomers α and β at a 70:30 ratio. Endosulfan is rapidly eliminated and does not readily accumulate.

Endosulfan metabolites were reported in water and sediments near areas of use in 2005–2006 (Bishop et al. 2000). Low concentrations of endosulfan sulphate were reported in eggs of tree swallows (*Tachycineta bicolor*) from an orchard environment in the 1990s (Bishop et al. 1998).

13.3.2 HEXACHLOROBENZENE

Hexachlorobenzene (HCB) was registered and applied in many jurisdictions as a fungicidal seed treatment. It was used in a number of industrial manufacturing processes, and was found as a

---

TABLE 13.7
A Summary of Published Critical Values for Other Organochlorines

Compound	Endpoint	Tissue	Concentration (µg/g ww)	Comments	Reference
Chlordane	Lethality	Brain	5	As oxychlordane	Stickel et al. 1979c
Heptachlor	Lethality	Brain	8	As heptachlor epoxide	Stickel et al. 1979c, 1983
	Reproduction	Egg	1.5	As heptachlor epoxide	Henny et al. 1983
HCB	Chick Survival	Egg	6.2	Chick survival reduced in chicks hatched from eggs produced by Japanese quail (*Coturnix coturnix japonica*) fed 20 ppm for 90 days	Schwetz et al. 1974
	Reproduction	Egg	0.35–4.4	NOAEL from Japanese quail fed 1 µg/g for 90 days	Vos et al. 1971
Mirex	Lethality	Brain	200	If carcass lipid < 2.4%	Stickel et al. 1973
	Lethality	Brain	177		Stickel et al. 1973
	Reproduction	Egg	20	NOAEL from mallards fed 10 µg/g in diet for 3 months	Heath and Spann 1973
	Sperm number	Liver	1.6	Liver residues in kestrels fed 8 µg/g in diet and tested for sperm quality	Bird et al. 1983
Toxaphene	Duckling growth	Egg	45.4 (based on area under 2 peaks)	Mean egg residues from long term study of reproduction of black ducks (*Anas rubripes*) fed 50 µg/g in diet. Ducklings from that group had depressed growth	Mehrle et al..1979, Haseltine et al. 1980
	Reproduction	Egg	13.9 (based on 3 peaks)	NOAEL in chickens (*Gallus domesticus*) fed 100 µg/g	Bush et al. 1977

contaminant in pesticides such as dacthal and pentachloronitrobenzene. It is a bioaccumulative and persistent compound.

Japanese quail fed 20 µg/g of HCB in the diet and the survived had a mean of 35.5 µg/g HCB (range 13.8–94 µg/g) in liver at sacrifice (Vos et al. 1968). Residues in liver ranging from 4.1 to 16.7 µg/g HCB were associated with increased liver weight and evidence of porphyria in Japanese quail fed 5 ppm of HCB (Vos et al. 1971). Livers of quail fed 1 µg/g in diet for 90 days contained 0.35–4.4 µg/g and were not associated with any evident effects. Chicks of Japanese quail fed 20 µg/g of HCB for 90 days exhibited a significant reduction in survival; mean HCB concentrations in eggs collected toward the end of the study were 6.2 µg/g wet weight (Schwetz et al. 1974).

Reproduction was apparently normal in both Canada geese with 2.97 µg/g HCB in eggs (Blus et al. 1984) and American kestrels with 2.4 µg/g in eggs (Henny et al. 1983). Recently, Bustnes et al. (2005) reported in a field study of the glaucous gull (*Larus hyperboreus*) from the Norwegian arctic that early chick growth was negatively related to maternal levels of four OCs, including HCB. Whole blood concentrations of HCB were in the range of 12.5–20 ng/g; DDE and particularly PCB concentrations were much higher. Helberg et al. (2005) reported that HCB and other contaminants in blood of female great black-backed gulls (*Larus marinus*) were associated with laying date. It is difficult to evaluate the significance of those results given the lack of experimental or other field data for comparison. However, the concentrations are much lower than would be expected in association with the no effect egg levels of 0.35–4.4 µg/g reported in quail by Vos et al. (1971).

13.3.3 HEXACHLOROCYCLOHEXANE

Hexachlorocyclohexane (1,2,3,5,6-hexachlorocyclohexane) also known as benzene hexachloride occurs as eight different isomers, but only three, α, β, and γ, are of environmental interest (Willett et al. 1998). The γ isomer is also known as lindane and is the most active insecticide (USDA 1980). The three major uses were on seed, on hardwood lumber, and on livestock (USDA 1980, Wiemeyer 1996). Lindane is readily metabolized and excreted in birds, and minimally accumulated in tissues (Blus et al. 1984), which was confirmed by Clark et al. (1987) who estimated a half-life for γ-hexachlorocyclohexane in herring gulls of 2.3 days. Thus, it is not surprising that residues of HCH, especially the γ-isomer, are generally found at low, for example, <0.1 µg/g concentrations in eggs and other tissues of wild birds. However, high concentrations of ΣHCHs, mainly β-HCH, regularly exceeding 1 µg/g wet weight and as high as 8.8 µg/g were reported in samples of resident and migratory birds from India (Tanabe et al. 1998).

There is a lack of data on toxicity of HCH to birds to form a basis for criteria or guidelines. Rock doves fed a high dose 72 mg/kg γ-HCH daily had liver residues of 4.2–37 µg/g γ-HCH when sacrificed after 15 days (Turtle et al. 1963). A dosing study with γ-HCH in pheasants reported no effects on hatching of eggs with residues of 10 µg/g (Ash and Taylor 1964).

13.3.4 METHOXYCHLOR

Methoxychlor (1,1″-(2,2,2-trichloroethylidene)-bis(4-methoxybenzene) is the p,p'-methoxy analogue of DDT and has been used to control a wide spectrum of insects. In Canada, it was used to control biting flies (Gardner and Bailey 1975). Methoxychlor is readily metabolized in birds and tends not to bioaccumulate (Gardner and Bailey 1975).

There has been some recent interest in the potential of methoxychlor as an endocrine-disrupting compound in wildlife. Ottinger et al. (2005) reported evidence of physiological effects, such as reduced circulating levels of sex steroids, in Japanese quail fed a low dose of 0.5 µg/g methoxychlor. Feeding of hens dosed with methoxychlor prior to egg laying resulted in concentrations of 2–6 ng/g methoxychlor in egg yolks. Gee et al. (2004) dosed zebra finches (*Taeniopygia guttata*) with methoxychlor and reported effects on egg quality and some other reproductive parameters. They considered that the dose was within the range potentially encountered by seed-eating birds in the field. Overall, monitoring studies have rarely detected methoxychlor residues in fish or wildlife (e.g., Schmitt et al. 1985). But recently, methoxychlor was detected intermittently but at concentrations up to 112 ng/g in chorioallantoic membranes of Morelet's crocodiles (*Crocodylus moreletti*) from Belize (Pepper et al. 2004).

13.3.5 Mirex

Mirex (1,1a,2,2,3,3a,4,5,5,5a,5b,6-dodechlorooctahydro-1,3,4-metheno-*H*-cyclo-buta[*cd*]pentalene) was used for control of fire ants in the southeastern United States, and as a flame retardant, while manufacturing residues contaminated Lake Ontario (Norstrom et al. 1980). All uses were cancelled in 1978 (Eisler 1985).

Stickel et al. (1973) concluded that lethal brain residues were greater than 117 µg/g if carcass lipids were depleted to less than 2.4% and that greater than 200 µg/g in brain was certainly associated with lethality. Mallard reproduction was not affected in birds given 10 ppm of mirex for 3 months, while egg levels on that dose contained 20 µg/g, which could be considered a no effect level (NOEL) when compared to other reproductive studies.

13.3.6 Toxaphene

Toxaphene is a complex mixture of chlorinated camphenes including 177 chemicals with a wide range of toxicities, most of which have not been identified (Eisler and Jacknow 1985). This has made chemical analysis problematic and interpretation of analytical results difficult and is compounded by changes in the composition of the residues found in tissues. It is one of the less frequently reported OC pesticides in wildlife tissues, although that may partly be due to methodology factors. It was one of the most widely used OC insecticides in the United States, but most registrations were cancelled in 1982 (Wiemeyer 1996; see also Pollock and Kilgore 1978). American black ducks were fed 0, 10, and 50 µg/g of toxaphene through two breeding seasons, and essentially experienced no reproductive effects, even with total toxaphene residues in the 50 µg/g group at 50 µg/g in eggs after two cycles (Haseltine et al. 1980). Mehrle et al. (1979) did find significant effects on growth of ducklings produced in that study from the 50 µg/g group.

Some studies have recently reported toxaphene residues in wildlife samples. Toxaphene was one of the OCs measured at high tissue concentrations and potentially implicated in mortality of waterbirds on restored Lake Apopka lands (Rauschenberger 2007). Total toxaphene residue concentrations greater than 1 µg/g were also measured in both eggs and nestling plasma of osprey nesting in remote alpine lakes in western Canada (Elliott et al. 2007).

13.4   AN ADDITIVE MODEL FOR TOXICITY OF OC MIXTURES

For both the DDTs and the cyclodienes, the essential mode of action is neurotoxicity, with similarities and differences in the apparent specific mechanisms, and in the appearance of intoxicated animals. Both groups of chemicals affect neuronal reflex arcs, which in the simplest form includes a peripheral afferent or sensory neuron connecting to the CNS, which in turn connects and interacts with efferent motor neurons leading back to muscle tissues. Signals are conducted along the neuron by changes in the membrane electric potential, involving small movements of cations, sodium and potassium, across the membrane. As the signal is propagated along the neuron, the electric potential returns to a normal or resting state. This process takes place against the gradient of ion concentrations and thus requires energy. This involves active ion transport via the "sodium pump" mediated by a specific enzyme $Na^+ K^+$–ATPase (Matsumura 1975).

DDT-poisoned animals exhibit intermittent periods of continuous tremoring and seizures, consistent with neurons discharging repetitively. The animal is also hyper-sensitive to stimuli. DDT effects on the nervous system are initially observed in the peripheral sensory tract. In contrast, toxicity of the cyclodienes, including aldrin, dieldrin, endrin, and the chlordanes is located in the CNS. Superficially at least, symptoms of cyclodiene toxicity resemble the DDTs, and there are similar effects on ion-transport systems (Matsumura 1975). Like DDTs, they also affect neuronal $Na^+-K^+-ATPase$ and $Ca^+-Mg^+-ATPase$ (Matsumura 1975, Wafford et al. 1989).

The problems associated with simultaneous exposure to more than one toxicant have been recognized for some time. Early researchers attempted to classify mixture responses, and several empirical studies determined that additive models were largely adequate for predicting the toxicity at least of paired exposure to range of toxic agents (see, e.g., Finney 1952, Pozzani et al. 1959, Smyth et al. 1969). In recent years there have been further efforts to address the challenges posed by exposure to both simple and complex mixtures (Groten et al. 2001). Although there are some differences in responses and mechanisms by which DDTs and cyclodienes cause intoxication of animals, they do impact the same target organ system and share some commonality of mechanism. That would appear to adequately address the requirements for application of an additive model proposed by Ferona et al. (1995). Additionally, Ludke's (1976) findings provide some empirical confirmation that two cyclodienes, chlordane and endrin, act in an additive manner. That was determined by comparing brain residue concentrations in bobwhite quail (*Colinus virginianus*) dosed only with endrin to those given chlordane followed by endrin. An earlier study reported additive effects of dosing with dieldrin and chlordane in rats (Keplinger and Deichmann 1967).

The acute toxicity of both cyclodiene and DDT compounds to birds and mammals has been evaluated by dose–response experiments, and by measurements of brain residues. Given that we have reasonably good information on what constitutes a lethal brain concentration, we believe there is adequate support to apply a simple additive model at least for purposes of interpreting brain residue data.

Stickel et al. (1970) devised an additive DDT equivalents model based on toxic brain residue concentrations. In their scheme, brain residues of 1 µg/g DDT equal 5 µg/g DDD and 15 µg/g DDE. They set a lower lethal limit of 10 µg/g DDT equivalents in brain. Blus (1996) expressed reservations about the conservative lethal brain residues recommended by Stickel et al. (1970). We concur with those reservations, and have used higher values of 25 µg/g for DDT and 300 µg/g for DDE based on concentrations of DDT and DDE in birds which died in dosing studies summarized by Blus (1996).

Using a similar approach and the data in their studies of cyclodiene and DDT toxicity to birds, we propose the use of a toxic equivalence scheme for both groups of OC compounds, as shown in Table 13.8. We have set dieldrin at unity, as the contaminant for which there is the most extensive data base on lethal residues in birds, and then set the toxic equivalence factors (TEFs) accordingly.

Thus, to determine the dieldrin toxic equivalents in a sample, we take the product of the concentration of the chemical and its TEF to obtain the toxicity equivalent concentration (TEQ). The

TABLE 13.8

A Dieldrin Toxic Equivalents Scheme for Interpreting Concentrations of Organochlorine Insecticide Residues in Brains of Birds or Mammals

Compound	Lethal Brain Concentration	Toxic Equivalence Factor	Reference
Dieldrin	4	1	Stickel et al. 1969
Endrin	1	4	Stickel et al. 1979b
Oxychlordane	5	0.8	Stickel et al. 1979c
Heptachlor	8	0.5	Stickel et al. 1979c
DDT	25	0.16	Stickel et al. 1970, Blus 1996
DDD	50	0.08	Stickel et al. 1970, Blus 1996
DDE	300	0.0133	Stickel et al. 1970, Blus 1996
DDT equivalents	25	0.16	Stickel et al. 1970, Blus 1996

concentration of a mixture of OC chemicals is then estimated by adding the converted concentrations to derive a TEQ for the mixture. Therefore,

$$TEQ_{mixture} = \Sigma(TEF_i * c_i)$$

where, c_i is the concentration of an individual compound and TEF_i is the corresponding TEF.

This provides, therefore, a relatively simple and reasonably defensible scheme for interpreting data on mixtures of cyclodiene and DDT compounds measured in brain of birds. A number of authors have previously devised (Sileo et al. 1977) or discussed the appropriateness of an additive model at least of cyclodiene toxicity (Wiemeyer 1996, Stansley and Roscoe 1999). Interestingly, for many of the cases, concentrations of the individual compounds are less than the putative lethal levels (Table 13.6). However, when the dieldrin toxic equivalence was calculated from the values reported for the individual cyclodienes, mean concentrations in brain exceeded 4 μg/g in all instances.

13.5 EXTRAPOLATION OF DIELDRIN EQUIVALENTS TO EGGS

As also discussed in the section on sublethal effects, we believe that dieldrin equivalents could be calculated for egg data, using only cyclodiene residue measurements (not DDTs), and used to infer toxicity. Egg values provide a reasonable metric for exposure of the female. For example, wet weight liver:egg ratios for oxychlordane, *trans*-nonachlor, heptachlor epoxide and dieldrin were all one or very close to one in a study of herring gulls (Braune and Norstrom 1989). Vermeer and Reynolds (1970) reported a liver: egg ratio of 1.5 in California gulls (*Larus californicus*) though r values were low. There is, however, a marked lack of data comparing brain residues with other tissues, especially eggs, in birds. Enderson and Berger (1970) experimentally dosed wild prairie falcons (*Falco*) with dieldrin and reported residues in eggs and brain, but it is not possible to determine ratios from those results. Liver: brain ratios in glaucous gulls (*Larus hyperborus*) of dieldrin and oxychlordane were remarkably close to one, although the poor body condition of those birds may have influenced tissue ratios (Gabrielsen et al. 1995). In California, gulls brain:egg ratios were 0.27 with an r = 0.937 (Vermeer and Reynolds 1970). Given the lack of data and the variability, the extrapolation from egg to brain has to be done with some caution and awareness that confidence intervals are large. Many factors could influence the relationships in any given situation, particularly the condition of the breeding birds and life history factors such as breeding strategies, income versus capital breeding for example, and diet.

As discussed in the sublethal effects section, dieldrin, and possibly other cyclodienes, could affect reproduction of wild birds via mortality of adult males. The male bird may not only have a higher residue burden as, unlike the female, he does not deposit some portion to the eggs, but also in many species the male must forage to provision himself, the female and the growing young (Newton 1998). Thus there is a risk of the male bird experiencing either overt toxicity, particularly if lipid stores are depleted during foraging efforts, or the more subtle anorexic effects, leading eventually lethal feedback process of reduced feeding and further mobilization of lipid stores and contaminants. Concentrations of total dieldrin equivalents in eggs exceeding 1 μg/g, therefore, may be an indirect indicator of the exposure of male birds to brain levels sufficient to cause some anorexia, and thus affect foraging activity and potentially survival.

Summary

In birds, a diagnosis of poisoning can be made with confidence when concentrations of dieldrin in brain are ≥10 μg/g. In a number of laboratory studies a concentration of 4 μg/g dieldrin in the brain is associated with death of a proportion of birds, and given the additional stresses of a field situation, it is a more protective criterion, and is recommended here (Table 13.5). Evidence from a number of lab and field studies suggests that dieldrin can cause cessation of feeding or anorexic behavior at concentrations as low as 1 µg/g in brain. Although we stress that value is a guideline as a cause for concern, but may be of particular value for interpretation of residues found in birds collected during migration or other stressful periods including reproduction. That rationale also can be applied to interpretation of egg residues, as the finding of greater than 1 µg/g dieldrin in eggs indicates a similar range of exposure of adult birds, and suggests that male birds in particular could be at risk given the stress of additional foraging (Table 13.5). Measurement of 1 µg/g dieldrin in plasma also indicates a similar value in brain and should also be considered as cause for concern. In juvenile owls as little as 0.1 µg/g dieldrin was associated with reduced survival at fledging.

Given the commonality of mechanism for the cyclodiene compounds and the overlap with the DDT compounds, we suggest use of a dieldrin toxic equivalence scheme for interpretation of data on mixtures of those compounds. The TEFs (Table 13.8) were derived from data on lethal brain residues and would be most applicable as an aid to diagnosing cause of death when a mixture of OC residues are measured in the brain. However, we suggest that calculation of dieldrin toxic equivalents using only cyclodiene residues (*excluding DDTs* as shell quality mechanisms are likely to dominate in the case of eggs) also could prove useful for interpretation of egg residues. The egg provides a valid metric of adult exposure and, therefore, whether adults were exposed to residues exceeding our suggested anorexic level, thus impairing foraging or survival of adult birds. We draw the readers' attention again to the recommended criteria of 1 µg/g dieldrin or dieldrin toxic equivalents in eggs as concentration potentially indicative of effects on reproduction via impaired foraging or reduced adult survival.

We have provided a summary of available criteria for other OCs (Table 13.7), which may be of value to interpretation of residues of HCB, mirex, and toxaphene in avian tissues.

ACKNOWLEDGMENTS

S. Lee assisted with drafting of figures and A. Potvin assisted in the compilation of tables. Useful comments on an earlier version of the paper were provided by B. Braune and C. Walker.

REFERENCES

Ash, J. S., and A. Taylor. 1964. Further trials on the effects of gamma BHC seed dressing on breeding pheasants. *Game. Res. Assoc. Annu. Rep.* 4:14–20.

Atkins, T. D., and R. L. Linder. 1967. Effects of dieldrin on reproduction of penned hen pheasants. *J. Wildl. Manage.* 31:746–753.

Babcock, K. M., and E. L. Flickinger. 1977. Dieldrin mortality of lesser snow geese in Missouri. *J. Wildl. Manage.* 41:100–103.

Bairlein, F., and E. Gwinner. 1994. Nutritional mechanisms and temporal control of migratory energy accumulation in birds. *Annu. Rev. Nutr.* 14:187–215.

Baxter, R. L., G. Linder, and R. B. Dahlgren. 1969. Effects in two generations of penned hen pheasants. *J. Wildl. Manage.* 33:96–102.

Bedford, C. T., D. H. Hudson, and I. L. Natof. 1975. The acute toxicity of endrin and its metabolites to rats. *Toxicol. Appl. Pharmacol.* 33:115–121.

Bird, D. M., P. H. Tucker, G. A. Fox, and P. C. Laguë. 1983. Synergistic effects of Aroclor® 1254 and mirex on the semen characteristics of American Kestrels. *Arch. Environ. Toxicol. Chem.* 12:633–639.

Bishop, C. A., S. L. Ashpole, A. M. Edwards, G. VanAggelen, and J. E. Elliott. 2010. Hatching success and pesticide exposures in amphibians living in agricultural habitats of the South Okanagan valley, British Columbia, Canada (2004–2006). *Environ. Toxicol. Chem.* 29:1593–1603.

Bishop, C. A., B. Collins, P. Mineau, N. M. Burgess, W. F. Read, and C. Risley. 2000. Reproduction of cavity-nesting birds in pesticide-sprayed apple orchard in southern Ontario, Canada (1988–1994). *Environ. Toxicol. Chem.* 19:588–599.

Bloomquist, J. R., and D. M. Soderlund. 1985. Neurotoxic insecticides inhibit GABA-dependent chloride uptake by mouse brain vesicles. *Biochem. Biophys. Res. Commun.* 133:37–43.

Blus, L. J. 1996. DDT, DDD, DDE in Birds. In *Environmental Contaminants in Wildlife*, eds. W. N. Beyer, G. H. Heinz, and A. W. Redmon-Norwood, pp. 49–71. Boca Raton, FL: CRC Press.

Blus, L. J. 2003. Organochlorine pesticides. In *Handbook of Ecotoxicology*, eds. D. J. Hoffman, B. A. Rattner, G. A. Burton, and J. Cairns, pp. 313–339. Boca Raton: CRC Press.

Blus, L. J., E. Cromartie, L. McNease, and T. Joanen. 1979. Brown pelican: population status, reproductive success, and organochlorine residues in Louisiana, 1971–1976. *Bull. Environ. Contam. Toxicol.* 22:128–135.

Blus, L. J, C. J. Henny, T. E. Kaiser, and R. A. Grove. 1983. Effects on wildlife from use of endrin in Washington State orchards. *Trans. N. Am. Nat. Res. Conf.* 48:159–174.

Blus, L. J., C. J. Henny, D. J. Lenhart, and T. E. Kaiser. 1984. Effects of heptachlor- and lindane-treated seed on Canada geese. *J. Wildl. Manage.* 48:1097–1111.

Bogan, J. A., and J. Mitchell. 1973. Continuing dangers to peregrines from dieldrin. *Br. Birds.* 66:437–439.

Both, C., and M. E. Visser. 2001. Adjustment to climate change is constrained by arrival date in a long-distance migrant bird. *Nature* 411:296–298.

Braune, B. M., and R. J. Norstrom. 1989. Dynamics of organochlorine compounds in herring gulls: III. Tissue distribution and bioaccumulation in Lake Ontario gulls. *Environ. Toxicol. Chem.* 8:957–968.

Busbee, E. L. 1977. The effects of dieldrin on the behaviour of young loggerhead shrikes. *Auk.* 94:28–35.

Bush, P. B., J. T. Kiker, R. K. Page, N. H. Booth, and O. J. Fletcher. 1977. Effects of graded levels of toxaphene on poultry residue accumulation, egg production, shell quality and hatchability in white leghorns. *J. Agric. Food Chem.* 25:928–932.

Bustnes, J. O., O. Miland, M. Fjeld, K. E. Erikstad, and J. U. Skaare. 2005. Relationships between ecological variables and four organochlorine pollutants in an arctic glaucous gull (*Larus hyperboreus*) population. *Environ. Pollut.* 136:175–185.

Clark, D. R. 1975. Effects of stress on dieldrin toxicity to male redwinged blackbirds (*Agelaius phoeniceus*). *Bull. Environ. Contam. Toxicol.* 14:250–256.

Clark, T. P., R. J. Norstrom, G. A. Fox, and H. T. Won. 1987. Dynamics of organochlorine compounds in herring gulls (*Larus argentatus*): II. A two-compartment model and data for ten compounds. *Environ. Toxicol. Chem.* 6:547–559.

Coats, J. R. 1990. Mechanisms of toxic action and structure-activity relationships for organochlorine and synthetic pyrethroid insecticides. *Environ. Health Perspect.* 87:255–262.

Dahlgren, R. B., and R. L. Linder. 1974. Effects of dieldrin in penned pheasants through the third generation. *J. Wildl. Manage.* 38:320–330.

Davison, K. L., J. L. Sell, and R. J. Rose. 1971. Dieldrin poisoning of chickens during severe dietary restriction. *Bull. Environ. Contam. Toxicol.* 5:493–501.

Deviche, P. 1995. Androgen regulation of avian pre-migratory hyperphagia and fattening: from eco-physiology to neuroendocrinology. *Am. Zool.* 35:234–245.

Ecobichon, D. J. 1991. Toxic effects of pesticides. In: *Casarett and Doull's Toxicology*, eds. M. O. Amdur, J. Doull, and C. D. Klaassen, pp. 565–622. New York: Pergamon.

Eisler, R. 1985. Mirex hazards to fish, wildlife and invertebrates: a synoptic review. *US Fish Wildl. Serv. Biol. Rep.* 85(1.1):42 pp.

Eisler, R. 1995. Chlordane hazards to fish, wildlife and invertebrates: a synoptic review. *US Fish Wildl. Serv. Biol. Rep.* 85(1.21):49 pp.

Eisler, R., and J. Jacknow. 1985. Toxaphene hazards to fish, wildlife and invertebrates: a synoptic review. *US Fish Wildl. Serv. Biol. Rep.* 85(1.4):26 pp.

Eldefrawi, M. E. S., S. M. Sherby, I. M. Abalis, and A. T. Eldefrawi. 1985. Interactions of pyrethroid and cyclodiene insecticides with nicotinic acetylcholine and GABA receptors. *Neurotoxicol.* 6:47–62.

Elliott, J. E., and M. L. Harris. 2001. An ecotoxicological assessment of chlorinated hydrocarbon effects on bald eagle populations. *Rev. Toxicol.* 4:1–60.

Elliott, J. E., and P. A. Martin. 1994. Chlorinated hydrocarbons and shell thinning in eggs of (*Accipiter*) hawks in Ontario, 1986–1989. *Environ. Pollut.* 86:189–200.

Elliott, J. E., M. M. Machmer, L. K. Wilson, and C. J. Henny. 2000. Contaminants in Ospreys from the Pacific Northwest: II. Organochlorine pesticides, polychlorinated biphenyls, and mercury, 1991–1997. *Arch. Environ. Contam. Toxicol.* 38:93–106.

Elliott, J. E., C. A. Morrissey, M. Wayland, L. K. Wilson, and D. C. G. Muir. 2007. Bioaccumulation of chlorinated hydrocarbons in relation to productivity and dietary factors in ospreys breeding in alpine regions of western Canada. Annual meeting of the Society of Environmental Toxicologists and Chemists. Milwaukee, WI, November 11–15.

Enderson, J. H., and D. D. Berger. 1970. Pesticides, eggshell thinning and lowered production of young in prairie falcons. *Bioscience* 20:355–356.

Fairchild, H. E. 1976. Heptachlor in relation to man and environment. EPA-540/4-76-007. US EPA, Washington DC, 65 pp.

Ferona, V. J., J. P. Grotena, J. A. van Zorgeb, F. R. Casseea, D. Jonkera,and P. J. van Bladerena. 1995. Toxicity studies in rats of simple mixtures of chemicals with the same or different target organs. *Toxicol. Lett.* 82:505–512.

Finney, D. J. 1952. *Probit Analysis*, 2nd Ed. London: Cambridge University Press.

Fleming, W. J. M., M. A. R. McLane, and E. Cromartie. 1982. Endrin decreases screech owl productivity. *J. Wildl. Manage.* 46:462–468.

Flickinger, E. L. 1979. Effects of aldrin exposure on snow geese in Texas rice fields. *J. Wildl. Manage.* 43:94–101.

Flickinger, E. L., and K. A. King. 1972. Some effects of aldrin-treated rice on Gulf Coast wildlife. *J. Wildl. Manage.* 36:706–777.

Flickinger, E. L., C. A. Mitchell, and A. J. Krynitsky. 1986. Dieldrin and endrin residues in fulvous whistling ducks in Texas in 1983. *J. Field Ornithol.* 57:85.

Fowler, J. F., L. D. Newsom, J. B. Graves, F. L. Bonner, and P. E. Schilling. 1971. Effect of dieldrin on egg hatchability, chick survival and eggshell thickness in purple and common gallinules. *Bull. Environ. Contam. Toxicol.* 6:495–501.

Frank, R. A., and R. S. Lutz. 1999. Productivity and survival of Great Horned Owls exposed to dieldrin. *Condor* 101:331–339.

Friend, M., and D. O. Trainer. 1974. Experimental dieldrin–duck hepatitis virus interaction studies. *J. Wildl. Manage.* 38:896–902.

Fuchs, P. 1967. Death of birds caused by application of seed dressings in the Netherlands. *Meded. Rijsfaculteit. Landbouwwet. Gent.* 32:855–859.

Gabrielsen, G. W., J. U. Skaare, A. Polder, and V. Bakken. Chlorinated hydrocarbons in glaucous gulls (*Larus hyperboreus*) in the southern part of Svalbard. *Sci. Tot. Environ.* 160–161:337–346.

Gardner, D. R., and J. R. Bailey. 1975. Methoxychlor: its effects on environmental quality. NRCC No. 14102. National Research Council of Canada, Ottawa, 164 pp.

Gee, J. M., C. B. Craig-Veit, and J. Millam. 2004. Posthatch methoxychlor exposure adversely affects reproduction of adult zebra finches. *Taeniopygia guttata* 73:607–612.

Genelly, R. E., and R. I. Rudd. 1956. Effects of DDT, toxaphene, and dieldrin on pheasant reproduction. *Auk.* 73:529–539.

Gesell, G. G., and R. J. Robel. 1979. Effects of dieldrin on operant behavior of bobwhites. *J. Environ. Sci. Health.* 14:153–170.

Graves, J. B., F. L. Bonner, W. F. McKnight, A. B. Watts, and E. A. Epps. 1969. Residues in eggs, preening glands, liver and muscle from feeding dieldrin-contaminated rice bran to hens and its effects on egg production, egg hatch, and chick survival. *Bull. Environ. Contam. Toxicol.* 4:375–383.

Groten, J. P., V. J. Feron, and J. Suhnel. 2001. Toxicology of simple and complex mixtures. *Trends Pharmacol. Sci.* 22(6):316–322.

Guyton, A. C. 1991. *Textbook of Medical Physiology*, 8th Ed. Philadelphia: Harcourt Brace Jovanovich, Inc.

Haseltine, S. D., M. T. Finley, and E. Cromartie. 1980. Reproduction and residue accumulation in black ducks fed toxaphene. *Arch. Environ. Contam. Toxicol.* 9:461–471.

Hayes, W. J. 1974. Distribution of dieldrin following a single oral dose. *Toxicol. Appl. Pharmacol.* 28:485–492.

Hayes, W. J., and E. R. Laws. 1991. *Handbook of Pesticide Toxicology, No. 2, Classes of Pesticides*. San Diego: Academic Press.

Heath, E. G., and J. W. Spann. 1973. Reproduction and related residues in birds fed mirex. In *Pesticides and the Environment: A Continuing Controversy*, ed. W. B. Deichmann, pp. 421–435. New York, NY: Intercontinental Medical Book Corp.

Heinz, G. H., and R. W. Johnson. 1979. Elimination of endrin by mallard ducks. *Toxicology* 12:189.

Heinz, G. H., and R. W. Johnson. 1981. Diagnostic brain residues of dieldrin. Some new insights. In *Avian and Mammalian Wildlife Toxicology. Second Conference ASTM STP 75 7*, eds. D. W. Lam and E. E. Kenaga, pp. 72–92. Philadelphia: American Society for Testing and Materials.

Helberg, M., J. O. Bustnes, K. E. Erikstad, K. O. Kristiansen, and J. U. Skaare. 2005. Relationships between reproductive performance and organochlorine contaminants in great black-backed gulls (*Larus marinus*). *Environ. Pollut.* 134:475–483.

Henny, C. J. 1997. DDE still high in white-faced Ibis eggs from Carson River, Nevada. *Col. Waterbirds* 20:478–484.

Henny, C. J., L. J. Blus, and C. J. Stafford. 1983. Effects of heptachlor on American kestrels in the Columbia basin. *J. Wildl. Manage*. 47:1080–1087.

Henriksen, E. O., G. W. Gabrielsen, S. Trudeau, J. Wolkers, K. Sagerup, and J. U. Skaare. 2000. Organochlorines and possible biochemical effects in glaucous gulls (*Larus hyperboreus*) from Bjornoya, the Barents Sea. *Arch. Environ. Contam. Toxicol*. 38:234–243.

Hudson, R. H., R. K. Tucker, and M. A. Haegele. 1984. Handbook of toxicity of pesticides to wildlife. US Fish and Wildlife Service, Washington DC, Resource Publ 153, 90 pp.

Jefferies, D. J., and M. C. French. 1972. Changes induced in the pigeon thyroid by *p,p'*-DDE and dieldrin. *J. Wildl. Manage*. 36:24–30.

Jefferies, D. J., and I. Prestt. 1966. Post-mortems of peregrines and lanners with particular reference to rogano-chlorine residues. *Br. Birds* 59:49–64.

Jones, D. M. D., D. Bennett, and K. E. Elgar. 1978. Deaths of owls traced to insecticide-treated timber. *Nature (Lond.)*. 272:52.

Keplinger, M. L., and W. B. Deichmann. 1967. Acute toxicity of combinations of pesticides. *Toxicol. Appl. Pharmacol*. 10:586–595.

Koeman, J. H., A. A. G. Oskamp, J. Veen, E. Brouwer, J. Rooth, P. Zwart, E. Van den Brock, and J. van Genderen J. 1967. Insecticides as a factor in the mortality of the sandwich tern (*Sterna sandvicensis*). A preliminary communication. *Meded. Rijsfaculteit. Land-bouwwet. Gent*. 32:841–854.

Linder, R. L., R. B. Dahlgren, and Y. A. Greichus. 1970. Residues in the brain of adult pheasants given dieldrin. *J. Wildl. Manage*. 34:954–956.

Lockie, J. D., D. A. Ratchliffe, and R. Balharry. 1969. Breeding success and organochlorine residues in golden eagles in West Scotland. *J. Appl. Ecol*. 6:381–389.

Ludke, J. L. 1976. Organochlorine pesticide residues associated with mortality: additivity of chlordane and endrin. *Bull. Environ. Contam. Toxicol*. 16:253–260.

Maier-Bode, H. 1968. Properties, effect, residues and analytics of the insecticide endosulfan. *Residue Rev*. 22:1–44.

Marra, P. P., K. A. Hobson, and R. T. Holmes. 1998. Linking winter and summer vents in a migratory bird by using stable-carbon isotopes. *Science*. 282:1884–1886.

Matsumura, F. 1975. *Toxicology of Insecticides*. New York, NY: Plenum Press.

Matsumura, F., and S. M. Ghiasudding. 1983. Evidence for similarities between cyclodiene type insecticides and picrotoxin in their action mechanisms. *J. Environ. Sci. Health B* 18:1–14.

McEwen, L. C., and R. L. Brown.1966. Acute toxicity of dieldrin and malathion to wild sharp-tailed grouse. *J. Wildl. Manage*. 30:604–611.

Mehrle, P. M., M. T. Finley, J. L. Ludke, F. L. Mayer, and T. E. Kaiser. 1979. Bone development in black ducks as affected by dietary toxaphene. *Pest. Biochem. Physiol*. 10:168–173.

Mendenhall, V. M., E. E. Klass, and A. R. McLean. 1983. Breeding success of barn owls (*Tyto alba*) fed low levels of DDE and dieldrin. *Arch. Environ. Contam. Toxicol*. 12:235–240.

Moore,F., and P. Kerlinger. 1987. Stopover and fat deposition by North American wood-warblers (*Parulinae*) following spring migration over the Gulf of Mexico. *Oecologia* 74:47–54.

Muralidharan, S. 1993. Aldrin poisoning of Sarus cranes (*Grus antigone*) and a few granivorous birds in Keoladeo National Park, Bharatpur, India. *Ecotoxicology* 2:196–202.

NRCC (National Research Council of Canada). 1974. Chlordane: its effects on Canadian ecosystems and its chemistry. Publ. NRCC 14094. Environmental Secretariat, Ottawa.

Newton, I. 1986. *The Sparrowhawk*. Carlton, UK: T&AD Poyser.

Newton, I. 1988. Determination of critical pollutant levels in wild populations, with examples from organo-chlorine insecticides in birds of prey. *Environ. Pollut*. 55:29–40.

Newton, I. 1998. *Population Limitation in Birds*. London: Academic Press.

Newton, I., and J. Bogan. 1974. Organochlorine residues, eggshell thinning and hatching success in British sparrowhawks. *Nature* 249:582–583.

Newton, I., and J. Bogan.1978. The role of different organochlorine compounds in the breeding of British spar-rowhawks. *J. Appl. Ecol*. 15:105–116.

Newton, I., and E. A. Galbraith. 1991. Organochlorines and mercury in the eggs of golden eagles *Aquila chrysaetos* from Scotland (UK). *Ibis 133:15–120*.

Newton, I., and M. B. Haas. 1988. Pollutants in Merlin eggs and their effects on breeding. *Br. Birds* 81:258–269.

Newton, I., I. Wyllie, and A. Asher. 1991. Mortality causes in British Barn Owls *Tyto alba*, with a discussion of aldrin-dieldrin poisoning. *Ibis*. 133:162–169.

Newton, I., I. Wylie, and A. Asher. 1992. Mortality from the pesticides aldrin and dieldrin in British sparrowhawks and kestrels. *Ecotoxicology* 1:31–44.

Nisbet, I. 1989. Organochlorine, reproductive impairment, and declines in bald eagle *Haliaeetus leucocephalus* populations: mechanisms and dose-response relationships. In *Raptors in the Modern World*, eds. B. U. Meyburg and R. D. Chancellow, pp. 483–489. Berlin: World Working Group on Birds of Prey.

Nisbet, I. C. T. 1998. The relative importance of DDE and dieldrin in the decline of peregrine falcon populations. In *Peregrine Falcon Populations: Their Management and Recovery*, eds. T. J. Cade, J. H. Enderson, C. G. Thelander, and C. M. White, pp. 351–375. Boise: Peregrine Fund.

Norstrom, R. J., D. J. Hallett, F. I. Onuska, and M. E. Comba. 1980. Mirex and its degradation products in Great Lakes herring gulls. *Environ. Sci. Technol.* 14:860–866.

Okoniewski, J. C., and E. Novesky. 1993. Bird poisonings with cyclodienes in suburbia: links to historic use on turf. *J. Wildl. Manage.* 57:630–639.

Ottinger, M. A., et al. 2005. Assessing the consequences of the pesticide methoxychlor: neuroendocrine and behavioural measures as indicators of biological impact of an estrogenic environmental chemical. *Brain Res. Bull.* 65:199–209.

Peakall, D. B. 1996. Dieldrin and other cyclodiene pesticides in wildlife. In *Environmental Contaminants in Wildlife: Interpreting Tissue Concentrations*, eds W. N. Beyer, G. H. Heinz, and A. W. Redmon-Norwood, pp. 73–97. Boca Raton: CRC Press..

Pepper, C. B., T. R. Rainwater, S. G. Platt, J. A. Dever, T. A. Anderson, and S. T. McMurry. 2004. Organochlorine pesticides in chorioallantoicmembranes of morelet's crocodile eggs from Belize. *J. Wildl. Dis.* 40:493–500.

Pollock, G. A., and W. W. Kilgore. 1978. Toxaphene. *Residue Rev.* 69:87–140.

Pozzani, U. C., C. S. Weil, and C. P. Carpenter. 1959. The toxicological basis of threshold limit values. 5. The experimental inhalation of vapor mixtures by rats, with notes upon the relationship between single dose inhalation and single dose oral data. *Am. Ind. Hyg. Asssoc. J.* 20:364–369.

Pu, S., M. R. Jain, T. L. Horvath, S. Diano, P. S. Kalra, and S. P. Kalra. 1999. Interactions between neuropeptide Y and y-aminobutyric acid in stimulation of feeding: a morphological and pharmacological analysis. *Endocrinol.* 140:933–940.

Ratcliffe, D. A. 1973. Studies of the recent breeding success of the peregrine, *Falco peregrinus*. *J. Reprod. Fertil.* 19(Suppl.):377–389.

Ratcliffe, D. 1988. *The Peregrine Falcon*. Vermillion: Buteo Books.

Rauschenberger, H. 2007. The Lake Apopka agreement. *Endanger. Species Bull.* 32:16–17.

Reading, C. M., G. H. Arscott, and I. J. Tinsley. 1976. Effect of dieldrin and calcium on the performance of adult Japanese quail (*Coturnix coturnix japonica*). *Poultry Sci.* 55:212–219.

Reichel, W. L., L. N. Locke, and R. M. Prouty. 1974. Peregrine falcon suspected of pesticide poisoning. *Avian Dis.* 18:487–489.

Robinson, J., and A. N. Crabtree. 1969. The effect of dieldrin on homing pigeons (*Columba livia var.*). *Mededlelingin van de Rijksfaculteit Landbouwwetenschappen te Gent* 34:413–427.

Robinson, J., A. Richardson, and V. K. H. Brown. 1967. Pharmacodynamics of dieldrin in pigeons. *Nature* 213:734–736.

Ronis, M. J. J., and C. H. Walker. 1989. The microsomal monooxygenases of birds. *Rev. Biochem. Toxicol.* 10:301–384.

Roy, R. 1997. Results from the American Kestrel (*Falco sparvenus*) biomonitoring study at the Rocky Mountain Arsenal National Wildlife Refuge, 1993–1996. USFWS. Dept. of Interior, 54 pp.

Rudd, R. L., and R. E. Genelly. 1956. Pesticides: their use and toxicity to wildlife. *Calif. Dep. Fish Game Bull.* No. 7, 209 pp.

Safe, S. 1990. Polychlorinated biphenyls (PCBs), dibenzo-p-dioxins (PCDDs), dibenzofurans (PCDFs), and related compounds: environmental and mechanistic considerations which support the development of toxic equivalency factors (TEFs). *Crit. Rev. Toxicol.* 21:51–88.

Schaub, M., and L. Jenni. 2001. Variation of fuelling rates among sites, days and individuals in migrating passerine birds. *Funct. Ecol.* 15:584–589.

Schmitt, C. J., J. L. Zajicek, and M. A Ribick. 1985. National Pesticide Monitoring Program: Residues of organochlorine chemicals in freshwater fish, 1980–81. *Arch. Environ. Contam. Toxicol.* 14:225–260.

Schwetz, B. A., J. M. Norris, R. J. Kocidba, P. A. Keeler, R. F. Cornier, and P. J. Gehrin. 1974. Reproduction study in Japanese quail fed hexachlorobutadiene for 90 days. *Toxicol. Appl. Pharmacol.* 30:255–265.

Scott, I., P. I. Mitchell, and P. R. Evans. 1994. Seasonal changes in body mass, body composition and food requirements in wild migratory birds. *Proc. Nutr. Soc.* 53:521–531.

Sharma, R. P., D. S. Winn, and J. B. Low. 1976. Toxic, neurochemical and behavioural effects of dieldrin exposure in mallard ducks. *Arch. Environ. Contam. Toxicol.* 5:43–53.

Sheail, J. 1985. *Pesticides and Nature Conservation the British Experience 1950–1975.* Oxford: Clarendon Press.

Sibly, R. M., I. Newton, and C. H. Walker. 2000. Effects of dieldrin on population growth rates of sparrowhawks 1963–1986. *J. Appl. Ecol.* 37:540–546.

Sileo, L., L. Karstad, R. Frank, M. V. H. Holdrinet, E. Addison, and H. E. Braun. 1977. Organochlorine poisoning of ring-billed gulls in southern Ontario. *J. Wildl. Dis.* 13:313–322.

Smyth, H. F., C. S. Weil, J. S. West, and C. P. Carpenter. 1969. An exploration of joint toxic action: twenty-seven industrial chemicals intubated issn rats in all possible pairs. *Toxicol. Appl. Pharmacol.* 14:340–347.

Stanley, P. I., and P. J. Bunyan. 1979. Hazards to wintering geese and other wildlife from the use of dieldrin, chlorfenviphos, and carbophenothion as wheat seed treatments. *Proc. R. Soc. London. Ser. B. Biol. Sci.* 205:31–45.

Stanley, P. I., and M. R. Fletcher. 1981. A review of the wildlife incidents investigated from October 1978 to September 1979. *Pest. Sci.* 252:55–63.

Stansley, W., and D. E. Roscoe. 1999. Chlordane poisoning of birds in New Jersey. *USA. Environ. Toxicol. Chem.* 18:2095.

Stansley, W., D. E. Roscoe, E. Hawthorne, and R. Meyer. 2001. Food chain aspects of chlordane poisoning in birds and bats. *Arch. Environ. Contam. Toxicol.* 40:285–291.

Stickel, W. H., J. A. Galyen, T. A. Dyrland, and D. L. Hughes. 1973. Toxicity and persistence of mirex in birds. In *Pesticides and the Environment: A Continuing Controversy*, ed. W. B. Deichman, pp. 437–467. New York, NY: Intercontinental Medical Book Corp.

Stickel, W. H., T. E. Kaiser, and W. L. Reichel. 1979a. Endrin vs. 12-ketoendrin in birds and rodents. In *Avian and mammalian wildlife toxicology*, ed. E. E. Kenaga, pp. 61–68. ASTM STP 693.

Stickel, W. H., W. L. Reichel, and D. L. Hughes. 1979b. Endrin in birds: lethal residues and secondary poisoning. In *Toxicology and Occupational Medicine*, ed. W. B. Deichman, pp. 397–406. North Holland: Elsevier.

Stickel, W. H., L. F. Stickel, and J. W. Spann. 1969. Tissue residues of dieldrin in relation to mortality in birds and mammals. In *Chemical Fallout*, eds. W. W. Morton and G. G. Berg, pp. 174–204. Springfield: Charles C. Thomas.

Stickel, W. H., L. F. Stickel, and F. B. Coon. 1970. DDE and DDD residues correlated with mortality of experimental birds. In *Pesticides Symposia, 7th Int. Am. Conf. Toxicol. Occ Med.*, ed. W. B. Deichman, pp. 287–294. Miami: Helios and Ass.

Stickel, L. F., W. H. Stickel, R. A. Dyrland, and D. L. Hughes. 1983. Oxychlordane, HCS-3260, and nonachlor in birds: lethal residues and loss rates. *J. Toxicol. Environ. Health* 12:611–622.

Stickel, L. F., W. H. Stickel, R. D. S. McArthur, and D. L. Hughes. 1979c. Chlordane in birds: a study of lethal residues and loss rates. In *Toxicology and Occupational Medicine*, ed. W. B. Deichman, pp. 387–396. North Holland: Elsevier.

Stone, W. B., and J. C. Okoniewski. 1988. Organochlorine pesticide-related mortalities of raptors and other birds in New York 1982–1986. In *Peregrine Falcon Populations: Their Management and Recovery*, eds. T. J. Cade, J. H. Enderson, C. G. Thelander, and C. M. White, pp. 429–438. Boise, ID: Peregrine Fund.

Tanabe, S., K. Senthilkumar, K. Kannan, and N. Subramanian. 1998. Accumulation features of polychlorinated biphenyls and organochlorine pesticides in resident and migratory birds from South India. *Arch. Environ. Contam. Toxicol.* 34:387–397.

Turtle, E. E., et al. 1963. The effects on birds of certain chlorinated insecticides used in seed dressings. *J. Sci. Food Agric.* 14:567–577.

US Department of Agriculture 1980. The biologic and economic assessment of lindane. *US Dept. Agric. Tech. Bull.* 1647:196 pp.

Vander Lee, B., and R. S. Lutz. 2000. Dose-tissue relationships for dieldrin in nestling black-billed magpies. *Bull. Environ. Contam. Toxicol.* 65:427–434.

Verboven, N., and M. E. Visser. 1998. Seasonal variation in local recruitment of great tits: the importance of being early. *Oikos* 81:511–524.

Vemeer, K., and L. M. Reynolds. 1970. Organochlorine residues in aquatic birds in the Canadian prairie provinces. *Can. Field Nat.* 84:117–130.

Vos, J. G., H. A. Breeman, and H. Benschop. 1968. The occurrence of the fungicide hexachlorobenzene in wild birds and its toxicological importance. A preliminary communication. *Meded. Rijksfac. Landbouwwet. Gent.* 33:1263–1269.

Vos, J. G., H. L. van der Maas, A. Musch, and E. Tam. 1971. Toxicity of hexachlorobenzene in Japanese quail with special reference to porphyria, liver damage, reproduction, and tissue residues. *Toxicol. Appl. Pharmacol.* 18:944–957.

Wafford, K. A., D. B. Satelle, D. B. Gant, A. T. Eldefrawi, and M. E. Eldefrawi. 1989. Non-competitive inhibition of GABA receptors in insect and vertebrae CNS by endrin and lindane. *Pest. Biochem. Physiol.* 33:213–219.

Walker, A. I., C. H. Neill, D. E. Stevenson, and J. Robinson. 1969. The toxicity of dieldrin (heod) to Japanese quail (*Coturnix japonica*). *Toxicol. Appl. Pharmacol.* 15:69–73.

Walker, C. H., and I. Newton. 1998. Effects of cyclodiene insecticides on the sparrowhawk (*Accipiter nisus*) in Britain—correction and updating of an earlier paper by Walker and Newton, *Ecotoxicology* 7, 185–189 (1998). *Ecotoxicology* 8:425–429.

Walker, C. H., and I. Newton. 1999. Effects of cyclodiene insecticides on raptors in Britain—correction and updating of an earlier paper by Walker and Newton, *Ecotoxicology* 7, 185–189 (1998). *Ecotoxicology* 8:425–429.

WHO. 1989. *World Health Organization Report on Aldrin/Dieldrin*. Environmental Health Criteria 91.

Wiemeyer, S. N. 1996. Other organochlorine pesticides in birds. In *Environmental Contaminants in Wildlife*, eds. W. N. Beyer, G. H. Heinz, and A. W. Redmon- Norwood, pp. 99–115. Boca Raton: CRC Press.

Wiemeyer, S. N., C. M. Bunck, and C. J. Stafford. 1993. Environmental contaminants in bald eagle eggs—1980–1984—and further interpretations of relationships to productivity and shell thickness. *Arch. Environ. Contam. Toxicol.* 24:213–227.

Wiemeyer, S. N., R. D. Porter, G. L. Hensler, and J. R. Maestrelli. 1986. DDE DDT and dieldrin: residues in American Kestrels and relations to reproduction. US Dept. Interior Fish and Wildlife Service. Fish and Wildlife Technical Report 6. Washington DC. 33 pp.

Wiese, I. H., N. C. J. Basson, J. H. Van der Vyver, and J. H. Van der Merwe. 1969. Toxicology and dynamics of dieldrin in the crowned guinea-fowl *Numida meleagris*. *Phytophylarctica* 1:161–176.

Wikelski, M., E. M. Tarlow, A. Raim, R. H. Diehl, R. P. Larkin, and G. H. Visser. 2003. Costs of migration in free-flying songbirds. *Nature* 423:704.

Willett, K. L., E. M. Ulrich, and R. A. Hites. 1998. Differential toxicity and environmental fate of hexachloro-cyclohexane isomers. *Environ. Sci. Technol.* 32:2197–2207.

Winker, K., D. W. Warner, and A. R. Weisbrod. 1992. Daily mass gains among woodland migrants at an inland stopover site. *Auk* 109:853–862.

Woodrey, M. S., and F. R. Moore. 1997. Age-related differences in the stopover of fall landbird migrants on the coast of Alabama. *Auk.* 114:695–707.

Common Cormorant

From *The Royal Natural History*, edited by Richard Lydekker, Frederick Warne & Co.,
London, 1893–94.

14 Effects of Polychlorinated Biphenyls, Dibenzo-*p*-Dioxins and Dibenzofurans, and Polybrominated Diphenyl Ethers in Wild Birds

Megan L. Harris
John E. Elliott

CONTENTS

14.1 Introduction 478
14.2 Avian Exposure to and Metabolism of Dioxin-Like Contaminants 480
    14.2.1 Measured Variability in PCB and PCDD Metabolism...480
    14.2.2 Routes of Exposure and Congener Patterns in Avian Tissues................................483
14.3 Effects on Reproduction 484
    14.3.1 Reproductive Effort 485
        14.3.1.1 Parental Behavior..485
        14.3.1.2 Clutch Size...490
        14.3.1.3 Secondary Sexual Characteristics 491
    14.3.2 Early Development 491
        14.3.2.1 Egg Quality...494
        14.3.2.2 Hatching Success 494
        14.3.2.3 Rates of Overt Morphological Deformities 498
        14.3.2.4 Nestling Mortality and Pre- and Posthatching Growth 501
        14.3.2.5 Other Measures of Nestling Fitness...502
    14.3.3 Productivity 505
14.4 Adult Survival..507
14.5 Possible Influence of PBDEs on Reproduction and Survival of Wild Birds 510
    14.5.1 Effects on Reproduction 511
    14.5.2 Effects on Nestling Fitness 511
Summary...512
Acknowledgments...517
References...517

14.1 INTRODUCTION

Polychlorinated biphenyls (PCBs), polychlorinated dibenzo-*p*-dioxins (PCDDs), and polychlorinated dibenzofurans (PCDFs) are structurally related families of persistent organic pollutants with
an extensive history of bioaccumulation and toxicity in wildlife around the world. PCBs are chlorinated aromatic hydrocarbons consisting of 209 congener variants of chlorine atoms around a biphenyl ring. PCBs were manufactured from the early 1930s to the mid-1980s in the developed world,
up to the 1990s in some countries. A global ban on their use was instated in 2004, in accordance
with the Stockholm Convention on Persistent Organic Pollutants (http://www.pops.int). PCDDs and
PCDFs were never intentionally produced for commercial purposes. Rather, they appeared in environmental media as by-products of industrial processes, such as manufacturing of chlorophenol-
based biocides and chlorine bleach pulping of wood, or during combustion, especially of plastics.
Once their presence in the environment was identified and their toxicological significance known,
steps were taken by industry to reduce their incidental production.

PCBs, PCDDs, and PCDFs are hydrophobic and lipophilic, thus tending to sequester to sediments
and accumulate in fatty tissues of organisms. Environmental persistence increases with the degree
of chlorination and toxicity is most often associated with those congeners that have a "co-planar"
conformation (for PCBs, these are the non-*ortho* substituted congeners, Table 14.1). The three families produce similar toxicity responses in aquatic organisms, with the dibenzodioxin 2,3,7,8-TCDD
considered the most potent compound (Peterson et al. 1993, Safe 1994, van den Berg et al. 1998). In
very recent egg-injection research, it appears that this may not hold true for wild birds, as some of the
dibenzofurans proved more embryo-lethal (Cohen-Barnhouse et al. 2010). PCBs that are not coplanar produce responses that are not dioxin-like in their mode of action, but those responses have yet
to be thoroughly evaluated in birds, and are treated only incidentally in this review. In this chapter,
"dioxin-like" refers to those effects initiated through the aryl hydrocarbon receptor (AhR) (detailed
in Section 14.2).

Following bans on use in North America and Europe (1977–1982), environmental monitoring
programs reported declines in bioavailability of PCBs throughout the 1990s (Gilbertson et al.
1991, Harris et al. 1993, Bowerman et al. 1995, van den Berg et al. 1995, Hario et al. 2000).
PCDDs and PCDFs were initially less prevalent globally in terms of environmental concentrations. Widespread process improvements limiting their production and environmental release
were initiated during the mid- to late-1990s. Past reviews of PCB toxicity to avian wildlife
described correlative associations between PCBs and reduced reproductive success found in field
monitoring studies, along with some supporting evidence for effects on embryonic development
from egg-injection studies mainly with chicken models (Bosveld and van den Berg 1994, Barron
et al. 1995, Hoffman et al. 1996a). Since that time, monitoring programs have continued to record
declines in tissue burdens of PCBs and PCDDs in wild birds (Helander et al. 2002, Norstrom and
Hebert 2006).

As discussed by previous reviewers, it has proven difficult to determine the relative contribution of PCBs to field observations of toxicity, where PCBs were and are almost always highly correlated with a number of other contaminants, notably PCDDs, PCDFs, and organochlorine (OC)
pesticides such as *p,p'*-dichlorodiphenyl trichloroethylene (-DDE), hexachlorobenzene (HCB), and
oxychlordane (Dawson 2000, Elliott and Harris 2001, Bosveld and van den Berg 2002, Rice et al.
2004, Bustnes 2006). Injection and *in vitro* studies, recently supported by data on the structure of
the AhR, have shown that the domestic chicken (*Gallus gallus*) is a sensitive outlier to dioxin-like
effects associated with exposure to PCDDs, PCDFs, and coplanar PCBs compared to most tested
species of wild birds (Hoffman et al. 1995, 1998, Sanderson and Bellward 1995, Powell et al. 1998,
Head et al. 2008). Previous reviewers did not find strong evidence that non-*ortho* PCBs were historically the main contributors to poor reproductive success via embryotoxicity or other tested mechanisms in populations of wild birds (Dawson 2000, Bosveld and van den Berg 2002). That weak
strength of association suggested by Dawson (2000) and Bosveld and van den Berg (2002) was in

TABLE 14.1

Proportionally or Toxicologically Dominant PCB, PCDD, and PCDF Congeners in Tissues of Wild Birds[a]

Congener	IUPAC Number	Toxic Equivalency Factor (TEF)	Metabolic Pathway
Non-*ortho* PCBs			
3,3',4,4'-tetraCB	77	0.05	CYP1A4-inducer
3,4,4',5-tetraCB	81	0.1	CYP1A4-inducer
3,3',4,4',5-pentaCB	126	0.1	CYP1A4-inducer
3,3',4,4',5,5'-hexaCB	169	0.001	CYP1A4-inducer
Mono-*ortho* PCBs			
2,3,3',4,4'-pentaCB	105	0.0001	CYP1A5-inducer[b]
2,3,4,4',5-pentaCB	114	0.0001	unknown
2,3',4,4',5-pentaCB	118	0.00001	CYP1A5-inducer[b]
2',3,4,4',5-pentaCB	123	0.00001	unknown
2,3,3',4,4',5-hexaCB	156	0.0001	unknown
2,3,3',4,4',5'-hexaCB	157	0.0001	unknown
2,3',4,4',5,5'-hexaCB	167	0.00001	unknown
2,3,3',4,4',5,5'-heptaCB	189	0.00001	unknown
Di-*ortho* PCBs			
2,2',4,4',5-pentaCB	99	—	unknown
2,2',3,4,4',5'-hexaCB	138	—	unknown
2,2',3,3',4,4'-hexaCB	153	—	Possible CYP1A antagonist[c]
2,2',3,3',4,4'-heptaCB	170	—	unknown
2,2',3,4,4',5,5'-heptaCB	180	—	unknown
PCDDs			
2,3,7,8-tetraCDD	—	1.0	CYP1A4-inducer
1,2,3,7,8-pentaCDD	—	1.0	CYP1A4-inducer
1,2,3,4,7,8-hexaCDD	—	0.05	CYP1A4-inducer
1,2,3,6,7,8-hexaCDD	—	0.01	CYP1A4-inducer
1,2,3,7,8,9-hexaCDD	—	0.1	CYP1A4-inducer
PCDFs			
2,3,7,8-tetraCDF	—	1.0	CYP1A4-inducer
1,2,3,7,8-pentaCDF	—	0.1	CYP1A4-inducer
2,3,4,7,8-pentaCDF	—	1.0	CYP1A4-inducer
1,2,3,4,7,8-hexaCDF	—	0.1	CYP1A4-inducer
1,2,3,6,7,8-hexaCDF	—	0.1	CYP1A4-inducer
1,2,3,7,8,9-hexaCDF	—	0.1	CYP1A4-inducer
2,3,4,6,7,8-hexaCDF	—	0.1	CYP1A4-inducer

[a] IUPAC = International Union of Pure and Applied Chemists; Toxic equivalency factors are those recommended for birds by van den Berg et al. 1998.

[b] Kennedy et al. 1998, Head and Kennedy 2007.

[c] Bosveld et al. 2000, Verhallen et al. 1997.

part due to a lack of mechanistic studies in birds, and the complexity of ecological moderators (food quality, food quantity, thermal stress, etc.) on effects expressed in the field.

Since those earlier reviews, several studies have sought to clarify the influence of PCBs and other chlorinated hydrocarbon contaminants on the health, and specifically, the survival and productivity of wild birds. Retrospective analyses of historical data using new statistical approaches, dietary exposure studies, an expansion of field studies to encompass altricial passerine species as well as semiprecocial water-birds, and attempts to find sources of the large interspecies variability in sensitivity to dioxin-like contaminants have contributed to a better understanding of PCB, PCDD, and PCDF effects in wild birds. This review attempts to collate the more recent findings of PCB, PCDD, and PCDF toxicity in wild birds, to propose effect thresholds and to identify high risk species groups. Although we incorporate discussions of some key findings published prior to 1996, we refer the reader to earlier reviews (Eisler 1986, Peterson et al. 1993, Bosveld and van den Berg 1994, Barron et al. 1995, Hoffman et al. 1996a) for a comprehensive analysis of earlier work. In addition, the emphasis in this chapter is placed on wild birds and the response complexities encountered in the field, thus the literature pertaining to dioxin-like toxicity in chickens has not been exhaustively referenced.

Polybrominated diphenyl ethers (PBDEs) are a group of polycyclic aromatic hydrocarbons of more recent concern (de Wit 2002, Ikonomou et al. 2002). Widespread use as flame retardants in plastics, textiles, and electronics has caused a global distribution approaching that of PCBs. PBDEs share many chemico-structural elements with PCBs; key differences are the replacement of chlorines with bromines on the benzene rings and an oxygen link between the rings. Dispersal, persistence, and bioaccumulation patterns appear similar to PCBs (Manchester-Neesvig et al. 2001), although there is some indication that they do not biomagnify with trophic level in marine wildlife to the same degree as PCBs (Kelly et al. 2008, Elliott et al. 2009). Data on temporal trends of PBDEs in wild birds have followed usage patterns in various jurisdictions, first increasing until the mid-1980s and then decreasing in Baltic guillemots (*Uria algae*) (Sellström et al. 2003), and increasing rapidly into the early 2000s in North American and arctic wildlife with some signs of possible decline since 2006 (Norstrom et al. 2002, Rayne et al. 2003, Elliott et al. 2005, de Wit et al. 2006, Gauthier et al. 2008). Although published data is limited, there is sufficient information on avian toxicity to suspect that PBDEs may have consequences for survival and/or reproduction of wild populations. PBDEs are reviewed in a separate section (Section 14.5) and tentative criteria suggested.

14.2   AVIAN EXPOSURE TO AND METABOLISM OF DIOXIN-LIKE CONTAMINANTS

14.2.1   MEASURED VARIABILITY IN PCB AND PCDD METABOLISM

It has long been recognized that the variability in response to PCB or TCDD exposure is considerable, both within and among bird species. Avian toxic equivalency factors (TEFs; van den Berg et al. 1998) were established to quantify the magnitude of dioxin-like effects using data for the domestic chicken, yet we know that the chicken is considerably more sensitive to dioxin-like contaminants than most wild birds (Sanderson and Bellward 1995, Kennedy et al. 1996, Hoffman et al. 1998, Head et al. 2008). Also, metabolic activation within individual birds of a single species is apparently triggered across a wide range of hepatic tissue values (Sanderson et al. 1998). Although those elements of variability were recognized some time ago, mechanistic studies exploring the root causes of variable sensitivity only began in earnest in the late 1990s with an expanded application of molecular techniques.

Studies of metabolic pathways for dioxin-like contaminants in birds have focused on actions mediated by the AhR. PCDDs, PCDFs, non-*ortho* and some mono-*ortho* PCBs produce dioxin-like toxicity, presumably through cellular activity subsequent to binding with the AhR, the first (phase I) step in TCDD metabolism (Whitlock 1999). AhR binding alters transcription rates of dioxin-responsive genes, and the most studied of those, the cytochrome P4501A (CYP1A) genes, have been applied as markers of PCB, PCDD, and PCDF exposure. Avian CYP1A isoforms sensitive to dioxin-like compounds have been classified as CYP1A4 and CYP1A5, distinct from mammalian isoforms (Gilday et al. 1996, Mahajan and Rifkind 1999). CYP1A4 induction is commonly measured as hepatic ethoxyresorufin-*O*-deethylase (EROD) or aryl hydrocarbon hydroxylase (AHH) activities, whereas CYP1A5 induction may be measured as arachidonic acid metabolism or as a cellular accumulation of porphyrins (via uroporphyrinogen oxidation).

There is some evidence that the relative potential of CYP1A to be induced in wild birds may also be indicative of species' sensitivity to PCB toxicity. One study suggests that the magnitude of *in vitro* EROD induction is predictive of *in ovo* CB-77 embryotoxicity in several species of Galiiformes and herring gulls (Kennedy et al. 1996). Another showed two orders of magnitude variation in hepatic *in ovo* EROD-inducing potency of TCDD among hatchlings of chicken, pigeon (*Columba livia*), great blue heron (*Ardea herodias*), and double-crested cormorant (*Phalacrocorax auritus*), which was predictive of hatchling viability (Sanderson and Bellward 1995). Interspecies variability in metabolic response to PCB exposure may be partially related to differences in the AhR molecular structure and to differences in preferential induction of CYP1A isoforms (Head 2006, Karchner et al. 2006, Head and Kennedy 2007, Yasui et al. 2007).

DNA cloning and sequencing work has identified amino acid differences in the ligand binding domain (LBD) of the AhR that were broadly predictive of species' sensitivity to dioxin-like contaminants (Head 2006, Karchner et al. 2006, Head et al. 2008). Karchner et al. (2006) showed that amino acids isoleucine-324 and serine-380 in the high affinity chicken receptor are replaced by valine-325 and alanine-381 in the lower affinity (by 7-fold) common tern (*Sterna hirundo*) receptor, and that those substitutions caused reductions in *in vitro* transactivation by the tern AhR. Further sequencing studies of the AhR LBD in other species confirm that those amino acid differences are predictive of binding affinity and species' sensitivity differences in a number of wild birds (Table 14.2). Of the species sequenced, the Galiiformes (primarily upland game birds), eastern bluebird (*Sialia sialis*), and black-footed albatross (*Phoebastria nigripes*) show an intermediate sensitivity to dioxin-like compounds, while other wild birds appear relatively insensitive (Head 2006, Yasui et al. 2007, Head et al. 2008).

Other recent DNA sequencing of AhRs in black-footed albatross and common cormorant (*Phalacrocorax carbo*) suggest that birds also have two AhR isoforms, a mammalian-like AhR1, which appears to be the dominant form, and a fish-like AhR2 (Yasui et al. 2007). Variability in induction of AhR2 may further explain species' response variability to PCBs.

Differential sensitivity of species to dioxin-like toxicity may also be linked to preferential induction of the two CYP1A isoforms. Kennedy et al. (2003) showed that hepatic EROD activity was not significantly correlated with total PCB concentrations in herring gull livers, and identified studies with other gull species that also found no relationship between PCBs and EROD (glaucous gulls *Larus hyperboreus*, Henriksen et al. 2000; black-headed gulls *L. ridibundus*, Fossi et al. 1986; black-tailed gulls *L. crassirostris*, Yamashita et al. 1993). *In vitro* investigations of CYP1A4 and CYP1A5 mRNA expression indicate that the apparent insensitivity to EROD induction by PCBs in gulls may relate to preferential induction of the non-EROD-inducing isoform, CYP1A5. Head and Kennedy (2007) found that TCDD-related induction and stability of the two CYP1A isoforms are differentially regulated in chicken and herring gull, and that CYP1A5 is preferentially induced in herring gull. This suggests that CYP1A5-regulated toxicity pathways, including porphyria, may be the most relevant for herring gull exposed to PCBs. The relative expression of CYP1A isoforms in mRNA of wild bird species is, therefore, another means of predicting high-risk species and the species-specific metabolic pathways of key interest.

Studies of CYP1A induction in PCB-exposed adult wild birds have frequently reported extreme variability in response among individuals of the same population and among subpopulations (Davis et al. 1997, Hilscherova et al. 2003, Head 2006). An analysis of EROD activity in primary hepatocytes of chicken, herring gull, ring-billed gull, double-crested cormorant and Forster's tern (*Sterna forsteri*) exposed to TCDD found that differences in sensitivity among individuals of the same species could be as large as the average differences among the tested species (Sanderson et al. 1998). Head (2006) found 20-fold and 125-fold variability in basal expression of CYP1A4 and CYP1A5 among 55 herring gull embryo hepatocyte cultures. In addition, a proportion of *in vitro* embryo

TABLE 14.2
Genotypes for the AhR Ligand Binding Domain (LBD) in Wild Birds[a]

Order and Species	Common Name	Genotype	Predicted Sensitivity to Dioxin-Like Toxicity
Anseriformes			
Aix sponsa	Wood duck	TTTVKA	Insensitive
Somateria mollissima	Common eider	TTTVKA	Insensitive
Charadriiformes			
Larus argentatus	Herring gull	ATTVKA	Insensitive
L. delawarensis	Ring-billed gull	ATTVKA	Insensitive
Sterna hirundo	Common tern	ATTVKA	Insensitive
Uria lomvia	Thick-billed murre	ATTVKA	Insensitive
Ciconiiformes			
Ardea herodias	Great blue heron	AATVKA	Insensitive
Falconiformes			
Falco sparverius	American kestrel	ATTVKA	Insensitive
Haliaeetus leucocephalus	Bald eagle	ATTVKA	Insensitive
Pandion haliaetus	Osprey	ATTVKA	Insensitive
Galiiformes			
Gallus gallus	Chicken	AATIKS	Highly sensitive
Coturnix japonica	Japanese quail	AATVKA	Insensitive
Phasianus colchicus	Ring-necked pheasant	TAIIKA	Intermediate sensitivity
Meleagris gallopavo	Turkey	TAIIKA	Intermediate sensitivity
Lagopus lagopus	Willow ptarmigan	AAIIKA	Intermediate sensitivity
L. mutus	Rock ptarmigan	AAVIKA	Intermediate sensitivity
Bonasa umbellus	Ruffed grouse	AAIIKA	Intermediate sensitivity
Passeriformes			
Tachycineta bicolor	Tree swallow	ATTVEA	Intermediate sensitivity
Pelecaniformes			
Phalacrocorax auritus	Double-crested cormorant	ATTVKA	Insensitive
P. carbo	Common cormorant	ATTVKA	Insensitive
Procellariiformes			
Phoebastria nigripes	Black-footed albatross	ATTIKA	Intermediate sensitivity

[a] Genotype refers to the 6 amino acids in the LBD: A = Ala (alanine); E = Glu (glutamic acid); I = Ile (isoleucine); K = Lys (lysine); S = Ser (serine); T = Thr (threonine); V = Val (valine); residue sequence positions are 256, 257, 297, 324, 341 and 380 of the chicken LBD.

Sources: Reproduced from Head, J.A. Variation in the cytochrome P4501A response to dioxin-like compounds in avian species. PhD dissertation, University of Ottawa, Ottawa, ON, 2006.
From Yasui, T. et al., *Toxicol. Sci.,* 99, 101–117, 2007.
Reproduced from Head, J.A. et al., *Environ. Sci. Technol.,* 42, 7535–7541, 2008.

cultures appear nonresponsive in all cases, ranging from 8% to 40% of individuals tested (Davis et al. 1997, Sanderson et al. 1998, Head 2006).

Sources of variation in CYP1A response among individuals of a species may be related to previous maternal or direct exposure to inducing contaminants, or to genetic differences associated with acquired resistance. Head et al. (2006) found that chicken *in ovo* exposure to CB-126 increased the CYP1A mRNA response to TCDD *in vitro* at concentrations of 0.8 and 1.6 µg/kg CB-126, but not at the environmentally relevant concentration of 0.4 µg/kg. That suggests there is a threshold for embryonic exposure, above which subsequent responses of the individual to dioxin-like contaminants would be enhanced. However, Bosveld et al. (1995) did not find any evidence for pre-induction of CYP1A *in ovo* in two colonies of common tern, a species that appears relatively insensitive to dioxin-like contaminants (Hoffman et al. 1998). Although there is a suggestion that history of exposure may have been a factor in individual responses when environmental contamination with dioxin-like contaminants was higher, it is unlikely that this source of variability is relevant today for wild birds, except those inhabiting highly contaminated sites.

Head (2006) also found limited evidence for genetic differences in CYP1A mRNA expression among subpopulations of herring gulls. A reproductively isolated and relatively uncontaminated Atlantic population of herring gulls expressed significantly higher basal levels of CYP1A4 and CYP1A5 than two Great Lakes colonies, while also containing a greater proportion of individuals that were effectively nonresponsive to TCDD exposure. It is still unclear whether this apparent differential expression of CYP1A would equate to genetic resistance in historically contaminated subpopulations of herring gull.

Papp et al. (2005) similarly identified genetic variability in basal CYP1A expression among tree swallows from a single breeding population along the northern shore of Lake Erie. They found a positive correlation between hatching date and EROD activity, independent of the significant associations between PCBs and both biological parameters. Laying date is used as a measure of parental fitness in wild birds (Wardrop and Ydenberg 2003).

Gender of breeding birds also appears to explain some of the variability in PCB metabolism among cohorts. Breeding females transfer some of their tissue contaminant burden to their eggs, and recent work with glaucous gulls suggests that maternal transfer favors the less persistent congeners (Verreault et al. 2006a). In a small passerine, the blue tit (*Cyanistes caeruleus*), however, when whole body burdens of PCBs, rather than maternal blood measurements, were compared to eggs, maternal transfer favored the more bioaccumulative and persistent compounds (Van den Steen et al. 2009). In predatory birds, there may also be differences in foraging patterns between the sexes that leads to variable congener patterns. For instance, male gulls may prey more on other seabirds and their eggs than females who show a greater reliance on fish (Verreault et al. 2006b). Despite the recognition that gender may alter the expression and regulation of CYP1A enzymes (Sanderson et al. 1997) we are not aware of published works that evaluated that source of variability.

There is some evidence that toxicokinetic interactions among PCB congeners may alter CYP1A induction in birds. Bosveld et al. (2000) fed common tern hatchlings a diet contaminated with CB-126 and found that the lowest-observed-effect-level (LOEL) for CYP1A induction was increased when an additional dosage of the di-*ortho* CB-153 was added. They suggested that CB-153 may act as an antagonist of CYP1A induction in terns and possibly other species. van den Berg et al. (1994) had also previously reported similar interactions between CB-153 and TCDD.

14.2.2 ROUTES OF EXPOSURE AND CONGENER PATTERNS IN AVIAN TISSUES

The availability of some PCB congeners appears to have changed since restrictions on use were imposed, as has the relative tissue burden of total PCBs compared to DDE, PCDDs, and PCDFs. Two retrospective analyses of long-term egg monitoring datasets suggest that higher chlorinated congeners have gradually increased as a proportion of total PCBs. Eggs of white-tailed sea eagle (*Haliaeetus albicilla*) collected in Sweden from the 1970s to the 1990s contained increasingly greater contributions from the higher chlorinated PCBs, with the inflection point occurring around CB-138 (Helander et al. 2002). Similarly, reanalyses of archived herring gull (*Larus argentatus*) eggs collected in the Great Lakes from 1971 to 1982 found a progressive shift in the congener pattern toward higher chlorinated PCBs (Norstrom and Hebert 2006). However, the relative contribution of CB-126 in both instances did not change over time, and the toxic potency of tissue burdens, as measured by TEQs, decreased in parallel with total PCBs. Hence, those long-term monitoring programs found no evidence for selective enrichment of PCBs during environmental weathering as had been suggested previously (Jones et al. 1993, Giesy et al. 1994, 1995), other than that associated with variable, congener-specific rates of biotic metabolism.

Species' differences in PCB congener pattern appear to be strongly influenced by the source of PCBs more so than varying metabolic capacities (Custer and Read 2006). Congeners 153, 138, and 180 dominated avian tissues collected from diverse locations and taxonomic groupings: tree swallows (*Tachycineta bicolor*) from Massachusetts (Custer and Read 2006), dippers from Wales (*Cinclus cinclus*; Ormerod et al. 2000) and western Canada (*C. mexicanus*; Morrissey et al. 2005), starlings (*Sturnus vulgaris*) from Illinois (Arenal et al. 2004), albatrosses (Procellariiformes) in the North and South Pacific (Guruge et al. 2001), various pelagic seabirds from the North Pacific (Elliott 2005); cormorants (*Phalacrocorax aristotelis*) from Norway (Murvoll et al. 2006a), black guillemots (*Cepphus grylle*) from Labrador (Kuzyk et al. 2003), and terns, shorebirds, and passerines from India and Russia (Kunisue et al. 2003) were among others. Other congeners such as CBs 118, 99, 105, 187, 170, and 31 vary in dominance according to source (e.g., Aroclor 1254, 1260, Clophen A50, Sovol or other; see Table 14.1 for a list of common congeners) and possibly, to a lesser extent, also species and trophic properties (Elliott et al. 1996a, 2000, Hebert et al. 1999, Kunisue et al. 2003, Custer and Read 2006).

PCDD and PCDF congener patterns in wild birds are broadly indicative of the dominant sources for those by-products in the environment. In intensively industrialized regions where combustion is the prevalent source, and at sites of chlorophenol manufacturing, the avian tissue pattern is dominated by TCDD, OCDD, and OCDF (Hebert et al. 1994, Custer et al. 2005). Conversely, in wild birds that forage in pulp mill discharge zones, the pattern is dominated by PnCDD and HxCDD congeners (Elliott et al. 1996a, 1998, 2001a, 2001b, Harris et al. 2003).

Stable isotope studies have proven useful in identifying ecological factors that influence exposure to PCBs, particularly food web structure and migration patterns (Sagerup et al. 2002, Morrissey et al. 2004, Hebert and Weseloh 2006). Retrospective analyses of stable nitrogen isotopes and fatty acids in herring gull eggs from the Great Lakes demonstrated the key role of food web structural changes on exposure patterns (Hebert and Weseloh 2006, Fox et al. 2007a). The food web, altered by declining populations of native invertebrate prey of lake trout (*Salvelinus namaycush*) and increasing numbers of exotic bivalves and fishes, influenced the true rate of decline of PCBs in the herring gull population such that it was slower than previously calculated. Carbon and nitrogen isotope analyses have also been used to examine intraspecies variability in contaminant levels associated with feeding ecology. Individual American dipper eating a higher proportion of salmon fry than invertebrates (Morrissey et al. 2004) and glaucous gull eating a higher proportion of seabird eggs than fish (Sagerup et al. 2002) contained greater tissue burdens of PCBs than their cohorts. In the case of the breeding glaucous gulls in the western Barents Sea, individual food preference determined from stable isotope signatures could explain up to 18% of the variation in OC concentrations (Sagerup et al. 2002).

Application of new statistical approaches, and molecular and ecological methodologies have refined our knowledge of metabolic and exposure pathways for wild birds. Further application of those techniques should help to identify species, populations, and even subpopulations most at risk from exposure to point sources of PCBs, PCDDs, and PCDFs.

14.3 EFFECTS ON REPRODUCTION

Most studies on the effects of PCBs and dioxins in wild birds have focused on reproductive outcomes. However, most study designs did not permit the discrimination of mechanism for effects observed, whether via embryotoxicity, alterations in parental reproductive effort, compromised nestling fitness or a complexity of multiple mechanisms (Barron et al. 1995, Hoffman et al. 1996a). Here, we examine the weight of evidence for which effects were and continue to be most clearly associated with the

PCB, PCDD, or PCDF content in environmental contaminant mixtures. Where the weight of evidence allows, we suggest threshold values for different tissues and endpoints in the section Summary.

14.3.1 Reproductive Effort

14.3.1.1 Parental Behavior

A few early studies suggested a parental behavior component to PCB effects on wild birds, but the response patterns were inconsistent among species, and logistical limitations often precluded robust field studies (Peakall 1996). Ringed turtledoves (*Streptopelia risoria*) dosed chronically with 10 ppm Aroclor 1254 in their diet (resulting in brain residues of 5.5 µg/g) exhibited reduced nest attentiveness and subsequent increased egg mortality from thermal stress (Peakall and Peakall 1973). However, other dietary dosing studies with captive mallards (*Anas platyrhynchos*) and wild Atlantic puffin (*Fratercula arctica*) did not find similar behavioral responses at comparable egg concentrations (dove egg = 16 µg/g, mallard egg = 23 µg/g, puffin egg = >10 µg/g; Custer and Heinz 1980, Haseltine and Prouty 1980, Harris and Osborn 1981). In the field, reduced nest attendance was considered to contribute to poor reproductive success in herring gull and Forster's tern populations in the Great Lakes during the 1970s and 1980s (Fox et al. 1978, Peakall et al. 1980, Kubiak et al. 1989). In those early studies in the Great Lakes, it was very difficult to isolate and identify critical variables in terms of both contaminants of greatest concern and principal pathways for effect, because the environmental contamination was so extreme and multifaceted. It is highly probable that multiple mechanisms were in play during that time.

An expansion of research to include more species of altricial birds, where young are more dependent on parents over a longer period of development, has enabled analysis of a greater range of behavioral endpoints. Those findings are summarized in Table 14.3. The exact mechanisms driving behavioral responses remain unclear (endocrine-disrupting, neurological or generic stress), but altered parenting behavior has been observed and statistically supported during courtship, egg-laying and incubation phases with PCB exposure in some birds. Asynchrony in parental duties and a preponderance of abnormal behaviors in the male birds were seen in a number of studies.

Prenesting behaviors, including nest-building and courtship interactions among males and females may be affected by PCB exposure in some species. Captive American kestrel (*Falco sparverius*) males and females fed on cockerels spiked with Aroclors 1248:1254:1260 (1:1:1 by weight) from one month prior to pairing through until hatching (7 mg/kg body weight/day or 3.8 µg/g in plasma at dose termination) exhibited a suite of altered breeding behaviors (Fisher et al. 2001, 2006, Fernie et al. 2001b, 2003a). Exposed males exhibited a significant increase in the number of aerial display and sexual behaviors compared to control males, with no corresponding increase in copulation or mounting frequencies (Fisher et al. 2001). In a quarter of exposed pairs (6 of 24 pairs), aggressive courtship interactions or a lack of any courtship behaviors were reported (Fernie et al. 2003a). When aggressive birds were replaced with other exposed birds, the majority of the new pairs (96%) went on to lay eggs.

During laying and incubation phases, those same kestrels displayed a number of differences in the frequency, duration, and timing of normal behaviors. Once again, the majority of disruptions occurred in males. Clutch initiation was significantly delayed in PCB-exposed pairs, the incubation period was extended, and 8% of pairs (vs. no controls) abandoned the clutch before hatching (Fernie et al. 2003a, Fisher et al. 2006). PCB-exposed males contributed significantly less time to incubation duties than control males (Fisher et al. 2006). Coordination of incubation recesses was also affected, as males were more likely to be late in returning to the nest for a switch of duties. Some of those effects were reported again in second generation breeding kestrels exposed *in ovo*. Delays in clutch initiation of about 5 or 3 days were seen in pairs with an *in ovo*-exposed female or male, respectively (Fernie et al. 2001a). One quarter of *in ovo*-exposed females failed to lay a clutch of eggs (Fernie et al. 2001a).

Captive zebra finch (*Taeniopygia guttata*) females exposed to a total 40 µg Aroclor 1248 through oral gavage in 4 pulse doses administered over 16 days prior to pairing laid eggs with 0.2 µg/g

TABLE 14.3
Significant Effects Observed in Parental Breeding Behavior or Physiology with Exposure to PCBs

Species	Exposure	Effect	ΣPCBs Tissue Residue	Reference
American kestrel	Captive, diet Aroclor 1248:1254:1260 7 mg/kg body weight/day 100 d	Abnormal male courtship behaviors Duller male plumage (courtship phase) Bright female plumage (courtship phase) Delay of clutch initiation Extended incubation time Nest abandonment (incubation phase) Reduced male incubation time Poorly coordinated incubation recesses	3.8 μg/g wet wt (male plasma)	Bortolloti et al. 2003, Fernie et al. 2003a, Fisher et al. 2001, 2006
	Captive, in ovo Aroclor 1248:1254:1260	Delay of clutch initiation Failure to initiate clutch Reduced median clutch size	34 μg/g wet wt (egg)	Fernie et al. 2001a, 2001b
	Captive, diet Aroclor 1242, 60 mg/kg ~150 d	Reduced circulating levels of thyroxine	Unmeasured	Quinn et al. 2002
Ringed turtle-doves	Captive, diet Aroclor 1254, 10 mg/kg	Reduced nest site attentiveness	5.5 μg/g wet wt (brain)	Peakall and Peakall 1973
	Captive, diet Aroclor 1254	Reduced brain dopamine and norepinephrine	2.8 μg/g wet wt (brain)	Heinz et al. 1980

Species	Study	Effect	Concentration	Reference
Zebra finch	Captive, oral Aroclor 1248, 40 µg 16 d	Increased number of nests constructed Increased number of clutches Extended incubation time	0.20 µg/g wet wt (egg)	Hoogesteijn et al. 2005
European starling	Field	Nest abandonment (construction phase) Reduced nest site attentiveness Reduced avoidance behavior	9 µg/g wet wt (whole body)	Arenal et al. 2004
Forster's tern	Field	Reduced nest site attentiveness	23 µg/g wet wt (egg) [+37 pg/g TCDD]	Kubiak et al. 1989
Common tern	Field + artificial incubation	Extended incubation time	~3.5 ng C-TEQs (egg yolk sac)	Bosveld et al. 1995
Glaucous gull	Field	Reduced nest site attentiveness	0.05–1.1 µg/g wet wt (blood)	Bustnes et al. 2001
	Field	Increased progesterone in males	0.15–2.18 µg/g wet wt (plasma)	Verreault et al. 2006b
Herring gull	Field	Reduced nest site attentiveness	142 µg/g wet wt (egg)	Gilman et al. 1977, Fox et al. 1978
Tree swallow	Field	Abnormal nest construction Burial of eggs (incubation phase) Nest abandonment (incubation phase) Super-normal clutch sizes Precocial plumage in sub-adult females	114 µg/g wet wt (whole body)	McCarty and Secord 1999a, 1999b, 2000
	Field	Nest abandonment (incubation phase) Reduced median clutch size	9 µg/g wet wt (whole body)	Neigh et al. 2006

ΣPCBs (Hoogesteijn et al. 2005). Those birds produced a greater number of clutches and nests constructed. Finches also exhibited an extended (by 7 days on average) incubation phase, similar to the dietary dosed kestrels.

Recent field studies have also attempted to assess effects of PCBs on parental breeding behavior during the nest-building and incubation phases. Nest box studies of tree swallows breeding along a PCB-contaminated stretch of the Hudson River (New York) reported abnormal nest-building, with small, low quality nests predominant (McCarty and Secord 1999a). Other tree swallows nesting near a sewage treatment plant in Vancouver (British Columbia) also built smaller nests with looser weaves than cohorts at a reference site (Dods et al. 2005). European starlings using nest boxes at a PCB-contaminated superfund site in the Crab Orchard National Wildlife Refuge in Illinois frequently failed to complete nest-building. Only 9% of the nests initiated were completed to full cup stage, as compared to a completion rate of 57–69% at other sites (Arenal et al. 2004). Incidences of nest abandonment and egg burial during incubation were elevated in the Hudson River (NY) population of tree swallows previously shown to exhibit abnormal nest-building behavior (McCarty and Secord 1999a, 1999b). However, site differences were apparent in only one of two study years. Tree swallows from a broadly contaminated stretch of the Kalamazoo River (Michigan) showed higher rates of nest abandonment than swallows from a reference location; however, egg burial was witnessed at both Michigan sites and was not always a precursor to abandonment (Neigh et al. 2006), as had been seen in the New York populations. Tree swallows nesting downstream of pulp mills and contaminated with PCDDs and PCDFs exhibited increased rates of nest abandonment during a one-season study along the Fraser River (British Columbia), but sample sizes were insufficient to distinguish between ecological and contaminant effects (Harris and Elliott 2000). Starlings breeding at a superfund site also experienced an increased rate of nest abandonment during incubation in some but not all years (McKee 1995, Arenal et al. 2004).

Male and female glaucous gulls from breeding colonies in the Norwegian north Atlantic showed significant associations between time absent from the nest and blood concentrations of total PCBs (Bustnes et al. 2001). Incubation time was not noticeably affected by PCB contamination, but attendance at the nest site by the nonincubating partner was reduced as circulating PCBs increased; incubating glaucous gull pairs take turns guarding the nest site against threats from predators and con-specifics. Female absences were longer and more frequent than male absences, but those gender differences could be at least partly explained by gender-specific foraging patterns. DDE and other persistent contaminants were present in gull blood, although PCB concentrations were five times greater than those of DDE (Bustnes et al. 2001, 2003).

Great black-backed gulls (*Larus marinus*) also from the Norwegian north Atlantic showed delays in clutch initiation associated with female blood levels of several contaminants, including PCBs but with HCB concentration being the best predictor of laying date (Helberg et al. 2005). Females with higher contaminant levels, particularly oxychlordane and DDE, also appeared at greater risk of nest predation from corvids. Higher contaminant concentrations in eggs were also related to longer incubation periods or delayed hatching in south polar skua (Bustnes et al. 2007) and in artificially incubated common tern (Bosveld et al. 1995).

Attempts to evaluate parental attentiveness during the chick rearing phase provided limited evidence for a PCB effect. Tree swallows feeding young along a contaminated section of the Kalamazoo River (Michigan) made significantly fewer visits to nest boxes than their counterparts in a reference area, but only during one of three consecutive breeding seasons (Neigh et al. 2006). In one breeding season, starlings rearing chicks in boxes within a PCB-contaminated superfund site (Illinois) made fewer foraging trips than parents in a nearby reference population, but only during part of the rearing phase. During days 1–5 after hatch, there were no differences among nest box sites in starling adult provisioning behavior; however, provisioning dropped significantly at the PCB site during days 6–15 after hatch (Arenal et al. 2004). Those starling adults from the contaminated site also exhibited poor avoidance behavior, frequently sitting on the roofs of the nest boxes and easily captured by researchers.

The comparison of results between laboratory and field studies is exacerbated by a lack of consistency in tissue residues analyzed. In a few instances, one can compare whole egg or adult plasma values. Parental behavioral effects observed in laboratory studies on kestrels and zebra finches occurred where subsequent egg PCB concentrations ranged between 0.2 and 34 µg/g wet weight (Hoogesteijn et al. 2005, Fisher et al. 2006). Similar observations in the field in Forster's tern, tree swallows and starlings occurred where eggs contained 5.1–29.5 µg/g ΣPCBs (Kubiak et al. 1989, McCarty and Secord 1999a, Arenal et al. 2004, Neigh et al. 2006). Effects observed in captive adult kestrels occurred at plasma concentrations of 3.8 µg/g ΣPCBs (Fisher et al. 2001, 2006), whereas those observed in wild gulls occurred at 0.2–2.2 µg/g ΣPCBs (Verreault et al. 2006b). In all other instances, residues cannot be directly compared using published values. Despite those logistic restrictions in comparative capacity, the concentration ranges described in field and laboratory studies appear quite similar. In future studies, consistent measurement of residues in adult blood, brain, and liver tissues for examination of parental behavior effects would greatly increase the capacity for cross-study comparisons.

The root causes for the aberrant parenting behaviors identified during laboratory and field studies remain unclear; many of the authors speculated effects on endocrine, neurological, or generic physiological stress pathways. The evidence for endocrine disruption provided by a small number of corresponding circulating hormone analyses is equivocal and hampered by design difficulties associated with the normal, rapid cycling of steroid hormones during the prenesting and incubation phases of breeding.

The adult male kestrels exposed to food contaminated with Aroclors 1248:1254:1260 prior to and during pairing (that resulted in plasma concentrations of 3.8 µg/g ΣPCBs, Fernie et al. 2001b) contained circulating total androgen levels similar to control males on day 7 after pairing (Fisher et al. 2001). Similarly, the exposed females in the study contained circulating 17β-estradiol levels similar to control females. Since the timing of those analyses corresponded to the peak period of behavioral anomalies, there was little evidence that the anomalies were the direct result of hormonal disruption. Also, Quinn et al. (2002) found no differences in circulating levels of estradiol in female and male American kestrels orally dosed with Aroclor 1242; however, the timing of their assessments, from egg-laying onward, did not correspond to the normal peaks in circulating hormones (during courtship activities) and basal levels in individuals were so low that many were not quantifiable.

Verreault et al. (2006b) measured the circulating levels of testosterone, 17β-estradiol, and progesterone in incubating glaucous gulls from the Norwegian population previously shown to exhibit reduced nest site attentiveness (Bustnes et al. 2001; though hormone tests were conducted during a different breeding season). They found no differences in testosterone or estradiol associated with varying contaminant burdens in blood; that was not unexpected, as blood concentrations of those hormones rapidly return to basal levels at the onset of incubation. However, progesterone levels in males were strongly positively related to a number of contaminants, including PCBs, DDTs, and PBDEs, suggesting potential interference with steroidogenesis in male but not female birds (Verreault et al. 2006b).

There has also been speculation that altered nest attentiveness may have a physiological basis in hyperthyroidism (Hoogesteijn et al. 2005). Quinn et al. (2002) detected depressed circulating levels of thyroxine (T_4) in female American kestrels dosed with Aroclor 1242 from the egg-laying phase throughout breeding; male T_4 was also lower in exposed birds relative to controls, but not significantly so. Behavioral markers were not assessed in those birds. Effects of PCBs on the developing thyroid system in embryos and nestlings have been reasonably well studied (see Section 14.3.2.5), but changes in adult thyroid function have not been comprehensively assessed (see Section 14.4) with the possible exception of herring gull populations in the Great Lakes. PCB-associated reductions in T_4 were observed in thyroid gland tissue of adult herring gulls, but not in plasma (McNabb and Fox 2003). Nonetheless, studies that measure thyroid function relative to breeding behavior in adult birds were not found in the published literature, and thus the potential mechanistic link between thyroid and nest attentiveness cannot be adequately evaluated.

Early laboratory studies drew a tentative link between abnormal parenting behavior and central nervous system effects of PCBs, but little follow-up of this neurological mechanism has occurred. Brain dopamine and norepinephrine were reduced in ringed turtle doves exposed to Aroclor 1254 and correlated to PCB residues in the brain (Heinz et al. 1980). Affected doves contained brain tissue PCBs of 2.8 µg/g wet weight.

Offspring of the zebra finch study reported earlier (dosed orally with 4 doses totalling 40 µg Aroclor 1248 and resulting in 0.19 µg ΣPCBs/g ww in eggs) also exhibited changes in the brain as a result of their *in ovo* exposure. Male and female 50-day-old offsprings had significantly smaller robustus arcopallialis centres in the brain, an area that is critical for song production in birds (Hoogesteijn et al. 2008). The authors suggested that PCBs were producing hormonally mediated neurological damage similar to that reported for robins exposed to DDT (Iwaniuk et al. 2006).

Where genders were evaluated independently, observed effects were frequently male-biased, suggesting male physiological processes may predispose them to PCB-induced behavior alterations. Bird et al. (1983) reported that sperm numbers declined by up to 27% per ejaculate in male kestrels exposed to 9–10 mg/kg body weight/day Aroclor 1254 in their diet (resulting in PCB concentrations in the testes and liver of 128 and 92 µg/g lipid weight, respectively). That implies a physiological problem with fertility in males that may or may not be related to male behavior patterns. Delays in clutch initiation may also be linked to effects in males if they failed to stimulate their mate's hormonal triggers for egg-laying (Fisher et al. 2006). Alternatively, delays may be related to effects on female ovulation or follicular development (Rattner et al. 1984) or the result of a complex combination of effects on both males and females (Fernie et al. 2001b).

Discrepancies in expression of behavioral effects among studies may be related in part to species differences in sensitivity and variation in exposure to specific congeners. Expression of behavioral breeding abnormalities in exposed kestrels during the season of exposure but not in the same individuals during their subsequent breeding year lead Fernie et al. (2003a) to postulate that the rapidly metabolized, less persistent PCB congeners (Drouillard et al. 2001) may have been responsible for the majority of those effects. That theory draws some support from studies of osprey (*Pandion haliaetus*) and bald eagle (*Haliaeetus leucocephalus*) which showed that exposure to PCDDs and PCDFs (similar in toxicity to persistent PCBs) was not directly associated with poor parental provisioning (Woodford et al. 1998, Gill and Elliott 2003). Although increased rates of nest abandonment have been seen in great blue herons and tree swallows exposed to elevated PCDDs and PCDFs (Harris and Elliott 2000, Elliott et al. 2001b, Gill and Elliott 2003), the evidence for cause–effect could not be established given the compounding presence of ecological stressors such as limited food supply and bad weather.

14.3.1.2 Clutch Size

Clutch sizes typically show a broadly modal distribution in populations of wild birds, and changes in the median can be difficult to detect in a small to moderate sample of the population. Captive breeding and nest box study designs help to overcome some of the logistical difficulties encountered with this measure of parental reproductive effort.

Results of captive breeding trials suggest that PCB exposure of adults early in breeding may both increase the normal variability in clutch size and lead to modest reductions in median size. Clutch size ranged from extreme lows to extreme highs in American kestrel exposed to an Aroclor mixture that produced egg concentrations of 34 µg/g ΣPCBs (one to eight eggs; Fernie et al. 2003a) and in zebra finch exposed to Aroclor 1248 that produced egg concentrations of 0.2 µg/g ΣPCBs (zero to eighteen eggs compared to a control range of 4 to 15 eggs; Hoogesteijn et al. 2005). Clutch sizes were smaller on average for kestrel pairs directly exposed to Aroclors and for pairs with one second generation *in ovo*-exposed individual (Fernie et al. 2001a, 2001b). Only the effect in *in ovo*-exposed pairs was significant.

Most field studies reporting clutch sizes did not find relationships with tissue burdens of PCBs or other dioxin-like contaminants (glaucous gulls, Bustnes et al. 2003; great black-backed gulls,

Helberg et al. 2005; wood ducks, White and Seginak 1994; European dippers, Ormerod et al. 2000; tree swallows, Custer et al. 1998, 2003; European starlings, Arenal et al. 2004). Tree swallow clutch sizes were smaller in a contaminated Michigan population in two of three seasons relative to reference values (Neigh et al. 2006), whereas McCarty and Secord (1999b) reported incidences of supernormal clutches in a contaminated New York population of tree swallows (Table 14.3).

14.3.1.3 Secondary Sexual Characteristics

Plumage color in mature birds has been evaluated as another marker of effects on reproductive effort in adults. Plumage in wild birds is a secondary sexual characteristic and brightness is correlated with health (low levels of pathogens and parasites) as well as high fertility of males (Palokangas et al. 1994, Wiehn 1997). One study of American kestrels found no effect of PCBs on plumage, while another found age- and gender-specific effects that varied seasonally. Where male and female adult kestrels were dosed with Aroclor 1242 (6–60 mg/kg) in feed from 1 week before egg-laying until the end of their autumn molt, there was no effect on color of secondary tail and wing flight feathers at the time of molt (Quinn et al. 2002). Where juvenile and adult kestrels of both sexes were dosed with Aroclor 1254:1248:1260 (7 mg/kg body weight/day, 3.8 μg/g ΣPCBs in male plasma) in feed from 1 month prior to pairing until hatching of young by adult pairs, there was a complexity of significant effects on plumage color and carotenoid concentrations (Bortolotti et al. 2003). Color of ceres and lores and plasma carotenoids were evaluated at pairing and courtship for adults, at fledging for juveniles, and again in winter for both age groups. PCB-exposed males were duller in color than control males during the critical period of courtship. Control females lost color during courtship, as expected when carotenoids are being diverted to developing ovaries; however, PCB-exposed females maintained color over the same period, suggesting that they may not have been physiologically prepared for breeding (Bortolotti et al. 2003). During winter months, the brightest individuals were PCB-exposed juveniles. The authors contend that the significant effects on plumage and carotenoids were consistent with endocrine disruption, although the mechanism of action remains unclear. Certainly, timing of exposure appears to be critical for the expression of altered pigmentation.

In field studies, subadult female tree swallows from nest box sites along the PCB-contaminated Hudson River (New York) showed color plumage patterns similar to adult females (Table 14.3; McCarty and Secord 2000). Tree swallow subadult females normally exhibit a dull brown plumage in comparison to the blue–green colors of mature females; 46% of female subadults from Hudson River exhibited blue–green plumage on the back, compared to a normal rate of approximately 15%. Again, the authors contended that precocious pigmentation was consistent with endocrine disruption, but the underlying mechanism could not be identified.

14.3.2 Early Development

A key point of debate regarding PCB, PCDD, and PCDF toxicity has been the relative contribution of intrinsic *in ovo* effects to the overall reproductive success of wild birds. Factors intrinsic to the egg may relate to nutritional quality or hormonal triggers for development, as well as acute toxicity to developing embryos. This section discusses the evidence for effects of PCBs and dioxin-like contaminants on egg quality, hatching success, rate of deformity in embryos and hatchlings, and nestling fitness prior to fledging. Table 14.4 provides a summary of the findings.

There also remains the complex issue of delayed effects on mature wild birds resulting from early exposure to TCDD-like compounds *in ovo* or during chick growth. It has been posited independently by a number of researchers that exposure of embryos or even chicks to xenobiotics during critical developmental windows may have consequences on later development of neural and reproductive systems, where there are timed releases of hormones. This "delayed response" scenario might occur at doses much lower than are toxic from direct exposure of adult animals (Peterson et al. 1993, Dawson 2000, Brunström et al. 2003, Giesy et al. 2003). There is evidence from mammalian studies

TABLE 14.4
Effects Observed in Early Development of Wild Birds with Exposure to PCBs or PCDDs

Species	Exposure	Effect	Tissue Residue[a]	Reference
American kestrel	Captive, *in ovo* Aroclor 1248:1254:1260	Egg yolk: albumin ratio increased	34 µg/g wet wt (egg)	Fernie et al. 2000, 2001b, 2003a
		More infertile eggs		
		Embryo & hatchling deformities		
		Increased mortality of pippers		
		Altered gender-specific nestling growth		
	Captive, egg injection PCB 77 or 126	Embryolethality	2.3–65 ng/g CB-126	Hoffman et al. 1998
		Embryo edema & beak defects	316 ng/g CB-77	
Zebra finch	Captive, oral Aroclor 1248, 40 µg, 16 d	Mortality of pippers	0.20 µg/g wet wt (egg)	Hoogesteijn et al. 2005
Common tern	Egg injection PCB 126	Embryolethality	44–104 ng/g CB-126	Hoffman et al. 1998
		Embryo edema & beak defects		
	Captive, dietary CB-126 + CB-153	Reduced bursa mass	25 ng/g lipid wt TEQ (nestling liver)	Bosveld et al. 2000
		Reduced plasma thyroxine		
	Field	Molar ratio retinol-retinyl palmitate increased		Murk et al. 1996
Double-crested cormorant	Egg injection PCB 126 or 2,3,7,8-TCDD	Embryolethality	158–177 ng/g CB-126 or 4–26 ng/g TCDD (egg)	Powell et al. 1997a, 1998
	Field, artificially incubated	Brain asymmetry	0.034 ng/g TCDD (egg)	Sanderson et al. 1994a, 1994b, Henshel 1998
Black-backed gull	Field	Reduced hatchling mass	13.6 µg/g wet wt (egg)	Custer et al. 1999
Glaucous gull	Field	Increased mortality of hatchlings	8–21 µg/g wet wt (nestling liver)	Hario et al. 2004
		Hatching success reduced	289 ng/g wet wt (adult female blood)	Bustnes et al. 2003
		Reduced hatchling mass		
Herring gull	Field	Altered vitamin A status in young	14–27 µg/g wet wt (egg)	Grasman et al. 1996
Black-crowned night heron	Field	Reduced hatchling mass	4.1 µg/g wet wt (egg)	Hoffman et al. 1986
Black guillemot	Field	Enlarged liver in female nestlings	0.073 µg/g wet wt (nestling liver)	Kuzyk et al. 2003
		Reduced vitamin A in storage and circulating in nestlings		

Great blue heron	Field, artificially incubated	Shortened limbs / Brain asymmetry	0.021 ng/g TCDD (egg)	Powell et al. 1997a, 1998
Gray heron	Field	Altered vitamin A status in young / Reduced T_3	1–6 µg/g wet wt (egg)	Champoux et al. 2006
	Field	Bone abnormalities / Nestling mortality elevated	619 ng/g lipid wt TEQ (nestling)	Thompson et al. 2006
European shag	Field	Reduced vitamin A in circulation	1.1 µg/g wet wt (egg yolk sac)	Murvoll et al. 2006a
Caspian tern	Field	Immunosuppression in nestlings	5.7 µg/g wet wt (egg)	Grasman and Fox 2001, Grasman et al. 1996
Forster's tern	Field, egg-exchange or artificially incubated	Hatching success reduced / Enlarged liver in nestlings	23 µg/g wet wt PCBs + 0.1 ng/g wet wt PCDDs (egg)	Kubiak et al. 1989
White-tailed sea eagle	Field	Embryolethality	500 µg/g lipid wt (egg) or 0.32 ng/g wet wt TEQs (egg)	Helander et al. 2002
	Field	Reduced brood size	5–9 ng/g CB-118 (blood)	Olsson et al. 2000
Osprey	Field, egg-exchange	Reduced nestling growth	0.03–0.16 ng/g wet wt TCDD (egg)	Woodford et al. 1998
	Field	Elevated stores of vitamin A	5–10 µg/g wet wt (egg)	Elliott et al. 2001a
Wood duck	Field, nest box	Hatching success reduced / More infertile eggs	0.19 ng/g wet wt PCDD/Fs (egg)	White and Seginak 1994
Tree swallow	Field, nest box	Hatching success reduced / More infertile eggs	9–30 µg/g wet wt (egg)	McCarty and Secord 1999b, 2000
	Field, nest box	Hatching success reduced	0.4–0.45 ng/g wet wt PCDD/Fs (egg)	Custer et al. 2003
	Field, nest box	Hatching success reduced	0.3–1.0 ng/g wet wt PCDD/Fs (egg)	Custer et al. 2005
	Field, nest box	Heart deformities	>4.4 µg/g wet wt (nestling)	DeWitt et al. 2006
House wren	Field, nest box	Hatching success reduced / Reduced nestling growth	6 µg/g wet wt (egg)	Neigh et al. 2007
Eastern bluebirds	Field, nest box	Heart deformities	1.6 µg/g wet wt (nestling)	DeWitt et al. 2006
Carolina chickadee	Field, nest box	Heart deformities	7.7 µg/g wet wt (nestling)	DeWitt et al. 2006
European starling	Field, nest box	Heart deformities	1.4–3.9 µg/g wet wt (nestling)	DeWitt et al. 2006
	Field, nest box	Increased mortality of hatchlings / Reduced nestling growth	6 µg/g wet wt (nestling)	Arenal et al. 2004

[a] ΣPCBs unless specified otherwise.

that *in utero* or lactational exposure to TCDD or PCBs affects neuro-endocrine and sexual development (Peterson et al. 1993, Dickerson and Gore 2007, Steinberg et al. 2008). Investigations on birds have focused primarily on the possible estrogenic or antiandrogenic effects of early exposure to DDE (Fry and Toone 1981, Helander et al. 2002, Holm et al. 2006, Iwaniuk et al. 2006) rather than delayed responses to PCBs or PCDDs. The few available studies (Fernie et al. 2001a, Hart et al. 2003, Hoogesteijn et al. 2008) are discussed in this section, but there is a lack of concerted experimentation in this area.

14.3.2.1 Egg Quality

Physical and nutritive constraints in incubated eggs have been examined in a number of field studies as potential mechanisms for contaminant effects on early embryos. The findings from those studies are equivocal, while support from laboratory experimentation is limited. Infertile American kestrel eggs containing an average 34 µg/g wet weight ΣPCBs from dietary exposure of the captive parents showed significantly altered nutrient composition. *In ovo*-exposed eggs had heavier yolks and less wet and dry albumin than control eggs (Fernie et al. 2000). Assuming those eggs were representative of the clutch, the embryos of exposed parents would have more lipid and less protein available for development than those of unexposed parents. Although survival to hatch may be enhanced by greater availability of yolk nutrients, albumin content is more often associated with hatchling size (Finkler et al. 1998). Thus, positive effects *in ovo* may be countered by negative effects on nestling viability. Incidences of heavier yolks with PCB exposure were also observed in European cormorants from the Netherlands (van den Berg et al. 1992), whereas the opposite relationship of lighter yolks with increasing TCDD-toxic equivalents (derived largely from PCDDs) was observed in double-crested cormorants in the Great Lakes and Canadian west coast (Sanderson et al. 1994a).

A number of studies have examined but found no relationship between PCBs and egg volume, shape, or mass in glaucous gulls, house wrens (*Troglodytes aedon*), American kestrels, osprey, and white-tailed sea eagle (Fernie et al. 2000, Elliott et al. 2001a, Helander et al. 2002, Bustnes et al. 2003, Neigh et al. 2007). Egg volume and embryo weight of double-crested cormorants from Green Bay (Lake Michigan) were negatively correlated with PCBs (Custer et al. 1999), but when these eggs were considered as part of a larger regional population that included uncontaminated colonies, the volume and mass associations with PCBs were no longer detected (Custer et al. 2001). That suggests other factors unrelated to PCB exposure may have had a stronger influence on egg morphology. Bosveld et al. (1995) reported that egg volume of common tern eggs collected from colonies in the Netherlands and Belgium and artificially incubated was negatively correlated with yolk-sac TEQs. Helberg et al. (2005) found that female great black-backed gulls with higher tissue burdens of persistent contaminants, particularly PCBs, produced smaller second and third eggs in a three-egg clutch relative to females with low tissue contaminant burdens.

In an assessment of a common tern population exhibiting incidences of ovotestes in male embryos at pipping, the concentrations of steroid hormones in eggs were measured as a potential causative factor (French et al. 2001). There were no relationships between total PCBs in eggs and concentrations of 17β-estradiol, 5α-dihydrotestosterone, testosterone, or androstenedione. In an egg-injection study with TCDD, Janz and Bellward (1996b) also found no effect of *in ovo* exposure to TCDD (0.1–2 µg/kg egg) on hepatic estrogen receptor or plasma estradiol concentrations in female hatchlings of chicken, domestic pigeon, or great blue heron.

14.3.2.2 Hatching Success

Hatching success is a measure of reproductive outcome that integrates a number of factors relating to both parental fitness and embryo viability during early development. Influences related to fertility and parenting behavior are classified as egg extrinsic and were discussed above. Conversely, influences of maternally deposited contaminants on embryonic development, growth and survival are classified as egg intrinsic. A clear distinction between these two mechanistic pathways for effects of PCBs and PCDDs in wild birds has been difficult to achieve in the field, and the most robust dataset indicating the presence of egg intrinsic effects is derived from laboratory egg-injection studies. Although the injection route of exposure does not exactly mimic that for field-exposed embryos, the studies do indicate some potential for direct embryotoxicity in birds.

Egg-injection methodology evolved through the 1970s and 1980s, with later research indicating that early injection was essential for evaluating the developmental stages most sensitive to embryolethality. Therefore, here we discuss only those studies that applied an early injection protocol. Recent research also suggests that hatching frequency is reduced by vertical incubation in standard incubators, as opposed to the natural horizontal positioning (McKernan et al. 2007). However, we could not distinguish on this basis, as most studies do not report incubation orientation.

An extensive suite of studies on chicken embryos provide a baseline for lethality in wild species, but comparative studies indicate the chicken is at least 30-fold more sensitive to embryotoxicity than the majority of tested species (Figure 14.1). A recent compilation of the primary literature determined average values for chicken embryo LD50s of 0.18 ng TCDD/g wet weight (ww), and 1.1 ng CB-126/g ww (Head et al. 2008). Other congener-specific analyses reported chicken LD50s of 2.6 ng CB-77/g ww (Hoffman et al. 1998), 3326 ng CB-105/g ww (Hoffman et al. 1995), and >14,000 ng CB-153/g ww (Hoffman et al. 1995). An early study using the formulation Aroclor 1248 reported 50% mortality in the range 1–5 μg/g ww PCBs (Brunström and Orberg 1982).

In a comparative developmental study of chicken, American kestrel and common tern, Hoffman et al. (1998) injected CB-126 or CB-77 into 4-day-old eggs (or the equivalent kestrel stage 6-day-old) and evaluated their development through hatching. They estimated LD50s for chicken embryolethality at 0.4 and 2.6 ng/g for CBs 126 and 77, respectively. American kestrels were less sensitive than chicken, with LD50s of 65 and 316 ng/g for CB-126 and 77, respectively. Common tern was the least sensitive species tested, with an LD50 of 104 ng/g for CB-126 (Hoffman et al. 1998). Similarly

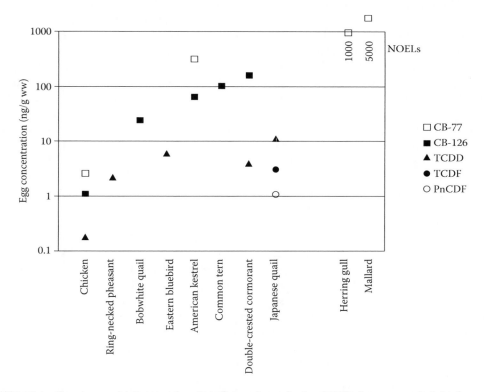

FIGURE 14.1 Species sensitivity to embryolethality, estimated using LD50s from egg early injection protocols. No-observed-effect-level (NOEL) concentrations for herring gull and mallard are included for broad comparative purposes, but should not be equated to the other LD50 values. Refer to Table 14.8 for data sources.

injected double-crested cormorant eggs exhibited a LD50 for CB-126 of 158–177 ng/g (Powell et al. 1997a, 1998); these eggs were injected prior to the start of incubation. A comparable lethal measure for TCDD was estimated at 4–26 ng/g TCDD. Terrestrial passerines and game birds appeared to be more sensitive than raptors and water-birds although still not as sensitive as the chicken. Nosek et al. (1993) reported an LD50 for ring-necked pheasant of 2.2 ng TCDD/g whereas Thiel et al. (1988) determined 50% mortality in eastern bluebirds occurred in the range 1–10 ng TCDD/g ww. Conversely, LD50s could not be attained in studies with herring gull, wood duck and mallard, with upper no-effect concentrations of 1000 ng CB-77/g, 4.6 ng TCDD/g, and 5000 ng CB-77/g, respectively (Brunström 1988, Augspurger et al. 2008).

In one early semicontrolled field injection study of herring gulls, contaminants extracted from Lake Ontario eggs were injected into reference eggs from Kent Island (New Brunswick) and left to incubate naturally by unexposed parents (Gilman et al. 1978). In that case, there were no elevations in rates of embryo or chick mortality, suggesting that the low hatching success observed in Lake Ontario colonies was not the result of embryotoxicity but rather egg-extrinsic factors such as parental care or infertility.

A very recent comparative egg-injection study with supportive *in vitro* experimentation suggests that some bird species may be more sensitive to furans than dioxins (Cohen-Barnhouse et al. 2010, Herve et al. 2010), thereby challenging the ranking currently used for Toxic Equivalency Factors (van den Berg et al. 1998). Using a relatively insensitive species, Japanese quail (*Coturnix japonica*), Cohen-Barnhouse et al. (2010) reported embryo LD50s for TCDD, 2,3,7,8-TCDF and 2,3,4,7,8-PentaCDF of 11.3 ng/g, 3.1 ng/g, and 1.1 ng/g, respectively. Those translate to potencies of 3.6 for TCDF and 10.3 for PeCDF, relative to TCDD. Their results have ramifications for past and present field studies which used conventional, chicken-derived TEQs to evaluate avian embryotoxicity at sites where furan contamination occurred as part of the environmental mixture.

Aside from egg-injection studies, there are few laboratory studies that examine embryotoxicity in birds other than chickens. Dietary exposures of captive breeding American kestrel to an Aroclor mixture reported some effects on hatching, at total PCB concentrations in eggs averaging 34 µg/g wet weight (Fernie et al. 2001a, 2001b). However, there was no evidence that embryolethality contributed significantly to the reduced hatching success; rather, PCB-exposed pairs laid more completely infertile clutches (7 from 25 pairs vs. 2 from 25 pairs, $p = .04$) and experienced delayed clutch initiation (average lag of 6.3 d compared to controls, $p = .003$) that was associated with poor hatching (Fernie et al. 2001b). These effects suggest more of a parental physiological element to the causative mechanism instead of a direct *in ovo* response to PCBs. Effects on fertility and hatching success were not observed one season after exposure in the same birds, nor in the fledged young that had been exposed *in ovo* during their first year breeding (Fernie et al. 2001a, 2001b). Since eggs in the first and second year of breeding contained comparable concentrations of total PCBs (34 vs. 29 µg/g wet weight) it was possible that fertility was related to effects of nonpersistent congeners (Drouillard et al. 2001).

These studies of captive birds suggest that embryolethality can occur with *in ovo* exposure to dioxin-like PCBs and PCDDs, but only at concentrations that are increasingly unlikely to occur in all but the most heavily contaminated sites. It seems probable that hatching success in wild birds is more often impacted through indirect mechanisms such as low fertility of the male or delayed clutch initiation by the female parent. These effects may be more the result of rapidly metabolized congeners (or their metabolites) rather than the non-*ortho* dioxin-like forms.

Several early field studies did not report on the nature of egg contents, such as the presence or not of a visible embryo (Tillitt et al. 1992, Larson et al. 1996), thus raising questions about mechanisms controlling variable hatching success in wild birds contaminated with a complex mixture of chemicals. It has proven difficult to establish cause–effect linkages between embryolethality and PCBs or PCDDs in the field. A protocol for distinguishing between infertility and dead early embryos has recently been developed (Birkhead et al. 2008) and may help to further resolve such questions in the future.

Two broad assessments of the long-term impact of contaminants on reproduction in wild raptors found that PCBs were negatively associated with success in a Swedish Baltic population of white-tailed sea eagle (1988–1997), and in a Great Lakes shoreline population of bald eagle (1986–2000). In both assessments, eggs with a visible embryo present contained higher concentrations of PCBs and other chlorinated hydrocarbons than eggs without a discernable embryo (Helander et al. 2002, Best et al. 2010). When Helander et al. (2002) corrected for lipid metabolism, CB-138 in undeveloped sea eagle eggs (~55 µg/g lipid weight, $n = 17$) remained significantly lower than in eggs containing embryos (~95 µg/g lipid weight, $n = 9$), whereas DDE did not differ among the two egg classes. They suggested that PCBs may have caused embryo lethality in sea eagles when whole egg values were above 500 µg/g lipid weight ΣPCBs, 160 ng/g lipid weight CB-126, or 320 pg/g fresh weight TEQs. In the same populations of sea eagle, researchers monitoring nests during a relatively short 3-year period (1996–1998) found a negative relationship between brood size (number of chicks hatched) and PCB 118 (Olsson et al. 2000).

Best et al. (in preparation) also attempted to compare contaminant concentrations in bald eagle eggs having a visible embryo with those in visibly undeveloped eggs. However, statistical analysis with Akaike's Information Criterion (designed to extract biologically relevant variables from large, noisy data sets) found that, for the whole dataset, chlordane compounds, and for the Great Lakes shoreline dataset, DDE, had the strongest association with presence of a visible embryo. PCBs were only significant as an interaction term and not as a main factor in the three statistical models used. Further, the low correlation coefficients produced for all of the measured contaminants suggested that other ecological or genetic factors were also influencing embryonic survival. In the full dataset of collected eggs, the majority (123 of 196 eggs or 63%) were visibly undeveloped and, therefore, the dominant direct determinant of hatching success in that population of bald eagles was early embryonic death and/or infertility.

Other field studies conducted well after the peaks in environmental contamination, from the early-1990s onward, also provide mixed messages about the influence of PCBs and PCDDs on hatching success of wild birds. Studies of dippers in Wales (Ormerod et al. 2000), double-crested cormorants and tree swallows in Green Bay, Lake Michigan (Custer et al. 1998, 1999), double-crested cormorants in Lakes Huron and Superior (Hilscherova et al. 2003), osprey and bald eagle in British Columbia (Elliott et al. 1996b, 2001a), American robins (*Turdus migratorius*) in Massachusetts (Henning et al. 2003), and starlings in Illinois (Arenal et al. 2004) found no significant associations between hatching success and exposure to PCBs or TEQs. Tissue concentrations of ΣPCBs were less on average than 560 µg/g lipid weight in bald eagle yolk sacs (TCDD < 3000 ng/kg lipid weight), 13.6 µg/g wet weight in cormorant eggs, 8.4 µg/g lipid weight in dipper eggs, 150 µg/g lipid weight in osprey yolk sacs (TCDD < 3600 ng/kg lipid weight), 3.3 µg/g wet weight in tree swallow pippers, 84 µg/g wet weight in robin eggs, and 9 µg/g wet weight in starling eggs.

Conversely, hatching success was reported to be significantly correlated with PCBs or PCDDs in glaucous gulls from the Norwegian Arctic (Bustnes et al. 2003), house wrens from Michigan (Neigh et al. 2007), and tree swallows from New York, Massachusetts, and Rhode Island (McCarty and Secord 1999b, Custer et al. 2003, 2005, Table 14.4). Bustnes et al. (2003) reported a significant positive relationship between the presence of unhatched eggs in a glaucous gull nest and the corresponding female blood concentrations of PCBs (289 ng/g ww) and other chlorinated hydrocarbons (including chlordanes); however, they did not report the condition of egg contents. Similarly, Neigh et al. (2007) did not report what proportions of house wren eggs that failed to hatch were infertile or contained dead embryos; average ΣPCB concentrations in eggs from the contaminated site were not particularly elevated at 6.3 µg/g wet weight, but within the range of concentrations shown to alter parental behavior in other passerines (see Section 14.3.1.1). Eastern bluebirds nesting at the same locations did not show associations between hatching success and contaminants, although their eggs contained 2 µg/g PCBs more on average than the wrens (Neigh et al. 2007). The majority (~60%) of tree swallow eggs that failed to hatch along the Hudson River in New York were undeveloped/infertile; eggs from this population contained 9.3–29.5 µg/g PCBs (McCarty and Secord 1999b).

That result again implies a physiological effect on the parent birds rather than an embryotoxic effect, and that is consistent with the possible aberrant nesting behavior also reported during their study (McCarty and Secord 1999a, 1999b).

Other tree swallow studies along the Housatonic (MS) and Woonasquatucket (RI) Rivers identified strong associations between hatching success and PCDDs (Custer et al. 2003, 2005). Housatonic River tree swallow pippers contained up to 101 μg/g wet weight PCBs on average, which is the highest reported for the species (Custer et al. 2003); however, a correlative model identified dioxin/furan TEQs and not PCB TEQs as being significantly associated with hatching success. On the Woonasquatucket River, TCDD concentrations of 300 to over 1000 pg/g wet weight in eggs were associated with 50–70% hatching success; total hatching failures in clutches became common at TCDD egg concentrations above 180 pg/g (Custer et al. 2005). One earlier study of wood ducks observed reduced hatching success with exposure to PCDDs and PCDFs, when total concentrations were higher than ~190 pg/g wet weight in eggs (White and Seginak 1994). In that case, most of the eggs that failed to hatch did not contain a visible embryo (35% with embryo, 65% without). Thus, a fertility response mechanism again seems more probable in those exposed birds. An embryotoxic sensitivity to TCDD would not be consistent with other studies reporting a relative insensitivity of duck species in general (Brunström 1988, Jin et al. 2001), and more specifically with a wood duck egg-injection study that reported no effect on embryolethality at 4600 pg/g wet weight in eggs (Augspurger et al. 2008). In addition, the field study was conducted downstream of a 2,4,5-T manufacturing plant, where other contaminants may have been present yet unmeasured.

In a few instances, egg-exchange designs have been used in the field to try and differentiate between intrinsic and extrinsic factors influencing embryonic development. The egg-exchange design involves switching incubated eggs between nests from contaminated and uncontaminated sites. Kubiak et al. (1989) used this technique with Forster's terns in Green Bay (Lake Michigan) and reported an apparent intrinsic effect of PCBs on hatching success (Table 14.4). However, the poorest hatching by a marked margin occurred where incubation was by "dirty" adults (those at contaminated sites), and the authors concluded that the over-riding influence on hatching success related to an extrinsic effect of parental inattendance. Woodford et al. (1998) found no direct or indirect effects of TCDD on hatching rates of osprey foraging in watersheds impacted by outflows from bleached kraft pulp mills (exposed egg concentrations ranged from 29 to 162 pg/g wet wt TCDD). Artificial incubation methods have also been used to eliminate some extrinsic factors (such as parental attentiveness), and have found no intrinsic effects of dioxin-like contaminants on hatching success in bald eagle (NOEL 303 pg TEQs/g ww, Elliott et al. 1996b, Elliott and Harris 2001), osprey (NOEL 136 pg TEQs/g ww, Woodford et al. 1998) and cormorants (NOEL 175 pg TEQs/g ww, Hilscherova et al. 2003). The afore-mentioned study of Forster's tern in Lake Michigan also applied artificial incubation methods and detected some reduction in hatching success related to intrinsic effects (Kubiak et al. 1989); however, as previously mentioned, the dominant effect occurred where extrinsic factors were acting alone (egg-exchange and incubation by "dirty" parents).

The more recent field studies corroborate the laboratory findings, in that the majority have not been able to identify intrinsic effects on bird embryos at the concentrations of PCBs and PCDDs prevalent in environments today. It seems likely that intrinsic effects were occurring in the 1970s in hotspots like the Great Lakes, but inconsistencies in reporting of the condition of unhatched eggs (with or without visible embryos) largely prevent the distinction between intrinsic and extrinsic effects.

14.3.2.3 Rates of Overt Morphological Deformities

The toxicity of dioxin-like compounds to birds first became apparent with the appearance in the 1950s of chick edema disease in the poultry industry. Poultry diets were found to be contaminated by PCDDs and PCDFs from feed fats collected from animal hides treated with chlorophenolics (Firestone 1973). That disease etiology was later applied to symptoms observed in wild birds during the 1970s and called Great Lakes Embryo Mortality, Edema and Deformities Syndrome (GLEMEDS)

(Gilbertson et al. 1991, Hoffman et al. 1996a). GLEMEDS was characterized in embryos and hatchlings by elevated rates of subcutaneous pericardial and peritoneal edemas, embryonic weight loss and mortality, and skeletal abnormalities including bill defects. Retrospective analyses of herring gull eggs collected from some Great Lakes colonies in the 1970s detected elevated concentrations of TCDD, in some cases over 1000 ng/g (Hebert et al. 1994), that could have contributed to the symptoms expressed in wild birds (Gilbertson 1983). However, PCBs also were suggested as a potential causative factor, because of their wider distribution, more elevated environmental levels, and the dioxin-like toxicity of coplanar congeners (Gilbertson et al. 1991).

Laboratory studies suggest that water-birds and raptors exposed to PCBs *in ovo* can exhibit deformities consistent with GLEMEDS, although there is considerable species variability in the form of expression of deformities and in sensitivity. Chicken and American kestrel eggs injected with CB-126 or CB-77 on day 4–6 of incubation produced embryos with edema and beak defects generally consistent with GLEMEDS (Hoffman et al. 1998). However, whereas chickens frequently expressed crossed bills, kestrels instead expressed shortened lower mandibles. Significant elevations in deformity incidence occurred at 0.3 ng/g CB-126 in chickens, and at 23 ng/g CB-126 in kestrels, a difference in sensitivity of almost two orders of magnitude. Although common tern embryos exhibited crossed bills, they appeared only at embryo-lethal concentrations (240 ng/g CB-126, Hoffman et al. 1998); presumably, those would not be seen in a wild tern population, because the affected individuals would die before hatching. Double-crested cormorant eggs injected with TCDD or CB-126 did not show a dose-response with rates of deformity at doses up to 11.7 and 698 ng/g, respectively, and appear to be relatively insensitive to the expression of developmental physical defects compared to kestrel and chicken (Powell et al. 1997a, 1997b, 1998).

Both dietary and *in ovo* exposures to a PCB mixture (29–34 µg/g ww ΣPCBs in eggs) produced embryo and hatchling deformities in American kestrel (Fernie et al. 2003a). The appearance of dwarfed limbs, shortened lower mandibles, craniofacial defects and hematomas only during dietary exposure and not one breeding year after exposure suggested that these particular abnormalities were induced by readily cleared congeners (Drouillard et al. 2001, Fernie et al. 2003a). Conversely, multiple deformities per exposed pair and underdevelopment and edema in embryos and hatchlings occurred with dietary exposure, 1 year after such exposure and with *in ovo* exposure, which implies that these abnormalities were induced by persistent congeners or unidentified stress factors (Fernie et al. 2003a). The overall incidence of abnormalities was not associated with total PCBs in sibling eggs even though on average the frequencies in PCB-exposed pairs were significantly greater than control pairs. The appearance of shortened lower mandibles, but not crossed bills, in that species is consistent with the egg-injection study of Hoffman et al. (1998).

The colonial water-birds of the Great Lakes have been the most extensively studied for the presence of deformities, which were considered a dramatic diagnostic of poor fitness in wildlife and of contamination of the Great Lakes ecosystem. The deformity frequency, however, was localized and not accompanied by population declines (Price and Weseloh 1986, Ryckman et al. 1998). Clustered incidences of bill deformities in Lake Michigan colonies of double-crested cormorants, particularly on Spider Island during the 1980s through to the early 1990s were regularly reported; Ludwig et al. (1996) and Giesy et al. (1994) considered that the correlative evidence was sufficient to implicate PCBs as a chemical driver of GLEMEDS disease in cormorants of the Great Lakes. Species other than double-crested cormorants that exhibited bill deformities at low levels in wild populations were herring gulls and common tern from Lake Ontario in the 1970s (Gilman et al. 1977, Gilbertson 1983), Forster's tern from Lake Huron (Yamashita et al. 1993), wood ducks from PCDD- and PCDF-contaminated wetlands in Arkansas (White and Seginak 1994; 6 of 250, all from one nest), bald eagles from the Great Lakes (Bowerman et al. 1994, Best et al. in preparation), white-tailed sea eagles from Sweden (Helander et al. 2002), and great blue herons from PCDD- and PCDF-contaminated coastal colonies in western Canada (Hart et al. 1991). Although edema often was reported more frequently than bill deformities, it is not in and of itself diagnostic of GLEMEDS or a developmental etiology (Peterson et al. 1993).

Retrospectively, the cause–effect relationship between developmental abnormalities and PCBs has been questioned, after more extensive field sampling and as a result of difficulties in replicating the syndrome with laboratory exposures of birds other than chickens. A lack of association between deformity rates and total PCB concentrations was evident in field studies where experimental design allowed correlation by nest rather than with colony averages. Incidences of bill deformities were observed in double-crested cormorants from Lake Michigan in 1988–1990 and 1994–1998 (Larson et al. 1996, Custer et al. 1999, Hilscherova et al. 2003); however, there were no correlations between embryonic deformities and PCBs on a nest/sibling basis, despite a reasonably wide range in PCB concentrations (1–72 µg/g ww). Re-analysis of historical data on deformity rates in the Great Lakes bald eagle population did not establish a clear statistical link with PCBs or other contaminants (Best et al. in preparation). In a sample of 73 bald eagle eggs having a visible embryo (of a total of 197 eggs salvaged during banding visits), eight deformities were observed and two of these were skewed bills (2.7%). Using Akaike's Information Criterion statistical models, there was no support for cause–effect relationships between any contaminant concentrations and eagle abnormalities (Best et al. in preparation).

PCBs and dioxins were recently linked to elevated incidences of skeletal deformities in grey heron (*Ardea cinerea*) nestlings from the United Kingdom (Thompson et al. 2006). Nestlings prone to multiple fractures of the tarsus, tibia and metacarpal bones contained a combined PCB—dioxin TEQ of 619 ng/g lipid weight in fat deposits. Heronries with high incidences of deformities also had elevated chick mortality. Previous studies reported correlations between shortened limbs and PCDD and PCDF exposure in wild great blue herons (Sanderson et al. 1994b) and captive PCB-exposed kestrels (Fernie et al. 2003a). The GLEMEDS etiology was not present, however, and there were inconsistencies among equally contaminated colonies in the expression of abnormalities.

The incidence of GLEMEDS-type symptoms appears to have decreased significantly since the mid-1990s in affected wild populations, and conclusive cause–effect relationships remain elusive. In the Canadian Great Lakes during the early 1990s, the prevalence of bill deformities in double-crested cormorant colonies was less than 3 per 10,000, down from as high as 12 per 10,000 in the 1980s (Ryckman et al. 1998). In Green Bay, Lake Michigan the prevalence of bill deformities in cormorants was 6 per 10,000 in the years 1997–2001 (Stromborg, unpublished data), down from 76 per 10,000 in the late 1980s (Larson et al. 1996). Although the full suite of OC tissue burdens has coincidentally declined, the major reductions in contamination occurred earlier, during the 1970s.

It remains difficult to separate the role of contaminants from contributions of extraneous factors such as genetics and disease. Elevated incidence of deformities in wild Great Lakes populations paralleled periods of rapid population growth in severely depleted populations of wild birds, which makes the argument for genetic contributions a plausible one (Briskie and Mackintosh 2004). Also, during the period 1988 to 1995, deformed hatchlings were routinely removed from Great Lakes populations for chemical analysis, which could have in effect removed a genetic mutation or an infectious agent (parasite, etc.) from the wild. Additionally, incidences of malformations documented in heron colonies outside of the Great Lakes were frequently colony-specific, and potentially having a genetic component (Sanderson et al. 1994b, Thompson et al. 2006). The documentation of crossed bills in captive, uncontaminated double-crested cormorants as a result of nutritional deficiencies, most probably vitamin D deficiencies, also reintroduces the importance of nutrition as a contributing variable in expression of developmental deformities (Kuiken et al. 1999).

For the purposes of diagnosing current impacts of PCBs, PCDDs, and PCDFs on wild birds, we do not find deformity rates useful as a definitive marker. Some wild bird species do not express the most visibly extreme symptoms of GLEMEDS (crossed bills) with laboratory exposure, and there is sufficient evidence from laboratory and field studies that incidences in the wild were not exclusively the result of dioxin-like toxicity. In the absence of a standardized scoring protocol and in the face of necessarily very large sample sizes (in the 1000s), the diagnosis of GLEMEDS is also logistically difficult to achieve.

14.3.2.4 Nestling Mortality and Pre- and Posthatching Growth

Nestlings may experience effects related to alterations *in ovo*, in parental attendance to them or in their own physiology when exposed to PCBs, PCDDs, and PCDFs. As such, they tend to show a diversity of responses that can be species- and gender-specific. Altricial and precocial chicks may show widely varying response patterns, because of their varying dependence on the parents for food and thermal protection. Significant effects on nestling viability are summarized in Table 14.4.

A number of cases of significant mortality shortly after hatching, usually within 3 days, have been reported in laboratory studies. Hoogesteijn et al. (2005) found many chicks of zebra finch hens exposed to Aroclor 1248 (laying eggs with 0.2 µg/g ΣPCBs) died immediately after pipping due to inanition or hypothermia. Similarly, 52% of kestrel hatchlings of PCB-exposed parents (laying eggs with 34 µg/g ΣPCBs) died within 3 days of hatching, compared to a rate of 14% for controls (Fernie et al. 2003b).

In a northern Baltic colony of lesser black-backed gulls suffering from unexplained disease, a high proportion of nestling deaths occurred within 3 days of hatching. Those individuals lost around 16% of their body mass during the time before death, and had hepatic concentrations of ΣPCBs from 4 to 36 µg/g wet weight (CB-77 + CB-126 = ~31 – 380 ng/g ww) and DDE from 2 to 30 µg/g wet weight (Hario et al. 2004). Starvation was evaluated as a possible cause for gull chick mortality and was rejected. Similarly, second clutches of starlings nesting in nest boxes at a PCB-exposed site in Illinois experienced 100% mortality, with most nestlings dying within 3 or 4 days of hatch (Arenal et al. 2004).

The mortalities of young nestlings may be related to thermoregulatory or food stress, either as a result of reduced parental attentiveness or physiological defects in the nestlings themselves. Since the species reportedly experiencing this early death phenomenon are altricial or semi-altricial, it is also possible that the hatchlings are being exposed to a more concentrated dose of contaminants as they absorb the last of the egg yolk within a few days of hatching (D. J. Hoffman, personal communication).

Nestling growth has been examined as both daily weight gain and terminal (fledgling) weight, but few definitive effects of PCB or PCDD exposure have been shown. The growth of captive American kestrel nestlings exposed *in ovo* to PCBs was affected in a complex and gender-specific manner (Fernie et al. 2003b, 2003c). First generation (exposed to 34 µg/g) and maternally exposed second generation (exposed to 29 µg/g) female nestlings grew at faster rates than control females and attained greater asymptotic mass. Exposed first generation male nestlings also grew at faster rates than control males and had stimulated bone growth, but showed lesser asymptotic mass (Fernie et al. 2003b). Maternally exposed second generation males grew at faster rates than control males, had greater asymptotic mass but shorter bones (Fernie et al. 2003c). Paternally exposed nestlings grew slower than controls and the females were smaller at fledging (Fernie et al. 2003c). Two sets of first generation nestlings were affected, in the year of parental exposure and also 1 year after exposure. Those observations on captive kestrels suggest that growth effects of PCBs may be sex-specific and hormonally mediated, and thus, overlooked in field studies where the gender of chicks was unknown or not distinguished.

In an artificial incubation experiment, Bosveld et al. (2000) could not find growth effects of CB-126 or CB-153 in reared common tern chicks exposed through food spiked with concentrations ranging 0.01–1.2 ng/g TEQ; this is in agreement with field studies of European colonies (Bosveld et al. 1995). Hatchling weights were also not affected by *in ovo* exposure to PCBs or TCDDs in field or egg-injection experiments with cormorants (double-crested, Sanderson et al. 1994a, Sanderson and Bellward 1995, Powell et al. 1997a, 1998, European, van den Berg et al. 1994, 1995), wood ducks (White and Seginak 1994), or American kestrels (Hoffman et al. 1998).

Where nestling growth has been statistically linked to dioxin-like contamination in field studies, the alterations have frequently been subtle and the long-term ramifications for survival consequently difficult to ascertain. In the field, exposure presumably occurs through a combination of *in ovo* and dietary routes. In egg-exchange experiments, Woodford et al. (1998) found that osprey nestlings from reference areas accrued mass significantly faster than nestlings transported to contaminated areas from reference areas as incubating eggs; contamination in that study was largely the result of PCDDs and PCDFs released from pulp mills (Table 14.4).

In nest box studies of the PCB-contaminated Kalamazoo River (Lake Michigan), Neigh et al. (2007) found that house wren and eastern bluebird nestlings showed reduced daily mass gain rates at contaminated versus reference sites in some but not all years. Nest-box tree swallow studies found no effect of *in ovo* or dietary exposure to PCBs or PCDDs on growth (Custer et al. 2005, Neigh et al. 2006) or an increased growth rate with PCB exposure (McCarty and Secord 1999b). A nest box study of starlings found reduced body weights in 3- and 15-day-old chicks from a PCB-contaminated site, but no difference in 9-day-old chicks (Arenal et al. 2004).

South polar skuas (*Catharacta maccormicki*) exhibited an eight percent reduction in hatchling body mass between adult females carrying low and high tissue burdens of a chlorinated hydrocarbon mix (Bustnes et al. 2007). Although total PCBs were elevated in the female blood of this species, significant contamination with HCB and mirex was also present. Similar reductions in hatchling mass were observed in northern colonies of glaucous gulls (Bustnes et al. 2003). In field studies on the Great Lakes, double-crested cormorant embryo mass was negatively correlated with total PCBs ($r = -0.28$, $p = 0.04$; Custer et al. 1999), while subdued growth curves for nestling Forster's terns were deemed suggestive of a wasting syndrome (Kubiak et al. 1989, Harris et al. 1993). Also in the Great Lakes, Hoffman et al. (1993) reported significant reductions in the femur length-to-body weight ratio of day-old common tern and black-crowned night heron nestlings that were negatively correlated with PCB concentrations in eggs ($r = -0.7$, $p \leq .05$). Similarly, a significant association between growth and total PCBs was detected in pipping black-crowned night herons (*Nycticorax nycticorax*) from San Francisco Bay ($r = -0.61$, $p < .05$); the relative reduction in mass in this case was ~15% (Hoffman et al. 1986).

14.3.2.5 Other Measures of Nestling Fitness

Long-term viability of nestlings is difficult to assess in wild birds, particularly those species that migrate and are logistically difficult to track after fledge. It is common to apply a weight of evidence approach to evaluating fitness, using a suite of morphological, physiological, and biochemical indicators. Although the majority of the applied indicators are not specific to xenobiotic response patterns, they have value when considered together for assessing the overall vigour of young birds (Fox et al. 2007a, 2007b).

A number of studies have reported relationships between exposure to PCBs or PCDDs and altered internal morphology, particularly brain asymmetry and weights of liver, heart, and bursa of Fabricius. Many of these markers are not specific to dioxin-like contamination. In general, measures taken of hatchlings have reported few significant effects (Elliott et al. 1996b, 2001a, Custer et al. 2001, Murvoll et al. 2006a), whereas more instances of altered morphology were reported with measures taken close to the time of fledge (Bosveld et al. 2000, Kuzyk et al. 2003, Papp et al. 2005, DeWitt et al. 2006). In the following text, the majority of studies were conducted in the field.

Brain asymmetry in response to TCDD exposure has been reported in great blue heron, double-crested cormorant and bald eagle nestlings, and was reviewed in detail by Henshel (1998). Unlike the majority of dioxin-like toxicity studies, chicken did not appear to be more sensitive to brain asymmetry effects than wild birds; Henshel (1998) reported ED50s for chicken, double-crested cormorant and great blue heron of 37, 34, and 21 pg/g TCDD in embryos, respectively. Unfortunately, we could not find any studies that measured brain asymmetry and then related findings to measured behavior responses, indicative of a whole-body effect. Custer et al. (2001) suggested that brain asymmetries in pipping embryos may at times be an artifact of physical constraints imposed by egg size, based on a correlation between asymmetry and egg volume in double-crested cormorants.

Liver enlargement and necrosis may be symptomatic of a generalized enzymatic response to xenobiotics. In American kestrel nestlings exposed to CB-126 through their diet, these hepatic responses increased in severity as doses increased from 50 to 1000 ng/g body weight (resulting in hepatic concentrations of 156–1098 ng/g ww, Hoffman et al. 1996b).

Kuzyk et al. (2003) observed 36% liver enlargement in female black guillemot nestlings exposed to PCBs in the field, which is similar to the altered size reported in wild birds of the Great Lakes during the 1980s (Kubiak et al. 1989). However, male nestlings did not exhibit comparable enlargement, with an average 5% increase. Despite correlations with ΣPCBs, there were no differences between genders in hepatic tissue burdens (0.17–6.2 µg/g ww), suggesting an enhanced susceptibility in female young. Hence, difficulty in differentiating gender during early development in other species may mask morphological effects. For instance, studies of tree swallows typically found no changes to organ weights despite elevated PCB and PCDD tissue burdens; however, results for males and females were often considered together (Bishop et al. 1999, Custer et al. 2005).

A suite of altricial passerines, namely tree swallow, house wren, Carolina chickadee (*Parus carolinensis*), eastern bluebird and red-winged blackbird (*Agelaius phoeniceus*) exhibited negative associations between heart weight or shape and total PCB concentrations in nestlings (DeWitt et al. 2006). Sensitivities to heart malformation appeared similar to those observed for chicken with exposure to TCDD or dioxin-like PCBs (Powell et al. 1996, Walker et al. 1997). Cheung et al. (1981) suggested a threshold for heart abnormalities in 14-day-old chicken embryos of 11 pg/g (Cheung et al. 1981). The lowest-observed-adverse-effect (LOAEL) levels for the wild passerines ranged from 0.01 to 10 µg/g PCBs (DeWitt et al. 2006). Once again, studies in wild birds generally failed to evaluate associations between heart malformation and whole body-effects such as cardiac fitness or first year return rates in migratory species.

In a companion study to that examining heart defects, Henshel and Sparks (2006) assessed a number of organ somatic indices in the same five passerine species. Several associations with PCB tissue burdens were detected, predominantly with brain and bursa weights; also of significance was the finding that tree swallows appeared less sensitive to morphological alterations than the other altricial species, and may not be representative of those passerines if used as a surrogate monitoring species. The findings of that study were, however, equivocal due to a lack of consistency in effects observed in various species. That may have been in part due to the use of site averages for correlation analysis rather than nest-by-nest correlations, and the masking of true effects by coincidental environmental variability. The results of such regional or site averaging on assessments of cause–effect were clearly evident among the varying findings of hatching success and bill deformities in Great Lakes species (see Sections 14.3.2.2 and 14.3.2.3).

Bursa of fabricius is an avian organ with important immune functions that has occasionally been associated with PCB contaminant effects in wild birds. Common tern chicks exposed to CB-126 and CB-153 in their food exhibited a dose-dependent decrease in bursa weight, which the authors suggested was consistent with immune dysfunction (Bosveld et al. 2000). Effects on bursa weight were evident at the LOEL for EROD induction and ED50 for plasma total T_4 reduction, namely 25 ng TEQ/g liver lipid. Reduced bursa weight was also evident in American kestrel nestlings similarly exposed to CB-126 (380–1000 ng/g ww in liver, Hoffman et al. 1996b).

Whereas organ weight changes are assumed to reflect potential physiological responses to xenobiotic insults in the avian body, biochemical markers are often used to infer endocrine disruption mechanisms in birds exposed to chlorinated hydrocarbons. Unfortunately, those measures too often are applied incompletely in field studies and in the absence of a clear understanding of the mechanistic actions of PCBs and PCDDs. There is sufficient evidence that avian and mammalian endocrine systems respond differently to these contaminants and there are few comprehensive mechanistic studies with birds. The remainder of this section describes what is known of potential endocrine disruption, measured as alterations in thyroid, vitamin A, and immune function.

The development of the thyroid system in nestling birds is still not well characterized and the use of these markers in young birds exposed to PCBs has produced equivocal results. Rolland (2000) provides a comprehensive review of the mammalian and avian literature on both thyroid and vitamin A function. Captive nestling American kestrels exposed *in ovo* to 34 µg/g wet weight

ΣPCBs exhibited lower plasma triiodothyronine (T_3) concentrations than controls, but no difference in T_4 levels (Smits et al. 2002). Nestlings of those kestrels that were exposed *in ovo* (the F_1 generation, that contained 29 μg/g wet weight ΣPCBs as eggs) exhibited the opposite pattern of effect. Chicken studies have suggested that the timing of exposure is important, as embryos from eggs injected prior to incubation exhibited reduced plasma T_4 while later injections produced no effect (Gould et al. 1997, 1999). Kestrel nestlings exposed to CB-126 through their diet for their first 10 days displayed lymphoid depletion of the spleen and bursa, reduced thyroid tissue and degenerative lesions, increasing in severity with dose (ranging from 156 to 1098 ng/g ww in liver, Hoffman et al. 1996b). Thyroid function is believed to be influenced by PCBs and PCDDs via induction of phase II biotransformation enzymes during exposure (specifically, uridine diphosphate glucuronosyltransferase) or via competitive inhibition binding to transthyretin (McNabb and Fox 2003). Altered thyroid function may influence growth, expression of developmental abnormalities or immune function.

There is a history of reported contaminant-correlated alterations in thyroid function in wild colonial birds of the Great Lakes. Pipping herring gull embryos from PCB-contaminated sites had significantly decreased yolk-sac stores of T_4, as little as one-tenth that of reference embryos, but half of these affected embryos had no corresponding decrease in plasma concentrations (McNabb and Fox 2003). Prefledgling herring gulls exhibited significant reductions in plasma T_4 as well as larger thyroid glands, indicative of hypothyroidism (McNabb et al. 2001, McNabb and Fox 2003). Those field studies of herring gulls suggest thyroid hormone responses to PCBs may be strongly age-specific, with compensation mechanisms developing as young birds age. A reduced capacity to respond to stresses associated with poor weather or limited food resources may occur as a result of those changes to thyroid function in developing herring gulls. Thyroid studies in the Great Lakes also may be confounded by natural iodine deficiencies present in the region (Rolland 2000, Fox et al. 2007b). Hatchling European cormorants exhibited a negative correlation between free T_4 and yolk-sac mono-*ortho* PCBs, but no similar relationships for plasma T_4 or T_3 (van den Berg et al. 1994). T_3 was negatively correlated with PCBs in great blue heron nestlings from the St. Lawrence River (Champoux et al. 2006).

Immune function is a highly complex system to evaluate with biochemical markers; toxicological studies with PCBs and PCDDs have, not surprisingly, reported species-specific, age-specific and gender-specific effects (Grasman 2002). Lavoie and Grasman (2007) measured a twofold suppression in antibody titers of 28-day-old chickens exposed *in ovo* to CB-126 (0.25 ng/g in eggs) or CB-77 (0.64 ng/g in eggs), but no phytohemagglutinin (PHA) skin response or altered blood chemistry. American kestrel nestlings exposed *in ovo* to PCBs (34 μg/g in eggs) exhibited a greater antibody response (as measured by the PHA skin test) than control nestlings, indicating no signs of immunosuppression (Smits et al. 2002). Nonetheless, immunosuppression was reported during the 1990s in Great Lakes populations of Caspian terns and herring gulls and related to PCBs (Grasman et al. 1996). Young Caspian terns collected in the late 1990s from Lake Huron continued to show T-lymphocyte-mediated immunosuppression, as measured by the PHA skin test and differential white blood cell counts (Grasman and Fox 2001). Those later incidences of immunosuppression were associated with elevated concentrations of PCBs (3.5–8.5 μg/g ww in eggs) and DDE (1.0–1.5 μg/g ww in eggs).

Stores of vitamin A (hepatic retinyl palmitate) and circulating concentrations (plasma retinol) have also been suggested as a marker of PCB and PCDD effects in young wild birds (Spear et al. 1986, 1990). Vitamin A is important for growth and development processes in wild birds. Laboratory studies with young Japanese quail exposed to PCBs in their diet suggested that liver stores of vitamin A were negatively affected by the exposure (Cecil et al. 1973). Similarly, CB-77 (10, 20, or 40 μg/g body weight) injected into juvenile ring doves reduced retinol concentrations (Spear et al. 1986). Suggested mechanisms for the action of dioxin-like contaminants on vitamin A status include accelerated metabolism, breakdown and clearance of vitamin A by interference with carrier proteins like retinol-binding protein and transthyretin (Brouwer et al. 1998).

Vitamin A in wild birds may be strongly influenced by dietary uptake which confounds the analysis of contaminant effects in the field. However, Spear et al. (1990) found strong positive associations between the molar ratio of retinol: retinyl palmitate and TCDD or TCDD-TEQs in herring gull eggs from the Great Lakes. Similar findings were reported for common tern (Murk et al. 1996) and great blue heron (Champoux et al. 2006).

Where nestlings have been monitored rather than eggs, the general response pattern appears to be one of reduced circulating concentrations with increasing PCB or PCDD contamination. Significant negative correlations between nestling plasma retinol and PCBs were reported in great blue heron from the St. Lawrence River (Canada, Champoux et al. 2006), European shag (*Phalacrocorax aristotelis*) from Norway (Murvoll et al. 2006a), black guillemots from Labrador (Kuzyk et al. 2003), and herring gulls and Caspian terns from the Great Lakes (Grasman et al. 1996), but not for bald eagle nestlings from British Columbia (Cesh et al. 2010). The study of black guillemots also reported a sex-specific reduction in liver stores with PCB exposure; the negative association was evident in female nestlings but not male nestlings (Kuzyk et al. 2003). In addition, hydroxylated metabolites of PCBs, OH-PCBs, were negatively correlated with plasma retinol in a 1998 study of bald eagle chicks from British Columbia (Newson et al. 2000), but not in a 2003 study (Cesh et al. 2010). Positive relationships between PCBs and/or TEQs with hepatic dehydro-retinol, retinyl-ester 2 or retinyl palmitate were reported in osprey and bald eagle hatchlings from British Columbia (Elliott et al. 1996b, 2001a). Martinovic et al. (2003a) found a positive relationship between PCDDs and the molar ratio of retinol: retinyl palmitate in kidneys of tree swallow nestlings from the St. Lawrence River; plasma concentrations were not measured. Those tree swallows also showed a negative correlation between basal plasma corticosterone levels and PCDFs (ranging from 5 to 121 pg/g ww), suggesting they were responding to a chronic stress (Martinovic et al. 2003b).

Thyroid and vitamin A status are the most widely applied markers of dioxin-like physiological effects in young and adult wild bird studies; however, there is still a need for better understanding of how age of birds influences the natural functioning of these systems. Also, studies that measure only one component of the complex feedback mechanisms, for instance only circulating levels of vitamin A and not storage forms, run the risk of missing subtle changes in the functioning of these important and highly complex systems. A further discussion of responses in adult birds is provided in Section 14.4.

14.3.3 PRODUCTIVITY

Productivity and adult survival (in Section 14.4) are the key determinants of population survival in wild birds. Productivity is more critical for short-lived species, whereas adult survival is more critical for long-lived species (Lebreton and Clobert 1991). These two gross measures of fitness express the cumulative impacts of ecological and toxicological factors, and thus present the greatest challenges for researchers trying to pinpoint cause–effect associations in complex ecosystems.

Early studies of colonial water-birds and raptors in the Great Lakes and Europe reported poor productivity associated with regional contamination from OC contaminants (Gilman et al. 1977, Price and Weseloh 1986, Kozie and Anderson 1991, Elliott and Harris 2001, Helander et al. 2002). Typically, PCBs were present in eggs of affected species at higher concentrations than other OCs, such as DDE and HCB, and were thus implicated as a probable dominant cause of poor productivity. However, the strength of correlations among OCs in these environments was very strong, and regional averages were largely insufficient to provide statistical evidence of cause–effect.

These confounding associations among contaminants continued to exert an influence in more recent productivity assessments (Table 14.5). The bulk of studies suggest that the strongest chemical determinants of productivity often were not PCBs, but instead DDE or PCDDs. For instance, in a retrospective analysis of white-tailed sea eagles from Sweden, productivity over the period 1964–1999 was most strongly associated with an eggshell desiccation index that was in turn strongly

TABLE 14.5
Evidence for Associations between PCBs or PCDDs and Productivity Measures in Wild Birds in the Field

Species	Effect/No Effect	Response Variable	Tissue Burden[a]	Other Significant Contaminants/ Variables	Reference
American robin	No effect	Productivity	84 μg/g (egg) 12 μg/g (whole nestling)		Henning et al. 2003
Bald eagle	No effect	5-year productivity	7–100 ng/g or 2.6–13 pg/g TEQs (nestling plasma)	Weak association with DDE	Elliott and Norstrom 1998, Gill and Elliott 2003
Dipper	No effect	Post-fledge survival	0.5–1.3 μg/g (egg)		Ormerod et al. 2000
House wren	No effect	3-year productivity	6 μg/g (egg) 0.8 μg/g (whole nestling)		Neigh et al. 2007
Great horned owl	No effect	Fledge success	0.5–40 μg/g (egg)		Strause et al. 2007
Osprey	No effect	Fledge rate	0.5–1.7 μg/g (egg) 4–119 pg/g PCDD-TEQs		Woodford et al. 1998
Tree swallow	No effect	Fledge success	5.1 μg/g (egg) 3.1 μg/g (whole nestling)		Papp et al. 2005, Neigh et al. 2006
Black-backed gull	Effect	Fledge rate	8–21 μg/g (nestling liver)	Chick disease (unknown etiology); association with DDE	Hario et al. 2000, 2004
Eastern bluebird	Effect	3-year productivity	8.3 μg/g (egg) 1.3 μg/g (whole nestling)	Interspecies competition for nest boxes and prey; associations with DDE	Neigh et al. 2007
European starling	Effect	Fledge rate	6 μg/g (whole nestling)		Arenal et al. 2004
Herring gull	Effect	Productivity	142 μg/g (egg)	Many co-contaminants	Gilman et al. 1977
Tree swallow	Effect	Fledge rate	9.3–29.5 μg/g (egg) 3.7–62 μg/g (nestling)		McCarty and Secord 1999b
	Effect	Nestling success	0.57 ng/g TCDD (nestling)	Year-to-year variability	Custer et al. 2005
White-tailed sea eagle	Effect	Productivity	380–750 μg/g lipid wt (egg)	Strong association with DDE	Helander et al. 2002
Wood duck	Effect	Productivity	0.02–0.05 ng/g PCDD-TEQs (egg)		White and Seginak 1994

[a] Units for ΣPCBs, wet weight unless stated otherwise.

associated with DDE (Helander et al. 2002). LOELs for reduced productivity were 120 μg/g lipid weight DDE and 500 μg/g lipid weight PCBs in eggs (Helander et al. 2002). A re-analysis of Great Lakes bald eagle productivity data for the period 1986–2000 reported associations on an individual nest basis for both PCBs and DDE, with a lower threshold value for DDE (Best et al. 2010). For the entire Great Lakes bald eagle sample, and for a subset of birds nesting on the Great Lakes shoreline, ∑PCBs had the clearest relationship with productivity, and the best relationship was dose-dependent and included ∑PCBs and DDE as factors. Reductions in productivity below a threshold level for population sustainability were associated with 6.5 µg/g wet weight DDE and 26 µg/g wet weight PCBs (Best et al. in preparation), which is similar to critical values estimated from North American regional productivity averages for the species (6 µg/g ww DDE and 20 µg/g ww PCBs; Elliott and Harris 2001).

Nest box studies with tree swallows and wood ducks reported associations between productivity, measured as fledge success, and TCDD or PCDDs in general. Tree swallows nesting along the TCDD-contaminated Woonasquatucket River (Rhode Island) were less likely to reach 12 days old than nestlings along a reference river system (Custer et al. 2005); however, the greatest influence on that fledge rate was poor hatching success, and some compensation during the nestling stage seemed to occur. Similarly, wood ducks nesting along the Bayou Meto in Arkansas showed reduced productivity at PCDD concentrations equivalent to 0.05 ng/g PCDD-TEQs (White and Seginak 1994), although it is unlikely that embryo survival would have been significantly affected at those exposure levels based on egg-injection data for that species (Augspurger et al. 2008).

A few nest box studies of passerine species recently reported reductions in productivity or fledge success statistically associated with PCBs (Table 14.5). Studies with starlings, bluebirds, and tree swallows suggest that productivity is reduced at threshold whole nestling concentrations in the range of 1–6 µg/g wet weight PCBs (McCarty and Secord 1999b, Arenal et al. 2004, Neigh et al. 2007).

Not surprisingly, many recent studies have had increasing difficulty quantifying the magnitude of influence of diminishing concentrations of OCs and PCDDs on productivity, relative to ecological variables such as food or thermal stress (Hario et al. 2000, Gill and Elliott 2003, Neigh et al. 2007). In long-term, extensive monitoring of herring gull populations in the Great Lakes, Hebert et al. (2002) determined that diet quality was significantly correlated with productivity, egg size and female body condition. These ecological determinants will remain relevant and require consideration in future field studies, necessitating individual nest correlative contaminant analyses rather than regional averaging techniques, and combined toxicological/ecological study designs.

14.4   ADULT SURVIVAL

Adult survival is a critical determinant of population growth rate in long-lived birds. PCBs and PCDDs have not been strongly associated with this parameter in the past, possibly because incidences of direct poisoning of adults were infrequent even during the 1960s and 1970s when bioavailability was relatively high. However, there is some evidence that OCs, PCBs, and PCDDs may have exerted indirect impacts on adult survival rates by exacerbating the effects of natural stressors such as low food availability (Elliott et al. 1996c, Gervais and Anthony 2003, Gill and Elliott 2003, Bustnes et al. 2006a, 2006b, Breton et al. 2008).

Table 14.6 summarizes the evidence for direct lethal effects of PCBs on wild bird adults. Laboratory studies on a variety of mostly passerine species indicated that the most reliable diagnostic tissue residue for PCB poisoning was the concentration in the brain (Prestt et al. 1970, Dahlgren et al. 1972, Stickel et al. 1984). Lethal brain residues in laboratory tests typically fell within the range 300–800 µg/g wet weight, depending on the species (Table 14.6). One study of wild, captured and dietary-exposed great cormorant reported a lower brain threshold of 76–180 µg/g; however, that study used Clophen A60 for dosing, a formulation that contained PCDF impurities (Koeman et al. 1973). The PCDFs may have contributed to a lower lethality threshold in the species, or alternatively the species may exhibit greater sensitivity. Coinciding field studies reporting 190 µg/g PCBs in cormorants found dead would lend support for the latter hypothesis (Koeman et al. 1973).

Only two other reports of suspected poisoning in wild populations measured brain residues within the reported laboratory range. One was a study of ring-billed gulls on the Great Lakes in the early 1970s (Sileo et al. 1977), whereas the other was more recent, the finding of an individual

TABLE 14.6
Effects of PCBs on Survival of Adult Wild Birds

Species	Exposure	Effect	Tissue Residue[a]	Other Significant Variables	Reference
			Laboratory Studies		
Pheasant	Aroclor 1254 Diet, up to 8 months	Poisoning	300–400 µg/g (brain)	Tests on immature females	Dahlgren et al. 1972
Bengalese finch	Aroclor 1254 Diet, 254 µg/g/d, 8 weeks	Poisoning	290 µg/g (brain)		Prestt et al. 1970
Great cormorant	Clophen A60 Diet, 2.3–9.1 g, 55–124 days	Poisoning	76–180 µg/g (brain)	Residual PCDFs	Koeman et al. 1973
Common grackle	Aroclor 1254 Diet, 1500 µg/g/d dw, 8 days	Poisoning	579 µg/g (brain)	Tests on immature males	Stickel et al. 1984
Red-winged blackbird	Aroclor 1254 Diet, 1500 µg/g/d dw, 6 days	Poisoning	579 µg/g (brain)	Tests on immature females	Stickel et al. 1984
Brown-headed cowbird	Aroclor 1254 Diet, 1500 µg/g/d dw, 7 days	Poisoning	579 µg/g (brain)	Tests on immature males	Stickel et al. 1984
European starling	Aroclor 1254 Diet, 1500 µg/g/d dw, 4 days	Poisoning	439 µg/g (brain)	Tests on immature females	Stickel et al. 1984
Japanese quail	Aroclor 1260 Diet, 1000 µg/g/d	Poisoning	780 µg/g (brain)		Call et al. 1973
			Field Studies		
Ring-billed gull	Great Lakes colony	Poisoning	300+ µg/g (brain)	Some individuals with high dieldrin as well	Sileo et al. 1977
Great cormorant	The Netherlands	Poisoning	190 µg/g (brain)	HCB, DDE, dieldrin and mercury also present	Koeman et al. 1973
Bald eagle	Michigan	Poisoning	190–235 µg/g (brain)	Additional high DDE	Belisle et al. 1972, Cromartie et al. 1975
Red-tailed hawk	New York	Poisoning	760 µg/g (brain)	Additional high dieldrin	Stone and Okoniewski 2000
Great blue heron	Wisconsin	Poisoning	220 µg/g (brain)	Possibly other contributors	Heinz et al. 1984
Caspian tern	Great Lakes colonies	Poor adult return rate	2.5–3.5 µg/g (plasma)	Strongest association with young-of-the-year	Mora et al. 1993
Glaucous gull	Scandinavian arctic	Poor adult return rate	0.1–1 µg/g (blood)	Stronger association with oxychlordane	Bustnes et al. 2003

[a] ΣPCBs unless specified otherwise.

debilitated and moribund red-tailed hawk (*Buteo jamaicensis*) in 1996 near a contaminated transformer manufacturing site in New York (Stone and Okoniewski 2000). The other cases listed in Table 14.6 were potential diagnoses of poisoning, where other chemical or ecological factors also likely contributed significantly to the observed mortalities.

Although the focus of many studies of PCB and PCDD effects in wild birds has been reproduction, a few have also attempted to measure chronic effects on adult survival in the field by investigating adult return rates in species exhibiting strong nest- or natal-site fidelity. In two instances, significant associations were reported between PCB tissue burdens and poor return rates (Table 14.6). In 1990, Mora et al. (1993) reported that return rates of Caspian tern to their natal region was negatively associated with regional averages of PCBs in blood of the adults. In particular, of those birds banded as chicks, the return rate at the two most contaminated colonies in Green and Saginaw Bays (Lakes Huron and Michigan, respectively) was 20–23%, compared to rates around 70% for colonies in northern Lake Huron (Mora et al. 1993). During the period 1997–2000, Bustnes et al. (2003) similarly found associations between blood PCBs and decreased adult return rates for glaucous gull in the Norwegian Arctic. Overall, a 10-fold increase in blood concentrations of OCs corresponded to a reduction in survival probability of up to 29% for females and up to 16% for males. There was a significant negative correlation between adult survivorship and both oxychlordane ($p < .01$) and PCBs ($p < .05$). A few other studies reported finding no associations between return rates and PCBs: in south polar skuas (Bustnes et al. 2007), in herring gull from Lake Ontario over the period 1993–1998 (Breton et al. 2008), and in tree swallow from the PCB-contaminated Housatonic River over the period 2000–2004 (Custer et al. 2007). Breton et al. (2008) considered that PCBs may have contributed to lower adult survival of herring gull during the 1990s or earlier, but the overwhelming explanatory variable of survival at Lake Ontario colonies appeared to be changing prey availability and nutritional value, as measured by the abundance of alewife. In the tree swallow study, Custer et al. (2007) actually found higher concentrations of PCBs in females that returned versus those that did not, and considered that higher lipid reserves (and consequently higher lipophilic contaminants) were the best predictor of survival between breeding seasons.

In short, there is little evidence that PCB and PCDD tissue burdens in wild birds are sufficient to impact on survival rates on their own. However, there are cases where they were suggested as potential contributors to reduced survival during periods of food stress or disease stress. In addition to the case of herring gull described earlier, Gill and Elliott (2003) suggested that PCDD and PCDF contamination from a pulp mill effluent exacerbated food availability-related reduced productivity in bald eagle from British Columbia over the period 1992–1998. Bustnes et al. (2006b) did not find that OCs in tissues influenced the survival of parasite-infested adult glaucous gulls, but tissue burdens did exacerbate the reproductive impacts of the infestation on those birds. Episodic food, thermal and disease stresses in wild birds will continue to occur, possibly more frequently with climate change, and it is therefore important to better understand the synergistic effects exerted by PCBs and PCDDs under such conditions.

The traditionally monitored biochemical markers of PCB and PCDD impacts, namely porphyria, vitamin A status, thyroid function, and immune function, may provide some indications of the vulnerability of adult wild birds to ecologically mediated stress events. They also provide some information on the potential for these contaminants to be impacting through endocrine disruption mechanisms. As discussed for nestlings, there is a need for further mechanistic laboratory studies to better understand contaminant impacts on endocrine function in avian species.

Immune dysfunction is one process by which PCBs, PCDDs, and PCDFs may reduce adult survival rates (Grasman 2002, Møller and Saino 2004). The thymus is an important organ for immune function and hence many studies link hematology, antibody-related immunity, and thymus function together to investigate immune function.

Captive studies with breeding adult American kestrel fed Aroclor 1248:1254:1260 (7 mg/ kg body weight/day for 120 days) reported higher white blood cell counts, principally related to lymphocyte proliferation (Smits et al. 2002). Lymphocytes (B and T cells) are involved in an adaptive immune response, proliferating once the body recognizes a specific antigen and often where a chronic infectious stress is present (Janeway et al. 1999). The response in kestrels was significant only for males, although the pattern of proliferation was consistent in both sexes. Those males also showed significantly depressed circulating corticosterone levels (Love et al. 2003), an increased PHA skin response under immune challenge, reduced plasma T_3 concentrations (Smits et al. 2002), and suppressed antibody production (Smits and Bortolloti 2001). In adult females, antibody production was enhanced rather than suppressed (Smits and Bortolloti 2001). Those results suggest chronic impacts on immune function by PCBs, mediated possibly by hormonal (sex-specific) mechanisms.

Earlier laboratory studies using mallards exposed by oral gavage to Aroclor 1254 reported fewer significant effects on immune and thyroid function (Fowles et al. 1997). This is consistent with ducks showing low sensitivity overall to PCB toxicity. Male mallards did not exhibit an effect on the PHA skin response with exposure up to 500 µg/g twice weekly for 5 weeks. However, liver and thyroid weights were increased in treatments above 100 µg/g, while plasma T_3 concentrations were decreased, with a LOEL of 20 µg/g (Fowles et al. 1997).

Consistencies exist between the few laboratory studies outlined above and field-based studies of immunity in breeding adult wild birds. Lymphocyte proliferation, in particular, was correlated to PCB (and other OC) exposure in herring gulls from the Great Lakes (Grasman et al. 2000), black-footed albatross from Midway Atoll (Finkelstein et al. 2007), and glaucous gulls from the Norwegian Arctic (Bustnes et al. 2004). Plasma T_4 but not T_3 concentrations were reduced in both the herring gulls (Fox et al. 2007a) and the glaucous gulls (Verreault et al. 2004). Moccia et al. (1986) and Fox et al. (2007b) also reported enlarged thyroids and microfollicular hyperplasia in thyroids of Great Lakes herring gulls, correlated with PCB tissue burdens.

There is insufficient evidence from these immune studies to propose effect levels from immune dysfunction in wild birds exposed to PCBs, but the consistencies between field and laboratory findings suggest that different species may be responding similarly and in a chronic fashion. Møller and Saino (2004) suggest that mounting an inappropriate immune response, either via suppression or hypersensitivity, can reduce the long-term survival of wild birds. Further research on parasite loads in the Arctic-breeding glaucous gulls described earlier reported that the more contaminated birds carried greater infestations of intestinal nematodes (Sagerup et al. 2000), supporting the idea that OC contaminants in general are associated with compromised immunity.

14.5 POSSIBLE INFLUENCE OF PBDES ON REPRODUCTION AND SURVIVAL OF WILD BIRDS

The recent addition of PBDEs to the environmental mix of similarly structured, persistent, and bio-accumulative contaminants creates further potential for additive effects of whole tissue burdens on avian wildlife. PBDEs have a similar chemical structure to PCBs, but have chlorines replaced with bromines and an oxygen bonding the two rings. Since the 1980s, these flame retardants have been produced in technical formulations that contain predominantly penta-, octa-, or deca-congeners (de Wit 2002). Early evidence of mammalian toxicity associated with lower brominated congeners has led to restrictions on manufacture of penta-technical mixtures. Three congeners in particular which are associated with the penta-formulations, BDE-47 (2,2′,4,4′-tetrabromoDE), BDE-99 (2,2′,4,4′,5-pentabromoDE), and BDE-100 (2,2′,4,4′,6-pentabromoDE), have been widely reported in eggs of water-birds and raptors (Norstrom et al. 2002, Elliott et al. 2005, McKinney et al. 2006, Murvoll et al. 2007, Gauthier et al. 2008).

Although the avian literature pertaining to effects of PBDEs is still very limited, a few recent studies suggest that PBDEs and their hydroxylated metabolites may alter reproduction and physiology in ways consistent with PCBs. Here we describe these early findings.

14.5.1 Effects on Reproduction

A small number of laboratory studies have explored the effects of the formulation DE-71 on parenting behavior, embryolethality, hatching and fledging success in wild birds and the chicken. In broad terms, PBDEs appear to invoke similar responses as PCBs, but the relative sensitivity of species does not follow the predicted pattern described earlier in Section 14.2.1 on measured variability in PCB and TCDD metabolism.

Captive adult American kestrels exposed to DE-71 in their diet for 75 days, starting 21 days prior to breeding, exhibited altered courtship behaviors. Individuals in both low (0.289 μg/g ww in eggs) and high (1.13 μg/g ww in eggs) exposure groups displayed fewer pair-bonding behaviors, spent less time in nest boxes and copulated less than control birds (Fernie et al. 2008). Exposed females also ate less, which may have impacted their capacity to produce high-quality eggs.

These same birds went on to show significant delays in clutch initiation (by 4 and 10 days for the low and high exposure groups) and smaller eggs (Fernie et al. 2009). Exposed females laid eggs with less volume, length, width, and mass than control eggs. Eggshells were also significantly thinned and embryos lost more weight during incubation when compared to control embryos. As a result, hatching success was significantly reduced (from 69% in controls to 49% in the high exposure group) and fledging success was nonsignificantly reduced (from 70% in controls to 56% in the high exposure group).

In an independent comparison of the kestrel with chicken and mallard, the kestrel was found to be more sensitive to embryotoxic effects of PBDEs than the chicken, unlike previous species comparisons with PCBs. Egg injection with DE-71 resulted in only 18% transfer of the dose to kestrel embryos and 29% transfer to chicken embryos (McKernan et al. 2009). Thus, the concentration affecting pipping and hatching success in kestrels was estimated to be as low as 1.8 μg ΣPBDE/g (from administered doses of 10 and 20 μg/g egg). That is comparable to the high exposure group described earlier in the adult oral dosing study (1.13 μg ΣPBDE/g, Fernie et al. 2008). In contrast, neither chicken or mallard embryo survival was altered by these dosage levels (McKernan et al. 2009). Chicken hatchlings did exhibit EROD induction and altered histology of the bursa, symptoms not observed in the other species.

The apparent effect level of 1–2 μg ΣPBDE/g from laboratory studies is supported by a single field study of osprey in the US northwest. Henny et al. (2009) found a negative relationship between osprey productivity and PBDE burdens in eggs from 30 nests along the Columbia and Willamette Rivers in 2006 and 2007. Productivity was markedly reduced where egg concentrations exceeded 1 μg/g (up to a maximum of 1.88 μg/g). Hence, PBDEs may not be present in environmental media at levels as high as PCBs even today, but they may nonetheless impact reproductive success through multiple egg intrinsic and extrinsic mechanisms at concentrations much lower than PCBs.

14.5.2 Effects on Nestling Fitness

A number of recent laboratory and field studies have focused on the potential for PBDEs and their hydroxylated metabolites to alter thyroid function in birds. Hydroxylated metabolites of both PCBs and PBDEs have been shown to bind more effectively to the transport protein, transthyretin, than either of the natural hormones, T_4 or T_3, in herring and glaucous gulls (Ucán-Marín et al. 2009). In theory, the transformed contaminants may compete with thyroid hormones and vitamin A (which also uses transthyretin as a means of transport), and disrupt their distribution within the body.

Captive American kestrel nestlings exposed *in ovo* and post-hatch for 29 days to a mix of PBDE congeners (BDE-47, -99, -100, and -153) showed no statistically significant ($p = .05$) differences from control birds in circulating T_4, T_3, and retinol (vitamin A) or in hepatic stores of retinol and retinyl palmitate (Fernie et al. 2005a). However, all but the last were negatively correlated to concentrations of one or more PBDE congeners. Dosed nestlings had whole body burdens of 86 ng ΣPBDEs/g wet weight on average. In addition to thyroid and vitamin A function, that study suggested some symptoms of induced oxidative stress (Fernie et al. 2005a) and indications of altered immune function

(Fernie et al. 2005b). However, in those PBDE-exposed nestlings, the PHA skin response, antibody-mediated response, and changes in the structure of the spleen, bursa and thymus were not significantly different from controls at $p = .05$, with the exception of the splenic somatic index.

Circulating and hepatic levels of vitamin A and vitamin E were also evaluated in mallard hatchlings exposed to PBDE-99 via a single egg injection of 0.1, 1, or 10 ng/g wet weight (Murvoll et al. 2005). While vitamin A was not affected by exposure, hepatic concentrations of tocopherol (vitamin E) were negatively correlated with PBDE-99 in yolk sacs. The authors suggested that tocopherol may be impacted through oxidative stress mechanisms, consistent with findings noted above for kestrel nestlings.

Despite the evidence for thyroid hormone and vitamin A and E effects during laboratory study, there is little suggestion that wild birds in the field experience similar altered physiology. Field-hatched Brünnich's guillemot (*Uria lomvia*), common eider (*Somateria mollissima*), North Atlantic kittiwake (*Rissa tridactyla*) and bald eagle showed no changes in vitamin status or T_4 levels associated with PBDEs in the range of 2–530 ng/g lipid weight in yolk sacs or 0.6–31 ng/g wet weight in nestling plasma (McKinney et al. 2006, Murvoll et al. 2006b, 2007, Cesh et al. 2009). Circulating T_3 and retinol were weakly and positively associated with hydroxylated metabolites of PBDEs in bald eagle nestlings from California and British Columbia (Cesh et al. 2009).

As with similar evaluations made with PCBs (Section 14.3.2.5), variations in the measure of tissue burdens among laboratory and field studies make comparison of these data difficult; however, a reporting of wet weight and lipid weight values for kittiwake and European shag by Murvoll et al. (2006b) suggests that at least some of these field hatchlings were more contaminated with PBDEs than those exposed in laboratory studies. It may be that the impacts on these functional systems in nestlings can be compensated for during development and are not expressed as widely in natural environments as in captive colonies.

Summary

PCBs, PCDDs, and PCDFs are legacy contaminants that continue to occur in environmental media with the potential to cause toxicological effects in wild birds. PBDEs are similarly structured more recent additions to the environment, and limited research indicates that they may also exert effects on reproduction and early development. Ongoing research has produced significant advances in our understanding of dioxin-like toxicity in birds since the last edition of this book was published in 1996. These advances include the following:

- Differences in the AhR-mediated response mechanisms between mammals and birds affect the extrapolation of mechanistic results from laboratory mammalian studies to wild birds. Birds have unique forms of the Cytochrome P450 enzymes responsible for mounting a response to tissue burdens of PCBs, PCDDs, or PCDFs (detailed discussion in Section 14.2.1).
- Different bird species exhibit varying sensitivity to dioxin-like contaminants in the field and those sensitivities may be broadly categorized according to amino acid differences in the structure of the AhR ligand binding domain. Whereas the chicken is the only species tested which exhibits high sensitivity, some upland game birds, passerines and albatross exhibit moderate sensitivity. All other birds (raptors, colonial water-birds, and ducks) appear relatively insensitive to dioxin-like contaminants. Considerable response variability remains within and among species in the latter category, driven by other factors, some of which may still be related to the AhR (detailed discussion in Sections 14.2.1 and 14.2.2).
- Evaluations of long-term datasets suggest that elevated PCBs in hotspots such as the Great Lakes and the Baltic region most probably impacted productivity of wild birds,

though the mechanisms are complex. In addition, then and now there is increasing evidence that PCBs, PCDDs, and PCDFs were eliciting responses in stressful ecological conditions, most frequently where food quality or quantity was reduced (detailed discussion in Section 14.3).

- There are indications from early development studies that TCDD may not be the most toxic congener in the suite of dioxin-like chemicals for wild birds. The furans, notably TCDF and PeCDF, are reported to be more toxic in some instances, thereby implying that correlative field assessments might have missed cause–effect relationships in the past (detailed discussion in Section 14.3.2.2).
- Accumulating evidence of behavioral drivers for PCB toxicity, through hormonally triggered mechanisms in maturing birds, is compelling. Although difficult to quantify in field studies, parenting behavior may be a sensitive endpoint for continuing low-level PCB and PBDE exposure, and likely to alter the outcomes of responses mounted to ecological stressors. There is increasing evidence from mammalian studies and limited evidence in birds, although with other chlorinated hydrocarbons such as DDE, that altered adult behaviors are the result of neuro-endocrine changes which occur principally as an outcome of exposure to contaminants during critical early stages of development (detailed discussion in Section 14.3.1).

Tables 14.7 and 14.8 provide some suggested critical thresholds for PCB and TCDD toxicity in birds, based on weight of evidence from laboratory and field studies.

The EROD activity provides the most consistently monitored endpoint for an AhR-mediated response in birds, although there are clearly large differences in sensitivity among species. Since the AhR mechanism is integral to the initiation of biochemical and physiological responses to dioxin-like contaminants, we suggest EROD be considered as a cross-species comparable LOEL. Although EROD is often considered a measure of exposure rather than effect, it does indicate a redirection of energy allocation in the individual away from processes critical for survival and reproduction. As such, it may be the most consistently measured indicator of low-grade effect. EROD induction has occurred in laboratory studies of wild bird hatchlings (not domesticated fowl such as the chicken) at concentrations ranging 0.3–5 ng TCDD/g wet weight and 25–230 ng CB-126/g ww (Table 14.7). In the field, where EROD is induced in response to a mix of AhR-active contaminants, corresponding threshold concentrations ranged 0.4–1.6 ng TEQs/g ww (Table 14.7). Although data exists for EROD induction at other life stages, the hatchling is the only age group where sufficient information exists to suggest these critical threshold values.

Although embryotoxicity may not be as sensitive to PCBs as behavioral alterations, it is easier to standardize and quantify this measure using egg-injection protocols. Therefore, we suggest embryolethality thresholds listed in Table 14.8 and Figure 14.1 be used as an additional comparison of expected variability in sensitivity among the various bird functional groups. The range in concentration thresholds described for wild species are 1–3 ng furans/g egg ww, ~1–11 ng TCDD/g egg ww, 24–158 ng CB-126/g egg ww, and 1–2 μg ΣPBDEs/g egg ww (Table 14.8). Hatching success, which incorporates factors that are both intrinsic and extrinsic to the egg (e.g., embryolethality and parental attendance), has been affected in species of intermediate sensitivity at egg concentrations ranging 6–50 μg ΣPCBs/g ww and 1–10 ng TCDD/g ww. Less sensitive species (raptors and water-birds) show reduced hatching success at egg concentrations ranging 23–142 μg ΣPCBs/g ww (Table 14.8).

Behavioral endpoints listed in Table 14.8 may be the most sensitive to PCB exposure and most relevant for scenarios likely in today's environment; however, there is a need for further mechanistic study in this area before they can be widely used to evaluate the effects of PCB exposure in the field. To date, the data suggests critical thresholds for altered parental care of 1–30 μg ΣPCBs/g ww in eggs or 2–4 μg ΣPCBs/g ww in adult plasma (Tables 14.3 and 14.8).

TABLE 14.7

Lowest-Observed-Effect-Levels (LOELs) of PCBs, PCDDs, and PCDFs for Hatchling Birds, as Estimated by Induction of the AhR-Mediated Cytochrome P450 (EROD) Enzyme System[a]

Species	Inducing Compound(s)	Exposure/Test Conditions	LOEL (ng/g ww)	Fold Induction over Control/Reference	Reference
			Laboratory Studies		
Chicken	TCDD	Egg injection	~0.03	5	Sanderson and Bellward 1995, Janz and Bellward 1996a
	CB-126	Egg injection	0.3	17	Hoffman et al. 1998
	CB-77	Egg injection	1.2	4	Hoffman et al. 1998
Pigeon	TCDD	Egg injection	0.3	5	Sanderson and Bellward 1995, Janz and Bellward 1996a
Double-crested cormorant	TCDD	Egg injection	1.3–3	5	Sanderson and Bellward 1995, Janz and Bellward 1996a, Powell et al. 1998
	CB-126	Egg injection	70	7	Powell et al. 1998
Great blue heron	TCDD	Egg injection	~2	3	Sanderson and Bellward 1995, Janz and Bellward 1996a
Common tern	CB-126	Egg injection	44	2	Hoffman et al. 1998
	CB-126	Dietary (to chicks)	25 ng TEQ/g lipid wt	2.4	Bosveld et al. 2000
Kestrel	CB-126	Egg injection	233	12	Hoffman et al. 1998
	CB-77	Egg injection	1000	2	Hoffman et al. 1998
Wood duck	TCDD	Egg injection	4.6	12	Augspurger et al. 2008
			Field Studies		
Common tern	Mix	Artificial incubation	16 ng TEQ/g lipid wt	4	Bosveld et al. 1995
Double-crested cormorant	Mix	Artificial incubation	~1.6 ng TEQ/g ww	4	Sanderson et al. 1994a
Great cormorant	ΣPCBs	Artificial incubation	0.1 ng/g lipid wt	? (30% increase)	Sanderson et al. 1994a
Bald eagle	Mix	Artificial incubation	0.4 ng TEQ/g ww	6	Elliott et al. 1996b, Elliott and Harris 2001
European starling	ΣPCBs	Field-collected	5900 (nestling)	?	Arenal et al. 2004
Tree swallow	ΣPCDDs + PCDFs	Field-collected	0.6 ng TEQ/g ww	?	Custer et al. 2005

[a] Chemicals were measured in eggs, while EROD was measured in hatchlings except for: wood ducks = late stage embryos; black guillemot, tree swallow, European starling = older chicks (12- to 15-days old).

continued

TABLE 14.8

Threshold Criteria for Survival or Reproduction in Birds Exposed to PCBs, PBDEs, PCDDs, or PCDFs

Endpoint	Species	Sensitivity	Critical Threshold	Reference
Adult poisoning	Galliformes (pheasant)	Intermediate	300–400 µg/g ww ΣPCBs (brain)	Dahlgren et al. 1972
	Passerines (finch, grackle, blackbird, cowbird, starling)	Intermediate	290–579 µg/g ww ΣPCBs (brain)	Prestt et al. 1970, Stickel et al. 1984
	Great cormorant	Low (?)	76–180 µg/g ww ΣPCBs (brain)	Koeman et al. 1973
	Japanese quail	Low	780 µg/g ww ΣPCBs (brain)	Call et al. 1973
Embryolethality	Chicken	High	0.18 ng/g ww TCDD (egg)	Hoffman et al. 1998, Head et al. 2008
			1.1 ng/g ww CB-126 (egg)	
			2.6 ng/g ww CB-77 (egg)	
	Galliformes (pheasant)	Intermediate	2.2 ng/g ww TCDD (egg)	Nosek et al. 1993
	Bobwhite quail	Intermediate (?)	24 ng/g ww CB-126 (egg)	Hoffman et al. 1995
	Passerines (E. bluebird)	Intermediate	1–10 ng/g ww TCDD (egg)	Thiel et al. 1988
	Raptors (American kestrel)	Low	65 ng/g ww CB-126 (egg)	Hoffman et al. 1998
			316 ng/g ww CB-77 (egg)	
	Terns (common)	Low	104 ng/g ww CB-126 (egg)	Hoffman et al. 1998
	Cormorants (double-crested)	Low	4 ng/g ww TCDD (egg)	Powell et al. 1997a, 1998
			158 ng/g ww CB-126 (egg)	
	Japanese quail	Low	11.3 ng/g ww TCDD (egg)	Cohen-Barnhouse et al. 2010
			3.1 ng/g ww TCDF (egg)	
			1.1 ng/g ww PeCDF (egg)	

TABLE 14.8 (continued)
Threshold Criteria for Survival or Reproduction in Birds Exposed to PCBs, PBDEs, PCDDs, or PCDFs

Endpoint	Species	Sensitivity	Critical Threshold	Reference
Reproductive success (hatching or fledge success)	Chicken	High	1–5 µg/g ww ΣPCBs (egg)	Hoffman et al. 1996a
	Passerines (E. bluebird, house wren, starling)	Intermediate	6–9 µg/g ww ΣPCBs (egg)	Arenal et al. 2004, Neigh et al. 2006
	Passerines (tree swallow)	Intermediate (?)	9–50 µg/g ww ΣPCBs (egg)	McCarty and Secord 1999b, 2000, Custer et al. 2005
	Passerines (E. bluebird)	Intermediate	1–10 ng/g ww TCDD (egg)	Thiel et al. 1988
	Passerines (tree swallow)	Intermediate (?)	1.7 ng/g ww TCDD (egg)	Custer et al. 2005
	Raptors (American kestrel)	Low	35 µg/g ww ΣPCBs (egg)	Fermie et al. 2000, 2001a, 2001b, 2003a, 2003b, 2009, McKernan et al. 2009
			1–2 µg/g ww ΣPBDEs (egg)	
	Terns & Gulls (Forster's, herring)	Low	23–142 µg/g ww ΣPCBs (egg)	Kubiak et al. 1989, Gilman et al. 1977
Productivity (3- or 5-year)	Passerines (E. bluebird)	Intermediate	8 µg/g ww ΣPCBs (egg)	Neigh et al. 2007
	Raptors (sea eagle, bald eagle)	Low	25 µg/g ww ΣPCBs (egg)	Helander et al. 2002, Best et al. in preparation, Elliott & Harris 2001
Parental care	Ringed turtle dove	Intermediate (?)	5.5 µg/g ww ΣPCBs (brain)	Peakall and Peakall 1973
	Passerines (E. bluebird, starling)	Intermediate	9 µg/g ww ΣPCBs (whole nestling)	Arenal et al. 2004, Neigh et al. 2006
	Raptors (American kestrel)	Low	4 µg/g ww ΣPCBs (adult male blood)	Fisher et al. 2001, 2006
	Terns & Gulls (Forster's, herring)	Low	23–142 µg/g ww ΣPCBs (egg)	Kubiak et al. 1989, Fox et al. 1978

ACKNOWLEDGMENTS

We greatly appreciate the access to unpublished information and valuable insights into recent developments provided by Jessica Head, Sean Kennedy, and Steve Bursian. This manuscript was greatly improved through review suggestions from David Hoffman and Mace Barron.

REFERENCES

Arenal, C. A., R. S. Halbrook, and M. J. Woodruff. 2004. European starling (*Sturnus vulgaris*): avian model and monitor of polychlorinated biphenyl contamination at a superfund site in southern Illinois, USA. *Environ.Toxicol. Chem.* 23:93–104.

Augspurger, T. P., D. E. Tillitt, S. J. Bursian, S. D. Fitzgerald, D. E. Hinton, and R. T. Di Giulio. 2008. Embryo toxicity of 2,3,7,8-tetrachlorodibenzo-*p*-dioxin to the wood duck (*Aix sponsa*). *Arch. Environ. Contam. Toxicol.* 55:659–669.

Barron, M. G., H. Galbraith, and D. Beltman. 1995. Comparative reproductive and developmental toxicology of PCBs in birds. *Comp. Biochem. Physiol.* 112C:1–14.

Belisle, A. A., et al. 1972. Residues of organochlorine pesticides, polychlorinated biphenyls, and mercury and autopsy data for bald eagles, 1969 and 1970. *Pest Monit. J.* 6:133–138.

Best, D. A., K. H. Elliott, W. W. Bowerman, M. Shieldcastle, S. Postupalsky, T. J. Kubaik, D. E. Tillitt, and J. E. Elliott. 2010. Productivity, embryo and eggshell characteristics and contaminants in bald eagles from the Great Lakes, USA, 1986–2000. *Environ. Toxicol. Chem.* 29:1581–1592.

Bird, D. M., P. H. Tucker, G. A. Fox, and P. C. Lague. 1983. Synergistic effects of Aroclor 1254 and mirex on the semen characteristics of American kestrels. *Arch. Environ. Contam. Toxicol.* 12:633–640.

Birkhead, T. R., J. Hall, E. Schut, and N. Hemmings. 2008. Unhatched eggs: methods for discriminating between infertility and early embryo mortality. *Ibis* 150:508–517.

Bishop, C. A., N. A. Mahony, S. Trudeau, and K. E. Pettit. 1999. Reproductive success and biochemical effects in tree swallows (*Tachycineta bicolor*) exposed to chlorinated hydrocarbon contaminants in wetlands of the Great Lakes and St. Lawrence River basin, USA and Canada. *Environ. Toxicol. Chem.* 18:263–271.

Bortolotti, G. R., K. J. Fernie, and J. E. Smits. 2003. Carotenoid concentration and coloration of American kestrels (*Falco sparverius*) disrupted by experimental exposure to PCBs. *Func. Ecol.* 17:651–657.

Bosveld, A. T. C., and M. van den Berg. 1994. Effects of polychlorinated biphenyls, dibenzo-*p*-dioxins, and dibenzofurans on fish-eating birds. *Environ. Rev.* 2:147–166.

Bosveld, A. T. C., and M. van den Berg. 2002. Reproductive failure and endocrine disruption by organohalogens in fish-eating birds. *Toxicology* 181:155–159.

Bosveld, A. T. C., et al. 1995. Effects of PCDDs, PCDFs, and PCBs in common tern (*Sterna hirundo*) breeding in estuarine and coastal colonies in the Netherlands and Belgium. *Environ. Toxicol. Chem.* 14:99–115.

Bosveld, A. T. C., et al. 2000. Biochemical and developmental effects of dietary exposure to polychlorinated biphenyls 126 and 153 in common tern chicks (*Sterna hirundo*). *Environ. Toxicol. Chem.* 19:719–730.

Bowerman, W. W., J. P. Giesy, D. A. Best, and V. J. Kramer. 1995. A review of factors affecting productivity of bald eagles in the Great Lakes region: implications for recovery. *Environ. Health Perspect.* 103(Suppl 4):51–59.

Bowerman, W. W., T. J. Kubiak, J. B. Holt, D. E. Evans, R. J. Eckstein, and C. R. Sindelar. 1994. Observed abnormalities in mandibles of nesting bald eagles. *Bull. Environ. Contam. Toxicol.* 53:450–457.

Breton, A. R., G. A. Fox, and J. W. Chardine. 2008. Survival of adult herring gulls (*Larus argentatus*) from a Lake Ontario colony over two decades of environmental change. *Waterbirds* 31:15–23.

Briskie, J. V., and M. Mackintosh. 2004. Hatching failure increases with severity of population bottlenecks in birds. *Proc. Nat. Acad. Sci.* 101:558–561.

Brouwer, A., et al. 1998. Interactions of persistent environmental organohalogens with the thyroid hormone system: mechanisms and possible consequences for animal and human health. *Toxicol. Ind. Health* 14:59–84.

Brunström, B. 1988. Sensitivity of embryos from duck, goose, herring gull, and various chicken breeds to 3,3′,4,4′-tetrachlorobiphenyl. *Poult. Sci.* 67:52–57.

Brunström, B., and J. Orberg. 1982. A method for studying embryotoxicity of lipophilic substances experimentally introduced into hens' eggs. *Ambio* 11:209–211.

Brunström, B., J. Axelsson, and K. Halldin. 2003. Effects of endocrine modulators on sex differentiation in birds. *Ecotoxicology* 12:287–295.

Bustnes, J. O. 2006. Pinpointing potential causative agents in mixtures of persistent organic pollutants in observational field studies: a review of glaucous gull studies. *J. Toxicol. Environ. Health* 69A:97–108.

Bustnes, J. O., T. Tveraa, O. Varpe, J. A. Henden, and J. U. Skaare. 2007. Reproductive performance and organochlorine pollutants in an Antarctic marine top predator: the south polar skua. *Environ. Int.* 33:911–918.

Bustnes, J. O., M. Helberg, K. B. Strann, and J. U. Skaare. 2006a. Environmental pollutants in endangered vs. increasing subspecies of the lesser black-backed gull on the Norwegian Coast. *Environ. Pollut.* 144:893–901.

Bustnes, J. O., K. E. Erikstad, S. A. Hanssen, T. Tveraa, I. Folstad, and J. U. Skaare. 2006b. Anti-parasite treatment removes negative effects of environmental pollutants on reproduction in an Arctic seabird. *Proc.R. Soc. B.* 273:3117–3122.

Bustnes, J. O., S. A. Hanssen, I. Folstad, K. E. Erikstad, D. Hasselquist, and J. U. Skaare. 2004. Immune function and organochlorine pollutants in arctic breeding glaucous gulls. *Arch. Environ. Contam. Toxicol.* 47:530–541.

Bustnes, J. O., K. E. Erikstad, J. U. Skaare, V. Bakken, and F. Mehlum. 2003. Ecological effects of organochlorine pollutants in the Arctic: a study of the glaucous gull. *Ecol. Applic.* 13:504–515.

Bustnes, J. O., V. Bakken, K. E. Erikstad, F. Mehlum, and J. U. Skaare. 2001. Patterns of incubation and nest-site attentiveness in relation to organochlorine (PCB) contamination in glaucous gulls. *J. Appl. Ecol.* 38:791–801.

Call, D. J., Y. A. Greichus, and J. J. Worman. 1973. Changes in a polychlorinated biphenyl (Aroclor 1260) standard within the bodies of Japanese quail. *Proc. SD. Acad. Sci.* 52:266–267.

Cecil, H. C., S. J. Harris, J. Bitman, and G. F. Fries. 1973. Polychlorinated biphenyl-induced decrease in liver Vitamin A in Japanese quail and rats. *Bull. Environ. Contam. Toxicol.* 9:179–185.

Cesh, L., et al. 2010. Polyhalogenated aromatic hydrocarbons and metabolites: relation to circulating thyroid hormone and retinol in nestling bald eagles (*Haliaeetus leucocephalus*). *Environ. Toxicol. Chem.* 29:1301–1310.

Champoux, L., J. Rodrigue, S. Trudeau, M. H. Boily, P. A. Spear, and A. Hontela. 2006. Contamination and biomarkers in the great blue heron, an indicator of the state of the St. Lawrence River. *Ecotoxicology* 15:83–96.

Cheung, M., E. F. Gilbert, and R. E. Peterson. 1981. Cardiovascular teratogenicity of 2,3,7,8-tetrachlorodibenzo-*p*-dioxin in the chick embryo. *Toxicol. Appl. Pharmacol.* 61:197–204.

Cohen-Barnhouse, A. M., et al. 2010. Sensitivity of Japanese quail (*Coturnix japonica*), common pheasant (*Phasianus colchicus*) and white leghorn chicken (*Gallus gallus domesticus*) embryos to in ovo exposure to 2,3,7,8-tetrachlorodibenzo-pdioxin (TCDD), 2,3,4,7,8-pentachlorodibenzofuran (PeCDF) and 2,3,7,8-tetrachlorodibenzofuran (TCDF). *Toxicol. Sci.* (accepted).

Cromartie, E., et al. 1975. Residues of organochlorine pesticides and polychlorinated biphenyls and autopsy data for bald eagles, 1971–72. *Pest. Monit. J.* 9:11–14.

Custer, C. M., and L. B. Read. 2006. Polychlorinated biphenyl congener patterns in tree swallows (*Tachycineta bicolor*) nesting in the Housatonic River watershed, western Massachusetts, USA, using a novel statistical approach. *Environ. Pollut.* 142:235–245.

Custer, T. W., and G. H. Heinz. 1980. Reproductive success and nest attentiveness of mallard ducks fed Aroclor 1254. *Environ. Pollut.* 21:313–318.

Custer, C. M., T. W. Custer, J. E. Hines, J. D. Nichols, and P. M. Dummer. 2007. Adult tree swallow (*Tachycineta bicolor*) survival on the polychlorinated biphenyl-contaminated Housatonic River, Massachusetts, USA. *Environ. Toxicol. Chem.* 26:1056–1065.

Custer, C. M., T. W. Custer, C. J. Rosiu, and M. J. Melancon. 2005. Exposure and effects of 2,3,7,8-tetrachlorodibenzo-*p*-dioxin in tree swallows (*Tachycineta bicolor*) nesting along the Woonasquatucket River, Rhode Island, USA. *Environ. Toxicol. Chem.* 24:93–109.

Custer, C. M., T. W. Custer, P. M. Dummer, and K. L. Munney. 2003. Exposure and effects of chemical contaminants on tree swallows nesting along the Housatonic River, Berkshire County, Massachusetts, USA, 1998–2000. *Environ. Toxicol. Chem.* 22:1605–1621.

Custer, C. M., T. W. Custer, P. D. Allen, K. L. Stromborg, and M. J. Melancon. 1998. Reproduction and environmental contamination in tree swallows nesting in the Fox River drainage and Green Bay, Wisconsin, USA. *Environ. Contam. Toxicol.* 17:1786–1798.

Custer, T. W., et al. 2001. Organochlorine contaminants and biomarker response in double-crested cormorants nesting in Green Bay and Lake Michigan, Wisconsin, USA. *Arch. Environ. Contam. Toxicol.* 40:89–100.

Custer, T. W., et al. 1999. Organochlorine contaminants and reproductive success of double-crested cormorants from Green Bay, Wisconsin, USA. *Environ. Toxicol. Chem.* 18:1209–1217.

Dahlgren, R. B., R. J. Bury, R. L. Linder, and R. F. Reidinger. 1972. Residue levels and histopathology in pheasants given polychlorinated biphenyls. *J. Wildl. Manage.* 36:524–533.

Davis, J. A., D. M. Fry, and B. W. Wilson. 1997. Hepatic ethoxyresorufin-*O*-deethylase activity and inducibility in wild populations of double-crested cormorants (*Phalacrocorax auritus*). *Environ. Toxicol. Chem.* 16:1441–1449.

Dawson, A. 2000. Mechanisms of endocrine disruption with particular reference to occurrence in avian wildlife: a review. *Ecotoxicology* 9:59–69.

de Wit, C. A. 2002. An overview of brominated flame retardants in the environment. *Chemosphere* 46:583–624.

de Wit, C. A., M. Mehran Alaee, and D. C. G. Muir. 2006. Levels and trends of brominated flame retardants in the Arctic. *Chemosphere* 64:209–233.

DeWitt, J. C., D. S. Millsap, R. L. Yeager, S. S. Heise, D. W. Sparks, and D. S. Henshel. 2006. External heart deformities in passerine birds exposed to environmental mixtures of polychlorinated biphenyls during development. *Environ. Toxicol. Chem.* 25:541–551.

Dickerson, S. M., and A. C. Gore. 2007. Estrogenic environmental endocrine-disrupting chemical effects on reproductive neuroendocrine function and dysfunction across the life cycle. *Rev. Endocr. Metab. Disord.* 8:143–159.

Dods, P. L., E. M. Birmingham, T. D. Williams, M. G. Ikonomou, D. T. Bennie, and J. E. Elliott. 2005. Reproductive success and contaminants in tree swallows (*Tachycineta bicolor*) breeding at a wastewater treatment plant. *Environ. Toxicol. Chem.* 24:3106–3112.

Drouillard, K. G., K. J. Fernie, J. E. Smits, G. R. Bortolotti, D. M. Bird, and R. J. Norstrom. 2001. Bioaccumulation and toxicokinetics of 42 polychlorinated biphenyl congeners in American kestrels (*Falco sparverius*). *Environ. Toxicol. Chem.* 20:2514–2522.

Eisler, R. 1986. Polychlorinated biphenyl hazards to fish, wildlife, and invertebrates: a synoptic review. *U.S. Fish Wildl. Serv. Biol. Rep.* 85(1.7), 72 pp.

Elliott, J. E. 2005. Chlorinated hydrocarbon contaminants and stable isotope ratios in pelagic seabirds from the North Pacific Ocean. *Arch. Environ. Contam. Toxicol.* 49:89–96.

Elliott, J. E., and M. L. Harris. 2001. An ecotoxicological assessment of chlorinated hydrocarbon effects on bald eagle populations. *Rev. Toxicol.* 4:1–60.

Elliott, J. E., and R. J. Norstrom. 1998. Chlorinated hydrocarbon contaminants and productivity of bald eagle populations on the Pacific coast of Canada. *Environ. Toxicol. Chem.* 17:1142–1153.

Elliott, J. E., M. L. Harris, L. K. Wilson, P. E. Whitehead, and R. J. Norstrom. 2001b. Monitoring temporal and spatial trends in polychlorinated dibenzo-*p*-dioxins (PCDDs) and dibenzofurans (PCDFs) in eggs of great blue heron (*Ardea herodias*) on the coast of British Columbia, Canada, 1983–1998. *Ambio* 30:416–428.

Elliott, J. E., M. M. Machmer, C. J. Henny, P. E. Whitehead, and R. J. Norstrom. 1998. Contaminants in ospreys from the Pacific Northwest: I. Trends and patterns in polychlorinated dibenzo-*p*-dioxins and dibenzofurans in eggs and plasma. *Arch. Environ. Contam. Toxicol.* 35:620–631.

Elliott, J. E., et al. 1996b. Biological effects of polychlorinated dibenzo-*p*-dioxins, dibenzofurans, and biphenyls in bald eagle (*Haliaeetus leucocephalus*) chicks. *Environ. Toxicol. Chem.* 15:782–793.

Elliott, J. E., R. J. Norstrom, and G. E. J. Smith. 1996a. Patterns, trends and toxicological significance of chlorinated hydrocarbons and mercury in bald eagle eggs. *Arch. Environ. Contam. Toxicol.* 31:354–367.

Elliott, J. E., L. K. Wilson, C. J. Henny, and M. M. Machmer. 2000. Contaminants in ospreys from the Pacific Northwest: II. Patterns and trends in organochlorine pesticides, polychlorinated biphenyls and mercury in eggs and plasma. *Arch. Environ. Contam. Toxicol.* 38:93–106.

Elliott, J. E., L. K. Wilson, C. J. Henny, S. F. Trudeau, F. A. Leighton, S. W. Kennedy, and K. M. Cheng. 2001a. Assessment of biological effects of chlorinated hydrocarbons in osprey chicks. *Environ. Toxicol. Chem.* 20:866–879.

Elliott, J. E., L. K. Wilson, K. E. Langelier, and R. J. Norstrom. 1996c. Bald eagle mortality and chlorinated hydrocarbons in liver samples from the Pacific coast of Canada, 1988–1995. *Environ. Pollut.* 94:9–18.

Elliott, J. E., L. K. Wilson, and B. Wakeford. 2005. Polybrominated diphenyl ether trends in eggs of marine and freshwater birds from British Columbia, Canada, 1979–2002. *Environ. Sci. Technol.* 39:5584–5591.

Elliott, K. H., L. S. Cesh, J. A. Dooley, R. J. Letcher, and J. E. Elliott. 2009. PCBs and DDE, but not PBDEs, increase with trophic level and marine input in nestling bald eagles. *Sci. Tot. Environ.* 407:3867–3875.

Fernie, K. J., J. L. Shutt, R. J. Letcher, I. J. Ritchie, and D. M. Bird. 2009. Environmentally relevant concentrations of DE-71 and HBCD alter eggshell thickness and reproductive success of American kestrels. *Environ. Sci. Technol.* 43:2124–2130.

Fernie, K. J., J. L. Shutt, R. J. Letcher, I. J. Ritchie, K. Sullivan, and D. M. Bird. 2008. Changes in reproductive courtship behaviors of adult American kestrels (*Falco sparverius*) exposed to environmentally relevant levels of the polybrominated diphenyl ether mixture, DE-71. *Toxicol. Sci.* 102:171–178.

Fernie, K. J., et al. 2005a. Exposure to polybrominated diphenyl ethers (PBDEs): changes in thyroid, vitamin A, glutathione homeostasis, and oxidative stress in American kestrels (*Falco sparverius*). *Toxicol. Sci.* 88:375–380.

Fernie, K. J., et al. 2005b. Evidence of immunomodulation in nestling American kestrels (*Falco sparverius*) exposed to environmentally relevant PBDEs. *Environ. Pollut.* 138:485–493.

Fernie, K., G. Bortolotti, and J. Smits. 2003a. Reproductive abnormalities, teratogenicity, and developmental problems in American kestrels (*Falco sparverius*) exposed to polychlorinated biphenyls. *J. Toxicol. Environ. Health* 66A:2089–2103.

Fernie, K., J. Smits, and G. Bortolotti. 2003b. Developmental toxicity of *in ovo* exposure to polychlorinated biphenyls: I. Immediate and subsequent effects on first-generation nestling American kestrels (*Falco sparverius*). *Environ. Toxicol. Chem.* 22:554–560.

Fernie, K., G. Bortolotti, K. Drouillard, J. Smits, and T. Marchant. 2003c. Developmental toxicity of *in ovo* exposure to polychlorinated biphenyls: II. Effects of maternal or paternal exposure on second-generation nestling American kestrels. *Environ. Toxicol. Chem.* 22:2688–2694.

Fernie, K. J., J. E. Smits, G. R. Bortolotti, and D. M. Bird. 2001a. *In ovo* exposure to polychlorinated biphenyls: reproductive effects on second-generation American kestrels. *Arch. Environ. Contam. Toxicol.* 40:544–550.

Fernie, K. J., J. E. Smits, G. R. Bortolotti, and D. M. Bird. 2001b. Reproduction success of American kestrels exposed to dietary polychlorinated biphenyls. *Environ. Toxicol. Chem.* 20:776–781.

Fernie, K. J., G. R. Bortolotti, J. E. Smits, J. Wilson, K. G. Drouillard, and D. M. Bird. 2000. Changes in egg composition of American kestrels exposed to dietary polychlorinated biphenyls. *J. Toxicol. Environ. Health* 60A:291–303.

Finkelstein, M. E., et al. 2007. Contaminant-associated alteration of immune function in black-footed albatross (*Phoebastria nigripes*), a north Pacific predator. *Environ. Toxicol. Chem.* 26:1896–1903.

Finkler, M. S., J. B. Van Orman, and P. R. Sotherland. 1998. Experimental manipulation of egg quality in chickens: influence of albumen and yolk on the size and body composition of near-term embryos in a precocial bird. *J. Comp. Physiol. B* 168:17–24.

Firestone, D. 1973. Etiology of chick-edema disease. *Environ. Health Perspect.* 5:59–66.

Fisher, S. A., G. R. Bortolotti, K. J. Fernie, D. M. Bird, and J. E. Smits. 2006. Behavioral variation and its consequences during incubation for American kestrels exposed to polychlorinated biphenyls. *Ecotoxicol. Environ. Saf.* 63:226–235.

Fisher, S. A., et al. 2001. Courtship behavior of captive American kestrels (*Falco sparverius*) exposed to polychlorinated biphenyls. *Arch. Environ. Contam. Toxicol.* 41:215–220.

Fossi, C., C. Leonzio, and S. Focardi. 1986. Mixed function oxidase activity and cytochrome P-450 forms in black-headed gulls feeding in different areas. *Mar. Pollut. Bull.* 17:546–548.

Fowles, J. R., A. Fairbrother, K. A. Trust, and N. I. Kerkvliet. 1997. Effects of Aroclor 1254 on the thyroid gland, immune function, and hepatic cytochrome P450 activity in mallards. *Environ. Res.* 75:119–129.

Fox, G. A., A. P. Gilman, D. B. Peakall, and F. W. Anderka. 1978. Behavioral abnormalities of nesting Lake Ontario herring gulls. *J. Wildl. Manage.* 42:477–483.

Fox, G. A., K. A. Grasman, and G. D. Campbell. 2007b. Health of herring gulls (*Larus argentatus*) in relation to breeding location in the early 1990s. II Cellular and histopathological measures. *J. Toxicol. Environ. Health A* 70:1471–1491.

Fox, G. A., D. A. Jeffrey, K. S. Williams, S. W. Kennedy, and K. A. Grasman. 2007a. Health of herring gulls (*Larus argentatus*) in relation to breeding location in the early 1990s. I Biochemical measures. *J. Toxicol. Environ. Health A* 70:1443–1470.

French, J. B., I. C. T. Nisbet, and H. Schwabl. 2001. Maternal steroids and contaminants in common tern eggs: a mechanism of endocrine disruption? *Comp. Biochem. Physiol.* 128C:91–98.

Fry, D. M., and C. K. Toone. 1981. DDT-induced feminization of gull embryos. *Science* 213:922–924.

Gauthier, L. T., C. E. Hebert, D. V. C. Weseloh, and R. J. Letcher. 2008. Dramatic changes in the temporal trends of polybrominated diphenyl ethers (PBDEs) in herring gull eggs from the Laurentian Great Lakes: 1982–2006. *Environ. Sci. Technol.* 42:1524–1530.

Gervais, J. A., and R. G. Anthony. 2003. Chronic organochlorine contaminants, environmental variability, and the demographics of a burrowing owl population. *Ecol. Appl.* 13:1250–1262.

Giesy, J. P., L. A. Feyk, P. D. Jones, K. Kannan, and T. Sanderson. 2003. Review of the effects of endocrine-disrupting chemicals in birds. *Pure Appl. Chem.* 75:2287–2303.

Giesy, J. P., et al. 1995. Contaminants in fishes from Great Lakes-influenced sections and above dams of three Michigan rivers: III. Implications for health of bald eagles. *Arch. Environ. Contam. Toxicol.* 29:309–321.

Giesy, J. P., J. P. Ludwig, and D. E. Tillitt. 1994. Dioxins, dibenzofurans, PCBs and colonial, fish-eating water birds. In: *Dioxins and health*, ed. A. Schecter, pp. 249–307. Plenum Press, NY.

Gilbertson, M. 1983. Etiology of chick edema disease in herring gulls in the lower Great Lakes. *Chemosphere* 12:357–370.

Gilbertson, M., T. Kubiak, J. Ludwig, and G. Fox. 1991. Great Lakes embryo mortality, edema, and deformities syndrome (GLEMEDS) in colonial fish-eating birds: similarity to chick-edema disease. *J. Toxicol. Environ. Health* 33:455–520.

Gilday, D., M. Gannon, K. Yutzey, D. Bader, and A. B. Rifkind. 1996. Molecular cloning and expression of two novel avian cytochrome P450 1A enzymes induced by 2,3,7,8-tetrachlorodibenzo-*p*-dioxin. *J. Biol. Chem.* 271:33054–33059.

Gill, C. E., and J. E. Elliott. 2003. Influence of food supply and chlorinated hydrocarbon contaminants on breeding success of bald eagles. *Ecotoxicology* 12:95–111.

Gilman, A. P., G. A. Fox, D. B. Peakall, S. M. Teeple, T. R. Carroll, and G. T. Haymes. 1977. Reproductive parameters and egg contaminant levels of Great Lakes herring gulls. *J. Wildl. Manage.* 41:458–468.

Gilman, A. P., D. J. Hallett, G. A. Fox, L. J. Allan, W. J. Learning, and D. B. Peakall. 1978. Effects of injected organochlorines on naturally incubated herring gull eggs. *J. Wildl. Manage.* 42:484–493.

Gould, J. C., K. R. Cooper, and C. G. Scanes. 1997. Effects of polychlorinated biphenyl mixtures and three specific congeners on growth and circulating growth-related hormones. *Gen. Comp. Endocrinol.* 106:221–230.

Gould, J. C., K. R. Cooper, and C. G. Scanes. 1999. Effects of polychlorinated biphenyls on thyroid hormones and liver type I monodeiodinase in the chick embryo. *Ecotoxicol. Environ. Saf.* 43:195–203.

Grasman, K. A. 2002. Assessing immunological function in toxicological studies of avian wildlife. *Integ. Comp. Biol.* 42:34–42.

Grasman, K. A., and G. A. Fox. 2001. Associations between altered immune function and organochlorine contamination in young Caspian terns (*Sterna caspia*) from Lake Huron, 1997–1999. *Ecotoxicology* 10:101–114.

Grasman, K. A., G. A. Fox, P. F. Scanlon, and J. P. Ludwig. 1996. Organochlorine-associated immunosuppression in prefledgling Caspian terns and herring gulls from the Great Lakes: an ecoepidemiological study. *Environ. Health. Perspect.* 104(Suppl 4):829–842.

Grasman, K. A., P. F. Scanlon, and G. A. Fox. 2000. Geographic variation in haematological variables in adult and prefledgling herring gulls (*Larus argentatus*) and possible associations with organochlorine exposure. *Arch. Environ. Contam. Toxicol.* 38:244–253.

Guruge, K. S., H. Tanaka, and S. Tanabe. 2001. Concentration and toxic potential of polychlorinated biphenyl congeners in migratory oceanic birds from the North Pacific and the Southern Ocean. *Mar. Environ. Res.* 52:271–288.

Hario, M., K. Himberg, T. Hollmén, and E. Rudbäck. 2000. Polychlorinated biphenyls in diseased lesser black-backed gull (*Larus fuscus fuscus*) chicks from the Gulf of Finland. *Environ. Pollut.* 107:53–60.

Hario, M., J. P. Hirvi, T. Hollmén, and E. Rudbäck. 2004. Organochlorine concentrations in diseased vs. healthy gull chicks from the northern Baltic. *Environ. Pollut.* 127:411–423.

Harris, H. J., T. C. Erdman, G. T. Ankley, and K. B. Lodge. 1993. Measures of reproductive success and PCB residues in eggs and chicks of Forster's tern on Green Bay, Lake Michigan–1988. *Arch. Environ. Contam. Toxicol.* 25:304–314.

Harris, M. L., and J. E. Elliott. 2000. Reproductive success and chlorinated hydrocarbon contamination in tree swallows (*Tachycineta bicolor*) nesting along rivers receiving pulp and paper mill effluent discharges. *Environ. Pollut.* 110:307–320.

Harris, M. P., and D. Osborn. 1981. Effect of a polychlorinated biphenyl on the survival and breeding of puffins. *J. Appl. Ecol.* 18:471–479.

Harris, M. L., L. K. Wilson, R. J. Norstrom, and J. E. Elliott. 2003. Egg concentrations of polychlorinated dibenzo-*p*-dioxins and dibenzofurans in double-crested (*Phalacrocorax auritus*) and pelagic (*P. pelagicus*) cormorants from the Strait of Georgia, Canada, 1973–1998. *Environ. Sci. Technol.* 37:822–831.

Hart, C. A., I. C. T. Nisbet, S. W. Kennedy, and M. E. Hahn. 2003. Gonadal feminization and halogenated environmental contaminants in common terns (*Sterna hirundo*): evidence that ovotestes in male embryos do not persist to the prefledgling stage. *Ecotoxicology* 12:125–140.

Hart, L. E., et al. 1991. Dioxin contamination and growth and development in great blue heron embryos. *J. Toxicol. Environ. Health* 32:331–344.

Haseltine, S. D., and R. M. Prouty. 1980. Aroclor 1242 and reproductive success of adult mallards (*Anas platy-rhynchos*). *Environ. Res.* 23:29–34.

Head, J. A. 2006. Variation in the cytochrome P4501A response to dioxin-like compounds in avian species. PhD dissertation, University of Ottawa, Ottawa, ON.

Head, J. A., and S. W. Kennedy. 2007. Differential expression, induction, and stability of CYP1A4 and CYP1A5 mRNA in chicken and herring gull embryo hepatocytes. *Comp. Biochem. Physiol.* 145C:617–624.

Head, J. A., M. E. Hahn, and S. W. Kennedy. 2008. Key amino acids in the aryl hydrocarbon receptor predict dioxin sensitivity in avian species. *Environ. Sci. Technol.* 42:7535–7541.

Head, J. A., J. O'Brien, and S. W. Kennedy. 2006. Exposure to 3,3′,4,4′,5-pentachlorobiphenyl during embryonic development has a minimal effect on the cytochrome P4501A response to 2,3,7,8-tetrachlorodibenzo-*p*-dioxin in cultured chicken embryo hepatocytes. *Environ. Toxicol. Chem.* 25:2981–2989.

Hebert, C. E., and D. V. C. Weseloh. 2006. Adjusting for temporal change in trophic position results in reduced rates of contaminant decline. *Environ. Sci. Technol.* 40:5624–5628.

Hebert, C. E., J. L. Shutt, and R. O. Ball. 2002. Plasma amino acid concentrations as an indicator of protein availability to breeding herring gulls (*Larus argentatus*). *Auk* 119:185–200.

Hebert, C. E., R. J. Norstrom, M. Simon, B. M. Braune, D. V. Weseloh, and C. R. Macdonald. 1994. Temporal trends and sources of PCDDs and PCDFs in the Great Lakes: herring gull egg monitoring, 1981–1991. *Environ. Sci. Technol.* 28:1268–1277.

Hebert, C. E., R. J. Norstrom, J. Zhu, and C. R. Macdonald. 1999. Historical changes in PCB patterns in Lake Ontario and Green Bay, Lake Michigan, 1971 to 1982, from herring gull egg monitoring data. *J. Great Lakes Res.* 25:220–233.

Heinz, G. H., E. F. Hill, and J. F. Contrera. 1980. Dopamine and norepinephrine depletion in ring doves fed DDE, dieldrin, and Aroclor 1254. *Toxicol. Appl. Pharmacol.* 53:75–82.

Heinz, G. H., D. M. Swineford, and D. E. Katsma. 1984. High PCB residues in birds from the Sheboygan River, Wisconsin. *Environ. Monit. Assess.* 4:155–161.

Helander, B., A. Olsson, A. Bignert, L. Asplund, and K. Litzén. 2002. The role of DDE, PCB, coplanar PCB and eggshell parameters for reproduction in the white-tailed sea eagle (*Haliaeetus albicilla*) in Sweden. *Ambio* 31:386–403.

Helberg, M., J. O. Bustnes, K. E. Erikstad, K. O. Kristiansen, and J. U. Skaare. 2005. Relationships between reproductive performance and organochlorine contaminants in great black-backed gulls (*Larus marinus*). *Environ. Pollut.* 134:475–483.

Henning, M. H., S. K. Robinson, K. J. McKay, J. P. Sullivan, and H. Brucker. 2003. Productivity of American robins exposed to polychlorinated biphenyls, Housatonic River, Massachusetts, USA. *Environ. Toxicol. Chem.* 22:2783–2788.

Henny, C. J., J. L. Kaiser, R. A. Grove, B. L. Johnson, and R. J. Letcher. 2009. Polybrominated diphenyl ether flame retardants in eggs may reduce reproductive success of ospreys in Oregon and Washington, USA. *Ecotoxicology Online First*. 18:802–813.

Henriksen, E. O., G. W. Gabrielsen, J. Trudeau, J. Wolkers, K. Sagerup, and J. U. Skaare. 2000. Organochlorines and possible biochemical effects in glaucous gulls (*Larus hyperboreus*) from Bjørnøya, the Barents Sea. *Arch. Environ. Contam. Toxicol.* 38:234–243.

Henshel, D. S. 1998. Developmental neurotoxic effects of dioxin and dioxin-like compounds on domestic and wild avian species. *Environ. Toxicol. Chem.* 17:88–98.

Henshel, D. S., and D. W. Sparks. 2006. Site specific PCB-correlated interspecies differences in organ somatic indices. *Ecotoxicology* 15:9–18.

Hervé, J. C., et al. 2010. Cytochrome P4501A induction by 2,3,7,8-tetrachlorodibenzo-p-dioxin and two chlorinated dibenzofurans in primary hepatocyte cultures of three avian species. *Toxicol. Sci.* 113:380–391.

Hilscherova, K., et al. 2003. Oxidative stress in laboratory-incubated double-crested cormorant eggs collected from the Great Lakes. *Arch. Environ. Contam. Toxicol.* 45:533–546.

Hoffman, D. J., M. J. Melancon, P. N. Klein, J. D. Eisemann, and J. W. Spann. 1998. Comparative developmental toxicity of planar polychlorinated biphenyl congeners in chickens, American kestrels, and common terns. *Environ. Toxicol. Chem.* 17:747–757.

Hoffman, D. J., M. J. Melancon, P. N. Klein, C. P. Rice, J. D. Eisemann, R. K. Hines, J. W. Spann, and G. W. Pendleton. 1996b. Developmental toxicity of PCB 126 (3,3′,4,4′,5-pentachlorobiphenyl) in nestling American kestrels (*Falco sparverius*). *Fundam. Appl. Toxicol.* 34:188–200.

Hoffman, D. J., M. J. Melancon, J. D. Eisemann, and P. N. Klein. 1995. Comparative toxicity of planar PCB congeners by egg injection. *SETAC Abstracts* 16:207.

Hoffman, D. J., B. A. Rattner, C. M. Bunck, A. Krynitsky, H. M. Ohlendorf, and R. W. Lowe. 1986. Association between PCBs and lower embryonic weight in black-crowned night herons in San Francisco Bay. *J. Toxicol. Environ. Health* 19:383–391.

Hoffman, D. J., C. P. Rice, and T. J. Kubiak. 1996a. PCBs and dioxins in birds. In: *Environmental contaminants in wildlife: interpreting tissue concentrations*, eds. W. N. Beyer, G. H. Heinz, and A. W. Redmon-Norwood. Boca Raton, FL.: CRC Press.

Hoffman, D. J., G. J. Smith, and B. A. Rattner. 1993. Biomarkers of contaminant exposure in common terns and black-crowned night herons in the Great Lakes. *Environ. Toxicol. Chem.* 12:1095–1103.

Holm, L., A. Blomqvist, I. Brandt, B. Brunstrom, Y. Ridderstale, and C. Berg. 2006. Embryonic exposure to *o,p'*-DDT causes eggshell thinning and altered shell gland carbonic anhydrase expression in the domestic hen. *Environ. Toxicol. Chem.* 25:2787–2793.

Hoogesteijn, A. L., T. J. DeVoogd, F. W. Quimby, T. De Caprio, and G. V. Kollias. 2005. Reproductive impairment in zebra finches (*Taeniopygia guttata*). *Environ. Toxicol. Chem.* 24:219–223.

Hoogesteijn, A. L., G. V. Kollias, F. W. Quimby, A. P. De Caprio, D. W. Winkler, and T. J. DeVoogd. 2008. Development of a brain nucleus involved in song production in zebra finches (*Taeniopygia guttata*) is disrupted by Aroclor 1248. *Environ. Toxicol. Chem.* 27:2071–2075.

Ikonomou, M. G., E. Rayne, M. Fischer, M. P. Fernandez, and W. Cretney. 2002. Occurrence and congener profiles of polybrominated diphenyl ethers (PBDEs) in environmental samples from coastal British Columbia, Canada. *Chemosphere* 46:649–663.

Iwaniuk, A. N., et al. 2006. The effects of environmental exposure to DDT on the brain of a songbird: changes in structures associated with mating and song. *Behav. Brain Res.* 173:1–10.

Janeway, C. A., P. Travers, M. Walport, and J. D. Capra. 1999. *Immunobiology: the immune system in health and disease,* 4th ed. London: Taylor and Francis.

Janz, D. M., and G. D. Bellward. 1996a. *In ovo* 2,3,7,8-tetrachlorodibenzo-*p*-dioxin exposure in three avian species. I. Effects on thyroid hormones and growth during the perinatal period. *Toxicol. Appl. Pharmacol.* 139:281–291.

Janz, D. M., and G. D. Bellward. 1996b. *In ovo* 2,3,7,8-tetrachlorodibenzo-*p*-dioxin exposure in three avian species. II. Effects on estrogen receptor and plasma sex steroid hormones during the perinatal period. *Toxicol. Appl. Pharmacol.* 139:292–300.

Jin, X., S. W. Kennedy, T. Di Muccio, and T. W. Moon. 2001. Role of oxidative stress and antioxidant defense in 3,3′,4,4′,5-pentachlorobiphenyl-induced toxicity and species-differential sensitivity in chicken and duck embryos. *Toxicol. Appl. Pharmacol.* 172:241–248.

Jones, P. D., et al. 1993. Biomagnification of bioassay-derived 2,3,7,8-tetrachlorodibenzo-*p*-dioxin equivalents. *Chemosphere* 26:1203–1212.

Karchner, S. I., D. G. Franks, S. W. Kennedy, and M. E. Hahn. 2006. The molecular basis for differential dioxin sensitivity in birds: role of the aryl hydrocarbon receptor. *Proc. Natl. Acad. Sci. USA* 103:6252–6257.

Kelly, B. C., M. G. Ikonomou, J. D. Blair, and F. A. P. C. Gobas. 2008. Bioaccumulation behaviour of polybrominated diphenyl ethers (PBDEs) in a Canadian Arctic marine food web. *Sci. Total Environ.* 401:60–72.

Kennedy, S. W., G. A. Fox, S. P. Jones, and S. F. Trudeau. 2003. Hepatic EROD activity is not a useful biomarker of polychlorinated biphenyl exposure in the adult herring gull (*Larus argentatus*). *Ecotoxicology* 12:153–161.

Kennedy, S. W., G. A. Fox, S. Trudeau, L. J. Bastien, and S. P. Jones. 1998. Highly carboxylated porphyrin concentration: a biochemical marker of PCB exposure in herring gulls. *Mar. Environ. Res.* 46:65–69.

Kennedy, S. W., A. Lorenzen, S. P. Jones, M. E. Hahn, and J. J. Stegeman. 1996. Cytochrome P4501A induction in avian hepatocyte cultures: a promising approach for predicting the sensitivity of avian species to toxic effects of halogenated aromatic hydrocarbons. *Toxicol. Appl. Pharmacol.* 141:214–230.

Koeman, J. H., H. C. W. Van Velzen-Blad, R. De Vries, and J. G. Vos. 1973. Effects of PCB and DDE in cormorants and evaluation of PCB residues from an experimental study. *J. Reprod. Fert.* 1973 (Suppl 19):353–364.

Kozie, K. D., and R. K. Anderson. 1991. Productivity, diet, and environmental contaminants in bald eagles nesting near the Wisconsin shoreline of Lake Superior. *Arch. Environ. Contam. Toxicol.* 20:41–48.

Kubiak, T. J., et al. 1989. Microcontaminants and reproductive impairment of the Forster's tern on Green Bay, Lake Michigan–1983. *Arch. Environ. Contam. Toxicol.* 18:706–727.

Kuiken, T., G. A. Fox, and K. L. Danesik. 1999. Bill malformations in double-crested cormorants with low exposure to organochlorines. *Environ. Toxicol. Chem.* 18:2908–2913.

Kunisue, T., M. Watanabe, A. Subramanian, A. M. Titenko, and S. Tanabe. 2003. Congener-specific patterns and toxic assessment of polychlorinated biphenyls in resident and migratory birds from southern India and Lake Baikal in Russia. *Arch. Environ. Contam. Toxicol.* 45:547–561.

Kuzyk, Z. Z. A., N. M. Burgess, J. P. Stow, and G. A. Fox. 2003. Biological effects of marine PCB contamination on black guillemot nestlings at Saglek, Labrador: liver biomarkers. *Ecotoxicology* 12:183–197.

Larson, J. M., et al. 1996. Reproductive success, developmental anomalies, and environmental contaminants in double-crested cormorants (*Phalacrocorax auritus*). *Environ. Toxicol. Chem.* 15:553–559.

Lavoie, E. T., and K. A. Grasman. 2007. Effects of *in ovo* exposure to PCBs 126 and 77 on mortality, deformities and post-hatch immune function in chickens. *J. Toxicol. Environ. Health A* 70:547–558.

Lebreton, J-D., and J. Clobert. 1991. Bird population dynamics, management and conservation: the role of mathematical modelling. In *Bird population studies: relevance to conservation and management*, eds. C. M. Perrins, J-D. Lebreton, and G. J. M. Hirons, pp. 105–121. London: Oxford University Press.

Love, O. P., L. J. Shutt, J. S. Silfies, G. R. Bortolotti, J. E. G. Smits, and D. M. Bird. 2003. Effects of dietary PCB exposure on adrenocortical function in captive American kestrels (*Falco sparverius*). *Ecotoxicology* 12:199–208.

Ludwig, J. P., et al. 1996. Deformities, PCBs and TCDD-equivalents in double-crested cormorants (*Phalacrocorax auritus*) and Caspian terns (*Hydroprogne caspia*) of the upper Great Lakes 1986–1991: testing a cause-effect hypothesis. *J. Great Lakes Res.* 22:172–197.

Mahajan, S. S., and A. B. Rifkind. 1999. Transcriptional activation of avian CYP1A4 and CYP1A5 by 2,3,7,8-tetrachlorodibenzo-*p*-dioxin: differences in gene expression and regulation compared to mammalian CYP1A1 and CYP1A2. *Toxicol. Appl. Pharmacol.* 155:96–106.

Manchester-Neesvig, J. B., K. Valters, and W. C. Sonzogni. 2001. Comparison of polybrominated diphenyl ethers (PBDEs) and polychlorinated biphenyls (PCBs) in Lake Michigan salmonids. *Environ. Sci. Technol.* 35:1071–1077.

Martinovic, B., D. R. S. Lean, C. A. Bishop, E. Birmingham, A. Secord, and K. Jock. 2003a. Health of tree swallow (*Tachycineta bicolor*) nestlings exposed to chlorinated hydrocarbons in the St. Lawrence River Basin. Part I. Renal and hepatic vitamin A concentrations. *J. Toxicol. Environ. Health* 66A:1053–1072.

Martinovic, B., D. R. S. Lean, C. A. Bishop, E. Birmingham, A. Secord, and K. Jock. 2003b. Health of tree swallow (*Tachycineta bicolor*) nestlings exposed to chlorinated hydrocarbons in the St. Lawrence River Basin. Part II. Basal and stress plasma corticosterone concentrations. *J. Toxicol. Environ. Health* 66A:2015–2029.

McCarty, J. P., and A. L. Secord. 1999a. Nest-building behavior in PCB-contaminated tree swallows. *Auk* 116:55–63.

McCarty, J. P., and A. L. Secord. 1999b. Reproductive ecology of tree swallows (*Tachycineta bicolor*) with high levels of polychlorinated biphenyl contamination. *Environ. Toxicol. Chem.* 18:1433–1439.

McCarty, J. P., and A. L. Secord. 2000. Possible effects of PCB contamination on female plumage color and reproductive success in Hudson River tree swallows. *Auk* 117:987–995.

McKee, M. J. 1995. Biological monitoring of chemical contamination at Crab Orchard National Wildlife Refuge. Final Report. Cooperative Wildlife Research Laboratory, Southern Illinois University, Carbondale, IL.

McKernan, M. A., B. A. Rattner, R. C. Hale, and M. A. Ottinger. 2007. Egg orientation affects toxicity of air cell administered polychlorinated biphenyl 126 (3,3′,4,4′,5-pentachlorobiphenyl) in chicken (*Gallus gallus*) embryos. *Environ. Toxicol. Chem.* 26:2724–2727.

McKernan, M. A., B. A. Rattner, R. C. Hale, and M. A. Ottinger. 2009. Toxicity of polybrominated diphenyl ethers (DE-71) in chicken (*Gallus gallus*), mallard (*Anas platyrhynchos*), and American kestrel (*Falco sparverius*) embryos and hatchlings. *Environ. Toxicol. Chem.* 28:1007–1017.

McKinney, M. A., L. S. Cesh, J. E. Elliott, T. D. Williams, D. K. Garcelon, and R. J. Letcher. 2006. Brominated flame retardants and halogenated phenolic compounds in North American west coast bald eaglet (*Haliaeetus leucocephalus*) plasma. *Environ. Sci. Technol.* 40:6275–6281.

McNabb, F. M. A., and G. A. Fox. 2003. Avian thyroid development in chemically contaminated environments: is there evidence of alterations in thyroid function and development? *Evol. Dev.* 5:76–82.

McNabb, F. M. A., R. J. R. McCleary, L. A. Fowler, C. M. Parsons, K. A. Grasman, and G. A. Fox. 2001. Thyroid function in polychlorinated biphenyl (PCB) exposed embryos and chicks. In *Perspectives in comparative endocrinology: unity and diversity*, eds. H. J. T. Goos, R. K. Rastogi, H. Vaudry, and R. Pierantoni, pp. 275–280. Bologna, Italy:Monduzzi Editore.

Moccia, R. D., G. A. Fox, and A. Britton. 1986. A quantitative assessment of the thyroid histopathology of herring gulls (*Larus argentatus*) from the Great Lakes and a hypothesis on the causal role of environmental contaminants. *J. Wildl. Dis.* 22:60–70.

Møller, A. P., and N. Saino. 2004. Immune response and survival. *Oikos* 104:299–304.

Mora, M. A., H. J. Auman, J. P. Ludwig, J. P. Giesy, D. A. Verbrugge, and M. E. Ludwig. 1993. Polychlorinated biphenyls and chlorinated insecticides in plasma of Caspian terns: relationships with age, productivity, and colony site tenacity in the Great Lakes. *Arch. Environ. Contam. Toxicol.* 24:320–331.

Morrissey, C. A., L. I. Bendell-Young, and J. E. Elliott. 2004. Linking contaminant profiles to the diet and breeding location of American dippers using stable isotopes. *J. Appl. Ecol.* 41:502–512.

Morrissey, C. A., L. I. Bendell-Young, and J. E. Elliott. 2005. Identifying sources and biomagnification of persistent organic contaminants in biota from mountain streams of southwestern British Columbia, Canada. *Environ. Sci. Technol.* 39:8090–8098.

Murk, A. J., et al. 1996. Effects of polyhalogenated aromatic hydrocarbons and related contaminants on common tern reproduction: integration of biological, biochemical, and chemical data. *Arch. Environ. Contam. Toxicol.* 31:128–140.

Murvoll, K. M., B. M. Jenssen, and J. U. Skaare. 2005. Effects of pentabrominated diphenyl ether (PBDE-99) on vitamin status in domestic duck (*Anas platyrhynchos*) hatchlings. *J. Toxicol. Environ. Health* 68A:515–533.

Murvoll, K. M., J. U. Skaare, E. Anderssen, and B. M. Jenssen. 2006a. Exposure and effects of persistent organic pollutants in European shag (*Phalacrocorax aristotelis*) hatchlings from the coast of Norway. *Environ. Toxicol. Chem.* 25:190–198.

Murvoll, K. M., J. U. Skaare, H. Jensen, and B. M. Jenssen. 2007. Associations between persistent organic pollutants and vitamin status in Brünnich's guillemot and common eider hatchlings. *Sci. Tot. Environ.* 381:134–145.

Murvoll, K. M., J. U. Skaare, B. Moe, E. Anderssen, and B. M. Jenssen. 2006b. Spatial trends and associated biological responses of organochlorines and brominated flame retardants in hatchlings of North Atlantic kittiwakes (*Rissa tridactyla*). *Environ. Toxicol. Chem.* 25:1648–1656.

Neigh, A. M., et al. 2006. Productivity of tree swallows (*Tachycineta bicolor*) exposed to PCBs at the Kalamazoo River Superfund site. *J. Toxicol. Environ. Health* 69A:395–415.

Neigh, A. M., et al. 2007. Reproductive success of passerines exposed to polychlorinated biphenyls through the terrestrial food web of the Kalamazoo River. *Ecotoxicol. Environ. Saf.* 66:107–118.

Newson, S. C., C. D. Sandau, J. E. Elliott, S. B. Brown, and R. J. Norstrom. 2000. PCBs and hydroxylated metabolites in bald eagle plasma—comparison to thyroid hormone and retinol levels. Paper presented in the Proceedings of the International Conference in Environmental Chemistry, May 7–11, Ottawa, ON.

Norstrom, R. J., and C. E. Hebert. 2006. Comprehensive re-analysis of archived herring gull eggs reconstructs historical temporal trends in chlorinated hydrocarbon contamination in Lake Ontario and Green Bay, Lake Michigan, 1971–1982. *J. Environ. Monit.* 8:835–847.

Norstrom, R. J., M. Simon, J. Moisey, B. Wakeford, and D. V. C. Weseloh. 2002. Geographical distribution (2000) and temporal trends (1981–2000) of brominated diphenyl ethers in Great Lakes herring gull eggs. *Environ. Sci. Technol.* 36:4783–4789.

Nosek, J. A., J. R. Sullivan, S. R. Craven, A. Gendron-Fitzpatrick, and R. E. Peterson. 1993. Embryotoxicity of 2,3,7,8-tetrachlorodibenzo-*p*-dioxin in the ring-necked pheasant. *Environ. Toxicol. Chem.* 12:1215–1222.

Olsson, A., K. Ceder, A. Bergman, and B. Helander. 2000. Nestling blood of the white-tailed sea eagle (*Haliaeetus albicilla*) as an indicator of territorial exposure to organohalogen compounds—an evaluation. *Environ. Sci. Technol.* 34:2733–2740.

Ormerod, S. J., S. J. Tyler, and I. Juttner. 2000. Effects of point-source PCB contamination on breeding performance and post-fledging survival in the dipper *Cinclus cinclus*. *Environ. Pollut.* 110:505–513.

Palokangas, P., E. Korpimaki, H. Hakkarainen, E. Huhta, P. Tolonen, and R. V. Alatalo. 1994. Female kestrels gain reproductive success by choosing brightly ornamented males. *Anim. Behav.* 47:443–448.

Papp, Z., G. R. Bortolotti, and J. E. G. Smits. 2005. Organochlorine contamination and physiological responses in nestling tree swallows in Point Pelee National Park, Canada. *Arch. Environ. Contam. Toxicol.* 49:563–568.

Peakall, D. B. 1996. Disrupted patterns of behavior in natural populations as an index of ecotoxicity. *Environ. Health Perspect.* 104(Suppl 2):331–335.

Peakall, D. B., G. A. Fox, A. D. Gilman, D. J. Hallet, and R. J. Norstrom. 1980. Reproductive success of herring gulls as an indicator of Great Lakes water quality. In *Hydrocarbons and halogenated hydrocarbons in the aquatic environment*, eds. B. K. Afghan, and D. MacKay, pp. 337–344. Plenum Press, NY.

Peakall, D. B., and M. L. Peakall. 1973. Effects of a polychlorinated biphenyl on the reproduction of artificially and naturally incubated dove eggs. *J. Appl. Ecol.* 10:863–868.

Peterson, R. E., H. M. Theobald, and G. L. Kimmel. 1993. Developmental and reproductive toxicity of dioxins and related compounds: cross-species comparisons. *CRC Rev.* 23:283–335.

Powell, D. C., R. J. Aulerich, K. L. Stromborg, and S. J. Bursian. 1996. The effects of 3,3′,4,4′-tetrachlorobiphenyl, 2,3,3′,4,4′-pentachlorobiphenyl, and 3,3′,4,4′,5-pentachlorobiphenyl on the developing chicken embryo when injected prior to incubation. *J. Toxicol. Environ. Health* 49:319–338.

Powell, D. C., et al. 1997a. Effects of 3,3′,4,4′,5-pentachlorobiphenyl (PCB 126), 2,3,7,8-tetrachlorodibenzo-*p*-dioxin (TCDD), or an extract derived from field-collected cormorant eggs injected into double-crested cormorant (*Phalacrocorax auritus*) eggs. *Environ. Toxicol. Chem.* 16:1450–1455.

Powell, D. C., et al. 1997b. Organochlorine contaminants in double-crested cormorants from Green Bay, Wisconsin: II. Effects of an extract derived from cormorant eggs on the chicken embryo. *Arch. Environ. Contam. Toxicol.* 32:316–322.

Powell, D. C., et al. 1998. Effects of 3,3′,4,4′,5-pentachlorobiphenyl and 2,3,7,8-tetrachlorodibenzo-*p*-dioxin injected into the yolks of double-crested cormorant (*Phalacrocorax auritus*) eggs prior to incubation. *Environ. Toxicol. Chem.* 17:2035–2040.

Prestt, I., D. J. Jefferies, and N. W. Moore. 1970. Polychlorinated biphenyls in wild birds in Britain and their avian toxicity. *Environ. Pollut.* 1:3–26.

Price, I. M., and D. V. Weseloh. 1986. Increased numbers and productivity of double-crested cormorants, *Phalacrocorax auritus* on Lake Ontario. *Can. Field-Nat.* 100:474–482.

Quinn, M. J., J. B. French, F. M. A. McNabb, and M. A. Ottinger. 2002. The effects of polychlorinated biphenyls (Aroclor 1242) on thyroxine, estradiol, molt, and plumage characteristics in the American kestrel (*Falco sparverius*). *Environ. Toxicol. Chem.* 21:1417–1422.

Rattner, B. A., V. P. Eroschenko, G. A. Fox, D. M. Fry, and J. Gorsline. 1984. Avian endocrine responses to environmental pollutants. *J. Exp. Zool.* 232:683–689.

Rayne, S., M. G. Ikonomou, and B. Antcliffe. 2003. Rapidly increasing polybrominated diphenyl ether concentrations in the Columbia River system from 1992 to 2000. *Environ. Sci. Technol.* 37:2847–2854.

Rice, C. P., P. O'Keefe, and T. Kubiak. 2004. Sources, pathways and effects of PCBs, dioxins and dibenzo-furans. In *Handbook of ecotoxicology*, eds. D. J. Hoffman, B. A. Rattner, G. A. Burton Jr, and J. Cairns Jr, pp. 501–573. London: Lewis Publishers.

Rolland, R. M. 2000. A review of chemically-induced alterations in thyroid and vitamin A status from field studies of wildlife and fish. *J. Wildl. Dis.* 36:615–635.

Ryckman, D. P., D. V. Weseloh, P. Hamr, G. A. Fox, B. Collins, P. J. Ewins, and R. J. Norstrom. 1998. Spatial and temporal trends in organochlorine contamination and bill deformities in double-crested cormorants (*Phalacrocorax auritus*) from the Canadian Great Lakes. *Environ. Monit. Assess.* 53:169–195.

Safe, S. 1994. Polychlorinated biphenyls (PCBs): environmental impact, biochemical and toxic responses, and implications for risk assessment. *Crit. Rev. Toxicol.* 24:87–149.

Sagerup, K., E. O. Henriksen, J. U. Skaare, and G. W. Gabrielsen. 2002. Intraspecific variation in trophic feeding levels and organochlorine concentrations in glaucous gulls (*Larus hyperboreus*) from Bjørnøya, the Barents Sea. *Ecotoxicology* 11:119–125.

Sagerup, K., E. O. Henriksen, A. Skorping, J. U. Skaare, and G. W. Gabrielsen. 2000. Intensity of parasitic nematodes increases with organochlorine levels in the glaucous gull. *J. Appl. Ecol.* 37:532–539.

Sanderson, J. T., and G. D. Bellward. 1995. Hepatic microsomal ethoxyresorufin *O*-deethylase-inducing potency *in ovo* and cytosolic Ah receptor binding affinity of 2,3,7,8-tetrachlorodibenzo-*p*-dioxin: comparison of four avian species. *Toxicol. Appl. Pharmacol.* 132:131–145.

Sanderson, J. T., D. M. Janz, G. D. Bellward, and J. P. Giesy. 1997. Effects of embryonic and adult exposure to 2,3,7,8-tetrachlorodibenzo-*p*-dioxin on hepatic microsomal testosterone hydroxylase activities in great blue herons (*Ardea herodias*). *Environ. Toxicol. Chem.* 16:1304–1310.

Sanderson, J. T., S. W. Kennedy, and J. P. Giesy. 1998. *In vitro* induction of ethoxyresorufin-*O*-deethylase and porphyrins by halogenated aromatic hydrocarbons in avian primary hepatocytes. *Environ. Toxicol. Chem.* 17:2006–2018.

Sanderson, J. T., R. J. Norstrom, J. E. Elliott, L. E. Hart, K. M. Cheng, and G. D. Bellward. 1994a. Biological effects of polychlorinated dibenzo-*p*-dioxins, dibenzofurans, and biphenyls in double-crested cormorant chicks (*Phalacrocorax auritus*). *J. Toxicol. Environ. Health* 41:247–265.

Sanderson, J. T., et al. 1994b. Monitoring biological effects of polychlorinated dibenzo-*p*-dioxins, dibenzofurans, and biphenyls in great blue heron chicks (*Ardea herodias*) in British Columbia. *J. Toxicol. Environ. Health* 41:435–450.

Sellström, U., A. Bignert, A. Kierkegaard, L. Haggberg, C. A. de Wit, and M. Olsson. 2003. Temporal trend studies on tetra- and pentabrominated diphenyl ethers and hexabromocyclododecane in guillemot egg from the Baltic Sea. *Environ. Sci. Technol.* 37:5496–5501.

Sileo, L., L. Karstad, R. Frank, M. V. H. Holdrinet, E. Addison, and H. E. Braun. 1977. Organochlorine poisoning of ring-billed gulls in southern Ontario. *J. Wildl. Dis.* 13:313–322.

Smits, J. E. G., and G. R. Bortolotti. 2001. Antibody-mediated immunotoxicity in American kestrels (*Falco sparverius*) exposed to polychlorinated biphenyls. *J. Toxicol. Environ. Health* 62A:217–226.

Smits, J. E., K. J. Fernie, G. R. Bortolotti, and T. A. Marchant. 2002. Thyroid hormone suppression and cell-mediated immunomodulation in American kestrels (*Falco sparverius*) exposed to PCBs. *Arch. Environ. Contam. Toxicol.* 43:338–344.

Spear, P. A., D. H. Bourbonnais, R. J. Norstrom, and T. W. Moon. 1990. Yolk retinoids (vitamin A) in eggs of the herring gull and correlations with polychlorinated dibenzo-*p*-dioxins and dibenzofurans. *Environ. Toxicol. Chem.* 9:1053–1061.

Spear, P. A., T. W. Moon, and D. B. Peakall. 1986. Liver retinoid concentrations in natural populations of herring gulls (*Larus argentatus*) contaminated by 2,3,7,8-tetrachlorodibenzo-*p*-dioxin and in ring doves (*Streptopelia risoria*) injected with a dioxin analogue. *Can. J. Zool.* 64:204–208.

Steinberg, R. M., D. M. Walker, T. E. Juenger, M. J. Woller, and A. C. Gore. 2008. Effects of perinatal polychlorinated biphenyls on adult female rat reproduction: development, reproductive physiology and second generational effects. *Biol. Reprod.* 78:1091–1101.

Stickel, W. H., L. F. Stickel, R. A. Dyrland, and D. L. Hughes. 1984. Aroclor 1254 residues in birds: lethal levels and loss rates. *Arch. Environ. Contam. Toxicol.* 13:7–13.

Stone, W. B., and J. C. Okoniewski. 2000. PCB poisoning in a red-tailed hawk (*Buteo jamaicensis*) near a site of terrestrial contamination in New York State. *Bull. Environ. Contam. Toxicol.* 64:81–84.

Strause, K. D., et al. 2007. Risk assessment of great horned owls (*Bubo virginianus*) exposed to polychlorinated biphenyls and DDT along the Kalamazoo River, Michigan, USA. *Environ. Toxicol. Chem.* 26:1386–1398.

Thiel, D. A., S. G. Martin, J. W. Duncan, M. J. Lemke, W. R. Lance, and R. E. Peterson. 1988. Evaluation of the effects of dioxin-contaminated sludges on wild birds. In: Proceedings of the 1988 Technical Association of Pulp and Paper Environmental Conference, 145–148. Charleston, SC.

Thompson, H. M., A. Fernandes, M. Rose, S. White, and A. Blackburn. 2006. Possible chemical causes of skeletal deformities in grey heron nestlings (*Ardea cinerea*) in North Nottinghamshire, UK. *Chemosphere* 65:400–409.

Tillitt, D. E., et al. 1992. Polychlorinated biphenyl residues and egg mortality in double-crested cormorants from the Great Lakes. *Environ. Toxicol. Chem.* 11:1281–1288.

Ucán-Marín, F., A. Arukwe, A. Mortensen, G. W. Gabrielsen, G. A. Fox, and R. J. Letcher. 2009. Recombinant transthyretin purification and competitive binding with organohalogen compounds in two gull species (*Larus argentatus* and *Larus hyperboreus*). *Toxicol. Sci.* 107:440–450.

van den Berg, M., L. H. J. Craane, T. Sinnige, I. J. Lutke-Schipholt, B. Spenkelink, and A. Brouwer. 1992. The use of biochemical parameters in comparative toxicological studies with the cormorant (*Phalacrocorax carbo*) in The Netherlands. *Chemosphere* 25:1265–1270.

van den Berg, M., L. H. J. Craane, S. Van Mourik, and A. Brouwer. 1995. The (possible) impact of chlorinated dioxins (PCDDs), dibenzofurans (PCDFs) and biphenyls (PCBs) on the reproduction of the cormorant *Phalacrocorax carbo*—an ecotoxicological approach. *Ardea* 83:299–313.

van den Berg, M., et al. 1994. Biochemical and toxic effects of polychlorinated biphenyls (PCBs), dibenzo-*p*-dioxins (PCDDs) and dibenzofurans (PCDFs) in the cormorant (*Phalacrocorax carbo*) after *in ovo* exposure. *Environ. Toxicol. Chem.* 13:803–816.

van den Berg, M., et al. 1998. Toxic equivalency factors (TEFs) for PCBs, PCDDs, PCDFs for humans and wildlife. *Environ. Health Perspect.* 106:775–792.

Van den Steen, E., V. L. B. Jaspers, A. Covaci, H. Neels, M. Eens, and R. Pinxten. 2009. Maternal transfer of organochlorines and brominated flame retardants in blue tits (*Cyanistes caeruleus*) *Environ. Int.* 35:69–75.

Verhallen, E. Y., M. van den Berg, and A. T. C. Bosveld. 1997. Interactive effects on the EROD inducing potency of polyhalogenated aromatic hydrocarbons in the chicken embryo hepatocyte assay. *Environ. Toxicol. Chem.* 16:277–282.

Verreault, J., R. J. Letcher, E. Ropstad, E. Dahl, and G. W. Gabrielsen. 2006b. Organohalogen contaminants and reproductive hormones in incubating glaucous gulls (*Larus hyperboreus*) from the Norwegian Arctic. *Environ. Toxicol. Chem.* 25:2990–2996.

Verreault, J., J. U. Skaare, B. M. Jenssen, and G. W. Gabrielsen. 2004. Effects of organochlorine contaminants on thyroid hormone levels in Arctic breeding glaucous gulls, *Larus hyperboreus. Environ. Health Perspect.* 112:532–537.

Verreault, J., R. A. Villa, G. W. Gabrielsen, J. U. Skaare, and R. J. Letcher. 2006a. Maternal transfer of organohalogen contaminants and metabolites to eggs of Arctic-breeding glaucous gulls. *Environ. Pollut.* 144:1053–1060.

Walker, M. K., R. S. Pollenz, and S. M. Smith. 1997. Expression of the aryl hydrocarbon receptor (AhR) and AhR nuclear translocator during chick cardiogenesis is consistent with 2,3,7,8-tetrachlorodibenzo-*p*-dioxin-induced heart defects. *Toxicol. Appl. Pharmacol.* 143:407–419.

Wardrop, S. L., and R. Ydenberg. 2003. Date and parental quality effects in the seasonal decline in reproductive performance of the tree swallow *Tachycineta bicolor*: interpreting results in light of potential experimental bias. *Ibis* 145:439–447.

White, D. H., and J. T. Seginak. 1994. Dioxins and furans linked to reproductive impairment in wood ducks. *J. Wildl. Manage.* 58:100–106.

Whitlock, J. P. 1999. Induction of cytochrome P4501A1. *Ann. Rev. Pharmacol. Toxicol.* 39:103–125.

Wiehn, J. 1997. Plumage characteristics as an indicator of male parental quality in the American kestrel. *J. Avian Biol.* 28:47–55.

Woodford, J. E., W. H. Karasov, M. W. Meyer, and L. Chambers. 1998. Impact of 2,3,7,8-TCDD exposure on survival, growth, and behaviour of ospreys breeding in Wisconsin, USA. *Environ. Toxicol. Chem.* 17:1323–1331.

Yamashita, N., S. Tanabe, J. P. Ludwig, H. Kurita, M. E. Ludwig, and R. Tatsukawa. 1993. Embryonic abnormalities and organochlorine contamination in double-crested cormorants (*Phalacrocorax auritus*) and Caspian terns (*Sterna caspia*) from the upper Great Lakes in 1988. *Environ. Pollut.* 79:163–173.

Yasui, T., E-Y. Kim, H. Iwata, D. G. Franks, S. I. Karchner, M. E. Hahn, and S. Tanabe. 2007. Functional characterization and evolutionary history of two aryl hydrocarbon receptor isoforms (AhR1 and AhR2) from avian species. *Toxicol. Sci.* 99:101–117.

Malayan Bear

By G. Mutzel, from *The Royal Natural History*, edited by Richard Lydekker, Frederick Warne & Co., London, 1893–94.

15 Toxicological Implications of PCBs, PCDDs, and PCDFs in Mammals

Matthew Zwiernik
Frouke Vermeulen
Steven Bursian

CONTENTS

15.1 Introduction 531
15.2 PCBs/PCDDs/PCDFs 532
    15.2.1 PCBs 532
    15.2.2 PCDDs and PCDFs 534
15.3 Spatial and Temporal Trends of PCBs/PCDDs/PCDFs in Mammals 534
15.4 Mechanism of Action of TCDD-Like PCBs/PCDDs/PCDFs 535
15.5 The Toxic Equivalency Approach 536
15.6 Field and Feeding Studies 537
15.7 Terrestrial Mammals 537
    15.7.1 Field Studies 537
    15.7.2 Feeding Studies 538
15.8 Marine Mammals 538
    15.8.1 Field Studies 539
15.9 Freshwater Aquatic Mammals 540
    15.9.1 Field Studies 541
    15.9.2 Feeding Studies 544
        15.9.2.1 Effects of Commercial PCB Mixtures 545
        15.9.2.2 Effects of Environmentally Derived PCBs, PCDDs, and PCDFs 545
15.10 Threshold Tissue Residue Concentrations 548
    15.10.1 Terrestrial Mammal Threshold Tissue Residue Concentrations 549
    15.10.2 Marine Mammal Threshold Tissue Residue Concentrations 549
    15.10.3 Aquatic Mammal Threshold Tissue Residue Concentrations 551
Summary 553
References 554

15.1 INTRODUCTION

Polychlorinated biphenyls (PCBs), polychlorinated dibenzo-*p*-dioxins (PCDDs), and polychlorinated dibenzofurans (PCDFs) constitute a group of structurally related halogenated aromatic compounds that are ubiquitous, persistent, and toxic (Safe 1990, 1998a, Van den Berg et al. 1994, Huwe 2002, Mandal 2005, Schecter et al. 2006). Because of their persistence and lipid solubility, these compounds can bioaccumulate in individual organisms and biomagnify up through higher trophic layers of the food chain. Mammalian characteristics, including food intake rate, life span, trophic status, and lactation in support of young, result in measurable residues of these compounds in virtually every species studied to date. Humans and mammalian wildlife share air, water, food chains, and biochemical and molecular processes. Thus, select species of wildlife are often used as sentinels, providing important real-world information on bioavailability and effects of environmental factors on toxicological responses. Episodic mortalities and declines of local wildlife populations that include physiological responses that are similar to those noted in controlled laboratory studies have raised scientific, public, and regulatory concern related to potential adverse environmental and human health effects associated with PCB/PCDD/PCDF exposure. These concerns have led to institutional regulation of PCB/PCDD/PCDF use and release in most developed countries. The good news is that since regulations have been imposed, mammalian exposure to these environmental contaminants is, in general, on the decline.

This chapter provides a review of literature pertaining to the exposure of mammalian wildlife species to PCBs, PCDDs, and PCDFs and aims to interpret those data in terms of potential for adverse effects as they relate to tissue concentrations. An overview of the chemistry, uses, and sources of PCBs/PCDDs/PCDFs is followed by a description of the spatial and temporal trends of these chemicals in mammalian wildlife species. Because the majority of toxic effects associated with exposure to PCBs/PCDDs/PCDFs are presumed to result from interaction with the aryl hydrocarbon receptor (AhR), an overview of the mechanism of action of these chemicals is provided. Based on this shared mechanism, combined with assumptions of toxicological additivity, the TCDD-like toxicity of complex mixtures of individual congeners can be predicted with some accuracy based on a toxic equivalency factor (TEF) approach. The strengths and limitations of such an approach are discussed. Field and feeding studies involving terrestrial (such as shrews, voles, and mice), marine (such as whales, dolphins, seals, sea otters, and polar bears), and aquatic (mink and river otters) mammalian species are reviewed. Where possible, tissue-based no observed adverse effect concentrations (NOAECs) and lowest observable adverse effect concentrations (LOAECs) are identified using ecologically relevant endpoints such as reproductive impairment and reduced offspring survivability and/or growth. In rare cases, NOAECs and LOAECs are based on measurement endpoints that are more difficult to interpret in terms of individual, or population level effects, such as alterations in vitamin A and hormone concentrations, immunological parameters, and tissue morphology. Finally, key studies are chosen that are used to derive threshold tissue residue concentrations (geometric means of the relevant NOAECs and LOAECs) for terrestrial, marine, and aquatic mammalian wildlife species. The uncertainties associated with data interpretations, including extrapolations across compounds and across species, are discussed.

15.2   PCBs/PCDDs/PCDFs

15.2.1   PCBs

PCBs are a nonpolar class of 209 individual compounds referred to as congeners, in which one to ten hydrogen atoms of a biphenyl molecule can be substituted for chlorine (Figure 15.1). The physical and chemical properties of PCBs that were advantageous for many industrial purposes, such as high stability, low flammability, inertness, and dielectric properties, led to the international use of PCBs in large quantities (Tanabe 1988). Commercial production of PCBs, primarily by the Monsanto Corporation, began in the United States in 1929 and lasted until 1977 (Kimbrough 1987, 1995, Tanabe 1988, Headrick et al. 1999). PCBs were used in closed-use systems including electrical transformers, capacitors, and heat transfer and hydraulic systems. For a period of time, PCBs also had a number of applications with direct environmental exposure including their use in paints, polymers, and adhesives, as lubricants, plasticizers, fire retardants, immersion oils, vehicles for

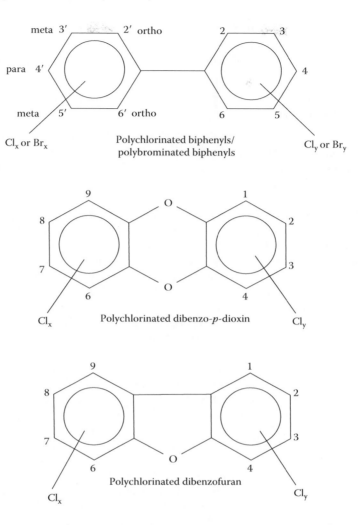

FIGURE 15.1 Structures and numbering of generic polychlorinated/polybrominated biphenyl (PCB/PBB), polychlorinated dibenzo-*p*-dioxin, and polychlorinated dibenzofuran molecules.

pesticide application and as agents for the suspension of pigments in carbonless copy paper (Safe 1990; Headrick et al. 1999). The PCB products that were manufactured by Monsanto in the United States had the trade name Aroclor followed by four digits that identified the particular mixture. Commercial PCB products similar to Monsanto's Aroclor mixtures were produced by manufacturers in Germany (Clophens, Bayer), France (Pheoclors and Pyralenes, Prodelec), Italy (Fenclors, Caffro), and Japan (Kanechlors, Kanegafuchi) (Kimbrough 1987, 1995, Safe 1994).

The estimated cumulative production of PCBs in the United States between 1930 and 1975 was 700,000 tons with an estimated 1.2 million tons produced worldwide. Domestic sales of PCBs in the United States during this time period totaled 627,000 tons (Kimbrough 1987, 1995, Tanabe 1988). As a result of extensive use, PCBs were identified in environmental media and biota as early as the 1960s. After the discovery of their widespread contamination of the environment, PCB production decreased and eventually ceased (Tanabe 1988). In 1971, Monsanto voluntarily stopped production of PCBs for open-ended uses, and subsequently only the lower chlorinated biphenyls were produced (Aroclor 1242 and 1016). In 1977, Monsanto ceased production entirely (Kimbrough 1987, 1995).

Although PCBs are no longer used commercially, because of their persistence about 31% of all the PCBs produced (370,000 tons) is present in coastal sediments and open ocean water. It is estimated that the remaining portion (780,000 tons) is either still in use or deposited in landfills or dumps or is in storage (Tanabe 1988).

15.2.2 PCDDs AND PCDFs

PCDDs and PCDFs are formed as by-products of commercial and natural processes (Safe 1990). PCDDs are composed of two benzene rings connected by two oxygen atom bridges and contain four to eight chlorines, for a total of 75 congeners (Figure 15.1). PCDFs are also composed of two benzene rings fused to a central furan ring, having four chlorine binding sites available on each ring (Figure 15.1). There are 135 different PCDF congeners (Di Carlo et al. 1978, Safe 1990, 1998a, Headrick et al. 1999, Huwe 2002, Mandal 2005, Schecter et al. 2006).

Some of the significant industrial sources of PCDDs and PCDFs included their formation as by-products in the production of PCBs, chlorinated phenols, and chlorinated phenol-derived chemicals, hexachlorobenzene, technical hexachlorocyclohexanes, and chlorides of iron, aluminum, and copper. PCDDs and PCDFs have also been identified in effluents, wastes, and pulp samples from the pulp and paper industry, and in finished paper products. Emissions from municipal and hazardous waste incinerators as well as home heating systems that use wood and coal, diesel engines, forest and grass fires, and agricultural and backyard burning contain PCDDs and PCDFs. Another source might be naturally formed PCDDs and PCDFs, which have been detected in deep soils and clays from the southern United States and Germany (Safe 1990, Huwe 2002).

Historical data suggest that the majority of anthropogenic PCDDs and PCDFs began entering the environment in the 1930s and 1940s with releases peaking in the 1970s. Since the 1970s, emissions have been decreasing (Bhavsar et al. 2008). The US Environmental Protection Agency (U.S. EPA) estimated that annual emissions of PCDDs and PCDFs decreased from 13.5 to 2.8 kg 2,3,7,8-tetrachlorodibenzo-p-dioxin (TCDD) toxic equivalents (TEQ)/year between 1987 and 1995 primarily due to improvements in incinerator performance and removal of incinerators that could not meet emission standards. Other regulations, including bans or restrictions on the production and use of chemicals such as the wood preservative pentachlorophenol (PCP), the phase-out of leaded gasoline that contained halogenated additives, and the elimination of chlorine bleaching in the pulp industry also contributed to reducing concentrations of PCDDs and PCDFs (Huwe 2002). There has been a further 50% decline in emissions between 1995 and 2000 from known sources in the United States (Bhavsar et al. 2008).

15.3 SPATIAL AND TEMPORAL TRENDS OF PCBs/PCDDs/PCDFs IN MAMMALS

Concentrations of PCBs/PCDDs/PCDFs in tissues collected from wildlife species, including mammalian wildlife species, have been used to assess temporal and spatial trends of environmental exposure (Addison et al. 1986, Muir et al. 1992, 1996, 1999, Mossner and Ballschmiter 1997, Muir and Norstrom 2000, Addison and Sobo 2001, Tanabe et al. 2003, Riget et al. 2004, 2006, Aguilar and Borrell 2005, Braune et al. 2005, Verreault et al. 2005, Martin et al. 2006a), thus providing important information on their presence and bioavailability in the environment and the potential for human exposure (Basu and Head 2010). Numerous researchers have reported greater tissue concentration for marine animals foraging around coastal areas associated with large human population centers, heavy industrial activity or in known areas of contamination. Bioaccumulation and biomagnification of these contaminants have been documented in marine mammal species like the harbor seal (*Phoca vitulina*) (Ross et al. 2004). Trends associated with global distribution have also been noted, including differences between hemispheres (Iwata et al. 1993, Muir et al. 1999), which are likely associated with disproportionate production and use of PCBs, cross-continental patterns associated with atmospheric release and deposition of contaminants (Muir and Norstrom 2000,

Anderson et al. 2001), and climate-related patterns thought to be associated with the "cold condensation hypothesis" (Weis and Muir 1997). Mammalian demographic patterns have been reported, including those based on age, sex, and reproductive history. For instance a number of researchers have noted that postreproduction, females of some marine mammalian species have lesser body burdens of PCBs/PCDDs/PCDFs than their similar aged male counterparts (Stern et al. 1994, Wells et al. 2005). The disparity is thought to be due to placental transfer of the chemical(s) to the fetus during pregnancy (Aguilar and Borrel 1994a) and more significantly, a reduction in maternal body burdens during lactation (Tanabe et al. 1994, Debier et al. 2003a, 2003b).

In terms of temporal changes, after an increase in tissue concentrations of PCBs and related chemicals in arctic marine mammals, such as polar bears (*Ursus maritimus*) and ringed seals (*Phoca hispida*) in the 1970s and 1980s, the most commonly noted trend has been a general decline in PCB/PCDD/PCDF tissue concentrations over the last 30 years (Muir and Norstrom 2000). Riget et al. (2004) noted a 78% decline in PCB concentrations in tissue samples of adipose collected from polar bears in Greenland between 1990 and 2000 while Henriksen et al. (2001) noted a 40% decline for 2,2′4,4′,5,5′-hexachlorbiphenyl (PCB 153) in fat of polar bears from Svalbard during the same time period. Similarly, temporal declines of PCBs in tissues of striped dolphins (*Stenella coeruleoalba*) in the Mediterranean Sea (Aguilar and Borrell 2005), ringed seals from the Canadian Arctic (Riget et al. 2004), and gray seals (*Halichoerus grypus*) of the Baltic Sea (Bergman et al. 2003) have all been noted. Since the 1980s, PCB concentrations in Eurasian river otters (*Lutra lutra*) have decreased by 6% to 11% annually (Mason and Madsen 1993, Roos et al. 2001). Martin and associates (2006a) compared contaminant concentrations in liver tissue of American mink (*Mustela vison*) trapped from 1998 to 2003 in the Lake Erie and Lake St. Clair basins to those of mink similarly obtained in 1978/1979. They reported that while concentrations of PCBs and other chlorinated hydrocarbons in mink generally decreased over the past two decades, PCB concentrations tended to increase in western Lake Erie mink over the same time period. In general, the declines in PCBs/PCDDs/PCDFs appear to be greatest in areas previously noted for elevated exposure and are likely the result of institutional controls, including the ban of production and open use of PCBs that occurred in the 1970s and the reduction in the allowable concentrations of PCDDs and PCDFs in smoke stack and waste water emissions. A temporal decline in tissue concentrations is not as apparent in areas that previously had low exposure as well as in areas proximal to countries with lesser institutional controls.

15.4 MECHANISM OF ACTION OF TCDD-LIKE PCBs/PCDDs/PCDFs

Certain approximate stereoisomers of PCBs/PCDDs/PCDFs as well as similarly structured planar polyhalogenated chemicals are known to induce a common suite of effects through a share mechanism of action. This mechanism is mediated by binding of the planar ligand to a specific high-affinity cytosolic protein, known as the aryl hydrocarbon receptor (AhR), which is differentially expressed in target tissues and organs. Mechanistically, the AhR is a ligand activated transcription factor that transduces extracellular signaling via DNA-binding dependent and independent mechanisms. There is a correlation between the AhR binding affinity of these chemicals and their structure. This structure–activity relationship supports the concept that the AhR is involved in the mediation of common responses induced by select PCDD, PCDF, and PCB congeners (Okey et al. 1994, Hahn 1998, 2002, Safe 1998a, Denison et al. 2002, Denison and Nagy 2003, Mandal 2005). While there are other halogenated and nonhalogenated polycyclic aromatic compounds that activate the AhR, the most commonly recognized chemicals in this group includes seven PCDD congeners, 10 PCDF congeners, and 12 PCB congeners. Toxicity and persistence of AhR-active chemicals are determined by structure, with lateral halogen substitutions on two coupled aromatic rings resulting in the most planar configuration and highest degree of toxicity. For the PCDDs and PCDFs, congeners with chlorines in the 2, 3, 7, and 8 positions fall into this category. The TCDD-like PCB congeners are the non-*ortho*- and mono-*ortho*-substituted compounds with no chlorines and one chlorine, respectively, in the *ortho* 2,2′, 6 or 6′ position (Safe 1990, Headrick et al. 1999, Huwe 2002, Mandal 2005, Schecter et al. 2006).

15.5   THE TOXIC EQUIVALENCY APPROACH

The common mechanism of action of TCDD and related compounds facilitates the use of the TEF approach to estimate TCDD-like toxicity of complex mixtures containing chemicals that resemble TCDD. TEFs are used to assess the additive toxicity of PCDDs, PCDFs, and TCDD-like PCB congeners that act through interaction with the AhR (Giesy and Kannan 2002). The TEF value for a TCDD-like congener is defined as the potency of the individual congener relative to TCDD. Using the TEF concept, a TCDD toxic equivalent (TEQ), which is the sum of the product of the concentration of each congener and its respective TEF, can be calculated for any complex mixture containing TCDD-like chemicals to provide an estimation of the total TCDD-like toxicity (Safe 1998a, Van den Berg et al. 1998, 2006, Huwe 2002). While it is likely that environmental mixtures containing PCBs, PCDDs, and PCDFs elicit toxicological responses by multiple mechanisms, it is generally assumed that the most sensitive and detrimental responses are associated with activation of the AhR (Blankenship et al. 2008).

There are some limitations with the TEQ approach. Blankenship et al. (2008) point out that TEFs are consensus values of relative potencies of the various PCB/PCDD/PCDF congeners rather than precise values. They may vary depending upon species, measurement endpoints and the relative proportions of individual congeners in complex mixtures. The assumption that toxic responses to TCDD-like chemicals are additive and that other classes of contaminants do not modify or add to the toxicity clearly is not always correct Safe (1998a, 1998b). In addition, some of the mammalian TEFs are based on *in vitro* studies. These types of studies do not account for whole animal exposure characteristics including absorption and distribution, nor does it account for response mechanisms including the effects of metabolism and elimination on target tissue exposure. Other TEFs are based on quantitative structure–activity relationships because of a lack of toxicity data for some congeners, which can introduce an additional degree of uncertainty into the determination of TEQs. Another limitation of the TEF approach is that it does not consider potential adverse effects of *ortho*-substituted PCB congeners. *Ortho*-substituted PCBs have been shown to elicit a variety of non-AhR-mediated toxic responses in experimental animals including neurobehavioral, neurotoxic, carcinogenic, and endocrine changes. However, studies reviewed by Giesy and Kannan (2002) suggest that it is unlikely that the effects of *ortho*-substituted PCBs are critical to survival at concentrations typically associated with the AhR-mediated toxic effects of PCBs for a number of reasons. In short, Giesy and Kannan (2002) suggest that relatively great concentrations of lesser-chlorinated di-*ortho*-substituted congeners need to accumulate in the brain to cause the observed effects; congeners that are active do not tend to accumulate in brains of animals exposed to complex mixtures of PCBs in the environment; and the most neurotoxic congeners with the greatest potency are less chlorinated and subsequently more easily degraded in the environment, less bioconcentrated and more readily metabolized and excreted.

TEFs are consensus values that intentionally overestimate the relative potency of congeners across a taxonomic class for the express purpose of risk assessment. TEF values are designed to be protective, rather than predictive of thresholds of effects (Blankenship et al. 2008).

In instances where data from the same or related species are available, relative potency factors (RPFs) from the scientific literature may be used in place of TEFs to reduce uncertainty (U.S. EPA 2003, Blankenship et al. 2008). A definition of relative potency is the dose of a reference compound required to cause a particular incidence of a specific toxic response in a specific species divided by the dose of a test compound needed to cause an equal incidence of that same effect in the same or closely related species. In this manner, toxicological assessments for a chemical of concern can be made in terms of another compound about which much is known (Jones 1995).

15.6 FIELD AND FEEDING STUDIES

A number of studies have been conducted pertaining to PCBs, PCDDs, and/or PCDFs involving aquatic, marine, and terrestrial mammalian species. These studies can be categorized as field or laboratory feeding studies. Field studies usually involve the assessment of tissue concentrations of PCBs, PCDDs, and PCDFs, or in some cases, TEQs with subsequent comparison of these concentrations to tissue concentrations associated with deleterious effects as determined in laboratory feeding studies. In a few cases, animals have been sampled for assessment of biochemical, morphological, and/or histological alterations, which have then been correlated to tissue concentrations of contaminants. The limitations inherent in field studies make it difficult to establish a relationship between contaminant concentrations in tissues and deleterious effects. These limitations include: exposure of the animal to confounding contaminants in addition to the specific contaminant(s) of concern; unknown impact of other stressors such as habitat alteration; unknown health and physiological status of the animal; possible deterioration of the animal at the time of tissue collection; possible deterioration of the tissue sample before analysis (Kamrin and Ringer 1996). Laboratory feeding studies offer distinct advantages over field studies in determining cause and effect relationships, but they may not accurately reflect environmental exposure to contaminants and other stressors.

15.7 TERRESTRIAL MAMMALS

Terrestrial mammals generally are used as indicators of pollution near industries and landfills where they have been shown to have greater concentrations of pollutants compared to less contaminated areas (Bremle et al. 1997). Different terrestrial species offer unique characteristics that may make them a preferred sentinel species in a particular location. Such considerations include abundance, home range, diet, and rate of metabolism.

15.7.1 FIELD STUDIES

A number of field studies examining PCB/PCDD/PCDF concentrations in tissues of wild terrestrial mammalian species have been conducted (see later). Of the field studies examined, none correlated specific reproductive or physiological endpoints with tissue concentrations. Instead, general observations were made, including the possibility that perceived population declines or conversely, the lack of observed deleterious effects, were expected based on comparison of field tissue concentrations with results from a laboratory feeding study. The various species examined include the short-tailed shrew (*Blarina brevicauda*), masked shrew (*Sorex cinerus*), meadow vole (*Microtus pennsylvanicus*), meadow jumping mouse (*Zapus hudsonicus*), white-footed deer mouse (*Peromyscus leucopus*), eastern chipmunk (*Tamius striatus*), red squirrel (*Tamiasciurus hudsonicus*) (Blankenship et al. 2005), Japanese field mouse (*Apodemus speciosus*), lesser Japanese mole (*Mogera imaizumii*), raccoon dog (*Nyctereutes procyonoides*) (Kunisue et al. 2006), raccoon (*Procyon lotor*) (Smith et al. 2003), pine martin (*Martes martes*) (Bremle et al. 1997), hedgehog (*Erinaceus europaeus*) (D'Havé et al. 2006), weasel (*Mustela nivalis*), stoat (*Mustela erminea*), polecat (*Mustela putorius*) (Leonards et al. 1998), arctic fox (*Alopex logopus*) (Fuglei et al. 2007), red fox (*Vulpes vulpes*) (Corsolini et al. 1999), wolf (*Canis lupus*) (Carril González-Barros et al. 1997, Gamberg and Braune 1999, Shore et al. 2001), ocelot (*Felis pardalis*) (Mora et al. 2000), European roe deer (*Capreolus capreolus*) (Naso et al. 2004), grizzly bear (*Ursus arctos horribilis*) (Christensen et al. 2007), New World fruit bat (*Pteropus marianus*), Indian pipistrelle bat (*Pipistrellus pipistrellus*), and short-nosed fruit bat (*Cynopterus brachyotis*) (Senthilkumar et al. 2001). It should be pointed out that none of these species has been studied extensively in terms of spatial and/or temporal changes in tissue contaminant concentrations.

15.7.2 Feeding Studies

Although no controlled feeding studies with wild terrestrial mammals were identified, there are laboratory rodent studies that can be used to establish tissue concentrations of a commercial PCB mixture that correspond to ecologically relevant endpoints such as reduced litter size. Linder et al. (1974) conducted a two-generation reproduction study with Sherman rats fed diets containing 0, 1, 5, 20, or 100 mg Aroclor 1254/kg feed. Rats exposed to Aroclor 1254 at dietary concentrations of 20 mg/kg diet or greater had fewer pups per litter compared to controls. Because tissue concentrations were not assessed in this study, a similar chronic rat study by Grant et al. (1974), in which rats were exposed to the same dietary doses of 0, 2, 20, or 100 mg Aroclor 1254/kg for 246 days, was used for hepatic PCB concentrations. In this study, the average hepatic concentration of Aroclor 1254 residues in the rats fed 20 mg Aroclor 1254/kg feed for 246 days was 4.8 mg/kg, wet weight. Thus, the dietary lowest observed adverse effect level (LOAEL) in the Linder et al. (1974) study of 20 mg Aroclor 1254/kg feed corresponds to a tissue-based LOAEC of 4.8 mg Aroclor 1254 residues/kg liver, wet weight. The no observed adverse effect level (NOAEL) in the Linder et al. (1974) study was 5 mg Aroclor 1254/kg feed. Tissue residue concentrations of Aroclor 1254 were not measured in Grant et al. (1974) study at a concentration corresponding to the dietary-based NOAEL in the Linder et al. (1974) study. However, Grant et al. (1974) reported a hepatic concentration of 1.1 mg Aroclor 1254/kg liver, wet weight for a dietary concentration of 2 mg Aroclor 1254/kg feed, which can be used as a reasonable approximation of the NOAEC (Table 15.1).

Similarly, there are laboratory rodent studies that can be used to establish tissue concentrations of TEQs that correspond to deleterious effects on reproduction and offspring survivability. Murray and associates (1979) conducted a three-generation reproduction study to evaluate the effects of chronic ingestion of TCDD in Sprague-Dawley rats. A daily dose of 0.01 μg TCDD/kg body weight/ day was considered to be the LOAEL based on decreases in fertility, litter size at birth, gestational survival and neonatal survival and growth. The NOAEL was identified as 0.001 μg TCDD/kg body weight/day. These daily doses corresponded to dietary concentrations of approximately 200 and 20 ng TCDD/kg feed, respectively, or 200 and 20 ng TEQs/kg feed, using a TEF value of 1.0 for TCDD (Van den Berg et al. 2006). While hepatic concentrations of TCDD were not determined in this study, they were determined in a 2-year chronic toxicity and oncogenicity study in Sprague-Dawley rats that reported the same NOAEL and LOAEL (Kociba et al. 1978). The concentration of TCDD in the liver corresponding to the NOAEL was 510 ng TCDD or TEQs/kg liver, wet weight, and the hepatic concentration corresponding to the LOAEL was 5100 ng TCDD or TEQs/kg liver, wet weight (Table 15.1).

15.8 MARINE MAMMALS

Marine mammals have some of the greatest reported tissue concentrations of PCBs/PCDDs/PCDFs as compared to other mammals (Ross 2000). Predators like seals, dolphins, whales, and polar bears can consume highly contaminated prey, such as fish or seals, making them some of the most highly exposed marine animal species. Marine mammals are long lived, consume great amounts of prey, and transfer their body burdens from one generation to the next through lactation, similar to their terrestrial and aquatic counterparts. In addition, marine mammals often require lipid-rich blubber for buoyancy, insulation, and nutrition to help them survive in the marine environment. The lipid-rich blubber is an ideal matrix for storage and transport of lipophilic contaminants such as PCBs/PCDDs/PCDFs. As a result, the greater relative proportion of lipids on a body mass basis associated with an effective contaminant transport system results in elevated exposures and body burdens. This unique characteristic may exacerbate target tissue contaminant exposure during times of physiological challenge such as illness, extreme environmental conditions, nutritional stress, or pregnancy as blubber stores are metabolized for survival.

TABLE 15.1
No Observed Adverse Effect Concentrations (NOAEC) and Lowest Observed Adverse Effect Concentrations (LOAEC) from Select Toxicity Studies

Study	Critical Effect	NOAEC ΣPCBs	NOAEC TEQ	LOAEC ΣPCBs	LOAEC TEQ
	Rat				
Linder et al. (1974)/Grant et al. (1974)[a]	↓ Litter size	1.1	—	4.8	—
Kociba et al. (1978)/Murray et al. (1979)[a]	Reproduction/survivability/ growth	—	510	—	5100
	Harbor Seal				
Boon et al. (1987)/Brouwer et al. (1989)[b]	↓ Plasma thyroid hormones/↓ plasma retinol	5.2	—	25	—
de Swart et al. (1994)/Ross et al. (1995, 1996a, 1996b)[c]	↓ Immune function	—	90[3]	—	286
	Otter				
Smit et al. (1996)/Murk et al. (1998)[d]	↓ Hepatic retinol/↓ hepatic retinyl palmitate	0.17	42	0.46	210
	Mink				
Restum et al. (1998)	↓ Kit weights	0.07	—	0.98	—
Halbrook et al. (1999)[a]	↓ Kit weights	5.9	—	7.3	—
Heaton et al. (1995a)/Tillitt et al. (1996)[a]	↓ Kit survivability	0.09	17	2.2	226
Bursian et al. (2006a)[a]	Reproduction/survivability/ growth	18	77	—	—
Bursian et al. (2006b, 2006c)[a]	↓ Kit survivability	3.1	50	3.1	189
Bursian et al. (2006a)[e]	Jaw lesion–juveniles	8.1	20	16	52
Bursian et al. (2006c)[e]	Jaw lesion–juveniles	0.73	16	1.7	32
Martin et al. (2006b)[f]	↓ Vitamin A–kits	11	120	19	237
Martin et al. (2006b)[e]	↓ Vitamin A–juveniles	16	52	18	77

[a] NOAECs and LOAECs are based on maternal hepatic concentrations (mg/kg, wet weight).
[b] NOAEC and LOAEC are based on plasma PCB concentrations (mg/kg lipid).
[c] NOAEC and LOAEC are based on blubber TEQ concentrations (ng/kg lipid).
[d] NOAEC and LOAEC (EC_{90}) are based on hepatic PCB (mg/kg, wet weight) and TEQ (ng/kg, wet weight) concentrations.
[e] NOAECs and LOAECs are based on juvenile hepatic concentrations (mg/kg, wet weight).
[f] NOAECs and LOAECs are based on kit hepatic concentrations (mg/kg, wet weight).

15.8.1 FIELD STUDIES

Studies related to marine mammal PCB/PCDD/PCDF exposure and effects are generally surveys of wild populations that have been initiated in response to mortality events or population declines. Only a small number of controlled or semifield exposure studies have been completed to date, largely due to legal, ethical, logistical, and practical limitations. PCBs, PCDDs, and PCDFs have been shown to cause a number of adverse effects in laboratory animals. These include reductions or failure to reproduce and challenges to survival including wasting syndrome, endocrine disruption, reduced immune function, and others. Although an unequivocal causal link has not been established to date, similar effects have been noted in a number of highly exposed marine mammal populations in the wild.

Poor reproduction in dolphins, whales, seals, sea lions, and sea otters have been noted. Reddy et al. (2001) identified significant differences in mean ΣPCB (based on the sum of 10 congeners) blubber concentrations for semicaptive bottlenose dolphins (*Tursiops truncatus*) whose calves survived beyond 6 months as compared to those who were stillborn or died within 12 days. Females with lesser reproductive success had mean ΣPCB concentrations of 14 mg/kg lipid, which was 2.5 fold greater than the more successful group (p = .076). In an examination of demographics of beluga whales (*Delphinapterus leucas*) from the St. Lawrence Estuary, Martineau and colleagues (1987) noted a lesser than expected number of calves and juveniles and suggested that a decreased birth rate or decreased juvenile survival could be the cause because ΣPCB blubber concentrations were as great as 576 mg/kg, wet weight. Helle et al. (1976) reported that an unusually great percentage of wild seals that annually failed to conceive also had significantly greater ΣPCB concentrations (77 mg/kg, wet weight) in their blubber as compared to their pregnant cohorts (56 mg/kg, wet weight). Similarly, a study of common or harbor seals fed fish containing substantial concentrations of PCBs (1.5 mg/kg, wet weight) from the Wadden Sea indicated sharp declines in pup production that appeared to be due to early abortion as compared to seals on a diet containing a lesser PCB concentration (0.22 mg/kg, wet weight) (Reijnders 1986). After a number of researchers noted early termination of California sea lion (*Zalophus californias*) pregnancies on San Miguel Island (California, U.S.), Delong and colleagues (1973) reported mean ΣPCB residues in the blubber (112 mg/kg, wet weight) and liver (5.7 mg/kg, wet weight) of pre-parturient females that were two to eight times greater than ΣPCB concentrations in the same tissues of full-term parturient females.

More specifically, PCBs/PCDFs/PCDDs may be bioactive in the endocrine system or other key biochemical pathways associated with reproduction and homeostasis. Polar bears and seals exposed to elevated concentrations of PCBs and related compounds in the wild are reported to have abnormal concentrations of reproductive hormones (Haave et al. 2003, Oskam et al. 2003). Bone-related effects in highly exposed seals may be associated with changes in vitamin D and thyroid hormones that in turn appear to be correlated to PCB exposure (Routti et al. 2008). In addition, researchers have noted correlations between the prevalence of disease and altered immune function with increased tissue concentrations of PCBs and related compounds in some of these same animals. The correlation was first noticed in studies investigating apparent pathogenically derived mass mortalities in dolphins (Kuehl et al. 1991, Aguilar and Borrell 1994b), and more recently, studies utilizing live free-ranging animals and nondestructive sampling have shown relationships between biochemical measures of immune response and PCB body burdens (Lahvis et al. 1995, Skaare et al. 2000). It should be noted that the earlier studies of wild populations all indicated the presence of cocontaminants, including, in some cases, greatly elevated concentrations of DTT and metabolites that confound these results. However, given the weight-of-evidence, it is highly probable that PCBs and related compounds have played at least some part in noted population declines and select mass mortalities of highly exposed marine mammals.

15.9 FRESHWATER AQUATIC MAMMALS

Many of the field studies and most of the laboratory feeding studies for aquatic mammals have focused on Mustelids. Several species of this family have been used in ecological risk assessments for sites involving aquatic habitats with elevated concentrations of PCBs, PCDDs, PCDFs, and related compounds (Blankenship et al. 2008) and are considered to be an important sentinel species. Because humans and other vertebrate species that share the same environment often have similar responses from exposure to toxic substances, certain of these animal species can be used as surrogates to monitor environmental contaminant exposure and effects. These sentinel species were defined by Grove et al. (2009) as those that are used to evaluate environmental contamination and its implications on environmental health based on their chemical sensitivity, position in the biotic community, exposure potential, and geographic distribution or abundance. A sentinel species must meet certain requirements as presented in Basu et al. (2007). These criteria include (1) widespread distribution, (2) high trophic status, (3) ability to accumulate contaminants, (4) maintained and studied in captivity, (5) captured in sufficient numbers, (6) restricted home range, (7) well-known biology, and (8) sensitive to contaminants.

The identification of the mink as a potential sentinel species can be traced back to 1968 when coho salmon (*Oncorhynchus kisutch*) that were collected from Lake Michigan tributaries during 1967 were incorporated into mink feed that was fed prior to and during the breeding and whelping periods on commercial fur farms, resulting in abnormally great kit mortality (Aulerich et al. 1971, 1973). Studies conducted at Michigan State University by Richard Aulerich and associates confirmed that Lake Michigan coho salmon, when fed to breeder mink at 30% of the diet, caused either complete reproductive failure or increased kit mortality. Chlorinated pesticide (DDT, DDT isomers, and dieldrin) contamination was ruled out by mink feeding experiments that indicated no reproductive effects at dietary concentrations in excess of what was detected in the fish (Aulerich et al. 1971, 1973). Further analysis of the fish indicated the presence of PCBs at concentrations up to 15 mg/kg whole body, wet weight. Subsequent feeding studies by Aulerich and colleagues demonstrated that mink were very sensitive to PCBs. Thirty milligram per kilogram of commercial PCB mixtures in feed was lethal to adult breeders within 4 months, 15 mg/kg feed caused reproductive failure and some adult mortality, and diets containing 10 mg of a commercial PCB mixture/kg feed resulted in reduced weight gain in growing mink. The clinical signs and gross lesions of mink fed diets containing the commercial PCB mixture were very similar to the signs noted in mink fed feed containing coho salmon (Aulerich et al. 1973, Aulerich and Ringer 1977).

15.9.1 FIELD STUDIES

The majority of PCB-related field studies involving aquatic mammalian species have focused on mink (Osowski et al. 1995, Haffner et al. 1998, Harding et al. 1999, Millsap et al. 2004, Beckett et al. 2005, Kay et al. 2005, Martin et al. 2006a, Zwiernik et al. 2008a, Haynes et al. 2009) and the Eurasian and river otter (*Lutra canadensis*) (Kruuk and Conroy 1991, Leonards et al. 1997, Simpson 1997, Sjoasen et al. 1997, Murk et al. 1998, Harding et al. 1999, Roos et al. 2001, Grove and Henny 2008).

In the study by Murk and associates (1998), the relationship between hepatic retinoids and corresponding ΣPCB concentrations (based on the sum of PCBs 26, 52, 101, 118, 138, 153, and 180) and TEQ concentrations (based on concentrations of non-*ortho* and mono-*ortho* PCBs using TEF values described by Ahlborg et al. 1994) in both captive and free-living Eurasian otters was determined. For environmentally exposed otters, a significant, negative correlation was found between hepatic retinol and retinyl palmitate when compared to TEQ concentrations. When expressed on a lipid weight basis, the authors estimated a NOAEC of 4 mg ΣPCBs/kg lipid (1 μg TEQs/kg lipid) and the concentration affecting 90% of the animals (EC_{90}) as 11 mg ΣPCBS/kg lipid (2 μg TEQs/kg lipid). The average hepatic lipid content was reported to be 4.2%, which results in a NOAEC of 0.17 mg ΣPCBs/kg liver, wet weight (42 ng TEQs/kg liver, wet weight) and an EC_{90} of 0.46 mg ΣPCBs/kg liver, wet weight (210 ng TEQs/kg liver, wet weight) (Table 15.1). Hepatic retinoid concentrations were suggested as a sensitive indicator for AhR-related PCB toxicity in otters because of the relationship between a deficiency of vitamin A and an increased risk of infection (Murk et al. 1998).

Harding et al. (1999) reported a significant correlation between hepatic ΣPCB concentrations (as Aroclor 1260) and baculum length in juvenile mink (but not river otter) collected on the Columbia and Frazer river systems in British Columbia, Canada. However, alterations of baculum length were not noted in subsequent laboratory or field studies with similar or greater ΣPCB exposures. Hepatic ΣPCB concentrations in animals in the Harding et al. (1999) study ranged from an estimated 0.01–0.18 mg/kg, wet weight. Feeding 12-week-old male mink a diet containing 3.2 mg Aroclor 1254/kg feed for approximately 5 months did not adversely affect baculum length or mass

(Aulerich et al. 2000). However, while there was a strong correlation between baculum length and mass in control animals, the correlation was weak in PCB-exposed animals suggesting subtle developmental alterations. Similarly, field studies of mink on the Kalamazoo (Millsap et al. 2004) and Tittabawasee (Zwiernik et al. 2008a) rivers (MI, U.S.) found no correlation between hepatic ΣPCB and TEQ concentrations and baculum length or mass, nor did they find significant differences in these parameters between exposed and unexposed mink.

Zwiernik and associates (2008a) assessed a number of endpoints in mink residing in the Tittabawassee River basin (Midland, MI, U.S.). Significant concentrations of PCDFs and PCDDs have been measured in sediments, soil, and biota collected from this area. The resident mink survey (26 animals in the reference area and 22 animals in the study area) indicated no observable adverse effects based on abundance, morphological measurements, sex ratios, population age structure, and gross and histological tissue examination. Hepatic concentrations of PCDDs, PCDFs, and total TEQs were significantly greater in animals collected from the study area compared to the reference animals. The mean total hepatic TEQ concentration in reference area mink was 20 ng/kg, wet weight when compared to 400 ng/kg, wet weight in study area mink. PCDDs, PCDFs, non-*ortho* PCBs, and mono-*ortho* PCBs contributed 21, 290, 80, and 5.8 ng TEQs/kg liver, wet weight, respectively. Of the contaminants analyzed, 2,3,4,7,8-pentachlorodibenzofuran (PeCDF) accounted for 56% of the total TEQs and occurred at concentrations approximately 86-fold greater in animals collected from the study area.

Recently, a lesion that had been described only in ranch mink fed diets containing PCB and PCDD congeners or environmentally derived PCBs/PCDDs/PCDFs was reported in wild-caught animals. The lesion was described as a mandibular and maxillary squamous epithelial proliferation with subsequent osteolysysis (Figures 15.2–15.4). These findings were similar to jaw and skull lesions previously observed in Baltic gray seals (*Halichoerus grypus*) that were heavily contaminated with DDTs and PCBs (Bergman et al. 1992) and thus has attracted attention as a potential endpoint indicative of compromised health. Initially, Render and associates (2000a) observed maxillary and mandibular osteoinvasive squamous epithelial proliferation in 12-week-old mink fed a diet containing 24 µg of 3,3′4,4′5-pentachlorobiphenyl (PCB 126)/kg feed. Gross examination revealed mandibular and maxillary lesions consisting of swollen tissue of the lower and upper jaws with nodular proliferation of the gingiva and loose teeth with increased gingival surface area. Radiographs indicated osteolysis of the maxilla and mandible, and histologic examination documented extensive osteoinvasion by squamous epithelial cells. A subsequent study by Render et al. (2001), in which 6- and 12-week-old mink were fed 24 µg PCB 126 or 2.4 µg TCDD/kg feed, verified induction of maxillary

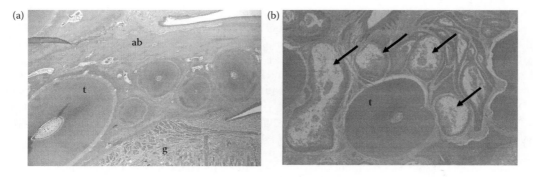

FIGURE 15.2 Mandibular and maxillary squamous epithelial proliferation in mink exposed to PCB 126. (a) Cross section through the mandible of a control mink shown for reference. Note the solid appearing alveolar bone (ab) surrounding the teeth (t), which are above the gingiva (g). (b) Cross section through the mandible of a mink fed a diet containing PCB 126. Note the infiltration of cysts of squamous cells throughout the alveolar bone (arrows), resulting in loosening of teeth.

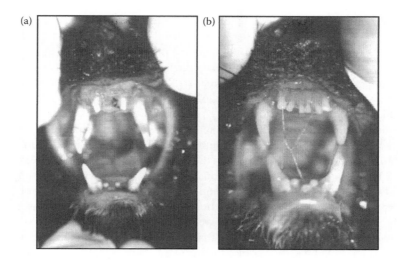

FIGURE 15.3 Dental malalignment and increased interdental spacing in mink induced by consumption of a diet containing 24.0 µg PCB 126/kg feed. (a) Inflammation of gingival tissue and moderate spreading of incisors. (b) Extended canines and missing incisors (mandibular and maxillary); gums are bloody and swollen.

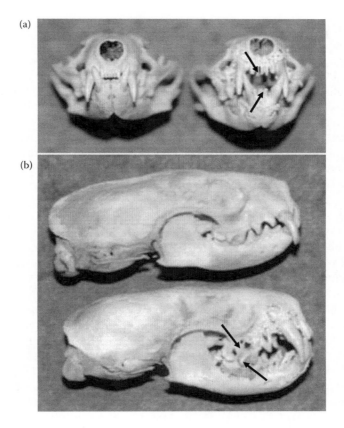

FIGURE 15.4 Osteolysis (arrows) induced by exposure to PCB 126. (a) Front view of skulls from a control mink (left) and a mink fed a diet containing PCB 126 (right). (b) Side view of skulls from a control mink (top) and a mink fed a diet containing PCB 126 (bottom).

and mandibular osteoinvasive squamous epithelial cell proliferation by PCB 126 and demonstrated its induction by TCDD. The latter study also showed that the lesion could be detected histologically after only 2 weeks of dietary exposure. Adult female mink that were fed 5.0 μg TCDD/kg feed for 6 months developed proliferation of squamous epithelial cells (Render et al. 2000b) that resulted in focal loss of alveolar bone or osteolysis. Studies by Render and associates (2000a, 2000b, 2001) suggested that juvenile mink exposed to PCB 126 or TCDD are more susceptible to proliferation of squamous epithelia than adult mink.

In a recent field study, Beckett et al. (2005) reported histological evidence of a jaw lesion characterized by hyperplasia of squamous epithelium in the mandible and maxilla in wild mink collected during an assessment of PCB exposure in the Kalamazoo River basin located in southwestern Michigan (U.S.) (Millsap et al. 2004, Kay et al. 2005). Lesion severity was positively correlated with hepatic ΣPCB (based on 100 congeners) and TEQ concentrations in mink collected from the study area. Hepatic ΣPCB and TEQ concentrations in mink having the lesion ranged from 2.9 to 6.0 mg/kg, wet weight and 0.21–1.3 ng/kg, wet weight, respectively.

Most recently, Haynes et al. (2009) reported that a mink collected near the shore of Lake Ontario within the Rochester Embayment area of concern (New York, U.S.) had histological and gross evidence of mandibular and maxillary squamous epithelial proliferation. This animal, which constitutes the first report of the lesion being detectable grossly in a naturally exposed mink, had a hepatic ΣPCB (based on 155 congeners/coeluters) concentration of 5.9 mg/kg, wet weight and a TEQ concentration based on PCDDs and PCDFs of 21.3 ng/kg, wet weight.

15.9.2 Feeding Studies

Mammalian wildlife-based laboratory feeding studies have focused on mink for a number reasons discussed previously. Mink laboratory feeding studies can be divided into those involving administration of commercial PCB mixtures (Aulerich and Ringer 1977, 1980, Jensen et al. 1977, Bleavins et al. 1980, Aulerich et al. 1985, Hornshaw et al. 1986, Wren et al. 1987a, 1987b, Kihlstrom et al. 1992, Brunstrom et al. 2001, Kakela et al. 2002a, 2002b), administration of discrete PCB, PCDD, or PCDF congeners (Aulerich et al. 1985, 1987, 1988, Hochstein et al. 1988, 1998, 2001, Render et al. 2000a, 2000b, 2001, Beckett et al. 2008, Zwiernik et al. 2008b, Moore et al. 2009, Zwiernik et al. 2009), or administration of environmentally derived PCB/PCDD/PCDF congener mixes by the incorporation of contaminated food items into the test diet (Aulerich et al. 1971, 1973, Platonow and Karstad 1973, Hornshaw et al. 1983, Heaton et al. 1995a, 1995b, Tillitt et al. 1996, Restum et al. 1998, Halbrook et al. 1999, Bursian et al. 2006a, 2006b, 2006c).

While the use of commercial PCB mixtures diminishes the potential of exposure to other contaminants, this exposure scenario does not accurately reflect the compositions of the mixtures that humans and wildlife are exposed to because of differences in metabolism and/or biodegradation rates of individual congeners comprising commercial PCB mixtures. An extensive discussion of this issue can be found in Giesy and Kannan (2002).

Similarly, the use of individual PCB/PCDD/PCDF congeners or small groups of congeners can provide important information related to the toxicokinetics and toxicodynamics of the specific congener. However, exposures do not accurately reflect what the animal encounters in the environment.

Incorporation of contaminated food items collected from the environment into the test diet alleviates the concerns about environmental relevancy, but cocontaminants such as chlorinated pesticides complicate the identification of causality. In addition, the contaminated food items are usually collected from a single site, thus results from a feeding study of this nature tend to be site-specific and may not be applicable to other locations because of differences in congener profiles. However, if a congener-specific analysis of the feed as well as tissues is conducted, then the relative contribution of individual congeners to the toxic potency of the mixture can be accounted for by using the TEQ approach, keeping in mind the limitations associated with the TEQ approach that were discussed previously.

15.9.2.1   Effects of Commercial PCB Mixtures

In an early study by Aulerich and colleagues (1973), three male and 12 female mink were fed diets containing 30% ocean fish (control), 30% Lake Michigan coho salmon, and 30% ocean fish plus 30 mg of commercial PCB mixtures/kg feed (10 mg/kg feed each provided by Aroclors 1242, 1248, and 1254) for 180 days beginning 2 months prior to breeding (January 1). None of the females on the coho salmon diet and Aroclor-supplemented diet whelped. Six of 15 mink on the coho salmon diet and all of the animals on the 30 mg Aroclor/kg feed treatment died between the beginning of breeding (March 1) and the end of the whelping period (mid-May). The clinical signs and lesions in mink that died on diets containing coho salmon or supplemental PCBs were similar and consisted of anorexia, bloody stools, fatty infiltration, and degeneration of the liver and kidneys, and hemorrhagic gastric ulcers. Selected tissues from animals that died on trial were analyzed for PCB residues. Tissue concentrations were equivalent in animals fed diets containing 30% Lake Michigan coho salmon and 30 mg Aroclors/kg feed. The greatest average ΣPCB concentration occurred in the brain (11 mg/kg, wet weight), which was approximately twofold greater than concentrations reported for the other tissues including the liver.

In a subsequent mink feeding study utilizing commercial PCB mixture (Aulerich and Ringer 1980), three male and 12 female mink were assigned to one of four treatment groups each (control or 2.0, 10, or 25 mg Aroclor 1016/kg feed) beginning 2 months prior to the breeding season through a second breeding season 18 months later. Aroclor 1016 (made by redistilling Aroclor 1242 to remove some of the more highly chlorinated PCB congeners) was introduced in 1970 as an environmentally acceptable replacement for Aroclor 1242. Reproduction was not adversely affected although kit body weights at 4 weeks of age were significantly less than control body weights in the 25 mg/kg feed group. Tissue concentrations of Aroclor 1016 in the adult females were greatest in adipose tissue (8.0 mg/kg, wet weight) with liver concentrations averaging 0.73 mg/kg, wet weight. Adipose and hepatic residue concentration in the females fed 10 ppm Aroclor 1016 averaged 5.2 and 0.41 mg/kg, wet weight, respectively. The authors noted that PCB residues in mink fed Aroclor 1016 were considerably less than those from mink fed comparable concentrations of a combination of Aroclors 1242, 1248, and 1254 (Aulerich et al. 1973, Aulerich and Ringer 1977), perhaps explaining their relative tolerance of dietary Aroclor 1016.

Adult mink (four males and 12 females) were fed diets containing 1.0 mg Aroclor 1254/kg feed alone or in combination with 1.0 mg methyl mercury/kg feed for a period of 184 days that included reproduction and lactation (Wren et al. 1987a, 1987b). A third treatment group was fed a diet containing 0.50 mg Aroclor 1254 plus 0.50 mg methyl mercury/kg feed. The diet containing 1 mg Aroclor 1254/kg feed had no significant effect on reproductive parameters, but growth rates of kits were significantly reduced at 3 and 5 weeks of age compared to controls. The NOAEC and LOAEC based on maternal hepatic PCB concentrations in the control and 1.0 mg Aroclor 1254/kg feed groups are thus 0.13 and 3.1 mg/kg, wet weight, respectively.

Kihlstrom et al. (1992) conducted a study in which a group of adult female mink was dosed with 1.64 mg Aroclor 1254/kg body weight/day beginning approximately 1 month prior to breeding until 5 days postwhelping. If a food consumption rate of 0.15 kg/day and body weight of 1.0 kg is assumed (U.S. EPA 1995), this dose is equivalent to an adult female mink consuming feed containing approximately 11 mg Aroclor 1254/kg. These animals had a significantly reduced whelping rate with no live kits being born. The concentration of Aroclor 1254 in the adipose tissue averaged 74 mg/kg, wet weight.

15.9.2.2   Effects of Environmentally Derived PCBs, PCDDs, and PCDFs

Halbrook et al. (1999) conducted a study in which adult female mink were fed diets containing 25%, 50%, or 75% fish collected from Poplar Creek (Tennessee, U.S.) 2 months prior to breeding through weaning of kits at 6 weeks of age. The diets provided 0.52, 1.01, or 1.86 mg ΣPCBs/kg feed with

Aroclor 1260 being the dominant Aroclor detected in both the fish and resulting feed. Because a congener-specific analysis was not conducted on feed and tissue samples, dietary and tissue TEQ concentrations are not available. Six-week-old male kits in the 1.86 mg ΣPCB/kg feed group had body weights that were significantly less than control body weights, resulting in a dietary NOAEL and LOAEL of 1.01 and 1.86 mg ΣPCBs/kg feed, respectively. The LOAEC was 7.3 mg ΣPCBs (as Aroclor 1260)/kg liver, wet weight. In this particular study, hepatic ΣPCB concentrations were not determined in the other treatment groups, but concentrations in adipose tissue were. The hepatic concentration of ΣPCBs in the 1.86 mg ΣPCB/kg feed group was 5.6% of the adipose tissue concentration. If this relationship is applied to adult females in the 1.01 mg ΣPCBs/kg feed group (the dietary NOAEL), then the NOAEC approximates 5.9 mg ΣPCBs/kg liver, wet weight (Table 15.1).

Restum and associates (1998) administered male and female mink (eight males and 16 females per group) diets containing fish collected from Saginaw Bay (Lake Huron, Michigan, U.S.) that provided ΣPCB concentrations (based on a multiple regression pattern analysis program used to quantify PCBs as a linear combination of Aroclors as described in Heaton et al. [1995a]) of 0.25, 0.50, or 1.0 mg/kg feed beginning approximately 2 months prior to breeding. Figure 15.5 provides a timeline for this two-generation reproduction study in which a portion of the animals were switched from their respective PCB-containing diets to the control diet. Because a congener-specific analysis was not conducted on feed and tissue samples, dietary and tissue TEQ concentrations are not available. Table 15.2 summarizes the critical effects that were reported in this study. The authors identified a dietary LOAEL of 0.25 mg ΣPCBs/kg feed based on reduced F_1-1 kit body weights at three and six weeks of age. Plasma and hepatic ΣPCB concentrations in these animals averaged 0.04 and 0.64 mg/kg, wet weight, respectively. Concentrations of ΣPCBs in plasma and liver of parental females continuously exposed to 0.25 mg ΣPCBs/kg feed for 18 months were 0.06 and 0.98 mg/kg, wet weight, respectively. Hepatic ΣPCB concentrations in the control F_1-1 kits and parental females averaged 0.07 mg/kg, wet weight (Table 15.1).

In a study reported by Heaton et al. (1995a, 1995b) and Tillitt et al. (1996), adult mink (three males and 12 females per group) were fed diets containing 10%, 20%, or 40% carp collected from Saginaw Bay (Lake Huron, Michigan, U.S.), which provided 0.72, 1.53, or 2.56 mg ΣPCBs/kg feed (based on a multiple regression pattern analysis program used to quantify PCBs as a linear combination of Aroclors), respectively, or 16.8, 32.8, or 65.7 ng TEQs/kg feed (recalculated using TEF

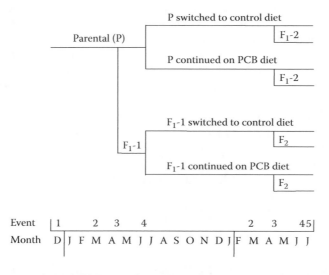

FIGURE 15.5 Exposure regimen for P, F_1-1, F_1-2, and F_2 generation mink exposed to various concentrations of PCBs derived from Saginaw Bay carp. Event numbers: 1 = study initiation (12/23/1991); 2 = breeding; 3 = whelping; 4 = weaning; 5 = study termination (7/1/1993). (From Restum, J. C. et al., J. Toxicol. Environ. Health Part A, 54, 343–375, 1998. With permission.)

TABLE 15.2
Summary of Critical Effects in Restum et al. (1988) Study

Generation[a]	Treatment[b]	Critical Effect
F_1-1	1.0	↓ Kit survivability
F_1-2	1.0 → 0	↓ Kit survivability
F_1-2	0.5 → 0.5	↓ Kit survivability
F_1-2	1.0 → 1.0	↓ Kit survivability
F_2	0.5 → 0.5	↓ Kit survivability
F_2	1.0 → 1.0	↓ Kit survivability
F_1-1	0.25	↓ Kit body weight
F_1-1	0.5	↓ Kit body weight
F_1-1	1.0	↓ Kit body weight
F_1-2	1.0 → 1.0	↓ Kit body weight
F_2	0.5 → 0.5	↓ Kit body weight
F_2	1.0 → 1.0	↓ Kit body weight

[a] F_1-1 animals were whelped (mid-April to mid-May, 1992) by adult females (P) fed diets containing PCBs derived from fish collected from Saginaw Bay, MI, U.S. At weaning of F_1-1 kits at 6 weeks of age, half of the P and F_1-1 animals were switched to the control diet and the remaining half were maintained on their respective diets containing 0.25, 0.50, or 1.0 mg ΣPCBs/kg diet. During the next breeding season (March, 1993), animals were bred within their respective age and treatment groups. F_1-2 kits were the second set of animals whelped by P females and F_2 kits were whelped by F_1-1 females.

[b] Numbers refer to parental treatment groups. For example, 1.0 → 0 refers to the situation where the parent was switched from 1.0 mg ΣPCBs/kg feed to the control diet prior to whelping while 1.0 → 1.0 signifies that the parent was in the 1.0 mg ΣPCBs/kg feed treatment group for the duration of the feeding trial.

values presented in Van den Burg et al. [2006]), respectively. A dietary concentration of 0.72 mg ΣPCBs/kg feed (16.8 ng TEQs/kg feed) resulted in a decrease in kit body weights at 3 and 6 weeks of age as well as reduced kit survivability at the two time points. The greatest feed concentration resulted in no viable kits within 24 h of whelping. The dietary LOAEL in this study is 0.72 mg ΣPCBs/kg feed (16.8 ng TEQs/kg feed) and the LOAEC is 2.19 mg ΣPCBs/kg liver, wet weight (226 ng TEQs/kg liver, wet weight) based on tissue concentrations in adult females. The authors reported a NOAEL and NOAEC based on ΣPCB and TEQ concentrations in control feed and livers from adult females in the control group. The control diet contained 40% ocean fish scraps that resulted in 0.015 mg ΣPCBs/kg feed or 0.7 ng TEQs/kg feed. Hepatic ΣPCB and TEQ concentrations in adult female controls were 0.09 mg/kg liver, wet weight and 17.4 ng/kg liver, wet weight, respectively (Table 15.1).

A subsequent study by Bursian et al. (2006a) utilized fish collected from the mouth of the Saginaw River (Michigan, U.S.), which empties into Saginaw Bay (Lake Huron, Michigan, U.S.), post sediment-remediation and 14 years after collection of fish used in the Heaton et al. (1995a, 1995b)/Tillitt et al. (1996) and Restum et al. (1998) studies. In this study, adult female mink (10 per treatment) were fed diets containing 10%, 20%, or 30% carp that provided 0.83 1.1 or 1.7 mg ΣPCBs/kg feed, respectively, and 22, 36, or 57 ng TEQs/kg feed (recalculated using TEF values presented in Van den Burg et al. [2006]), respectively. Animals were exposed beginning 3 weeks prior to breeding through weaning of their offspring at 6 weeks of age (119 days). Eight kits from each treatment group were continued on their respective diets through 27 weeks of age (an additional 147 days). Consumption of diets containing up to 1.7 mg ΣPCBs/kg feed (57 ng TEQs/kg feed) had no significant effect on mink reproduction and kit survivability. Although maternal hepatic ΣPCB and TEQ concentrations were not assessed in this study, if it is assumed that the hepatic concentrations in the adult females were similar to concentrations in the 27-week-old juveniles as has been seen in other studies (Bursian et al. 2006c), then the NOAEC would be 18 mg ΣPCBs/kg liver, weight wet and 77 ng TEQs/kg liver, wet weight (Table 15.1). Juvenile mink developed histologically identified maxillary and mandibular squamous epithelial proliferation at the two greatest dietary concentrations (1.1 and 1.7 mg ΣPCBs/kg feed or 36 and 57 ng TEQs/kg feed). Corresponding liver concentrations were 16 and 18 mg ΣPCBs/kg, wet weight or 52 and 77 ng TEQs/kg, wet weight. The dietary no observed effect level (NOEL) based on the histologically identified jaw lesion is 0.83 mg ΣPCBs/kg feed (22 ng TEQs/kg feed) and the no observed effect concentration (NOEC) is 8.1 mg ΣPCBs/kg liver, wet weight (20 ng TEQs/kg liver, weight wet) (Table 15.1).

Mink kits and juveniles from the aforementioned Saginaw River feeding study were assessed for changes in thyroid and vitamin A status (Martin et al. 2006b). No evidence of overt thyroid toxicity was reported, but plasma retinol and total ester concentrations were decreased in both kits and juveniles exposed to 1.7 mg ΣPCBs/kg feed (57 ng TEQs/kg feed). This results in a dietary-based NOEL and LOEL of 1.1 and 1.7 mg ΣPCBs/kg feed or 36 and 57 ng TEQs/kg feed. The NOEC and LOEC for kits are 11 and 19 mg ΣPCBs/kg, wet weight or 120 and 237 ng TEQs/kg, wet weight, respectively. The NOEC and LOEC for the juveniles are 16 and 18 mg ΣPCBs/kg, wet weight or 52 and 77 ng TEQs/kg, wet weight (Table 15.1).

A study similar to the Saginaw River study (Bursian et al. 2006a) was conducted utilizing fish collected from the Housatonic River (Massachusetts, U.S.) (Bursian et al. 2006b, 2006c). Twelve adult female mink per group were fed diets containing varying percentages of fish that provided dietary concentrations of ΣPCBs (based on the sum of concentrations of 100 congeners) ranging from 0.34 to 3.7 mg/kg feed or 2.5 to 50.5 ng TEQs/kg feed (recalculated using TEF values presented in Van den Burg et al. [2006]). Exposure of adults began approximately two months prior to breeding through weaning of kits at six weeks of age (160 days). Twelve kits per group were maintained on their respective treatments through 31 weeks of age (an additional 168 days). Mink kits exposed to 3.7 mg ΣPCBs/kg feed (50.5 ng TEQs/kg feed) *in utero* and during lactation had reduced survivability between 3 and 6 weeks of age. This dietary concentration resulted in a maternal hepatic concentration of 3.1 mg ΣPCBs/kg, wet weight or 189 ng TEQs/kg, wet weight). Using kit survivability as an environmentally relevant endpoint, the dietary NOAEL is 1.6 mg ΣPCBs/kg feed (12.1 ng TEQs/kg feed) and the NOAEC is 3.1 mg ΣPCBs/kg liver, wet weight (50.0 ng TEQs/kg liver, wet weight) (Table 15.1). Mandibular and maxillary squamous cell proliferation was apparent in 31-week-old juveniles exposed to as low as 0.96 mg ΣPCBs/kg feed (6.6 ng TEQs/kg feed). Animals in this exposure group had hepatic concentrations of 1.7 mg ΣPCBs/kg, wet weight and 32.2 ng TEQs/kg, wet weight. The dietary NOEL based on the jaw lesion is 0.61 mg ΣPCBs/kg feed (4.2 ng TEQs/kg feed) and the NOEC is 0.73 mg ΣPCBs/kg liver, wet weight (15.7 ng TEQs/kg liver, wet weight) (Table 15.1).

15.10 THRESHOLD TISSUE RESIDUE CONCENTRATIONS

Threshold tissue residue concentrations are presented in the following sections for terrestrial, marine and aquatic mammals on both a ΣPCB and TEQ basis. The threshold tissue residue concentration is the geometric mean of the NOAEC and LOAEC for a specific effect from a specific study or from a group of closely related studies (Tillitt et al. 1996, Kannan et al. 2000). This value is a conservative endpoint between the NOAEC, which is over-protective, and the LOAEC, which is under-protective and is assumed to represent a tissue concentration below which the specific effect in the species of concern will not occur. Depending upon the specific application, some of the values presented here could be used as a toxicity reference values (TRVs).

A TRV is a chemical concentration expressed as an administered dose or as a media concentration that is used in conjunction with an exposure prediction to estimate health hazard or ecological risk

(USACHPPM 2000). There are several possible approaches to derive tissue-based TRVs (Sample et al. 1996, Blankenship and Giesy 2002, U.S. EPA 2003). These include the use of a single definitive study when one study is significantly more scientifically defendable than all others available, or the use of multiple studies to define a range or develop a multistudy dose–response curve. Ideally, TRVs are derived from chronic toxicity studies in which a dose–response relationship has been observed for ecologically relevant endpoints in the species of concern, or a closely related species (Sample et al. 1996, U.S. EPA 1997). Furthermore, the duration of exposure should include sensitive life stages to assess potential developmental and reproductive effects and there should be minimal impact of cocontaminants (Blankenship et al. 2008).

In deriving a TRV, it is essential to critically evaluate the applicability of the data to the site-specific receptors of concern and exposure pathways. TRVs derived in the same species are generally not available for the majority of wildlife receptors. Thus, it is often necessary to derive TRVs using toxicological data for surrogate species in combination with uncertainty factors (UFs) to account for uncertainty associated with factors such as species differences, different laboratory endpoints and differences in experimental design (Blankenship et al. 2008). For this reason, only the threshold tissue residue concentrations based on mink offspring survivability could be used as TRVs for this species without further modification.

15.10.1 Terrestrial Mammal Threshold Tissue Residue Concentrations

Due to its insectivorous diet, relatively great ingestion rate and metabolism per unit of body weight (Lawlor 1979), relatively small home range (varies from 0.39 to 0.96 ha; Buckner 1966, Faust et al. 1971), and its relatively short lifespan (~1 year; Russell 1998), the short-tailed shrew has been recommended as a sentinel species of contaminant exposure based on its trophic level and value as a biomonitor of the local biome (Talmage and Walton 1991). The shrew is often selected as a surrogate species for terrestrial mammalian wildlife that would be expected to have maximal exposure to ΣPCBs in the soils. A threshold tissue residue concentration for shrews was developed from the two rodent studies by Linder et al. (1974) and Grant et al. (1974) in which a hepatic Aroclor 1254 concentration of 4.8 mg/kg liver, wet weight was the LOAEC and 1.1 mg Aroclor 1254/kg liver, weight wet was the NOAEC based on reduced litter size. The geometric mean of the NOAEC and LOAEC resulted in a threshold tissue concentration of 2.3 mg Aroclor 1254/kg liver, wet weight. When expressed on a lipid weight basis (assuming the liver contains 5% lipid [Poole et al. 1995]), the threshold tissue residue concentration is 46 mg Aroclor 1254/kg lipid (Table 15.3). A TEQ-based threshold tissue residue concentration of 1613 ng TEQ/kg liver, wet weight was derived using a NOAEC of 510 ng TEQ/kg liver, wet weight and a LOAEC of 5100 ng TEQ/kg liver, wet weight from the two rodent studies by Kociba et al. (1978) and Murray et al. (1979). The TEQ-based threshold tissue residue concentration expressed on lipid weight basis is 32 μg TEQ/kg lipid (Table 15.3).

15.10.2 Marine Mammal Threshold Tissue Residue Concentrations

Despite calls for controlled studies on captive marine mammals (Marine Mammal Commission 1999), ethical, logistical, and practical difficulties have resulted in a limited data set for only a few individuals. Thus, threshold tissue residue concentrations must rely largely on other mammalian data sets with a focus on AhR associated responses.

In a review by Kannan et al. (2000), threshold tissue residue concentrations were derived for marine mammal exposure to PCBs and related compounds in terms of ΣPCBs and TEQs using three data sets: (1) data associated with semicontrolled exposure studies of marine mammals, (2) exposure studies in Eurasian otters, and (3) PCB exposure studies in mink. TEQs used by Kannan et al. (2000) are those presented in the original citations and do not reflect the most recent TEF values reported by Van den Berg et al. (2006). Controlled studies of mink and otter were selected to

TABLE 15.3
Threshold Tissue Residue Concentrations of PCBs (mg/kg lipid) and Toxic Equivalents (TEQ; ng/kg lipid) Based on the Geometric Mean of the NOAEC and LOAEC

Mammal Type	Tissue Type	ΣPCBs	TEQs
Terrestrial	Liver	46[a]	32,255[b]
Marine	Liver and blood	8.6[c]	1292[c]
	Blubber	17[d]	160[e]
Aquatic	Liver	41[f]	1540[f]

[a] NOAEC and LOAEC were used to derive the geometric mean based on liver wet weight concentrations in Grant et al. (1974) and associated effect concentrations noted in Linder et al. (1974). The lipid content of the liver is assumed to be 5%.

[b] The NOAEC and LOAEC used to derive the geometric mean were based on effects data from Murray et al. (1979) and tissue residue concentration data from Kociba et al. (1978) who examined the effects of TCDD on rats. The lipid content of the liver is assumed to be 5%.

[c] Geometric mean of NOAECs and LOAECs for seals, otter and mink based on lipid normalized blood and liver.

[d] The threshold tissue residue concentration ΣPCBs for marine liver in blood was multiplied by 2 based on Reddy et al. (1998) from Kannan et al. (2000).

[e] Value based on seal data derived by Kannan et al. (2000).

[f] Geometric mean of NOAECs and LOAECs are based on lipid normalized mink hepatic concentrations from Heaton et al. (1995a)/Tillitt et al. (1996) and Bursian et al. (2006a, 2006b, 2006c).

supplement the limited marine mammal data because these species are piscivorous mammals with a similar reproductive physiology. NOAEC and LOAEC values for toxic effects of PCBs and related compounds in seals, dolphins, otter, and mink were compared and used to derive threshold tissue residue concentrations. A seal blood-based threshold tissue residue concentration was derived using data from a study that assessed the effects of feeding captive harbor seals PCB-contaminated fish on plasma retinol and thyroid hormone concentrations (Boon et al. 1987, Brouwer et al. 1989) (Table 15.1). Kannan et al. (2000) identified a NOAEC of 5.2 mg ΣPCBs/kg lipid and a LOAEC of 25 mg ΣPCBs/kg lipid, resulting in a threshold concentration of 11 mg ΣPCBs/kg lipid. In a semifield study conducted with harbor seals that assessed the effects of feeding PCB-contaminated fish on immunological parameters (de Swart et al. 1994, Ross et al. 1995, 1996a, 1996b) (Table 15.1), a blubber-based threshold tissue residue concentration of 160 ng TEQs/kg lipid (Table 15.3) was derived using a NOAEC of 90 ng TEQs/kg lipid and a LOAEC of 286 ng TEQs/kg lipid. A semifield study that examined hepatic retinoids and corresponding ΣPCB concentrations in environmentally exposed feral and captive otter (Smit et al. 1996, Murk et al. 1998) (Table 15.1) was used by Kannan et al. (2000) to derive tissue-based threshold ΣPCB and TEQ concentrations in otters. The NOAEC, based on a reduction in hepatic retinol concentrations, was 4 mg ΣPCBs/kg lipid (1000 ng TEQs/ kg/lipid) and the LOAEC was 11 mg ΣPCBs/kg lipid (2000 ng TEQs/kg lipid). The geometric mean of the NOAEC and LOAEC resulted in a threshold tissue residue concentration of 6.6 mg ΣPCBs/ kg lipid or 1400 ng TEQs/kg lipid. The Heaton et al. (1995a)/Tillitt et al. (1996) mink feeding study (Table 15.1) was used by Kannan et al. (2000) to derive a threshold tissue residue concentration for this species. The NOAEC and LOAEC, based on reduced kit survivability, were 1.8 and 44 mg ΣPCBs/kg lipid (220 and 6480 ng TEQ/kg lipid), respectively, resulting in threshold tissue residue concentrations of 8.9 ΣPCBs/kg lipid and 1193 ng TEQ/kg lipid. The geometric mean of the threshold tissue residue concentrations for seal blood, seal blubber, otter liver, and mink liver was then calculated to derive an overall threshold tissue concentration for aquatic animals. This value is 8.6 mg ΣPCBs/kg lipid or 644 ng TEQ/kg lipid. However, if separate threshold tissue residue concentrations are derived for blood/liver and blubber, then the values for blood/liver would be 8.6 ΣPCBs/kg lipid and 1292 ng TEQs/kg lipid and the values for blubber would be 17 mg ΣPCBs kg lipid (8.6 mg ΣPCBs/kg lipid in blood/liver × 2 based on the twofold difference in lipid normalized concentrations for PCBs in blood and blubber of bottlenose dolphins reported by Reddy et al. 1998) and 160 ng TEQs/kg lipid (Table 15.3).

Schwacke et al. (2002) provided a probabilistic risk assessment of reproductive effects on bottlenose dolphins. The authors incorporated the Kannan et al. (2000) threshold tissue residue concentrations to assess hazard and compared it to a probabilistic approach whereby measured tissue concentrations were compared with dose–response data from mink studies. For this approach, a threshold value (EC_{10}) was estimated on the basis of a fitted dose–response model and resulted in a threshold tissue residue concentration of 15 mg ΣPCBs/kg lipid with a 95% confidence limit (CL) of 11–18 mg/kg lipid.

Both the Kannan et al. (2000) and Schwacke et al. (2002) threshold tissue residue concentrations relied heavily on the Heaton et al. (1995a)/Tillitt et al. (1996) mink feeding study described in detail above. Since 2002, two additional mink feeding studies of high quality have been published (Bursian et al. 2006a, 2006b, 2006c) that are also described earlier. If these data are incorporated into the Kannan et al. (2000) and Schwacke et al. (2002) analyses, a slightly less conservative blubber-specific threshold tissue residue concentration would result.

15.10.3 AQUATIC MAMMAL THRESHOLD TISSUE RESIDUE CONCENTRATIONS

As stated previously, ecologically relevant endpoints are generally those considered to relate to reproductive performance, survival, and growth (Blankenship et al. 2008). Using kit survivability as the most ecologically relevant endpoint, there are three mink feeding studies that provide data related to this endpoint: the Heaton et al. (1995a)/Tillitt et al. (1996) Saginaw Bay fish feeding study, the Bursian et al. (2006a) Saginaw River fish feeding study and the Bursian et al. (2006b, 2006c) Housatonic River fish feeding study. The designs of the Saginaw River (Bursian et al. 2006a) and the Housatonic River (Bursian et al. 2006b, 2006c) studies are essentially identical and the two studies were conducted only a year apart, whereas the Saginaw Bay study (Heaton et al. 1995a/ Tillitt et al. 1996) was conducted 14 years previously using analytical procedures appropriate for the time. In comparing the Saginaw River study to the Housatonic River study, the latter had both NOAEC and LOAEC values while the greatest dose in the Saginaw River study was the NOAEC. Thus, if the Housatonic River study is used as the critical study, the threshold tissue residue concentration (calculated as the geometric mean of the NOAEC and LOAEC [Tillitt et al. 1996]) for reduced mink kit survivability is 3.1 mg ΣPCB/ kg liver, wet weight or 97 ng TEQs/kg liver, wet weight. Expressed on a lipid weight basis (assuming a liver lipid content of 5% [Poole et al. 1995]), the corresponding values are 62 mg ΣPCBs/kg lipid and 1940 ng TEQs/kg lipid. If a threshold tissue residue concentration is calculated using the NOAECs and LOAECs from all three studies, then this value would be 2.0 mg ΣPCBs/kg liver (77 ng TEQs/kg liver), or, if expressed on a lipid weight basis, 41 mg ΣPCBs/kg lipid (1540 ng TEQs/kg lipid) (Table 15.3).

Blankenship et al. (2008) recommended tissue-based TRVs based on the Saginaw River and Housatonic River studies (Bursian et al. 2006a, 2006b, 2006c). These values are 50–70 ng TEQs/kg liver, wet weight for the NOAEC and 189 ng TEQs/kg liver, wet weight for the LOAEC. The geometric mean of these values is 90 ng TEQs/kg liver or 1800 ng TEQs/kg lipid, assuming a hepatic lipid concentration of 5%.

A few studies have modeled data from existing mink feeding studies to derive median effect concentrations (EC_{50}) for reproductive effects. Leonards et al. (1995) used a one-compartment bioaccumulation model to estimate whole-body concentrations of ΣPCBs and TEQs in mink in relation to changes in litter size and kit survival. The authors proposed a critical body residue (EC_{50}) for reduced litter size expressed on a ΣPCB basis of 1.2 mg/kg body weight. When expressed on a TEQ basis (using TEFs proposed by Safe [1987, 1993]), the EC_{50} value for reduced litter size was 160 ng TEQs/kg body weight and 200 ng TEQs/kg body weight for kit survival. Fuchsman et al. (2008) conducted a quantitative analysis of more than 50 mink feeding studies that assessed the reproductive effects of PCBs to determine how effectively different methods of measuring and assessing PCB concentrations approximated a toxicologically relevant dose for this endpoint. They reported EC_{50} values of 1.5 mg ΣPCBs/kg whole body (corresponding to 1.3 mg ΣPCBs/kg liver, wet weight), 25 ng TEQs/kg whole body (using TEF values in Van den Berg et al. [2006]) and 19 ng TEQs/kg whole body (using TEFs applicable to internal doses of PCBs as reported by DeVito et al. [1997, 2000] and Tillitt et al. [1996]). Blankenship et al. (2008) cautioned against combining information from more than one study, in that the approach assumes that the studies are of equal quality and of similar methodology.

It is difficult to choose a single tissue concentration of ΣPCBs or TEQs at which deleterious effects would be expected. Studies involving administration of commercial PCB mixtures or pure PCB/PCDD/PCDF congeners do not accurately reflect environmental exposures. On the other hand, studies evaluating the effects of environmentally weathered mixtures of PCBs, PCDDs, and PCDFs have inherent limitations that need to be considered. Environmental mixtures of PCBs, PCDDs, and PCDFs have congener profiles that tend to be site-specific, thus the potency of the mixture can change from site to site or temporally. For example, the NOAEC for the Saginaw Bay study (Heaton et al. 1995a/Tillitt et al. 1996) was 0.09 mg ΣPCBs/kg liver, wet weight and the NOAEC in the Saginaw River study (Bursian et al. 2006a), which was conducted 14 years later, was eightfold greater at 18 mg ΣPCBs/kg liver, wet weight (Table 15.1). The NOAEC expressed as TEQs was almost fivefold greater in the latter study (77 ng TEQs/kg liver, wet weight vs. 17 ng TEQs/kg liver, weight wet) (Table 15.1). Examination of hepatic congener profiles in both studies indicates that while the percentage of PCDD and mono-*ortho* PCB congeners were similar between the two studies (14% and 3%), the percentage of non-*ortho* PCB congeners in the liver decreased from 54% to 34%, and the percentage of PCDF congeners increased from 29% to 50% (Table 15.4). Thus, in the earlier Saginaw Bay study (Heaton et al. 1995a/Tillitt et al. 1996), non-*ortho* PCB congeners contributed a majority of the hepatic TEQs, while in the latter Saginaw River study (Bursian et al. 2006a), PCDF congeners contributed a majority of the hepatic TEQs. The comparison of the NOAECs and hepatic TEQ profiles in these two studies suggests that the PPCB/PCDD/PCDF congener profile is influencing toxicity in a way that is not accounted for using the TEF approach. In addition, Beckett et al. (2008) reported that exposure of adult female mink to dietary concentrations of 24 and 2.4 μg PCB 126/kg feed resulted in complete reproductive failure (2.4 and 0.24 μg TEQs/kg feed, respectively) while a dietary concentration of 0.24 μg PCB 126/kg feed (0.024 μg TEQs/kg feed) was the NOAEL. In contrast, Zwiernik et al. (2009) reported that an equivalent TEQ concentration (24.2 μg/kg feed) provided by 2,3,7,8-tetrachlorodibenzofuran (TCDF) had no effect on mink reproduction and survivability of offspring, indicating that for mink, the TEF value for TCDF is inaccurate.

An additional limitation of studies utilizing environmentally weathered mixtures of PCBs, PCDDs, and PCDFs is the presence of cocontaminants that may be influencing toxicity. As an example, in the Saginaw Bay mink study reported by Heaton et al. (1995a)/Tillitt et al. (1996), the total dietary TEQ concentration calculated from the identified TCDD-like compounds was approximately half the TEQ concentration based on the H4IIE bioassay, suggesting synergism of compounds present in the mixture or the presence of AhR-active chemicals that were not measured yet contributed to the TCDD-like potency of the mixture (Tillitt et al. 1996). Furthermore, Kannan et al. (2002) reported perfluorooctanesulfonate (PFOS) at concentrations of 240–300 μg/kg, wet weight in carp used in the Heaton et al. (1995a)/Tillitt et al. (1996) study, resulting in dietary concentrations of 105–160 μg PFOS/kg feed. Bursian et al. (2006a) reported hepatic Σpolybrominated diphenyl ether (PBDE) concentrations up to 23 μg/kg, wet weight in 6-week-old kits fed diets containing carp collected from the Saginaw Bay. It is not known how these cocontaminants may be influencing toxicity that is attributed to PCBs.

TABLE 15.4

Toxic Equivalent (TEQ) Contribution by PCDDs, PCDFs, Non-*Ortho* PCBs and Mono-*Ortho* PCBs in Congener-Based Mink Feeding Studies

Study	Medium	Chemical Group	% of Total TEQs	Predominant Congener	% of Total TEQs
Heaton et al. (1995a)/	Feed	PCDDs	18	TCDD	11
Tillitt et al. (1996)		PCDFs	9	PeCDF	6
(Saginaw Bay)		Non-*ortho* PCBs	64	PCB126	62
		Mono-*ortho* PCBs	9	PCB118	6
	Maternal liver	PCDDs	14	TCDD	8
		PCDFs	29	PeCDF	24
		Non-*ortho* PCBs	54	PCB126	53
		Mono-*ortho* PCBs	3	PCB118	2
Bursian et al. (2006a)	Feed	PCDDs	35	TCDD	16
(Saginaw River)		PCDFs	25	PeCDF	17
		Non-*ortho* PCBs	35	PCB126	33
		Mono-*ortho* PCBs	5	PCB118	4
	Juvenile liver	PCDDs	14	TCDD	8
		PCDFs	50	PeCDF	44
		Non-*ortho* PCBs	34	PCB126	34
		Mono-*ortho* PCBs	2	PCB118	1
Bursian et al. (2006b,	Feed	PCDDs	2	TCDD	1
2006c) (Housatonic		PCDFs	4	PeCDF	3
River)		Non-*ortho* PCBs	87	PCB126	81
		Mono-*ortho* PCBs	7	PCB123	2
	Maternal liver	PCDDs	1	HxCDD	1
		PCDFs	8	PeCDF	7
		Non-*ortho* PCBs	89	PCB126	85
		Mono-*ortho* PCBs	2	PCB118	1
	Juvenile liver	PCDDs	1	TCDD	<1
		PCDFs	7	PeCDF	6
		Non-*ortho* PCBs	87	PCB126	83
		Mono-*ortho* PCBs	5	PCB118	3

Summary

All mammalian wildlife species, including those animals inhabiting the most remote regions of the globe, are exposed to PCBs, PCDDs, and PCDFs. Historically, localized observations of poor reproduction and episodic mortalities have been noted among mammalian wildlife with high tissue concentrations of PCBs, PCDDs, and PCDFs. Fortunately, the general global trend appears to be toward lesser exposure, likely due to institutional controls adopted by most developed countries. Despite this general downward trend, areas remain where mammalian wildlife species continue to experience elevated exposures to PCBs and related compounds as well as areas where mammals remain under the threat of future exposure. After 40 years of research on PCBs, PCDDs, and PCDFs, much is known about their environmental presence but there remain many uncertainties associated with predicting what concentrations of these environmental contaminations will illicit detrimental effects. This poor predictability is derived from fundamental knowledge gaps including specific mechanisms of action, interspecies response sensitivity and inaccuracies associated with predicting the potency of congener mixtures.

There have been a great number of field-based studies addressing wildlife exposure and effects. These studies are important for integrating environmental parameters. However, the results are, by definition, correlations of exposures and effects, and do not establish causation. For terrestrial-based mammalian wildlife, essentially all of the studies compared field measured tissue concentrations to those associated with deleterious effects as determined in laboratory feeding studies with other species. A similar situation existed for marine mammals, although in a few cases, semicaptive animals were sampled for assessment of biochemical, morphological, and/or histological alterations, which were then correlated to tissue concentrations of contaminants. To date, the majority of laboratory studies that systematically relate tissue concentrations with effects have been conducted with ranch mink that are less than a century removed from their wild counterparts. Thus, for aquatic, piscivorous mammals, mink dose–response data are some of the most robust for predicting toxicological effects. As noted earlier, mink also have a number of desirable species characteristics including noted sensitivity to PCBs and related compounds that add utility to the data set. A number of laboratory feeding studies using mink have assessed the effects of neat commercial PCB mixtures and single congeners. However, an advantage of laboratory mink is the ability to expose them to actual site-specific environmental prey (fish), thus mimicking the exposure of their wild counterparts.

Out of necessity, scientists need to be able to know at what point exposure to environmental contaminants pose an unacceptable risk to human and wildlife populations. Based on the available data, a number of threshold tissue residue concentrations for PCBs and related compounds have been proposed. It should be noted that there is a variable amount of uncertainty associated with these estimates. Observed potency differences based on unpredicted shifts in dose–response curves for various environmental mixtures highlight the significant uncertainties associated with expressing concentrations on a ΣPCB basis when congener mixtures vary from site to site. Even when PCB/PCDD/PCDF concentrations in environmental mixtures and tissues are expressed on a TEQ basis, unexpected differences in toxicity occur, suggesting uncertainties associated with TEFs for specific congeners in specific species. Ideally, threshold tissue residue concentrations for PCBs/PCDDs/PCDFs should be site-specific and species-specific. Since this currently is not realistic, it is suggested that the following tissue-based threshold values be used with caution: 46 mg ΣPCBs (3.2×10^4 ng TEQs)/kg lipid in the liver for terrestrial mammals; 8.6 mg ΣPCBs (1.3×10^3 ng TEQs)/kg lipid in the liver and 17 mg ΣPCBs (160 ng TEQs)/kg lipid in the blubber for marine mammals; 41 mg ΣPCBs (1.5×10^3 ng TEQs)/kg lipid for aquatic mammals.

REFERENCES

Addison, R. F., and W. T. Sobo. 2001. Trends in organochlorine residue concentrations and burdens in grey seals (*Halichoerus grypus*) from Sable Is., NS, Canada, between 1974 and 1994. *Environ. Pollut.* 112:505–513.

Addison, R. F., M. E. Zinck, and T. G. Smith. 1986. PCBs have declined more than DDT-group residues in Arctic ringed seals (*Phoca hispida*) between 1972 and 1981. *Environ. Sci. Technol.* 20:253–256.

Aguilar, A., and A. Borrell. 1994a. Reproductive transfer and variation of body load of organochlorine pollutants with age in fin whales (*Balaenoptera physalus*). *Arch. Environ. Contam. Toxicol.* 27:546–554.

Aguilar, A., and A. Borrell. 1994b. Abnormally high polychlorinated biphenyl levels in striped dolphins (*Stenella coeruleoalba*) affected by the 1990–92 Mediterranean epizootic. *Sci. Total Environ.* 154:237–247.

Aguilar, A., and A. Borrell. 2005. DDT and PCB reduction in the western Mediterranean from 1987 to 2002, as shown by levels in stripped dolphins (*Stenella coeruleoalba*). *Mar. Environ. Res.* 59:391–404.

Ahlborg, U. G., et al. 1994. Toxic equivalency factors for dioxin-like PCBs-Report on a WHO-ECEH and IPCS consultation, December 1993. *Chemosphere* 28:1049–1067.

Anderson, M., et al. 2001. Geographical variation of PCB congeners in polar bears (*Ursus maritimus*) from Svalbard east to Chukchi Sea. *Polar Biol.* 24:321–238.

Aulerich, R. J., et al. 1971. Effects of feeding coho salmon and other Great Lakes fish on mink reproduction. *Can. J. Zool.* 49:611–616.

Aulerich, R. J., et al. 1987. Toxicity of 3,4,5,3′,4′,5′-hexachlorobiphenyl to mink. *Arch. Environ. Contam. Toxicol.* 16:53–60.

Aulerich, R. J., S. J. Bursian, W.J. Breslin, and R. K. Ringer. 1985. Toxicological manifestations of 2,4,5,2′,4′,5′-, 2,3,6,2′,3′,6′-, and 3,4,5,3′,4′,5′-hexachlorobiphenyl and Aroclor 1254 in mink. *J. Toxicol. Environ. Health* 15:63–79.

Aulerich, R. J., S. J. Bursian, and A. C. Napolitano. 1988. Biological effects of epidermal growth factor and 2,3,7,8-tetrachlorodibenzo-*p*-dioxin on developmental parameters of neonatal mink. *Arch. Environ. Contam. Toxicol.* 17:27–31.

Aulerich, R. J., S. J. Bursian, A. C. Napolitano, and T. Oleas. 2000. Feeding growing mink (*Mustela vison*) PCB Aroclor 1254 does not affect baculum (os-penis) development. *Bull. Environ. Contam. Toxicol.* 64:443–447.

Aulerich, R. J., and R. K. Ringer. 1977. Current status of PCB toxicity to mink, and their effect on their reproduction. *Arch. Environ. Contam. Toxicol.* 6:279–292.

Aulerich, R. J., and R. K. Ringer. 1980. *Toxicity of the polychlorinated biphenyl Aroclor 1016 to mink.* Environmental Research Laboratory, Office of Research and Development, US Environmental Protection Agency. EPA-600/3-80-033. Duluth, MN, USA.

Aulerich, R. J., R. K. Ringer, and S. Iwamoto. 1973. Reproductive failure and mortality in mink fed on Great Lakes fish. *J. Reprod. Fertil. Suppl.* 19:365–376.

Basu, N., and J. Head. 2010. Mammalian wildlife as complementary models in environmental neurotoxicology. *Neurotoxicol. Teratol.* 32:114–119.

Basu, N., A. M. Scheuhammer, S. J. Bursian, K. Rouvinen-Watt, J. Elliott, and H. M. Chan. 2007. Mink as a sentinel species in environmental health. *Environ. Res.* 103:130–144.

Beckett, K. J., S. D. Millsap, A. L. Blankenship, M. J. Zwiernik, J. P. Giesy, and S. J. Bursian. 2005. Squamous epithelial lesion of the mandibles and maxillae of wild mink (*Mustela vison*) naturally exposed to polychlorinated biphenyls. *Environ. Toxicol. Chem.* 24:674–677.

Beckett, K. J., B. Yamini, and S. J. Bursian. 2008. The effects of 3,3′,4,4′,5-pentachlorobiphenyl (PCB 126) on mink (*Mustela vison*) reproduction and kit survivability and growth. *Arch. Environ. Contam. Toxicol.* 54:123–129.

Bergman, A., M. Olsson, and S. Relland. 1992. Skull-bone lesions in the Baltic gray seal (*Halichoerus grypus*). *Ambio* 21:517–519.

Bergman, A., A. Bignert, and M. Olsson. 2003. Pathology in Baltic grey seals (*Halichoerus grypus*) in relation to environmental exposure to endocrine disruptors. In *Toxicology of marine mammals*, eds. J. G. Vos, et al., pp. 507–533. Boca Raton, FL: Taylor and Francis.

Bhavsar, S. P., et al. 2008. Temporal trends and spatial distribution of dioxins and furans in lake trout or lake whitefish from the Canadian Great Lakes. *Chemosphere* 73:S158–S165.

Blankenship, A. L., and J. P. Giesy. 2002. Use of biomarkers of exposure and vertebrate tissue residues in the characterization of PCBs at contaminated sites: applications to birds and mammals. In *Environmental analysis of contaminated sites: Toxicological methods and approaches*, eds. G. I. Sunahra, A. Y. Renoux, C. Thellen, C. L. Gaudet, and A. Pilon, pp. 153–180. New York, NY: Wiley.

Blankenship, A. L., et al. 2005. Differential accumulation of polychlorinated biphenyl congeners in the terrestrial food web of the Kalamazoo River Superfund Site, Michigan. *Environ. Sci. Technol.* 39:5954–5963.

Blankenship, A. L., et al. 2008. Toxicity reference values for mink exposed to 2,3,7,8-tetrachlorodibenzo-*p*-dioxin (TCDD) equivalents (TEQs). *Ecotoxicol. Environ. Saf.* 69:325–349.

Bleavins, M. R., R. J. Aulerich, and R. K. Ringer. 1980. Polychlorinated biphenyls (Aroclors 1016 and 1242): Effects on survival and reproduction in mink and ferrets. *Arch. Environ. Contam. Toxicol.* 9:627–635.

Boon, J. P., P. J. H. Reijnders, J. Dols, P. Wensvoort, and M. T. J. Hillebrand. 1987. The kinetics of individual polychlorinated biphenyl congeners in female harbour seals (*Phoca vitulina*), with evidence for structure-related metabolism. *Aquat. Toxicol.* 10:307–324.

Braune, B. M., et al. 2005. Persistent organic pollutants and mercury in marine biota of the Canadian Arctic: an overview of spatial and temporal trends. *Sci. Total Environ.* 351–352:4–56.

Bremle, G., P. Larsson, and J. O. Helldin. 1997. Polychlorinated biphenyls in a terrestrial predator, the pine marten (*Martes martes* L.). *Environ. Toxicol. Chem.* 16:1779–1784.

Brouwer, A., P. J. H. Reijnders, and J. H. Koeman. 1989. Polychlorinated biphenyl (PCB)-contaminated fish induces vitamin A and thyroid hormone deficiency in the common seal (*Phoca vitula*). *Aquat. Toxicol.* 15:99–106.

Brunstrom, B., et al. 2001. Reproductive toxicity in mink (*Mustela vison*) chronically exposed to environmentally relevant polychlorinated biphenyl concentrations. *Environ. Toxicol. Chem.* 20:2318–2327.

Buckner, C. H. 1966. Populations and ecological relationships of shrews in tamarack bogs of southeastern Manitoba. *J. Mammal.* 47:181–194.

Bursian, S. J., et al. 2006a. Assessment of effects in mink caused by consumption of carp collected from the Saginaw River, Michigan, USA. *Arch. Environ. Contam. Toxicol.* 50:614–623.

Bursian, S. J., et al. 2006b. Dietary exposure of mink (*Mustela vison*) to fish from the Housatonic River, Berkshire County, Massachusetts, USA: Effects on reproduction, kit growth, and survival. *Environ. Toxicol. Chem.* 25:1533–1540.

Bursian, S. J., et al. 2006c. Dietary exposure of mink (*Mustela vison*) to fish from the Housatonic River, Berkshire County, Massachusetts, USA: Effects on organ weights and histology and hepatic concentrations of polychlorinated biphenyls and 2,3,7,8-tetrachlorodibenzo-*p*-dioxin toxic equivalence. *Environ. Toxicol. Chem.* 25:1541–1550.

Carril González-Barros, S. T., M. E. Alvarez Piñeiro, J. Simal Lozano, and M. A. Lage Yusty. 1997. PCBs and PCTs in wolves (*Canis lupus,* L) in Galicia (N.W. Spain). *Chemosphere* 35:1243–1247.

Christensen, J. R., M. MacDuffee, M. B. Yunker, and P. S. Ross. 2007. Hibernation-associated changes in persistent organic pollutant (POP) levels and patterns in British Columbia grizzly bears (*Ursus arctos horribilis*). *Environ. Sci. Technol.* 41:1834–1840.

Corsolini, S., S. Focardi, C. Leonzio, S. Lovari, F. Monaci, and G. Romeo. 1999. Heavy metals and chlorinated hydrocarbon concentrations in the red fox in relation to some biological parameters. *Environ. Monit. Assess.* 54:87–100.

D'Havé, H., J. Scheirs, A. Covaci, P. Schepens, R. Verhagen, and W. De Coen. 2006. Nondestructive pollution exposure assessment in the European hedgehog (*Erinaceus europaeus*): III. Hair as an indicator of endogenous organochlorine compound concentrations. *Environ. Toxicol. Chem.* 25:158–167.

de Swart, R. L., et al. 1994. Impairment of immune function in harbor seals (*Phoca vitulina*) feeding on fish from polluted waters. *Ambio* 23:155–159.

Debier, C., et al. 2003a. Quantitative dynamics of PCB transfer from mother to pup during lactation in UK grey seals, *Halichoerus grypus. Mar. Ecol. Prog. Ser.* 247:237–248.

Debier, C., et al. 2003b. Dynamics of PCB transfer from mother to pup during lactation in UK grey seals *Halichoerus grypus*: Differences in PCB profile between compartments of transfer and changes during the lactation period. *Mar. Ecol. Prog. Ser.* 247:249–256.

Delong, R. L., W. G. Gilmartin, and J. G. Simpson. 1973. Premature births in California sea lions: associations with organochlorine pollutant residue levels. *Science* 181:1168–1170.

Denison, M. S., and S. R. Nagy. 2003. Activation of the aryl hydrocarbon receptor by structurally diverse exogenous and endogenous chemicals. *Annu. Rev. Pharmacol. Toxicol.* 43:309–334.

Denison, M. S., A. Pandini, S. R. Nagy, E. P. Baldwin, and L. Bonati. 2002. Ligand binding and activation of the Ah receptor. *Chem. Biol. Interact.* 141:3–24.

DeVito, M. J., J. J. Diliberto, D. G. Ross, M. G. Menache, and L. S. Birnbaum. 1997. Dose-response relationships for induction for polyhalogenated dioxins and dibenzofurans following subchronic treatment in mice. 1. CYP1A1 and CYP1A2 enzyme activity in liver, lung and skin. *Toxicol. Appl. Pharmacol.* 147:267–280.

DeVito, M. J., M. G. Menache, J. J. Diliberto, D. G. Ross, and L. S. Birnbaum. 2000. Dose-response relationships for induction of CYP1A1 and CYP1A2 enzyme activity in liver, lung, and skin in female mice following subchronic exposure to polychlorinated biphenyls. *Toxicol. Appl. Pharmacol.* 167:157–172.

Di Carlo, F. J., J. Seifter, and V. J. DeCarlo. 1978. Assessment of the hazards of polybrominated biphenyls. *Environ. Health Perspect.* 23:351–365.

Faust, B. F., M. H. Smith, and W. B. Wray. 1971. Distances moved by small mammals as an apparent function of grid size. *Acta Theriol.* 11:161–177.

Fuchsman, P. C., T. R. Barber, and M. J. Bock. 2008. Effectiveness of various exposure metrics in defining dose-response relations for mink (*Mustela vison*) exposed to polychlorinated biphenyls. *Arch. Environ. Contam. Toxicol.* 54:130–144.

Fuglei, E., J. O. Bustnes, H. Hop, T. Mørk, H. Björnfoth, and B. van Bavel. 2007. Environmental contaminants in arctic foxes (*Alopex lagopus*) in Svalbard: Relationships with feeding ecology and body condition. *Environ. Pollut.* 146:128–138.

Gamberg, M., and B. M. Braune. 1999. Contaminant residue levels in arctic wolves (*Canis lupus*) from the Yukon Territory, Canada. *Sci. Total Environ.* 243/244:329–338.

Giesy, J. P., and K. Kannan. 2002. Dioxin-like and non-dioxin-like effects of polychlorinated biphenyls: Implications for risk assessment. *Lakes Reserv. Res. Manag.* 7:139–181.

Grant, D. L., C. A. Moodie, and W. E. J. Phillips. 1974. Toxicodynamics of Aroclor 1254 in the male rat. *Environ Physiol. Biochem.* 4:214–225.

Grove, R. A., and C. J. Henny. 2008. Environmental contaminants in male river otters from Oregon and Washington, USA, 1994–1999. *Environ. Monit. Assess.* 145:49–73.

Grove, R. A., C. J. Henny, and J. L. Kaiser. 2009. Osprey: Worldwide sentinel species for assessing and monitoring environmental contamination in rivers, lakes, reservoirs, and estuaries. *J. Toxicol. Environ. Health, Part B Crit. Rev.* 12:25–44.

Haave, M., et al. 2003. Polychlorinated biphenyls and reproductive hormones in female polar bears at Svalbard. *Environ. Health Perspect.* 111:431–436.

Haffner, G. D., V. Glooschenko, C. A. Straughan, C. A. Herbert, and R. Lazar. 1998. Concentrations and distribution of polychlorinated biphenyls, including non-*ortho* congeners, in mink populations from southern Ontario. *J. Gt. Lakes Res.* 24:880–888.

Hahn, M. E. 1998. The aryl hydrocarbon receptor: a comparative perspective. *Comp. Biochem. Physiol. Part C Toxicol. Pharmacol.* 121C:23–53.

Hahn, M. E. 2002. Aryl hydrocarbon receptors: Diversity and evolution. *Chem. Biol. Interact.* 141:131–160.

Halbrook, R. S., R. J. Aulerich, S. J. Bursian, and L. Lewis. 1999. Ecological risk assessment in a large river-reservoir: 8. Experimental study of the effects of PCBs on reproductive success in mink. *Environ. Toxicol. Chem.* 18:649–654.

Harding, L. E., M. L. Harris, C. R. Stephen, and J. E. Elliot. 1999. Reproductive and morphological condition of wild mink (*Mustela vison*) and river otters (*Lutra canadensis*) in relation to chlorinated contamination. *Environ. Health Perspect.* 107:141–147.

Haynes, J. M., S. T. Wellman, K. J. Beckett, J. J. Pagano, S. D. Fitzgerald, and S. J. Bursian. 2009. Field and literature-based confirmations that histological lesions in mink jaws are a highly sensitive biomarker of effect after exposure to TCDD-like pollutants. *Arch. Environ. Contam. Toxicol.* 57:803–807.

Headrick, M. L., K. Hollinger, R. A. Lovell, and J. C. Matheson. 1999. PBBs, PCBs, and dioxins in food animals, their public health implications. *Vet. Clin. N. Am.: Food Anim. Pract.* 15:109–131.

Heaton, S. N., et al. 1995a. Dietary exposure of mink to carp from Saginaw Bay, Michigan. 1. Effects on reproduction and survival, and the potential risks to wild mink populations. *Arch. Environ. Contam. Toxicol.* 28:334–343.

Heaton, S. N., et al. 1995b. Dietary exposure of mink to carp from Saginaw Bay, Michigan. 2. Hematology and liver pathology. *Arch. Environ. Contam. Toxicol.* 29:411–417.

Helle, E., M. Olsson, and S. Jensen. 1976. DDT and PCB levels and reproduction in ringed seals from the Bothnian Bay. *Ambio* 5:188–198.

Henriksen, E. O., O. Wiig, J. U. Skaare, G. W. Gabrielsen, and A. E. Derocher. 2001. Monitoring PCBs in polar bears: lessons learned from Svalbard. *J. Environ. Monit.* 3:493–498.

Hochstein, J. R., R. J. Aulerich, and S. J. Bursian. 1988. Acute toxicity of 2,3,7,8-tetrachloro-dibenzo-*p*-dioxin to mink. *Arch. Environ. Contam. Toxicol.* 17:33–37.

Hochstein, J. R., S. J. Bursian, and R. J. Aulerich. 1998. Effects of dietary exposure to 2,3,7,8-tetrachlorodibenzo-*p*-dioxin in adult female mink (*Mustela vison*). *Arch. Environ. Contam. Toxicol.* 35:348–353.

Hochstein, J. R., J. A. Render, S. J. Bursian, and R. J. Aulerich. 2001. Chronic toxicity of dietary 2,3,7,8-tetrachlorodibenzo-*p*-dioxin to mink. *Vet. Hum. Toxicol.* 43:134–139.

Hornshaw, T. C., R. J. Aulerich, and H. E. Johnson. 1983. Feeding Great Lakes fish to mink: Effects on mink and accumulation and elimination of PCBs by mink. *J. Toxicol. Environ. Health* 11:933–946.

Hornshaw, T. C., J. Safronoff, R. K. Ringer, and R. J. Aulerich. 1986. LC$_{50}$ test results in polychlorinated biphenyl-fed mink: age, season, and diet comparisons. *Arch. Environ. Contam. Toxicol.* 15:717–723.

Huwe, J. K. 2002. Dioxins in food: A modern agricultural perspective. *J. Agric. Food Chem.* 50:1739–1750.

Iwata, H., S. Tanabe, N. Sakai, and R. Tatsukawa. 1993. Distribution of persistent organochlorines in the oceanic air and surface seawater and the role of the ocean on their global transport and fate. *Environ. Sci. Technol.* 27:1080–1098.

Jensen, S., J. E. Kihlstrom, M. Olsson, C. Lundberg, and J. Orberg. 1977. Effects of PCB and DDT on mink (*Mustela vison*) during the reproductive season. *Ambio* 6:239.

Jones, T. D. 1995. Toxicological potency of 2,3,7,8-tetrachlorodibenzo-*p*-dioxin relative to 100 other compounds: a relative potency analysis of *in vitro* and *in vivo* test data. *Arch. Environ. Contam. Toxicol.* 29:77–85.

Kakela, A., R. Kakela, H. Hyvarnen, and J. Asikainen. 2002a. Vitamins A1 and A2 in hepatic tissues and subcellular fractions in mink feeding on fish-based diets and exposed to Aroclor 1242. *Environ. Toxicol. Chem.* 21:397–403.

Kakela, R., I. Jokinen, A. Kakela, and H. Hyvarinen. 2002b. Effects of gender, diet, exogenous melatonin and subchronic PCB exposure on plasma immunoglobin G in mink. *Comp. Biochem. Physiol. Part C Toxicol. Pharmacol.* 132:67–74.

Kamrin, M. A., and R. K. Ringer. 1996. Toxicological implications of PCB residues in mammals. In *Environmental contaminants in wildlife*, eds. W. N. Beyer, G. H. Heinz, and A. W. Redmon-Norwood, pp. 153–163. New York, NY: Lewis Publishers.

Kannan, K., A. L. Blankenship, P. D. Jones, and J. P. Giesy. 2000. Toxicity reference values for the toxic effects of polychlorinated biphenyls to aquatic mammals. *Hum. Ecol. Risk. Assess.* 6:181–201.

Kannan, K., J. Newsted, R. S. Halbrook, and J. P. Giesy. 2002. Perfluorooctane and related fluorinated hydrocarbons in mink and river otters from the United States. *Environ. Sci. Technol.* 36:2566–2571.

Kay, D. P., et al. 2005. Differential accumulation of polychlorinated biphenyl congeners in the aquatic food web at the Kalamazoo River superfund site, Michigan. *Environ. Sci. Technol.* 39:5964–5974.

Kihlstrom, J. E., M. Olsson, S. Jensen, A. Johansson, J. Ahlbom, and A. Bergman. 1992. Effects of PCB and different fractions of PCB on the reproduction of mink (*Mustela vison*). *Ambio* 21:563–569.

Kimbrough, R. D. 1987. Human health effects of polychlorinated biphenyls (PCBs) and polybrominated biphenyls (PBBs). *Annu. Rev. Pharmacol. Toxicol.* 27:87–111.

Kimbrough, R. D. 1995. Polychlorinated biphenyls (PCBs) and human health: An update. *Crit. Rev. Toxicol.* 25:133–163.

Kociba, R. J., et al. 1978. Results of a two-year chronic toxicity and oncogenicity study of 2,3,7,8-tetrachlorodibenzo-*p*-dioxin in rats. *Toxicol. Appl. Pharmacol.* 46:279–303.

Kruuk, H., and J. W. H. Conroy. 1991. Mortality of otters (*Lutra lutra*) in Shetland. *J. Appl. Ecol.* 28:83–94.

Kuehl, D. W., R. Haebler, and C. Potter. 1991. Chemical residues in dolphins from the U.S. Atlantic coast including Atlantic bottlenose obtained during the 1987/88 mass mortality. *Chemosphere* 22:1071–1084.

Kunisue, T., et al. 2006. PCDDs, PCDFs, and coplanar PCBs in wild terrestrial mammals from Japan: Congener specific accumulation and hepatic sequestration. *Environ. Pollut.* 140:525–535.

Lahvis, G. P., R. S. Wells, D. W. Kuehl, J. L. Stewart, H. L. Rhinehart, and C. S. Vis. 1995. Decreased lymphocyte responses in free-ranging bottlenose dolphins (*Tursiops truncates*) are associated with increased concentrations of PCBs and DDT in peripheral blood. *Environ. Health. Perspect.* 103:67–72.

Lawlor, T. E. 1979. *Handbook to the orders and families of living mammals*. Eureka, CA: Mad River Press.

Leonards, P. E. G., et al. 1995. Assessment of experimental data on PCB-induced reproduction inhibition in mink, based on an isomer- and congener-specific approach using 2,3,7,8-tetrachlorodibenzo-*p*-dioxin toxic equivalency. *Environ. Toxicol. Chem.* 14:639–652.

Leonards, P. E. G., Y. Zierikzee, U. A. T. Brinkman, W. P. Cofino, N. M. van Straalen, and B. van Hattum. 1997. The selective dietary accumulation of planar polychlorinated biphenyls in the otter (*Lutra lutra*). *Environ. Toxicol. Chem.* 16:1807–1815.

Leonards, P. E. G., et al. 1998. Studies of bioaccumulation and biotransformation of PCBs in mustelids based on concentration and congener patterns in predators and preys. *Arch. Environ. Contam. Toxicol.* 35:654–665.

Linder, R. E., T. B. Gaines, and R. D. Kimbrough. 1974. The effect of polychlorinated biphenyls on rat reproduction. *Fd. Cosmet. Toxicol.* 12:63–77.

Mandal, P. K. 2005. Dioxin: a review of its environmental effects and its aryl hydrocarbon receptor biology. *J. Comp. Physiol. B Biochem. Syst. Environ. Physiol.* 175:221–230.

Marine Mammal Commission. 1999. *Marine Mammals and Persistent Ocean Contaminants: Proceedings of the Marine Mammal Commission Workshop*, Keystone, Colorado, October 12–15, 1998, 150 pp. Bethesda, MD: Marine Mammal Commission.

Martin, P. A., T. V. McDaniel, and B. Hunter. 2006a. Temporal and spatial trends in chlorinated concentrations of mink in Canadian Lakes Erie and St. Clair. *Environ. Monit. Assess.* 113:245–263.

Martin, P. A., G. J. Mayne, S. Bursian, V. Palace, and K. Kannan. 2006b. Changes in thyroid and vitamin A status in mink fed polyhalogented-aromatic-hydrocarbon-contaminated carp from the Saginaw River, Michigan, USA. *Environ. Res.* 101:53–67.

Martineau, D., P. Beland, C. Desjardins, and A. Lagace. 1987. Levels of organochlorine chemicals in tissues of beluga whales (*Delphinapterus leucas*) from the St. Lawrence Estuary, Quebec, Canada. *Arch. Environ. Contam. Toxicol.* 16:137–147.

Mason, C. F., and A. B. Madsen. 1993. Organochlorine pesticide residues and PCBs in Danish otters (*Lutra lutra*). *Sci. Total. Environ.* 133:73–81.

Millsap, S. D., et al. 2004. Comparison of risk assessment methodologies for exposure of mink to PCBs on the Kalamazoo River, Michigan. *Environ. Sci. Technol.* 38:6451–6459.

Moore, J. N., et al. 2009. Relationships between P450 enzyme induction, jaw histology and tissue morphology in mink (*Mustela vison*) exposed to polychlorinated dibenzofurans (PCDFs). *Arch. Environ. Contam. Toxicol.* 57:416–425.

Mora, M. A., et al. 2000. Environmental contaminants in blood, hair, and tissues of ocelots from the lower Rio Grande Valley, Texas, 1986–1997. *Environ. Monit. Assess.* 64:477–492.

Mossner, S., and K. Ballschmiter. 1997. Marine mammals as global pollution indicators for organochlorines. *Chemosphere* 34:1285–1296.

Muir, D. C. G., and R. J. Norstrom. 2000. Geographical differences and time trends of persistent organic pollutants in the Arctic. *Toxicol. Lett.* 112–113:93–101.

Muir, D. C. G., R. Wagemann, B. T. Hargrave, D. J. Thomas, D. B. Peakall, and R. J. Norstrom. 1992. Arctic marine ecosystem contamination. *Sci. Total Environ.* 122:75–134.

Muir, D. C. G., K. Koczanski, B. Rosenberg, and P. Beland. 1996. Persistent organochlorines in Beluga whales (*Delphinapterus leucas*) from the St Lawrence River Estuary-II. Temporal trends, 1982–1994. *Environ. Pollut.* 93:235–245.

Muir, D., et al. 1999. Spatial and temporal trends and effects of contaminants in the Canadian Arctic marine ecosystem: a review. *Sci. Total Environ.* 230:83–144.

Murk, A. J., P. E. G. Leonards, B. van Hattum, R. Luit, M. E. J. van der Weiden, and M. Smit. 1998. Application of biomarkers for exposure and effect of polyhalogenated aromatic hydrocarbons in naturally exposed European otters (*Lutra lutra*). *Environ. Toxicol. Pharmacol.* 6:91–102.

Murray, F. J., F. A. Smith, K. D. Nitschke, C. G. Humiston, R. J. Kociba, and B. A. Schwetz. 1979. Three-generation reproduction study of rats given 2,3,7,8-tetrachlorodibenzo-*p*-dioxin (TCDD) in the diet. *Toxicol. Appl. Pharmacol.* 50:241–252.

Naso, B., et al. 2004. Organochlorine pesticides and polychlorinated biphenyls in European roe deer *Capreolus capreolus* resident in a protected area in Northern Italy. *Sci. Total Environ.* 328:83–93.

Okey, A. B., D. S. Riddick, and P. A. Harper. 1994. The Ah receptor: Mediator of the toxicity of the toxicity of 2,3,7,8-tetrachlorodibenzo-*p*-dioxin (TCDD) and related compounds. *Toxicol. Lett.* 70:1–22.

Oskam, I. C., et al. 2003. Organochlorines affect the major androgenic hormone, testosterone, in male polar bears (*Ursus maritimus*) at Svalbard. *J. Toxicol. Environ. Health Part A* 66:2119–2139.

Osowski, S. L., L. W. Brewer, O. E. Baker, and G. P. Cobb. 1995. The decline of mink in Georgia, North Carolina, and South Carolina: the role of contaminants. *Arch. Environ. Contam. Toxicol.* 29:418–423.

Platonow, N. S., and L. H. Karstad. 1973. Dietary effects of polychlorinated biphenyls on mink. *Can. J. Comp. Med.* 37:391–400.

Poole, K. G., B. T. Elkin, and R. W. Bethke. 1995. Environmental contaminants in wild mink in the Northwest Territories, Canada. *Sci. Total Environ.* 160:473–486.

Reddy, M. L., S. Echols, B. Finklea, D. Busbee, J. Reif, and S. Ridgway. 1998. PCBs and chlorinated pesticides in clinically healthy *Tursiops truncatus*: Relationships between levels in blubber and blood. *Mar. Pollut. Bull.* 36:892–903.

Reddy, M. L., J. S. Reif, A. Bachand, and S. H. Ridgway. 2001. Opportunities for using Navy marine mammals to explore associations between organochlorine contaminants and unfavorable effects on reproduction. *Sci. Total Environ.* 274:171–182.

Reijnders, P. J. H. 1986. Reproductive failure in common seals feeding from polluted coastal waters. *Nature* 324:456–457.

Render, J. A., R. J. Aulerich, S. J. Bursian, and R. F. Nachreiner. 2000a. Proliferation of maxillary and mandibular periodontal squamous cells in mink fed 3,3′,4,4′,5-pentachlorobiphenyl (PCB 126). *J. Vet. Diagn. Invest.* 12:477–479.

Render, J. A., J. R. Hochstein, R. J. Aulerich, and S. J. Bursian. 2000b. Proliferation of periodontal squamous epithelium in mink fed 2,3,7,8-tetrachlorodibenzo-*p*-dioxin (TCDD). 2000. *Vet. Hum. Toxicol.* 42:85–86.

Render, J. A., S. J. Bursian, D. S. Rosenstein, and R. J. Aulerich. 2001. Squamous epithelial proliferation in the jaws of mink fed diets containing 3,3′,4,4′,5-pentachlorobiphenyl (PCB 126) or 2,3,7,8-tetrachlorodibenzo-*p*-dioxin (TCDD). *Vet. Hum. Toxicol.* 43:22–26.

Restum, J. C., et al. 1998. Multigenerational study of the effects of consumption of PCB-contaminated carp from Saginaw Bay, Lake Huron, on mink. 1. Effects on mink reproduction, kit growth and survival, and selected biological parameters. *J. Toxicol. Environ. Health Part A* 54:343–375.

Riget, F., R. Dietz, K. Vorkamp, P. Johansen, and D. Muir. 2004. Levels and spatial and temporal trends of contaminants in Greenland biota: an updated review. *Sci. Total Environ.* 331:29–52.

Riget, F., K. Vorkamp, R. Dietz, and S. C. Rastogi. 2006. Temporal trends studies on polybrominated diphenyl ethers (PBDEs) and polychlorinated biphenyls (PCBs) in ringed seals from East Greenland. *J. Environ. Monit.* 8:1000–1005.

Roos, A., E. Greyerz, M. Olsson, and F. Sandegren. 2001. The otter (*Lutra lutra)* in Sweden—population trends in relation to ∑DDT and total PCB concentrations during 1968–99. *Environ. Pollut.* 111:457–469.

Ross, P. S. 2000. Marine mammals as sentinels in ecological risk assessment. *Hum. Ecol. Risk. Assess.* 6:29–46.

Ross, P. S., R. L. de Swart, P. J. H. Reijnders, H. van Loveren, J. G. Vos, and A. D. M. E. Osterhaus. 1995. Contaminant-related suppression of delayed-type hypersensitivity and antibody responses in harbor seals fed herring from the Baltic Sea. *Environ. Health Perspect.* 103:162–167.

Ross, P., R. de Swart, R. Addison, H. Van Loveren, J. Vos, and A. Osterhaus. 1996a. Contaminant-induced immunotoxicity in harbor seals: wildlife at risk? *Toxicology* 112:157–169.

Ross, P. S., 1996b. Suppression of natural killer cell activity in harbour seals (*Phoca vitulina*) fed Baltic Sea herring. *Aquat. Toxicol.* 34:71–84.

Ross, P. S., S. J. Jefferies, M. B. Yunker, R. E. Addison, M. G. Ikonomou, and J. C. Calambokidis. 2004. Harbor seals (*Phoca vitulina*) in British Columbia, Canada, and Washington State, USA, reveal a combination of local and global polychlorinated biphenyl, dioxin, and furan signals. *Environ. Toxicol. Chem.* 23:157–165.

Routti, H., M. Nyman, B. M. Jenssen, C. Backman, J. Koistinen, and G. W. Gabrielsen. 2008. Bone-related effects of contaminants in seals may be associated with vitamin D and thyroid hormones. *Environ. Toxicol. Chem.* 27:873–880.

Russell, J. S. 1998. Evaluation of liver cytochromes P450 induction in the short-tailed shrew and white-tailed mouse. M.S. Thesis, Southern Illinois University.

Safe, S. 1987. Determination of 2,3,7,8TCDD toxic equivalence factors (TEFs)—support for the use of the in vitro AHH induction assay. *Chemosphere* 16:791–802.

Safe, S. 1990. Polychlorinated biphenyls (PCBs), dibenzo-*p*-dioxins (PCDDs), dibenzofurans (PCDFs), and related compounds: environmental and mechanistic considerations which support the development of toxic equivalency factors (TEFs). *Crit. Rev. Toxicol.* 21:51–88.

Safe, S. 1993. Polychlorinated biphenyls—toxicology and risk assessment. Proceedings, Dioxin'93, 13th International Symposium on Dioxins and Related Compounds, Vienna, Austria, September 20–24, p. 53.

Safe, S. 1994. Polychlorinated biphenyls (PCBs): Environmental impact, biochemical and toxic responses, and implications for risk assessment. *Crit. Rev. Toxicol.* 24:87–149.

Safe, S. 1998a. Development validation and problems with the toxic equivalency factor approach for risk assessment of dioxins and related compounds. *J. Anim. Sci.* 76:134–141.

Safe, S. 1998b. Limitations of the toxic equivalency factor approach for risk assessment of TCDD and related compounds. *Teratog. Carcinog. Mutagen.* 17:285–304.

Sample, B. E., D. M. Opresko, and G. W. Suter II. 1996. *Toxicological Benchmarks for Wildlife*. 1996 Revision, ES/ER/TM86-/R3. Oak Ridge National Laboratory, Health Sciences Research Division, Oak Ridge, TN.

Schecter, A., L. Birnbaum, J. J. Ryan, and J. D. Constable. 2006. Dioxins: an overview. *Environ. Res.* 101:419–428.

Schwacke, L. H., E. O. Voit, L. J. Hansen, R. S. Wells, G. B. Mitchum, A. E. Hohn, and P. A. Fair. 2002. Probabilistic risk assessment of reproductive effects of polychlorinated biphenyls on bottlenose dolphins (*Tursiops truncatus*) from the southeast United States coast. *Environ. Toxicol. Chem.* 21:2752–2764.

Senthilkumar, K., K. Kannan, A. Subramanian, and S. Tanabe. 2001. Accumulation of organochlorine pesticides and polychlorinated biphenyls in sediments, aquatic organisms, birds, bird eggs and bat collected from South India. *Environ. Sci. Pollut. Res.* 8:35–47.

Shore, R., et al. 2001. Organochlorine pesticide, polychlorinated biphenyl and heavy metal concentrations in wolves (*Canis lupus* L. 1758) from north-west Russia. *Sci. Total Environ.* 280:45–54.

Simpson, V. R. 1997. Health status of otters (*Lutra lutra*) in south-west England based on postmortem findings. *Vet. Rec.* 141:191–197.

Sjoasen, T., J. Ozolins, E. Greyerz, and M. Olsson. 1997. The otter (*Lutra lutra*) situation in Latvia and Sweden related to PCB and DDT levels. *Ambio* 26:196–201.

Skaare, J. U., et al. 2000. Organochlorines in top predators at Svalbard-occurrence, levels and effects. *Toxicol. Lett.* 112–113:103–109.

Smit, M. D., P. E. G. Leonards, A. J. Murk, A. W. J. J. de Jongh, and B. van Hattum. 1996. *Development of Otter-Based Quality Objectives for PCBs*. Institute for Environmental Studies. Vrije Universiteit, Amsterdam, The Netherlands.

Smith, P. N., K. A. Johnson, T. A. Anderson, and S. T. McMurry. 2003. Environmental exposure to polychlorinated biphenyls among raccoons (*Procyon lotor*) at the Paducah gaseous diffusion plant, Western Kentucky, USA. *Environ. Toxicol. Chem.* 22:406–416.

Stern, G. A., D. C.G. Muir, M. D. Segstro, R. Dietz, and M. P. Heide Jørgensen. 1994. PCBs and other organochlorine contaminants in white whales (*Delphinapterus leucas*) from west Greenland: Variations with age and sex. *Medd. om Grønland* 39:243–257.

Talmage, S. S., and B. T. Walton. 1991. Small mammals as monitors of environmental contaminants. *Rev. Environ. Contam. Toxicol.* 119:48–143.

Tanabe, S. 1988. PCB problems in the future: Foresight from current knowledge. *Environ. Pollut.* 50:5–28.

Tanabe, S., H. Iawata, and R Tatsukawa. 1994. Global contamination by persistent organochlorines and their ecotoxicological impact on marine mammals. *Sci. Total Environ.* 154:163–177.

Tanabe, S., S. Niimi, T. B. Minh, N. Miyazaki, and E. A. Petrov. 2003. Temporal trends of persistent organochlorine contamination in Russia: a case study of Baikal and Caspian Seal. *Arch. Environ. Contam. Toxicol.* 44:533–545.

Tillitt, D. E., et al. 1996. Dietary exposure of mink to carp from Saginaw Bay. 3. Characterization of dietary exposure to planar halogenated hydrocarbons, dioxin equivalents, and biomagnification. *Environ. Sci. Technol.* 30:283–291.

USACHPPM. 2000. *Standard Practice for Wildlife Toxicity Reference Values.* USACHPPM Technical Guide 254. Aberdeen Proving Ground, MD, USACHPPM.

U.S. EPA. 1995. *Great Lakes Water quality initiative criteria documents for the protection of wildlife: DDT, Mercury, 2,3,7,8-TCDD, PCBs.* EPA-820-B-95-0083. Washington, DC, USEPA.

U.S. EPA. 1997. *Ecological risk assessment guidance for superfund: Process for designing and conducting ecological risk assessments interim final.* EPA 540-R-97-006. Washington, DC, USEPA.

U.S. EPA. 2003. Attachment 4-5. *Ecological soil screening levels (Eco-SSLs) standard operating procedure (SOP) 34: Derivation of wildlife toxicity reference value (TRV).* OWSER Directive 9285.7-55. Washington, DC, USEPA.

Van den Berg, M., J. De Jongh, H. Poiger, and J. R. Olson. 1994. The toxicokinetics and metabolism of polychlorinated dibenzo-*p*-dioxins (PCDDs) and dibenzofuran (PCDFs) and their relevance for toxicity. *Crit. Rev. Toxicol.* 24:1–74.

Van den Berg, M., et al. 1998. Toxic equivalency factors (TEFs) for PCBs, PCDDs, PCDFs for humans and wildlife. *Environ. Health Perspect.* 106:775–792.

Van den Berg, M., et al. 2006. The 2005 World Health Organization reevaluation of human and mammalian toxic equivalency factors for dioxins and dioxin-like compounds. *Toxicol. Sci.* 93:223–241.

Verreault, J., et al. 2005. Chlorinated hydrocarbon contaminants and metabolites in polar bears (*Ursus maritimus*) from Alaska, Canada, East Greenland, and Svalbard: 1996–2002. *Sci. Total Environ.* 351–352:369–390.

Weis, I. M., and D. C. G. Muir. 1997. Geographical variation of persistent organochlorine concentrations in blubber of ringed seal (*Phoca hispida*) from the Canadian arctic: Univariate and multivariate approaches. *Environ. Pollut.* 96:321–333.

Wells, R. S., et al. 2005. Integrating life-history and reproductive success data to examine potential relationships with organochlorine compounds for bottlenose dolphins (*Tursiops truncates*) in Sarasota Bay, Florida. *Sci. Total Environ.* 349:106–119.

Wren, C. D., D. B. Hunter, J. F. Leatherland, and P. M. Stokes. 1987a. The effects of polychlorinated biphenyls and methylmercury, singly and in combination, on mink. I: Uptake and toxic responses. *Arch. Environ. Contam. Toxicol.* 16:441–447.

Wren, C. D., D. B. Hunter, J. F. Leatherland, and P. M. Stokes. 1987b. The effects of polychlorinated biphenyls and methylmercury, singly and in combination, on mink. II: Reproduction and kit development. *Arch. Environ. Contam. Toxicol.* 16:449–454.

Zwiernik, M. J., et al. 2008a. Exposure and effects assessment of resident mink (*Mustela vison*) exposed to polychlorinated dibenzofurans and other dioxin-like compounds in the Tittabawassee River basin, Midland, Michigan, USA. *Environ. Toxicol. Chem.* 27:2076–2087.

Zwiernik, M., et al. 2008b. Toxicokinetics of 2,3,7,8-TCDF and 2,3,4,7,8-PeCDF in mink (*Mustela vison*) at ecologically relevant exposures. *Toxicol. Sci.* 105:33–43.

Zwiernik, M. J., et al. 2009. Chronic effects of polychlorinated dibenzofurans on mink in laboratory and field environments. *Integr. Environ. Assess. Manage.* 5:291–301.

King-Vulture

By G. Mutzel, from *The Royal Natural History,* edited by Richard Lydekker, Frederick Warne & Co.,
London, 1893–94.

16 Lead in Birds

J. Christian Franson
Deborah J. Pain

CONTENTS

16.1 Introduction 563
16.2 Notes on Terminology 565
16.3 Lead Distribution among Avian Tissues 565
16.4 Factors Influencing Concentrations of Lead in Tissues..567
16.5 Tissue Lead Concentrations and Sublethal Effects on Enzyme Systems...........................568
16.6 Lead and Immunosuppression..569
16.7 Background Lead Concentrations in Avian Tissues..569
16.8 Recommendations for the Interpretation of Tissue Lead Concentrations in Birds 570
    16.8.1 Individual Birds 570
    16.8.2 Populations..572
    16.8.3 Species 573
16.9 Lead Residues Reported in Selected Field and Laboratory Studies in Birds 574
    16.9.1 Anseriformes (Ducks, Geese, and Swans)..574
    16.9.2 Falconiformes (Hawks, Falcons, and Vultures)..576
    16.9.3 Columbiformes (Doves and Pigeons) 578
    16.9.4 Galliformes (Pheasants, Grouse, Quail, and Partridge) 579
    16.9.5 Passeriformes (Perching Birds) 580
    16.9.6 Charadriiformes (Shorebirds, Gulls, and Alcids)..580
    16.9.7 Gruiformes (Cranes and Their Allies)..580
    16.9.8 Ciconiiformes (Herons and Their Allies) 581
    16.9.9 Gaviiformes (Loons)..581
    16.9.10 Strigiformes (Owls) 581
    16.9.11 Procellariiformes (Tubenoses) 582
    16.9.12 Pelecaniformes (Pelicans and Their Allies) 582
    16.9.13 Phoenicopteriformes (Flamingos) 582
    16.9.14 Piciformes (Woodpeckers)...582
Summary...582
References..583

16.1 INTRODUCTION

Lead is a highly toxic heavy metal that acts as a nonspecific poison affecting all body systems and has no known biological requirement. Absorption of low concentrations may result in a wide range of sublethal effects in animals, and higher concentrations may result in mortality (Demayo et al. 1982).

    Lead has been mined and smelted by humans for centuries, but the use of lead-based products increased greatly following the Industrial Revolution. Consequently, lead today is ubiquitous in air, water, and soil, in both urban and rural environments (Eisler 2000). Vertebrates are exposed to lead mainly via inhalation and ingestion. A proportion of lead entering the body is absorbed into the bloodstream and subsequently becomes distributed among body tissues, primarily the blood, liver, kidney, and bone. As a result of anthropogenic activities, most animals have higher tissue lead concentrations than in preindustrialized times. Although even very low tissue lead concentrations have some measurable physiological effects, the concentrations usually encountered in the wider environment (i.e., distant from lead emission sources) have not generally been considered to directly affect survival of most wildlife.

However, significant numbers of wild birds may be exposed to large amounts of lead through the ingestion of spent lead from ammunition (i.e., shotgun pellets, bullets, and fragments thereof). This source of exposure primarily affects birds in two groups. The first includes birds that ingest lead shotgun pellets, presumably as grit or food particles, in areas that are hunted over, including wetlands, farmland, terrestrial game shooting areas, and shooting ranges. The second includes birds that prey upon or scavenge the flesh of game species, or other hunted species, such as pests or predators. This latter group consists primarily of hawks, eagles, falcons, owls, vultures, and other scavenging birds that ingest lead from ammunition in carcasses of wildlife that have been killed but unretrieved or shot but survived, or that ingest lead bullet fragments embedded in the offal and other tissues of field-dressed game animals. Ingestion of lead from ammunition has resulted in widespread avian mortality in the United States, Europe, and elsewhere for over a century (Bellrose 1959, Sanderson and Bellrose 1986, Locke and Friend 1992, Pain 1992, Kendall et al. 1996, Kurosawa 2000, Fisher et al. 2006, Kreager et al. 2008, Mateo 2009, Pain et al. 2009). Consequently, the use of lead ammunition has been restricted in at least 29 countries around the world (Avery and Watson 2009). Lead restrictions have been introduced primarily due to concerns over lead poisoning in waterfowl or avian scavengers. The most common restriction is a ban on the use of lead gunshot for shooting waterfowl and/or over wetlands, although a few countries have banned the use of lead shot for all hunting (Avery and Watson 2009). The ingestion of lead fishing weights also has resulted in lead exposure and mortality in waterfowl, particularly swans, common loons (*Gavia immer*), and a variety of other waterbirds (Sears 1988, Blus et al. 1989, Franson et al. 2003, Scheuhammer et al. 2003b, Sidor et al. 2003, Pokras et al. 2009).

Nonparticulate lead exposure is frequently the result of the ingestion of sediments containing lead, particularly at locations surrounding mines and smelters, but also occurs from the consumption of lead-contaminated prey (Beyer et al. 1985, Henny et al. 1994, Johnson et al. 1999, Sileo et al. 2001, Beyer et al. 2004, Bostan et al. 2007). In addition, lead exposure of birds has occurred in urban environments from airborne sources (Ohi et al. 1974, Hutton 1980, Hutton and Goodman 1980, Ohi et al. 1981, Grue et al. 1986, Nam and Lee 2005, Scheifler et al. 2006, Roux and Marra 2007) and from the ingestion of paint chips near buildings painted with lead-based paint (Sileo and Fefer 1987, Work and Smith 1996, Finkelstein et al. 2003).

The primary focus of this chapter is the interpretation of tissue lead concentrations resulting from exposure to metallic lead, including lead shotgun pellets, bullets, bullet fragments, and fishing weights. However, the guidelines provided can also be applied to evaluating the effects of other sources of lead exposure in birds, such as contaminated sediments, airborne emissions, and lead-based paints.

Research on lead poisoning of birds, under both experimental and field conditions, has been conducted since the 1950s, when Bellrose (1959) completed the first comprehensive study of lead poisoning in waterfowl. More recently, lead poisoning has been investigated in many additional species of aquatic and terrestrial birds. This chapter draws upon this research, covering the distribution of lead within the body, factors that influence tissue lead concentrations, and the effects of lead poisoning. The concentrations of lead found in tissues of unexposed and lead-poisoned birds are discussed, along with the threshold tissue concentrations that are considered to indicate excessive lead absorption or poisoning. The chapter concludes with suggested thresholds for tissue lead concentrations indicative of different levels of exposure and poisoning and discusses factors associated with their interpretation at both individual and population levels.

16.2 NOTES ON TERMINOLOGY

1. "Natural" environmental concentrations no longer exist, because lead resulting from anthropogenic emissions is ubiquitous. Concentrations in the wider environment far from lead emission sources have consequently been described as "background." The magnitude of such "background exposure" in birds varies according to the specific geographic areas used throughout the life cycle. "Elevated exposure" is that resulting from exposure to above-background concentrations.

2. Lead exposure can be characterized according to duration as acute (exposure over a short period of time) or chronic (sustained exposure), and magnitude, depending on the amount of lead the animal is exposed to. However, "acute" is generally used to describe a high magnitude of exposure over a short time period, and "chronic" to describe exposure over a more protracted time period, often at a lower level.

3. Residues from the literature that were reported as parts per million (ppm) were converted to milligram per kilogram and expressed on a wet weight or dry weight basis when specified as such. Blood lead data reported in the literature as ppm, µg/g, or mg/kg were assumed to be on a wet weight basis. For reference, 1 µg/g = 1 mg/kg and is equal to 1 ppm. In blood, 1 ppm lead is approximately equal to 1 µg/mL, 1 µg/g, or 100 µg/dL. Interpretive threshold concentrations for soft tissues are given on wet weight basis, as is commonly done, although reporting on a dry weight basis is often recommended to control for variation in moisture content (Adrian and Stevens 1979). Unless otherwise noted, conversions between wet and dry weight concentrations were calculated using moisture levels reported for mallard (*Anas platyrhynchos*) tissues by Scanlon (1982). Thus, 1 µg/g wet weight equals approximately 4.6 µg/g dry weight for blood, 3.1 µg/g dry weight for liver, 4.3 µg/g dry weight for kidney, and 1.2 µg/g dry weight for bone.

16.3 LEAD DISTRIBUTION AMONG AVIAN TISSUES

Ingested lead shot, bullets, fishing weights, or fragments from these are subject to dissolution by acids in the avian stomach, as well as mechanical grinding in those species that retain grit in their muscular gizzards. The resultant toxic lead salts are absorbed into the bloodstream. However, retention of lead particles, such as shot, is variable. They may be evacuated immediately or soon after ingestion with little or no absorption of lead, retained and partially eroded/dissolved with significant lead absorption, or retained until they are completely eroded/dissolved and absorbed. For example, all black ducks (*Anas rubripes*) that died 4–6 days after being dosed retained the lead shot that was administered (Pain and Rattner 1988). However, if waterfowl that have ingested shot survive long enough, most lead shot disappear from the gizzard within about 20 days, either because it has passed through the gastrointestinal tract or has been eroded (Franson et al. 1986, Sanderson and Bellrose 1986). In mourning doves (*Zenaida macroura*), 33–100% of administered lead shot were voided in 2 days (Schulz et al. 2007), while another study reported that up to 95% of shot were voided within 3 weeks (Marn et al. 1988). In a 211-day study of lead poisoning in turkey vultures (*Cathartes aura*), some of the birds were redosed because lead shot was regurgitated or defecated, whereas regurgitation occurred only once in Andean condors (*Vultur gryphus*) and they retained the administered lead shot for 39–49 days (Carpenter et al. 2003, Pattee et al. 2006).

Once absorbed, some of the lead in the bloodstream is deposited rapidly into soft tissues, primarily the liver and kidney, bone, and in growing feathers. Relative concentrations in different tissues depend on time postexposure and absorption. However, in general, the highest lead concentrations are found in bone, followed by kidney and liver, with intermediate concentrations in brain and blood, and the lowest concentrations in muscle (Longcore et al. 1974b, Johnson et al. 1982, Custer et al. 1984, Garcia-Fernandez et al. 1995). In lead-exposed birds, lead concentrations are often greater in kidney than liver of many species (Longcore et al. 1974b, Custer et al. 1984, Beyer et al. 1988,

Marn et al. 1988, Carpenter et al. 2003, Beyer et al. 2004, Pattee et al. 2006, Schulz et al. 2006). However, the reverse has been reported for eagles and some other raptors (Pattee et al. 1981, Wayland et al. 1999, Iwata et al. 2000, Kenntner et al. 2001, Krone et al. 2004, Battaglia et al. 2005, Martin et al. 2008).

In cases of acute exposure, lead in blood and soft tissues retains a fairly mobile equilibrium and usually remains elevated from several weeks to several months following exposure, in relation to the initial amount absorbed. When exposure is chronic blood lead will remain elevated for proportionately longer. Lead in bone is relatively immobile, loss is very slow, and lead accumulates in bone throughout the lifetime. However, the dynamics of lead absorption and release appear to be accelerated in the medullary bone of egg-laying females as described later. Bone lead concentrations in bird populations generally increase with age, although they may be very high in young individuals subject to high levels of exposure (Stendell et al. 1979, Clausen et al. 1982, García-Fernández et al. 1997, Scheuhammer et al. 1999, Pain et al. 2005). The use of feathers for monitoring of metals in birds is predicated on the fact that metals are deposited during the period of feather growth, but a variety of factors can affect concentrations (Burger 1993). Lead concentrations in feathers of juvenile birds have been shown to be a reliable indicator of dietary exposure at the time of feather formation and to be correlated with lead in some tissues, particularly kidney and bone (Golden et al. 2003). However, the use of adult feathers for assessing lead exposure can be problematic because they are subject to the deposition of atmospheric lead for a longer time, and must be cleaned thoroughly before analysis. Pain et al. (2005) found that, even after cleaning, lead concentrations in the feathers of Spanish imperial eagles (*Aquila adalberti*) from museums were positively correlated with specimen age, suggesting that externally deposited lead can be difficult to remove. Fry and Maurer (2003) and Fry (2004) have proposed using lead analysis of portions of feathers along their length as indicators of lead circulating in the feather pulp at the time of growth in California condors (*Gymnogyps californianus*), relating the findings to feather growth rates to evaluate lead exposure over time.

The chronicity of exposure to lead has an important influence upon the concentrations of lead in various tissues of birds. In cases of chronic exposure, the highest lead concentrations are generally found in bone, with lower concentrations in soft tissues such as liver, kidney, and blood (Custer et al. 1984, Pattee 1984, Mautino and Bell 1986, Mautino and Bell 1987). However, when birds die following acute exposure after the ingestion and absorption of large amounts of lead, concentrations in kidney and/or liver may exceed those in bone. For example, in a die-off of lesser scaup (*Aythya affinis*), recent lead exposure and rapid poisoning were suspected as body weights were positively correlated with the number of shot in the gizzard, indicating that birds exposed to the most lead died before the effects of lead poisoning resulted in weight loss (Anderson 1975). Tissue lead concentrations were consistent with the rapid absorption of large amounts of lead, as mean lead concentrations were 46 mg/kg wet weight (c. 141 mg/kg dry weight) in liver, 66 mg/kg wet weight (c. 282 mg/kg dry weight) in kidney, and 40 mg/kg dry weight in bone (Anderson 1975). Concentrations of lead in liver and kidney of Canada geese (*Branta canadensis*) found during a lead poisoning die-off were more than twice as high as those in bone (Szymczak and Adrian 1978). In a study with bald eagles (*Haliaeetus leucocephalus*), three of four birds that died did so within 20 days after dosage with 10 or more No. 4 lead shot and lead concentrations in tissues, on a dry weight basis were approximately 2.5–10 times higher in liver and kidney than in bone (Pattee et al. 1981). Zebra finches (*Taeniopygia guttata*) that received lead acetate in drinking water for 38 days and mourning doves dosed with lead shot and held for 3 weeks also had higher mean concentrations of lead in kidney than in bone (Marn et al. 1988, Snoeijs et al. 2005). In another study with mourning doves, birds that died within 5 weeks after receiving 1–4 lead shot had kidney lead concentrations that were 1.3–2 times greater than concentrations in bone (Buerger et al. 1986).

Because lead is lost far more slowly from bone than from soft tissues, few birds sampled from wild populations have higher lead concentrations in soft tissues than bone. Clausen et al. (1982) found only 10% of mute swans (*Cygnus olor*) had high liver and low bone lead concentrations, whereas 31% had low liver and high bone lead concentrations. Franson et al. (2009) found that in mourning doves shot by hunters, birds both with and without ingested shot had higher lead concentrations in bone than in liver. In 12 species of raptors, mean lead concentrations were consistently higher in bone than in liver (Martin et al. 2008).

Thus, bone lead concentration is generally considered the best indicator of lead exposure over the total lifetime of the bird, but the least useful indicator of recent lead exposure and absorption. The tissues usually chosen to evaluate recent exposure are blood, liver, and occasionally kidney.

16.4 FACTORS INFLUENCING CONCENTRATIONS OF LEAD IN TISSUES

It is often difficult to relate lead exposure directly to tissue lead concentrations in birds, as many factors influence lead retention and its absorption and deposition into body tissues. The influence of the magnitude and duration (chronic or acute) of lead exposure on tissue lead concentrations was discussed earlier. Both the absorption of lead and its deposition in tissues are affected by a variety of additional factors, including gender, breeding condition, age, stomach type, and diet, as discussed later.

Finley et al. (1976a) found that laying mallards dosed with one No. 4 lead shot (193 mg) accumulated significantly higher liver, kidney, and bone lead concentrations than did males. Although lead concentrations remained relatively low in soft tissues of both genders (maximum means of 1.2 mg/kg wet weight in liver and 3.5 mg/kg wet weight in kidney), mean bone lead concentrations were 112 mg/kg dry weight in females compared with 10 mg/kg dry weight in males. Subsequent studies revealed a similar result and additionally found lead concentrations in femurs to be 4 times higher in laying than nonlaying females (Finley and Dieter 1978). As females mobilize calcium from medullary bone for eggshell formation, the intestinal absorption of calcium (and concurrently lead) increases, resulting in higher bone lead concentrations in females (Krementz and Ankney 1995, Scheuhammer 1996). Lead also is deposited to a greater extent in bones of high medullary content (e.g., femur, sternum, and tibia) than in those with low medullary content (e.g., humerus, ulna, and radius) (Finley and Dieter 1978, Pattee 1984, Rocke and Samuel 1991). Medullary bone is of particular significance in females because of its accumulation, before egg laying, as a source of calcium storage for eggshells, followed by its destruction during eggshell formation (Simkiss 1961). Although lead concentrations of wing bones (low medullary bone content) of waterfowl collected in the autumn did not differ by gender (White and Stendell 1977), a study of American woodcock (*Scolopax minor*) collected in the autumn showed that females had higher lead concentrations in wing bones than males (Scheuhammer et al. 1999).

Although, as mentioned previously, lead concentrations in bones generally increase with age, evidence also exists that large amounts of lead are deposited in bones of immature birds during the time when calcium absorption and deposition are rapid. For example, 70% of American woodcock chicks collected from nests or broods had bone lead concentrations of >20 mg/kg dry weight, compared with 43% of hatch year woodcock with >20 mg/kg collected in the autumn (Strom et al. 2005) and Anderson (1975) reported mean bone lead concentrations of 56 mg/kg dry weight in immature lesser scaup versus 26 mg/kg dry weight in adults during a lead poisoning mortality event. Mateo et al. (2001) also reported higher bone lead concentrations in duckling (<9-day old) than fully grown marbled teal (*Marmaronetta angustirostris*).

The anatomical characteristics of the avian ventriculus (stomach) vary by species and can influence the retention, and thus to some extent the absorption, of metallic lead objects. Stomachs of carnivores, scavengers, and fish-eating birds are adapted for a relatively soft diet, whereas birds that consume grain, vegetation, and insects have a more muscular stomach (or gizzard) adapted for hard diets (Denbow 2000). The acidic conditions in both types of stomachs facilitate the dissolution of lead, but ingested lead may be retained with grit and mechanically ground down in the muscular gizzard of some birds, particularly waterfowl. Although passerines and columbiforms also have muscular gizzards, lead shot dosing experiments suggest that these groups of birds may void shot relatively quickly (Vyas et al. 2001, Schulz et al. 2007). On the other hand, raptors and many other carnivorous birds have a thin-walled stomach and share the particular characteristic of forming and regurgitating pellets, or casts, consisting of undigested bones, hair, and feathers of prey (Duke 1986). Field studies confirm that particulate lead may sometimes be egested with the casts (Platt 1976, Nelson et al. 1989, Mateo et al. 1999, 2007). However, the significance of such regurgitation on total lead absorption is variable, as one experimental study found that bald eagles regurgitated lead from 12 h to 48 days after dosage (Pattee et al. 1981).

Diet, one of the most important factors influencing lead absorption and deposition in tissues, has been extensively studied in waterfowl and to a lesser extent in other species (Jordan and Bellrose 1951, Longcore et al. 1974a, Sanderson and Irwin 1976, Koranda et al. 1979, Sanderson and Bellrose 1986, Marn et al. 1988, Scheuhammer 1996, Vyas et al. 2001). Findings indicate that nutritional, chemical, and physical characteristics of diet are important. In general, nutritionally balanced species-appropriate diets high in protein and calcium tend to mitigate the effects of lead exposure. Much of the mitigating effect of nutrient-rich diets appears to occur in the digestive system, with high calcium and protein levels reducing the gastrointestinal absorption of lead and lowering the total body burden (Koranda et al. 1979, Sanderson 1992, Scheuhammer 1996). Furthermore, when lead is ingested along with food, certain chemical groups in food components have a ligand effect, binding lead in an insoluble form in the intestine (Morton et al. 1985). In waterfowl, there is evidence that the selection of larger grit size is positively related to the prevalence of lead shot ingestion (Pain 1990, Mateo et al. 2000, Figuerola et al. 2005).

Finally, when evaluating reports of lead concentrations in tissues, one should consider the quality assurance/quality control data for the analytical technique used, such as lead concentrations in procedural blanks, spiked samples, duplicates, and standard reference materials. Although standard practice today, these parameters were not reported in many of the older studies, when inadequate sensitivity and contamination issues may have resulted in less precise analyses.

16.5   TISSUE LEAD CONCENTRATIONS AND SUBLETHAL EFFECTS ON ENZYME SYSTEMS

Tissue lead "threshold" concentrations for lead poisoning can be defined according to the tissue lead concentrations at which measurable effects occur. Because of the dynamics of lead uptake and retention, lead concentrations in blood and soft tissues are more easily related to effects than are bone lead concentrations. The effects of blood lead upon hematological parameters and enzyme systems have been extensively studied in birds. Blood lead studies provide useful information, as blood lead concentrations can be related to effect over time, with the extent and duration of lead exposure experimentally controlled.

Lead inhibits the activities of several enzymes necessary for the synthesis of heme, for example, delta-aminolevulinic acid dehydratase (ALAD) and heme synthetase. Heme is incorporated into hemoglobin and mitochondrial cytochromes and is part of cytochrome P-450, which is required in the liver for certain detoxification processes (Sassa et al. 1975, Dieter and Finley 1979). Heme synthetase is responsible for the incorporation of ferrous iron into protoporphyrin IX (PPIX) at the last stage of heme formation. Inhibition of heme synthetase activity results in an accumulation of PPIX in the blood. PPIX fluoresces when exposed to specific wavelengths, and this fluorescence has been used for the quantitative estimation of the PPIX concentration in birds and, consequently, as an indicator of lead exposure (Roscoe et al. 1979, Beyer et al. 1988, O'Halloran et al. 1988a, 1988b, Franson et al. 1996a, Vyas et al. 2000, Carpenter et al. 2003, Pattee et al. 2006). However, the first measurable biochemical change resulting from lead absorption appears to be the inhibition of erythrocyte ALAD activity (Hernberg et al. 1970, Tola et al. 1973). The inhibition of erythrocyte ALAD activity by blood lead is described in some detail later and illustrates the difficulty of interpreting the biological significance of sublethal effects, even when they can be directly related to tissue lead concentrations.

Inhibition of erythrocyte ALAD activity persists for several weeks to several months following lead absorption, in relation to elevated blood lead concentrations (Finley et al. 1976b, Dieter and Finley 1978, Pain 1987, Redig et al. 1991, Carpenter et al. 2003, Pattee et al. 2006). Inhibition of

ALAD activity in avian blood has been reported at blood lead concentrations of <5 μg/dL (c. 0.05 mg/kg) (Pain 1989, Martínez-López et al. 2004). Blood lead concentrations of 15–20 μg/dL (c. 0.15–0.2 mg/kg), 30–80 μg/dL (c. 0.3–0.8 mg/kg), and ≥100 μg/dL (c. 1 mg/kg) have been reported to result in inhibition in ALAD of 50%, 60–80%, and 75% to nearly 100%, respectively (Finley et al. 1976b, 1976c, Dieter and Finley 1978, Hoffman et al. 1981, Pain 1989, Redig et al. 1991, Work and Smith 1996, Franson et al. 2002, Beyer et al. 2004, Pattee et al. 2006).

Birds appear to be able to tolerate some reduction of erythrocyte ALAD activity without showing signs of reduced hematocrit or hemoglobin concentration, although anemia may occur following sustained low level ALAD inhibition. Rapid decreases in hematocrit after exposure to a large amount of lead may be associated with hemolytic anemia, as well as severe (e.g., >75%) ALAD inhibition (Pain and Rattner 1988, Mateo et al. 2003). However, although the significance of reduced erythrocyte ALAD activity is not always easily determined, ALAD activity is also inhibited in other body tissues. One month after administering doses of one lead shot to mallards, Dieter and Finley (1979) recorded a 75% reduction in erythrocyte ALAD activity (blood lead, 98 μg/dL, c. 0.98 mg/kg), a 42% inhibition in liver ALAD activity (liver lead, 2.24 mg/kg of wet weight), a 50% reduction in ALAD activity in the cerebellum, and a 35% reduction in the cerebral hemisphere (brain lead, 0.43 mg/kg of wet weight). ALAD activity was correlated with the lead concentration in all tissues but was more sensitive to lead in the brain than in the liver, where some lead may possibly be bound in a biologically inactive form. The authors also recorded a significant increase in butylcholinesterase (a marker enzyme for glial or supportive cells) activity in the brain, suggesting possible brain damage. The results of this study suggest that blood lead concentrations of 100 μg/dL (c. 1 mg/kg) may be associated with pathological changes in waterfowl brains, but the histologic lesions found in the central nervous system of lead-exposed birds have been minor (Wobeser 1997).

16.6 LEAD AND IMMUNOSUPPRESSION

Several studies have addressed the immunosuppressive effects of lead exposure in birds. Mallards dosed with one No. 4 lead shot exhibited depressed hemagglutination titers to sheep red blood cells, indicating an effect on antibody-mediated immunity by day 7 after treatment, and titers remained low until the end of the 3-week experiment (Trust et al. 1990). Three ducks died during the experiment, exhibiting clinical signs and lesions consistent with lead poisoning. In Japanese quail (*Coturnix coturnix*), lead acetate in drinking water suppressed antibody-mediated response, but only at dosages that also caused clinical lead poisoning (Grasman and Scanlon 1995). In two other studies with Japanese quail, lead exposure did not affect any of the immune system parameters measured (Morgan et al. 1975, Fair and Ricklefs 2002). Western bluebird (*Sialia mexicana*) nestlings dosed with three No. 9 lead shot showed a reduction in cell-mediated immune response, although those receiving one or two lead shots were not affected (Fair and Myers 2002). Redig et al. (1991) reported depressed cell-mediated immunity in red-tailed hawks (*Buteo jamaicensis*) that received increasing amounts of lead acetate over a 10-week period. As discussed previously, the distribution of lead among tissues may be different between females and males during the prebreeding and breeding seasons. The storage of higher proportions of lead in bone prior to and during breeding by females could result in gender differences in the immune response of birds exposed to a similar amount of lead during this period. For example, Rocke and Samuel (1991) noted an increased immunosuppressive effect of lead in male compared with female mallards during the prebreeding season. In nonbreeding zebra finches, however, antibody-mediated immunity was suppressed by lead in females compared with males, but only in birds on a low calcium diet (Snoeijs et al. 2005).

16.7 BACKGROUND LEAD CONCENTRATIONS IN AVIAN TISSUES

As discussed earlier, there are difficulties associated with relating exposure to tissue lead concentrations and with relating tissue lead concentrations to effect. However, for managing wildlife on contaminated areas, it is important to provide recommendations for interpreting tissue concentrations. These recommendations are based on the tissue lead concentrations in unexposed wild birds and the concentrations at which clinical effects and mortality may occur.

Background blood lead concentrations in birds are generally low, usually <20 µg/dL (c. 0.2 mg/kg) and frequently well below 10 µg/dL (c. 0.1 mg/kg) (Dieter et al. 1976, Szymczak and Adrian 1978, Birkhead 1983, Franson et al. 1986, 2004, Pain 1989, Trust et al. 1990, Redig et al. 1991, Carpenter et al. 2003, Pattee et al. 2006). Birds with no history of lead poisoning usually have liver and kidney lead concentrations of <2 mg/kg wet weight and frequently of <1 mg/kg wet weight (Bagley and Locke 1967, Irwin 1975, Clausen and Wolstrup 1979, Custer et al. 1984, Spray and Milne 1988, Kingsford et al. 1989, Franson et al. 1995a, Beyer et al. 1998, Carpenter et al. 2003, Kenntner et al. 2003, Martin et al. 2008). Bone lead concentrations are more difficult to interpret, and concentrations tend to be higher because of accumulation. Different values for background concentrations have been proposed, but concentrations of <10–20 mg/kg dry weight are generally considered as background or are reported in unexposed birds (Moore 1978, Szymczak and Adrian 1978, Pain et al. 1992, Martin et al. 2008, Franson et al. 2009).

16.8 RECOMMENDATIONS FOR THE INTERPRETATION OF TISSUE LEAD CONCENTRATIONS IN BIRDS

The interpretation of tissue lead concentrations is facilitated by the availability of information on exposure and clinical signs of poisoning. In live birds, useful information may include the presence of lead in the casts regurgitated by raptors, the presence of lead particles in the gastrointestinal tracts of birds identified by x-ray, and clinical signs of lead poisoning. At necropsy, pathological observations and the presence of ingested lead particles are useful indicators, along with history about the bird before death. However, detailed history information is not always available. In these situations, a prediction can be made regarding the degree to which both live and dead birds were affected by lead poisoning by evaluating tissue residues. We categorize lead residues in the blood, liver, and kidney of Anseriformes, Falconiformes, and Columbiformes according to increasing severity of effects (Table 16.1): (1) subclinical poisoning, a range of residues reported to cause physiological effects that are insufficient to severely impair normal biological functioning, resulting in no external signs of poisoning, and from which the bird would probably recover if lead exposure were terminated; (2) clinical poisoning, an approximate threshold level marking the initiation of clinical signs (pathological manifestations of physiological effects) such as anemia, microscopic lesions in tissues, weight loss, muscular incoordination, green diarrhea, and anorexia leading to probable death if lead exposure were to continue; and (3) severe clinical poisoning, an approximate threshold value at which the effects may be directly life threatening in field, captive, and/or experimental cases of lead poisoning. We consider residues below the subclinical poisoning range as "background," that is, evidence of environmental exposure distant from any specific source of lead contamination. Table 16.1 is meant to provide general guidance in the assessment of residues, with an awareness that numerous factors, some of which are discussed later, may contribute to an overlap of residue values in the three categories. It is also important to note that birds in experimental studies may accumulate higher levels of lead in tissues than wild birds subject to stressors in the environment.

16.8.1 INDIVIDUAL BIRDS

It is important to note that for lead, as for many other contaminants, toxic effects may depend upon factors other than simply the concentrations in tissues. These factors include the level and duration of lead exposure, previous history of exposure, species variability in response to exposure, the overall health of the bird, the extent of damage already done, and the potential interactions between lead and other disease agents. The biological significance of a tissue lead concentration may, therefore, be difficult to determine if the history of the bird is unknown. The chronicity of exposure is

TABLE 16.1
Suggested Interpretations of Tissue[a] Lead Concentrations in Three Orders of Birds

Order	Blood (μg/dL)[b]	Liver[c] (mg/kg ww)	Kidney (mg/kg ww)	Reference
Anseriformes				
Subclinical poisoning	20 < 50	2 < 6	2 < 6	Dieter and Finley (1979), Degernes (1991)
Clinical poisoning	50–100	6–10	6–15	Longcore et al. (1974b), Degernes (1991), Beyer et al. (2000)
Severe clinical poisoning	>100	>10	>15	Cook and Trainer (1966), Longcore et al. (1974b), Mautino and Bell (1986), Beyer et al. (1988, 2000), Pain and Rattner (1988), Pain (1989), Blus et al. (1991, 1999), Degernes (1991), Kelly et al. (1998), Nakade et al. (2005), Degernes et al. (2006)
Falconiformes				
Subclinical poisoning	20 < 50	2 < 6	2 < 4[d]	Custer et al. (1984), Henny et al. (1991), Kramer and Redig (1997)
Clinical poisoning	50–100	6–10	4–6[d]	Lumeij et al. (1985), Kramer and Redig (1997)
Severe clinical poisoning	>100	>10	>6[d]	Redig et al. (1980), Hoffman et al. (1981), Pattee et al. (1981, 2006), Langelier et al. (1991), Kramer and Redig (1997)
Columbiformes				
Subclinical poisoning	20 < 200	2 < 6	2 < 15	Ohi et al. (1974), Cory-Slechta et al. (1980), Kendall et al. (1982), Kendall and Scanlon (1982), DeMent et al. (1987), Scheuhammer and Wilson (1990)
Clinical poisoning	200–300	6–15	15–30	Cory-Slechta et al. (1980), Anders et al. (1982), Boyer et al. (1985)
Severe clinical poisoning	>300	>15	>30	Locke and Bagley (1967), Barthalmus et al. (1977), Cory-Slechta et al. (1980), Anders et al. (1982), Boyer et al. (1985), Schulz et al. (2006)

[a] Lead concentrations in bone reflect lifetime accumulation and chronic low exposure to lead may result in similar concentrations in bone as acute exposure to higher levels. If evidence of acute exposure exists, we recommend that bone lead concentrations (dry weight basis) of <10 μg/g be considered background, 10–20 μg/g be considered evidence of subclinical to clinical poisoning, and >20 μg/g be considered evidence of severe clinical poisoning.

[b] Divide μg/dL by 100 for an approximate conversion to mg/kg.

[c] In general, a diagnosis of lead poisoning in an individual bird can be reached if necropsy observations are consistent with lead poisoning and the lead concentration in the liver is ≥6 mg/kg wet weight.

[d] Although in many species lead concentrations are often higher in kidney than in liver, the reverse has been found in lead-exposed and poisoned eagles and some species of hawks; thus, for Falconiformes, we have suggested a conservative lead threshold concentration in kidney.

particularly important. Birds exposed to relatively low lead levels on a sustained basis may suffer similar effects, but with lower soft tissue lead concentrations, than birds acutely exposed to higher levels of lead for a short period of time. Also, exposure and tissue lead concentrations are not always associated in individual birds because of the varying retention time of shot in the gizzard and the uptake/retention dynamics of lead in tissues. However, in a live bird, sequential blood lead analyses from an individual give a much clearer picture of the significance of contamination as chronicity can be established. In addition, hematological measurements such as ALAD activity, PPIX concentration, hematocrit, and hemoglobin concentrations will indicate biochemical damage.

A wide range of clinical signs may be observed in lead-poisoned birds following chronic exposure, including anorexia, emaciation, anemia, lethargy, wing droop, ataxia, green diarrhea staining the vent, and neurological signs such as leg paralysis or convulsions (Locke and Thomas 1996, Wobeser 1997, Friend 1999, Eisler 2000, Pattee and Pain 2003). Some of the gross and microscopic lesions in waterfowl with lead poisoning (Coburn et al. 1951, Cook and Trainer 1966, Karstad 1971, Clemens et al. 1975, Forbes and Sanderson 1978, Hunter and Wobeser 1980, Wobeser 1997) have been reported for other avian species (Locke and Bagley 1967, Hunter and Haigh 1978, Pattee et al. 1981, 2006, Beyer et al. 1988, Langelier et al. 1991, Vyas et al. 2001, Carpenter et al. 2003). Muscle wasting and loss of fat reserves are some of the most consistent lesions associated with lead poisoning across avian taxa. Other gross lesions include impactions of the esophagus or proventriculus, distended gallbladder, dark discolored gizzard lining, light areas (gross evidence of necrosis) in heart or gizzard muscle, wasting of internal organs, pale flabby heart, pale internal organs and muscle tissue, and atrophied internal organs. Particulate lead may or may not be present in the stomach. The presence of acid-fast intranuclear inclusion bodies in kidney tubular cells are indicative of lead poisoning, but are not present in all cases. Additional microscopic lesions include hemosiderosis in the liver, necrosis of myocardial and gizzard muscle, and degenerative changes in the brain and peripheral nerves. In cases where birds die rapidly following acute exposure to high levels of lead, many of these lesions may be absent. It should be noted that while acid-fast intranuclear inclusion bodies are a relatively specific indicator of lead poisoning, other signs are nonspecific and may be observed in association with a variety of other conditions. In waterfowl, Beyer et al. (1998) found impactions of the upper alimentary tract, submandibular edema, myocardial necrosis, and biliary discoloration of the liver to be the gross lesions most reliably associated with lead poisoning.

In individual birds, lead poisoning as a cause of death should be distinguished from the observation that the bird simply has been exposed to lead, based solely on tissue residues. The confidence with which a diagnosis of lead poisoning as the cause of death can be made will increase with the amount of information available. Ideally, a diagnostic evaluation will include exposure history data, necropsy observations, and pathological findings. Necropsy and pathological findings consistent with lead poisoning, in conjunction with lead residues of ≥6 mg/kg wet weight in liver, for example, support a diagnosis of lead poisoning as the cause of death. On the other hand, if necropsy observations and pathological findings are not available, a more conservative approach is recommended. Beyer et al. (1998) studied liver lead concentrations in waterfowl diagnosed as lead-poisoned, finding that 95% of 421 ducks and geese diagnosed with lead poisoning had lead concentrations of at least 38 mg/kg dry weight (10 mg/kg wet weight), but less than 1% of birds that died of other causes had a level that high. The authors concluded that 38 mg/kg dry weight in liver is a defensible criterion for identifying waterfowl suffering from lead poisoning, although a definitive diagnosis requires the finding of concurrent lesions compatible with lead toxicosis. Single bone lead concentrations are the least useful indices of poisoning due to accumulation over time. However, high liver and low bone lead concentrations may suggest severe clinical poisoning following recent exposure to a large amount of lead.

16.8.2 POPULATIONS

The determination of lead concentrations in blood is a useful, and nonlethal, way to study lead exposure in populations of birds. Flint et al. (1997) collected blood samples from a population of threatened spectacled eiders (*Somateria fischeri*) in Alaska, finding that the probability of lead exposure (≥20 µg/dL blood lead, or c. 0.2 mg/kg) increased through the breeding season. In a follow-up study with spectacled eiders, adult females with ≥20 µg/dL blood lead before hatching their eggs were found to survive at a much lower rate than females with blood lead concentrations of <20 µg/dL before hatch (Grand et al. 1998). Sympatrically nesting common eiders (*Somateria mollissima*) were found to have much lower frequencies of elevated lead than spectacled eiders, probably as a result of differing foraging behavior and brood rearing strategies (Flint et al. 1997, Wilson et al. 2007). In canvasbacks (*Aythya valisineria*), Hohman et al. (1995) found that winter survival rates for immatures with ≥0.2 mg/kg blood lead were lower than in those with <0.2 mg/kg during two of three winters. DeStefano et al. (1991) measured blood lead concentrations and studied lead exposure in the Eastern Prairie Population of Canada geese, reporting increasing levels of exposure as geese moved from breeding to wintering grounds. Pain et al. (1997) found a temporal variation in the proportion of marsh harriers (*Circus aeruginosus*) with elevated blood lead concentrations, with a far higher incidence during than outside the hunting season. Blood lead concentrations have been used in recent years to monitor California condor populations for lead exposure, and to form the basis for decisions regarding chelation therapy in individuals of this highly endangered species (Hall et al. 2007, Parish et al. 2007).

The activity of ALAD is well correlated with blood lead concentrations, particularly up to about 100–150 µg/dL (c. 1–1.5 mg/kg). As absolute levels of ALAD vary among individuals, the ratio of lead inhibited ALAD activity to reactivated ALAD activity, with the lead displaced, has been recommended for monitoring lead exposure in populations (Dieter 1979, Pain 1989, Scheuhammer 1989). Although less useful than other tissues in individual cases, bone lead concentrations have been used to determine geographical patterns of lead poisoning in populations (Stendell et al. 1979, Scheuhammer and Dickson 1996). In addition, at a population level, bone and liver lead concentrations may be correlated (Anderson 1975), and tissue lead concentrations may sometimes be correlated with exposure, as measured by the presence of shot in gizzards (White and Stendell 1977). Pain et al. (1992) found a positive correlation between exposure to shot (measured as the percentage of shot ingestion in eight species of waterfowl and shore birds) and liver and bone lead concentrations.

Concentrations of lead in tissues will sometimes be elevated via pathways other than metallic lead from ammunition or angler's weights, such as exposure to mining or smelting wastes, lead-based paint, or industrial effluents. An evaluation of the distribution of blood or liver lead concentrations may help identify possible sources of exposure. Distributions of blood and soft tissue lead concentrations from populations in which some birds have ingested metallic lead tend to be very skewed or have distinct outliers (Dieter 1979, Beyer et al. 1998), whereas concentrations resulting from a more general source of lead, for example, in water or the atmosphere, may be more normally distributed. The determination of stable lead isotope ratios, Pb[204,206,207,208], in bird tissues may also assist in determining potential sources of lead exposure (Scheuhammer and Templeton 1998, Meharg et al. 2002, Scheuhammer et al. 2003a, Pain et al. 2007). For example, Church et al. (2006) found lead isotope values in free-flying California condors with relatively high blood lead concentrations to be consistent with lead from ammunition and distinct from those of pre-release condors with low blood lead concentrations.

16.8.3 SPECIES

Lead concentrations in tissues associated with poisoning in birds vary among orders, with Galliformes and Columbiformes often having higher residues than others (Pattee et al. 1981, Kendall et al. 1983, Gjerstad and Hanssen 1984, Reichel et al. 1984, Beyer et al. 1988, 1998, Vyas et al. 2001, Lewis et al. 2001, Schulz et al. 2006). Within-order differences among species also have been noted, as lead-poisoned Anseriformes of lighter weight tended to have higher lead concentrations in the liver than species of greater weight (Beyer et al. 1998). Within Falconiformes, one study of birds similarly dosed with lead acetate found that the blood lead concentration of a turkey vulture was nearly 8 times greater than in two red-tailed hawks after 6 weeks of dosage and that the vulture developed clinical signs of lead poisoning, but the hawks did not (Reiser and Temple 1981). In a more recent study with turkey vultures, Carpenter et al. (2003) reported that, although four of six birds died or were euthanized because of severe clinical signs, survival time, even with frequent redosing because of regurgitated or defecated lead shot, was quite long and lead pellet dosage was high compared with other avian studies. The authors concluded that turkey vultures are relatively tolerant to lead exposure and would not be a good model for evaluating the risk of lead exposure in California condors (Carpenter et al. 2003).

Although nontoxic shot regulations have been introduced in many countries, with a few exceptions these primarily restrict the use of lead gunshot over wetlands or for waterfowl hunting. While, where compliance is good, this should reduce lead ingestion in waterfowl and their main predators, other avian predators and scavengers, a range of other terrestrial birds continue to be subject to lead exposure and poisoning because of their feeding habits. Lead poisoning in the California condor, which feeds almost exclusively on carrion, is now considered a major obstacle to the recovery of the species (Meretsky et al. 2000, Cade 2007, Snyder 2007). California condors ingest lead fragments from ammunition in carcasses of animals or offal left in the field (Church et al. 2006, Hunt et al. 2006, Cade 2007), and considerable effort is devoted to monitoring and attempts to reduce lead exposure (Hall et al. 2007, Parish et al. 2007, Sullivan et al. 2007). Lead poisoning of wintering bald eagles and golden eagles (*Aquila chrysaetos*) was long thought to be primarily due to the ingestion of shotgun pellets in dead or crippled waterfowl, but Kramer and Redig (1997) found that the prevalence of lead poisoning in eagles did not decrease after the implementation of the 1991 nontoxic shot regulations for hunting waterfowl and American coots (*Fulica americana*) in the United States. The authors suggested that carcasses of animals still hunted with lead ammunition, such as small mammals and birds, and deer, could be additional sources of exposure (Kramer and Redig 1997). Lead exposure and poisoning has been reported in a variety of eagle species and other raptors throughout the world (Pain et al. 2009).

16.9   LEAD RESIDUES REPORTED IN SELECTED FIELD AND LABORATORY STUDIES IN BIRDS

16.9.1 ANSERIFORMES (DUCKS, GEESE, AND SWANS)

Lead concentrations in the liver and kidney of sick and dead Canada geese in an early field report ranged from 9 to 27 mg/kg and 12 to 57 mg/kg (wet weight presumed but not stated) (Adler 1944). One of the early experimental studies of lead exposure in waterfowl that reported concentrations in tissues was that of Coburn et al. (1951) who dosed mallards with lead nitrate. Birds that died in that study had average bone and liver lead concentrations of 469 and 208 mg/kg dry weight, respectively (Coburn et al. 1951). Canada geese that died of experimental lead shot exposure had lead concentrations in the liver of 5–32 mg/kg (wet weight presumed but not stated) (Cook and Trainer 1966). Longcore et al. (1974b) summarized results of seven experimental lead exposure studies in Canada geese and mallards, in which birds both died and were euthanized, finding that mean liver lead concentrations ranged from 12 to 51 mg/kg wet weight. Mallards that died after being dosed with one No. 4 lead buckshot had mean liver and kidney lead concentrations (mg/kg wet weight) of 51–64 and 158–259, respectively (Longcore et al. 1974b). In wild Canada geese, tundra swans (*Cygnus columbianus*), and mallards picked up during lead poisoning die-offs, mean liver lead concentrations were 12–28 mg/kg wet weight (Longcore et al. 1974b). In another lead poisoning event, the mean lead concentration in livers of Canada geese was 16 mg/kg wet weight (Bagley et al. 1967). Lesser scaup found sick or dead during a lead poisoning mortality event had average lead concentrations in tissues of 46 mg/kg wet weight in liver, 66 mg/kg wet weight in kidney, and 40 mg/kg dry weight in wing bones (Anderson 1975). Average lead concentrations in liver and kidney (mg/kg wet weight assumed but not stated) of lead poisoned mallards and mute swans were 40 and 58, and 33 and 105, respectively (Clausen and Wolstrup 1979). Mean liver lead concentrations in four species of waterfowl picked up during a lead poisoning die-off ranged from 15 to 31 mg/kg wet weight, or 86–131 mg/kg dry weight (Zwank et al. 1985). Liver lead concentrations in four spectacled eiders and a common eider found dead or moribund and diagnosed with lead poisoning ranged from 26 to 52 mg/kg wet weight (Franson et al. 1995b). The lead concentration in a blood sample from one of the spectacled eiders was 8.5 mg/kg (Franson et al. 1995b). In a field study of lead poisoning in tundra swans and trumpeter swans (*Cygnus buccinator*), Degernes et al. (2006) reported mean liver lead concentrations of 42–203 mg/kg dry weight in lead-poisoned swans, and 0.5–2.1 mg/kg dry weight in swans not poisoned by lead. Whooper swans (*Cygnus cygnus*) that were diagnosed as lead poisoned had liver lead concentrations of 5.5–44.3 mg/kg wet weight (Ochiai et al. 1992). The mean and fifth percentiles of liver lead concentrations in waterfowl diagnosed with lead poisoning were 115 and 38 mg/kg dry weight, respectively (Beyer et al. 1998). Mallards dosed with lead shot, half of which died, had liver and kidney lead concentrations of 78 and 256 mg/kg dry weight (Kelly et al. 1998). When given a chronic dose of lead acetate, mallards died with lower tissue lead concentrations than when given a large dose of lead from lead pellets (Beyer et al. 1988).

In experimental studies of lead exposure in waterfowl, blood lead concentrations can reach 7 mg/kg or greater. After one week, concentrations of lead in blood of ring-necked ducks (*Aythya collaris*) dosed with one No. 4 lead shot and in mallards dosed with two No. 4 lead shot were 7.6 and 7.8 mg/kg (Mautino and Bell 1986, 1987). Three and six days after canvasbacks were dosed with one No. 4 lead shot, blood lead concentrations were 7.4 mg/kg and about 4.5 mg/kg, respectively (Franson et al. 1986). In Canada geese dosed with multiple No. 4 lead shot, peak blood lead concentrations at days 3 and 10 were 16.8 mg/kg and 6.7 mg/kg, respectively (Cook and Trainer 1966). Concentrations of lead in blood of ducks that survive experimental lead shot exposure generally decline to about 1 mg/kg or less after 1 month (Dieter and Finley 1978, 1979, Franson et al. 1986, Mautino and Bell 1986, 1987). Concentrations of lead in blood associated with clinical signs in field cases of lead-poisoned waterfowl tend to be lower than in experimental studies. Trumpeter swans with 1.0–1.99 mg/kg lead in their blood had pronounced clinical signs, including weakness and neurological abnormalities (Degernes 1991). With treatment, 6 of 10 birds in this group survived (Degernes 1991). Concentrations of lead in blood of six whooper swans that were diagnosed with lead poisoning were 300–630 µg/dL (c. 3–6.3 mg/kg) (Ochiai et al. 1992). Six whooper swans and two tundra swans that were captured exhibiting signs of weakness, green feces, and pale conjunctiva had blood lead concentrations ranging from 2.5 to 6.7 mg/kg (Nakade et al. 2005). Seven of the eight swans died on the second day after capture, lead shot were recovered from gizzards, and liver and kidney lead concentrations (wet weight presumed but not stated) ranged from 14.0 to 30.4 mg/kg and 30.2 to 122 mg/kg, respectively. The single whooper swan that survived had an original blood lead concentration of 2.9 mg/kg, which declined to about 0.6 mg/kg at 30 days (Nakade et al. 2005). Sick Canada geese captured during a lead poisoning die-off had a mean blood lead concentration of 5.2 mg/kg (Szymczak and Adrian 1978).

Tissue residues described above are primarily the result of lead shot ingestion. Poisoning of waterfowl from the ingestion of lead in sediments can result in similar tissue concentrations. Mortality of tundra swans near an area of lead mining and smelting was reported as early as 1924 and early analysis of liver samples revealed lead concentrations of 18–37 mg/kg wet weight (Chupp and Dalke 1964). Later studies in the same area reported mean liver lead concentrations of 10–24 mg/kg wet weight and mean blood lead concentration of 3.3 mg/kg in swans and up to 8 mg/kg in blood and 14 mg/kg wet weight in liver of wood ducks (*Aix sponsa*) (Blus et al. 1991, 1993, 1999). The mean and fifth percentile blood lead concentrations in tundra swans moribund from lead poisoning were 3.6 and 1.9 mg/kg, respectively (Beyer et al. 2000). Sileo et al. (2001) diagnosed lead poisoning in 219 waterfowl found sick or dead, none of which had metallic lead in their gizzards. Liver lead concentrations in 216 birds in that study ranged from 6.3 to 90 mg/kg wet weight. The three others were diagnosed with lead poisoning as follows: a tundra swan and a Canada goose had <6 mg/kg wet weight lead in liver, but had inclusion bodies in the kidney, and a mallard had 38.7 mg/kg wet weight lead in the kidney (Sileo et al. 2001). Heinz et al. (1999) conducted a series of experiments feeding mallards 3–48% sediment (containing lead concentrations of 3400 and 4000 mg/kg dry weight) from the mining and smelting area. Lead concentrations in blood, liver, and kidney were as high as 6.9, 38, and 31 mg/kg wet weight, respectively (Heinz et al. 1999).

16.9.2 FALCONIFORMES (HAWKS, FALCONS, AND VULTURES)

A wide range of liver and kidney residues have been reported from raptors dying of lead poisoning, but levels are often in the range of 5–40 mg/kg wet weight. In lead-exposed or poisoned eagles, lead concentrations are often higher in liver than in kidney. Wayland et al. (1999) studied relationships of lead concentrations among tissues of bald and golden eagles, finding that when the lead concentration in liver was 6 mg/kg dry weight, the predicted renal concentration was 4.6 mg/kg, but when the concentration in the liver was 30 mg/kg dry weight, the predicted renal concentration was 18 mg/kg. The published records of 37 wild bald eagles that died of lead poisoning report liver lead residues of 5–61 mg/kg wet weight and of 5 and 12 mg/kg wet weight in the two kidneys tested (Mulhern et al. 1970, Kaiser et al. 1980, Reichel et al. 1984, Frenzel and Anthony 1989, Craig et al. 1990, Langelier et al. 1991, Gill and Langelier 1994). The mean lead concentration in kidneys of five bald eagles that showed evidence of lead poisoning was 34 mg/kg dry weight (Elliott et al. 1992). Of three captive bald eagles that died of lead poisoning, two had liver lead residues of 23 and 15 mg/kg and kidney residues of 11 mg/kg each (wet weight presumed but not stated) (Jacobson et al. 1977, Redig et al. 1980). The third had 26 mg/kg wet weight of lead in the liver and 9 mg/kg wet weight in the kidney (Janssen et al. 1979). In an experimental study of lead shot poisoning, four eagles died with mean lead residues of 17 mg/kg wet weight in the liver, 6 mg/kg wet weight in the kidney, 1.4 mg/kg wet weight in the brain, and 10 mg/kg dry weight in the bone (Pattee et al. 1981). A fifth eagle in that study that became blind and was euthanized had lead concentrations of 3 mg/kg wet weight in the liver and kidney, 2 mg/kg wet weight in the brain, and 13 mg/kg dry weight in the bone.

A Steller's sea eagle (*Haliaeetus pelagicus*) and a white-tailed sea eagle (*Haliaeetus albicilla*) with lead fragments in the intestine and gizzard, respectively, had liver lead concentrations of 139 and 79 mg/kg dry weight and kidney lead concentrations of 67 and 58 mg/kg dry weight (Kim et al. 1999). Mean lead concentrations in liver and kidney of two Steller's sea eagles and three white-tailed sea eagles with lead bullet fragments in their stomachs were 92 and 156 mg/kg dry weight, respectively (Iwata et al. 2000). Five white-tailed sea eagles diagnosed with lead poisoning had mean liver and kidney lead concentrations of 22.4 and 9.6 mg/kg wet weight, respectively (Krone et al. 2004, 2006, 2009).

Liver lead concentrations in three California condors that died of lead poisoning were 6, 23, and 35 mg/kg wet weight, respectively (Wiemeyer et al. 1988). The relatively low liver lead concentration in the first bird was probably the result of chelation therapy administered before it died. The liver and kidney lead concentrations (mg/kg wet weight) in Andean condors dosed with lead shot, and that died or were euthanized, ranged from 45 to 110 and 124 to 179, respectively (Pattee et al. 2006). A wild griffon vulture (*Gyps fulvus*) that died after admittance to a rehabilitation center had a piece of lead in its gizzard and a liver lead concentration of 52 mg/kg dry weight (Mateo et al. 1997b). In a lead shot dosing study with turkey vultures, lead concentrations (mg/kg wet weight) in liver and kidney of birds that died or were euthanized with severe clinical signs ranged from 6.8 to 34 and 180 to 246, respectively (Carpenter et al. 2003). Reports of lead poisoning mortality in captive condors and vultures include an Andean condor with a liver lead level of 38 mg/kg wet weight (Locke et al. 1969) and two captive king vultures (*Sarcoramphus papa*) with liver lead concentrations of 63 and 7 mg/kg and kidney residues of 71 and 25 mg/kg wet weight (Decker et al. 1979).

A wild red-tailed hawk that died of lead poisoning had a liver lead concentration of 71 mg/kg (wet weight presumed but not stated) and eight ingested lead shot (Sikarskie 1977). Two red-tailed hawks found emaciated, but with no lead in their stomachs, had lead concentrations in liver of 4.3 and 10 mg/kg wet weight (Franson et al. 1996b) and a third emaciated red-tailed hawk with green staining around the vent had lead concentrations of 22.3 mg/kg dry weight in liver and 8.2 mg/kg dry weight in kidney (Martin et al. 2008). Two additional red-tailed hawks had necropsy lesions consistent with lead poisoning and lead concentrations in liver or kidney of >30 or >20 mg/kg dry weight, respectively (Clark and Scheuhammer 2003). A common buzzard (*Buteo buteo*) with paresis of the legs and contracted talons had lead concentrations (all on dry weight basis) of 47.7 mg/kg in liver, 6.6 mg/kg in kidney, and 42.0 mg/kg in bone (Battaglia et al. 2005). MacDonald et al. (1983) reported lead poisoning mortality, probably caused by the presence of metallic lead in food items, in captive Falconiformes including a common buzzard, Eurasian sparrowhawk (*Accipiter nisus*), two peregrine falcons (*Falco peregrinus*), and a lagger falcon (*Falco jugger*). Liver lead concentrations were 36–175 mg/kg dry weight, and kidney residues were 31–221 mg/kg dry weight. Two prairie falcons (*Falco mexicanus*) that died of lead poisoning after being fed parts of animals shot by hunters had lead residues of 17 and 57 mg/kg in the liver and 6 and 78 mg/kg (wet weight presumed but not stated) in the kidney (Benson et al. 1974, Redig et al. 1980). Three raptors that died after receiving daily doses of lead acetate for several weeks had lead residues (dry weight basis) of 2.8–19.7 mg/kg in the liver, 4.2–17.4 mg/kg in the kidney, and 33.4–41 mg/kg in bone (Reiser and Temple 1981). Nestling American kestrels (*Falco sparverius*) dosed with metallic lead powder exhibited reduced ALAD activity and anemia with liver and kidney lead concentrations of 4 and 7 mg/kg wet weight, while survivors of a dosage that killed 40% of the treatment group had lead residues of 6 and 16 mg/kg wet weight in the liver and kidney (Hoffman et al. 1985a, 1985b). American kestrels fed a diet containing about 450 mg/kg dry weight of biologically incorporated lead had mean tissue residues of 10 mg/kg dry weight in the liver, 15 in the kidney, 2 in the brain, and 18 in bone with no effects on body weight, hematocrit, hemoglobin concentration, or red blood cell count (Custer et al. 1984). The authors concluded that their study provided further evidence that lead poisoning of raptors is probably due to the ingestion of metallic lead, not biologically incorporated lead.

Raptors exhibited anemia, anorexia, and bile-stained feces when blood lead residues reached 5–8 mg/kg (Reiser and Temple 1981). After 14 days of exposure to lead shot, five bald eagles had a mean blood lead concentration of 5.4 mg/kg and exhibited 80% ALAD depression and a 20–25% reduction in hematocrit and hemoglobin concentration (Hoffman et al. 1981). A California condor that was captured in a weakened condition and later died had a blood lead concentration of 420 µg/dL (c. 4.2 mg/kg) (Janssen et al. 1986). Of 437 blood samples collected from wild California condors in Arizona, 137 had lead concentrations of 15–59 µg/dL (c. 0.15–0.59 mg/kg), 25 were >100 µg/dL (c. 1 mg/kg), 10 were >200 µg/dL (c. 2 mg/kg), and 5 were >400 µg/dL (c. 4 mg/kg) (Parish et al. 2007). Among 214 blood samples from 44 individual California condors in southern California, 95 had lead concentrations >20 µg/dL (c. 0.2 mg/kg), 18 had 60–99 µg/dL (c. 0.6–0.99 mg/kg), and 7 had >100 µg/dL (c. 1 mg/kg) (Hall et al. 2007). Lead concentrations in 126 blood samples from 33 wild Big Sur California condors were >20 µg/dL (c. 0.2 mg/kg) in 27, 60–99 µg/dL (c. 0.6–0.99 mg/kg) in 4, and >100 µg/dL (c. >1 mg/kg) in 2 (Sorenson and Burnett 2007). In captive Andean condors dosed with lead shot, blood lead concentrations reached 16 and 17 mg/kg wet weight (Pattee et al. 2006). Two turkey vultures with clinical signs consistent with lead toxicosis had blood lead residues of 320 µg/dL (c. 3.2 mg/kg) and 11 mg/kg, but both recovered (Janssen et al. 1979, Reiser and Temple 1981). Turkey vultures that died after exposure to lead shot had 6–30 mg/kg wet weight lead in the heart blood clot (Carpenter et al. 2003). A prairie falcon with weakness, weight loss, and anemia had 11 mg/kg of lead in its blood but recovered after chelation therapy (Redig et al. 1980). A honey buzzard (*Pernis apivorus*) with a blood lead concentration of 80 µg/dL (c. 0.8 mg/kg) exhibited clinical signs of green diarrhea, muscle wasting, and weakness and had one lead shot in its stomach (Lumeij et al. 1985). Two weeks after removal of the lead shot when the blood lead concentration was 16 µg/dL (c. 0.16 mg/kg), the bird was released to the wild.

American kestrels fed lead-contaminated diets had blood lead levels of 1.69 mg/kg with no resultant anemia (Custer et al. 1984). Redig et al. (1991) reported no mortality or clinical signs of toxicity, based on gross observations and body weight, in red-tailed hawks with blood lead concentrations of up to 1.58 µg/mL (c. 1.58 mg/kg). However, Kramer and Redig (1997) stated that in bald and golden eagles, blood lead concentrations >1.2 mg/kg were invariably associated with mortality. In a survey of blood lead concentrations in 162 wild golden eagles Pattee et al. (1990) considered 36% of the birds to have been exposed to lead, but the overall mean blood lead concentration was 0.25 mg/kg. Harmata and Restani (1995) found ≥1.0 mg/kg lead in blood of 5.4% of bald eagles and 2.3% of golden eagles. Ospreys (*Pandion haliaetus*) from a mining and smelting area had a mean blood lead concentration of 0.20 mg/kg and inhibited ALAD activity, but there was no effect on hemoglobin or hematocrit (Henny et al. 1991).

16.9.3   COLUMBIFORMES (DOVES AND PIGEONS)

Soft tissue lead residues associated with mortality in doves and pigeons tend to be in the range of 20–60 mg/kg wet weight or even higher, as in the case of a wild mourning dove that died of lead poisoning with 72 mg/kg wet weight of lead in the liver (Locke and Bagley 1967). Clausen and Wolstrup (1979) reported liver and kidney lead residues of 48 and 200 mg/kg (wet weight presumed but not stated), respectively, in a wood pigeon (*Columba palumbus*) that died of lead poisoning. Mourning doves that died after dosage with lead shot had mean lead residues of 80–93 mg/kg dry weight in the liver, 230–300 mg/kg in the kidney, and 116–192 mg/kg in bone (Buerger et al. 1986). In another lead shot dosing study, a mourning dove that died had lead residues of 267 mg/kg dry weight in the liver, 1901 mg/kg in the kidney, 11 mg/kg in the brain, and 403 mg/kg in bone (Kendall et al. 1983). Mourning doves in that study that were euthanized after 9 days of lead shot exposure had microscopic lesions of lead poisoning and lead concentrations of 150–179 mg/kg dry weight in the liver, 1182–1298 mg/kg in the kidney, 11–12 mg/kg in the brain, and 473–528 mg/kg in bone. Castrale and Oster (1993) reported that blood lead concentrations in mourning doves dosed with 4 #8 lead shot reached 10 mg/kg (wet weight presumed but not stated) after about 5 days, and eight of ten birds receiving this dose survived to 28 days, when blood lead levels had approached those in control birds. Ringed turtle-doves (*Streptopelia risoria*) given doses of lead shot and euthanized 9 days later had liver lead residues of 24–128 mg/kg and kidney residues of 633–2384 mg/kg dry weight (Kendall et al. 1981). Several of these birds had seizures, and all had microscopic lesions in the kidney and liver, but none died during the course of the experiment. Ringed turtle-doves that received lead acetate in drinking water for 90 days had mean kidney and liver residues of 900 and 8 mg/kg dry weight, respectively (Kendall and Scanlon 1981). Although no birds died, cellular necrosis and lead inclusions were noted in the kidneys. Two clinically normal feral pigeons (*Columba livia*) with lead shot in their gizzards had liver lead concentrations of 5 and 13 mg/kg and kidney residues of 14 and 81 mg/kg wet weight (DeMent et al. 1987).

Reduced ALAD activity occurred in ringed turtle-doves that had liver and kidney lead residues of 1 and 6 mg/kg wet weight, respectively (Scheuhammer and Wilson 1990). In other studies with ringed turtle-doves, ALAD activity was inhibited with tissue lead levels of 4–9 mg/kg dry weight in the brain, 9–19 mg/kg in the liver, and 84–839 mg/kg in the kidney (Kendall and Scanlon 1982, Kendall et al. 1982). The hemoglobin concentration was reduced when tissue residues reached 28 mg/kg dry weight in the liver, 12 mg/kg in the brain, and 457 mg/kg in the kidney (Kendall and Scanlon 1982). ALAD activity was reduced in feral pigeons with lead residues as low as 1.7 mg/kg wet weight in kidney and 16.5 mg/kg wet weight in bone (Ohi et al. 1974). Domestic pigeons that became anemic after treatment with lead acetate had lead residues of 8–20 mg/kg wet weight in the liver, 33–603 mg/kg in the kidney, 0.9–2.3 mg/kg in the brain, and 57–501 mg/kg in bone (Anders et al. 1982).

Blood lead concentrations associated with death in Columbiformes can be extremely high. Two domestic pigeons treated with lead acetate had blood lead residues of 2320 and 4000 µg/dL

(c. 23.2 and 40 mg/kg) shortly before they died (Dietz et al. 1979, Anders et al. 1982). Other pigeons that were treated with lead acetate developed blood lead concentrations of 250–440 µg/dL (c. 2.5–4.4 mg/kg), resulting in anemia and lead inclusions in the kidneys, but survived the dosage regimen for up to 64 weeks (Anders et al. 1982). Domestic pigeons that died after experimental lead acetate exposure had blood lead levels of 569–1235 and 1245 µg/dL (c. 5.69–12.35, and 12.45 mg/kg), crop stasis occurred at 450–1100 µg/dL (c. 4.5 and 11 mg/kg), but there were no observable clinical signs when blood lead was less than 200 µg/dL (c. 2 mg/kg) (Barthalmus et al. 1977, Cory-Slechta et al. 1980, Boyer et al. 1985). ALAD activity was inhibited in ringed turtle-doves with blood lead concentrations of 21, 81–122, and 142–245 µg/dL (c. 0.21, 0.81–1.22, and 1.42–2.45 mg/kg), and hemoglobin concentration was reduced with mean blood lead concentrations of 395 µg/dL (c. 3.95 mg/kg) (Kendall and Scanlon 1982, Kendall et al. 1982, Scheuhammer and Wilson 1990). Depression of ALAD in feral pigeons has been associated with lead residues as low as 15 µg/dL (c. 0.15 mg/kg) in the blood (Ohi et al. 1974). Two urban pigeons with lead shot in their gizzards had blood lead concentrations of 95 and 1870 µg/dL (c. 0.95 and 18.7 mg/kg), but they were not anemic or emaciated (DeMent et al. 1987).

16.9.4 GALLIFORMES (PHEASANTS, GROUSE, QUAIL, AND PARTRIDGE)

Data suggest that Galliformes are relatively less sensitive to lead than some other groups as evidenced by high tissue lead residues in individuals poisoned by lead. A wild turkey (*Meleagris gallopavo*) that died of lead poisoning had a liver lead concentration of 17 mg/kg wet weight (Stone and Butkas 1978). A wild ring-necked pheasant (*Phasianus colchicus*) found dead with 29 lead shot in its gizzard had 168 mg/kg (wet weight presumed but not stated) of lead in the liver (Hunter and Rosen 1965). Two female ring-necked pheasants from shooting estates with ingested lead shot had lead concentrations of 378 and 220 mg/kg dry weight in wing bones (Butler et al. 2005). A northern bobwhite (*Colinus virginianus*) observed in the field with partial paralysis one day and found dead the next had two lead shot pellets in its gizzard and a liver lead concentration of 399 mg/kg wet weight (Lewis and Schweitzer 2000). Keymer and Stebbings (1987) reported lead poisoning as the cause of death in a gray partridge (*Perdix perdix*) with 40 mg/kg wet weight of lead in the liver and 100 mg/kg wet weight in the kidney. An emaciated gray partridge had lead residues of 130 mg/kg in the liver and 440 in the kidney (wet weight presumed but not stated) with 34 lead pellets in the gizzard (Clausen and Wolstrup 1979). Gjerstad and Hanssen (1984) administered doses to willow ptarmigan (*Lagopus lagopus*) of one, three, or six lead shot. Three ptarmigan that died had liver lead residues of 64, 134, and 274 mg/kg wet weight. Birds given doses of one lead shot survived with no clinical signs and had mean liver lead residues of about 3 mg/kg wet weight 15 days after dosing. Thirty-three greater sage-grouse (*Centrocercus urophasianus*) diagnosed with West Nile virus or trauma as cause of death had liver lead concentrations of ≤0.66 mg/kg wet weight (Dailey et al. 2008). When northern bobwhite were fed increasing amounts of lead acetate until half of the birds died, lead residues in the liver and kidney were 21–277 and 85–500 mg/kg wet weight, respectively (Beyer et al. 1988). Chickens (*Gallus gallus*) that died after being fed lead-containing grit had liver lead residues of up to 54 mg/kg wet weight, while liver lead in chickens that exhibited no clinical signs and were euthanized was 1–6 mg/kg wet weight (Salisbury et al. 1958). ALAD was inhibited but there was no effect on hemoglobin or packed cell volume in Japanese quail that had liver lead concentrations of about 0.5, 4, and 6 mg/kg and kidney concentrations of 6, 20, and 30 mg/kg wet weight (Stone and Soares 1976, Stone et al. 1977, 1979). Northern bobwhite given doses of lead shot had a pooled blood lead concentration of 43 mg/kg 9 days post exposure (McConnell 1967). Weakness, lethargy, and loss of weight were seen in 19% of the quail and 10% died. Chickens fed lead acetate exhibited severely inhibited ALAD activity, weight loss, and mild anemia with blood lead concentrations of 322–832 µg/dL (c. 3.22–8.32 mg/kg) (Franson and Custer 1982). Mean liver and kidney lead residues were 17 and 56 mg/kg dry weight after 28 days. Japanese quail given doses of lead shot had lead residues of 7 mg/kg wet weight in the blood, 3 mg/kg in the liver, and

5 mg/kg in the kidney (Yamamoto et al. 1993). ALAD was inhibited, but no other clinical signs were reported.

16.9.5 PASSERIFORMES (PERCHING BIRDS)

Two yellow-billed cuckoos (*Coccyzus americanus*) collected near a zinc smelter had liver lead concentrations of 18 and 25 mg/kg wet weight and kidney residues of 21 and 14 mg/kg wet weight (Beyer et al. 1985). ALAD was inhibited, but there was no anemia or gross or microscopic lesions of lead poisoning. Two American robins (*Turdus migratorius*), a northern cardinal (*Cardinalis cardinalis*), and a brown thrasher (*Toxostoma rufum*) sampled in a area with a history of mining and smelting lead and zinc had liver lead concentrations of 12–94 mg/kg dry weight, kidney lead concentrations of 25–150 mg/kg dry weight, and ALAD activities of <10% of reference values (Beyer et al. 2004). Getz et al. (1977) measured tissue lead residues in songbirds collected in urban and rural areas. House sparrows (*Passer domesticus*), starlings (*Sturnus vulgaris*), common grackles (*Quiscalus quiscula*), and American robins had maximum liver lead residues of 10–16 mg/kg dry weight and kidney residues of 14–98 mg/kg dry weight. The concentration of lead in the liver of a dark-eyed junco (*Junco hyemalis*) collected at a trap and skeet range was 9.3 mg/kg dry weight (Vyas et al. 2000). Three brown-headed cowbirds (*Molothrus ater*) that died after being dosed with lead shot had liver lead concentrations of 71–137 mg/kg dry weight (Vyas et al. 2001). Starlings given doses of trialkyl lead died within 6 days of treatment, with mean liver and kidney residues of 40 and 20 mg/kg wet weight, respectively (Osborn et al. 1983). Beyer et al. (1988) fed lead acetate to red-winged blackbirds (*Agelaius phoeniceus*), brown-headed cowbirds, and common grackles until half of the birds in each group died. Median liver lead concentrations were 20–50 mg/kg wet weight (range, 4–97 mg/kg), and median kidney residues were 22–160 mg/kg wet weight (range, 2–740 mg/kg).

16.9.6 CHARADRIIFORMES (SHOREBIRDS, GULLS, AND ALCIDS)

Locke et al. (1991) reported a liver lead concentration of 52 mg/kg wet weight in a marbled godwit (*Limosa fedoa*) that died of lead poisoning after ingesting lead shot. Seven herring gull (*Larus argentatus*) chicks that received a one-time injection of lead nitrate survived 45 days but exhibited slower growth than did controls, with mean lead concentrations of 197 µg/dL (c. 1.97 mg/kg) in the blood, 21 mg/kg dry weight in the liver, and 41 mg/kg dry weight in the kidney (Burger and Gochfeld 1990). Apparently healthy adult laughing gulls (*Larus atricilla*) collected in a lead-contaminated area had mean liver lead residues of 4–5 mg/kg and kidney residues of 2 mg/kg wet weight (Munoz et al. 1976, Hulse et al. 1980). Adult royal terns (*Thalasseus maximus*) collected in the same region had liver and kidney residues of 0.4 and 1.1 mg/kg wet weight, while sandwich terns (*Thalasseus sandvicensis*) had liver and kidney residues of 0.5 and 0.8 mg/kg wet weight (Maedgen et al. 1982). Herring gulls feeding at a dump site had liver and kidney lead residues of 3 and 13 mg/kg dry weight, respectively (Leonzio et al. 1986). Black-headed gulls (*Larus ridibundus*) collected at the same location had liver and kidney residues of 8 and 31 mg/kg dry weight, respectively.

16.9.7 GRUIFORMES (CRANES AND THEIR ALLIES)

Windingstad et al. (1984) reported on two wild sandhill cranes (*Grus canadensis*) that died of lead poisoning with parts of fishing sinkers in their gizzards. One bird had liver and kidney lead concentrations of 23 and 30 mg/kg wet weight, while the other had liver and kidney lead residues of 259 and 113 mg/kg dry weight. A captive sandhill crane that died after ingesting two 0.22-caliber rifle cartridges had 30 mg/kg wet weight of lead in its liver, and another dead captive sandhill crane with an ingested copper-coated penny had a liver lead concentration of 24 mg/kg wet weight

(Windingstad et al. 1984). A sandhill crane that died after exposure to lead-based paint in a zoo facility had liver and kidney lead concentrations of 29 and 19 mg/kg (wet weight presumed but not stated), respectively (Kennedy et al. 1977). Four additional cranes in this facility developed clinical signs of lead poisoning with blood lead concentrations of 146–378 µg/100 mL (c. 1.46–3.78 mg/kg), but they recovered following chelation treatment. Lead concentrations in kidneys of two houbara bustards (*Chlamydotis undulata maqueenii*) that died on a farm and were diagnosed with lead poisoning resulting from the ingestion of paint flakes were 47 and 5.5 mg/kg, wet weight (Bailey et al. 1995). A whooping crane (*Grus americana*) had a blood lead concentration of 5.6 mg/kg shortly before it died and was found to have liver and kidney concentrations of 24 and 10 mg/kg wet weight (Snyder et al. 1991). An emaciated Mississippi sandhill crane (*Grus canadensis pulla*) found with an unidentified lead object in its stomach had 70 mg/kg wet weight lead in liver tissue (Franson and Hereford 1994). Two Japanese cranes (*Grus japonensis*) found dead had liver lead concentrations of 32 and 62 mg/kg dry weight and kidney lead concentrations of 31 and 34 mg/kg dry weight (Teraoka et al. 2007). Apparently healthy soras (*Porzana carolina*) collected with lead shot in their gizzards had mean liver lead residues of up to 3 mg/kg wet weight and up to 11 mg/kg dry weight in bone (Stendell et al. 1980).

16.9.8 CICONIIFORMES (HERONS AND THEIR ALLIES)

A black-crowned night heron (*Nycticorax nycticorax*) with a lead jig head in its stomach had a liver lead concentration of 26 mg/kg wet weight (Franson et al. 2003), but a number of surveys of lead in herons have reported liver concentrations of <6.7 mg/kg dry weight (see Custer 2000). Apparently healthy cattle egrets (*Bubulcus ibis*) collected in an industrialized area had liver lead concentrations of 0.07–1.3 mg/kg and kidney concentrations of 0.08–3.5 mg/kg wet weight (Hulse et al. 1980).

16.9.9 GAVIIFORMES (LOONS)

Locke et al. (1982) reported liver lead residues of 21–38 mg/kg wet weight in three common loons that died of lead poisoning. Two of the three had lead fishing sinkers in gizzard contents. Liver lead concentrations were 5–41 mg/kg wet weight in 36 additional cases of lead-poisoned loons, 33 of which had ingested lead fishing weights (Franson and Cliplef 1992, Pokras and Chafel 1992, Stone and Okoniewski 2001). Of 522 common loons examined from New England, Sidor et al. (2003) confirmed lead poisoning in 68 and assigned a diagnosis of suspected lead poisoning in 50 more. The average lead concentration in the livers of confirmed lead poisoning cases in that study was 17.6 mg/kg wet weight. Franson et al. (2003) found liver lead concentrations of 8.0–16.9 mg/kg wet weight in four common loons with ingested lead fishing weights. In two other common loons with ingested lead, blood and liver lead levels (mg/kg wet weight) were 4.2 and 16.6, and 6.0 and 12.8, respectively, whereas one loon with an ingested jig head had 0.30 mg/kg lead in blood (liver was not tested) (Franson et al. 2003). In a summary of Canadian data, the average lead concentrations in liver (*n* = 12) and kidney (*n* = 25) of common loons with ingested lead artifacts were 59 and 218 mg/kg dry weight, respectively (Scheuhammer et al. 2003b). Kidney lead concentrations were 15–167 mg/kg wet weight in six common loons that died of lead poisoning, four of which were determined to have remnants of lead fishing weights in their gizzards (Daoust et al. 1998). Although lead fishing weights account for most of the reported lead poisonings in loons, a Pacific loon (*Gavia pacifica*) found in Alaska had three ingested lead shot, and a liver lead concentration of 31 mg/kg wet weight (Wilson et al. 2004).

16.9.10 STRIGIFORMES (OWLS)

Two captive snowy owls (*Nyctea scandiaca*) that died of lead poisoning had 45 and 204 mg/kg dry weight of lead in the liver and 68 and 146 mg/kg dry weight in the kidney (MacDonald et al. 1983). Exposure was thought to have resulted from consumption of bullet fragments in food items. Eastern screech owls (*Otus asio*) fed lead acetate until half the birds died had median liver and kidney lead concentrations of 22 and 33 mg/kg wet weight (Beyer et al. 1988). In a great-horned owl (*Bubo virginianus*) with necropsy lesions consistent with lead poisoning, the lead concentration in the liver or kidney was >30 or >20 mg/kg dry weight, respectively (Clark and Scheuhammer 2003).

16.9.11 PROCELLARIIFORMES (TUBENOSES)

Lead poisoning from consumption of paint chips in association with buildings contributed to epizootic mortality in Laysan albatross (*Diomedea immutabilis*) chicks on Midway Atoll (Sileo and Fefer 1987). Some of the sick birds were unable to retract their wings, resulting in a "droop-wing" appearance, and in 10 of these birds diagnosed with lead poisoning, lead was detected in the blood of each (maximum of 4.8 mg/kg wet weight), in the liver of 8 (maximum 110 mg/kg dry weight), and in the kidney of 1 (44 mg/kg dry weight) (Sileo and Fefer 1987). In later studies on Midway, droop-wing albatross chicks had mean lead concentrations of 14.4 and 14.0 mg/kg wet weight in liver and kidney, respectively, and 410 µg/dL (c. 4.10 mg/kg) in blood (Burger and Gochfeld 2000, Finkelstein et al. 2003). Work and Smith (1996) found that ALAD in albatross chicks decreased at blood lead concentrations >0.05 µg/mL (c. >0.05 mg/kg).

16.9.12 PELECANIFORMES (PELICANS AND THEIR ALLIES)

Seven brown pelicans (*Pelecanus occidentalis*) with ingested lead fishing weights, sampled when still alive, had blood lead concentrations of 0.04–13.9 mg/kg wet weight (Franson et al. 2003).

16.9.13 PHOENICOPTERIFORMES (FLAMINGOS)

Liver lead concentrations of 17 Caribbean flamingos (*Phoenicopterus ruber ruber*) that died of lead poisoning after consumption of lead shot ranged from 128 to 771 mg/kg dry weight (Schmitz et al. 1990). In two lead poisoning events in greater flamingos (*Phoenicopterus ruber*), dead birds had eight or more ingested lead shot and minimum liver lead concentrations were 12.6 mg/kg wet weight and 77.2 mg/kg dry weight (Ramo et al. 1992, Mateo et al. 1997a).

16.9.14 PICIFORMES (WOODPECKERS)

Liver and kidney lead residues were 20 and 76 mg/kg wet weight, respectively, in an emaciated gray-headed woodpecker (*Picus canus*), 9.4 and 1.4 mg/kg wet weight, respectively in a second gray-headed woodpecker that died of traumatic injuries, and 26.2 and <0.02 mg/kg, respectively, in a white-backed woodpecker (*Dendrocopus leucotos*) that died during a translocation program (Mörner and Petersson 1999).

Summary

Lead is a nonessential, highly toxic heavy metal that affects all body systems. Birds are exposed to lead from direct consumption of spent lead shot, bullets, or fragments thereof in prey items, ingestion of lead fishing weights and chips of lead-based paints, and environmental contamination of urban and industrial areas. However, the majority of cases of poisoning result from the ingestion of lead from spent ammunition or ammunition fragments, or angler's lead weights. First recognized more than a century ago, lead poisoning of birds has been reported in many areas of the world.

When lead is absorbed into the bloodstream, the first measurable effects are the inhibition of heme-biosynthetic enzymes. There does not appear to be a no-effect level for lead, because the activities of certain enzymes have been reported to be inhibited at blood lead concentrations of <5 µg/dL (c. <0.05 mg/kg). From the blood, lead is deposited into soft tissues, particularly liver and kidney, growing feathers, and bone. Lead has different retention times in these tissues, with blood and other soft tissue concentrations reflecting recent exposure, lead in feathers reflecting exposure at the time of feather growth, and bone lead concentrations reflecting long-term absorption and accumulation over the lifetime of the bird. Many factors influence lead absorption and distribution within the body, including age, gender, physiological condition, diet, and exposure level and duration. Tissue lead residues associated with physiological injury, clinical signs, and death due to lead poisoning also vary among species, and assessments of toxicosis should be done by comparison with data from phylogenetically related groups whenever possible.

We consider background concentrations of lead to be <20 µg/dL (c. <0.2 mg/kg) in blood, <2 mg/kg wet weight in liver and kidney, and <10 mg/kg dry weight in bone of birds. The suggested thresholds of increasing severity of effects for Anseriformes are: 20–<50 µg/dL (c. 0.2–<0.5 mg/kg) in blood, 2 < 6 mg/kg wet weight in liver and kidney (subclinical poisoning); 50–100 µg/dL (c. 0.5–1 mg/kg) in blood, 6–10 mg/kg wet weight in liver, 6–15 mg/kg wet weight in kidney (clinical poisoning); >100 µg/dL (c. >1 mg/kg) in blood, >10 mg/kg wet weight in liver, >15 mg/kg wet weight in kidney (severe clinical poisoning). Suggested thresholds for Falconiformes: 20–<50 µg/dL (c. 0.2–<0.5 mg/kg) in blood, 2 < 6 mg/kg wet weight in liver, 2 < 4 mg/kg wet weight in kidney (subclinical poisoning); 50–100 µg/dL (c. 0.5–1 mg/kg) in blood, 6–10 mg/kg wet weight in liver, 4–6 mg/kg wet weight in kidney (clinical poisoning); >100 µg/dL (c. >1 mg/kg) in blood, >10 mg/kg wet weight in liver, >6 mg/kg wet weight in kidney (severe clinical poisoning). Suggested thresholds for Columbiformes: 20 < 200 µg/dL (c. 0.2 < 2 mg/kg) in blood, 2 < 6 mg/kg wet weight in liver, 2 < 15 mg/kg wet weight in kidney (subclinical poisoning); 200–300 µg/dL (c. 2–3 mg/kg) in blood, 6–15 mg/kg wet weight in liver, 15–30 mg/kg wet weight in kidney (clinical poisoning); >300 µg/dL (c. >3 mg/kg) in blood, >15 mg/kg wet weight in liver, >30 mg/kg wet weight in kidney (severe clinical poisoning). For birds in general, liver lead concentrations within the clinical poisoning range (≥6 mg/kg wet weight) support a lead poisoning diagnosis where necropsy observations are consistent with lead poisoning.

Bone lead concentrations of >20 mg/kg dry weight are considered to suggest excessive exposure. Because of the rapid uptake and slow release of lead from bone, bone lead concentrations are the least useful index of recent lead poisoning, although bone lead concentrations can be used to determine geographical patterns of poisoning in populations.

Care should be exercised when comparing tissue lead residues from experimental studies with field data. Wild birds with inadequate diets and exposed to ambient environmental conditions may be more susceptible to lead toxicosis than are birds in controlled situations. A clinical evaluation of health and the determination of lead residues in sequential blood samples will enhance the assessment of lead exposure and toxicity in live birds. Although conclusions can be drawn regarding the severity of lead poisoning from tissue concentrations alone, a definitive diagnosis of lead poisoning as the cause of death in an individual bird requires interpretation of tissue lead residues in dead birds accompanied by an examination of carcasses for gross and microscopic lesions of lead poisoning.

REFERENCES

Adler, F. E. W. 1944. Chemical analyses of organs from lead-poisoned Canada geese. *J. Wildl. Manag.* 8:83–85.

Adrian, W. J., and M. L. Stevens. 1979. Wet versus dry weights for heavy metal toxicity determinations in duck liver. *J. Wildl. Dis.* 15:125–126.

Anders, E., et al. 1982. Morphological, pharmacokinetic, and hematological studies of lead-exposed pigeons. *Environ. Res.* 28:344–363.

Anderson, W. L. 1975. Lead poisoning in waterfowl at Rice Lake, Illinois. *J. Wildl. Manag.* 39:264–70.

Avery, D., and R. T. Watson. 2009. Regulation of lead-based ammunition around the world. In *Ingestion of lead from spent ammunition: Implications for wildlife and humans*, eds. R. T. Watson, M. Fuller, M. Pokras, and W. G. Hunt. Boise: The Peregrine Fund. DOI 10.4080/ilsa.2009.0115.

Bagley, G. E., and L. N. Locke. 1967. The occurrence of lead in tissues of wild birds. *Bull. Environ. Contam. Toxicol.* 2:297–305.

Bagley, G. E., L. N. Locke, and G. T. Nightingale. 1967. Lead poisoning in Canada geese in Delaware. *Avian Dis.* 11:601–608.

Bailey, T. A., J. H. Samour, J. Naldo, and J. C. Howlett. 1995. Lead toxicosis in captive houbara bustards (*Chlamydotis undulata maqueenii*). *Vet. Rec.* 137:193–194.

Barthalmus, G. T., J. D. Leander, D. E. McMillan, P. Mushak, and M. R. Krigman. 1977. Chronic effects of lead on schedule-controlled pigeon behavior. *Toxicol. Appl. Pharmacol.* 42:271–284.

Battaglia, A., S. Ghidini, G. Campanini, and R. Spaggiari. 2005. Heavy metal contamination in little owl (*Athene noctua*) and common buzzard (*Buteo buteo*) from northern Italy. *Ecotoxicol. Environ. Saf.* 60:61–66.

Bellrose, F. C. 1959. Lead poisoning as a mortality factor in waterfowl populations. *Ill. Nat. Hist. Surv. Bull.* 27:235–288.

Benson, W. W., B. Pharaoh, and P. Miller. 1974. Lead poisoning in a bird of prey. *Bull. Environ. Contam. Toxicol.* 11:105–108.

Beyer, W. N., O. H. Pattee, L. Sileo, D. J. Hoffman, and B. M. Mulhern. 1985. Metal contamination in wildlife living near two zinc smelters. *Environ. Pollut. Ser. A* 38:63–86.

Beyer, W. N., J. W. Spann, L. Sileo, and J. C. Franson. 1988. Lead poisoning in six captive avian species. *Arch. Environ. Contam. Toxicol.* 17:121–130.

Beyer, W. N., J. C. Franson, L. N. Locke, R. K. Stroud, and L. Sileo. 1998. Retrospective study of the diagnostic criteria in a lead-poisoning survey of waterfowl. *Arch. Environ. Contam. Toxicol.* 35:506–512.

Beyer, W. N., D. J. Audet, G. H. Heinz, D. J. Hoffman, and D. Day. 2000. Relation of waterfowl poisoning to sediment lead concentrations in the Coeur d'Alene River Basin. *Ecotoxicology* 9:207–218.

Beyer, W. N., et al. 2004. Zinc and lead poisoning in wild birds in the tri-state mining district (Oklahoma, Kansas, and Missouri). *Arch. Environ. Contam. Toxicol.* 48:108–117.

Birkhead, M. 1983. Lead levels in the blood of mute swans *Cygnus olor* on the River Thames. *J. Zool. (Lond.)* 199:59–73.

Blus, L. J., R. K. Stroud, B. Reiswig, and T. McEneaney. 1989. Lead poisoning and other mortality factors in trumpeter swans. *Environ. Toxicol. Chem.* 8:263–271.

Blus, L. J., C. J. Henny, D. J. Hoffmann, and R. A. Grove. 1991. Lead toxicosis in tundra swans near a mining and smelting complex in northern Idaho. *Arch. Environ. Contam. Toxicol.* 21:549–555.

Blus, L. J., C. J. Henny, D. J. Hoffman, and R. A. Grove. 1993. Accumulation and effects of lead and cadmium on wood ducks near a mining and smelting complex in Idaho. *Ecotoxicology* 2:139–154.

Blus, L. J., C. J. Henny, D. J. Hoffman, L. Sileo, and D. J Audet. 1999. Persistence of high lead concentrations and associated effects in tundra swans captured near a mining and smelting complex in Northern Idaho. *Ecotoxicology* 8:125–132.

Bostan, N., M. Ashraf, A. S. Mumtaz, and I. Ahmad. 2007. Diagnosis of heavy metal contamination in agro-ecology of Gujranwala, Pakistan using cattle egret (*Bubulcus ibis*) as bioindicator. *Ecotoxicology* 16:247–251.

Boyer, I. J., D. A. Cory-Slechta, and V. DiStefano. 1985. Lead induction of crop dysfunction in pigeons through a direct action on neural or smooth muscle components of crop tissue. *J. Pharmacol. Exp. Ther.* 234:607–615.

Buerger, T. T., R. E. Mirarchi, and M. E. Lisano. 1986. Effects of lead shot ingestion on captive mourning dove survivability and reproduction. *J. Wildl. Manag.* 50:1-8.

Burger, J. 1993. Metals in avian feathers: Bioindicators of environmental pollution. *Rev. Environ. Toxicol.* 5:203–311.

Burger, J., and M. Gochfeld. 1990. Tissue levels of lead in experimentally exposed herring gull (*Larus argentatus*) chicks. *J. Toxicol. Environ. Health* 29:219–233.

Burger, J., and M. Gochfeld. 2000. Metals in Laysan albatrosses from Midway Atoll. *Arch. Environ. Contam. Toxicol.* 38:254–259.

Butler, D. A., R. B. Sage, R. A. H. Draycott, J. P. Carroll, and D. Potts. 2005. Lead exposure in ring-necked pheasants on shooting estates in Great Britain. *Wildl. Soc. Bull.* 33:583–589.

Cade, T. J. 2007. Exposure of California condors to lead from spent ammunition. *J. Wildl. Manag.* 71:2125–2133.

Carpenter, J. W., et al. 2003. Experimental lead poisoning in turkey vultures (*Cathartes aura*). *J. Wildl. Dis.* 39:96–104.

Castrale, J. S., and M. Oster. 1993. Lead and delta-aminolevulinic acid dehydratase in the blood of mourning doves dosed with lead shot. *Proc. Indiana Acad. Sci.* 102:265–272.

Chupp, N. R., and P. D. Dalke. 1964. Waterfowl mortality in the Coeur d'Alene River Valley, Idaho. *J. Wildl. Manag.* 28:692–702.

Church, M. E., et al. 2006. Ammunition is the principal source of lead accumulated by California condors re-introduced to the wild. *Environ. Sci. Technol.* 40:6143–6150.

Clark, A. J., and A. M. Scheuhammer. 2003. Lead poisoning in upland-foraging birds of prey in Canada. *Ecotoxicology* 12:23–30.

Clausen, B., and C. Wolstrup. 1979. Lead poisoning in game from Denmark. *Dan. Rev. Game Biol.* 11:1–22.

Clausen, B., K. Elvestad, and O. Karlog. 1982. Lead burden in mute swans from Denmark. *Nord. Veterinaermed.* 34:83–91.

Clemens, E. T., L. Krook, A. L. Aronson, and C. E. Stevens. 1975. Pathogenesis of lead shot poisoning in the mallard duck. *Cornell Vet.* 65:248–285.

Coburn, D. R., D. W. Metzler, and R. Treichler. 1951. A study of absorption and retention of lead in wild waterfowl in relation to clinical evidence of lead poisoning. *J. Wildl. Manag.* 15:186–192.

Cook, R. S., and D. O. Trainer. 1966. Experimental lead poisoning of Canada geese. *J. Wildl. Manag.* 30:1-8.

Cory-Slechta, D. A., R. H. Garman, and D. Seidman. 1980. Lead-induced crop dysfunction in the pigeon. *Toxicol. Appl. Pharmacol.* 52:462–467.

Craig, T. H., J. W. Connelly, E. H. Craig, and T. L. Parker. 1990. Lead concentrations in golden and bald eagles. *Wilson Bull.* 102:130–133.

Custer, T. W. 2000. Environmental contaminants. In *Heron conservation,* eds. J.A. Kushlan and H. Hafner, pp. 251–267. San Diego: Academic Press.

Custer, T. W., J. C. Franson, and O. H. Pattee. 1984. Tissue lead distribution and hematologic effects in American kestrels (*Falco sparverius* L.) fed biologically incorporated lead. *J. Wildl. Dis.* 20:39–43.

Dailey, R. N., M. F. Raisbeck, R. S. Siemion, and T. E. Cornish. 2008. Liver metal concentrations in greater sage-grouse (*Centrocercus urophasianus*). *J. Wildl. Dis.* 44:494–498.

Daoust, P. Y., G. Conboy, S. McBurney, and N. Burgess. 1998. Interactive mortality factors in common loons from Maritime Canada. *J. Wildl. Dis.* 34:524–531.

Decker, R. A., A. M. McDermid, and J. W. Prideaux. 1979. Lead poisoning in two captive king vultures. *J. Am. Vet. Med. Assoc.* 175:1009.

Degernes, L. A. 1991. The Minnesota trumpeter swan lead poisoning crisis of 1988–89. In *Proceedings and papers of the 12th trumpeter swan society conference*, pp. 114–118. Minneapolis: Trumpeter Swan Society.

Degernes, L., et al. 2006. Epidemiologic investigation of lead poisoning in trumpeter and tundra swans in Washington state, USA, 2000–2002. *J. Wildl. Dis.* 42:345–358.

Demayo, A., M. C. Taylor, K. W. Taylor, and P. V. Hodson. 1982. Toxic effects of lead and lead compounds on human health, aquatic life, and wildlife plants, and livestock. *CRC Crit. Rev. Environ. Control* 12:257–305.

DeMent, S. H., J. J. Chisolm, Jr., M. A. Eckhaus, and J. D. Strandberg. 1987. Toxic lead exposure in the urban rock dove. *J. Wildl. Dis.* 23:273–278.

Denbow, D. M. 2000. Gastrointestinal anatomy and physiology. In *Sturkie's avian physiology*, 5th ed., ed. G. C. Whittow, pp. 299–325. San Diego: Academic Press.

DeStefano, S., C. J. Brand, D. H. Rusch, D. L. Finley, and M. M. Gillespie. 1991. Lead exposure in Canada geese of the eastern prairie population. *Wildl. Soc. Bull.* 19:23–32.

Dieter, M. P. 1979. Blood delta-aminolevulinic acid dehydratase (ALAD) to monitor lead contamination in canvasback ducks (*Aythya valisineria*). In *Animals as monitors of environmental pollutants*, eds. S. W. G. Nielsen, G. Migaki, and D. G. Scarpelli, pp. 177–191. Washington, D.C.: National Academy of Sciences.

Dieter, M. P., and M. T. Finley. 1978. Erythrocyte delta-aminolevulinic acid dehydratase activity in mallard ducks: duration of inhibition after lead shot dosage. *J. Wildl. Manage.* 42:621–625.

Dieter, M. P., and M. T. Finley. 1979. Delta-aminolevulinic acid dehydratase enzyme activity in blood, brain, and liver of lead-dosed ducks. *Environ. Res.* 19:127–135.

Dieter, M. P., M. C. Perry, and B. M. Mulhern. 1976. Lead and PCB's in canvasback ducks: relationship between enzyme levels and residues in blood. *Arch. Environ. Contam. Toxicol.* 5:1–13.

Dietz, D. D., D. E. McMillan, and P. Mushak. 1979. Effects of chronic lead administration on acquisition and performance of serial position sequences by pigeons. *Toxicol. Appl. Pharmacol.* 47:377–384.

Duke, G. E. 1986. Raptor physiology. In *Zoo and wild animal medicine*, 2nd ed., ed. D. Pedersen, pp. 370–376. Philadelphia: W.B. Saunders Company.

Eisler, R. 2000. *Handbook of chemical risk assessment*, Vol. 1, Metals. Boca Raton, FL: Lewis Publishers.

Elliott, J. E., K. M. Langelier, A. M. Sheuhammer, P. H. Sinclair, and P. E. Whitehead. 1992. Incidence of lead poisoning in bald eagles and lead shot in waterfowl gizzards from British Columbia, 1988–91. *Canadian Wildlife Service Progress Notes No.* 200:1–7.

Fair, J. M., and O. B. Myers. 2002. The ecological and physiological costs of lead shot and immunological challenge to developing western bluebirds. *Ecotoxicology* 11:199–208.

Fair, J. M., and R. E. Ricklefs. 2002. Physiological, growth, and immune responses of Japanese quail chicks to the multiple stressors of immunological challenge and lead shot. *Arch. Environ. Contam. Toxicol.* 42:77–87.

Figuerola, J., R. Mateo, A. J. Green, J. Y. Mondain-Monval, H. LeFranc, and G. Mentaberre. 2005. Grit selection in waterfowl and how it determines exposure to ingested lead shot in Mediterranean wetlands. *Environ. Conserv.* 32:226–234.

Finkelstein, M. E., R. H. Gwiazda, and D. R. Smith. 2003. Lead poisoning of seabirds: environmental risks from leaded paint at a decommissioned military base. *Environ. Sci. Technol.* 37:3256–3260.

Finley, M. T., M. P. Dieter, and L. N. Locke. 1976a. Lead in tissues of mallard ducks dosed with two types of lead shot. *Bull. Environ. Contam. Toxicol.* 16:261–269.

Finley, M. T., M. P. Dieter, and L. N. Locke. 1976b. Delta-aminolevulinic acid dehydratase: inhibition in duck dosed with lead shot. *Environ. Res.* 12:243–249.

Finley, M. T., M. P. Dieter, and L. N. Locke. 1976c. Sublethal effects of chronic lead ingestion in mallard ducks. *J. Toxicol. Environ. Health* 1:929–937.

Finley, M. T., and M. P. Dieter. 1978. Influence of laying on lead accumulation in bone of mallard ducks. *J. Toxicol. Environ. Health* 4:123–129.

Fisher, I. J., D. J. Pain, and V. G. Thomas. 2006. A review of lead poisoning from ammunition sources in terrestrial birds. *Biol. Conserv.* 131:421–432.

Flint, P. L., M. R. Petersen, and J. B. Grand. 1997. Exposure of spectacled eiders and other diving ducks to lead in western Alaska. *Can. J. Zool.* 75:439–443.

Forbes, R. M., and G. C. Sanderson. 1978. Lead toxicity in domestic animals and wildlife. In *The biogeochemistry of lead in the environment. Part B. Biological effects*, ed. J. O. Nriagu, pp. 225–277. Amsterdam: Elsevier/North-Holland Biomedical Press.

Franson, J. C., and T. W. Custer. 1982. Toxicity of dietary lead in young cockerels. *Vet. Hum. Toxicol.* 24:421–423.

Franson, J. C., G. M. Haramis, M. C. Perry, and J. F. Moore. 1986. Blood protoporphyrin for detecting lead exposure in canvasbacks. In *Lead poisoning in wild waterfowl—a workshop*, eds. J. S. Feierabend, and A. B. Russell, pp. 32–37. Washington, D.C.: National Wildlife Federation.

Franson, J., and D. J. Cliplef. 1992. Causes of mortality in common loons. In *Proceedings from the 1992 conference on the loon and its ecosystem: status, management, and environmental concerns*, eds. L. Morse, S. Stockwell, and M. Pokras, pp. 2–12. Concord: U.S. Fish and Wildlife Service.

Franson, J. C., and S. G. Hereford. 1994. Lead poisoning in a Mississippi sandhill crane. *Wilson Bull.* 106:766–768.

Franson, J. C., P. S. Koehl, D. V. Derksen, T. C. Rothe, C. M. Bunck, and J. F. Moore. 1995a. Heavy metals in seaducks and mussels from Misty Fjords National Monument in southeast Alaska. *Environ. Monit. Assess.* 36:149–167.

Franson, J. C., M. R. Petersen, C. U. Meteyer, and M. R. Smith. 1995b. Lead poisoning of spectacled eiders (*Somateria fischeri*) and of a common eider (*Somateria mollissima*) in Alaska. *J. Wildl. Dis.* 31:268–271.

Franson, J. C., W. L. Hohman, J. L. Moore, and M. R. Smith. 1996a. The efficacy of protoporphyrin as a predictive biomarker for lead exposure in canvasback ducks: effect of sample storage time. *Environ. Monit. Assess.* 43:181–188.

Franson, J. C., N. J. Thomas, M. R. Smith, A. H. Robbins, S. Newman, and P. C. McCartin. 1996b. A retrospective study of postmortem findings in red–tailed hawks. *J. Raptor. Res.* 30:7–14.

Franson, J. C., T. Hollmén, M. Hario, M. Kilpi, and D. L. Finley. 2002. Lead and delta-aminolevulinic acid dehydratase in blood of common eiders (*Somateria mollissima*) from the Finnish archipelago. *Ornis Fenn.* 79:87–91.

Franson, J. C., et al. 2003. Lead fishing weights and other fishing tackle in selected waterbirds. *Waterbirds* 26:345–352.

Franson, J. C., T. E. Hollmén, P. L. Flint, J. B. Grand, and R. B. Lanctot. 2004. Contaminants in molting long-tailed ducks and nesting common eiders in the Beaufort Sea. *Mar. Pollut. Bull.* 48:504–513.

Franson, J. C., S. P. Hansen, and J. H. Schulz. 2009. Ingested shot and tissue lead concentrations in mourning doves. In *Ingestion of lead from spent lead ammunition: Implications for wildlife and humans*, eds. R.T. Watson, M. Fuller, M. Pokras, and W.G. Hunt. Boise: The Peregrine Fund. DOI 10.4080/ilsa.2009.0202.

Frenzel, R. W., and R. G. Anthony. 1989. Relationship of diets and environmental contaminants in wintering bald eagles. *J. Wildl. Manag.* 53:792–802.

Friend, M. 1999. Lead. In *Field manual of wildlife diseases, general field procedures and diseases of birds*, eds. M. Friend and J. C. Franson, pp. 317–334. U.S. Geological Survey Information and Technology Report 1999-001. Washington, D. C.: U.S. Department of the Interior.

Fry, D. M. 2004. Analysis of lead in California condor feathers: determination of exposure and depuration during feather growth. California Department of Fish and Game, Species Conservation and Recovery Program Report, 2004-02, Sacramento.

Fry, D. M., and J. R. Maurer. 2003. Assessment of lead contamination sources exposing California condors. Final report to the California Department of Fish and Game, Sacramento.

Garcia-Fernandez, A. J., J. A. Sanchez-Garcia, P. Jimenez-Montalban, and A. Luna. 1995. Lead and cadmium in wild birds in southeastern Spain. *Environ. Toxicol. Chem.* 14:2049–2058.

García-Fernández, A. J., M. Motas-Guzmán, I. Navas, P. María-Mojica, A. Luna, and J. A. Sánchez-García. 1997. Environmental exposure and distribution of lead in four species of raptors in southeastern Spain. *Arch. Environ. Contam. Toxicol.* 33:76–82.

Getz, L. L., L. B. Best, and M. Prather. 1977. Lead in urban and rural song birds. *Environ. Pollut.* 12:235–238.

Gill, C. E., and K. M. Langelier. 1994. Acute lead poisoning in a bald eagle secondary to bullet ingestion. *Can. Vet. J.* 35:303–304.

Gjerstad, K. O., and I. Hanssen. 1984. Experimental lead poisoning in willow ptarmigan. *J. Wildl. Manag.* 48:1018–1022.

Golden, N. H., B. A. Rattner, J. B. Cohen, D. J. Hoffman, E. Russek-Cohen, and M. A. Ottinger. 2003. Lead accumulation in feathers of nestling black-crowned night herons (*Nycticorax nycticorax*) experimentally treated in the field. *Environ. Toxicol. Chem.* 22:1517–1524.

Grand, J. B., P. L. Flint, M. R. Petersen, and C. L. Moran. 1998. Effect of lead poisoning on spectacled eider survival rates. *J. Wildl. Manag.* 62:1103–1109.

Grasman, K. A., and P. F. Scanlon. 1995. Effects of acute lead ingestion and diet on antibody and T-cell-mediated immunity in Japanese quail. *Arch. Environ. Contam. Toxicol.* 28:161–167.

Grue, C. E., D. J. Hoffman, W. N. Beyer, and L. P. Franson. 1986. Lead concentrations and reproductive success in European starlings *Sternus vulgaris* nesting within highway roadside verges. *Environ. Pollut. Ser. A* 42:157–182.

Hall, M., J. Grantham, R. Posey, and A. Mee. 2007. Lead exposure among reintroduced California condors in southern California. In *California condors in the 21st century*. Series in ornithology, no. 2, eds. A. Mee and L. S. Hall, pp. 139–162. Cambridge, MA: Nuttall Ornithology Club and Washington, D.C.: The American Ornithologists' Union.

Harmata, A. R., and M. Restani. 1995. Environmental contaminants and cholinesterase in blood of vernal migrant bald and golden eagles in Montana. *Intermountain J. Sci.* 1:1–15.

Heinz, G. H., D. J. Hoffman, L. Sileo, D. J. Audet, and L. J. LeCaptain. 1999. Toxicity of lead-contaminated sediment to mallards. *Arch. Environ. Contam. Toxicol.* 36:323–333.

Henny, C. J., L. J. Blus, D. J. Hoffman, R. A. Grove, and J. S. Hatfield. 1991. Lead accumulation and osprey production near a mining site on the Coeur d'Alene River, Idaho. *Arch. Environ. Contam. Toxicol.* 21:415–424.

Henny, C. J., L. J. Blus, D. J. Hoffman, and R. A. Grove. 1994. Lead in hawks, falcons and owls downstream from a mining site on the Coeur D'Alene River, Idaho. *Environ. Monit. Assess.* 29:267–288.

Hernberg, S., J. Nikkanen, G. Mellin, and H. Lilius. 1970. Delta-aminolevulinic acid dehydrase as a measure of lead exposure. *Arch. Environ. Health* 21:140–145.

Hoffman, D. J., O. H. Pattee, S. N. Wiemeyer, and B. Mulhern. 1981. Effects of lead shot ingestion on delta-aminolevulinic acid dehydratase activity, hemoglobin concentration, and serum chemistry in bald eagles. *J. Wildl. Dis.* 17:423–431.

Hoffman, D. J., J. C. Franson, O. H. Pattee, C. M. Bunck, and A. Anderson. 1985a. Survival, growth, and accumulation of ingested lead in nestling American kestrels (*Falco sparverius*). *Arch. Environ. Contam. Toxicol.* 14:89–94.

Hoffman, D. J., J. C. Franson, O. H. Pattee, C. M. Bunck, and H. C. Murray. 1985b. Biochemical and hematological effects of lead ingestion in nestling American kestrels (*Falco sparverius*). *Comp. Biochem. Physiol.* 80C:431–439.

Hohman, W. L., J. L. Moore, and J. C. Franson. 1995. Winter survival of immature canvasbacks in inland Louisiana. *J. Wildl. Manag.* 59:384–392.

Hulse, M., J. S. Mahoney, G. D. Schroder, C. S. Hacker, and S. M. Pier. 1980. Environmentally acquired lead, cadmium, and manganese in the cattle egret, *Bubulcus ibis*, and the laughing gull, *Larus atricilla*. *Arch. Environ. Contam. Toxicol.* 9:65–78.

Hunt, W. G., W. Burnham, C. N. Parish, K. K. Burnham, B. Mutch, and J. L. Oaks. 2006. Bullet fragments in deer remains: Implications for lead exposure in avian scavengers. *Wildl. Soc. Bull.* 34:167–170.

Hunter, B., and J. C. Haigh. 1978. Demyelinating peripheral neuropathy in a guinea hen associated with subacute lead intoxication. *Avian Dis.* 22:344–349.

Hunter, B. F., and M. N. Rosen. 1965. Occurrence of lead poisoning in a wild pheasant (*Phasianus colchicus*). *Calif. Fish Game* 51:207.

Hunter, B., and G. Wobeser. 1980. Encephalopathy and peripheral neuropathy in lead-poisoned mallard ducks. *Avian Dis.* 24:169–178.

Hutton, M. 1980. Metal contamination of feral pigeons *Columba livia* from the London area: Part 2—Biological effects of lead exposure. *Environ. Pollut. Ser. A* 22:281–293.

Hutton, M., and G. T. Goodman. 1980. Metal contamination of feral pigeons *Columba livia* from the London area: Part 1—Tissue accumulation of lead, cadmium, and zinc. *Environ. Pollut. Ser. A* 22:207–217.

Irwin, J. C. 1975. Mortality factors in whistling swans at Lake St. Clair, Ontario. *J. Wildl. Dis.* 11:8–12.

Iwata, H., et al. 2000. Contamination by chlorinated hydrocarbons and lead in Steller's sea eagle and white-tailed sea eagle from Hokkaido, Japan. In *First symposium on Steller's and white-tailed sea eagles in east Asia*, eds. M. Ueta, and M. J. McGrady, pp. 91–106. Tokyo: Wild Bird Society of Japan.

Jacobson, E., J. W. Carpenter, and M. Novilla. 1977. Suspected lead toxicosis in a bald eagle. *J. Am. Vet. Med. Assoc.* 171:952–954.

Janssen, D. L., P. T. Robinson, and P. K. Ensley. 1979. Lead toxicosis in three captive avian species. In *Proc. 1979 Annu. Meet. Am. Assoc. Zoo Vet.*, 40–42.

Janssen, D. L., J. E. Oosterhuis, J. L. Allen, M. P. Anderson, D. G. Kelts, and S. N. Wiemeyer. 1986. Lead poisoning in free-ranging California condors. *J. Am. Vet. Med. Assoc.* 189:1115–1117.

Johnson, G. D., et al. 1999. Lead exposure in passerines inhabiting lead-contaminated floodplains in the Coeur D'Alene River Basin, Idaho, USA. *Environ. Toxicol. Chem.* 18:1190–1194.

Johnson, M. S., H. Pluck, M. Hutton, and G. Moore. 1982. Accumulation and renal effects of lead in urban populations of feral pigeons, *Columba livia. Arch. Environ. Contam. Toxicol.* 11:761–767.

Jordan, J. S., and F. C. Bellrose. 1951. Lead poisoning in wild waterfowl. *Ill. Nat. Hist. Surv. Biol. Notes* No. 26.

Kaiser, T. E., et al. 1980. Organochlorine pesticide, PCB, and PBB residues and necropsy data for bald eagles from 29 states—1975–77. *Pestic. Monit. J.* 13:145–149.

Karstad, L. 1971. Angiopathy and cardiopathy in wild waterfowl from ingestion of lead shot. *Conn. Med.* 35:355–360.

Kelly, M. E., et al. 1998. Acute effects of lead, steel, tungsten-iron, and tungsten-polymer shot administered to game-farm mallards. *J. Wildl. Dis.* 34:673–687.

Kendall, R. J., and P. F. Scanlon. 1981. Chronic lead ingestion and nephropathy in ringed turtle doves. *Poult. Sci.* 60:2028–2032.

Kendall, R. J., and P. F. Scanlon. 1982. The toxicology of ingested lead acetate in ringed turtle doves *Streptopelia risoria. Environ. Pollut. Ser. A* 27:255–262.

Kendall, R. J., P. F. Scanlon, and R. T. Di Giulio. 1982. Toxicology of ingested lead shot in ringed turtle doves. *Arch. Environ. Contam. Toxicol.* 11:259–263.

Kendall, R. J., P. F. Scanlon, and H. P. Veit. 1983. Histologic and ultrastructural lesions of mourning doves (*Zenaida macroura*) poisoned by lead shot. *Poult. Sci.* 62:952–956.

Kendall, R. J., H. P. Veit, and P. F. Scanlon. 1981. Histological effects and lead concentrations in tissues of adult male ringed turtle doves that ingested lead shot. *J. Toxicol. Environ. Health* 8:649–658.

Kendall, R. J., et al. 1996. An ecological risk assessment of lead shot exposure in non-waterfowl avian species: upland game birds and raptors. *Environ. Toxicol. Chem.* 15:4–20.

Kennedy, S., J. P. Crisler, E. Smith, and M. Bush. 1977. Lead poisoning in sandhill cranes. *J. Am. Vet. Med. Assoc.* 171:955–958.

Kenntner, N., O. Krone, R. Altenkamp, and F. Tataruch. 2003. Environmental contaminants in liver and kidney of free-ranging northern goshawks (*Accipiter gentilis*) from three regions of Germany. *Arch. Environ. Contam. Toxicol.* 45:128–135.

Kenntner, N., F. Tataruch, and O. Krone. 2001. Heavy metals in soft tissue of white-tailed eagles found dead or moribund in Germany and Austria from 1993 to 2000. *Environ. Toxicol. Chem.* 20:1831–1837.

Keymer, I. F., and R. St. J. Stebbings. 1987. Lead poisoning in a partridge (*Perdix perdix*) after ingestion of gunshot. *Vet. Rec.* 120:276–277.

Kim, E. Y., R. Goto, H. Iwata, Y. Masuda, S. Tanabe, and S. Fujita. 1999. Preliminary survey of lead poisoning of Steller's sea eagle (*Haliaeetus pelagicus*) and white-tailed sea eagle (*Haliaeetus albicilla*) in Hokkaido, Japan. *Environ. Toxicol. Chem.* 18:448–451.

Kingsford, R. T., J. Flanjak, and S. Black. 1989. Lead shot and ducks on Lake Cowal. *Aust. Wildl. Res.* 16:167–172.

Koranda, J., K. Moore, M. Stuart, and C. Conrado. 1979. Dietary effects on lead uptake and trace element distribution in mallard ducks dosed with lead shot. UCID-18044. Livermore: Lawrence Livermore Laboratory, Environmental Sciences Division.

Kramer, J. L., and P. T. Redig. 1997. Sixteen years of lead poisoning in eagles, 1980-95: an epizootilogic view. *J. Raptor Res.* 31:327–332.

Kreager, N., B. C. Wainman, R. K. Jayasinghe, and L. J. S. Tsuji. 2008. Lead pellet ingestion and liver-lead concentrations in upland game birds from southern Ontario, Canada. *Arch. Environ. Contam. Toxicol.* 54:331–336.

Krementz, D. G., and C. D. Ankney. 1995. Changes in total body calcium and diet of breeding house sparrows. *J. Avian Biol.* 26:162–167.

Krone, O., A. Berger, and R. Schulte. 2009. Recording movement and activity pattern of a white-tailed sea eagle (*Haliaeetus albicilla*) by a GPS datalogger. *J. Ornithol.* 150:273–280.

Krone, O., T. Stjernberg, N. Kenntner, F. Tataruch, J. Koivusaari, and I. Nuuja. 2006. Mortality factors, helminth burden, and contaminant residues in white-tailed sea eagles (*Haliaeetus albicilla*) from Finland. *Ambio* 35:98–104.

Krone, O., F. Wille, N. Kenntner, D. Boertmann, and F. Tataruch. 2004. Mortality factors, environmental contaminants, and parasites of white-tailed sea eagles from Greenland. *Avian Dis.* 48:417–424.

Kurosawa, N. 2000. Lead poisoning in Steller's sea eagles and white-tailed sea eagles. In *First symposium on Steller's and white-tailed sea eagles in east Asia*, eds. M. Ueta and M. J. McGrady, pp. 107–109. Tokyo: Wild Bird Society of Japan.

Langelier, K. M., C. E. Andress, T. K. Grey, C. Wooldridge, R. J. Lewis, and R. Marchetti. 1991. Lead poisoning in bald eagles in British Columbia. *Can. Vet. J.* 32:108–109.

Leonzio, C., C. Fossi, and S. Focardi. 1986. Lead, mercury, cadmium, and selenium in two species of gull feeding on inland dumps, and in marine areas. *Sci. Total Environ.* 57:121–127.

Lewis, L. A., and S. H. Schweitzer. 2000. Lead poisoning in a northern bobwhite in Georgia. *J. Wildl. Dis.* 36:180–183.

Lewis, L. A., R. J. Poppenga, W. R. Davidson, J. R. Fischer, and K. A. Morgan. 2001. Lead toxicosis and trace element levels in wild birds and mammals at a firearms training facility. *Arch. Environ. Contam. Toxicol.* 41:208–214.

Locke, L. N., and G. E. Bagley. 1967. Lead poisoning in a sample of Maryland mourning doves. *J. Wildl. Manage.* 31:515–518.

Locke, L. N., G. E. Bagley, D. N. Frickie, and L. T. Young. 1969. Lead poisoning and aspergillosis in an Andean condor. *J. Am. Vet. Med. Assoc.* 155:1052–1056.

Locke, L. N., and M. Friend. 1992. Lead poisoning of avian species other than waterfowl. In *Lead poisoning in waterfowl*, ed. D. J. Pain, pp. 19–22. IWRB Spec. Publ. 16. Slimbridge: International Waterfowl and Wetlands Research Bureau.

Locke, L. N., S. M. Kerr, and D. Zoromski. 1982. Lead poisoning in common loons (*Gavia immer*). *Avian Dis.* 26:392–396.

Locke, L. N., M. R. Smith, R. M. Windingstad, and S. J. Martin. 1991. Lead poisoning of a marbled godwit. *Prairie Nat.* 23:21–24.

Locke, L. N., and N. J. Thomas. 1996. Lead poisoning of waterfowl and raptors. In *Noninfectious diseases of wildlife*, 2nd ed., eds. A. Fairbrother, L. N. Locke, and G. L. Hoff, pp. 108–117. Ames, IA: Iowa State University Press.

Longcore, J. R., R. Andrews, L. N. Locke, G. E. Bagley, and L. T. Young. 1974a. Toxicity of lead and proposed substitute shot to mallards. *U.S. Fish Wildl. Serv. Spec. Sci. Rep. Wildl.* No. 183. Washington, D.C.: U.S. Department of the Interior.

Longcore, J. R., L. N. Locke, G. E. Bagley, and R. Andrews. 1974b. Significance of lead residues in mallard tissues. *U.S. Fish Wildl. Serv. Spec. Sci. Rep. Wildl.* No. 182. Washington, D.C.: U.S. Department of the Interior.

Lumeij, J. T., W. T. C. Wolvekamp, G. M. Bron-Dietz, and A. J. H. Schotman. 1985. An unusual case of lead poisoning in a honey buzzard (*Pernis apivorus*). *Vet. Q.* 7:165–168.

MacDonald, J. W., C. J. Randall, H. M. Ross, G. M. Moon, and A. D. Ruthven. 1983. Lead poisoning in captive birds of prey. *Vet. Rec.* 113:65–66.

Maedgen, J. L., C. S. Hacker, G. D. Schroder, and F. W. Weir. 1982. Bioaccumulation of lead and cadmium in the royal tern and sandwich tern. *Arch. Environ. Contam. Toxicol.* 11:99–102.

Marn, C. M., R. E. Mirarchi, and M. E. Lisano. 1988. Effects of diet and cold-exposure on captive female mourning doves dosed with lead shot. *Arch. Environ. Contam. Toxicol.* 17:589–594.

Martin, P. A., D. Campbell, K. Hughes, and T. McDaniel. 2008. Lead in the tissues of terrestrial raptors in southern Ontario, Canada, 1995–2001. *Sci. Total Environ.* 391:96–103.

Martínez-López, E., et al. 2004. Lead in feathers and delta-aminolevulinic acid dehydratase activity in three raptor species from an unpolluted Mediterranean forest (southeastern Spain). *Arch. Environ. Contam. Toxicol.* 47:270–275.

Mateo, R. 2009. Lead poisoning in wild birds in Europe and the regulations adopted by different countries. In *Ingestion of lead from spent ammunition: implications for wildlife and humans*, eds. R. T. Watson, M. Fuller, M. Pokras, and W. G. Hunt. Boise: The Peregrine Fund. DOI 10.4080/ilsa.2009.0107.

Mateo, R., W. N. Beyer, J. W. Spann, D. J. Hoffman, and A. Ramis. 2003. Relationship between oxidative stress, pathology, and behavioral signs of lead poisoning in mallards. *J. Toxicol. Environ. Health Part A* 66:1371–1389.

Mateo, R., J. C. Dolz, J. M. Aguilar Serrano, J. Belliure, and R. Guitart. 1997a. An epizootic of lead poisoning in greater flamingos (*Phoenicopterus ruber roseus*) in Spain. *J. Wildl. Dis.* 33:131–134.

Mateo, R, A. J. Green, C. W. Jeske, V. Urios, and C. Gerique. 2001. Lead poisoning in the globally threatened marbled teal and white-headed duck in Spain. *Environ. Toxicol. Chem.* 20:2860–2868.

Mateo, R., A. J. Green, H. Lefranc, R. Baos, and J. Figuerola. 2007. Lead poisoning in wild birds from southern Spain: a comparative study of wetland areas and species affected, and trends over time. *Ecotoxicol. Environ. Saf.* 66:119–126.

Mateo, R., R. Guitart, and A. J. Green. 2000. Determinants of lead shot, rice and grit ingestion in ducks and coots. *J. Wildl. Manage.* 64:939–947.

Mateo, R., R. Molina, J. Grífols, and R. Guitart. 1997b. Lead poisoning in a free ranging griffon vulture (*Gyps fulvus*). *Vet. Rec.* 140:47–48.

Mateo, R., et al. 1999. Lead shot ingestion by marsh harriers Circus aeruginosus from the Ebro delta, Spain. *Environ. Pollut.* 104:435–440.

Mautino, M., and J. U. Bell. 1986. Experimental lead toxicity in ring-necked duck. *Environ. Res.* 41:538–545.

Mautino, M., and J. U. Bell. 1987. Hematological evaluation of lead intoxication in mallards. *Bull. Environ. Contam. Toxicol.* 38:78–85.

McConnell, C. A. 1967. Experimental lead poisoning of bobwhite quail and mourning doves, pp. 208–219. New Orleans: Proc. 21st Annu. Conf. Southeast. Assoc. Game Fish Comm.

Meharg, A. A., et al. 2002. Isotopic identification of the sources of lead contamination for white storks (*Ciconia ciconia*) in a marshland ecosystem (Doñana, S. W. Spain). *Sci. Total Environ.* 300: 81–86.

Meretsky, V. J., N. F. R. Snyder, S. R. Beissinger, D. A. Clendenen, and J. W. Wiley. 2000. Demography of the California condor: implications for reestablishment. *Conserv. Biol.* 14:957–967.

Moore, K. C. 1978. Investigations of lead poisoning in waterfowl in California. In *Trans. West. Sec. Wildl. Soc. Annu. Mtg.* 209–220.

Morgan, G. W., F. W. Edens, P. Thaxton, and C. R. Parkhurst. 1975. Toxicity of dietary lead in Japanese quail. *Poult. Sci.* 54:1636–1642.

Mörner, T., and L. Petersson. 1999. Lead poisoning in woodpeckers in Sweden. *J. Wildl. Dis.* 35:763–765.

Morton, A. P., S. Partridge, and J. A. Blair. 1985. The intestinal uptake of lead. *Chem. Br.* Oct:923–927.

Mulhern, B. M., et al. 1970. Organochlorine residues and autopsy data from bald eagles 1966-68. *Pestic. Monit. J.* 4:141–144.

Munoz, R. V., Jr., C. S. Hacker, and T. F. Gesell. 1976. Environmentally acquired lead in the laughing gull, *Larus atricilla. J. Wildl. Dis.* 12:139–142.

Nakade, T., et al. 2005. Lead poisoning in whooper and tundra swans. *J. Wildl. Dis.* 41:253–256.

Nam, D. H., and D. P. Lee. 2005. Possible routes for lead accumulation in feral pigeons (*Columba livia*). *Environ. Monit. Assess.* 121:355–361.

Nelson, T. A., C. Mitchell, and C. Abbott. 1989. Lead-shot ingestion by bald eagles in western Arkansas. *Southwest. Nat.* 34:245–249.

Ochiai, K., et al. 1992. Pathological study of lead-poisoning in whooper swans (*Cygnus cygnus*) in Japan. *Avian Dis.* 36:313–323.

O'Halloran, J., P. F. Duggan, and A. A. Myers. 1988a. Biochemical and haematological values for mute swans (*Cygnus olor*): effects of acute lead poisoning. *Avian Pathol.* 17:667–678.

O'Halloran, J., A. A. Myers, and P. F. Duggan. 1988b. Blood lead levels and free red blood cell protoporphyrin as a measure of lead exposure in mute swans. *Environ. Pollut.* 52:19–38.

Ohi, G., H. Seki, K. Akiyama, and H. Yagyu. 1974. The pigeon, a sensor of lead pollution. *Bull. Environ. Contam. Toxicol.* 12:92–98.

Ohi, G., H. Seki, K. Minowa, M. Ohsawa, I. Mizoguchi, and F. Sugimori. 1981. Lead pollution in Tokyo—the pigeon reflects its amelioration. *Environ. Res.* 26:125–129.

Osborn, D., W. J. Every, and K. R. Bull. 1983. The toxicity of trialkyl lead compounds to birds. *Environ. Pollut. Ser. A* 31:261–275.

Pain, D. J. 1987. Lead poisoning in waterfowl: an investigation of sources and screening techniques. Ph.D. Thesis. Oxford Univ. Oxford, UK.

Pain, D. J. 1989. Haematological parameters as predictors of blood lead and indicators of lead poisoning in the black duck (*Anas rubripes*). *Environ. Pollut.* 60:67–81.

Pain, D. J. 1990. Lead shot ingestion by waterbirds in the Camargue, France: an investigation of levels and interspecific differences. *Environ. Pollut.* 66:273–285.

Pain, D. J. 1992. Lead poisoning of water fowl: a review. In *Lead poisoning in waterfowl*, ed. D. J. Pain, pp. 7–13. IWRB Spec. Publ. 16. Slimbridge: International Waterfowl and Wetlands Research Bureau.

Pain, D. J., and B. A. Rattner. 1988. Mortality and hematology associated with the ingestion of one number four lead shot in black ducks, *Anas rubripes. Bull. Environ. Contam. Toxicol.* 40:159–164.

Pain, D. J., C. Amiard-Triquet, and C. Sylvestre. 1992. Tissue lead concentrations and shot ingestion in nine species of waterbird from the Camargue (France). *Ecotoxicol. Environ. Saf.* 24:217–233.

Pain, D. J., C. Bavoux, and G. Burneleau. 1997. Seasonal blood lead concentrations in marsh harriers *Circus aeruginosus* from Charente-Maritime, France: relationship with the hunting season. *Biol. Conserv.* 81:1–7.

Pain, D. J., A. A. Meharg, M. Ferrer, M. Taggart, and V. Penteriani. 2005. Lead concentrations in bones and feathers of the globally threatened Spanish imperial eagle. *Biol. Conserv.* 121:603–610.

Pain, D. J., et al. 2007. Lead contamination and associated disease in captive and reintroduced red kites *Milvus milvus* in England. *Sci. Total Environ.* 376:116–127.

Pain, D. J., I. J. Fisher, and V. G. Thomas. 2009. A global update of lead poisoning in terrestrial birds from ammunition sources. In *Ingestion of lead from spent ammunition: implications for wildlife and humans*, eds. R. T. Watson, M. Fuller, M. Pokras, and W. G. Hunt. Boise: The Peregrine Fund. DOI 10.4080/ilsa.2009.0108.

Parish, C. N., W. R. Heinrich, and W. G. Hunt. 2007. Lead exposure, diagnosis, and treatment in California condors released in Arizona. In *California condors in the 21st century*. Series in ornithology, no. 2, eds. A. Mee and L. S. Hall, pp. 97–108. Cambridge, MA: Nuttall Ornithology Club and Washington, D.C.: The American Ornithologists' Union.

Pattee, O. H. 1984. Eggshell thickness and reproduction in American kestrels exposed to chronic dietary lead. *Arch. Environ. Contam. Toxicol.* 13:29–34.

Pattee, O. H., S. N. Wiemeyer, B. M. Mulhern, L. Sileo, and J. W. Carpenter. 1981. Experimental lead-shot poisoning in bald eagles. *J. Wildl. Manage.* 45:806–810.

Pattee, O. H., P. H. Bloom, J. M. Scott, and M. R. Smith. 1990. Lead hazards within the range of the California condor. *Condor* 92:931–937.

Pattee, O. H., and D. J. Pain. 2003. Lead in the Environment. In *Handbook of ecotoxicology*, 2nd ed., eds. D. J. Hoffman, B. A. Rattner, G. A. Burton, Jr., and J. Cairns, Jr., pp. 373–408. Boca Raton: Lewis Publishers.

Pattee, O. H., et al. 2006. Lead poisoning in captive Andean condors (*Vultur gryphus*). *J. Wildl. Dis.* 42:772–779.

Platt, J. B. 1976. Bald eagles wintering in a Utah desert. *Am. Birds* 30:783–788.

Pokras, M. A., and R. Chafel. 1992. Lead toxicosis from ingested fishing sinkers in adult common loons (*Gavia immer*) in New England. *J. Zoo Wildl. Med.* 23:92–97.

Pokras, M. A., M. R. Kneeland, A. Major, R. Miconi, and R. H. Poppenga. 2009. Lead objects ingested by common loons in New England. Extended abstract. In *Ingestion of lead from spent ammunition: implications for wildlife and humans*, eds. R. T. Watson, M. Fuller, M. Pokras, and W. G. Hunt. Boise: The Peregrine Fund. DOI 10.4080/ilsa.2009.0116.

Ramo, C., C. Sánchez, and L. H. Saint-Aubin. 1992. Lead poisoning of greater flamingos *Phoenicopterus ruber. Wildfowl* 43:220–222.

Redig, P. T., C. M. Stowe, D. M. Barnes, and T. D. Arent. 1980. Lead toxicosis in raptors. *J. Am. Vet. Med. Assoc.* 177:941–943.

Redig, P. T., E. M. Lawler, S. Schwartz, J. L. Dunnette, B. Stephenson, and G. E. Duke. 1991. Effects of chronic exposure to sublethal concentrations of lead acetate on heme synthesis and immune function in red-tailed hawks. *Arch. Environ. Contam. Toxicol.* 21:72–77.

Reichel, W. L., et al. 1984. Pesticide, PCB, and lead residues and necropsy data for bald eagles from 32 states—1978-81. *Environ. Monit. Assess.* 4:395–403.

Reiser, M. H., and S. A. Temple. 1981. Effects of chronic lead ingestion on birds of prey. In *Recent advances in the study of raptor diseases*, eds. J. E. Cooper and A. G. Greenwood, pp. 21–25. West Yorkshire: Chiron Publications, Ltd.

Rocke, T. E., and M. D. Samuel. 1991. Effects of lead shot ingestion on selected cells of the mallard immune system. *J. Wildl. Dis.* 27:1–9.

Roscoe, D. E., S. W. Nielsen, A. A. Lamola, and D. Zuckerman. 1979. A simple quantitative test for erythrocytic protoporphyrin in lead-poisoned ducks. *J. Wildl. Dis.* 15:127–136.

Roux, K. E., and P. P. Marra. 2007. The presence and impact of environmental lead in passerine birds along an urban to rural land use gradient. *Arch. Environ. Contam. Toxicol.* 53:261–268.

Salisbury, R. M., E. L. J. Staples, and M. Sutton. 1958. Lead poisoning of chickens. *N. Z. Vet. J.* 6:2–7.

Sanderson, G. C. 1992. Lead poisoning mortality. In *Lead poisoning in waterfowl,* ed. D. J. Pain, pp. 14–18. IWRB Spec. Publ. 16. Slimbridge: International Waterfowl and Wetlands Research Bureau.

Sanderson, G. C., and F. C. Bellrose. 1986. A review of the problem of lead poisoning in waterfowl. *Ill. Nat. Hist. Surv. Spec. Publ.* No. 4. Champaign, IL: Illinois Natural History Survey.

Sanderson, G. C., and J. C. Irwin. 1976. Effects of various combinations and numbers of lead:iron pellets dosed in wild-type captive mallards. Final Report Contract No. 14-16-0008-914. U. S. Fish and Wildlife Service and Illinois Natural History Survey.

Sassa, S., S. Granick, and A. Kappas. 1975. Effect of lead and genetic factors on heme biosynthesis in the human red cell. *Ann. N. Y. Acad. Sci.* 244:419–440.

Scanlon, P. F. 1982. Wet and dry weight relationships of mallard (*Anas platyrhynchos*) tissues. *Bull. Environ. Contam. Toxicol.* 29:615–617.

Scheifler, R., et al. 2006. Lead concentrations in feathers and blood of common blackbirds (*Turdus merula*) and in earthworms inhabiting unpolluted and moderately polluted urban areas. *Sci. Total Environ.* 371:197–205.

Scheuhammer, A. M. 1989. Monitoring wild bird populations for lead exposure. *J. Wildl. Manage.* 53:759–765.

Scheuhammer, A. M. 1996. Influence of reduced dietary calcium on the accumulation and effects of lead, cadmium, and aluminum in birds. *Environ. Pollut.* 94:337–343.

Scheuhammer, A. M., and K. M. Dickson. 1996. Patterns of environmental lead exposure in waterfowl in eastern Canada. *Ambio* 25:14–20.

Scheuhammer, A. M., and D. M. Templeton. 1998. Use of stable isotope ratios to distinguish sources of lead exposure in wild birds. *Ecotoxicology* 7:37–42.

Scheuhammer, A. M., and L. K. Wilson. 1990. Effects of lead and pesticides on delta-aminolevulinic acid dehydratase of ring doves (*Streptopelia risoria*). *Environ. Toxicol. Chem.* 9:1379–1386.

Scheuhammer, A. M., C. A. Rogers, and D. Bond. 1999. Elevated lead exposure in American woodcock (*Scolopax minor*) in eastern Canada. *Arch. Environ. Contam. Toxicol.* 36:334–340.

Scheuhammer, A. M., D. E. Bond, N. M. Burgess, and J. Rodrigue. 2003a. Lead and stable lead isotype ratios in soil, earthworms, and bones of American woodcock (*Scolopax minor*) from eastern Canada. *Environ. Toxicol. Chem.* 22:2585–2591.

Scheuhammer, A. M., S. L. Money, D. A. Kirk, and G. Donaldson. 2003b. Lead fishing sinkers and jigs in Canada: review of their use patterns and toxic impacts on wildlife. Occasional Paper #108. Ottawa: The Canadian Wildlife Service.

Schmitz, R. A., A. A. Aguirre, R. S. Cook, and G. A. Baldassarre. 1990. Lead poisoning of Caribbean flamingos in Yucatan, Mexico. *Wildl. Soc. Bull.* 18:399–404.

Schulz, J. H., et al. 2006. Acute lead toxicosis in mourning doves. *J. Wildl. Manage.* 70:413–421.

Schulz, J. H., X. Gao, J. J. Millspaugh, and A. J. Bermudez. 2007. Experimental lead pellet ingestion in mourning doves (*Zenaida macroura*) *Am. Midl. Nat.* 158:177–190.

Sears, J. 1988. Regional and seasonal variations in lead poisoning in the mute swan *Cygnus olor* in relation to the distribution of lead and lead weights, in the Thames area, England. *Biol. Conserv.* 46:115–134.

Sidor, I. F., M. A. Pokras, A. R. Major, R. H. Poppenga, K. M. Taylor, and R. M. Miconi. 2003. Mortality of common loons in New England, 1987 to 2000. *J. Wildl. Dis.* 39:306–315.

Sikarskie, J. 1977. The case of the red-tailed hawk. *Intervet* 8:4.

Sileo, L., and S. I. Fefer. 1987. Paint chip poisoning of Laysan albatross at Midway Atoll. *J. Wildl. Dis.* 23:432–437.

Sileo, L., et al. 2001. Lead poisoning of waterfowl by contaminated sediment in the Coeur d'Alene River. *Arch. Environ. Contam. Toxicol.* 41:364–368.

Simkiss, K. 1961. Calcium metabolism and avian reproduction. *Biol. Rev.* 36:321–367.

Snoeijs, T., T. Dauwe, R. Pinxten, V. M. Darras, L. Arckens, and M. Eens. 2005. The combined effect of lead exposure and high or low dietary calcium on health and immunocompetence in the zebra finch (*Taeniopygia guttata*). *Environ. Pollut.* 134:123–132.

Snyder, N. F. R. 2007. Limiting factors for wild California condors. In *California condors in the 21st century.* Series in ornithology, no. 2, eds. A. Mee, and L. S. Hall, pp. 9–33. Cambridge, MA: Nuttall Ornithological Club and Washington, D.C.: The American Ornithologists' Union.

Snyder, S. B., M. J. Richard, R. C. Drewien, N. Thomas, and J. P. Thilsted. 1991. Diseases of whooping cranes seen during annual migration of the Rocky Mountain flock. In *1991 Proc. Am. Assoc. Zoo Vet.*, ed. R. E. Junge, pp. 74–80. Cambridge, MA: The Nutthall Ornithological Club and Washington, D.C.: The American Orithologists' Union.

Sorenson, K. J., and L. J. Burnett. 2007. Lead concentrations in the blood of Big Sur California condors. In *California condors in the 21st century*. Series in ornithology, no. 2, eds. A. Mee and L. S. Hall, pp. 185–195. Cambridge, MA: Nuttall Ornithological Club and Washington, D.C.: The American Ornithologists' Union.

Spray, C. J., and H. Milne. 1988. The incidence of lead poisoning among whooper and mute swans *Cygnus cygnus* and *C. olor* in Scotland. *Biol. Conserv.* 44:265–281.

Stendell, R. C., R. I. Smith, K. P. Burnham, and R. E. Christensen. 1979. Exposure of waterfowl to lead: a nationwide survey of residues in wing bones of seven species, 1972–73. *U.S. Fish Wildl. Serv. Spec. Sci. Rep. Wildl.* No. 223. Washington, D.C.: U.S. Department of the Interior.

Stendell, R. C., J. W. Artmann, and E. Martin. 1980. Lead residues in sora rails from Maryland. *J. Wildl. Manage.* 44:525–527.

Stone, W. B., and S. A. Butkas. 1978. Lead poisoning in a wild turkey. *N. Y. Fish Game J.* 25:169.

Stone, C. L., and J. H. Soares, Jr. 1976. The effect of dietary selenium level on lead toxicity in the Japanese quail. *Poult. Sci.* 55:341–349.

Stone, C. L., M. R. S. Fox, A. L. Jones, and K. R. Mahaffey. 1977. Delta-aminolevulinic acid dehydratase—a sensitive indicator of lead exposure in Japanese quail. *Poult. Sci.* 56:174–181.

Stone, C. L., K. R. Mahaffey, and M. R. S. Fox. 1979. A rapid bioassay system for lead using young Japanese quail. *J. Environ. Pathol. Toxicol.* 2:767–779.

Stone, W. B., and J. C. Okoniewski. 2001. Necropsy findings and environmental contaminants in common loons from New York. *J. Wildl. Dis.* 37:178–184.

Strom, S. M., K. A. Patnode, J. A. Langenberg, B. L. Bodenstein, and A. M. Scheuhammer. 2005. Lead contamination in American woodcock (*Scolopax minor*) from Wisconsin. *Arch. Environ. Contam. Toxicol.* 49:396–402.

Sullivan, K., R. Sieg, and C. Parish. 2007. Arizona's efforts to reduce lead exposure in California condors. In *California condors in the 21st century*. Series in ornithology, no. 2, eds. A. Mee and L. S. Hall, pp. 109–121. Cambridge, MA: Nuttall Ornithology Club and Washington, D.C.: The American Ornithologists' Union.

Szymczak, M. R., and W. J. Adrian. 1978. Lead poisoning in Canada geese in southwest Colorado. *J. Wildl. Manage.* 42:299–306.

Teraoka, H., et al. 2007. Heavy metal contamination status of Japanese cranes (*Grus japonensis*) in east Hokkaido, Japan—extensive mercury pollution. *Environ. Toxicol. Chem.* 26:307–312.

Tola, S., S. Hernberg, S. Asp, and J. Nikkanen. 1973. Parameters indicative of absorption and biological effect in new lead exposure: a prospective study. *Br. J. Ind. Med.* 30:134–141.

Trust, K. A., M. W. Miller, J. K. Ringelman, and I. M. Orme. 1990. Effects of ingested lead on antibody production in mallards (*Anas platyrhnchos*). *J. Wildl. Dis.* 26:316–322.

Vyas, N. B., J. W. Spann, G. H. Heinz, W. N. Beyer, J. A. Jaquette, and J. M. Mengelkoch. 2000. Lead poisoning of passerines at a trap and skeet range. *Environ. Pollut.* 107:159–166.

Vyas, N. B., J. W. Spann, and G. H. Heinz. 2001. Lead shot toxicity to passerines. *Environ. Pollut.* 111:135–138.

Wayland, M., and T. Bollinger. 1999. Lead exposure and poisoning in bald eagles and golden eagles in the Canadian prairie provinces. *Environ. Pollut.* 104:341–350.

White, D. H., and R. C. Stendell. 1977. Waterfowl exposure to lead and steel shot on selected hunting areas. *J. Wildl. Manage.* 41:469–475.

Wiemeyer, S. N., J. M. Scott, M. P. Anderson, P. H. Bloom, and C. J. Stafford. 1988. Environmental contaminants in California condors. *J. Wildl. Manage.* 52:238–247.

Wilson, H. M., J. L. Oyen, and L. Sileo. 2004. Lead shot poisoning of a pacific loon in Alaska. *J. Wildl. Dis.* 40:600–602.

Wilson, H. M., P. L. Flint, and A. N. Powell. 2007. Coupling contaminants with demography: effects of lead and selenium in pacific common eiders. *Environ. Toxicol. Chem.* 26:1410–1417.

Windingstad, R. M., S. M. Kerr, L. N. Locke, and J. J. Hurt. 1984. Lead poisoning of sandhill cranes (*Grus canadensis*). *Prairie Nat.* 16:21–24.

Wobeser, G. A. 1997. *Diseases of wild waterfowl*, 2nd ed. New York: Plenum Press.

Work, T. M., and M. R. Smith. 1996. Lead exposure in Laysan albatross adults and chicks in Hawaii: prevalence, risk factors, and biochemical effects. 1996. *Arch. Environ. Contam. Toxicol.* 31:115–119.

Yamamoto, K., M. Hayashi, M. Yoshimura, H. Hayashi, A. Hiratsuka, and Y. Isii. 1993. The prevalence and retention of lead pellets in Japanese quail. *Arch. Environ. Contam. Toxicol.* 24:478–482.

Zwank, P. J., V. L. Wright, P. M. Shealy, and J. D. Newsom. 1985. Lead toxicosis in waterfowl on two major wintering areas in Louisiana. *Wildl. Soc. Bull.* 13:17–26.

Spotted Cuscus

By G. Mutzel, from *The Royal Natural History*, edited by Richard Lydekker, Frederick Warne & Co.,
London, 1893–94.

17 Lead in Mammals

Wei-Chun Ma

CONTENTS

17.1 Introduction 595
17.2 Interpreting Lead Levels in Blood...596
17.3 Interpreting Lead Levels in Soft Tissues...598
17.4 Interpreting Lead Levels in Hard Tissue..600
17.5 Lead Levels in Association with Subclinical Targets...601
Summary 602
References...604

17.1 INTRODUCTION

Lead has no known essential biochemical function but nevertheless is ubiquitously present at measurable concentrations in mammals and humans, due to global and regional anthropogenic contamination of air, water, and soil. The health hazards of lead are known since ancient times and it is a widely accepted view that everyday low-level exposure to lead has significantly contributed to the collapse of the Roman Empire. Lead is banned from consumer products such as rifle shot, gasoline, and paints and the emission from smelters and other industrial activities is also now much better regulated. However, although environmental lead contamination is globally and locally decreasing it will continue to be a serious problem for the foreseeable future.

Common pathways of exposure to mammals and humans include the inhalation of wind-blown lead-laden dust and the oral intake of lead in contaminated water and food. Soil ingestion is a risk factor for grazing animals and for ground-living species of small mammals as well as for young children with hand-to-mouth behavior. Lead is able to pass through the placenta and expose the developing fetus and, after delivery, lead is transferred from the maternal blood to the milk to expose the weanling (Hallén et al. 1995). Apart from the amount and bioavailability of lead exposure, the health risk for mammals is dependent on a multitude of confounding variables, including dietary composition, nutritional status, age, and metabolism.

As a multitargeted toxicant, lead affects a range of different physiological systems, including the central nervous system (CNS), the renal system, the hematopoietic system, the cardiovascular system, and the gastrointestinal system. As a consequence, lead induces a continuum of such general phenomena as anemia, hypertension, nephropathy (kidney damage), encephalopathy (brain injury), peripheral neuropathy (injury to peripheral nerves), excessive salivation, vomiting, and intestinal colic, as well as abnormal behavior patterns such as insomnia, ataxia (uncoordinated movement), loss of appetite, and lassitude. At extremely high dose levels lead may cause acute death. Since most of these clinical symptoms are nonspecific, causal relations with lead exposure must be verified by demonstrating elevated lead in the blood or in other body tissues.

A large body of human toxicological literature is available on lead (reviewed in WHO 1995 and ATSDR 2007). The information can be perused to better understand and anticipate the risk of lead to mammalian wildlife. Conversely, mammals are commonly used in toxicological studies as experimental models to assess the risk of toxic exposure in humans (Casteel et al. 2006, Smith et al. 2008). Outcomes of occupational lead exposure and of exposure in experimental animals were therefore included in this chapter. The focus of this chapter was the interpretation of internal lead levels as diagnosis criteria of suspected lead toxicity in mammals. Body tissues that will be discussed as biomarkers of lead dose include the blood, the kidneys and skeletal bone. In addition, some specific lead-sensitive biochemical and histological markers are briefly discussed.

17.2 INTERPRETING LEAD LEVELS IN BLOOD

Lead in blood is a reliable biomarker of acute exposure to a lead dose in the most recent few months. Lead in whole blood (PbB) is for about 99% confined to the erythrocytes, where it is to a large part bound to hemoglobin. Lead in whole blood samples can be determined by using graphite furnace atomic absorption spectroscopy (GFAAS) and has an analytical detection limit of 1 μg of Pb per 100 mL (1 μg/dL) (Hu et al. 2007). Mammals and humans in supposedly uncontaminated environments generally show a background or baseline blood lead level <10 μg/dL, but mostly <5 μg/dL. Examples include 5–7 μg/dL in grazing cattle (Neuman and Dollhopf 1992), 4 μg/dL in laboratory rats (Barrett and Livesey 1985) and 5 μg/dL in monkeys (Laughlin et al. 1983). In wild populations of small mammals in a natural duneland area background blood lead levels of 4 μg/dL were measured in bank voles (*Clethrionomys glareolus*) and 8 μg/dL in wood mice (*Apodemus sylvaticus*) (Ma 1994). Background blood lead in humans generally show mean values in the range 2–5 μg/dL (Piomelli et al. 1980, CDC 2005) and thus are similar as those in mammals. However, background blood lead levels can widely differ between countries, such as a mean background value of 8 μg/dL with a range of 4–16 μg/dL reported in children (He et al. 2009). By way of comparison, the U.S. childhood blood lead level of concern has been set at a 10 μg/dL standard (CDC 2005).

The concentration blood lead in mammals may rise to a steady state after one to several months of continuous exposure to an external lead source (Azar et al. 1973). Once the animal is removed from the source, blood lead of acutely exposed animals may show a half-life in the order of weeks or months. A half-life of blood lead of 10 days was thus estimated for blood lead in cattle suffering from acute lead poisoning after accidental ingestion of feed with high amounts of lead shot (Baars et al. 1990). Half-lives of 68–266 days were measured in cattle that had licked a discarded lead battery (Miranda et al. 2007). During chronic cumulative lead exposure, however, blood lead is in equilibrium with delivery into the blood of lead that has been sequestered in skeletal bone. The time-course of lead clearance from blood, once the animal is removed from the external lead source, is then considerably extended and may show a half-life of one or more years. Such conditions of chronic exposure apply to wild mammals exposed in the field to environmental lead. They may also apply to human occupational exposures. A median half-life of blood lead of 619 days independent of initial values has been reported for chronic occupationally exposed workers (Hryhorczuk et al. 1985).

Lead is able to pass through the blood-brain barrier and cause neurobehavioral disturbances and other neurotoxic effects in mammals. The young stages are especially vulnerable as even very low blood lead levels are associated with a wide range of neurological deficits. Blood lead levels >15–20 μg/dL were associated in rats with behavioral deficits and impaired visuospatial discrimination capabilities (Altmann et al. 1993). Rats with the same blood lead levels showed inhibition of isolation-induced aggression behavior (Ogilvie and Martin 1982, Cory-Schlechta et al. 1985). Rats with a mean blood lead level of 26 μg/dL at 50 days of age showed deficits in extinction learning of a conditioned fear response to a tone stimulus (McGlothan et al. 2008).

There is no evidence to indicate that rats may differ with other mammals in their susceptibility to the neurological effects of lead. Blood lead levels of 15–25 μg/dL in newly born macaque monkeys were associated with impaired ability to perform motor discrimination reversal tasks at 3 years of age (Rice 1985). By that time blood lead had decreased to 11–13 μg/dL, but was still higher than the baseline blood lead of 3 μg/dL of the controls. This example illustrates the fact that neurobehavioral effects may persist in mammals for many years beyond the termination of lead exposure, depending on the elimination rate of lead from the blood. Learning deficits may even be permanent as was shown in rhesus monkeys that had been exposed to blood lead levels of 40–60 µg/dL from birth to 1 year of age. The animals otherwise continued to show a normal food consumption and body weight gain (Laughlin et al. 1983).

During the period of pregnancy and lactation the mobilization of lead from the skeleton into the blood circulation is greatly enhanced (Gulson et al. 1997, Manton et al. 2003). The developing fetus may thus be at an increased risk (Barltop 1969, Lidsky and Schneider 2003). Maternal blood lead levels of 21–70 µg/dL were associated with prolonged deficits in learning and motor function in squirrel monkeys after these had been exposed *in utero* only (Newland et al. 1996). The deficits were detected by comparison with offspring of unexposed controls with a baseline blood lead of 4–9 µg/dL. Impaired visual discrimination was found in lambs after *in utero* exposure to a maternal blood lead level of 35 µg/dL (Seppäläinen et al. 1983). Both maternal and paternal exposures may have consequences for the viability of the offspring. Maternal blood lead levels in rats as low as 11 µg/dL were linked to neurobehavioral effects in offspring when the exposure was combined with other forms of stress (Virgolini et al. 2008). Paternal exposure of rabbits to blood lead levels >40 µg/dL was associated with behavioral deficits in the offspring (Nelson et al. 1997). Initial genomic expression was affected in embryos fathered by male rats with blood lead levels down to 15–23 µg/dL (Gandley et al. 1999).

The neonate is exposed to lead in the maternal milk, the concentration of which shows a linear relationship with the concentration of lead in the maternal blood (Gulson et al. 1998). In addition, the efficiency of lead absorption is very high at this early stage. Lead absorption in newborn rats is 90%, decreasing to 15% only after 20–30 days of age (Forbes and Reina 1972). Concentrations of lead in milk of lead-poisoned cows can be as high as 125 µg Pb per liter (Baars et al. 1990). Mean concentrations of 25 µg Pb per liter with a maximum of 35 µg Pb per liter have been found in the breast milk of women living in areas near metal smelters and showing a mean blood lead level of 46 µg/dL with a maximum of 99 µg/dL (Namihira et al. 1993). Children may already suffer from neurobehavioral deficits at blood lead <10 µg/dL, the CDC blood lead level of concern, and it even seems doubtful whether any safe threshold would exist at all in this regard (Jusko et al. 2008). For mammals a blood lead level of 18 µg/dL has been suggested as a safe level for regulatory purposes to protect mammalian wildlife against the risk of toxic lead exposure (Buekers et al. 2009). However, neurotoxic effects were not included in the dataset used. As discussed earlier, neurobehavioral impairments and deficits in visuospatial ability and motor functions appear to correlate with blood lead levels <10 µg/dL.

Lead-induced neurotoxicity includes peripheral neuropathy with demyelination and axonal degeneration of motor nerves and impaired peripheral nerve conduction velocities. Such effects were seen in chronic occupationally exposed workers with blood lead levels >30 µg/dL (Seppäläinen et al. 1983, Chia et al. 1996). Altered auditory perceptual processing was observed in monkeys with a mean blood lead level of 56 µg/dL. The animals were compared with controls with a baseline blood lead of 9 µg/dL (Lilienthal et al. 1990). Lead in blood has been associated with the impairment of sensory modalities including hearing and vision. Monkeys with life-time exposure to blood lead levels between 50 and 170 µg/dL tended to exhibit an elevated frequency threshold for pure tones (Rice 1997). Blood lead levels >7 µg/dL were associated with hearing loss in steel workers when this was tested at sound frequencies of 3000 through 8000 Hz (Hwang et al. 2009). Lead-induced visual system pathology includes amblyopia (poor transmission of the visual image to the brain), optic neuritis (damage to the optic nerve) and peripheral and central scotomas (areas with impaired visual acuity). Impaired visual acuity due to scotopic deficits was found in rhesus monkeys with mean blood lead level of 85 µg/dL after postnatal exposure (Bushnell et al. 1977). Such effects appear to be related to apoptosis (programmed cell death) in the rod and retinal cells caused by sustained overload of Pb^{2+} and Ca^{2+} ions after inhibition of cGMP phosphodiesterase (Fox et al. 1997).

Compared with neurotoxicity effects, less information is available on the effect of lead on the mammalian reproductive system. Blood lead levels >30–40 µg/dL in male rats and mice were associated with prostatic hyperplasia, impaired sperm maturation and motility, reduced testicular weight, seminiferous tubular damage, and spermatogenic cell arrest (Hildebrand et al. 1973, Johansson and Wide 1986). Semen characteristics such as sperm count and motility and sperm forms may be affected as well. Blood lead levels >24 µg/dL in male rabbits and >40–50 µg/dL in human males were associated with deficits in semen quality (WHO 1995). Male fertility of rats was reduced at blood lead levels in the range 27–60 µg/dL (Gandley et al. 1999). Relatively little is known about the effect of chronic lead exposure on the female reproductive system, which may include infertility, miscarriage, and premature delivery. Possible hormonal effects that are linked to ovarian activity and fertility include the observation of suppressed luteal function in female rhesus monkeys with mean blood lead of 70 µg/dL. The animals showed no significant differences in body weight, hematocrit, or general health in comparison with controls with blood lead <10 µg/dL (Franks et al. 1989). A blood lead level >35 µg/dL has been associated with the suppression of the luteinizing and follicle stimulating hormone and estradiol in female monkeys (Foster 1993).

17.3 INTERPRETING LEAD LEVELS IN SOFT TISSUES

Lead in the kidneys, liver, and brain can be determined using routine GFAAS techniques. Concentrations reported in the literature are given on either a wet weight (ww) or a dry weight (dw) basis. Since the water content of the biomarker tissues is mostly unknown, the original data of each reference cited are given here. However, for sake of standardization concentrations reported as wet weight values are also given as dry weight values by applying an approximate conversion factor of 6.5 for kidney tissue and 4.0 for liver (Ma 1994), whereas a factor of 4.4 was used for brain tissue (McIlwain and Bachelard 1985).

Lead concentrations reported in the soft tissues of individual mammals generally decrease in the following order: kidneys > liver > brain ≥ muscle. Muscle is a poor lead accumulator, whereas lead levels in the brain are defined by boundaries of very high toxicity. In contrast, lead has a large potential of bioaccumulation in the kidneys. Lead in the kidneys and liver has similar dynamics as in blood, whereas lead in the brain is much more persistent with half-lives that can be several years (Barry 1975). Concentrations of lead in the brain may therefore not correlate well with levels of lead in the other soft tissues.

The kidneys are a significant target organ and biomarker for lead exposure. Metals in the kidneys of life-time exposed mammals generally reach steady state before the animals have grown into adults. Kidney metal concentrations in life-time exposed shrews (*Sorex araneus*) appear to reach steady states when the animals are half a year of age and in the later subadult stage of their development (Ma and van der Voet 1993). Table 17.1 shows an example of the age relation of lead accumulation in the kidneys of mammals. The kidney lead concentration tended to reach steady state in the

TABLE 17.1
The Effect of Age on the Concentration of Lead in Badgers (*Meles meles*)

Area	Pb Kidney (µg/g dw)	Age Class (year)				
		1	2	3	4	5
A	Geom.mean	1.7	2.6	3.2	4.6	3.0
	Range	0.9–2.8	1.3–4.8	2.4–5.6	2.3–11.3	2.2–3.5
	N	4	12	6	6	3
B	Geom.mean	0.6	2.3	2.1	1.7	1.8
	Range	—	2.3–2.4	0.7–5.6	0.9–3.9	—
	N	1	3	5	5	1
C	Geom.mean	—	4.6	4.3	3.1	4.5
	Range	—	2.9–7.7	2.2–4.6	2.1–5.4	2.8–6.3
	N	0	4	4	5	4

Note: The samples were obtained from populations of three different areas in the Netherlands. *N* is the number of animals of each group.

Source: Ma, W. C., and S. Broekhuizen, *Lutra*, 32, 151, 1989. With permission.

second year when the animals were still in the subadult stage of their development. Sex was not a significant confounding variable in the age relation (Ma and Broekhuizen 1989).

The risk of exposure of wild mammals to lead present in polluted soils is greatly dependent on the bioavailability for uptake in vegetation and invertebrate soil organisms utilized as food. Kidney lead levels in shrews varied between 3 and 11 μg/g dw in natural areas on clay or peat and between 13 and 19 μg/g dw in natural areas on sandy soil (Ma et al. 1992). Mice and voles in the respective areas had a kidney lead level of 0.2–0.6 μg/g dw and 0.4–1.5 μg/g dw. These background levels of kidney lead are not expected to correlate with any toxicity effect in mammals. Birth rate and litter survival were found to remain unaltered in mice with a mean kidney lead of 8.5 μg/g dw. The animals had a lead concentration in the liver of 3.3 and 0.84 μg/g dw in the brain (Mierau and Favara 1975). The risk of exposure for rodents, however, is less great than for insectivores because of the specific dietary requirements and higher metabolic rate of the latter (Ma et al. 1991, Ma and Talmage 2001).

The importance of bioavailability for the exposure of wild mammals to lead-polluted soil is illustrated in Table 17.2 for two smelter areas on a similar type of sandy soil with a similar total amount of lead but different availability. The kidney lead burden of the moles (adult stage) showed on average an almost twenty times difference between the areas. The animals with a kidney lead level >200 μg/g dw were likely to suffer from severe lead intoxication, although nothing was known about demographics. A median kidney lead of 225 μg/g dw in shrews, acquired by feeding the animals in the laboratory with lead-contaminated earthworms, was associated with reduced body weight and increased mortality rate (Pankakoski et al. 1994).

Cattle with mean lead in kidney of 50 μg/g ww (~325 dw) and 26 μg/g ww (~104 dw) in liver and horses with 16 μg/g ww (~104 dw) in kidney and 18 μg/g ww (~72 μg/g dw) in liver have been diagnosed with lead poisoning (Osweiler and van Gelder 1983). Dogs with kidney lead concentrations of 32 μg/g ww (~208 dw) by feeding on a lead-enriched diet showed clinical symptoms of lead poisoning. The animals also had a lead concentration in the liver of 23 μg/g ww (~92 dw) and 1.2 μg/g ww (~5.2 dw) in the brain (Forbes and Sanderson 1978). Rats with a kidney lead level of 120 μg/g dw suffered from loss of body weight (Goyer et al. 1970). Cattle with mean kidney lead concentrations of 87 μg/g dw and liver lead 32 μg/g dw have actually died from lead poisoning (Baars et al. 1990).

The somatic kidney index (SKI), defined as the ratio of the total fresh weight of the kidneys to the total fresh body weight (Goyer et al. 1970, Bankowska and Hine 1985) is a useful marker of advanced lead-induced nephrotoxicity in mammals. An increased SKI value, indicative of renal edema, is specifically induced by lead, other nephrotoxic metals such as arsenic and cadmium are much less effective in this regard (Mahaffey and Fowler 1977). The association between the SKI response and kidney lead may thus be interpreted as a causal relation. In a systematic study of a sympatric population of small mammals in a lead-shot polluted area, the SKI was found to be increased in shrews with a mean kidney-Pb level of 270 μg/g dw (Ma 1989). However, an increased SKI can be linked to kidney lead levels that are considerably lower, as shown in bank voles with mean kidney lead to 16 μg/g dw. It was shown also in wood mice with mean kidney lead of 47 μg/g dw (Roberts et al. 1978). The SKI was only found to remain unaltered in wood mice that had a low kidney lead level of 5.9 μg/g dw (Ma 1989).

TABLE 17.2
The Effect of Lead Availability in Soil on the Concentration of Lead in Moles (*Talpa europea*)

Ecotope Type	Pb Soil (μg/g dw)	SOM (%)	Soil pH	Pb Kidney (μg/g dw) Geom. Mean	Range
Grassland	135	10.2	6.5	18	8–35 ($N = 4$)
Heathland	149	2.0	4.1	338	238–438 ($N = 5$)

Note: SOM is the soil organic matter content. N is the number of samples.
Source: Ma, W. C., *Bull. Environ. Contam. Toxicol.*, 39, 933–938, 1987. With permission.

Lead in the brain is assumed to alter the expression of particular genes and proteins that make up the hippocampal NMDA receptors, which play an important role in brain development and cognition. This change is associated with impairments of nerve communication in the brain. Concentrations of lead in the brain of 0.07 μg/g ww (~0.3 dw), compared with 0.03 μg/g ww (~0.1 dw) in unexposed controls, were associated with behavioral deficits in rats when exposed as weanlings to lead in drinking water (Cory-Schlechta et al. 1985). Similarly, deficits in learning and memory, and spatial and visual discrimination were demonstrated in adult rats with brain-Pb concentrations of 0.09–0.16 μg/g ww (~0.4–0.7 dw) (Altmann et al. 1993).

17.4   INTERPRETING LEAD LEVELS IN HARD TISSUE

Lead in bone reflects cumulative dose and is a reliable biomarker of long-term lead exposure (Hryhorczuk et al. 1998). Lead in bone samples, for which the tibia and femur are most commonly used, can be determined with adapted GFAAS methods. It is, however, also possible to measure bone lead with noninvasive *in vivo* K-shell X-ray fluorescence (K-XRF) techniques (Chettle 2005). Concentrations of lead in cortical bone are reported in the literature on the basis of dry weight (dw) or ash dry weight (adw). For purpose of standardization to dry weight concentrations, it can be assumed that adw is 40% of dry weight (McHugh et al. 2003).

The adult skeleton consists of about 80% of cortical (compact) bone and 20% of trabecular (spongy) bone and is a major depository of lead in the body. Up to 95% of the total pool of lead is stored in the skeleton of adults and about 75% in that of children (Barry 1975). Lead in the skeleton has a long half-life in the order of 10–30 years. Concentrations of lead in bone do not bear a significant relation with concentrations present in the soft tissues, although lead in trabecular bone is relatively more exchangeable with surrounding fluids than lead in cortical bone (Hu et al. 2007). Skeletal bone is continuously resorbed and rebuilt by the activity of the osteoclast and osteoblast bone cells. Mobilization of skeletal lead stores during bone resorption and its release into the blood circulation are especially important during pregnancy and lactation when normal bone turnover is enhanced (Manton et al. 2003). The rates of release, however, of lead from bone into the blood and thence into the milk are to some extent reduced by the dietary intake of calcium (Ettinger et al. 2006). Bone lead levels decrease upon advanced age when bone turnover is increased, the decline is accompanied by an increase of lead concentrations in the brain, liver, and kidneys (Cory-Schlechta and Schaumburg 2000).

Lead sequestered in bone is biomarker of chronic exposure. Concentrations in bone may reveal historical incidences of acute exposures to lead that are no longer reflected in current blood lead levels. Background concentrations of bone lead of mice and voles living in natural sites are in the range 2–3 μg/g dw (Jefferies and French 1972, Kisseberth et al. 1984, Ma 1989). The lowest level of bone lead that have been reported for wild populations of shrews is 12 μg/g dw (Ma and Talmage 2001). In humans, background bone lead levels include 1.5 μg/g dw in the age group of 0–11 months and 3 μg/g dw in the age group of 12–19 years (Samuels et al. 1989).

Lead in bone is a relevant exposure metric for wild mammals with life-time exposure to environmental lead. Lead measured in femur bone of shrews living in lead-polluted sites have shown mean levels of 550 μg/g dw, with an upper boundary of approximately 1500 μg/g dw in some individuals (Ma 1989). Mean bone lead concentrations of 352 μg/g dw were found in wood mice living in a mining area (Roberts et al. 1978). An increase of the lead level in tibia bone of occupationally exposed workers with a mean bone lead of 57 μg/g dw was inversely linked to a decline of cognitive performance (Khalil et al. 2009). Dogs with bone lead of 735 μg/g dw showed symptoms of lead poisoning (Forbes and Sanderson 1978). Clinical signs of lead poisoning in mice at 20 weeks of age were associated with a mean bone lead of 1326 μg/g adw (~530 dw) in females and a mean level of 227 μg/g adw (~91 dw) in males (Donald et al. 1987).

Bank voles with a mean bone lead level of 26 μg/g dw showed advanced nephropathy as indicated by increased SKI values (Ma 1989). The SKI was unaltered in wood mice with a mean bone lead level of 14 μg/g dw. Birth rate and litter survival also remained unaltered in a population of deer mice showing a mean bone lead level of 52 μg/g dw (Mierau and Favara 1975). However, population endpoints of lead exposure are likely to be less sensitive than endpoints that are assessed at the individual level.

17.5 LEAD LEVELS IN ASSOCIATION WITH SUBCLINICAL TARGETS

Biochemical and histopathological changes in certain subclinical markers may provide supporting evidence for lead exposure. Routinely used biochemical markers with a high specificity for lead include key enzymes that are involved in the heme synthetic pathway. Among the most sensitive ones is δ-aminolevulinic acid dehydratase (ALAD), which converts δ-aminolevulinic acid (ALA) to porphobilinogen. The inhibition of erythrocyte ALAD can already be demonstrated at blood lead levels down to 5–10 μg/dL and is almost complete at blood lead levels >70–80 μg/dL (Schutz and Skerfving 1976, Angle et al. 1982).

A less sensitive but nevertheless frequently used enzymatic marker of lead exposure is the inhibition of ferrochelatase, which catalyzes the incorporation of a ferrous ion into protoporphyrin IX to form heme. Inhibition of this blood enzyme results in the incorporation of zinc instead of iron, resulting in the formation of zinc-protoporphyrin (ZPP). However, it may take a prolonged exposure time in order for ZPP and free erythrocyte protoporphyrin (FEP) to rise to significantly elevated levels in erythrocytes (Cory-Schlechta and Schaumburg 2000). A detectable rise in erythrocyte ZPP or FEP may occur at blood lead levels of 15–25 μg/dL in young children and at 20–35 μg/dL in adults (Hammond et al. 1985). On the other hand, the results appear to be rather variable across studies, probably because of the vulnerability of measurements to the blood iron status of the animal. ZPP and FEP were found to remain unaltered in mice treated with lead acetate in their drinking water for 6 months during which blood lead increased from a baseline of 4.5 μg/dL to a level of 60 μg/dL and ALAD activity decreased from 21.5 units per liter to 6.8 units per liter (Torra et al. 1989). Behavior reactions and decision times of lead workers, which did correlate inversely with blood lead in the range <20–80 μg/dL, nevertheless failed to show an association with ZPP levels (Stollery 1996).

Hematological changes are also frequently used as markers of lead exposure, although they are less sensitive than the enzymatic parameters. Changes in hematological parameters were not observed in cattle with blood lead levels between 48 and 76 μg/dL (Miranda et al. 2007). Mice living in a lead-polluted smelter area were not found to have altered hematocrit and hemoglobin values, whereas they did show a strongly reduced ALAD activity (Beyer et al. 1985). It is important that anemia is demonstrated in animals when the reduction of heme synthesis occurs concomitantly with a shortened life span of erythrocytes. The risk of lead-induced anemia as indicated by a reduction of hematocrit with more than 65% was 2% at blood lead levels of 20–39 μg/dL, 18% at 40–59 μg/dL, and 40% at levels >60 μg/dL (Schwartz et al. 1990). The demonstration of anemia thus appears to be restricted to cases of severe lead poisoning. Blood lead levels >70 μg/dL in children and >80 μg/dL in adults were associated with decreased hemoglobin levels with anemia (Goyer et al. 1970). Similarly, a reduced hemoglobin content was found at blood lead levels >40 μg/dL in children and >50 μg/dL in adults, and with a reduced erythrocyte survival time at >60 μg/dL (Hernberg 1980).

Finally, microscopic demonstration of renal edema, interstitial nephritis, and tubular and glomerular damage in the kidneys provides important supporting evidence of a lead exposure diagnosis. Lead-induced histopathological changes in the kidneys include the appearance of acid-fast-staining intranuclear inclusion bodies in the proximal convoluted tubular epithelial cells (Goyer et al. 1970, Kisseberth et al. 1984). Acid-fast intranuclear inclusion bodies were found in the kidneys of a shrew with a kidney lead level of 280 μg/g dw (Beyer et al. 1985). Kidney lead levels >20 μg/g ww (~130 dw) in raccoons were associated with acid-fast intranuclear inclusions and microscopical lesions consisting of tubular degeneration (Hamir et al. 1999). Inclusion bodies and mitochondrial alterations were found in rats with a kidney lead level of 30 μg/g dw (Goyer et al. 1970).

Summary

Evidence has been reviewed for the possibility of interpreting internal lead doses in terms of their association with a range of different endpoints of toxic lead exposure. Critical methodological problems of this approach include the understanding of the dynamics of lead in biomarker tissues as well as the dynamics of the toxicity progress. Age relations thus contribute to the uncertainty of associations made and require to distinguish between acute and chronic lead exposure and between the exposure of prenatal, postnatal and adult stages. Lead in the blood and soft tissues is useful as a biomarker of acute lead poisoning, whereas lead in bone is a more relevant exposure metric for evaluating health effects that have a long latency period.

Table 17.3 summarizes the mean or lowest reported level of blood lead that has been associated with some indicator of adverse health effects observed after chronic exposure. Biochemical responses were excluded from the dataset. There was no evidence of a difference in susceptibility between mammals and humans, suggesting a mutual predictiveness of their lead dose–effect relations. A progression can be seen Table 17.3 within the continuum of lead-induced health effects across the range of reported blood lead levels. Blood lead levels >5 µg/dL were associated with neurobehavioral deficits and other neurotoxic effects. Moderate blood lead levels >20 µg/dL were associated with adverse reproductive effects, whereas still higher levels of >40 µg/dL were associated with nephrotoxic and hematological changes. Very high blood lead levels >80 µg/dL were associated with loss of body weight and death.

TABLE 17.3
Summary of Lead Concentrations in the Blood of Chronically Exposed Mammals and Humans in Association with Various Toxicity Effects

Species	Exposure Stage	Health Effect	Pb Blood (µg/dL)	Reference (Abbreviated)
Humans	Adult	Impaired auditory function	7	Hwang 09
Rats	Prenatal	Neurotoxic effects	11	Virgolini et al. 08
Rats	Adult	Neurotoxic effects	15	Ogilvie 82
Monkeys	Postnatal	Neurotoxic effects	15	Rice 85
Rats	Pre/postnatal	Neurotoxic effects	15	Altmann 93
Rats	Adult	Impaired offspring	15	Gandley 99
Humans, monkeys, rats	Postnatal	Impaired visual function	20	Fox 97
Monkeys	Prenatal	Neurotoxic effects	21	Newland 96
Rabbits	Adult	Impaired male function	24	WHO 95
Rats	Postnatal	Neurotoxic effects	26	McGlothan 08
Rats	Adult	Impaired male fertility	27	Gandley 99
Humans	Adult	Neurotoxic effects	30	Seppalainen 83
Rats, mice	Adult	Impaired male function	32	Hildebrand 73
Cattle	Adult	Clinical lead poisoning	35	Hammond 64, etc.
Monkeys	Adult	Impaired hormonal system	35	Foster 92
Lambs	Prenatal	Impaired visual function	35	Seppalainen 83
Monkeys	Postnatal	Neurotoxic effects	40	Laughlin 83
Humans	Adult	Impaired male function	40	WHO 95
Rabbits	Adult	Impaired offspring	40	Nelson 97
Mammals, humans	Adult	Neurotoxic effects	40	WHO 95
Humans	Subadult	Anemia	40	Hernberg 80
Humans	Adult	Anemia	50	Hernberg 80
Monkeys	Life-time	Impaired auditory function	50	Rice 97

TABLE 17.3 (continued)
Summary of Lead Concentrations in the Blood of Chronically Exposed Mammals and Humans in Association with Various Toxicity Effects

Species	Exposure Stage	Health Effect	Pb Blood (µg/dL)	Reference (Abbreviated)
Monkeys	Adult	Impaired auditory function	56	Lilienthal 90
Rats, humans	Adult	Impaired kidney function	60	WHO 77
Monkeys	Adult	Impaired hormonal system	70	Franks 89
Humans	Adult	Gastrointestinal colic	60	WHO 95
Humans	Subadult	Anemia	70	Goyer 70
Humans	Adult	Anemia	80	Goyer 70
Monkeys	Postnatal	Impaired visual function	85	Bushnell 77
Rats	Adult	Loss of body weight	98	Azar et al. 73
Pigs	Adult	Death	120	Link 66
Dogs	Adult	Anemia, weight loss	170	Forbes 78

Note: Values represent the mean or lowest reported blood lead concentration of a range that was associated with an observed effect.

TABLE 17.4
Summary of Lead in the Kidneys and Liver of Mammals

Species	Exposure Stage	Health Effect	Pb Kidney (µg/g dw)	Pb Liver (µg/g dw)	Reference (Abbreviated)
Bank voles	Life-time	Kidney damage	16	5	Ma 89
Raccoons	Adult	Kidney damage	20	—	Hamir 99
Rats	Adult	Kidney damage	30	—	Goyer 70
Wood mice	Life-time	Kidney damage	47	12	Roberts 78
Cattle	Adult	Various symptoms	87	32	Baars 90
Horses	Adult	Various symptoms	104	72	Osweiler 83
Rats	Adult	Body weight loss	120	—	Goyer 70
Dogs	Adult	Various symptoms	208	92	Forbes 78
Shrews	Adult	Weight loss, death	225	19	Pankakoski 94
Shrews	Life-time	Kidney damage	270	16	Ma 89
Shrews	Life-time	Kidney damage	280	—	Beyer 85
Cattle	Adult	Weight loss, death	325	104	Osweiler 83

Note: Values represent the mean or lowest reported lead concentration of a range that was associated with an observed effect. Wet weight concentrations were converted to dry weight values, when applicable.

The kidneys are a strong bioindicator of lead exposure as well as a significant target tissue. Kidney lead levels >15 µg/g dw were associated with structural and functional kidney damage (Table 17.4). Kidney lead levels >80 µg/g dw, indicative of either acute or chronic exposure, were associated with body weight loss and death. Lead in the liver seems to be a less suitable biomarker of lead exposure because of a greater variability of the outcomes.

Lead in hard tissues, such as bone and teeth, is a suitable biomarker of chronic or cumulative lead exposure. Table 17.5 summarizes the concentrations of lead in skeletal bone that have been associated with toxicity effects. Bone lead levels >25 µg/g dw were associated with kidney damage.

It may be concluded that the toxicity endpoint that is most sensitively affected by lead are related to neurotoxic effects due to exposure of the early developmental stages. It almost appears that the mere presence of lead in the brain is indicative of functional brain damage. Since lead in the brain has a long half-life in the order of years, lead-induced brain injuries are most likely to

TABLE 17.5
Summary of Bone Lead Concentrations in Mammals, Represented as the Mean or Lowest Reported Lead Concentration of a Range that Was Associated with an Observed Effect

Species	Exposure Stage	Health Effect	Pb in Bone (μg/g dw)	Reference (Abbreviated)
Bank voles	Life-time	Kidney damage	26	Ma 89
Humans	Adult	Neurobehavioral	57	Khalil 09
Mice (males)	Subadult	Various symptoms	91	Donald 87
Wood mice	Life-time	Kidney damage	352	Roberts 78
Mice (females)	Subadult	Various symptoms	530	Donald 87
Shrews	Life-time	Kidney damage	550	Ma 89
Dogs	Adult	Various symptoms	735	Forbes 78

Note: Ash dry weight concentrations were converted to dry weight values, when applicable.

be permanent. The CNS effects of chronic exposure to low levels of environmental lead are especially considered relevant for humans because of cognitive and behavioral concomitants. However, neurotoxicological effects of lead may also apply to wild mammals, although more in terms of ecological relevance. A great impediment to our potential of risk evaluation for wild mammals is the present inability to tie neurotoxic effects to the performance of individuals in the field. Almost nothing is known about a combination effect of neurotoxicity with other forms of stress in animals. More insight is desirable into the impact that the lead-induced deficits in sensory and motor functions are likely to have on the competitiveness and survival fitness of mammals.

REFERENCES

Altmann, L., F. Weinsberg, K. Sveinsson, H. Lilienthal, H. Wiegand, and G. Winneke. 1993. Impairment of long-term potentiation and learning following chronic lead exposure. *Toxicol. Lett.* 66:105–112.

Angle, C. R., M. S. McIntire, M. S. Swanson, and S. J. Stohs. 1982. Erythrocyte nucleotides in children-increased blood lead and cytidine triphosphate. *Pediatr. Res.* 16:331–334.

ATSDR (Agency for Toxic Substances and Disease Registry). 2007. *Toxicological profile for lead.* US Department of Health and Human Services. Public Health Services, Atlanta.

Azar, A., A. A. Trochimowics, and M. E. Maxfield. 1973. Review of lead studies in animals carried out at Haskell laboratory. *Proc. Int. Symp., Amsterdam. CEC, Luxemburg*, pp.199–210.

Baars, A. J., et al. 1990. Lead intoxication in cattle in the northern part of the Netherlands. *Tijdschr. Diergeneesk.* 115:882–890 (in Dutch).

Bankowska, J., and C. Hine. 1985. Retention of lead in the rat. *Arch. Environ. Contam. Toxicol.* 14: 621–629.

Barltop, D. 1969. Transfer of lead to the human fetus. In *Mineral metabolism in pediatrics*, eds. D. Barltop and W. L. Barland, pp. 135–151. Philadelphia: Davis.

Barrett, J., and P. J. Livesey. 1985. Low level lead effects on activity under varying stress conditions in the developing rat. *Pharmacol. Biochem. Behav.* 22:107–118.

Barry, P. S. I. 1975. A comparison of concentrations of lead in human tissues. *Brit. J. Ind. Med.* 32:119–139.

Beyer, W. N., O. H. Pattee, L. Sileo, D. J. Hoffman, and B. M. Mulhern. 1985. Metal contamination in wildlife living near two zinc smelters. *Environ. Pollut. Ser. A.* 38:63–86.

Buekers, J., E. Steen Redeker, and E. Smolders. 2009. Lead toxicity to wildlife: derivation of a critical blood concentration for wildlife monitoring based on literature data. *Sci. Total Environ.* 407:3431–3438.

Bushnell, P. J., J. R. Allan, and R. J. Marlar. 1977. Scotopic vision deficits in young monkeys exposed to lead. *Science* 196:333–335.

Casteel, S. W., C. P. Weis, G. M. Henningsen, and W. J. Brattin. 2006. Estimation of relative bioavailability of lead in soil and soil-like materials using young Swine. *Environ. Health Perspect.* 114:1162–1171.

CDC (Centers for Disease Control and Prevention). 2005. Preventing lead poisoning in young children. US Department of Health and Human Services, Atlanta.

Chettle, D. R. 2005. Three decades of *in vivo* x-ray fluorescence of lead in bone. *X-Ray Spectrometry* 34:446–450.

Chia, S. E., H. P. Chia, C. N. Ong, and J. Jeyaratnam. 1996. Cumulative blood lead levels and nerve conduction parameters. *Occup. Med.* 46:59–64.

Cory-Schlechta, D. A., B. Weiss, and C. Cox. 1985. Performance and exposure indices of rats exposed to low concentrations of lead. *Toxicol. Appl. Pharmacol.* 71:342–352.

Cory-Schlechta, D. A., and H. H. Schaumburg. 2000. Lead, inorganic. In *Experimental and clinical neurotoxicology*, eds. P. S. Spencer, H. H. Schaumburg, and A. C. Ludolph, pp. 708–720. New York: Oxford University Press.

Donald, J. M., M. G. Cutler, and M. R. Moore. 1987. Effects of lead in the laboratory mouse. Development and social behaviour after lifelong exposure to 12 μM lead in drinking fluid. *Neuropharmacology* 26:391–399.

Ettinger, A. S., et al. 2006. Influence of maternal bone lead burden and calcium intake on levels of lead in breast milk over the course of lactation. *Am. J. Epidemiol.* 163:48–56.

Forbes, R. M., and J. C. Reina. 1972. Effect of age on gastrointestinal absorption (Fe, Sr, Pb) in the rat. *J. Nutr.* 102:647–659.

Forbes, R. M., and G. C. Sanderson. 1978. Lead toxicity in domestic animals and wildlife. In *The biogeochemistry of lead in the environment. Part B. Biological effects*, ed. J. O. Nriagu, pp. 225–277. Amsterdam: Elsevier/North Holland Biomedical Press.

Foster, W. G. 1993. Reproductive toxicity of chronic lead exposure in the female cynomologus monkey. *Reprod. Toxicol.* 6:123–131.

Fox, D. A., M. L. Campbell, and Y. S. Blocker. 1997. Functional alterations and apoptotic cell death in the retina following developmental or adult lead exposure. *Neurotoxicology* 18:645–664.

Franks, P. A., N. K. Laughlin, D. J. Dierschke, R. E. Bowman, and P. A. Meller. 1989. Effect of lead on luteal function in rhesus monkeys. *Biol. Reprod.* 41:1055–1062.

Gandley, R., L. Anderson, and E. K. Silbergeld. 1999. Lead: male-mediated effects on reproduction and development in the rat. *Environ. Res.* 80:355–363.

Goyer, R. A., D. L. Leonard, J. F. Moore, B. Rhyne, and M. R. Krigman. 1970. Lead dosage and the role of the intranuclear inclusion body. *Arch. Environ. Health* 20:705–711.

Gulson, B. L., C. W. Jameson, K. R. Mahaffey, K. J. Mizon, M. J. Korsch, and G. Vimpani. 1997. Pregnancy increases mobilization of lead from maternal skeleton. *J. Lab. Clin. Med.* 130:51–62.

Gulson, B. L., et al. 1998. Relationships of lead in breast milk to lead in blood, urine, and diet of the infant and mother. *Environ. Health Perspect.* 106:667–674.

Hallén, I. P., L. Jorhem, and A. Oskarsson. 1995. Placental and lactational transfer of lead in rats: a study on the lactational process and effects on offspring. *Arch. Toxicol.* 69:596–602.

Hamir, A. N., B. Lehmann, N. Raju, J. G. Ebel, K. L. Manzell, and C. E. Ruprecht. 1999. Experimental lead toxicosis of raccoons (*Procyon lotor*). *J. Comp. Pathol.* 120:147–154.

Hammond, P. B., R. L. Bornschein, and P. Succop. 1985. Dose-effect and dose-response relationships of blood lead to erythrocytic protoporphyrin in young children. *Environ. Res.* 38:187–196.

He, K., S. Wang, and J. Zhang. 2009. Blood lead levels of children and its trend in China. *Sci. Total Environ.* 407:3986–3993.

Hernberg, S. 1980. Biochemical and clinical effects and responses as indicated by blood lead concentration. In *Lead toxicity*, eds. R. L. Singhal and J. Thomas, pp. 367–399. Baltimore, MD: Schwarzenberg.

Hildebrand, D. C., H. Der, W. Griffin, and F. S. Fahim. 1973. Effect of lead acetate on reproduction. *Am. J. Obstret. Gynecol.* 115:1058–1065.

Hryhorczuk, D. O., et al. 1985. Elimination kinetics of blood lead in workers with chronic lead intoxication. *Am. J. Ind. Med.* 8:33–42.

Hryhorczuk, D. O., et al. 1998. Bone lead as a biological marker in epidemiologic studies of chronic toxicity: conceptual paradigms. *Environ. Health Perspect.* 106:1–8.

Hu, H., R. Shih, S. Rothenberg, and B. S. Schwartz. 2007. The epidemiology of lead toxicity in adults: measuring dose and consideration of other methodologic issues. *Environ. Health Perspect.* 115:455–462.

Hwang, Y.-H., H-Y. Chiang, M-C. Yen-Jean, and J-D. Wang. 2009. The association between low levels of lead in blood and occupational noise-induced hearing loss in steel workers. *Sci. Total Environ.* 408:43–49.

Jefferies, D. J., and M. C. French. 1972. Lead concentrations in small mammals trapped on roadside verges and field sites. *Environ. Pollut.* 3:147–156.

Johansson, L., and M. Wide. 1986. Long-term exposure of the male mouse to lead: effects on fertility. *Environ. Res.* 41:481–487.

Jusko, T. A., C. R. Henderson Jr., B. P. Lanphear, D. A. Cory-Schlechta, P. J. Parsons, and R. L. Canfield. 2008. Blood lead concentrations <10 μg/dl and child intelligence at 6 years of age. *Environ. Health Perspect.* 116:243–248.

Khalil, N., L. A. Morrow, and H. Needleman. 2009. Association of cumulative lead and neurocognitive function in an occupational cohort. *Neuropsychol.* 23:10–19.

Kisseberth, W. C., J. P. Sundberg, R. W. Nyboer, J. D. Reynolds, S. C. Kasten, and V. R. Beasley. 1984. Industrial lead contamination of an Illinois wildlife refuge and indigenous small mammals. *J. Am. Vet. Med. Assoc.* 185:1309–1313.

Laughlin, N. K., E. E. Bowman, E. D. Levin, and P. J. Bushnell. 1983. Neurobehavioral consequences of early exposure to lead in rhesus monkeys: effects on cognitive behaviors. In *Reproductive and developmental toxicity of metals*, eds. T. W. Clarkson, G. F. Nordberg, and P. R. Sager, pp. 497–515. New York: Plenum Press.

Lidsky, T. I., and J. S. Schneider. 2003. Lead neurotoxicity in children: basic mechanisms and clinical correlates. *Brain* 126:5–19.

Lilienthal, H., G. Winneke, and T. Ewert. 1990. Effects of lead on neurophysiological and performance measures: animal and human data. *Environ. Health Perspect.* 89:21–25.

Link, R. P., and R. R. Pensinger. 1966. Lead toxicosis in swine. *Am. J. Vet. Res.* 27:759–763.

Ma, W. C. 1987. Heavy metal accumulation in the mole, *Talpa europea*, and earthworms as an indicator of metal bioavailability in terrestrial environments. *Bull. Environ. Contam. Toxicol.* 39:933–938.

Ma, W. C. 1989. Effect of soil pollution with metallic lead pellets on lead bioaccumulation and organ/body weight alterations in small mammals. *Arch. Environ. Contam. Toxicol.* 18:617–622.

Ma, W. C. 1994. Methodological principles of using small mammals for ecological hazard assessment of chemical soil pollution, with examples on cadmium and lead. In *Ecotoxicology of soil pollution*, eds. M. H. Donker, H. Eijsackers, and F. Heimbach, pp. 357–371. Boca Raton, FL: Lewis Publishers.

Ma, W. C., and S. Broekhuizen. 1989. Heavy metals in badgers (*Meles meles*): possible effect of the contaminated flood plains of the river Meuse. *Lutra* 32:151 (in Dutch, with English summary).

Ma, W. C., and S. Talmage. 2001. Chapter 4: Insectivora. In *Ecotoxicology of wild mammals*, eds. R. F. Shore, and B. A. Rattner, pp. 123–158. Chichester, UK: John Wiley & Sons Ltd.

Ma, W. C., and H. van der Voet. 1993. A risk-assessment model for toxic exposure of small mammalian carnivores to cadmium in contaminated natural environments. *Sci. Total Environ. Suppl.* 1993:1701–1714.

Ma, W. C., W. Denneman, and J. Faber. 1991. Hazardous exposure of ground-living small mammals to cadmium and lead in contaminated terrestrial ecosystems. *Arch. Environ. Contam. Toxicol.* 20:266–270.

Ma, W. C., H. van Wezel, and D. van den Ham. 1992. Background concentrations of fifteen metal elements in soil, vegetation, and soil fauna of twelve nature areas in the Netherlands (in Dutch). Rep. No. 92/11. Institute for Forestry and Nature Research, The Netherlands.

Mahaffey, K. R., and B.A. Fowler. 1977. Effects of concurrent administration of dietary lead, cadmium and arsenic in the rat. *Environ. Health Perspect.* 19:165.

Manton, W. I., C. R. Angle, K. L. Stanek, D. Kuntzelman, Y. R. Reese, and T. J. Kuehnemann. 2003. Release of lead from bone in pregnancy and lactation. *Environ. Res.* 92:139–151.

McGlothan, J. L., M. Karcz-Kubicha, and T. R. Guilare. 2008. Developmental lead exposure impairs extinction of conditioned fear in young adult rats. *Neurotoxicology* 29:1127–1130.

McHugh, N. A., M. Haydee, R. W. E. Vercesi, and J. A. Hey. 2003. Receptor activator of NF-κB ligand arrests bone growth and promotes cortical bone resorption in growing rats. *J. Appl. Physiol.* 95:672–676.

McIlwain, H., and H. S. Bachelard. 1985. *Biochemistry and the central nervous system.* Edinburgh: Churchill Livingstone.

Mierau, G. W., and B. E. Favara. 1975. Lead poisoning in roadside populations of deer mice. *Environ. Pollut.* 8:55–64.

Miranda, M., M. López-Alonso, P. García-Partida, J. Velasco, and J. L. Benedito. 2007. Long-term follow-up of blood lead levels and haematological and biochemical parameters in heifers that survived an accidental lead poisoning episode. *J. Vet. Med. Series A* 53:305–310.

Namihira, D., L. Saldivar, N. Pustilnik, G. J. Carreon, and M. E. Salinas. 1993. Lead in human blood and milk from nursing women living near a smelter in Mexico City. *J. Toxicol. Environ. Health* 38:225–232.

Nelson, B. K., W. J. Moorman, S. M. Schrader, P. B. Shaw, and E. F. Krieg Jr. 1997. Paternal exposure of rabbits to lead: behavioral deficits in offspring. *Neurotoxicol. Teratol.* 19:191–198.

Neuman, D. R., and D. J. Dollhopf. 1992. Lead levels in blood from cattle near a lead smelter. *J. Environ. Qual.* 21:181–184.

Newland, M. C., Y. Sheng, B. Lögdberg, and M. Berlin. 1996. *In utero* lead exposure in squirrel monkeys: motor effects seen with schedule-controlled behavior. *Neurotoxicol. Teratol.* 18:33–40.

Ogilvie, D. M., and A. H. Martin. 1982. Aggression and open-field activity of lead-exposed mice. *Arch. Environ. Contam. Toxicol.* 11:249–252.

Osweiler, G. D., and G. A. van Gelder. 1983. Epidemiology of lead poisoning in animals. In *Toxicity of heavy metals in the environment. Part I*, ed. F. W. Oehme, pp. 143–177. New York, NY: Marcel Dekker.

Pankakoski, E., I. Koivisto, H. Hyvärinen, J. Terhivuo, and K. M. Tähkä. 1994. Experimental accumulation of lead from soil through earthworms to common shrews. *Chemosphere* 29:1639–1649.

Piomelli, S. L., M. B. Corash, C. Seaman, P. Mushak, B. Glover, and R. Padgett. 1980. Blood lead concentrations in a remote Himalayan population. *Science* 210:1135–1137.

Rice, D. C. 1985. Chronic low-lead exposure from birth produces deficits in discrimination reversal in monkeys. *Toxicol. Appl. Pharmacol.* 77:201–210.

Rice, D. C. 1997. Effects of lifetime lead exposure in monkeys on detection of pure tones. *Fund. Appl. Toxicol.* 36:112–118.

Roberts, R. D., M. S. Johnson, and M. Hutton. 1978. Lead contamination of small mammals from abandoned metalliferous mines. *Environ. Pollut.* 15:61–69.

Samuels, E. R., J. C. Meranger, B. L. Tracy, and K. S. Subramanian. 1989. Lead concentrations in human bones from the Canadian population. *Sci. Total Environ.* 89:261–269.

Schutz, A., and S. Skerfving. 1976. Effect of a short, heavy exposure to lead dust upon blood lead level, erythrocyte delta-aminolevulinic acid, and coproporphyrin. *Scand. J. Work Environ. Health* 3:176–184.

Schwartz, J., P. J. Landrigan, E. L. Baker Jr, W. A. Orenstein, and I. H. von Lindern. 1990. Lead-induced anemia: dose-response relationship and evidence for a threshold. *Am. J. Public Health* 80:165–168.

Seppäläinen, A. M., S. Hernberg, R. Vesanto, and B. Kock. 1983. Early neurotoxic effects of occupational lead exposure: a prospective study. *Neurotoxicology* 4:181–192.

Smith Jr, D. M., H. W. Mielke, and J. B. Heneghan. 2008. Subchronic lead feeding study in male rats. *Arch. Environ. Contam. Toxicol.* 55:518–528.

Stollery, B. T. 1996. Reaction times changes in workers exposed to lead. *Neurotoxicol. Teratol.* 18:477–483.

Torra, M., M. Rodamilans, J. To-Figueras, I. Hornos, and J. Corbella. 1989. Delta aminolevulinic dehydratase and ferrochelatase activities during chronic lead exposure. *Bull. Environ. Contam. Toxicol.* 42:476–481.

Virgolini, M. B., A. Rossi-George, R. Lisek, D. D. Weston, M. Thiruchelvan, and D. A. Cory-Schlechta. 2008. CNS effects of developmental Pb exposure are enhanced by combined maternal and offspring stress. *Neurotoxicol.* 29:812–827.

WHO. 1995. Environmental health criteria. 165. Inorganic lead. World Health Organization, Geneva.

African Adjutant

By F. Specht, from *The Royal Natural History,* edited by Richard Lydekker, Frederick Warne & Co., London, 1893–94.

18 Mercury in Nonmarine Birds and Mammals

Richard F. Shore
M. Glória Pereira
Lee A. Walker
David R. Thompson

CONTENTS

18.1 Introduction 609
18.2 Aims and Approach.. 610
18.3 Mercury–Selenium Interactions 611
18.4 Indicative Dietary Concentrations of Mercury.. 612
    18.4.1 Dietary Methyl Mercury Toxicity–Lethality... 612
    18.4.2 Dietary Methyl Mercury Toxicity–Reproductive Effects............................ 613
18.5 Indicative Tissue Mercury Concentrations.. 614
    18.5.1 Tissue Concentrations in Birds Associated with Lethality 614
    18.5.2 Tissue Concentrations in Mammals Associated with Lethality................... 616
    18.5.3 Tissue Concentrations in Birds Associated with Impaired Reproduction.............. 616
    18.5.4 Tissue Concentrations in Mammals Associated with Impaired Reproduction 617
    18.5.5 Indicative Egg Mercury Concentrations.. 617
Summary 618
References.. 621

18.1 INTRODUCTION

Mercury (Hg) is a highly toxic nonessential heavy metal emitted into the environment from natural sources such as volcanic emissions, continental particulate and volatile matter, and fluxes from the marine environment (Nriagu 1989). Hg also has a long association with Man that dates back to the ancient Greeks and includes incidents of human poisoning, such as the relatively recent and well documented mass poisoning incidents around Minamata Bay in Japan (Kurland et al. 1960) and in Iraq (Bakir et al. 1973). Currently, there is a wide range of anthropogenic emissions to the environment. These include metal mining/smelting, chlor-alkali production, biomedical waste, consumer products, fossil fuel combustion, waste incineration, and sewage sludge (Nriagu and Pacyna 1988), which, in sum, are larger than natural emissions (Nriagu 1989). Temporal trends in anthropogenic emissions vary between countries but global Hg emissions are predicted to increase, largely driven by the expansion of coal-fired electricity generation in the developing world, particularly Asia (Streets et al. 2009).

Hg occurs in the environment both in inorganic and organic forms. In the atmosphere, elemental Hg (Hg^0) is most abundant at about 98% (Jackson 1997). It is volatile and undergoes long-range transport around the globe, including to remote locations. Before being deposited, Hg^0 is oxidized to inorganic Hg [mainly Hg^{2+} and to a less extent Hg^+ (Jackson 1997)] which is the most common form of Hg in soils and surface waters. Hg can also be converted by sulfate-reducing bacteria in sediments and water to methyl mercury (MeHg) (Westcott and Kalff 1996). Methylation can be enhanced in aquatic environments by high concentrations of organic carbon, low pH and/or low alkalinity (Spry and Wiener 1991).

MeHg is thought to typically be the most stable, bioavailable and toxic form of Hg to wildlife. It is absorbed efficiently from the diet (the main route of exposure) and biomagnified through the food web, although the exact mechanisms are poorly understood (Pelletier 1986). Apex predators can therefore be exposed to relatively high dietary concentrations and it has been shown that up to 100% of the total Hg in the liver, kidney and muscle of birds can be in the methylated form (Thompson and Furness 1989, Thompson et al. 1991). Hg is also incorporated into hair and feathers, and moult is a significant elimination pathway; for example 70–93% of the body burden has been found to be the plumage of some species (Burger et al. 1993).

MeHg is a neurotoxin, an endocrine-disrupter and a teratogen (Clarkson and Magos 2006). Hg in vertebrates may cause mortality (Scheuhammer 1987, Heinz 1996) and severe pathology in tissues, especially nerve tissues (Eisler 1987, Longcore et al. 2007). The most common acute neurological effects are the loss of motor skills, coordination, and reduction in motivation (Wolfe et al. 1998). At sublethal concentrations Hg can cause brain lesions, demyelination of neurons, and changes in hematology, hormone levels, neurology, and histology (Heinz and Locke 1976, Spalding et al. 2000). Hg intoxication can affect reproduction indirectly by altering parental behavior such as nest attendance (Barr 1986), and by direct toxicity to the embryo, which seems to be highly sensitive (Scheuhammer 1987, Wiener et al. 2003). Maternal transfer of Hg can disrupt early development, resulting in reduced embryonic survival, increased mortality in developing young, and neurological impairment and behavioral abnormalities in surviving offspring (Heinz, 1976b, 1979, 1996).

18.2 AIMS AND APPROACH

The previous edition of this chapter (Thompson 1996) described dietary and tissue (liver, kidney, and egg) Hg concentrations that were associated with adverse effects on reproduction and/or survival. The values were determined on the basis of review of published data and associated expert judgment. Our aim in this revised edition is to update, refine and expand on the previous assessment by incorporating new data and, where possible, defining on a statistical basis those dietary and tissue Hg concentrations that are likely to be indicative of adverse effects on the generic endpoints of survival and reproduction. We did not relate diet/tissue concentrations to milder endpoints (such as blood chemistry, subtle behavioral effects) because their possible impact on individual reproductive success and survival, two key parameters affecting population numbers, are unclear.

Two key difficulties in describing a single diet/tissue concentration that is associated with adverse effects are: (i) how to summarize data from multiple studies on the same species and (ii) to account for interspecies variation in susceptibility to Hg. To overcome the first problem, we took the most sensitive value when there were multiple effects measures for the same generic endpoint; for example, when there were measures for hatching success and chick survival, both of which we classed under the generic endpoint of impaired reproduction. We also found multiple studies that measured the same specific endpoint (such as hatching success) in the same species. In such cases, differences between datasets may reflect variation between studies in a number of factors (e.g., strain of test species, form of MeHg administered, duration of dosing). When we encountered multiple datasets for the same endpoint and species, we took the median value. To address the second issue of interspecies variation in response to Hg, we used a statistical approach where possible. This was done by describing species sensitivity distributions for lethality and reproduction endpoints by fitting cumulative Gaussian percentage distributions (Prism, v5, Graphpad, www.graphpad.com) to concentration data associated with adverse effects; where such data were derived from experimental studies, these data were effectively Lowest Observed Adverse Effect Concentrations (LOAECs). For each distribution, we calculated the fifth percentile of the distribution. Assuming that the distribution of the available data corresponds to that for all species, the fifth percentile can be taken as an approximation of the concentration below which only 5% of species are likely to suffer adverse effects; in other words, this concentration should be protective for 95% of species. This is analogous to the HC5 (fifth percentile of species NOEC values) approach (Aldenberg and Slob 1993) that has been used previously for assessing secondary poisoning risk in mammals and birds from metals (Spurgeon and Hopkin 1996), and more recently to assess the significance of tissue lead residues in wildlife (Buekers et al. 2009). However, this approach can be overly conservative when the number of species is small, although there is no clear agreement on the minimum number (Erik Smolders personal communication). We only calculated the fifth percentile of the species sensitivity distribution when there were data for eight or more species.

In this chapter, we initially describe the available data for different dietary, tissue, and egg concentrations. Concentrations are described on a wet weight (wet wt.) basis for total Hg, unless stated otherwise. We summarize the adverse effect concentrations obtained for different species as geometric means and ranges. The geometric rather than arithmetic mean is used because the data usually follow a log-normal distribution. However, the geometric mean is higher than the LOAEC values for some test species and most likely for some nontest species, and so is not a protective measure. Therefore, where possible, we also present the calculated fifth percentile of the species distribution of LOAECs as an indicative value that may approximate to the concentration protective of most (95%) species. For the purposes of this chapter we refer to this value as the HC5, although we recognize that it is based on LOAECs, not on NOAECs.

18.3 MERCURY–SELENIUM INTERACTIONS

Hg and selenium (Se) can occur together at high concentrations, especially in marine mammals (Koeman et al. 1973), and in some birds (Norheim 1987). Hg and Se interactions are normally antagonistic (see Yang et al. 2008, Cuvin-Aralar and Furness 1991 for detailed descriptions) and a general detoxification mechanism for MeHg in seabirds and marine mammals has been proposed (Ikemoto et al. 2004, Yang et al. 2008). Birds and mammals with the highest liver–Hg concentrations often have the lowest MeHg fraction, despite being primarily exposed to MeHg (Ikemoto et al. 2004). This is thought to be because a proportion of the MeHg is demethylated to inorganic Hg by reactive oxygen species (Suda and Takahashi 1992), gut microflora (Rowland 1988) and selenium (Cuvin-Aralar and Furness 1991, Yang et al. 2008). Demethylation occurs primarily in the liver, but also in the kidneys and spleen (Kim et al. 1996, Henny et al. 2002), and a high molecular weight Hg–Se protein compound, that contains Hg and Se at an equimolar ratio, is formed (Cuvin-Aralar and Furness 1991, Palmisano et al. 1995). This Hg–Se complex subsequently undergoes lysosomal degradation to form an insoluble Hg–Se compound that accumulates in the liver (Arai et al. 2004, Ikemoto et al. 2004). Wildlife species vary in their capacity to demethylate Hg and this is believed to affect their sensitivity to MeHg (Scheuhammer et al. 2007).

While the formation of nontoxic Hg–Se complexes appears to be the most accepted mechanism to explain the antagonistic relationship between these two elements, additive and synergistic effects have been reported (Cuvin-Aralar and Furness 1991). Heinz and Hoffman (1998) showed that selenomethionine had opposite toxicological interactions with MeHg in mallards (*Anas platyrhynchos*) depending on whether ducklings or adults were exposed. Furthermore, the chemical form of both Hg and Se may also determine the type of reaction between these two elements (Cuvin-Aralar and Furness 1991); for instance it has been suggested that Se could be methylated within organisms to form dimethyl selenide, which then acts synergistically with Hg (Wilber 1980). The toxic interactions and the additive and synergistic effects of Hg and Se appear to be complex and need further study. Physiological variability in the handling of Hg and Se, together with variation in Hg:Se dietary ratios, is likely to contribute to intra and interspecies variability in Hg toxicity. This can make interpreting the significance of Hg residues in biota problematic.

18.4 INDICATIVE DIETARY CONCENTRATIONS OF MERCURY

18.4.1 Dietary Methyl Mercury Toxicity–Lethality

There have been various studies in which organic Hg has been fed to avian laboratory test species and lethality recorded. Dietary concentrations of Hg of between 12.5 and 21 mg/kg, administered in organic form, caused ≥90% mortality in adult pheasants (*Phasianus colchicus*) (Spann et al. 1972), while dietary MeHg chloride concentrations, equivalent to 16 mg/kg Hg, caused between 55% and 80% mortality in Japanese quail (*Coturnix coturnix japonica*) and bobwhite quail (*Colinus virginianus*) (Ganther et al. 1972, Spann et al. 1986). In mallards, one of 12 adult males fed a dietary concentration of 10 mg/kg Hg (as MeHg) died and, by the end of the study (77 days), eight others suffered paralysis of the legs, likely to prove lethal under nonlaboratory conditions (Heinz and Hoffman 1998). In studies on nonstandard test species, dietary concentrations of 40 mg/kg Hg, administered as MeHg dicyandiamide in Morsodren, were fed to grackles (*Quiscalus quiscula*), starlings (*Sternus vulgaris*), cowbirds (*Molothrus ater*), and red-winged blackbirds (*Agelaius phoeniceus*) and killed all dosed individuals (*n* = 5 per species) within 11 days (Finley et al. 1979). There have also been a small number of experimental studies on predatory birds. Kestrels (*Falco tinnunculus*), red tailed hawks (*Buteo jamaicensis*), and goshawks (*Accipiter gentiles*) all died after being fed mice or chicks that had been dosed with MeHg dicyandiamide. The total Hg concentrations in the diet ingested ranged between 10 and 14 mg/kg (Borg et al. 1970, Fimreite and Karstad 1971, Koeman et al. 1971).

The data for the dietary concentrations of Hg associated with lethality in mammals are more limited than that for birds. There has been a number of Hg feeding studies with mink (*Mustela vision*). Dietary concentrations of 5 mg/kg of Me–Hg (Aulerich et al. 1974), and of 1.1 and 0.7 mg/kg Hg, incorporated into the diet as MeHg chloride (Wobeser et al. 1976b, Wren et al. 1987a) all caused mortality or overt signs of poisoning, whereas a mean Hg concentration of 0.44 mg/kg in a fish diet (presumably largely MeHg) did not have detectable detrimental effects (Wobeser et al. 1976a). The median value of 1.1 mg/kg Hg has been taken as the value for mink for this assessment, even though Jernelov et al. (1976) found no detectable adverse effects of feeding mink fish containing 5.8 mg/kg Hg, almost all of which was in methylated form; Se may have been an ameliorating factor in that study as Se concentrations increased with those of Hg in the liver and kidneys of dosed animals.

Dietary Hg concentrations associated with mortality or signs of poisoning in other mammals include 1.9 mg/kg Hg (incorporated as Me–Hg in the diet) in river otters (*Lutra canadensis*) (O'Connor and Nielsen 1980, Eisler 2004), 5 mg/kg Hg in ferret-polecats (*Mustela putorius*) fed chickens contaminated through MeHg-dressed wheat feed (Hanko et al. 1970), and 6 mg/kg Hg in cats (*Felis catus*) fed fish homogenate containing Hg as MeHg (Albanus et al. 1972). Studies on laboratory rodents which report mortality include studies to assess carcinogenicity and dietary Hg concentrations of 14 and 7 mg/kg, administered as MeHg chloride, caused lethality in laboratory mice and rats, respectively (U.S. EPA 1995).

Analysis of the data from the earlier studies indicates that, when Hg is in an organic (typically methyl) form in the diet, the geometric mean dietary Hg concentration associated with lethality is 20 mg/kg (range 10–40) and 4.3 mg/kg (range 1.1–14) in birds and mammals, respectively. These values are not LOAECs but rather dietary concentrations associated with substantial mortality, as typically between 50% and 100% of test animals died at the doses given. The calculated dietary HC5 for mortality in birds, based on data for 11 species, is 6 mg/kg (Figure 18.1). It was not possible to calculate an HC5 for mammals as there were too few data. However, a generic approach to dietary risk assessment for pesticides in birds and mammals recently advocated by the European

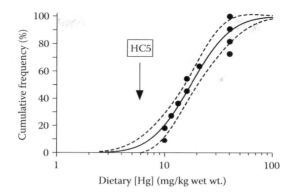

FIGURE 18.1 Cumulative frequency of Hg concentrations (mg/kg wet wt.) in diet associated with lethality in birds. See text for details of the data. Mercury in the diet was typically as MeHg. (From Oehme, G., *Hercynia*, 18, 353–364, 1981. With permission.)

Commission (European Food Safety Authority 2008) is to use the geometric mean dietary No Observed Effect Concentration (NOEC) for when there are toxicity data for multiple species and an assessment factor of 5 for chronic studies. This means that dietary exposure that is one fifth or more of the geometric mean NOAEC at the first tier of assessment would trigger a requirement for more detailed assessment to refine the risk further. Applying this approach to the calculated geometric mean dietary Hg for lethality in mammals gives a value of 0.9 mg/kg, which is similar to the bottom of the range of values available for mammals. This approach, is of course, less conservative than for pesticide risk assessment as it is based in this case on dietary concentrations associated with adverse effect, rather than NOAEC data.

18.4.2 Dietary Methyl Mercury Toxicity–Reproductive Effects

Laboratory and field studies have shown that adverse effects on reproduction in birds can be caused by dietary Hg concentrations that are one to two orders of magnitude below those associated with mortality. Laboratory studies in which pheasants and black ducks (*Anas rubripes*) were fed MeHg dicyandiamide contaminated diet found that the lowest total Hg dietary concentrations that caused adverse effects were 2.25 mg/kg [eggs without shells, reduced hatchability, unfertilized eggs (Fimreite 1971)] and 3 mg/kg [reduced duckling survival; (Finley and Stendell 1978)], respectively. The lowest dietary concentration of Hg (incorporated as MeHg chloride) observed to adversely affect reproduction (egg weight, fertility, and hatchability) was a little higher in Japanese quail and chickens (*Gallus domesticus*) at 10 mg/kg (Scott et al. 1975, Scott 1977). However, chick survival one week posthatching was reduced when adult mallards were exposed long-term to diet contaminated with MeHg dicyandiamide that was equivalent to only 0.1 mg/kg Hg in natural diet (Heinz 1976a). In nonlaboratory species, it has been estimated that productivity among loons (*Gavia immer*) was reduced to half the maximum when Hg concentrations in fish prey were 0.21 mg/kg (Burgess and Meyer 2008).

On the basis of these data, the geometric mean dietary Hg concentration associated with adverse effects on reproduction in birds can be calculated to be 1.2 mg/kg diet (range 0.1–10); this is when dietary Hg is predominantly in the methyl form. Thus, impairment of reproduction appears to occur at dietary concentrations at least an order of magnitude lower than those associated with mortality. There were insufficient data to calculate an HC5 for avian reproduction (data for seven species only) but, using the EU approach of using a fifth of the geometric mean value derived from multiple species (European Food Safety Authority 2008), then a dietary Hg concentration exceeding 0.25 mg/kg (when Hg was in a methylated form) would trigger a requirement to refine the risk assessment further.

Data on adverse effects of MeHg on reproduction in mammals are sparse. Adverse effects on reproduction (including reduction in litter size and fetal malformations) have been reported in laboratory rodents and guinea pigs but have often been the result of short-term, high dose studies (U.S. EPA 1995). Tubular atrophy of the testes was observed in male laboratory mice fed approximately 7 mg/kg Hg (as MeHg chloride) for 104 weeks. Studies on mink reported reduced kit survival in mink associated with dietary Hg concentrations (administered as MeHg chloride) of 0.7–1 mg/kg (Wren et al. 1987b, Dansereau et al. 1999). From these scant data, it would appear that adverse effects on reproduction appear to be associated with dietary Hg concentrations that are within the range of those associated with lethality.

18.5   INDICATIVE TISSUE MERCURY CONCENTRATIONS

18.5.1   TISSUE CONCENTRATIONS IN BIRDS ASSOCIATED WITH LETHALITY

Tissue Hg concentrations (measured as total Hg) associated with death have been reported in individual birds that have been found dead and were diagnosed as being poisoned by Hg, or in dying animals exhibiting symptoms consistent with chronic/acute Hg poisoning. Published data are more numerous for birds than mammals and the largest datasets are for the liver and kidney. Often, Hg has been measured in one or other organ but both liver and kidney Hg residues have been measured in the same animals in some studies. Comparison of these data suggests that there is a linear relationship between liver and kidney Hg concentrations across species for the Hg concentration range associated with adverse effects (Figure 18.2). We used this relationship to predict organ Hg concentrations for species where residue data were missing for liver but available for kidney, and vice versa. By using such an approach, it was possible to increase the number of species for which data were available so that an HC5 approach could be calculated for liver and kidney concentrations in birds that were associated with lethality.

We found direct measures of liver Hg concentrations associated with death for six avian species, and used measured kidney concentrations to estimate the liver concentrations in two more species. In free-living birds, liver Hg concentrations have been directly measured in sea eagles (*Haliaaetus albicilla*) in various studies (Henriksson et al. 1966, Koeman et al. 1972, Oehme 1981, Falandysz 1984, 1986, Falandysz et al. 1988) and four median or midpoint values (where ranges

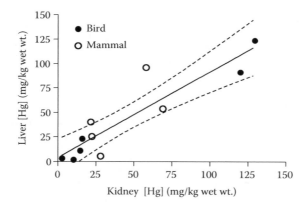

FIGURE 18.2   Relationship between liver and kidney total Hg concentrations (mg/kg wet wt.) in birds and mammals. The regression line is described by the equation: liver [Hg] = 0.872 kidney [Hg] + 4.1, R^2 = 0.82, $F_{(1,9)}$ = 41.9, $p < .0001$. Dashed lines are the 95% confidence limits for the equation. Data are for: black duck (Finley and Stendell 1978), cat (Eaton et al. 1980), ferret-polecat (Hanko et al. 1970), goshawk (Borg et al. 1970), mallard (Heinz 1976b), mink (Wobeser et al. 1976b), river otter (Wren 1985), rat (Endo et al. 2005), starling (Nicholson and Osborn 1984), and white tailed sea eagle (Oehme 1981).

were reported) can be derived; these range between 16 and 91 mg/kg. The median of these four values is 36.3 mg/kg and is the value we used for sea eagles. Koeman et al. (1969) measured kidney Hg concentrations of 85 and 93 (median 89) mg/kg in two buzzards (*Buteo buteo*), and 68 mg/kg in a long-eared owl (*Asio otus*) which were either dead or exhibiting symptoms of Hg poisoning; these kidney concentrations can be estimated (Figure 18.2) to be equivalent to liver concentrations of 82 (buzzard) and 63 mg/kg (owl). There have also been a number of controlled feeding studies in which liver residues were measured in birds that died from Hg intoxication. Hepatic Hg concentrations of 16.7–20 (midpoint 18.4), 54.5, 49–122 (midpoint 85.5), 103–144 (midpoint 123.5) and 126.5 mg/kg were measured in experimentally poisoned red tailed hawks, grackles, kestrels, goshawks, and red-winged blackbirds, respectively (Borg et al. 1970, Fimreite and Karstad 1971, Koeman et al. 1971, Finley et al. 1979). Based on these data, the geometric mean liver Hg concentration associated with death is 63 mg/kg (range 18.4–127) and the estimated HC5 value is 22 mg/kg (Figure 18.3), close to the minimum for the range of species measured.

A number of studies have directly measured kidney residues in birds diagnosed as having been fatally poisoned by Hg. These include the kidney residues described earlier for buzzard and long-eared owl and measured kidney concentrations in experimentally poisoned goshawks and kestrels of 121–138 (midpoint 130) and 73–125 (midpoint 99) mg/kg, respectively (Koeman et al. 1969, Borg et al. 1970). Kidney concentrations in Hg-intoxicated white tailed sea eagles have been reported to be 48.6–123 (midpoint 85.8) and 120 mg/kg (Henriksson et al. 1966, Oehme 1981), giving a median value of 103 mg/kg. Kidney concentrations of 21, 55, and 120 mg/kg can also be estimated from the liver concentrations reported earlier for experimentally poisoned red-tailed hawks, grackles, and red-winged blackbirds. These data indicate a geometric mean kidney Hg concentration associated with lethality of 76 mg/kg and an HC5 value of 40 mg/kg (Figure 18.3).

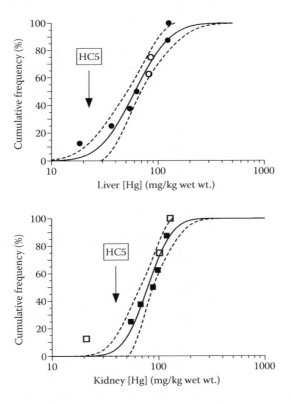

FIGURE 18.3 Cumulative frequency of total Hg concentrations (mg/kg wet wt.) in liver and kidney associated with lethality in birds. Open symbols indicate predicted values. See text for details of data sources.

Data on tissue Hg concentrations in other tissues associated with death appear scanty. Mercury (as MeHg) brain concentrations of >15 mg/kg in adult birds and >3 mg/kg in developing bird embryos are associated with lethality (Scheuhammer et al. 2007).

18.5.2 TISSUE CONCENTRATIONS IN MAMMALS ASSOCIATED WITH LETHALITY

Studies on tissue residues associated with lethality in mammals have focused on carnivores. There have been various studies on mink and Hg concentrations associated with lethality were approximately 20–60 (median: 44) mg/kg for the liver and 20–40 (median: 28) mg/kg for the kidney (Aulerich et al. 1974, Wobeser and Swift 1976, Wobeser et al. 1976b, Wren et al. 1987a). Liver and kidney residues of 33 and 39 mg/kg, respectively in river otters (Hanko et al. 1970, O'Connor and Nielsen 1980) and of 54 and 69 mg/kg, respectively in polecat-ferrets (Hanko et al. 1970, O'Connor and Nielsen 1980) were associated with death in feeding trials. In cats (*Felis catus*), two separate feeding studies both reported mean terminal liver Hg concentrations of approximately 40 mg/kg (Albanus et al. 1972, Eaton et al. 1980); mean kidney concentrations ranged between 21 and 31 mg/kg. Roelke et al. (1991) recorded a liver Hg concentration of 110 mg/kg in the liver of a Florida panther (*Felis concolor coryi*) found dead and presumed to have died from Hg poisoning, but it is notable that this concentration is at least twice that associated with lethality in the other carnivorous species. Based on the experimental studies on mink, otter, polecats-ferret, and cats, the geometric mean Hg concentration associated with lethality is 42 mg/kg (range 33–54) for the liver and 37.5 (range 26–69) mg/kg for the kidney. Based on the ranges of these data, it would appear that liver and kidney residues exceeding approximately 25–30 mg/kg in both organs may be associated with lethality in carnivorous mammals, and perhaps other mammal groups.

Brain Hg concentrations associated with lethality have also been measured in a small number of mammal species that have been experimentally poisoned. Brain Hg concentrations of approximately 12, 15, 19, and 20 mg/kg (median: 17 mg/kg) have been associated with death in mink (Aulerich et al. 1974, Wobeser and Swift 1976, O'Connor and Nielsen 1980, Wren et al. 1987a), and concentrations of 21 and 30 mg/kg have likewise been associated with lethality in river otters (O'Connor and Nielsen 1980, Wren 1985). Eaton et al. (1980) observed a brain Hg concentration of 11.3 mg/kg in cats that had been experimentally poisoned, while blindness, and seizures, likely to lead to death in the wild, is associated with brain Hg concentrations of 12–20 (midpoint 16) mg/kg in small mammals (Burbacher et al. 1990). Using median values for multiple measures for the same species, it can be calculated that a geometric mean brain concentration associated with lethality in the mammal species tested is 16 mg/kg (range 11–21), and concentrations above 10 mg/kg may be associated with lethality.

18.5.3 TISSUE CONCENTRATIONS IN BIRDS ASSOCIATED WITH IMPAIRED REPRODUCTION

Data on liver concentrations associated with impaired reproduction can be derived from both laboratory and field studies. Feeding adult ring-neck pheasants 2–3 mg/kg of MeHg as methyl mercury dicyandiamide for 12 weeks resulted in an increase in eggs without shells, a decrease in egg weight, decreased hatchability, and an increase in unfertilized eggs; liver Hg concentrations in adult pheasants were approximately 2 mg/kg. In two studies on mallards fed Hg in the diet over long periods of time, hens fed Hg produced fewer 1-week-old ducklings and duckling growth rate was lower compared with the productivity and growth of controls; average liver Hg concentrations in adult hens were 0.9 mg/kg (Heinz 1976a) and 11mg/kg (Heinz 1976b). Clutch size, egg production, hatchability, and duckling survival were all lower in black ducks experimentally fed Hg and that had an average liver concentration of 23 mg/kg. In field studies, liver Hg concentrations of tree swallows (*Tachycineta bicolor*) and house wrens (*Troglodytes aedon*) of 4.1 and 3.0 mg/kg were associated with poor hatchability (70–74% compared with a nationwide average for the two species of ~85%) (Custer et al. 2007). Fimreite (1974) also reported a complete absence of loon chicks in Ball Lake, northwestern Ontario, where adult loons had average liver Hg concentrations of 52 mg/kg fledging success in common terns (*Sterna hirundo*) was also poor (estimated at 10%) at this site compared with another area, Lake Wabigoon (although the influence of Hg upon hatching success was not clear-cut when Hg concentrations were compared among clutches of one, two, or three eggs); average liver Hg concentrations in adult terns from Ball Lake and Lake Wabigoon were 21 and 9 mg/kg, respectively. Based on the experimental and field data available for different species, the geometric mean adult liver Hg concentration associated with adverse effects on reproduction is 8.7 (range 2–52 mg/kg). This is approximately an order of magnitude lower that that associated with lethality, and is consistent with dietary intakes associated with reproduction also being roughly ten fold lower than those associated with lethality in birds. The range of liver concentrations that have been associated with impaired reproduction is relatively wide, and overlaps that associated with lethality, but the data suggest that liver concentrations in adult birds exceeding 2 mg/kg may be associated with adverse effects on reproduction.

There have been small numbers of studies on other tissue concentrations associated with impaired reproduction in birds. Adult kidney Hg concentrations of 1–15 mg/kg in mallards (Heinz 1976a, 1976b) and 11–16 mg/kg in black ducks (Finley and Stendell 1978) have been associated with impaired reproduction. Adult brain concentrations in birds from the same studies were 0.44–4.6 mg/kg (mallard) and 2–4mg/kg (black ducks). Blood and feather concentrations of 3 and 40 mg/kg, respectively have been associated with adverse effects on multiple endpoints in common loons, including impaired reproductive output (Evers et al. 2008), but concentrations of 6.7 (blood) and 40 (feather) mg/kg were not apparently associated with impaired reproduction in bald eagles *Haliaeetus leucocephalus* (Weech et al. 2006). The feather concentrations associated with the adverse reproductive effects suffered by mallards and black ducks fed experimental Hg diets were 9–70 mg/kg and approximately 60 mg/kg, respectively (Heinz 1976a, 1976b, Finley and Stendell 1978).

18.5.4 TISSUE CONCENTRATIONS IN MAMMALS ASSOCIATED WITH IMPAIRED REPRODUCTION

There appear to be few data for tissue residues associated with impaired reproduction in mammals. Liver and kidney Hg residues of 30 and 36 mg/kg have been associated with impaired reproduction in mink (Wren et al. 1987b), within the range associated with lethality for this species. It was also considered that Hg accumulation from nonungulate prey by female panthers resulted in reduced kitten survival when Hg concentrations in the blood exceeded 0.5 mg/kg (mean, 0.17 kittens/female year) compared with blood levels ≤0.25 mg/kg [mean, 1.46 kittens/female year; (Roelke et al. 1991)].

18.5.5 INDICATIVE EGG MERCURY CONCENTRATIONS

There are a relatively large number of field and a smaller number of laboratory studies in which no adverse effects and/or adverse effects on reproduction have been associated with total Hg concentrations in eggs (Table 18.1). Some species (such as common tern), although associated with marine environments, feed extensively in fresh waters and so are included. Laboratory derived values are effectively NOAECs and LOAECs, whereas field-derived adverse effect values are simply the Hg concentrations in (often failed) eggs from nests or colonies where there is evidence of impaired reproduction. Interestingly, the laboratory LOAECs fall within the range of values derived from field studies (Figure 18.4).

Overall, the geometric mean egg Hg concentration associated with no adverse effect is 0.4 mg/kg (range 0.07–1.6) and the equivalent concentration associated with adverse effect is 1.9 mg/kg (range 0.8–5.1). There are data for sufficient species, including those such as American kestrel and snowy egret, which are thought to be highly sensitive to the embryotoxic effects of Hg (Heinz et al. 2009), to estimate an HC5 for adverse effects; the predicted value is 0.56 mg/kg (Figure 18.4). This suggests that reproduction may be impaired in most (95%) species by egg Hg concentrations of ≥0.6 mg/kg. This is higher than the experimentally derived median lethal concentration (LC50)

TABLE 18.1
Egg Hg Concentrations (mg/kg wet wt.) Associated Either with No Adverse Effects or with Adverse Effects on Reproductive Output in Different Species

Species	[Hg] mg/kg	Reference
Concentrations Associated with No Adverse Effects		
Prairie Falcon *Falco mexicanus*	0.07	Fyfe et al. 1976
Richardson's merlin *Falco columbarius richardsonii*	0.13	Fyfe et al. 1976
White-faced ibis *Plegadis chihi*	0.2	Henny and Herron 1989
Peregrine falcon *Falco peregrinus*	0.25	Newton et al. 1989
Forster's tern *Sterna forsteri*	0.4	King et al. 1991
Marsh harrier *Circus aeruginosus*	0.4	Odsjo and Sondell 1977
Black Skimmer *Rynchops niger*	0.46	King et al. 1991
Herring Gull *Larus argentatus*	0.51	Gilman et al. 1977
Tree swallow *Tachycineta bicolor*	0.77	Custer et al. 2007
Black-crowned night heron *Nycticorax nycticorax*	0.8	Hill et al. 2008
Common loon *Gavia immer*	1	Barr 1986
Common tern *Sterna hirundo*	1	Eisler 2004, Fimreite 1974
White-tailed sea eagle *Haliaeetus albicilla*	1.59	Mid value from Helander et al. 1982, Koivusaari et al. 1980
Mallard[a] *Anas platyrhynchos*	0.895	Midpoint of values from Heinz 1974, 1976b
American kestrel[a] *Falco sparverius*	0.08	Albers et al. 2007
Concentrations Associated with Adverse Effects		
Snowy egret *Egretta thula*	0.8	Hill et al. 2008
Mallard[a] *A. platyrhynchos*	1	Midpoint of values from Heinz 1974, 1976a, 1976b, Heinz and Hoffman 1998, 2003
White-tailed eagle *H. albicilla*	1.15	Helander et al. 1982
Common Loon *G. immer*	1.3	Evers et al. 2003
Tree swallow *Tachycineta bicolor*	1.58	Custer et al. 2007
Ring-necked pheasant[a] *Phasianus colchicus*	1.62	Midpoint of values from Fimreite 1971, Borg et al. 1969, Spann et al. 1972
American kestrel[a] *Falco sparverius*	2	Albers et al. 2007
Common tern *Sterna hirundo*	3.5	Midpoint of values from Fimreite 1974, Eisler 2004
Chicken[a] *Gallus domesticus*	5	Tejning 1967
Black duck[a] *Anas rubripes*	5.1	Finley and Stendell 1978

[a] Lab study.

for 12 out of 23 species for which egg toxicity was tested by injecting methyl mercury, but MeHg appears to induce greater toxicity when injected than when derived from natural maternal transfer (Heinz et al. 2009).

Summary

Dietary, organ, and egg Hg concentrations that may be indicative of adverse effects on survival and reproduction are summarized in Table 18.2. Mercury concentrations in other organs that

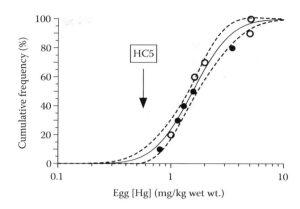

FIGURE 18.4 Cumulative frequency of total Hg concentrations (mg/kg wet wt.) in eggs associated with reproductive impairment. Open and closed symbols indicate data derived from laboratory and field studies, respectively. Data and sources are described in Table 18.1.

have been associated with adverse effects have been highlighted in the text, but are derived only from one or two species and so not included in the table. However, it should also be remembered that the values in Table 18.2 are also only derived from a limited number of species, and the mammalian data are from studies in just one taxonomic group, the Carnivora.

The overall intra and interspecies variation in pharmacokinetic handling of, and sensitivity to, Hg makes deriving single indicative values for all species problematic. The values in Table 18.2 should be used only as a tool in an initial assessment of potential risk from Hg. Exceedence of these values should not be taken as evidence of adverse effects. However, these indicative values are of real value as they enable initial assessment of potential risk for the myriad species for which there are no specific data on Hg toxicity. When used for such purposes, it should be noted that the indicative values are likely to be conservative to some extent as they are based on either an HC5 type approach, an uncertainty factor applied to a geometric mean concentration, or they represent the lowest value in a small dataset. Consequently, the LOAECs for most individual species are likely to be higher than the indicative values. This is apparent from the species sensitivity distributions presented in this chapter, and may particularly be true for species that have coevolved with relatively high environmental levels of Hg. Given this, when appropriate taxonomic-specific data are available, they most likely will prove a better indicator of risk than the indicative values.

The review of toxicity data conducted here suggests, at least superficially, that birds may be less sensitive than mammals to dietary Hg concentrations. The geometric mean dietary concentration associated with lethality is 20 mg/kg (95% geometric confidence limits 13.9–29.8, $n = 11$ species) for birds and 4.3 mg/kg (1.6–11.3, $n = 6$ species) for mammals. Although the difference between the avian and mammalian data is statistically significant (Mann–Whitney U test, $p < .01$), such comparison is problematic, partly because the duration of exposure, the speciation of ingested Hg, feeding rate and severity of effect (proportion of animals dying) all potentially differed between studies. If there is a real difference in sensitivity between mammals and birds to dietary Hg, this does not appear to be reflected in the liver concentrations associated with adverse effects. For example, the indicative values for liver concentrations associated with death in birds were broadly similar to that for mammals. This may suggest that the internal distribution of Hg between body tissues varies markedly between birds and mammals, at least when toxic doses are ingested. It is also notable that dietary concentrations and liver residues associated with adverse effects on reproduction in birds were approximately an order of magnitude lower than those associated with lethality, but this does not appear to be the case from the

TABLE 18.2

Summary of Dietary and Tissue Total Hg Concentrations (mg/kg wet wt.) Associated with Adverse Effects on Survival and Reproduction in Nonmarine Birds and Mammals and Proposed Concentrations That May Be Indicative of Adverse Effects

| | No/ Species[a] | Hg Concentration (mg/kg wet wt.) Associated with Adverse Effect | | Description of Proposed Indicative Value |
		Geometric Mean (Range)	Proposed Indicative Value	
		Birds-Death		
Diet	11	20 (10–40)	>6	Predicted HC5 value, based on dietary Hg being MeHg
Liver	8	63 (18.4–127)	>20	Predicted HC5 value
Kidney	8	76 (21–130)	>40	Predicted HC5 value
Brain			>15 (adults) >3 (developing young)	Based on Scheuhammer et al. 2007
		Birds-Reproduction		
Diet	7	1.2 (0.1–10)	>0.25	Geometric mean divided by five[b]
Liver	5	8.7 (2–52)	>2	Lower end of range
Egg	10	1.9 (0.8–5.1)	>0.6	Predicted HC5 value
		Mammals-Death[c]		
Diet	6	4.3 (1.1–14)	>0.9	Geometric mean divided by five[b]
Liver	4	42 (33–54)	>25–30	Based on the lower end of ranges for liver and kidney Hg. Values derived from data for carnivorous mammals only
Kidney	4	38 (26–69)	>25–30	
Brain	4	16 (11–21)	>10 (adults)	Bottom of range

[a] Number of species for which data were available and from which geometric mean, range, and proposed diagnostic Hg concentrations are derived. The concentrations are associated with adverse effects but are not necessarily derived from LOAEC or LOAEL data.

[b] Approach analogous to the generic approach to dietary risk assessment for pesticides in birds and mammals recently advocated by the European Commission (European Food Safety Authority 2008) where the geometric mean dietary No Observed Effect Concentration (NOEC) and an assessment factor of five are used as the trigger value at the first tier of assessment.

[c] Diet and tissue concentrations associated with impaired reproduction in carnivorous mammals were in approximately the same range as those associated with lethality.

limited data available for mammals. There may, therefore, also be taxonomic differences in the relative sensitivity of different endpoints to Hg, although data for more species are needed.

In conclusion, the indicative values in Table 18.2 are somewhat lower and more precisely defined than those given in the previous edition of this chapter. This is partly due to the approaches used. The biggest differences is probably that adverse effects on reproduction in birds were previously associated with a rather poorly defined dietary Hg concentration of "up to 3" (Thompson 1996), whereas the indicative value given here is almost an order of magnitude lower. However, the overall datasets for Hg toxicity in birds and mammals remain sparse and future improvements in assessing likely risk will require toxicity data for additional species. If sufficiently large datasets are obtained, it may become possible to refine indicative values on the basis of ecological and physiological traits. Any new data derived from field studies need to couple observational data on adverse effects with measurements of dietary and tissue residues, preferable across a range of sites that vary in their Hg (but no other) contamination, so that evidence for dose–response relationships can be determined.

REFERENCES

Albanus, L., et al. 1972. Toxicity for cats of methylmercury in contaminated fish from Swedish lakes and of methylmercury hydroxide added to fish. *Environ. Res.* 5:425–442.

Albers, P. H., et al. 2007. Effects of methylmercury on reproduction in American kestrels. *Environ. Toxicol. Chem.* 26:1856–1866.

Aldenberg, T., and W. Slob. 1993. Confidence-limits for hazardous concentrations based on logistically distributed NOEC toxicity data. *Ecotox. Environ. Safe.* 25:48–63.

Arai, T., et al. 2004. Chemical forms of mercury and cadmium accumulated in marine mammals and seabirds as determined by XAFS analysis. *Environ. Sci. Technol.* 38:6468–6474.

Aulerich, R., R. Ringer, and S. Iwamoto. 1974. Effects of dietary mercury on mink. *Arch. Environ. Contam. Toxicol.* 2:43–51.

Bakir, F., et al. 1973. Methylmercury poisoning in Iraq. *Science* 181:230–241.

Barr, J. 1986. Population dynamics of common loon (*Gavia immer*) associated with mercury -contaminated waters in northwestern Ontario. *Canadian Wildlife Service Occasional Paper* 56:23.

Borg, K., K. Erne, E. Hanko, and H. Wanntorp. 1970. Experimental secondary methyl mercury poisoning in the goshawk (*Accipiter G. Gentilis L.*). *Environ. Pollut.* 1:91–104.

Borg, K., H. Wanntorp, K. Erne, and E. Hanko. 1969. Alkyl Mercury poisoning in Swedish wildlife. *Viltrevy (Stokh)* 6:301–379.

Buekers, J., E. Steen Redeker, and E. Smolders. 2009. Lead toxicity to wildlife: derivation of a critical blood concentration for wildlife monitoring based on literature data. *Sci. Total Environ.* 407:3431–3438.

Burbacher, T. M., P. M. Rodier, and B. Weiss. 1990. Methylmercury developmental neurotoxicity—a comparison of effects in humans and animals. *Neurotoxicol. Teratol.* 12:191–202.

Burger, J., J. A. Rodgers, and M. Gochfeld. 1993. Heavy metal and selenium levels in endangered wood storks *Mycteria americana* from nesting colonies in Florida and Costa Rica. *Arch. Environ. Contam. Toxicol.* 24:417–420.

Burgess, N. M., and M. W. Meyer. 2008. Methylmercury exposure associated with reduced productivity in common loons. *Ecotoxicology* 17:83–91.

Clarkson, T., and W. L. Magos. 2006. The toxicology of mercury and its chemical compounds. *Crit. Rev. Toxicol.* 36:609–662.

Custer, C. M., T. W. Custer, and E. F. Hill. 2007. Mercury exposure and effects on cavity-nesting birds from the Carson River, Nevada. *Arch. Environ. Contam. Toxicol.* 52:129–136.

Cuvin-Aralar, M. L. A., and R. W. Furness. 1991. Mercury and selenium interaction—a review. *Ecotoxicol. Environ. Saf.* 21:348–364.

Dansereau, M., N. Lariviere, D. Du Tremblay, and D. Belanger. 1999. Reproductive performance of two generations of female semidomesticated mink fed diets containing organic mercury contaminated freshwater fish. *Arch. Environ. Contam. Toxicol.* 36:221–226.

Eaton, R., D. Secord, and P. Hewitt. 1980. An experimental assessment of the toxic potential of mercury in ringed-seal liver for adult laboratory cats. *Toxicol. Appl. Pharmacol.* 55:514–521.

Eisler, R. 1987. Mercury hazards to fish, wildlife, and invertebrates: a synoptic review. *Biological Reports*, pp. 63. U.S. Fish and Wildlife Service.

Eisler, R. 2004. Mercury hazards from gold mining to humans, plants, and animals. *Rev. Environ. Contam. Toxicol.* 181:139–198.

Endo, T., Y. Hotta, K. Haraguchi, and M. Sakata. 2005. Distribution and toxicity of mercury in rats after oral administration of mercury-contaminated whale red meat marketed for human consumption. *Chemosphere* 61:1069–1073.

European Food Safety Authority. 2008. Scientific opinion of the panel on plant protection products and their residues on a request from the EFSA PRAPeR Unit on risk assessment for birds and mammals. *EFSA J.* 734:1–181.

Evers, D. C., et al. 2008. Adverse effects from environmental mercury loads on breeding common loons. *Ecotoxicology* 17:69–81.

Evers, D. C., K. M. Taylor, A. Major, R. J. Taylor, R. H. Poppenga, and A. M. Scheuhammer. 2003. Common loon eggs as indicators of methylmercury availability in North America. *Ecotoxicology* 12:69–81.

Falandysz, J. 1984. Metals and organochlorines in a female white-tailed eagle from Uznam Island, southwestern Baltic Sea. *Environ. Conserv.* 11:262–263.

Falandysz, J. 1986. Metals and organochlorines in adult and immature males of white-tailed eagle. *Environ. Conserv.* 13:69–70.

Falandysz, J., B. Jakuczun, and T. Mizera. 1988. Metals and organochlorines in 4 female white-tailed eagles. *Mar. Pollut. Bull.* 19:521–526.

Fimreite, N. 1971. Effects of dietary methylmercury on ring necked pheasants with special references to reproduction. *Can. Field Nat.* 9:39.

Fimreite, N. 1974. Mercury contamination of aquatic birds in northwestern Ontario. *J. Wildl. Manage.* 38:120–131.

Fimreite, N., and L. Karstad. 1971. Effects of dietary methylmercury on red tailed hawks. *J. Wildl. Manage.* 35:293–300.

Finley, M. T., and R. C. Stendell. 1978. Survival and reproductive success of black ducks fed methyl mercury. *Environ. Pollut.* 16:51–64.

Finley, M. T., W. H. Stickel, and R. E. Christensen. 1979. Mercury residues in tissues of dead and surviving birds fed methylmercury *B. Environ. Contam. Toxicol.* 21:105–110.

Fyfe, R. W., R. W. Riseborough, and W. Walker. 1976. Pollutant effects on the reproduction of the Prairie Falcons and Merlins of the Canadian prairies. *Can. Field Nat.* 90:345–355.

Ganther, H. E., et al. 1972. Selenium: relation to decreased toxicity of methylmercury added to diets containing tuna. *Science* 175:1122–1124.

Gilman, A. P., G. A. Fox, D. B. Peakall, S. M. Teeple, T. R. Carroll, and G. T. Haymes. 1977. Reproductive parameters and egg contaminant levels of Great Lakes herring gulls. *J. Wildl. Manage.* 41:458–468.

Hanko, E., K. Erne, H. Wanntorp, and K. Borg. 1970. Poisoning in ferrets by tissues of alkyl mercury-fed chickens. *Acta Vet. Scand.* 11:268–82.

Heinz, G. H. 1974. Effects of low dietary levels of methyl mercury on mallard reproduction. *B. Environ. Contam. Toxicol.* 11:386–392.

Heinz, G. H. 1976a. Methyl-mercury-second generation reproductive and behavioral effects on mallard ducks. *J. Wildl. Manage.* 40:710–715.

Heinz, G. H. 1976b. Methylmercury—2nd-year feeding effects on mallard reproduction and duckling behavior. *J. Wildl. Manage.* 40:82–90.

Heinz, G. H. 1979. Methylmercury—reproductive and behavioral effects on 3 generations of mallard ducks. *J. Wildl. Manage.* 43:394–401.

Heinz, G. H. 1996. Mercury poisoning in wildlife. In *Noninfectious diseases of wildlife,* eds. A. Fairbrother, L. N. Locke, and G. L. Hoff, pp. 118–127. Ames, IA: Iowa State University Press.

Heinz, G. H., and D. J. Hoffman. 1998. Methylmercury chloride and selenomethionine interactions on health and reproduction in mallards. *Environ. Toxicol. Chem.* 17:139–145.

Heinz, G. H., and D. J. Hoffman. 2003. Embryotoxic thresholds of mercury: estimates from individual mallard eggs. *Arch. Environ. Contam. Toxicol.* 44:257–264.

Heinz, G. H., D. J. Hoffman, J. D. Klimstra, K. R. Stebbins, S. L. Kondrad, and C. A. Erwin. 2009. Species differences in the sensitivity of avian embryos to methylmercury. *Arch. Environ. Contam. Toxicol.* 56:129–138.

Heinz, G. H., and L. N. Locke. 1976. Brain lesions in mallard ducklings from parents fed methylmercury. *Avian Dis.* 20:9–17.

Helander, B., M. Olsson, and L. Reutergardh. 1982. Residue levels of organochlorine and mercury-compounds in unhatched eggs and the relationships to breeding success in white-tailed sea eagles *Haliaeetus albicilla* in Sweden. *Holarctic Ecol.* 5:349–366.

Henny, C. J., and G. B. Herron. 1989. DDE, selenium, mercury, and white-faced ibis reproduction at Carson Lake, Nevada. *J. Wildl. Manage.* 53:1032–1045.

Henny, C. J., E. F. Hill, D. J. Hoffman, M. G. Spalding, and R. A. Grove. 2002. Nineteenth century mercury: Hazard to wading birds and cormorants of the Carson River, Nevada. *Ecotoxicology* 11:213–231.

Henriksson, K., E. Karppanen, and M. Helminen. 1966. High residue of mercury in Finnish white tailed eagles. *Ornis Fennica* 43:38–45.

Hill, E. F., C. J. Henny, and R. A. Grove. 2008. Mercury and drought along the lower Carson River, Nevada: II. Snowy egret and black-crowned night-heron reproduction on Lahontan Reservoir, 1997–2006. *Ecotoxicology* 17:117–131.

Ikemoto, T., T. Kunito, H. Tanaka, N. Baba, N. Miyazaki, and S. Tanabe. 2004. Detoxification mechanism of heavy metals in marine mammals and seabirds: interaction of selenium with mercury, silver, copper, zinc, and cadmium in liver. *Arch. Environ. Contam. Toxicol.* 47:402–413.

Jackson, T. A. 1997. Long-range atmospheric transport of mercury to ecosystems, and the importance of anthropogenic emissions—a critical review and evaluation of the published evidence. *Environ. Rev.* 5:99–120.

Jernelov, A., A. H. Johansson, L. Sorensen, and A. Svenson. 1976. Methylmercury degradation in mink. *Toxicology* 6:315–321.

Kim, E. Y., T. Murakami, K. Saeki, and R. Tatsukawa. 1996. Mercury levels and its chemical form in tissues and organs of seabirds. *Arch. Environ. Contam. Toxicol.* 30:259–266.

King, K. A., T. W. Custer, and J. S. Quinn. 1991. Effects of mercury, selenium, and organochlorine contaminants on reproduction of Forster's terns and black skimmers nesting in a contaminated Texas bay. *Arch. Environ. Contam. Toxicol.* 20:32–40.

Koeman, J., R. Garssen-Hoekstra, E. Pels, and J. De Goeij. 1971. Poisoning of birds of prey by methylmercury compounds. *Meded Rijksfac Landbouwwet gent* 36:43–49.

Koeman, J., R. Hadderingh, and M. Bijlveld. 1972. Persistent pollutants in the white-tailed eagle (*Haliaeetus albicilla*) in the Federal Republic of Germany. *Biol. Conserv.* 4:373–377.

Koeman, J. H., W. H. M. Peeters, C. H. M. Koudstaal, P. S. Tjioe, and J. Goeij. 1973. Mercury-selenium correlations in marine mammals. *Nature* 245:385–386.

Koeman, J., J. Vink, and J. Goeij. 1969. Causes of mortality in birds of prey and owls in the Netherlands in the winter of 1968–1969. *Ardea* 57:67–76.

Koivusaari, J., I. Nuuja, R. Palokangas, and M. Finnlund. 1980. Relationships between productivity, eggshell thickness and pollutant contents of addled eggs in the population of white-tailed eagles *Haliaetus albicilla* L. in Finland during 1969–1978. *Environ. Pollut. A* 23:41–52.

Kurland, L. T., S. N. Faro, and H. Siedler. 1960. Minamata disease: the outbreak of a neurological disorder in Minamata, Japan, and its relationship to the ingestion of seafood contaminated by mercuric compounds. *World Neurol.* 1:370–395.

Longcore, J. R., R. Dineli, and T. A. Haines. 2007. Mercury and growth of tree swallows at Acadia National Park, and at Orono, Maine, USA. *Environ. Monit. Assess.* 126:117–127.

Newton, I., J. A. Bogan, and M. B. Haas. 1989. Organochlorines and mercury in the eggs of British peregrines *Falco peregrinus. Ibis* 131:355–376.

Nicholson, J. K., and D. Osborn. 1984. Kidney lesions in juvenile starlings *Sturnus vulgaris* fed on a mercury-contaminated synthetic diet. *Environ. Pollut. A* 33:195–206.

Norheim, G. 1987. Levels and interactions of heavy metals in sea birds from Svalbard and the Antarctic. *Environ. Pollut.* 47:83–94.

Nriagu, J. O. 1989. A global assessment of natural sources of atmospheric trace metals. *Nature* 338:47–49.

Nriagu, J. O., and J. M. Pacyna. 1988. Quantitative assessment of worldwide contamination of air, water and soils by trace-metals. *Nature* 333:134–139.

O'Connor, D., and S. Nielsen. 1980. Environmental survey of methylmercury levels in wild mink (*Mustela vison*) and otter (*Lutra canadensis*) from the northeastern United States and experimental pathology of methylmercurialism in the otter. In *First Worldwide Furbearer Conference*, eds. J. A. Chapman and D. Pursley, pp. 1728–1745. Frostburg, MD.

Odsjo, T., and J. Sondell. 1977. Population development and breeding success in the marsh harrier Cirus aeruginosus in relation to levels of DDT, PCB, and mercury. *Var Fagelvarld* 36:152–160.

Oehme, G. 1981. Zur Quecksilberruckstandsbelastunt tot aufgefundener Seeadler, Haliaeetus albicilla, in den jahren 1976–1978. *Hercynia* 18:353–364.

Palmisano, F., N. Cardellicchio, and P. G. Zambonin. 1995. Speciation of mercury in dolphin liver—a two stage mechanism for the demethylation accumulation process and role of selenium. *Mar. Environ. Res.* 40:109–121.

Pelletier, E. 1986. Mercury-selenium interactions in aquatic organisms—a review. *Mar. Environ. Res.* 18:111–132.

Roelke, M., A. Schutz, C. Facemire, and S. Sundlof. 1991. Mercury contaminations in the free-ranging endangered Florida panther (felis concolor coryi). *Proc. Am. Assoc. Zoo Veterin.* 20:277–283.

Rowland, I. R. 1988. Interactions of the gut microflora and the host in toxicology. *Toxicol. Pathol.* 16:147–153.

Scheuhammer, A. M. 1987. The chronic toxicity of aluminum, cadmium, mercury, and lead in birds—a review. *Environ. Pollut.* 46:263–295.

Scheuhammer, A. M., M. W. Meyer, M. B. Sandheinrich, and M. W. Murray. 2007. Effects of environmental methylmercury on the health of wild birds, mammals, and fish. *Ambio* 36:12–18.

Scott, M. L. 1977. Effects of PCBs, DDT, and mercury compounds in chickens and Japanese quail. *Fed. Proc. Amer. Soc. Exper. Biol.* 36:1888–1893.

Scott, M. L., J. R. Zimmermann, S. Marinsky, P. A. Mullenhoff, G. L. Rumsey, and R. W. Rice. 1975. Effects of PCBs, DDT, and mercury compounds upon egg-production, hatchability and shell quality in chickens and Japanese-quail. *Poultry Sci.* 54:350–368.

Spalding, M. G., et al. 2000. Histologic, neurologic, and immunologic effects of methylmercury in captive great egrets. *J. Wildl. Dis.* 36:423–435.

Spann, J. W., et al. 1986. Differences in mortality among bobwhite fed methylmercury chloride dissolved in various carriers. *Environ. Toxicol. Chem.* 5:721–724.

Spann, J. W., J. F. Kreitzer, R. G. Heath, and L. N. Locke. 1972. Ethyl mercury para toluene sulfonanilide—lethal and reproductive effects on pheasants. *Science* 175:328–331.

Spry, D. J., and J. G. Wiener. 1991. Metal bioavailability and toxicity to fish in low-alkalinity lakes—a critical-review. *Environ. Pollut.* 71:243–304.

Spurgeon, D. J., and S. P. Hopkin. 1996. Risk assessment of the threat of secondary poisoning by metals to predators of earthworms in the vicinity of a primary smelting works. *Sci. Total Environ.* 187:167–183.

Streets, D. G., Q. Zhang, and Y. Wu. 2009. Projections of global mercury emissions in 2050. *Environ. Sci. Technol.* 43:2983–2988.

Suda, I., and H. Takahashi. 1992. Degradation of methyl and ethyl mercury into inorganic mercury by other reactive oxygen species besides hydroxyl radical. *Arch. Toxicol.* 66:34–39.

Tejning, S. 1967. Biological effects of methyl mercury dicyandiamide-treated grain in the domestic fowl *Gallus gallus* L. *Oikos* (Suppl 8):116 pp.

Thompson, D. R. 1996. Mercury in birds and terrestrial mammals. In *Environmental contaminants in wildlife: interpreting tissue concentrations*, eds. W. N. Beyer, G. H. Heinz, and A. W. Redmon-Norwood, pp. 341–356. Boca Raton, FL: CRC Lewis Publishers.

Thompson, D. R., and R. W. Furness. 1989. The chemical form of mercury stored in south Atlantic seabirds. *Environ. Pollut.* 60:305–317.

Thompson, D. R., K. C. Hamer, and R. W. Furness. 1991. Mercury accumulation in great skuas *Catharacta skua* of known age and sex, and its effects upon breeding and survival. *J. Appl. Ecol.* 28:672–684.

U.S. EPA. 1995. Methylmercury (MeHg) (CASRN 22967-92-6). *US EPA Integrated Risk Information System*, www.epa.gov/iris/subst/0073.htm.

Weech, S. A., A. M. Scheuhammer, and J. Elliott. 2006. Mercury exposure and reproduction in fish-eating birds breeding in the Pinchi Lake region, British Columbia, Canada. *Environ. Toxicol. Chem.* 25:1433–1440.

Westcott, K., and J. Kalff. 1996. Environmental factors affecting methyl mercury accumulation in zooplankton. *Can. J. Fish. Aquat. Sci.* 53:2221–2228.

Wiener, J. G., D. P. Krabbenhoft, and A. Scheuhammer. 2003. Ecotoxicology of mercury. In *Handbook of ecotoxicology,* eds. D. J. Hoffman, B. A. Rattner, G. V. Burton, and J. J. Cairns, pp. 409–463. Boca Raton, FL: Lewis.

Wilber, C. 1980. Toxicology of selenium: a review. *Clin. Toxicol.* 17:171–230.

Wobeser, G., N. O. Nielsen, and B. Schiefer. 1976a. Mercury and mink. I. Mercury contaminated fish as food for ranch mink. *Can. J. Comp. Med.* 40**,** 30–33.

Wobeser, G., N. O. Nielsen, and B. Schiefer. 1976b. Mercury and mink. II. Experimental methyl mercury intoxication. *Can. J. Comp. Med.* 40:34–45.

Wobeser, G., and M. Swift. 1976. Mercury poisoning in a wild mink. *J. Wildl. Dis.* 12:335–340.

Wolfe, M. F., S. Schwarzbach, and R. A. Sulaiman. 1998. Effects of mercury on wildlife: a comprehensive review. *Environ. Toxicol. Chem.* 17:146–160.

Wren, C. D. 1985. Probable case of mercury-poisoning in a wild otter, *Lutra canadensis,* in northwestern Ontario. *Can. Field Nat.* 99:112–114.

Wren, C. D., D. B. Hunter, J. F. Leatherland, and P. M. Stokes. 1987a. The effects of polychlorinated-biphenyls and methylmercury, singly and in combination, on mink. I. Uptake and toxic responses. *Arch. Environ. Con. Toxicol.* 16:441–447.

Wren, C. D., D. B. Hunter, J. F. Leatherland, and P. M. Stokes. 1987b. The effects of polychlorinated-biphenyls and methylmercury, singly and in combination, on mink. II. Reproduction and kit development. *Arch. Environ. Con. Toxicol.* 16:449–454.

Yang, D.-Y., Y.-W Chen, J. M. Gunn, and N. Belzile. 2008. Selenium and mercury in organisms: interactions and mechanisms. *Environ. Rev.* 16:71–92.

European Squirrel

By G. Mutzel, from *The Royal Natural History,* edited by Richard Lydekker, Frederick Warne & Co.,
London, 1893–94.

19 Cadmium in Small Mammals

John A. Cooke

CONTENTS

19.1 Introduction 627
19.2 Metabolism and Toxicity 628
19.3 Cadmium in Terrestrial Habitats 629
19.4 Tissue Concentrations in Wild Species 631
19.5 Critical Tissue Concentrations... 636
Summary 638
Acknowledgment 638
References.. 639

19.1 INTRODUCTION

There has been much research into the toxicity of cadmium over the last four decades. This chapter assesses the available information, and updates a previous review (Cooke and Johnson 1996). It focuses on the relationship between the toxicology of cadmium and critical tissue concentrations to provide an assessment of environmental risk to wild small mammals. The main sources of information are human epidemiological studies, chronic studies of experimental exposure of laboratory rodents and captive wild species, and field studies on wild small mammals.

Controlled experimental studies involving laboratory rats and mice, in particular, have provided relevant data concerning the effects of cadmium in wild small mammals. However, many of these studies have involved the administration of soluble cadmium salts and metallothionein (MT)-bound cadmium through injection and these, especially at high doses, lack relevance to the long-term oral exposure and gut absorption that characterizes the more realistic environmental conditions. Most of the field-based investigations of wild populations of small mammals over the last 25 years have focused on the cadmium concentrations in tissues, especially in kidney and liver. These have provided a substantial knowledge base of concentrations of cadmium in animals from uncontaminated habitats and from localized, often highly, contaminated habitats. Some of these field studies have provided data relating tissue concentrations of cadmium to estimated dietary concentrations, dose–response characteristics and pathological effects. However, very few have yielded results that show the ecotoxicological effects at the population level and of the effects on the life history of animals in the wild. Most of the field studies of contaminated sites have been at sites that contain elevated levels of other metals, especially copper, lead, and zinc. Because physiological interactions between these metals and cadmium are likely, the ecological and toxicological significance of such interactions are an important consideration in the assessment of risk of environmental cadmium. Over the last 15 years there have been a number of studies that have attempted to bridge the gap between laboratory and field approaches, by conducting controlled experiments using laboratory populations of widely distributed wild species of small mammals commonly found in Europe.

19.2   METABOLISM AND TOXICITY

Cadmium has no known biological function. Its toxicity may originate through exposure via respiration or ingestion. The incidence of acute toxicity is rare, although the inhalation of cadmium fume is a well-known industrial hazard (Fielder and Dale 1983). Fifty percent lethal dose values for acute toxicity of about 100–300 mg of cadmium per kilogram have been obtained for a range of soluble cadmium compounds in the mouse and rat although lower lethal doses have been recorded in a few studies (Fielder and Dale 1983, WHO 1992a, ATSDR 1999).

Environmental cadmium exposure is predominantly a chronic problem reflected in the gradual accumulation of the metal from the diet in target organs with eventual tissue dysfunction. The metabolism of cadmium is broadly similar for most mammalian species. Absorption from the diet is low, <5% of the ingested cadmium, but diets low in calcium, zinc, and iron can significantly increase how much dietary cadmium is absorbed and the percentage retained in the body (Reeves and Chaney 2002, 2004). Low protein, copper, and vitamin D may have similar effects on absorption and retention (Bremner 1979). Excretion of cadmium in feces will include intestinal epithelium with bound cadmium, sloughed into the intestine many weeks after initial absorption (McLellan et al. 1978, Reeves and Chaney 2004). However, this bound cadmium may also be reabsorbed from the gut lumen and lead to delayed absorption long after the cadmium was originally ingested (McLellan et al. 1978). Biliary excretion and urinary excretion of assimilated cadmium can occur, but the amounts are normally very small (<1% of absorbed cadmium), leading to long biological half-lives (100–300 days in rats) (Friberg et al. 1974, Bremner 1979).

Absorbed cadmium is largely bound to protein, particularly MT, and the main sites of accumulation are the intestinal mucosa, kidney, and liver (Bremner 1979, Zalups and Ahmad 2003). The cadmium enters the intestinal mucosa enterocytes via transporter proteins and can induce MT synthesis. Cadmium is released slowly into the blood from the intestinal mucosa where the majority goes to the liver where it is bound to liver MT. It is not clear how much cadmium can go directly to the kidney from the blood and how much is transported via the liver to the kidney as cadmium bound to MT. The binding of intracellular cadmium to MT in the kidney protects against the toxicity of cadmium, and it is probably the "free" cadmium that initiates tissue injury. Renal cadmium, which is not bound to MT, may occur because a "critical concentration" of MT-bound cadmium has been exceeded because of either high doses or age accumulation, both may result in there being insufficient MT to detoxify the unbound cadmium (WHO 1992a, Liu et al. 1998, Habeebu et al. 2000, Liu et al. 2001, Schnellmann 2001). Because of the binding of cadmium to MT and low excretion rates, the biological half-life is long. Retention times of 10–30 years in humans, result in age accumulation whereby the concentrations in the kidney renal cortex have been shown to increase up to the age of 50 years and then to decline (Ryan et al. 1982, WHO 1992a). This decline may be associated with renal tubule dysfunction, which leads to increased urinary excretion of cadmium.

The effects on the kidney are characterized by tubular dysfunction and tubular cell damage, although glomerular dysfunction may also occur. The most usual early sign of cadmium nephrotoxicity is the impairment of the reabsorptive function of the proximal tubules manifested as the presence of low molecular weight proteins, enzymes, amino acids, and other compounds and minerals in urine. Hyperexcretion of calcium and phosphorus, and other important bone minerals, can lead to osteomalacia, osteoporosis and decreased mineral density and bone strength. "Itai-itai" disease is the Japanese name for this syndrome among people (mostly multiparous women over 50 years) living on a rice diet high in cadmium. In a recent study, Alfven et al. (2002) have shown, for a large sample of people from cadmium-exposed and nonexposed groups in southern Sweden, a strong positive relationship between blood cadmium, and tubular proteinuria, and a strong negative relationship between bone mineral density and blood cadmium in their older age (>60 years) group. However, it is likely that "itai-itai" disease is also associated with multiple dietary deficiencies (Friberg et al. 1986), especially low dietary zinc and iron. Deficiencies in iron and zinc can increase the absorption and toxicity of cadmium (Reeves and Chaney 2004).

Exposure to cadmium (usually as cadmium chloride) through the diet has been extensively investigated in experimental studies with laboratory rats, rabbits, and mice, and in the last 10 years a number of European research groups have used laboratory populations of bank voles (*Myodes glareolus*—the former genus name *Clethrionomys* is used in this chapter) and wood mice (*Apodemus sylvaticus*). The major effects of cadmium include the following: reduction of food and water intake; growth depression; renal dysfunction (proteinuria); histological lesions in the kidney; reduced calcification of bone and osteoporosis; hypertension (increased blood pressure); anemia and bleaching of incisors; and cancers (Samarawickrama 1979, Fielder and Dale 1983, WHO 1992a). Hepatotoxicity has been noted in rats but usually occurs only at dose levels higher than those producing nephrotoxicity (Fielder and Dale 1983, ATSDR 1999). Testicular effects can occur, but the oral doses required are likely to be high (Fielder and Dale 1983, ATSDR 1999). However, cadmium-induced testicular necrosis generally results in permanent infertility (WHO 1992a).

The understanding of cadmium exposure and absorption, even in laboratory experiments, and the standardization of dose–response data is difficult because of the influence of body mass, sex, dietary components other than cadmium, the length of the experiment (exposure), animal age, and whether given in drinking water or food. The definitions of Critical Concentrations or the Lowest Observable Adverse Effects Levels (LOAELs) are dependent in turn on the dose–effects relationship (WHO 1992a). Thus the definition and judgment of what are the critical effects and whether the magnitude of these effects in a particular case represent adverse functional changes are crucial. For example, early effects of cadmium on the kidney can be measured as proteinuria but what is the cut-off point (i.e., the concentration of low molecular proteins such as β_2-microglobulin in the urine usually standardized against the creatinine concentration) above which the effect is deemed to be a marker of "an adverse functional change"? One review of dose–response data from over one hundred laboratory rat experiments has attempted to categorize adverse effects and LOAELs as it "believes that there is sufficient merit in this approach to warrant an attempt at distinguishing between 'less serious' and 'serious' effects" (ATSDR 1999). Serious effects "are those that evoke failure in a biological system and can lead to morbidity or mortality" and less serious effects "are those that are not expected to cause significant dysfunction or death." Inevitably there is an overlap between these two categories ("less serious" and "serious") in the LOAELs so derived (ATSDR 1999).

Also important in this judgment of critical concentrations or LOAELs are the individual responses within a treated experimental group, for example was the adverse effect shown in 50% (an EC_{50}, or some other defined effects percentage) of individuals in the group (WHO 1992a). This population information within an experimental group is usually not defined in laboratory rodent cadmium experiments investigating chronic adverse effects.

It would seem that with increasing chronic oral exposure there would be an increase in the severity: of renal dysfunction (proteinuria and tubular necrosis) from a LOAEL of about 5 mg of cadmium per kilogram body weight per day; of reduction of >20% in body weight gain from about 7 mg/kg body weight; of reproductive effects (necrosis and atrophy of seminiferous tubule epithelium) from about 10 mg/kg body weight (ATSDR 1999). In an analysis of nine laboratory studies Shore and Douben (1994) concluded that the LOAEL for chronic oral exposure can be estimated to be in the range 3.5–7.5 mg/kg body weight. These values were based on proteinuria, disturbed bone metabolism and decreased body weight and growth. These data suggests that a reasonable estimate of a no observed adverse effects level (NOAEL) for laboratory rats is a dose of between 3.5 and 5 mg/kg body weight for oral exposure over an exposure period of 15–300 days.

19.3 CADMIUM IN TERRESTRIAL HABITATS

Cadmium concentrations tend to be very low in most environmental media. Typical levels in uncontaminated soils are <1 mg/kg, although naturally higher values can occur when the soils are derived from the weathering of parent materials high in cadmium, such as black shales (Jackson and Alloway 1992). Environmental contamination through atmospheric deposition can lead to much higher soil concentrations. Comparison of the major natural atmospheric source of cadmium, volcanic action (Nriagu 1979), with the major anthropogenic sources, nonferrous metal production, iron and steel making, fuel combustion (coal, oil, and gas), and refuse incineration, for the European Community gave relative annual emission figures of 20, 33, 34, 8.5, and 31 metric tons (t)/year, respectively (Hutton 1983). Thus about 85–90% of the total airborne cadmium in global cycles arises from anthropogenic sources (WHO 1992b) and indicates the importance of human activity in the global cycling of cadmium. Such emissions over the last 100 years have increased cadmium levels in soils of urban, rural, and industrial sites (Korte 1982).

The interception and retention of cadmium-bearing atmospheric particulates by leaves, other trapping surfaces (bark, stems, surface litter), and soil with its subsequent uptake through plant roots (Martin et al. 1982) will lead to circulation of the element within the terrestrial ecosystem (Hughes et al. 1980). In this way, many plant-derived substrates with high cadmium concentrations will be provided for animal feeding, leading to elevated concentrations in most invertebrates and small mammals within the contaminated ecosystems. Martin and Coughtrey (1982), although primarily concerned with the biological monitoring of heavy metal pollution, review many examples of the concentrations of cadmium in both invertebrates (especially earthworms, wood lice, snails, and slugs) and small mammals from many different polluted habitats. Hunter et al. (1987a, 1987b, 1987c, 1989) described, in considerable detail, the distribution of cadmium in grassland ecosystems near a copper–cadmium refinery in northwest England. Some of their results are summarized in Table 19.1 and are both illustrative and representative of the ecological consequences of environmental cadmium contamination. Aerial deposition of cadmium induced proportional increases in most biotic components of the grassland ecosystem studied. Moreover, cadmium is generally regarded as one of the most mobile toxic metals, because its transfer occurs along most terrestrial food chains, and because its transference potential is much greater than those of other metals like lead and zinc (Roberts and Johnson 1978). Cadmium can often be measured in equal or higher concentrations in animal consumers relative to the concentrations in their diet. For example, earthworms had higher concentrations of cadmium than those from surface soil, and isopods had concentrations greater

TABLE 19.1
Cadmium (mg/kg Dry Weight) in Grasslands from a Reference (Clean) Site and from Sites Near a Copper–Cadmium Refinery in Northwest England

	No. of Samples Analyzed from Each Site	Reference Site	1-km Site	Refinery Site
Surface soil	40	0.8 ± 0.08[a]	6.9 ± 0.6	15.4 ± 2.3
Creeping bent grass (leaf) (*Agrostis stolonifera*)	100			
Live		0.63 ± 0.06	1.32 ± 0.07	3.3 ± 0.36
Senescent		0.68 ± 0.06	2.06 ± 0.13	10.4 ± 1.7
Earthworms (Oligochaeta)	60	4.1 ± 0.3	34.0 ± 2.8	107 ± 24.6
Wood lice (Isopoda)	60	14.7 ± 1	130 ± 45.9	231 ± 131
Spiders (Lycosidae)	400	2.6 ± 0.3	34.5 ± 5	102 ± 7.5
Beetles (Carabidae)	400	0.7 ± 0.1	5.6 ± 0.9	15.1 ± 1.1
Field vole (kidney) (*Microtus agrestis*)	19–23	1.7 ± 0.2	23.9 ± 5.6	88.8 ± 23.3
Common shrew (kidney) (*Sorex araneus*)	20–25	20.5 ± 1.6	156 ± 25	253 ± 75

Sources: From Hunter, B. A. et al., *J. Appl. Ecol.,* 24, 573–586, 1987a; Hunter, B. A. et al., *J. Appl. Ecol.,* 24, 587–599, 1987b; Hunter, B. A. et al., *J. Appl. Ecol.,* 26, 89–99, 1989. With permission.

[a] Mean ± standard error.

than those in litter (senescent leaf tissue; Table 19.1). Significant accumulation can occur in wild small mammal kidney tissue. This is evident from a comparison between the values for the herbivore, field vole (*Microtus agrestis*), and those for live creeping bent grass (*Agrostis stolonifera*) leaf material (Table 19.1). Insectivorous small mammals, such as the common shrew (*Sorex araneus*), usually contain higher concentrations in the kidney than do herbivorous or omnivorous species in the same habitat (Table 19.1).

It is also clear that the application of phosphatic fertilizers and sewage sludges and other organic wastes to agricultural land can increase soil cadmium levels considerably. A survey of sludge-treated soils in the United Kingdom showed cadmium concentrations ranging from 0.27 to 158.7 mg/kg (Jackson and Alloway 1992). Thus, extremely high local concentrations can occur because of the high variability in the cadmium content of sludges and sludge application rates. Dried sludge on vegetation and soil surfaces can also be directly ingested by grazing animals (Chaney et al. 1987). It is possible for cadmium in sludge-treated grasslands to accumulate in the kidneys and livers of small mammals as field experiments with the meadow vole (*Microtus pennsylvanicus*) in Ohio have shown (Anderson et al. 1982, Maly 1984).

Nonferrous metal mines, especially those exploiting lead–zinc ores, and the activities involved in processing the ores can be a significant source of environmental contamination through the disposal of cadmium-rich wastes (Roberts and Johnson 1978). Wind dispersal of waste from spoil tips at a derelict mine complex was shown to lead to considerable increases in the cadmium concentrations in herbivorous and carnivorous invertebrates and indigenous small mammals (Roberts and Johnson 1978). Considerable mobility of cadmium within the soil-plant-animal system can also occur after the rehabilitation of industrial waste disposal sites to grassland (Andrews and Cooke 1984).

19.4   TISSUE CONCENTRATIONS IN WILD SPECIES

In a review that combined data from 13 studies between 1974 and 1987, the average concentrations (on a dry weight basis) from reference ("clean") sites for the whole body, liver, and kidney were 0.1–1.4, 0.2–1.5, and <0.1–5.6 mg/kg, respectively (Talmage and Walton 1991). This data set included mice, rats, and voles but excludes moles and shrews that accumulate higher cadmium levels from their insectivorous diet. Cadmium concentrations values for shrews (both subfamilies, the soricine and crocidurine) from uncontaminated sites have been found to be in the range 1.2–4 mg/kg (whole body), 2.3–25.4 mg/kg (liver), and 4.1–25.7 mg/kg (kidney) (Talmage and Walton 1991, Sanchez-Chardi et al. 2007). However, cadmium kidney values from uncontaminated sites can show large intraspecific variability with individual high values; for example, up to 60 mg/kg in shrews and moles (*Talpa europea*) in the Netherlands (Ma and Talmage 2001) and with similar values for moles in Austria (Komarnicki 2000), with the higher concentrations being associated with the older animals. High cadmium levels in insectivores are because of food chain accumulation where food chains involve earthworms and other surface-soil dwelling invertebrates. It is interesting to note that a study of badgers (*Meles meles*) in the Netherlands where badgers often forage on earthworms, also recorded high concentrations of cadmium in kidney tissue of older animals (Van den Brink and Ma 1998). Unusually high values were measured by Pedersen and Lierhagen (2006) of up to 220 mg/kg dry weight with a mean of 107 mg/kg in kidney of adult arctic hares (*Lepus arcticus*) in Nunavut, Canada, which the authors thought was because of high cadmium inputs from the natural local geology as there was no source of contamination.

When small mammals are found in cadmium-contaminated sites, the increased dietary concentrations lead to high concentrations in a range of tissues and organs, particularly the kidney and liver, regardless of whether the contamination is primarily derived from polluted soil or atmospheric deposition (Table 19.2). Lesser increases in tissue concentrations (but often statistically significantly higher than for animals caught at reference sites) are found in the heart, femur, muscle, pancreas, and hair (Andrews et al. 1984, Hunter et al. 1989). There seems little evidence of increased cadmium in the brain, lung, and testis, even at highly contaminated sites (Hunter et al. 1989).

TABLE 19.2
Kidney and Liver Cadmium Concentrations[a] in Various Species of Small Mammals from Contaminated Habitats

Species	Contaminated Site	Concentration (mg/kg dry wt.) Kidney	Liver	Ratio Kidney/Liver	Reference
Cotton-tail rabbit (*Sylvilagus floridanus*)	Zinc Smelter, Pennsylvania, USA	381	61	6.2	Beyer and Storm (1995)
Wood mouse (*Apodemus sylvaticus*)	Cu/Cd refinery, England	41.7	18.2	2.3	Hunter et al. (1989)
	Minera, Pb/Zn mine, Wales	39.7	9.8	4.1	Johnson et al. (1978)
	Smelter waste, Wales	18.0	5.5	3.3	Johnson et al. (1978)
	Y Fan, Pb/Zn mine, Wales	10.3	2.49	4.1	Johnson et al. (1978)
	Fluorspar waste, England	1.78	0.71	2.5	Cooke et al. (1990)
Deer mouse (*Peromyscus maniculatus*)	Zinc Smelter, Pennsylvania, USA	42	13	3.2	Beyer and Storm (1995)
	Superfund site	3.85	1.05	3.7	Pascoe et al. (1996)
White-footed mouse (*Peromyscus leucopus*)	Zinc Smelter, Pennsylvania, USA	70	19	3.7	Beyer and Storm (1995)
	Reclaimed wetland near smelter	6.3	3.8	1.7	Levengood and Heske (2008)
	Wastewater-irrigated site, USA	2.3	0.5	5.0	Anthony and Kozlowski (1982)
Short-tailed field vole (*Microtus agrestis*)	Cu/Cd refinery, England	88.8	22.7	3.9	Hunter et al. (1989)
	Y Fan, Pb/Zn mine, Wales	8.91	1.06	8.4	Johnson et al. (1978)
	Fluorspar waste, England	5.3	1.8	2.9	Andrews et al. (1984)
	Budel, the Netherlands	2.7	0.57	4.7	Ma et al. (1991)
Meadow vole (*Microtus pennsylvanicus*)	Sludge-treated fields, USA	23[a]	7.9[a]	2.9	Anderson et al. (1982)
	Superfund site	2.45	1.75	1.4	Pascoe et al. (1996)
Bank vole (*Clethrionomys glareolus*)	Dulowa Forest, Poland	29.6	12.8	2.3	Sawicka-Kapusta et al. (1990)
	Zn–Cd smelter	27	5.84	4.6	Sawicka-Kapusta et al. (1999)
	Y Fan, Pb/Zn mine, Wales	16.8	5.1	3.3	Johnson et al. (1978)
	Pb/Zn mine	1.9	0.3	6.3	Milton et al. (2003)
Common shrew (*Sorex araneus*)	Cu/Cd refinery, England	253	578	0.43	Hunter et al. (1989)
	Budel, the Netherlands (February–March)	200	268	0.75	Ma et al. (1991)
	Near smelter, Poland	176	288	0.61	Swiergosz-Kowalewska et al. (2005)

TABLE 19.2 (continued)
Kidney and Liver Cadmium Concentrations[a] in Various Species of Small Mammals from Contaminated Habitats

Species	Contaminated Site	Concentration (mg/kg dry wt.)		Ratio Kidney/Liver	Reference
		Kidney	Liver		
	Fluorspar waste, England	158	236	0.67	Andrews et al. (1984)
	Pb/Zn Smelter, England	142	452	0.31	Read and Martin (1993)
	Budel, the Netherlands (October–November)	126	180	0.7	Ma et al. (1991)
Pygmy shrew (*Sorex minutus*)	Pb/Zn Smelter, England	49.9	63.7	0.78	Read and Martin (1993)
European mole (*Talpa europea*)	Budel, the Netherlands	224	227	0.99	Ma (1987)

[a] Values changed from wet weight as necessary (dry weight = wet weight × 3.5).

TABLE 19.3
Estimated Dietary Concentrations and Daily Cadmium Intake, as Well as Concentrations in Kidney and Liver, for Three Species of Small Mammals[a]

	Site[a]	Dietary Concentration (mg/kg dry wt.)	Cadmium Dose (mg/kg Body Weight Per Day)	Kidney Concentration (mg/kg dry wt.)	Liver Concentration (mg/kg dry wt.)	Kidney– Liver Ratio
Wood mouse	1	0.89	0.22	2.0	0.4	5
(*Apodemus*	2	1.4		8.5	1.8	4.7
sylvaticus)	3	3.01	0.94	41.7	18.2	2.3
Short-tailed field	1	0.67	0.35	1.7	0.7	2.4
vole (*Microtus*	2	1.3		23.9	8.7	2.7
agrestis)	3	3.3	2.11	88.8	22.7	3.9
Common shrew	1	1.78	0.58	20.5	13.6	1.5
(*Sorex araneus*)	2	16.0		156	245	0.63
	3	55.0	25.0	253	578	0.43

Sources: From Hunter, B. A. et al., *J. Appl. Ecol.,* 24, 601–614, 1987c; Hunter, B. A. et al., *J. Appl. Ecol.,* 26, 89–99, 1989. With permission.

[a] Site 1, reference site; site 2, 1 km from Cu/Cd refinery; site 3, close to Cu/Cd refinery.

Nearly all measurements of cadmium in the tissues of rodents caught in the wild have shown that the kidney has the highest concentrations, with the liver being the next most important in terms of concentration (Table 19.2). Table 19.2 also shows that, for most species, the ratios of kidney/liver concentration are between 2 and 8. There are obvious differences for shrews and the European mole (both insectivores), where the kidney/liver ratio is below 1.0, with both the kidney and liver concentrations considerably higher than those for the rodent species listed in Table 19.2.

It is important to relate the dietary concentrations and dose (on a body weight basis) to these tissue concentrations. Table 19.3 summarizes some data and the response of three sympatric species found in contaminated grasslands near a copper–cadmium refinery, compared with a clean reference site (Hunter et al. 1987c, 1989). Two species, the woodmouse (*A. sylvaticus*) and the short-tailed field vole (*M. agrestis*), in the contaminated site were exposed to a increase in dietary concentrations (between the range 0.7 and 3.3 mg/kg dry weight) or daily dose (between 0.2 and 2 mg/kg body weight), and the response in both was a considerable increase in their kidney and liver concentrations. Shrews (*S. araneus*) received a much higher level of dietary cadmium than did the other two species occupying the same grasslands in terms of both concentration and estimated dose (Table 19.3). The difference in the estimated dietary concentration between shrews and the other two species increased from about 2 times at the clean reference site to 18 times at the site closest to the refinery. This is because of the contrasting nature of feeding habits between the species and in the relatively high concentrations reached in earthworms and ground-dwelling invertebrates, the prey of the common shrew (Hunter et al. 1987b; Table 19.1). The response by the shrew to the higher cadmium intake rates is to accumulate higher concentrations in the liver compared with those in the kidney, so that the kidney/liver ratios are above one at the reference site but well below one at the two contaminated sites (Table 19.3).

A similar pattern in the relative magnitudes of the cadmium in kidney has been found in a number of studies of sympatric species in contaminated habitats allowing the comparison of shrews (*Sorex* spp.), voles (*Microtus* spp. and *Clethrionomys glareolus*) and mice (*Peromyscus* spp. and *Apodemus* spp.). Using the median values for the kidney cadmium ratios of shrew/vole or shrew/mouse as a comparative measure shrews had kidney cadmium concentrations between 8 and 24 times that of voles and mice (based on data in Hunter and Johnson 1982, Andrews et al. 1984, Hunter et al. 1989, Cooke et al. 1990, Ma et al. 1991, Nickleson and West 1996, Swiergosz-Kowalewska et al. 2005). Shore (1995), using published data, has taken this approach further and shown, for cadmium in kidney and liver tissue, that interspecific relationships occur between shrews, voles and mice in the same habitat such that the cadmium residues in any one species could be predicted from the values for one of the other species. Such interspecific extrapolation if combined with the knowledge of the kidney or liver critical concentrations for a species, especially for population parameters like impaired reproductive success, would be a powerful risk assessment tool for a rare or declining species.

In an analysis of published field studies, Shore and Douben (1994) showed a significant positive relationship (regression) between the calculated cadmium dose (daily intake) and cadmium residue magnitude in kidney and liver for field voles and shrews but not for wood mice. In a recent study of mine sites in Wales and Ireland there was no relationship between estimated dietary intakes of cadmium and kidney or liver concentrations in wood mice (Milton et al. 2004). An explanation is that the estimation of diet of wood mice in the field is probably more difficult than for shrews or voles especially as wood mice are likely to be less granivorous and more strongly omnivorous, selecting a wide variety of food items that vary in cadmium content, in disturbed and contaminated grassland ecosystems where the productivity is low. However, there is also evidence from a laboratory experiment with captive wood mice that there can be a surprisingly wide variation in cadmium absorption and accumulation in kidney and liver from feeding on a relatively uniform diet (Walker et al. 2002). These intraspecific differences in cadmium absorption could have a genetic basis, possible via differences in inducible tissue MT, and be applicable to most species of small mammals that have a relatively high level of genetic variation.

The long biological half-lives of cadmium in the kidney and liver mean that cadmium tissue concentrations do depend on the age of the animal and so are dependent on both dose and the duration of exposure. Age accumulation has been demonstrated in field-caught animals by comparing the often large significant differences in the kidney and liver cadmium concentrations between juveniles and adults, for example, for the greater white toothed shrew *Crocidura russula* (Sanchez-Chardi and Nadal 2007) and for *S. araneus and S. minutus* (Read and Martin 1993, Ma and Talmage 2001). It has been also been inferred by a significant positive relationship between body weight and kidney or liver concentration within a population sample, for example, for *S. araneus* (Hunter et al. 1989) and *Clethronomys glareolus* (Milton et al. 2003), and between total body burden (TBB) and body weight, for example, for the shrew, *Blarina brevicauda* (Schlesinger and Potter 1974); and for *S. araneus*, *M. agrestis*, and *A. sylvaticus* from a reference site and a restored mine site (Cooke et al. 1990). A few studies have aged the animals into year classes. For moles who live for about 4 years, Komarnicki (2000) was able to show an increase of 12 mg/kg dry weight per year in kidney cadmium. In badgers tooth wear was used to determine age and significant increases in kidney cadmium were shown with age leading to substantial concentrations in 5-year-old females with a mean kidney concentration of 203 mg/kg dry weight (Van den Brink and Ma 1998).

The relation between age and cadmium concentrations must be recognized when interpreting tissue concentrations, such as those given in Table 19.2, where sample populations were not separated into age groups. Further, age accumulation of cadmium could have significant ecotoxicological consequences: for predators who preferentially take adult prey (Walker et al. 2002); and for breeding success where this depends on older mature animals who territorially exclude young animals from breeding (Van den Brink and Ma 1998).

The absorption and retention of cadmium is clearly dose and duration dependent in wild small mammals. In young animals in the majority of contaminated sites, cadmium in the diet will be gradually accumulated in the kidney and liver and the majority of the total absorbed body burden will be in these two organs. It should be noted that the gut and its contents may make up a considerable percentage of the cadmium TBB available to a predator especially in a young juvenile prey animal; however, very little of this gut cadmium is likely to be absorbed and bound in the intestinal mucosa (Walker et al. 2002). Thus of the absorbed cadmium the kidney will have higher concentrations than does the liver and can be regarded as the critical target organ. However, this may change with very high doses of dietary cadmium, with kidney concentrations tending to reach a plateau level with the liver concentrations still increasing. It has been shown in field-caught shrews (*S. araneus*) that, with increasing TBBs of cadmium, the kidney/liver concentration ratios decrease, because the proportion of the body burden in the kidney decreases while the proportion in the liver increases as the TBB rises (Andrews and Cooke 1984, Hunter et al. 1989, Ma et al. 1991). This has been confirmed in feeding trials of a laboratory population of *S. araneus* fed cadmium chloride-contaminated diets, where the proportion of the TBB (over the range of 60–400 µg of cadmium) declined in the kidney from 17% to 8% and increased in the liver from about 60% to 80% (Dodds-Smith et al. 1992). Similar experimental findings are available for the laboratory rat and suggest that the saturation of the kidney is a general phenomenon in small mammals (Mitsumori et al. 1998).

As well as being a response to the magnitude of dietary cadmium, it may be that species differences highlighted in Tables 19.2 and 19.3 reflect differences in the cadmium bioavailability between the different components of the diet of different species of small mammals. This is still a difficult area to review and come to any firm conclusions. Differences in dietary cadmium speciation (e.g., metallothionein-bound, bound to other compounds as in ingested soil, or as soluble inorganic cadmium) could lead to differences both in the tissue levels of cadmium and in the relative tissue distribution in relation to the dietary dose of cadmium (Jackson and Alloway 1992, Walker et al. 2002). However, the similarities in the tissue levels of cadmium and in the relative distribution between the kidney and the liver in field-caught shrews feeding on invertebrates compared with shrews from laboratory experiments using cadmium chloride diets might suggest that dose together with basic physiological differences in the absorption of cadmium between species are as important as the dietary form of cadmium (Dodds-Smith et al. 1992). Further, the dietary levels of Zn, Fe, and Ca seem to have a considerable effect on cadmium retention and in laboratory experiments with the provision of only marginal nutritional status of these elements, cadmium retention increased by up to 10 fold (Reeves and Chaney 2002, 2004). An argument has been made that there is considerably higher cadmium accumulation in small mammals when dietary zinc is low and where the zinc/cadmium ratio is significantly lower than the natural ratio in soils and small mammal diets (Brown et al. 2002). The very high Cd levels in animals caught near the Cu–Cd Alloy Plant in northern England by Hunter et al. (see Tables 19.2 and 19.3) where high cadmium contamination was not accompanied by increased zinc contamination, as it is in most other contaminated sites and thus the cadmium/zinc ratio is low, is given as an example of this (Brown et al. 2002).

19.5 CRITICAL TISSUE CONCENTRATIONS

It is widely accepted that the kidney is the critical organ in chronic mammalian cadmium toxicity in that it has the highest cadmium concentrations and it is the first organ in which damage is observed or adverse functional changes start to occur. In humans, one third to one half of the TBB of cadmium may be in the kidney, with the concentration in the cortex being 1.25 times that of the whole organ (Friberg et al. 1986). The critical concentration of cadmium in the human kidney cortex at which tubular dysfunction and/or morphological kidney changes occur has been given as 200 mg/kg wet weight (Friberg et al. 1974). This threshold has been re-evaluated using *in vivo* neutron activation analysis and various metabolic models and is thought to remain the best estimate for the critical concentration at which renal dysfunction is likely to occur in 10% of the exposed human population (Friberg et al. 1986). In reviews of critical cadmium concentrations for renal dysfunction in experimental animals (mainly rats) values between 100 and 200 mg/kg wet weight have usually been cited (Fielder and Dale 1983, Nomiyama 1986, WHO 1992a, ATSDR 1999). Many experimental studies over 30 years using rats have supported this threshold of 100 mg/kg wet weight which equates to about 350 mg/kg on a dry weight basis (e.g., Tohyama et al. 1987, Liu et al. 1998, Mitsumori et al. 1998, Noel et al. 2004).

However, there are other experimental studies on laboratory rats and mice that suggest cadmium toxicity can be found at lower kidney concentrations. Values as low as 30–60 mg/kg wet weight (105–210 mg/kg dry weight) in the rat kidney have been associated with proteinuria (Prigge 1978), changes in urinary excretion of trace elements (Chmielnicka et al. 1989), and cell necrosis and degenerative changes in the proximal tubules (Itokawa et al. 1978, Aughey et al. 1984). At similar kidney concentrations in mice (110–260 mg/kg dry weight) severe tissue damage has been shown and although the animals were described by the authors as "outwardly healthy," there were changes in the proximal tubules including cell necrosis, nuclear pyknosis, and mitochondrial swelling (Nicholson et al. 1983). These studies have led to a number of authors using much lower kidney cadmium values of 100–120 mg/kg dry weight as the critical threshold to compare with wildlife mammalian data (e.g., Ma et al. 1991, Shore and Douben 1994).

It is clearly important in the use of a critical threshold kidney concentration that cadmium-induced changes are *actually* indicative of real adverse permanent damage and of significant physiological risks of cadmium toxicity. It is certainly the case that the measurement, for example, using Proton Nuclear Magnetic Resonance, of early biochemical changes in kidney tissue, such as changes in lipid concentrations and cellular acidosis, can be detected at very low tissue cadmium concentrations in bank voles (Griffin et al. 2000) and rats (Griffin et al. 2001). However, these changes are probably an adaptive tissue response to cadmium rather than the beginning of a toxic response leading to physiological (functional) kidney damage (Jones et al. 2007). Further, it is necessary to assess whether the magnitude of an adverse effect in a particular study as indicated by a biomarker of renal dysfunction (e.g., proteinuria), or even histopathological changes in kidney tissue is or is not of sufficient magnitude to lead to serious physiological dysfunction (ATSDR 1999). This perspective needs to be considered when evaluating the studies of cadmium toxicity in field studies of wild mammals. For example, in two studies of wild caught bank voles, where Leffler and Nyholm (1996) found increased proteinuria (but not an increase in creatinine clearance) associated with about 20 mg/kg dry weight (6 mg/kg wet weight) and Damek-Poprawa and Sawicka-Kapusta (2004) observed that the contraction of the proximal tubules occurred at kidney concentrations of 33 mg/kg dry weight.

Kidney cadmium levels and pathological injury are strongly affected by dose and duration (length of exposure), form of cadmium, and the route of exposure. One challenge with comparing subcutaneous injection (e.g., the studies of Chmielnicka et al. 1989 and Nicholson et al. 1983 above) with oral exposure is that the doses quoted for different exposure routes are not comparable in terms of the "absorbed dose," and the form of the cadmium reaching the liver and kidney may be different. Injection circumvents gut absorption and in particular whether cadmium is MT-bound in the tissue of the intestinal tract before it is transported to the kidney. This can affect the accumulation and toxic effects of cadmium in the kidney (Ohta et al. 2000, Liu et al. 2001).

If one focuses on longer term experimental studies of cadmium oral exposure, it is necessary to distinguish between acute (high dose usually short exposure) and chronic (lower dose with longer exposure). The magnitude of the dose can affect the toxicokinetics of absorption, storage, accumulation and toxic effects, for example, through the effects on the inducibility and rates of synthesis of metallothionein. A long-term experiment (60 weeks) using oral administration by Ohta et al. (2000), showed that with relatively low doses (2–10 mg/kg) more cadmium accumulated in the kidney than the high dose groups (20–60 mg/kg) where more accumulated in the liver. Further, in the high dose group renal dysfunction (proteinuria) was associated with kidney values of 30–60 mg/kg wet weight whereas there was no renal injury in the 2 mg/kg dose group with cadmium concentrations of over 120 mg/kg wet weight. The high dose groups also showed liver damage and a decrease in femur bone mineral density. Thus, it can be concluded that the rate of cadmium accumulation, and the cadmium saturation of, and subsequent release from, the intestinal mucosa, affect the distribution to and accumulation in the kidney, and the mechanisms of toxicity.

In one of the most comprehensive electron microscopy studies of wild animals, *S. araneus* from a polluted smelter site exhibited widespread kidney and liver damage (Hunter et al. 1984). Over the range of 150–560 mg/kg dry weight in the kidney, the degree of ultrastructural damage in proximal tubule cells, including enlargement of apical cytoplasmic vesicles and scattered cell necrosis, was found to correlate with the tissue concentration of cadmium. Limited glomerular damage also occurred throughout the full age range of the population; that is, this damage did not correlate with the kidney concentration. Corresponding liver concentrations of cadmium (300–1000 mg/kg) were associated with damage to hepatocytes that showed disrupted rough endoplasmic reticulum, swollen mitochondria, dilation of the smooth endoplasmic reticulum, and invagination of nuclei. Hepatocytes from adults contained numerous electron-dense cytoplasmic inclusion bodies with much of the cadmium bound to metallothionein (Hunter et al. 1984). Notwithstanding the tissue damage to both kidneys and liver, no evidence for clinical renal dysfunction could be found from analysis of urine, and the animals were seemingly in good condition when caught in the field (Hunter 1984). This was confirmed by experiments, which showed that kidney concentrations of 1000 mg/kg could be found with no adverse effects (Dodds-Smith et al. 1992). Thus, as shown in Tables 19.2 and 19.3, although shrews and other insectivorous species, such as the moles (*Talpidae*), often have the highest exposure to cadmium they are able to tolerate very high concentrations of cadmium in their kidneys and livers and may not be at the greatest toxicological risk.

Having considered the difficulties in assessing what are the significant physiological risks of cadmium to the health of the animal the next fundamental consideration is understanding whether kidney damage (and that to liver and other tissues) is related to ecological fitness. The kidney does have spare functional capacity, and its regenerative capacity is great (Nicholson and Osborn 1983). Similarly, cadmium-induced proteinuria, although it may persist throughout life, may be a condition that can be tolerated and is not necessarily indicative of progression to a more serious condition, such as renal failure. However, it can be argued with equal validity that tissue damage in wild small mammals subject to predation, food shortages and marginal nutrition, low temperatures, etc., can be of much greater significance than equivalent damage in relatively inactive, disease-free, well-fed laboratory rodents. Because of these uncertainties, it has been concluded, in terms of environmental exposure, that cadmium is an element "looking for a disease to cause" (Davies 1992) and that "there has been a tendency in the ecotoxicological literature to exaggerate the toxicity of cadmium" (Beyer 2000).

If one summarizes the complex dose–effect aspects of cadmium accumulation in the kidney, it would seem that at chronic doses commonly encountered by wild small mammals in contaminated habitats say up to about 10 mg/kg body weight, then because of the normal toxicodynamics, renal accumulation will dominate and the previously proposed critical limits of serious adverse effects would seem to still be appropriate (Cooke and Johnson 1996). Thus, 100 mg of cadmium per kilogram wet weight or 350 mg of cadmium per kilogram dry weight in the whole kidney is taken as an appropriate critical cadmium concentration for serious adverse effects in wild small mammals. However, at higher doses (which might be regarded as high chronic or subacute dose) and higher rates of absorption, serious nephrotoxicity can occur at lower kidney cadmium concentrations. Here the critical limits which have commonly been quoted in the range of 30–60 mg/kg wet weight and 105–210 mg/kg dry weight may be appropriate. In this latter situation the ratio of kidney/liver cadmium concentrations will be lower and generally less than 1. This latter range would also characterize the band of tissue concentrations where wild small mammals receiving a chronic long-term dose would show less serious (ATSDR 1999) adverse toxic effects which generally do not affect survivorship or reproduction.

Summary

Cadmium is a widespread element in the environment, but at low concentrations. Increased cadmium in terrestrial habitats of small mammals is largely derived from anthropogenic sources including atmospheric deposition, the application of phosphatic fertilizers and sewage sludges to land, and disused mine waste. Field studies have shown that cadmium, whether derived from the atmosphere or soil, is commonly found in most biotic components within a terrestrial ecosystem.

Concentrations measured in wild small mammals caught in contaminated sites show elevated cadmium in many tissues and organs, but most of the body burden is in the kidney and liver. Kidney concentrations ranged up to 70 mg of cadmium per kilogram dry weight in mice and 89 mg/kg in voles where the corresponding liver values were 19 and 23 mg/kg dry weight respectively. In shrews, kidney values were higher ranging up to 250 mg of cadmium per kilogram dry weight and liver concentrations were higher than kidney ranging up to 578 mg of cadmium per kilogram. This pattern of organ accumulation in shrews can be explained by the high dietary exposure because of food chain accumulation, higher absorbed dose, and the saturation of the kidney as the primary site of accumulation and the subsequent increase in liver concentration. Shrews would seem to be high cadmium accumulators without serious symptoms of cadmium toxicity.

Using primarily laboratory experiments with rats, mice, captive wild species, at chronic oral doses equivalent to those encountered by wild small mammals in contaminated habitats, 100 mg of cadmium per kilogram wet weight or 350 mg of cadmium per kilogram dry weight could be considered as the critical kidney concentration on a whole-organ basis for serious adverse effects that can effect survivorship and ecological fitness. However, at higher (subacute) cadmium doses serious nephrotoxicity can occur at lower kidney cadmium concentrations. Here, critical limits in the range of 30–60 mg/kg wet weight and 105–210 mg/kg dry weight may be more appropriate. This latter range also characterizes the band of critical kidney cadmium concentrations where wild small mammals receiving chronic oral doses typical of contaminated habitats, show less serious adverse biochemical and physiological effects including renal tubule damage and proteinuria.

ACKNOWLEDGMENT

I thank Alicia Nadesan for the assistance with the literature search for this chapter.

REFERENCES

Alfven, T., L. Jarup, and C. Elinder, 2002. Cadmium and lead in blood in relation to low bone mineral density and tubular proteinuria. *Environ. Health Perspect.*110:699–702.

Anderson, T. J., G. W. Barren, C. S. Clark, V. J. Eiia, and V. A. Majeti. 1982. Metal concentrations in tissues of meadow voles from sewage sludge-treated fields. *J. Environ. Qual.* 11:272–277.

Andrews, S. M., and J. A. Cooke. 1984. Cadmium within a contaminated grassland ecosystem established on metalliferous mine waste. In *Metals in animals. Inst. Terrestrial Ecol. Publ. No. 12,* ed. D. Osborn, pp. 11–15. Cambridge, U.K.: Natural Environment Research Council.

Andrews, S. M., M. S. Johnson, and J. A. Cooke. 1984. Cadmium in small mammals from grassland established on metalliferous mine waste. *Environ. Pollut. Ser. A. Ecol. Bio.* 33:153–162.

Anthony, R. G., and R. Kozlowski. 1982. Heavy metals in tissues of small mammals inhabiting waste-water-irrigated habitats. *J. Environ. Qual.* 11:20–22.

ATSDR. 1999. Toxicological profile for cadmium. Agency for Toxic Substances and Disease Registry, Atlanta, GA, USA.

Aughey, E., G. S. Fell, R. Scott, and M. Black. 1984. Histopathology of early effects of oral cadmium in the rat kidney. *Environ. Health Perspect.* 54:153–161.

Beyer, W. N. 2000. Hazards to wildlife from soil-borne cadmium reconsidered. *J. Environ. Qual.* 29:1380–1384.

Beyer, W. N., and G. Storm. 1995. Ecotoxicological damage from Zinc Smelting at Palmerton, Pennsylvania. In *Handbook of ecotoxicology*, eds. D. J. Hoffman, B. A. Rattner, G. A. Burton and J. Cairns, pp. 596–608. Boca Raton: Lewis.

Bremner, I. 1979. Mammalian absorption, transport, and excretion of cadmium. In *The chemistry, biochemistry, and biology of cadmium*, ed. M. Webb, pp. 175–193. Amsterdam: Elsevier/North-Holland Biomedical Press.

Brown, S., R. L. Chaney, M. Sprenger, and H. Compton. 2002. Assessing impact to wildlife at biosolid remediated sites. *Biocycle* August:50–58.

Chaney, R. L., R. J. Bruins, D. E. Baker, R. F. Korcak, J. E. Smith, and D. Cole. 1987. Transfer of sludge-applied trace elements to the food chain. In *Land application of sludge: Food chain implications*, eds. A. L. Page, T. J. Logan, and J. A. Ryan. Chelsea, MI: Lewis Publishers.

Chmielnicka, J., T. Halatek, and U. Jedlinska. 1989. Correlation of cadmium-induced nephropathy and the metabolism of endogenous copper and zinc in rats. *Ecotoxicol. Environ. Saf.* 18:268–276.

Cooke, J. A., and M. S. Johnson. 1996. Cadmium in small mammals. In *Environmental contamination in wildlife*, eds. W. N. Beyer, G. H. Heinz, and A. W. Redmond-Norwood, pp. 377–388. Boca Raton: Lewis Publishers.

Cooke, J. A., S. M. Andrews, and M. S. Johnson. 1990. Lead, zinc, cadmium, and fluoride in small mammals from contaminated grassland established on fluorspar tailings. *Water Air Soil Pollut.* 51:43–54.

Damek-Poprawa, M., and K. Sawicka-Kapusta. 2004. Histopathological changes in the liver, kidneys, and testes of bank voles environmentally exposed to heavy metal emissions from the steelworks and zinc smelter in Poland. *Environ. Res.* 96:72–78.

Davies, B. E. 1992. Trace metals in the environment: retrospect and prospect. In *Biogeochemistry of trace metals*, ed. D. C. Adriano, pp. 1–17. Boca Raton, FL: Lewis Publishers.

Dodds-Smith, M. E., M. S. Johnson, and D. J. Thompson. 1992. Trace metal accumulation by the shrew, *Sorex araneus.* II. Tissue distribution in kidney and liver. *Ecotoxicol. Environ. Saf.* 24:118–130.

Fielder, R. J., and E. A. Dale. 1983. Cadmium and its compounds. *Toxicity Review No. 7. Health and Safety Executive.* Her Majesty's Stationery Office, London.

Friberg, L., T. Kjellstrom, and G. F. Nordberg. 1986. Cadmium. In *Handbook on the toxicology of metals,* 2nd ed, eds. L. Friberg, G. F. Nordberg, and V. Vouk. Amsterdam: Elsevier Science Publishers.

Friberg, L., M. Piscator, G. F. Nordberg, and T. Kjellstrom. 1974. *Cadmium in the environment*, 2nd ed. Cleveland, OH: CRC Press.

Griffin, J. L., L. A. Walker, J. Troke, D. Osborn, R. F. Shore, and J. K. Nicholson. 2000. The initial pathogenesis of cadmium induced renal toxicity. *FEBS Lett.* 478:147–150.

Griffin, J. L., L. A. Walker, R. F. Shore, and J. K. Nicholson. 2001. Metabolic profiling of chronic cadmium exposure in the rat. *Chem. Res. Toxicol.* 14:1428–1434.

Habeebu, S. S., J. Liu, Y. Liu, and C. D. Klaassen. 2000. Metallothionein-null mice are more sensitive than wild-type mice to liver injury induced by repeated exposure to cadmium. *Toxicol. Sci.* 55:223–232.

Hughes, M. K., N. W. Lepp, and D. A. Phipps. 1980. Aerial heavy metal pollution in terrestrial ecosystems. *Adv. Ecol. Res.* 11:218–327.

Hunter, B. A. 1984. The Ecology and Toxicology of Trace Metals in Contaminated Grasslands. Ph.D. thesis. University of Liverpool, Liverpool, England.

Hunter, B. A., and M. S. Johnson. 1982. Food chain relationships of copper and cadmium in contaminated grassland ecosystems. *Oikos* 38:108–117.

Hunter, B. A., M. S. Johnson, and D. J. Thompson. 1984. Cadmium induced lesions in tissues of Sorex araneus from metal refinery grasslands. In *Metals in animals. Inst. Terrestrial Ecol. Publ. No.12,* ed. D. Osborn, pp. 39–44. Cambridge, UK: Natural Environment Research Council.

Hunter, B. A., M. S. Johnson, and D. J. Thompson. 1987a. Ecotoxicology of copper and cadmium in a contaminated grassland ecosystem. I. Soil and vegetation contamination. *J. Appl. Ecol.* 24:573–586.

Hunter, B. A., M. S. Johnson, and D. J. Thompson. 1987b. Ecotoxicology of copper and cadmium in a contaminated grassland ecosystem. II. Invertebrates. *J. Appl. Ecol.* 24:587–599.

Hunter, B. A., M. S. Johnson, and D. J. Thompson. 1987c. Ecotoxicology of copper and cadmium in a contaminated grassland ecosystem. III. Small mammals. *J. Appl. Ecol.* 24:601–614.

Hunter, B. A., M. S. Johnson, and D. J. Thompson. 1989. Ecotoxicology of copper and cadmium in a contaminated grassland ecosystem. IV. Tissue distribution and age accumulation in small mammals. *J. Appl. Ecol.* 26:89–99.

Hutton, M. 1983. Sources of cadmium in the environment. *Ecotoxicol. Environ. Saf.* 7:9–24.

Itokawa, Y., K. Nishino, M. Takashima. T. Nakata, H. Kaito, E. Okamoto, K. Daijo, and J. Kawamura. 1978. Renal and skeletal lesions in experimental cadmium poisoning of rats. Histology and renal function. *Environ. Res.* 15:206–217.

Jackson, A. P., and B. J. Alloway. 1992. The transfer of cadmium from agricultural soils to the human food chain. In *Biogeochemistry of trace metals,* ed. D. C. Adriano, pp. 109–158. Boca Raton, FL: Lewis Publisher.

Johnson, M. S., R. D. Roberts, M. Hutton, and M. J. Inskip. 1978. Distribution of lead, zinc, and cadmium in small mammals from polluted environments. *Oikos* 30:153–159.

Jones, O. A. H, L. A. Walker, J. K. Nicholson, R. F. Shore, and J. L. Griffin. 2007. Cellular acidosis in rodents exposed to cadmium is caused by adaptation of the tissue rather than an early effect of toxicity. *Comp. Biochem. Physiol. Part D* 2:316–321.

Komarnicki, G. J. K. 2000. Tissue, sex and age specific accumulation of heavy metals (Zn, Cu, Pb, Cd) by populations of the mole (*Talpa europeae* L.) in a central urban area. *Chemosphere* 41:1593–1602.

Korte, F. 1982. Ecotoxicology of cadmium. *Regul. Toxicol. Pharmacol.* 2:184–208.

Leffler, P. E., and N. E. I. Nyholm. 1996. Nephrotoxicity effects in free-living bank voles in a heavy metal polluted environment. *Ambio* 25:471–420.

Levengood, J. M., and E. J. Heske. 2008. Heavy metal exposure, reproductivity, activity, and demographic patterns in white-footed mice (*Peromyscus leucopus*) inhabiting a contaminated floodplain wetland. *Sci. Total Environ.* 389:320–328.

Liu, J., S. L. Habeebu, Y. Liu, and C. D. Klaasen 1998. Acute CdMT nephropathy: comparison of chronic CdCl2 and CdMT exposure with acute CdMT injection in Rats. *Toxicol. Appl. Pharmacol.* 153:48–58.

Liu, Y., J. Liu, and C. D. Klaasen. 2001. Metallothionein-null and wild-type mice show similar cadmium absorption and tissue distribution following oral cadmium administration. *Toxicol. Appl. Pharmacol.* 175:253–259.

Ma, W. C. 1987. Heavy metal accumulation in the mole *Talpa europea* and earthworms as an indicator of metal bioavailability in terrestrial environments. *Bull. Environ. Contam. Toxicol.* 39:933–938.

Ma, W. C., W. Denneman, and J. Faber. 1991. Hazardous exposure of ground-living small mammals to cadmium and lead in contaminated terrestrial ecosystems. *Arch. Environ. Contam. Toxicol.* 20:266–270.

Ma, W. C., and S. S. Talmage. 2001. Insectivora. In *Ecotoxicology of wild mammals, ecotoxicological & environmental toxicology series,* eds. R. F. Shore and B. A Rattner, pp. 123–158. New York, NY: John Wiley and Sons Ltd.

Maly, M. S. 1984. Survivorship of meadow voles *Microtus pennsylvanicus* from sewage sludge-treated fields. *Bull. Environ. Contam. Toxicol.* 32:724–731.

Martin, M. H., and P. J. Coughtrey. 1982. *Biological monitoring of heavy metal pollution.* London: Applied Science Publishers.

Martin, M. H., E. M. Duncan, and P. J. Coughtrey. 1982. The distribution of heavy metals in a contaminated woodland ecosystem. *Environ. Pollut. Ser. B. Chem. Phys.* 3:147–157.

McLellan, J. S., P. R. Flanagan, M. J. Chamberlain, and L. S. Valberg. 1978. Measurement of dietary cadmium absorption in humans. *J. Toxicol. Environ. Health* 4:131–138.

Milton, A., J. A. Cooke, and M. S. Johnson. 2003. Accumulation of lead, zinc, and cadmium in a wild population of *Clethrionomys glareolus* from an abandoned lead mine. *Arch. Environ. Contam. Toxicol.* 44:405–411.

Milton, A., J. A. Cooke, and M. S. Johnson. 2004. A Comparison of cadmium in ecosystems on metalliferous mine tailings in Wales and Ireland. *Water Air Soil Pollut.* 153:157–172.

Mitsumori, K., M. Shibutani, S. Satoh, H. Onodera, J. Nakawa, Y. Hayashi, and M. Ando. 1998. Relationship between the development of hepato-renal toxicity and cadmium accumulation in rats given minimum to large amounts of cadmium chloride in the long term: preliminary study. *Arch. Toxicol.* 72:545–552.

Nicholson, J. K., and D. Osborn. 1983. Kidney lesions in pelagic seabirds with high tissue levels of cadmium and mercury. *J. Zool. (Lond.)* 200:99–118.

Nicholson, J. K., M. D. Kendall, and D. Osborn. 1983. Cadmium and mercury nephrotoxicity. *Nature (Lond.)* 304:633–635.

Nickelson, S. A., and S. D. West. 1996. Renal cadmium in mice and shrews collected from forest lands treated with biosolids. *J. Environ. Qual.* 25:86–91.

Noel, L., T. Guerin, and M. Kolf-Clauw. 2004. Subchronic dietary exposure of rats to cadmium alters the metabolism of metals essential to bone health. *Food Chem. Toxicol.* 421:1203–1210.

Nomiyama, K. 1986. The chronic toxicity of cadmium: influence of environmental and other variables. In *Handbook of experimental pharmacology, Vol. 80*, ed. E. C. Foulkes, pp. 101–133. Berlin: Springer-Verlag.

Nriagu, J. O. 1979. Global inventory of natural and anthropogenic sources of metals in the atmosphere. *Nature (Lond.)* 279:409–411.

Ohta, H., Y. Yamauchi, M. Nakakita, H. Tamaka, S. Asami, Y. Seki, and H. Yoshikawa. 2000. Relationship between renal dysfunction and bone mineral disorder in male rats after long-term oral quantitative cadmium administration. *Ind. Health* 38:339–355.

Pascoe, G. A., R. J. Blancher, and G. Linder. 1996. Food chain analysis of exposures and risks to wildlife at a metal-contaminated wetland. *Arch. Environ. Contam. Toxicol.* 30:306–318.

Pedersen, S., and S. Lierhagen. 2006. Heavy metal accumulation in arctic hares (*Lepus arcticus*) in Nunavut, Canada. *Sci. Total Environ.* 368:951–955.

Prigge, E. 1978. Early signs of oral and inhalative cadmium uptake in rats. *Arch. Toxicol.* 40:231–247.

Read, H. J., and M. H. Martin. 1993. The effect of heavy metals on populations of small mammals from woodlands in Avon (England) with particular emphasis on metal concentrations in *Sorex araneus* L. and *Sorex minutus* L. *Chemosphere* 27:2197–2211.

Reeves, P. G., and R. L. Chaney. 2002. Nutritional status affects the absorption and whole-body and organ retention of cadmium in rats fed rice-based diets. *Environ. Sci. Technol.* 36:2684–2692.

Reeves, P. G., and R. L. Chaney. 2004. Marginal nutritional status of zinc, iron, and calcium increases cadmium retention in the duodenum and other organs of rats fed rice-based diets. *Environ. Res.* 96:311–322.

Roberts, R. D., and M. S. Johnson. 1978. Dispersal of heavy metals from abandoned mine workings and their transference through terrestrial food chains. *Environ. Pollut.* 16:293–310.

Ryan, J. A., H. R. Pahren, and J. B. Lucas. 1982. Controlling cadmium in the human food chain: a review and rationale based on health effects. *Environ. Res.* 28:251–302.

Samarawickrama, G. P. 1979. Biological effects of cadmium in mammals. In *The chemistry, biochemistry, and biology of cadmium,* ed. M. Webb, pp. 341–421. Amsterdam: Elsevier/North-Holland Biomedical Press.

Sanchez-Chardi, A., C. C. Marques, J. Nadal, and M. da Luz Mathias. 2007. Metal bioaccumulation in the greater white-toothed shrew, *Crocidura russula*, inhabiting an abandoned pyrite mine site. *Chemosphere* 67:121–130.

Sanchez-Chardi, A., and J. Nadal 2007. Bioaccumulation of metals and effects of landfill pollution in small mammals. Part I. The greater white-toothed shrew, *Crocidura russula*. *Chemosphere* 68:703–711.

Sawicka-Kapusta, K., R. Swiergosz, and M. Zakrzewska. 1990. Bank voles as monitors of environmental contamination by heavy metals. A remote wilderness area in Poland imperilled. *Environ. Pollut.* 67:315–324.

Sawicka-Kapusta, K., M. Zakrzewska, and T. Orzechowski. 1999. Seasonal changes of metal concentration in small mammals' populations from different contaminated areas. In *Proc. SECOTOX 99. Fifth Eur. Conf. Ecotoxicol. Environ. Safety*, eds. A. Ketrup and K.-W. Schramm. March 15–17, Munich.

Schlesinger, W. H., and G. L. Potter. 1974. Lead, copper, and cadmium concentrations in small mammals in the Hubbard Brook Experimental Forest. *Oikos* 25:148–152.

Schnellman, R. G. 2001. Toxic response of the kidney. In *Casarett and Doull's Toxicology—the basic science of poisons*, ed. C. D. Klaassen. New York: McGraw-Hill.

Shore, R. F. 1995. Predicting cadmium, lead and fluoride levels in small mammals from soil residues and by species–species extrapolation. *Environ. Pollut.* 88:333–340.

Shore, R. F., and P. E. T. Douben. 1994. The ecotoxicological significance of cadmium intake and residues in terrestrial small mammals. *Ecotoxicol. Environ. Saf.* 29:101–112.

Swiergosz-Kowalewska, R., M. Gramatyka, and W. Reczynski. 2005. Metals distribution and interactions in tissues of shrews (*Sorex* spp.) from copper- and zinc-contaminated areas in Poland. *J. Environ. Qual.* 34:1519–1529.

Talmage, S. S., and B. T. Walton. 1991. Small mammals as monitors of environmental contaminants. *Rev. Environ. Contam. Toxicol.* 119:47–145.

Tohyama C., N. Sugihira, and H. Saito. 1987. Critical concentration of cadmium for renal toxicity in rats. *J. Toxicol. Environ. Health* 22:255–259.

Van den Brink, N. W., and W. C. Ma. 1998. Spatial and temporal trends in levels of trace metals and PCBs in the European badger Meles meles (L., 1758) in The Netherlands: implications for reproduction. *Sci. Total Environ.* 222:107–118.

Walker, L. A., L. J. Bailey, and R. F. Shore. 2002. The importance of the gut and its contents in prey as a source of cadmium to predators. *Environ. Toxicol. Chem.* 21:76–80.

WHO. 1992a. *Environmental Health Criteria 134 Cadmium.* Geneva: WHO.

WHO. 1992b. *Environmental Health Criteria135 Cadmium-Environmental Aspects.* Geneva: WHO.

Zalups, R. K., and S. Ahmad. 2003. Molecular handling of cadmium in transporting epithelia. *Toxicol. Appl. Pharmacol.* 186:163–188.

Eider Ducks

By G. Mutzel, from *The Royal Natural History,* edited by Richard Lydekker, Frederick Warne & Co., London, 1893–94.

20 Cadmium in Birds

Mark Wayland
Anton M. Scheuhammer

CONTENTS

20.1   Introduction..645
20.2   Uptake and Elimination 646
20.3   Metallothionein Induction..647
20.4   Tissue Distribution and Accumulation...648
20.5   Toxic Effects 651
20.6   Intestinal Damage and Nutrient Uptake...651
20.7   Kidney Damage and Vitamin D Metabolism 652
20.8   Skeletal Effects 652
20.9   Effects on Osmoregulation...653
20.10 Effects on Energy Metabolism...653
20.11 Effects on Reproductive Organs and Reproduction...................................654
20.12 Endocrine Disruption 654
20.13 Anemia 655
20.14 Behavioral Alterations..655
20.15 Association between Tissue Concentrations and Effects 655
Summary 659
References...660

20.1   INTRODUCTION

Cadmium is a toxic metal with no known nutritional value. In the northern hemisphere cadmium emissions to the atmosphere rose steeply following the onset of the Industrial Revolution and peaked during the 1960s and 1970s. By the 1990s cadmium emissions had declined to about one-half of peak values (Boutron 1995). Human activities were estimated to contribute 2.3 times more cadmium to the atmosphere than natural emissions by the mid-1990s (Pacyna and Pacyna 2001). Moreover, the anthropogenic inputs of cadmium into the aquatic environment were estimated to greatly exceed those into the atmosphere (Nriagu and Pacyna 1988).

Despite recent declines in anthropogenic cadmium releases, human activities continue to contribute to cadmium pollution at local and regional scales (Rice 1999, Müller et al. 2000, Mahler et al. 2006). Consequently, cadmium levels are often higher in biota near heavily urbanized and industrialized sites or near sites with historical evidence of mining pollution than in biota in more pristine areas (Schmitt and Brumbaugh 1990, Beyer and Storm 1995, Blus et al. 1995, García-Fernández et al. 1996, Besser et al. 2007, Peltier et al. 2008).

Birds accumulate cadmium in their tissues, primarily through their diets (Scheuhammer 1987, García-Fernández et al. 1995, Burger 2008). Thus, there is concern that environmental cadmium pollution could result in avian exposure to potentially toxic levels of the element (Anderson and Van Hook, Jr. 1973, Bryan and Langston 1992). That concern has created a need for assessment of the toxicological risk that cadmium poses to birds. One approach to assessing risk from cadmium exposure is to identify critical tissue thresholds as determined in laboratory and field studies and to relate such concentrations to those known to occur in free-living birds.

In an earlier edition of this book, Furness (1996) reviewed the relationships between tissue cadmium concentrations and toxic effects in birds to identify critical tissue thresholds. He suggested that 40 mg/kg in liver or 100 mg/kg in kidneys (wet weight) should be considered tentative threshold concentrations and further indicated that such thresholds could be higher for pelagic seabirds. Our purpose is to update Furness' work by reviewing the older literature together with more recent research that has appeared since 1996. The following major topics are considered in the chapter: uptake and elimination, metallothionein (MT) induction, tissue distribution and accumulation, toxic effects and relationships between tissue concentrations and toxic effects.

20.2   UPTAKE AND ELIMINATION

Absorption of cadmium via the respiratory route (7–40%) is higher than intestinal absorption (<1–7%) the latter of which is dose-dependent with greater proportions being absorbed at higher doses (Sell 1975, Koo et al. 1978, Friberg et al. 1986, Scheuhammer 1988). Although cadmium is more effectively taken up via the respiratory route than via the diet, inhalation is unimportant in contributing to cadmium body burdens in free-living animals except possibly in areas of severe air pollution (Blanco et al. 2003, Burger 2008).

Dietary essential elements and micronutrients affect intestinal absorption of cadmium. For example, low levels of dietary calcium enhanced intestinal uptake of cadmium (Koo et al. 1978) which could explain why higher levels of cadmium were found in kidneys and livers of birds on low-calcium diets than of birds on high-calcium diets (Silver and Nudds 1995, Scheuhammer 1996). Cholecalciferol supplementation and low levels of dietary phosphorus were reported to increase cadmium absorption (Koo et al. 1978) while iron (II) and ascorbic acid supplementation decreased cadmium absorption (Fox et al. 1980). Maintenance of moderate iron stores appears to be important in minimizing cadmium absorption. The effect of ascorbic acid on cadmium absorption may depend primarily on its influence in improving iron absorption (Fox et al. 1980). Cadmium uptake was inversely proportional to dietary zinc concentration (Fox et al. 1984, McKenna et al. 1992) while copper- and iron-deficient diets increased cadmium levels in kidney, presumably due to effects of deficiencies of these essential elements on cadmium absorption in the small intestine (Fox et al. 1984).

Once absorbed, cadmium enters the bloodstream bound to high or low molecular weight proteins and is transported to the two main organs in which it is stored, the liver and the kidneys where it binds mainly with MT (Friberg et al. 1986). Radiolabeled [109]Cd was not detected in blood of laying hens (*Gallus domesticus*) later than 48 h after injection, indicating the rapidity with which cadmium is removed from circulation (Sell 1975). In ducks that had originally received various levels of dietary cadmium for 90 days, blood concentrations of cadmium declined by 15–70% following a period of 30 days on a "clean" diet (White and Finley 1978).

After being deposited in kidneys, cadmium is eliminated very slowly. Cadmium levels in kidneys, liver, and gonads of ducks that had been fed diets containing various levels of cadmium for 90 days, did not decline following a 30-day period on a "clean" diet (White and Finley 1978). The whole-body half-life of cadmium in adult chipping sparrows (*Spizella passerina*) was 99 days, a period that was judged to be sufficiently long as to allow for accumulation of cadmium to levels exceeding those in the diet (Anderson and Van Hook, Jr. 1973). In contrast, the half-life of cadmium was only 30 days in kidneys of ducklings that had been fed diets containing toxic levels of cadmium (Mayack et al. 1981). However, those authors speculated that the relatively rapid rate of cadmium elimination was likely the result of cadmium-induced kidney damage, which may have allowed cadmium to be excreted as a low-molecular weight protein complex. The major route of elimination of ingested cadmium is by gut clearance (Anderson and Van Hook, Jr. 1973). Feather growth, uropygial gland and, in seabirds, salt gland secretions are routes of elimination of absorbed cadmium (Burger 1993, Pilastro et al. 1993, Bennett et al. 2000, Burger et al. 2000). Eggs are not considered to be an important source of cadmium elimination (Sell 1975, White and Finley 1978, Sato et al. 1997). The importance of urinary excretion has not been examined in birds.

20.3 METALLOTHIONEIN INDUCTION

MTs are low molecular weight (~6500 daltons), cysteine-rich metal-binding proteins, first identified as Cd/Zn-binding proteins in horse kidney tissue (Margoshes and Vallee 1957, Kagi and Vallee 1960), and later shown to be synthesized in a number of tissues by a wide variety of animals, both vertebrate and invertebrate (Chatterjee and Maiti 1990). The synthesis of MT can be induced in tissues (especially liver and kidney) by exposure to metals, the most effective inducers being the essential trace metals zinc and copper, and the nonessential metals cadmium and (inorganic) mercury. MT functions in the normal storage, distribution, and metabolic activity of zinc and copper (Carpene et al. 2007) and also serves to protect cells from the damaging effects of cadmium by binding cadmium in a stable complex with a long biological half-life (Klaassen et al. 1999). In birds (and mammals), a majority of the body burden of cadmium is localized in the liver and kidneys bound to MT; and the long biological half-life of the Cd–MT complex largely accounts for the often-reported tendency for cadmium to accumulate with age. Scheuhammer and Templeton (1990) compared liver and kidney of ring doves (*Streptopelia risoria*) with respect to MT production in response to chronic, environmentally realistic dietary cadmium exposure, reporting that kidney was apparently only 35% as responsive as liver to tissue-cadmium accumulation. However, when all potentially relevant metals (Cd + Zn + Cu) were taken into account, liver and kidney were found to be almost equally responsive to dietary cadmium accumulation. The apparently greater responsiveness of liver was due to a greater coaccumulation of zinc with cadmium in liver, which was associated with additional MT synthesis, compared to kidney.

The kidney is widely recognized as the major target organ of chronic cadmium exposure and accumulation. As cadmium is taken up by the kidney, it induces the synthesis of renal MT to which it subsequently binds; through time, cadmium and MT concentrations of the renal cortex increase. Bound to cytosolic MT, intracellular cadmium is rendered largely nontoxic; however, as cadmium and MT concentrations continue to increase, a threshold is reached above which the ability of MT to protect cells against cadmium toxicity declines, and nephropathy characterized by proximal tubular necrosis, proteinuria, increased urinary cadmium, decreased kidney-cadmium, and the appearance of MT in plasma occurs. Although there is uncertainty regarding the levels of cadmium and MT that are associated with a commencement of toxicity (see Section 20.15), without the presence of MT, toxic thresholds would almost certainly be achieved at much lower levels of tissue cadmium. Klaassen et al. (1999) suggested that approximately 7% of the human population currently has some degree of renal dysfunction from cadmium exposure; thus, if humans did not synthesize MT, normal dietary cadmium exposure would be sufficiently high to be nephrotoxic for a large proportion of the human population. The same conclusion can probably be made for numerous wild bird species, some of which demonstrate quite elevated kidney cadmium concentrations.

MT concentrations have been measured in livers and/or kidneys of at least 19 free-living avian species, primarily sea birds and sea ducks but also one terrestrial passerine species, the Great Tit (*Parus major*) (Elliott et al. 1992, Stewart et al. 1996, Elliott and Scheuhammer 1997, Trust et al. 2000, Debacker et al. 2001, Barjaktarovic et al. 2002, Braune and Scheuhammer 2008, Vanparys et al. 2008). In virtually all of these studies, renal concentrations of cadmium and MT were positively and significantly correlated, although sometimes, consistent with Scheuhammer and Templeton (1990), the correlation with MT was improved by accounting for all MT-inducing metals (Cd + Zn + Cu) rather than cadmium alone (e.g., Trust et al. 2000).

Based on the known protective role of MT against cadmium toxicity, it might be asked, for studies on wild birds where both cadmium and MT were reported, whether measured MT concentrations were generally sufficient to bind all the cadmium, and thus protect the tissue from toxicity. *In vivo*, vertebrate MTs typically bind metals in two clusters, one mol MT binding a total of 7 g-atoms metal in a 4-metal α-cluster, plus a 3-metal β-cluster (Furey et al. 1986); thus, a theoretical maximum of 7 g-atoms cadmium can be sequestered per mol MT. However, native MTs are usually heterogeneous in metal composition, with zinc, cadmium, and copper occurring in varying proportions. In ring doves chronically exposed to different levels of dietary cadmium, approximately 0.5 μmol zinc and 0.3 μmol copper are accumulated in kidney tissue per μmol cadmium (Scheuhammer and Templeton 1990); and Braune and Scheuhammer (2008) reported that both cadmium and zinc concentrations were positively correlated with MT in kidneys of Arctic seabirds. In mixed Cd/Zn-MT, the 4-metal α-cluster is generally considered to be an obligate Zn-containing cluster. Thus, it is probably incorrect to assume that MT synthesized in response to cadmium exposure *in vivo* will be fully saturated with cadmium; it is more likely that a maximum of 4–5 g-atoms cadmium will be sequestered per mol MT in free-living birds exposed to dietary cadmium, with other binding sites occupied by zinc and/or copper. The highest concentrations of cadmium and MT reported in individual free-living seabirds (Leach's storm petrel (*Oceanodroma leucorhoa*), thick-billed murre, and black-legged kittiwake (Elliott et al. 1992, Braune and Scheuhammer 2008) were approximately 75–100 and 1000–1500 μg/g, respectively. On a molar basis, these concentrations correspond to an average of about 3–4 g-atoms Cd per mol MT, indicating sufficient MT for full sequestration of accumulated intracellular cadmium and little expectation of renal toxicity at these, or lower, concentrations of cadmium.

20.4 TISSUE DISTRIBUTION AND ACCUMULATION

Throughout this chapter, cadmium concentrations are presented on a wet weight basis unless otherwise indicated, following Furness (1996). When literature values were reported on a dry weight basis, we converted them to approximate wet weight values using mean values for tissue moisture content that were provided by the cited publications. Alternatively, when publications did not provide moisture contents, we used mean tissue moisture contents from Scanlon (1982).

Liver and kidneys account for approximately 67–97% of the total body burden of cadmium in birds (White and Finley 1978, Scheuhammer 1987, García-Fernández et al. 1996, Nam et al. 2005) while feathers contain 1–28% of the body burden (Burger 1993, Agusa et al. 2005, Nam et al. 2005). However, concentrations of cadmium in feathers are likely affected not only by internal biological sequestering but also by external contamination (see later). The latter source would lead to inflated estimates of the proportion of the body burden in feathers. Concentrations of cadmium are normally greatest in kidney. Mean cadmium concentrations in liver ranged from only 10% to 70% of those in kidney, while mean concentrations in pancreas and gonads ranged from only 13% to 23% and from 8% to 17%, respectively, of those in kidney. Most other tissues including brain, muscle, blood, feathers, and eggs contained very low, sometimes nondetectable, levels of cadmium (Osborn et al. 1979, Hutton 1981, Burger and Gochfeld 1996, García-Fernández et al. 1996, Kim et al. 1998, Mora 2003, Nam et al. 2005, Orlowski et al. 2007).

High exposure to cadmium can impair the reabsorption capability of kidneys, resulting in excretion of cadmium and a consequent decrease in kidney cadmium concentration (Friberg et al. 1986). After cadmium-induced renal toxicity has occurred, concentrations of cadmium in kidneys decline, and concentrations in liver may exceed those in kidney (Hughes et al. 2000).

Liver and kidneys are the two organs most frequently used for monitoring cadmium levels in free-living birds probably because they account for most of the cadmium body burden. Liver was considered to be a better indicator of chronic cadmium exposure than kidneys because the cadmium content of the liver is relatively stable and resistant to the toxic effects of cadmium, unlike the kidneys in which cadmium concentrations decline after the production of renal tubular dysfunction (Scheuhammer

1987). However, Scheuhammer (1987) further indicated that kidneys should provide a superior sample for monitoring cadmium exposure in bird populations characterized by background exposure to cadmium in which renal toxicity is unlikely. Renal and hepatic cadmium levels are highly correlated (Figure 20.1), suggesting that in many instances, either organ would suffice for monitoring.

Feathers are often used as a nondestructive means of assessing relative cadmium exposure in wild birds. Concentrations in feathers have sometimes been correlated with those in kidneys or liver (Pilastro et al. 1993, Agusa et al. 2005) although such correlations were not found in other studies (Stock et al. 1989, Nam et al. 2005, Orlowski et al. 2007). While Goede and de Voogt (1985) considered feathers to be of dubious value for monitoring exposure to cadmium, Burger (1993) concluded that cadmium is likely sequestered to growing feathers from internal sources, and therefore feathers are useful for monitoring cadmium levels in birds (assuming a minimal degree of external feather contamination). Some of the cadmium in feathers were reported to originate from uropygial gland secretions, which were transferred to feathers during preening (Pilastro et al. 1993). However, Dauwe et al. (2003) and Jaspers et al. (2004) concluded that external contamination is the dominant source of cadmium in feathers while Ek et al. (2004) reported that cadmium is primarily sequestered from internal sources with a few externally attached particles of high concentration. Thus, despite the wide use of feathers in monitoring studies, uncertainty continues to exist about their value for accurately assessing cadmium exposure in birds.

Blood has been used to assess cadmium exposure in birds. Levels of cadmium in blood reflect recent dietary exposure whereas levels in kidneys and liver integrate chronic exposure (Friberg et al. 1986). Nevertheless, blood cadmium levels were correlated with levels in liver or kidneys both within and among species of birds (García-Fernández et al. 1995, 1996, Wayland et al. 2001a). Furthermore, blood cadmium levels increased in a dose-dependent manner in birds chronically exposed to various levels of cadmium in their diets (White and Finley 1978, Świergosz and Kowalska 2000). Together, these lines of evidence indicate that blood is a valid medium for assessing cadmium exposure in birds. A major difficulty with using blood has been that concentrations of cadmium in blood are often below the detection limits of standard analytical methods. However, this problem is not insurmountable as more sensitive analytical methods have been developed (Stoeppler and Brandt 1980, García-Fernández et al. 1995).

Cadmium concentrations in tissues of wild birds vary widely according to ecosystem use, diet, age, and physiological status. The highest levels of cadmium in kidneys and livers were found in

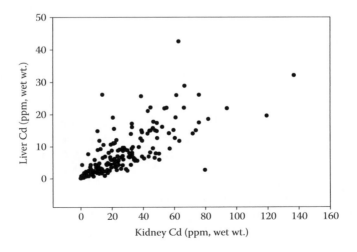

FIGURE 20.1 Correlation between concentrations of cadmium in kidney and liver based on mean or median values for individual species from studies shown in Table 20.1.

TABLE 20.1
**Summary of Background Cadmium Concentrations (μg/g wet wt.) in
Livers and Kidneys of Birds According to Habitat Type**[a]

Habitat	Kidney	Liver
Terrestrial	0.4 (0.02–57.2), 26	0.1 (0.1–9.3), 43
Freshwater[b]	0.4 (0.1–4.9), 19	0.1 (0–7.6), 27
Shoreline	1.1 (0.1–7.8), 17	1.5 (0–3.5), 6
Marine-coastal	12.7 (0.2–79.8), 57	2.9 (0.1–17.1), 71
Marine-pelagic	31.3 (1.2–137.0), 110	8.2 (0.4–42.5), 129

Note: Values are medians (ranges) and sample sizes for individual species and age class
(adult, subadult, juvenile—where age class data were available) mean or median values
from 30 studies.

[a] Bull et al. 1977, Osborn et al. 1979, Hutton 1981, Muirhead and Furness 1988, Stock et al.
1989, Honda et al. 1990, Henny et al. 1991, 1995, Ohlendorf et al. 1991, Elliott et al. 1992,
Lock et al. 1992, García-Fernández et al. 1995, Dietz et al. 1996, Elliott and Scheuhammer
1997, Kim et al. 1998, Trust et al. 2000, Wayland et al. 2001b, Barjaktarovic et al. 2002,
Mochizuki et al. 2002, Stoudt et al. 2002, Savinov et al. 2003, Braune and Simon 2004,
Campbell et al. 2005, Alleva et al. 2006, Pedersen et al. 2006, Horai et al. 2007, Rodrigue
et al. 2007, Teraoka et al. 2007, Braune and Scheuhammer 2008, Braune and Noble 2009.

[b] Refers to species that primarily use freshwater habitats, although they may use estuarine or
coastal habitats during the winter.

pelagic seabirds and the lowest in terrestrial birds (Table 20.1). Coastal birds also contained relatively high levels of cadmium while shorebirds and species that use primarily freshwater habitats,
including piscivores, contained lower concentrations than coastal birds and seabirds but slightly
higher concentrations than terrestrial birds . Diet is probably one of the two most important determinants of cadmium levels in birds, the other being the age of the birds. Certain marine invertebrates,
such as hyperiid amphipods and sea-skaters (*Halobates*), have been shown to contain high (~10–40
μg/g dry weight) concentrations of cadmium (Macdonald and Sprague 1988, Schulz-Baldes 1989).
Among adult seabirds, insectivores such as albatrosses, petrels, and storm-petrels tended to have the
highest levels of cadmium, sometimes exceeding 100 μg/g in kidney and 25 μg/g in liver. Seabirds
that eat mollusks or fish or that scavenge other species of marine birds and mammals tended to have
lower, but nevertheless elevated (compared with most terrestrial species) levels of cadmium in their
kidneys and livers. In the Canadian Arctic, the highest renal cadmium concentrations were found
in an individual thick-billed murre (*Uria lomvia*) (108 μg/g) and an individual black-legged kittiwake (*Rissa tridactyla*) (102 μg/g) (Braune and Scheuhammer 2008). While most terrestrial species
contain low levels of cadmium (<2 μg/g), a notable exception is the willow ptarmigan (*Lagopus
lagopus*) in which mean renal and hepatic levels of up to 57 and 9 μg/g, respectively, have been
reported (Pedersen et al. 2006, Rodrigue et al. 2007). Willow, a cadmium-accumulating plant species, is the major food eaten by willow ptarmigan (Rodrigue et al. 2007). In blood, cadmium levels
varied according to geographical location (González-Solís et al. 2002, Wayland et al. 2008a) and
diet (García-Fernández et al. 1995, Burger and Gochfeld 1997). Cadmium levels in blood samples of
wild birds are normally <50 ng/mL (García-Fernández et al. 1995, Thompson and Dowding 1999,
Wayland et al. 2001a, 2008b, Martínez-López et al. 2005, Finkelstein et al. 2007, Tsipoura et al.
2008), although in some instances concentrations as high as 990 ng/mL were reported (Burger and
Gochfeld 1997, Wilson et al. 2004).

Age is an important determinant of cadmium levels in birds. Prefledged birds have very low
levels of cadmium in their tissues. Renal and hepatic cadmium levels increased during the first

year of life in willow ptarmigan and tended to level off after that (Myklebust and Pedersen 1999). Renal cadmium levels were approximately 2–30 times higher in adult birds than in hatch-year birds (Hutton 1981, Blomqvist et al. 1987, Stock et al. 1989, Henny et al. 1991, Stewart et al. 1994, Trust et al. 2000, Kojadinovic et al. 2007, Orlowski et al. 2007, Rodrigue et al. 2007, Braune and Noble 2009). In a high cadmium environment, adult white-tailed ptarmigan (*Lagopus leucurus*) accumulated cadmium to levels that were approximately 100-fold higher than those in immatures (Larison et al. 2000). Higher levels of cadmium have been reported in adults than in subadults of some species (Hutton 1981, García-Fernández et al. 1995, Larison et al. 2000, Trust et al. 2000, Nam et al. 2005), but not others (Blomqvist et al. 1987, Stock et al. 1989). While some studies have reported continued accumulation of renal, hepatic, or muscle cadmium with age in known-age adult birds (Hutton 1981, Donaldson et al. 1997, Larison et al. 2000), other studies found no such relationships (Stewart and Furness 1998, Agusa et al. 2005). In blood samples, cadmium levels were reported to be higher in adults than in immatures (García-Fernández et al. 1996) while there was no relationship between cadmium levels and age of known-age adult birds (González-Solís et al. 2002).

Body condition and physiological status also influence levels of cadmium in tissues of wild birds. Renal and hepatic cadmium levels increased by a factor of 1.3–2.2 between the prebreeding and nesting periods in common guillemots (*Uria aalge*) and common eiders (*Somateria mollissima*) (Stewart et al. 1994, Wayland et al. 2005). Following moult, cadmium levels in common guillemots declined to about one-half of levels recorded during the nesting period (Stewart et al. 1994). Such changes may be related to normal patterns of seasonal change in lipid and protein content of organs (Osborn 1979) or possibly to changes in diet or availability of MT-binding sites (Stewart et al. 1994).

20.5 TOXIC EFFECTS

We were unable to find reports of acute cadmium poisoning of wild birds in the scientific literature. In humans, most acute poisoning cases have resulted from inhalation of cadmium dust in industrial settings (Hammond and Beliles 1980), a scenario that would be extremely unlikely to occur in wild birds. Instead, it is the chronic dietary exposure of birds to cadmium that, if sufficiently elevated, has the potential to harm their health and fitness. Although the kidney has long been thought to be the primary target of cadmium accumulation and toxicity in vertebrates, a number of other physiological systems may also be at risk. Below, we describe the various toxic effects of cadmium reported in birds. We then summarize relationships between toxic effects and tissue cadmium concentrations.

20.6 INTESTINAL DAMAGE AND NUTRIENT UPTAKE

Quail (*Coturnix coturnix japonica*) chicks fed cadmium at 75 µg/g in their diets had dilated and thin-walled small intestines similar to those observed in cases of human malabsorption syndrome. Villi were short and thickened with evidence of damage to absorptive cells (Richardson and Fox 1974). Similarly, Bokori et al. (1996) observed intestinal lesions characterized by enlarged villi and mononuclear cell infiltration of the mucosa in chickens fed cadmium at 25 µg/g in their diets for 274 days beginning at 2 weeks of age. Hughes et al. (2000) reported that ducks treated with both salt water and dietary cadmium at 300 µg/g experienced hypertrophy of small intestines. Berzina et al. (2007) also observed hypertrophy of small intestines as well as an increase in the number of goblet cells and granular lymphocytes in intestinal mucosa of chicks fed a diet containing 50 µg/g cadmium for 30 days.

Cadmium was found to reduce intestinal absorption of some essential elements most commonly calcium, zinc, and iron (Freeland and Cousins 1973, Fullmer et al. 1980, Korének et al. 2000). In Japanese quail, some of the effects of dietary cadmium exposure resembled iron deficiency (anemia, bone-marrow hyperplasia, and cardiac hypertrophy) or zinc deficiency (testicular hypoplasia) (Richardson et al. 1974). Reduced absorption of calcium may have been caused, in part, by a decline in concentrations of vitamin D-dependent calcium-binding protein in intestinal mucosa

(Fullmer et al. 1980). Whether this effect was related to cadmium-induced cytotoxicity remains unclear (Corradino 1979, Fullmer et al. 1980).

In chickens, cadmium was reported by two studies to impair the secretion of digestive enzymes from the pancreas into the small intestine. In the first study, cadmium was fed at four dietary concentrations (control, 75, 300, and 600 µg/g) (Kósa et al. 2004). In the second it was administered at 0.12 mg/day in drinking water (Korének et al. 2007). Co-administration of zinc overcame the cadmium effect on trypsin and chymotrypsin, suggesting that cadmium may impair absorption of zinc, and/or replace zinc in zinc-dependent enzymes, thereby impairing their activity (Korének et al. 2007). Co-administration of vitamin D_3 reduced the effect of cadmium on chymotrypsin activity, perhaps by enhancing uptake of calcium with a consequent reduction in cadmium uptake (Korének et al. 2000).

20.7 KIDNEY DAMAGE AND VITAMIN D METABOLISM

The kidney is the primary target for cadmium toxicity. The time required for kidney lesions to occur varies according to the dietary cadmium concentration, and other factors such as age. Thus, for example, a diet containing 200 µg/g cadmium was required to produce kidney lesions in adult mallards within 8 weeks (White et al. 1978), whereas only 20 µg/g dietary cadmium produced similar toxic effects in mallard ducklings within 12 weeks (Cain et al. 1983). A hallmark of chronic cadmium toxicity is damage to renal proximal tubules. Numerous studies of birds exposed to elevated levels of dietary cadmium have reported inflammatory reactions, fibrosis, necrosis, and/or proteinaceous and mineralized material in proximal tubules, and some have also reported damage to Bowman's capsule and glomeruli and/or impairment of glomerular filtration rate (White et al. 1978, Mayack et al. 1981, Cain et al. 1983, Nicholson and Osborn 1983, Prasada-Rao et al. 1989a, 1989b, Bennett et al. 1993, Bokori et al. 1996, Bennett et al. 2000, Hughes et al. 2000, Larison et al. 2000, Świergosz and Kowalska 2000).

Renal tubular damage can lead to renal dysfunction characterized by proteinuria, glycosuria, and increased urinary cadmium (Scheuhammer 1987). Furthermore, when proximal tubules are damaged, re-uptake of minerals into the bloodstream may decline, thereby reducing their availability for essential functions (Friberg et al. 1986, Bennett et al. 2000). Damage to glomeruli reduced the glomerular filtration rate, which had adverse consequences for kidney function and osmoregulation (Bennett et al. 2000). Proximal tubules are a primary site for conversion of vitamin D_3 to its active form, 1,25-dihydroxy-vitamin D_3. Damage to proximal tubules can impair vitamin D_3 metabolism (WHO Task Group on Environmental Health Criteria for Cadmium 1992). Such an effect was demonstrated in chick kidneys, *in vitro* (Feldman and Cousins 1973) but not in chicks fed diets containing cadmium at 80 µg/g (Fullmer et al. 1980). Thus, whether or not cadmium induced-kidney damage impairs vitamin D metabolism in birds, as it does in mammals, remains uncertain.

20.8 SKELETAL EFFECTS

In humans and experimental mammals, one of the most serious consequences of excessive cadmium exposure is the damage caused to bones. Such damage may come in the form of osteomalacia, osteoporosis, or osteopenia. These conditions are likely to occur in subjects whose kidneys have been damaged by cadmium because such damage enhances excretion of essential minerals like calcium and phosphorus and interferes with the synthesis of the active form of vitamin D, a hormone that plays a vital role in intestinal absorption of calcium (WHO Task Group on Environmental Health Criteria for Cadmium 1992). Cadmium may also affect the metabolism of other calciotropic hormones (Brzóska and Moniuszko-Jakoniuk 2005).

There is little information on the effects of cadmium on bone metabolism in birds. *In vitro* studies on embryonic chick femurs indicated that cadmium increases resorption of calcium and phosphorus, inhibits bone matrix formation and decreases bone mineral content (Miyahara et al. 1980, 1983, Kaji et al. 1988). However, these studies cannot indicate whether cadmium affects bone metabolism in free-living birds indirectly by causing kidney damage and associated changes in vitamin D metabolism. This is unfortunate because a study of white-tailed ptarmigan showed that calcium content in bones of female birds with renal cadmium levels that were high enough to cause kidney damage ($\geq$100 µg/g) averaged 8–10% lower than that in birds whose kidneys contained lower levels of cadmium (<100µg/g) (Larison et al. 2000). Furthermore, while the authors did not consider that the birds' ages may have affected bone calcium content independently of cadmium, they postulated that the skeletal effects may have contributed to the relatively high mortality rates among adult female ptarmigan in the study area.

20.9 EFFECTS ON OSMOREGULATION

High salt content of food and drinking water in marine environments represents an osmoregulatory challenge to seabirds, especially species of sea ducks, which migrate between freshwater and salt-water environments. Major organs involved in osmoregulation are kidneys, intestines and, in seabirds, salt glands. As mentioned earlier, cadmium caused intestinal hypertrophy, kidney damage, and reduced glomerular filtration rate in ducks that were acclimated to salt water over time (Bennett et al. 2000, Hughes et al. 2000). Cadmium exposure also produced other signs and symptoms of osmoregulatory stress in these birds. Specifically, it increased interstitial fluid space at the expense of intracellular fluid, which would be expected to stimulate salt gland secretion; lowered plasma osmolality and sodium levels; delayed and raised the threshold for activation of salt gland secretion (Bennett et al. 2000); increased mass of salt glands (Hughes et al. 2000); and, during the period of salt water acclimation, increased plasma volume relative to packed cell volume and suppressed the increase in angiotensin II, a hormone that regulates sodium balance and resists extracellular fluid volume shrinkage (Hughes et al. 2003). These effects were clearly evident at a dietary cadmium concentration of 300 µg/g while some, but not all of these effects were also evident at a dietary concentration of 50 µg/g. Some of these effects were sex-specific. The authors concluded that cadmium ingestion caused a redistribution of body water that affected the ability of kidneys and salt glands to effectively eliminate excess sodium chloride. However, the relatively high dietary and tissue cadmium levels at which these effects occurred coupled with the presumed lower salt tolerance of the experimental birds compared to wild seabirds created some uncertainty as to the likelihood that cadmium exposure poses a significant osmoregulatory challenge in seabirds.

20.10 EFFECTS ON ENERGY METABOLISM

Cadmium did not affect body weight of fully grown birds (White et al. 1978, Di Giulio and Scanlon 1985, Leonzio et al. 1992, Bennett et al. 2000, Congiu et al. 2000), except at the highest doses (450 µg/g in diet, Di Giulio and Scanlon 1984; 3 mg/kg body weight by intraperitoneal injection, Rahman et al. 2007). In young, growing birds, dietary or aqueous cadmium was reported to reduce growth rates or body weight gain in some studies (Richardson et al. 1974, Bokori et al. 1996, Świergosz and Kowalska 2000, Erdogan et al. 2005) but not in others (Mayack et al. 1981, Cain et al. 1983). Differences among these studies in dosing regimes cannot explain the different results.

While cadmium exposure does not easily induce birds to lose or gain weight, some studies reported that it can affect energy metabolism in less obvious ways. Plasma triglyceride levels were elevated in Japanese quail fed dietary cadmium at 100 µg/g (Leonzio et al. 1992). In mallards (*Anas platyrhynchos*) that were fed cadmium at 450 µg/g, aldolase activity increased and gluconeogenesis was enhanced, the latter as evidenced by increased mass of adrenal glands, total adrenal corticosterone, and plasma uric acid concentration. In addition, lipids were mobilized as evidenced by an increase in plasma nonesterified fatty acids and a decrease in liver glycogen concentrations (Di Giulio and Scanlon 1984). Di Giulio and Scanlon (1985) observed that dietary cadmium at 50 µg/g enhanced energy drain in starved ducks, suggesting that birds whose access to food is restricted may be at greater risk to cadmium exposure than birds with abundant food supplies.

20.11 EFFECTS ON REPRODUCTIVE ORGANS AND REPRODUCTION

Male birds experimentally exposed to cadmium had small or atrophied testes (Richardson et al. 1974, White et al. 1978, Bokori et al. 1996). Lesions, including minimal germinal cell maturation, fibrous tissue and inflammatory cells scattered throughout interstitial tissue, an absence of spermatozoa and narrow seminiferous tubules with damaged epithelia were observed (Richardson et al. 1974, White et al. 1978, Hughes et al. 2000, Świergosz and Kowalska 2000). Spermatogenesis was reduced or absent (White et al. 1978, Bokori et al. 1996), semen volume and sperm concentration were reduced, sperm cells exhibited morphological abnormalities and the activity of zinc-dependent angiotensin converting enzyme (ACE) in sperm cells was reduced by daily intramuscular injections of cadmium chloride (Mohan et al. 1990). Cadmium may have replaced zinc in ACE, thus interfering with the enzyme's role in sperm production.

It was reported that cadmium damaged the vasculature of hens' ovaries, resulting in hemorrhagic necrosis (Nolan and Brown 2000). Although not statistically significant, dietary exposure to cadmium at 300 µg/g reduced mean ovary mass to 36% of that in birds fed a control diet (Hughes et al. 2000).

Egg production was reduced in mallards and chickens by relatively high exposure to dietary cadmium or by intraperitoneal cadmium injection (White and Finley 1978, Leach et al. 1979, Nolan and Brown 2000, Rahman et al. 2007). In contrast, exposure to cadmium for 3 months at 1.5 ng/L in drinking water did not affect egg production by pheasants (*Phasianus colchicus*) (Toman et al. 2005).

Eggshell thinning has been observed in eggs of birds injected with cadmium. Quail experienced a transitory decrease in eggshell thickness following a single 0.3 mg/kg body weight intraperitoneal injection of cadmium (Rahman et al. 2007). Shell thickness declined by an average of 6% and 12% in two experiments in which laying hens were fed a diet containing cadmium at 48 µg/g (Leach et al. 1979). However, eggshell thinning was not observed in eggs of ringed doves fed a diet containing 20 µg/g cadmium plus lead and aluminum (Scheuhammer 1996).

Egg hatchability was unaffected in experiments in which cadmium was provided to birds in their diets or in drinking water (Scheuhammer 1996, Toman et al. 2005). Mixed results were obtained when cadmium was injected into fertile eggs. Injection of 0.5 ng cadmium as $CdSO_4$ into mallard and chicken eggs reduced hatchability by 40–70% compared to vehicle-injected eggs (Kertész and Hlubik 2002, Kertész and Fáncsi 2003). In contrast, injections of ≤1.2 µg cadmium did not reduce hatchability of chicken eggs but injections of 4.8 and 19.2 µg did (Sato et al. 1997). The value of these egg injection studies in interpreting the toxicological significance of cadmium levels in eggs of wild birds is questionable since even the higher effect levels were likely similar to those which have been recorded in eggs of wild birds with no known reproductive problems (range of concentrations from 0.002 to 0.9 µg/g) (Burger 1994).

20.12 ENDOCRINE DISRUPTION

Very high dietary cadmium concentrations (450 µg/g for 42 days) suppressed levels of plasma tri-iodothyronine (T_3), increased mass of adrenal glands and increased total corticosterone content of adrenal glands in mallards, effects that may have been secondary to cadmium-induced alterations in energy metabolism in the experimental birds (Di Giulio and Scanlon 1984). Dietary cadmium at 50 µg/g augmented the decline in plasma T_3 levels and the increase in total adrenal corticosterone that were attributed to food restriction, suggesting that starved birds may be more sensitive to cadmium than well-fed birds (Di Giulio and Scanlon 1985). Decreased T_3 levels in chickens were related to a cadmium-induced, dose-dependent depression in type-I iodothyronine 5'-monodeiodinase activity, which may have resulted from a corresponding increase in the generation of free radicals (Gupta et al. 2000).

Cadmium, injected intraperitoneally at 0.1 mg/kg body weight, was reported to mimic estrogen function in male Japanese quail, whereas at 10 mg/kg body weight it was reported to suppres estrogen-responsive gene transcription (Rahman et al. 2008). In laying hens fed a diet containing cadmium at 153 µg/g, mean vitellogenin and estrogen levels were reduced to 64% and 58%, respectively, of levels in birds fed a control diet (Nolan and Brown 2000).

20.13 ANEMIA

In chicks fed dietary cadmium at 75 µg/g, intestinal absorption of iron was reduced by an average of 39–62% of control values (Freeland and Cousins 1973). Reduced absorption of iron may result in lower levels of stored iron in kidneys and livers of birds as demonstrated by several experimental studies in which cadmium was fed to birds at dietary levels from 25 to 80 µg/g (Freeland and Cousins 1973, Prasada Rao et al. 1989a, Bokori et al. 1996, Świergosz and Kowalska 2000). Anemia was observed in birds that were fed cadmium (Jacobs et al. 1969, Richardson et al. 1974, Bokori et al. 1996). In immature birds fed cadmium at dietary concentrations ranging from 20 to 75 µg/g, mean hematocrits ranged from 8% to 48% lower and mean hemoglobin ranged from 6% to 25% lower than in controls (Jacobs et al. 1969, Freeland and Cousins 1973, Cain et al. 1983). However, the diet containing 20 µg/g reduced hematocrits and hemoglobin in young birds only at 8 weeks of age but not at 4 or 12 weeks of age (Cain et al. 1983). Hematocrits and/or hemoglobin were unaffected in adult birds fed cadmium at levels from 20 to 200 µg/g (White and Finley 1978, Scheuhammer 1996). From these studies, it appears that cadmium is more likely to cause anemia in immature birds than in fully grown or adult birds.

20.14 BEHAVIORAL ALTERATIONS

Behavioral alterations have been observed in birds fed cadmium at relatively low concentrations. In a controlled study, 1-week-old ducklings fed a diet containing cadmium at 4 µg/g ran farther from a frightening object than did control ducklings or ducklings fed cadmium at 40 µg/g (Heinz et al. 1983). It was postulated that the hyper-responsiveness in ducklings fed the lower cadmium diet could be just as harmful to birds in the wild as a failure to respond might be. Black ducks (*Anas rubripes*) fed a diet containing 4 µg/g cadmium were significantly more active than ducks on a control diet but activity levels quickly returned to normal after cadmium was removed from the diet (Silver and Nudds 1995). In 1 of 2 years of the study, the effect of cadmium was heightened in birds simultaneously fed low levels of calcium. Free-living, brood-rearing willow ptarmigan dosed with cadmium by osmotic minipumps exhibited weaker distraction displays and flushed further from their broods than saline-dosed birds (Pedersen and Saether 1999). Survival of chicks of the cadmium-contaminated birds was lower than that of the saline-treated birds, and the authors suggested that cadmium negatively affected the ability of parents to care for their young.

A physiological mechanism by which cadmium affects behavior has not been elucidated. In the case of poor parental behavior it was suggested that cadmium exposure may inhibit prolactin secretion (Pedersen and Saether 1999), while in the case of increased activity it was postulated that cadmium may alter the secretion or metabolism of catecholamines (Silver and Nudds 1995). Both hypotheses are partially supported by *in vitro* evidence (Winstel and Callahan 1992, Yamagami et al. 1994) but we are unaware of *in vivo* evidence that secretion of prolactin or catecholamines is altered in animals exposed to low levels of cadmium. The lack of a unifying, proven explanation as to the physiological mechanisms by which low levels of cadmium alter behavior remains a serious shortcoming to accepting that such exposure can indeed affect behavioral patterns.

20.15 ASSOCIATION BETWEEN TISSUE CONCENTRATIONS AND EFFECTS

Certain aspects of cadmium toxicity should be considered when attempting to relate tissue concentrations to effects. First, young, growing birds are more sensitive to cadmium than fully grown birds (Scheuhammer 1987). Thus, when evaluating tissue concentrations in relation to toxic effects, these two age groups should be considered separately. Second, it is not known if sensitivity to cadmium ranges widely among species. This makes it difficult to interpret the results of experimental feeding studies, which mostly have used mallards, quail, and chickens, in terms of the likelihood of effects on species that accumulate high levels of cadmium in the wild, for example seabirds and ptarmigan. Third, the current state of knowledge about cadmium toxicity to birds does not permit different endpoints to be confidently ranked according to their relative sensitivity. Fourth, kidneys and liver are the two tissues for which there is the most information available to examine tissue concentration–effects relationships. Finally, experimental dosing studies, on which much of the toxicological information is based, have typically exposed birds to high levels of cadmium for short periods of time (weeks to months) and have reported tissue concentrations and effects associated with those dosing regimes. It has not been demonstrated whether toxic effects associated with such dosing regimes (and resultant tissue concentrations) would also occur under more natural conditions in which concentrations in the diet are likely to be substantially lower but high levels of cadmium can nevertheless gradually accumulate in tissue because of chronic exposure (years).

In fully grown birds, the lowest, study-based mean or median renal concentrations at which structural/functional damage to tissues, including kidneys, testes, liver, gut, and salt glands, were observed ranged from approximately 25 to 178 µg/g (White and Finley 1978, White et al. 1978, Nicholson et al. 1983, Prasada Rao et al. 1989a, 1989b, Bokori et al. 1996, Bennett et al. 2000, Hughes et al. 2000, 2003). Severe damage to proximal tubules of starlings (*Sturnus vulgaris*) subcutaneously injected with cadmium chloride occurred at concentrations approximately ≥37 µg/g (Nicholson and Osborn 1983). Energy metabolism was altered in such a way that would likely cause energy stores to decline over the long term in experimental mallards with mean renal cadmium concentrations of 124 µg/g (Di Giulio and Scanlon 1984). Experimental starlings that experienced hepatotoxicity in the form of enlarged livers and reduced activity of the antioxidant enzyme, glutathione peroxidase, had a mean renal cadmium concentration of 72 µg/g (Congiu et al. 2000). Effects on structural/functional damage to tissues were not observed in wild birds or birds experimentally exposed to cadmium with mean or median renal cadmium concentrations ranging from approximately 22 to 56 µg/g (White and Finley 1978, White et al. 1978, Elliott et al. 1992, Trust et al. 2000, Wayland et al. 2001b). Energy metabolism was unaffected in experimental mallards with a mean kidney cadmium concentrations of 44 and 77 µg/g (Di Giulio and Scanlon 1984, 1985). Nor was there evidence of hepatotoxicity in experimental starlings with a mean kidney cadmium concentration of 27 µg/g (Congiu et al. 2000). Using cadmium concentrations from the above studies in a logistic regression, we estimated there would be a 50% probability of adverse effects on some individual birds at a renal cadmium concentration of 66 µg/g (Figure 20.2). However, confidence limits on the estimate were very wide, indicating substantial uncertainty about the relationship between renal cadmium levels and adverse effects in birds based on current scientific knowledge.

Unfortunately, field studies confirming that accumulation of 66 µg/g in kidney poses a risk to birds are lacking. One study (Larison et al. 2000) showed that accumulation of >100 µg/g in kidneys of white-tailed ptarmigan was associated with renal damage which may have resulted in bone decalcification and ultimately in reduced survivorship of adult birds. While the relationships between cadmium exposure, bone decalcification and survivorship were correlative and the possible unifying effect of age on the patterns of the observed relationships was not considered, this study remains the only example of a free-living population of birds possibly experiencing toxic effects from exposure to cadmium. Thus, the threshold of 66 µg/g should be viewed as a conservative but uncertain risk level at which some individuals may experience minor but significant toxicity, whereas 100 µg/g should be viewed as a more liberal level above which substantial adverse effects are likely to occur.

Logistic regression analysis of liver data failed to yield a maximum likelihood estimate because of complete separation in liver cadmium concentrations between affected and nonaffected birds. In livers of fully developed birds, which did not exhibit any signs of adverse effects, mean cadmium levels ranged from 7 to 45 µg/g (median = 17.5 µg/g) (White and Finley 1978, White et al. 1978,

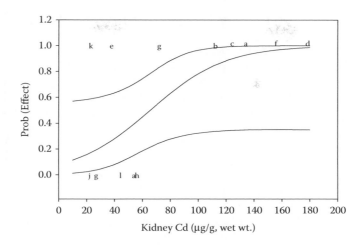

FIGURE 20.2 Probability of an adverse effect in relation to kidney cadmium concentration in fully grown birds. Top and bottom lines are the 95% confidence limits of the estimates. An adverse effect is defined as an alteration in energy metabolism or structural and/or functional damage to tissues. Letters show lowest study-based effect concentrations or highest study-based no-effect concentrations. (a) White and Finley 1978, White et al. 1978; (b) Bennett et al. 2000, Hughes et al. 2000, 2003; (c) Di Giulio and Scanlon 1984; (d) Prasada-Rao et al. 1989a, 1989b; (e) Nicholson and Osborn 1983 (starlings); (f) Bokori et al. 1996; (g) Congui et al. 2000; (h) Elliott et al. 1992 (Leach's Storm-Petrel); (i) Wayland et al. 2001b (King Eider); (j) Trust et al. 2000 (Spectacled Eider); (k) Nicholson et al. 1983 (seabirds); (l) Di Giulio and Scanlon 1985 (birds fed *ad-lib*).

Di Giulio and Scanlon 1984, Elliott et al. 1992, Congui et al. 2000, Trust et al. 2000, Wayland et al. 2001b). At the lowest effect concentration (68 µg/g), the livers of starlings were enlarged and hepatic activity of the antioxidant enzyme, glutathione peroxidase, was inhibited (Congiu et al. 2000). At a slightly higher mean hepatic concentration (73 µg/g), chickens experienced structural damage to their livers and kidneys, enlargement of liver and heart, and atrophy of testes (Bokori et al. 1996). At a mean hepatic cadmium concentration of 110 µg/g, the kidneys of mallards were structurally damaged, their testes were atrophied, their kidneys were enlarged, and egg production was suppressed (White and Finley 1978, White et al. 1978). At a mean concentration of 144 µg/g, experimental mallards lost weight and liver mass but had enlarged kidneys and adrenals and experienced changes in carbohydrate metabolism consistent with a reduction in stored energy (Di Giulio and Scanlon 1984). Finally, at approximately 220 µg/g, gut hypertrophy and changes in some aspects of osmoregulation and damage to kidneys, testes, and adrenal chromaffin cells were observed in experimental mallards (Hughes et al. 2000, 2003). Based on results from these studies, we estimate that the threshold effect level for hepatic, renal, or testicular toxicity lies between approximately 45 and 70 µg/g in liver.

Birds that experience prolonged food shortages may be more sensitive to cadmium exposure than birds that are not food-stressed. Energy drain appeared to be exacerbated in cadmium-exposed experimental mallards with mean hepatic and renal concentrations of approximately 15 and 32 µg/g, respectively (Di Giulio and Scanlon 1985).

The renal and hepatic thresholds we have suggested are based on studies of few species. It is possible that some species may be affected at lower levels of cadmium while others may be affected only at levels substantially higher than these thresholds. The threshold for cadmium damage to seabird tissues remains uncertain. Although one study reported tissue damage in seabirds whose renal cadmium levels were as low as approximately 23 µg/g (Nicholson et al. 1983), other studies of seabirds have failed to find tissue damage at mean renal cadmium levels of approximately 2, 7, 18, 20, 22, 37, and 57 µg/g (Elliott et al. 1992, Trust et al. 2000, Wayland et al. 2001b). Furthermore, maximum concentrations in studies of species whose kidneys were not damaged by cadmium ranged from approximately 32 to 120 µg/g (Elliott et al. 1992, Trust et al. 2000, Wayland et al. 2001b), which are higher than concentrations reported by Nicholson et al. (1983) to harm seabird kidneys. Notably, Nicholson et al. (1983) concluded that the kidney damage they observed in seabirds was probably due to the birds having high cadmium burdens, although they also recognized that mercury accumulation may have been sufficiently high to cause some of the damage.

In young, growing birds, whole organism responses or structural and/or functional damage to tissues were observed over a wide range of renal and hepatic cadmium concentrations. Pheasant and Japanese quail chicks and mallard and wood duck (*Aix sponsa*) ducklings exhibited many effects, including reduced growth rates, reduced feeding efficiency, oxidative stress, anemia and/or damage to liver, kidney, testes, intestines, or other organs at mean hepatic concentrations of 6, 37, 24, and 209 µg/g (Richardson et al. 1974, Richardson and Fox 1974, Mayack et al. 1981, Cain et al. 1983, Świergosz and Kowalska 2000). In contrast, mallard and wood duck ducklings were unaffected by cadmium dosing that resulted in mean hepatic cadmium levels of 12 and 27 µg/g (Mayack et al. 1981, Cain et al. 1983). Mean renal cadmium concentrations as low as 11 µg/g, were associated with lesions in kidneys, liver, and testes of growing pheasants (Świergosz and Kowalska 2000) and with lipid peroxidation, reduced growth rates and reduced feed conversion efficiency in broiler chickens (Erdogan et al. 2005). Kidneys of wood duck ducklings were damaged when their mean renal cadmium concentrations reached 132 µg/g, but their growth rates were not impaired nor were their kidneys, livers, guts, or brains damaged at a mean renal cadmium concentration of 62 µg/g (Mayack et al. 1981). Because of this wide range in effects-levels and the broad overlap in cadmium concentrations between unaffected and affected birds, we cannot recommend a specific threshold-effects concentration which would apply to young, growing individuals of all or even most species of birds. Nevertheless, in the wild, tissues of young, growing birds contain levels of cadmium that are lower than even the lowest effects-levels identified earlier, making it unlikely that cadmium would adversely affect young, growing birds under natural conditions.

Although reproduction may be adversely affected by exposure to cadmium (Nolan and Brown 2000, Rahman et al. 2007), there are few studies relating such effects to tissue cadmium concentrations. Egg production was reduced to 10–75% of control values at mean hepatic concentrations of approximately 110 and 26 µg/g and at mean renal concentrations of 134 and 87 µg/g (White and Finley 1978, Leach et al. 1979). In comparison, egg production was unaffected at mean hepatic and renal concentrations ranging from approximately 4 to 12 and 10 to 168 µg/g, respectively (Leach et al. 1979, Toman et al. 2005). Hatchability of eggs was not affected by dietary exposure to cadmium at mean hepatic and renal cadmium levels ranging from approximately 4 to 35 and 10 to 115 µg/g (Scheuhammer 1996, Toman et al. 2005). Based on this information, it is not possible to recommend broadly applicable tissue thresholds associated with reproductive effects in birds.

Bone calcium content was significantly lower in white-tailed ptarmigan whose kidneys contained >100 µg/g cadmium than in those whose kidneys contained <100 µg/g, and this may have contributed to the low survival rates of adult birds in the Colorado ore belt, an area where ptarmigan diets contain potentially toxic levels of cadmium (Larison et al. 2000).

Behavioral alterations were observed in experimentally dosed birds whose mean hepatic and renal cadmium concentrations ranged from approximately 4 to 8 and 17 to 33 µg/g, respectively (Silver and Nudds 1995, Pedersen and Saether 1999). In comparison, control birds, which presumably exhibited normal parental behavior, had mean hepatic and renal concentrations of 4 and 22 µg/g (Pedersen and Saether 1999). However, in the study by Silver and Nudds (1995), behavioral effects were transitory, disappearing after cadmium was removed from the experimental diet. This may indicate that levels of cadmium in long-term storage organs such as liver and kidneys are inappropriate for evaluating its effects on behavior.

There is little information about cadmium toxicity thresholds in tissues other than liver or kidneys. Mean testicular concentrations of cadmium in chickens and mallards that exhibited structural and functional damage to testes were 3.7 and 8.5 µg/g, respectively (White et al. 1978, Bokori et al. 1996). Adverse effects were not observed in mallard testes containing <1 µg/g cadmium (White et al. 1978). In blood samples, the lowest mean cadmium concentration associated with numerous types of adverse effects in adult mallards was 0.26 µg/g (White and Finley 1978) while in 4-week-old pheasant chicks, this value may have been as low as 0.017 µg/g, although a more liberal interpretation of results of this study is that effects were most pronounced when blood concentrations reached 0.048–0.095 µg/g in 42-day-old chicks (Świergosz and Kowalska 2000). Cadmium concentrations of approximately 1 µg/g in long bones have been associated with adverse effects in several organs of young, growing birds, and fully developed birds (Cain et al. 1983, Bokori et al. 1996, Świergosz and Kowalska 2000).

Summary

Cadmium is a toxic metal with both natural and anthropogenic sources. Despite recent global decreases in anthropogenic emissions of cadmium, human activities continue to contribute to cadmium pollution at local and regional scales.

Birds are exposed to cadmium primarily through their diets. Intestinal absorption of cadmium is low (<1–7%). Once absorbed, cadmium is transported by blood and deposited primarily in the liver and kidneys, which together contain approximately 67–97% of the total body burden. Kidneys are long-term storage organs for cadmium and contain higher concentrations of cadmium than other organs. Concentrations in liver are normally about 10–70% of those in kidneys. Concentrations of cadmium in blood, which is normally ≤50 ng/mL, reflect recent exposure but can also be correlated with those in long-term storage organs like kidneys and liver.

Cadmium levels in birds vary according to diet, ecosystem use, age, and physiological status. They are higher in seabirds, especially pelagic species, than in most terrestrial birds, freshwater birds, and shorebirds. They are also from 2 to 100 times higher in adult birds than in immature birds.

Once absorbed, cadmium binds mainly to MT in liver and kidneys. Under normal circumstances of dietary exposure in wild birds, cadmium induces the synthesis of MT in quantities sufficient to bind and detoxify nearly all of the cadmium present in these organs. Nevertheless, chronic dietary exposure to elevated concentrations of cadmium has the potential to harm the health and fitness of birds. Although the kidney has long been thought to be the primary target of cadmium accumulation and toxicity in vertebrates, a number of other physiological systems may also be at risk. Cadmium can damage absorptive cells in the digestive tract and impair the uptake of essential minerals like calcium, zinc, and iron. In kidneys, damage to proximal tubules, Bowman's capsule and glomeruli may occur. Such damage can lead to renal dysfunction and can impair the re-uptake of minerals into the bloodstream, thereby reducing their availability for essential functions. Whether cadmium induced-kidney damage impairs vitamin D metabolism in birds, as it does in mammals, remains uncertain. In humans and experimental mammals, one of the most serious consequences of excessive cadmium exposure is the damage caused to bones. Birds with toxic levels of cadmium in their kidneys were found to have lower levels of bone calcium than birds whose kidneys contained normal levels of cadmium. Exposure to high levels of dietary cadmium has the potential to affect osmoregulation in seabirds by causing a redistribution of body water that affects the ability of kidneys and salt glands to eliminate effectively excess sodium chloride. Exposure to excessive dietary cadmium can alter energy metabolism in birds, cause testicular damage, disrupt the synthesis or metabolism of certain hormones, suppress egg production, contribute to hyperactivity, impair parental care and reduce the absorption and storage of iron, resulting in anemia.

In deriving threshold tissue concentrations, it is important to take into consideration that (a) immature, growing birds are likely more sensitive to cadmium than fully grown birds, (b) there is likely to be variability among species in sensitivity to cadmium, (c) there is an adequate amount of information on which to base threshold concentrations in kidneys and liver of adult birds but insufficient information for other tissues and age classes, and (d) our understanding of tissue concentrations associated with cadmium toxicity to birds is based primarily on short-term, high-dose studies but the applicability of such studies to natural conditions of lower dietary exposure over much longer periods of time remain unknown.

We calculated that a cadmium concentration of approximately 65 µg/g wet weight in kidneys would be associated with a 50% probability (95% CI: 22–79%) of alterations in energy metabolism or structural/functional damage to tissues, including kidneys, testes, liver, gut, or salt glands and suggest this concentration to be a conservative, yet uncertain, risk level for cadmium-induced damage in adult birds. A more liberal level above which adverse effects are likely to occur is 100 µg/g. For liver tissue, a threshold effect level in adult birds may lie between 45 and 70 µg/g wet weight. Testicular cadmium concentrations ≥4 µg/g wet weight may cause damage to testes. Blood cadmium levels ≥0.26 µg/g (~260 ng/mL) and bone cadmium levels ≥1 µg/g were associated with adverse effects including structural damage to kidneys and testes, intestinal lesions, or absence of spermatogenesis in adult, experimental birds. Birds with blood or bone cadmium concentrations exceeding those levels should be considered at risk of suffering toxic effects.

At present, it is not possible to suggest tissue threshold effect concentrations that would be applicable to immature, growing birds; however, in the wild, such birds usually have very low levels of cadmium in their tissues. Even the lowest effect levels documented in experimental studies of immature birds were higher than tissue concentrations that occur in young birds in the wild, suggesting that cadmium is not likely to pose a risk to free-living immature, growing birds.

Finally, behavioral changes of a type that may have important implications for reproductive success or long-term survival may be induced by exposure to cadmium at concentrations that are lower than those required to cause other toxic effects. Since some behavioral changes are reversible, cadmium concentrations in long-term storage organs like liver and kidneys are probably inappropriate for judging whether or not cadmium exposure is high enough to alter behavior. Blood may be a better tissue for this purpose. Experimental studies that link behavioral changes in cadmium-exposed birds to blood cadmium levels are needed.

REFERENCES

Agusa, T., et al. 2005. Body distribution of trace elements in black-tailed gulls from Rishiri Island, Japan: age-dependent accumulation and transfer to feathers and eggs. *Environ. Toxicol. Chem.* 24:2107–2120.

Alleva, E., N. Francia, M. Pandolfi, A. M. DeMarinis, F. Chiarotti, and D. Santucci. 2006. Organochlorine and heavy-metal contaminants in wild mammals and birds of Urbino-Pesaro Province, Italy: an analytic overview for potential biomarkers. *Arch. Environ. Contam. Toxicol.* 51:123–134.

Anderson, S. H., and R. I. Van Hook, Jr. 1973. Uptake and biological turnover of [109]Cd in chipping sparrows, *Spizella passerina*. *Environ. Physiol. Biochem.* 3:243–247.

Barjaktarovic, L., J. E. Elliott, and A. M. Scheuhammer. 2002. Metal and metallothionein concentrations in scoter (*Melanitta* spp.) from the Pacific Northwest of Canada, 1989–1994. *Arch. Environ. Contam. Toxicol.* 43:486–491.

Bennett, D. C., M. R. Hughes, J. E. Elliott, A. M. Scheuhammer, and J. E. Smits. 2000. Effect of cadmium on Pekin duck total body water, water flux, renal filtration, and salt gland function. *J. Toxicol. Environ. Health*, Pt A. 59:43–56.

Bennett, J. M., D. N. Prashad, R. O. Blackburn, and R. Deane. 1993. The effect of cadmium on glomerular basement membrane thickness in the domestic fowl. *J. Physiol.* 473:219P.

Berzina, N., J. Markovs, S. Isajevs, M. Apsite, and G. Smirnova. 2007. Cadmium-induced enteropathy in domestic cocks: a biochemical and histological study after subchronic exposure. *Basic Clin. Pharmacol.* 101:29–34.

Besser, J. M., W. G. Brumbaugh, T. W. May, and C. J. Schmitt. 2007. Biomonitoring of lead, zinc and cadmium in streams draining lead-mining and non-mining areas, southeast Missouri, USA. *Environ. Monitor. Assess.* 129:227–241.

Beyer, W. N., and G. Storm. 1995. Ecotoxicological damage from zinc smelting at Palmerton, Pennsylvania. In *Handbook of ecotoxicology,* eds. D. J. Hoffman, B. A. Rattner, G. A. Burton, Jr., and J. Cairns, Jr., pp. 596–608. Boca Raton, FL: CRC Press.

Blanco, G., O. Frías, B. Jiménez, and G. Gómez. 2003. Factors influencing variability and potential routes of heavy metals in black kites exposed to emissions from a solid waste incinerator. *Environ. Toxicol. Chem.* 22:2711–2718.

Blomqvist, S., A. Frank, and L. R. Petersson. 1987. Metals in liver and kidney tissues of autumn-migrating dunlin *Calidris alpina* and curlew sandpiper *Calidris ferruginea* staging at the Baltic Sea. *Mar. Ecol. Prog. Ser.* 35:1–13.

Blus, L. J., C. J. Henny, D. J. Hoffman, and R. A. Grove. 1995. Accumulation in and effects of lead and cadmium on waterfowl and passerines in northern Idaho. *Environ. Pollut.* 89:311–318.

Bokori, J., S. Fekete, R. Glávits, I. Kádár, J. Koncz, and L. Kövári. 1996. Complex study of the physiological role of cadmium. IV. Effects of prolonged dietary exposure of broiler chickens to cadmium. *Acta. Vet. Hung.* 4:57–74.

Boutron, C. F. 1995. Historical reconstruction of the earth's past atmospheric environment from Greenland and Antarctic snow and ice cores. *Environ. Rev.* 3:1–28.

Braune, B. M., and D. G. Noble. 2009. Environmental contaminants in Canadian shorebirds. *Environ. Monitor. Assess.* 148:185–204.

Braune, B. M., and A. M. Scheuhammer. 2008. Trace element and metallothionein concentrations in seabirds from the Canadian Arctic. *Environ. Toxicol. Chem.* 27:645–651.

Braune, B. M., and M. Simon. 2004. Trace elements and halogenated organic compounds in Canadian Arctic seabirds. *Mar. Pollut. Bull.* 48:986–992.

Bryan, G. W., and W. J. Langston. 1992. Bioavailability, accumulation and effects of heavy metals in sediments with special reference to United Kingdom estuaries: a review. *Environ. Pollut.* 76:89–112.

Brzóska, M. M., and J. Moniuszko-Jakoniuk. 2005. Effect of low-level lifetime exposure to cadmium on calciotropic hormones in aged female rats. *Arch. Toxicol.* 79:636–646.

Bull, K. R., R. K. Murton, D. Osborn, and P. Ward. 1977. High levels of cadmium in Atlantic seabirds and sea-skaters. *Nature* 269:507–509.

Burger, J. 1993. Metals in avian feathers: bioindicators of environmental pollution. *Rev. Environ. Toxicol.* 5:203–311.

Burger, J. 1994. Heavy metals in avian eggshells: another excretion method. *J. Toxicol. Environ. Health* 41:207–220.

Burger, J. 2008. Assessment and management of risk to wildlife from cadmium. *Sci. Total Environ.* 389:37–45.

Burger, J., and M. Gochfeld. 1996. Heavy metal and selenium levels in Franklin's gull (*Larus pipixcan*) parents and their eggs. *Arch. Environ. Contam. Toxicol.* 30:487–491.

Burger, J., and M. Gochfeld. 1997. Age differences in metals in the blood of herring (*Larus argentatus*) and Franklin's (*Larus pipixcan*) gulls. *Arch. Environ. Contam. Toxicol.* 33:436–440.

Burger, J., C. D. Trivedi, and M. Gochfeld. 2000. Metals in herring and great black-backed gulls from the New York Bight: the role of salt gland excretion. *Environ. Monitor. Assess.* 64:569–581.

Cain, B. W., L. Sileo, J. C. Franson, and J. Moore. 1983. Effects of dietary cadmium on mallard ducklings. *Environ. Res.* 32:286–297.

Campbell, L. M., R. J. Norstrom, K. A. Hobson, D. C.G. Muir, S. Backus, and A. T. Fisk. 2005. Mercury and other trace elements in a pelagic Arctic marine food web (Northwater Polynya, Baffin Bay). *Sci. Total Environ.* 351–352:247–263.

Carpene, E., G. Andreani, and G. Isani. 2007. Metallothionein functions and structural characteristics. *J. Trace Elem. Med. Biol.* 21 (Suppl. 1):35–39.

Chatterjee, A., and I. B. Maiti. 1990. Comparison of the immunological properties of mammalian (rodent), bird, fish, amphibian (toad), and invertebrate (crab) metallothioneins. *Mol. Cell Biochem.* 94:175–181.

Congiu, L., M. Chicca, A. Pilastro, M. Turchetto, and L. Tallandini. 2000. Effects of chronic dietary cadmium on hepatic glutathione levels and glutathione peroxidase activity in starlings (*Sturnus vulgaris*). *Arch. Environ. Contam. Toxicol.* 38:357–361.

Corradino, R. A. 1979. Cadmium inhibition of vitamin D-mediated responses in organ-cultured embryonic chick duodenum. *Toxicol. Appl. Pharmacol.* 48:257–261.

Dauwe, T., L. Bervoets, R. Pinxten, R. Blust, and M. Eens. 2003. Variation of heavy metals within and among feathers of birds of prey: effects of molt and external contamination. *Environ. Pollut.* 124:429–436.

Debacker, V., L.-S. Schiettecatte, T. Jauniax, and J.-M. Bouquegneau. 2001. Influence of age, sex and body condition on zinc, copper, cadmium and metallothioneins in common guillemots (*Uria aalge*) stranded at the Belgian coast. *Mar. Environ. Res.* 52:427–444.

Di Giulio, R. T., and P. F. Scanlon. 1984. Sublethal effects of cadmium ingestion on mallard ducks. *Arch. Environ. Contam. Toxicol.* 13:765–771.

Di Giulio, R. T., and P. F. Scanlon. 1985. Effect of cadmium ingestion and food restriction on energy metabolism and tissue metal concentrations in mallard duck (*Anas platyrhynchos*). *Environ. Res.* 37:433–444.

Dietz, R., F. Riget, and P. Johansen. 1996. Lead, cadmium, mercury and selenium in Greenland marine animals. *Sci. Total Environ.* 186:67–93.

Donaldson, G. M., B. M. Braune, A. J. Gaston, and D. G. Noble. 1997. Organochlorine and heavy metal residues in breast muscle of known-age thick-billed murres (*Uria lomia*) from the Canadian Arctic. *Arch. Environ. Contam. Toxicol.* 33:430–435.

Ek, K. H., G. M. Morrison, P. Lindberg, and S. Rauch. 2004. Comparative tissue distribution of metals in birds in Sweden using ICP-MS and laser ablation ICP-MS. *Arch. Environ. Contam. Toxicol.* 47:259–269.

Elliott, J. E., and A. M. Scheuhammer. 1997. Heavy metal and metallothionein concentrations in seabirds from the Pacific Coast of Canada. *Mar. Pollut. Bull.* 34:794–801.

Elliott, J. E., A. M. Scheuhammer, F. E. Leighton, and P. A. Pearce. 1992. Heavy metal and metallothionein concentrations in Atlantic Canadian seabirds. *Arch. Environ. Contam. Toxicol.* 22:63–73.

Erdogan, Z., S. Erdogan, S. Celik, and A. Unlu. 2005. Effects of ascorbic acid on cadmium-induced oxidative stress and performance of broilers. *Biol. Trace Elem. Res.* 104:19–31.

Feldman, S. L., and R. J. Cousins. 1973. Influence of cadmium on the metabolism of 25-hydoxycholecalciferol in chicks. *Nutr. Rep. Internat.* 8:251–259.

Finkelstein, M. E., et al. 2007. Contaminant-associated alteration of immune function in black-footed albatross (*Phoebastria nigripes*), a North Pacific predator. *Environ. Toxicol. Chem.* 26:1896–1903.

Fox, M. R.S., R. M. Jacobs, A. O.L. Jones, B. E. Fry, Jr., and C. L. Stone. 1980. Effects of vitamin C and iron on cadmium metabolism. *Ann. N. Y. Acad. Sci.* 355:249–261.

Fox, M. R.S., S.-H. Tao, C. L. Stone, and B. E. Fry, Jr. 1984. Effects of zinc, iron and copper deficiencies on cadmium in tissues of Japanese quail. *Environ. Health Perspect.* 54:57–65.

Freeland, J. H., and R. J. Cousins. 1973. Effect of dietary cadmium on anemia, iron absorption, and cadmium binding protein in the chick. *Nutr. Rep. Internat.* 8:337–347.

Friberg, L., T. Kjellstrom, and G. F. Nordberg. 1986. Cadmium. In *Handbook on the toxicology of metals,* eds. L. Friberg, G. F. Nordberg, and V. Vouk, pp. 130–184. Amsterdam: Elsevier.

Fullmer, C. S., T. Oku, and R. H. Wasserman. 1980. Effect of cadmium administration on intestinal calcium absorption and vitamin D-dependent calcium-binding protein. *Environ. Res.* 22:386–399.

Furey, W. F., A. H. Robbins, L. L. Clancy, D. R. Winge, B. C. Wang, and C. D. Stout. 1986. Crystal structure of Cd, Zn metallothionein. *Science* 231:704–710.

Furness, R. W. 1996. Cadmium in birds. In *Environmental contaminants in wildlife: interpreting tissue concentrations,* eds. W. N. Beyer, G. H. Heinz, and A. W. Redmon-Norwood, pp. 389–404. Boca Raton, FL: CRC Lewis Publishers.

García-Fernández, A. J., J. A. Sánchez-García, P. Jiménez-Montalbán, and A. Luna. 1995. Lead and cadmium in wild birds in southeastern Spain. *Environ. Toxicol. Chem.* 14:2049–2058.

García-Fernández, A. J., J. A. Sánchez-Garcíia, M. Gómez-Zapata, and A. Luna. 1996. Distribution of cadmium in blood and tissues of wild birds. *Arch. Environ. Contam. Toxicol.* 30:252–258.

Goede, A. A., and P. deVoogt. 1985. Lead and cadmium in waders from the Dutch Wadden Sea. *Environ. Pollut. (Ser. A)* 37:311–322.

González-Solis, J., C. Sanpera, and X. Ruiz. 2002. Metals and selenium as bioindicators of geographic and trophic segregation in giant petrels *Macronectes* spp. *Mar. Ecol. Prog.* Ser. 244:257–264.

Gupta, P., C. C. Chaurasia, P. K. Maiti, and A. Kar. 2000. Cadmium induced inhibition of type-I iodothyronine 5′-monodeiodinase in young cockerel: the possible involvement of free radicals. *Wat. Air Soil Poll.* 117:245–250.

Hammond, P. B., and R. P. Beliles. 1980. Metals. In *Casarett and Doull's toxicology: the basic science of poisons,* eds. J. Doull, C. D. Klaassen, and M. O. Amdur, pp. 409–467. New York, NY: Macmillan Publ. Co.

Heinz, G. H., S. D. Heseltine, and L. Sileo. 1983. Altered avoidance behavior of young black ducks fed cadmium. *Environ. Toxicol. Chem.* 2:419–421.

Henny, C. J., D. D. Rudis, T. J. Roffe, and E. Robinson-Wilson. 1995. Contaminants and sea ducks in Alaska and the circumpolar region. *Environ. Health Perspect.* 103:41–49.

Henny, C. J., L. J. Blus, R. A. Grove, and S. P. Thompson. 1991. Accumulation of trace elements and organochlorines by surf scoters wintering in the Pacific Northwest. *Northwest Nat.* 72:43–60.

Honda, K., J. E. Marcovecchio, S. Kan, R. Tatsukawa, and H. Ogi. 1990. Metal concentrations in pelagic seabirds from the North Pacific Ocean. *Arch. Environ. Contam. Toxicol.* 19:704–711.

Horai, S., C. Watanabe, H. Takada, Y. Iwamizu, T. Hayashi, S. Tanabe, and K. Kuno. 2007. Trace element concentrations in 13 avian species collected from the Kanto area, Japan. *Sci. Total Environ.* 373:512–525.

Hughes, M. R., D. C. Bennett, D. A. Gray, J. Sharp, A. M. Scheuhammer, and J. E. Elliott. 2003. Effects of cadmium ingestion on plasma and osmoregulatory hormone concentrations in male and female Pekin ducks. *J. Toxicol. Environ. Health* Pt A. 66:565–579.

Hughes, M. R., J. E. Smits, J. E. Elliott, and D. C. Bennett. 2000. Morphological and pathological effects of cadmium ingestion on Pekin ducks exposed to saline. *J. Toxicol. Environ. Health, Part A* 61:591–608.

Hutton, M. 1981. Accumulation of heavy metals and selenium in three seabird species from the United Kingdom. *Environ. Pollut. (Ser. A)* 26:129–145.

Jacobs, R. M., M. R.S. Fox, and M. H. Aldridge. 1969. Changes in plasma proteins associated with the anemia produced by dietary cadmium in Japanese quail. *J. Nutr.* 99:119–128.

Jaspers, V., T. Dauwe, R. Pinxten, L. Bervoets, R. Blust, and M. Eens. 2004. The importance of exogenous contamination on heavy metal levels in bird feathers. A field experiment with free-living great tits, *Parus major*. *J. Environ. Monitor.* 6:356–360.

Kagi, J. H. R., and B. L. Vallee. 1960. Metallothionein: a cadmium- and zinc-containing protein from equine renal cortex. *J. Biol. Chem.* 235:3460–3465.

Kaji, T., R. Kawatani, M. Takata, T. Hoshino, T. Miyahara, H. Kozuka, and F. Koizumi. 1988. The effects of cadmium, copper or zinc on formation of embryonic chick bone in tissue culture. *Toxicology* 50:303–316.

Kertész, V., and T. Fáncsi. 2003. Adverse effects of (surface water pollutants) Cd, Cr, and Pb on the embryogenesis of the mallard. *Aquat. Toxicol.* 65:425–433.

Kertész, V., and I. Hlubik. 2002. Plasma ALP and blood PCV value changes in chick fetuses due to exposure of the egg to different xenobiotics. *Environ. Pollut.* 117:323–327.

Kim, E. Y., R. Goto, S. Tanabe, H. Tanaka, and R. Tatsukawa. 1998. Distribution of 14 elements in tissues and organs of oceanic seabirds. *Arch. Environ. Contam. Toxicol.* 35:638–645.

Klaassen, C. D., J. Liu, and S. Choudhuri. 1999. Metallothionein: an intracellular protein to protect against cadmium toxicity. *Annu. Rev. Pharmacol. Toxicol.* 39:267–294.

Kojadinovic, J., M. LeCorre, R. P. Cosson, and P. Bustamante. 2007. Trace elements in three marine birds breeding on Reunion Island (western Indian Ocean): Part 1: factors influencing their bioaccumulation. *Arch. Environ. Contam. Toxicol.* 52:418–430.

Koo, S. I., C. S. Fullmer, and R. H. Wasserman. 1978. Intestinal absorption and retention of [109]Cd: effects of cholecalciferol, calcium status and other variables. *J. Nutr.* 108:1812–1822.

Korének, M., et al. 2000. Effect of vitamin D3 on chymotrypsin activity in the droppings of laying hens after an exposure to cadmium. *Bull. Vet. Inst. Pulawy.* 44:227–233.

Korének, M., et al. 2007. Effects of cadmium and zinc on the activity of chymotrypsin and trypsin in droppings of Japanese quails. *Internat. J. Environ. Stud.* 64:221–227.

Kósa, E., Z. Rigó, I. Csutorás, and Á. Szakács. 2004. Laboratorial investigations on effects of different concentrations of cadmium in broiler chickens. *Trace Elem. Electro.* 21:211–214.

Larison, J. R., G. E. Likens, J. W. Fitzpatrick, and J. G. Crock. 2000. Cadmium toxicity among wildlife in the Colorado Rocky Mountains. *Nature* 406:181–183.

Leach, R. M., Jr., K. W. Wang, and D. E. Baker. 1979. Cadmium and the food chain: the effect of dietary cadmium on tissue composition in chicks and laying hens. *J. Nutr.* 109:437–443.

Leonzio, C., M. C. Fossi, L. Lari, and S. Focardi. 1992. Influence of cadmium on polychlorobiphenyl uptake, MFO activity and serum lipid levels in Japanese quail. *Arch. Environ. Contam. Toxicol.* 22:238–241.

Lock, J. W., D. R. Thompson, R. W. Furness, and J. A. Bartle. 1992. Metal concentrations in seabirds of the New Zealand region. *Environ. Pollut.* 75:289–300.

Macdonald, C. R., and J. B. Sprague. 1988. Cadmium in marine invertebrates and arctic cod in the Canadian Arctic. Distribution and ecological implications. *Mar. Ecol. Prog. Ser.* 47:17–30.

Mahler, B. J., P. C. Van Metre, and E. Callender. 2006. Trends in metals in urban and reference lake sediments across the United States, 1970 to 2001. *Environ. Toxicol. Chem.* 25:1698–1709.

Margoshes, M., and B. L. Vallee. 1957. A cadmium kidney protein from equine kidney cortex. *J. Amer. Chem. Soc.* 79:4813–4814.

Martínez-López, E., P. María-Mojica, J. E. Martínez, J. F. Calvo, D. Romero, and A. J. García-Fernández. 2005. Cadmium in feathers of adults and blood of nestlings of three raptor species from a nonpolluted Mediterranean forest, southeastern Spain. *Bull. Environ. Contam. Toxicol.* 74:477–484.

Mayack, L. A., P. B. Bush, O. J. Fletcher, R. K. Page, and T. T. Fendley. 1981. Tissue residues of dietary cadmium in wood ducks. *Arch. Environ. Contam. Toxicol.* 10:637–645.

McKenna, I. M., R. L. Chaney, S.-H. Tao, R. M. Leach, Jr., and F. M. Williams. 1992. Interactions of plant zinc and plant species on the bioavailability of plant cadmium to Japanese quail fed lettuce and spinach. *Environ. Res.* 57:73–87.

Miyahara, T., M. Miyakoshi, Y. Saito, and H. Kozuka. 1980. Influence of poisonous metals on bone metabolism. III. The effect of cadmium on bone resorption in tissue culture. *Toxicol. Appl. Pharmacol.* 55:477–483.

Miyahara, T., Y. Oh-e, E. Takaine, and H. Kozuka. 1983. Interaction between cadmium and zinc, copper, or lead in relation to the collagen and mineral content of embryonic chick bone in culture. *Toxicol. Appl. Pharmacol.* 67:41–48.

Mochizuki, M., R. Hondo, K. Kumon, R. Sasaki, H. Matsuba, and F. Ueda. 2002. Cadmium contamination in wild birds as an indicator of environmental pollution. *Environ. Monitor. Assess.* 73:229–235.

Mohan, J., R. P. Moudgal, and N. B. Singh. 1990. Effects of cadmium salt administration on the angiotensin converting enzyme and other semen parameters in domestic fowl (*Gallus domesticus*). *Med. Sci. Res.* 18:799–801.

Mora, M. A. 2003. Heavy metals and metaloids in egg contents and eggshells of passerine birds from Arizona. *Environ. Pollut.* 125:393–400.

Muirhead, S. J., and R. W. Furness. 1988. Heavy metal concentrations in the tissues of seabirds from Gough Island, South Atlantic Ocean. *Mar. Pollut. Bull.* 19:278–283.

Müller, J., H. Ruppert, Y. Muramatsu, and J. Schneider. 2000. Reservoir sediments—a witness of mining and industrial development (Matler Reservoir, eastern Erzegebirge, Germany). *Environ. Geol.* 39:1341–1351.

Myklebust, I., and H. C. Pedersen. 1999. Accumulation and distribution of cadmium in willow ptarmigan. *Ecotoxicol.* 8:457–465.

Nam, D.-H., Y. Anan, T. Ikemoto, Y. Okabe, E.-Y. Kim, A. Subramanian, K. Saeki, and S. Tanabe. 2005. Specific accumulation of 20 trace elements in great cormorants (*Phalacrocorax carbo*) from Japan. *Environ. Pollut.* 134:503–514.

Nicholson, J. K., and D. Osborn. 1983. Kidney lesions in pelagic seabirds with high tissue levels of cadmium and mercury. *J. Zool. Lond.* 200:99–118.

Nicholson, J. K., M. D. Kendall, and D. Osborn. 1983. Cadmium and mercury nephrotoxicity. *Nature* 304:633–635.

Nolan, T. D., and D. Brown. 2000. The influence of elevated dietary zinc, selenium, and their combination on the suppressive effect of dietary and intraperitoneal cadmium on egg production in laying hens. *J. Toxicol. Environ. Health Pt A* 60:549–565.

Nriagu, J. O., and J. M. Pacyna. 1988. Quantitative assessment of worldwide contamination of air, water and soils by trace metals. *Nature* 333:134–139.

Ohlendorf, H. M., K. C. Marois, R. W. Lowe, T. E. Harvey, and P. R. Kelly. 1991. Trace elements and organochlorines in surf scoters from San Francisco Bay, 1985. *Environ. Monitor. Assess.* 18:105–122.

Orlowski, G., R. Polechonski, W. Dobicki, and Z. Zawada. 2007. Heavy metal concentrations in the tissues of the black-headed gull *Larus ridibundus* L. nesting in the Dam Reservoir in south-western Poland. *Pol. J. Ecol.* 55:783–793.

Osborn, D. 1979. Seasonal changes in the fat, protein and metal concentrations of the liver of the starling *Sturnus vulgaris*. *Environ. Pollut.* 19:145–154.

Osborn, D., M. P. Harris, and J. K. Nicholson. 1979. Comparative tissue distribution of mercury, cadmium and zinc in three species of pelagic seabirds. *Comp. Biochem. Physiol.* 64C:61–67.

Pacyna, J. M., and E. G. Pacyna. 2001. An assessment of global and regional emissions of trace metals to the atmosphere from anthropogenic sources worldwide. *Environ. Rev.* 9:269–298.

Pedersen, H. C., and M. Saether. 1999. Effects of cadmium on parental behaviour in free-living willow ptarmigan. *Ecotoxicol.* 8:1–7.

Pedersen, H. C., F. Fossøy, J. A. Kålås, and S. Lierhagen. 2006. Accumulation of heavy metals in circumpolar willow ptarmigan (*Lagopus l. lagopus*) populations. *Sci. Total Environ.* 371:176–189.

Peltier, G. L., J. L. Meyer, C. H. Jagoe, and W. A. Hopkins. 2008. Using trace element concentrations in *Corbicula fluminea* to identify potential sources of contamination in an urban river. *Environ. Pollut.* 154:283–290.

Pilastro, A., L. Congiu, L. Tallandini, and M. Turchetto. 1993. The use of bird feathers for the monitoring of cadmium pollution. *Arch. Environ. Contam. Toxicol.* 24:355–358.

Prasada Rao, P. V. V., S. A. Jordan, and M. K. Bhatnager. 1989a. Combined nephrotoxicity of methylmercury, lead and cadmium in Pekin ducks: metallothionein, metal interactions and histopathology. *J. Toxicol. Environ. Health* 26:327–348.

Prasada Rao, P. V. V., S. A. Jordan, and M. K. Bhatnager. 1989b. Ultrastructure of kidney of ducks exposed to methylmercury, lead and cadmium in combination. *J. Environ. Pathol. Toxicol. Oncol.* 9:19–44.

Rahman, M. S., M. Mochizuki, and M. Mori. 2008. Cadmium disrupts the diethylstilbestrol effect on very-low density apolipoprotein II gene transcription in the liver of Japanese quail (*Coturnix japonica*). *J. Poult. Sci.* 45:62–66.

Rahman, M. S., T. Sasanami, and M. Mori. 2007. Effects of cadmium administration on reproductive performance of Japanese quail (*Coturnix japonica*). *J. Poult. Sci.* 44:92–97.

Rice, K. C. 1999. Trace element concentrations in streambed sediment across the conterminous United States. *Environ. Sci. Technol.* 33:2499–2504.

Richardson, M. E., and M. R.S. Fox. 1974. Dietary cadmium and enteropathy in the Japanese quail: histochemical and ultrastructural studies. *Lab. Invest.* 31:722–731.

Richardson, M. E., M. R.S. Fox, and B. E. Fry, Jr. 1974. Pathological changes produced in Japanese quail by ingestion of cadmium. *J. Nutr.* 104:323–338.

Rodrigue, J., L. Champoux, D. Leclair, and J.-F. Duchense. 2007. Cadmium concentrations in tissues of willow ptarmigan (*Lagopus lagopus*) and rock ptarmigan (*Lagopus muta*) in Nunavik, northern Quebec. *Environ. Pollut.* 147:642–647.

Sato, S., M. Okabe, T. Emoto, M. Kurasaki, and Y. Kojima. 1997. Restriction of cadmium transfer to eggs from laying hens exposed to cadmium. *J. Toxicol. Environ. Health* 51:15–22.

Savinov, V. M., G. W. Gabrielsen, and T. N. Savinova. 2003. Cadmium, zinc, copper, arsenic, selenium and mercury in seabirds from the Barents Sea: levels, inter-specific and geographical differences. *Sci. Total Environ* 306:133–158.

Scanlon, P. F. 1982. Wet and dry weight relationships in mallard (*Anas platyrhynchos*) tissues. *Bull. Environ. Contam. Toxicol.* 29:615–617.

Scheuhammer, A. M. 1987. The chronic toxicity of aluminium, cadmium, mercury and lead in birds: a review. *Environ. Pollut.* 46:263–295.

Scheuhammer, A. M. 1988. The dose-dependent deposition of cadmium into organs of Japanese quail following oral administration. *Toxicol. Appl. Pharmacol.* 95:153–161.

Scheuhammer, A. M. 1996. Influence of reduced dietary calcium on the accumulation and effects of lead, cadmium, and aluminum in birds. *Environ. Pollut.* 94:337–343.

Scheuhammer, A. M., and D. M. Templeton. 1990. Metallothionein production: similar responsiveness of avian liver and kidney to chronic cadmium administration. *Toxicology* 60:151–159.

Schmitt, C. J., and W. G. Brumbaugh. 1990. National Contaminant Biomonitoring Program: concentrations of arsenic, cadmium, copper, lead, mercury, selenium and zinc in U.S. freshwater fish, 1976–1984. *Arch. Environ. Contam. Toxicol.* 19:731–747.

Schulz-Baldes, M. 1989. The sea-skater *Halobates micans*: an open ocean bioindicator for cadmium distribution in Atlantic surface waters. *Mar. Biol.* 102:211–215.

Sell, J. L. 1975. Cadmium and the laying hen: apparent absorption, tissue distribution and virtual absence of transfer to eggs. *Poult. Sci.* 54:1674–1678.

Silver, T. M., and T. D. Nudds. 1995. Influence of low-level cadmium and reduced calcium intake on tissue Cd concentrations and behaviour of American black ducks. *Environ. Pollut.* 90:153–161.

Stewart, F. M., and R. W. Furness. 1998. The influence of age on cadmium concentrations in seabirds. *Environ. Monit. Assess.* 50:159–171.

Stewart, F. M., D. R. Thompson, R. W. Furness, and N. Harrison. 1994. Seasonal variation in heavy metal levels in tissues of common guillemots, *Uria aalgae* from northwest Scotland. *Arch. Environ. Contam. Toxicol.* 27:168–175.

Stewart, F. M., R. W. Furness, and L. R. Monteiro. 1996. Relationships between heavy metal and metallothionein concentrations in lesser black-backed gulls, *Larus fuscus*, and Cory's shearwater, *Calonectris diomedea*. *Arch. Environ. Contam. Toxicol.* 30:299–305.

Stock, M., R. F. Herber, and H. M.A. Gerón. 1989. Cadmium levels in oystercatcher *Haematopus ostralegus* from the German Wadden Sea. *Mar. Ecol. Prog. Ser.* 53:227–234.

Stoeppler, M., and K. Brandt. 1980. Contributions to automated trace analysis. V. Determination of Cd in whole blood and urine by electrothermal AAS. Fresenius Z. *Anal. Chem.* 300:372–380.

Stoudt, J. H., K. A. Trust, J. F. Cochrane, R. S. Suydam, and L. T. Quakenbush. 2002. Environmental contaminants in four eider species from Alaska and arctic Russia. *Environ. Pollut.* 119:215–226.

Świergosz, R., and A. Kowalska. 2000. Cadmium accumulation and its effects on growing pheasants *Phasianus colchicus* (L.). *Environ. Toxicol. Chem.* 19:2742–2750.

Teraoka, H., et al. 2007. Heavy metal contamination status of Japanese cranes (*Grus japonensis*) in East Hokkaido, Japan—extensive mercury pollution. *Environ. Toxicol. Chem.* 26:307–312.

Thompson, D. R., and J. E. Dowding. 1999. Site-specific heavy metal concentrations in blood of South Island pied oystercatchers *Haematopus ostralegus finschi* from the Auckland Region, New Zealand. *Mar. Pollut. Bull.* 38:202–206.

Toman, R., P. Massányi, N. Lukáč, L. Ducsay, and J. Golian. 2005. Fertility and content of cadmium in pheasant (*Phasianus colchicus*) following cadmium intake in drinking water. *Ecotoxicol. Environ. Safe.* 62:112–117.

Trust, K. A., K. T. Rummel, A. M. Scheuhammer, I. L. Brisbin, and M. J. Hooper. 2000. Contaminant exposure and biomarker responses in spectacled elders (*Somateria fischeri*) from St. Lawrence Island, Alaska. *Arch. Environ. Contam. Toxicol.* 38:107–113.

Tsipoura, N., et al. 2008. Metal concentrations in three species of passerine birds in the Hackensack Meadowlands of New Jersey. *Environ. Res.* 107:218–228.

Vanparys, C., et al. 2008. Metallothioneins (MTs) and δ–aminolevulinic acid dehydratase (ALAd) as biomarkers of metal pollution in great tits (*Parus major*) along a pollution gradient. *Sci. Total Environ.* 401:184–193.

Wayland, M., A. J. García-Fernández, E. Neugebauer, and H. G. Gilchrist. 2001a. Concentrations of cadmium, mercury and selenium in blood, liver and kidney of common eider ducks from the Canadian Arctic. *Environ. Monitor. Assess.* 71: 255–267.

Wayland, M., et al. 2001b. Trace elements in king eiders and common eiders in the Canadian Arctic. *Arch. Environ. Contam. Toxicol.* 41:491–500.

Wayland, M., H. G. Gilchrist, and E. Neugebauer. 2005. Concentrations of cadmium, mercury and selenium in common eider ducks in the eastern Canadian Arctic: influence of reproductive stage. *Sci. Total Environ.* 351:323–332.

Wayland, M., R. T. Alisauskas, D. K. Kellett, and K. R. Mehl. 2008a. Trace element concentrations in blood of nesting king eiders in the Canadian Arctic. *Arch. Environ. Contam. Toxicol.* 55:683–690.

Wayland, M., et al. 2008b. Survival rates and blood metal concentrations in two species of North American sea ducks. *Environ. Toxicol. Chem.* 27:698–704.

White, D. H., and M. T. Finley. 1978. Uptake and retention of dietary cadmium in mallard ducks. *Environ. Res.* 17:53–59.

White, D. H., M. T. Finley, and J. F. Ferrell. 1978. Histopathologic effects of dietary cadmium on kidneys and testes of mallard ducks. *J. Toxicol. Environ. Health* 4:551–558.

WHO Task Group on Environmental Health Criteria for Cadmium. 1992. Cadmium: Environmental Health Criteria 134. World Health Organization, Geneva.

Wilson, H. M., M. R. Petersen, and D. Troy. 2004. Concentrations of metals and trace elements in blood of spectacled and king eiders in northern Alaska, USA. *Environ. Toxicol. Chem.* 23:408–414.

Winstel, C., and Callahan, P. 1992. Cadmium exposure inhibits the prolactin secretory response to thyrotrophin releasing hormone (THR) *in vitro*. *Toxicology* 74:9–17.

Yamagami, K., S. Nishimura, and M. Sorimachi. 1994. Cd^{2+} and Co^{2+} at micromolar concentrations stimulate catecholamine secretion by increasing the cytosolic free Ca^{2+} concentration in cat adrenal chromaffin cells. *Brain Res.* 646:295–298.

Avocets

By C.G. Specht, from *The Royal Natural History*, edited by Richard Lydekker, Frederick Warne & Co., London, 1893–94.

21 Selenium in Birds

Harry M. Ohlendorf
Gary H. Heinz

CONTENTS

21.1 Introduction 669
21.2 Dietary Requirements versus Toxicity 671
21.3 Egg and Tissue Concentrations..676
    21.3.1 Eggs 676
        21.3.1.1 Laboratory Studies..677
        21.3.1.2 Field Studies...681
    21.3.2 Liver..683
        21.3.2.1 Laboratory Studies..683
        21.3.2.2 Field Studies...685
    21.3.3 Kidney..686
    21.3.4 Muscle..687
    21.3.5 Blood..687
    21.3.6 Integument/Feathers 689
21.4 Biomarkers...690
    21.4.1 Biochemical 690
    21.4.2 Morphological..690
21.5 Interactions 691
21.6 Hormesis 693
Summary 693
Acknowledgments...696
References...696

21.1 INTRODUCTION

Selenium (Se) is a metalloid trace element that birds and other wildlife need in small amounts for good health. The main purpose of this chapter is to interpret tissue concentrations of Se. However, because food is the main source of Se accumulation for birds and other wildlife, and because dietary concentrations for effects on bird reproduction have been reported, we also provide interpretive information on Se in the diet.

Se deficiencies in domestic poultry and livestock occur in some parts of the world and must be corrected by additions of Se to the diet. However, the range of dietary concentrations that provides adequate but nontoxic amounts of Se is narrow compared with the ranges for most other essential trace elements.

In the 1930s, grains grown on seleniferous soils in South Dakota caused reproductive failure when fed to chickens (*Gallus domesticus*) (Poley and Moxon 1938). The most drastic incident of Se poisoning in wild birds occurred at Kesterson Reservoir (located on the Kesterson National Wildlife Refuge) in California during the early and mid-1980s (Ohlendorf et al. 1986a, 1988, Ohlendorf 1989, 2002, Ohlendorf and Hothem 1995). Water used to irrigate crops in the San Joaquin Valley of California dissolved naturally occurring Se salts from the soil, and when the Se-laden subsurface water was drained from agricultural fields into Kesterson Reservoir, levels of Se that were toxic to birds accumulated in plants and animals used as foods by the birds. Reproductive failure and adult mortality occurred. The findings at Kesterson Reservoir received extensive publicity and led to a series of laboratory and field studies (summarized in this chapter) that provide one of the best case studies in ecotoxicology during the past 30 years. The integrated field studies at Kesterson and related laboratory studies have been recognized as a "gold standard" in the field of ecotoxicology (Suter 1993). Similar problems of impaired bird reproduction were subsequently discovered elsewhere in the western United States, most notably in the Tulare Basin in California (Skorupa and Ohlendorf 1991, Skorupa 1998a).

High concentrations of Se in foods of wildlife are not limited to areas where soils are naturally high in Se. They also can be the result of the disposal of sewage sludge or fly ash, mining activity, or emissions from metal smelters (Robberecht et al. 1983, Wadge and Hutton 1986, Cappon 1991, Skorupa 1998a, Ratti et al. 2006, Wayland and Crosley 2006).

An assessment of the toxicity of Se is complicated by its occurrence in many different chemical forms, some differing greatly in their toxicity to birds. The four common oxidation states are selenide (-2), elemental Se (0), selenite (+4), and selenate (+6). Elemental Se is virtually insoluble in water and presents little risk to birds. Both selenite and selenate are toxic to birds, but organic selenides pose the greatest hazard. Among the organic selenides, selenomethionine has been shown to be highly toxic to birds and seems to be the form most likely to harm wild birds because it results in high bioaccumulation of Se in their eggs.

Much has been learned about Se toxicity to birds during the last 25 years; some of that information was summarized in the earlier edition by Heinz (1996). Other reviews in relation to exposure and effects of Se in birds are provided by Skorupa (1998a), O'Toole and Raisbeck (1998), U.S. DI (1998), Eisler (2000), Hoffman (2002), and Ohlendorf (2003). The purpose of this chapter is to identify the concentrations of Se in avian diets and in avian eggs and other tissues that are toxic, and to discuss how different chemical forms of Se and their interactions with other environmental contaminants can alter toxicity. We also present what are considered background (or no-effect) concentrations of Se from Se-normal areas, when available.

Background and reference area concentrations can be very useful for interpreting the possible toxic thresholds of a contaminant, especially when it is known with some certainty that the reference area has no known source of the contaminant in question. However, because some "background" concentrations of contaminants such as Se are reported from areas where the Se input is unknown, and may not, in fact, be what might be called "normal," "baseline," or "uncontaminated," they should be referred to as "reference area" samples, and a certain degree of caution must be exercised when using those concentrations as being synonymous with safe levels. The rigorous identification of safe levels of Se, or other contaminants, can really come only from the findings of controlled laboratory dosing studies and carefully designed field studies. In other words, merely because a contaminant like Se is at a level that has been reported from what are believed to be Se-normal areas does not, in itself, prove that the levels are safe.

The manner in which different authors present Se concentrations can be confusing, so it is important to understand the various ways results can be presented. Se concentrations typically are reported as micrograms per liter (μg/L) in most fluids (but sometimes μg/g or μg/dL in blood) and milligrams per kilogram (mg/kg), or micrograms per gram (μg/g) in soil, sediment, plant or animal tissues, and diets. Concentrations in soil, sediment, tissues, and diets can be expressed either on a wet-weight (or fresh-weight basis, which is considered to be synonymous) or a dry-weight basis. Although moisture loss during sample processing can be controlled fairly well in the laboratory, it is sometimes difficult to do so under field conditions. Therefore, reporting results on dry-weight basis helps ensure comparability of values.

Conversion from one basis to the other is a function of the moisture content in the sample (which should be reported regardless of which basis is used), as follows:

$$\text{Dry-weight conc.} = \text{wet-weight conc.} \times \frac{100}{\left(100 - \text{Moisture percentage}\right)}.$$

In this chapter, we preferentially provide Se concentrations in diets and tissues on dry-weight basis (unless otherwise noted), and provide typical moisture content of eggs and tissues to enable readers to make conversions. When results were originally reported on wet-weight basis, the original concentrations are given in parentheses following the approximate dw concentration.

Se's ability to interact with other nutrients and environmental contaminants, especially other elements, also sometimes complicates an interpretation of toxic thresholds in tissues of birds. Although we do not attempt a comprehensive review to interpret critical levels of Se in the presence of elevated levels of other pollutants, we include a brief section on interactions, and the reader should be aware that such interactions exist.

21.2 DIETARY REQUIREMENTS VERSUS TOXICITY

In general, the diet is the most important exposure pathway for birds and, whenever possible, dietary concentrations should be included when reporting results or evaluating the effects observed in experimental or field studies. With the previously stated caution about "background" levels of Se in mind, mean background concentrations in diets of freshwater and terrestrial avian species are typically <3 mg/kg, with thresholds for reproductive impairment in the range of 3–8 mg/kg (Table 21.1).

TABLE 21.1
Published Assessment Values for Effects of Dietary or Tissue Concentrations of Se on Birds

Medium and Level/Status[a]	Concentration (mg Se/kg, dw)	Effects	Comments	Reference
		Diet[b]		
Adequate	0.30–1.1	Nutritional needs are met for poultry	Lower dietary concentrations are marginal or deficient, and diets must be fortified	Puls 1988
High	3.0–5.0	Levels are excessive but not considered toxic to poultry	Poultry are relatively sensitive to effects of selenium	Puls 1988
Toxic	>5.0	Reduced egg hatchability and teratogenic effects in embryos/chicks	Poultry are relatively sensitive to effects of selenium	Puls 1988
Background	<3.0	None	Deficiencies associated with lower concentrations have not been reported in wild birds	U.S. DI 1998, Eisler 2000
Reproductive impairment	3–8	Reduced egg hatchability; potential deformities in embryos/chicks at upper end of range	Sensitivity varies by species and chemical form of Se in diet	U.S. DI 1998, Eisler 2000

continued

TABLE 21.1 (continued)
Published Assessment Values for Effects of Dietary or Tissue Concentrations of Se on Birds

Medium and Level/Status[a]	Concentration (mg Se/kg, dw)	Effects	Comments	Reference
Reproductive impairment	4.0 (95% CI = <0.5–7.3)	EC_{10} for reduced egg hatchability	Based on studies of mallard, American kestrel, chicken, black-crowned night-heron, eastern screech-owl and ring-necked pheasant using logistic regression analysis	Wayland et al. 2007
Reproductive impairment	4.4 (95% CI = 3.8–4.8)	EC_{10} for reduced egg hatchability	Based on results of six laboratory studies with mallards, using hockey-stick regression analysis	Adams (personal communication; see Ohlendorf 2007)
Reproductive impairment	4.9 (95% CI = 3.6–5.7)	EC_{10} for reduced egg hatchability	Based on results of six laboratory studies with mallards, using logistic regression analysis	Ohlendorf 2003
Eggs[c]				
Adequate	0.66–5.0 (0.20–1.5 ww)	Nutritional needs are met for poultry	Lower dietary concentrations are marginal or deficient, and diets must be fortified	Puls 1988
High	5.0–16 (1.5–5.0 ww)	Levels are excessive and upper end of range may be toxic to poultry	Poultry are relatively sensitive to effects of selenium	Puls 1988
Toxic	>8.2 (>2.5 ww)	Reduced egg hatchability and teratogenic effects in embryos/chicks	Poultry are relatively sensitive to effects of selenium	Puls 1988
Background	Mean < 3.0 (typically 1.5–2.5); individual eggs <5	None	Concentrations may be higher in some marine birds (Ohlendorf and Harrison 1986, Braune et al. 2002)	Ohlendorf and Harrison 1986, Skorupa and Ohlendorf 1991, U.S. DI 1998, Eisler 2000
Reproductive impairment	6–7 (about 1.8–2.1 ww)	EC_{10} on a clutch-wise (or hen-wise) basis and EC_{03} on egg-wise basis	Based on results of extensive field studies of black-necked stilts	Skorupa 1998b, 1999
Reproductive impairment	7.7 (about 2.3 ww)	EC_{10} for reduced egg hatchability	Based on results of one laboratory study with mallards, assuming hormetic effects	Beckon et al. 2008
Reproductive impairment	9.0	$EC_{8.2}$ for impaired clutch viability	Based on results of one laboratory study with mallards, using linear regression analysis	Lam et al. 2005
Reproductive impairment	12 (95% CI = 6.4–16)	EC_{10} for reduced egg hatchability	Based on results of six laboratory studies with mallards, using logistic regression analysis	Ohlendorf 2003

TABLE 21.1 (continued)
Published Assessment Values for Effects of Dietary or Tissue Concentrations of Se on Birds

Medium and Level/Status[a]	Concentration (mg Se/kg, dw)	Effects	Comments	Reference
Reproductive impairment	12 (95% CI = 9.7–14)	EC_{10} for reduced egg hatchability	Based on results of six laboratory studies with mallards, using hockey-stick analysis	Adams (personal communication; see Ohlendorf 2007)
Reproductive impairment	14	$EC_{11.8}$ for reduced egg hatchability	Based on results of extensive field studies of black-necked stilts	Lam et al. 2005
Teratogenicity	13–24	Threshold for teratogenic effects on population level	Sensitivity varies widely by species	Skorupa and Ohlendorf 1991
Teratogenicity	23	EC_{10} for teratogenic effects in mallard	Mallard is considered a "sensitive" species	Skorupa 1998b, U.S. DI 1998
Teratogenicity	37	EC_{10} for teratogenic effects in stilt	Stilt is considered an "average" species	Skorupa 1998b, U.S. DI 1998
Teratogenicity	74	EC_{10} for teratogenic effects in American avocet	Avocet is considered a "tolerant" species	Skorupa 1998b, U.S. DI 1998
Liver[d]				
Adequate	1.2–3.3 (0.35–1.0 ww)	Nutritional needs are met	Lower liver concentrations are marginal or deficient, and diets must be fortified	Puls 1988
High	6.6–20 (2.0–6.0 ww)	Levels are excessive but not considered toxic to poultry	Poultry are relatively sensitive to effects of selenium	Puls 1988
Toxic	13–76 (4.0–23 ww)	Reduced egg hatchability and teratogenic effects in embryos/chicks	Poultry are relatively sensitive to effects of selenium	Puls 1988
Background for freshwater and terrestrial species	<10	None	Deficiencies associated with lower concentrations have not been documented in wild birds	U.S. DI 1998, Eisler 2000
Background for marine species	20–75 in some species (see text)	None	Found in livers of several species from uncontaminated areas	Elliott et al. 1992, Dietz et al. 1996, Trust et al. 2000, Grand et al. 2002, Mallory et al. 2004, Campbell et al. 2005, Elliott 2005
Elevated and potentially toxic	10–20	Considered suspicious of selenium toxicosis when accompanied by symptoms listed for toxic effects	Sensitivity varies by species	Ohlendorf et al. 1988, Albers et al. 1996, O'Toole and Raisbeck 1997, 1998

continued

TABLE 21.1 (continued)
Published Assessment Values for Effects of Dietary or Tissue Concentrations of Se on Birds

Medium and Level/Status[a]	Concentration (mg Se/kg, dw)	Effects	Comments	Reference
Toxic	20–25	Diagnostic when accompanied by emaciation, poor quality of shed nails, bilaterally symmetrical alopecia of the head and neck, hepatic lesions, and necrosis of maxillary nails	Based on field observations and laboratory studies with mallards	Ohlendorf et al. 1988, Albers et al. 1996, O'Toole and Raisbeck 1997, 1998
Toxic	351–735	Many effects on liver and other tissues	Common eiders seem to be more tolerant of selenium in tissues than are mallards	Franson et al. 2007
Kidney[e]				
Adequate	2.2–5.2 (0.50–1.2 ww)	Nutritional needs are met in poultry	Similar to wild birds, concentrations tend to be higher than in liver	Moksnes 1983, Puls 1988
High	6.4–22 (1.5–5.2 ww)	Levels are excessive but not considered toxic to poultry	Similar to wild birds, concentrations tend to be equal to or lower than in liver	Moksnes 1983, Puls 1988
Muscle[f]				
Adequate	0.49–4.9 (0.13–1.3 ww)	Nutritional needs are met	Lower muscle concentrations are marginal or deficient, and diets must be fortified	Puls 1988
High	1.5–21 (0.40–5.5 ww)	Levels are excessive but may not be toxic to poultry	Wide range of concentrations that overlaps with toxic level	Puls 1988
Toxic	4.9 (1.3 ww)	Toxic level is below the midpoint of the "high" range	Concentrations in muscle are not very useful for diagnosing current exposure because of long lag in reaching equilibrium	Puls 1988
Background	1–3	None	Accumulation in muscle varies by bird species and chemical form of selenium; concentrations above background in muscle more useful for assessing human health risks than diagnosing toxic effects in birds	U.S. DI 1998, Eisler 2000
Blood[g]				
Adequate	0.62–0.96 (0.13–0.20 ww)	Nutritional needs are met	Lower blood concentrations are marginal or deficient, and diets must be fortified	Puls 1988

TABLE 21.1 (continued)
Published Assessment Values for Effects of Dietary or Tissue Concentrations of Se on Birds

Medium and Level/Status[a]	Concentration (mg Se/kg, dw)	Effects	Comments	Reference
Background	0.48–1.9 (0.10–0.40 ww)	None	Deficiencies associated with lower concentrations have not been documented in wild birds	U.S. DI 1998, Eisler 2000
Provisional threshold warranting further study	4.8 (1.0 ww)	Interpretive relationship to effects is limited, but elevated levels associated with effects on reproduction or survival	Blood selenium concentrations are good indicator of current/recent exposure, and especially important for sampling when animals should not be sacrificed	Heinz et al. 1990, Heinz and Fitzgerald 1993a, O'Toole and Raisbeck 1997, U.S. DI 1998, Yamamoto et al. 1998, Santolo et al. 1999, Eisler 2000
Feathers[h]				
Background	1–4 (typically 1–2)	None	Based on breast feathers; concentrations in feathers vary by type and reflect exposure at the time feathers were grown, rather than current exposure	Burger 1993, Ohlendorf 1993, U.S. DI 1998, Eisler 2000
Provisional threshold warranting further study	5	Interpretive relationship to effects is limited, but elevated levels associated with exposure when the feathers were developing	Feather selenium concentrations are not good indicator of current/recent exposure, but may be useful if limitations are understood (see text)	Burger 1993, Ohlendorf 1993, U.S. DI 1998, Eisler 2000

[a] Typical moisture content (%) and approximate conversion factor are shown in footnotes for each medium. Values that are shaded are based on domestic poultry rather than wild species.

[b] Variable moisture; laboratory diet typically ~10%, but natural diet varies widely (<10–>90%).

[c] 65–80% moisture, varying with species and incubation stage; use 70% (i.e., factor of 3.3) for approximate conversion.

[d] 70% moisture; use factor of 3.3 for approximate conversion.

[e] 76–78% moisture, based on limited data; use factor of 4.3 for approximate conversion.

[f] 74% moisture; use factor of 3.8 for approximate conversion.

[g] 79% moisture in lab studies, variable under field conditions; use factor of 4.8 for approximate conversion.

[h] 10% moisture assumed (not well defined); use factor of 1.1 for approximate conversion.

For birds, as for most other animals, dietary Se requirements appear to be between 0.05 and 0.5 mg/kg (NAS-NRC 1976, 1983, Combs and Combs 1986, Oldfield 1990, 1998, Eisler 2000). Excess Se in the diet of female birds during the period just before egg-laying can result in the transfer of Se to the eggs or other tissues at harmful levels, although sensitivity to Se varies among species (Skorupa and Ohlendorf 1991, Ohlendorf 1996, Skorupa 1998a, 1998b). Detwiler (2002) analyzed field-collected eggs and conducted laboratory studies with chickens to determine partitioning of Se in eggs (to albumen, yolk, and embryo) and to identify toxicokinetic causes of species variability in sensitivity to Se. As expected, differences among species, as well as those due to form of Se in the diet, are complex. Those complexities are not described in detail here, but readers may wish to read about them in Detwiler's (2002) work.

Ohlendorf (2003) used data from six laboratory studies with mallards (*Anas platyrhynchos*) (Heinz et al. 1987, 1989, Stanley et al. 1994, 1996, Heinz and Hoffman 1996, 1998) to calculate an EC_{10} (i.e., the "effective concentration" that caused a 10% effect; in this case, the dietary concentration that reduced hatching of eggs 10% below that of the control group in the same study) along with 95% confidence intervals (95% CI) for the mean Se concentration in the diet. The dietary EC_{10} was calculated to be 4.9 mg Se/kg, with 95% CI of 3.6–5.7 mg Se/kg.

The EC_{10} of 4.9 mg Se/kg was estimated by fitting a logistic regression model to the available data. It should be noted, however, that the mallard studies used a "dry" diet that had about 10% moisture. Ohlendorf (2003) used the reported dietary Se concentrations without adjustment for that moisture content, but an upward adjustment of the values (by 11%; to about 5.4 mg/kg) would be appropriate to account for the moisture content of the duck diet.

Adams et al. (2003) used hockey-stick regression on data for egg Se concentrations and adverse effects in mallards to derive toxicity thresholds, such as EC_{10} values. On further analyses (as described in Ohlendorf 2007), they found a threshold to exist when dietary Se was plotted against egg inviability and duckling mortality (which incorporated the cumulative effects of fertilization success and hatchability plus survival of ducklings for 6, 7, or 14 days after hatching, as reported for the different studies). The inflection point occurred at a dietary Se concentration of 3.9 mg/kg. The predicted EC_{10} was 4.4 mg Se/kg (just slightly above the inflection point), and the 95% CI around the predicted EC_{10} ranged from 3.8 to 4.8 mg Se/kg.

Wayland et al. (2007) used logistic regression to calculate EC_{10} values based on experimental studies of six species (mallard, American kestrel [*Falco sparverius*], domestic chicken, black-crowned night-heron [*Nycticorax nycticorax*], eastern screech-owl [*Megascops asio*] and ring-necked pheasant [*Phasianus colchicus*]). The EC_{10} was 4.0 mg Se/kg with 95% CI from <0.5 to 7.3 mg Se/kg. The effect of including several species was to widen the confidence limits substantially (compared to mallard EC_{10}), indicating a high degree of difference among species in sensitivity to Se.

Information on forms of Se in invertebrates (as potential diets for birds) is limited, but Andrahennadi et al. (2007) found variability in the Se speciation among aquatic insects that included mayflies (Ephemeroptera), stoneflies (Plecoptera), caddisflies (Trichoptera), and craneflies (Diptera) from streams in Alberta, Canada. Higher percentages of inorganic Se were found in primary consumers, detritivores, and filter feeders than in predatory insects. Among the organic forms, organic selenides constituted a major fraction in most organisms. A form of selenide, believed to represent selenomethionine, varied widely among aquatic insects (from 36% to 98% of the total Se), indicating a high degree of variability in bioaccumulation potential from diet to eggs. Nevertheless, the chemical forms of Se in aquatic foods of birds have received little study. It is likely that varying chemical forms of Se are present to some degree in plants and animals eaten by birds, yet the toxic concentrations of few Se compounds have been determined in birds.

Interpretive guidelines that have resulted from extensive testing with poultry are provided by Puls (1988). The Se concentrations for diet (as well as those for eggs and other tissues) are helpful guidelines for wild birds as well as domestic poultry. Dietary Se concentrations of less than 0.30 mg/kg are considered to be below the range adequate for good adult health and reproduction, 3.0–5.0 mg/kg are high, and above 5.0 mg/kg are toxic (Table 21.1).

21.3　EGG AND TISSUE CONCENTRATIONS

21.3.1　Eggs

Mean background Se concentrations in eggs of freshwater and terrestrial birds are <3 mg/kg dw (typically 1.5–2.5 mg/kg dw); concentrations lower than about 0.66 mg/kg dw may indicate inadequate Se in the diet, and maximums for individual eggs are <5 mg/kg dw (Table 21.1). Moisture content of eggs varies by stage of incubation (decreasing throughout incubation) and by species, but typical moisture content of field-collected eggs is usually 65–80% (Ohlendorf and Hothem 1995). Fresh mallard eggs, such as those collected from laboratory studies, have about 70% moisture (Stanley et al. 1996). The latter value provides a reasonable conversion factor (3.3) for estimating from one basis to the other and, except where noted, is used in this chapter when Se concentrations in eggs were originally reported on wet-weight basis, but the moisture content of samples was not reported.

21.3.1.1 Laboratory Studies

In a wide variety of species, if one expresses both the diet and eggs on a dry-weight basis, Se concentrations in bird eggs range from roughly equal to about three or four times the concentrations in the diet of the female at the time of egg-laying (Heinz et al. 1987, 1989, Smith et al. 1988, Ohlendorf 1989, Stanley et al. 1994, 1996, Wiemeyer and Hoffman 1996, Santolo et al. 1999). However, Se transfer from diet to egg varies by species and the chemical form of Se in the diet.

When birds fed on Se-contaminated diets during the laying season, the exposure was quickly reflected in elevated levels of Se in eggs (Heinz 1993b, Latshaw et al. 2004, DeVink et al. 2008a). Similarly, when the birds were switched to a clean diet, Se concentrations in eggs declined quickly. When mallard hens were fed a diet containing 15 mg Se/kg (as selenomethionine), levels peaked in eggs (to about 43–66 mg Se/kg dw; 13–20 mg Se/kg ww) after about 2 weeks on the treated diet and leveled off at a relatively low level (<16 mg Se/kg dw; <5 mg Se/kg ww) about 10 days after switching to an untreated diet (Heinz 1993b). The findings of this study and two others with ring-necked pheasants (*Phasianus colchicus*) (Latshaw et al. 2004) or lesser scaup (*Aythya affinis*) (DeVink et al. 2008a) summarized later have important implications for evaluation of field exposures, such as how quickly and for what duration Se exposure may adversely affect bird reproduction. Concentrations of Se in eggs are especially important because they provide the best samples for evaluating potential adverse reproductive effects (Skorupa and Ohlendorf 1991). Knowing Se concentrations in food items available to wild birds at a site also can be useful in assessing risks of reproductive effects, but relationships between the available food and concentrations that occur in eggs can vary widely on the basis of physiology and feeding ecology of the birds. Se speciation in the diet also may be important in this regard (i.e., plant vs. animal diets).

When ring-necked pheasants received feed that contained 9.3 mg Se/kg because of a feed mixing problem, severe effects occurred within 4 days (Latshaw et al. 2004). The rate of egg production decreased and bird aggression increased. About 12% of the hens died within a week; necropsy results were consistent with Se toxicity. After 8 days, the toxic feed was removed and replaced with fresh feed. Egg production, which had dropped by 50%, returned to normal within 10 days of feed replacement. Hatchability of eggs laid from days 8 to 14 after the pheasants received the toxic feed dropped to 35%, and more than 50% of the embryos that survived to the point where they could be examined had deformed beaks and abnormal eyes. Hatchability of eggs laid 21–28 days after the hens had received the toxic feed (i.e., 13–20 days after it was replaced by new feed) was almost 80%. Similar to the study with mallards, this incident showed a rapid onset of effects and a rapid recovery in response to dietary Se concentrations.

To assess the possible effects of Se on reproduction and fitness (measured as body mass) of lesser scaup, captive scaup were fed a control diet or one supplemented with Se at 7.5 or 15 mg/kg for 30 days to simulate dietary exposure to Se during late spring migration (DeVink et al. 2008a). The treated feed was removed after 30 days, just before the birds began laying. There was no effect of Se on body mass, breeding probability, or clutch initiation dates. Se concentrations in the first eggs laid by these birds were 25–30 mg/kg in the 7.5-mg/kg and 30–35 mg/kg in the 15-mg/kg treatment groups. Egg Se concentrations of both treatment groups decreased rapidly after the Se-supplemented feed was removed, and within 8 and 12 days, respectively, the egg Se concentration was less than 9 mg/kg dw. There was no significant intraclutch variation in egg Se deposition.

The embryo is the avian life stage most sensitive to Se (Poley et al. 1937, Poley and Moxon 1938, Heinz et al. 1987, 1989, Hoffman and Heinz 1988). Because it is the Se in the egg, rather than in the parent bird, that causes developmental abnormalities and death of avian embryos, Se in the egg gives the most sensitive measure for evaluating hazards to birds (Skorupa and Ohlendorf 1991). Given the rapid accumulation and loss patterns of Se in birds (Heinz et al. 1990, Heinz 1993b, Heinz and Fitzgerald 1993b, Latshaw et al. 2004), Se concentrations in eggs also probably best represent contamination of the local environment. Additional advantages of measuring Se in eggs are that eggs are frequently easier to collect than adult birds, the loss of one egg from a nest probably has little effect on a population, and the egg represents an integration of exposure of the adult female during the few days or weeks before egg-laying.

The concentration detected in eggs and the toxicity of that concentration seem to depend on the chemical form of the ingested Se. Organoselenium compounds are believed to be major forms in plants and animals. One organoselenium compound, selenomethionine, when fed to breeding mallards was more toxic to embryos than was selenocystine or sodium selenite (Heinz et al. 1989). Selenomethionine is a major form of Se in wheat seeds and soybean protein (Olson et al. 1970, Yasumoto et al. 1988). Hamilton et al. (1990) found selenomethionine to be an excellent model for Se poisoning in Chinook salmon (*Oncorhynchus tshawytscha*) when compared with the toxicity of Se that was biologically incorporated into mosquitofish (*Gambusia affinis*) collected at Kesterson Reservoir in California. Yamamoto et al. (1998) measured Se concentrations in blood and excreta of American kestrels fed either a selenomethionine-fortified diet or animals from Kesterson. They found no significant differences in concentrations or in accumulation and depuration of Se among experimental groups that received Se as selenomethionine or naturally incorporated in tissue of animals from Kesterson.

When mallards were fed a diet containing 10 mg Se/kg as selenomethionine (and about 10% moisture), reproductive success was significantly lower in the treated ducks than in controls, and a small sample of five eggs from the treated birds contained a mean of about 15 mg Se/kg dw (4.6 mg Se/kg ww) (Heinz et al. 1987). Because mallards were fed only one dietary concentration of Se in the form of selenomethionine, no safe level was established in this experiment. All that can be said is that the safe level in eggs was below about 15 mg Se/kg dw.

In a subsequent study, mallards were fed a diet containing about 10% moisture and 0, 1, 2, 4, 8, or 16 mg/kg of added Se as selenomethionine (Heinz et al. 1989). The reproductive success of the groups fed 1, 2, or 4 mg Se/kg did not significantly differ from that of controls; mean Se concentrations in a sample of 15 eggs from each of these groups were about 2.7, 5.3, and 11 mg/kg dw (0.83, 1.6, and 3.4 mg/kg ww). The group fed 8 mg Se/kg produced 57% as many healthy ducklings as the controls; the reduction in numbers was caused mainly by hatching failure and the early death of those that did hatch. A sample of 15 eggs from this group contained about 36 mg Se/kg dw (11 mg Se/kg ww). The group fed 16 mg Se/kg failed to produce any healthy young, and a sample of 10 of their eggs contained an average of about 59 mg Se/kg dw (18 mg Se/kg ww). Therefore, based on this study, the highest mean Se concentration in eggs not associated with reproductive impairment was about 11 mg/kg dw (3.4 mg/kg ww), and the lowest mean toxic concentration was 36 mg/kg dw (11 mg/kg ww).

Lam et al. (2005) subjected the data from this study with mallards (Heinz et al. 1989) to statistical analyses to estimate the threshold for effects on clutch viability. They normalized treatment response for control response and subjected the data to linear regression analysis, and then used a stepwise increment of 0.5-mg Se/kg concentration units followed by a one-tailed, one-sample t-test comparing the percentage of impairment of clutch viability (±95% CI) with zero to derive threshold effect levels of Se in eggs associated with impaired hatchability. They determined that 9 mg Se/kg was the lowest concentration in eggs at which clutch viability was significantly different than zero, and that the value represented an $EC_{8.2}$ for effects. A recent paper by Beckon et al. (2008) used the mean response data from the same laboratory study with mallards (Heinz et al. 1989) to evaluate potential hormetic effects exhibited by the treatment groups, and found an EC_{10} of 7.7 mg Se/kg (see later section on Hormesis).

In another study, Heinz and Hoffman (1996) compared the toxicity of three forms of selenomethionine. In nature, selenomethionine occurs almost exclusively in the L form, which is one of the two stereoisomer forms it can take (Cukierski et al. 1989). The other stereoisomer is the D form, and in many feeding studies with birds a mixture of the two forms (seleno-DL-methionine) has been fed. In yeast, most of the Se is in the form of seleno-L-methionine (Beilstein and Whanger 1986), and in addition to being in the naturally occurring form, it is biologically incorporated into the yeast. Pairs of breeding mallards were fed 10 mg Se/kg in each of the three forms. The results suggested that seleno-DL-methionine and seleno-L-methionine were of similar toxicity and both were more toxic than the Se in selenized yeast, but the lower toxicity of selenized yeast may have been due to a lower bioavailability of the selenomethionine in the yeast. A sample of eggs from the pairs fed seleno-L-methionine contained a mean of about 30 mg Se/kg dw (8.9 mg Se/kg ww), which resulted in a severe reduction in reproductive success (6.4% hatching of fertile eggs compared to 41.3% for controls). Eggs from pairs fed the seleno-DL-methionine contained a mean of about 31 mg Se/kg dw (9.2 mg Se/kg ww), and hatching of fertile eggs was 7.6%. Eggs from the pairs fed the selenized yeast contained a mean of only about 22 mg Se/kg dw (6.6 mg Se/kg ww), and hatching success was 27.0%. Because even the 22 mg Se/kg derived from the selenized yeast had a profound effect on reproductive success a toxic threshold was not established, but was obviously well below 22 mg Se/kg. Three studies were conducted to evaluate the interactive effects of Se with arsenic (As) (Stanley et al. 1994), boron (B) (Stanley et al. 1996), or mercury (Hg) (Heinz and Hoffman 1998), which are described in a later section (Interactions).

Using the same approach as that described earlier for the dietary values associated with reduced egg hatchability in mallards, Ohlendorf (2003) found the EC_{10} in eggs was 12 mg Se/kg dw, with 95% CIs of 6.4–16 mg Se/kg dw. The EC_{10} of 12 mg Se/kg was estimated by fitting a logistic regression model to the results of the six laboratory studies with mallards mentioned earlier.

The EC_{10} for mallard duckling mortality, as reported in Adams et al. (2003), ranged from 12 to 16 mg Se/kg dw in eggs. These EC_{10} values are based on a synthesis of the same six laboratory studies as mentioned earlier, but using the final endpoint of duckling mortality (the same effects data used in the dietary EC_{10} evaluation with hockey-stick regression above); the range of EC_{10} values reflects different statistical approaches for analyzing the data. Based on further analyses of those data, Adams (personal Communication; see Ohlendorf 2007) determined that the inflection point of the hockey stick occurred at an egg Se concentration of 9.8 mg/kg dw, with a predicted EC_{10} of about 12 mg/kg dw, which was comparable to that derived by Ohlendorf (2003). The 95% CI using hockey-stick regression was much narrower (9.7–14 mg/kg dw) than that derived by Ohlendorf using logistic regression (6.4–16 mg/kg dw). Given that there is a clear egg–Se threshold at which effects begin to be observed, a unimodal model, such as logistic regression, may result in exaggerated confidence intervals, particularly in the tails.

In a laboratory study designed to measure the lingering effects of an overwinter exposure to selenomethionine on reproduction, mallards were fed a diet containing 15 mg Se/kg for 21 weeks before the onset of laying (Heinz and Fitzgerald 1993b). Females began laying after various lengths of time off treatment. This experimental design was not ideal for determining the lowest concentration of Se in eggs associated with reproductive impairment, but the authors were able to make some general conclusions. Some eggs hatched when Se in eggs was as high as about 20–30 mg/kg dw (6–9 mg/kg ww), but other eggs failed to hatch when Se concentrations were estimated to be between 9.9 and 16 mg/kg dw (3 and 5 mg/kg ww). The authors concluded that the most logical reason why some embryos die while others survive when exposed to a given concentration of Se is that mallard embryos vary in their individual sensitivity to Se.

When black-crowned night-herons were fed a diet containing 10 mg Se/kg as selenomethionine (on close to a dry-weight basis) hatching success of fertile eggs was not reduced (Smith et al. 1988). The eggs of treated herons contained a mean concentration of about 11 mg Se/kg dw (3.3 mg Se/kg ww). The results from this study must be taken with some caution, however, because sample sizes were small ($n = 5$ pairs per group) and hatching success of fertile eggs of the control group was poor (32%).

Martin (1988) fed Japanese quail (*Coturnix coturnix japonica*) diets containing 5 or 8 mg Se/kg and chickens 10 mg Se/kg as selenomethionine, respectively. At 5 mg Se/kg, the hatching success of fertile quail eggs (56.4%) was lower than that of controls (76.4%); eggs from treated females contained about 23 mg Se/kg dw (7.1 mg Se/kg ww). At 8 mg Se/kg, the hatching of quail eggs was further decreased to 10.4% (compared with 75.1% for controls in that trial), and Se in eggs averaged about 40 mg/kg dw (12 mg/kg ww). The hatching success of the chickens fed 10 mg Se/kg also was depressed (23.2% compared with 84.5% for controls), and Se in eggs averaged about 36 mg/kg dw (9.6 mg/kg ww; the conversion from ww to dw [3.8] was based on the contents of chicken eggs containing about 73.6% water [Romanoff and Romanoff 1949]). No-effect concentrations in the diet or eggs were not determined.

In another study with chickens, diets were supplemented with seleniferous grains in amounts to produce dietary concentrations of 2.5, 5, and 10 mg Se/kg (Moxon and Poley 1938, Poley and Moxon 1938). Modern statistical techniques were not applied to these data, and chemical analyses were different from those used today, but at 2.5 mg Se/kg in the diet, the hatching success of fertile eggs was no different from that of controls, and a sample of eggs contained Se at about 15 mg/kg dw in albumen and 3.2 mg/kg dw in yolk 1.75 mg/kg and 1.67 mg/kg ww, respectively; conversions from ww to dw here and below (multiply ww concentrations by 8.3 for albumen and by 1.9 for yolk) were based on the fact that chicken eggs are composed of about 55.8% albumen, 31.9% yolk, and 12.3% shell, and that the moisture content of albumen is about 87.9% while that of yolk is 48.7% (Romanoff and Romanoff 1949). At 5 mg Se/kg in the diet, the hatching of eggs was "slightly reduced," and Se in egg albumen and yolks averaged about 24 and 5.2 mg/kg dw (2.95 and 2.73 mg/kg ww), respectively. At 10 mg Se/kg, hatching decreased to zero, and albumen and yolks contained about 53 and 7.4 mg Se/kg dw (6.40 and 3.92 mg Se/kg ww), respectively. Based on the percentages of albumen and yolk in chicken eggs and the respective percentages of water in albumen and yolk, a Se threshold of about 10 mg/kg dw (3 mg/kg ww) in whole eggs was associated with reproductive impairment in the study where chickens were fed 5 mg Se/kg; this threshold is similar to the findings of more rigorous recent studies with mallards.

Harmful concentrations of Se in eggs may be of a different magnitude when another chemical form of Se, sodium selenite, is fed to birds. A diet containing 7 mg Se/kg as sodium selenite caused reproductive impairment in chickens but resulted in only about 7.2 and 3.8 mg Se/kg dw (0.87 and 2.02 mg Se/kg ww) in egg albumen and yolk (Ort and Latshaw 1978).

In another study with chickens, a diet containing 8 mg Se/kg as sodium selenite impaired reproduction, and whole eggs contained from about 5.5 to 7.1 mg/kg dw (1.46–1.86 mg/kg ww) of Se (Arnold et al. 1973). The chemical form of Se in chicken eggs seems to be different when sodium selenite rather than selenomethionine is fed (Latshaw 1975, Latshaw and Osman 1975).

In mallards, a dietary concentration of 25 mg Se/kg as sodium selenite impaired reproduction but resulted in a mean of only about 4.3 mg/kg dw (1.3 mg/kg ww) of Se in eggs (Heinz et al. 1987). Therefore, although higher dietary concentrations of sodium selenite than selenomethionine must be fed to mallards to harm reproduction, lower concentrations of Se in eggs are associated with harm.

Selenium also may affect egg fertility in some species, but egg fertility is not always reported from field or laboratory studies. Lack of reporting on fertility effects in some studies of Se effects in birds may be due in part to a general practice of simply including infertile eggs as inviable eggs (i.e., "infertility" effects may not be separated from "embryotoxic" effects in the overall measurement of hatchability). Failure to measure infertility as a separate endpoint may be due to the difficulty often associated with distinguishing infertile eggs from those containing embryos that have died very early in development. Nevertheless, decreased fertility is a distinct effect from embryotoxicity, particularly in that it can indicate a mechanism acting on adult, rather than embryonic, physiology. In American kestrels fed selenomethionine at 12 mg Se/kg, egg fertility was significantly reduced (by over 14%) compared to kestrels fed 6 mg Se/kg (Santolo et al. 1999). Results obtained in kestrels suggest that infertility may be an important factor contributing to the overall reproductive impairment in some species. However, in mallards (Heinz et al. 1987, Heinz and Hoffman 1996, 1998) and black-crowned night-herons (Smith et al. 1988) fed 10 mg Se/kg as selenomethionine, egg fertility was not reduced compared with controls. Similarly, fertility was not affected in mallards fed diets containing Se at 7 mg/kg (Stanley et al. 1996) or 16 mg/kg (Heinz et al. 1989) as selenomethionine, but hatchability of fertile eggs was significantly reduced. Thus, effects on egg fertility in mallards and night-herons are not likely to be as ecologically significant as reduced hatchability.

21.3.1.2 Field Studies

Selenium concentrations in the eggs of marine species are variable, but may be higher than in freshwater or terrestrial birds, even in remote areas (Ohlendorf 1989). For example, eggs of three species (wedge-tailed shearwater [*Puffinus pacificus*], red-footed booby [*Sula sula*], and sooty tern [*Sterna fuscata*]) were sampled at four locations throughout the Hawaiian Archipelago, from Oahu to Midway (Ohlendorf and Harrison 1986). Mean Se concentrations varied only slightly by location, from about 4.4 to 5.3 mg/kg dw (1.1–1.4 mg/kg ww) for shearwaters, 5.0–6.1 mg/kg (0.76–0.92 mg/kg ww) for boobies, and 4.1–5.1 mg/kg (1.1–1.4 mg/kg ww) for terns, but all were higher than typical of freshwater species. Henny et al. (1995) predicted egg concentrations (21.3 or 29.2 mg Se/kg dw, based on different regressions) from liver concentrations in white-winged scoters (*Melanitta fusca*) (mean of 54 mg Se/kg dw for combined males and females; concentration not given separately for females) based on established liver–egg relationships for freshwater species (Henny and Herron 1989, Ohlendorf et al. 1990, Ohlendorf and Hothem 1995). However, they found that Se concentrations in eggs were only about 10% of the predicted concentrations, from 2.7 to 4.7 mg/kg dw.

Braune et al. (2002) analyzed eggs of glaucous gulls (*Larus hyperboreus*), black-legged kittiwakes (*Rissa tridactyla*), thick-billed murres (*Uria lomvia*), and black guillemots (*Cepphus grylle*) from the Canadian Arctic. Mean Se concentrations varied somewhat by species and location, with all means between 1.1 and 2.7 mg/kg dw except for kittiwakes (with means of 4.4 mg/kg at two locations), so kittiwakes were the only species with means greater than typical of freshwater and terrestrial birds.

Eggs of common eiders (*Somateria mollissima*) collected from five locations in the Baltic Sea near coastal Finland also had median Se concentrations (0.55 mg/kg ww; about 1.65 mg/kg dw) that were similar to background for freshwater and terrestrial birds (Franson et al. 2000). Thus, there seems to be no consistent difference between marine and other birds.

Using the results of extensive field studies of black-necked stilts (*Himantopus mexicanus*), Skorupa (1998a, 1999) found a threshold of 6–7 mg Se/kg in eggs to be associated with impaired egg hatchability. That concentration is about equivalent to the EC_{10} on a clutch-wise (or hen-wise) basis and the EC_{03} on an egg-wise basis. Lam et al. (2005) used the same statistical approach as described earlier for the laboratory study with mallards to estimate the threshold for effects on stilt clutch viability. They derived an $EC_{11.8}$ of 14 mg Se/kg at which clutch viability was significantly impaired (i.e., greater than zero impairment). It should be noted that the background rate of clutch inviability (when Se concentrations in eggs are <6 mg/kg) is estimated at 8.7% (U.S. DI 1998).

Studying birds at Kesterson Reservoir in California, Ohlendorf et al. (1986b) used logistic regression to estimate a 50% chance of embryo death or deformity in American coots (*Fulica americana*) when Se concentrations in eggs were about 18 mg/kg dw. The estimated Se concentration causing the same effect in black-necked stilts was 24 mg/kg. The value for eggs of eared grebes (*Podiceps nigricollis*) could not be calculated because even the lowest Se concentration detected in eggs (44 mg/kg) was embryotoxic. The logistic approach is best suited to estimate the 50% effect concentration, not the concentrations of Se in eggs at which embryo deaths and deformities begin for each species. These concentrations would obviously be somewhat lower than the 50% effect levels.

Skorupa and Ohlendorf (1991) examined the relation between Se concentrations in eggs of various aquatic bird species and reproductive impairment at the population level. Embryo deformities were detected in only 3 of 55 populations of birds that had a mean Se concentration of less than 3 mg/kg in eggs (and these deformities were not characteristic of those induced by Se); this is a concentration of Se judged to represent a background level (Figure 21.1). However, as discussed earlier, reference area concentrations may not always be the same as concentrations from known uncontaminated areas and, therefore, are not necessarily always synonymous with safe levels. Deformities were detected in nine of ten populations of aquatic birds in which the mean Se concentration in eggs exceeded about 48 mg/kg. Their data suggested that a teratogenic threshold at the population level existed between about 13 and 24 mg Se/kg, as illustrated in the figure.

The nature of Se-related deformities makes them a good measure for characterizing the dose–response relation between Se concentrations in eggs and the incidence of severe reproductive impairment in avian populations because (1) the embryo is either deformed or normal (a presence/absence indicator), and (2) the deformities resulting from Se toxicosis are diagnostic of Se toxicosis. It should be noted, however, that the data plotted in Figure 21.1 represent a population-level analysis and can not be used to infer probability of teratogenesis in individual eggs of known Se content.

Using data on Se in eggs from the Tulare Basin (southern San Joaquin Valley), combined with data from several other western sites where elevated Se was found, Skorupa (1998a, 1998b; also in U.S. DI 1998) documented a detailed exposure-response relationship. Statistically distinct teratogenesis response functions were delineated for ducks, stilts, and American avocets (*Recurvirostra americana*) using the Tulare Basin data. The Tulare curves were used to estimate expected frequencies of teratogenesis for ducks, stilts, and avocets using other sites, and the predicted levels were tested against the observed frequencies from the sites. The predicted and observed frequencies of teratogenesis were not significantly different, so the data were combined to generate final response curves. Using these data, Skorupa (1998b) developed species-specific response curves for stilts and avocets and a composite duck curve (using combined data from gadwalls [*Anas strepera*], mallards, pintails [*A. acuta*], and redheads [*Aythya americana*]).

Based on the response coefficients and their standard errors, the teratogenesis function for ducks, stilts, and avocets were significantly different (Skorupa 1998b). Within this data set, these responses represent "sensitive" (duck), "average" (stilt), and "tolerant" (avocet) species. The probability of overt teratogenesis in stilts increased markedly when Se concentrations in eggs were greater than 40 mg/kg, with an EC_{10} for teratogenic effects of 37 mg/kg. In contrast, the thresholds for teratogenesis (expressed as an EC_{10}) were 23 mg Se/kg in mallards and 74 mg Se/kg in avocets. Sensitivity

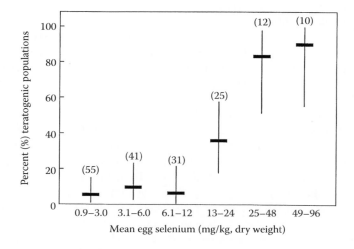

FIGURE 21.1 Dose–response relation between mean egg Se and teratogenic classification of aquatic bird populations (from Skorupa and Ohlendorf 1991, with kind permission of Springer Science and Business Media). For each dose interval, the observed percentage of populations classified as teratogenic is plotted along with 95% binomial confidence intervals. Sample sizes (number of populations assessed) for each dose interval are listed above the response plots.

of these species to effects of Se on egg hatchability followed a similar pattern, with mallards being more sensitive than stilts, which are more sensitive than avocets (U.S. DI 1998).

21.3.2 Liver

Background Se concentrations in livers of freshwater and terrestrial birds are <10 mg/kg dw (Table 21.1), while livers of marine birds from uncontaminated areas tend to have considerably higher Se concentrations (often 20 mg/kg or more; Dietz et al. 1996, Trust et al. 2000, Grand et al. 2002, Mallory et al. 2004, Elliott 2005). Typical moisture content is about 70% (Ohlendorf et al. 1990, Stanley et al. 1996).

21.3.2.1 Laboratory Studies

In a manner similar to that for eggs, Se concentrations in the liver respond quickly when birds are placed on or taken off a Se-contaminated diet (Heinz et al. 1990). When mallards were fed a diet containing 10 mg Se/kg, Se concentrations in liver were predicted to reach 95% of equilibrium in 7.8 days; the rate of loss from liver also was rapid, with half-time of 18.7 days. Thus, Se concentrations measured in the livers of birds sampled outside the breeding season are not good predictors of potential reproductive effects. In laboratory studies of reproductive effects, livers of male mallards had higher concentrations of Se than those of females, probably because females excreted part of the Se they had accumulated through egg-laying (e.g., Heinz et al. 1987, 1989, Heinz and Hoffman 1998). Nevertheless, analysis of livers of either male or female field-collected birds can provide a useful indication of the relative level of exposure experienced by the population.

Laboratory studies have been conducted with mallards to determine the kinds of lesions and other measurements that can be used for diagnosis of Se toxicosis in birds (Albers et al. 1996, Green and Albers 1997, O'Toole and Raisbeck 1997, 1998). Dietary concentrations of added Se ranged from 10 to 80 mg/kg in these studies. Various hepatic lesions were associated with dietary exposures greater than 10 mg Se/kg, and Se concentrations in livers increased in response to the dietary levels. In general, ducks that received diets containing more than 20 mg Se/kg developed a number of lesions of the liver, and those receiving 40 mg/kg or more Se in their diets lost weight and had abnormal changes in the integument (described later) in addition to the liver. Lesions of the integument and liver, and weight loss, when corroborated by elevated Se concentrations in tissues (especially the liver), can be diagnostic of Se toxicosis in birds. It should be noted, however, that some birds died without exhibiting any significant morphological lesions even though they were emaciated. Although a clear threshold Se concentration in livers (or other tissues) for diagnosis of Se toxicity could not be defined, concentrations greater than 10 mg/kg were considered suspicious of Se toxicosis, particularly when accompanied by emaciation, poor quality (and sloughing) of nails, bilaterally symmetrical alopecia of the head and neck, toxic hepatic lesions, and necrosis of maxillary nails.

In laboratory studies with birds fed diets containing selenomethionine, when Se concentrations in the diet and in livers of mallards, night-herons, and eastern screech-owls were expressed on a dry-weight basis, liver concentrations ranged from roughly equal to the dietary concentrations to about three times the dietary levels (Heinz et al. 1987, 1989, Smith et al. 1988, Stanley et al. 1994, 1996, Wiemeyer and Hoffman 1996). At Kesterson Reservoir, Se concentrations in livers of European starling (*Sturnus vulgaris*) nestlings (7.5 mg/kg) were only slightly higher than those in the invertebrates being fed to the chicks (6.2 mg/kg) by adults (Santolo 2007).

In a laboratory study, surviving mallard ducklings fed 40 mg Se/kg as selenomethionine had a mean Se concentration of about 224 mg/kg dw (68 mg/kg ww) in the liver, whereas ducklings that died had a mean of about 198 mg/kg dw (60 mg/kg ww) (Heinz et al. 1988). In another laboratory study, this time with adult male mallards fed 100 mg Se/kg as selenomethionine, the livers of survivors contained a mean of about 142 mg Se/kg dw (43 mg Se/kg ww), and the livers of birds that died contained a mean of about 125 mg Se/kg dw (38 mg Se/kg ww) (Heinz 1993a).

When adult male mallards were fed 32 mg Se/kg as selenomethionine, they accumulated an average of about 96 mg Se/kg dw (29 mg Se/kg ww) in their livers (Hoffman et al. 1991). One of 10 birds fed 32 mg Se/kg died, and others had hyperplasia of the bile duct and hemosiderin pigmentation of the liver and spleen. Various other sublethal effects, such as elevated plasma alkaline phosphatase activity and a change in the ratio of hepatic oxidized glutathione to reduced glutathione, were observed in ducks with lower hepatic concentrations. At a dietary concentration of 8 mg Se/kg, which caused several of the physiological effects mentioned above, the mean concentration of Se in the liver was about 41 mg/kg dw (12.5 mg/kg ww).

Based on these laboratory studies, in which Se was present as selenomethionine in the diet and was the only element fed at toxic concentrations, mortality of young and adult mallards could occur when hepatic concentrations of Se reach roughly 66 mg/kg dw (20 or more mg/kg ww), and important sublethal effects are likely when the concentrations exceed about 33 mg/kg dw (10 mg/kg ww).

Using Se concentrations in adult female livers to predict when reproductive impairment occurs in birds is not nearly as good as using Se concentrations in eggs, because it is the Se in the egg that actually harms the embryo (Skorupa and Ohlendorf 1991). Extrapolating from liver to egg will introduce additional uncertainty above that already existing for the egg. However, in a controlled laboratory study, the correlation between Se concentrations in eggs and in the livers of laying females was demonstrated by feeding mallards selenomethionine (Egg $Se_{mg/kg\ ww}$ = $-1.10 + 2.6$ (Liver $Se_{mg/kg\ ww}$); $R^2 = 0.83$; $p < .01$; Heinz et al. 1989). Therefore, when Se concentrations in eggs are not available, the concentrations in the livers of females during the breeding season can be used to estimate whether reproduction might be impaired. When Se concentrations are known for both the eggs and livers of breeding females, judgments on the hazards of Se to reproduction should be based on Se in the egg.

In laboratory studies of reproduction, the livers of male mallards contained more Se than did the livers of females fed the same diets (Heinz et al. 1987, 1989). Because females may use the egg as a route of Se excretion unavailable to males, one would expect that, in the field, the lowest reproductive effect threshold of Se would be in the livers of laying females and that the livers of males would be less useful in predicting effects on reproduction, even if the males were collected during the breeding season and from the area where reproduction is of concern. The advantage of sampling laying females, however, may be more academic than practical. In nature, it is easier and more likely that a female would be collected before or after egg-laying, at which time the concentration of Se in her liver should be the same as in the liver of a male. If one collects breeding males in the wild or has reason to believe that the collected females were not collected during egg-laying, a 10-mg/kg dw (3-mg/kg ww) threshold concentration of Se in the liver would be on the low side (and would represent the upper end of background conditions); a value of about 13–20 mg Se/kg dw (4–6 mg Se/kg ww) might be more appropriate for freshwater birds. However, some marine species typically have higher hepatic Se concentrations even in remote areas (as noted previously), so these values would not be appropriate for those species.

Female mallards that were fed 10 mg Se/kg as selenomethionine had reduced reproductive success and a mean of about 16 mg Se/kg dw (4.7 mg Se/kg ww) in their livers (Heinz et al. 1987). Because no dietary concentrations below 10 mg/kg were used, a no-effect level of Se in the liver was not determined in this study.

A dietary concentration of 8 mg Se/kg as selenomethionine significantly reduced reproductive success of mallards, and livers of the treated females contained a mean of about 12 mg Se/kg dw (3.5 mg Se/kg ww) (Heinz et al. 1989). In the same study, reproductive success was not significantly different between females fed 4 mg Se/kg and controls, and livers contained a mean of about 7.9 mg Se/kg dw (2.4 mg Se/kg ww). Based on a regression equation of Se concentrations in female livers versus their eggs (Heinz et al. 1989), the threshold Se concentration of 10 mg/kg dw (3 mg/kg ww) in eggs corresponds to a Se value of about 5.3 mg/kg dw (1.6 mg/kg ww) in the liver. However, we do not know whether the data for this regression were linear in the lower end of the Se range. If the data were curvilinear, a value of 10 mg Se/kg dw (3 mg/kg ww) in eggs may correspond to a value of roughly 10 mg Se/kg dw (3 mg/kg ww) for the liver.

In these laboratory studies with mallards, between 16 and 31 eggs were laid before each female was sacrificed. Depletion of Se through egg-laying, therefore, may have been greater in the laboratory than in nature where birds lay fewer eggs. If depletion of Se is greater by females in a laboratory study, the Se concentrations in the liver associated with reproductive impairment could be on the low side.

Separate studies were conducted to evaluate the interactive effects of Se with As (Stanley et al. 1994), B (Stanley et al. 1996), and Hg (Heinz and Hoffman 1998). The results of the interactions are described in more detail in a later section (Interactions); here we discuss only the effects of the Se treatment by itself. When Se was fed alone at dietary concentrations of 3.5 or 7.0 mg/kg in the B study, the mean Se concentration in livers of females was about 11 mg/kg dw (3.5 mg/kg diet) or 17 mg/kg (7 mg/kg diet) (3.2 and 5.1 mg/kg ww in liver). Hatching success was reduced in the 7-mg Se/kg treatment group when compared to controls and the 3.5-mg Se/kg treatment group. No embryonic deformities were found in that study; although Se reduced duckling weight, it did not affect duckling survival. When ducks were fed Se at 10 mg/kg in both the As and Hg studies, Se accumulated significantly in eggs and livers, reduced hatching success and duckling survival (or production per pair), and was teratogenic. In the As study, the mean Se concentration in livers of ducks receiving the 10-mg/kg diet was 31 mg/kg in females and 34 mg/kg in males. In the Hg study, the mean Se concentration in livers of hens receiving the 10-mg/kg diet was about 20 mg/kg dw (6.0 mg/kg ww), and in males it was about 32 mg/kg dw (9.6 mg/kg ww).

Franson et al. (2007) fed common eiders a diet containing 20 mg Se/kg as seleno-L-methionine or a diet that was started at 20 mg Se/kg and increased over time to 60 mg Se/kg. Among the ducks fed the 20-mg Se/kg diet, 57% exhibited lipidosis and hypertrophy of Kupffer cells in the liver. Among the ducks fed the 60-mg Se/kg diet, 83% exhibited cellular lipidosis and 100% had hypertrophy of Kupffer cells. One duck in the 60-mg Se/kg group died after 30 days and another was euthanized on day 32 after developing a staggering gait and a 35% weight loss. Selenium concentrations in livers averaged 351 mg/kg dw in the 20-mg/kg dietary group and 735 mg/kg dw in the 60-mg/kg dietary group. The authors of that study stated that the effects of Se generally were comparable to those seen in mallards fed similar dietary concentrations of selenomethionine; however, the eiders accumulated more Se in their livers than did the mallards. For example, in one study (O'Toole and Raisbeck 1997) mallards fed 60 mg Se/kg accumulated about 200 mg Se/kg dw (60.6 mg Se/kg ww) in liver versus the 735 mg Se/kg dw for the eiders fed 60 mg Se/kg in the Franson et al. (2007) study, leading the authors of the eider study to conclude that eiders, and probably other sea ducks, apparently have a higher adverse effects threshold of Se in tissues than do freshwater species.

21.3.2.2 Field Studies

Selenium concentrations in the liver have been used to estimate both exposure and effects on birds. For example, livers of adult birds (coots, stilts, and ducks) collected from Kesterson Reservoir and reference areas showed time-period differences related to collection site and duration of exposure (Ohlendorf et al. 1990). In addition, Se concentrations in prefledging juvenile birds of some species were generally similar to those in livers of late-season adults. Geometric means for Se in adult stilts in 1983 were as follows: Kesterson Reservoir—41.8 mg/kg early, 94.4 mg/kg late nesting season; Volta Wildlife Area—10.7 mg/kg early, 5.41 mg/kg late nesting season. Selenium concentrations in juveniles were 94.6 mg/kg at Kesterson and 4.10 mg/kg at the Volta Wildlife Area.

Although accumulation in the liver is dose-dependent (Hoffman et al. 1991), the hepatic concentration is only an imprecise estimator of the pathological condition of a bird. The cutoff is not clear between Se concentrations in the livers of birds killed by Se poisoning and others exposed to high concentrations but collected alive. The livers of birds found dead at the Kesterson Reservoir contained 26–86 mg Se/kg, whereas the livers of birds shot there contained 38–85 mg Se/kg (Ohlendorf et al. 1988).

Selenium toxicosis effects in several species of aquatic birds found at Kesterson Reservoir in 1984–1986 were described previously (Ohlendorf et al. 1988, 1990, Ohlendorf 1989, 1996, Ohlendorf and Hothem 1995). Those birds exhibited many of the same signs of selenosis as those later found in mallards (as described above), including hepatic lesions, alopecia, necrosis of the beak, and weight loss.

Livers of diving ducks (such as scoters [*Melanitta* spp.] and scaups [*Aythya* spp.]) from estuarine habitats have been found to contain higher concentrations of Se than other aquatic birds in the same habitats (Ohlendorf et al. 1986c, 1989, 1991, Henny et al. 1991). One possible reason for the higher concentrations of Se in these diving ducks is that they forage on benthic organisms, which bioaccumulate Se to a higher degree than foods of some other aquatic birds. However, many species of marine birds, including some that feed on planktonic crustaceans or other near-surface organisms, also tend to have higher hepatic Se concentrations than typical of freshwater birds (Elliott et al. 1992, Dietz et al. 1996, Campbell et al. 2005, Elliott 2005). Those include species such as Leach's storm-petrel (*Oceanodroma leucorhoa*), northern fulmar (*Fulmarus glacialis*), black-footed albatross (*Diomedea nigripes*), and black-legged kittiwake that have mean Se concentrations up to 75 mg/kg.

Based on field data, a very high risk of embryonic deformity exists when the mean Se concentration in the livers of a population of birds using nonmarine habitats (both sexes included and females not necessarily laying) exceeded about 30 mg/kg dw (U.S. Fish and Wildlife Service 1990). Populations with means below about 10 mg Se/kg dw generally did not have many deformed embryos. Some species of marine birds can accumulate high concentrations of Se in their livers without correspondingly high concentrations in their eggs (e.g., Henny et al. 1995, Braune et al. 2002, Campbell et al. 2005, DeVink et al. 2008b).

21.3.3 KIDNEY

Background Se concentrations in bird kidneys have not been clearly defined, and there is no consistent trend regarding liver/kidney ratios. Selenium concentrations in kidneys of birds from Se-normal areas were somewhat higher than those in the liver (liver/kidney ratios of less than 1), but concentrations in the two tissues were similar in birds from the Se-contaminated Kesterson Reservoir (Ohlendorf et al. 1988, 1990) and in the Imperial Valley of California (Koranda et al. 1979). Selenium concentrations in liver and kidneys of American coots from Kesterson Reservoir and the reference site (Volta Wildlife Area) were significantly correlated ($r = 0.98$). The average moisture content of kidneys was 76–78%, so a conversion factor of 4.3 can be used to estimate from wet-weight to dry-weight concentrations.

When chickens were fed 0.1 mg Se/kg as selenomethionine for 18 weeks, Se concentrations in kidneys (about 3.3 mg/kg dw; 0.77 mg/kg ww) were higher than those in the liver (about 2.0 mg/kg dw; 0.60 mg/kg ww), but when the diet contained 6 mg Se/kg the kidney and liver Se concentrations were essentially equal (both about 22 mg/kg dw; 5.2 and 6.6 mg/kg ww, but with different moisture contents assumed for kidney and liver) (Moksnes 1983).

In a study to determine body distribution of trace elements in black-tailed gulls (*Larus crassirostris*) nesting on Rishiri Island in Hokkaido Prefecture, Japan, Se concentrations in kidneys of both adults (6.9 mg/kg) and juveniles (6.5 mg/kg) were significantly ($p < .001$) higher than in livers (adults, 4.5 mg/kg; juveniles, 5.3 mg/kg) (Agusa et al. 2005).

In a laboratory study with mallards (Albers et al. 1996), Se concentrations in livers of surviving ducks were consistently higher than those in kidneys when the ducks were fed diets supplemented with Se at 0 (control), 10, 20, or 40 mg/kg. However, concentrations in the two tissues were more similar among the birds that died during the exposure period. When expressed on a dry-weight basis, Se concentrations in livers were about two or three times the dietary concentration, whereas those in kidneys averaged less than twice the dietary concentration.

Although concentrations of Se in kidneys representative of those diagnostic of harm to adult health or reproductive success are poorly understood, if one had no other information on Se values in tissues other than in kidneys, one could assume a roughly one-to-one correspondence between the concentration of Se in kidney and liver. In this way one could make a preliminary assessment of possible harm to birds, but this assessment would be weak compared to those based on concentrations in eggs or livers.

21.3.4 MUSCLE

Background Se concentrations in muscle tissues of birds are 1–3 mg/kg (Table 21.1). Average moisture content of mallard muscle in a laboratory study was 74% (Heinz et al. 1987).

As in eggs and liver, Se concentrations in muscle increase and decrease in response to changes in dietary exposure, but the changes occur more slowly (Heinz et al. 1990) and diagnostic concentrations for effects are not readily available. Heinz et al. (1990) fed female mallards 10 mg Se/kg as selenomethionine for 6 weeks, followed by 6 weeks off treatment, and measured Se in the liver and breast muscle. By 6 weeks, Se in breast muscle averaged about 24 mg/kg dw (6.3 mg/kg ww). Selenium in the liver had nearly peaked after about 1 week, whereas muscle was projected to reach a peak of about 30 mg Se/kg dw (8 mg Se/kg ww) after 81 days. Likewise, Se was eliminated faster from the liver than from breast muscle, indicating that the two tissues may contain similar concentrations of Se, but only after both reach equilibrium. This difference in accumulation and loss rates between tissues helps explain the variability observed in the muscle–liver relationships at Kesterson Reservoir and the reference site described below (Ohlendorf et al. 1990).

Selenium concentrations in breast muscle from juvenile ducks (*Anas* spp.) at Kesterson Reservoir and a reference site (Volta Wildlife Area) were measured because of concern about human consumption of ducks harvested in the vicinity of Kesterson (Ohlendorf et al. 1990). Mean Se concentrations were higher at Kesterson than the reference site, and were only slightly lower than those in livers of these birds. However, the relationship between muscle and liver ($R^2 = 0.69$) of the ducks was considerably more variable than that between kidneys and livers of American coots from the two sites ($R^2 = 0.97$). The predictive equation was:

$$\text{Log Se in muscle} = 0.22 + 0.65 \text{ log Se in liver.}$$

When mallards were fed 10 mg Se/kg as selenomethionine in a laboratory study, females had similar concentrations of Se in the liver (about 16 mg/kg dw; 4.7 mg/kg ww) and breast muscle (about 19 mg/kg dw; 4.9 mg/kg ww), whereas males had much higher concentration in the liver (about 28 mg/kg dw; 8.6 mg/kg ww) than in breast muscle (about 12 mg/kg dw; 3.1 mg/kg ww) (Heinz et al. 1987). Because the females were laying eggs, they may have been using stores of Se from the liver to incorporate into eggs.

Fairbrother and Fowles (1990) reported more Se in breast muscle (about 22 mg/kg) than in the liver (about 16 mg/kg) of male mallards given drinking water containing 2.2 mg Se/L (as selenomethionine) for 12 weeks. When chickens were fed 0.1 mg Se/kg as selenomethionine for 18 weeks, Se concentrations in breast muscle (about 1.1 mg/kg dw; 0.29 mg/kg ww) were about half of those in the liver (about 1.9 mg/kg dw; 0.60 mg/kg ww), but when fed 6 mg Se/kg in the diet nearly equal Se concentrations were reported in the breast muscle and liver (20 and 22 mg/kg dw; 5.4 and 6.6 mg/kg ww) (Moksnes 1983).

As was the case with liver, much more Se was accumulated in muscle when ducks received an organic form of Se (selenomethionine) at 10 mg/kg than when fed a diet supplemented with an equivalent concentration of inorganic Se (selenite, which is used routinely, but at much lower concentrations, in poultry diets) (Heinz et al. 1987). Also, females that received the organic Se during the reproductive study accumulated significantly more Se in breast muscle than the males receiving the same treatment.

21.3.5 BLOOD

Background Se concentrations in whole blood of nonmarine birds are 0.1–0.4 mg/L on a wet-weight basis (Table 21.1). However, marine birds inhabiting unpolluted areas often have higher

Se concentrations in their blood (e.g., Franson et al. 2000, Wayland et al. 2001, 2008, Grand et al. 2002), and similar findings were observed at Great Salt Lake, UT (Conover and Vest 2009).

Under uniform sampling conditions, the moisture content of blood is fairly uniform, but under field conditions the moisture content can vary substantially. For example, when mallard blood was sampled over a period of about 3 months by exsanguination in a laboratory study, the dry-weight content of blood averaged 21.70 ± 0.21% (mean ± SE) (Scanlon, 1982). In a laboratory study with kestrels (Yamamoto et al. 1998, Santolo et al. 1999, G. M. Santolo, personal communication), the dry-weight content of blood averaged 21.40 ± 0.11% (mean ± SE) with a range from 14% to 25%. However, when kestrels and other raptors were sampled in the field (Santolo and Yamamoto 1999, G. M. Santolo, personal communication), the dry-weight content of blood averaged 19.30 ± 0.14% (mean ± SE) with a range from 9% to 32%. In both the laboratory and field studies of kestrels (and other raptors), blood samples were taken in a consistent manner from the birds by the same investigators. However, there was much greater variability in moisture content of birds collected in the field (variance = 8.3) and than in the lab (variance = 2.2).

In experimental studies, Se concentrations in blood of mallards (Heinz et al. 1990, Heinz and Fitzgerald 1993a, O'Toole and Raisbeck 1997) and American kestrels (Yamamoto et al. 1998, Santolo et al. 1999) reflected dietary exposure levels. Mallards receiving Se (as selenomethionine) at dietary concentrations of 10, 25, or 60 mg/kg had blood–Se concentrations of about 22, 43, or 77 mg/L dw (4.5, 8.9, or 16 mg/L ww) (O'Toole and Raisbeck 1997). The concentration of Se in blood increased in a time- and dose-dependent manner and reached a plateau after 40 days.

When female mallards were fed increasingly high dietary concentrations of Se as selenomethionine (from 10 to 160 mg/kg over a period of 31 days), birds began to die at the end of the 31-day exposure (Heinz et al. 1990). Survivors contained means of about 60 mg Se/kg dw (12 mg Se/kg ww) in the blood on day 31, when their diet was switched to an untreated diet. Half-time for loss of Se from blood was 9.8 days, which was much faster than for muscle (23.9 days). In another study (Heinz and Fitzgerald 1993a), adult male mallards were fed 10, 20, 40, or 80 mg Se/kg as selenomethionine. Mortality began in the 40- and 80-mg Se/kg treatment groups during the third week on treatment, when samples of blood from surviving ducks in the same pens contained means of about 25 or 70 mg Se/kg dw (5 or 14 mg Se/kg ww). Blood–Se concentrations of the ducks fed lower-Se diets plateaued after 8 weeks at about 42 mg/kg dw (8.4 mg/kg ww) for the 10-mg/kg treatment group and 70 mg/kg dw (14 mg/kg ww) for the 20-mg/kg dietary concentration. However, samples of blood were not taken from any of the birds that died. Therefore, comparisons of Se concentrations between the dead and the survivors were not possible.

In American kestrels (Yamamoto et al. 1998), maximal blood concentrations, when expressed on a dry-weight basis, were about the same as those in the selenomethionine-supplemented diet. The Se concentration in blood after 77 days on treatment was 5.0 mg/kg for kestrels receiving the 5 mg/kg diet and 8.9 mg/kg for those receiving the 9 mg/kg dietary concentration. Selenium concentrations in blood returned to near the control concentrations in 28 days after the experimental diets were removed. Selenium concentrations in excreta of the kestrels were higher than those in blood during the treatment period, indicating that they excrete a substantial amount of the ingested Se.

To assess the possible effects of Se on reproduction and fitness (measured as body mass) of lesser scaup, captive scaup were fed a control diet or one supplemented with Se at 7.5 or 15 mg/kg for 30 days to simulate late spring migration (DeVink et al. 2008a). The treated feed was removed after 30 days, before the birds began laying. There was no effect of Se on body mass, breeding probability, or clutch initiation dates. Blood–Se concentrations differed between the treatment groups in proportion to dose, with mean Se concentrations in blood after 30 days on treatment (16.3 and 30.8 mg/kg) about twice the concentration in the diet. The half-lives for Se concentrations in blood were 22 days for the 7.5-mg/kg treatment group and 16 days for the 15-mg/kg treatment group.

When Franson et al. (2007) fed common eiders a diet containing 20 mg Se/kg as seleno-L-methionine or a diet that was started at 20 mg Se/kg and increased over time to 60 mg Se/kg (as described in Liver section), the eiders accumulated high concentrations of Se in their blood.

Within 35 days on the high-Se diet the eiders lost about 30% of their body mass and mean blood Se concentration was about 88 mg/kg (17.5 mg/kg ww). Body mass of the eiders on the 20-mg Se/kg diet was similar to that of controls, although mean blood Se in the 20-mg/kg group was about 70 mg/kg (14 mg/kg ww), which was higher than that of controls (about 2 mg/kg; <0.4 mg/kg ww).

Differences in the relationship between blood and liver Se concentrations may be attributed to more rapid initial elimination from liver than blood (Heinz et al. 1990, Wayland et al. 2001) and to binding of Se to inorganic mercury (IoHg) forming an inert Hg–Se protein with a long half-life (Scheuhammer et al. 1998).

Selenium concentrations in wild-trapped birds can be measured in blood as a nonlethal approach for assessing exposure and, when combined with laboratory findings, can be interpreted as to whether exposures are potentially harmful. For example, Se concentrations were measured in terrestrial birds of several species from Kesterson Reservoir, the area surrounding that site, and several reference areas in California from 1994 to 1998 (Santolo and Yamamoto 1999). Except for loggerhead shrikes (*Lanius ludovicianus*), blood-Se was higher in birds from within Kesterson than in birds from other areas. For shrikes, the mean Se concentrations for birds from Kesterson (13 mg/kg dw) were not significantly different than those from nearby surrounding areas (8.5 mg/kg), although the maximum Se concentration at Kesterson (38 mg/kg) was more than twice the maximum for the surrounding area (16 mg/kg). Among species at Kesterson Reservoir, blood–Se concentration was higher in loggerhead shrikes and northern harriers (*Circus cyaneus*) than in the other species (hawks and owls) sampled. This difference among species is likely due to the differing sizes of foraging ranges of the various species (nesting harriers and young were sampled). Adult starlings collected from nest boxes within Kesterson had a mean Se concentration of 16 mg/kg in blood, and concentrations in eggs were significantly correlated with those in blood (Santolo 2007).

Based on the information available, we conclude that Se concentrations in blood can indicate recent dietary exposures of birds, but relationships vary among species, and concentrations in blood can not be clearly related to effects on reproduction or individual health and fitness.

21.3.6 INTEGUMENT/FEATHERS

Background concentrations of Se in feathers are 1–4 mg/kg, and are typically less than 2 mg/kg (Table 21.1), with moisture content of about 10%. As is the case for liver and other tissues, Se concentrations may be higher in the feathers of birds from areas with elevated levels of Hg, because of the interactions between these two elements. Analyses of feathers may provide useful information concerning exposures of birds to Se if they are considered carefully. It is important to recognize that the Se may have been deposited into the feathers at the time they were formed (which may have been months earlier and thousands of miles away from the sampling time and location), or the Se may be the result of external contamination (Goede and de Bruin 1984, 1985, 1986, Goede et al. 1989, Burger 1993). Concentrations also may have been reduced through leaching. Different kinds of feathers from the same bird may contain different concentrations, depending partly on when and where the feathers were grown during the molt cycle.

Overall, feathers are not very useful for diagnosing potential harm in birds, especially because Se concentrations in them are not good indicators of current or recent exposure (unless, perhaps, while the feathers are growing) (Burger 1993, Ohlendorf 1993, U.S. DI 1998, Eisler 2000). However, a Se concentration of 5 mg/kg was identified as a threshold warranting further study (U.S. DI 1998).

Feather loss (bilateral alopecia) is one of the signs of chronic selenosis in birds that may be observed in the field when dietary concentrations are high (Ohlendorf et al. 1988, Ohlendorf 1996). As mentioned earlier, laboratory studies have been conducted with mallards to determine the kinds of lesions and other measurements that can be used for diagnosis of Se toxicosis in birds (Albers et al. 1996, Green and Albers 1997, O'Toole and Raisbeck 1997, 1998). In general, ducks that received diets containing more than 20 mg Se/kg developed a number of lesions of the integument.

Those receiving 40 mg/kg or more Se in their diets lost weight and had abnormal changes in the integument that involved structures containing hard keratin, such as feathers (alopecia/depterylation [i.e., feather loss]), beaks (necrosis), and nails (onychoptosis [sloughed or broken]). When corroborated by elevated Se concentrations in tissues (especially the liver), the observed integumentary and hepatic lesions, as well as weight loss, can serve for diagnosis of Se toxicosis in birds. It should be noted, however, that some birds died without exhibiting any significant morphological lesions even though they were emaciated.

In conclusion, Se concentrations in feathers can indicate exposure of birds at the time the feathers grew, but concentrations that may be diagnostic of problems have not been developed.

21.4 BIOMARKERS

21.4.1 BIOCHEMICAL

A number of studies have described physiological changes that are associated with Se exposure in field-collected or laboratory-exposed birds (Ohlendorf et al. 1988, Hoffman et al. 1989, 1991, 1998, Hoffman and Heinz 1998). These generally involved changes in measurements associated with liver pathology and glutathione metabolism (e.g., glycogen, protein, total sulfhydryl and protein-bound sulfhydryl concentrations; and glutathione peroxidase activity). In lesser scaup, results of a field study suggested that corticosterone release may be influenced by complex contaminant interactions in relation to body condition and body size (Pollock and Machin 2009). When cadmium concentrations were high and birds were in good body condition, there was a negative relationship between liver Se and corticosterone, but not in birds with poor body condition. The overall mean Se concentration in livers was 4.3 mg/kg, with no apparent difference between the two groups.

Wayland et al. (2002) found an inverse association between stress response (measured as corticosterone concentrations following capture) and Se in common eiders nesting in the Canadian Arctic in 1999. Following capture and blood sampling, the birds were placed in a flight pen on-site for 8 days to examine immune function. Cell-mediated immunity was positively related to hepatic Se (geometric means were 14.1 mg/kg in females, 32.1 mg/kg in males). The heterophil: lymphocyte ratio was inversely related to hepatic Se. In 1998, hepatic Se (geometric mean of 17.2 mg/kg in females) was positively related to body mass, abdominal fat mass, kidney mass, and liver mass.

Hoffman (2002) and Spallholz and Hoffman (2002) provide discussions of the mechanisms and role of Se toxicity and oxidative stress in aquatic birds. As dietary and tissue concentrations of Se increase, increases in plasma and hepatic glutathione peroxidase activities occur, followed by dose-dependent increases in the ratio of hepatic oxidized to reduced glutathione, and ultimately hepatic lipid peroxidation. At a given tissue (or egg) Se concentration, one or more of these oxidative effects were associated with teratogenesis (at about 15 mg Se/kg dw [4.6 mg Se/kg ww] in eggs), reduced growth of ducklings (at about 50 mg Se/kg dw [15 mg Se/kg ww] in liver), diminished immune system (at about 16 mg Se/kg dw [5 mg Se/kg ww] in liver), and histopathological lesions (at about 96 mg Se/kg dw [29 mg Se/kg ww] in liver) in adults. These effects have been documented in field and laboratory studies, as reviewed by Hoffman (2002).

21.4.2 MORPHOLOGICAL

The characteristic reproductive effects of Se observed in both field and laboratory studies include reduced hatchability of eggs (due to embryo mortality) and a high incidence of embryo deformities (teratogenic effects) (Ohlendorf 1996, 2003). Selenium-induced abnormalities are often multiple and include defects of the eyes (microphthalmia = abnormally small eyes; possible anophthalmia = missing eyes), feet or legs (amelia = absence of legs; ectrodactylia = absence of toes), beak (incomplete development of the lower beak, spatulate narrowing of the upper beak), brain (hydrocephaly = a swelling of the skull due to fluid accumulation in the brain; exencephaly = an opening in the skull that exposes the brain), and abdomen (gastroschisis = an opening of the gut wall, exposing the intestines and other internal organs). Most of these abnormalities are illustrated through photographs that have been published elsewhere (e.g., Ohlendorf et al. 1986a, 1988, Ohlendorf 1989, 1996, Ohlendorf and Hothem 1995, O'Toole and Raisbeck 1998).

Morphological changes in adult birds as a result of chronically consuming diets with excessive Se have been documented in field and laboratory studies, as described in earlier sections and other reviews (e.g., Ohlendorf 1989, 1996, 2003, O'Toole and Raisbeck 1998, Eisler 2000). They include poor body condition (i.e., weight loss and loss of body lipids), feather loss, and histopathological changes in tissues. Tissue concentrations that cause these changes are not clear-cut, but effects are sometimes observed when hepatic Se is >10 mg/kg. American kestrels fed a diet containing Se at a concentration of 12 mg/kg lost lean body mass, suggesting that they were burning muscle mass as a result of this exposure (not seen in the lower treatment group fed 6 mg/kg); this may be the cause of wasting seen in other species (Yamamoto and Santolo 2000).

21.5 INTERACTIONS

The most studied interactions of Se with other environmental contaminants are between Se and Hg, where each may counteract the toxicity of the other (Cuvin-Aralar and Furness 1991) but also may increase bioaccumulation in tissues (e.g., Furness and Rainbow 1990, Heinz and Hoffman 1998). However, Se toxicity has also been reported to be reduced by elevated levels of lead (Donaldson and McGowan 1989), copper and cadmium (Hill 1974), silver (Jensen 1975), and arsenic (Thapar et al. 1969, Stanley et al. 1994). Despite their common occurrence, biological effects of metal contaminant mixtures are poorly understood and difficult to predict.

Interactions between Se and vitamins A, C, and E, as well as sulfur-containing amino acids also have been documented (NAS-NRC 1976, 1983, Kishchak 1998, Eisler 2000). The interactions may be synergistic or antagonistic in terms of effects on uptake and metabolism, and the degree of interaction is affected by numerous factors. Thus, the topic of interactions is too complex to be addressed in detail in this review, and only a few examples of recent studies are discussed. Nevertheless, some of the interactions of Se with other chemicals can be important factors in the design of field or laboratory studies and in the evaluation of results, and they should be taken into consideration.

After adverse effects characteristic of Se toxicosis were observed in field studies at Kesterson Reservoir, California (described earlier), a series of laboratory studies was conducted, primarily with mallards, to help interpret the potential toxicity of different forms of Se, dietary sources of Se, and interactions with other dietary components including methionine, protein, and various trace elements that might be encountered in nature. Hamilton and Hoffman (2003) provide a review of the findings from the various laboratory studies, including Se concentrations in diets or tissues associated with the effects.

Here we summarize only the laboratory studies conducted to assess interactions with As (Stanley et al. 1994), B (Stanley et al. 1996), and Hg (Heinz and Hoffman 1998) in addition to relevant field studies. Each of the laboratory studies involved varying levels of dietary exposures of breeding mallards to Se alone, one of the other elements alone, and Se in combination with the other chemical. In each study, Se and the other chemical caused significant adverse effects on reproduction when present alone in the diet at higher treatment levels, but the interactions varied by chemical. Antagonistic interactions between As and Se occurred whereby As reduced Se accumulation in duck livers and eggs, and reduced the effects of Se on hatching success and embryo deformities when dietary As concentrations were 100 or 400 mg/kg. As the authors noted, however, the importance of the observed As–Se interaction in the environment is unknown because As may not be present in bird food items at contaminated sites in the form used in the study (sodium arsenate).

There was little evidence of interaction between B and Se when ducks were fed the two chemicals in combination. When the diet contained 10 mg Se/kg plus 10 mg Hg/kg, the effects on reproduction were worse than for either Se or Hg alone, even though Se concentrations in eggs were elevated only modestly by the presence of Hg. The 10-mg Se/kg diet produced a mean of about 25 mg Se/kg on a dw basis (7.6 mg Se/kg ww) in eggs, and reduced the hatching success of fertile eggs to 24.0% compared to 44.2% for controls. When 10 mg Hg/kg was fed along with the 10 mg Se/kg, Se concentrations in eggs rose only to about 31 mg/kg dw (9.3 mg/kg ww), but hatching success dropped to 1.4%. Either the embryotoxicity of the Se had been increased by the presence of Hg, the embryotoxicity of the Hg was added to that of the Se, or some combination of these synergistic effects had occurred. In any case, the 31 mg Se/kg measured in eggs was associated with a greater-than-expected level of embryonic death were one to focus only on the Se in the eggs. In addition to the number of young produced per female being significantly reduced in the earlier study, the frequency of teratogenic effects was significantly increased by the combination of Hg and Se in the diet, and Hg enhanced the storage of Se in duck tissues. Female mallards fed the combination diet had about 1.5 times higher hepatic Se concentrations than those fed the Se-only diet, and male mallards fed the combination diet had almost 12 times the Se concentration of those fed the Se-only diet. In contrast to the synergistic effects on reproduction, the combined Se plus Hg diet was less toxic to adult male mallards than either Se or Hg alone. In male mallards fed only the 10 mg Se/kg diet, livers contained a mean of about 32 mg Se/kg dw (9.6 mg Se/kg ww), but when 10 mg Hg/kg was also in the diet, male livers contained a mean of about 380 mg Se/kg (114 mg Se/kg ww). A value of 380 mg Se/kg in the liver of ducks would almost certainly be equated with severe harm, but the coexistence of about 217 mg Hg/kg (65 mg Hg/kg ww) in the livers seemingly nullified the toxicity of the Se. Likewise, the 217 mg Hg/kg is well above the level normally associated with harm in birds; in this study a level of about 237 mg Hg/kg (71 mg Hg/kg ww) was reported in the male mallards fed only the 10 mg Hg/kg, and Hg-induced toxicity and mortality were observed in this group of males. Obviously, the Hg and Se had conferred a mutually antagonistic effect on each other, but only as far as the adult birds were concerned.

Mercury and Se concentrations in the livers of various free-living carnivorous mammals often are highly correlated in a molar ratio of 1:1 (Scheuhammer 1987, Furness and Rainbow 1990, Cuvin-Aralar and Furness 1991, Eisler 2000). However, there is no consistent pattern for such a correlation in the livers of birds. For example, in diving ducks from San Francisco Bay, hepatic Hg and Se were correlated, but Se concentrations exceeded Hg concentrations by 6- to 15-fold on molar basis (Ohlendorf et al. 1986c, 1991). Elsewhere, Hg and Se concentrations were positively correlated in some bird livers, but not in others, or they were negatively correlated (see review by Ohlendorf 1993). These relationships may change as birds remain at the sampling location (due to differential accumulation and loss rates for Hg and Se), they may vary because of differing relative concentrations of the two elements, and other factors (such as the chemical forms present) also may complicate the patterns of bioaccumulation.

When there is a low concentration of Hg, a lower molar ratio is observed; however, at high Hg and Se concentrations in the liver, most Se binds Hg resulting in a Hg:Se ratio greater than 1.0 (Kim et al. 1996). For example, livers of black-footed albatross that contained total mercury (THg) concentrations over 100 mg/kg had an equivalent molar ratio of 1:1 between THg and Se, but such a relationship was unclear when birds had relatively low Hg levels. Studies by Henny et al. (2002) and Spalding et al. (2000) have shown high correlations of Se with IoHg on a molar basis in livers of fish-eating birds. As the THg concentration increased, the percentage present as methylmercury (MeHg) decreased. Those authors suggested that Se may contribute to the sequestration of IoHg, thereby reducing its toxicity. This conclusion would be consistent with the results of a Se–Hg inter-action study with mallards by Heinz and Hoffman (1998) described earlier.

Recent work by Eagles-Smith et al. (2009) provides a useful understanding of Se–Hg relation-ships. They assessed the role of Se in demethylation of MeHg in the livers of adults and chicks of four waterbird species that commonly breed in San Francisco Bay (American avocets, black-necked stilts, Caspian terns [*Hydroprogne caspia*; formerly *Sterna caspia*], and Forster's terns [*Sterna forsteri*]). In adults (all species combined) there was strong evidence for a threshold model where demethylation of MeHg occurred above a hepatic THg concentration threshold of 8.51 ± 0.93 mg/kg, and there was a strong decline in percent MeHg values as THg concentrations increased above 8.51 mg/kg. Conversely, there was no evidence for a demethylation threshold in chicks, and they found that percent MeHg values declined linearly with increasing THg concentrations. For adults, they also found taxonomic differences in the demethylation responses, with avocets and stilts showing a higher demethylation rate than terns when concentrations exceeded the threshold, whereas terns had a lower demethylation threshold (7.48 ± 1.48 mg/kg) than avocets and stilts (9.91 ± 1.29 mg/kg). Selenium concentrations were positively correlated with IoHg in livers of birds above the demethylation threshold, but not below, suggesting that Se may act as a binding site for demethylated Hg and may reduce the potential for secondary toxicity.

Similar findings were reported by Scheuhammer et al. (2008) for common loons (*Gavia immer*) and bald eagles (*Haliaeetus leucocephalus*), although the thresholds were very different. In liver, both species had a wide range of THg concentrations, substantial demethylation of MeHg, and coaccumulation of Hg and Se. There were molar excesses of Se over Hg up to about 50–60 mg Hg/kg, above which there was an approximate 1:1 molar ratio of Hg:Se in both species. Thus, the amount of Se bound to Hg at any given concentration of THg is likely to vary among species, suggesting that the 8.5 mg Hg/kg threshold described earlier is not a universal one.

At this time it is not possible to enumerate what concentrations of Se need to be in eggs or tissues to cause harm when certain concentrations of other contaminants such as Hg are also present in the samples. Likewise, the concentrations of combinations of Se and other chemicals that would lead one to conclude that no harm from Se, or the other chemical, is likely to occur are unknown. However, when elevated concentrations of other contaminants, especially Hg, are found along with Se in eggs or tissues, caution should be exercised in interpreting the significance of the Se (and the other contaminant). When warranted and feasible (due to time and resource constraints), this caution would translate into conducting careful field studies at the contaminated site to determine if reproduction and adult health are normal, compared to an uncontaminated reference area.

21.6 HORMESIS

Selenium is an essential trace element for bird diets, as described earlier, and inadequate dietary levels of bioavailable Se may result in low Se in eggs. When poultry diets contain Se concentrations of less than 0.30 mg/kg and eggs contain less than about 0.66 mg/kg dw (0.20 mg/kg ww), they are considered to be below the "adequate" range (Puls 1988).

Consideration of the hormetic effects of Se may result in lowering of thresholds for diet and eggs described above. A recent paper by Beckon et al. (2008) used the mean response data for the control and five treatment levels from the mallard study by Heinz et al. (1989) to evaluate potential hormetic effects exhibited by the treatment groups. They concluded that the EC_{10} from that study was 7.7 mg Se/kg (although their Figure 5 says 7.3 mg Se/kg). Because Se concentrations in bird eggs may be used in setting site-specific water quality standards for Se (e.g., Great Salt Lake; State of Utah 2008), the difference in conclusions between the Ohlendorf (2003) and Beckon et al. (2008) results are important from a regulatory as well as scientific standpoint. Consequently, further analyses of the available data from the six studies with mallards (Heinz et al. 1987, 1989, Heinz and Hoffman 1996, 1998, Stanley et al. 1994, 1996) are underway by the authors of this chapter.

Summary

Selenium is an essential nutrient for birds, with a narrow range of concentrations between what is a beneficial diet (< 3 mg/kg dw) and what represents a threshold for reproductive impairment (in the range of 3–8 mg/kg, depending on species and the form of Se in the diet). When birds eat a high-Se diet, Se levels in the diet are quickly reflected in concentrations in eggs, liver, and blood, but more slowly in muscle. Similarly, when birds are switched from a high-Se diet to one with a lower concentration (or when they migrate from a high-Se area to a Se-normal area), the eggs, liver, and blood adjust relatively quickly to the lower concentrations.

Kidneys are not as useful as livers for diagnosing Se status of birds, although the concentrations in kidneys and livers are highly correlated. In Se-normal areas, concentrations in kidneys tend to be higher than those in the liver, but concentrations in the two tissues are similar in birds from high-Se areas. Feathers can be useful under some circumstances, but it is important to recognize that Se concentrations in feathers reflect the exposure of the bird when the feathers were developing, not their current exposure.

Background, elevated, and various effect levels of Se in bird diets, eggs, and various tissues are summarized in Table 21.1. We present there a range of effect concentrations because different techniques have been used to develop them, and the reader can select from the range of values those that are appropriate for the degree of protectiveness (conservatism) desired under a particular set of circumstances.

Based on our experience and review of the literature, we recommend the values presented in Table 21.2 as diagnostic levels for Se concentrations in eggs, livers, and diet to evaluate the probability that Se may be causing adverse effects in birds. Se concentrations in eggs and livers should be considered the primary diagnostic levels, complemented by Se levels in the diet and observed effects on egg hatchability or signs of toxicosis such as those described for liver or other tissues. As stated previously, when Se concentrations are known for both the eggs and livers of breeding females, judgments on the hazards of Se to reproduction should be based on Se in the egg.

Short of doing a time-consuming study of reproductive success, analysis of eggs is by far the best way to determine status of a population with respect to potential reproductive impairment. No single criterion is available for diagnosis of Se toxicosis in young or adult birds, but Se toxicosis is indicated when elevated Se concentrations in tissues (especially when greater than 20 mg/kg in the liver) are accompanied by emaciation, poor quality of shed nails, bilaterally symmetrical alopecia of the head and neck, toxic hepatic lesions, and necrosis of maxillary nails.

Regardless of which kind of sample is being analyzed (diet, egg, or other tissue), we highly recommend measuring moisture content of the samples and reporting those values along with the Se concentration. The literature contains a mixture of wet-weight and dry-weight concentrations in different media, and it is difficult to relate concentrations on one basis to the other without knowing the moisture content of the samples. This is important because moisture content varies by sample type and handling procedures.

Physiological changes associated with Se exposure in field-collected or laboratory-exposed birds generally involve changes in measurements associated with liver pathology and glutathione metabolism (e.g., glycogen, protein, total sulfhydryl and protein-bound sulfhydryl, concentrations; glutathione peroxidase activity). As dietary and tissue concentrations of Se increase, increases in plasma and hepatic glutathione peroxidase activities occur, followed by dose-dependent increases in the ratio of hepatic oxidized to reduced glutathione, and ultimately hepatic lipid peroxidation. At a given tissue (or egg) Se concentration, one or more of these oxidative effects were associated with teratogenesis (when Se concentrations in eggs reached about 15 mg/kg dw = 4.6 mg/kg ww), reduced growth of ducklings (at about 50 mg Se/kg dw = 15 mg Se/kg ww in liver), diminished immune system (at about 16 mg Se/kg dw = 5 mg Se/kg ww in liver) and histopathological lesions (about 96 mg Se/kg dw = 29 mg Se/kg ww in liver) in adults.

The characteristic reproductive effects of Se observed in both field and laboratory studies include reduced hatchability of eggs (due to embryo mortality) and high incidence of developmental abnormalities (due to teratogenesis). Se-induced abnormalities are often multiple and include defects of the eyes (microphthalmia and possible anophthalmia [i.e., abnormally small or missing eyes]), feet or legs (amelia and ectrodactylia [absence of legs or toes]), beak (incomplete development of the lower beak, spatulate narrowing of the upper beak), brain (hydrocephaly and

TABLE 21.2
Recommended Assessment Values for Effects of Dietary or Tissue Concentrations of Se on Birds

Medium and Level/ Status/Effects[a]	Concentration (mg Se/kg, dw)	Comments
Diet[b]		
Background	<3.0	Typical concentrations in diet items for birds; deficiencies associated with low concentrations have not been reported in wild birds
Reproductive impairment	<4.0	Low probability for reduced egg hatchability; near background concentration for many aquatic food-chain items (i.e., field diet, typically <3.0 mg/kg); value based on studies of multiple species
Reproductive impairment	>5.0	Elevated probability for reduced egg hatchability in sensitive species; effects down to this concentration may be measurable in the laboratory but unlikely to be detectable in the field unless dietary concentrations are considerably higher
Eggs[c]		
Background	Mean < 3.0 (typically 1.5–2.5); individual eggs < 5	Concentrations may be higher in some marine birds
Reproductive impairment	<8.0	Low probability for reduced egg hatchability, including effects in sensitive species
Reproductive impairment	>12	Elevated probability for reduced egg hatchability in sensitive and moderately sensitive species
Teratogenicity	<20	Low probability for teratogenic effects in most species, and threshold for statistically discernable incidence in sensitive species such as mallard
Teratogenicity	>35	Probability for teratogenic effects in species of "average" sensitivity such as black-necked stilt
Liver[d]		
Background for freshwater and terrestrial species	<10	Low probability of adverse effects in these species
Background for some marine species	20–75 in some species (see text)	Low probability of adverse effects in these species; must consider species differences compared to freshwater and terrestrial species
Elevated and potentially toxic in freshwater and terrestrial species	10–20	Considered suspicious of selenium toxicosis when accompanied by symptoms listed for toxic effects (see text); sensitivity varies by species
Toxic	>20	Diagnostic of Se toxicosis when accompanied by emaciation, poor quality of shed nails, bilaterally symmetrical alopecia of the head and neck, hepatic lesions, and necrosis of maxillary nails; based on field observations and laboratory studies with mallards

Notes: No specific recommendations are made for kidney, muscle, blood, or feathers, although each of them can indicate levels of exposure. Kidney concentration is generally correlated with liver; muscle responds more slowly than eggs, liver, or blood in reflecting current exposure; and feathers reflect exposure at the time they were growing rather than the time of sampling.

[a] Typical moisture content (%) and approximate conversion factor are shown in footnotes for each medium.

[b] Variable moisture; laboratory diet typically ~10%, but natural diet varies widely (<10–>90%).

[c] 65–80% moisture, varying with species and incubation stage; use 70% (i.e., factor of 3.3) for approximate conversion.

[d] 70% moisture; use factor of 3.3 for approximate conversion.

exencephaly [fluid accumulation in the brain and exposure of the brain]), and abdomen (gastroschisis [an open fissure of the abdomen]).

Selenium interacts with a number of other environmental contaminants and nutrients of interest for birds. The interactions of Se with Hg have been studied most extensively, but interactions with As also may be important. Se and Hg each may counteract or increase the toxicity of the other but also may increase bioaccumulation in tissues. Dietary Hg and Se together were more harmful to mallard reproduction than either element was alone, while they were less toxic to adult birds in combination than they were alone. Consequently, where Hg may be elevated, both Se and Hg should be evaluated. In a similar study of Se and inorganic As, interactions between As and Se were antagonistic, whereby As reduced Se accumulation in duck livers and eggs, and reduced the effects of Se on hatching success and embryo deformities.

Recent work on Se–Hg interactions has shown strong evidence for a threshold above which demethylation of MeHg occurred, and there was a strong decline in percent MeHg values as THg concentrations increased above the threshold. Conversely, there was no evidence for a demethylation threshold in chicks, and percent MeHg values declined linearly with increasing THg concentrations. For adults, there were taxonomic differences in the demethylation responses, with avocets and stilts showing a higher demethylation rate than terns when concentrations exceeded the threshold, whereas terns had a lower demethylation threshold than avocets and stilts. Selenium concentrations were positively correlated with IoHg in livers of birds above the demethylation threshold, but not below, suggesting that Se may act as a binding site for demethylated Hg and may reduce the potential for secondary toxicity.

In summary, the ecotoxicology of Se is complex, because of the variable chemical forms in which it occurs in the environment, its interactions with other environmental contaminants, and large differences in species sensitivity to the adverse effects of Se. The most likely effects to be observed in the field are reproductive impairment, which has been documented at a number of locations during the past 25 years or so. However, Se toxicosis and mortality of adult birds also has been observed and may occur when exposures are higher than those causing reproductive impairment. The assessment values for diet, eggs, and other tissues presented in Table 21.1 can be used to evaluate risks of adverse effects in birds.

ACKNOWLEDGMENTS

We appreciate the assistance of G. M. Santolo in providing some of the material for this chapter through our previous work, and for his helpful review of the draft. C. M. and T. W. Custer, M. Wayland, and W. N. Beyer also reviewed the manuscript and provided useful comments.

REFERENCES

Adams, W. J., K. V. Brix, M. Edwards, L. M. Tear, D. K. DeForest, and A. Fairbrother. 2003. Analysis of field and laboratory data to derive selenium toxicity thresholds for birds. *Environ. Toxicol. Chem.* 22:2020–2029.

Agusa, T., et al. 2005. Body distribution of trace elements in black-tailed gulls from Rishiri Island, Japan: Age-dependent accumulation and transfer to feathers and eggs. *Environ. Toxicol. Chem.* 24:2107–2120.

Albers, P. H., D. E. Green, and C. J. Sanderson. 1996. Diagnostic criteria for selenium toxicosis in aquatic birds: Dietary exposure, tissue concentrations, and macroscopic effects. *J. Wildl. Dis.* 32:468–485.

Andrahennadi, R., M. Wayland, and I. J. Pickering. 2007. Speciation of selenium in stream insects using X-ray absorption spectroscopy. *Environ. Sci. Technol.* 41:7683-7687.

Arnold, R. L., O. E. Olson, and C. W. Carlson. 1973. Dietary selenium and arsenic additions and their effects on tissue and egg selenium. *Poult. Sci.* 52:847-854.

Beckon, W. N., C. Parkins, A. Maximovich, and A. V. Beckon. 2008. A general approach to modeling biphasic relationships. *Environ. Sci. Technol.* 42:1308-1314.

Beilstein, M. A., and P. D. Whanger. 1986. Deposition of dietary organic and inorganic selenium in rat erythrocyte proteins. *J. Nutr.* 116:1701–1710.

Braune, B. M., G. M. Donaldson, and K. A. Hobson. 2002. Contaminant residues in seabird eggs from the Canadian Arctic. II. Spatial trends and evidence from stable isotopes for intercolony differences. *Environ. Pollut.* 117:133–145.

Burger, J. 1993. Metals in avian feathers: Bioindicators of environmental pollution. *Rev. Environ. Toxicol.* 5:203–311.

Campbell, L. M., R. J. Norstrom, K. A. Hobson, D. C. G. Muir, S. Backus, and A. T. Fisk. 2005. Mercury and other trace elements in a pelagic Arctic marine food web (Northwater Polynya, Baffin Bay). *Sci. Total Environ.* 351/352:247–263.

Cappon, C. J. 1991. Sewage sludge as a source of environmental selenium. *Sci. Total Environ.* 100:177–205.

Combs, G. F., Jr., and S. B. Combs. 1986. *The role of selenium in nutrition.* Orlando, FL: Academic Press, Inc.

Conover, M. R., and J. L. Vest. 2009. Selenium and mercury concentrations in California gulls breeding on the Great Salt Lake, Utah, USA. *Environ. Toxicol. Chem.* 28:324–329.

Cukierski, M. J., et al. 1989. 30-day oral toxicity study of L-selenomethionine in female long-tailed macaques (*Macaca fascicularis*). *Fund. Appl. Toxicol.* 13:26–39.

Cuvin-Aralar, M. L. A., and R. W. Furness. 1991. Mercury and selenium interaction: A review. *Ecotoxicol. Environ. Saf.* 21:348–364.

Detwiler, S. J. 2002. Toxicokinetics of selenium in the avian egg: Comparisons between species differing in embryonic tolerance. PhD diss., University of California, Davis.

DeVink, J.-M. A., R. G. Clark, S. M. Slattery, and T. M. Scheuhammer. 2008a. Effects of dietary selenium on reproduction and body mass of captive lesser scaup. *Environ. Toxicol. Chem.* 27:471–477.

DeVink, J.-M. A., R. G. Clark, S. M. Slattery, and M. Wayland. 2008b. Is selenium affecting body condition and reproduction in boreal breeding scaup, scoters, and ring-necked ducks? *Environ. Pollut.* 152:116–122.

Dietz, R., F. Riget, and P. Johansen. 1996. Lead, cadmium, mercury and selenium in Greenland marine animals. *Sci. Total Environ.* 186:67–93.

Donaldson, W. E., and C. McGowan. 1989. Lead toxicity in chickens: Interaction with toxic dietary levels of selenium. *Biol. Trace Elem. Res.* 20:127–133.

Eagles-Smith, C. A., J. T. Ackerman, J. Yee, and T. L. Adelsbach. 2009. Mercury demethylation in waterbird livers: Dose-response thresholds and differences among species. *Environ. Toxicol. Chem.* 28:568–577.

Eisler, R. 2000. *Handbook of chemical risk assessment: Health hazards to humans, plants, and animals*, Vol. 3, pp. 1649–1705. Boca Raton, FL: Lewis Publishers.

Elliott, J. E. 2005. Trace metals, stable isotope ratios, and trophic relations in seabirds from the North Pacific Ocean. *Environ. Toxicol. Chem.* 24:3099–3105.

Elliott, J. E., A. M. Scheuhammer, F. A. Leighton, and P. A. Pearce. 1992. Heavy metal and metallothionein concentrations in Atlantic Canadian seabirds. *Arch. Environ. Contam. Toxicol.* 22:63–73.

Fairbrother, A., and J. Fowles. 1990. Subchronic effects of sodium selenite and selenomethionine on several immune functions in mallards. *Arch. Environ. Contam. Toxicol.* 19:836–844.

Franson, J. C., T. Hollmén, R. H. Poppenga, M. Hario, M. Kilpi, and M. R. Smith. 2000. Selected trace elements and organochlorines: Some findings in blood and eggs of nesting common eiders (*Somateria mollissima*) from Finland. *Environ. Toxicol. Chem.* 19:1340–1347.

Franson, J. C., et al. 2007. Effects of dietary selenium on tissue concentrations, pathology, oxidative stress, and immune function in common eiders (*Somateria mollissima*). *J. Toxicol. Environ. Health (Part A)* 70:861–874.

Furness, R. W., and P. S. Rainbow. 1990. *Heavy metals in the marine environment.* Boca Raton, FL: CRC Press.

Goede, A. A., and M. de Bruin. 1984. The use of bird feather parts as a monitor for metal pollution. *Environ. Pollut. (Ser. B)* 8:281–298.

Goede, A. A., and M. de Bruin. 1985. Selenium in a shore bird, the dunlin, from the Dutch Waddenzee. *Mar. Pollut. Bull.* 16:115–117.

Goede, A. A., and M. de Bruin. 1986. The use of bird feathers for indicating heavy metal pollution. *Environ. Monit. Assess.* 7:249–256.

Goede, A. A., T. Nygard, M. de Bruin, and E. Steinnes. 1989. Selenium, mercury, arsenic and cadmium in the lifecycle of the dunlin, *Calidris alpina*, a migrant wader. *Sci. Total Environ.* 78:205–218.

Grand, J. B., J. C. Franson, P. L. Flint, and M. R. Petersen. 2002. Concentrations of trace elements in eggs and blood of spectacled and common eiders on the Yukon-Kuskokwim Delta, Alaska, USA. *Environ. Toxicol. Chem.* 21:1673–1678.

Green, D. E., and P. H. Albers. 1997. Diagnostic criteria for selenium toxicosis in aquatic birds: Histologic lesions. *J. Wildl. Dis.* 33:385–404.

Hamilton, S. J., and D. J. Hoffman. 2003. Trace element and nutrition interactions in wildlife. In *Handbook of ecotoxicology*, 2nd ed., eds. D. J. Hoffman, B. A. Rattner, G. A. Burton, Jr., and J. Cairns, Jr., pp. 1197–1235. Boca Raton, FL: Lewis Publishers.

Hamilton, S. J., K. J. Buhl, N. L. Faerber, R. H. Wiedmeyer, and F. A. Bullard. 1990. Toxicity of organic selenium in the diet to chinook salmon. *Environ. Toxicol. Chem.* 9:347–358.

Heinz, G. H. 1993a. Re-exposure of mallards to selenium after chronic exposure. *Environ. Toxicol. Chem.* 12:1691–1694.

Heinz, G. H. 1993b. Selenium accumulation and loss in mallard eggs. *Environ. Toxicol. Chem.* 12:775–778.

Heinz, G. H. 1996. Selenium in birds. In *Environmental contaminants in wildlife: Interpreting environmental contaminants in animal tissues*, eds. W. N. Beyer, G. H. Heinz, and A. W. Redmon-Norwood, pp. 447–458. Boca Raton, FL: Lewis Publishers.

Heinz, G. H., and M. A. Fitzgerald. 1993a. Overwinter survival of mallards fed selenium. *Arch. Environ. Contam. Toxicol.* 25:90–94.

Heinz, G. H., and M. A. Fitzgerald. 1993b. Reproduction of mallards following overwinter exposure to selenium. *Environ. Pollut.* 81:117–122.

Heinz, G. H., and D. J. Hoffman. 1996. Comparison of the effects of seleno-L-methionine, seleno-DL-methionine, and selenized yeast on reproduction of mallards. *Environ. Pollut.* 91:169–175.

Heinz, G. H., and D. J. Hoffman. 1998. Methylmercury chloride and selenomethionine interactions on health and reproduction in mallards. *Environ. Toxicol. Chem.* 17:139–145.

Heinz, G. H., D. J. Hoffman, A. J. Krynitsky, and D. M. G. Weller. 1987. Reproduction in mallards fed selenium. *Environ. Toxicol. Chem.* 6:423–433.

Heinz, G. H., D. J. Hoffman, and L. G. Gold. 1988. Toxicity of organic and inorganic selenium to mallard ducklings. *Arch. Environ. Contam. Toxicol.* 17:561–568.

Heinz, G. H., D. J. Hoffman, and L. G. Gold. 1989. Impaired reproduction of mallards fed an organic form of selenium. *J. Wildl. Manage.* 53:418–428.

Heinz, G. H., G. W. Pendleton, A. J. Krynitsky, and L. G. Gold. 1990. Selenium accumulation and elimination in mallards. *Arch. Environ. Contam. Toxicol.* 19:374–379.

Henny, C. J., and G. B. Herron. 1989. Selenium, mercury, and white-faced ibis reproduction at Carson Lake, Nevada. *J. Wildl. Manage.* 53:1032–1045.

Henny, C. J., L. J. Blus, R. A. Grove, and S. P. Thompson. 1991. Accumulation of trace elements and organochlorines by surf scoters wintering in the Pacific Northwest. *Northwest Nat.* 72:43–60.

Henny, C. J., D. D. Rudis, T. J. Roffe, and E. Robinson-Wilson. 1995. Contaminants and sea ducks in Alaska and the circumpolar region. *Environ. Health Perspect.* 103:41–49.

Henny, C. J., E. F. Hill, D. J. Hoffman, M. G. Spalding, and R. A. Grove. 2002. Nineteenth century mercury: Hazard to wading birds and cormorants of the Carson River, Nevada. *Ecotoxicology* 11:213–231.

Hill, C. H. 1974. Reversal of selenium toxicity in chicks by mercury, copper, and cadmium. *J. Nutr.* 104:593–598.

Hoffman, D. J. 2002. Role of selenium toxicity and oxidative stress in aquatic birds. *Aquat. Toxicol.* 57:11–26.

Hoffman, D. J., and G. H. Heinz. 1988. Embryotoxic and teratogenic effects of selenium in the diet of mallards. *J. Toxicol. Environ. Health* 24:477–490.

Hoffman, D. J., and G. H. Heinz. 1998. Effects of mercury and selenium on glutathione metabolism and oxidative stress in mallard ducks. *Environ. Toxicol. Chem.* 17:161–166.

Hoffman, D. J., G. H. Heinz, and A. J. Krynitsky. 1989. Hepatic glutathione metabolism and lipid peroxidation in response to excess dietary selenomethionine and selenite in mallard ducklings. *J. Toxicol. Environ. Health* 27:263–271.

Hoffman, D. J., G. H. Heinz, L. J. LeCaptain, C. M. Bunck, and D. E. Green. 1991. Subchronic hepatotoxicity of selenomethionine ingestion in mallard ducks. *J. Toxicol. Environ. Health* 32:449–464.

Hoffman, D. J., H. M. Ohlendorf, C. M. Marn, and G. W. Pendleton. 1998. Association of mercury and selenium with altered glutathione metabolism and oxidative stress in diving ducks from the San Francisco Bay region, USA. *Environ. Toxicol. Chem.* 17:167–172.

Jensen, L. S. 1975. Modification of a selenium toxicity in chicks by dietary silver and copper. *J. Nutr.* 105:769–775.

Kim, E. Y., K. Saeki, S. Tanabe, H. Tanaka, and R. Tatsukawa. 1996. Specific accumulation of mercury and selenium in seabirds. *Environ. Pollut.* 94:261–265.

Kishchak, I. T. 1998 Supplementation of selenium in the diets of domestic animals. In *Environmental chemistry of selenium*, eds. W. T. Frankenberger, Jr., and R. A. Engberg, pp. 143–152. New York, NY: Marcel Dekker.

Koranda, J. J., M. Stuart, S. Thompson, and C. Conrado. 1979. Biogeochemical studies of wintering waterfowl in the Imperial and Sacramento Valleys, Report UCID-18288, Lawrence Livermore Laboratory, University of California, Livermore, California.

Lam, J. C. W., S. Tanabe, M. H. W. Lam, and P. K. S. Lam. 2005. Risk to breeding success of waterbirds by contaminants in Hong Kong: Evidence from trace elements in eggs. *Environ. Pollut.* 135:481–490.

Latshaw, J. D. 1975. Natural and selenite selenium in the hen and egg. *J. Nutr.* 105:32–37.

Latshaw, J. D., and M. Osman. 1975. Distribution of selenium in egg white and yolk after feeding natural and synthetic selenium compounds. *Poult. Sci.* 54:1244–1252.

Latshaw, J. D., T. Y. Morishita, C. F. Sarver, and J. Thilsted. 2004. Selenium toxicity in breeding ring-necked pheasants (*Phasianus colchicus*). *Avian Dis.* 48:935–939.

Mallory, M. L., B. M. Braune, M. Wayland, H. G. Gilchrist, and D. L. Dickson. 2004. Contaminants in common eiders (*Somateria mollissima*) of the Canadian Arctic. *Environ. Rev.* 12:197–218.

Martin, P. F. 1988. The toxic and teratogenic effects of selenium and boron on avian reproduction. M.S. Thesis, University of California, Davis.

Moksnes, K. 1983. Selenium deposition in tissues and eggs of laying hens given surplus of selenium as selenomethionine. *Acta Vet. Scand.* 24:34–44.

Moxon, A. L., and W. E. Poley. 1938. The relation of selenium content of grains in the ration to the selenium content of poultry carcass and eggs. *Poult. Sci.* 17:77–80.

National Academy of Sciences–National Research Council (NAS-NRC). 1976. Selenium. Committee on Medical and Biologic Effects of Environmental Pollutants, NRC. National Academy Press, Washington, D.C.

National Academy of Sciences–National Research Council (NAS-NRC). 1983. Selenium in Nutrition. Subcommittee on Selenium, Committee on Animal Nutrition, Board on Agriculture, NRC. National Academy Press, Washington, D.C.

Ohlendorf, H. M. 1989. Bioaccumulation and effects of selenium in wildlife. In *Selenium in agriculture and the environment,* ed. L. W. Jacobs, pp. 133–177. Special Publication 23. American Society of Agronomy and Soil Science Society of America, Madison, WI.

Ohlendorf, H. M. 1993. Marine birds and trace elements in the temperate North Pacific. In *The status, ecology, and conservation of marine birds of the North Pacific,* eds. K. Vermeer, K. T. Briggs, K. H. Morgan, and D. Siegel-Causey, pp. 232–240. Can. Wildl. Serv. Spec. Publ., Ottawa.

Ohlendorf, H. M. 1996. Selenium. In *Noninfectious diseases of wildlife, Second Edition,* eds. A. Fairbrother, L. N. Locke, and G. L. Hoff, pp. 128–140. Ames, IA: Iowa State University Press.

Ohlendorf, H. M. 2002. The birds of Kesterson Reservoir: A historical perspective. *Aquatic Toxicol.* 57:1–10.

Ohlendorf, H. M. 2003. Ecotoxicology of selenium. In *Handbook of ecotoxicology,* 2nd., eds. D. J. Hoffman, B. A. Rattner, G. A. Burton Jr., and J. C. Cairns Jr., pp. 465–500. Boca Raton, FL: Lewis Publishers.

Ohlendorf, H. M. 2007. Threshold values for selenium in Great Salt Lake: Selections by the science panel. Final Technical Memorandum. Prepared by CH2M HILL for the Great Salt Lake Science Panel. February 28; available at http://www.deq.utah.gov/Issues/GSL_WQSC/selenium.htm

Ohlendorf, H. M., and C. S. Harrison. 1986. Mercury, selenium, cadmium and organochlorines in eggs of three Hawaiian seabird species. *Environ. Pollut. (Series B)* 11:169-191.

Ohlendorf, H. M., and R. L. Hothem. 1995. Agricultural drainwater effects on wildlife in central California. In *Handbook of ecotoxicology,* eds. D. J. Hoffman, B. A. Rattner, G. A. Burton, Jr., and J. Cairns, Jr., pp. 577–595. Boca Raton, FL: Lewis Publishers.

Ohlendorf, H. M., D. J. Hoffman, M. K. Saiki, and T. W. Aldrich. 1986a. Embryonic mortality and abnormalities of aquatic birds: Apparent impacts of selenium from irrigation drainwater. *Sci. Total Environ.* 52:49–63.

Ohlendorf, H. M., R. L. Hothem, C. M. Bunck, T. W. Aldrich, and J. F. Moore. 1986b. Relationships between selenium concentrations and avian reproduction. *Trans. N. Am. Wildl. Nat. Resour. Conf.* 51:330–342.

Ohlendorf, H. M., R. W. Lowe, P. R. Kelly, and T. E. Harvey. 1986c. Selenium and heavy metals in San Francisco Bay diving ducks. *J. Wildl. Manage.* 50:64–71.

Ohlendorf, H. M., R. W. Kilness, J. L. Simmons, R. K. Stroud, D. J. Hoffman, and J. F. Moore. 1988. Selenium toxicosis in wild aquatic birds. *J. Toxicol. Environ. Health* 24:67–92.

Ohlendorf, H. M., K. C. Marois, R. W. Lowe, T. E. Harvey, and P. R. Kelly. 1989. Environmental contaminants and diving ducks in San Francisco Bay. In *Selenium and agricultural drainage: Implications for San Francisco Bay and the California environment,* Proceedings of the Fourth Selenium Symposium, Berkeley, CA, March 21, 1987, ed. A. Q. Howard, pp. 60–69. Sausalito, CA: The Bay Institute of San Francisco.

Ohlendorf, H. M., R. L. Hothem, C. M. Bunck, and K. C. Marois. 1990. Bioaccumulation of selenium in birds at Kesterson Reservoir, California. *Arch. Environ. Contam. Toxicol.* 19:495–507.

Ohlendorf, H. M., K. C. Marois, R. W. Lowe, T. E. Harvey, and P. R. Kelly. 1991. Trace elements and organochlorines in surf scoters from San Francisco Bay, 1985. *Environ. Monit. Assess.* 18:105–122.

Oldfield, J. E. 1990. Selenium: Its uses in agriculture, nutrition & health, and environment. Special Publication of Selenium-Tellurium Development Association, Inc., Darien, Connecticut.

Oldfield, J. E. 1998. Environmental implications of uses of selenium with animals. In *Environmental chemistry of selenium*, eds. W. T. Frankenberger, Jr. and R. A. Engberg, pp. 129–142. New York: Marcel Dekker.

Olson, O. E., E. J. Novacek, E. I. Whitehead, and I. S. Palmer. 1970. Investigations on selenium in wheat. *Phytochemistry (Oxf.)* 9:1181–1188.

Ort, J. F., and J. D. Latshaw. 1978. The toxic level of sodium selenite in the diet of laying chickens. *J. Nutr.* 108:1114–1120.

O'Toole, D., and M. F. Raisbeck. 1997. Experimentally induced selenosis of adult mallard ducks: Clinical signs, lesions, and toxicology. *Vet. Pathol.* 34:330–340.

O'Toole, D., and M. F. Raisbeck. 1998. Magic numbers, elusive lesions: Comparative pathology and toxicology of selenosis in waterfowl and mammalian species. In *Environmental chemistry of selenium*, eds. W. T. Frankenberger, Jr. and R. A. Engberg, pp. 355–395. New York: Marcel Dekker.

Poley, W. E., and A. L. Moxon. 1938. Tolerance levels of seleniferous grains in laying rations. *Poult. Sci.* 17:72–76.

Poley, W. E., A. L. Moxon, and K. W. Franke. 1937. Further studies of the effects of selenium poisoning on hatchability. *Poult. Sci.* 16:219–225.

Pollock, B., and K. L. Machin. 2009. Corticosterone in relation to tissue cadmium, mercury and selenium concentrations and social status of male lesser scaup (*Aythya affinis*). *Ecotoxicology* 18:5–14.

Puls, R. 1988. *Mineral levels in animal health: Diagnostic data.* Clearbrook, British Columbia, CAN: Sherpa International.

Ratti, J. T., A. M. Moser, E. O. Garton, and R. Miller. 2006. Selenium levels in bird eggs and effects on avian reproduction. *J. Wildl. Manage.* 70:572–578.

Robberecht, H., H. Deelstra, D. Vanden Berghe, and R. Van Grieken. 1983. Metal pollution and selenium distributions in soils and grass near a non-ferrous plant. *Sci. Total Environ.* 29:229–241.

Romanoff, A. L., and A. J. Romanoff. 1949. *The avian egg.* New York, NY: John Wiley & Sons, Inc.

Santolo, G. M. 2007. Selenium accumulation in European starlings nesting in a selenium-contaminated environment. *Condor* 109:863–870.

Santolo, G. M., and J. T. Yamamoto. 1999. Selenium in blood of predatory birds from Kesterson Reservoir and other areas of California. *J. Wildl. Manage.* 63:1273–1281.

Santolo, G. M., J. T. Yamamoto, J. M. Pisenti, and B. W. Wilson. 1999. Selenium accumulation and effects on reproduction in captive American kestrels fed selenomethionine. *J. Wildl. Manage.* 63:502–511.

Scanlon, P. F. 1982. Wet and dry weight relationships of mallard (*Anas platyrhynchos*) tissues. *Bull. Environ. Contam. Toxicol.* 29:615–617.

Scheuhammer, A. M. 1987. The chronic toxicity of aluminium, cadmium, mercury, and lead in birds: A review. *Environ. Pollut.* 46:263–295.

Scheuhammer, A. M., A. H. K. Wong, and D. Boyd. 1998. Mercury and selenium accumulation in common loons (*Gavia immer*) and common mergansers (*Mergus merganser*) from eastern Canada. *Environ. Toxicol. Chem.* 17:197–201.

Scheuhammer, A. M., et al. 2008. Relationships among mercury, selenium, and neurochemical parameters in common loons (*Gavia immer*) and bald eagles (*Haliaeetus leucocephalus*). *Ecotoxicology* 17:93–101.

Skorupa, J. P. 1998a. Selenium poisoning of fish and wildlife in nature: Lessons from twelve real world experiences. In *Environmental chemistry of selenium*, eds. W. T. Frankenberger, Jr. and R. A. Engberg, pp. 315–354. New York, NY: Marcel Dekker.

Skorupa, J. P. 1998b. Risk assessment for the biota database of the National Irrigation Water Quality Program. Prepared for the National Irrigation Water Quality Program, U.S. Department of the Interior, Washington, DC. April.

Skorupa, J. P. 1999. Beware of missing data and undernourished statistical models: Comment on Fairbrother et al.'s critical evaluation. *Hum. Ecol. Risk Assess.* 5:1255–1262.

Skorupa, J. P., and H. M. Ohlendorf. 1991. Contaminants in drainage water and avian risk thresholds. In *The economics and management of water and drainage in agriculture*, eds. A. Dinar and D. Zilberman, pp. 345–368. Norwell, MA: Kluwer Academic Publishers.

Smith, G. J., G. H. Heinz, D. J. Hoffman, J. W. Spann, and A. J. Krynitsky. 1988. Reproduction in black-crowned night-herons fed selenium. *Lake Reservoir Manage.* 4:175–180.

Spalding, M. G., P. C. Frederick, H. C. McGill, S. N. Bouton, and L. R. McDowell. 2000. Methylmercury accumulation in tissues and effects on growth and appetite in captive great egrets. *J. Wildl. Dis.* 36:411–422.

Spallholz, J. E., and D. J. Hoffman. 2002. Selenium toxicity: Cause and effects in aquatic birds. *Aquat. Toxicol.* 57:27–37.

Stanley, T. R., Jr., J. W. Spann, G. J. Smith, and R. Rosscoe. 1994. Main and interactive effects of arsenic and selenium on mallard reproduction and duckling growth and survival. *Arch. Environ. Contam. Toxicol.* 26:444–451.

Stanley, T. R., Jr., G. J. Smith, D. J. Hoffman, G. H. Heinz, and R. Rosscoe. 1996. Effects of boron and selenium on mallard reproduction and duckling growth and survival. *Environ. Toxicol. Chem.* 15:1124–1132.

State of Utah. 2008. Utah Administrative Code; Rule R317–2, Standards of Quality for Waters of the State, and Rule R317–2-14, Numeric Criteria, October 22.

Suter II, G. W. 1993. *Ecological risk assessment.* Boca Raton, FL: Lewis Publishers.

Thapar, N. T., E. Guenthner, C. W. Carlson, and O. E. Olson. 1969. Dietary selenium and arsenic additions to diets for chickens over a life cycle. *Poult. Sci.* 48:1988–1993.

Trust, K. A., K. T. Rummel, A. M. Scheuhammer, I. L. Brisbin Jr., and M. J. Hooper. 2000. Contaminant exposure and biomarker responses in spectacled eiders (*Somateria fischeri*) from St. Lawrence Island, Alaska. *Arch. Environ. Contam. Toxicol.* 38:107–113.

U.S. Department of the Interior (U.S. DI). 1998. Guidelines for interpretation of the biological effects of selected constituents in biota, water, and sediment. National Irrigation Water Quality Program Information Report No. 3. USDI, Denver, CO. November.

U.S. Fish and Wildlife Service. 1990. Summary report: Effects of irrigation drainwater contaminants on wildlife, pp. 1–38. U.S. Fish and Wildlife Service, Patuxent Wildlife Research Center, Laurel, MD.

Wadge, A., and M. Hutton. 1986. The uptake of cadmium, lead, and selenium by barley and cabbage grown on soils amended with refuse incinerator fly ash. *Plant Soil.* 96:407–412.

Wayland, M., and R. Crosley. 2006. Selenium and other trace elements in aquatic insects in coal mine-affected streams in the Rocky Mountains of Alberta, Canada. *Arch. Environ. Contam. Toxicol.* 50:511–522.

Wayland, M., A. J. Garcia-Fernandez, E. Neugebauer, and H. G. Gilchrist. 2001. Concentrations of cadmium, mercury and selenium in blood, liver and kidney of common eider ducks from the Canadian Arctic. *Environ. Monit. Assess.* 71:255–267.

Wayland, M., H. G. Gilchrist, T. Marchant, J. Keating, and J. E. Smits. 2002. Immune function, stress response, and body condition in Arctic-breeding common eiders in relation to cadmium, mercury, and selenium concentrations. *Environ. Res.* 90:47–60.

Wayland, M., R. Crosley, and E. Woodsworth. 2007. A dietary assessment of selenium risk to aquatic birds on a coal mine affected stream in Alberta, Canada. *Hum. Ecol. Risk Assess.* 13:823–842.

Wayland, M., K. L. Drake, R. T. Alisauskas, D. K. Kellett, J. Traylor, C. Swoboda, and K. Mehl. 2008. Survival rates and blood metal concentrations in two species of free-ranging North American sea ducks. *Environ. Toxicol. Chem.* 27:698–704.

Wiemeyer, S. J., and D. J. Hoffman. 1996. Reproduction in eastern screech-owls fed selenium. *J. Wildl. Manage.* 60:332–341.

Yamamoto, J. T., and G. M. Santolo. 2000. Body condition effects in American kestrels fed selenomethionine. *J. Wildl. Dis.* 36:646–652.

Yamamoto, J. T., G. M. Santolo, and B. W. Wilson. 1998. Selenium accumulation in captive American kestrels (*Falco sparverius*) fed selenomethionine and naturally incorporated selenium. *Environ. Toxicol. Chem.* 17:2494–2497.

Yasumoto, K., T. Suzuki, and M. Yoshido. 1988. Identification of selenomethionine in soybean protein. *J. Agric. Food Chem.* 36:463–467.

Continental Field-Vole

By C. G. Specht, from *The Royal Natural History*, edited by Richard Lydekker, Frederick Warne & Co., London, 1893–94.

22 Radionuclides in Biota

Bradley E. Sample
Cameron Irvine

CONTENTS

22.1 Introduction 703
22.2 Basic Concepts 704
22.3 Units of Radiation Measurement 705
22.4 Sources of Radiation 705
22.5 Effects of Ionizing Radiation on Biological Tissue 707
    22.5.1 Molecular and Cellular Effects 707
    22.5.2 Whole-Body Effects 708
    22.5.3 Relative Sensitivity to Ionizing Radiation 708
    22.5.4 Levels of Ecological Organization 710
22.6 Databases and Tools for Evaluation of Radionuclide Exposure-Effects 710
22.7 Studies of Exposure to and Effects from Ionizing Radiation to Biota 713
    22.7.1 Radiation Effects on Mammals 713
        22.7.1.1 Molecular-Level and Genetic Effects 713
        22.7.1.2 Reproduction 714
        22.7.1.3 Immune/Endocrine Responses 715
        22.7.1.4 Mortality 715
    22.7.2 Radiation Effects on Birds 716
        22.7.2.1 Molecular and Hematological Responses 716
        22.7.2.2 Growth and Development 717
        22.7.2.3 Reproduction 718
        22.7.2.4 Mortality 719
    22.7.3 Insects and Other Terrestrial Invertebrates 719
    22.7.4 Fish and Aquatic Invertebrates 721
22.8 Ecological Radiation Effects Thresholds 723
Summary 726
References 728

22.1 INTRODUCTION

Radioactive elements do not differ chemically from nonradioactive or stable elements and can be accumulated and integrated into tissue of plants and animals. Once integrated into tissues, emissions from radioactive elements have the potential to produce deleterious effects. Radioactive elements differ from conventional elements in that the radiation they emit can also exert biological effects at a distance without directly contacting the organism.

Quantification of tissue-based exposures to radioactive elements differs from chemical exposure. In contrast to chemical exposures which can be evaluated simply through direct measurement or estimation of concentrations in target tissues, both internal (e.g., due to radioactive elements retained in tissues) and external (e.g., due to emissions from radioactive elements in the environment) exposure must be considered for radioactive elements.

Effects of radioactive elements on biological tissue are qualitatively similar, differing only in magnitude, regardless of which radioactive element is considered. As a consequence, exposure to multiple radioactive elements is additive. Total radiation exposure therefore is the sum of exposure to all radioactive elements retained in tissue and in the immediate environment (i.e., the sum of internal and external exposure).

The objectives of this chapter are threefold: First we present a brief introduction to radiation physics, summarize how exposure to radioactive elements is measured, and how radiation can affect biological tissue. Second, we provide an overview of the considerable research that has been conducted to quantify environmental exposure of and effects on biota from both naturally occurring and man-made radioactive elements since the 1950s. Finally, we summarize the history of development of radiation effect thresholds for protection of nonhuman biota.

22.2 BASIC CONCEPTS*

All matter is comprised of atoms, which in turn are made up of subatomic particles (i.e., protons, neutrons, electrons, etc.). The nucleus of atoms of the same element all have the same number of protons, but may have varying numbers of neutrons. These different atoms of the same element are known as isotopes. If the number of protons and neutrons in the nucleus are balanced, the atom is stable and is referred to as a stable isotope. However, if the nucleus has too few or too many neutrons, the atom is unstable and is known as a radioisotope or radionuclide.

Radionuclides release energy or decay to form either new radionuclides (daughters) or stable elements. Some radionuclides decay through a series of daughter products before reaching a stable state. This process is known as a decay chain. The duration of time required for the activity of a specific radionuclide to decay by 50% is known as a half-life. Radionuclide half-lives are a measure of their persistence. Radionuclides with long half-lives are generally of greater environmental concern than are those with short half-lives.

Release of energy through the decay process may occur via emission of particles (alpha [α], beta [β], or neutrons), or by electromagnetic radiation (gamma [γ] or x-rays). Because of their ability to produce charged particles (ions) in matter through which they pass, these emissions are collectively referred to as ionizing radiation. Alpha, beta, and gamma radiation are the primary forms of ionizing radiation of concern. A key characteristic of ionizing radiation is the amount of linear energy transfer (LET) they exhibit. LET characterizes the amount of energy lost by a particle or radiation, and its ability to create ions, as it passes through tissue. High-LET radiation can therefore be more damaging than low-LET radiation.

Alpha particles have an electrical charge of +2 and are composed of two protons and two neutrons. Because of their large size, alpha particles have low penetration capacity, can be stopped by a sheet of paper, and cannot pass through skin. However, alpha particles are high-LET radiation that cause considerable damage (i.e., ionization; discussed in more detail later in this chapter) if they, or radionuclides that emit them, are ingested or integrated into tissue. Energy from alpha particles is dissipated in a few centimeters of air or up to 0.04 mm of tissue.

Beta particles are generally negatively charged electrons ejected from the nucleus of an atom. Less frequently they consist of positrons (positively charged electrons). Beta particles are lighter and have higher penetrating capacity, but lower LET than alpha particles. At least 3 mm of metal, or 6 mm of wood, or 5 mm of tissue are needed to shield from beta radiation.

---

* This text is derived from ATSDR 2001, Eisler 1994, Copplestone et al. 2001, Sazykina et al. 2003, and Talmage and Meyers-Schone 1995.

Gamma and X-ray radiation are comprised of high energy photons with a short wavelength. While gamma radiation originates from the nucleus of an atom, X-rays are derived from the electron shell. Gamma radiation has a lower LET than either alpha and beta particles, and although it can travel deeper into tissue, it causes comparatively less damage (ionization). Shielding of 5–10 cm of lead or 30–60 cm of concrete are needed to stop gamma radiation.

22.3 UNITS OF RADIATION MEASUREMENT

Quantification of radiation was originally described by Curies (Ci), Röentgens (R), rad, and REM (Röentgen equivalent man) (Eisler 1994, Talmage and Meyers-Schone 1995). These units have subsequently been replaced by the International System (SI) of Units, Becquerel (Bq), Gray (Gy), and Sievert (Sv). Bequerels and Curies are measures of radioactive decay, where 1 Bq equals one atomic disintegration per second, which equals 27 pCi (Table 22.1).

The activity per unit mass of each radionuclide (Bq/g or pCi/g) and the energy it releases determines the dose. Absorbed dose (Gy or rad) is quantified by the amount of energy absorbed by a receptor, whereas 1 Gy = 1 joule/kg = 100 rad = 10000 erg/g (Table 22.1). It should be noted that the absorbed dose is not a good indicator of the likely biological effect. For example, 1 Gy of alpha radiation would be much more biologically damaging than 1 Gy of photon (gamma) radiation because of its high-LET.

Radiation weighting factors (wR) can be applied to absorbed doses to reflect the varying magnitude of biological effects associated with equal doses of different radiation types. The equivalent dose (H_T) is the product of absorbed radiation dose to tissue and wR, in an attempt to allow for the different relative biological effects (RBEs) of different types of ionizing radiation. Equivalent dose is therefore a less fundamental quantity than radiation absorbed dose, but is more biologically significant. Equivalent dose is expressed as either sieverts (SI units) or REM. Dose-wR for different types of radiation require extensive experimentation. Recommended wR for human exposure are 1 for beta and gamma or X-ray radiation and 20 for alpha particles (ICRP 1991). Weighting factors for nonhuman biota are poorly developed and lack consensus. Values range from 5 to 20 for alpha and 1 to 3 for beta (Sazykina et al. 2003). For this reason most doses for nonhuman biota are simply reported as the absorbed dose.

22.4 SOURCES OF RADIATION

Ionizing radiation can be both naturally occurring and anthropogenic in origin. Natural background sources of ionizing radiation include sunlight, cosmic rays, and naturally occurring radionuclides

TABLE 22.1
Units of Radiation Measurement

Type of Measurement	SI Unit	Standard Unit	Description
Activity	1 Bq	2.7E-11 Ci (27 pCi)	1 Bq = 1 disintegration/sec
	3.7E + 10 Bq	1 Ci	1Ci = 3.7E + 10 disintegration/sec
Absorbed Dose	1 Gy	100 rad	1Gy = 1 J/kg tissue
	0.01 Gy	1 rad	1 rad = 100 erg/g
Exposure		1 R	2.58E-04 coulomb/kg air
Dose equivalent	1 Sv	100 rem	1 Gy = 1 J/kg tissue
	0.01 Sv	1 rem	1 rem = 1.07 R

Bq = Bequerel; Ci = Curie; Gy = Gray; R = Röentgen; Sv = Sievert.

Röentgen is defined as unit of radiation exposure equal to the quantity of ionizing radiation that will produce one electrostatic unit of electricity in one cubic centimeter of dry air at 0°C and standard atmospheric pressure.

Source: Adapted from Talmage, S. S., and L. Meyers-Schone. 1995. *Handbook of Ecotoxicology*, Lewis Publishers, Boca Raton, FL, pp. 469–491; Eisler, R., Radiation hazards to fish, wildlife, and invertebrates: A synoptic review. Biological Report 26, Contaminant Hazard Reviews Report 29. December 1994. With permission.

from geologic sources (e.g., uranium and radon gas in soil). It has been estimated that the total annual dose to humans from natural background sources is 3.11 mGy/year (0.0085 mGy/d), of which 17.4% is external (10.6% from cosmic radiation and 6.7% from terrestrial sources) and 82.6% is from internal sources (68% from radon and 14.6% from other radionuclides; NCRP 2009). Some areas, such as Ramsar in northern Iran, have naturally elevated levels of radionuclides (primarily ^{226}Ra and daughters). Background dose levels as high as 0.71 mGy/d have been measured there (Ghiassi-nejad et al. 2002). For comparison, Beresford et al. (2008) estimated that doses to representative nonhuman biota in Great Britain range from 0.0017 to 0.015 mGy/d (Table 22.2).

Anthropogenic sources of radionuclides in the environment are diverse and include, in no particular order: uranium mining operations, nuclear weapons production and testing, nuclear power plants (including disposal of waste fuel), and accidental releases. Although above-ground testing of nuclear weapons and nuclear power production are major sources of anthropogenic radionuclides in the environment (Eisler 1994), areas of very significant contamination have resulted from accidental releases from power and weapons production plants. These accidental releases have provided invaluable opportunities to observe fate and effects of radionuclides on natural ecosystems. Key examples of accidental radionuclide releases are summarized below:

In September 1957, a tank containing liquid radioactive waste exploded at the "Mayak" nuclear complex in the southern Ural Mountains, in what is now Russia. This event, which occurred at the height of the Cold War and was unknown outside the Soviet Union until the 1970s, ultimately became known as the Kyshtym accident (named after the nearest population center). The Kyshtym explosion released 7.4E + 17 Bq of radiation, spreading contamination 300 km downwind, and contaminating an area of 15,000–23,000 km^2 (Sazykina and Kryshev 2006, Jones 2008). Maximum ^{90}Sr concentrations, the dominant long-lived radionuclide that was released, were 1.5E + 08 Bq/m^2. The human population was evacuated from a 1000 km^2 area with ^{90}Sr concentrations >7.4E + 02 Bq/m^2.

A month after the Kyshtym accident (October 1957), a fire in the graphite core of a plutonium production reactor at the Windscale nuclear complex in Great Britain resulted in the release of 1.86E + 16 Bq of radiation into the air, surrounding countryside and the Irish Sea (Talmage and Meyers-Schone 1995, Jones 2008). Principal isotopes released included ^{131}I, ^{137}Cs, ^{210}Po, ^{89}Sr, and ^{90}Sr.

In 1979, a loss of cooling water resulted in a partial meltdown of the core at one of two reactors at the Three Mile Island nuclear power plant (TMI) in Pennsylvania. Although significant radiation was

TABLE 22.2
Estimated Total Weighted Absorbed Dose Rates to Reference Terrestrial Animals and Plants Due to Background Concentrations of ^{40}K, ^{238}U Series, and ^{233}Th Series Radionuclides in England and Wales

Taxa	Absorbed Dose (mGy/d)
Wild grass	0.012
Pine tree	0.0017
Earthworm	0.015
Bee	0.012
Frog	0.007
Duck	0.006
Deer	0.0017
Rat	0.003

Source: Beresford, N. A. et al., *J. Environ. Radioact.*, 99, 1430–1439, 2008. With permission.

released (>1.59E + 15 Bq), it was predominantly as radioactive krypton gas, with ^{131}I comprising less than 0.05% of total released radiation (Walker 2004).

The 1986 explosion and fire in Reactor 4 at the Chernobyl nuclear plant was the largest known accidental release of radionuclides into the environment, 1.85E + 18–8E + 19 Bq of radiation (Talmage and Meyers-Schone 1995, Geras'kin et al. 2008). Because the fire carried contamination into the atmosphere, global dispersion of radionuclides occurred. Beta and gamma radiation in Germany, 1200 km from Chernobyl, ranged from less than 1.5E + 3–2.4E + 5 Bq/m² (Talmage and Meyers-Schone 1995). Soil concentrations of 2.00E + 4–6.00E + 4 Bq/m² of ^{137}Cs, 6.00E + 3–1.00E + 5 Bq m² of ^{90}Sr, and 370–3700 Bq/m² of plutonium, existed shortly after the accident in areas over 30 km from the reactor (UNAS 1997). Lower levels of contamination spread over a large area in the Ukraine, Belarus, and Russia, and the world.

22.5 EFFECTS OF IONIZING RADIATION ON BIOLOGICAL TISSUE

The mechanism by which ionizing radiation produces effects in biological tissue is both simple and complex. At its very base, it is initiated by a simple transfer of energy from the decaying radionuclide to neutral molecules and atoms, thereby producing electrically charged ions. Complexity arises in the cascade of responses and subsequent reactions that occurs as a result of the creation of ionized molecules and atoms within biological tissue.

Effects from exposure to ionizing radiation have been categorized as both deterministic and stochastic (Copplestone et al. 2001, Delistraty 2008). Deterministic effects increase in severity above a defined threshold level, and can be clearly associated with an exposure, both in time and causation. Organ or tissue damage due to cell death are examples of deterministic effects. Stochastic effects increase in probability (but not severity) with increasing dose, do not display a threshold for effects, and exhibit a delayed and uncertain association with exposure. Cancer and heritable genetics effects represent stochastic effects.

22.5.1 MOLECULAR AND CELLULAR EFFECTS

The chemical form and biological function of molecules and cells can be changed by exposure to ionizing radiation. Chemically active free radicals and reactive oxygen species (ROS) are created through radiolysis of water within biological material (Pollard 1969, Sazykina et al. 2003, O'Neill and Wardman 2009). The ROS and free radicals initiate numerous reactions with biological molecules (i.e., enzymes, lipids, proteins, DNA, etc.) and inorganic substances within cells, generally through oxidation and peroxidation. The resulting reactions can impair or otherwise deactivate these biological molecules. When they lose their biological function, these molecules become waste products that must be eliminated. In genetic material, the primary target of ionizing radiation, direct effects are expressed in the form of single- and double-strand breaks (ICRP 2008). If not repaired, damage to DNA can be perpetuated, leading to mutations that may be passed to offspring. O'Neill and Wardman (2009) provide an extensive review and discussion of the mechanisms for radiation-induced molecular/cellular damage and repair.

At the cellular level, ionizing radiation can damage membranes of both organelles in the cell (i.e., nucleus, mitochondria, etc.) and the cell itself. This damage to cell membranes disrupts osmotic regulation and ion balance, leading to cell death (Sazykina et al. 2003).

The effects of ionizing radiation at the molecular and cellular levels is somewhat mitigated by the inherent capabilities of cells to repair damage. Repair mechanisms are most effective for low to moderate levels of damage. In addition, damage from beta particles and gamma radiation is more easily repaired than is that from alpha particles. Repair at the molecular and cellular level is an ongoing process. As a consequence, the same absorbed dose received as multiple smaller exposures is generally less harmful than when it is received as a single large exposure. However, because repairs may not always be correct or complete, or if some damage is subtle and not recognized by the cell, cellular and genetic damage can accumulate over time.

All cell types are not equally sensitive to radiation exposure. In general, mitotically active and functionally undifferentiated cells are more radiosensitive than are specialized cells (Environment Canada and Health Canada 2003, Delistraty 2008). This generalization that young rapidly growing tissues are more radiosensitive is known as the Law of Bergonie and Tribondeau, and was first postulated in 1906 (Bergonie and Tribondeau 1906). For example, mammalian nerve and muscle cells (which are highly specialized) exhibit lower radiosensitivity than do reproductive germ cells, dermal and gastrointestinal stem cells, and erythroblasts (all of which are mitotically active and comparatively undifferentiated). In addition to these cell types, bone marrow and the cells in the eye lens are also radiosensitive (ICRP 2008). As a result of the radiosensitivity of these cell types, the organ systems of which they are a component (e.g., the reproductive, digestive, and immune systems) are also radiosensitive. Likewise, meristematic tissues in stems and root tips are the most sensitive cells in plants.

22.5.2 Whole-Body Effects

When taken up by an organism, radionuclides are integrated into tissue and follow the same biochemical pathways as their stable analogs. For example, ^{131}I accumulates in the thyroid, ^{90}Sr in bones, and ^{137}Cs in muscle. As a consequence, exposure to ionizing radiation from internally incorporated radionuclides is not uniform. Depending on the radionuclide, different tissues experience different doses, and therefore develop specific pathologies.

The immune system in vertebrates is complex and is especially radiosensitive (Sazykina et al. 2003). Lymphocytes are readily killed by radiation exposure. Further, radiation-induced damage to cell membranes facilitates infection by microbial, viral, and parasitic agents. Secondary infections can therefore become established and contribute to further effects when the immune system is compromised.

Reproductive tissues and rapidly dividing cells are also particularly radiosensitive. Radiation doses that have no effect on survival of an animal may produce total sterility. Sazykina et al. (2003) pointed out that reproductive effects are dependant on exposure of one or both parents. Irradiation of breeding adults can result in embryonic effects due to genetic damage, cellular damage to sexual organs, or impairment of the adults' immune system resulting in embryonic exposure to pathogens. Radiation-induced reproductive effects may be expressed a significant time postexposure for the adults and can be evident even if no embryonic exposure occurred (UNSCEAR 1996).

22.5.3 Relative Sensitivity to Ionizing Radiation

Radiosensitivity varies between different taxonomic groups, and among species (Eisler 1994, Harrison and Anderson 1996, Sazykina et al. 2003). Although there is considerable overlap among taxa, more primitive biota are more resistant to acute radiation exposure and more complex organisms are more sensitive (Figure 22.1). Acute[*] radiation doses lethal to higher plants and vertebrates are generally lower than doses lethal to invertebrates, which in turn are lower than doses lethal to algae and microbes. Among plants, Real et al. (2004) report that gymnosperms are the most sensitive and cryptogams[†] the most radioresistant under chronic exposure regimes.

When effects of acute and chronic radiation exposure on reproductive success are considered, similar taxonomic trends in radiosensitivity are evident (Figure 22.2). Mammals are most sensitive and invertebrates least sensitive. Fish sensitivity to ionizing radiation tends to be intermediate.

Radiosensitivity can also vary between individuals of the same species exposed to the same doses at different life stages (Willard 1963, Mole 1990). Organisms chronically exposed in the environment may also develop radioresistance so that effects to surviving individuals are not as pronounced as those exposed acutely (Rodgers and Baker 2000).

---

[*] *Acute* exposures typically occur over a short time period (relative to the development of a biological response), whereas a chronic exposure continues over a large fraction of the natural lifespan of the organism.

[†] Cryptogams are plants which reproduce using spores rather than seeds (e.g., algae, lichens, mosses, and ferns).

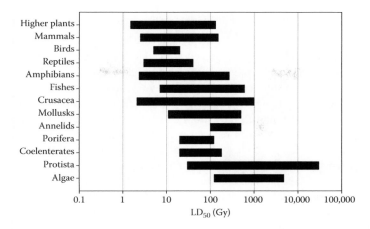

FIGURE 22.1 Summary of the range of lethal doses (LD_{50}) for various taxa following single, acute exposures (data from Eisler, 1994 and Harrison and Anderson, 1996).

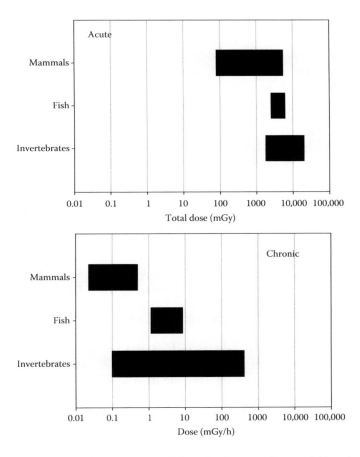

FIGURE 22.2 Comparison of relative sensitivity of fertility in invertebrates, fish, and mammals to acute and chronic exposures to radiation. Data represent thresholds for effects on fertility from Harrison and Anderson (1996).

22.5.4 Levels of Ecological Organization

Similar to other environmental contaminants, expression of radiological effects occurs over multiple biological levels of organization. Biological and ecological processes are driven by mechanisms that are one level below, and effects are expressed at the next higher level (Caswell 1996). For example, adverse effects to bone marrow stem cells or reproductive germ cells due to radionuclide exposure may adversely affect the immune or reproductive systems. Effects to the immune or reproductive systems in turn may result in impaired individual survival and reproductive success. If sufficient individuals are affected, individual-level responses may ultimately be expressed as population or community level effects. Although effects observed at the individual or suborganismal level may have an effect on higher levels of organization, that is not necessarily true (e.g., Anderson and Wild 1994). Some effects to individual survival are reported to be masked at the population level by immigration or environmental influences (Cooley and Miller 1971, Trabalka and Allen 1977). Likewise, population responses are not always affected by the same doses that negatively affect individuals, if the individual response does not reduce fitness (Marshall 1962, 1963, 1966, O'Farrell et al. 1972, Benjamin et al. 1998).

22.6 DATABASES AND TOOLS FOR EVALUATION OF RADIONUCLIDE EXPOSURE-EFFECTS

Within the past 10 years, multiple organizations have undertaken efforts to assemble, review, and interpret the extensive volume of literature available concerning radionuclides in the environment. Most of these activities have been undertaken to better define exposure-response relationships and develop tools to facilitate the evaluation of exposure to and risks from radionuclides experienced by nonhuman biota. Jones and Gilek (2004) reviewed and compared processes, assumptions, and tools associated with existing risk assessment approaches for both ionizing radiation and chemical contaminants. Although the general structures of the assessment frameworks were comparable, site-specificity, level of detail, and definitions of acceptable exposure/effects differed among frameworks. Summaries of key efforts for evaluation of ionizing radiation in the United States and Europe are presented later.

The U.S. Department of Energy (U.S. DOE) manages and maintains a number of facilities that have become contaminated by radionuclides as part of historical research, development, and fabrication of nuclear weapons, research on nuclear power generation, and other nuclear energy applications. The Graded Approach for Evaluating Radiation Doses to Aquatic and Terrestrial Biota (U.S. DOE 2002) was developed in support of the effort to assess the ecological risks and guide remediation of radionuclide contaminated sites. This approach is a 4-tiered process that progresses from simple comparisons of radionuclide concentrations in abiotic environmental media to Biota Concentration Guides (BCGs*), to more detailed analyses requiring field sampling and exposure modeling with site-specific data. BCGs represent radionuclide concentrations in soil, sediment, or surface water that correspond to radionuclide effect thresholds from IAEA (1992; 10 mGy/d for aquatic animals and terrestrial plants, and 1 mGy/d for terrestrial animals). Conservative exposure assumptions that integrate both internal and external exposure for four biota types (aquatic, riparian, and terrestrial animals and terrestrial plants) are integral to the BCGs (Higley et al. 2003). In the first tier of the Graded Approach, environmental radionuclide concentrations are divided by the medium-specific BCG to generate a ratio. Because radionuclide effects are additive, these ratios are summed if multiple radionuclides are present. If the sum of the ratios is less than unity, then it can be assumed that radionuclide exposure is insufficient to present a risk to resident biota. However, if the sum of the ratios exceeds unity, the Graded Approach proceeds to the next tier and additional

---

* BCGs are radionuclide concentrations in environmental media that would not exceed recommended dose rate guidelines (U.S. DOE 2002, 2009).

evaluation is necessary to more definitively define exposure and risk. Higher tiers include the use of site-specific data (e.g., bioaccumulation factors), a kinetic/allometric modeling tool to assess internal exposure (i.e., assumes first-order kinetics and allometric relationships, which relate body size to physiological and life history traits), and site-specific dose assessment (i.e., collection and analysis of representative biota samples, use of tissue radionuclide data with dosimetric models to estimate internal dose to a site-specific receptor). The overall process outlined by U.S. DOE (2002) is intended to protect biota at the population level and is consistent with ecological risk assessment guidance from the U.S. Environmental Protection Agency (U.S. EPA 1997, 1998). RESRAD-Biota (http://web.ead.anl.gov/resrad/home2/) is a software application developed by Argonne National Laboratory to facilitate the implementation of the Graded Approach.

Whereas exposure estimation and risk screening were the focus of the development of the Graded Approach by the U.S. DOE, a more comprehensive effort has been initiated in Europe. This effort consists of a consecutive series of programs that developed databases of radiation exposure and effects, modeling tools for radiation exposure estimation and risk evaluation, and effects thresholds for evaluation of potential impacts of estimated exposures.

The first program, "Framework for ASSessment of Environmental impacT" (FASSET), was initiated by the European Commission in 2000 (FASSET 2004, Larsson 2004). It focused on potential impacts of radionuclides on nonhuman biota within seven ecosystems (forests, seminatural pastures and heathlands, agricultural ecosystems, wetlands, freshwaters, marine, and brackish waters). Biota were represented by approximately 30 reference organisms indicative of functional components important to each ecosystem. One critical component of the FASSET program was the FASSET Radiation Effects Database (FRED). The FRED database included approximately 25,000 entries from over 1000 references, dating from 1945 to 2001, representing 16 wildlife groups, and focusing on four umbrella effect types: morbidity (effects on growth, immune systems, behavior, etc.), mortality, reproductive success, and mutation. The database includes raw dose/effect data from laboratory and field experiments, of which 80% are acute exposures and 20% are chronic exposures. Real et al. (2004) evaluated the effects data in the FRED database in an effort to determine its adequacy for threshold development. They concluded that significant data gaps existed, particularly in relation to chronic studies of exposure to alpha and beta radiation. Nonetheless, the available data indicated the threshold for statistically significant effects was approximately 100 uGy/h (2.4 mGy/d).

The Environmental Protection from Ionizing Contaminants (EPIC) database was developed concurrently with FRED, with the objective of compiling and analyzing published and unpublished dose-effects data from the arctic, with an emphasis on data from Russia and areas of the former Soviet Union (Sazykina et al. 2003). Results were compiled from 435 sources and included approximately 1600 records primarily describing chronic effects from relatively low radiation doses, but also from high doses, ranging from <1.0E-05 Gy/d to >1 Gy/d. Data in the EPIC database are classified according to seven effect categories (morbidity, reproduction, mortality/life shortening, cytogenic, ecological, stimulation, and adaptation) and five taxonomic groups (terrestrial animals, aquatic animals, terrestrial plants, soil fauna, and microorganisms). Based on these data, Sazykina (2005) concluded that the threshold for effects for arctic wildlife from chronic exposure to low-LET radiation was in the 0.5–1 mGy/d range. However, it was noted that survival of species with high reproductive rates, such as mice and some fish species, were unaffected by exposures up to 10 mGy/d, although other radiation-induced effects were evident. Ranges of exposures and associated effects observed among arctic biota are summarized in Table 22.3.

The Environmental Risks from Ionizing Contaminants: Assessment and Management (ERICA) project was initiated in 2004 with the intent to further develop the tools and databases originated under FASSET. Under the ERICA program, risk characterization tools and decision-support elements not addressed under FASSET were developed. The ERICA assessment tool (http://www.project.facilia.se/erica/download.html) is a software program developed to facilitate estimation of radionuclide exposure and risks to nonhuman biota. Another key function was the integration of

TABLE 22.3
Summary of Chronic Dose–Response Relationships for Arctic Biota Exposed to Low-LET Radiation

Dose Rate Range (mGy/d)	Effects Observed
0.001–0.01	• Natural range of background radiation exposure.
0.01–0.1	• No data on radiation effects above background.
0.1–0.5	• Minor cytogenetic effects. Stimulation of the most sensitive vertebrate species.
0.5–1	• Threshold for minor effects on morbidity in sensitive vertebrate animals.
2–5	• Threshold for effects on reproductive organs of vertebrate animals.
5–10	• Threshold for life shortening of vertebrate animals. • Threshold for effects in invertebrate animals. • Threshold for effects on growth of coniferous plants.
10–100	• Life shortening of vertebrate animals; chronic radiation sickness. • Considerable damage to coniferous trees.
100–1000	• Acute radiation sickness of vertebrate animals. • Death of coniferous plants. • Damage to eggs and larvae of invertebrate animals.
>1000	• Acute radiation sickness of vertebrate animals. • Lethal dose is received within several days. • Increased mortality of eggs and larvae of invertebrate animals. • Mortality in coniferous plants. • Damage to deciduous plants.

Source: EPIC database, Sazykina 2005.

the FRED and EPIC databases to produce the FREDERICA database (http://87.84.223.229/fred/mainpage.asp). This combined database contains 1509 references from 1945 to 2004 and 29,400 data entries.

In 2006 the PROTECT program (http://www.ceh.ac.uk/PROTECT/) was initiated as a successor to FASSET and ERICA. The purpose of the ongoing PROTECT program is to evaluate different approaches for assessment and management of ionizing radiation risks to nonhuman biota and to compare these approaches to those employed for chemical contaminants. The goal is to develop the scientific foundation on which to base numerical standards for protection of nonhuman biota from ionizing radiation.

Most recently, the system of Reference Animals and Plants was introduced by the ICRP (2008). Reference Animals and Plants are an extension of the Reference Person employed to evaluate human exposure and effects. Reference Animals and Plants are representative of biota that occupy multiple ecological niches and may occur in many habitats and environments. Taxa represented include deer, rat, duck, frog, trout, flatfish, bee, crab, earthworm, pine tree, wild grass, and brown seaweed. Although these specific taxa may not be present at all sites, information available for these reference species can be used to develop an approximate estimate of exposure and effects for species that do occur.

Because taxa-specific data are lacking for most ecological receptors, the Reference Animals and Plants are intended to facilitate initial evaluations of radioecological exposures and effects. ICRP (2008) not only provides parameters needed to estimate internal and external exposure to 75 radio-nuclides (internal and external dose conversion factors, body weights, and assumptions on body geometries), but also provides extensive summaries of available radionuclide effects data for each Reference Animal and Plant.

22.7 STUDIES OF EXPOSURE TO AND EFFECTS FROM IONIZING RADIATION TO BIOTA

As has been described in general terms previously in this chapter, the magnitude of effects displayed varies according to the duration of exposure, radioisotope and type of ionizing radiation, organ system considered, and taxonomy of biota exposed. We reviewed a limited selection of studies extracted from current publications and the extensive databases developed by others, to provide an overview of the current understanding of exposure and effects of radionuclides in mammals, birds, terrestrial invertebrates, fish, and aquatic invertebrates. Studies were selected to represent multiple types of responses to ionizing radiation including molecular-level and genetic effects, immune and endocrine responses, reproductive effects, and mortality. Because exposure data have been reported in multiple units, chronic doses were converted to mGy/d to aid the comparison of data from various studies. Acute exposures typically occurred over periods much less than 1 day and were converted to Gy to allow comparisons across studies.

22.7.1 RADIATION EFFECTS ON MAMMALS

Considerable published and gray literature is available describing the effects of ionizing radiation on mammals. Most published studies have been under laboratory conditions, with fewer studies evaluating effects in the field. Real et al. (2004) reported only 7 of 183 mammalian study references in the FASSET database corresponding with field exposures and effects.

22.7.1.1 Molecular-Level and Genetic Effects

Radiation exposures that produce molecular-level and genetic effects in mammals are variable and overlap with doses that have not been shown to produce effects. Real et al. (2004) reported that genetic effects may occur at dosages as low as 10 mGy/d, while exposures up to 90 mGy/d produced significant genetic effects among wild bank voles (*Clethrionomy glareolus*) within the Chernobyl exclusion zone (Rodgers and Baker 2000, Wickliffe et al. 2002). However, different populations of the same species, and different species exposed in the same areas, have been unaffected.

For example, Wiggins et al. (2002) observed intrapopulation variation in the chromosomal positions and abundance of chromatin among resident voles (*Microtus* sp.) living near Chernobyl Reactor #4 and exposed to 19–53 mGy/d, but these differences were similar to those observed among voles from reference areas, and therefore could not be attributed to radiation exposure. Likewise, bank voles from Chernobyl have had among the highest cesium body burdens and dose rates reported for a mammal (mean [134]Cs and ^{137}Cs burdens and dose rates were 2.8E + 04 Bq/g and 20 mGy/d, respectively [Baker et al. 2001]). No detrimental effects to vole populations were observed despite these high body burdens and elevated mitochondrial DNA diversity. Wickliffe et al. (2003) investigated mitochondrial DNA (mtDNA) abnormalities in embryos from a female bank vole to test the hypothesis that chronic irradiation posed a genetic risk to offspring. A pregnant female bank vole was captured in the Red Forest with a ^{137}Cs burden of 73000 Bq/g and an estimated external dose of 50 mGy/d. No significant differences in mtDNA mutations were found in embryos from the exposed female when compared to a control female and embryos. Although the sample size was small, chronic exposure to ambient irradiation did not increase genetic risks for this exposed individual. Genetic effects may not be long lasting in some species or individuals, as intracellular repair mechanisms act to correct genetic damage. In addition, populations exposed to elevated levels of ionizing radiation over multiple generations may select for genetic repair mechanisms and develop radioresistance (Wiggins et al. 2002). Radioresistance was postulated as the explanation for the lack of difference in frequency of micronuclei[*] among bank voles from the Chernobyl exclusion

---

[*] Chromosomal fragments that have not been integrated into the daughter nuclei at cell division. The abundance of these fragments are widely used as a measure of chromosomal damage (Sugg et al. 1996).

zone and that among voles from uncontaminated reference areas. To test this, Rodgers et al. (2001b) placed naïve bank voles in exposures for 30 days. Despite an exposure to 36.2 mGy/d, no increase in the incidence of micronuclei in red blood cells was observed. The authors concluded that bank voles are inherently radioresistant. Similarly, no increases in point mutations were observed in a similar study with naïve Big Blue® mice chronically exposed in the Ukraine's Red Forest to 33 mGy/d from ^{137}Cs for 90 days (Wickliffe et al. 2003). Recent work by Meeks et al. (2007) further supports the hypothesis that some small mammals are inherently radioresistant.

Contrary to the lack of genetic effects reported by Rodgers et al. (2001b) and Wickliffe et al. (2003), Rodgers et al. (2001a) reported that house mice (*Mus musculus*) from the Red Forest exposed to 40 mGy/d for 30 and 40 days exhibited a significant increase in the frequency of micronuclei. Ryabokon and Goncharova (2006) also reported genetic effects in mice exposed to very low dose levels at Chernobyl. Among the resident small mammal population at the Chernobyl site, they observed that a consistently elevated level of genetic damage and an increased frequency of embryonic lethality had developed over 22 generations. At two of the more contaminated sampling sites, whole-body doses decreased from 0.61 and 0.087 mGy/d in 1986, shortly after the accident, to 0.041 and 0.007 mGy/d in 1996. The frequency of genetically aberrant bone marrow cells was relatively constant (1.21% and 1.99% in 1986, and 2.04% and 1.95% in 1996, at the two sites, respectively). The frequency of genetically aberrant bone marrow cells at control sites was less than 0.64%. Ryabokon and Goncharova (2006) also observed that offspring of wild-caught gravid females that had been raised under uncontaminated laboratory conditions continued to display elevated levels of chromosomal abnormalities. Their findings were inconsistent with the assumption that the radiation-induced genetic effects should dissipate as the whole-body absorbed dose rate decreases, and the authors propose that the consistent elevated level of genetic anomalies reflects a transgenerational transmission and accumulation of genetic damage. Green (1964) observed similar transgenerational transmission of effects from radiation exposure among laboratory mice. Reproductive fitness was impaired among nonirradiated offspring of mice acutely exposed to 0.5 and 1 Gy X-ray doses.

22.7.1.2 Reproduction

Reproduction and reproductive organs can be particularly sensitive to ionizing radiation. This is because DNA in developing tissue and stem cells are easily damaged by ionizing radiation. This damage may not be repaired before it is replicated and distributed through entire organisms or tissues, resulting in physical malformations, developmental delays, and increased mortality (Sazykina et al. 2003). Based on data in the FASSET database, primarily from laboratory or domestic species, Real et al. (2004) concluded that the threshold for chronic reproductive effects in mammals was 2.4 mGy/d with effects clearly evident at doses of 24 mGy/d. Other data however, suggest that wild mammal populations may be less sensitive. For example, Sugahara (1964) evaluated the chronic effects of gamma radiation on laboratory mice. Mice received 4.7 mGy/d over three generations. Statistically significant increases in sterility were observed in the second and third generations, but not in the parent or first generations. Juvenile mortality also increased to double that in the controls.

French et al. (1974) conducted a multiyear field study of populations of desert rodents exposed to low-level gamma radiation. Three 9 ha enclosures were established, two of which served as controls and one was fitted with a suspended ^{137}Cs source. Monthly censuses of small mammals, plants and invertebrates were conducted for 5 years. Dosimeters were fitted to some small mammals to quantify the doses they experienced. Measured dose rates for individual long-tailed pocket mice (*Perognathus formosus*) ranged from 0.0005 to 0.02 mGy/d. Although the chronic exposures reduced survival, particularly in the 1–6 month age groups, no negative reproductive effects were observed. Rather reproduction was somewhat greater among irradiated rodents. The authors note that a similar response, attributed to stimulatory effects of low-level radiation on fertility, has been observed among laboratory rat populations. Although no negative reproductive effects were observed for rodents, sterility was observed among the four lizard species that also occurred in the field enclosures, with the most severe effects observed for the longest lived species.

Mihok (2004) exposed wild, free-ranging meadow voles (*Microtus pennsylvanicus*) to a ^{137}Cs source under natural conditions for 1–1.5 years. The maximum dose measured using dosimeters attached to individual voles was 44 mGy/d. No significant reproductive or population-level effects were observed over the three generation study. Although the lack of effects on reproduction or survival was clear, the author cautions that their results are limited to external exposure from low-LET gamma radiation.

Acute radiation exposures have been shown to adversely affect reproduction among wild rodents. O'Farrell et al. (1972) exposed free-ranging Great Basin pocket mice (*Perognathus parvus*) to one-time ^{60}Co doses ranging from 5 to 9 Gy. Animals were recaptured for 3 years postexposure and reproductive status was evaluated. Although sterility was not evident among males at any dose level, females at all doses were infertile. The authors stated that the reproductive responses they observed in pocket mice were comparable to those observed in laboratory studies with domestic mice. Green (1964) observed reproductive effects following acute X-ray doses of as low as 0.5 Gy.

22.7.1.3 Immune/Endocrine Responses

The responses of the immune and endocrine systems may be affected by chronic high-level irradiation. Bustad et al. (1955) fed sheep diets containing beta-emitting ^{131}I at doses ranging from 4 to 43000 mGy/d for 4 years. Thyroid activity was reduced at exposures of 214 mGy/d or greater. Significant reductions in leukocyte counts, plus extensive inflammation and necrosis of the thyroid were observed at doses of 5000 and 43000 mGy/d. Similar but much less severe effects were observed at doses of 3700 mGy/d. No significant effects on thyroid function or blood chemistry were observed at the 4 mGy/d dose.

Gridley et al. (2002) observed effects on the immune system of mice following acute alpha radiation exposures. Mice were acutely exposed to 0.1, 0.5, or 2 Gy of [56]Fe, or 2 Gy of [28]Si. Mass of the spleen and thymus were reduced in a dose-dependant manner at 4 days postexposure, as were numbers of all major leukocyte cell types. At 113 days postexposure, abundance of T cells and natural killer cells were still reduced at the 2 Gy exposures. Based on these results the authors concluded that the immune system is sensitive to relatively low-level exposures of high-LET radiation.

In contrast to the adverse effects observed at higher doses, low-level exposures may be beneficial. Evidence of potentially beneficial immune and endocrine responses from low radiation doses (i.e., hormesis) was reported in meadow voles (*Microtus pennsylvanicus*) by Boonstra et al. (2005) in association with the ecological studies of Mihok (2004) that showed no evidence for population effects. Chronic ^{137}Cs gamma radiation exposures over 4 years ranged from 0.005 to 0.01 mGy/d (control) to 0.54 mGy/d (low dose) and 92.2 mGy/d (high dose). Corticosterone concentrations and abundance of neutrophils were significantly elevated in low dose individuals as compared to both controls and high dose voles. Elevated corticosterol is a beneficial response that protects an individual against excessive immune and inflammatory responses. Male and female voles responded differently, with females producing higher corticosterone levels than males. The only negative response observed in this study was a slight reduction in hematocrit at both doses. However, the authors stated that this should be viewed with caution given the highly variable nature of hematocrits in wild rodents.

Nonetheless, Calabrese and Baldwin (2004) suggested the enhanced longevity observed in association with low-level gamma exposure is due to stimulatory effects on the hematopoietic and immune systems that occur following low-level chronic bone marrow injury. This is an example of hormesis, where low-level exposures can stimulate beneficial responses. Hormesis appears to be a general biological phenomenon that may represent an over-compensation to an alteration in homeostasis. Hormetic responses have been reported in diverse species, for many biological endpoints, and from a wide range of stressors (both nonradionuclides and ionizing radiation).

22.7.1.4 Mortality

Most mortality data in the literature are for small mammals. Golley et al. (1965) compared responses to acute gamma radiation exposure of wild rodents to that of laboratory mice. Oldfield mice (*Peromyscus polionotus*), cotton mice (*P. gossypinus*), harvest mice (*Reithrodontomys humulis*), and wild and laboratory strains of house mice were all acutely exposed to gamma radiation from a ^{60}Co source. Significant taxonomic differences were observed. Whereas the $LD_{50/30}$ (median acute lethal dose, killing animals within 1 month) for oldfield, cotton, and harvest mice were 11.25, 11.3, and 12 Gy, respectively, wild and laboratory strains of house mice were more sensitive with $LD_{50/30}$ values of 7 and 7.25 Gy.

The survival of wild Great Basin pocket mice (*Perognathus parvus*) exposed to radiation was compared to the survival of laboratory populations of the same species under the similar exposure conditions (O'Farrell et al. 1972). Both field-collected and laboratory-raised mice received acute doses ranging from 5 to 10 Gy. Field-collected animals were returned to the study area and periodically recaptured for evaluation. Wild mice ($LD_{50/30}$ of 8.3 Gy) were no more or less sensitive to the acute effects of ionizing radiation than laboratory populations ($LD_{50/30}$ of 8.6 Gy). In a follow-on study, O'Farrell et al. (1973) observed wild pocket mice to be more sensitive to acute gamma exposure than laboratory pocket mice. $LD_{50/30}$ were 7.8 and 8.8 Gy for wild and laboratory mice, respectively. They attributed this difference to a synergism between radiation-induced and environmental mortality factors.

Mortality is also associated with chronic radiation exposure. In their review of the data in the FASSET database, Real et al. (2004) observed that little mortality occurred in mammals at chronic gamma and beta radiation exposures lower than 2.4 mGy/d. At 24–120 mGy/d more overt signs of toxicity and mortality can occur. Real et al. (2004) concluded that reduced survival is likely above 240 mGy/d. Tanaka et al. (2007) documented the lifespans and causes of mortality for several thousand mice exposed to dose rates of 21, 1.1, and 0.05 mGy/d (^{137}Cs source) from 8 weeks old to approximately 400 days old. Neoplasms accounted for 86.7% of all deaths. The frequency of myeloid leukemia in males, soft tissue neoplasms and malignant granulosa cell tumors in females, and hemangiosarcoma in both sexes exposed to 21 mGy/d were significantly increased over controls. Total counts of multiple primary neoplasms per mouse were significantly increased in mice irradiated at 21 mGy/d. The authors concluded that early death from chronic irradiation was typically due to a variety of neoplasms and not specific increased neoplasms.

Caratero et al. (1998) evaluated the effects of chronic low-dose gamma irradiation on longevity in laboratory mice. They exposed mice to lifetime exposures of 0.19 or 0.38 mGy/d. Whereas median life span for control mice was 549 days, that of both irradiated treatments was 673 days. These findings are consistent with those of other researchers (e.g., Calabrese and Baldwin 2004). In contrast to these laboratory findings, French et al. (1974) observed reduced survival, particularly in the 1–6 month age groups among long-tailed pocket mice with measured dose rates ranging from 0.0005 to 0.02 mGy/d.

22.7.2 RADIATION EFFECTS ON BIRDS

Fewer studies on the effects of ionizing radiation to birds are available in the published literature compared to small mammals. Birds are generally more difficult to study than mammals, resulting from birds' longer life spans, difficulties rearing large numbers of birds in a lab, and birds' greater mobility hampering controlled field experiments. Several studies providing examples from the literature are provided here.

22.7.2.1 Molecular and Hematological Responses

Few studies were identified that evaluated molecular-level effects in birds or on their hematopoietic system, although these two endpoints can be evaluated together when they are studied. One reported no-effect to mitochondrial DNA and mutation rates in wood ducks (*Aix sponsa*) exposed to ambient ionizing irradiation (primarily ^{137}Cs) at the Savannah River site (Johnson et al. 1999). DNA sequencing found the mutation rate was low for both the control and the experimental groups. The frequency of mutations observed in offspring was similar to their parents.

Genetic effects have also been documented among barn swallows (*Hirundo rustica*) breeding in the vicinity of the Chernobyl site (Ellegren et al. 1997, Moller et al. 2007, Bonisoli-Alquati et al. 2009). Increased mutation rates, two to tenfold higher than uncontaminated areas, were recorded as were reduced fitness and an increased frequency of albinism (Ellegren et al. 1997). DNA damage in red-blood cells was greater in barn swallows found in areas around Chernobyl than in low-level irradiation areas (Bonisoli-Alquati et al. 2009). These genetic indicators have been linked to reduced viability (Moller et al. 2007). These authors concluded that chronic exposures to radioactive contaminants at Chernobyl continued to affect wildlife living there (Moller et al. 2008).

However, others (e.g., Smith 2008) argue the "apparent impacts on birds" reported by Moller et al. (2007) may be due to land use changes following abandonment of contaminated land near Chernobyl, weak dosimetry, and inappropriate grouping of "Chernobyl" study sites.

One study demonstrated the recovery and cellular level repair mechanisms in response to ionizing radiation exposures. Newly hatched white leghorn chicks were acutely exposed to 2.25 Gy from a gamma source (Malhotra et al. 1990). Hematological changes over the following 28 days included a sudden decline and gradual recovery of total red-blood cells (RBCs) and white blood cells (WBCs), the level of hemoglobin and hematocrit. Corresponding increases in the mean corpuscular hemoglobin (MCH), mean corpuscular volume (MCV), and the MCH concentration (MCHC) were also noted. Recovery from these adverse effects, and stimulation of increased RBC and WBC counts, occurred within 28 days of exposure.

Compensatory and repair mechanisms can allow some exposed species to recover from relatively minor effects of chronic and acute exposures. Whole-body exposure of white leghorn chicks, 15-days posthatch, to a single acute dose of 2.1 Gy or two doses of 2.1 Gy from a gamma radiation source caused reversible blood pathologies (Malhotra et al. 1990). The chickens recovered from depressed total red-blood cell counts, hemoglobin content, hematocrit, MCV, and MCH within 60 days. The next higher dose tested (6.6 Gy) caused mortality to all exposed birds within 7 days. Adverse effects begin to manifest when compensatory or repair mechanisms no longer keep up with the damage caused by irradiation.

22.7.2.2 Growth and Development

Environmental exposures provide realistic conditions that integrate spatial and temporal variability, but there are few experimental controls. A study by Millard and Whicker (1990) reported reduced growth rates and body weights in barn swallows exposed to radionuclides at the Test Reactor Area of the Idaho National Engineering Laboratory, but the author could not conclude that the observed differences from controls were due to the radiation. During the breeding season, barn swallows nested near radioactive leaching ponds. These birds fed on arthropods from the leaching ponds and built nests with radioactive contaminated mud (Millard and Whicker 1990). The radionuclide accumulated in greatest abundance was ^{51}Cr, with tissue concentrations averaging 16.1 Bq/g. Thyroids contained mean ^{131}I concentrations of 3330 Bq/g, resulting in individual thyroid doses of 4.3 mGy/d or 0.45 Gy over the entire breeding season. Eggs and nestlings showed no increased mortality at whole-body doses 0.84 and 2.2 mGy/d, respectively. Mean growth rates and body weights of exposed birds were significantly lower than controls, by 5% and 9%. However, the authors concluded that the cause of differing growth rates could not be definitively attributed to the radiation exposure, citing weather, food shortage, predation, and other possible stressors in addition to radiation.

Freshly hatched tree swallows (*Tachycineta bicolor*) were collected in the field and returned to the lab for acute gamma radiation exposures of 0.9, 2.7, and 4.5 Gy from a ^{60}Co source (Zach and Mayoh 1984). The exposure doses did not induce mortality, but body mass was reduced in the 2.7 and 4.5 Gy exposure groups in a dose-dependent manner (5% and 13% reduction in growth). Some measures of development and timing were also affected by irradiation—the timing of the primary feather development was half a day later in the two highest exposure groups—while others were not. Primary feather length at emergence did not differ between exposed tree swallows and controls. In another laboratory exposure, Zach and Mayoh (1986a) acutely irradiated tree swallows nests at

0.4, 0.8, 1.6, or 3.2 Gy at 7–8 days into incubation. Similar effects were observed as when irradiating hatchlings. Hatching and fledging success was not reduced at doses up to 3.2 Gy. Growth was significantly reduced by 4–14% in eggs acutely exposed at 1.6–3.2 Gy, respectively. Incubation time increased, while foot and primary feather lengths at fledging were also significantly depressed at the highest dose. These data show that for birds, growth can be a more sensitive endpoint for measuring effects of ionizing irradiation than hatching or fledgling success.

22.7.2.3 Reproduction

Many avian studies have investigated reproduction, as this is a relatively simple endpoint to measure, and it incorporates subtle toxicological effects on multiple organ systems.

Zach et al. (1993) chronically exposed tree swallows to 0.144 mGy/d, equivalent to 45 times background radiation levels. This was done during the breeding season to investigate the potential for reproductive effects from the gamma radiation of spent uranium fuel pellets. Reproduction, measured by hatching and fledging success, the number of eggs that resulted in fledglings (breeding success), and chick growth measures (chick body weight, primary feather length, and foot length) was not significantly different in exposed tree swallows as compared to controls. Although no physical effects were observed at low-level irradiation of wild tree swallows, both swallows and house wrens (*Troglodytes aedon*) avoided nesting in areas with radiation 100 times greater than background from a ^{137}Cs source (from 0.7 to 6 mGy/d) (Zach and Mayoh 1982). Increased human activity and reduced prey abundance in areas near irradiation sources could have contributed to the avoidance behavior in addition to the radiation. Breeding season exposures in this study also did not cause any significant reproduction effects to breeding success and chick growth. Avoidance behavior may have protected these birds from adverse effects.

Field observations of wild birds exposed to low-level nuclear waste at the Savannah River Site in South Carolina found relatively high body burdens up to 55.6 Bq ^{90}Sr/g of bone, and 667 Bq ^{137}Cs/g in muscle (Willard 1963). In addition to these body burdens, eastern bluebirds (*Sialia sialis*) and great crested flycatchers (*Myiarchus crinitus*) were exposed to ambient radiation from waste disposal area soil. Wild birds did not appear to have any obvious physical effects from the ambient radiation at this site. The effects of single acute ^{60}Co gamma doses to 9-day-old and 2-day-old chicks (both bluebirds and flycatchers) were investigated for 2 weeks after exposures ranging from 3 to 30 Gy. Significant reductions in body weight (up to 63%) were observed at all doses; primary feather growth was reduced (up to 68%) at all but the lowest (3 Gy) dose. Two-day-old chicks were also more sensitive to radiation exposures than the 9-day-old chicks, and showed greater differences in weight and feather length relative to controls. Thus, earlier life stages can be more sensitive to irradiation than older individuals.

Acute doses should have greater effects than lower level chronic doses, but this is not always the case. In a comparison between the effects of acute and chronic irradiation to the same species, chronic gamma irradiation to tree swallow nests at ~1 Gy/d reduced hatching by 32% (56% compared to 82% controls) while single acute doses up to 3.2 Gy did not (Zach and Mayoh 1986a). Fledging success, measured as the number of hatched eggs that successfully fledged, was not reduced in acutely or chronically exposed hatchlings. Similar levels of effects were reported for house wrens (Zach and Mayoh 1986b). Survival of freshly hatched house wrens was not reduced over the 15-day nesting period after acute whole-body exposures ranging from 0.9 to 6 Gy of (^{60}Co) gamma radiation. Significantly reduced growth (e.g., body mass, primary feathers, and foot length) was observed in a linear dose-dependent manner, and birds that received the highest dose exhibited stunted growth. Dose fractionation, which refers to the separation of radiation exposures with periods of rest, allows time for DNA repair. Likewise, acute irradiation damage can sometimes be repaired, while the additive effects from continuous exposures may not. The early life-stages receiving acute exposures in this study (7–8 days posthatch) could also have helped minimize the irradiation due to the relatively low sensitivity of undifferentiated cells.

Another study evaluated the differences between acute and chronic exposures of the same total gamma radiation dose to chicken and black-headed gull eggs (*Larus ridibundus*) (Phillips and

Coggle 1988). Eggs were exposed to a ^{60}Co source for total doses of 1.9–28.8 Gy during incubation as 100–2000 mGy/d for 20 days, or to single acute doses of 1.92–28.8 Gy on day 10 of incubation. Cumulative chronic exposures above 9.6 Gy and acute exposures above 4.8 Gy reduced hatchability and development to full-term. The frequency of foot and limb deformities also increased in chicks exposed to cumulative chronic doses of 9.6 Gy.

Zakaria (1991) reported reduced hatchability in broiler chicken eggs following exposures at doses below those that had no effect on house wrens or tree swallows. Eggs were acutely exposed to ^{60}Co gamma doses ranging from 0.05 to 2.1 Gy prior to incubation. Acute exposures below 1.6 Gy did not affect hatchability or chick body weight, although hatchability and chick weight were reduced at 2.1 Gy. Results of this study, which tested over 23,000 eggs, were consistent with previously reported effect levels for other avian species.

How the rate of exposure affects the responses to acute doses of ionizing radiation is poorly studied. Leghorn chicken (*Gallus domestica*) hens were acutely exposed to varying dose rates to determine effects on reproductive performance of exposed hens, egg hatchability, and reproductive performance of progeny (Mraz and Woody 1973). Each exposure group received 8 Gy from a ^{60}Co source at 0.01, 0.05, 0.25, or 0.43 Gy/min. Egg production in the exposed hens was reduced in all but the lowest exposure, and this effect lasted for 5–7 weeks. Egg hatchability in the first 14 days after irradiation was reduced only in the two highest dose rates. Egg production was not reduced in any of the surviving offspring. Half of the hens were again exposed to 8 Gy for 23 weeks after the initial dose. The second exposure reduced egg production below that observed after the initial dose. All groups recovered from these second exposures by the end of the 8-week observation period, except for the group previously exposed at the highest dose rate (0.43 Gy/min). These results indicate that the first exposure had a residual effect on the hens, making them more sensitive to additional irradiation.

A similar study investigated the effects of gamma radiation on egg production of 21-week-old Japanese quail (*Corturnix corturnix japonica*) (Baumgartner 1985). Quail were acutely exposed to doses ranging from 3 to 30 Gy from a ^{60}Co source. Doses above 6 Gy caused a dose-dependent reduction in egg production by 30 days postirradiation. Egg weight likewise decreased in exposures above 6 Gy.

22.7.2.4 Mortality

Birds are about as sensitive to acute radiation exposure as mammals, with $LD_{50/30}$ ranging from 5 to 20 Gy (Woodhead 1998). Earlier life stages are also more radiosensitive than adults. Mellinger and Schultz (1975) performed one of the initial reviews of the effects of ionizing radiation in birds. Effects documentation from various radiation sources was quite limited and focused on tissue concentrations as a measure of exposure. The general conclusion from review of these studies was that "minor population effects [to birds] that might results from ionizing radiation probably would be masked by more consequential environmental effects."

Although few chronic exposure studies showing avian mortality were available, acute thresholds had been identified for multiple species. Reported $LD_{50/30}$s from exposure to ^{137}Cs for adult green-winged teal (*Anas crecca*), blue-winged teal (*Anas discors*), and northern shoveler (*Anas clypeata*) were 4.85, 7.15, and 8.94 Gy, respectively (Tester et al. 1968). Immature, 4-month-old mallards exposed to gamma radiation had an $LD_{50/30}$ of 7.04 Gy (Abraham 1972). The 24-h LD_{50} for white leghorn chicks (15-days posthatch) was 6.6 Gy (Malhotra et al. 1990). Baumgartner (1985) observed an $LD_{50/30}$ of 22.5 Gy for 21-week-old Japanese quail exposed to gamma radiation from a ^{60}Co source.

22.7.3 Insects and Other Terrestrial Invertebrates

Insects and other terrestrial invertebrates are generally recognized as among the least sensitive taxa to ionizing radiation. A review conducted by Wood et al. (2005) indicated that the levels of chronic radiation exposure that produced effects in insects and soil invertebrates ranged from 2 to 1000 mGy/d (Table 22.4). These data were subsequently used to identify radiation exposure levels for a

TABLE 22.4

Effects Observed among Insects and Soil Invertebrates Resulting from Chronic Radiation Exposure

Dose Rate (mGy/d)	Taxa	Radiation Type	Effect
2	Earthworm	Alpha	Reduced abundance, plus reproductive and histological effects
2	Insect larvae	Alpha	Reduced abundance
2	Scorpion	Gamma	Increased chromosome aberations
2	Midge	Mixed	Chromosome aberations
20–100	Earthworm		Reduced abundance
100–200	Myriapods, spiders, earthworms	Beta	Reduced abundance
200–1000	Bark beetle		Reduced pupal survival
200–1000	Soil invertebrates		Population reduction
1000	Ants		Behavioral changes in colony

Source: Adapted from Wood, M. D. et al., *Radioprotection*, Suppl 1, 40, S229–S234, 2005. With permission.

study on previously untested taxa, the terrestrial isopod, *Porcellio scaber*. Morbidity, mortality and reproductive capacity were evaluated in terrestrial isopods exposed to gamma radiation for 96 days (Hingston et al. 2005). No adverse effects were observed following a chronic exposure of up to 192 mGy/d.

Hertel-Aas et al. (2007) exposed earthworms (*Eisenia foetida*) to five levels (4.32, 41, 96, 264, and 1032 mGy/d) of gamma radiation from a ^{60}Co source for two generations. At the highest dose, hatchability was reduced to 60% of controls. The lowest dose rate at which reproductive effects were observed was 96 mGy/d in the F_0 generation and 264 mGy/d in the F_1 generation. The authors concluded that although chronic gamma exposure reduces reproductive capacity in earthworms, extensive exposure (13 weeks) was necessary for effects to be expressed.

Effects of acute gamma irradiation on the collembolan, *Folsomia candida*, were investigated by Nakamori et al. (2008). Growth and survival of these soil insects was measured following acute gamma radiation exposures ranging from 35 to 1770 Gy. Reproductive effects were also measured in response to gamma doses ranging from 4 to 110 Gy. Acute ED_{50}s for survival, growth, and reproduction for these insects were 1356, 144, and 21.9 Gy, respectively.

Field data from Chernobyl indicated that immediately following the accident, soil invertebrates experienced exposures ranging from 9 to 29 Gy. These dose levels resulted in 20- to 30-fold reductions in density of the soil invertebrate community (Geras'kin et al. 2008). One year after the accident, abundance of soil invertebrates had recovered to 45% of that at control sites. By 2.5 years, soil invertebrate abundance was generally equal to that present prior to the accident, but diversity was still reduced by 20%. This reduction in diversity was still evident up to 10 years after the accident. Field observations from several locations in the Chernobyl Exclusion Zone in 2001, 15 years following the accident, indicated that whereas diversity of flying insects and soil invertebrates declined as contamination increased, overall invertebrate biomass did not (Jackson et al. 2005). Although data from this study were limited, the authors suggested the reduced diversity was due to chronic impacts on more sensitive taxa. Most recently, Moller and Mousseau (2009) evaluated the abundance of several insect and arachnid taxa (bumblebees, butterflies, grasshoppers, dragonflies, and spiders) in the Chernobyl exclusion zone, 20 years after the accident. Abundance was reduced for all taxa in a dose-dependent manner, among sites with dose levels ranging from less than 0.0024 mGy/d to over 2.4 mGy/d. The authors suggested that these results imply that the ecological effects of radiation from Chernobyl on terrestrial invertebrates are greater than previously assumed.

22.7.4 FISH AND AQUATIC INVERTEBRATES

Although fish are generally among the least radiosensitive vertebrates, they are more sensitive than aquatic invertebrates. Based on studies summarized in the EPIC database (Sazykina et al. 2003), eggs of fish are twice as sensitive to acute gamma exposure as are eggs of snails (Table 22.5). When chronically exposed to gamma radiation, fish eggs are approximately 70 times more sensitive than are snail eggs (Table 22.6). Among fish species, salmon are more sensitive to chronic gamma exposures than are pike. Whereas 130–330 mGy/d resulted in 100% egg mortality in salmon, 940 mGy/d was needed to produce the same effect in pike eggs.

Real et al. (2004) summarized the dose-response data for fish available in the FASSET database. Most data were for gamma irradiation in laboratory settings. General trends from these data are

TABLE 22.5
Responses Observed among Arctic Fish and Invertebrates in Exposure to Acute Gamma Radiation

Dose (Gy)	Taxa	Effect
3	Salmon eggs	LD_{50} for eggs exposed in initial developmental stage
5	Salmon eggs	~100% mortality for eggs exposed in initial developmental stage
2	Pike eggs	30% survival reduction considerable embryo abnormality
4	Pike eggs	~100% mortality for eggs exposed in initial developmental stage
3	Pond snail (*Limnea stagnalis*) eggs	Slight increase in egg mortality
6	Pond snail (*L. stagnalis*) eggs	~50% mortality for eggs
12–15	Pond snail (*L. stagnalis*) eggs	~100% mortality for eggs
2	Daphnid	Slight increase in mortality
20	Community of 6 Cladocerans	Compensatory increase in fecundity
200	Zooplankton community	66% decrease in community richness

Source: From Sazykina, T. G. et al., Report of dose-effects relationships for reference (or related) Arctic biota. EPIC database "Radiation effects on biota." A deliverable report for EPIC (Environmental Protection from Ionizing Contaminants) Project CIA2-CT-2000-10032, 2003. With permission.

TABLE 22.6
Responses Observed among Arctic Fish and Invertebrates in Exposure to Chronic Gamma Radiation

Dose Rate (mGy/d)	Taxa	Effect
1–5	Salmon eggs	30–40% decrease in egg survival
30	Salmon eggs	~50% mortality of eggs
130–330	Salmon eggs	~100% egg mortality
5	Pike eggs	Decreased egg development time Increased abnormalities
30	Pike eggs	Increased frequency of chromosomal aberrations
300–470	Pike eggs	Decreased egg and larval survival
940	Pike eggs	~100% egg mortality
17	Pond snail (*Limnea stagnalis*)	Reduced adult size increased radioresistance
2,100	Pond snail (*L. stagnalis*)	Decreased egg survival increased abnormalities
0.0055	Daphnids	Slight decrease in fertility

Source: From Sazykina, T. G. et al., Report of dose-effects relationships for reference (or related) Arctic biota. EPIC database "Radiation effects on biota." A deliverable report for EPIC (Environmental Protection from Ionizing Contaminants) Project CIA2-CT-2000-10032, 2003. With permission.

TABLE 22.7
Summary of Chronic Irradiation of Fish from the FASSET Database[a] Unless Otherwise Noted

Dose Rate (mGy/d)	Taxa	Effect
2.4–24	Plaice, medaka, roach	Reduction in testes mass and sperm production; lower fecundity; delayed spawning
24–120	Plaice, eelpout, medaka, guppy, rainbow trout	Reduction in testes mass and sperm content; severe depletion of spermatogonia; reduced fertility or complete infertility; reduce fecundity; reduced male courtship
120–240	Medaka	Depletion of spermatogonia
350	Medaka	Reduced number of eggs/female; reduced egg viability; reduced percentage of larvae from viable eggs[b]
240–1200	Medaka, guppy	Sterility; reduction in larval survival; increase in vertebral abnormalities
>1200	Guppy	No impact on offspring survival following parental irradiation

Sources: [a] Real, A. et al., *J. Radiol. Prot.*, 24, A123–A137, 2004; [b] Hinton, T. G. et al., *J. Environ. Radioactiv.*, 74, 43–55, 2004a. 2004. With permission.

summarized in Table 22.7. The overall conclusions were that effects on the immune system of fish may occur at exposures of less than 0.2 mGy/d (but results were inconsistent), and that although one study indicated that effects on gametogenesis may occur at 5.5 mGy/d, most data indicated effects occurred when exposure was greater than 24 mGy/d. Data were too few to identify chronic radiation exposure thresholds for mortality in fish (Real et al. 2004).

As can be seen in the data summarized earlier, most available data on the effects of ionizing radiation in fish pertains to effects of gamma radiation. Data with which to determine the RBE of alpha radiation relative to gamma radiation on an ecologically important endpoint such as reproduction have been lacking. To address this issue, Knowles (2002) chronically exposed zebrafish (*Danio rerio*) to both alpha and gamma radiation and evaluated reproduction. Fish were exposed to three gamma radiation doses (7.2, 24, and 178 mGy/d) from a ^{137}Cs source, and four alpha radiation doses (0.23, 0.456, 2, and 5.1 mGy/d from ^{210}Po added to their water) for over 33 weeks. Among the gamma-irradiated fish, reproductive effects were observed only at the 178 mGy/d dose. The number of nonviable eggs was significantly greater, and by the 20th week of the study, all pairs of fish had ceased laying eggs. In contrast, no reproductive effects were observed for any of the alpha radiation doses. Because no reproductive effects from exposure to alpha radiation were observed, the RBE of alpha radiation relative to gamma radiation could not be precisely measured from this study. However, the author suggested a conservative estimate of the RBE of less than 35 based on the highest no-effect alpha radiation dose.

A similar series of studies was also conducted as part of the FASSET-ERICA projects to investigate the chronic effects of alpha and gamma radiation on aquatic invertebrates. Alonzo et al. (2006) evaluated growth, physiology, and reproduction of *Daphnia magna* exposed to the alpha-emitting radionuclide, ^{241}Am, at doses of 0.48, 2.64, and 23.8 mGy/d for 23 days. Although food ingestion was not affected at any dose rate, growth was reduced at the two highest doses. In addition, although fecundity was not affected at any dose rate, egg masses, and neonates were significantly smaller. In a subsequent study, Alonzo et al. (2008) evaluated effects of ^{241}Am on *Daphnia magna* at doses of 7.2, 36, and 360 mGy/d over three generations (70 days). At the highest dose rate, juvenile mortality was 38–90%, with survivors displaying delayed reproduction and reduced fecundity in the F_1 and F_2 generations. At the two lower doses, juvenile mortality ranged from 31% to 38%, but was only evident in the F_1 generation. Whereas the proportion of breeding females was reduced to 62% and 69% in the 36 and 7.2 mGy/d exposures, respectively, this reproductive effect was evident in the F_1 generation at the 36 mGy/d dose, but did not occur until the F_2 generation at the 7.2 mGy/d dose. In comparison to alpha radiation, chronic effects from gamma radiation on *Daphnia magna* were less severe. Gilbin et al. (2008) exposed *Daphnia magna* to gamma radiation from a ^{137}Cs source at a dose rate of 9.8, 100.8, and 744 mGy/d for 23 days. None of the doses affected growth or survival. Reproduction (a 21% reduction in fecundity) was affected only at the 744 mGy/d dose.

In a laboratory study of marine polychaetes, Knowles and Greenwood (1996) observed significant reductions in larvae following chronic exposure to both beta (from ^{3}H in water) and gamma (^{137}Cs) radiation at doses of 175 mGy/d. Whereas egg production but not larval survival was reduced among worms exposed to gamma radiation, survival of eggs to larvae but not egg production was reduced among beta radiation-exposed worms.

Subtle effects on expression of reproductive traits in invertebrate populations in the field have been observed at much lower exposure rates. Tsytsugina and Polikarpov (2003) evaluated the radiation effects on populations of aquatic oligochaetes chronically exposed in the environment near the Chernobyl facility. They studied two lakes that were similar in terms of water chemistry and chemical contamination, but differing in radionuclide contamination (primarily ^{90}Sr). Whereas doses to worms in the control lake were determined to be 0.017 mGy/d, doses in the contaminated lake were 0.336 mGy/d. The incidence of chromosomal aberrations was significantly greater among worms from the contaminated pond. The degree of chromosomal damage was highly correlated with the proportion of individuals in the worm populations with developing germ cells (an indication that the worms had changed from the asexual to the sexual mode of reproduction). The authors postulate that unfavorable environmental conditions (i.e., radiation exposure) had initiated the transition to sexual reproduction, which provides a greater potential for adaptive responses within the population.

22.8 ECOLOGICAL RADIATION EFFECTS THRESHOLDS

Interpretation of doses, whether from chemical or radiological contaminants, requires identification of effect thresholds. Understanding of the nature and magnitudes of effects associated with a given dose permits risk management decisions to be made such that potential adverse effects may be mitigated. To that end, dose-response data, such as that summarized earlier and in the radionuclide databases have been used by a number of organizations as the basis for ecological radiation effects thresholds. Examples of effects thresholds for ionizing radiation are summarized in Table 22.8.

Prior to development of the extensive reviews of radiation dose-response literature, the protection of nonhuman biota from exposure to ionizing radiation was guided by the International Council for Radiation Protection (ICRP) paradigm that if humans were adequately protected, then plants and animals would also be protected (ICRP 1977). Hinton et al. (2004b) questioned the validity of this paradigm because it does not consider habitats where humans are absent, nor consider instances where protection of individuals is important (i.e., threatened or endangered species). It should also be noted that nonhuman biota may experience unique exposure pathways (e.g., root uptake), possess increased radiosensitivity (e.g., fish eggs, plant root hairs), or experience secondary effects (e.g., reduced food supply) that would not affect humans, and therefore they may not be protected under the ICRP paradigm. In contrast, Smith (2005) argued that the paradigm is valid because effects from regulated releases of ionizing radiation are relatively insignificant and the environmental benefit gained by additional regulation to further protect nonhuman biota is not commensurate with the additional economic cost that would be required. Smith (2005) however, did recommend that analyses continue to test the ICRP paradigm.

In one of the first efforts to evaluate the validity of the ICRP paradigm, the National Council for Radiation Protection (NCRP) reviewed available laboratory and field data concerning dose–response relationships for freshwater and marine biota, including both fish and invertebrates (NCRP 1991). Response types that were considered included genetic and cellular effects, physiology, reproduction, growth and development, and mortality. They observed a 28-fold difference between the maximum chronic environmental exposure that produced no effects (0.36 mGy/d) and the lowest dosage from laboratories studies at which effects were observed (10 mGy/d). They therefore

TABLE 22.8
Summary of Chronic Effects Thresholds (mGy/d) for Protection of Nonhuman Biota Recommended by Multiple Organizations

Taxa	NCRP (1991)[a]	IAEA (1992)[a]	UNSCEAR (1996)[a]	Environment Canada and Health Canada (2003)[b]	Garnier-Laplace and Gilbin (2006)[c]	Andersson et al. (2008)[d]
Generic						0.24
Terrestrial						
Plants		10	10	2.7	0.24	1.7
Animals		1	1		0.24	
Invertebrates				5.5		4.8
Birds						0.05
Mammals				2.7		0.05
Aquatic						
Freshwater Organisms	10	10	10		0.24	
Algae				2.7		
Macrophytes				2.7		
Benthic invertebrates				5.5		4.8
Amphibian						0.05
Fish				0.55		0.05
Marine organisms	10		10		0.24	
Deep sea organisms		24			0.24	

[a] Chronic exposures of this level or less are unlikely to produce adverse effects in exposed populations.
[b] Represents an estimated no-effects value (ENEV) for protection of populations, derived by use of uncertainty factors.
[c] Default predicted no-effect level for all ecosystems and organisms—derived from species sensitivity distributions for chronic exposures.
[d] Predicted no-effect dose rates.

concluded that a chronic dose rate to the maximally exposed individual that did not exceed 10 mGy/d would be protective of endemic populations. They acknowledged that some minor effects from radiation exposure at 10 mGy/d are possible, but in the context of other natural compensatory mechanisms, the significance of these effects is likely trivial. Based on their analyses, NCRP (1991) also concluded that protection of humans from radionuclide exposure via aquatic pathways would also be protective of aquatic biota.

A broader review of radiation effects on nonhuman biota, also with the purpose of determining if available data supported the ICRP paradigm, was conducted by the International Atomic Energy Agency (IAEA 1992). This review had two objectives: (1) to determine whether the ICRP paradigm was supported by the available data and (2) to determine whether standards for protection of aquatic and terrestrial biota were warranted. IAEA (1992) considered acute and chronic data relating to terrestrial plants, mammals, birds, reptiles, amphibians, and invertebrates. A detailed evaluation of data for aquatic biota was not performed because the existing comprehensive reviews (i.e., NCRP 1991) of these data were considered adequate. Reproduction was identified as the most limiting endpoint for survival of populations for both acute and chronic exposures. Acute doses of 0.1 Gy/d or less were considered unlikely to produce persistent, significant adverse effects on populations or communities of terrestrial plants or animals. A chronic exposure of 1 mGy/d was identified as a no-effect level for radiosensitive terrestrial animal populations. For aquatic biota and terrestrial plants, a chronic dose rate of 10 mGy/d was determined to be unlikely to have adverse population-level effects. IAEA cautioned that further consideration should be considered for long-lived species with low reproductive rates. Based on their analysis, IAEA (1992) concurred with the ICRP paradigm, but noted that there may be circumstances, such as the presence of rare or endangered species, in which the generic effects thresholds may be insufficient and site-specific analyses may be required.

UNSCEAR (1996) reviewed effects of ionizing radiation on terrestrial plants (crops, forest trees, shrubs, herb communities, and lichen), terrestrial animals (mammals, birds, reptiles, amphibians, and invertebrates), and aquatic biota. Ionizing radiation doses of 10–1000 Gy were identified as acutely lethal to plants. Chronic effects were evident among sensitive plants at exposures of 0.024–72 mGy/d. Based on the available data, UNSCEAR (1996) recommended that 10 mGy/d be considered a threshold protective of plants because although it may have effects on sensitive plants, only trivial effects would be expected on the broader plant community. Among terrestrial animals, most available data was for mammals. Acute lethal doses ranged from 6 to 10 Gy for small mammals, to 1.5–2.5 Gy for livestock. When chronic data were considered, the beagle was identified as the most sensitive animal. Whereas an ionizing radiation dose of 4 mGy/d resulted in sterility in a few months, a lifetime dose of 0.9 mGy/d had no effect. UNSCEAR (1996) concluded that an exposure of 1 mGy/d for the most exposed individual in a population would be unlikely to have an effect on reproduction at the population level. Because the available data indicated that birds were approximately as sensitive as small mammals, reptiles and amphibians were less sensitive than mammals, and invertebrates are quite insensitive to ionizing radiation, protection of mammals effectively protects other terrestrial animals. Fish were found to be the most sensitive taxa among aquatic biota, with fish embryos being the most sensitive lifestage. Whereas 10–25 Gy was acutely lethal to marine fish, an acute dose of 0.16 Gy was lethal to salmon embryos. Chronic exposures of 240–720 mGy/d had no-effect on marine invertebrates (snails, scallops, clams, and crabs). Chronic exposures of 24–40 mGy/d resulted in sterility among some freshwater fish species. Because it was concluded that significant effects to reproductive capacity in fish were unlikely at doses of less than 24 mGy/d, UNSCEAR (1996) identified the dose of 10 mGy/d as a protective threshold for aquatic biota.

In preparation for developing radiological standards for the protection of aquatic and terrestrial biota, the U.S. DOE convened a workshop to evaluate the data presented in IAEA (1992). The purpose was to evaluate if these data were adequate to support development of regulatory criteria and to determine whether the conclusions were still valid. The results of this workshop are documented in Barnthouse (1995). The consensus of participants was that the thresholds identified in IAEA (1992) were valid, but that better guidance on their application was needed. They also concluded that the ICRP paradigm was supported except when human access is restricted but not for other biota, unique exposure pathways exist, rare or endangered species are present, or other stresses are significant. Based on the conclusions outlined in Barnthouse (1995), the IAEA thresholds were selected to form the basis for the BCGs developed by U.S. DOE (2002) and incorporated into RESRAD-BIOTA (U.S. DOE 2009).

Estimated no-effect values (ENEVs) for exposure of biota to radionuclides were developed by Environment Canada and Health Canada (2003) as part of their effort to assess environmental impacts of releases from nuclear facilities. The ENEVs represent chronic low or no effect-levels applicable to the survival of populations of the most sensitive species extracted from the literature and evaluated in accordance with Environment Canada (1997). The IAEA values assume that although some individuals in an exposed population may be affected, compensatory mechanisms will preclude expression of any population-level effects. In contrast, the ENEVs from Environment Canada and Health Canada (2003) represent levels where the probability of effects is extremely low such that populations with limited home ranges or small population sizes will not be affected. Environment Canada and Health Canada (2003) focused on reproduction and survival as the limiting endpoints for population effects. Although data were reviewed for birds, mammals, amphibians, reptiles, fish, terrestrial and aquatic invertebrates, and terrestrial and aquatic plants, data were considered insufficient to support development of ENEVs for birds, amphibians, and reptiles. ENEVs for other taxonomic groups ranged from 0.55 mGy/d for fish to 5.5 mGy/d for aquatic and terrestrial invertebrates (Table 22.8).

Radionuclide dose-response data were extracted from the FRED database as part of the ERICA project and used to develop acute ED50 (doses in Gy that produced a 50% effect) and chronic effect dose rates (EDR10) (doses in uGy/h that produced a 10% effect) values for multiple taxonomic groups and effect endpoints (Garnier-Laplace and Gilbin 2006, Garnier-Laplace et al. 2006). Geometric means of acute ED50s for aquatic biota ranged from 3.4 to 430 Gy for salmonid reproduction and alga mortality, respectively. For terrestrial biota, geometric means of acute ED50s ranged from 1.2 to 2061 Gy for mammalian reproduction and insect mortality, respectively. Comparable patterns were observed for chronic values; whereas the lowest geometric mean EDR10s were observed for fish and mammal reproduction (12.4 and 0.75 mGy/d, respectively), the highest geometric mean EDR10s were observe for aquatic invertebrate reproduction and morbidity in lichen (1.1E + 04 and 3.8E + 03 mGy/d, respectively). These data were then used to develop acute predicted no-effect doses (PNEDs) and chronic predicted no-effect dose rates (PNEDRs) for aquatic and terrestrial biota based on species sensitivity distributions (SSDs) in accordance with methods outlined by the European Commission (EC 2003). Based on their analyses, Garnier-Laplace and Gilbin (2006) recommend a chronic PNEDR of 0.24 mGy/d as suitable for all receptors in all ecosystems. For acute exposures in terrestrial and freshwater systems, a PNED of 900 mGy was determined, while a value of 300 mGy was calculated for acute exposures in marine systems. Subsequent work by Garnier-Laplace et al. (2006) identified PNEDRs ranging from 0.24 mGy/d for freshwater ecosystems to 1.6 mGy/d for terrestrial ecosystems.

The most recent effect thresholds for exposure to ionizing radiation were developed by the PROTECT program based on data extracted from the FREDERICA database (Andersson et al. 2008). Similar to the thresholds developed for ERICA (Garnier-Laplace and Gilbin 2006), the PROTECT thresholds were developed based on SSDs following EC (2003) and are referred to as predicted no-effect dose rates (PNEDRs). Whereas Garnier-Laplace and Gilbin (2006) based their SSDs on geometric mean EDR10s, the SSDs developed as part of the PROTECT program used the lowest EDR10 values for a given species. Using this approach, a generic PNEDR for all biota, and PNEDRs for vertebrates, invertebrates, and plants were developed (Table 22.8). It should be noted that although taxa-specific PNEDRs were developed, data were limited and therefore the confidence intervals (and therefore uncertainty) are large (upper 95% confidence limits were 80–200 times larger than lower limits). In contrast, a larger, more robust dataset was available for the generic PNEDR. Uncertainty associated with the generic value is therefore lower.

Summary

Radionuclides are unstable forms of elements that emit particles (alpha [α], beta [β], or neutrons) or electromagnetic radiation (gamma or X-rays) as they decay. These emissions are collectively referred to as ionizing radiation because of their ability to produce charged particles (ions) in matter that they pass through. Alpha, beta, and gamma radiation are the primary forms of ionizing radiation of concern.

Radioactive elements differ from conventional elements in that the radiation they emit can also exert biological effects at a distance without directly contacting the organism. On the other hand they are similar to stable elements because radionuclides can be accumulated and integrated into the tissues of plants and animals, where they can produce adverse effects. Because radionuclides emit radiation, effects may be induced by both internal and external exposure. Further, because effects from radionuclides are qualitatively similar (differing only in magnitude), exposure to multiple radionuclides is additive. Total radiation exposure is therefore the sum of exposure to all radioactive elements retained in tissue and in the immediate environment (i.e., the sum of internal and external exposure).

Ionizing radiation can be both naturally occurring and anthropogenic in origin. Total annual dose to humans or large terrestrial animals from natural background sources is estimated to be 3.11 mGy/year (0.0085 mGy/d), of which 17.4% is external (10.6% from cosmic radiation and 6.7% from terrestrial sources) and 82.6% is from internal sources (68% from radon and 14.6% from other radionuclides).

The mechanism by which ionizing radiation produces effects in biological tissue is both simple and complex. The initial effect is simply the transfer of energy from the decaying radionuclide in the form of ionizing radiation, to neutral molecules and atoms to produce electrically charged ions. This results in a cascade of responses and subsequent reactions that occur as the ionized molecules interact with biological tissue.

Radiosensitivity varies between different taxonomic groups, and among species. Although there is considerable overlap among taxa, more primitive biota are more resistant to acute radiation exposure and more complex organisms are more sensitive. Acute radiation doses lethal to higher plants and vertebrates are generally lower than doses lethal to invertebrates, which in turn are lower than doses lethal to algae and microbes. When effects of acute and chronic radiation exposure on reproductive success are considered, similar taxonomic trends in radiosensitivity are evident, with mammals being most sensitive, fish intermediate, and invertebrates least sensitive.

Mammals are among the most sensitive taxa, however radiation exposure levels that produce effects are variable and overlap. Genetic effects have been observed at chronic exposures as low as 10 mGy/d and the threshold for reproductive effects from chronic radionuclide exposures is 2.4 mGy/d. Acute exposures at which mortality is observed in small mammals range from 7 to 12 Gy.

Birds generally display radiosensitivity comparable to mammals. Whereas chronic gamma exposure at 0.144 mGy/d has no effect on reproductive success, hatching success is reduced at 1 Gy/d. Acute doses as low as 3 mGy/d have been shown to affect the growth of hatchling barn swallows. Acute mortality thresholds for birds have been reported to range from 4.9 to 22.5 Gy.

Insects and other terrestrial invertebrates are among the least sensitive taxa to ionizing radiation, with effects from chronic radiation exposure observed for doses ranging from 2 to 1000 mGy/d. Acute effects from gamma irradiation on survival, growth, and reproduction of collembola occur at doses of 1356, 144, and 21.9 Gy, respectively.

Although fish are generally among the least radiosensitive vertebrates, they are more sensitive than aquatic invertebrates. Eggs of fish are twice as sensitive to acute gamma exposure and 70 times more sensitive to chronic gamma exposure than are eggs of snails. Available data indicates effects on the immune system in fish occur at chronic gamma exposures less than 0.2 mGy/d, and effects on gametogenesis may occur at chronic doses 5.5 mGy/d to greater than 24 mGy/d. In zebrafish, reproductive effects occurred at chronic gamma doses of 178 mGy/d. No effects were observed at an alpha dose of 5.1 mGy/d.

Laboratory studies of chronic exposure of *Daphnia magna* to both alpha and gamma-emitting radionuclides observed a 100-fold difference in effect levels. Whereas reproductive effects from alpha radiation occurred at exposures as low as 7.2 mGy/d, reproductive effects were observed at a gamma radiation exposure of 744 mGy/d.

Since the 1970s multiple organizations have researched and proposed effects thresholds to protect nonhuman biota from chronic exposure to ionizing radiation. Initial values, such as those from the NCRP, IAEA, and UNSCEAR, ranged from 1 mGy/d for terrestrial animals to 10 mGy/d for plants and aquatic animals. Environment Canada and Health Canada developed values that ranged from 0.55 mGy/d for fish to 5.5 mGy/d for terrestrial invertebrates. The most recent values developed in Europe range from 0.05 mGy/d for terrestrial and aquatic vertebrates, to 1.7 mGy/d for plants, and 4.8 mGy/d for terrestrial and aquatic invertebrates. Given uncertainties associated with limited datasets, the most robust value is the generic PNEDR of 0.24 mGy/d.

REFERENCES

Agency for Toxic Substances and Disease Registry (ATSDR). 2001. Draft Toxicological Profile for Cesium. United States Department of Health and Human Services, Public Health Service, Agency for Toxic Substances and Disease Registry. July 2001. 264 pp.

Abraham, R. L. 1972. Mortality of mallards exposed to gamma radiation. *Rad. Res.* 49:322–327.

Alonzo, F., R. Gilbin, S. Bourrachot, M. Floriani, M. Morello, and J. Garnier-Laplace. 2006. Effects of chronic internal alpha irradiation on physiology, growth, and reproductive success of *Daphnia magna. Aquatic Toxicol.* 80:228–236.

Alonzo, F., R. Gilbin, F. A. Zeman, and J. Garnier-Laplace. 2008. Increased effects of internal alpha irradiation on *Daphnia magna* after chronic exposure over 3 successive generations. *Aquatic Toxicol.* 87:146–156.

Anderson, S. L., and G. C. Wild. 1994. Linking genotoxic responses and reproductive success in ecotoxicology. *Environ. Health Perspect.* 102 Suppl. 12:9–12.

Andersson, P., et al. 2008. Numerical benchmarks for protecting biota against radiation in the environment: proposed levels and underlying reasoning. Deliverable 5B, Protection of the Environment from Ionizing Radiation in a Regulatory Context (PROTECT). European Commission, 6th Framework, Contract N° 036425 (FI6R). September 2008. http://www.ceh.ac.uk/protect/

Baker, R. J., et al. 2001. Consequences of polluted environments on population structure: the bank vole (*Clethrionomys glareolus*). *Ecotoxicology*. 10:211–216.

Barnthouse, L. R. 1995. Effects of ionizing radiation on terrestrial plants and animals: a workshop report. Prepared by the Oak Ridge National Laboratory, Prepared for the US Department of Energy under Contract DE-AC05–84OR21400. 20 pp.

Baumgartner, J. 1985. The effect of total body gamma-irradiation on mortality, egg production and egg weight of Japanese quails. *Int. Radiat. Biol. Relat. Stud. Phys., Chem. Med.* 47:591–597.

Benjamin, S. A., A. C. Lee, G. M. Angleton, W. J. Saunders, T. J. Keefe, and C. H. Mallinckrodt. 1998. Mortality in beagles irradiated during prenatal and postnatal development. II. Contribution of benign and malignant neoplasia. . *Radiat. Res.* 150:330–348.

Beresford, N. A., et al. 2008. Background exposure rates of terrestrial wildlife in England and Wales. *J. Environ. Radioact.* 99:1430–1439.

Bergonie, J., and L. Tribondeau. 1906. De quelques résultats de la radiotherapie et essai de fixation d'une technique rationnelle. *Comptes-Rendus des Séances de l'Académie des Sciences* 143:983–985

Bonisoli-Alquati, A., A. Voris, T. A. Mousseau, A. P. Møller, N. Saino, and M. D. Wyatt. 2010. DNA damage in barn swallows (Hirundo rustica) from the Chernobyl region detected by use of the comet assay. *Comp. biochem. physiol. C Toxicol. pharmacol.* 151: 271–277.

Boonstra, R., R. G. Manzon, S. Mihok, and J. L. Helson. 2005. Hormetic effects of gamma radiation on the stress of natural populations of meadow voles (*Microtus pennslvanicus*). *Env. Toxicol. Chem.* 24(2):334–343.

Bustad, L. K., L. A. George, Jr., S. Marks, D. E. Warner, C. M. Barnes, K. E. Herde, and H. A. Kornberg. 1955. *Biological effects of* ^{131}I *chronically administered to sheep*. Hanford Atomic Products Operation, Hanford, WA. HW-38757.

Calabrese, E. J., and L. A. Baldwin. 2004. The effects of gamma rays on longevity. *Biogerontology.* 1:309–319.

Caratero, A., M. Courtade, L. Bonnet, H. Planel, and C. Caratero. 1998. Effect of continuous irradiation at a very low dose on life span of mice. *Gerontology.* 44:272–276.

Caswell, H. 1996. Demography meets toxicology: untangling the population-level effects of toxic substances. In *Ecotoxicology: A hierarchical approach,* eds. M. C. Newman and C. H. Jagoe, pp. 255–292. Boca Raton, FL: Lewis Publishers.

Cooley, J. L., and F. L. Miller. 1971. Effects of chronic irradiation on laboratory populations of the aquatic snail *Physa heterostropha. Radiat. Res.* 47:716.

Copplestone, D., S. Bielby, S. R. Jones, D. Patton, P. Daniel, and I. Gize. 2001. Impact Assessment of Ionizing Radiation on Wildlife. Environment Agency R&D Publication 128. 222pp.

Delistraty, D. 2008. Radioprotection of non-human biota. *J. Environ. Radioactiv.* 99:1863–1869.

Environment Canada. 1997. Environmental assessments of Priority Substances under the *Canadian Environmental Protection Act.* Guidance manual version 1.0—March 1997. Chemical Evaluation Division, Commercial Chemicals Evaluation Branch, Hull, Quebec (EPS 2/CC/3E).

EC (European Commission). 2003. Technical guidance document in support of Commission Directive 93/67/ EEC on risk assessment for new notified substances and Commission Regulation (EC) No 1488/94 on risk assessment for existing substances, Directive 98/8/EC of the European Parliament and of the Council concerning the placing of biocidal products on the market. Part II. Office for Official Publication of the European Communities, Luxembourg.

Eisler, R. 1994. Radiation hazards to fish, wildlife, and invertebrates: a synoptic review. Biological Report 26, Contaminant Hazard Reviews Report 29. December.

Ellegren, H., G. Lindgren, C. R. Primmer, and A. P. Moller. 1997. Fitness loss and germline mutations in barn swallows breeding in Chernobyl. *Nature*. 389:593–596.

Environment Canada and Health Canada. 2003. Releases of Radionuclides from Nuclear Facilities (Impact on Non-human Biota). Priority Substances List Report. Canadian Environmental Protection Act, 1999.

FASSET. 2004. Framework for assessment of environmental impact. Final Report. European Commission, 5th Framework, Contract N°FIGE-CT-2000-00102.

French, N. R., B. G. Maza, H. O. Hill, A. P. Achwanded, and H. W. Kaaz. 1974. A population study of irradiated desert rodents. *Ecol. Monogr*. 44:45–72.

Garnier-Laplace, J., and R. Gilbin. 2006. Derivation of predicted no effect dose rate values for ecosystems (and their sub-organizational levels) exposed to radioactive substances. Deliverable D5. European Commission, 6th Framework, Contract N°FI6RCT-2003-508847.

Garnier-Laplace, J., C. Della-Vedova, R. Gilbin, D. Copplestone, J. Hingston, and P. Ciffroy. 2006. First derivation of predicted-no-effect values for freshwater and terrestrial ecosystems exposed to radioactive substances. *Environ. Sci. Technol*. 40:6498–6505.

Geras'kin, S. A., S. V. Fesenko, and R. M. Alexakhin. 2008. Effects of non-human species irradiation after the Chernobyl NPP accident. *Environ. Int*. 34:880–897.

Ghiassi-nejad, M., S. M. J. Mortazavi, J. R. Cameron, A. Niroomand-rad, and P. A. Karam. 2002. Very high background radiation areas of Ramsar, Iran: preliminary biological studies. *Health Phys*. 82(1):87–93.

Gilbin, R., F. Alonzo, and J. Garnier-Laplace. 2008. Effects of chronic external gamma irradiation on growth and reproductive success of *Daphnia magna*. *J. Environ. Radioactiv*. 99:134–145.

Golley, F. B., J. B. Gentry, E. F. Menhinick, and J. L. Carmon. 1965. Response of wild rodents to acute gamma radiation. *Rad. Res*. 24:350–356.

Green, E. L. 1964. Reproductive fitness of irradiated populations of mice. *Genetics*. 50:423–434.

Gridley, D. S. M. J. Pecaut, and G. A. Nelson. 2002. Total-body irradiation with high-LET particles: acute and chronic effects on the immune system. A. *J. Physiol. Regul. Integr. Comp. Physiol*. 282:R677–R688.

Harrison, F. L., and S. L. Anderson. 1996. Taxonomic and developmental aspects of radiosensitivity. Lawrence Livermore National Laboratory. UCRL-JC-125920.

Hertel-Aas, T., D. H. Oughton, A. Jaworska, H. Bjerke, B. Salbu, and G. Brunborg. 2007. Effects of chronic gamma irradiation on reproduction in the earthworm *Eisenia fetida* (Oligochaeta). *Radiat. Res*. 168:515–528.

Higley, K. A., S. L. Domotor, and E. J. Antonio. 2003. A kinetic-allometric approach to predicting tissue radionuclide concentrations in biota. *J. Environ. Radioactiv*. 66:61–74

Hingston, J. L., M. D. Wood, D. Copplestone, and I. Zinger. 2005. Impact of chronic low-level ionising radiation exposure on terrestrial invertebrates. *Radioprotection* 40 (Suppl. 1):S145–S150.

Hinton, T. G., D. P. Coughlin, Y. Yi, and L. C. Marsh. 2004a. Low dose rate irradiation facility: initial study on chronic exposures to medaka. *J. Environ. Radioactiv*. 74:43–55.

Hinton, T. G., J. S. Bedford, J. C. Congdon, and F. W. Whicker. 2004b. Effects of radiation on the environment: a need to question old paradigms and enhance collaboration among radiation biologists and radiation ecologists. *Radiat. Res*. 162:332–338.

International Atomic Energy Agency (IAEA). 1992. Effects of ionizing radiation on plants and animals at levels implied by current radiation protection standards. Technical Reports Series 332. International Atomic Energy Agency, Vienna, Austria. 54 pp.

International Commission of Radiological Protection (ICRP). 1977. Recommendations of the International Commission of Radiological Protection. Ann. ICRP 1, No. 3, Pergamon Press, Oxford. Reprinted (with additions) in 1987. Superseded by ICRP Publication 60.

International Commission of Radiological Protection (ICRP). 1991. 1990 Recommendations of the International Commission of Radiological Protection. ICRP Publication 60. Ann. ICRP 21 (1–3), Oxford: Pergamon Press.

International Commission of Radiological Protection (ICRP). 2008. Environmental protection: the concept and use of reference animals and plants. Consultation draft from http://www.icrp.org (not to be cited as a finalized document).

Jackson, D., D. Copplestone, D. M. Stone, and G. M. Smith. 2005. Terrestrial invertebrate population studies in the Chernobyl exclusion zone, Ukraine. *Radioprotection* 40 (Suppl. 1):S857–S863.

Johnson, K. P., J. Stout, I. L. Brisbin, Jr., R. M. Zink, and J. Burger. 1999. Lack of demonstratable effects of pollutants on Cyt b sequences in wood ducks from a contaminated nuclear reactor cooling pond. *Environ. Res.* 81(A):146–150.

Jones, S. 2008. Windscale and Kyshtym: a double anniversary. *J. Environ. Radioactiv.* 99:1–6.

Jones, C., and M. Gilek. 2004. Overview of programmes for the assessment of risks to the environment from ionising radiation and hazardous chemicals. *J. Radiol. Prot.* 24:A157–A177.

Knowles, J. F. 2002. An investigation into the effects of chronic radiation on fish. Research and Development Technical Report P3–053/TR. Centre for Environment, Fisheries, and Aquaculture Science. Environment Agency, UK. Research and Development Technical Report P3-053/TR.

Knowles, J. F., and L. N. Greenwood. 1996. A comparison of the effects of long-term beta and gamma irradiation on the reproductive performance of a marine invertebrate *Ophryotrocha diadema* (Polychaeta, Dorvilleidae). *J. Environ. Radioactiv.* 34:1–7.

Larsson, C-M. 2004. The FASSET framework for assessment of environmental impacts of ionizing radiation in European ecosystems—an overview. *J. Radiol. Prot.* 24:A1–A12.

Malhotra, N., N. Rani, K. Rana, and R. K. Malhotra. 1990. Radiation induced blood pathology in chick erythrocytes and related parameters. *Exp. Path.* 38:241–248.

Marshall, J. S. 1962. The effects of continuous gamma radiation on the intrinsic rate of natural increase of *Daphnia pulex*. *Ecology* 43:598.

Marshall, J. S. 1963. The effect of continuous, sub-lethal gamma radiation on the intrinsic rate of natural increase and other population attributes of *Daphnia pulex*. In *Radioecology*, eds. V. Schultz and A. W. Klement, Jr, pp. 363–366. New York, NY: Reinhold.

Marshall, J. S. 1966. Population dynamics of *Daphnia pulex* as modified by chronic radiation stress. *Ecology* 47:561.

Meeks, H. N., J. K. Wickliffe, S. R. Hoofer, R. K. Chesser, B. E. Rodgers, and R. J. Baker. 2007. Mitochondrial control region variation in bank voles (*Clethrionomys glareolus*) is not related to Chernobyl radiation exposure. *Environ. Toxicol. Chem.* 26:361–369.

Mellinger, P. J., and V. Schultz. 1975. Ionizing radiation and wild birds: a review. *CRC Crit. Rev. Environ. Cont.* 5(3):397–421.

Mihok, S. 2004. Chronic exposure to gamma radiation of wild populations of meadow voles (*Microtus pennsylvanicus*). *J. Environ. Radiat.* 75:233–266.

Millard, J. B., and F. W. Whicker. 1990. Radionuclide uptake and growth of barn swallows nesting by radioactive leaching ponds. *Health Phys.* 58:429–439.

Mole, R. H. 1990. The effect of prenatal radiation exposure on the developing human brain. *Int. J. Radiat. Biol.* 57:647–663.

Moller, A. P., T. A. Mousseau, F. de Lope, and N. Saino. 2007. Elevated frequency of abnormalities in barn swallows from Chernobyl. *Biol. Lett.* 3:414–417.

Moller, A. P., T. A. Mousseau, F. de Lope, and N. Saino. 2008. Anecdotes and empirical research in Chernobyl. *Biol. Lett.* 4:65–66.

Moller, A. P., and T. A. Mousseau. 2009. Reduced abundance of insects and spiders linked to radiation at Chernobyl 20 years after the accident. *Biol. Lett.* 5:356–359.

Mraz, F. R., and M. C. Woody. 1973. The effect of dose rate upon reproductive performance of white leghorn hens and their progeny. *Radiat. Res.* 54:549–555.

Nakamori, T., S. Yoshida, Y. Kubota, T. Ban-nai, N. Kaneko, M. Hasegawa, and R. Itoh. 2008. Effects of acute gamma irradiation on *Folsomia candida* (Collembola) in a standard test. *Ecotox. Environ. Safe.* 71:590–596.

National Council on Radiation Protection (NCRP). 1991. Effects of ionising radiation on aquatic organisms, NCRP Report No 109. National Council on Radiation Protection and Measurement, Washington D.C.

National Council on Radiation Protection (NCRP). 2009. Ionising radiation exposure of the population of the United States, NCRP Report No 160. National Council on Radiation Protection and Measurement. Bethesda, MD.

O'Farrell, T. P., J. D. Hedlund, R. J. Olson, and R. O. Gilbert. 1972. Effects of ionizing radiation on survival, longevity, and reproduction in free ranging pocket mice, *Perognathus parvus*. *Rad. Res.* 49:611–623.

O'Farrell, T. P., J. D. Hedlund, R. J. Olson, and R. O. Gilbert. 1973. Radiation effects in free-ranging pocket mice, *Perognathus parvus*, during breeding season. *Science.* 179:291–293.

O'Neill, P., and P. Wardman. 2009. Radiation chemistry comes before radiation biology. *Int. J. Radiat. Biol.* 85:9–25.

Phillips, L. J., and J. E. Coggle. 1988. The radiosensitivity of embryos of domestic chickens and black-headed gulls. *Int. J. Radiat. Biol.* 53:309–317.

Pollard, E. C. 1969. The biological action of ionizing radiation. *Am. Sci.* 57:206–236.

Real, A., S. Sundell-Bergmann, J. K. Knowles, D. S. Woodhead, and I. Zinger. 2004. Effects of ionizing radiation exposure on plants, fish, and mammals: relevant data for environmental radiation protection. *J. Radiol. Prot.* 24:A123–A137.

Rodgers, B. E., and R. J. Baker. 2000. Frequencies of micronuclei in bank voles from zones of high radiation at Chornobyl, Ukraine. *Environ. Toxicol. Chem.* 19(6):1644–1648.

Rodgers, B. E., R. K. Chesser, J. K.Wickliffe, C. J. Phillips, and R. J. Baker. 2001a. Subchronic exposure of BALB/c and C57BL/6 strains of *Mus musculus* to the radioactive environment of the Chornobyl, Ukraine exclusion zone. *Environ. Toxicol. Chem.* 20(12):2830–2835.

Rodgers, B. E., J. K. Wickliffe, C. J. Phillips, R. K. Chesser, and R. J. Baker. 2001b. Experimental exposure of naïve bank voles (*Clethrionomys glareolus*) to the Chornobyl, Ukraine, environment; a test of radioresistance. *Environ. Toxicol. Chem.* 20(9):1936–1941.

Ryabokon, N. I., and R. I. Goncharova. 2006. Transgenerational accumulation of radiation damage in small mammals chronically exposed to Chernobyl fallout. *Radiat. Environ. Biophys.* 45:167–177.

Sazykina, T. G, A. Jaworska, and J. Brown. 2003. Report of dose-effects relationships for reference (or related) Arctic biota. EPIC database "Radiation effects on biota". A deliverable report for EPIC (Environmental Protection from Ionizing Contaminants) Project CIA2-CT-2000–10032.

Sazykina, T. G. 2005. A system of dose-effects relationships for the northern wildlife: radiation protection criteria. *Radioprotection* 40 (Suppl. 1):S889–S892.

Sazykina, T. G., and I. I Kryshev. 2006. Radiation effects in wild terrestrial vertebrates—the EPIC collection. *J. Environ. Radioactiv.* 88:11–48.

Smith, J. T. 2005. The case against protecting the environment from ionizing radiation. *Radioprotection* 40 (Suppl. 1):S967–S972.

Smith, J. T. 2008. Is Chernobyl radiation really causing negative individual and population-level effects on barn swallows? *Biol. Lett.* 4:63–64.

Sugahara, T. 1964. Genetic effects of chronic irradiation given to mice through three successive generations. *Genetics* 50:1143–1158.

Sugg, D. W., et al. 1996. DNA damage and radiocesium in channel catfish from Chernobyl. *Environ. Toxicol. Chem.* 15:1057–1063.

Talmage, S. S., and L. Meyers-Schone. 1995. Nuclear and thermal. In *Handbook of ecotoxicology,* eds. J. Hoffman, B. A. Rattner, G. A. Burton, and J. Cairns, Jr., pp. 469–491. Boca Raton, FL: Lewis Publishers.

Tanaka, I. B., III, S. Tanaka, K. Ichinohe, S. Matsushita, T. Matsumoto, H. Otsu, Y. Oghiso, and F. Sato. 2007. Cause of death and neoplasia in mice continuously exposed to very low dose rates of gamma rays. *Radiat. Res.* 167:417–437.

Tester, J. R., F. McKinney, and D. B. Siniff. 1968. Mortality of three species of ducks—*Anas discors, A. crecca* and *A. clypeata*—exposed to ionizing radiation. *Radiat. Res.* 33:364–370.

Trabalka, J. R., and C. P. Allen. 1977. Aspects of fitness of a Mosquitofish *Gambusia affinis* population exposed to chronic low-level environmental radiation. *Radiat. Res.* 70:198–211.

Tsytsugina, V. G., and G. G. Polikarpov. 2003. Radiological effects on populations of Oligochaeta in the Chernobyl contaminated zone. *J. Environ. Radioactiv.* 66:141–154.

Ukrainian National Academy of Sciences (UNAS). 1997. Organization and supervision of the efforts aimed to mitigate social and economical consequences of the Chernobyl catastrophe. In. Chernobyl Disaster. Kiev, Ukraine. pp. 51–120.

UNSCEAR. 1996. Sources and Effects of Ionizing Radiation. United Nations Scientific Committee on the effects of atomic radiation (UNSCEAR). Report to the General Assembly, United Nations, New York, 17 pp + Annexes.

U.S. Department of Energy (U.S. DOE). 2002. A graded approach for evaluating radiation doses to aquatic and terrestrial biota. DOE-STD-1153–2002.

U.S. Department of Energy (U.S. DOE). 2009. RESRAD-BIOTA for Windows version 1.5. Argonne National Laboratory, U.S. Department of Energy. Argonne, IL. May 11, 2006. http://web.ead.anl.gov/resrad/home2/ biota.cfm

U.S. Environmental Protection Agency (U.S. EPA). 1997. Ecological risk assessment guidance for superfund: process for designing and conducting ecological risk assessments, interim final. Office of Solid Waste and Emergency Response. EPA 540-R-97–006. June.

U.S. Environmental Protection Agency (U.S. EPA). 1998. *Final Guidelines for Ecological Risk Assessment, Risk Assessment Forum.* U.S. EPA, Washington, DC, EPA/630/R-95/002F. April.

Walker, J. S. 2004. *Three mile island: A nuclear crisis in historical perspective.* Berkeley, CA: University of California Press.

Wickliffe, J. K., R. K. Chesser, B. E. Rodgers, and R. J. Baker. 2002. Assessing the genotoxicity of chronic environmental irradiation by using mitochondrial DNA heteroplasmy in the bank vole (*Clethrionomys glareolus*) at Chornobyl, Ukraine. *Environ. Toxicol. Chem.* 21(6):1249–1254.

Wickliffe, J. K., et al. 2003. Exposures to chronic, low-dose rate gamma-radiation at Chornobyl does not induce point mutations in Big Blue® mice. *Environ. Molec. Mut.* 42:11–18.

Wiggins, L. E., R. A. Van den Bussche, M. J. Hamilton, R. K. Chesser, and R. J. Baker. 2002. Utility of chromosomal position of heterochromatin as a biomarker of radiation-induced genetic damage: a study of Chornobyl voles (*Microtus* sp.). *Ecotoxicology.* 11:147–154.

Willard, W. K. 1963. Relative sensitivity of nestlings of wild passerine birds to gamma radiation. In *Radioecology*, eds. V. Schultz and A. W. Klement, Jr., pp. 345–349. New York, NY: Reinhold.

Wood, M. D., J. L. Hingston, D. Copplestone, and I. Zinger. 2005. Development of experimental protocols for chronic irradiation studies: the application of a good practice guide framework. *Radioprotection* 40 (Suppl. 1):S229–S234.

Woodhead, D. S. 1998. The impact of radioactive discharges on native British wild-life and the implications for environmental protection. R&D Technical Report P135, Environment Agency, Bristol, 80pp.

Zach, R,. and K. R. Mayoh. 1982. Breeding biology of tree swallows and house wrens in a gradient of gamma radiation. *Ecology* 63(6):1720–1728.

Zach, R., and K. R. Mayoh. 1984. Gamma radiation effects on nestling tree swallows. *Ecology* 65(5):1641–1647.

Zach, R., and K. R. Mayoh. 1986a. Gamma irradiation of tree swallow embryos and subsequent growth and survival. *Condor* 88:1–10.

Zach, R., and K. R. Mayoh. 1986b. Gamma radiation effects on nestling house wrens: a field study. *Radiat. Res.* 105:49–57.

Zach, R., J. L. Hawkins, and S. C. Sheppard. 1993. Effects of ionizing radiation on breeding swallows at current radiation protection standards. *Environ. Toxicol. Chem.* 12:779–786.

Zakaria, A. H. 1991. Effect of low doses of gamma irradiation prior to egg incubation on hatchability and body weight of broiler chickens. *Brit. Poult. Sci.* 32:103–107.

Index

A

Absorbed dose, *see also* Gray (Gy)
    nonhuman biota, 706
    quantification, 705
Absorption, distribution, metabolism, and excretion
    (ADME), 386
AC, *see* Adobe Creek (AC)
ACE, *see* Angiotensin converting enzyme (ACE)
Acquired dose, 3
    internal, 258
Active pharmaceutical ingredients (APIs), 288, 313
    aquatic exposure, 293–294
    aquatic plants uptake, 314–315
    aquatic tissue levels, 297–300
    background, 289–295
    BCFs, 304
    binding to melanin, 292
    bioconcentration data, 317–324
    DBPs and metabolites, 314
    as environmental pollutants, 289
    general considerations, 324–325
    Huggett model, 331–333
    human plasma PTD values, 330
    hydrology, 325–326
    multianalyte studies, 315–317
    PBPK models, 331
    predictive modeling, 300–301
    properties, 289
    published data summary, 317
    residue data, 301
    select site-specific factors, 325
    site-specific pH and API pKa, 327
    sources, 295–297
    tissue levels, 335–337
    tissue sampling and surrogate monitoring, 328–329
    variables, 297
    wastewater treatment, 326–327
    in wild-caught fish, 316
Active toxicant fraction, 266
Acute exposure, 565, 708
    DDT, 58
    heptachlor, 77
    lead, 566, 596
    mirex, 84
    to OC pesticides, 448
    radiation, 719, 726
    toxaphene, 83
Additive model, 467
Additivity, 113, 270
Adenosine triphosphate (ATP), 268
ADME, *see* Absorption, distribution, metabolism, and
    excretion (ADME)
Administered dose, 3, 258

Adobe Creek (AC), 213
Adult survival, 505, 507
    PCB and PCDD effects on, 509
AE, *see* Assimilation efficiency (AE)
African adjutant, 608
Agelaius phoeniceus, see Red-winged blackbirds
Agency for Toxic Substances and Disease Registry
    (ATSDR), 365
Agricultural seeds, 18
Agrostis stolonifera, see Creeping bent grass
Ah, *see* Aryl hydrocarbon (Ah)
Ah-gene battery, 148
AHH, *see* Aryl hydrocarbon hydroxylase (AHH)
AhR, *see* Aryl hydrocarbon receptor (AhR)
AhR nuclear transporter (ARNT), 358
Air monitoring, 52
ALA, *see* δ-aminolevulinic acid (ALA)
ALAD, *see* δ-aminolevulinic acid dehydratase (ALAD)
Aldicarb sulfoxide, 416
Aldrin, 72, 449; *see also* Cyclodienes; Dieldrin;
    Organochlorine (OC); Organohalogen (OH)
    compounds
    in fish, 72
    invertebrate CBRs, 73
    properties, 50
    toxicity to birds, 448
    in United States, 49
Ameiurus nebulosus, see Brownbullhead
American avocets (*Recurvirostra americana*), 668
    EC10 for teratogenesis, 15
    teratogenesis function, 682
American mink (*Mustela vison*), 535
    PCB effects on, 13
American robin (*Turdus migratorius*), 451
    cyclodiene residues in, 461
    DDT, DDD, and DDE residues in, 430
    DDT and DDD effects, 426, 427
    dieldrin mortality, 453
    liver lead residues, 580
    PCBs and dioxins effects, 24, 506
AMG, *see* Autometallography (AMG)
Analysis of variance (ANOVA), 114
Anas platyrhynchos, see Mallards
Anemia, 601
    ALAD inhibition, 569
    cadmium induced, 655
    lead-induced, 601
Angiotensin converting enzyme (ACE), 654
Anomala orientalis, see Oriental beetle
ANOVA, *see* Analysis of variance (ANOVA)
Antagonistic interactions
    As and Se, 691
    Hg and Se interactions, 611
    Se and vitamins, 691

Antibiotics, 295, 310–311
  in aquaculture, 314
  fluoroquinolone, 327
APIs, *see* Active pharmaceutical ingredients (APIs)
Apodemus sylvaticus, see Wood mice
Aquatic exposure, 325
    APIs, 288, 289, 295–296, 310
    sources/origins of APIs, 295–296
    uses, 293–294
    variables, 297, 298–299
Aquatic wildlife, 18, 19
Arctic marine mammals, 535
Ardea herodias, see Great blue heron
ARNT, *see* AhR nuclear transporter (ARNT)
Aroclor 1254
    dietary dosing studies, 485, 490, 491, 538
    ecotoxicity data, 153–154
    exposure species sensitivity, 155
Aryl hydrocarbon (Ah), 24; *see also* Dioxins
Aryl hydrocarbon hydroxylase (AHH), 481
Aryl hydrocarbon receptor (AhR), 148, 356, 478, 532
    dimerization, 358
    dioxin toxicity, 356, 480
    DNA sequencing, 481
    function, 535
    LBD in wild birds, 482
    ligands, 148
Assimilation efficiency (AE), 235
    influencing factors, 58
    laboratory measurement, 243
    metals and food types, 235
    methylmercury, 171
ATP, *see* Adenosine triphosphate (ATP)
ATSDR, *see* Agency for Toxic Substances and Disease
        Registry (ATSDR)
Autometallography (AMG), 386
Aythya collaris, see Ring-necked ducks
Azocyclotin (ACT), 257

B

Background, 1, 565, 670; *see also* Lead
        (Pb)—exposure
    blood lead concentrations, 570
    levels of Se, 671
BAF, *see* Bioaccumulation factor (BAF)
Balaena mysticetus, see Bowhead whale
Bald eagle (*Haliaeetus leucocephalus*), 428
    DDE, 428, 434
    dieldrin, 459
    ecological variables, 26
    Hg exposure, 19
    lead, 566, 574, 576
    effect of PCB, 500, 505
    PCBs and dioxins effects, 23
Barbed hedgehog, 168
Barnacles (Cirripedia), 232
    Cu requirement, 240
    lethal body concentration, 244
    trace metal accumulation, 232
    Zn accumulation, 240, 241–242, 246
Bass, common, 192
BCF, *see* Bioconcentration factor (BCF)
Becquerel (Bq), 705

Beluga whale (*Delphinapterus leucas*), 378, 383
    Ag accumulation, 395
    Cd accumulation, 394
    contamination assessments studies, 387
    effects of PCB, 540
    Hg exposure, 385
    PBT exposure and CYP1A activity, 362
Benthic organisms, 66
Benzene hexachloride (BHC), 81, 465; *see also*
        Hexachlorocyclohexane (HCH);Lindane
BEST, *see* Biomonitoring of Environmental Status and
        Trends (BEST)
β-blockers, 302, 305–306
BFR, *see* Brominated flame retardant (BFR)
BHC, *see* Benzene hexachloride (BHC); γ-isomer of HCH
        (BHC)
Bioaccumulation, 57, 110; *see also* Bioconcentration
    in aquatic invertebrates, 245, 248
    BSAF, 289
    chemical structure and, 110
    ciprofloxacin, 311
    DDT, 58
    factors influencing, 417
    in fish, 61
    lead, 598
    methylmercury, 171
    OH compounds, 56
    organism physiology and, 418
    organotins, 263
    pesticide, 411
    selenium, 195
Bioaccumulation factor (BAF), 56, 109, 110, 410; *see also*
        Bioconcentration, 109
    data for APIs in aquatic tissues, 317–324
    diclofenac, 302, 305
    of fluoxetine, 303, 316
    predictive modeling, 300
Bioconcentration factor (BCF), 56, 109, 110, 410, 413
    for APIs, 333
    field-based Se, 195–196
    fluoxetine, 303
    for polychlorinated compounds, 110
    for selective serotonin reuptake inhibitor, 328
    TBT, 263
    2,3,7,8-TCDD and aroclors, 111
    unit, 288
Biodynamic modeling, 236
    application, 243
    bioaccumulation, 245
    explanatory power, 244, 246
Bioindicator, *see* Biomarker
Biological molecules, 707
    ionizing radiation effect, 707
    radiosensitive growing cell, 708
    ROS and free radicals effect, 707
Biomagnification, 60, 109
    DDT, 60
    veterinary drugs, 310
Biomagnification factor (BMF), 60
    calculation, 410
Biomarker, 290
    application, 15
    endpoint measurements, 15
    of exposure, 291

lead, 596, 598, 603
in marine mammals, 358
PBT exposure, 362
for PCB monitoring, 26
selenium, 198
vitellogenin, 62
Biomonitoring of Environmental Status and Trends
(BEST), 49
Biomonitors, 219, 220
aquatic invertebrate, 248
bryophytes, 315
harbor seal, 390
Biota-sediment accumulation factor (BSAF), 109, 110,
289, 313, 365
Biotransformation, 58; *see also* Detoxification
in aquatic animals, 413–415
components, 414
DDT, 58
elimination of contaminants, 358
Hg, 384, 390
insecticide, 414, 416
OH compounds, 57
via P450, 265
potential, 415
processes in aquatic animals, 413
as protective mechanism, 72
toxicological responses to, 416
Bisphenol A (BPA), 360
Blackbirds (*Icteridae*), 460
Black-crowned night herons (*Nycticorax nycticorax*), 439,
440, 502
Blubber-rich ringed seal (*Phoca hispida*), 354
Bluegill sunfish (*Lepomis macrochirus*), 62, 72, 194
concentration-response relationships, 217
DDT, 62
direct dietary Se, 209, 210, 211
exposure, 197, 209, 211
mirex, 81
Se effects, 217
Se exposure, 194, 196–197
Se maternal transfer, 201, 202, 204
survival, 225
toxicity test conduction, 211
BMF, *see* Biomagnification factor (BMF)
Bobwhite quail (*Colinus virginianus*), 612
DDD, 426
dieldrin lethal levels, 452
lead, 579
PCBs and dioxins, 24
pesticides toxicity, 448
Bottlenose dolphin (*Tursiops truncatus*), 313
CYP1A1 expression, 362
Hg concentrations, 382, 384
PAH toxicity, 359
PCB, 540
Bowhead whale (*Balaena mysticetus*), 382, 395
BPA, *see* Bisphenol A (BPA)
Bq, *see* Becquerel (Bq)
Brominated flame retardant (BFR), 107, 357;
see also Polybrominated diphenyl
ether (PBDE)
Brown pelican (*Pelecanus occidentalis*), 439
blood lead concentrations, 582
DDE critical dietary level, 439

DDE residues, 432, 435, 436, 437
endrin, 460
Brownbullhead (*Ameiurus nebulosus*), 303
BSAF, *see* Biota-sediment accumulation factor (BSAF)
Bursa of fabricus, 503
in aquatic organisms, 262
Butyltins, 265; *see also* Phenyltins
activators of RXR, 270
neurotoxic effects, 276
tissue concentration, 261

C

Cadmium (Cd), 393–394, 645, 659
absorption and retention, 635, 646
anemia, 655
behavioral alterations, 655
in birds, 645
in blood, 659
Cetaceans, 394
contamination, 649
critical concentration, 659
critical tissue concentrations, 636–638
dietary, 633
dietary exposure, 651
dose effects, 637
effects, 628, 629
elimination route, 646
endocrine disruption, 654–655
energy metabolism, 653
exposure, 629, 652, 659
in feathers, 648, 649
fume, 628
in grasslands, 630
interception and retention, 630
intestinal damage and nutrient uptake, 651–652
Itai-itai, 628
in kidney, 628, 632–633, 647, 648–650, 652
kidney damage and vitamin D metabolism, 652
in liver, 632–633, 649, 650
in mammals, 627
metabolism and toxicity, 628–629
metallothionein induction, 647–648
to MT binding, 628
Mysticetes, 395–396
nephrotoxicity, 628
Odontocetes, 394–395
osmoregulation, 653
renal and hepatic thresholds, 657, 660
reproduction, 654
seabirds, 648, 649–650
skeletal effects, 652–653
terrestrial habitats, 629–631
tissue concentration, 635
tissue concentrations, 655–659
tissue distribution and accumulation, 648–651
toxic effects, 651
toxicity, 636
uptake and elimination, 646–647
variability in crustaceans, 233
vitamin D3 and, 652
wild birds, 649
wild species, 631–635
willow, 650

Camphechlor, *see* Toxaphene
Canadian Council of Ministers of the Environment (CCME), 365
Canadian Environmental Protection Act (CEPA), 365
Capreolus capreolus, see European roe deer
Carbamazepine (CBZ), 302, 311
Carbon (C), 260
    isotope analyses, 382, 484
Carolina chickadee (*Parus carolinensis*), 503
Catfish (Siluridae), 58; *see also* Channel Catfish, 32
    DDT, 58
    mirex, 84
    selenium, 196
Cause-and-effect relation, 3, 425
CBR, *see* Critical body residue (CBR)
CBZ, *see* Carbamazepine (CBZ)
CCME, *see* Canadian Council of Ministers of the Environment (CCME)
Cd, *see* Cadmium (Cd)
CEE, *see* Conjugated equine estrogen (CEE)
Central nervous system (CNS), 383, 595
    cyclodienes, 449
    mechanism, 449
    OH compounds, 61
CEPA, *see* Canadian Environmental Protection Act (CEPA)
CF, *see* Contamination factor (CF)
Channel Catfish, 302
Chemical toxicity distribution (CTD), 329
Chick edema disease, 21
Chinook salmon (*Oncorhynchus tshawytscha*), 58, 211
    concentration-response relationships, 218
    DDT concentrations, 54
    PBDE, 127, 128
    Se toxicity, 210
Chlordane, 76, 460; *see also* Cyclodienes; Organohalogen (OH) compounds
    CBR ranges and medians, 65
    critical values, 464
    fish, 77
    fish CBRs, 79
    invertebrates, 77
    metabolism, 58
    metabolite half-lives, 450
    properties, products and uses, 50
    toxicity to birds, 448, 463
Chlorinated pesticide, 541
Chlorination, 314
Cholinesterase activity, 14
Chronic exposure, 565, 708, *see also* Acute exposure
    cadmium, 393
    DDT, 58
    lead, 566, 604
    OC pesticides, 448
    PAHs, 359
    radiation, 724–725
Ci, *see* Curies (Ci)
Cirripedia, *see* Barnacles
CL, *see* Confidence limit (CL)
CNS, *see* Central nervous system (CNS)
Cold condensation hypothesis, 534–535
Colinus virginianus, see Bobwhite quail
Common loons (*Gavia immer*), 613
    gaviiformes, 581
Confidence limit (CL), 551
    calculation, 3

Conjugated equine estrogen (CEE), 307
Contaminated prey, highly, 538
Contamination factor (CF), 86
Continental field-vole, 702
Copper (Cu), 239; *see also* Cadmium; Trace metals
    accumulation, 233, 241, 242
    amphipod crustacean, 242
    concentrations among crustaceans, 233
    detoxified component, 239
    dietary exposure, 242
    essential requirements, 239
    excretion in crustaceans, 236
    metabolic requirements, 238
    variability in crustaceans, 233
Cormorants, 484
    AhR isoforms, 481
    common, 476
    DDE residues in eggs, 432, 436
    PCB effects, 21, 23, 492, 508
Corticosterol, 715
Coturnix coturnix japonica, see Japanese quail
Creeping bent grass (*Agrostis stolonifera*), 631
    cadmium, 630
Critical body residue (CBR), 3, 49, 279, 288
    in *Chironomus dilutus*, 414
    concept, 325
    in fish, 412–413
    fish and invertebrate species, 412–413
    in fish and invertebrate species, 412–414
    lowest invertebrate, 90
    measurement, 411
    for OH compounds, 65
    organochlorine pesticides, 79, 80
Critical concentration, 2, 3, 629
    cadmium, 636
    dieldrin in brain, 451
    from field studies, 4
    for PCBs and DDE, 26
Critical organ, 2
Critical organ concentration, 2
CTD, *see* Chemical toxicity distribution (CTD)
Curies (Ci), 705
Cyclodienes, 49, 449
    action mechanism, 450
    acute toxicity, 467
    aldrin, 72
    in brains of birds, 461–463
    chlordane, 76–77, 460
    CNS mechanism, 449, 450
    dieldrin, 72, 449, 450–451
    endosulfan, 75–76
    endrin, 74–75, 460
    field studies, 461–463
    fish, 72–73, 75, 76, 77, 78, 79
    heptachlor, 77, 463
    heptachlor epoxide, 77
    in invertebrates, 73, 74, 75, 76, 77, 78, 79, 80
    pesticides, 61
    poisoning, 460
    residues in brains, 461–463
    structure, 449
    toxicity, 467
Cygnus columbianus, see Tundra swans
CYP1A, *see* Cytochrome P4501A (CYP1A)

CYP450, *see* Cytochrome P450 monoxygenases (CYP450)
Cytochrome P450 monoxygenases (CYP450), 414
    activity in crustacean species, 415
Cytochrome P4501A (CYP1A), 480
    avian isoforms, 480
    in birds, 27
    dioxin toxicity, 148
    expression, 483
    induction in PCB-exposed birds, 481
    PBT, 358, 362

D

Dall's porpoise (*Phocoenoides dalli*), 387
    butyltins maternal transfer, 265
    Cd concentration, 394
    Hg concentration, 388
DBP, *see* Disinfection By-Product (DBP)
DDD, *see* 1,1-dichloro-2,2-bis(4-chlorophenyl) ethane (DDD)
DDE, *see* 1,1'-(2,2-dichlor-ethenylidene]-bis[4–2,2-chlorobenzene] (DDE);
    Dichlorodiphenyldichloroethane (DDE)
DDT, *see* Dichlorodiphenyl trichloroethane (DDT)
Decabromodiphenyl ether (Deca-PBDE), 107; *see also*
    Polybrominated diphenyl ether (PBDE)
Deca-PBDE, *see* Decabromodiphenyl ether (Deca-PBDE)
DEHP, *see* Di(2-ethylhexyl) phthalate (DEHP)
δ-aminolevulinic acid (ALA), 601
δ-aminolevulinic acid dehydratase (ALAD), 601
    lead effect, 568–569, 573
Detoxification, 239
    cadmium, 393, 394
    lethal body concentration, 246
    mercury, 388
    methylmercury, 182
    selenium, 379
Di(2-ethylhexyl) phthalate (DEHP), 360
Diagnosis, 3
Diagnostic, 3
Diagnostic residue, 3
Diagnostician, 3
Dibutyltin (DBT), 257
    in aquatic environments, 259
    in aquatic species, 262, 267
    as biocide, 257
    imposex, 274
    lethal responses for, 278
    organotin properties, 257
    sublethal response for, 270
    tissue concentrations, 275, 278
    toxicity, 268, 273
1,1'-(2,2-Dichlor-ethenylidene]-bis[4-2,2-chlorobenzene] (DDE), 48
    in brain, 426, 427, 428
1,1-Dichloro-2,2-bis(4-chlorophenyl) ethane (DDD), 48; *see also* Dichlorodiphenyltrichloroethane (DDT)
    in aquatic invertebrates, 64
    in eggs, 432–433, 436
    elimination rate coefficient, 57
    in liver, 429, 430
    o,p'-DDD, 69
    p,p'-DDD, 70
    rothane, 48
    in sample eggs, 437–438

Dichlorodiphenyldichloroethane (DDE), 12. *See also*
    Dichlorodiphenyltrichloroethane (DDT)
    in aquatic invertebrates, 64
    in bird brain, 427
    in bird liver, 430
    CDF and CBRs, 89
    dietary concentration, 439
    effect on insect fecundity, 63
    in eggs, 432–433, 436
    eggshell strength, 434
    eggshell thinning, 431, 432, 433
    elimination rate coefficient, 57
    estimated residues, 435
    lethal body, 72
    in liver, 429, 430
    o, p'-DDE, 68
    p,p'-DDE, 68, 69
    residue level, 428
    in sample eggs, 437–438
Dichlorodiphenyltrichloroethane (DDT), 11, 48, 63, 356, 357, 425. *See also* Pesticide
    acute toxicity, 467
    bioconcentration, 305
    in brain, 426–429
    CBR, 72
    DDE estimated residues in eggs, 435
    discussion, 70
    effects, 11, 466–467
    eggshell strength, 434–435
    eggshell thinning, 431–434
    elimination rate coefficient, 57
    endocrine effects, 62
    environmental DDT concentrations, 54–56
    estrogenic properties, 358
    fate of, 59
    fish, 63–65, 67–71
    food, 439
    in hepatic organ, 59
    insecticidal formulation, 426
    invertebrates, 66, 68, 69, 70, 71–72
    lethal residues, 426
    liver, 429–431
    maternally transferred isomers, 67
    mean lethal residue, 428
    metabolites, 48
    o, p'-DDT, 67
    other tissues, 431, 440
    p,p'-DDT, 63
    productivity, 435–439
    sublethal residues, 431
    and transformation products, 59
    in wildlife tissues, 12
Diclofenac
    APIs, 302
    effects, 17
Dieldrin, 449, 450–451
    acute effects, 451
    concentration, 72, 459
    effect on feeding, 454–455
    effect zone, 458
    entire nest failure, 458
    extrapolation equivalents to eggs, 468–469
    on feeding, 454–455, 456
    field studies, 453

Dieldrin (*Cont.*)
    in goose brain, 457
    implications for migrants, 456–457
    lethal levels, 452
    mortality, 453, 454
    on nestling survival, 458
    population level assessments, 458–459, 460
    on reproduction, 457–458
    stereoisomer, 74
    studies, 452
    sublethal effects, 451
    TEFs, 467
    toxic equivalents, 448
    toxicity, 72
    wholebody residue, 72
Diethyltin (DET), 257
    in aquatic environments, 259
    organotin properties, 257
    sublethal response for, 270
Dihexyltin (DHT), 257
    lethal responses for, 278
    organotin properties, 257
    sublethal response for, 270
Dimethyltin (DMT), 257
    in aquatic species, 267
Dioctyltin (DOT), 257, 271
    lethal responses for, 278
    organotin properties, 257
Diorganotins, 259, 278
    lethal values, 269, 271
    toxic action, 270
    variability, 267
Dioxin, 21, 103, 107; *see also* Polychlorinated biphenyl (PCB)
    with "dioxin-like" toxicity, 105
    ecotoxicity data, 149–151
    ecotoxicological effects, 127, 148, 152
    environmental fate, 107
    highest mean, 117
    mode of action, 358
    PCDD, 356
    TCDD, 21
    TEQ, 115, 116, 117, 118
    tissue residues, 113–114, 117, 118
    toxic equivalency factors, 111–112, 113
    toxicity, 16, 148
    in wild birds, 23–24
Dioxin TEQ
    edible portion of fish, 118
    fillet with skin, 116
    fillet without skin, 116
    whole body, 115
Dioxin-like, 105, 478
    contaminants, 480–481, 504
Dioxin-like PCB congeners, 16, 106
    aquatic fate, 108–109
    atmospheric fate, 107–108
    bioaccumulation, 109, 110, 111
    characteristics, 104–105
    contamination, 21
    in edible portion of fish, 123
    embryolethality, 496
    mode of action, 148
    toxic potency, 356
    toxicity to birds, 498

Diphenyltin (DPT), 257
    in aquatic species, 267
    imposex, 274
    lethal responses for, 278
    organotin properties, 257
    sublethal response for, 270
Dipropyltin (DPrT), 257
    lethal responses for, 278
    organotin properties, 257
    sublethal response for, 270
Dirty adults, 498
Dirty Dozen chemicals, 357, 365
Disinfection By-Product (DBP), 314
Distribution coefficient
    dissociative system, 327
    liposome-water, 327
    water-particle, 108
Dose, 3, 256
Dose assessment, 711
Dose metric, 411
Doses in μGy that produced a 50% effect (ED50), 726
Doses in μGy/h that produced a 10% effect (EDR10), 726
DOT, *see* Dioctyltin (DOT)
Double-crested cormorant (*Phalacrocorax auritus*), 25, 481
    DDE residues, 432, 436
    egg thinning, 433
    PCB, 492, 494
    PCBs and dioxins, 23
Drins, *see* Endrin

E

E17α-ethinylestradiol (EE2), 290
    synthetic hormone, 295
    transformation products, 314
Eastern mosquitofish (*Gambusia holbrooki*), 200
Ebers papyrus, 9
Ecotoxicological effects, 127
    accumulated metal concentrations, 231
    benthic organisms, 66
    at the community level, 249
Ecotoxicologist, 3
Ecotoxicology, 1
    gold standard, 670
    of methylmercury in freshwater fish, 170
    of Se, 696
    of white-tailed sea-eagle, 26
ED50, *see* Doses in μGy that produced a 50% effect (ED50)
EDCs, *see* Endocrine-disrupting compounds (EDCs)
EDR10, *see* Doses in μGy/h that produced a 10% effect (EDR10)
EE2, *see* E17α-ethinylestradiol (EE2)
Effect ratio (ER), 332
Effective concentration, 258, 676
Effective residue, 258
Effects concentrations, 104
    for Aroclor-1254, 153–154
    Hg affecting fish, 183
    for PCDDs, 149–150
    for PCDFs, 151
Egg yolk precursor protein, *see* Vitellogenin
Egg-exchange design, 498

Egg-injection methodology, 495
Eggshell thinning, 435
Eider ducks, 644
Embryotoxic effects, 680
Embryotoxicity, 22
    chicken, 495, 496
    CYP1A induction, 27, 481
    effects of DDE, 448
    hatching success, 494–495
    Hg, 692
    nest failure, 22
    PCBs, 13, 478
Encephalopathy, 595
Endocrine disruptors, 62, 288
    AhR, 356
    Cd, 654
    DDE, 69, 70, 448
    DDT, 62, 63, 70
    dioxin-like chemicals, 27
    genotoxic, 362
    Hg, 378
    irradiation, 715
    methoxychlor, 465
    methylmercury, 178–179, 610
    obesogen/somatogen, 277
    OCs, 410
    OH pesticides, 61, 62, 91
    organotins, 270, 273
    PCB, 362, 489, 494, 503
    PCDDs, 503
    PFOS, 364
    TBT and TPT, 279
    in thyroid gland, 62
    WWTPs in removing, 326
Endocrine system, 62
    biomagnifications, 60
    CBR, 65, 79
    elimination, 58
    elimination rate coefficient, 57
Endocrine-disrupting compounds (EDCs), 301; *see also*
    Endocrine disruptors
Endosulfan, 75, 464; *see also* Organochlorine pesticides
    (OC pesticides)
    in aquatic invertebrates, 74
    CBR, 65
    CF, 86
    fish, 76
    invertebrates, 76
    metabolic products, 58
    properties, products and uses, 50
    toxicity, 448
Endrin, 74–75, 460; *see also* Cyclodienes; Dielderin
    in aquatic invertebrates, 73
    fish, 75, 78
    fish lethal CBRs, 89, 90
    invertebrates, 75
    metabolism, 460
    properties, products and uses, 50
    structure, 449
    toxicity, 448
ENEV, *see* Estimated no-effect value (ENEV)
Environmental Protection Agency (EPA), 49
Environmental Protection from Ionizing Contaminants
    (EPIC), 711

Environmental Residue Effects Database (ERED), 63, 118
Environmental Risks from Ionizing Contaminants:
    Assessment and Management (ERICA), 711
Enzymatic indicators, 14
EPA, *see* Environmental Protection Agency (EPA)
EPIC, *see* Environmental Protection from Ionizing
    Contaminants (EPIC)
Equilenin (Eqn), 307
Equilin (Eq), 307
Equivalent dose, 705
ER, *see* Effect ratio (ER)
ERED, *see* Environmental Residue Effects Database
    (ERED)
ERICA, *see* Environmental Risks from Ionizing
    Contaminants: Assessment and Management
    (ERICA)
EROD, *see* Ethoxyresorufin-*O*-deethylase (EROD)
Erythrocyte inhibition, 568, 601
Estimated no-effect value (ENEV), 725
17β-estradiol, 63
Estrogenic compounds, 86
Ethoxyresorufin-*O*-deethylase (EROD), 481
    CYP450, 415
    induction, 481, 513
    PCB, 481
EU, *see* European Union (EU)
Eurasian kestrels (*Falco tinnunculus*), 429, 612
European roe deer (*Capreolus capreolus*), 537
European squirrel, 626
European Union (EU), 365
EVOS, *see* Exxon Valdez Oil Spill (EVOS)
Exposure, 290
    sign of, 292
Exxon Valdez Oil Spill (EVOS), 359

F

Fabricius bursa, 503
Falco, *see* Falcons
Falco tinnunculus, *see* Eurasian kestrels
Falcons (*Falco*), 468
FASSET, *see* Framework for ASSessment of
    Environmental impacT (FASSET)
FASSET Radiation Effects Database (FRED), 711
FEP, *see* Free erythrocyte protoporphyrin (FEP)
FI, *see* Food ingestion (FI)
Field exposures, 677
Field vole (*Microtus agrestis*), 631
Fifth percentile hazard concentration (HC5), 17
    calculation requirement, 3
    for SSD, 309
    use, 3
Final sink, 359
5 hour-stop feeding, 454
Flat configuration, 107
Flavin-containing monooxygenases (FMOs), 414
Fleshing greases, 21
Flipper factor, 383
Flounder (*Paralichthys olivaceus*), 274
Fluoxetine, 303
Flying fish, 102
FMOs, *see* Flavin-containing monooxygenases (FMOs)
Food chain uptake, 460
Food ingestion (FI), 366

Forster's tern (*Sterna forsteri*), 25, 481
4 Ts, 297
Framework for ASSessment of Environmental impacT (FASSET), 711
FRED, *see* FASSET Radiation Effects Database (FRED)
Free erythrocyte protoporphyrin (FEP), 601
Fresh water shrimp, 408
Fucus-like sea horse, 254
Fungicides, 306
    bioconcentration data, 319–324
    organotins, 256

G

GABA, *see* Gamma-aminobutyric acid (GABA)
Gambusia holbrooki, *see* Eastern mosquitofish
Gamma-aminobutyric acid (GABA), 62, 276
γ-isomer of HCH (BHC), 81
Gas chromatography (GC), 22
Gavia immer, *see* Loon, common
GC, *see* Gas chromatography (GC)
Genyonemus lineatus, *see* White croaker
GFAAS, *see* Graphite furnace atomic absorption spectroscopy (GFAAS)
GLEMEDS, *see* Great Lakes Embryo Mortality, Edema and Deformities Syndrome (GLEMEDS)
GLFMP, *see* Great Lakes Fish Monitoring Program (GLFMP)
GLNPO, *see* Great Lakes National Program Office (GLNPO)
Glucosidation, 415; *see also* Biotransformation
Glutathione (GSH), 392
    activities of, 181
Glutathione-S-transferase (GST), 414
Golden shiners (*Notemigonus crysoleucas*), 178
Graded approach, 710
    software application, 711
Graphite furnace atomic absorption spectroscopy (GFAAS), 596
Grasshopping, 49
Gray (Gy), 705
Great blue heron (*Ardea herodias*), 428
Great Lakes Embryo Mortality, Edema and Deformities Syndrome (GLEMEDS), 25, 498, 498–499
    cause–effect relationship, 500
    colonial water-bird studies, 499
Great Lakes Fish Monitoring Program (GLFMP), 120
Great Lakes National Program Office (GLNPO), 120
Great pipe-fish, 286
GSH, *see* Reduced glutathione (GSH)
GSSG, *see* Oxidized glutathione (GSSG)
GST, *see* Glutathione-S-transferase (GST)
Gy, *see* Gray (Gy)

H

Half-life, 704
    of blood Pb, 596
    of Cd, 646
    γ-hexachlorocyclohexane, 465
    of pesticides, 418
Haliaeetus leucocephalus, *see* Bald eagle
Halobates, *see* Sea-skaters

Harbor seal (*Phoca vitulina*), 359
    captive studies, 363
    Hg biomonitoring, 390
    methylmercury concentration, 391
    PAHs, 359
    PCB, 355, 540
    PDV, 360
    in vitro research, 363
Hatchling, 513
    European cormorants, 504
    reduction in body mass, 502
HBCD, *see* Hexabromocyclododecane (HBCD)
HC5, *see* Fifth percentile hazard concentration (HC5)
HCB, *see* Hexachlorobenzene (HCB)
HCH, *see* Hexachlorocyclohexane (HCH)
Heavy metal, 379
Hemocyanin, 238
Henry's law constant (H), 107, 108
Hepatic retinoids, 541
Heptachlor, 76, 77, 463
Heptachlor epoxide, 77, 449
    lethal concentrations, 463
    in whole-body fish, 78
Herring gull (*Larus argentatus*), 13
    feeding, 580
    field injection study, 496
    monitoring program, 13–14
    retrospective analyses, 499
Heteropneustes fossilis, *see* Catfish
Hexabromocyclododecane (HBCD), 107
Hexachlorobenzene (HCB), 464, 465
Hexachlorocyclohexane (HCH), 49, 465
Hexachlorocyclohexane, *see* Benzene hexachloride (BHC)
Hg, *see* Mercury (Hg)
HgSe, *see* Mercuric selenide (HgSe)
Histopathologist, 2
Hormesis, 277, 693, 715
Horsethief Canyon State Wildlife Area (HT), 213
Horsethief east wetland (HTEW), 213
HT, *see* Horsethief Canyon State Wildlife Area (HT)
HT site, 215
HTEW, *see* Horsethief east wetland (HTEW)
Huggett Model, 331–333
Hydrophobic organics, 60
Hypertension, 629

I

ICRP, *see* International Council for Radiation Protection (ICRP)
Ictalurus punctatus, *see* Channel Catfish
Icteridae, *see* Blackbirds
Immune function, 504
    Bursa of fabricus, 503
    o,p'-DDE exposure on, 69
    thymus, 509
Imposex
    CBR for, 268
    in molluscs, 273–274, 279
    RXR for, 277
    threshold for, 274
Industrialized ecosystems, 22
Infertility effects, 680
Inorganic mercury (IoHg), 689

Insecticide, 410
  high molecular weight chlorinated, 49
  OC, 11, 416, 447
  OP, 416
  phenylpyrazole, 414
Intelligent ecotoxicity testing, 292
International Council for Radiation Protection (ICRP), 723
International System (SI), 705
Inviable eggs, 680
  effects on biota, 713
IoHg, see Inorganic mercury (IoHg)
Ionizing radiation, 704, 705, 727
  dose–response relationships, 712
  ecological organization levels, 710
  fish and aquatic invertebrates, 721–725
  genetic effects, 713–714, 716–717
  growth, 717–718
  immune/endocrine responses, 715
  insects and invertebrates, 719–720
  irradiation, 722
  on mammals, 713
  molecular responses, 716–717
  on molecules and cells, 707–708
  mortality, 715–716, 719
  radiosensitivity, 708
  relative sensitivity to ionizing radiation, 708
  reproduction, 714–715, 718–719
  units, 705
  whole-body effects, 708
Isotopes, 704
  analysis in biologic molecules, 382
  use, 17, 60
Itai-itai disease, 628

J

Japanese quail (*Coturnix coturnix japonica*), 24, 465, 680
  Cd exposure effect, 651
  dieldrin toxicity, 451
  dietary concentrations, 633
  dietary dieldrin, 457
  Hg, 385
  Pb exposure, 569
  PCB exposure, 504

K

Kidney
  Cd, 632–633, 636
  Pb in, 603
  residues, 615
  Se, 686–687
King vulture (*Sarcoramphus papa*), 562, 576
Kow, see Octanol–water partition coefficient (Kow)
K-shell X-ray fluorescence (K-XRF), 600
K-XRF, see K-shell X-ray fluorescence (K-XRF)

L

Lagopus lagopus, see Willow ptarmigan
Largemouth bass (*Micropterus salmoides*), 78, 200, 207
  exposure, 69, 73
  ovary Se EC10, 220–221
  whole-body CBR, 78

Larus argentatus, see Herring gull
Law of Bergonie and Tribondeau, 708
LBD, see Ligand binding domain (LBD)
LC50, see Lethal concentration affecting 50% (LC50)
Lead (Pb), 379, 563, 582
  age effect, 598
  from ammunition, 564
  in avian tissues, 565–567, 568, 569–570
  bioavailability, 599
  in blood, 596, 598, 602, 603
  in bone, 600, 602
  chronic exposure, 598
  concentration, 572, 580, 581, 602–603
  critical intake, 11
  diet, 568
  droop-wing appearance in birds, 582
  exposure, 565, 599, 601
  factors influencing, 567–568
  in hard tissue, 600–601
  health hazards of, 595
  immunosuppression, 569
  ingestion of, 564
  in kidney, 598, 599
  neonate exposure, 597
  poisoning, 577
  proportion of, 564
  in soft tissues, 598–600
  in subclinical targets, 601
  sublethal effects, 568–569
  thresholds of, 583
  tissue concentration, 568, 570–574
  toxic, 565, 596
Lead in whole blood (PbB), 596
Lead residues in birds, 574–582
Lepomis macrochirus, see Bluegill sunfish
LET, see Linear energy transfer (LET)
Lethal concentration affecting 50% (LC50), 617
  brain, 2
  heptachlor epoxide, 463
  organotins, 271
  of trace metals, 244, 246–247
  zinc, 242
Lethal tissue residue to 50% (LR50), 118, 119, 125
Lethality
  CBR, 414
  to pink shrimp, 83
  tissue Hg concentrations, 614, 616
Level Leading to Population Decline (LPD), 459
Ligand binding domain (LBD), 481
  genotypes for AhR LBD in wild birds, 482
Lindane, 49, 50, 81, 82, 465
  in fish, 81, 82, 84
  invertebrate CBR, 90–91
  median lethal, 84
  toxicity, 82, 86
Linear energy transfer (LET), 704
Lipid regulators, 305
Liquid scintillation counting (LSC), 51
Little owl, 446
LOAEC, see Lowest observed adverse effect concentration (LOAEC)
LOAEL, see Lowest observed adverse effect level (LOAEL)

LOEC, *see* Lowest-observed effect concentration
    (LOEC)
LOEL, *see* Lowest-observed-effect-level (LOEL)
LOER, *see* Lowest observable effect residue (LOER)
Loon, common (*Gavia immer*), 17
    DDD poisoning, 429
    DDE residues, 435
    egg Hg concentrations, 618
    lead poisoning, 581
Lossless model, 334
Lowest observable effect residue (LOER), 258
Lowest observed adverse effect concentration (LOAEC),
    532, 539, 611
    for individual species, 619
    for PCB toxic effects, 550
Lowest observed adverse effect level (LOAEL), 16, 629
    of Aroclor feed, 538
    dietary, 547
    and NOAEC, 550
    of ΣPCBs/kg feed, 546
    TDI calculation with, 366
    for wild passerines, 503
Lowest-observed effect concentration (LOEC), 20
    and NOEC, 548
    ovary, 201
    Se concentration, 200, 211
Lowest-observed-effect-level (LOEL), 483, 514
    PCBs, PCDDs, and PCDFs, 514
    productivity in wild birds, 506
LPD, *see* Level Leading to Population Decline (LPD)
LR50, *see* Lethal tissue residue to 50% (LR50)
LSC, *see* Liquid scintillation counting (LSC)
Lutra lutra, *see* River otters

M

Macrocyclic lactones, 306
Malachite green, 313–314
Malayan bear, 530
Mallards (*Anas platyrhynchos*), 574, 653
    lead shot, 575
    Se, 688
Mammals, 350, 352, 538, 708
    bone lead concentrations in, 604
    cadmium in, 633
    herbivorous, 353
    Hg in, 382, 391
    insectivorous, 631
    lactation duration, 354
    lead in kidneys, 603
    lead in liver, 603
    LR50 values, 271, 272
    mortality data, 715–716
    nonfur-bearing, 359
    PBTs, 355–356
    PCB, 355
    population effect, 359
    radiation exposure levels, 727
    SKI, 599
    studies, 360–362, 362–363
    toxicology, 396
    trophic levels, 352–353
    WHO TEF for, 112
Mass spectrometry (MS), 22

Maternal transfer, 85, 274
    DDT, 67
    Hg, 610
    OH compounds, 50, 61
    PCBs, 155, 483
    selenium, 197, 198, 200, 202, 215–217, 220, 222
    TBT, 265
Mayflies (Ephemeroptera), 676
MBT, *see* Monobutyltin (MBT)
MCH, *see* Mean corpuscular hemoglobin (MCH)
MCH concentration (MCHC), 717
MCHC, *see* MCH concentration (MCHC)
MCV, *see* Mean corpuscular volume (MCV)
MDL, *see* Method detection limit (MDL)
Mean corpuscular hemoglobin (MCH), 717
Mean corpuscular volume (MCV), 717
Mechanism of action (MeOA), 256
    of cyclodiene insecticides, 450
    for dioxin, 148
    of PCBs/PCDDs/PCDFs, 535–536
    toxic response, 269–270
Mechanism/mode of action (MOA), 288
MeHg, *see* Methylmercury (MeHg)
Meleagris gallopavo, *see* Wild turkey
MeOA, *see* Mechanism of action (MeOA)
MeO-PBDEs, *see* Methoxylated PBDEs (MeO-PBDEs)
Mercuric selenide (HgSe), 384
Mercury (Hg), 17, 382, 609; *see also* Metal toxicology
    aims and approach, 610–611
    cetaceans, 382–383
    concentrations of, 180, 184, 185
    dietary concentration, 613, 619
    dosing studies, 18
    elimination in mammals, 172
    in environment, 18, 378
    exposure, 17
    in fish fillets, 184
    hepatic concentration, 388
    impaired reproduction, 613–614, 616–617
    incidence of acute, 628
    indicative dietary concentrations, 612
    indicative egg concentrations, 617
    indicative tissue concentrations, 614
    lethality, 612–613, 614–616
    low concentration, 389
    in mammals, 382, 391
    maternal transfer, 610
    mercury–selenium interactions, 611–612
    mysticetes, 388–389
    odontocetes, 383
    pinnipeds, 389–391
    and Se, 391–393
    sensitive environments, 19–20
    sources of, 627
    in tissues, 170, 173–177
    total concentrations, 612
    toxicity, 619
MeSO2, *see* Methyl sulfone (MeSO2)
Metal toxicity, 380
Metal toxicology, 380–381
    genomic tools, 381
    oral intake/feeding ecology approaches, 381–382
    in vitro approaches, 381
Metalloids, 231, 379

Metallothionein (MT), 627, 646
Metallothionein (MTH), 380
    against cadmium toxicity, 648
    coaccumulation, 647
    synthesis, 647
Method detection limit (MDL), 261
Methoxychlor, 465–466
Methoxylated PBDEs (MeO-PBDEs), 125
    dietary and tissue total Hg, 620
    egg Hg concentrations, 618
    liver and kidney relationship in birds, 614
Methyl sulfone (MeSO2), 358
Methylmercury (MeHg), 610, 692
    bioaccumulation, 171–172
    consequences of, 178
    contamination, 170
    cumulative frequency, 613, 615, 619
    demethylation, 19
    dicyandiamide, 18
    dietary, 20, 180
    effects, 172, 178–180, 180–182
    passage of, 171
    percentage of, 391
    rate of, 172
    reproductive effects, 613–614
    in tissues, 182, 183–184
    toxicity–lethality, 612–613
    toxicological effects, 172
Micropterus salmoides, see Largemouth bass
Microtus agrestis, see Field vole
Migration, 456
Migratory restlessness, 456
Ministry of the Environment (MOE), 113
    dental malalignment in, 543
    exposure regimen, 546
    interdental spacing in, 543
    squamous epithelial proliferation, 542
Mink, 554
Mirex, 80, 84
    in carcass, 81
    nonlethal body residues, 81
    uses, 466
MMT, *see* Mono methyltin (MMT)
MOA, *see* Mechanism/mode of action (MOA)
Mode of action (MoOA), 61, 269
    DDT, 62
    immunotoxicity, 273
    for PBDEs, 127
MOE, *see* Ministry of the Environment (MOE)
Molecular weight (MW), 301
Moles (*Talpidae*), 637
    lead effect, 599
Monobutyltin (MBT), 257, 259
    in aquatic species, 262, 267
    imposex, 274
    organotin properties, 257
    tissue concentrations, 275, 278
Monomethyltin (MMT), 257
    in aquatic environments, 259
Monooctyltin (MOT), 257
Monophenyltin (MPT), 257
    in aquatic species, 267
    imposex, 274
    organotin properties, 257

Monopropyltin (MPrT), 257
Monosodium methanearsonate (MSMA), 16
MoOA, *see* Mode of action (MoOA)
Morone americana, see White perch
MS, *see* Mass spectrometry (MS)
MT, *see* Metallothionein (MT)
MTH, *see* Metallothionein (MTH)
Mustela vison, see American mink
MW, *see* Molecular weight (MW)

N

Narwhal, 376
National Contaminant Biomonitoring Program (NCBP), 49
National Council for Radiation Protection (NCRP), 723
National Pesticide Monitoring Program (NPMP), 49
National Pollutant Discharge Elimination System (NPDES), 325
National Rivers and Streams Assessment (NRSA), 333
National Water-Quality Assessment (NAWQA), 75–76, 225, 410
    runoff, 85, 410
    whole-body Se concentrations, 225
NAWQA, *see* National Water-Quality Assessment (NAWQA)
NCBP, *see* National Contaminant Biomonitoring Program (NCBP)
NCRP, *see* National Council for Radiation Protection (NCRP)
Nephropathy, 595
Nest box studies, 488
    PCB contamination, 502
    TCDD-contamination, 507
Nestling fitness measures, 502–505
    brain asymmetry, 502
    fabricius bursa, 503
    immune function, 504
    liver enlargement, 502–503
    thyroid function, 504
    Vitamin A influence, 504–505
Neurotoxicity, 466
    lead-induced, 597
Nitrogen (N), 380
N-methyl-D-aspartate receptor (NMDAR), 276
No adverse effect level (NOAEL), 16
No observable effect tissue residue (NOER), 118
    dioxins, 119
    geometric mean, 550
    PBDEs, 127
    PCBs, 124–125
No observed adverse effect concentration (NOAEC), 532, 539; *see also* Lowest observed adverse effect concentration (LOAEC)
    expressed as TEQs, 552
    for Saginaw Bay, 552
No observed adverse effect level (NOAEL), 68
    Aroclor, 538
    dietary, 548
    laboratory study, 629
    TDI calculation, 366
No observed effect concentration (NOEC), 17, 548, 613
    dietary, 204, 215
    and LOEC, 210, 548
    Se concentration, 200

No observed effect level (NOEL), 466, 548
    concentrations, 495
    dietary, 548
NOAEC, *see* No observed adverse effect concentration
    (NOAEC)
NOAEL, *see* No observed adverse effect level (NOAEL)
NOEC, *see* No observed effect concentration (NOEC)
NOEL, *see* No observed effect level (NOEL)
NOER, *see* No observable effect tissue residue (NOER)
Nonhuman biota, 723
    chronic effects thresholds, 724
    weighting factors for, 705
Nonpolar lipids, 60
Nonsteroidal anti-inflammatory drug (NSAID), 295,
    304–305
Notemigonus crysoleucas, see Golden shiners
NPDES, *see* National Pollutant Discharge Elimination
    System (NPDES)
NPMP, *see* National Pesticide Monitoring Program
    (NPMP)
NRSA, *see* National Rivers and Streams Assessment
    (NRSA)
Nycticorax nycticorax, see Black-crowned night herons

O

OA, *see* Oxolinic acid (OA)
Obesogens, 277
OC, *see* Organochlorine (OC)
Octabromodiphenyl ether (Octa-PBDE), 107
Octanol–water partition coefficient (Kow), 108, 259, 290
Octa-PBDE, *see* Octabromodiphenyl ether (Octa-PBDE)
Odontocetes, 383–384, 394–395
    elemental and MTH interactions, 384–385
    Hg toxicodistribution, 387–388
    liver and kidney mercury, 385–386
    in vitro, 387
    in vivo, 384
OH, *see* Organohalogen (OH)
Oil globule, 68
Oncorhynchus tshawytscha, see Chinook salmon
OP, *see* Organophosphate (OP)
Organic contaminants, 15, 182, 350, 359
    challenge, 363–364
    chemicals of emerging concern, 357–358
    exposure routes, 351–352
    marine mammal affected by, 351
    nonpersistent bioaccumulative, 359
    PAH, 359
    persistent bioaccumulative, 356
    PPCPs, 359–360
    regulations, 364–365, 365–367
    studies, 360–363
    toxic action, 358
    toxic effects, 356
    transportation, 87
    in vitro studies, 363
    weight of evidence, 364
Organic moiety, 259
Organobromines, 125
Organochlorine (OC), 27
    critical body residues, 79, 80
    critical values, 464
    cyclodienes, 449

    dieldrin TEFs, 467
    dieldrin toxic equivalents scheme, 467
    endosulfan, 464
    groups, 447
    HCB, 464, 465
    hexachlorocyclohexane, 465
    index, 429
    methoxychlor, 465–466
    mirex, 466
    neurotoxicity, 466
Organochlorine pesticides (OC pesticides)
    acute toxicity, 448
    pesticide toxicity, 448
    TEQ, 468
    toxaphene, 466
Organohalogen (OH), 48
    in air, 52
    analytical method considerations, 50–51
    bioaccumulation, 56–57, 58, 59
    biomagnification, 60
    CBR, 65
    concentration, 58
    DDT effects, 63
    environmental occurrence, 52
    in fish tissues, 53
    hormone activity disturbance, 62
    ke and 95% SS in invertebrates, 57
    life stage/tissue type, 87–88
    lindane effects, 81–83
    lipid dynamics, 60–61
    maternal transfer, 61
    mirex effects, 80
    mixtures, 85–86
    pesticides, 48, 91
    potency of, 61
    properties, 50
    resistance, 86–87
    in sediment, 52–53
    tendency of, 56
    tissue-residue, 63, 85
    toxaphene effects, 83–84
    toxicity, 61–63
    uptake and elimination, 57
    in water, 52
Organophosphate (OP), 410
Organotin, 256, 258, 259, 263
    active molecule causing, 258
    as agonists, 270
    alkyl, 259
    as antifoulants, 258
    applications, 256
    in aquatic biota, 256, 257
    bioaccumulation, 259, 263–265
    butyltins, 261
    carbon impact, 260
    critical triorganotin body residues, 268
    elimination, 265
    endocrine disruptors, 273
    environmental chemistry, 259–260
    ionizable, 259
    Kow, 259, 263
    legislation, 258
    lethal concentrations, 272
    lethal values, 269

obesogens, 277
parental transfer, 265
phenyltins, 261
properties, 257
reporting, 258
somatogen, 277
tissue concentration, 261
toxicity, 266–277
uptake, 264–265
in water and sediment, 260–261
Organotin toxicity, 266
from ambient exposure, 267, 268, 269
behavioral, 276, 277
fish reproduction, 274, 276
growth, 272, 273
immunotoxicity, 273
imposex, 273–274
mechanisms of, 269–270
mortality, 270–272
neurological, 276
obesogen/somatogen, 277
reproductive effects, 276
tissue-residue toxicity, 270
Oriental beetle (*Anomala orientalis*), 460
Osprey, 424
Osteolysis, 543
OTC, *see* Oxytetracycline (OTC)
Outwardly healthy, 636
Oxidative phosphorylation, 278, 412
Oxidized glutathione (GSSG), 212
Oxolinic acid (OA), 295
Oxychlordane, 449, 460
Oxytetracycline (OTC), 310

P

PAHs, *see* Polycyclic aromatic hydrocarbons (PAHs)
Palaemon serratus, 230
Paralichthys olivaceus, see Flounder
Paroxetine, 303
Parus carolinensis, see Carolina chickadee
Pb, *see* Lead (Pb)
PbB, *see* Lead in whole blood (PbB)
PBDEs, *see* Polybrominated diphenyl ethers (PBDEs)
PBHDs, *see* Polybrominated hexahydroxanthene derivatives (PBHDs)
PBT, *see* Persistent, Bioaccumulative, and Toxic (PBT)
PCB, *see* Polychlorinated biphenyl (PCB)
PCDFs, *see* Polychlorinated dibenzofurans (PCDFs)
PCP, *see* Pentachlorophenol (PCP)
PCPs, *see* Personal care products (PCPs)
PDV, *see* Phocine distemper virus (PDV)
PeCDF, *see* 2,3,4,7,8-pentachlorodibenzofuran (PeCDF)
Pelecanus occidentalis, see Brown pelican
Pentabromodiphenyl ether (Penta-PBDE), 107
2,3,4,7,8-pentachlorodibenzofuran (PeCDF), 542
Pentachlorophenol (PCP), 356, 412, 534
Penta-PBDE, *see* Pentabromodiphenyl ether (Penta-PBDE)
Peregrine falcon, 8
Perfluoroalkylated surfactants (PFAS), 357
Perfluorooctane sulfonate (PFOS), 17, 552
Perfluorooctane sulfonic acids (PFOS), 357
Perfluorooctanoic acid (PFOA), 358

Peroxisome proliferator-activated receptor (PPARγ), 270
health effects, 361
Persistent, Bioaccumulative, and Toxic (PBT), 103, 350, 357, 361
dirty dozen chemicals, 357
risk at trophic levels, 353
in water, 365
Persistent organic pollutants (POPs), 103, 290, 364
Personal care products (PCPs), 289
Pesticide, 11, 409, 419, 447
anticholinesterase, 14
background, 409–410
bioaccumulation, 411, 417–419
biotransformation, 411, 413–415, 416–417
CBR, 412–413
deformities, 16
dietary risk assessment, 612–613
endocrine disruptors, 62
environmental movement, 12–13
environmental occurrence, 52
exposure studies, 12
half-life, 359
hazards, 16
OH compounds, 48–51, 52, 447, 448
poisoning, 1, 11
pseudopersistence, 360
quantification, 13
residue, 91
resistance, 86
secondary, 418
Silent Spring, 12
Thirty Years' War, the, 15
in tissue, 12, 53, 59, 65, 411
toxicity, 61, 411
toxicological responses, 416–417
Petroleum spills, 14–15
PFAS, *see* Perfluoroalkylated surfactants (PFAS)
PFOA, *see* Perfluorooctanoic acid (PFOA)
PFOS, *see* Perfluorooctane sulfonate (PFOS)
PHA, *see* Phytohemagglutinin (PHA)
Pharmaceuticals and PCPs (PPCPs), 316
in aquatic ecosystems, 359–360
detection of, 360
Phenylpyrazole insecticide, 414, 416
Phenyltins, 261
in aquatic organisms, 262, 278
parental transfer, 265
Phoca hispida, see Blubber-rich ringed seal
Phoca vitulina, see Harbor seal
Phocine distemper virus (PDV), 360
delayed neuropathy, 15
elemental, 11
hazard of ingested, 10
organophosphorus pesticides, 11, 14
osteomalacia, 628, 652
Phocoena phocoena, see Porpoises
Phocoenoides dalli, see Dall's porpoise
Phosphorus
Cd absorption, 646
vitamin D synthesis, 652
white, 11
Physiological-Based Pharmaco(Toxico) Kinetic Models (PBPK Models), 331
Phytohemagglutinin (PHA), 504

Pinnipeds, 353, 358
　　carcinoma, 361
　　dioxin-like responses, 362
　　Hg in, 389–391
　　PAH exposure, 352, 356, 363
Plecoptera, *see* Stoneflies
PNEDRs, *see* Predicted no-effect dose rates (PNEDRs)
PNEDs, *see* Predicted no-effect doses (PNEDs)
POCIS, *see* Polar organic chemical integrative sampler (POCIS)
Polar bear (*Ursus maritimus*), 535
　　contaminated prey, 538
　　PBT exposure, 352, 354, 361
　　PCB exposure, 540
Polar lipids, 60
Polar organic chemical integrative sampler (POCIS), 328, 334
Polybrominated biphenyl (PBB), 107
　　atmospheric fate, 107
　　bioaccumulation, 109–111
　　disruption in thyroid, 364
　　generic structure, 533
　　in marine mammals, 357
　　on nestling, 511
　　as PBT, 103
Polybrominated diphenyl ether (PBDE), 17, 107; *see also* Polychlorinated biphenyls (PCBs)
　　aquatic fate, 108
　　properties, 106
　　on reproduction, 510
　　structure, 105
　　TEFs, 111–113
　　tissue residues, 123–127
　　toxicological effects, 155
Polybrominated diphenyl ethers (PBDEs), 107, 357, 480
　　in aquatic biota, 128, 130–146, 147
　　aquatic fate, 108–109
　　atmospheric fate, 107–108
　　bioaccumulation, 109, 110, 111
　　ecotoxicological effects, 155, 156
　　environmental fate, 107
　　threshold criteria, 515–516
　　tissue residues, 123, 125, 126, 127
　　toxic equivalency factors, 111–112, 113
　　on wild birds, 510–512
Polybrominated hexahydroxanthene derivatives (PBHDs), 125
Polychlorinated biphenyl (PCB), 21, 105, 106–107
　　on adult survival, 508, 509
　　AhR LBD, 481, 482
　　aquatic fate, 108–109
　　assessments, 497
　　atmospheric fate, 107–108
　　on avian reproduction, 22
　　bioaccumulation, 109, 110, 111
　　clutch size, 490–491
　　concentration, 22, 354
　　concentration variation, 355
　　congeners, 106, 479
　　contamination surveys, 22
　　critical effects, 547
　　CYP1A, 480, 481, 483
　　dietary exposures, 496
　　DNA cloning and sequencing, 481

ecotoxicological effects, 152, 155
　　egg quality, 494
　　egg-exchange design, 498
　　egg-injection methodology, 495, 496
　　embryolethality, 495
　　embryotoxicity, 22
　　environmental fate, 107
　　environmental importance of, 1
　　estimated cumulative, 533
　　in fish tissue, 121, 122
　　on hatching, 498
　　hatching success, 494–495, 497
　　isotope studies, 484
　　laboratory studies, 489
　　morphological deformity rates, 498–500
　　nestling fitness measures, 502–505
　　nestling growth, 501–502
　　nestling mortalities, 501
　　parental behavior, 485, 486–487, 488
　　parental breeding, 486–487
　　prenesting behaviors, 485
　　products, 533
　　properties, 106, 532
　　quantification, 22
　　secondary sexual characteristics, 491
　　structure, 356
　　threshold criteria, 515–516
　　tissue residues, 120, 122, 123, 126, 550
　　toxic equivalency factors, 111–112, 113
　　toxicity of, 106
　　variability, 480
　　on wild bird, 492–493
　　wild bird development, 491–494
　　wild birds studies, 23–24
　　wildlife toxicity, 21
Polychlorinated compound, 532–534
　　adult survival, 507–510
　　aquatic mammals, 540–541
　　avian development, 491–494
　　avian exposure, 480
　　clutch size, 490–491
　　congener patterns, 483
　　effects, 484–485, 510–512, 545–548
　　egg quality, 494
　　exposure routes, 483–484
　　field studies, 537, 538, 539, 541–544
　　hatching success, 485, 494–498, 501–502
　　marine mammals, 538
　　measured variability in metabolism, 480
　　mechanism of action, 535–536
　　morphological deformities, 498–500
　　nestling fitness measures, 502–505
　　parental behavior, 485–490
　　productivity, 505–507
　　secondary sexual characteristics, 491
　　terrestrial mammals, 537
　　threshold tissue residue concentrations, 548–553
　　toxic equivalency approach, 536
　　trends, 534–535
　　wild birds, 478
Polychlorinated Dibenzofurans (PCDFs), 21, 103, 356; *see also* Polychlorinated biphenyl (PCB)
　　congener, 479, 484
　　as contaminants, 23

effects of, 543, 545–548
in marine mammals, 538
mechanism of action, 535
properties, 106, 478, 531, 534
quantification, 22
TEQ, 553
threshold criteria, 515–516
Polychlorinated dibenzo-*p*-dioxins (PCDDs), 21, 103, 105, 356, 478, 531; *see also* Polychlorinated biphenyl (PCB)
AhR LBD, 481, 482
congener, 479, 484
congeners in tissues, 479
CYP1A, 480, 481, 483
DNA cloning and sequencing, 481
effects on wild bird, 492–493
in pesticides, 21
properties, 106
structure, 105
threshold criteria, 515–516
variability, 480
Polycyclic aromatic hydrocarbons (PAHs), 4, 359
crude oil contaminants, 352
Polyhalogenated chemicals, 535
Polyvinyl chloride (PVC), 257, 360
POPs, *see* Persistent organic pollutants (POPs)
Porpoises (*Phocoena phocoena*), 385
PPARγ, *see* Peroxisome proliferator-activated receptor (PPARγ)
PPCPs, *see* Pharmaceuticals and PCPs (PPCPs)
PPIX, *see* Protoporphyrin IX (PPIX)
Predators, 18, 538, 564
benthic invertebrates and, 359
concentrations in top, 351, 610
Hg poisonings, 18
metal for assimilation, 235
nontoxic Hg–Se complexes, 19
Pb poisoning, 574
Predicted no-effect dose rates (PNEDRs), 726
Predicted no-effect doses (PNEDs), 726
Probabilistic therapeutic distributions (PTDs), 330, 331
PROTECT program, 712
cadmium on kidney, 629, 637, 647, 652
cadmium toxicity, 636
Proteinuria, 636
Protoporphyrin IX (PPIX), 568
Pseudocrenilabrus sp, *see* Southern mouthbrooder
Pseudopersistence, 298, 360
PTDs, *see* Probabilistic therapeutic distributions (PTDs)
Pure chlordane, 76
PVC, *see* Polyvinyl chloride (PVC)

Q

Quantitative structure–activity relationship (QSAR), 56, 259, 263, 410

R

Radionuclides, 704
absorbed dose, 705, 706
arctic fish and invertebrates, 721
equivalent dose, 705
exposure, 704, 710–712
on fertility, 709, 714
graded approach, 710
half-life, 704
ionizing radiation, 704
isotopes, 704
lethal doses, 709
measurement units, 705
radiation effects thresholds, 723–726
radiosensitivity, 708
source, 705
weighting factors, 705
whole-body effects, 708
Radiotracer technique, 51
Razorback suckers (*Xyrauchen texanus*), 200
RBCs, *see* Red-blood cells (RBCs)
RBEs, *see* Relative biological effects (RBEs)
RC, *see* Reference concentrations (RC)
REACH, *see* Regulation on Registration, Evaluation, Authorization and Restriction of Chemicals (REACH)
Reactive oxygen species (ROS), 707
Recurvirostra americana, *see* American avocets
Red-blood cells (RBCs), 717
Reduced glutathione (GSH), 212
Red-winged blackbirds (*Agelaius phoeniceus*), 612
Reference area, 670, 682
Reference concentrations (RC), 366
Reference site, 687
Regulation on Registration, Evaluation, Authorization and Restriction of Chemicals (REACH), 365
Relative biological effects (RBEs), 705
Relative potency factors (RPFs), 536
Renal proximal tubules damage, 652
Renal tubular dysfunction, 648
Resident mink survey, 542
Retinoid X receptor (RXR), 270
Ringed seal, 390
Ring-necked ducks (*Aythya collaris*), 575
River otters (*Lutra lutra*), 535
Röentgens (R), 705
ROS, *see* Reactive oxygen species (ROS)
Rothane, 48
Route of uptake, 171
RPFs, *see* Relative potency factors (RPFs)
RXR, *see* Retinoid X receptor (RXR)

S

Saint Lawrence Estuary (SLE), 383
Sample egg technique, 437
Sampling, nondestructive, 16
Sandhill crane, 580, 581
Sarcoramphus papa, *see* King vulture
SDS, *see* Sodium dodecylsulfate (SDS)
Seal, common, 348; *see also* Harbor seal (*Phoca vitulina*)
Sea-perch, 192
Sea-skaters (*Halobates*), 650
Selective Serotonin Reuptake Inhibitors (SSRI), 302–304
Selenium (Se), 19
age, size, and trophic position, 196
aquatic food chain, 195, 196
assessment values, 671–675, 695

Selenium (Se) (*Cont.*)
  on avian embryo, 678
  bioaccumulation, 195, 196, 197–198
  biogeochemistry of, 194
  biomarkers, 690–691
  blood, 687–689
  bluegill, 201, 204, 209, 211
  chemical form effect, 678
  chinook salmon, 211–212
  concentrations, 197, 200, 207, 215, 219, 670–671, 676
  critical form, 196
  demethylation, 392
  diet *vs.* toxicity, 671–676
  dietary exposures, 209
  dietary TTFs, 216
  distribution thresholds, 223
  Dolly varden, 206
  dose–response relation, 682
  effects, 217, 222, 688
  in egg, 208, 209, 221, 676–677
  elevated levels, 198
  exposure route influence, 196–197
  fathead minnow, 206–207, 212
  in feathers, 689–690, 694
  field studies, 681–683, 685–686
  and fish deformities, 219
  50% effect concentration, 681
  hatching success, 680
  with Hg, 696
  hormesis, 693
  and hormetic effects, 678, 693
  inadequate, 676–677
  inflection point, 676
  interactions, 691–693
  in kidney, 686–687
  lab studies, 677–681, 683–685
  in largemouth bass, 207
  in liver, 683, 684, 690
  LOEC and NOEC, 222
  maternal intake, 205
  maternal transfer, 200, 220
  in MeHg demethylation, 692
  mosquitofish, 207–208
  muscle, 687
  northern pike, 208
  in ovary, 201, 217
  oxidation states, 670
  parent *vs.* offspring, 221
  in pinnipeds, 393
  poisoning, 195, 669–670
  in rainbow trout, 208, 212–213
  in razorback sucker, 208–209, 213, 214–215
  reproductive effects, 694
  safe level, 670
  selenomethionine, 196
  sodium selenite, 680
  source of, 669, 671
  stereoisomer, 679
  tissue thresholds, 220
  toxicity, 679, 681, 690
  toxicity thresholds, 223
  toxicokinetics, 216
  trout, 204–206
  Tulare curves, 682

  and vitamins, 691
  in white sturgeon, 215
  in white sucker, 209
Selenium (Se) threshold
  comparisons, 223, 224–225
  concentration-response relationships, 218, 219
  Egg hatchability, 15
  in egg, 220
  fish tissue threshold, 215, 222
  life stage considerations, 216
  maternal transfer and juvenile toxicity, 216, 217–218
  in ovary, 220, 221
  in tissue-based toxicity, 219
  in whole-body, 221, 222
Selenoamino acids, 196
Selenomethionine, 196, 678
Self-biasing detectability, 300
Semipermeable membrane devices (SPMD), 328
Serotonin-Norepinephrine Reuptake Inhibitors (SNRI), 302–304
Sex hormone-binding globulin (SHBG), 309
Sex hormones, 62
SHBG, *see* Sex hormone-binding globulin (SHBG)
Short-snouted sea-horse, 286
Shrews (*Sorex araneus*), 634
SI, *see* International System (SI)
Sievert (Sv), 705
Σpolybrominated diphenyl ether (PBDE), 552
Silent spring, 12
Silicone membrane equilibrator, 329
Siluridae, *see* Catfish
Silver (Ag), 394
Silvery light-fish, 168
Skeleton, adult, 600
SKI, *see* Somatic kidney index (SKI)
SLE, *see* St. Lawrence Estuary (SLE)
Slender-clawed crayfish, 46
Small mammals
  insectivorous, 631
Sn, *see* Tin (Sn)
Sodium dodecylsulfate (SDS), 393
Sodium pump, 466
Sodium selenite, 680
Soil ingestion, 595
Soil organic matter, 599
Solid-phase microextraction (SPME), 52, 316
Somatic kidney index (SKI), 599
Somatogen, 277
Sorex araneus, *see* Shrews
Source-to-effects continuum, 291
Southern mouthbrooder (*Pseudocrenilabrus* sp.), 76
Species sensitivity distribution (SSD), 118, 309, 726
SPMD, *see* Semipermeable membrane devices (SPMD)
SPME, *see* Solid-phase microextraction (SPME)
Spotted cuscus, 594
SSD, *see* Species sensitivity distribution (SSD)
St. Lawrence Estuary (SLE), 361
Starlings (*Sturnus vulgaris*), 460
  endosulfan, 75
  of PCBs/PCDDs/PCDFs effects, 535
  selenomethionine, 679
Stenella coeruleoalba, *see* Striped dolphins
Stereoisomer
  of dieldrin, 74

Sterna forsteri, see Forster's tern
Sterna hirundo, see Tern, common
Steroid, 306–310
    hormones, 78
Stone-bass, 192
Stoneflies (Plecoptera), 676
Striped dolphins (*Stenella coeruleoalba*), 535
Sturnus vulgaris, see Starlings
Subatomic particles, 704
Sublethal effects, 684
    DDE, 69, 89
    DDT, 66, 70, 90
    endocrine system disruption, 61
    heptachlor, 78
    ligands, 171
    lindane, 84
    OH compounds, 88
    of PCBs, 25
    Se impact, 694
    toxaphene, 83–84, 91
Sulfhydryls
    in Hg detoxification, 392
Sv, *see* Sievert (Sv)
Synthetic progestins, 333

T

T3, *see* Triiodothyronine (T3)
T4, *see* Thyroxine (T4)
Taeniopygia guttata, see Zebra finches
Talpidae, see Moles
TBB, *see* Total body burden (TBB)
TBBPA, *see* Tetrabromobisphenol A (TBBPA)
TBT, *see* Tributyltin (TBT)
TCDD, *see* Tetrachlorodibenzo-*p*-dioxin (TCDD)
TCDF, *see* Tetrachlorodibenzofuran (TCDF)
TDI, *see* Tolerable daily intake (TDI)
TeBT, *see* Tetrabutyltin (TeBT)
TeET, *see* Tetraethyltin (TeET)
TEF, *see* Toxic equivalence factor (TEF)
TeMT, *see* Tetramethyltin (TeMT)
TEQ, *see* Toxic equivalent (TEQ)
Tern, common (*Sterna hirundo*), 617
    Hg concentrations, 617, 618
    PCB effects, 25
Testosterone (T), 309
TET, *see* Triethyltin (TET)
Tetrabromobisphenol A (TBBPA), 107
Tetrabutyltin (TeBT), 257, 270
    in aquatic species, 267
Tetrachlorodibenzofuran (TCDF), 552
Tetrachlorodibenzo-*p*-dioxin (TCDD), 21
    concentration, 538
    tissue residue, 120
Tetraethyltin (TeET), 257, 261
    in aquatic species, 267
Tetramethyltin (TeMT), 257, 261
    in aquatic species, 267
Tetraphenyltin (TePT), 257
    in aquatic species, 267
Tetrapropyltin (TePrT), 257
    in aquatic species, 267
THg, *see* Total mercury (THg)
Thin layer chromatographic (TLC), 51

Thirty Years' War, the, 14
Threshold, 2
Thymus, 509
Thyroid, 362
    as biomarkers, 362, 505
    contaminated feed and, 363
    nest attentiveness, 489, 503
    PBT contaminants, 358, 362
    PCB and PCDD impact on, 504
    physiological functions, 62
    Thyroid function, 504
    Thyroid hormones, 358
Thyroxine (T4), 393
Tin (Sn), 257, 258, 259
    critical concentration, 2
    interpretation, 4, 16–17
    and mercury, 183
    organotin properties, 257, 259
    reporting concentrations, 258
    reporting unit, 258
    in TBT debutylation, 265
    terminology, 2–3
    toxicity, 267
Tissue residue concentration, 1, 548, *see*
        Acquired dose
    to assess exposure and effects, 9
    and toxic effects, 182–185
    toxicokinetic model, 17
Tissue Residue Guidelines (TRGs), 365
Tissue residue threshold, 548–549
    aquatic mammal, 551–553
    marine mammal, 549–551
    terrestrial mammal, 549
Tissue sampling, 316
TLC, *see* Thin layer chromatographic (TLC)
TnHT, *see* Tri-*n*-hexyltin (TnHT)
Tolerable daily intake (TDI), 366
TOT, *see* Trioctyltin (TOT)
Total body burden (TBB), 634
Total mercury (THg), 692
Totality of exposure, 334
Toxaphene, 50, 83–84, 466
    in aquatic invertebrates, 82
    ban, 49
    quantification, 51
    range of, 91
Toxic equivalence factor (TEF), 24, 104, 111–113, 467
    approach, 536
    approach limitations, 356
    TCDD-like toxicity, 532
Toxic slick, 351
Toxic unit (TU), 85
Toxicant, 2
    consumption criteria, 380–381
    DDT, DDD, and DDE, 64
    heptachlor, 78
    mirex, lindane, and toxaphene, 82
    multiple exposure, 467
    multitargeted, 595
    neurological, 276, 389
    reproductive, 279
    SLE on, 384
    toxicant-nutrient interaction, 388
    4Ts, 297

Toxicity
  cyclodienes, 73–74
  elemental interactions, 379
  predictors, 417
  to wildlife, 12
Toxicity equivalent (TEQ), 113, 467, 534, 536, 553
  approach, 536, 544
  correlative model, 498
  dioxin, 115–177
  embryotoxicity, 496
  for mixture, 468
  threshold concentration, 549, 550
Toxicity reference value (TRV), 365, 548
  tissue-based, 549, 551
Toxicity thresholds, 676
  cadmium, 658
  Se, 197, 218, 223
  tissue-based, 219
Toxicokinetic model, 17, 266
Toxicologist, 2, 9–10, 27
  aquatic, 2–3
  biochemical measurements, 15
  comparative, 378
  critical body residues, 412–413
  ecotoxicologist, 1
  ecotoxicology, 1, 411
  exposure, 290
  poison detection, 10
  Se, 199–200, 202, 210
Toxicology, 9; *see also* Wildlife toxicology
  adverse effects, 11
  weight of evidence in, 364
Trace metal, 231, 378, 379–380
  accumulation, 232, 236–237, 240, 241, 242, 243, 248
  aquatic invertebrate body metal content, 237
  in bioaccumulated metal, 247, 249
  bioavailabilities, 247
  biodynamic modeling, 243–244
  burrowing invertebrate, 234
  detoxified component, 239–240
  essential, 238, 239
  excretion, 236
  lethal body concentration, 244, 246–247
  metabolic requirements, 238
  toxicity, 14, 244
  transmembrane routes, 235
  uptake, 232, 233–236
Trace metal, essential, 234
  accumulated concentration, 236
  metabolic requirements for, 238
Transmembrane protein, 234–235
Transpiration stream concentration factors (TSCFs), 314
TRGs, *see* Tissue Residue Guidelines (TRGs)
Triazoles, 302
  fungicides, 306
Tributyltin (TBT), 257
  in aquatic species, 267
  chambering response, 273
  concentrations, 274, 275
  contamination, 274
  exposure, 273
  imposex, 274
  lethal concentrations, 272
  lethal toxicity for, 272

  lethality values, 278
  metabolism, 247
  organotin properties, 257
  in pesticides, 256
  sublethal response for, 270
  toxicity, 271
Tri-*c*-hexyltin (TcHT), 257
  lethal concentrations, 272
  organotin properties, 257
  sublethal response for, 270
  toxicity, 271
Triclocarban, 311
  in aquatic tissues, 322
  biocide, 302
  uptake, 297
Triclosan (TCS), 311
  accumulation, 312
  in aquatic tissues, 322, 323
  biocide, 301, 302
  in marine mammal, 313
  toxicity, 315
  trophic-level responses to, 298
Triethyltin (TET), 257
  in aquatic species, 267
  lethality values, 278
  organotin properties, 257
  sublethal response for, 270
Triiodothyronine (T3), 393
Trimethyltin (TMT), 257
  in aquatic environments, 259
  in aquatic species, 267
  lethal concentrations, 272
  lethality values, 278
  organotin properties, 257
  sublethal response for, 270
  toxicity, 266, 267
Tri-*n*-hexyltin (TnHT), 257
  toxicity, 271
Trioctyltin (TOT), 257
  membrane permeability, 271
Triorganotin, 271
  in aquatic environments, 259
  bioaccumulation and toxicity, 273
  body residues, 268
  Kow values, 263
  lethal values, 269, 271, 278
  lethality values, 278
  organotin properties, 257
  pesticides, 256
  sublethal response for, 270
  toxicity, 266, 267, 271, 272
Triphenyltin (TPT), 256
  in aquatic organisms, 262, 267
  aqueous, 273
  exposure, 267
  imposex, 274
  lethal concentrations, 272
  metabolism, 265
  metabolites, 261
  organotin properties, 257
  RXR activator, 270
  sublethal response for, 270
  tissue concentrations, 275, 278
  triorganotin body residues, 268

Tripropyltin (TPrT), 257
Trophic transfer factors (TTFs), 196
  risk at trophic levels, 353
  Se, 196
  Se dietary, 216
True seals, 353
TRV, *see* Toxicity reference value (TRV)
TSCFs, *see* Transpiration stream concentration factors (TSCFs)
TTFs, *see* Trophic transfer factors (TTFs)
TU, *see* Toxic unit (TU)
Tulare curves, 682
Tundra swans (*Cygnus columbianus*), 574
Turdus migratorius, see American robin
Turkey (*Melleagris gallopavo*), 24
Tursiops truncates, see Bottlenose dolphin

U

Uncertainty factor (UF), 366, 549
  chromosomal abnormalities in, 714
  herring gulls, 483
  reference lake, 209
  sites, 232
Uncontaminated, 670; *see also* Background; Reference area
  cadmium in, 629, 631
UNEP, *see* United Nations Environment Program (UNEP)
Unimodal model, 679
United Nations Environment Program (UNEP), 48, 357
United States Geological Survey (USGS), 225
Uptake from food, 235
  levels, 303–304
U.S. Department of Energy (USDOE), 710
U.S. Environmental Protection Agency (USEPA), 105, 534
  national study, 114
U.S. Fish and Wildlife Service (USFWS), 49
U.S. Geological Survey (USGS), 49

V

Venlafaxine, *see also* Serotonin-Norepinephrine Reuptake Inhibitors (SNRI)
  bioconcentration, 318
Vertebrate cytosolic protein, *see* Aryl hydrocarbon receptor (AhR)
Vitamin A, 362, 504
  AhR-related PCB toxicity, 541
  as biomarkers, 358, 504, 505, 509
  contaminant effect on, 363, 505
  PBDE-99 effect, 512
  ΣPCB effect, 548
  transformed contaminants, 511
Vitellogenin, 61
  estrogenic compound effect, 86
  plasma concentration, 62

W

Wastewater steroid effect, 314
Wastewater treatment plants (WWTPs), 307
  CEEs, 307
  removal efficiency of APIs, 296
  TCS, 311
  technologies, 326–327
WBCs, *see* White blood cells (WBCs)

Weight of evidence, 364
  on effects of PCBs, 484, 540
  in marine mammal toxicology, 364
  nestling fitness, 502
  PCB and TCDD thresholds, 513
  in regulatory guidelines, 367
Wet weight (ww), 118
White blood cells (WBCs), 717
White croaker (*Genyonemus lineatus*), 71
  PBDE levels, 129
White perch (*Morone americana*), 184
  methylmercury accumulation, 184
  SSRIs, 303
WHO, *see* World Health Organization (WHO)
Whooper swan (*Cygnus cygnus*), 575
  lead poisoning, 575
Wild griffon vulture, 576
Wild turkey (*Meleagris gallopavo*), 579
  AhR LBD genotype, 482
Wildlife toxicology, 10–11
  arsenic poisoning, 10
  contaminant exposure, 14, 15
  contaminant studies, 11
  dioxins and PCBs, 21
  history, 10
  interpreting exposure, 16
  mercury in wild birds, 17
  pesticides, 11–12
  petroleum spills, 14–15
  poisoning diagnosis, 14
  population level effects, 13
  productivity assessments, 505–506
  tissue residues in, 1
  use of stable isotopes, 17
Willow, 650
Willow ptarmigan (*Lagopus lagopus*), 579
  AhR LBD genotypes, 482
  cadmium effect, 650
  lead effect, 579
Wood mice (*Apodemus sylvaticus*), 596
  blood lead levels, 596
  cadmium effects, 629, 632
  dietary cadmium level, 633
World Health Organization (WHO), 111, 112
ww, *see* Wet weight (ww)
WWTPs, *see* Wastewater treatment plants (WWTPs)

X

Xenobiotic response elements (XRE), 358
Xenoestrogenic substances, 62
XRE, *see* Xenobiotic response elements (XRE)
Xyrauchen texanus, see Razorback suckers

Z

Zebra finches (*Taeniopygia guttata*), 465
  Aroclor 1248 effects, 485
  lead effect, 566
  methoxychlor, 465
Zinc (Zn)
  concentration, 247
  uptake and loss, 246
  variability in crustaceans, 233
Zinc-protoporphyrin (ZPP), 601